S0-DQW-252

CELL BIOLOGY

Neal O. Thorpe
Augsburg College

JOHN WILEY & SONS
New York
Chichester
Brisbane
Toronto
Singapore

Production supervised by Ruth Greif Schild.
Book and cover designed by Ann Marie Renzi.
Electron micrograph and computer reconstruction on the cover were kindly provided by Dr. Jürg P. Rosenbusch of the University of Basel, Switzerland.
Photo research by Kathy Bendo.
Illustrations by John Balbalis with the assistance of the Wiley Illustration department.
Manuscript edited by Brenda Griffing under the supervision of Susan Winick.

Library of Congress Cataloging in Publication Data

Thorpe, Neal O. (Neal Owen), 1938-
Cell biology.

Includes index.
1. Cytology. I. Title.
QH581.2.T47 1984 574.87 83-14785
ISBN 0-471-08278-3

Printed in the United States of America

10 9 8 7 6 5 4 3 2

To the cell's Chief Architect
and to four beautiful cell collections . . .
Kay, Lisa, Heather, and Peter

Preface

The introduction of a new text in cell biology needs little defense, for no single literary source can do justice to the cell. Several textbooks are now available on the subject, each with its own coverage and character. In addition, there are thousands of monographs dealing with various aspects of cell structure and function. In writing this text I have sought to condense a very complex and diversified field into a book that is easy to read and clearly focused. The central theme is the structure and function of the cell and its organelles.

Students of cell biology often have quite diverse academic backgrounds. Certainly introductory courses in biology and chemistry are prerequisite to a study of the cell, but beyond that some students may have little organic chemistry and others may be well grounded in both organic chemistry and biochemistry. This text is designed to speak to a heterogeneous audience, partly by the way in which material is presented and partly because of the introduction of marginal comments. Students with a minimum of organic chemistry or biochemistry will find in appropriate marginal locations structures and comments that will enable them to follow discussions that have a molecular or biochemical emphasis. Students with stronger chemical backgrounds will discover the value of these comments for recall. In addition, marginal comments are used to provide information to enrich a discussion or as a running glossary, especially when the information would interrupt the flow of the topic if thrust into the heart of the material.

The book is organized to examine the cell by starting with its environment, then moving to its surface components, and finally into its interior. Chapters dealing with cell components that are functionally related have been grouped into sections so that students can come to appreciate the way in which different cell structures and functions are interrelated and interdependent.

The first chapter surveys the field in a way that serves as a brief refresher course for all students, placing them on a common ground. Then that which is exterior to the plasma membrane—the cell environment and the cell wall—is discussed. These have a strong influence on the cell and in many instances are essential for its survival. The next section deals with the plasma membrane and the surface properties of antigenicity, reception, adhesion, and communication.

The cell interior is emphasized next, first by discussing organelles that have specialized biosynthetic and oxidative functions, the endoplasmic reticulum and microbodies. The mechanisms of transporting materials into the cell in bulk form by endocytosis and the subsequent digestive activities of lysosomes are the subjects of the next section. Within the cell, materials are moved through a pathway that involves the Golgi complex, a fascinating function of the cell that is covered in a separate chapter. Energy transduction, the activities of mitochondria and chloroplasts, constitutes the next discussion. This is followed by a section that describes the manner in which information is compartmentalized and directed in the cell. A section on the molecular anatomy of form and movement completes the main narrative.

Within each chapter the topics are developed to move from rather low to high resolution, such as from the microscopic to the molecular. Where appropriate, this approach is integrated with a chronological development of the subject, taking the student from the technically simpler past into the more sophisticated present state of the art.

Even though the cell is a highly integrated structure, the study surrounding any given component has developed into its own area of research. Each area has a certain emphasis and uniqueness. These important scientific flavors have been retained throughout the book by, for example, providing a brief description of each research area from a historical perspective. Our present understanding of the parts of a cell draws from the roots of each field of research, a fact that is important to acknowledge. In most cases, the past has set the tone for the development of the field and for its current status. In general, I sought to move a topic in the direction and as far as experts in the field have taken it, thus maintaining the scientific atmosphere in which the study of a particular cell feature has emerged. Beyond that, I point out questions that are not yet answered and work that still must be done before topics can be discussed with a high level of confidence.

The material of each chapter is referenced with two basic objectives in mind. One is to acknowledge the sources of particularly important advances or to point to sources that may shed light on specialized areas that are not necessarily expanded in the text. A second is to indicate sources of information that were especially helpful to me and should be useful for student research and enrichment, as well. Each chapter ends with a list of books and articles that are excellent starting points for additional study.

The text contains an abundance of electron micrographs, diagrams, and tables. In every case, I have tried to select the best figure available to illustrate a point, and wherever practical I have employed an original rather than a newly constructed drawing. Investigators who have kindly provided prints that are reproduced in figures are acknowledged in the figure legends. All other credits for tables and line drawings based on published works are collected by chapter in a credit section at the end of the text. Figure legends are not merely labels. They are instructive and have important content to add to the narrative.

Since the study of the cell depends on a variety of methods, some explanation of the techniques employed and the kinds of information derived from them is found at the end of the text. This is not a thorough theoretical treatment of methods but merely a survey to help the student understand the experimental approaches used to solve particular problems. Throughout the text, marginal notes direct the student to a specific method when it would be helpful, and the discussion of the method itself directs the student back to several examples of its use.

From the beginning to the end, the student will gain a feeling of the discipline of cell biology—a sense of its development and areas of current work and speculation. The subject of cell biology is extremely dynamic. To reflect this, I have avoided the use of dogmatic statements that might suggest that the final word is in. In many cases it simply is not, and it is good for the student to gain a sense of the drama of investigation and to catch a vision of opportunity to participate as a professional in the continuing development of the field. I have not avoided divergent points of view or interpretations of data or lines of speculation. These are all a part of the fabric of present-day cell biology. At the same time, I looked for and emphasized consensus, to give the subject a firm foundation.

To close these comments without thanking a number of people would be a serious omission, for many have influenced the writing of this text. First, to my colleague, John Holum, whose "Go—write!" was the shove I needed to leap into the chasm of textbook writing and who then frequently joined me down there with invaluable wisdom and encouragement, my heartfelt thanks. Then, to Jon Singer, who kindly provided an idyllic setting under the eucalyptus and Torrey pines where the writing was begun while snow and cold smothered the landscape of the Midwest, my deepest gratitude. The transfer from nearly illegible scrawl to typed perfection was efficiently executed by Tammi Trelstad, Carolyn Pratt, Kayla Polzin, and Judie Wester. To each, I offer my appreciation, as well as to Cathy Marlett for her beautifully drawn models. My departmental colleagues Ralph Sulerud, Robert Herforth, Roberta Lammers, and Erwin Mickelberg came to my rescue on many occasions when I stepped into puzzling scientific quagmires. Many thanks.

Scores of investigators from the four corners of the earth sent prints, advice, reviews, and permissions to help create a well-illustrated and scientifically accurate text. Their generosity was a heartening boost to the work, and to them I offer my sincere appreciation.

The combination of artistic talent and scientific understanding of John Balbalis is a treasure that has contributed greatly to the quality of this text. Ann Marie Renzi effectively applied her creative and artistic talents to the design of this text and Ruth Greif guided me through the intricacies of production with a wealth of skill and perseverance. Kathy Bendo significantly eased the effort in researching and procuring certain electron micrographs.

I am most grateful to Fred Corey, who skillfully struck the delicate balance between encouragement, guidance, and keeping me on track.

Finally, to my family, who daily helped me back out of the chasm with patience, understanding, and encouragement. To them I cannot adequately express my indebtedness and gratitude and love.

Neal O. Thorpe

Contents

1

Cell Biology: A Preview

There is no "typical person." *Homo sapiens,* although a single genus and species, is made up of an assortment of individuals differing widely in regard to color, shape, size, mass, and behavior, to name but a few people properties.

Groups of people do exist, of course, in which the individuals are somewhat similar. We may say, for example, that a person is a typical Norwegian or Italian. But even if the statement is made with caution, it is likely to be disputed, especially by the individual in question, for there is within either group about as much variation as is found between groups. No two Norwegians are exactly alike, or at least no two would ever admit to this. The same is true for the members of any other group.

In the same sense, to refer to a "typical cell" would not be proper. Cells come as wildly assorted in size, form, and function as do people. This is, indeed, the basis for the heterogeneity of people. It is also the basis for the different kinds of tissues found in a single organism. A red blood cell in an elephant bears little resemblance to a striated muscle cell in the same animal and even less to a giant motor neuron running from the pachyderm's spinal column to the end of its trunk.

But groups of cells that possess similar traits are often found together. This is generally the case for a tissue, which is basically a group of cells with common structures and functions. It therefore is appropriate to speak of a typical liver parenchymal cell or nerve tissue cell. Were this not the case, a pathologist examining a biopsy specimen would have no basis for distinguishing between malignant and benign cells, in this sense, atypical and typical cells.

Fortunately, liver cells, or cells of any other type for that matter, are recognizable regardless of species. Thus, the difference between the livers of an elephant and a shrew lies not in the form or size of the constituent liver cells but rather in cell number. It would even be safe to say that there would be no microscopic difference between the liver cells of Norwegians and Italians. This law of likeness is something for which researchers are eternally grateful, for without it, mice and frogs could not take their place in experiments.

But except for this kind of likeness, cells display variations with respect to almost every conceivable property and structure. Hence, cells exist that are every color of the rainbow, with or without walls, with or without certain organelles, ranging in size from macroscopic to the edge of microscopic, and of shapes that challenge the most creative of imaginations.

There is no "typical" cell. This fact must be kept in mind when discussing cells in a course in cell biology. Nevertheless, we want to focus on structural and functional traits that are most typically evidenced by cells. In doing so, it is well to remember that variants, even in traits, are not infrequent in the cellular kingdom.

1.1 THE COMPOSITE CELL

When describing cells in a comprehensive sense, the best we can do is to speak of "composite" rather than typical cells. These are hypothetical cells, ideals, that embody the features of all cells. Such composite cells and electron micrographs from which models of this sort are derived are depicted in Figure 1.1. We use these composites to sketch very briefly,

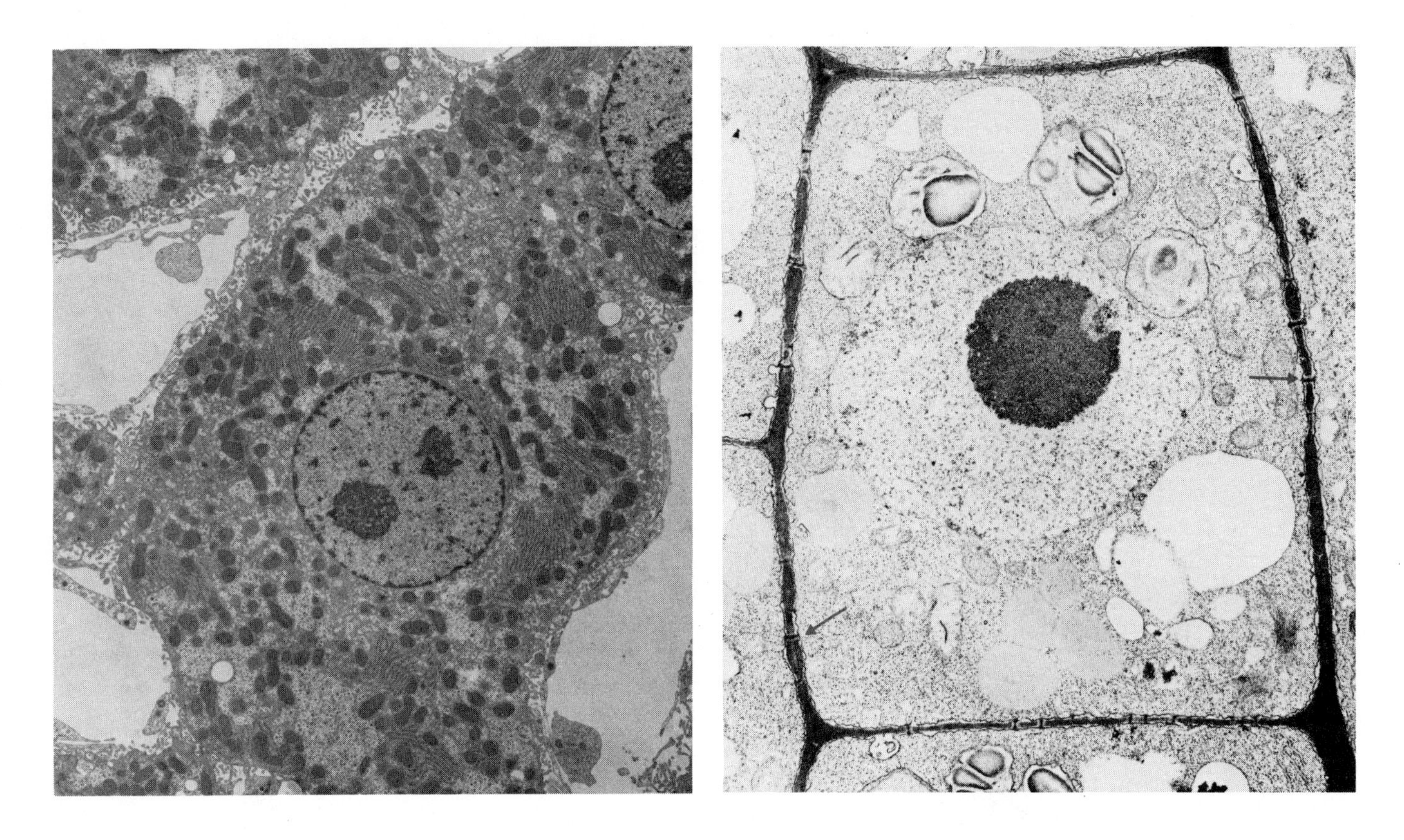

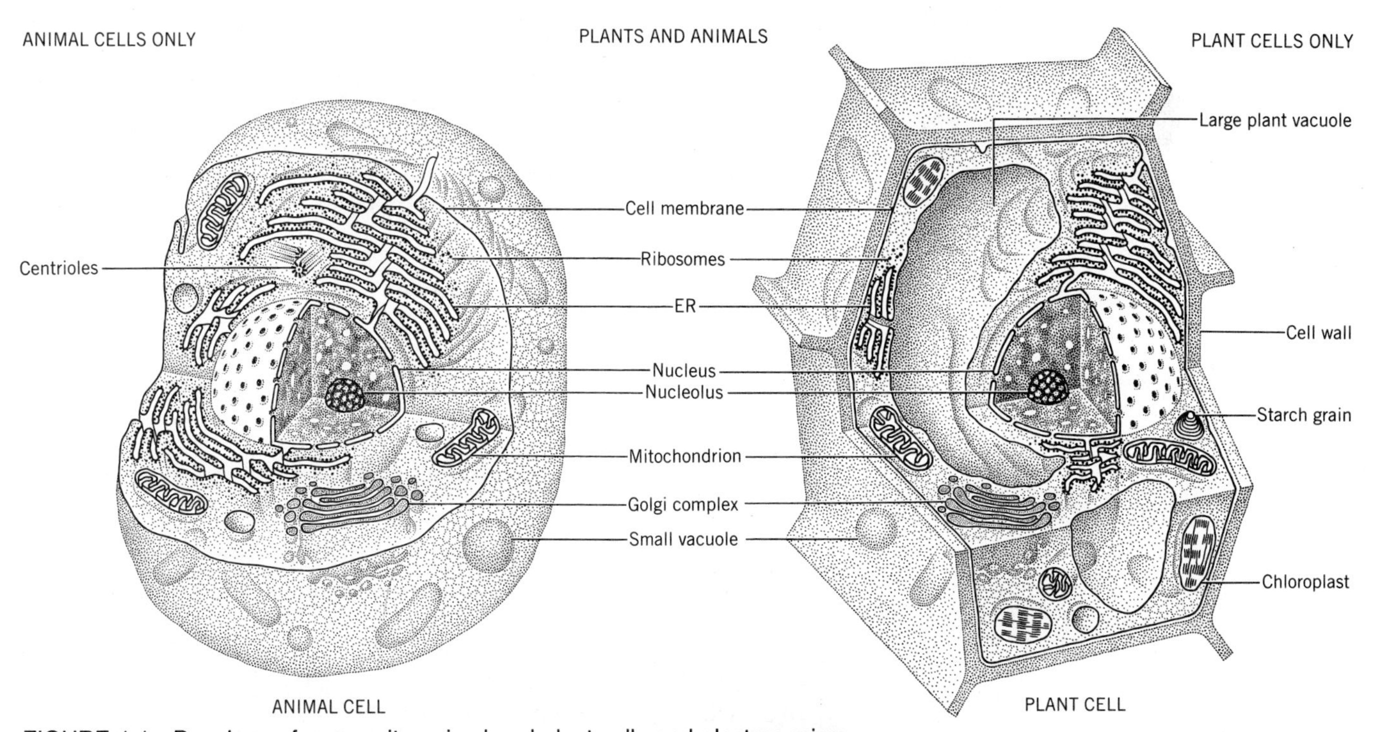

FIGURE 1.1 Drawings of composite animal and plant cells and electron micrographs illustrating some of the features depicted in the models. (Micrograph of animal cell courtesy of Dr. R. L. Wood; plant cell courtesy of Dr. Harry Horner.)

in preliminary form, the major structural and functional features of their components.

1.2 OUTSIDE THE PLASMA MEMBRANE

The Extracellular Environment

Cells of some sort inhabit virtually every niche in the biosphere. Some, therefore, are found in water, both fresh and salt, at its surface and at great depths. Some are found in air, in moist places, dry places, and hot and cold places. Others are found within masses of plant or animal tissue, where the environment may range from a fluid to a semisolid or solid matrix to that of neighboring cells. The environmental possibilities for cells are enormous.

The cell cannot ignore its environment. Its forces may desiccate it or squeeze it or impose strong osmotic pressures on its borders. Many characteristics of cells that we see displayed are there to cope with the difficult task of surviving in an adverse environment.

But the environment is not always harsh; in many cases it is beneficial. For example, it is by means of the environment that cells receive messages that influence their metabolism and provide for a stable cell economy. In addition, precisely controlled environmental conditions are absolutely essential for the proper differentiation of the cell.

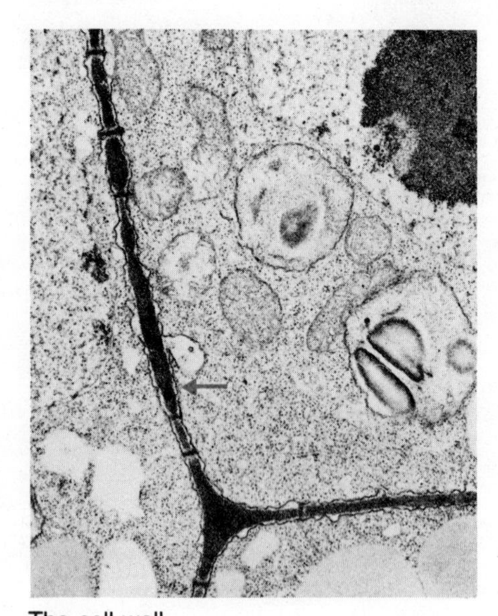

The cell wall.

The Cell Wall

Walls surround certain kinds of cells. Bacteria, blue-green algae (cyanobacteria), fungi, higher algae, and plants are contained by walls. Walls are always exterior to the plasma membrane, where they serve both a protective and a structural, or support, function. They enable cells to survive in hypotonic environments without rupture, they prevent dehydration, and they provide plants with the opportunity for an aerial existence.

Although cell walls possess many functions in common, the walls of cells of different types are far from identical in structure. There are major differences both in their composition and in their ultrastructural makeup. However, a common theme runs throughout the ultrastructure of walls: they contain high molecular weight linear polymers that are cross-linked to give three-dimensional structures that are extremely resistant to rupture and distortion. At the same time, these rigid barriers can be mobilized to respond to the growth and reproductive requirements of the cell. They are thus dynamic structures that have primarily a static role.

1.3 EXTERIOR MEMBRANES AND SURFACE COMPONENTS

The Plasma Membrane

In spite of the structural and functional diversity of cells emphasized at the beginning of this chapter, *all cells possess a plasma membrane*. It is the bag that holds together the contents. Walls cannot serve this function because they are too porous. *The plasma membrane is therefore the one common feature of all cells.*

The plasma membrane is a complex of lipids and proteins, often covered by carbohydrates, that is stabilized in its bilayer structure because of a molecular abhorrence for water and an affinity for self-type molecules. The membrane's content of amphipathic molecules makes this structure possible. In fact, membranes form spontaneously because of their unique chemical makeup.

The membrane functions as a selective barrier completely surrounding its contents. It is solely responsible for maintaining a difference in composition between the inside and the outside of the cell. Any defect in the membrane that neutralizes this difference spells a cellular struggle or death. Of all organelles, the plasma membrane is perhaps the most crucial to the second by second homeostasis of the cell. The mature human red blood cell serves as an example of this, for although devoid of all other internal organelles, it possesses a plasma membrane.

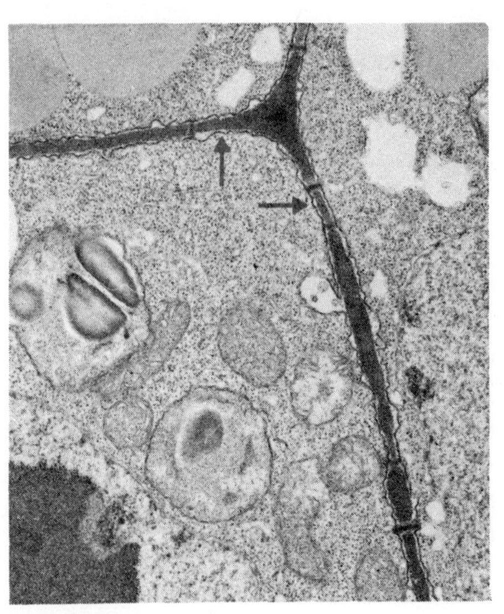

The plasma membrane.

The Cell Surface

In cells without walls, the cell surface is the plasma membrane. It is a dynamic mosaic of proteins, lipids, and carbohydrates.

Different components of the surface and patterns of components function as receptor sites for a variety of extracellular molecules that bring messages to the cell interior. By this means cells respond to their environments, are triggered to carry out various physiological activities, are stimulated to differentiate, and are controlled.

A specialized group of receptors is primarily concerned with intercellular adhesion. Surfaces interact to varying degrees, either directly or across short spaces, to maintain the integrity of multicellular systems.

1.4 INTERIOR MEMBRANES: SPECIALIZED BIOSYNTHESIS AND OXIDATION

The Endoplasmic Reticulum

Within the cell interior there is a network of membranes referred to as the endoplasmic reticulum. The forms in which the endoplasmic reticulum exists, and the amount present, are usually related to the function of the cell.

A cell that synthesizes large amounts of protein generally possesses an abundance of endoplasmic reticulum with ribosomes attached. Ribosomes, in turn, are nucleoprotein particles on which the reactions of protein synthesis take place. This type of endoplasmic reticulum is called rough endoplasmic reticulum (RER).

A cell that specializes in the synthesis of steroids has endoplasmic reticulum without attached ribosomes. This form is called smooth endoplasmic reticulum (SER). This type of membranous system not only lacks the attached ribosomes of rough endoplasmic reticulum but tends to have a vesicular rather than a layered conformation.

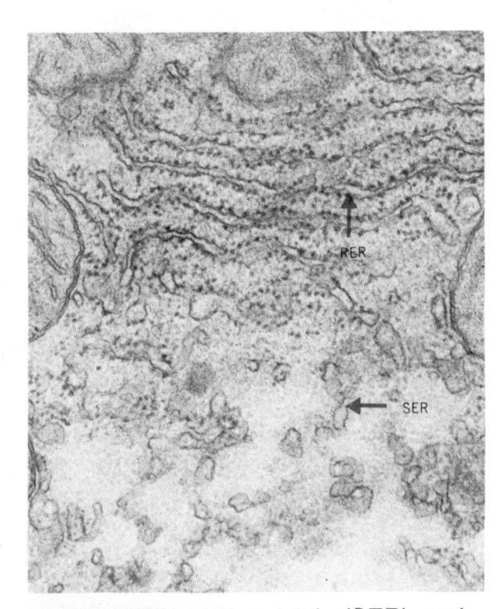

Rough (RER) and smooth (SER) endoplasmic reticula.

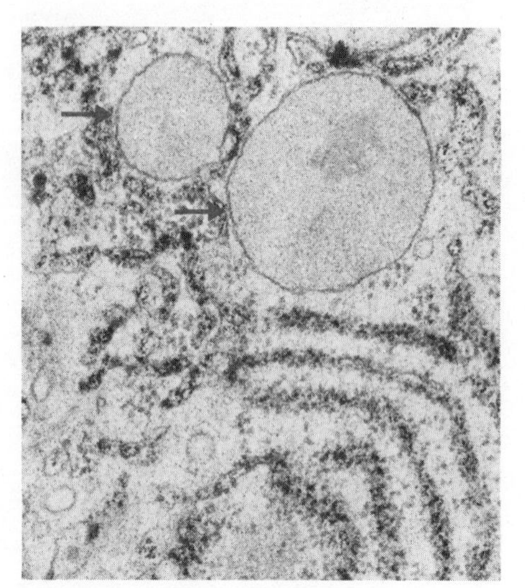

Peroxisomes.

Microbodies

Microbodies are membranous organelles that are often closely associated with the endoplasmic reticulum, as well as with mitochondria and chloroplasts. They are enzyme bags, noted for their unique content of oxidative enzymes that act on a wide variety of substrates.

Two classes of microbodies have been identified. The members of one class, called peroxisomes, contain catalases and oxidases. They are found in both plant and animal tissues. The others, called glyoxysomes, contain in addition to the enzymes above part or all of the enzymes of the glyoxylate cycle. Glyoxysomes are found in the endosperm of seeds, where they have a special function in germination.

Whereas the role of the membranes of the endoplasmic reticulum is largely biosynthetic, the role of microbodies is to break down certain materials with a special set of enzymes.

1.5 BULK TRANSPORT AND INTRACELLULAR DIGESTION

Endocytosis

It has been known for some time that certain cells, such as amoebae and white blood cells, engulf extracellular materials and digest them intracellularly. It is now generally believed that most cells carry out some form of engulfment or endocytosis. During this process, the plasma membrane invaginates and brings into the cell interior either particles (via phagocytosis) or liquid (via pinocytosis).

Although the act of endocytosis is a continuum of cellular responses and movements, it can be considered to consist of several stages. The cell must sense a signal to begin the pursuit of an engulfable material. Then the material and the receptor sites on the endocytic cell interact. Finally, engulfment proceeds.

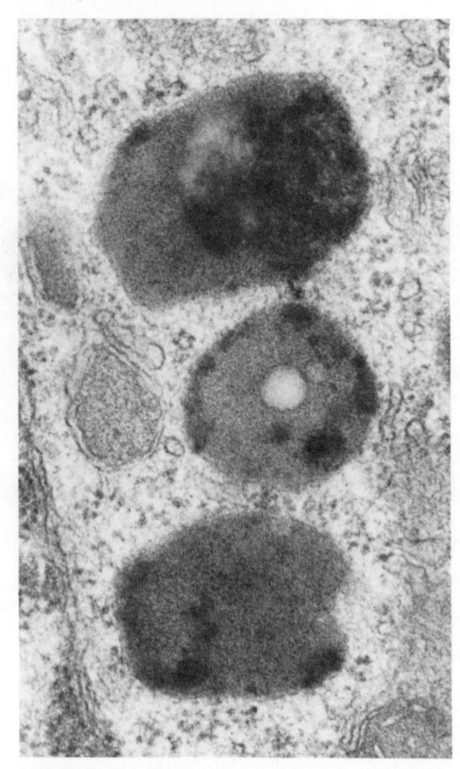

Lysosomes.

Lysosomes

The endocytic vacuole fuses with a lysosome, a membrane-bound collection of hydrolytic enzymes that can act on a large variety of substrates. The products released from this degradation are in turn made available to the cell as nutrients or vital building blocks for biosynthetic pathways.

Lysosomes may also act on other intracellular organelles and degrade them. This is apparently a mechanism to provide nutrients to the cell when they are not available from the outside. It is also a means of tissue mobilization such as that which takes place upon metamorphosis in the tadpole or tissue regression of the uterus after parturition.

Improperly functioning lysosomes often lead to "storage" diseases. Cells become saturated with materials they cannot break down, and the whole organism is adversely affected.

1.6 MODIFICATION AND EXPORT

The Secretory Pathway

The secretory pathway is a traffic route of transport through the cell interior. It begins on the ribosomes where secretory proteins are synthesized. These are then moved through the cisternae of the endoplasmic reticulum and they eventually traverse the Golgi complex region. Here they are modified and leave the complex as secretory granules, packaged with membrane covers.

Secretory granules may be stored for some time before they are released from the cell, or the pathway may function more or less continuously to provide a constant efflux of products to the immediate cell environment. Eventually they are expelled by a type of reverse endocytosis mechanism.

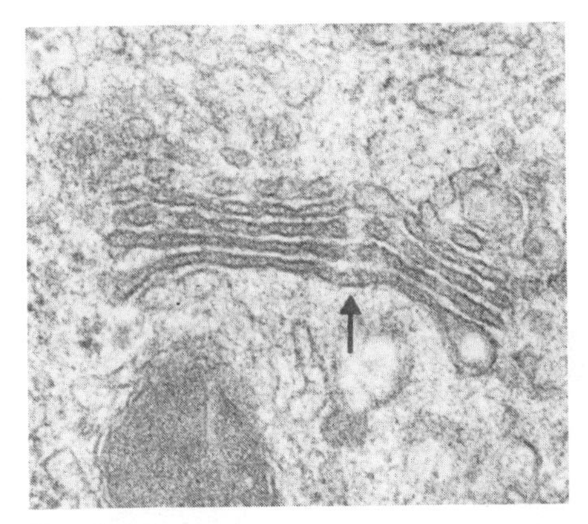

The Golgi complex.

The Golgi Complex

The Golgi complex is a system of membranes that ranges in structure from extremely amorphous, and hard to recognize, to highly organized, with a characteristic stacked arrangement of flattened sacs. It functions as a processing center for materials that are shipped to the cell exterior.

The chief function of the Golgi complex is to glycosylate proteins that are destined either for export or for incorporation into the plasma membrane. It also appears to modify membranes as they move along the secretory pathway. Thus it is a vital transformation center in the cell.

1.7 ENERGY TRANSDUCTION: MITOCHONDRIA AND CHLOROPLASTS

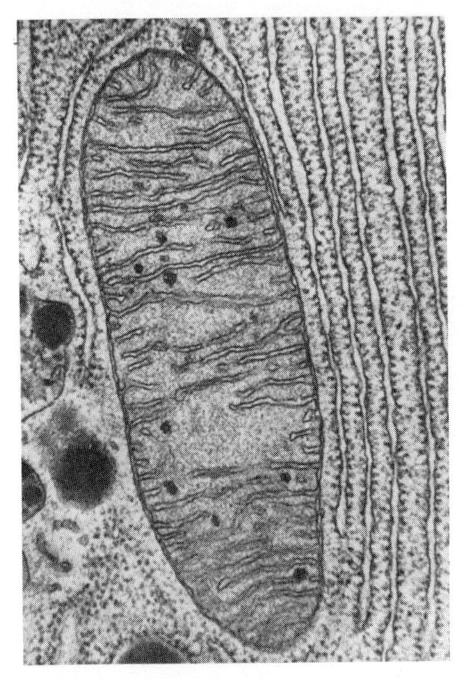

A mitochondrion.

Mitochondria and chloroplasts are double membrane-bound organelles of adenosine triphosphate (ATP) production. Both plant and animal cells contain mitochondria, but chloroplasts are found only in photosynthetic cells, almost exclusively plants.

Mitochondria are present in high numbers with specially designed interiors in those cells that have high energy demands. A variety of metabolic pathways are conducted within mitochondria, including the reactions of the tricarboxylic acid cycle, the β-oxidation of fatty acids, and oxidative phosphorylation.

Chloroplasts generate ATP and reduced coenzymes in higher plants. Once these two ingredients have been formed by molecular entrapment of radiant energy, carbon dioxide is fixed into organic compounds, a process that does not directly require light. Photophosphorylation, the production of ATP via radiant energy, is one of the most remarkable reactions to take place in the biosphere. It requires no nutrients, only a supply of electrons from the ubiquitous water molecule, and radiant energy. All living forms except for a small group of bacteria ultimately depend on this reaction.

Both mitochondria and chloroplasts contain their own genomes, but they are not sufficiently complete to allow these organelles an autonomous existence. Both organelles depend on the nuclear genome to become properly functional. Given this incentive, the rest of the cell obviously benefits enormously.

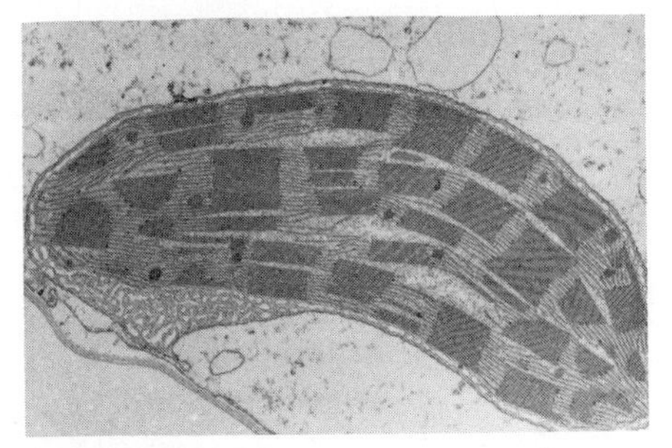

A chloroplast.

1.8 COMPARTMENTALIZATION AND INTRACELLULAR FLOW OF INFORMATION

Within the cell there is a dynamic pathway of information flow that originates in the genetic material and culminates in proteins. Although systems have been discovered in recent years whereby portions of this pathway can be reversed it is for the most part a unidirectional flow that is crucial both to cellular reproduction and to the overall economy of the cell.

Chromosomes

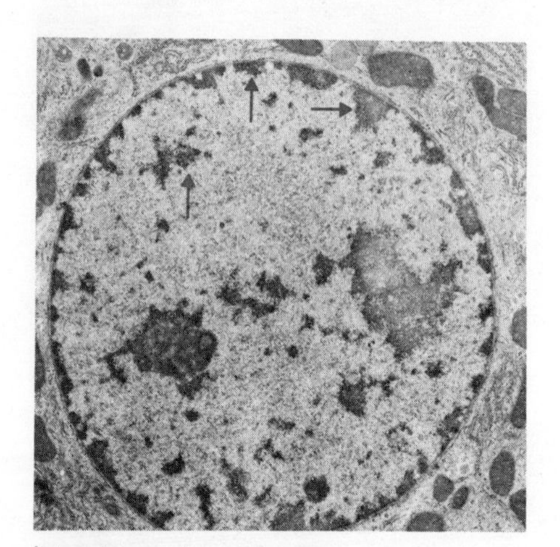

Interphase chromosomes.

Chromosomes in eucaryotic cells are protein–deoxyribonucleic acid (DNA) complexes. They are the primary seat of the genetic information of the cell, with much smaller genetic roles played by mitochondria and chloroplasts.

Chromosome morphology is synchronized with the cell cyle. During interphase, chromosomes are not visible by light microscopy because they exist as extended fibers of a diameter beyond the resolving power of the light microscope. During late prophase, metaphase, and early anaphase, the same chromosomes are visible because they take on a highly coiled conformation. The function of the extended form as seen in interphase is that of permitting DNA replication and the manufacture of RNA from DNA templates. The function of the coiled form is to permit a reliable separation of chromosomes into daughter cells during mitosis or meiosis and to retain the genetic information in a nonexpressed state.

Chromosomes are about half nucleic acid and half protein. The protein component is of two types: histone and nonhistone proteins. Histone proteins have an important structural role in the chromosome, whereas nonhistone proteins regulate the expression of genes by way of blocking or effecting the transcription of DNA.

A recent finding concerning chromosome ultrastructure is the presence of nucleosomes as elemental repeating units of the chromosome. The nucleosome is a histone–DNA complex that takes on the appearance of beads on a string when the chromosome fiber is extended. The fiber is shortened and thickened by a supercoiling of nucleosomes into a cylindrical solenoid.

The Nucleolus

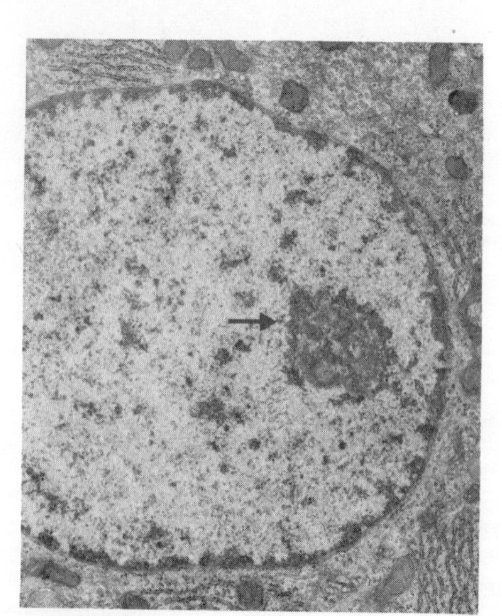

Nucleolus.

For many decades light microscopy has revealed dense areas in nuclei called nucleoli. Electron microscopy, with its greater resolving power, has shown these regions to be granular and fibrous, generally in physical association with particular chromosomes in the cell. Some very interesting experimental approaches have now unveiled the mystery of these regions. They are sites where ribosome subunits are assembled before they are exported to the cytoplasm.

The nucleolus, therefore, is not an organelle in the usual sense. It is rather a depot area where manufacture, assembly, and transient storage of products occur.

The Nuclear Envelope

For ribosomes to get to the cytoplasm, their natural habitat in the cell, they must pass across a formidable double membrane barrier, the nuclear envelope. The nuclear envelope forms a continuous membrane shell around the chromosomes, structurally isolating the nuclear contents from the rest of the cell and functionally permitting only a highly controlled movement of materials to and from the nucleus.

The control routes apparently involve pores that pockmark the surface of the nuclear envelope as seen by freeze-fracture electron microscopy. These are complex structures that permit the passage outward of messenger RNA (mRNA) and ribosome subunits and the passage inward of proteins essential for the replication and transcription of DNA and the assembly of ribosome subunits.

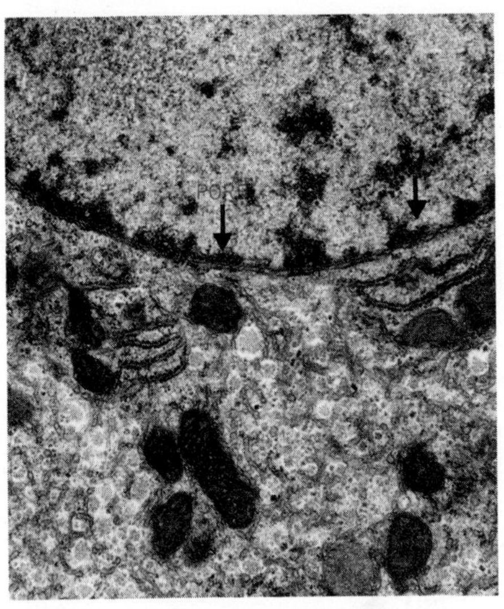

Nuclear envelope (NE) and pore.

The Nuclear Interior

The interior of the nucleus contains, besides chromosomes and nucleoli, a structural network called the matrix. It consists of only a few major proteins that aggregate to provide a nuclear skeleton. This structure is thought to have roles in DNA replication, in transcription, and in the posttranscriptional processing and transport of RNA products.

Within the nucleus, between the larger structures, there resides a heterogeneous population of granules and fibrils. In the region of the nucleolus these are ribosome subunits in the process of assembly. In other regions these particles and fibrils are early forms of RNA, referred to as heterogeneous nuclear ribonucleoproteins. These are processed and modified before they are exported to the cytoplasm as finished messengers.

Ribosomes

When ribosome subunits leave the nucleus to take up residence either on the surface of endoplasmic reticulum or in a free unattached state, they are apparently completely formed. From the point of view of composition, ribosomes are approximately half protein and half RNA. From the point of view of structure, they consist of two subunits that are reversibly dissociated and associated into the intact particle. From the point of view of their molecular anatomy, they consist of approximately 55 different proteins and 3 or 4 different RNA molecules.

Ribosomes are the sites of proteins synthesis when an appropriate complex forms between the ribosome, mRNA, transfer RNA (tRNA), and a variety of other factors essential to the process.

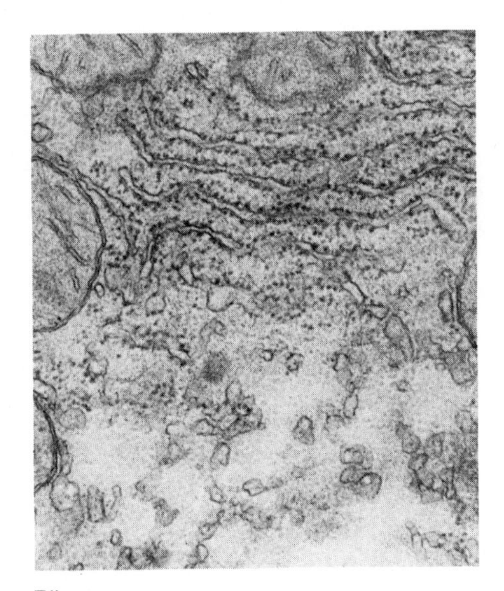

Ribosomes.

1.9 THE MOLECULAR ANATOMY OF FORM AND MOVEMENT

Distributed throughout the cell at different stages in its life cycle are systems of microtubules and microfilaments.

Microtubules

Microtubules are the largest of the intracellular fibrils, with an outside diameter of approximately 24 nm. They are hollow structures, like tubes, with a wall made up of a repeating protein dimer called tubulin.

Microtubules are present as core structures in cilia and flagella as well as in basal bodies and centrioles. Cilia and flagella are generated by basal bodies, and centrioles are concerned with the formation of the microtubules making up the mitotic apparatus.

Microtubule movements are the basis of beating in cilia and flagella and are responsible for the movements of chromosomes during anaphase. They also, along with microfilaments, provide a cytoskeleton for the cell.

Microfilaments

Microfilaments are multiple subunit protein strands having a diameter of near 7 nm. The proteins that make up the microfilament are contractile proteins with properties similar to those of the actin and myosin molecules of muscle. Functionally, microfilaments have roles in amoeboid cellular movement, in cytokinesis, and in other dynamic cellular processes.

Intermediate Filaments

Intermediate filaments are intracellular protein structures with a diameter of 10 nm and an indefinite length. Their functions are not as clearly understood as are those for the other fibrils, but their presence in bundles near the periphery of the cell suggests that they have a role as a component of the cytoskeleton in conferring and maintaining cell shape.

1.10 EUCARYOTIC AND PROCARYOTIC CELLS

The features of the composite cell as reviewed above are characteristic only of cells that possess a true nucleus. Plants, animals, fungi, algae, and protozoa are made up of this type of cell, called a *eucaryotic* cell. It is a participant in both plant and animal kingdoms, hence may be either photosynthetic or not, walled or not. The eucaryotic cell has an information center that is membrane surrounded, and the cell possesses a variety of organelles. But it is only one type of cell that we find in nature.

A second major cell type is about the size of an average mitochondrion in a eucaryotic cell. This type of cell provides the structural framework for the bacteria. It is generally surrounded by a wall and contains ribosomes but has no other membranous organelles. It has no nucleus wherein the DNA is confined but rather bears free DNA strands. This type of cell is called a *procaryotic* cell. The features of this type of cell as compared to those of eucaryotes are summarized in Table 1.1

Some procaryotic cells are photosynthetic, and, being surrounded by a rigid wall, are very plantlike even though they are too small to carry chloroplasts. This is true for three different groups of photobacteria, including the group that is often referred to as blue-green algae or cyanobacteria. Others are more animallike in that they have no photosynthetic apparatus and are highly motile. Neither of these are clearly plants nor animals, and yet they are certainly animal- and plantlike.

Organisms with eucaryotic cells are not always clearly plants or animals either. Fungi possess walls but are not photosynthetic; some protozoa have chloroplasts but no walls; some multicellular forms never have organs or other highly differentiated tissues. For many the entire organism consists of only a single cell. In spite of these anomalies, many taxonomists place these organisms into either kingdom Plantae or kingdom Animalia. But not all specialists agree.

Table 1.1 Principal Structural Differences between Procaryotic and Eucaryotic Cells

Characteristic	Procaryote	Eucaryote
Cell wall	Chemically heterogeneous, generally present	Chemically more homogeneous when present
Cell membrane	Generally lacks sterols	Sterols present
Membrane-bounded organelles	Absent	Numerous types present
Genetic center	No nuclear membrane, DNA naked	Nuclear envelope; DNA complexed with proteins
Nucleoli	Absent	Present
Ribosomes	One type—small	Two types—large and small
Flagella	Simple; an aggregate of proteins	Complex, with much ultrastructure

1.11 THE KINGDOM PROTISTA

Another approach was developed some time ago to solve the probelm of classification. In 1866 Haekel, a German zoologist, proposed a third kingdom, termed Protista. Into this kingdom are placed the protozoa, fungi, algae, and bacteria. Those with eucaryotic cells are called "higher" protists and those with procaryotic cells are referred to as "lower" protists.

Although these organisms appear to have little in common, it is convenient to keep them in a group by themselves. They do, however, generally possess certain features that provide a rationale for creating this kingdom. One trait most have is a lack of extensive cellular differentiation and thus no organs and few specialized tissues. Another is a general lack of cell arrangement. This, of course, is related to a lack of tissues wherein there is normally a special architecture for their constituent cells. Finally, the individual organism often cannot be seen without the assistance of a microscope. Even though some giant amoebae can be seen with the naked eye, most unicellular organisms cannot.

1.12 MASS AND MORPHOLOGICAL VARIATIONS AMONG CELLS

Individual cells of different species may vary in mass by several orders of magnitude. This is emphasized in the data of Table 1.2. There is a significant mass increment between eucaryotic and procaryotic cells (10^{-9} to 10^{-11}g), but within each category there is a wide range of weights as well.

Cell morphology is one of the most unique properties of cells. Bacteria have some of the simpler shapes, ranging from spheres to rods to spirals (see Figure 1.2).

Eucaryotic cells display an endless array of shapes, with the shape of a given type of cell generally tied to its function. Figure 1.3 illustrates several different cell shapes that are commonly seen among eucaryotic cells.

Mammalian red blood cells are biconcave, a feature that increases surface area for effective exchange of carbon dioxide and oxygen with

Table 1.2 Representative Cell Masses

Cell	Mass (gs)
Eggs of dinosaur, ostrich	10^3–10^2
Valonia (marine alga)	10^1–10^0
Vitella (green alga)	10^{-1}
Eggs of frog	10^{-2}–10^{-3}
Human muscle cell	10^{-4}
Human ovum	10^{-5}
Paramecium (protozoan)	10^{-6}
Human liver cell	10^{-7}
Entamoeba histolytica (protozoan)	10^{-8}
Human sperm	10^{-9}
Bacteria	10^{-11}–10^{-14}

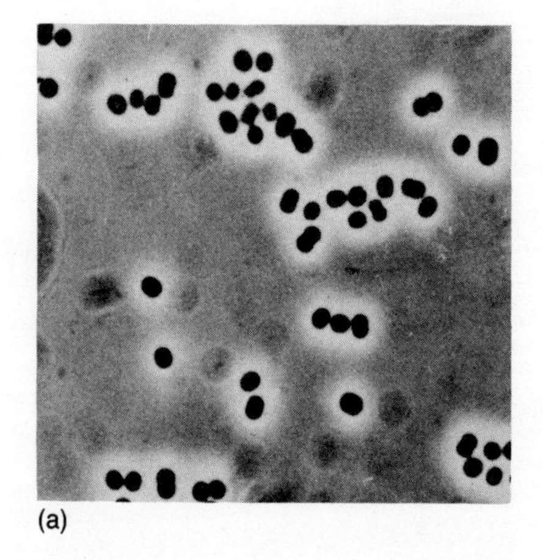
(a)

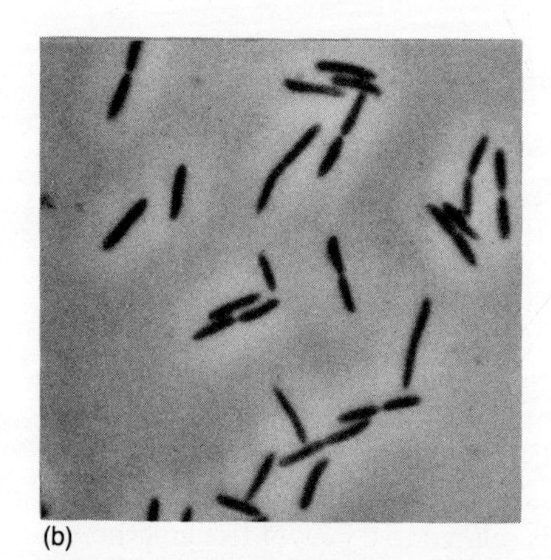
(b)

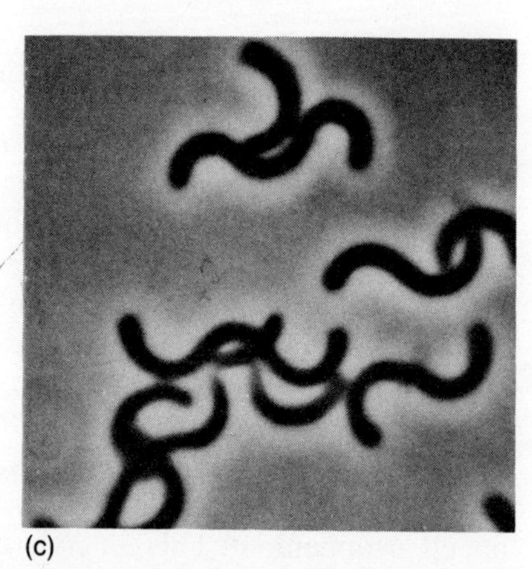
(c)

FIGURE 1.2 Micrographs illustrating three common shapes possessed by bacteria. The forms demonstrated here are called cocci (*a*), rods or bacilli (*b*), and spirilla (*c*). (Courtesy of Drs. R. Stanier, M. Doudoroff, and E. Adelberg.)

their environment and yet keeps them small and pliable enough to percolate through capillary beds. Epithelial cells of the skin are flat, and their geometry permits adjacent cells to fit together like pieces of a puzzle in two dimensions. In the third dimension they are layered, thus forming an effective protective surface to discourage injury, penetration, and dehydration.

Epithelial cells that line inner surfaces, such as the intestines or the respiratory tract, are elongated and polar. These cells are absorbers of nutrients or suppliers of lubricating materials for the lumen surface. They thus have two working surfaces: toward the lumen and opposite from the lumen. The sides are long, giving opportunity for extensive cell–cell adhesion. This is important in keeping the lumen surface intact, since there is only one layer of cells to depend on for this.

Cells that are clustered around tubules are wedge shaped or nearly

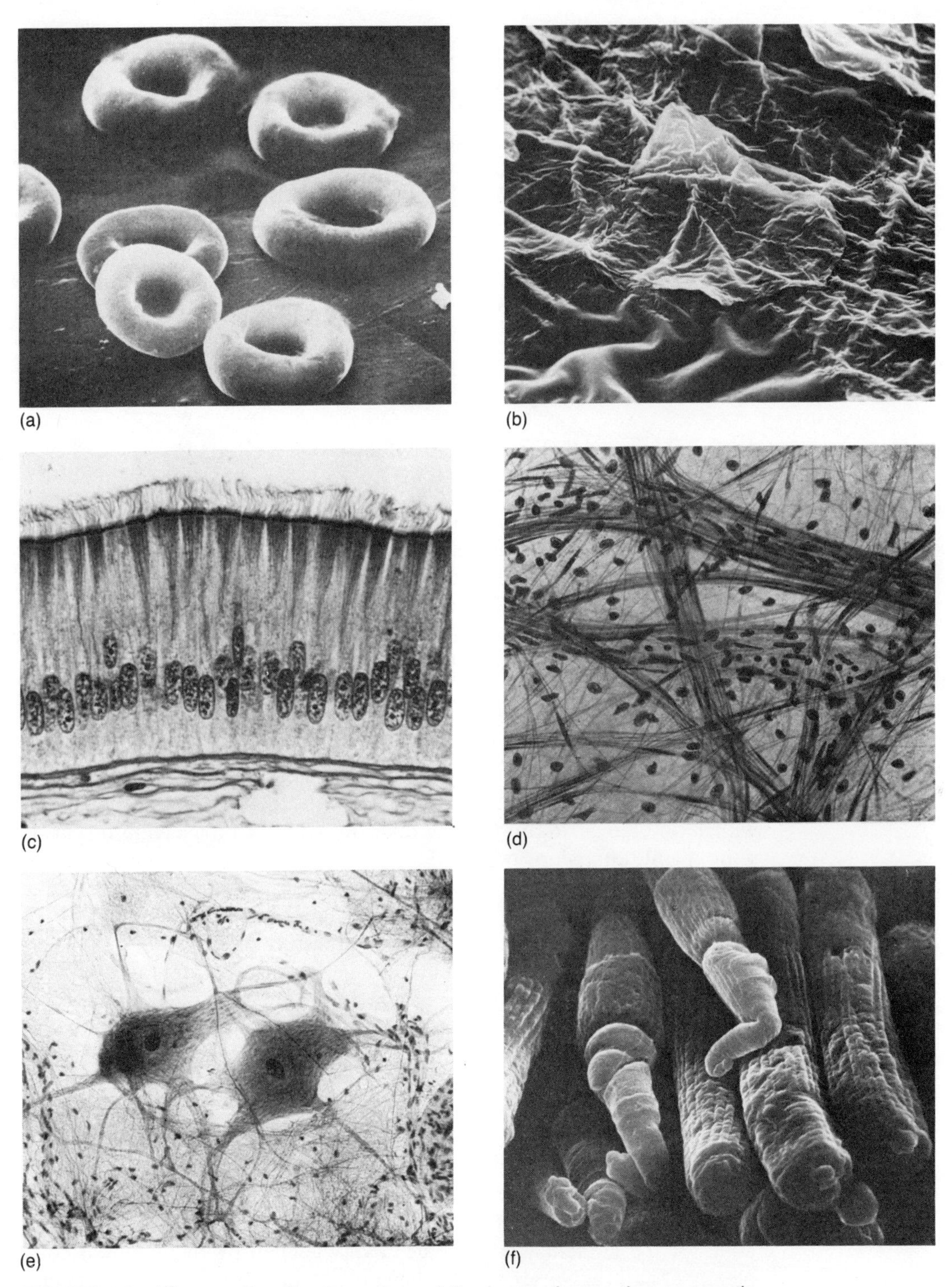

FIGURE 1.3 Micrographs showing a few of the many shapes that eucaryotic cells display. (*a*) Human red blood cells. (*b*) Human surface skin cells. (*c*) Ciliated columnar epithelium from the gut of a mussel. (*d*) Smooth muscle cells from a urinary bladder. (*e*) Spinal nerve cells. (*f*). Rods and cones in the retina of a mud puppy. (Micrographs courtesy of F. Morel, R. Baker & H. Wayland, *a*; David N. Menton, *b*; Don W. Fawcett, *c*; Manfred Kage and Peter Arnold, *d*; Carolina Biological Supply Company, *e*; Edwin R. Lewis, *f*.)

cuboidal, as is true for pancreas and kidney, respectively. Muscle cells are elongated or spindle shaped to enable contraction and expansion along their longitudinal axes. Nerve cells have long extensions, permitting information to be sent over long distances and to be coordinated among various parts of the organism. Many other examples could be given to demonstrate a correlation between morphology and function.

Thermodynamically, the most likely shape for a unicellular organism without a wall is the sphere, and some indeed are spherical. But amoebae, unicellular protozoans, are as dynamic and diverse as cells can get, suggesting that something is going on inside the cell to energize it and drive it into conformations that are not thermodynamically favorable. Indeed, as we shall see, cell shape is influenced not only by walls and membranes, but also by contractile organelles operating in the interior.

1.13 THE FUNCTIONAL UNITY OF CELLS

The protists are an important group of organisms for researchers because they can be cultivated and controlled better than most in the laboratory setting. They are therefore the object of many investigations. Since results obtained from the study of protists often are projected to multicellular organisms, an understanding of higher forms, especially the human, has come about through a study of simpler and more cooperative systems.

For example, the most thoroughly studied organism in the world is *Escherichia coli,* a bacterium. It can be grown easily to large numbers in the laboratory with a new generation produced every 17 to 20 min. It can be subjected to structural, genetic, and behavioral studies as well as to any conceivable biochemical approach. But even though it is procaryotic, most of the molecular information yielded by *E. coli* can be transposed with little change to any other organism whether procaryotic or eucaryotic, plant, animal, or protist. This is so because the molecular laws and principles in the biological world are generally universal. The result of this is that patterns, such as those for metabolism and protein synthesis, will see only slight variations in widely differing organisms.

Because of this functional unity among cells, not all types have to be studied to understand a biological principle. One cell can be a prototype, or an archetypal cell, for the rest. Thus, when cell walls or membranes are studied, investigators select a cell or tissue source from which walls and membranes can be most readily isolated and purified. When the mechanisms of cell secretion are the subject of the research, the cells that are the most profuse secreters and the most amenable to study are used. In each case, because of functional unity, the information obtained from the prototype has universal value.

This is a most fortunate break for the researcher and a wellspring of hope for the student who becomes overwhelmed by the apparent complexity of nature. Nature is indeed complex, but there are common themes that run throughout its dimensions. When the themes are discovered and understood, the variations on the themes as evidenced in individual species simply make the effect more interesting and enjoyable.

1.14 CELL BIOLOGY: A DYNAMIC DISCIPLINE

As for any other field of science, studies of the cell today are quite different from those carried out when the field was in its infancy. Although early statements concerning the cell theory are often attributed to Mathias Schleiden and Theodor Schwann in the 1800s, cells were certainly observed and studied earlier. Some of the earliest work on the cell probably took place in the mid-1500s by Conrad Gesner, who must have used some type of microscopic assistance in his work on Foraminifera.

But it was not until the technology of magnification was developed and refined that cells could be studied in earnest. This principle, of course, has generally been true in science: the level of technology limits the development of the discipline.

In the late 1600s Antony van Leeuwenhoek made a major technological breakthrough in grinding lenses that enabled him to observe a wide variety of cells with magnifications approaching 300×. Unfortunately, his skills were not shared widely with the scientific community of his time; hence the development of the field was temporarily impeded.

Robert Hooke was a contemporary of Leeuwenhoek. He also constructed microscopes and during his studies of cork coined the term "cell." But it took an additional 200 years before the light microscope was fully developed and the cell became an object of study around the world. Even then, the cell interior was largely beyond the reach of careful scrutiny because of the resolving limitations of light microscopy. The major organelles (nuclei, chloroplasts, mitochondria, and flagella) could be seen, but their ultrastructures were essentially undetermined. Beyond this, the functions of the various cell components were largely subjects for speculation.

So it took between 300 and 400 years before cell biology emerged as a discipline in its own right. This portion of the developmental curve of the field could be described as a lag phase. There was some growth, some advancement, but more than anything a time of tooling up that erupted into an exponential phase of dynamic growth in the 1940s. The timing of this eruption was largely due to the development of the electron microscope, as well as physical and chemical techniques that enabled researchers to break open and fractionate cells and to analyze molecular structures and events associated with all parts of the cell.

During the past four decades, an overwhelming amount of information has been generated on the cell. Furthermore, our present comprehension of cell structures and events is built on a foundation of chemistry, mathematics, and physics. Whereas at one time it was possible to learn all that was known about the cell with little academic background, now years must be spent studying the basic sciences before the cell can be viewed intelligently.

The amount of information and its level of sophistication are not the only special challenges in studying cell biology, however. In recent years, the field has become so dynamic that often what was observed and published yesterday must be reinterpreted in the light of new data today. Since this is not so readily apparent to the student, it is well to emphasize that the field is often both dynamic and transient. For the teacher this presents a particular challenge, for to keep up to date in a discipline as broad and complex as cell biology is impossible. For the student, transiency in science is generally frustrating. The textbook is viewed as the final word. Concepts outlined on those pages should certainly not be changed.

Nevertheless, one of the chief characteristics of cell biology is that it is still in exponential growth, and as a result dynamic and changing. Even statements that can be made with certainty for one cell system may have to be made with reservation for another system—not because cells violate physical and chemical principles, but rather because they seem to glory in being different from one another.

Thus, cells are structurally diverse yet functionally united, with an enormous number of variations on both themes. Cell biology reflects this condition, which must be kept up front as together we turn the page.

SECTION 1

OUTSIDE THE PLASMA MEMBRANE

Our consideration of the cell begins with the environment that surrounds it. For some cells the environment is of enormous depth and volume—the ocean, a lake, air—over which the cell appears to exert little control. Rather than in control, it is under control. For other cells, the environment may be a comparatively thin layer of semisolid or solid material—blood, collagens, a wall—that may be produced by the cell, hence largely under its control. In either case, the cell is scarcely a unit independent of its environment. Everything outside the plasma membrane is either a challenge with which the cell must deal or a vital substance for which the cell must strive, or it is placed there by the cell for survival, growth, and reproduction.

2

The Extracellular Environment

In the matter of science versus extracellular matrix, we believe we have established the prima facie *involvement of the accused with respect to its impact on gene expression. Several modes of action have been clearly identified and described, . . . the defense attorney has argued that the evidence is circumstantial, that no general mechanism has been established, . . . Under the circumstances, as prosecuting attorney, I am putting the case to the jury with the conviction that it will find no reasonable doubt that extracellular matrix does have a considerable impact on gene expression.*

C. GROBSTEIN, 1974

Clifford Grobstein, in summing up a sheaf of presentations made at the second international Santa Catalina Island Colloquium, was reflecting on the evidence presented that gene expression depends to some degree on the environment in which the cell is located. This is most certainly a conservative statement. The impact of the environment on gene expression is not simply whether the gene is expressed; we must consider the properties and functions of gene products that result from environmental influence as well. It would be quite apropos to broaden the concept to say that cells are what they are, structurally and functionally, because of the environment in which they are immersed.

Even though scientists are becoming convinced of this, the data are not as numerous and definitive as most would like. When cells are studied, they are usually taken from their natural environments and transported to foreign "*in vitro*" atmospheres wherein their behavior is always somewhat suspect. Even when strenuous efforts are made to simulate the natural extracellular environment, only an approximation generally is achieved. In many cases, it is simply impossible to fabricate the native environment.

Thus the real effect of the natural environment on the economy of a cell is difficult to ascertain. Where information is available, it has come largely from studies using contrived environments in the laboratory setting. But although most studies are conducted *in vitro,* an awareness and understanding of the dynamic interaction between the cell and its environment are taking form and assuming greater impact in considerations of the biology of the cell.

In this chapter we cannot deal exhaustively with the topic of the cell versus its environment. Rather, we will examine a select group of common extracellular environments to heighten our consciousness of the importance of the environment to the structure and expression of the cell.

2.1 THE ADAPTABILITY OF CELLS TO DIVERSE ENVIRONMENTS

If the sea constituted the prime seed bed for living things, early cellular life on the earth must have been somewhat drab. Cellular diversification would not have been necessary, at least initially, since the sea provides an enormous volume of habitat with a relatively fixed composition.

As cellular life began to probe the habitat potentials of new environments, the ability to adapt depended on the genetic machinery and on its ability to change. Change resulting in an advantage often was accompanied by the loss of certain abilities. Thus cells began to specialize according to the properties of their environments, with some tolerating water of lower salinity and others adjusting to different climates of temperature, pressure, oxygen, and pH.

As unicellular life realized an advantage in multicellular existence, further specialization took place. Cells began to interact not only with their environment, but with one another. Often this was accomplished by the creation of connecting materials, making feasible the grouping of specialized cells that could interact indirectly by way of extracellular matrices.

Thus, not only are cells governed in part by the extracellular material in which they are immersed, but they have become dependent on one another. They, in turn, either individually or collectively, frequently generate their own environment by synthesizing and secreting matrix material.

Procaryotic cells are more versatile than eucaryotic cells in adapting to new environments. They appear to be more versatile biochemically, and they have simpler genetic systems that are more readily mobilized to take advantage of mutations and the production of adaptive enzymes. They have remained largely unicellular and independent of other cells. Furthermore, they have penetrated and established themselves in almost every conceivable environmental niche.

Eucaryotic cells have developed highly sophisticated and specialized intracellular hardware, which has made it more difficult for them to adapt because of the complexity of their design. In addition, their use of specialized internal structures has made intercellular dependence an ultimate advantage. This type of specialization has promoted the development of tissues and organ systems. These features have given multicellular organisms an ability to survive in extremely inhospitable environments only as long as their individual cells are sheltered in a microenvironment compatible with life. Hence the necessity for protective coverings, circulatory systems, and mechanisms to maintain an osmotic balance among cells.

Every environment has physical or chemical factors that make it unique compared to any other environment. For an organism or a cell to adapt to a new environment, it must be able to handle a new set of factors, some of which may be lethal. The options available to the cell are few.

One option is to keep the new factor out. This means modifying walls or membranes or transport mechanisms. Another option is to modify the factor so that it is no longer a metabolic or structural threat. This means inducing enzymes to act on the factor. A third option is to become dependent on the factor. If the factor continues to be lethal for other organisms, it will give the dependent organism a competitive advantage in the environment.

Cells are neither structurally nor functionally independent of their environments. They are in one sense products of the environment in which they have successfully adapted, and in another sense creators of the environment in which they live.

2.2 ENVIRONMENTAL EXTREMES

An environment that is perfectly "normal" for one organism may be extreme to the point of lethality for another. Therefore extremes as we are viewing them are not absolutes, but relative differences in one or more environmental factors that pose hardships to the structure or activities of a given organism.

Several factors may contribute toward a condition of environmental extreme. We consider each briefly.

Factors that may vary in magnitude and thereby contribute toward conditions we call environmental extremes are:

Water activity
Salinity
Hydrostatic pressure
pH
Temperature
Oxygen
Nutrients
Radiations

Water Activity

All living things have high water contents and depend on the availability of water to carry out the numerous reactions of metabolism. If water is not readily available, metabolism is depressed and so are all the related activities of the organism, including growth and reproduction.

One way of viewing water availability is by the concept of water activity, a_w. Two factors commonly influence a_w. One is the degree to which water is adsorbed to surfaces, hence unavailable to the cell; and the second is the extent to which solutes are hydrated in aqueous systems. In either case, if water is not readily available to the organism, its activities will be depressed.

Two types of environment in the biosphere contain low water activities. One is the dry environment, where the small amounts of water present are tightly bound to soil particles or other surfaces. The other is the saline environment, where hydrated salts cut down on the availability of water to the organism.

Although most organisms grow best at high water activities ($a_w > 0.95$), many are *osmotolerant* and some even *osmophilic*. Those tolerating saline environments, where water activity is decreased, and those loving it, do much better than normal organisms in these environments, as is evident from their growth rates (Figure 2.1).

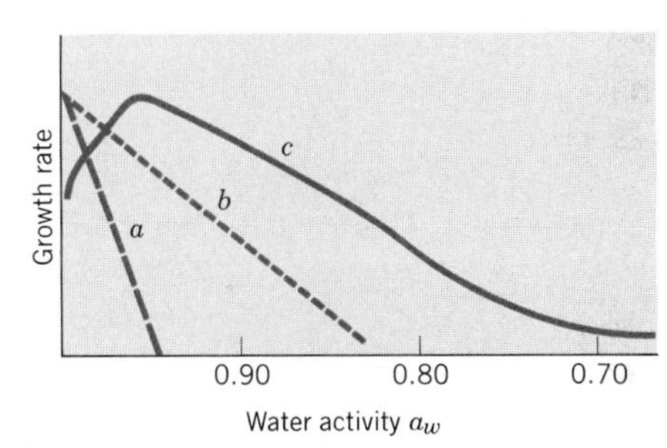

FIGURE 2.1 Effect of variation in water activity a_w on the growth of microorganisms. Curve *a*: a normal organism is markedly inhibited when water activity is reduced (as in saline environments). Curve *b*: an osmotolerant organism will tolerate reduced water activities better than a normal organism but still grows best at high a_w. Curve *c*: an osmophilic organism has a growth optimum at some reduced level of water activity and tolerates significantly reduced a_w.

As a point of reference, seawater, with a salt content of 3.5%, has a water activity of 0.980, whereas the a_w in the Great Salt Lake, Utah (23% NaCl) is 0.850. Most microorganisms cannot tolerate water activities below 0.60.

Salinity

The biological effect of increased salinity on an organism is related in part to a decrease in water activity. Although we usually view this as an osmotic phenomenon, it is more correctly a function of water activity. As the salt concentration in the environment increases, less water is available for the organism and more work is required to sequester water molecules from the environment.

But increased salinity may have other effects on the cell. Certain ions may be toxic and thus limit the forms that will tolerate them. Some organisms dependent on high concentrations of the ions likely to be found in saline environments. This is true for the organism *Halobacterium*, which requires relatively high concentrations of sodium ion (Na^+) for growth. It is an example of an extreme *halophile*.

The name *halogen* is derived from Greek words meaning "formers of salt." The halogens are fluorine, chlorine, bromine, and iodine. These form salts with Na^+ and are found in particular abundance in seawater. An organism that prefers a salt-containing environment is called a *halophile*. From this we can see how the extreme halophilic organism was given the name *Halobacterium*.

Hydrostatic pressure

Organisms that live in water live under a pressure exerted by the depth of water in which they are found. For each 10 m of water depth there is an approximate hydrostatic pressure of 1 atm.

Since the hydrostatic pressure in a large body of water is a continuum from top to bottom, it is difficult to say at which depth the pressure becomes extreme.

Some organisms are *barotolerant;* that is, they tolerate high pressures and will grow and reproduce in such environments. These organisms have generally adapted to this environment by possessing enzymes that are not pressure inactivated and by having membranes that are not severely affected by high pressures.

Even though there is some controversy regarding the existence of truly barophilic organisms, the term is often used when discussing living forms present at high pressures. We will use the term again later in this chapter.

It is not now clear whether any organisms are truly *barophilic*. Most barotolerant organisms do better at lower pressures. Thus, they tolerate high pressures, but prefer lower pressures.

pH

The conformation and charge of proteins, hence their enzymatic activities and structural roles, are highly pH dependent. Even a slight change in the pH, or hydrogen ion concentration, in the environment may adversely affect the metabolism of an organism.

In the biosphere acidic environments are more common than alkaline habitats. The ocean is slightly alkaline, pH 8, whereas rivers and lakes may have pH values ranging from near neutrality to pH 5 or 6. These are comfortable pH ranges for most organisms.

A number of environments are not as comfortable. In bogs and swamps and in volcanic and other geothermal habitats the environment can be highly acidic. Extreme examples of acid environments are solfataras, the sulfur-rich geothermal areas found in many parts of the world. These are basically sulfuric acid environments with pH values that are commonly less than 2. In Yellowstone National Park a solfatara soil with a pH of 0.05 was found to have a living algal organism.[1]

Different organisms have different lower pH limits (see Table 2.1). Fungi tolerate extremely low pH values, as do certain bacterial and algal species. Vascular plants may live at a pH as low as 3 and fish in waters as low as about 4.

Because of the relationship between pH and protein structure, it is obvious that aciduric organisms must accomplish one of two things for survival. Either they must have enzymes that have low pH optima or they must be able to maintain an internal cell environment that is optimum in spite of the external conditions. Both abilities probably are important in most cases.

Temperature

Microorganisms generally grow best at, or love, certain temperature ranges. Thermophiles, lovers of heat, grow well between 40 and 90°C, or even higher. Psychrophiles or cryophiles, lovers of the cold, grow best below 20°C. Mesophiles, those that love the middle, do best between 20 and 50°C.

Organisms, in particular certain microorganisms, are capable of living at enormous temperature extremes. Some live, metabolizing and reproducing, below 0°C, and others are found at the boiling temperatures of their aqueous environments. The crucial factor in both cases appears to be the availability of liquid water.

Table 2.1 Lower pH Limits for Different Groups of Organisms

Group	Approximate Lower pH Limit[a]	Examples of Species Found at Lower Limit
Animals		
Fish	4	Carp
Insects	2	Ephydrid flies
Protozoa	2	Amoebae, heliozoans
Plants		
Blue-green algae	4	*Mastigocladus, Synechococcus*
Vascular plants	2.5–3	*Eleocharis sellowiana*
		Eleocharis acicularis
		Carex sp.
		Ericacean plants (heather, blueberries, cranberries, etc.)
Mosses	3	*Sphagnum*
Eucaryotic algae	1–2	*Euglena mutabilis*
		Chlamydomonas acidophila
		Chlorella sp.
	0	*Cyanidium caldarium*
Fungi	0	*Acontium velatum*
Bacteria	0.8	*Thiobacillus thiooxidans*
		Sulfolobus acidocaldarius
	2–3	*Bacillus, Streptomyces*

[a]Lower pH limits are only approximate, and may vary depending on other environmental factors.

In addition, organisms found at high temperatures must possess thermostable enzymes, which the majority of organisms do not have. Thus even though hot springs are unsuitable to support most forms of life, for some species they are a normal habitat. An environment is extreme only for organisms that are not adapted to it.

It is often true that organisms may survive if one factor is extreme but not if several are extreme. For example, a combination of low pH and high temperature is lethal for almost all forms of life. Biological molecules are subject to hydrolysis under these conditions, and cell death results.

Other Factors

Additional environmental factors could be mentioned that, when high or low, can make the environment intolerable.

The concentration of oxygen markedly affects the complement of enzymes that an organism may use for metabolism. Some cannot tolerate its presence and others cannot survive without it, and there are all shades in between.

The level of essential nutrients can also contribute to environmental extremes. In deep aquatic environments growth is commonly substrate or nutrient limited. An organism requiring a complex diet would not do well in abyssal waters.

Finally, radiation can present organisms with extremely adverse environments. Microorganisms are especially subject to destruction by this means, especially when in or near the air.

2.3 AIR AS AN EXTRACELLULAR ENVIRONMENT

The air is not a natural habitat for any type of individual cell, but it is for a wide variety of terrestrial multicellular systems. Unicellular organisms, especially procaryotes like bacteria, derive some benefit from the air in being transported about by air currents to new and more favorable environments. But the major impact of air on the cell, if it is the only extracellular environment, is harsh and destructive. It can be viewed as an environmental extreme.

Desiccation

Perhaps the most harmful effect of air on a cell is to bring its water content down to levels at which metabolism is impossible. Very few unicellular organisms can combat the determined forces of desiccation or a decrease in water activity.

Several structural modifications are frequently employed by cells in response to this environmental problem. One is the construction of a cell wall that cuts down on evaporation. As an example, the bacterium that causes tuberculosis, *Mycobacterium tuberculosis,* is surrounded by a cell wall rich in waxy lipids that permits survival for months in dried sputum or dust. This wall can be viewed as a structural adaptation that buys survival time in the air while the organism is transported from person to person.

A second type of modification brings certain cells into another form of dormancy. In bacteria, a structure called an endospore is generated (Figure 2.2). Its thick coat and unique composition create an environment within which the cell's DNA and a few critical enzymes are held in a dormant state for an indefinite period.

Still another modification is demonstrated by eucaryotic cells, such as protozoa. They become encysted when under adverse environmental conditions by a protective wall (Figure 2.3). Cells that are thus protected can withstand the drying effects of air for long periods. When the environment is appropriate, a new vegetative cell emerges from the cyst.

Terrestrial multicellular organisms employ another mechanism to avoid desiccation. They assign specialized surface cells the task of synthesizing and secreting protective materials having antidesiccant properties.

But adaptation to the dehydrating effects of air is not sufficient for survival. There are other destructive effects of this extracellular environment on cells. One of these is radiation.

Both purines and pyrimidines may undergo photochemical alterations by ultraviolet light. The most significant and destructive is the photodimerization of adjacent thymines in DNA.

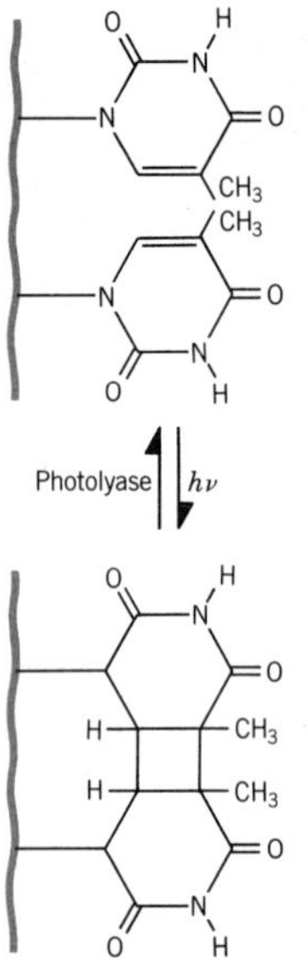

The resulting cyclobutane dimers block DNA replication. A photoreactivating enzyme, DNA photolyase, reverses this reaction.

Radiation

Mycobacterium tuberculosis, the same cell that can avoid desiccation for months, cannot survive in sunlight for more than a couple of hours. The sterilizing factor in sunlight is the ultraviolet component of the radiation that is selectively absorbed by nucleic acids. The most important detrimental effect is the photodimerization of thymine to form a cyclobutane dimer. This blocks the replication of DNA, accounting for the lethal and mutagenic effect of ultraviolet radiation on organisms.

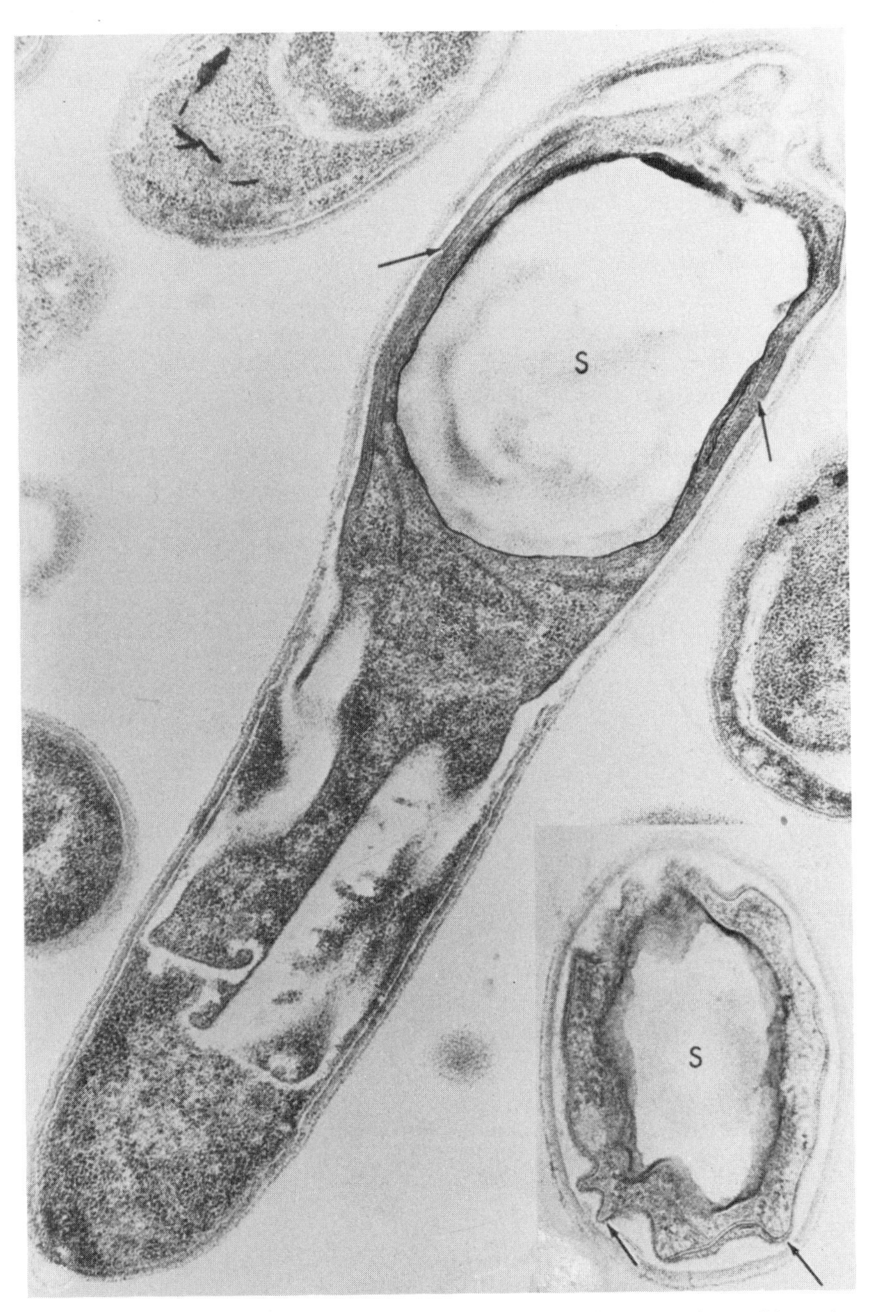

FIGURE 2.2 Bacterial endospore. During certain environmental conditions the DNA is sequestered into a compartment along with certain enzymes and is walled off. The resulting endospore (S), surrounded by a material called the exosporium (arrows), can survive under dehydrating conditions for thousands of years. When the climate is right, it will germinate a new bacterial cell. (Courtesy of Dr. Hilmer A. Frank.)

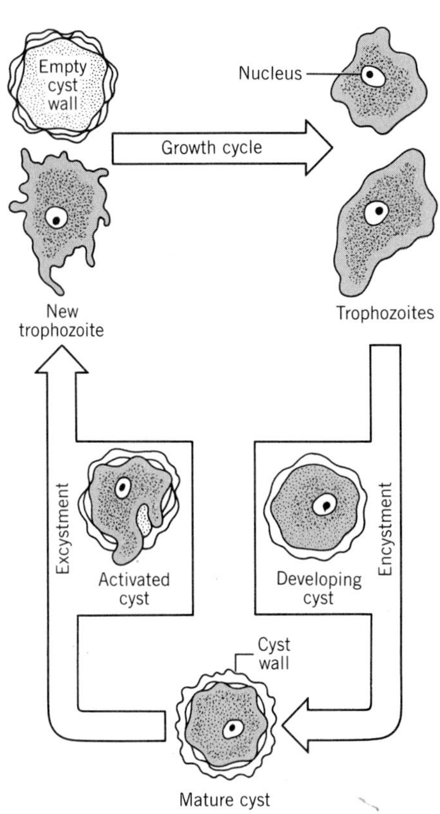

FIGURE 2.3 The formation of cysts from trophozoites, or vegetative cells, in protozoa.

Apparently destruction by this means is significant enough that mechanisms have been developed to combat it. Bacteria possess very low levels of an enzyme called photolyase. It absorbs light maximally at 380 nm, which is in the visible range. Light at this wavelength activates the enzyme to reverse the dimerization reaction and regenerate normal independent thymine residues. This process, termed *photoreactivation,* permits bacteria to survive better in spite of the destructive effects of ultraviolet light.

Bacteria, other unicellular organisms, and multicellular organisms made up of small numbers of cells are most susceptible to destruction by ultraviolet radiation. Some of these employ screens that protect them. Pigment formation is an example of an antiradiation screen.

Higher forms of life use a variety of protective screens, ranging from layers of outer skin cells to hair, feathers, shells, and exoskeletons. Obviously these structures have other functions too, but for the present discussion it is important to reflect on how cells have taken on special protective structures to deal with the life-threatening effects of their habitat.

In summary, an extracellular environment of air poses a special challenge for the cell. It is a powerful dehydrating medium penetrated with destructive ultraviolet radiation. Only cells with special abilities to combat these forces can survive, and only higher organisms covered by cells with these properties can tolerate air as a habitat.

2.4 WATER ENVIRONMENT

Hydrolysis is especially important in digestion, where polymerized molecules are hydrolyzed to their monomer units. The monomers are then taken into the cell. Proteins, polysaccharides, lipids, and nucleic acids are all subject to hydrolysis.

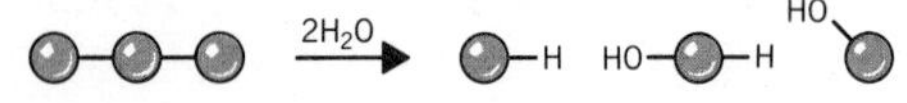

During photosynthesis, chlorophyll is excited by light and donates an electron to an acceptor.

$$\mathrm{Chl} \xrightarrow{h\nu} \mathrm{Chl}^* \xrightarrow{A} \mathrm{Chl}^+ + \mathrm{A}^-$$

The electron lost from chlorophyll is replaced from water.

$$2H_2O \rightarrow 4H^+ + O_2 + 4e^-$$

We will discuss this in more detail in Chapter 13.

Water is a basic requirement for all forms of life. Approximately 80% of the average cell is water, to which several functions can be attributed.

One function of water is to provide an environment in which nutrients and metabolic intermediates are dissolved. In this form they can be transported across membranes into the cell and can diffuse from one enzyme to another as metabolic reactions are carried out.

In addition, water is a direct participant in two fundamental molecular reactions: it has a role in hydrolysis and acts as an electron donor in oxidation–reduction reactions. Many examples could be given for both, but one example is found in the most important biological process occurring on the face of the earth, namely, photosynthesis. Photosynthesis depends on water as an electron donor to the photosynthetic machinery of the chlrochloroplast.

A third function of water is in maintaining the structure of the cell membrane. Amphipathic molecules that make up the membrane interact to form a bilayer structure only because of water. The membrane is in a minimum free energy state, which is assumed spontaneously. Without water there would be no membrane structure. Therefore, water is an absolute requirement for the cell, both internally and as an extracellular environment.

Fresh Water

When cells are immersed in pure water or water relatively free of salts, they immediately encounter an osmotic pressure problem. Most cells have an internal ionic environment that represents variations of the theme seen in seawater. Seawater is about 3% salt with a small collection of principal ions as seen in Table 2.2. An organism with an internal environment roughly reflecting this composition must be able to prevent an influx of water when in a freshwater environment. Without such a mechanism, cells will lyse.

Table 2.2 **The Principal Ions of Seawater Compared to the Ionic Composition in the Blood or Body Fluids of Various Animals**

Location of Seawater		Ion					
		Na	K	Ca	Mg	Cl	SO_4
Woods Hole, MA		100	2.74	2.79	13.94	136.8	7.10
Japan		100	2.14	2.28	11.95	119.0	5.95
		Ion					
Animal	Classification	Na	K	Ca	Mg	Cl	SO_4
Aurelia	Coelenterate	100	2.90	2.15	10.18	113.05	5.15
Strongylocentrotus	Echinoderm	100	2.30	2.28	11.21	116.1	5.71
Phascolosoma	Sipunculid	100	10.07	2.78	—	114.06	—
Venus	Mollusk	100	1.66	2.17	5.70	117.3	5.84
Carcinus	Crustacean	100	2.32	2.51	3.70	105.2	3.90
Cambarus	Crustacean	100	3.09	2.60	6.70	30.9	—
Hydrophilus	Insect	100	11.1	0.92	16.8	33.6	0.12
Lophius	Fish	100	2.85	1.01	1.61	71.9	—
Frog	Amphibian	100	2.40	1.92	1.15	71.4	—
Man	Mammal	100	3.99	1.78	0.66	83.97	1.73

*Calculated from data in Prosser. ed., 1973, *Comparative Animal Physiology,* 3rd ed. W. B. Saunders Co. Philadelphia. Used with permission.

Two major cell modifications allow unicellular organisms to survive in low salt, aqueous environments. One is the formation of a cell wall that can resist high osmotic pressure built up within the cell. Bacteria and cyanobacteria possess a wall that can withstand internal pressures of 20 to 30 atm, giving them the option of living quite comfortably in freshwater lakes and rivers, and even in a carboy of distilled water in the laboratory. Other freshwater organisms, such as eucaryotic algae and freshwater fungi, also possess walls permitting life to go on in spite of potential cell-splitting pressures.

A second mechanism to combat low salt environments is employed primarily by protozoans. These organisms have water pumps, internal water-concentrating organelles, which when they are loaded, dump their contents back into the extracellular environment. Protozoans thus keep from cellular lysis by constantly returning the water to the outside.

Seawater and Body Fluids

One solution to the osmotic pressure problem is to select or create an environment that contains salts at a concentration that approaches that of the cell interior. There are two such environments in the biosphere. One is the sea, the largest salt-containing environment available to cells. It is a natural environment, in some sense the original cell environment. It is not generated by cells, and its composition does not have to be monitored or maintained by cells.

The second salt-containing environment is blood or body fluids. This is a contrived extracellular environment that is constantly monitored and maintained by cells to keep it at a composition that is agreeable with life. It is thus a created environment and one on which certain cells depend completely.

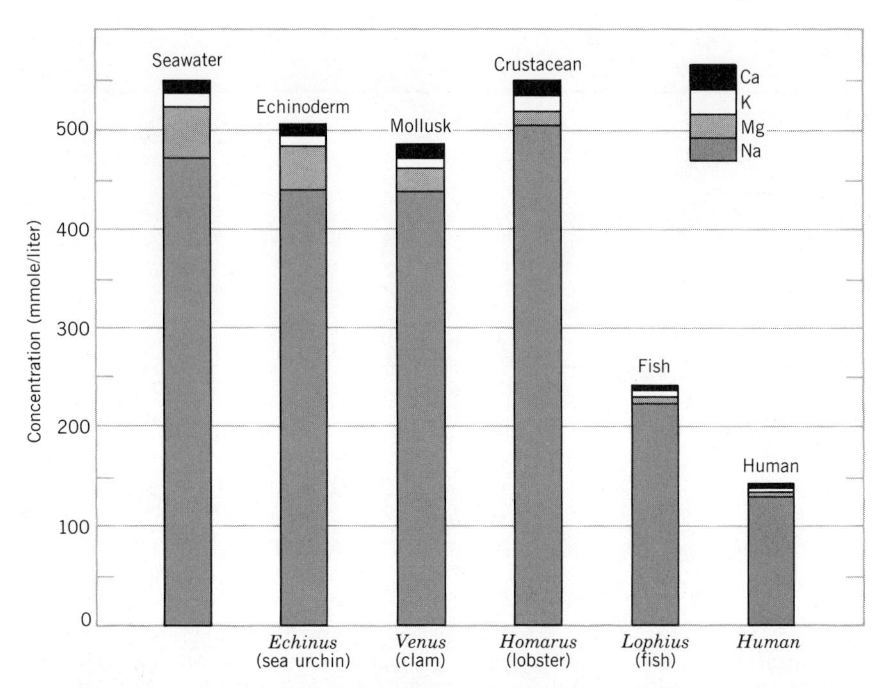

FIGURE 2.4 Cations in seawater, body fluid, and blood. Note how closely the total salt level and its relative composition in echinoderms (sea urchin) resemble these properties in seawater. In contrast, fish and the human depart strikingly from the pattern of ions in the sea.

The principal cations in seawater are sodium, magnesium, potassium and calcium in approximate decreasing order as seen from Table 2.2 and Figure 2.4. The relative proportions of these salts are reflected in most marine invertebrates, as the figure indicates. Echinoderms, which have no kidneys, have internal fluid environments most like the seawater in which they reside. Mollusks and crustaceans, possessors of kidneys, maintain salt compositions in their body fluids that differ from the sea, being relatively richer in potassium and poorer in magnesium.

All other higher forms of life have body fluids that are lower in salt concentration than the sea. Vertebrates range from 0.68 to 0.9% salt and fresh water and terrestrial invertebrates even lower (0.3–0.7%).

The physiological significance of the various levels of salt and the relative contributions made by different ions is beyond the scope of this text. Our brief discussion of these salt environments simply points out the range of aqueous environments with which the cell must cope for survival. Thus a unicellular or small multicellular organism with an internal salt concentration lower than the sea must have cellular mechanisms to prevent water loss and must perform work to extract water from the sea. This implies a specialization of membranes quite different from that needed by protozoans in fresh water, where the problem is to get rid of excess water.

Multicellular sea-dwelling organisms with internal environments lower in salt than the sea must depend on the specialized cellular functions of kidneys to enforce these homeostatic levels and to maintain relative compositions of salts that are most favorable for these complex organisms. In addition, the internal salt-containing environment must be protected

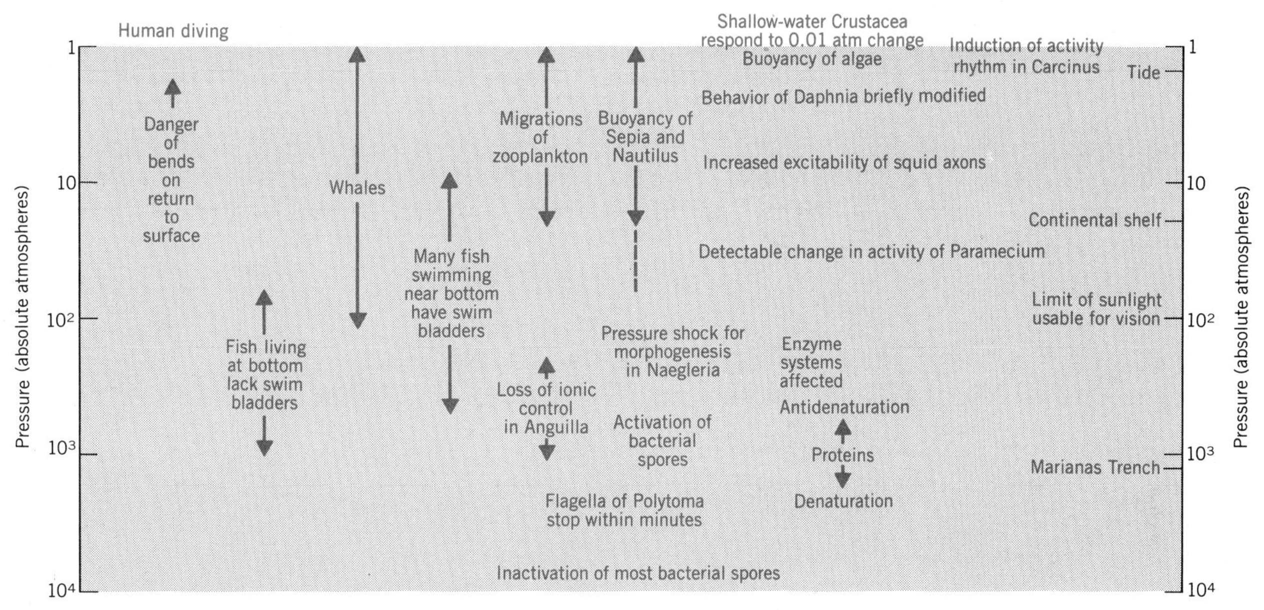

FIGURE 2.5 The ranges of biological pressure phenomena. In the vicinity of 10^3 atm enzyme activities may be altered. At these levels the polymer molecules of flagella are also disrupted, and at extreme depths even bacterial endospores may be inactivited.

from the external sea environment because they differ not only in concentration but in composition.

Thus, although the sea contains salts, eliminating or reducing the osmotic pressure problems encountered by organisms in fresh water, it presents its own set of extracellular influences that the cell must overcome for survival.

A small group of organisms can tolerate a salinity up to 30%, found in salt lakes and brines. Among these are bacteria, yeasts, protozoans, and crustaceans such as brine shrimp. The internal salt concentration of these organisms is generally higher than normal but not as high as their environment. These organisms survive the high salt levels by employing enzymes that are much more salt tolerant than normal enzymes.

Hydrostatic Pressures

Living forms are found in the deepest parts of the oceans where the high hydrostatic pressure and absence of light have demanded extraordinary cell creativity for the maintenance of life.

One atmosphere of pressure is the pressure of dry air at 0°C at sea level at 45° latitude. This is equal to 760 mm Hg or 14.696 pounds per square inch (psi).

The deepest part of the ocean reaches down to 10,860 m at the bottom of the Challenger Deep in the North Pacific. Here the hydrostatic pressure is 1160 atm, but the region is not devoid of life. Both unicellular and multicellular organisms are present. Photosynthetic organisms are of course absent, being confined to the uppermost 100 m of the ocean through which light can penetrate at sufficient energies to activate photosynthetic machinery.

The biological range of pressure is summarized in Figure 2.5. The thicker the layer of water, the more the cell must make adjustments at the molecular level so that its metabolism is not blocked. Barotolerant and bar-

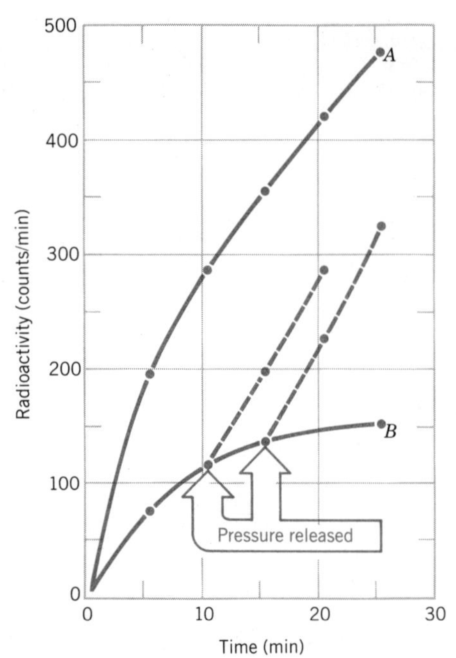

FIGURE 2.6 The rate of protein synthesis at atmospheric pressure (curve *A*) and at 666 atm hydrostatic pressure (curve *B*). Protein synthesis was measured in counts of radioactive emissions per minute of [^{14}C]-glycine taken up by *E. coli.* The rate of incorporation is given by the slope of the lines plotting counts against time. The effect is reversible, as indicated by an increase in protein synthesis upon the release of pressure.

Table 2.3 Hydrostatic Pressure that Inactivates Certain Enzymes in Barophobic Microbes at Temperatures Between 20 and 35°C

Enzyme	Whole Cells[a]	Pressure (atm)
Invertase	Yeast	8000–15,000
α-Amylase	*Bacillus subtilis*	4000–10,000
β-Galactosidase	*Escherichia coli*	300–680
Succinic dehydrogenase	Sulfate reducer	1200–1800
Succinic dehydrogenase	*E. coli*	200–800
Formic dehydrogenase	*E. coli*	200–1000
Malic dehydrogenase	*E. coli*	200–1000
Aspartase	*E. coli*	100–1000
Nitrate reductase	*Pseudomonas* sp.	1000–1800
Urease	*Micrococcus* sp.	250–800
Luciferase	*Photobacterium* sp.	330–680
Phosphatase	Various marine bacteria	200–1000

[a]Of the organisms listed, only yeast possesses a eucaryotic cell.

ophilic organisms, those tolerating and preferring high hydrostatic pressures, have adapted to this extracellular environment by using enzymes that are not adversely depressed by these conditions.

In general terms, hydrostatic pressure has a threefold effect on biological systems:[2] it inactivates enzymes, it affects physiological reaction rates, and it influences growth or reproduction.

Hydrostatic pressures of sufficient magnitude, on the order of 1000 to 3000 atm, denature many enzymes, whereas pressures lower than this cause enzyme inactivation (see Table 2.3). It is therefore apparent that organisms that operate successfully at these pressures must have isoenzymes that are not seriously affected by the pressures. One speculation is that barotolerant enzymes tend to be monomeric or dimeric, and their terrestrial counterparts polymeric. This is in line with the observation that high pressures depolymerize complex biological structures.

The effect of hydrostatic pressure on the rate of protein synthesis in the bacterium *Escherichia coli* is presented in Figure 2.6. Protein synthesis represents the result of the combined action of many enzymes, so although the data do not pinpoint the molecular location of the pressure effect, the impact on this obligatory cell process is apparent.

Whereas most enzyme activities are depressed at high pressures, certain enzymes are pressure activated. This is true for the enzyme fructose diphosphatase,[3] an important enzyme in gluconeogenesis (see Figure 2.7). The enzyme, in this case, does not appear to exist in a markedly special form for barophilic organisms. Both the surface-dwelling trout and the abyssal rat-tail possess quite similar pressure-activated fructose diphosphatases. However, certain kinetic parameters of the enzymes reveal differences that provide selective advantages for the environments in which these two organisms are found.

In contrast to fructose diphosphatase, pyruvate kinase is inhibited by pressure and lactate dehydrogenase is pressure independent. Other enzymes could be placed into these three categories as well. These are examples of molecular loci at which the environment exerts a controlling effect on the metabolic activity of the cell.

Fructose diphosphatase circumvents a thermodynamic barrier in glycolysis that cannot be reversed under the conditions extant in the cell. Its role is therefore essential if an organism must make glucose from precursors such as pyruvate, lactate, certain amino acids, and citric acid cycle intermediates.

Glucose ⇄ Glucose-6-P ⟶ Fructose-6-P
Fructose diphosphatase ⇅
Fructose-1,6-diP
↓

Pyruvate kinase operates toward the end of glycolysis in the forward direction.

⟶ Phosphoenolpyruvate + ADP ⟶ Pyruvate + ATP

In mammalian liver it is a regulatory enzyme, activated by fructose 1,6-diphosphate and phosphoenolpyruvate but inhibited by ATP, AMP, citrate, and alanine. In barophilic organisms it is also subject to pressure inhibition.

Hydrostatic pressure also has an effect on sol–gel transitions within cells. At the molecular level, these transitions are the expression of a change from monomer to polymer forms of proteins, respectively. The formation of hydrogen bonds is encouraged at increased pressures and ionic and hydrophobic interactions are discouraged.[4] Therefore, cellular processes such as cell division and amoeboid movement that depend on the formation of microtubules are inhibited at high pressures. At pressures greater than 400 atm, dividing sea urchin eggs revert to a one-cell stage and pseudopodia formation in *Amoeba proteus* is inhibited.

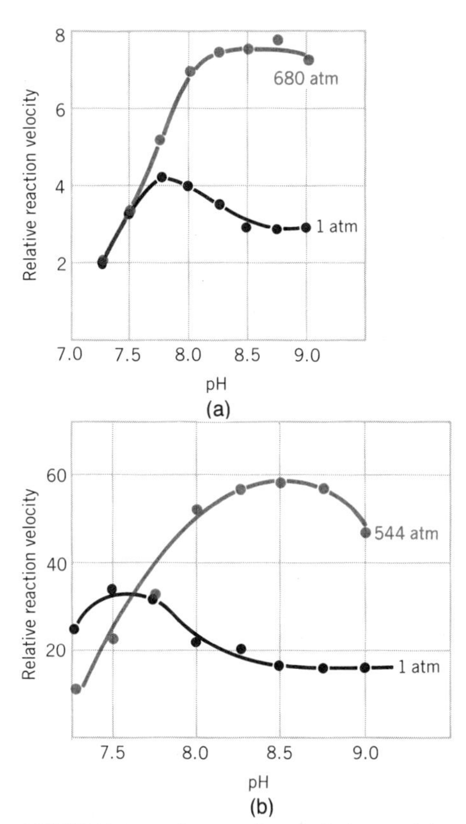

FIGURE 2.7 Comparison of the activity and pH profiles of liver fructose diphosphatase under conditions of saturating substrate concentrations for (*a*) rainbow trout at 3°C and (*b*) rat-tail at 2°C. The enzyme in the rat-tail (a barophile) responds to pressure as does the enzyme in the surface-dwelling trout. However, the reaction velocity of the enzyme from the bottom dweller is higher, thus giving this organism an advantage at high pressures.

Effect on Water Structure

There are now several theories concerning the structure of water, ranging from a complete absence of ordering to the existence of discrete, geometrically identifiable clusters of molecules. One point of view is that water structure is best described as a mixture model, a state dynamically balanced somewhere between these extremes.[5] For this discssion, we should be aware of the view that water near a solid interface, such as a cell surface, takes on a different form from that of bulk water. It is believed that this water is stabilized by interaction with the surface molecules.

The effect of high pressure on cells, as discussed above, may be due to an effect on water structure. One form of water, termed the clathrate hydrate, may be stabilized near proteins at higher pressures with structural voids in the proteins increasingly occupied. At still higher pressures, another form of water, the ice polymorph, may predominate. Thus the effect of pressure may have an impact on protein conformation by way of vicinal water structure. This, in turn, affects enzyme activities.

Effect on Cell Shape and Movement

The effect of high pressure on cell membranes has been studied by observing changes in cell shape and movement. The rationale for this is that the cell membrane is a contributing factor to cell shape and thus an effect on the membrane is transmitted to that of a shape change. Cell movement as observed, for example, in amoebae is also a membrane-related event. Therefore a change in properties of the membrane may indeed be expected to alter movement.

Changes in both shape and movement are thought to be related to the sol–gel transitions of the cell that are pressure sensitive.[6] Sol formation is preferred at higher pressures. Therefore a loss in shape or cell movement is reasonably attributable to sol formation in the protoplasm near the cell surface where, under normal circumstances, polymerized proteins make up a scaffolding structure governing membrane movements.

Sol–gel transitions represent a change from monomer to polymer.
A number of cell functions are dependent on the polymer form.

Sol (high pressure) Gel (low pressure)

Bacteria

At hydrostatic pressures greater than those of their normal habitats, bacteria become filamentous and lose their motility. The change in morphology from single cells to filaments apparently results from an inhibitory effect on cell wall synthesis, a polymerization reaction. A loss of motility can readily be explained by a direct effect on bacterial flagella, which are simply organized assemblies of polymerized proteins that are not covalently bound.

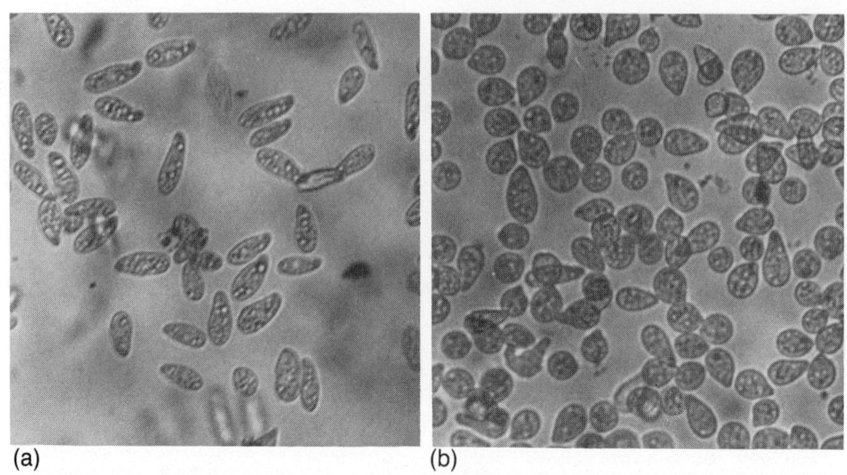

FIGURE 2.8 Effects of pressure on a protozoan ciliate *Tetrahymena pyriformis*. At atmospheric pressure the cells are elongated and motile (*a*) whereas at 10,000 psi for 10 minutes (*b*) they round up, lose their translational movement and display only uncoordinated ciliary activity. (Courtesy of Dr. Selma Zimmerman).

Protozoans

At pressures between 300 and 400 atm, amoebae become spherical and quiescent. The effect is reversible within a few minutes when the organisms are brought back to normal pressures, suggesting sol–gel transitions as likely candidates for the molecular target of this effect.

Pinocytosis, a membrane function, is also blocked at increased hydrostatic pressures.

High pressures cause both ciliates and flagellates to round up and lose their locomotive abilities (see Figure 2.8). Flagellates are somewhat more resistant to pressure-induced shape changes than ciliates, possibly because of the presence of the pellicle, a specialized surface structure possessed by these organisms. High pressures generally cause a loss of flagella as well, and thus an inhibition of motility.

Marine Eggs

Cytokinesis is the process whereby the cytoplasm of a cell is segregated and separated into two daughter cells.

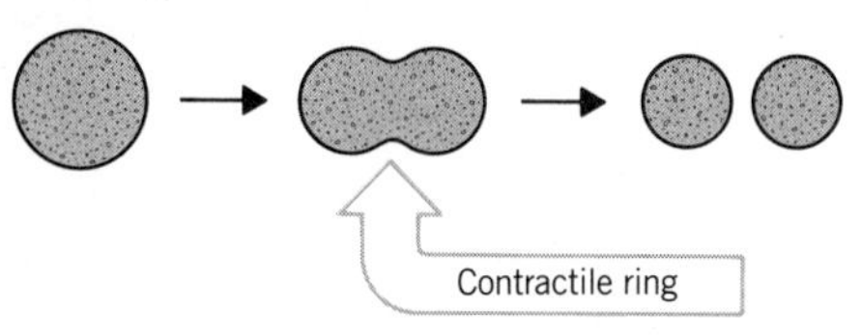

A contractile ring mechanism involving actin and myosin proteins in the cell cortex appears to cause the furrowing. Actin and myosin are both polymeric proteins.

Cell division is inhibited in sea urchin eggs at high pressures with apparently a twofold effect. The cortical plasmagel, which appears to be responsible for cytokinesis, is one part of the cell affected by the high pressures. The structure of the mitotic apparatus is also altered thereby affecting cell cleavage.

Cells of Higher Organisms

Although less is known about the effect of high pressure on more complicated organisms, some studies on human cells and embryonic chick heart tissues have revealed that these cells respond to pressure in a manner similar to that of the other systems discussed. In general, irregular cell shapes disappear under pressure and a spherical configuration is assumed.

The structures of the nuclei, the mitochondria, and the plasma membrane itself do not seem to be affected by high pressures (476–680 atm). Neither do the desmosomes, which are specialized intercellular points of contact between cells.

Table 2.4 Effects of Temperature on Physical and Chemical Properties of Water

Property	Effect of an Increase in Temperature
Physical	
Density	Decrease
Viscosity	Decrease
Surface tension	Decrease
Volume	Increase
Dielectric constant	Decrease
Vapor pressure	Increase
Heat capacity	Decrease
Compressibility	Decrease
Refractive index	Decrease
Diffusion	Increase
Chemical	
Ionization	Increase
pH	Decrease
Oxygen solubility	Decrease
Solubility of most organic and inorganic compounds	Increase

In a number of systems that have been studied, high pressure disrupts existing microtubules and prevents the formation of new microtubules. Many cellular activities depend on these pressure-sensitive polymeric structures (see Chapter 18).

Life at High Temperatures

When water temperature is increased, it undergoes a number of physical and chemical changes to which an organism must adjust for survival. These changes are summarized in Table 2.4. Certain of these changes are of more biological import than others.

Thomas Brock and a host of co-workers carried out an intensive investigation of thermophilic microorganisms primarily in Yellowstone National Park from 1965 through 1975.[1] They made many important discoveries concerning the presence and properties of life forms at temperatures and under conditions not normally considered to be compatible with life.

Hot springs differ widely in their chemical compositions, making it possible for investigators to examine the effects of a variety of factors in the geothermal environments on cells.

Upper Temperature Limits

Different organisms possess different upper temperature limits beyond which growth is inhibited or death realized. These upper limits for several groups of organisms are presented in Table 2.5. In general, eucaryotic organisms are more heat sensitive than organisms possessing procaryotic cells.

Procaryotic microorganisms have been found in habitats of superheated pools at 93.5 to 95.5° C, but not in fumaroles where temperatures are above boiling and liquid water does not exist. These observations led

Table 2.5 Upper Temperature Limits for Growth of Various Microbial Groups

Group	Approximate Upper Temperature(°C)
Animals	
Fish and other aquatic vertebrates	38
Insects	45–50
Ostracods (crustaceans)	49–50
Plants	
Vascular plants	45
Mosses	50
Eucaryotic microorganisms	
Protozoa	56
Algae	55–60
Fungi	60–62
Procaryotic microorganisms	
Blue-green algae (cyanobacteria)	70–73
Photosynthetic bacteria	70–73
Chemolithotrophic bacteria	>90
Heterotrophic bacteria	>90

Brock to conclude that the upper temperature limit for procaryotes is any temperature at which liquid water can be found.

On the other hand, eucaryotic organisms appear to possess definite upper temperature limits, even if liquid water is present. The most heat-resistant organisms are the fungi, which will grow at temperatures near 60° C. As is apparent from Table 2.5, protozoa and algae are more heat sensitive.

It is thought that differences in internal membrane structure are responsible for differences in heat tolerance between procaryotic and eucaryotic organisms. Eucaryotic organisms depend on a proper functioning of a variety of membrane-bound organelles, such as mitochondria, lysosomes, and nuclei. The composition of these membranes differs somewhat from that of the plasma membrane and, of course, the internal membranes have different functions with respect to transport. Speculation is that it may not be possible for a eucaryote to synthesize organelle membranes that are both functional and thermostable.

Thermostable Enzymes

Both procaryotes and eucaryotes living at high temperatures have enzymes that are stable under these conditions. Several thermostable enzymes have now been isolated and purified to homogeneity. Their amino acid sequences have been compared to those of comparable enzymes from mesophiles, and only minor differences were reported between the two groups, in many cases involving a few amino acids. However, these differences, even though minor, must markedly affect the three-dimensional stability of the molecules when challenged by heat.

Aldolase is an enzyme of glycolysis, breaking the six-carbon compound fructose 1,6-diphosphate into two three-carbon molecules: dihydroxyacetone phosphate and glyceraldehyde 3-phosphate.

The thermostability for the enzyme aldolase taken from *Thermus aquaticus* is demonstrated in Figure 2.9. This organism is a bacterium that is found in hot springs and in unnatural environments such as commercial hot water heaters, which are generally run at 70 to 80°C. At 97° C the

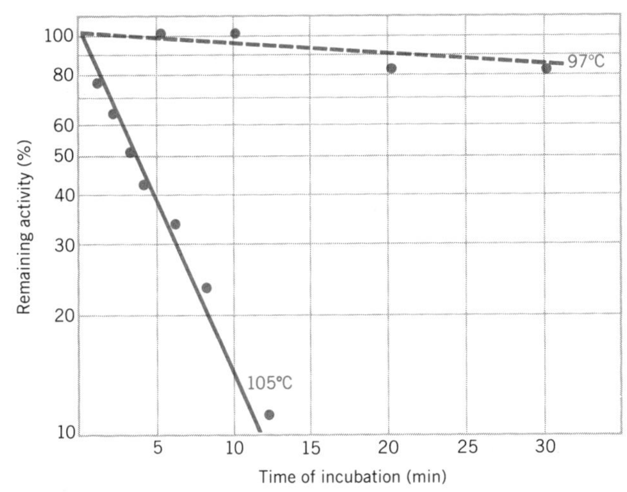

FIGURE 2.9 Heat inactivation of the enzyme aldolase from *Thermus aquaticus*.

enzyme loses little activity with time, but at 105° C it is inactivated in about 10 min. Below 60° C the enzyme has little activity.

Nucleic acids with higher G–C contents have higher melting temperatures; that is, it takes a higher temperature to separate the complementary strands. Magnesium ion may stabilize negative repulsion between regions. Thiolation is thought to increase the number of polarizable electrons in this position, thereby increasing stacking interactions.

Ribothymidine

5-Methyl-2-thiouridine

Thermostable Nucleic Acids

Protein synthesis at high temperatures requires not only thermostable enzymes but also messenger, transfer, and ribosomal RNAs that can withstand the higher temperatures.

One approach to assessing the thermal stability of nucleic acids has been to look at the nucleotide sequence of tRNA from a thermophilic organism and compare it with that obtained from a mesophile. In one study, comparing tRNA from *Thermus thermophilus* with *E. coli,* the former was found to differ in the following ways: (1) an increase in guanine–cytosine (G-C) content and tight stacking of base pairs in one region of the molecule, (2) Mg^{2+} ion protection from heat denaturation, and (3) thiolation of ribothymidine to a 5-methyl-2-thiouridine base at position 55 of tRNA.

These chemical changes, although minor, appear to change the physical properties of tRNA so that it is inherently more thermostable.

Thermostable Membranes

Protoplasts of mesophilic bacteria such as *Streptococcus faecalis,* when heated to 60° C, lyse. Their membranes are thermolabile. In contrast, lysozyme-induced spheroplasts of *T. aquaticus* do not lyse even when boiled.

Protoplasts are cells completely devoid of walls. They generally assume a spherical morphology. Spheroplasts are also spherical, but they may have small wall fragments attached. Functionally, protoplasts and spheroplasts are about the same.

The molecular basis of the thermostability of membranes is not yet clear, but it is probably related to the nature of both the proteins and lipids present in the plasma membrane. A major phospholipid, unique but not yet characterized, has been isolated from the membrane of *T. aquaticus*. Furthermore, the higher the temperature of the environment in which the organism is grown, the lower the content of monoenoic and branched-C_{17} fatty acids in the lipids.

The majority of membrane lipids have ester linkages, such as in the phospholipids:

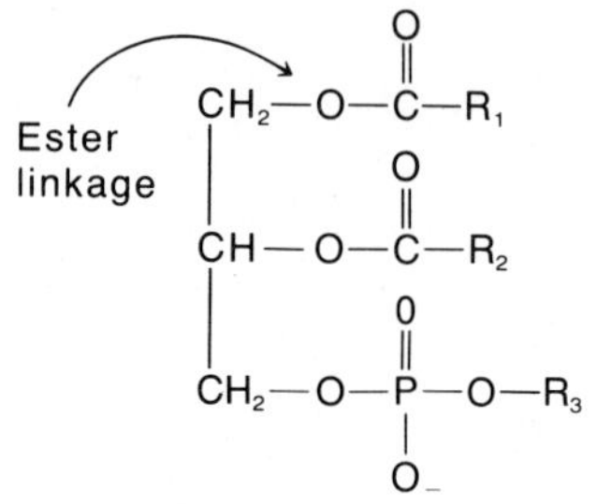

```
                      O
                      ‖
 Ester      CH2—O—C—R1
 linkage     |        O
             |        ‖
            CH—O—C—R2
             |        O
             |        ‖
            CH2—O—P—O—R3
                      |
                      O−
```

Thus, as the temperature of the environment is increased, the organism produces lipids that have higher melting temperatures.

Perhaps an extreme example of membrane modification is demonstrated by the organism *Thermoplasma acidophilum*. The normal habitat for this organism is coal refuse piles, which have temperatures ranging from 32 to 80° C and pH values from 1.2 to 5.2. The organism must be stable to both high temperatures and acids to survive. One reason for its stability appears to involve its membrane content of an unusual lipid shown in Figure 2.10. The structure is a tetraether, with two glycerol units cross-linked by two saturated C_{40} components that are based on the isoprenoid unit.

The diglycerol tetraether has three features that may contribute toward membrane stability. One, the absence of ester bonds, normally found in membrane lipids, renders it stable from acid hydrolysis. Two, the long chain saturated bridges melt only at high temperatures. Three, and perhaps most unique, the molecule has a dimension that enables it to completely span the plasma membrane. Normally, membranes have bilayers of lipids in which the hydrophobic ends reaching from opposite membrane sides interdigitate. The diglycerol tetraether provides a monolayer rather than a bilayer, which limits the lateral movement of lipids independently of one another on the two membrane surfaces. Apparently this strategy stabilizes the membrane in the presence of both heat and acid.

2.5 CELL-CREATED ENVIRONMENTS

The majority of cells in multicellular organisms have fixed positions relative to one another. Circulating red and white blood cells and wandering macrophages are examples of cells that are free from these constraints, but most others are cemented into a matrix that they themselves have created.

The amount of matrix between adjacent cells varies with different tissue and organ types. For example, the matrix width between cells in epithelial tissue is minimal. There are even points of intercellular membrane fusion, eliminating the extracellular environment completely.

In cartilage, cells are widely spaced in a matrix of much greater volume than that occupied by the cells.

In the material that follows we review briefly the chemical nature of some extracellular matrices, emphasizing the collagen-type matrix, which has been extensively studied.

Collagen

The most ubiquitous and abundant substance in matrix material is collagen. More than 70% of the dry weight of tendon, skin, and cornea is collagen. This protein is actually a family of gene products rather than a single type of protein molecule, so it is appropriate to use the plural, collagens, when discussing these systems.

The native collagen molecule consists of three polypeptide chains wound to form a triple helix. Figure 2.11 contains a model of the basic collagen molecule. At least five types of collagen molecule are recognized, each

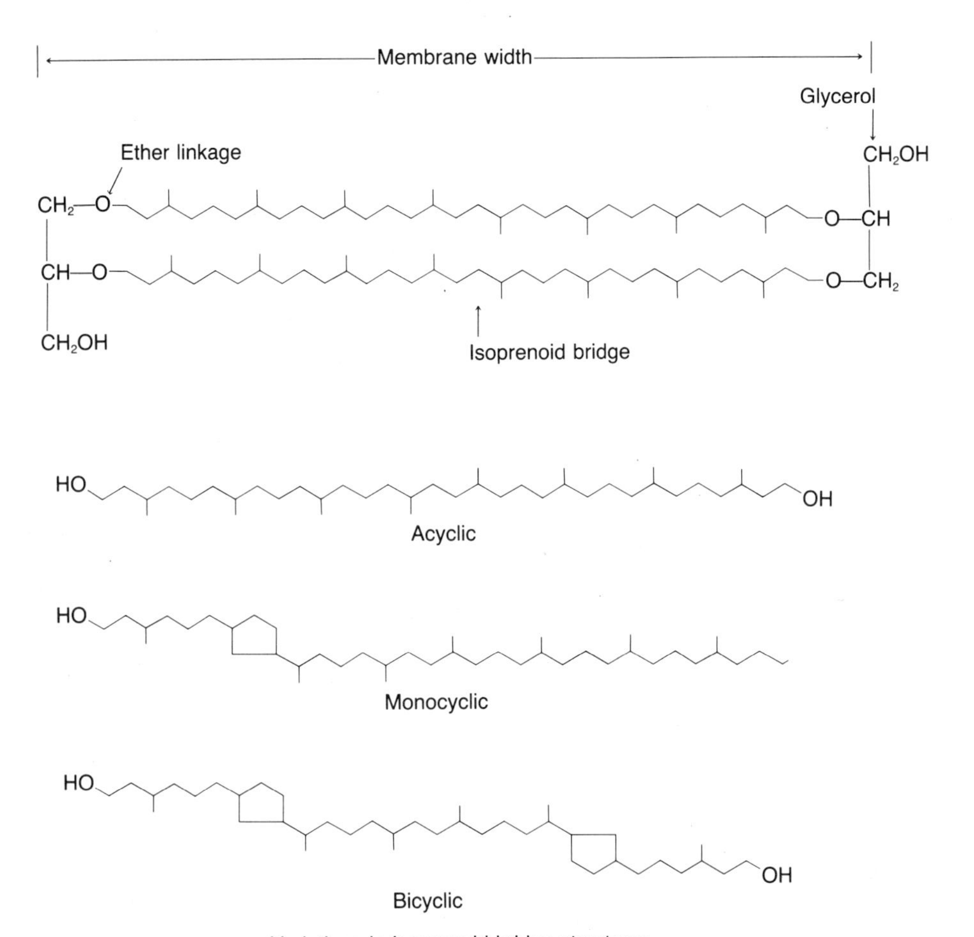

Variations in isoprenoid bridge structures

FIGURE 2.10 Unusual lipids in thermophilic organisms like *Thermoplasma.* These lipids in their membranes may account for both their thermostability and resistance to acid hydrolysis. Two glycerols are linked by isoprenoid polymers, which may appear in either acyclic or cyclic forms. The molecule spans the plasma membrane and the ether bonds render it stable in acidic environments.

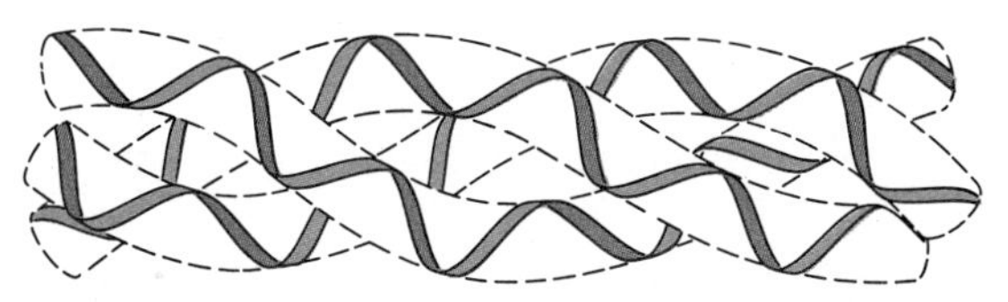

FIGURE 2.11 Model of collagen. Three left-handed polypeptide helices (solid) are intertwined to form a right-handed triple super helix.

Table 2.6 The Collagen Isotypes

Collagen type	Composition	Some Distinguishing Characteristics	Tissue Location	Cell Types that Synthesize Various Collagen Types
I	$[\alpha 1(I)]_2\alpha 2(I)$	Low carbohydrate, low hydroxylation of lysine	Skin, bone, tendon, cornea	Fibroblasts, osteoblasts, smooth muscle cells, epithelium
II	$[\alpha 1(II)]_3$	$>$10 hydroxylysines per chain	Cartilage, cornea, vitreous body	Chondrocytes, neural retinal cells, notochord cells
III	$[\alpha 1(III)]_3$	Contains cysteine, low hydroxylation of lysine	Fetal skin, blood vessels, organs	Fibroblasts, myoblasts
IV	$[\alpha 1(IV)]_3$ $[\alpha 2(IV)]_3$	Many lysines hydroxylated and glycosylated, sugars other than glucose and galactose, low alanine, high 3-hydroxyproline	Basement membrane	Endothelial and epithelial cells
V	$[\alpha 1(V)]_2\alpha 2(V)$ (formerly αA- and αB chains)	Elevated hydroxylysine, low alanine, contains 3-hydroxyproline	Blood vessels, smooth muscle	Smooth muscle cells, chondrocytes under certain conditions

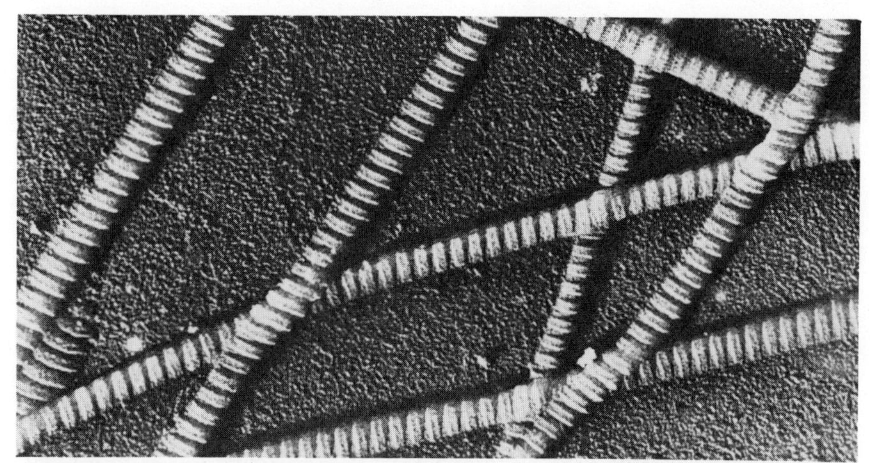

FIGURE 2.12 Electron micrograph demonstrating the classic 640-Å repeat unit of collagen fibrils. The basic structural unit of the fibril is the rod-shaped tropocollagen molecule, which is 3000 Å long and 15 Å in diameter. Tropocollagens are quarter-staggered with a 400-Å gap between their ends. (Courtesy of Dr. Jerome Gross.)

made up of a different combination of α -chains (see Table 2.6). Type I, for example, consists of two $\alpha 1$(I) chains and one $\alpha 2$(I) chain. This type of collagen appears in some of the harder tissues, such as bone and tendon. A second type of $\alpha 1$ chain, designated II, is found in triplicate in the collagens of cartilages.

Table 2.6 lists five genetically distinct α -chains in various types of collagen molecules. Although there is some degree of tissue specificity for the collagen types, a given tissue may synthesize more than one type of collagen.

Method I

Electron microscopic studies on the collagens reveal them to be striated fibers with repeat distances of 640 Å. The classic fibril ultrastructure is shown in Figure 2.12. This morphology is best explained by a staggered arrangement of end-to-end tropocollagen molecules, each tropocollagen being the elemental unit of the intact collagen fiber.

The amino acid composition of the collagens is quite unusual compared to other proteins. The α chain, for example, contains about 1050 amino acids, with glycine contributing about one third of the total residues. The primary structure is thus $(\text{—Gly—X—Y})_n$, where X and Y may be filled by a variety of amino acids. Most commonly, however, the X-position is taken by proline and the Y-position by hydroxyproline.

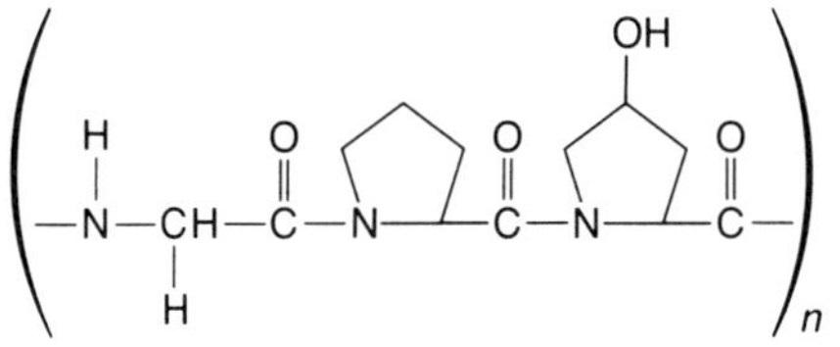

Collagens are synthesized by cells in precursor forms termed *procollagens*. The mechanism of cell synthesis and secretion follows the same pathway as that utilized for any other secretory protein, namely, synthesis on ribosomes of the rough endoplasmic reticulum followed by transport through the Golgi complex and exocytosis from the cell (see Chapter 10 for a more detailed discussion of the secretory pathway).

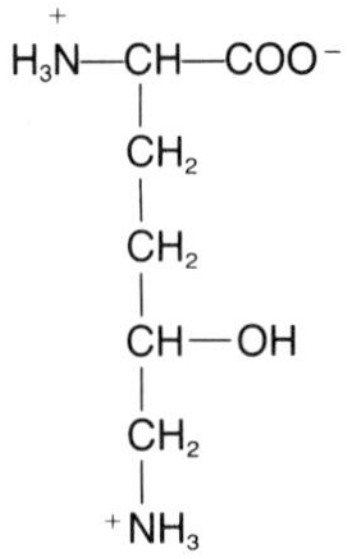

The procollagen molecule, when secreted, is already a hydroxylated and glycosylated triple helix molecule. Proline and lysine are both hydroxylated, with about 10% of the proline and 0.5 to 4% of the lysine in hydroxylated form. The hydroxylation reactions proceed by specific hydroxylase enzymes operating on the growing polypeptide chains as they are being formed on the membrane-bound ribosomes (see Figure 2.13). Carbohydrates are covalently attached to the collagens through the side chains of hydroxylysine. As shown in Figure 2.14, glucose and galactose are commonly found as a disaccharide. Glycosylation takes place in the cell at the level of the nascent peptides by UDP-galactosyl transferase and UDP-glucosyl transferase enzymes.

Thus there are four intracellular enzymes that effect a posttranslational modification of the collagens, two concerned with hydroxylations and two with glycosylations.

The assembly of the triple-helical configuration also takes place before secretion. It is now thought that the pairing of the α chains is promoted by extensions called *registration peptides*, which are eventually excised from the procollagen before the collagen fiber is formed. The "registration" stretches of the chains are depicted at both their amino- and carboxy-terminal ends in Figure 2.15. Concomitant with triple helix formation, disulfide bonds are formed between the chains, shown here toward the *C*-terminal ends of the chains. The helix is thereby stabilized by covalent linkages and in this form procollagen is secreted from the cell.

Figure 2.16 depicts the stepwise processing of procollagen that takes place in the extracellular matrix, after the molecule has been secreted from the cell. The amino-terminal registration peptides are cleaved off first, leaving a processed procollagen that still contains the carboxy-terminal stretches cross-linked by disulfide bonds. These stretches are subsequently excised to generate the *tropocollagen molecule*. The collagen fibril is then formed by a self-assembly process in which tropocollagen molecules aggregate. The fibril is finally strengthened by covalent linkages employing oxidized forms of lysine and hydroxylysine that condense.

Two adjacent tropocollagen molecules may be cross-linked by an aldol condensation between their oxidized lysines. Another molecule may be tied to the complex by forming a Schiff base between the free aldehyde and an amino group of lysine. Still another tropocollagen can be bound by a reaction between an imidazole group of histidine and the unsaturated carbon–carbon bond.

The repeating triad of collagen:

$$(\text{—Gly—X—Y—})_n$$

Hydroxylysine:

Monosaccharides, such as glucose and galactose, that are destined to be incorporated into polymers are generally activated and carried by UDP.

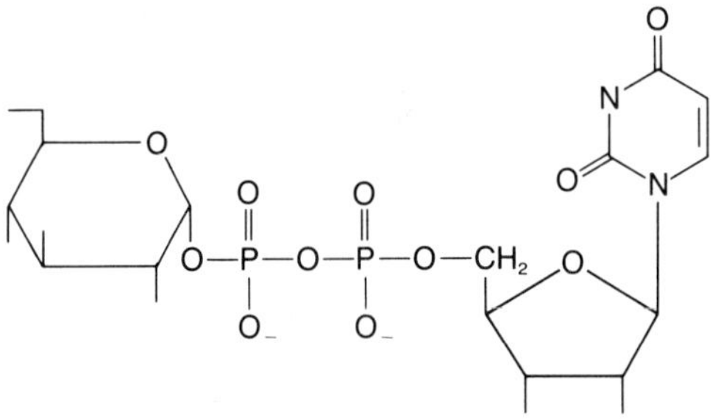

UDP-glucose

A copper-dependent enzyme, lysyl oxidase, oxidatively deaminates lysine or hydroxylysine residues. The aldehydes generated by this enzyme undergo a variety of reactions with other aldehydes, with lysine and its derivatives, or even with histidine.

lysyl oxidase

aldol condensation

Histidine

Lysine

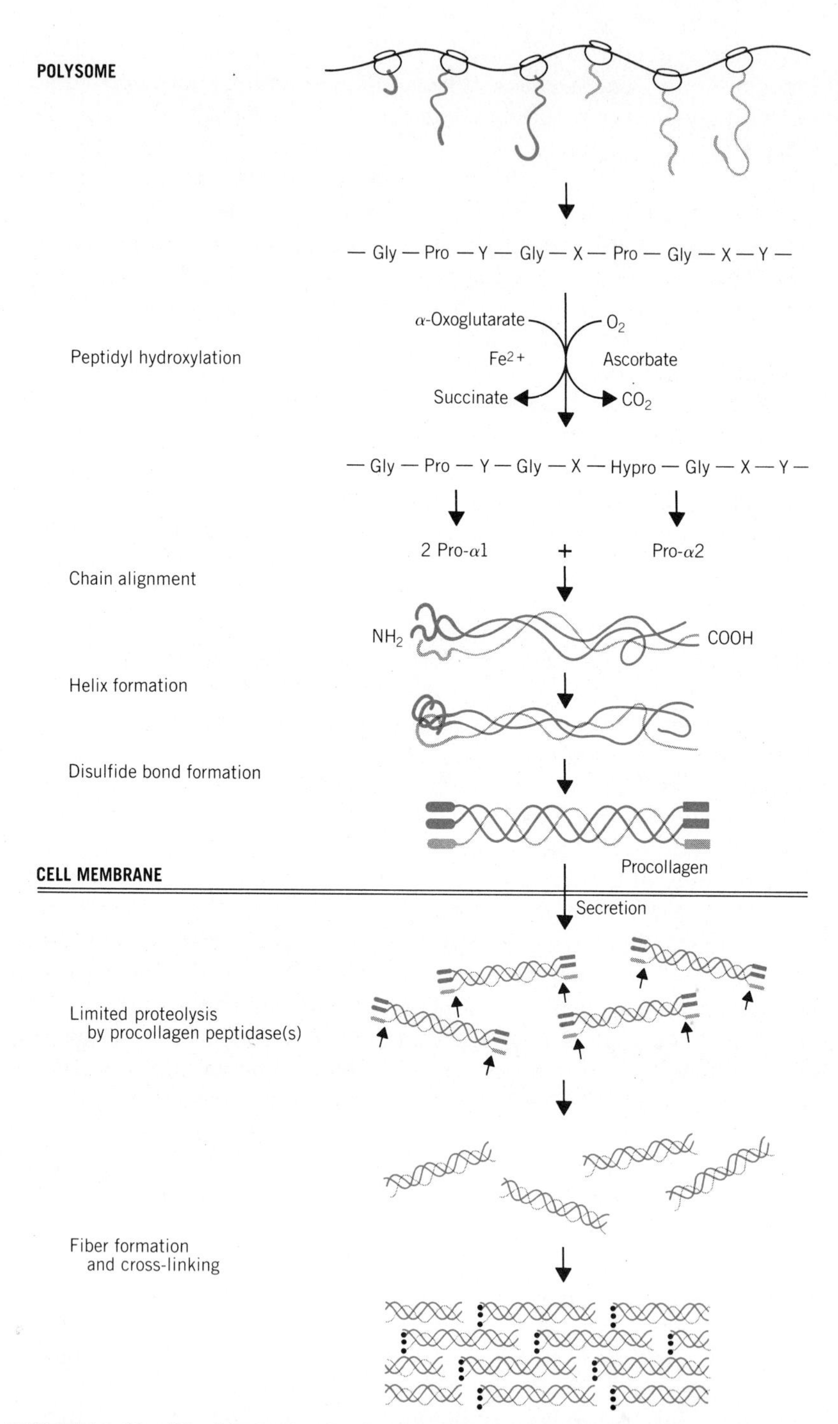

FIGURE 2.13 The biosynthesis of procollagen and its conversion to a collagen fibril. Hydroxylation takes place while the chains are being formed on the polysome. Procollagen is secreted from the cell, where it is enzymatically modified and condensed into fibril form.

One or more of these reactions may take place to increase the tensile strength of the collagen fiber. Although the number and variety of cross-linkages varies with the type of collagen, the linkages are usually grouped toward the ends of the tropocollagen molecules.

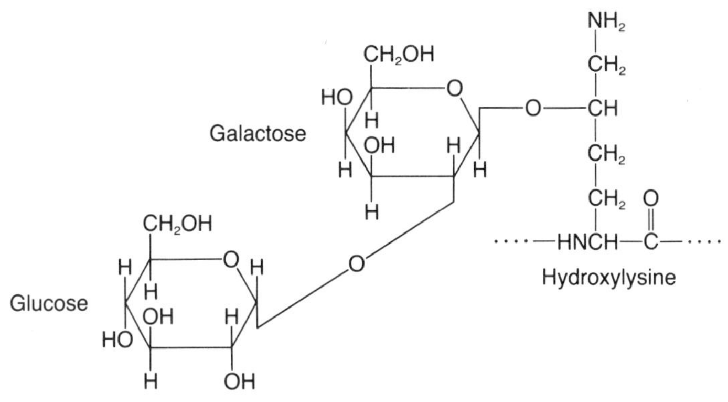

FIGURE 2.14 Glucose and galactose are commonly found as a disaccharide unit attached to collagen by way of the side chain of hydroxylysine. The temporal order during biosynthesis thus requires hydroxylation first, then glycosylation.

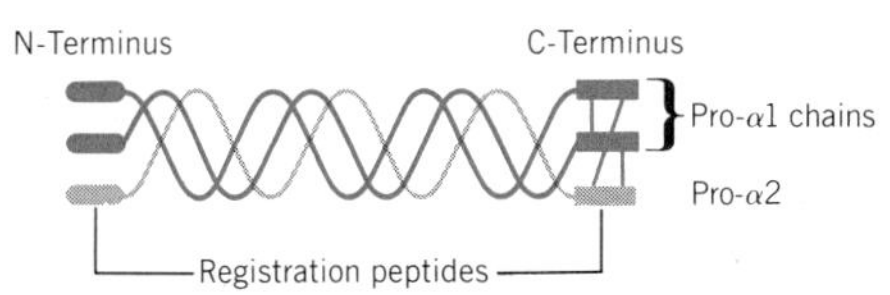

FIGURE 2.15 Schematic representation of type 1 procollagen as secreted from the cell. Both *N*-terminal and *C*-terminal ends contain registration peptides that assist in the proper alignment of $\alpha1$ and $\alpha2$ chains. These peptides are not present in the tropocollagen molecule.

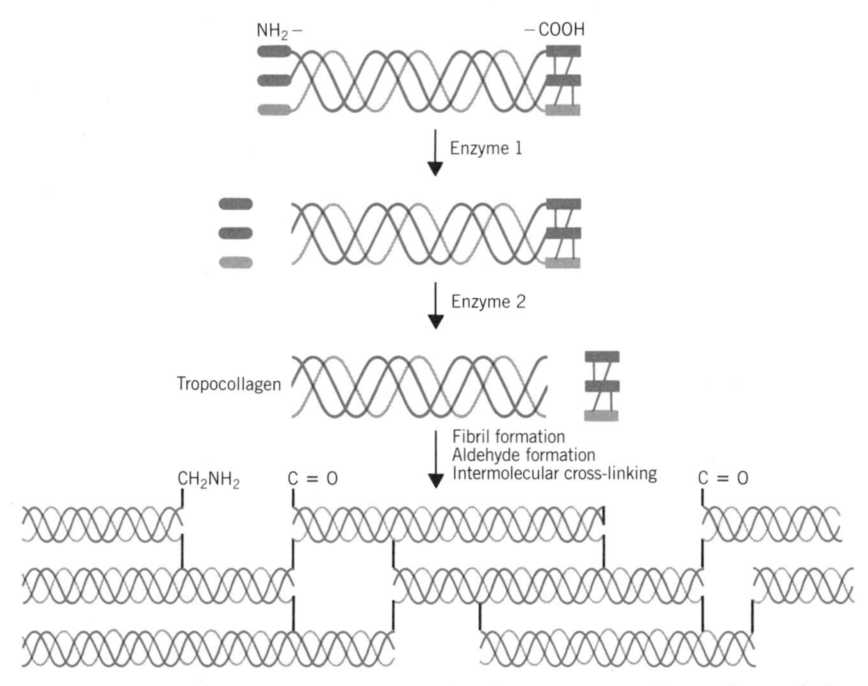

FIGURE 2.16 Summary of events that take place on secreted procollagen during its conversion to a collagen fibril. Registration peptides are enzymatically excised and the resultant tropocollagen molecules are chemically cross-linked to generate a fibril with high tensile strength.

These, then, are the types of molecule that make up a good deal of the material in the extracellular matrix in tissues and organs. The full impact of matrix collagen on the cell is not yet realized, but an extensive amount of research is currently underway to ascertain its effects upon cell properties.

It is known from *in vitro* studies that collagen enhances such cell properties as attachment, growth, and differentiation. It is likely that these cell activities are affected similarly *in vivo*. Furthermore, different types of collagen influence cells differently.

Epidermal cells attach best to type IV collagen, whereas fibroblasts

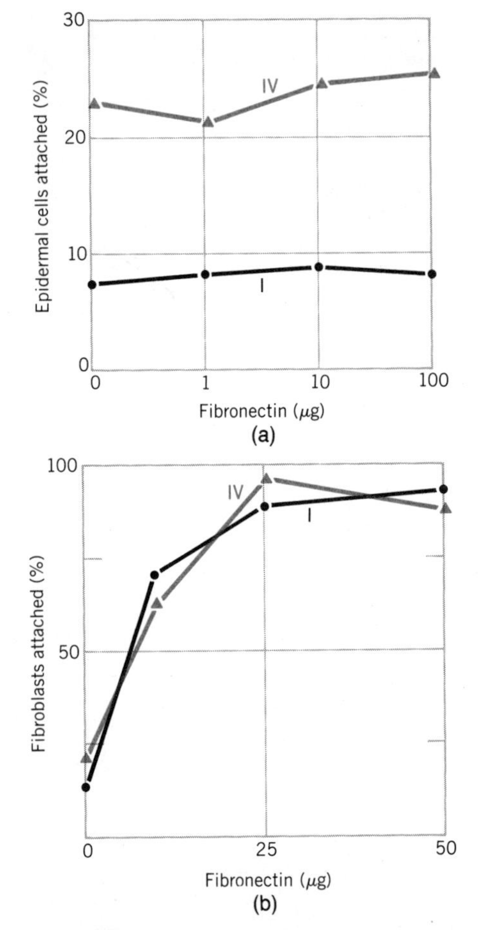

FIGURE 2.17 Effect of fibronectin on the attachment of (*a*) epidermal cells and (*b*) fibroblasts to collagen, types I and IV. This study employed collagen-coated bacteriological dishes to which the attachment of cells could be quantitated. The results show a preference of epidermal cells to type IV collagen over type I; fibroblasts adhere equally to both types. Type IV collagen is that found in basement membranes.

adhere equally to all collagen types.[7] The attachment of fibroblasts to collagen, although not influenced by collagen type, is enhanced by a protein that has been given the name *fibronectin*. This protein apparently acts as a mediator between the cell and the matrix. The attachment of epidermal cells, on the other hand, is not enhanced by fibronectin (see Figure 2.17). It therefore appears that matrix–cell adherence is governed by the type of matrix molecules and by mediators.

These observations lend support to the notion that during development cell localization and orientation may depend on specific interactions with matrix molecules. Epidermis, for example, is made up of multilayered squamous cells, the lower ones associated with the basement membrane, which is a matrix rich in type IV collagen. Experimentally it has been shown that when epidermal cells attach to type IV collagen they differentiate and form multilayered systems. Thus the evidence is becoming stronger that matrix–cell interactions underlie cell growth and differentiation and subsequent tissue formation.

Noncollagen Matrix Molecules

In addition to collagen, there are other molecules prominent in the matrix. Most abundant are giant aggregates referred to as *proteoglycans*. Proteoglycans consist of proteins and glycosaminoglycans (or mucopolysaccharides) linked both covalently and noncovalently to form large amorphous aggregates.

Table 2.7 lists the common names of the glycosaminoglycans present in extracellular matrices. These are high molecular weight molecules that have ordered helical structures according to X-ray diffraction studies on crystalline fiber and film preparations.[8]

The proteoglycan aggregate is presently viewed as having a structure like that depicted in Figure 2.18. The backbone of the aggregate is an extended hyaluronic acid polymer, highly acidic because of its content of carboxyl groups on every other sugar unit. Around this backbone proteoglycan subunits are clustered, held in place by noncovalent associations. The proteoglycan subunit appears to consist of three regions,[9] a globular end that binds noncovalently to the hyaluronic acid, an intermediate length of molecule to which keratan sulfate is covalently attached by way of glutamic acid residues, and an extended region of variable length where both keratan sulfate and chondroitan sulfate are covalently bound. Another protein, called a link protein, is present in these aggregates. Its function appears to be to stabilize the noncovalent association between the proteoglycan subunits and the hyaluronic acid backbone.

Articular cartilage is the cartilage of joints.

In articular cartilage, where the cells are widely separated by large amounts of extracellular matrix, the proteoglycan aggregate makes a very significant contribution to the extracellular material. In this cartilage, the matrix consists of approximately 70% collagen, 18% proteoglycan, and 10% additional noncollagenous protein.

Chondrogenesis: (Greek: *chondros* = cartilage + *genesis* = production), the formation of cartilage.

Proteoglycans interact with collagen molecules to varying degrees, depending on their composition and the composition of chains making up the collagens. An area of research that is now receiving concentrated effort is the relation between these components during development. New evidence is beginning to suggest that these substances and their interactions may have a regulatory role in cytodifferentiation and chondrogenesis.[10]

Table 2.7 Extracellular Glycosaminoglycans

Polysaccharide Name	Chemical Units	Distribution
Chondroitin	Glucuronic acid, *N*-acetylgalactosamine	Cornea
Chondroitin sulfate	Glucuronic acid, *N*-acetylgalactosamine (4 or 6)-sulfate	Cartilage
Dermatan sulfate	Iduronic acid, *N*-acetylgalactosamine 4-sulfate	Skin
Heparin	Glucosamine 6-sulfate, glucuronic acid-2-sulfate, iduronic acid	Lung, liver
Hyaluronate	Glucuronic acid, *N*-acetylglucosamine	Synovial fluid, intercellular cement
Keratan sulfate	Galactose, galactose 6-sulfate, *N*-acetylglucosamine 6-sulfate	Cornea, skin

Repeating unit of hyaluronic acid:

COO^- CH_2OH O O NH C=O CH_3 n

Glucuronic acid N-acetyl-glucosamine

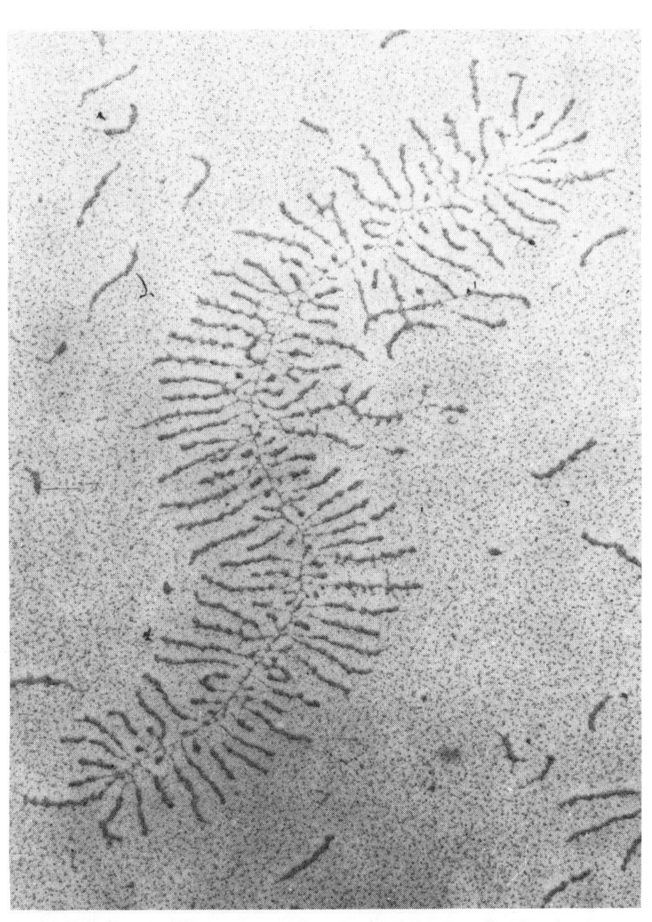

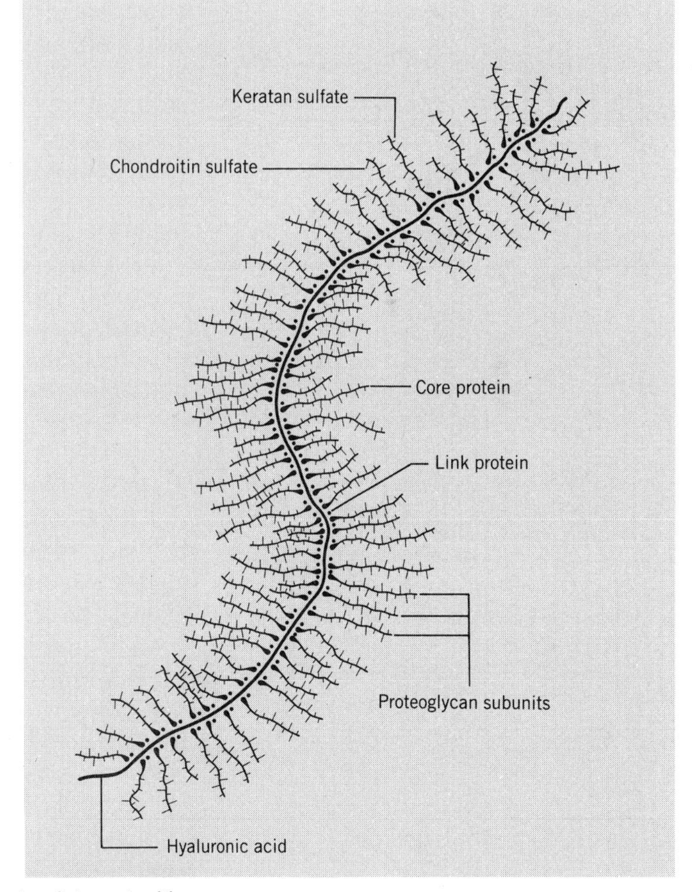

FIGURE 2.18 The proteoglycan aggregate. Such giant aggregates interact with collagens in the extracellular matrix. (Electron micrograph courtesy of Dr. Joseph A. Buckwalter.)

Basement Membranes

Basement membranes are not membranes at all. They are rather a specialized form of connective tissue made up of extracellular matrices rich in type IV collagen, not in any way related chemically or structurally to the plasma membrane.

Table 2.8 Sites of Basement Membranes

Membranes Adjacent to
Endothelial cell layer of blood vessels and capillaries.
Epithelial layers of skin viscera and glands.
Specialized tissues of the eye:
Lens capsule epithelium
Bowman's epithelium of the cornea
Endothelium of Descemet's membrane
Embryonic layers of the parietal and visceral yolk sac

Table 2.8 lists the basement membrane sites, which are widely distributed in most of the tissues and organs of the body.

It is now understood that basement membranes are formed by the cells with which they are located. For example, the glomerular basement membrane is made by the epithelial cell, and so is the lens capsule and basement membranes in the vicinity of respiratory, glandular and intestinal epithelium.[11] In like manner, Descemet's membrane is produced by endothelial cells and Reichert's membrane of the embryonic yolk sac by endodermal cells.

A glomerulus is a tuft or cluster, in this case used to describe the cluster of blood vessels in the kidney.

Descemet's (des -ĕ- māź) membrane is the posterior limiting layer of the cornea. It is named after Jean Descemet, a French anatomist (1732–1810). Reichert's membrane, named after the German anatomist Karl B. Reichert (1811–1884), is the anterior limiting layer of the cornea.

Basement membranes have served well as model systems for the study of the extracellular environment. They can be isolated, with those from the glomerulus and the lens more frequently isolated and studied than the others. Their amino acid composition and carbohydrate content have been explored, and certain physical properties, such as solubility, have been examined.

From these studies it can be concluded that basement membranes contain only peptide and carbohydrate. Basement membranes from different sources are qualitatively similar but quantitatively distinct, as can be seen in Table 2.9. There is characteristically a high content of proline and hydroxyproline, glycine, and hydroxylysine.

The carbohydrate content of these extracellular matrices is high, ranging between 11 and 14%. Hexoses are predominant, with the presence of glucose setting basement membranes apart from most other tissue and plasma proteins. Table 2.10 shows some characteristic carbohydrate compositions.

One of the proteins of basement membranes is clearly a type of collagen that contains glucose and galactose almost exclusively, which together form glucosylgalactose disaccharides distributed on the α chain.

In addition to the collagen, a high molecular weight glycoprotein is present in basement membranes.

Based on the studies above, it has also been concluded that basement membranes probably have at least two functions. One function is to provide an elastic form of support to epithelial and endothelial cell linings. It is a medium on which these cells are anchored, thereby enabling them to congregate into tissue form. A second function appears to be that of filtration. Molecules that move across the epithelial surface into the blood and ultimately across the capillary walls must make their way through basement membrane material. When considering the polymeric nature of the molecules in the basement membrane, one could anticipate that a certain amount of barrier activity for certain kinds of molecule would be inevitable.

Table 2.9a Amino Acid Composition of Anterior Lens Capsule (residues/1000 residues)

Amino Acid	Human	Canine	Bovine	Ovine	Chick
→Hydroxylysine	34.5	32.0	35.0	39.0	20.61
Lysine	19.4	11.0	13.2	12.3	17.55
Histidine	15.2	12.0	10.2	11.6	11.83
Arginine	39.5	38.3	43.0	37.0	47.70
→3-Hydroxyproline	21.3	19.5	15.0	19.0	12.43
→4-Hydroxyproline	85.0	80.0	85.0	104.0	81.73
Aspartic	57.0	57.8	55.0	55.0	64.41
Threonine	31.0	31.0	29.0	29.0	40.40
Serine	43.4	48.0	42.0	44.0	49.31
Glutamic	94.4	95.0	92.5	92.0	90.01
→Proline	67.3	66.0	68.0	68.8	71.99
→Glycine	260.0	288.0	275.0	266.0	223.95
Alanine	40.6	46.0	42.8	44.7	60.89
½-Cystine	21.0	18.0	28.0	12.0	29.92
Valine	33.2	35.0	30.0	30.0	42.20
Methionine	5.0	6.2	8.0	5.0	7.12
Isoleucine	32.0	30.0	28.8	26.0	28.21
Leucine	57.7	56.5	58.0	62.0	60.26
Tyrosine	13.0	11.0	10.0	11.0	12.13
Phenylalanine	29.8	29.0	32.0	28.0	27.26

Table 2.9b Amino Acid Composition of Glomerular Basement Membranes (residues/1000 residues)

Amino Acid	Human		Canine	Ovine	Rat	Reichert's Membrane, Rat
→Hydroxylysine	24.5	26.1	25.0	22.5	21.8	18.9
Lysine	26.4	19.5	26.0	27.0	40.0	32.3
Histidine	18.7	14.2	14.4	18.5	21.0	18.3
Arginine	48.3	38.0	48.2	50.5	52.6	46.0
→3-Hydroxyproline	12.0	22.2	8.5	7.0	7.0	6.6
→4-Hydroxyproline	53.0	81.3	56.5	56.0	50.0	45.3
Aspartic	70.0	67.7	70.0	72.0	70.0	83.0
Threonine	40.3	36.7	40.5	41.5	45.5	46.0
Serine	54.2	48.9	49.0	54.5	62.8	64.0
Glutamic	101.3	98.3	97.0	98.0	100.0	111.0
→Proline	64.1	57.9	69.8	66.0	62.0	62.2
→Glycine	225.2	221.0	229.0	220.0	200.0	180.0
Alanine	58.6	59.8	65.0	60.0	67.0	64.0
½-Cystine	22.0	22.8	22.7	20.0	20.4	19.2
Valine	36.0	35.6	36.0	38.0	43.0	44.0
Methionine	7.0	13.2	5.0	7.5	11.0	11.8
Isoleucine	28.6	31.3	28.1	29.5	30.0	32.0
Leucine	60.3	65.3	60.2	59.0	60.0	73.0
Tyrosine	20.5	15.3	22.0	23.0	17.0	17.5
Phenylalanine	28.3	24.9	26.8	28.0	19.0	31.0

Basement membranes from different sources and organisms, although qualitatively similar, are quantitatively distinct. For example, the content of leucine exceeds phenylalanine, which exceeds tyrosine in every case shown. However, each shows slight quantitative variations from all others. Note in particular the high levels of proline, hydroxyproline, glycine, and hydroxylysine (→).

Table 2.10 Carbohydrate Composition of the Collagens Isolated From Canine Basement Membranes and Tendon (g/100 g)

Carbohydrate	Glomerulus	Lens Capsule	Descemet's Membrane	Tendon
Hexose	10.4	10.4	10.0	0.60
Glucose	5.1	5.0	5.0	0.28
Galactose	5.25	5.2	5.2	0.40
Mannose	0.25	0.2	0.2	0.00
Hexosamine	0.15	0.1	0.1	0.00

The carbohydrate content of collagen isolated from three different types of basement membrane is contrasted with the collagen of tendon. About 10% of the dry weight is hexose, with glucose and galactose obviously the predominant monosaccharides. Except for collagen isolated from vitreous humor of the eye, no other collagen contains this high hexose level.

2.6 CELLULAR–EXTRACELLULAR INTERACTIONS

It should be quite obvious that cells do not function independently of their environments. The full impact of the environment on the cell is still somewhat beyond our reach, but some principles of interaction between the two are becoming clearer.

It is clear that cells do not tolerate an extracellular environment of air and must be protected from it. The interaction is thus one primarily of defense. The desiccating effects of air and the radiations of the atmosphere have severely limited the habitat for unicellular forms of life. These environmental factors have encouraged the generation of multicellular systems in which the exterior cells take on the role of protection for the rest. Cell wall structures have also been promoted in some organisms to protect them from the harshness of the air.

As vital as water is to life, it is also clear that environmental water can pose a problem for cells. An environment of fresh water requires defensive tactics to keep cells from bursting as a result of high internal osmotic pressures. The defense most frequently used is the cell wall, which is able to sustain extremely high internal pressures. In exchange for this protection, the cell has sacrificed flexibility and often freedom of movement.

Water at high pressures has given the cell another set of problems. The interaction in this case is between water and proteins, both those that have a structural role in the membrane and in microtubules, and those that have enzymatic functions. Protein conformation is changed at high hydrostatic pressures, depressing catalytic activities and inhibiting polymerization reactions. Some cells have successfully adapted to these conditions by developing and using proteins that are less sensitive to pressure denaturation and in some cases are even activated enzymatically by high pressures. This is another type of defensive tactic that enables survival.

The salinity of seawater is usually significantly higher than that of cells and is of such a composition that cells must make special compensations to survive. The effect of this extracellular environment is thus both dehydration and enzyme inhibition. This has required cells to modify their membranes so that water can be sequestered from the environment, and halotolerant isoenzymes are employed. This is yet another defensive ploy to cope with the environment.

Cells have taken offensive action toward their environments by the creation of their own extracellular matrices. Along with this comes multicellularity. The synthesis of matrix not only facilitates intercellular adherence but creates a cell community wherein cells are assigned specialized tasks.

In multicellular organisms the destructive effects of air are thwarted by surface cells that are antidesiccant and able to protect against radiation damage. The aqueous environment in which the cells are bathed is cell generated with optimal levels of ions of an appropriate composition. As long as there is an external source of vital materials, the extracellular environment is maintained. Armed with these offensive weapons, multicellular organisms can survive and indeed reproduce in environments that are strongly conducive to cell death rather than life.

The extracellular matrix created by cells is not nearly as static as it was once assumed to be. As suggested by the introductory comments in this chapter, the matrix does affect cell behavior. One effect that is coming into view through extensive research currently being conducted by many groups is a regulatory effect at the level of gene expression. No general scheme is recognized yet as a mechanism of control, yet it is quite apparent from the data that matrix molecules do not go unnoticed by the cell. In many instances, such as during cytodifferentiation, the destiny of the cell ultimately depends on its extracellular matrix.

Summary

The presence of cellular life throughout the diverse environments of the biosphere indicates that cells have a tremendous ability to adapt. Procaryotic cells in particular have adapted to extremes of temperature, pH, pressure, and salt concentration.

The environment makes an impact on the structure and activities of cells, especially on their membranes, polymer molecules, and enzyme activities.

Air is a harsh environment, bringing about desiccation and transmitting lethal radiation.

Fresh water, being hypotonic to cells, causes elevated internal osmotic pressures. Seawater is normally dehydrating. High hydrostatic pressures generally inhibit cell replication and metabolism by depolymerizing microfilaments and depressing enzyme activities. Cells exist at high temperatures only if they possess thermostable enzymes and membranes.

In tissues, cells create their own extracellular environments by synthesizing and secreting materials that are largely protein and polysaccharide. These materials assist cells in adhering to one another and protect cells from desiccation and osmotic pressure problems while keeping them bathed in the proper environment for optimal activity.

References

1. *Thermophilic Microorganisms and Life at High Temperatures,* T.D. Brock, Springer-Verlag, New York, 1978.
2. Effects of Deep-Sea Pressures on Microbial Enzyme Systems, C.E. Zobell and J. Kim, in *The Effects of Pressure on Organisms,* Symposia of the Society for Experimental Biology, No. XXVI, Academic Press, New York, 1972.

3. The Adaptation of Enzymes to Pressure in Abyssal and Midwater Fishes, P.W. Hochachka, T.W. Moon, and T. Mustafa, in *The Effects of Pressure on Organisms,* Symposia of the Society for Experimental Biology, No. XXVI, Academic Press, New York, 1972.
4. Effect of Pressure on Biopolymers and Model Systems, K. Suzuki and Y. Taniguchi, in *The Effects of Pressure on Organisms,* Symposia of the Society for Experimental Biology, No. XXVI, Academic Press, New York, 1972.
5. Effects of Pressure on the Structure of Water in Various Aqueous Systems, W. Drost-Hansen, in *The Effects of Pressure on Organisms,* Symposia of the Society for Experimental Biology, No. XXVI, Academic Press, New York, 1972.
6. The Effects of Hydrostatic Pressure on Cell Membranes, S.B. Zimmerman and A.M. Zimmerman, in *Mammalian Cell Membranes,* vol. 5, G.A. Jamieson and D.M. Robinson, eds., Butterworths, London, 1977.
7. Epidermal Cells Adhere Preferentially to Type IV (Basement Membrane) Collagen, J.C. Murray, G. Stingl, H.K. Kleinman, G.R. Martin, and S.I. Katz, *J. Cell Biol.* 80:197(1979).
8. Glycosaminoglycan Conformations, S. Arnott, J.M. Guss, and W.T. Winter, in *Extracellular Matrix Influences on Gene Expression,* H.C. Slavkin and R.C. Greulich, eds., Academic Press, New York, 1975.
9. Organization of Extracellular Matrix in Bovine Articular Cartilages, L. Rosenberg, R. Margolis, C. Wolfenstein-Todel, S. Pal, and W. Strider, in *Extracellular Matrix Influences on Gene Expression,* H.C. Slavkin and R.C. Greulich, eds., Academic Press, New York, 1975.
10. Proteoglycan–Collagen Interaction: Possible Developmental Significance, B.P. Toole and T.F. Linsenmayer, in *Extracellular Matrix Influences on Gene Expression,* H.C. Slavkin and R.C. Greulich, eds., Academic Press, New York, 1975.
11. Basement Membranes, N.A. Kefalides, in *Mammalian Cell Membranes,* vol. 2, G.A. Jamieson and D.M. Robinson, eds., Butterworths, London, 1977.

Selected Books and Articles

Books

Cell Biology of Extracellular Matrix, E.D. Hay, ed., Plenum Press, New York, 1981.

Extracellular Matrix Influences on Gene Expression, H.C. Slavkin and R.C. Greulich, eds., Academic Press, New York, 1975.

Mammalian Cell Membranes, vol. 5, G.A. Jamieson and D.M. Robinson, eds., Butterworths, London, 1977.

The Effects of Pressure on Organisms, Symposia of the Society for Experimental Biology, No. XXVI, Academic Press, New York, 1972.

Thermophilic Microorganisms and Life at High Tempratures, T.D. Brock, Springer-Verlag, New York, 1978.

Articles

Biosynthesis of Procollagen, J.H. Fessler and L.I. Fessler, *Annu. Rev. Biochem.* 47:129(1978).

High-Pressure Microbial Physiology, R.E. Marquis, *Adv. Microb. Physiol.* 14:159(1976).

Microbial Life in the Deep Sea, H.W. Jannasch and C.O. Wirsen, *Sci. Am.* 236:42(1977).

3

The Cell Wall

This is the forest primeval. The murmuring
pines and the hemlocks,
Bearded with moss, and in garments green,
indistinct in the twilight,
Stand like Druids of eld, with voices
sad and prophetic,
Stand like harper's hoar, with beards
that rest on their bosoms.
Loud from its rocky caverns, the deep-voiced
neighboring ocean
Speaks, and in accents disconsolate answers the
wail of the forest.

HENRY WADSWORTH LONGFELLOW, 1847

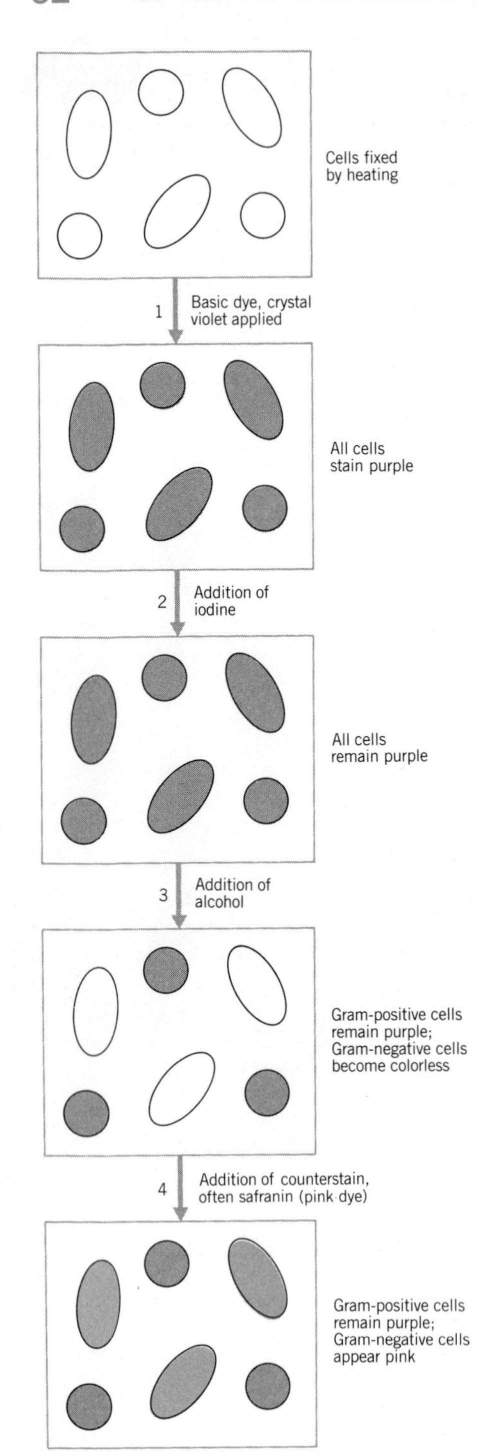

FIGURE 3.1 Steps in the Gram staining procedure.

The familiar lines from the tragic story of Evangeline and Gabriel quoted at the beginning of this chapter may seem altogether out of place in a textbook that deals primarily with the stark scientific realities in the life of a cell. Indeed, for the student with interests only for the cell and its ultrastructure, if such a person exists, this is subjective and sentimental verbiage that cannot be scrutinized with the tools of science. But perhaps these poetic lines can help us keep our study in proper perspective.

There is an important connection between the pleasant beauty of the forest primeval and the topic of this chapter, for it is only by virtue of the support given by cell walls that the murmuring pines and the hemlocks rise to their heights of grandeur.

Imagine for a moment the landscape without the cell wall. Aerial life would not exist. All plant forms would be restricted to a height of only a few centimeters. The giant redwood without a cell wall would be an enormous slug of protoplasm spread out on the mountain side, always creeping toward the valley. All other plants would be similarly amorphous. A walk through the forest would require pulling on hip boots or slipping into a new generation of air boat.

Of course, plant life without walls would dictate a completely different form for most all of the biosphere. For the poet, drinking in the landscape, if one could survive without sinking out of sight, there would be little of aerial beauty that pleases the eye and excites the imagination. And certainly, *Evangeline,* in its familiar form, would never have been written. We owe a good deal to cell walls.

3.1 THE WALLS OF PROCARYOTIC CELLS

Not all organisms that possess walls lend significant aesthetic qualities to the landscape. In fact, those possessing *procaryotic* cells are more readily identified with water pollution and spoilage.

There are two major classes of procaryotic cells: bacteria and blue-green algae (cyanobacteria). The general features of procaryotic cells have been discussed in Chapter 1. Since the cell walls of bacteria have been investigated much more intensely than those of the blue-green algae, our discussion centers on bacteria.

Blue-green algae are also called cyanobacteria. Except for being photosynthetic, they have many structural features in common with Gram-negative bacteria.

All bacteria possess cell walls except for two groups, the mycoplasmas and the L-forms. Neither group appears to have a major role in nature, and compared to other bacteria neither has been the center of attention for most microbiologists.

Therefore, for the discussion that follows we will view bacteria as cells encased by rigid walls.

The Gram Stain: A Detector of Wall Differences

Almost all bacteria can be placed in one of two categories depending on how they react to a staining procedure named after its originator, Hans Christian Gram (1853–1938). The steps of this procedure are outlined in Figure 3.1.

All bacteria take up the dye crystal violet and are intensely purple when emerging from the first step. Iodine functions as a mordant in the next step to fix the crystal violet more firmly within the cell.

The crucial step is step 3. When alcohol is flooded across a smear of bacteria after the first two steps, the crystal violet is extracted out of the cells of some species while other species retain the dye. The result is therefore either colorless cells or cells that remain intensely purple. Routinely, a counterstain of different color is then added to the smear. Safranin, for example, stains colorless cells pink but has no effect on cells that are already saturated with dye.

The molecular basis for the operation of the Gram stain is still controversial. Iodine appears to form a complex or precipitate with crystal violet within the cell and/or in conjunction with the wall. Alcohol dissolves the complex and removes it in Gram-negative bacteria.

The final result of the Gram stain is cells that are either purple or pink. Those that are purple because they retain the original dye are called *Gram positive*. Those that are pink are *Gram negative*.

The cellular or molecular basis by which the Gram reaction distinguishes bacterial types has been a point of controversy for many years. Several theories have been offered, but one that has received broad acceptance suggests that differences in the chemical composition and structure of the cell wall are ultimately responsible for the different results. We discuss these differences at length in this chapter.

Electron Microscopy of Bacterial Cell Walls

When bacteria are sectioned and stained, the cell wall is seen as a pronounced electron-dense layer outside the plasma membrane. This is apparent in the electron micrographs of Figure 3.2, which shows a difference in wall thickness and complexity between Gram-positive and Gram-negative bacteria. Gram-positive bacteria are surrounded by a single electron-dense layer, whereas the material encasing Gram-negative bacteria is thicker

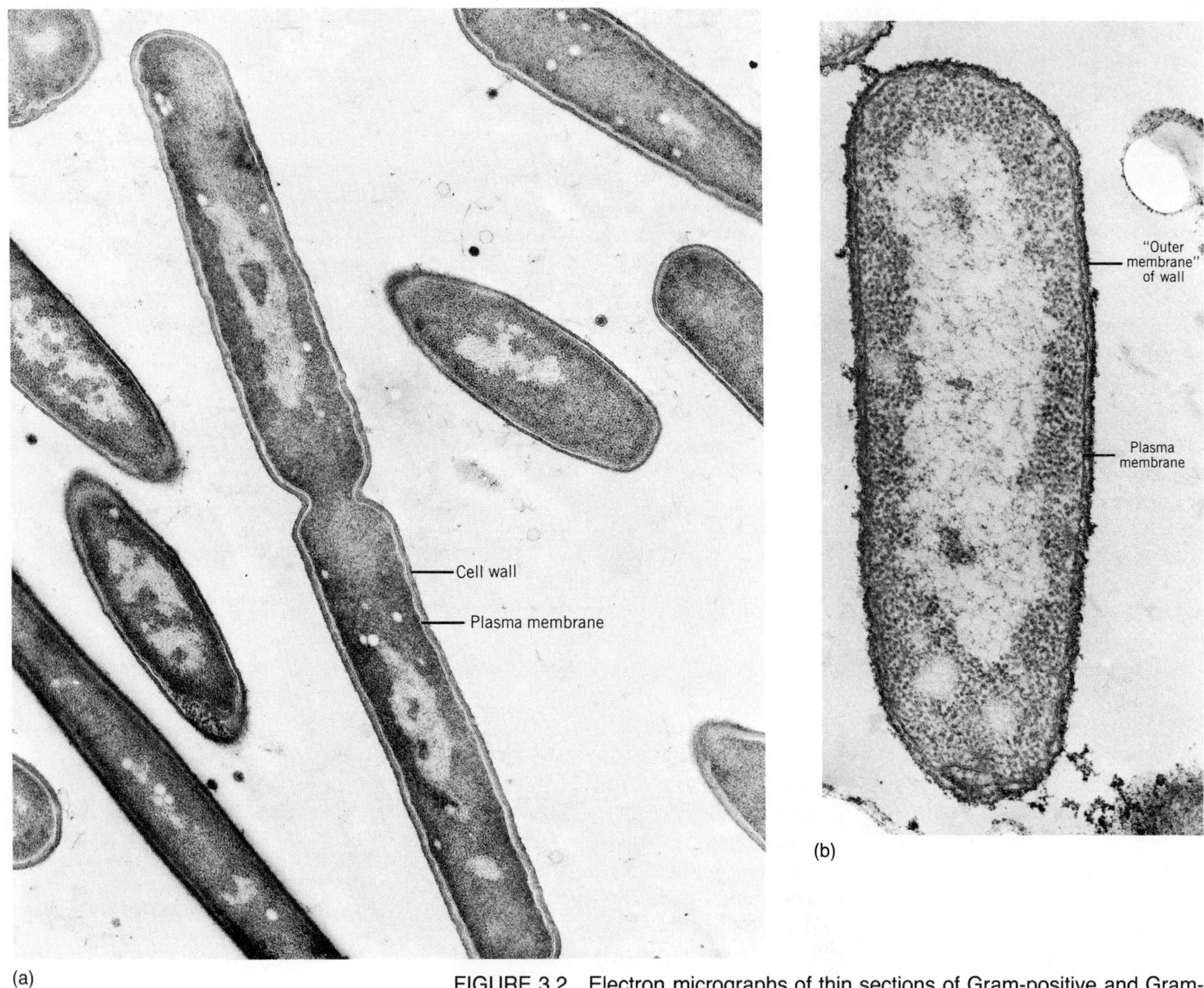

FIGURE 3.2 Electron micrographs of thin sections of Gram-positive and Gram-negative bacteria. The cell walls of Gram-positive bacteria (*a*) look thick but relatively homogeneous compared to those of Gram-negative bacteria (*b*). Both the plasma membrane and the "outer membrane" of the wall can be seen in the section of Gram-negative cells. (Courtesy of Dr. Dwight Anderson.)

and more complex. As we shall see, this difference is completely in accord with chemical studies that demonstrate the cell walls of Gram-negative bacteria to be more complex.

Several studies have suggested that the thickness of the wall seen in electron micrographs is significantly less than that found in the living cell. The measured thickness in sectioned material may be on the order of 18 nm for *Staphylococcus aureus,* a Gram-positive bacterium, but the use

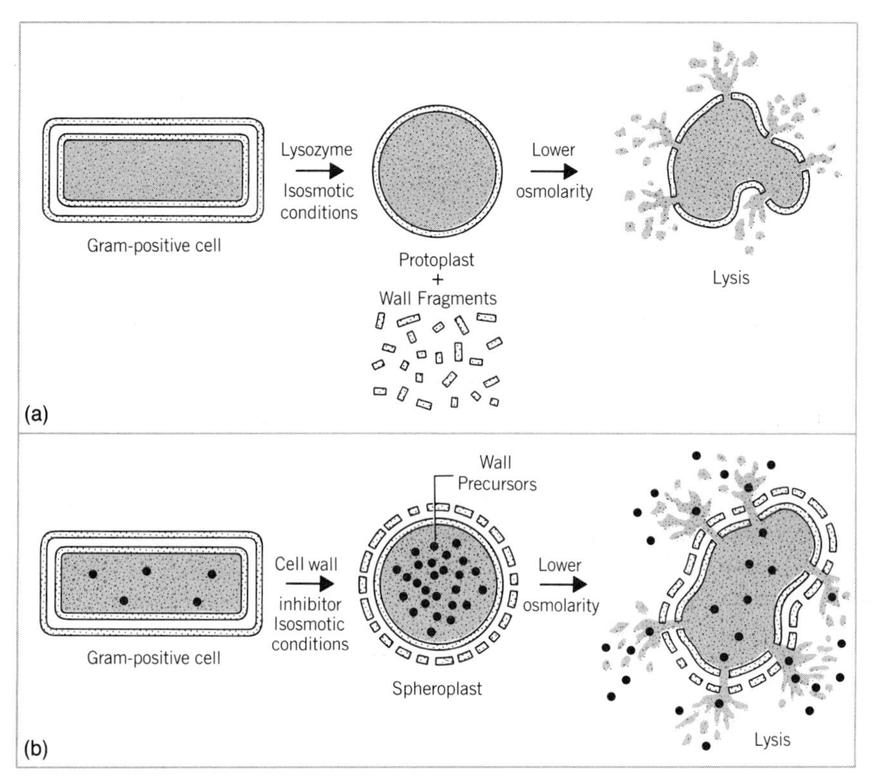

FIGURE 3.3 Production of protoplasts (*a*) and spheroplasts (*b*). Morphologically the two are identical (spherical under isosmotic conditions), but compositionally they are not. Spheroplasts retain wall fragments, whereas protoplasts are completely stripped of their walls. Functionally they are about the same.

of other techniques on living cells, such as light scattering and refractive index measurements, give a value of 86 nm.[1] The techniques required to fix and stain cells for electron microscopy modify the appearance of the walls considerably, limiting the value of this approach in defining the physical parameters of bacterial walls.

Isolation of Wall Material

Two major experimental approaches have been used to isolate cell wall materials for compositional and structural studies. One is to treat cells with enzymes or chemicals that remove the wall. The other is to force the accumulation of cell wall precursors by inhibiting wall biosynthesis.

Several different procedures can be used to strip wall material from procaryotic cells. The walls of Gram-positive bacteria in particular can be fragmented enzymatically by the use of lysozyme (Figure 3.3). Lysozyme hydrolyzes peptidoglycan, which is the most abundant material in the walls of this class of bacteria.

When this type of enzymatic treatment is carried out in a solution that is isosmotic with the cell interior, the "naked" cell normally assumes a spherical shape that is the lowest free energy configuration of the system.

Such a cell, when completely free of cell wall, is called a *protoplast*. When the osmolarity of the solution is lowered protoplasts lyse.

Method III

Differential centrifugation can be used to sediment intact protoplasts from cell wall fragments, leaving partially degraded cell wall material in the supernatant solution.

Another approach, which has been very successfully used in determining the structure of cell wall material, is to administer cell wall inhibitors to growing cells. Penicillin is one of several inhibitors.

In the presence of penicillin, under isosmotic conditions, growing cells accumulate high concentrations of cell wall precursors that cannot be incorporated into the final cell wall structure. These cells may also assume spherical forms, in this case usually retaining fragments of old wall material. Although morphologically similar to protoplasts these fragment-containing spheres are called *spheroplasts*. The precursor materials can be extracted from the cells with trichloroacetic acid and separated from the spheroplasts by differential sedimentation techniques. Spheroplasts will also lyse if they are exposed to a hypotonic environment.

The term *outer membrane* is used because it is similar to the plasma membrane in gross composition and structure. Its function is not the same, however, and it is considered to be wall material, not membrane material. The term is in some respects misleading.

The walls of Gram-negative bacteria, being more complex than those of Gram-positive cells, require special techniques for their isolation. Walls in these organisms consist of an outer layer, called the *outer membrane*, and an underlying layer of peptidoglycan. For detailed studies, these two layers are normally separated from each other and studied independently.

Although many procedural modifications are used, there are basically three approaches to isolating these walls and separating the layers. They are based on: (1) buoyant density differences of the layers shed from cells lysed after treatment with lysozyme–EDTA, or by use of a French pressure cell, (2) electric charge differences, since the outer membrane possesses a higher net negative charge than the peptidoglycan layer, and (3) selective solubilization of one or the other layer with detergents.[2]

Peptidoglycan: The Common Denominator of Bacterial Walls

G

NH
C=O
CH_3

N-Acetylglucosamine

M

O
CH_3—CH—COO^-
NH
C=O
CH_3

N-Acetylmuramic acid
(the 3-*O*-D-lactic acid ether of G)

Both Gram-positive and Gram-negative bacteria contain *peptidoglycan* in the cell wall. The walls of Gram-positive bacteria may contain between 40 and 90% peptidoglycan, whereas levels as low as 1% are found in Gram-negative bacteria.

Peptidoglycan is actually a closely related family of polymeric molecules that vary slightly between different species of bacteria. Nevertheless, they are fundamentally similar in that they comprise a giant macromolecular net that surrounds the entire bacterial cell.

Three variations of peptidoglycan structure are presented schematically in Figure 3.4, and one of these variants is illustrated in more detail in Figure 3.5.

Not only is peptidoglycan unique to procaryotic cell walls, but it is remarkably designed to resist stretch and thus prevent osmotic lysis of bacteria in hypotonic environments. In one dimension, the peptidoglycan chain consists of alternating units of *N*-acetylglucosamine (G) and *N*-acetylmuramic acid (M). Since the alternating glycan units are covalently linked, there is little stretch in this dimension.

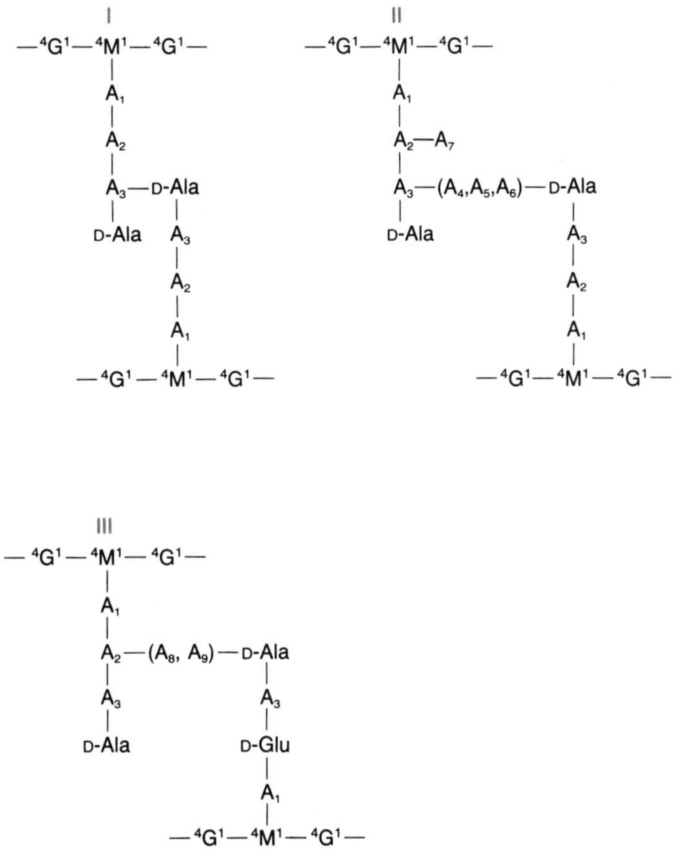

I

$—^4G^1—^4M^1—^4G^1—$
A_1
A_2
A_3—D-Ala
D-Ala A_3
A_2
A_1
$—^4G^1—^4M^1—^4G^1—$

II

$—^4G^1—^4M^1—^4G^1—$
A_1
$A_2—A_7$
$A_3—(A_4,A_5,A_6)$—D-Ala
D-Ala A_3
A_2
A_1
$—^4G^1—^4M^1—^4G^1—$

III

$—^4G^1—^4M^1—^4G^1—$
A_1
$A_2—(A_8, A_9)$—D-Ala
A_3 A_3
D-Ala D-Glu
A_1
$—^4G^1—^4M^1—^4G^1—$

FIGURE 3.4 Some general structures of peptidoglycans. The main differences between the types is the length and complexity of the peptide bridging system. In all cases shown, a peptide of four residues extends from *N*-acetylmuramic acid (M). In type I these are directly covalently linked. In types II and III additional bridging amino acids tie the peptides together. The main backbone of the system consists of alternating residues of *N*-acetylglucosamine (G) and *N*-acetylmuramic acid.

The second dimension of peptidoglycan is formed by parallel glycan chains cross-linked with peptide bridges. The muramic acid units are substituted by tetrapeptides, which in turn are cross-linked either directly or by an interpeptide bridge. In *Staphylococcus aureus,* the tetrapeptide is an (L-alanyl-D-isoglutaminyl-L-lysyl-D-alanine) sequence (see Figure 3.5). There is some variation in this sequence among different species, in terms of composition as well as connecting points to the interpeptide bridge. Also, in *S. aureus*, the interpeptide bridge is a sequence of five glycine units.

There are three unusual features of this peptide structure. One is the presence of D-amino acids, whereas in nature L-amino acids are the norm. A second is the presence of mesodiaminopimelic acid, which is a variant of the common amino acid lysine. A third unusual feature is the linkage between glutamic acid and its carboxyl neighbor. Normally the α-carboxyl group rather than the γ-carboxyl group of the amino acid side chain participates in peptide linkage in a polypeptide.

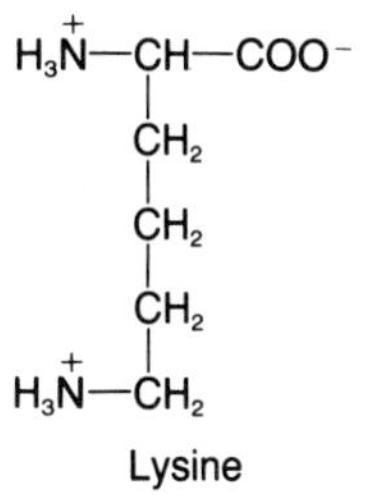

$H_3\overset{+}{N}—CH—COO^-$
CH_2
CH_2
CH_2
$H_3\overset{+}{N}—CH_2$
Lysine

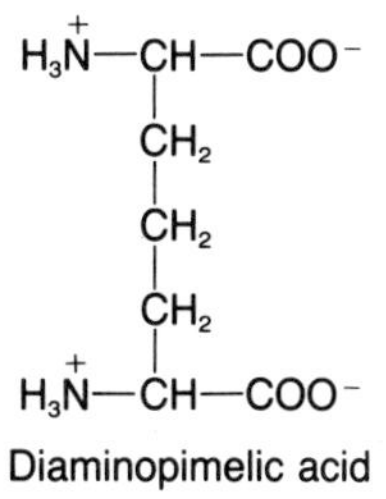

$H_3\overset{+}{N}—CH—COO^-$
CH_2
CH_2
CH_2
$H_3\overset{+}{N}—CH—COO^-$
Diaminopimelic acid

Since this second dimension of peptidoglycan is tied together by co-

FIGURE 3.5 Chemical structure of peptidoglycan. Unusual features in this expanded view of the type I peptidoglycan (Figure 3.4) are the presence of D-amino acids and the presence of isoglutamic acid in which a peptide bond is formed off the side chain carboxyl group. The peptides are tied together by a covalent bond between the carboxyl group of alanine and the ε-amino group of lysine.

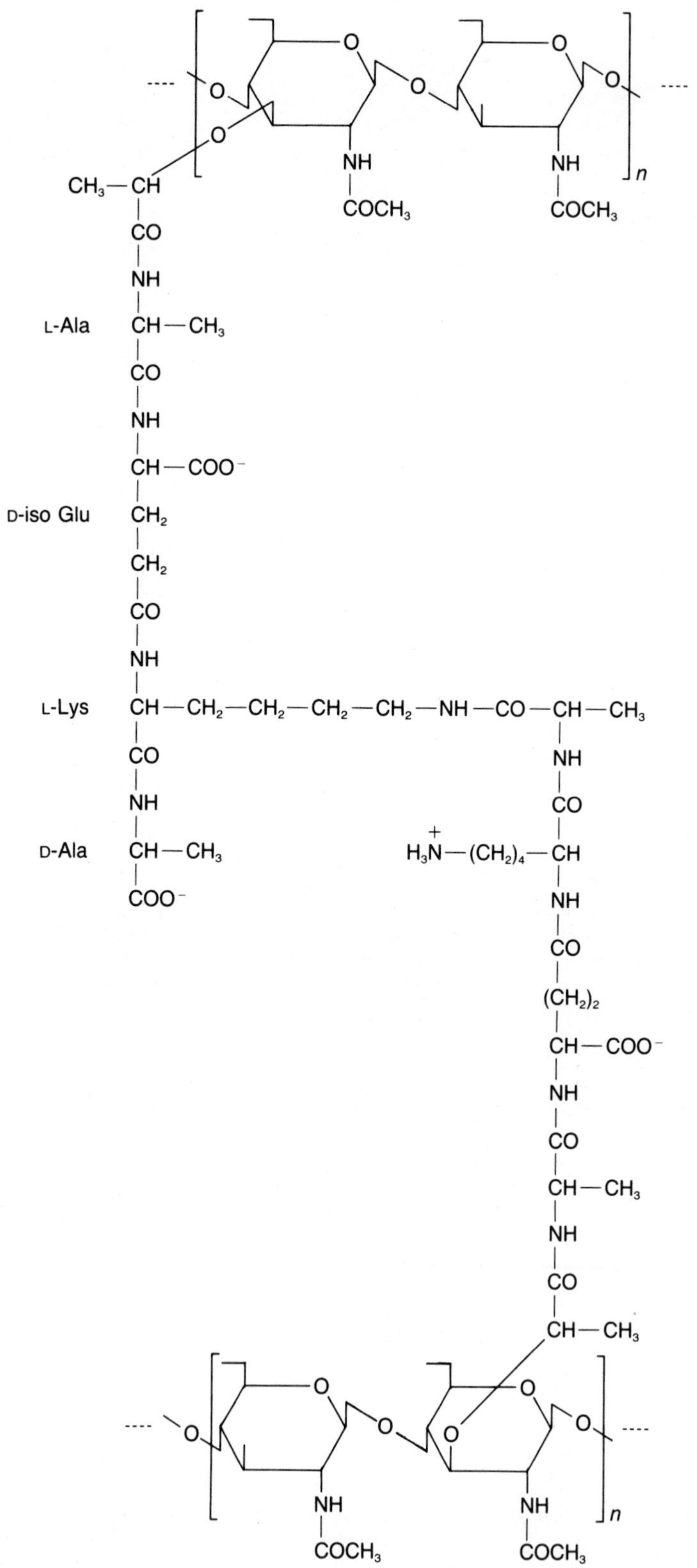

CH_3—CH
CO
NH
L-Ala CH—CH_3
CO
NH
CH—COO^-
D-iso Glu CH_2
CH_2
CO
NH
L-Lys CH—CH_2—CH_2—CH_2—CH_2—NH—CO—CH—CH_3
CO
NH
D-Ala CH—CH_3
COO^-

NH
CO
$H_3\overset{+}{N}$—$(CH_2)_4$—CH
NH
CO
$(CH_2)_2$
CH—COO^-
NH
CO
CH—CH_3
NH
CO
CH—CH_3

NH
$COCH_3$
n

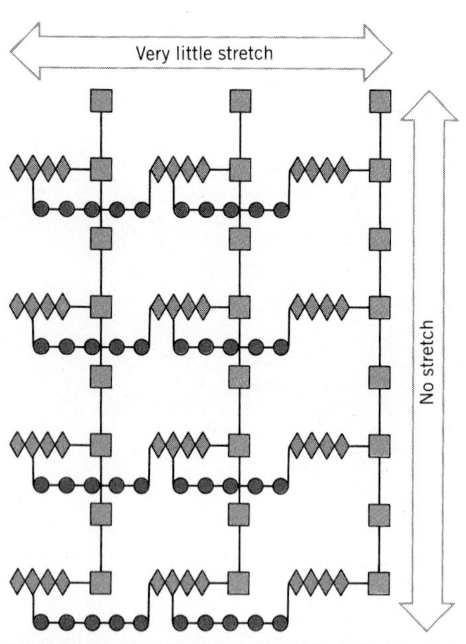

FIGURE 3.6 Schematic surface view of peptidoglycan sheet. The polysaccharides of alternating *N*-acetylglucosamine and *N*-acetylmuramic acid units are represented by colored squares. The peptide extending from muramic acid is shown as a series of diamonds. The bridge unit, which could be a pentaglycine run, is a string of spheres. The bacterial cell is completely encased by this molecular net.

valent bonds, it is a strong structure. One might envision a slight amount of stretch, but the molecular sheet is very resistant to rupture. A surface view of this sheet, drawn schematically, is presented in Figure 3.6.

The third dimension of peptidoglycan is not as well understood. It may consist of additional layers of peptidoglycan with connecting bridges or a single layer with peptide arms extending from a variety of points. It is generally thought that Gram-negative organisms contain from 1 to 3 peptidoglycan layers[3] and Gram-positive organisms may possess 30 to 40 such layers.[4]

Most workers in this field agree that the predominant role of peptidoglycan in bacterial cell walls is to provide strength to prevent lysis. Bacteria are normally found in hypotonic environments, which create internal pressures on the order of 20 to 30 atm. A secondary role, which is more controversial among cell wall experts, is that of maintaining the shape of the cell. It is likely that shape is a result of the contributions made by all the chemical species in the wall, not just the peptidoglycan. In some studies, protoplasts have maintained the particular morphology of the intact cell, even without wall material present.

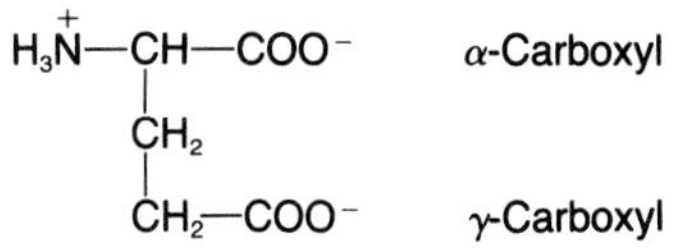

$H_3\overset{+}{N}-CH(-CH_2-CH_2-COO^-)-COO^-$ (α-Carboxyl; γ-Carboxyl)

Glutamic acid

Teichoic Acids

The second most abundant material in the cell walls of many Gram-positive bacteria is a group of polyolphosphate polymers called *teichoic acids*. Some of these compounds are present in both the wall and the cytoplasmic membranes. In walls, they frequently account for 20 to 50% of the dry weight of the wall.

Two classes of polyolphosphates predominate: ribitol phosphates and glycerol phosphates. The earliest and most extensively studied teichoic acid was of the ribitol type, obtained from the walls of *Staphylococcus aureus*. Its structure is shown in Figure 3.7. As extracted from cell walls,

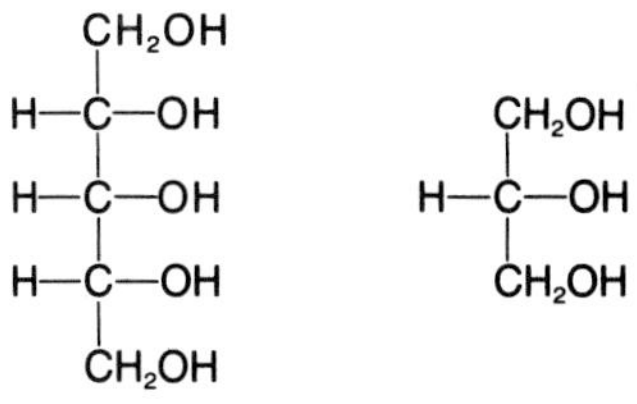

$CH_2OH-(H-C-OH)_3-CH_2OH$

Ribitol

$CH_2OH-(H-C-OH)-CH_2OH$

Glycerol

R = α- or β-*N*-acetylglucosamine
Ala = D-alanine

(*a*)

R = H or glycosyl residue
Ala = alanine

(*b*)

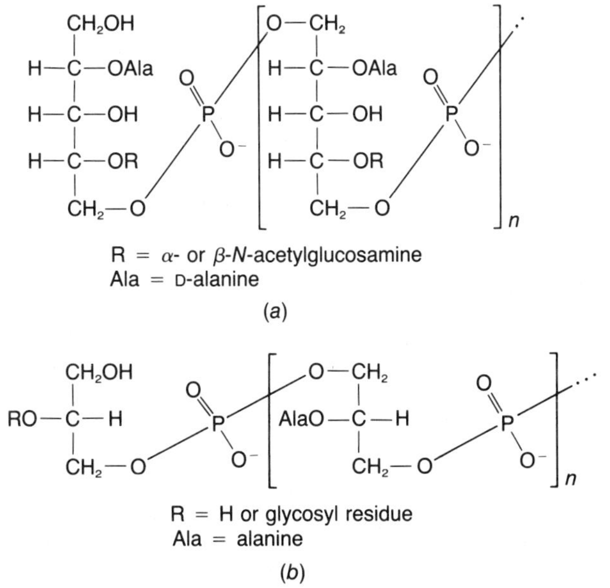

FIGURE 3.7 Teichoic acids: (*a*) ribitol type and (*b*) glycerol type.

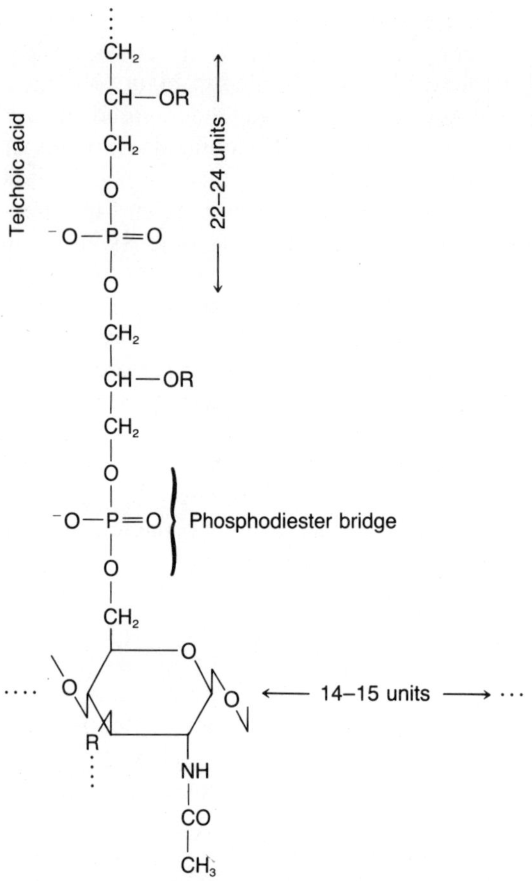

FIGURE 3.8 Relation between teichoic acid side chain and polysaccharide polymer of peptidoglycan. A phosphodiester bridge links the glycerol end of the teichoic acid to the 6-hydroxy group of muramic acid.

the polymer has an average chain length of 6 to 10 repeating units, with each unit commonly possessing a D-alanine ester residue and an *N*-acetylglucosaminyl residue as noted in the figure.

Glycerol phosphate teichoic acids are more widespread than the polyribitol phosphates, occurring in both membranes and walls. A typical glycerol teichoic acid is also shown in Figure 3.7. Alanine is esterified to the repeat unit with an occasional glycosyl group in the 2-hydroxy position of glycerol.

Determining the spatial relationship between the teichoic acids and peptidoglycan has been a difficult research problem, as can be imagined from the complexities of these polymers. Teichoic acids are extractable from walls by trichloroacetic acid or alkali, but the slow rate at which they are removed has long suggested a covalent linkage between the two macromolecules. A number of studies are now beginning to clarify the nature of the linkage, but there appear to be many variations on the theme.

One type of linkage appears to be a direct one between the teichoic acid and the 6-hydroxy group of the muramyl residues in the peptidoglycan.[5] The bridge point would be a phosphodiester, as shown in Figure 3.8.

Other linkages do not appear to be quite as direct or simple. In several organisms linker groups exist, bridging the teichoic acids to the peptidoglycan. These linker groups consist of two or three glycerol residues and a molecule of *N*-acetylglucosamine.

In one organism studied in detail, *Staphylococcus lactis*, one teichoic acid 22 to 24 residues long was found per 7.4 disaccharide units.[6] Furthermore, the teichoic acids appeared to be on the outside surface of the wall according to immunological studies. It is likely that these polymers project perpendicularly from the peptidoglycan surface. These results are beginning to bring into focus the three-dimensional picture of the walls of Gram-positive bacteria.

Method VII

Membrane teichoic acids are anchored to the outer surface of the plasma membrane through the glycolipids, which are a part of the bilayer structure of the membrane.

Teichuronic Acid

Another family of carbohydrate polymers found in certain bacteria is called *teichuronic acid*. The organism *Micrococcus lysodeikticus* has been a good source for this material because teichuronic acid is the main ingredient of its wall in addition to peptidoglycan.

The teichuronic acid of this organism has the repeat sequence illustrated in Figure 3.9. Between 10 and 40 repeating units make up the polysaccharide chain. As is true for the teichoic acids, teichuronic acid is covalently linked to peptidoglycan and can be removed from the latter by enzymatic hydrolysis.

The Outer Membrane of Gram-Negative Bacteria

Gram-negative bacteria have cell walls that are considerably more complex than the structure that has been described for Gram-positive bacteria.

$\longrightarrow \left[\text{ManNAcUA} \xrightarrow{\beta\text{-1,6}} \text{Glc} \right] \text{——} \cdots$

(a)

COO⁻ COCH₃ O—CH₂ NH O 10–40 repeats

N-Acetylmannosaminuronic acid (ManNAcUA) Glucose (Glc)

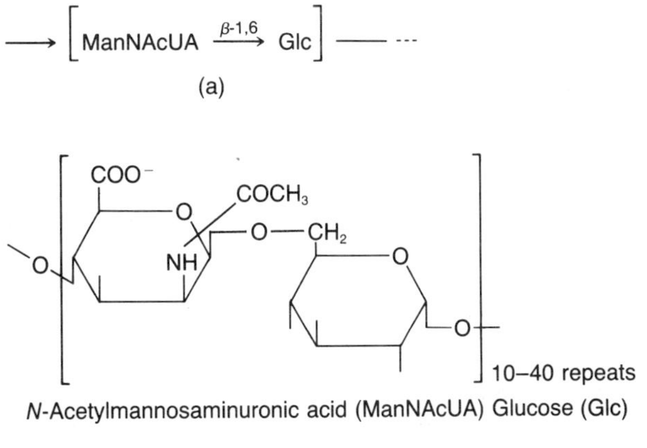

(b)

FIGURE 3.9 Teichuronic acid of *Micrococcus lysodeikticus:* (*a*) the repeating disaccharide and (*b*) its chemical structure. The linkage between teichuronic acid and peptidoglycan is not known with the degree of certainty posited for the teichoic acid–peptidoglycan link. The most likely candidate is the 6-hydroxy group of muramic acid with some type of bridge linking it to the teichuronic acid chain.

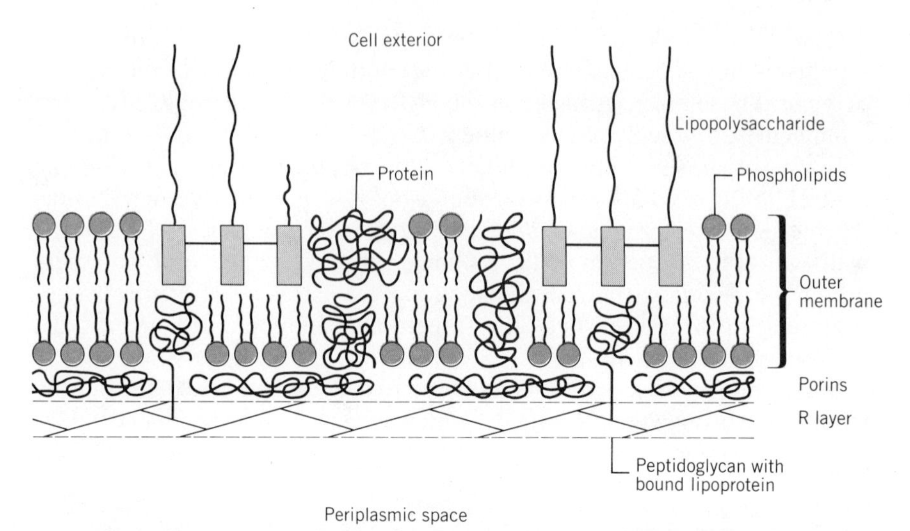

FIGURE 3.10 General model for the cell wall of Gram-negative bacteria. In addition to containing amphipathic lipids characteristic of other membranes, the outer membrane of the wall contains a complex lipid to which long polysaccharide chains are attached. Proteins, covalently linked to peptidoglycan, are also submerged in the lipid of the outer membrane. The two layers are thus effectively anchored to one another.

In addition to peptidoglycan, they have an outer cell wall layer, called the outer membrane, that contains some of the most complex biological molecules known. Figure 3.10 depicts in schematic form the cell wall of Gram-negative cells.

One of the main components of the outer membrane is an amphipathic molecule called *lipopolysaccharide* (LPS). The structure of this complex molecule is shown in Figure 3.11. The end of the molecule most distally situated from the cell surface contains a hydrophilic oligosaccharide stretch of three or four sugars repeated many times. This portion of the LPS is the most variable, accounting for the numerous serological types that are present even within a given species of bacteria. The oligosaccharide is often referred to as "O-antigen" in serological typing.

Adjacent to the oligosaccharide chain is a less variable run of a heterogeneous mix of molecules that are also hydrophilic. In fact, a high net negative charge is borne by this stretch because of the presence of ionized carboxyl and phosphate groups. This intermediate region of the LPS, called the "R core," has been arbitrarily divided into inner and outer core regions. The outer core region, closest to the O antigen, is composed of hexoses and *N*-acetylated hexosamines. The inner core region contains two types of compound that are unique to bacteria. One is 2-keto-3-deoxyoctonic acid (KDO) and the other is glyceromannoheptose (Hep).

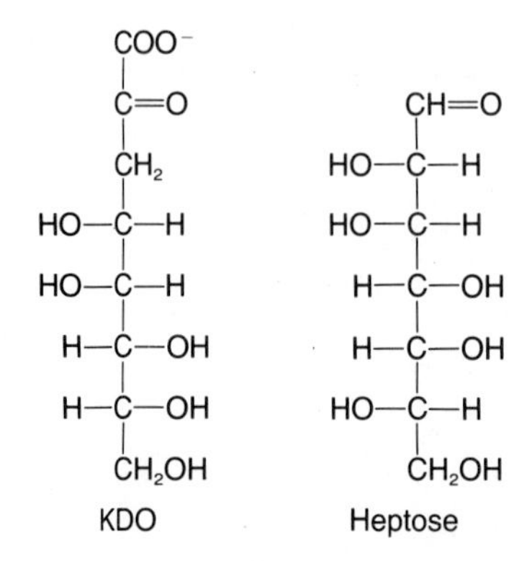

The remainder of LPS constitutes the hydrophobic end of the amphipathic complex. This portion is termed "lipid A." Its structure is shown in Figure 3.12. The chemical backbone of the structure is a glucosamine disaccharide to which fatty acids are attached through either amide or ester linkages. The aliphatic side chains of the fatty acids are saturated, thus imparting to this end of the molecule the property of low fluidity. Lipid A is attached to the R core by glycosidic linkage between the 3′-hydroxyl group of glucosamine units and KDO.

Compare this structure with its schematic illustration in Figure 3.10 to see how it is situated in the outer membrane.

The outer membrane and the peptidoglycan layer are very complex and are chemically distinct from one another. But physically they interact, so that the outer membrane is not completely independent of the layer beneath it.

In *E. coli,* for example, peptidoglycan is covalently linked to a lipoprotein through a trypsin-sensitive bond between diaminopimelic acid and lysine (Figure 3.13). The hydrophobic portion of the protein reaches up into the hydrophobic lipid bilayer of the outer membrane. Treatment with trypsin affords a good means of separating the two layers with the lipoprotein moving with the outer membrane. Statistically, every tenth to twelfth diaminopimelic acid residue has a lipoprotein attached.

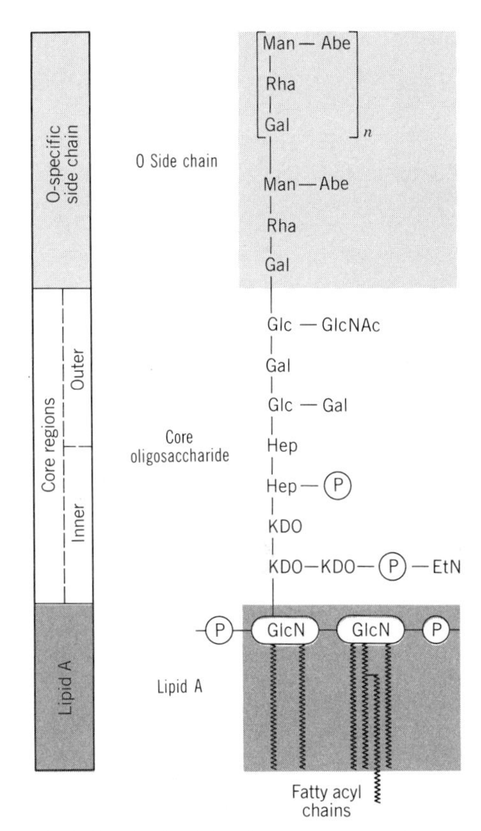

FIGURE 3.11 The lipopolysaccharide of the Gram-negative bacterium *Salmonella typhimurium.* Three regions are defined: lipid A, a core oligosaccharide with inner (proximal to the lipid) and outer regions, and the polysaccharide O side chain. Lipid A is submerged in the outer membrane, and the core and O side chain extend into the aqueous environment around the cell. Abe, abequose; EtN, ethanolamine; Gal, galactose; Glc, glucose; GlcN, glucosamine; GlcNAc, *N*-acetylglucosamine; Hep, heptose; KDO, 2-keto-3-deoxyoctonate; Man, mannose; Rha, rhamnose.

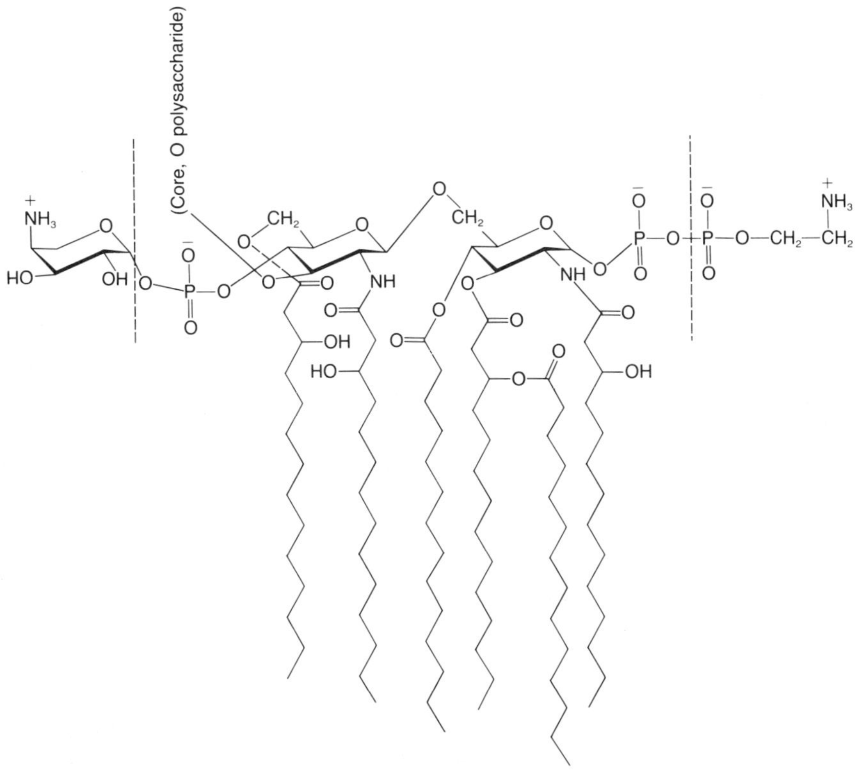

FIGURE 3.12 Structure of lipid A from *Salmonella minnesota.* In this organism, three fatty acyl residues [dodecanoyl, hexadecanoyl (palmitic), and 3-*O*-tetradecanoyl-tetradecanoyl residues] are linked to the hydroxyl groups of the glucosamine disaccharide, but their exact locations are unknown. Some units lack the arabinosamine residue, the phosphoryl-ethanolamine residue, or both (dotted lines).

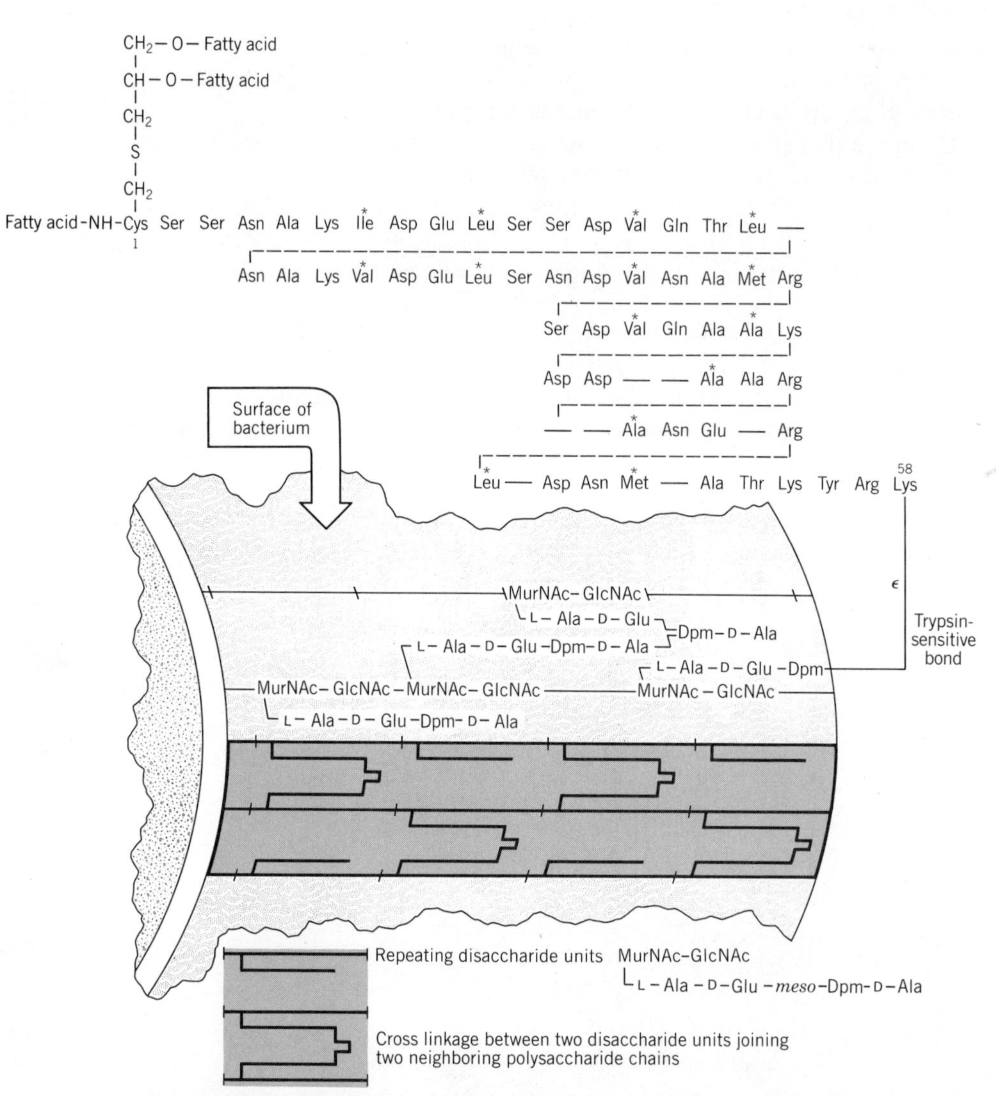

FIGURE 3.13 Structure of the lipoprotein–peptidoglycan complex in *E. coli.* The glycan chains are arbitrarily drawn parallel with the long axis of the organism. The peptidoglycan is composed of about 10^6 repeating units, to which approximately 10^5 lipoprotein molecules are covalently bound. The lipoprotein replaces a D-alanine on the diaminopimelate (Dpm) residue. Dashed lines represent hypothetical deletions of amino acids that may have occurred during evolution. A variety of fatty acids are bound to the glycerylcysteine by ester linkage or to the *N*-terminus by amide linkage, but about half of the fatty acids are palmitic acid. Asterisks indicate hydrophobic amino acids.

The outer membrane also contains a number of different proteins, most of which have been characterized only electrophoretically. Certain of these proteins are associated with the underlying peptidoglycan by covalent or noncovalent linkages. One type of peptidoglycan-associated protein, having a molecular weight of about 36,500 daltons, is called *porin* because when added to phospholipid–LPS mixtures it creates transmembrane diffusion pores.[7]

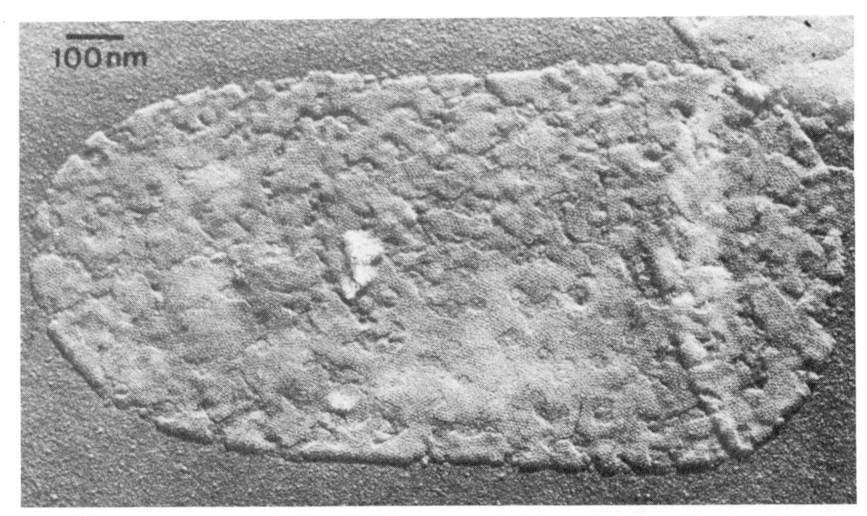

FIGURE 3.14 Electron micrograph of a cell wall specimen tungsten-shadowed after freeze-drying. The intact peptidoglycan–matrix protein complex retains the rodlike shape of the cell. The surface shows a regular hexagonal array of porins. (Courtesy of Dr. J.P. Rosenbusch.)

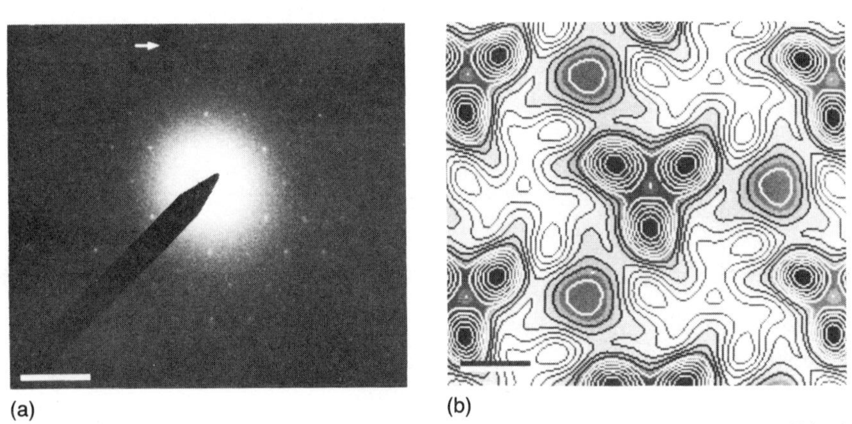

FIGURE 3.15 Electron diffraction pattern (*a*) and computer reconstruction (*b*) of a negatively stained reconstituted sheet of matrix porin. The reconstruction was calculated by combining phases, extracted from electron images, with amplitudes, obtained from electron diffraction. Stain-filled triplet indentations (in black with white contour lines) are roughly 2 to 4 nm deep (contour line intervals 0.15–0.3 nm). This estimate is inferred from electron scattering measurements, strongly suggesting that channels span the membrane. The scale bar in *b* represents 3 nm. (Courtesy of Dr. J.P. Rosenbusch.)

Bacteria can be prepared for electron microscopy in such a way that the outer membrane is removed with sodium dodecyl sulfate (SDS) and the peptidoglycan and associated proteins are left intact. These bacteria retain their morphology and have surfaces that show a regular hexagonal array of porins. This is illustrated in Figure 3.14. The porin molecules form a monolayer covering the outer surface of peptidoglycan.

The molecules are arranged in a threefold symmetry on a hexagonal lattice, a feature that is evident in Figure 3.15. The unit cell is thought to contain three molecules of porins, between which are indentation triplets with a center-to-center spacing of 3 nm. These indentations are penetrated by stain and are believed to demarcate channels across the protein layer.

Escherichia coli B possesses only one major species of porin protein, whereas wild type K-12 produces two.[8] *Salmonella typhimurium*, a closely related bacterium, contains three different porins. But regardless of the numbers produced and the species of organism, all the porins bear similarities chemically and functionally. All have molecular weights in the vicinity of 34,000 to 42,000 daltons. All remain noncovalently bound to peptidoglycan upon removal of the outer membrane with SDS plus Mg^{2+} extraction. Finally, all appear to allow permeation of simple sugars and amino acids and of uncharged molecules up to a molecular weight of about 600 daltons.

Sodium dodecyl sulfate is a detergent that disrupts most protein–protein and lipid–protein interactions. When it is used in conjunction with polyacrylamide gel electrophoresis, the technique is abbreviated SDS–PAGE.

$$CH_3—(CH_2)_{10}—CH_2OSO_3^-Na^+$$

SDS

Porins are important in transport. Mutants that lack certain porins show defective transport properties. They may have certain other roles as well, such as serving as phage receptors and even as antigenic determinants. But all the roles and structures must await more exhaustive research before they can be confidently stated.

The Functions of Wall Components

The specific functions of all the components of the bacterial cell wall are not yet known with certainty, but at this point some very well educated guesses are being made by cell wall investigators.

The major role of peptidoglycan appears to be as a macromolecular corset, keeping the cell from osmotic rupture. It may also contribute to the general rigidity and shape of the cell, but probably it is not solely responsible for the cell's morphological features.

The teichoic acids appear to have quite a different role, even though they are chemically linked to peptidoglycan. At present the most likely role is thought to involve the securing of magnesium from the environment for the metabolic reactions of the cell.[9] According to this notion, cell wall teichoic acids bind magnesium and plasma membrane teichoic acids then move the magnesium into the cell interior.

The outer membrane of Gram-negative bacteria has several discernible functions. An obvious function is that it is the outermost surface of the cell and in this capacity confers on the cell several biological properties.

One property of the outer membrane is negative charge. This is important in maintaining a unicellular existence for the organism. A second property is to confer hydrophilicity to the surface. This appears to be important in preventing engulfment by phagocytic cells. A third property is to provide a ready mechanism to generate surface variability. This may be essential for survival when hosts begin to develop antibodies against an organism of particular surface type. All these functions relate most directly to the oligosaccharide portion of LPS.

Another type of function of the outer membrane is to perform as a penetration barrier against several kinds of molecule. Gram-negative bacteria are generally more resistant than Gram-positive organisms to a variety of antibiotics, such as actinomycin D, erythromycin, novobiocin, and rifamycin SV. In addition, they resist penetration by dyes such as crystal violet and by bile salts and certain detergents.[2] LPS is probably the main ingredient of the outer membrane that functions in these instances as a barrier.

The monomolecular layer of porin molecules also functions in transport, serving as a secondary discriminator for the molecules that are able to penetrate the outer membrane.

With a better understanding of bacterial cell wall structure we can return to the Gram stain and its ability to distinguish between types of bacteria. Gram-positive bacteria retain the crystal violet–iodine complex, apparently because it is blockaded in the cell by multiple peptidoglycan layers. Alcohol does not appear to influence the porosity of Gram-positive cells.

Gram-negative bacteria lose the crystal violet–iodine complex by alcohol extraction. With much less peptidoglycan and more lipid and protein, Gram-negative cell walls become more porous, permitting the exit of purple dye.

The Biosynthesis of Wall Polymers

Detailed information about the chemistry of cell wall biosynthesis can be quite easily retrieved from appropriate reviews listed at the end of this chapter. We emphasize here the locus of these events and the time course of biosynthesis.

Peptidoglycan

During the past 25 years the steps in peptidoglycan synthesis have been unraveled. It has been a remarkably difficult, yet successful, research endeavor.

It is convenient to break down the process into the three stages illustrated in Figures 3.16, 3.17, and 3.18.

Stage 1 takes place within the interior of the cell requiring the soluble enzymes of the cytoplasmic matrix. The product is a uridine diphosphate *N*-acetylmuramyl pentapeptide. It is this material that accumulates when cells are administered the wall inhibitor penicillin.

The second stage consists of reactions that take place within or on the plasma membrane. The key participant in these reactions is a membrane C_{55}-isoprenyl phosphate called undecaprenyl phosphate. It picks up the product of the first stage, forms the glucosamine disaccharide, and transfers it to the exterior side of the plasma membrane. There it is given up to a growing point on a peptidoglycan chain. The chains are extended by adding these disaccharide units to the reducing ends of the polymers.

Stage 3 is a cross-linking reaction between adjacent peptidoglycan chains involving the interpeptide bridges. This takes place outside the plasma membrane at the site of cell wall growth. This reaction, called transpeptidation, results in the elimination of the terminal D-alanine and the formation of a peptide bond between the α-amino group of glycine and the α-carboxyl group of the penultimate D-alanine.

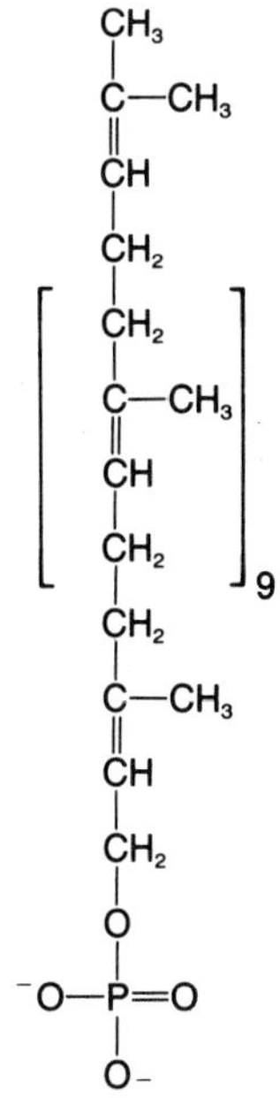

Undecaprenyl phosphate

Teichoic Acids

Teichoic acids are synthesized simultaneously along with peptidoglycan and are subsequently attached to peptidoglycan outside of the membrane. CDP-ribitol and CDP-glycerol provide the phosphoglycan units for polyolphosphate chain polymerization within the cell. The pathway is not entirely clear yet, but it may be that teichoic acid shares the undecaprenyl

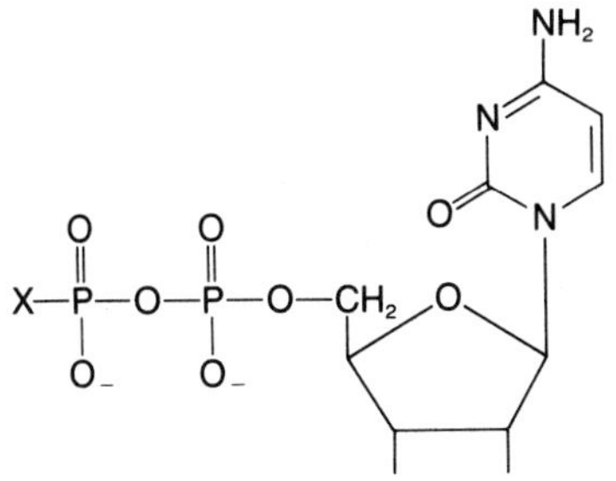

CDP-X (ribitol or glycerol)

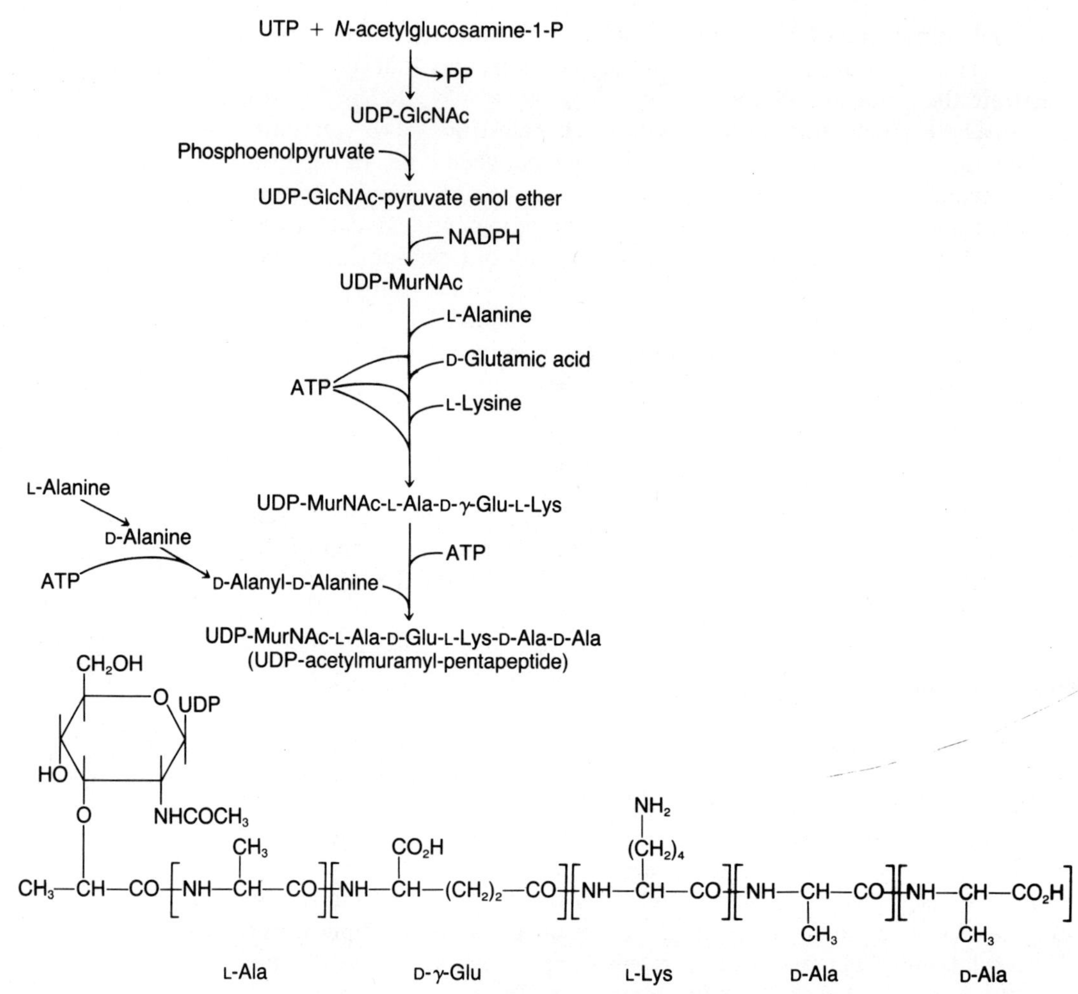

FIGURE 3.16 Stage one in the synthesis of peptidoglycan. Starting with UTP, a hexosamine phosphate, and appropriate amino acids, a complex UDP-product is formed. *These reactions take place within the cell.*

phosphate carrier with peptidoglycan, along with most of the other cell wall components that are made within the cell and assembled outside.

A scheme proposing the formation of a poly glucosyl glycerol phosphate teichoic acid is presented in Figure 3.19.

Teichuronic Acid

The biosynthesis of teichuronic acid, at least as worked out under *in vitro* conditions, occurs in two stages. The first stage uses the undecaprenol phosphate carrier to make an attached $(ManNAcUA)_2$-GlcNAc short polymer.[10] The first part of Figure 3.20 illustrates the steps in this stage. This is a rate-limiting stage in the formation of wall material and is carried out *in vitro* by a particulate enzyme fraction prepared from *Micrococcus lysodeikticus*.

The second stage, included in the same figure, involves the attachment

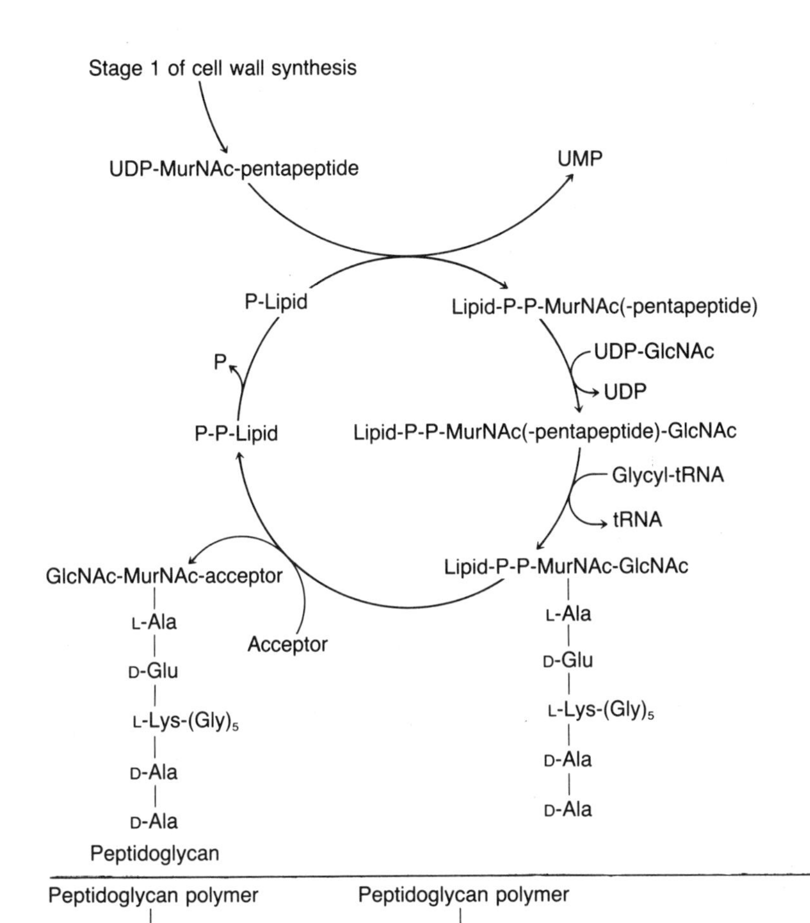

FIGURE 3.17 Stage two in the synthesis of peptidoglycan. Undecaprenyl phosphate (P-Lipid) picks up the UDP product, in which form *N*-acetylglucosamine and glycine residues are added (in this case the organism is *Staphylococcus aureus,* a Gram-positive bacterium). The final step is a transfer of the product to a growing point on peptidoglycan, whereupon the lipid carrier is released. *These reactions take place in conjunction with the plasma membrane.*

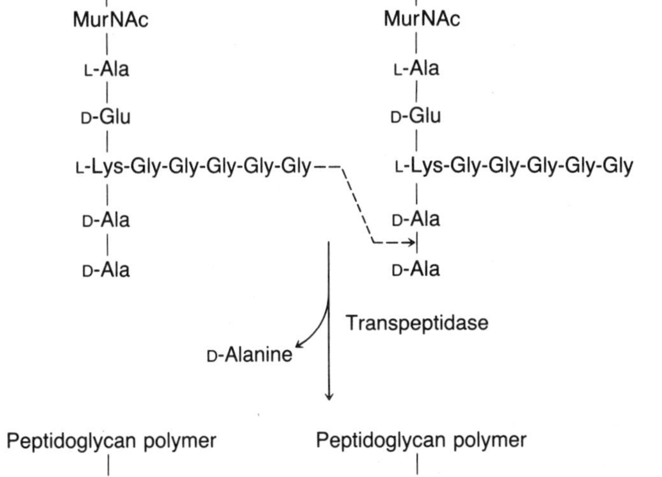

FIGURE 3.18 Stage three in the synthesis of peptidoglycan. The main feature of this stage is a transpeptidation in which the terminal D-alanine of the peptide run is eliminated and a peptide bond is formed between the α-carboxyl group of the penultimate D-alanine on one peptidoglycan polymer and the α-amino group of the pentaglycine bridge on an adjacent polymer. *These reactions take place outside the plasma membrane.*

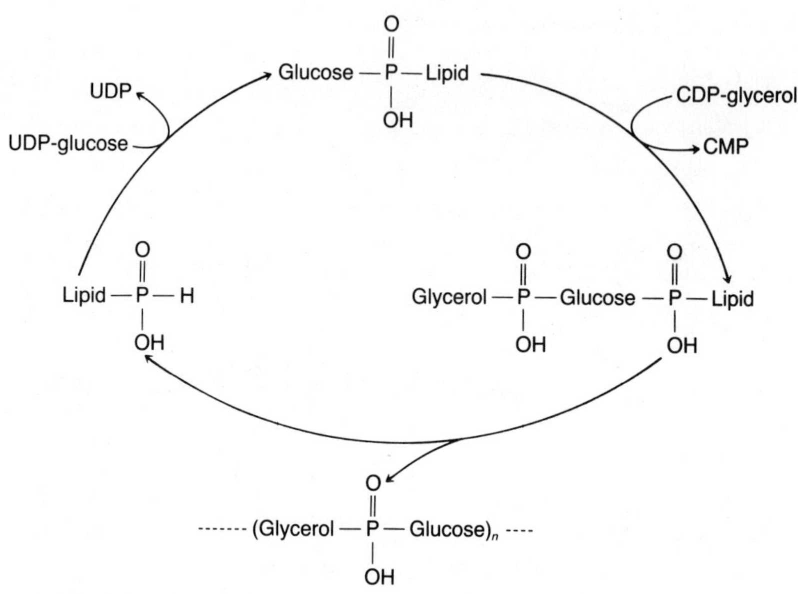

FIGURE 3.19 Scheme illustrating the formation of a glycerol teichoic acid as it has been studied in the Gram-positive bacterium *Bacillus licheniformis.* The alternating units in this teichoic acid are glycerol and glucose.

of a glucose residue from UDP-glucose to the short polymer.[11] With this residue in place another ManNAcUA is attached establishing a ManNAcUA-Glc sequence. Teichuronic acid is then built to its final high molecular weight form by a polymerization of this sequence from 10 to 40 times.

The teichuronic acid polymer is then transferred to peptidoglycan, but the structure of the resulting linkage between the two is not completely settled. One possibility is an attachment of teichuronic acid from a GlcNAc residue by means of phosphate to the 6-hydroxy group of an *N*-acetylmuramic acid unit of peptidoglycan.[12] This GlcNAc residue is the one shown next to the lipid carrier of component C in Figure 3.20.

Lipopolysaccharide

The topological aspects of LPS biosynthesis are rapidly being clarified. A scheme showing the proposed topology of *O*-side chain biosynthesis in the bacterium *Salmonella typhimurium* is presented in Figure 3.21. The side chain is first polymerized on the inner surface of the plasma membrane and then transferred to the outer surface where it is attached to the R core. The LPS is finished by connection to lipid A and transported to the outer membrane. There are obvious gaps in molecular detail, but the general course of the biosynthetic route is taking form.

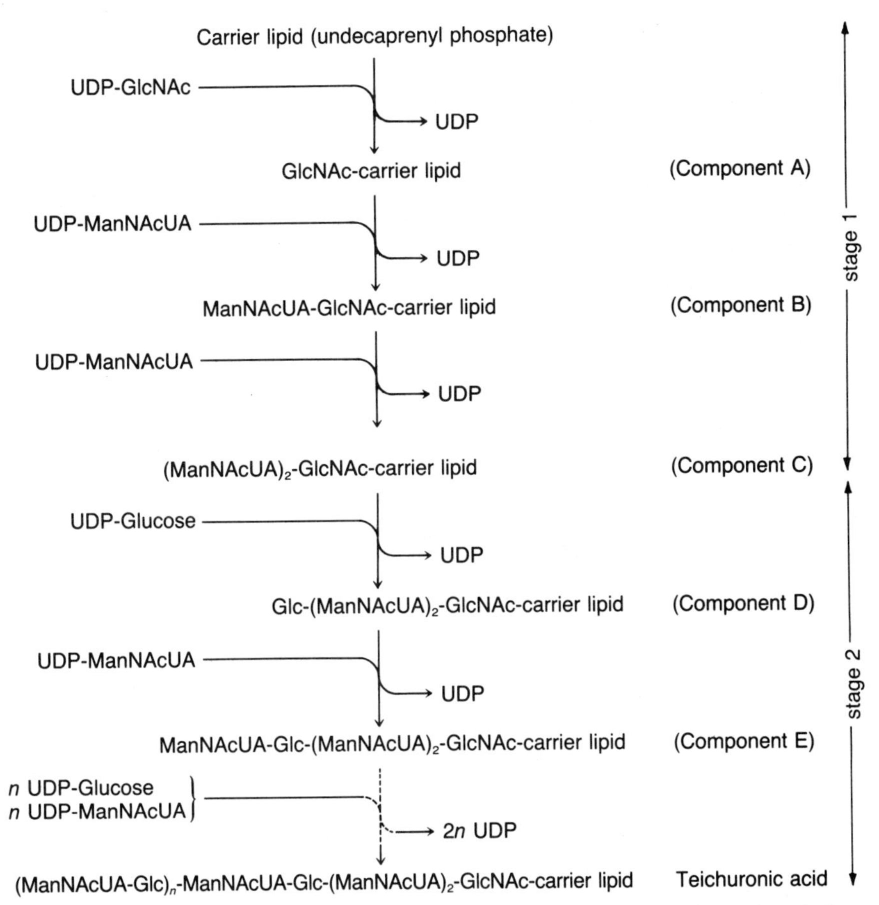

FIGURE 3.20 Proposed synthesis of teichuronic acid in *Micrococcus lysodeikticus.* Stage 1 culminates in the formation of component C, a short polymer attached to a lipid carrier. Stage 2 incorporates glucose, which becomes an alternating residue in the final polymer.

Antibiotics versus Cell Wall Growth

Several antibiotics inhibit bacterial cell growth by interfering with the assembly of new cell walls. Bacteria that reproduce with weakened walls as a result of this action are subject to osmotic rupture, which is the immediate cause of cell death. Even conditions less severe than rupture, as when a cell develops membrane leaks because of incompletely formed walls, may spell death for the organism.

The value of antibiotics in treating infections of various types needs no explanation. But antibiotics are also useful tools to govern *in vitro* reactions conducted by researchers. In particular, they have been used to force the accumulation of cell wall precursors that have then been analyzed and placed into proper metabolic pathways. The molecular bases for the inhibitory effect of several antibiotics have now been worked out in considerable detail.

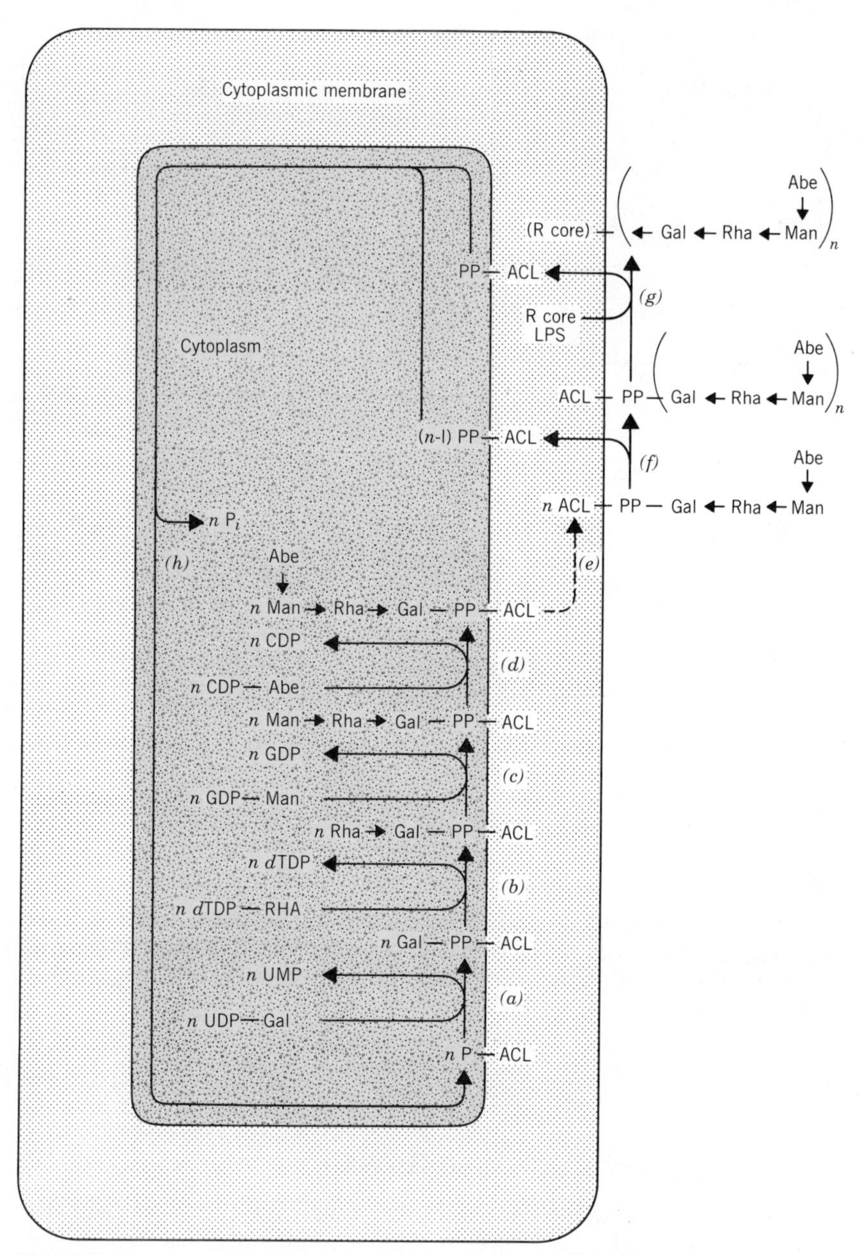

FIGURE 3.21 Lipid cycle in the biosynthesis of O side chains in *Salmonella typhimurium*, a Gram-negative organism. The tetrasaccharide repeating unit is synthesized on a carrier lipid (P-ACL) (reactions *a* through *d*), then the units are polymerized (*f*), and the O side chain is transferred to the R core (*g*). Reactions *a* through *d* are assumed to take place at the inner surface of the plasma membrane and reactions *f* through *g* at the outer surface. This requires the hypothetical transport of the carrier and its repeating unit across the membrane, step e.

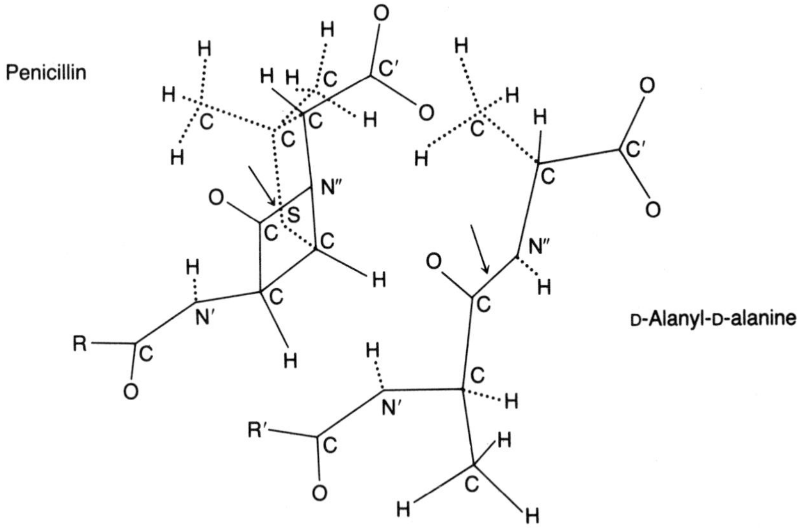

Penicillin

D-Alanyl-D-alanine

FIGURE 3.22 Dreiding stereomodels of penicillin (left) and of the D-alanyl-D-alanine end of the peptidoglycan strand (right). Arrows indicate the position of the CO—N bond in the β-lactam ring of penicillin and of the CO—N bond in D-alanyl-D-alanine at the end of the peptidoglycan strand. The two structures are strikingly similar, allowing their competition for the transpeptidation enzyme site.

Penicillins and Cephalosporins

Two families of antibiotics, the penicillins and the cephalosporins, inhibit the transpeptidation reaction, the last step in cell wall formation illustrated in Figure 3.18.

The terminal alanine residue in the strand to be cross-linked is eliminated when the interpeptide bridge is brought into place.

A reasonable theory as to how penicillin may interfer with this reaction has been proposed by Strominger and his co-workers.[13] The structure of penicillin is very similar to that of the terminal D-alanyl-D-alanine of the peptidoglycan strand (see Figure 3.22). Therefore, penicillin should fit the substrate binding site of the transpeptidation enzyme and acylate the enzyme because of the highly reactive —CO—N— bond in the β-lactam ring. The mechanism that describes this is illustrated in Figure 3.23.

The product is a penicillinoyl enzyme, an enzyme completely inactivated by the reaction. Reaction products have been isolated and identified that are consistent with this mechanism.

Bacitracins

The bacitracins are peptide antibiotics isolated from species of *Bacillus* bacteria. The proposed structure of bacitracin A is shown in Figure 3.24.

The bacitracins inhibit cell wall growth in a manner quite different from that of penicillin, exerting their effect on the transmembrane carrier undecaprenyl phosphate. Specifically, the bacitracins inhibit dephosphorylation of the pyrophosphate form of the carrier (see Figure 3.17), thereby blocking its ability to accept the MurNAc-pentapeptide and causing it to accumulate in this unreactive form in the membrane.

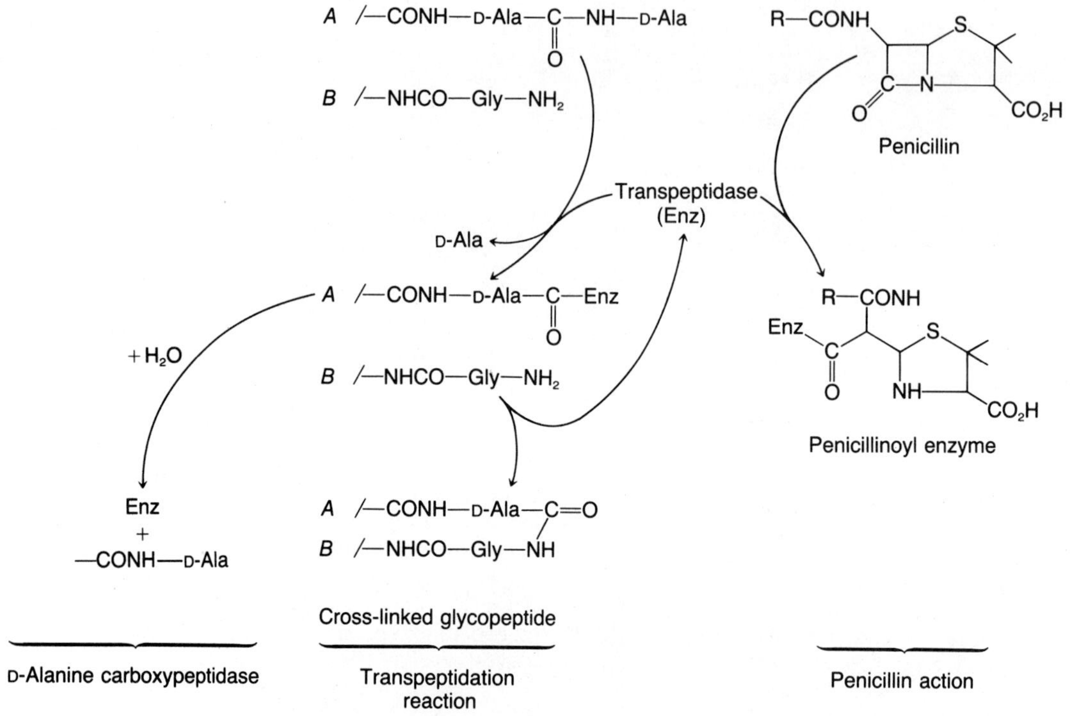

A /—CONH—D-Ala—C(=O)—NH—D-Ala

B /—NHCO—Gly—NH_2

R—CONH

Penicillin

Transpeptidase (Enz)

D-Ala

A /—CONH—D-Ala—C(=O)—Enz

B /—NHCO—Gly—NH_2

R—CONH

Enz

NH

CO_2H

Penicillinoyl enzyme

$+H_2O$

Enz + —CONH—D-Ala

A /—CONH—D-Ala—C=O

B /—NHCO—Gly—NH

Cross-linked glycopeptide

D-Alanine carboxypeptidase

Transpeptidation reaction

Penicillin action

FIGURE 3.23 Proposed mechanism of inhibition of transpeptidation by penicillins: *A* represents the end of the main peptide chain of the glycan strand; *B* represents the end of the pentaglycine substituent from an adjacent strand. If the acyl enzyme intermediate can react with water instead of the acceptor (left), the enzyme would be regenerated and the substrate released. The overall reaction would be the hydrolysis of the terminal D-alanine residue of the substrate. Such a reaction could be an "uncoupled transpeptidation" reaction, but it also seems likely that carboxypeptidation (left) and transpeptidation (center) reactions can be catalyzed by separate proteins, perhaps using similar mechanisms.

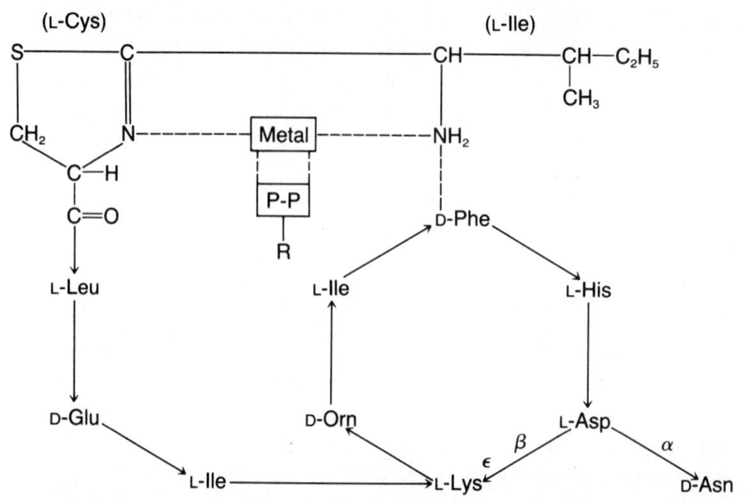

FIGURE 3.24 Proposed structure of bacitracin A. Arrows represent C—N bonds. Cysteine and isoleucine are shown in detail to illustrate the thiazoline ring. Dashed lines show likely chelation arrangements in the complex.

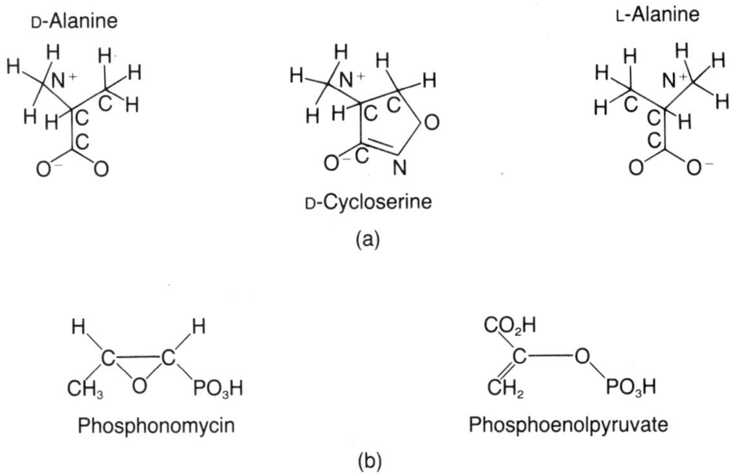

FIGURE 3.25 (*a*) Cycloserine compared to the zwitterionic forms of D- and L-alanine. Note its similarity to D-alanine. (*b*) Structure of phosphonomycin, showing it to be a structural analogue of phosphoenolpyruvate, a product of glycolysis.

In this case, the antibiotic does not react with and inactivate the enzyme. Rather, the pyrophosphate group of the lipid carrier fits into the pocket of bacitracin with a metal ion (Zn^{2+} being most effective) serving as a ligand connecting the carrier to the antibiotic. Thus, the process of wall building is stopped at the level of plasma membrane translocation.

Cycloserine and Phosphonomycin

The antibiotics cycloserine and phosphonomycin inhibit wall formation during the first stage of reactions taking place within the cell (see Figure 3.16).

Cycloserine is a structural analogue of alanine, as is evident in Figure 3.25. It is a true competitive inhibitor for two enzymes: alanine racemase and D-alanyl-D-alanine synthetase. In the presence of cycloserine, the completed UDP-MurNAc-pentapeptide cannot be formed in the cell and is therefore not attached to the translocating lipid carrier.

Phosphonomycin is a structural analogue of phosphoenolpyruvate. The enzyme normally forming the 3-*O*-lactic acid ether bond to make muramic acid reacts with phosphonomycin forming a stable C—S bond and becomes irreversibly inactivated.

Physical Aspects of Cell Wall Growth

In spite of an accumulating wealth of knowledge concerning the reactions of cell wall polymer formation, it is still difficult to imagine how the molecular events described above culminate in the expansion of a wall and the separation of one cell into two complete individuals.

An electron micrograph of a dividing coccus bacterium caught in the midexponential phase of growth is presented in Figure 3.26. It is apparent that an intruding septum has grown down to separate the cell into two compartments. In addition there must be surface growth.

Studies on *Streptococcus faecalis*, a Gram-positive organism, have pro-

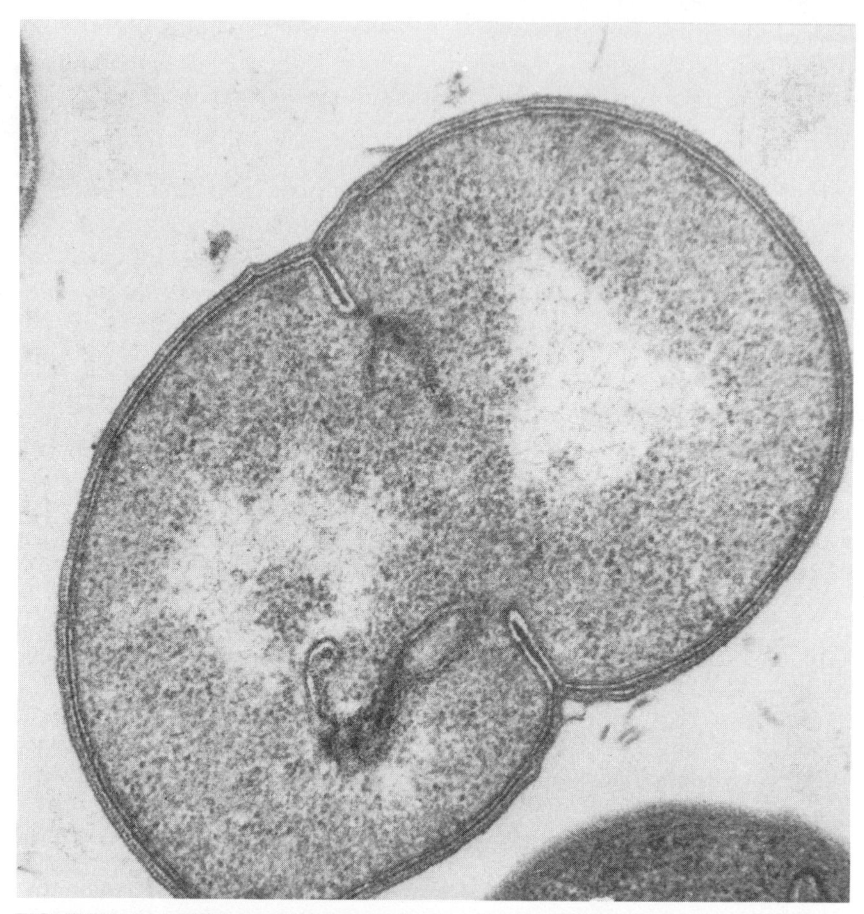

FIGURE 3.26 Electron micrograph showing the formation of a transverse septum in a dividing cell of *Sporosarcina ureae.* The septum grows down to separate the cells. The junction between new and old wall can be seen as a ridge (compare this with Figure 3.27). (Courtesy of Dr. T. J. Beveridge.)

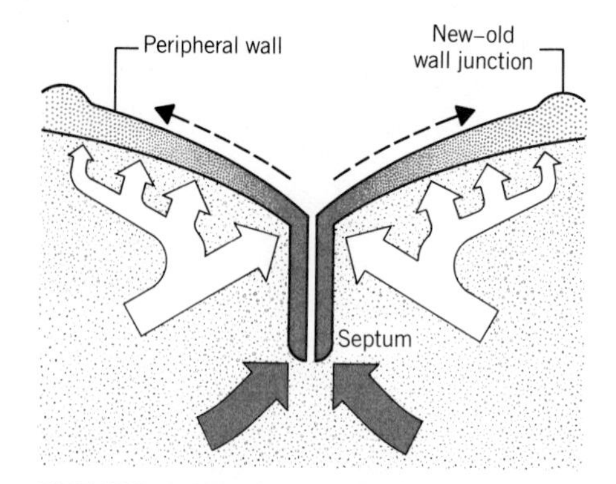

FIGURE 3.27 A model of cell wall expansion of the Gram-positive organism *Streptococcus faecalis.* New peripheral wall material grows out from the septum with the primary point of biosynthesis being the bottom of the septum. The two cell products can separate only after the septum has split.

vided some interesting insight into these ultrastructural maneuvers.[14] The septum seen in the electron micrograph appears to be the source of new wall material. The walls are fed out from the septum onto the surface. Eventually the septum splits, making it physically possible for the newly forming surfaces to move apart from one another. This is modeled in Figure 3.27.

The septum should be viewed as a ring encircling the entire cell. Therefore a new wall spreads over the entire surface of the cell beginning at a ring where new material meets preexisting wall. The septum does not close until the newly formed cells are complete.

Periplasmic Proteins

Outside the plasma membrane, but within the outer cell wall membrane, there resides a class of proteins called *periplasmic proteins*. One group of these proteins is made up of hydrolases. Their apparent function is to degrade large molecules that must be reduced in size before entry into the cell across the plasma membrane. Another function of the hydrolases may be protective, namely, degrading intruding foreign nucleic acids that may come from bacteriophages.

Another group of periplasmic proteins functions mainly with respect to transport. These proteins bind particular nutrient molecules, especially amino acids and sugars, and serve as the first step in active transport.

Still another group appears to function as receptor proteins. They may be involved in the chemotaxic activity of the organism.

3.2 THE WALLS OF EUCARYOTIC CELLS

The cells of fungi, algae, and higher plants are bounded by walls that differ considerably in composition and physical structure from those of procaryotic cells. Furthermore, major differences exist in wall structure among eucaryotic classes, especially when comparing photosynthetic and nonphotosynthetic organisms.

As is generally the case in the research world, certain organisms have been selected for study because of their availability and amenability to research techniques. Thus we have a substantial body of information on a limited number of species, and for the present these must serve to represent the rest.

The Fungi

Two distinct morphological varieties of fungi exist. One is a microscopic unicellular form, represented by common baker's yeast, *Saccharomyces cerevisiae*. A second is a filamentous form made up of branched, tubular structures called *hyphae*. A mass of hyphae containing reproductive structures constitutes macroscopic forms that are commonly called molds. Some fungi are dimorphic; that is, they take on one growth form in one environment and the other under different conditions. This is often true for the pathogenic fungi that grow as unicellular "yeasts" in the body and as "molds" in the laboratory.

Both forms have cell walls that provide support and impart structure to the cells. The prototype cell about which most is known is the yeast *Saccharomyces cerevisiae*.

Wall Architecture

The general picture of the fungal cell wall is one of a stratified structure, the degree of stratification depending on the maturity of the wall.

Figure 3.28 is a cross-sectional view of a growing hyphal tip of a commonly studied fungus, *Neurospora crassa*. The outer surface of the wall is smooth or slightly granular. The interior next to the plasma membrane contains crystalline components making up the skeletal portion of the wall.

Two high molecular weight polymers, *chitin* and the *β-glucans,* appear to contribute to the crystalline or fibrous properties of the inner wall. It is the presence of chitin, in particular, that sets the cell wall of fungi apart from other plants. The amorphous material of the wall, both coating and

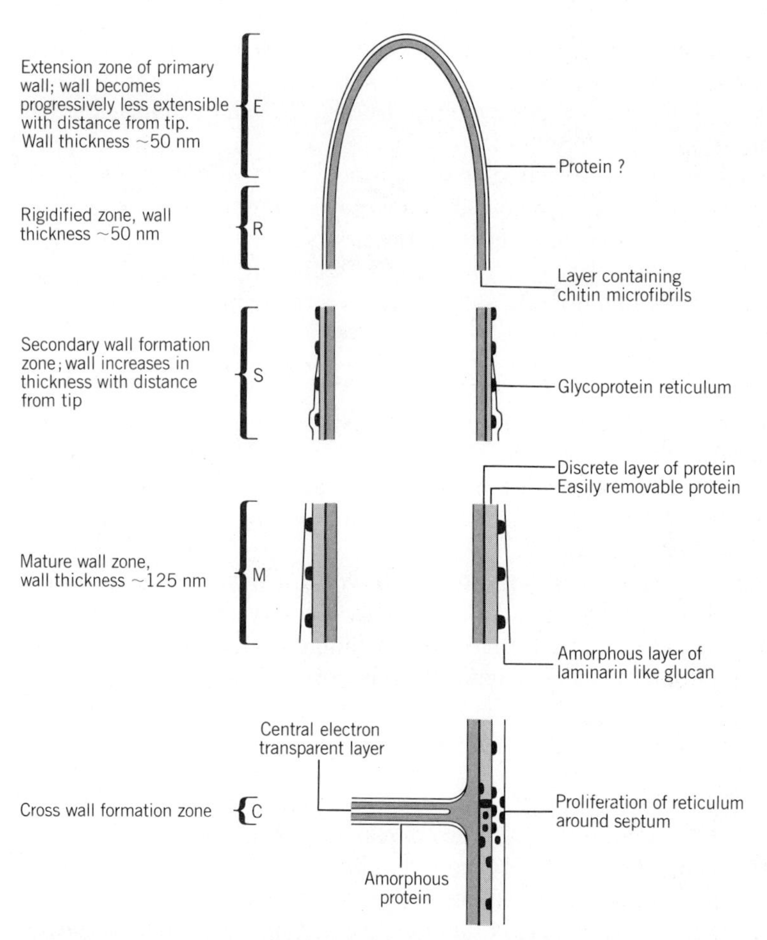

FIGURE 3.28 Wall and septal structure of a hyphal tip of *Neurospora crassa.*

penetrating the more highly structured skeleton, is largely a mixture of homo- and heteropolysaccharides often linked to proteins. Of these latter molecules, the mannan–protein molecule has been the most extensively investigated.

The repeating unit in chitin is:

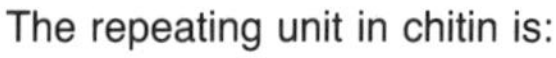

Chitin

Chitin is the major wall component in most of the filamentous fungi. It has a structural role in fungal cell walls comparable to that played by cellulose in the walls of other plants.

The structure of chitin is similar in certain respects to that of cellulose. It is a linear homopolysaccharide made up entirely of β-1,4-linked *N*-acetylglucosamine residues. These polymers are arranged in microscopic microfibrils 10 to 25 nm thick (Figure 3.29), much the same as is true for the cellulose in plant cells.

Glucans

Glucans are D-glucose polymers that exist in an array of forms in fungal cell walls. A limited amount of cellulose, a type of glucose polymer, is found in certain species, but its presence in fungi is more the exception than the rule. We defer a discussion of its structure until we consider the plant cell wall.

The β configuration is the most commonly observed interunit bond, with a mixing of β-1,3 and β-1,6 linkages. Figure 3.30 contains a partial structure of a yeast β-1,3-glucan molecule.

Two prominent physical properties of β-glucans contribute to their function as a skeletal framework for the cell: they have low solubility and a high degree of crystallinity.

FIGURE 3.29 Electron micrograph of internal surface of a sporangiophore wall of *Mucor rouxii* showing bed of chitin microfibrils. In this part of the wall the microfibrils run in directions approximately perpendicular to each other and oblique to the longitudinal axis of the sporangiophore. (Courtesy of Dr. S. Bartnicki-Garcia.)

The Mannan Component

Whereas chitin is a major ingredient in the walls of filamentous fungi, *mannan* is a major polysaccharide in most yeasts. In the wall, the mannan is covalently linked to protein, thus forming a stable glycoprotein complex.

The structure of the mannan component of the glycoprotein in the wall of *S. cerevisiae* is shown in Figure 3.31. Two species of mannan are shown, differing both in their structures and in their attachment to the protein.

One type of mannan has a branched outer chain and an inner core through which the polymer is linked by *N*-glycosidic bonds to an asparaginyl residue in the protein. In this species of mannan, the backbone linkage is α-1,6 throughout except for the last trisaccharide unit, where the 1,4-linkages are β. The branches are made up of a mixture of α-1,2 and α-1,3 glycosidic bonds. The linker region between the polymannose backbone and the asparagine is a diacetylchitobiose unit.

The other type of mannan is linear, with the polymannose backbone resembling the branch chains of the outer and inner core of the first species. The glycosidic bonds are α-1,2 and α-1,3. These polymers are directly attached to serine or threonine residues in the protein by an *O*-glycosidic bond that can be split by weak alkali.

Figure 3.32 is a schematic model of the mannan glycoprotein.

Attachments of mannan to protein:

O, H, O, N—C—CH_2—CH, NH, Protein, NH, $COCH_3$

Mannan–asparaginyl bond

O, O, C, O—CH_2—CH, NH, Protein, O

Mannan–serine bond
(configuration uncertain)

```
G—(1⟶3)—[G]a—(1⟶3)—G
                     |
                     1
                     ↓
                     6
                     |
G—(1⟶3)—[G]b—(1⟶3)—G—(1⟶3)—[G]c—(1⟶3)—G
                                        |
                                        1
                                        ↓
                                        6
                                        |
                    ····G—(1⟶3)—G—(1⟶3)—G····
```

FIGURE 3.30 Partial structure of a segment of yeast β-1,3-glucan. The exact lengths of *a, b,* and *c* are not known, but together they comprise about 60 glucose (G) residues. Note that all the linkages are β-1,3 except for branch points.

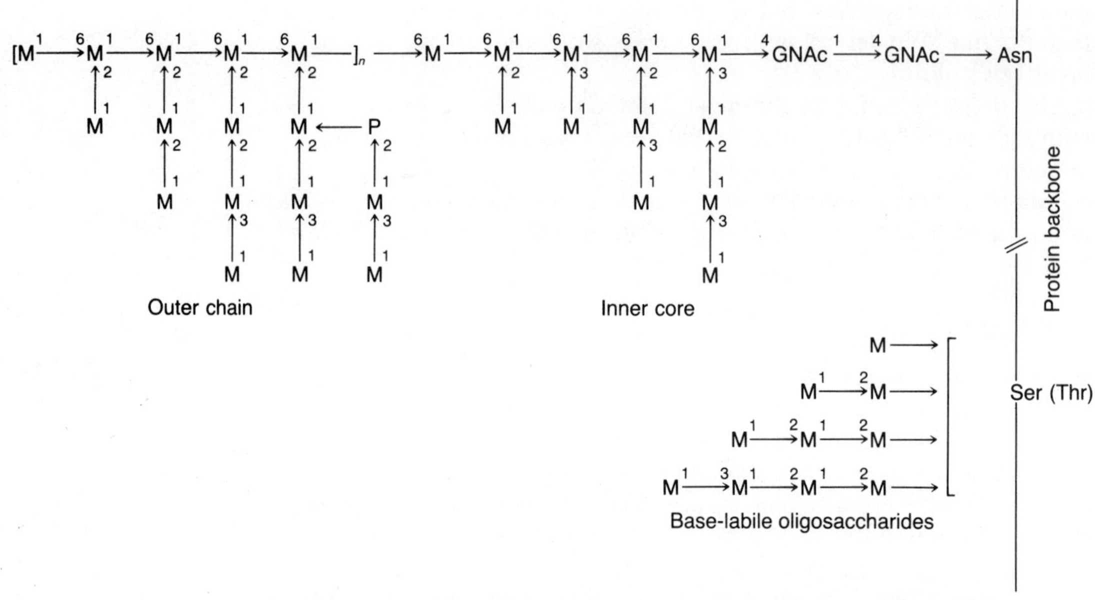

FIGURE 3.31 Detailed structure of the mannans from the yeast *Saccharomyces cerevisiae.* Two forms are shown, one linked to the protein via asparagine (Asn) and the other via serine (Ser) or threonine (Thr). M, mannose; GNAc, N-acetylglucosamine.

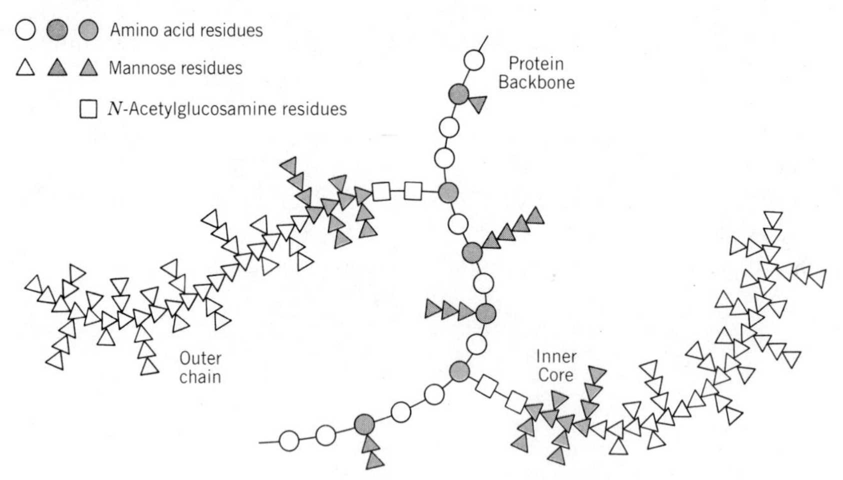

FIGURE 3.32 A schematic model of mannan emphasizing its highly branched nature. Mannose residues are distinguished as being attached to hydroxyamino acids (▲) or constituents of inner cores (▲) and outer chains (△). *N*-Acetylglucosamine bridges between polysaccharide chains and protein are represented by squares. Amino acid residues are circles, with ● indicating serine or threonine and ● asparagine.

The Biosynthesis of Fungal Walls

There are many unresolved questions concerning the mechanism of synthesis of fungal cell walls. But it is quite possible at this point to put the story together in general terms.

According to the foregoing discussions, two types of molecule are incorporated into cell walls. One type is a crystalline polysaccharide, represented by chitin and the β-glucans. The other is amorphous, made up

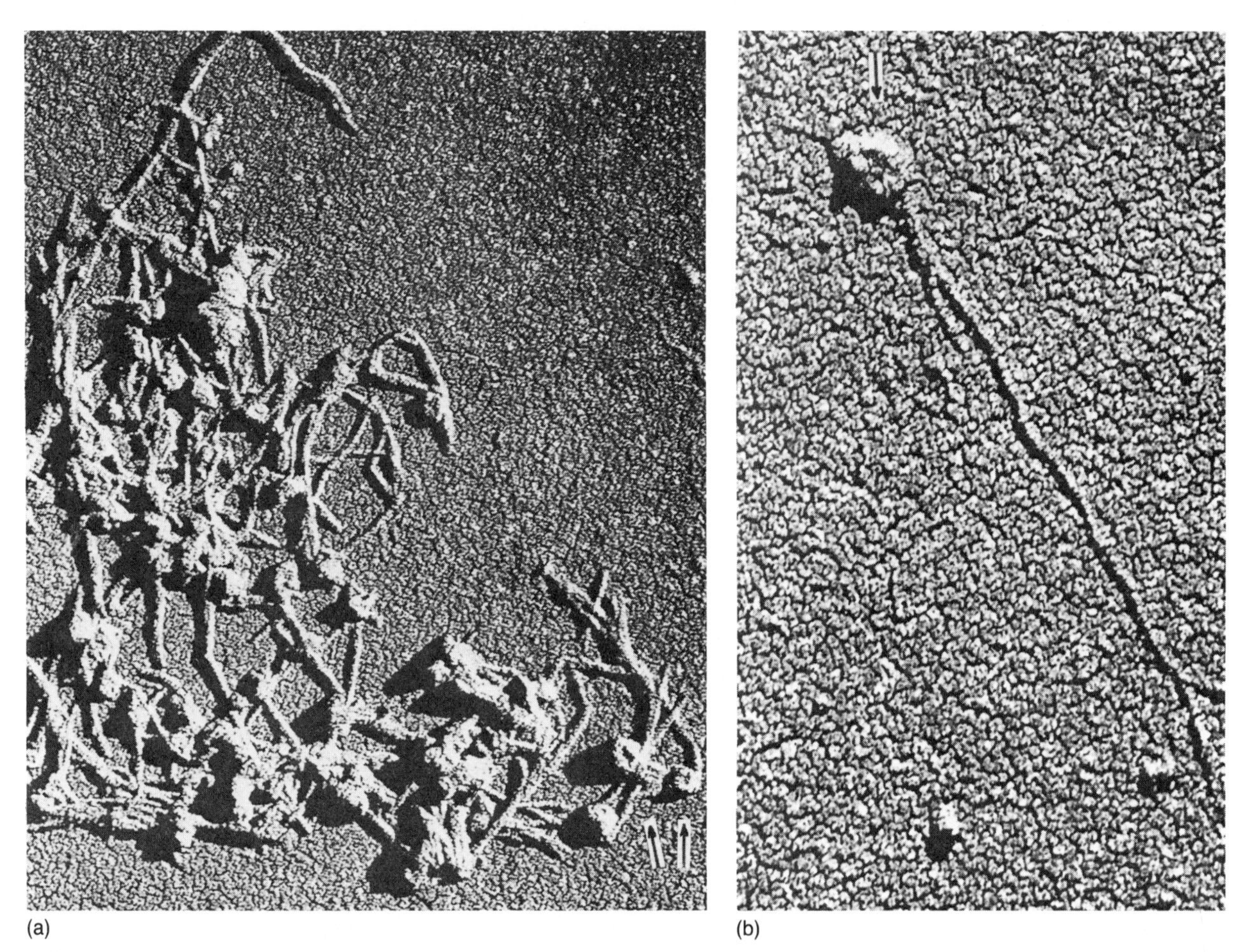

(a) (b)

FIGURE 3.33 Micrographs of isolated chitosomes. (*a*) A collection of chitosomes and associated microfibrils. × 33,300. (*b*) A single chitosome (arrow) showing an extended chitin microfibril that appears to be made up of two units wrapped around each other. × 56,600. (Courtesy of Dr. S. Bartnicki-Garcia.)

of polysaccharides and protein–polysaccharide complexes. These two types of wall component are assembled in significantly different ways.[15]

The crystalline skeletal polymers are synthesized by enzymes that are uniformly distributed in the plasma membrane. The general reaction carried out by chitin synthetase is as follows.

$$\text{UDP-GlcNAc} + \begin{bmatrix} \beta\text{-(1,4)-GlcNAc} \\ \text{primer} \end{bmatrix}_n \rightarrow \begin{bmatrix} \beta\text{-(1,4)-GlcNAc} \\ \text{product} \end{bmatrix}_{n+1} + \text{UDP}$$

A single enzyme, chitin synthetase, is apparently responsible for this polymerization reaction. As isolated from membranes, it appears in granules, which have been termed *chitosomes*. When incubated with UDP-GlcNAc, chitosomes spin out microfibrils of chitin chains, as shown in Figure 3.33. The microfibrils are made up of a large number of parallel chains of chitin, suggesting simultaneous synthesis of chains rather than aggregation of individually synthesized chains to form the microfibril.

The relationship between the chitosomes, seen in *in vitro* systems and the *in vivo* biosynthesis of chitin, is not certain. But what is clear is that chitosomes are too large (35–100 nm in diameter) to fit inside the plasma membrane (8–9 nm thick).

Less is known of the biosynthesis of the other skeletal polymers, the β-glucans. The enzyme system responsible for their biosynthesis appears to be associated with the plasma membrane, perhaps on its outer surface, and to utilize UDP-glucose as a substrate for the polymerization reaction.

An interesting property of both the chitin- and β-glucan-forming enzymes is that they exist in inactive and active forms. The mechanism of activation is currently being studied to gain insight into the regulation of fungal growth. Growth of the wall in hyphal cells is limited to the tip, where it is assumed the enzymes are activated for their biosynthetic work. In contrast, wall growth in spherical cells takes place uniformly over the entire surface area.

The amorphous polysaccharide and protein–polysaccharide components of the membrane are assembled by enzymes located in the endoplasmic reticulum of the cell. They are then packaged and moved to the cell periphery where they are expelled from the cell and added to growing wall points.

The biosynthetic route for the formation of the mannan molecule is quite complex, as could well be imagined from its highly branched structure. A scheme proposing the sequence of reactions in producing the inner core is presented in Figure 3.34: note that the mannosyl units are supplied by a GDP-mannose carrier. This carrier is sufficient for the direct attachment of mannosyl residues to growing outer and inner cores. However, the formation of the *O*- and *N*-glycosidic linkages to the protein requires a polyisoprene carrier called dolichol phosphate.

Dolichol phosphate is similar to undecaprenyl phosphate except that it contains 20 isoprene units instead of 11.

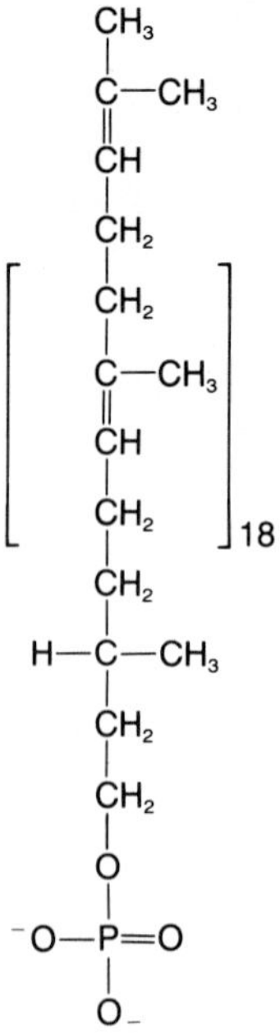

In summary, the sites of synthesis of the crystalline components of the wall, chitin and the β-glucans, are in or near the plasma membrane by constitutive enzymes located in these positions. The amorphous wall components are synthesized in the cell and exported to the surface in a manner similar to that for a secretory protein.

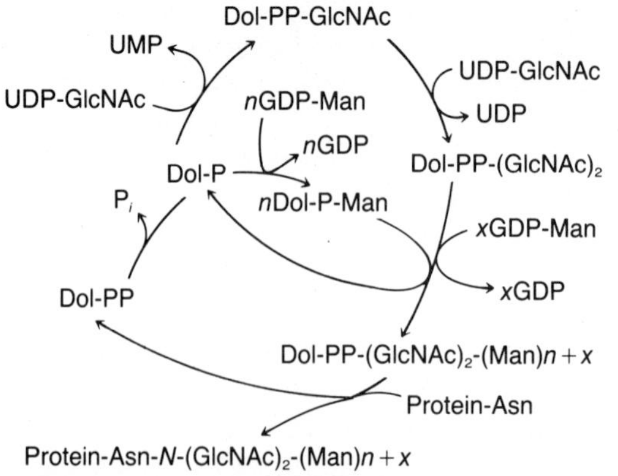

FIGURE 3.34 Scheme illustrating a proposed cycle of reactions for the synthesis of inner core material of yeast mannan. Two types of carrier are employed. Dolichol phosphate (Dol-P) carries sugar residues that become linked to protein via *O*- and *N*-glycosidic bonds. GDP carries mannosyl residues that merely increase the size of inner and outer cores by attachment to growing points on polysaccharide chains. $n + x = 0$ to 17 units.

A final question concerning cell wall synthesis is how new material can be inserted into a rigid wall structure. The sequence of events that is assumed to occur is as follows.

First, polysaccharide hydrolases begin to weaken the wall at a given site, thus causing a loss of contact between the wall and the plasma membrane. This loosening of the wall activates the constitutive polysaccharide synthetases located in the plasma membrane to begin production of skeletal polysaccharide material. This evokes formation of a bud on the wall, which marks the primary growth region.

Next, the appropriate cytoplasmic vesicles containing enzymes and wall material move from the cell interior into the primary growth region where their contents are discharged into the loosened zone. The wall continues to grow, and its morphological development depends on the activities and sites of postulated extracytoplasmic inhibitors operating on the polymerizing enzymes.

This is obviously a very superficial sketch of cell wall synthesis. There are still many unknowns and enormous amounts of difficult research that must be done before these biosynthetic mechanisms are completely sorted out.

Higher Plants

Perhaps the most distinguishing feature of plant cells, setting them apart from animal cells, is the cell wall. The walls of different species often contain unique geometrical patterns and, depending on the maturity of the cell, may be relatively thin and pliable or thick and rigid.

In aqueous environments walls serve the important function of protecting the protoplast from lysis as well as giving support to the plant. Plants growing in the air depend on walls to prevent dehydration and to make possible an aerial existence for the plant.

Electron Microscopy of Walls

An electron micrograph depicting a typical plant cell wall in cross section is shown in Figure 3.35. Even though the walls are thick and rigid, immobilizing the population of cells in these tissues, the walls are not absolute barriers. Cells still communicate by interconnecting pathways called *plasmadesmata*. Thus walls provide for functional interaction between cells, despite the impossibility of physical interaction.

Surface views of plant walls reveal that the wall is made up of microfibrils, similar in gross morphology to that which we saw in fungal cell walls. Figure 3.36 contains electron micrographs of two different regions in the cell wall.

One region is characterized by microfibrils that run in random orientation. These microfibrils are the basic morphological units of the cell wall. Their length is indefinite, but their widths are generally in the vicinity of 10 to 25 nm. A random orientation is characteristic of the *primary wall*, which is the first wall to form during cell replication and the only form that has plasticity.

A second region is characterized by microfibrils that run in parallel array. This more highly ordered structure is seen in *secondary walls*, which surround cells that are more mature.

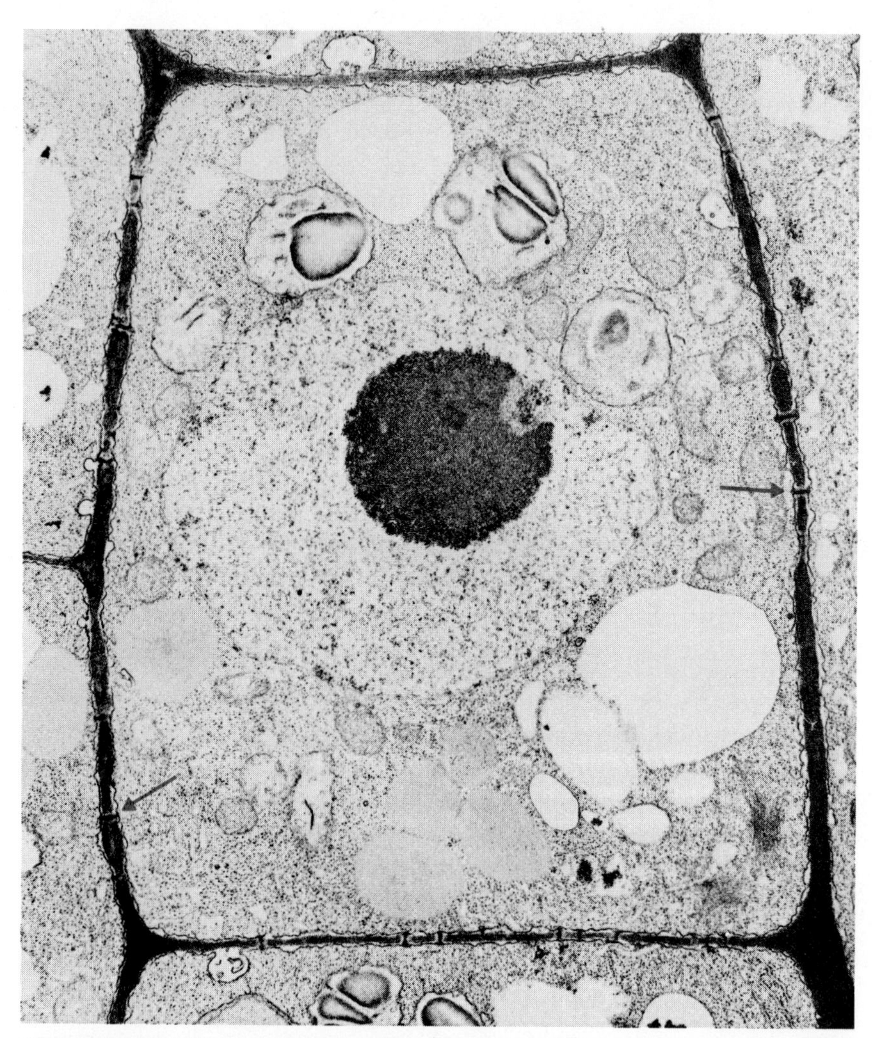

FIGURE 3.35 Electron micrograph showing cell walls of meristematic leaf of *Typha angustifolia*. Plasmadesmata connect the cytoplasm of adjacent cells (arrows). × 3,560. (Courtesy of Dr. Harry T. Horner.)

The Chemical Building Blocks of Wall Polymers

The principal chemical components of plant cell walls are polysaccharides. Walls are structurally biphasic: ordered microfibrils are embedded in an amorphous matrix. Both phases contain a variety of polysaccharides that are linked together to form a network of polysaccharides.

The polysaccharides of walls are made up of sugars linked by glycosidic bonds in ways that are characteristic of the particular polysaccharide of which they are a part. The most commonly observed monosaccharides are shown in Figure 3.37. Of the various conformations possible for these molecules, the predominant form is the pyranose chair (except for arabinose, which assumes a planar furanose ring form).

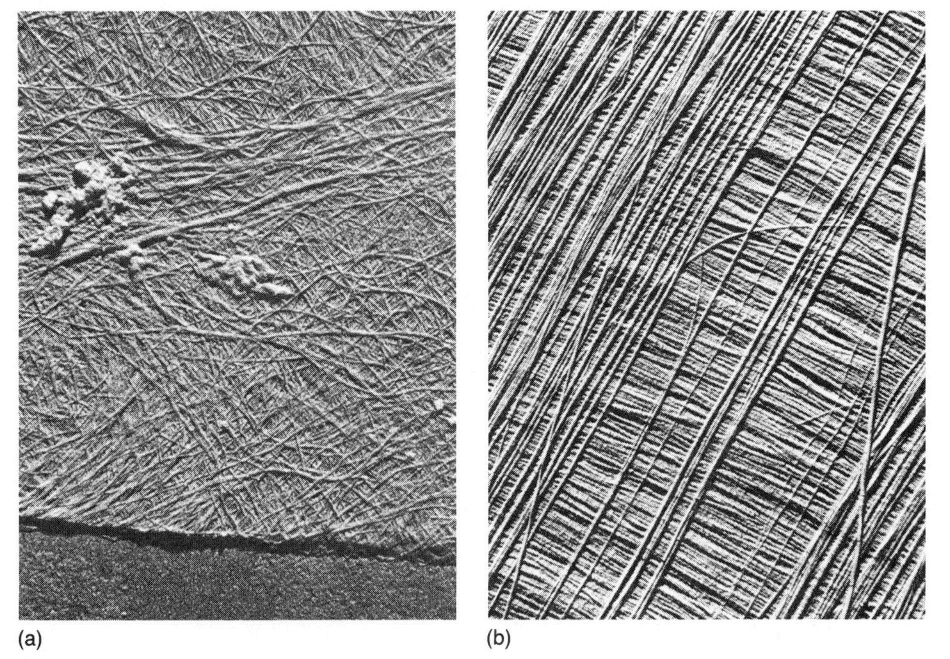

FIGURE 3.36 Surface views of primary (*a*) and secondary (*b*) cell walls illustrating the random cellulose microfibril arrangement in primary walls and the sheets of parallel microfibrils in secondary walls. (*a*) Parenchyma cells from a wheat coleoptile, metal shadowed. Microfibrils are 10 nm in diameter. × 20,800. (Electron micrograph courtesy of Dr. R. D. Preston.) (*b*) Two lamellae from the wall of the marine alga *Chaetomorpha melagonium,* metal shadowed. Microfibrils are 20 nm in diameter. × 18,150. (Courtesy of Dr. Eva Frei and Dr. R. D. Preston.)

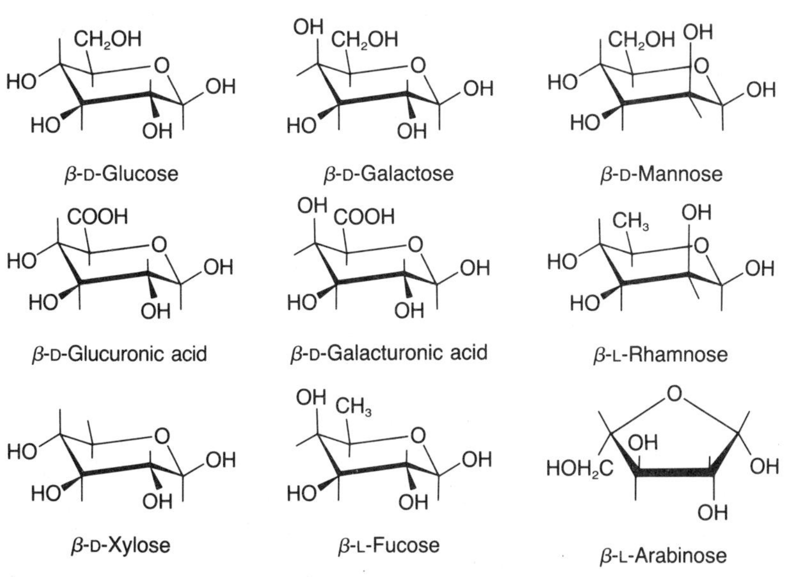

FIGURE 3.37 Common sugars of plant cell walls. The most stable conformation is the pyranose chair form except for arabinose, which prefers the planar furanose ring.

The Polysaccharides of Walls

Three types of polysaccharide are generally present in cell walls. These are *cellulose, hemicelluloses,* and *pectic polysaccharides.* In addition, structural protein is found in the primary wall, and biological plastics coat outer surfaces and impregnate the primary and secondary wall polysaccharides. The most common plant cell wall biopolymers are listed in Table 3.1.

Extraction procedures have been devised to selectively isolate the different polysaccharides present in the wall. Figure 3.38 is a scheme showing such an extraction sequence. Starting with whole dried cells, the major nonfibrillar polysaccharides can be removed and effectively separated from one another.

Cellulose. Cellulose is the most abundant polysaccharide product in nature. Of all the cell wall biopolymers formed by different cells, it has probably been the most extensively studied. Although it is present in both primary and secondary walls, most of the structural studies have been carried out on secondary wall cellulose. It is generally assumed that there is no significant difference in celluloses that may be related to their position within the cell wall.

Cellulose accounts for about 45% of the dry weight in typical woods and for 98% of the weight of the dry fiber of the cotton boll. The cotton boll has supplied the majority of chemical studies on cellulose structure.

Cellulose is a very high molecular weight linear polymer of glucose units linked with β-1,4-glycosidic bonds (see Figure 3.39). On the order of 8000 to 15,000 glucose units are present in a single molecule.

Some investigators believe that the glucan chains have no natural termination points but rather are of indefinite lengths. Termination during biosynthesis may simply be due to separation from biosynthetic enzymes.

Within the plant wall cellulose exists as fibers made up of 40 to 70 glucan chains. Actually, these measurements are very difficult to make, so the range of chain numbers is large. Fibers on the order of 4.5 × 8.5

Table 3.1 Carbohydrate Polymers of Plant Cell Walls

General Category	Structural Classification
Cellulose	β1,4D-Glucan
Pectic substances	Galacturonans and rhamnogalacturonans
	Arabinans
	Galactans and arbinogalactans I[a]
Hemicelluloses	Xylans [including arabinoxylans and (4-*O*-methyl)glucuronoxylans]
	Glucomannans and galactoglucomannans
	Xyloglucans
	β-D-Glucans (1,3 and 1,4)
Other polysaccharides	β1,3-linked D-glucans (callose)
	Arabinogalactans II[a]
	Glucuronomannans
Glycoproteins	

[a]Arabinogalactans of type I are characterized by essentially linear β1,4 D-galactan chains, whereas those of type II contain highly branched interior chains with 1,3 and 1,6 intergalactose linkages.

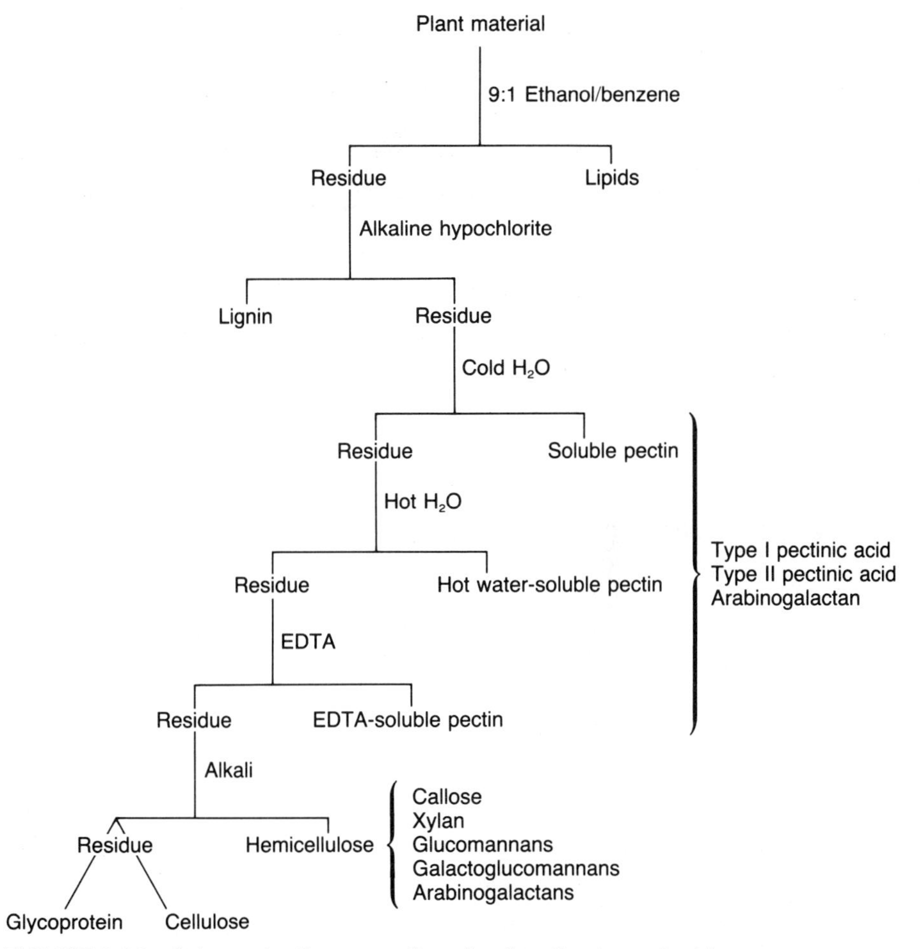

FIGURE 3.38 Scheme for the separation of cell wall polysaccharides.

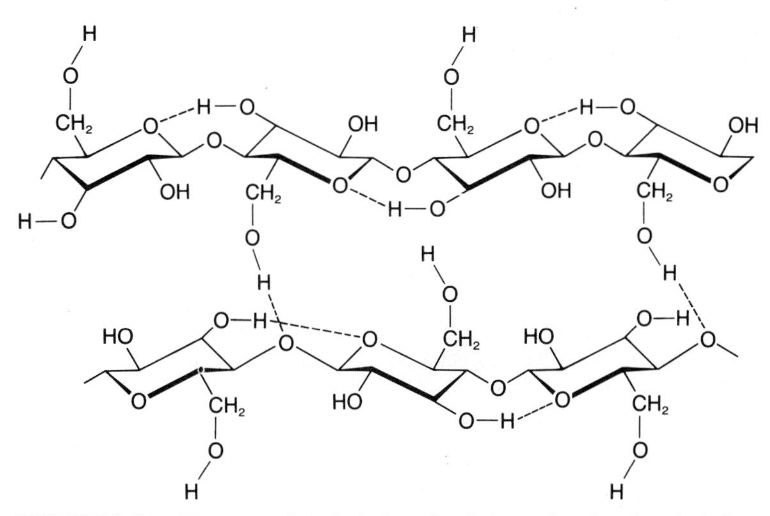

FIGURE 3.39 Two associated chains of cellulose showing the chair-form glucose units linked by β-1,4-glycosidic bonds. Both intrachain and interchain types of hydrogen bonding are present.

nm in diameter have been measured by electron microscopy, and these serve as a basis for the estimation of chain number.[16]

Based on the pioneering efforts in 1937 of Meyer and Misch applying X-ray diffraction to cellulose,[17] followed by additional studies by other research groups, a crystalline model of cellulose has been proposed. In this model alternating glucose residues lie in the same plane and adjacent cellulose chains run in an antiparallel fashion, stabilized by extensive interchain hydrogen bonding (see Figure 3.39). The repeating glucose residues are alternately inverted to minimize stearic hindrance and maximize hydrogen bonding between chains. The result is flat, ribbonlike structures that fit together so tightly that internal water is eliminated. It is an extremely stable, stiff crystalline rod.

In spite of the degree of order in cellulose and its crystalline nature, the run of the cellulose molecule in the microfibril is still a matter of differing opinions. Several models for the structure of the microfibril are shown in Figure 3.40. The first two models incorporate the idea of antiparallelism, but it is an intrachain rather than an interchain phenomenon. In the second two models the cellulose molecules are not folded back on themselves. The final two models propose the existence of crystalline cores around which are packed less ordered cellulose and hemicellulose molecules. In any event, it is clear that the last word in the crystalline structure of cellulose is not in. These different models are presented not to confuse the issue but to emphasize that this is a difficult experimental problem. These models have a useful function in serving as a frame of reference for the interpretation of results that will continue to come in for some time.

At this time, the antiparallel nature of the adjacent cellulose chains is a point of controversy. It is felt by many investigators that the biosynthesis of cellulose microfibrils would be most difficult to explain if the chains were antiparallel. It appears from the literature that a parallel orientation with the same polarity of chains may be the most common and accurate representation of fiber interior.

Hemicelluloses. The hemicelluloses are an extremely varied group of polysaccharides. They differ from cellulose in that they are heteropolymers and in that they are branched molecules (Figure 3.41).

Some of the common hemicelluloses go under the names xylans, arabinoxylans, glucomannans, galactomannans, and xyloglucans. The names do not tell us much about their structures, but do reflect their compositions.

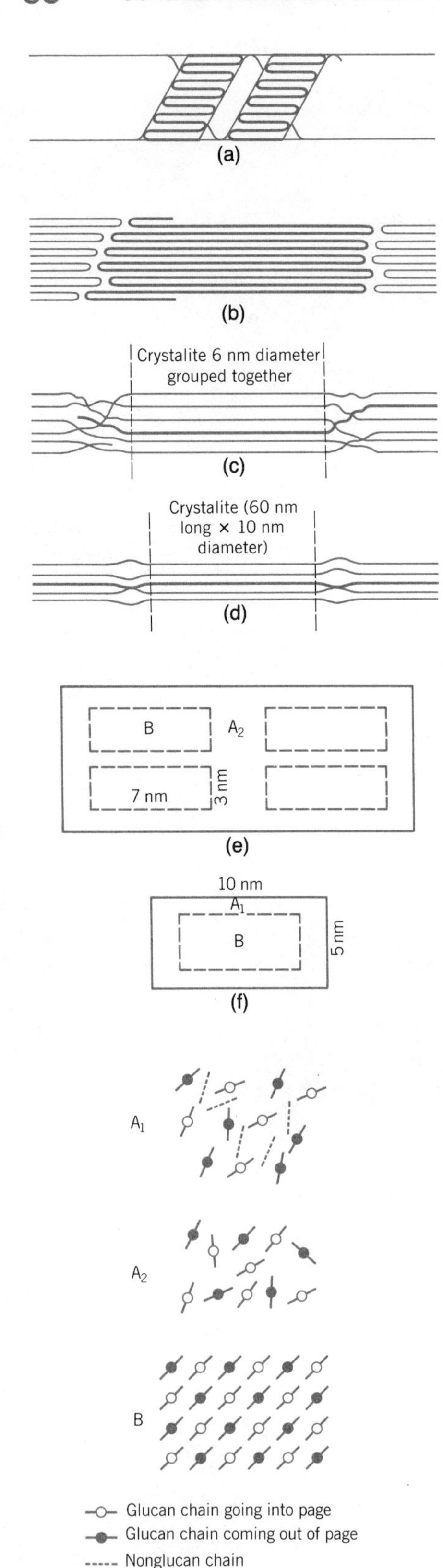

FIGURE 3.40 Models for the structure of the cellulose microfibril. In models (*a*) to (*d*) one cellulose chain is drawn heavily to clarify its run. (*a*) Each molecule is folded on itself in a plane and then twisted (model of Manley). (*b*) Each molecule is folded parallel to the fiber axis (model of Marx-Figini and Schulze). (*c*) Fibers run from crystallite to crystallite and start and end in intervening regions of lesser order (model of Hess, Mahl, and Gutter). (*d*) As in (*c*) except fibers start and end in crystallite (model of Ranby). (*e*) and (*f*) These models depict crystalline cores around which are packed less ordered polysaccharides (models of Frey-Wyssling and Preston and Cronshaw, respectively).

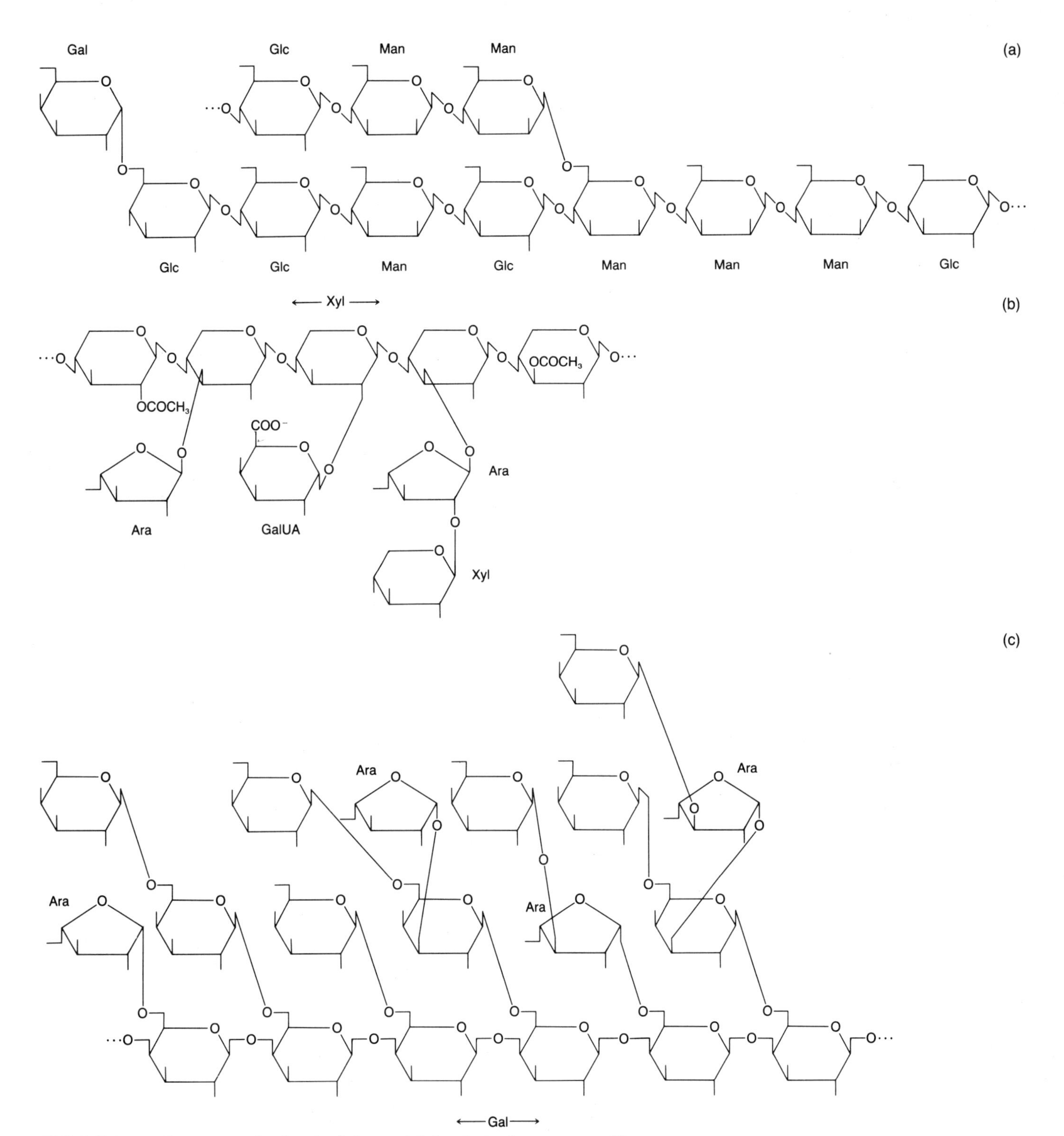

FIGURE 3.41 Representative hemicelluloses. (*a*) A galactoglucomannan. These hemicelluloses generally contain more mannose than glucose and are often more heavily galactosylated than shown here. Glucomannans are similar to this except they lack galactose. (*b*) A xylan. The core of this is a xylose polymer with short side chains containing arabinose and galacturonic acid. (*c*) An arabinogalactan. This complex hemicellulose consists of galactose and arabinose. Ara, arabinose; Gal, galactose; GalUA, galacturonic acid; Glc, glucose; Man, Mannose; Xyl, xylose.

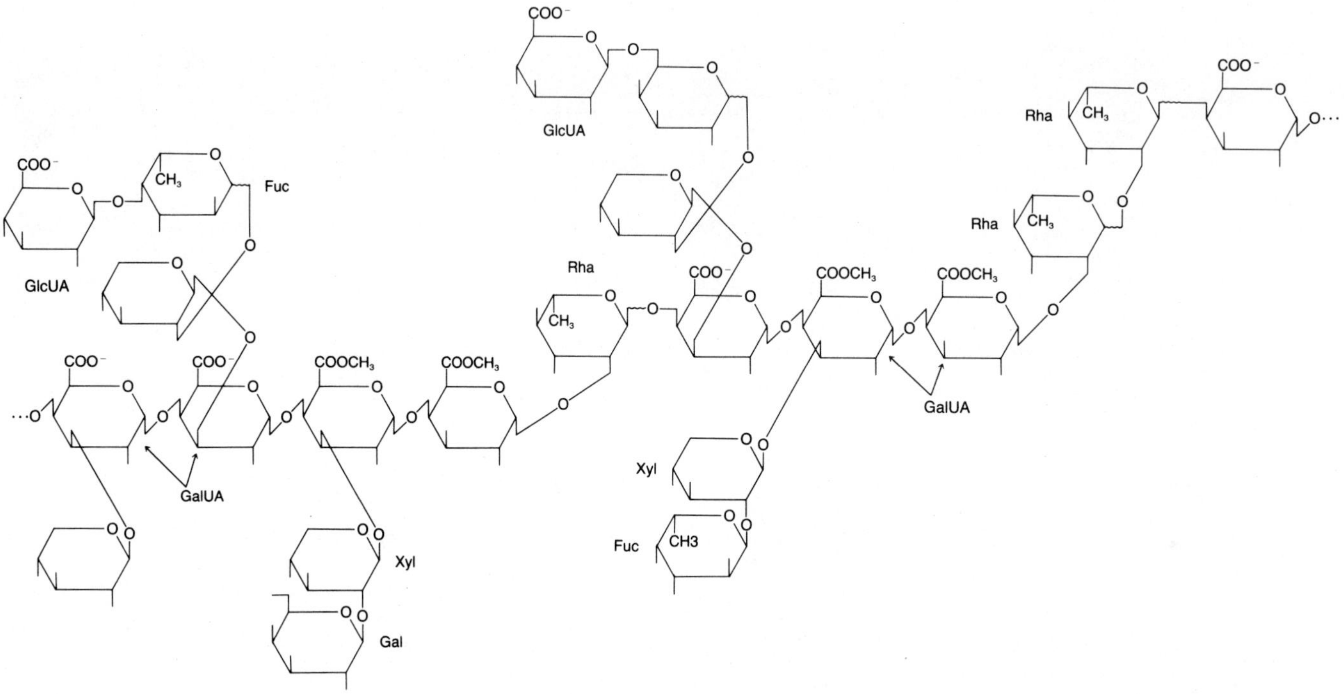

FIGURE 3.42 Detailed structure of a representative pectin. This is a short section of a galacturonorhamman chain to which are attached various types of branches. Many variations of these structures occur. Carboxyl groups exist in free form or are acetylated. Fuc, fucose; GalUA, galacturonic acid; GlcUA, glucuronic acid; Rha, rhamnose.

The various hemicelluloses have two structural features in common:[18] a flat β-1,4-linked backbone with short side chains, often just one sugar, and a three-dimensional structure that discourages self-aggregation into crystalline structures. Although these materials cannot self-crystallize readily, they do appear to cocrystallize with cellulose. A function of the hemicelluloses may therefore be to coat cellulose microfibrils, acting as a gluelike substance.

Pectins. Pectins are a very complex family of polysaccharides that show great structural variation. One common feature is the presence of acidic groups, contributed mainly by glucuronic and galacturonic acid residues.

An example of a pectin is shown in Figure 3.42. It is highly branched and acetylated and contains carboxyl groups that are negatively charged at physiological pH.

Structural Proteins and Plastics

Cell walls also contain nonpolysaccharide components. Protein fragments with very unusual features have been isolated from primary cell walls.

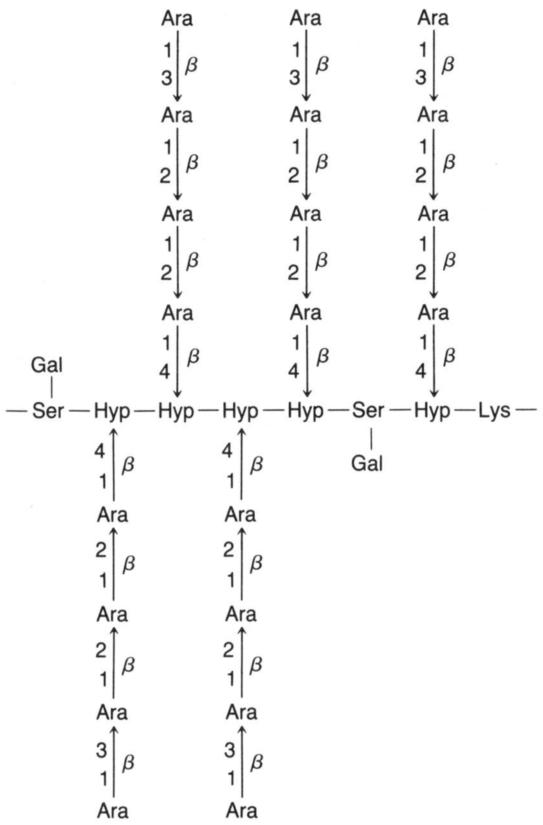

FIGURE 3.43 A model for a portion of the hydroxyproline-rich structural glycoprotein of primary cell walls. Note the sequence Ser–Hyp_4 in the protein. Ara, arabinose; Hyp, hydroxyproline; Gal, galactose; Ser, serine; Lys, lysine.

The protein is rich in hydroxyproline, containing over 25% hydroxyproline in its makeup. Tryptic peptides appear to commonly contain the sequence Ser-Hyp_4, with arabinose tetrasaccharides glycosidically linked to the hydroxyproline residues (Figure 3.43). In other instances the hydroxyproline residues carry glucose and galactose.

The three-dimensional structure of this protein is thought to be a rigid, extended rod that is quite resistant to proteolytic breakdown. Although its function in the wall is unknown, it is speculated to play a role in cell wall organization.

Another nonpolysaccharide material in the wall is a biological plastic called *lignin*. Lignins arise by a free radical polymerization of a family of lignin precursors shown in Figure 3.44. Lignin impregnates both the primary and secondary walls with a hydrophobic encrusting material, providing rigidity and resistance to deterioration by the environment.

Cutin is also a biological plastic found as a coating material on the cell walls of surface cells. Its structure has not been worked out, but it is known to contain C_{16}–C_{18} hydroxy fatty acids, which in turn are covalently linked to each other by way of ester bonds. These polymers are further coated and impregnated with waxes. This surface is very resistant to

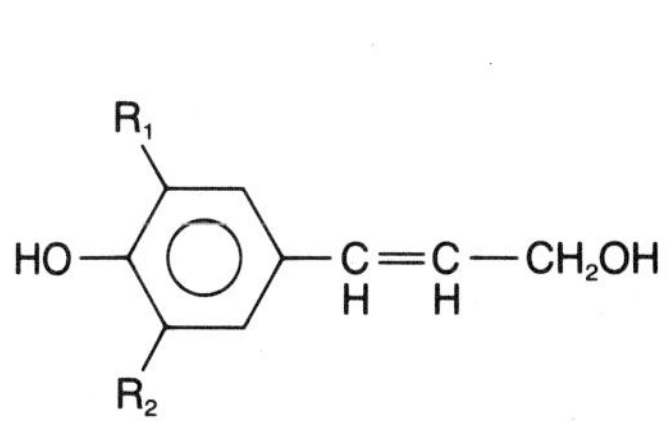

FIGURE 3.44 Lignins are nonpolysaccharide biological plastics that arise by a polymerization of free radicals of precursors based on this structure. Depending on the R groups, these precursors are called coumaryl alcohol (R_1 = H, R_2 = H), coniferyl alcohol (R_1 = H, R_2 = OCH_3), or synapyl alcohol (R_1 = OCH_3, R_2 = OCH_3).

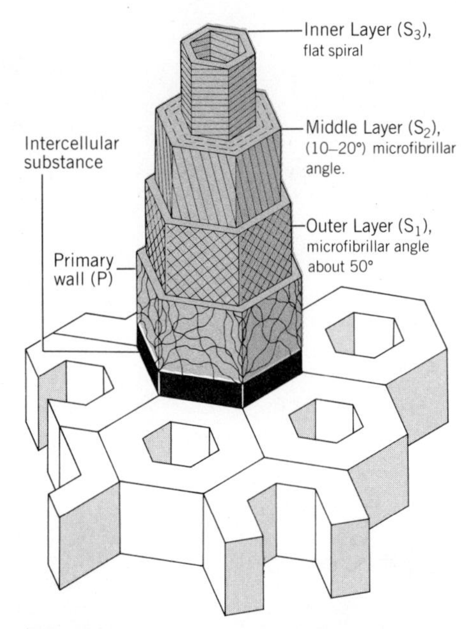

FIGURE 3.45 Model showing the secondary wall depicted as consisting of three regions, S_1, S_2, and S_3. Microfibrils are the most regularly oriented toward the interior of the wall.

dehydration and also serves as a protective barrier for the cell against injury or pathogens.

Primary and Secondary Walls

Plant cell walls are generally layered, or lamellar, structures. Two layers, quite arbitrarily defined, are commonly noted, termed primary and secondary walls. The characteristics of these layers are outlined in Table 3.2.

The primary cell wall is characterized structurally by having microfibrils dispersed in a noncrystalline matrix, and functionally by its ability to flex and extend along with that of the protoplast. The hemicellulose content is high and the cellulose content quite low. Pectic substances are also present in the primary wall. The primary wall is the first wall structure laid down next to the middle lamella, the mid region between cells.

In contrast, the secondary wall is characterized structurally by microfibrils arranged in parallel order and functionally is rigid and nonextensible. The hemicellulose content is lower than that found in primary walls and the cellulose content may represent a majority of molecular species in the wall.

The secondary wall is a multilamellar structure with three layers often seen. These are designated S_1, S_2, and S_3 (Figure 3.45). The microfibrils in these layers run parallel, but in different directions in the different layers.

Not all layers are present in all plants. The S_2 layer, for example, may be the predominant layer, as it is in wood, which has properties that dictate the overall structural features of the plant. Secondary walls often become encrusted with lignin and the hemicelluloses.

Both the primary and the secondary walls have been likened structurally to reinforced concrete. The cellulose microfibrils are analogous to the reinforcement rods, and the remaining noncrystalline materials are comparable to the concrete matrix. The strength provided by this kind of design is remarkable for living tissue. The tensile strength of the cell wall of flax fibers is about 110 kg/mm^2, compared to steel, which is on the order of 150 kg/mm^2. The analogy is good up to a point. It fails at the point of plant cell growth, where the primary wall is required to be extensible.

The primary wall may represent the limit of wall formation in many plants, and it is the required wall form at the growth points of plant tissue. Once a secondary wall has been deposited on a primary wall, wall extensibility is lost.

As a result of studying the primary wall of sycamore cells, which have three principal noncellulose polysaccharides, Peter Albersheim and his colleagues have formulated a model of the cell wall that places into perspective the relation between cellulose and the noncellulose polysaccharides (see Figure 3.46). In their model (Figure 3.47), the cellulose microfibrils, each made up of about 40 cellulose chains, are interconnected by polysaccharide bridges. Xyloglucan molecules coat the microfibrils and rhamnogalacturonan polysaccharides run parallel to the fibers but at a distance from them. Arabinogalactan bridges connect the two, thus tying all units together. In this model, cellulose is hydrogen bonded to xyloglucan, but the bonds between the other polysaccharides are covalent glycosidic linkages.

Table 3.2 Characteristics of Primary and Secondary Walls

Characteristic	Primary Walls	Secondary Walls
Flexibility and extensibility	High	Low
Thickness	Dynamic	Static
Microfibril order	Nonordered	Parallel layers
Growth	Multinet intussusception	Apposition
Crystallinity	~40%	~70%
Cellulose content	Low	High
Hemicellulose content	~50%	~25%
Lipid content	5–10%	Little or none
Protein content	~5%	Low

Cell Wall Growth

The physical problems of cell wall formation and extension, given the properties of the wall as discussed, are difficult ones that are not completely understood. Nevertheless, the picture is getting clearer and researchers are beginning to think more in terms of molecular models for growth.

First, it is important to recognize that plant growth involves two distinct processes: cell division, then cell elongation.

Cell division takes place in specialized regions of tissue called meristematic regions. These are found at growing points, such as the tips of roots and stems and in the buds that generate flowers and leaves. New daughter cells, formed in these tissues, are smaller than older cells, having arisen by a halving of parent cells.

Once a daughter cell has formed, cell elongation takes place, a process that involves extension of the primary wall. Figure 3.48 illustrates how the primary cell wall microfibrils reorient during cell elongation according to the *multinet growth theory*. The microfibrils are laid down on the inner face of the wall in a direction transverse to the length of the cell. As the wall elongates, the microfibrils are reoriented toward the longitudinal axis of the cell until they are essentially parallel with this axis.

It is apparent from this picture of wall extension that the microfibrils move relative to one another. Thus, molecular models of the cell wall must account for both static and dynamic wall features.

The dynamic feature of wall extension is explained by Albersheim and co-workers according to the model in Figure 3.49. In this highly speculative diagram, postulated enzymes cleave the cross-bridging polysaccharides. The stress of elongation then moves the microfibrils relative to one another. When new positions are found for the microfibrils, the polysaccharides again reform their covalent interactions, anchoring the microfibrils firmly in place.

The Formation of Cellulose Fibrils

Although the chemical composition and structure of the glucan polymer of cellulose is known with certainty, the physical placement of the chains within the polymer is uncertain. Even the matter of microfibril size is debatable. For many years there has been a search for the "elementary

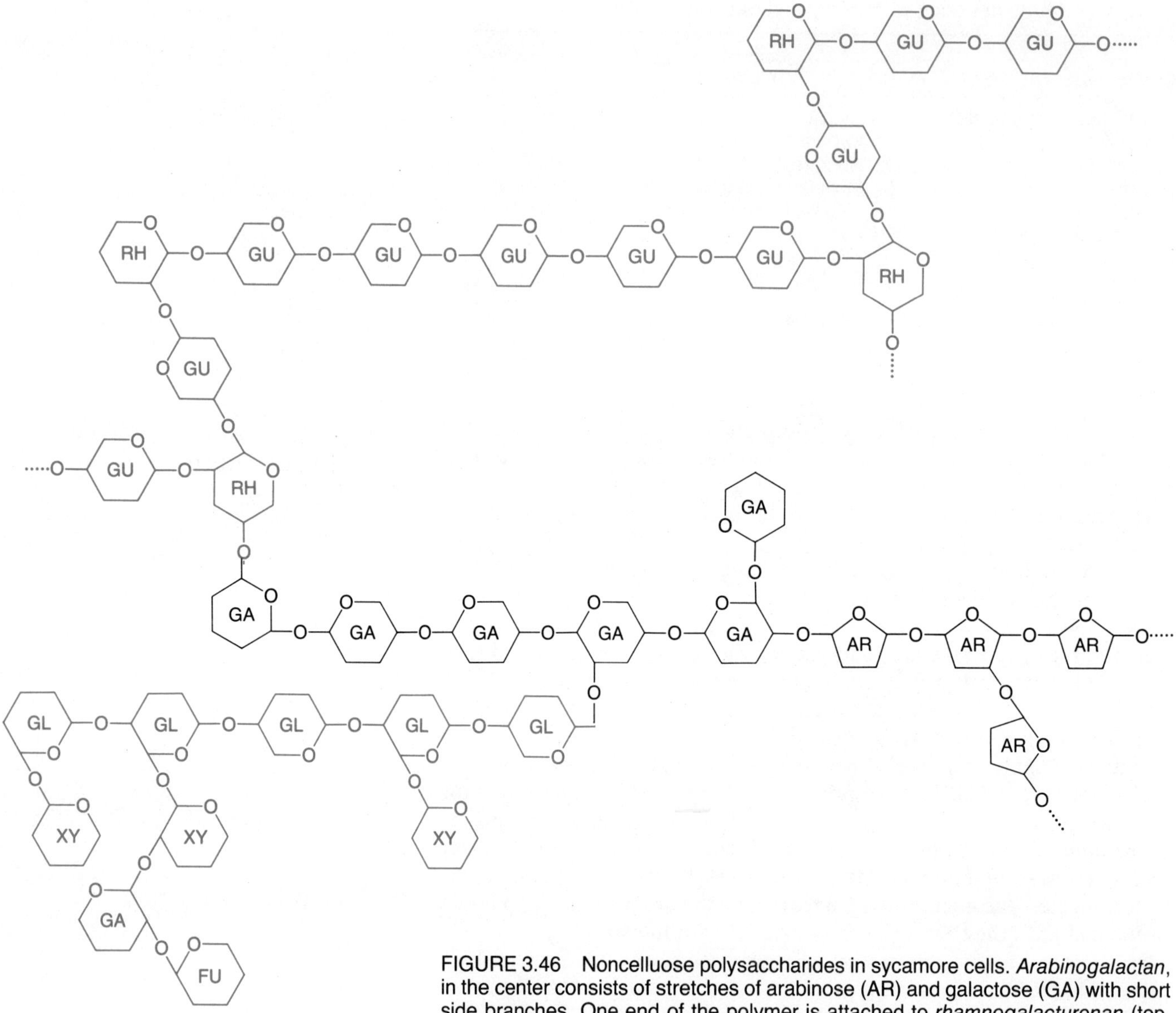

FIGURE 3.46 Noncelluose polysaccharides in sycamore cells. *Arabinogalactan*, in the center consists of stretches of arabinose (AR) and galactose (GA) with short side branches. One end of the polymer is attached to *rhamnogalacturonan* (top, color), which is a straight polymer of galacturonic acid (GU) interrupted by rhamnose (RH) kinks. *Xyloglucan* (bottom, color) is also linked directly to arabinogalactan. This polymer is made up of β-1,4-linked glucose (GL) units to which xylose (XY), galactose,and fucose (FU) are branched.

fibril,'' that is, the fundamental unit on which larger structures can be built. But the search appears not to have reached its conclusion. If, in fact, there is an elementary fibril, it may well vary with the species of cell producing the cellulose.

Since the physical form of the immediate product of biosynthesis is uncertain, the task of determining the mechanisms of biosynthesis has been especially difficult. Clearly, if the cellulose-containing plant cell uses mechanisms that parallel other wall-forming cells, there are three stages

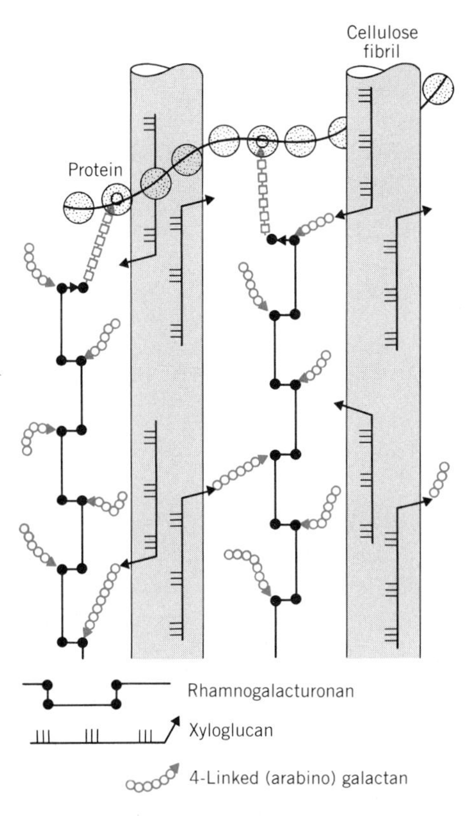

FIGURE 3.47 Model showing spatial arrangements between major cell wall components in sycamore primary walls. Xyloglucan, a hemicellulose, is shown cocrystallized with cellulose. The reducing ends of some of the xyloglucans are glycosidically linked to arabinogalactan side chains of the rhamnogalacturonan polymers. Thus rhamnogalacturonans are the main bridging element between cellulose fibrils through their side chains. In addition, certain of these side chains tie into structural proteins. Arrowheads indicate reducing ends of polysaccharides.

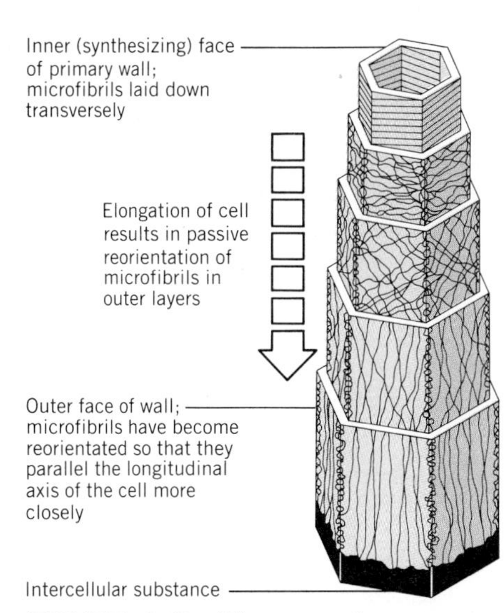

FIGURE 3.48 Diagrammatic representation of multinet growth. The transversely oriented microfibrils in the primary walls of elongated cells become reoriented as the cell extends.

of biosynthesis. First, the fundamental unit of the glucan chain, glucose, must be activated in the cell interior, followed, perhaps, by partial or complete polymerization into the glucan chain. Next, the intermediate or final glucan polymer must be transported across the plasma membrane to the cell exterior. Finally, microfibril formation and orientation must take place. Beyond this, there is the formation of cross-links between polymer molecules.

The Biochemistry of Cellulose Biosynthesis

The first stage in cellulose biosynthesis is the activation of glucose by the formation of a glucose nucleotide. This is a common reaction employed in biosynthesis to provide "active" glucose for polymerizations.

The form of nucleotide used in this step varies somewhat with species.[19] Some of the cellulose-forming bacteria and cotton use UDP-glucose; other

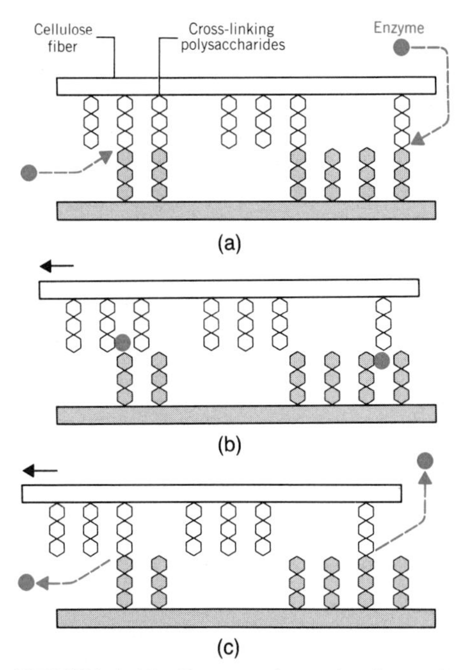

FIGURE 3.49 Proposed mechanism of cell wall extension. Enzymes break the linkage between two cell wall polysaccharides (*a*) and remain attached to one of the break points. The polysaccharides are then free to shift (*b*) and move until the enzyme reestablishes new linkages (*c*).

green plants use GDP-glucose. The activation reaction in either case is as follows.

$$\text{Glucose-1-phosphate} + \text{UTP (or GTP)} \longrightarrow \text{UDP-glucose (or GDP-glucose)} + PP_i$$

Glucose-1-phosphate

UTP (or GTP)

UDP-glucose (or GDP-glucose)

You will recall that during the biosynthesis of peptidoglycan in bacterial cells, the intermediate peptidoglycan disaccharide unit was transferred across the membrane by a lipid carrier (see Figure 3.17). There is evidence for a lipid carrier in cellulose biosynthesis also, but its role may not be limited to transport from cell interior to exterior.

A scheme summarizing the events of cellulose biosynthesis up to this point is presented in Figure 3.50. This cycle represents what is thought to occur mainly in bacterial systems. UDP-Glucose gives up its glucose to a lipid carrier that appears to contain 13 isoprene residues but is of unknown final structure. Upon reaction with another UDP-glucose, the carrier takes on another sugar unit, forming cellobiose. Cellobiose is then transferred off the carrier to an acceptor end of a soluble cellulose polymer.

It appears that green plants also use a lipid carrier in a scheme very similar to that of bacteria. But for green plants there appear to be two features that differ from the bacterial scheme: (1) the lipid carrier is different, having the properties of dolichol, and (2) a protein may be involved as a shuttle between the lipid carrier and the growing glucan polymer.

All the reactions up to this point, including the formation of glucan oligomers, take place within the cell. In bacteria, the reactions, particularly those involving the lipid carrier, probably take place near or on the plasma membrane. In green plants part of the process includes Golgi complex membranes and part polymerization on the plasma membrane.

Physical Mechanisms of Microfibril Formation

The second and third stages of cellulose fibril formation are transport across the plasma membrane and fibril growth and orientation, respectively. A good deal of speculation and controversy surrounds these two stages because the appropriate experimental evidence is difficult to obtain.

Bacteria. Cellulose synthesized by bacteria, such as the organism *Acetobacter xylinum,* is apparently formed without the aid of intact membranes or walls. This was demonstrated as early as 1958 by L. Glaser[20] who showed cellulose synthesis by a cell-free particulate enzyme system.

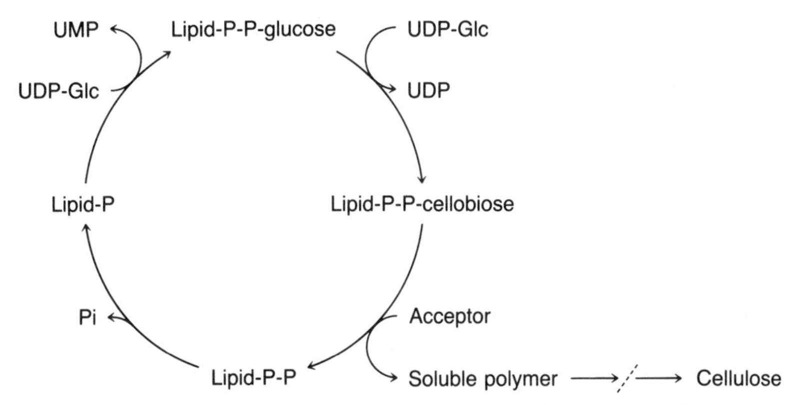

FIGURE 3.50 Summary of events in cellulose biosynthesis. Glucose is transferred from a UDP carrier to a lipid carrier on which a cellobiose disaccharide unit is created. Polymer length increases by the addition of cellobiose units.

Thus the biochemical mechanisms of polyglucan formation are sufficient to make the polymer, and an alignment of polyglucan chains to form the microfibril occurs without direct membrane or wall assistance.

More recent studies by J.R. Colvin and co-workers[19] have shown that cellulose polymers are extruded from the cell through funnel-shaped pores in the wall. These infant fibrils are small and pliable. In aqueous media they align spontaneously to form highly hydrated assemblies about 100 nm wide. With time the assemblies rearrange into a more compact and crystalline form characteristic of the mature microfibril.

These observations on bacterial systems have provided the important information that microfibril formation can be a purely physical process once the polyglucan chains have been synthesized.

Higher Green Plants. The physical process of microfibril formation in higher plants is largely unexplored. In some systems that have been examined by electron microscopy cellulose fibers have been observed on the outside of the plasma membrane. This suggests that the physical formation of the fibril takes place outside the protoplast.

In vitro synthesis of cellulose is most difficult to achieve in higher plants, and until systems have been devised to support this, the process will probably remain beyond the reach of good experimental analysis.

Green Algae. Some recent work on the unicellular green alga *Micrasterias denticulata* by Andrew Staehelin and co-workers has uncovered some ultrastructural features of the membranes of these organisms that appear to be related to microfibril formation.[21] These organisms lay down a primary cell wall during cell division and synthesize a secondary wall when more mature. Eventually the primary wall is shed and replaced by an amorphous layer outside the secondary wall.

Fracturing through the amorphous layer gives a view of the outer surface of the secondary wall (Figure 3.51). Parallel fibrils can be easily seen, occurring in bands of up to about 17 multiple fibrils. Close examination of these bands has revealed that the fibrils toward the center of the bundle are large and planar and those toward the periphery are smaller and rounded. The larger fibrils have diameters around 28 nm.

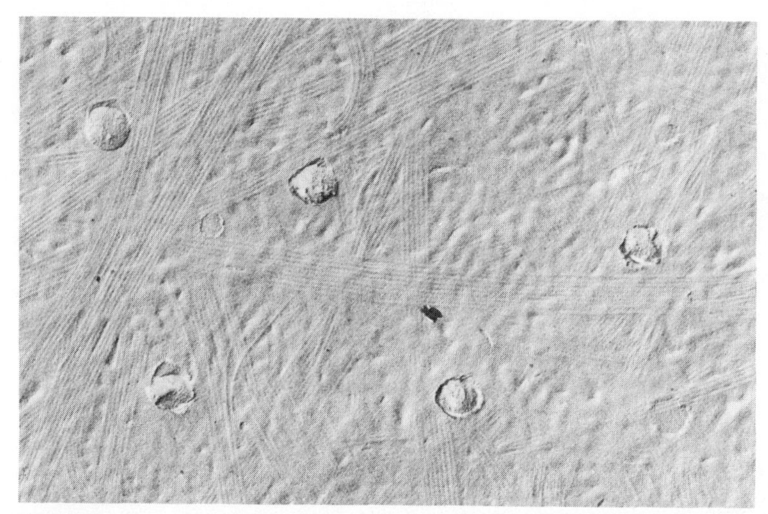

FIGURE 3.51 Electron micrograph showing the fibril packing, as revealed when a fracture is made through the amorphous cell wall layer of the green alga *Micrasterias denticulata.* The ridges correspond to the fibrils of the underlying secondary cell wall. × 18,000. (Courtesy of Dr. Thomas H. Giddings, Jr.)

Fractures through the plasma membrane reveal particles in the membrane that line up in hexagonal array (Figure 3.52). Each particle is composed of six subunits put together to appear as a rosette of 22 nm outer diameter. The central space between the subunits is 7 to 8 nm, about the same diameter as the individual subunits of the rosette.

It is of striking interest that the largest complexes have about 175 rosettes lined up in about 16 rows. In other words, there is a close correlation between the number of rows of rosettes in a complex and the number of fibrils in the bands of cellulose in secondary walls. Not only do numbers correspond, but so also do distances. The distance between fibrils in the band is the same as the distance between rosettes.

The conclusion drawn is that the widest fibrils in the band are made by the longest rows of rosettes and the narrower outer fibrils by the shorter outside rows.

Cells that are engaged in the synthesis of primary cell walls have membranes containing randomly distributed rosettes. This would be in accord with the observation that primary cell walls have randomly oriented microfibrils.

These observations are summarized in the models presented in Figures 3.53 and 3.54. Figure 3.53 shows a model of a rosette from which a microfibril of about 5 nm is being extruded. A microfibril of this width could contain about 50 polyglucan chains. This microfibril may be the elusive elementary fibril discussed earlier in this chapter. At least for this particular organism this would appear to be the case.

As the microfibril extends in length, the rosette is thought to move laterally in the plane of the biomolecular leaflet. The random movement would give rise to randomly oriented fibrils that run over and under one another.

Figure 3.54 is an interpretation of how aligned rosettes would give rise to fibrils of varying dimensions within a band in secondary walls. Each

(a)

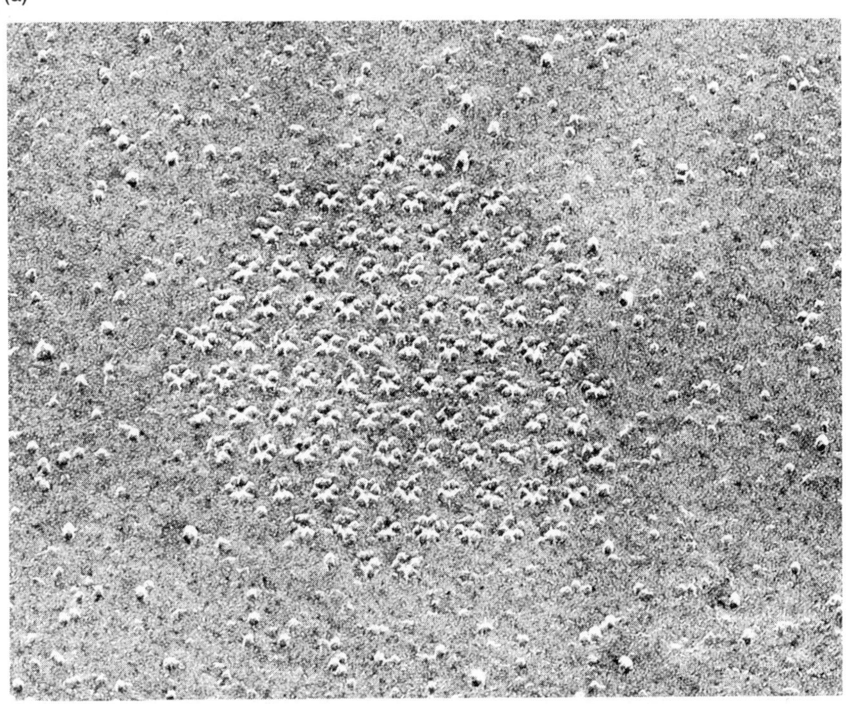

(b)

FIGURE 3.52 The P face of the plasma membrane at low (*a*) and high (*b*) magnification showing hexagonal arrays of rosettes. The largest complexes have about 16 rows of rosettes, each rosette consisting of six subunits. (*a*) ×45,000; (*b*) ×168,000. (Courtesy of Dr. Thomas H. Giddings, Jr.)

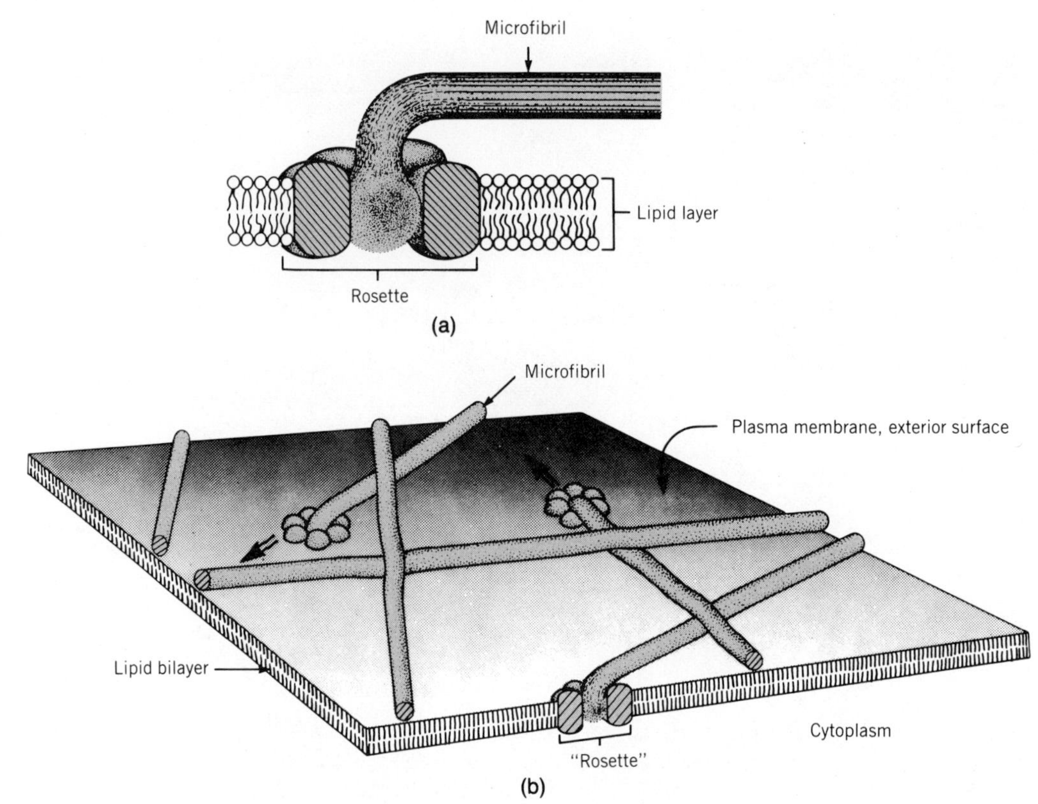

FIGURE 3.53 Model of microfibril deposition during primary wall formation in *Micrasterias*. (*a*) Side view of a rosette embedded in the unit membrane spinning out a microfibril. (*b*) Surface view of the plasma membrane exterior showing random movement of rosettes as they are pushed along in the plane of the lipid bilayer by extending microfibrils. (Courtesy of Dr. Thomas H. Giddings, Jr.)

rosette in a row extrudes one 5-nm microfibril that aggregates laterally with other microfibrils emerging from rosettes in the same row. The longer the row, the larger the fibril.

The universality of this phenomenon among cellulose-producing plants is not known, but intramembrane complexes with rosettelike structures have been detected in other algae and in higher plants as well. It may well be that these complexes constitute the enzymes that produce the cellulose unit microfibril.

Orientation of Microfibrils during Growth

The multinet theory of fiber orientation during growth illustrated in Figure 3.48 is primarily a passive model of fibril orientation that occurs because of cell wall stretch during growth. It undoubtedly is a factor in fibril alignment.

Another theory that invokes active means of fibril orientation is gaining ground and is most intriguing. It has arisen in part to explain the fact that layers of parallel fibrils form in walls of plant cells that are no longer growing. Wall stretch is therefore not sufficient to account for this continued deposition of parallel fibers. Tracheids are an example of cells in

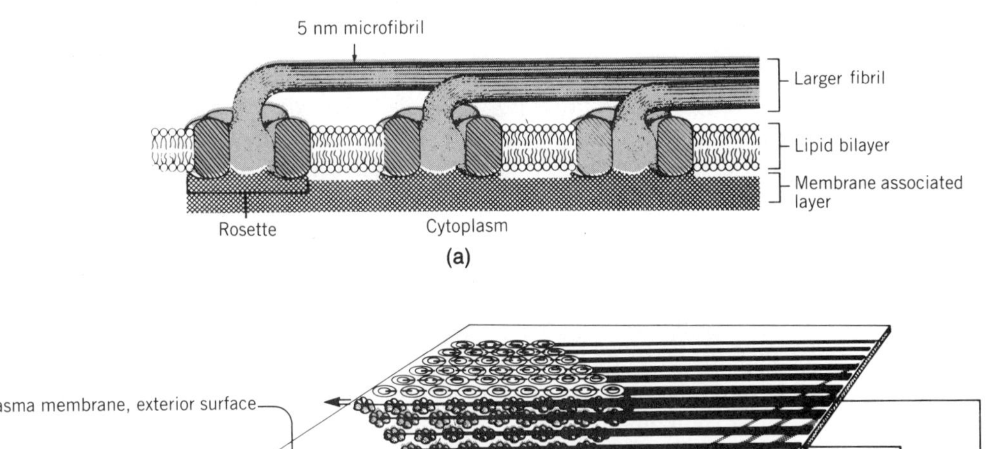

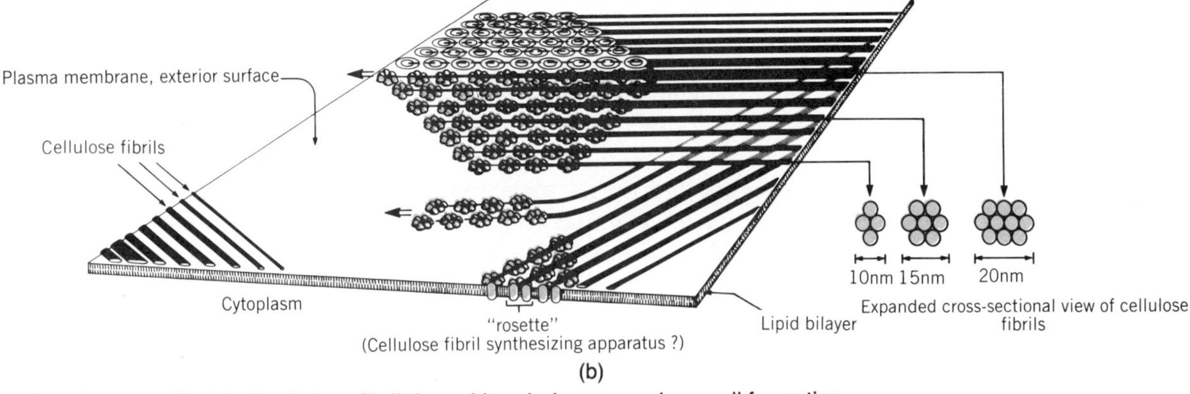

FIGURE 3.54 Model of cellulose fibril deposition during secondary wall formation in *Micrasterias.* Each rosette is proposed to form one 5-nm microfibril with a row of rosettes forming a set that aggregates to make the larger cellulose fibers. (*a*) Side view showing the presumed site of microfibril formation as a stippled area in the rosette center. The "membrane-associated layer" may hold rosettes in formation. (*b*) Surface view with expanded cross-sectional view of cellulose fibrils. (Courtesy of Dr. Thomas H. Giddings, Jr.)

which parallel fibrils of secondary wall are laid down without concomitant wall stretch.

The components of the *active orientation theory* are as follows. *Cyclosis* in the cell interior may be the driving force for fibril orientation on the cell exterior. Cyclosis may deform the plasma membrane by producing bulges or corrugations parallel to the direction of flow. Cellulose fibrils that are being extruded during this event would tend to orient on the membrane outer surface along the axes of the corrugations. With time the fibrils would crystallize and embed into position.

Cyclosis is a streaming phenomenon involving the cytoplasmic materials in the cells of plant leaves.

When the direction of cyclosis changes, which is a common occurrence in many plant cells, a new pattern of corrugations would be established, hence a new layer of parallel fibrils oriented in a direction at some angle from the first. Cells in which cyclosis does not occur or is weak or variable would be expected to have cell wall fibrils that are randomly oriented.

It should be apparent that although much has been accomplished in delineating cell wall structures, much remains to be done. The questions to be answered range all the way from the precise interaction of cellulose strands in the microfibril to the mechanisms of reorientation during cell wall extension. Therefore, we close this chapter realizing that the subject is far from closed. It will be an area of research to watch with interest in the coming years.

Summary

Procaryotic cells have walls made up of complex biological macromolecules. Peptidoglycan, a polymer of *N*-acetylated glucosamines with attached peptide chains, is present in both Gram-positive and Gram-negative bacteria. Some Gram-positive bacteria also have polyolphosphate polymers called teichoic and teichuronic acids. Gram-negative bacteria, although possessing less peptidoglycan in their walls, have an outer layer of wall called the outer membrane. It contains polar lipids and a material termed lipopolysaccharide. LPS consists of a lipid core to which is attached a long polysaccharide arm. LPS is antigenic and accounts for some of the surface properties of Gram-negative bacteria.

Peptidoglycan serves as a restraint on the protoplast to prevent osmotic lysis of the cells in hypotonic media.

Several antibiotics are effective as therapeutic agents because they inhibit specific reactions concerned with cell wall growth. When new walls are improperly formed, membranes will leak or growing cells will lyse.

Fungi have walls that consist of crystalline components embedded in noncrystalline material. The main crystalline materials are chitin and the β-glucans, occurring predominantly in fungal hyphae and yeasts, respectively. A major noncrystalline material is a protein–mannan complex. The polysaccharide portion of this complex consists of either highly branched or linear mannose polymers. The former is linked to the protein via asparaginyl residues and the latter through serine and threonine.

Green plants contain cell walls that consist largely of cellulose, hemicellulose, and pectic substances. Cellulose is a linear polymer of β-1,6-linked glucan residues that, along with other cellulose chains, takes on a crystalline character. Hemicelluloses are made up of β-1,4-linked glucans with short side chains. Hemicelluloses do not self-crystallize but can cocrystallize with cellulose. Pectins are varied in structure, containing acidic groups. They are branched and acetylated.

In the wall, the microfibril, made up of many cellulose chains, is coated with hemicellulose, which in turn is cross-linked to other hemicelluloses with pectins and other polysaccharides.

The primary cell wall is the first wall to be laid down by a growing cell. In this wall the cellulose microfibrils are randomly oriented. The secondary wall is formed after the cell has reached its maximum size. In this wall microfibrils run parallel in layers with the run of the microfibrils in a given layer at some angle to the run in an apposing layer.

The elementary microfibril appears to form within bacterial cells and is extruded from a funnellike pore. In eucaryotic cells the microfibril is extruded from rosettelike structures embedded in their membranes. The elementary microfibril is 5 nm in diameter. After extrusion it aggregates with other microfibrils of similar diameter, forming fibrils that vary in size depending on the number of elementary fibrils they contain.

Once microfibrils have been extruded, they are oriented in the wall in manners characteristic of primary and secondary walls. The multinet theory of orientation suggests that as the plant cell elongates, the microfibrils in the primary wall reorient from transverse or random to longitudinal with the cell axis. The theory of active orientation suggests that cyclosis promotes a furrowing of the plasma membrane. Microfibrils are aligned parallel with the furrow. When cyclosis changes direction, furrowing also changes, and so does the alignment of the next layer of microfibrils.

References

1. Cell Wall Thickness, Size Distribution, Refractive Index Ratio and Dry Weight Content of Living Bacteria, P.J. Wyatt, *Nature* 226:277(1970).
2. The Outer Membrane of Gram-negative Bacteria, H. Nikaido and T. Nakae, *Adv. Microb. Physiol.* 20:164(1979).
3. Model for the Structure of the Shape-maintaining Layer of the *Escherichia coli* Cell Envelope, V. Braun, H. Gnirke, V. Henning, and K. Rehn, *J. Bacteriol.* 114:1264(1973).
4. Three-Dimensional Molecular Models of Bacterial Cell Wall Mucopeptides (Peptidoglycans), M. V. Kelemen and H. J. Rogers, *Proc. Natl. Acad. Sci. U.S.* 68:992(1971).
5. Biogenesis of the Wall in Bacterial Morphogenesis, H. J. Rogers, *Adv. Microb. Physiol.* 19:1(1979).
6. Molecular Arrangement of Teichoic Acid in the Cell Wall of *Staphylococcus lactis,* A. R. Archibald, J. Baddiley, and J. E. Heckels, *Nature New Biol.* 241:29(1973).
7. Outer Membrane of *Salmonella,* T. Nakae, *J. Biol. Chem.* 251:2176(1976).
8. Chemical Heterogeneity of Major Outer Membrane Pore Proteins of *Escherichia coli,* D. Lee, C. A. Schnaitman, and A. P. Pugsley, *J. Bacteriol.* 138:861(1979).
9. Biochemistry of Bacterial Cell Envelopes, V. Braun and K. Hantke, *Annu. Rev. Biochem.* 43:89(1974).
10. Initial Reactions in Biosynthesis of Teichuronic Acid of *Micrococcus lysodeikticus* Cell Walls, T. E. Rohr, G. N. Levy, N. J. Stark, and J. S. Anderson, *J. Biol. Chem.* 252:3460(1977).
11. Reactions of Second Stage of Biosynthesis of Teichuronic Acid of *Micrococcus lysodeikticus* Cell Walls, N. J. Stark, G. N. Levy, T. E. Rohr, and J. S. Anderson, *J. Biol. Chem.* 252:3466(1977).
12. The Chemical Structure of a Fragment of *Micrococcus lysodeikticus* Cell Wall, Nasir-ud-Din and R. W. Jeanloz, *Carbohydr. Res.* 47:245(1976).
13. The Actions of Penicillin and Other Antibiotics on Bacterial Cell Wall Synthesis, J. L. Strominger, in *Biochemistry of Cell Walls and Membranes,* C. F. Fox, ed., Butterworths, London, 1975.
14. Study of a Cycle of Cell Wall Assembly in *Streptococcus faecalis* by Three-Dimensional Reconstruction of Thin Sections of Cells, M. L. Higgins and G. D. Shockman, *J. Bacteriol.* 127:1346(1976).
15. Biosynthesis of Cell Walls of Fungi, V. Farkăs, *Microbiol. Rev.* 43:117(1979).
16. *The Physical Biology of Plant Cell Walls,* R. D. Preston, Chapman and Hall, London, 1974.
17. Positions des Atomes dans le Nouveau Modèle Spatial de la Cellulose, K. H. Meyer and L. Misch, *Helv. Chim. Acta* 20:232(1937).
18. Plant Cell Walls, W. D. Bauer, in *The Molecular Biology of Plant Cells,* H. Smith, ed., University of California Press, Berkeley, 1977.
19. The Biosynthesis of Cellulose, J. R. Colvin, in *The Biochemistry of Plants,* vol. 3, J. Preiss, ed., Academic Press, New York, 1980.
20. The Synthesis of Cellulose in Cell-Free Extracts of *Acetobacter xylinum,* L. Glaser, *J. Biol. Chem.* 232:627(1958).
21. Visualization of Particle Complexes in the Plasma Membrane of *Micrasterias denticulata* Associated with the Formation of Cellulose Fibrils in Primary and Secondary Cell Walls, T. H. Giddings, D. L. Brower, and L. A. Staehelin, *J. Cell Biol.* 84:327(1980).

Selected Books and Articles

Books

Biogenesis of Plant Cell Wall Polysaccharides, F. Loewus, ed., Academic Press, New York, 1973.

Cellulose and Other Natural Polymer Systems, R. Malcolm Brown, Jr., ed., Plenum Press, New York, 1982.

Dynamic Aspects of Plant Ultrastructure, A. W. Robards, ed., McGraw-Hill, New York, 1974.

Surface Carbohydrates of the Prokaryotic Cell, I. Sutherland, ed., Academic Press, New York, 1977.

The Biochemistry of Plants, vol. 1, N. E. Tolbert, ed., Academic Press, New York, 1980.

The Fungi, vol. 1, G. C. Ainsworth and A. S. Sussman, eds., University of California Press, Berkeley, 1977.

The Physical Biology of Plant Cell Walls, R. D. Preston, Chapman and Hall, London, 1974.

Articles

Biogenesis of the Wall in Bacterial Morphogenesis, H. J. Rogers, *Adv. Microb. Physiol.* 19:1(1979).

Biosynthesis of Cell Walls of Fungi, V. Farkăs, *Microbiol. Rev.* 43:117(1979).

Plant Cell Walls, W. D. Bauer, in *The Molecular Biology of Plant Cells,* H. Smith, ed., University of California Press, Berkeley, 1977.

Structure and Biosynthesis of the Mannan Component of the Yeast Cell Envelope, C. Ballou, *Adv. Microb. Physiol.* 14:93(1976).

The Actions of Penicillin and Other Antibiotics on Bacterial Cell Wall Synthesis, J. L. Strominger, in *MTP International Review of Science,* C. F. Fox, ed., Butterworths, London, 1975.

The Biosynthesis of Cellulose, J. R. Colvin, in *The Biochemistry of Plants,* vol. 3, J. Preiss, ed., Academic Press, New York, 1980.

The Outer Membrane of Gram-negative Bacteria, H. Nikaido and T. Nakae, *Adv. Microb. Physiol.* 20:164(1979).

The Wall of the Growing Plant Cell: Its Three-Dimensional Organization, J.-C. Roland and B. Vian, *Int. Rev. Cytol.* 61:129(1979).

The Walls of Growing Plants, P. Albersheim, *Sci. Am.* 232:81(1975).

Visualization of Particle Complexes in the Plasma Membrane of *Micrasterias denticulata* Associated with the Formation of Cellulose Fibrils in Primary and Secondary Cell Walls, T. H. Giddings, D. L. Brower, and L. A. Staehelin, *J. Cell Biol.* 84:327(1980).

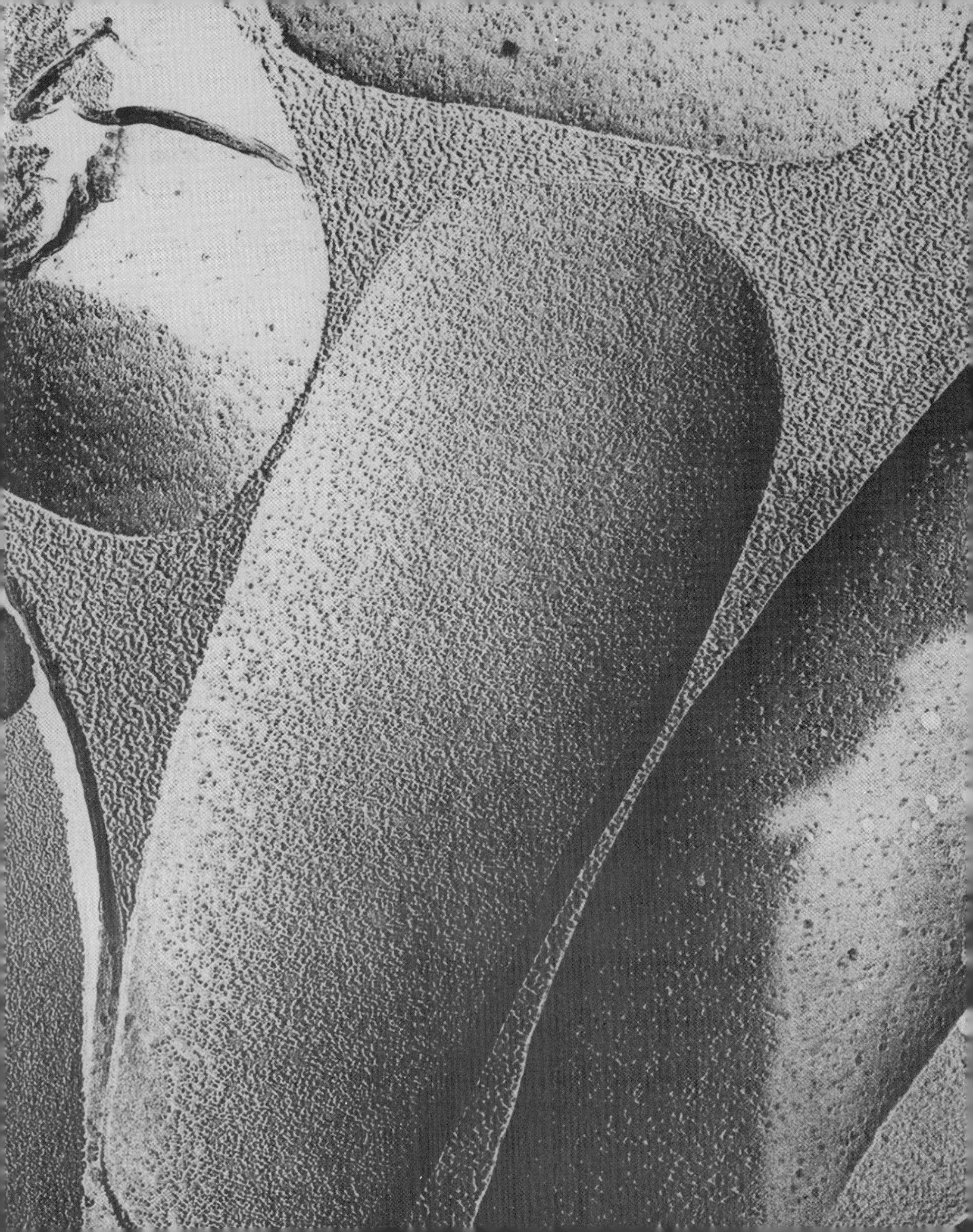

SECTION 2

EXTERIOR MEMBRANES AND SURFACE COMPONENTS

Not all cells possess walls or manufacture extracellular matrices. But all possess a plasma membrane that assumes for the cell a variety of vital functions. Among other things, it is a selective transporter–barrier, a gradient producer and maintainer, and a communicator with other cells and the cell environment. It is, for cells without walls, the cell surface.

In Chapter 4 we discuss the structure and function of the plasma membrane so that we can better appreciate its surface role as described in Chapter 5.

4

The Plasma Membrane

It now seems to be agreed that [the plasma membrane's] basic structure is that which I suggested in 1934, and it is highly probable that the same structure is present in many other intracellular membranes. So far as it is possible to predict at the present time, it is unlikely that this general picture will be substantially disturbed, and the focus of attention is likely to shift to other fields.

J. F. DANIELLI, 1962

Trends and developments in biological research are shaky subjects for prognostication. Only a decade after Danielli made the foregoing statement, the field of membrane research blossomed and proliferated as a major effort in hundreds of laboratories around the world. It continues as a focus of attention today for many research groups.

Because of the concentrated effort of thousands of scientists and supporting technicians, the basic concept of membrane structure has indeed been disturbed from that prevalent in 1934. True, some of the fundamentals have remained over the years, and the universality of the basic structure among different membrane systems is becoming well documented. So in this sense Danielli was absolutely correct. Nevertheless, the field has moved and changed rapidly, especially in the 1970s, and the newer concepts of membrane structure are making a major impact on our understanding of a variety of cellular behaviors.

The existence of the cell plasma membrane, or outermost boundary, was difficult to substantiate by direct examination before 1930 because of technological limitations. The membrane is beyond the resolution of the light microscope, rendering a morphological approach to its study unfeasible with this instrument. Thus most of the experimental approaches provided only indirect evidence for the existence of such a membrane, based, for example, on the relative permeativity of molecules into cells, or phenomena such as the plasmolysis of plant cells in hypertonic solutions.

Method I

Of course, the very fact that a cell, especially an animal cell, which has no wall, can exist as a physically defined entity suggests that it must have some sort of boundary. But the physical properties of this boundary eluded researchers for many decades. For some time it was thought of as a peripheral region of specialized molecules randomly clustered around the cell in a relatively diffuse cloud. The concept of an ordered molecular organization with high stability had to await the development and application of tools such as X-ray diffraction analysis and polarized and electron microscopy. Once these tools were employed, speculation and doubt regarding the plasma membrane were quickly removed.

Methods I and VIII

Electron microscopy suggested the outer cell barrier to be rigid, a point of information that occasionally presented more problems that it solved. Cells had long been understood as open-ended systems, taking up certain molecules and giving off others. Furthermore, they exhibit the properties of growth, contraction, fragmentation, and differentiation, and many other dynamic activities. Somehow, the static structure seen by electron microscopy had to be reconciled with these dynamic cellular traits.

The manner in which this problem has been attacked and is still unfolding represents one of the most intriguing and significant chapters in cellular biology. The story began nearly a century ago.

4.1 MEMBRANE ARCHITECTURE: THE EVOLUTION OF MEMBRANE MODELS

From a Diffuse Lipoid Layer to a Distinct Physical Boundary

One of the early studies suggesting the existence of a boundary layer around cells was conducted in 1899 by Overton[1]. After performing some 10,000 experiments with more than 500 different chemicals, he concluded that the peculiar osmotic properties of living protoplasts are due to a *selective solubility* mechanism. Hydrophobic compounds entered cells more rapidly than hydrophilic ones. Overton believed this was because of an outer *lipoid layer* in which hydrophobic compounds were more soluble, and he speculated that it might contain cholesterol, lecithin, and fatty oils. Today we know that these speculations were remarkably well on target, and these early notions contributed substantively to the concept of the lipid membrane.

Thirty-one years later, Plowe[2] showed convincingly that the outer boundary of plant protoplasts was an elastic layer distinct from the rest of the cytoplasm. By using microneedles she showed that the *plasmalemma,* as she preferred to call this layer, could be stretched and penetrated, causing only local, reversible injury to the protoplast. More extensive tearing resulted in cell death, confirming the idea of a protective role for the membrane. Plowe's work was a very significant milestone toward establishing the plasma membrane as a distinct physical entity having a molecular organization that set it apart from the rest of the cell.

Origins of the Smectic Lipid Bilayer

At nearly the same time as Plowe's work was reported, other investigators were carrying out some of the first direct analyses of membrane constituents. Gorter and Grendel[3] conducted a study on red blood cell membranes that has had a major and lasting effect on the construction of membrane models. They extracted the lipids from erythrocyte ghosts and spread them out on monolayers in the Langmuir trough apparatus. These investigators discovered that the area covered by the monolayer was twice that needed to cover the surface of the cells from which the lipid was retrieved. They concluded that red cells were covered by a layer of lipids two molecules thick, oriented with polar groups toward the inside and outside of the cell. This *smectic,* or extended, lipid bilayer form is still recognized today as the most likely predominant configuration of the lipid in the plasma membrane.

The Danielli–Davson Model

Membrane model building had its real genesis with the work of Danielli and Davson, and since their contributions the models have undergone constant evolution. The salient features of their concepts arose from surface tension studies, comparing the properties of oil drops in water with those in cell extracts. Interfacial tensions between oil drops and cell extracts were significantly lower than tensions between oil and water, a property that suggested the presence of a substance in the cell with the ability to lower interfacial tensions. This effect was explained in an interface model set forth by Danielli and Harvey (Figure 4.1). In this model, amphipathic lipid molecules are oriented with their hydrophobic domains toward the oil phase. Hydrated proteins act as a coating buffer between the hydrophilic lipid heads and the aqueous phase.

This concept became an important forerunner of the membrane model published by Danielli and Davson in 1935.[4] Depicted in Figure 4.2, the model incorporates the features of the oil–cell aqueous phase interface system, with hydrated protein adsorbed both to the interior side of the membrane and to its exterior surface. An unresolved feature of the model is the *undefined thickness* of the lipoid layer. This is the concept that Danielli thought probably would not be disturbed for some time.

Amphipathic molecules contain both a hydrophilic and a hydrophobic component. Although many kinds of molecule can meet these requirements, in membranes the major amphipathic molecules are phosphoglycerides, sphingomyelin, glycolipids, and cholesterol. These molecules are often symbolized as follows.

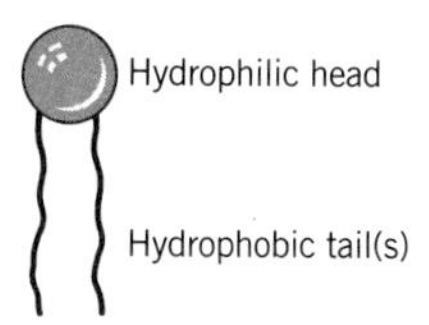

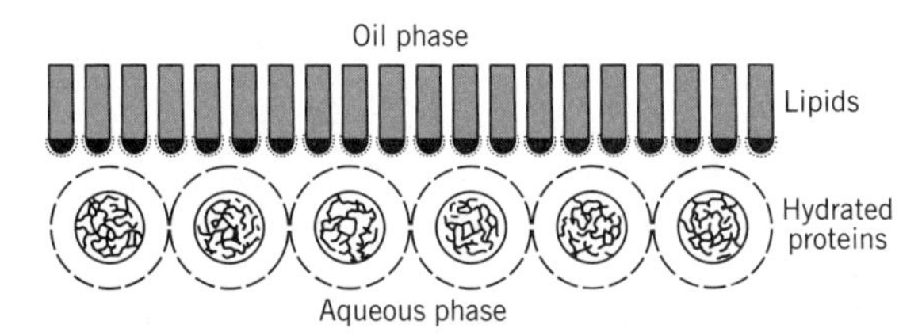

FIGURE 4.1 A model of the structure of the oil–aqueous phase interface.

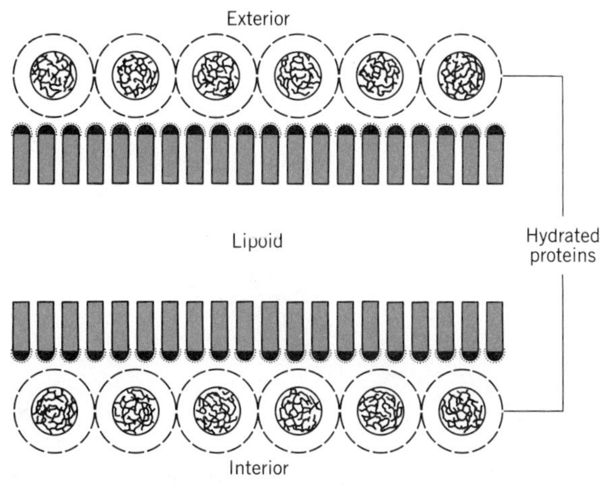

FIGURE 4.2 An early model of cell membrane structure depicting a lipoid layer of undefined thickness sandwiched between globular proteins. The concept of Figure 4.1 is directly applied in this model.

An important factor that enabled this model to be published with a reasonable degree of certainty was the development and increased use of X-ray diffraction and polarization and electron microscopy. For a time, these techniques were intensively applied to a study of the myelin sheath of nerve cell axones without realizing that the sheath is actually a concentrically layered membrane system of the Schwann cell (see Figure 4.3). Thus the information gained from its study could in many instances be directly applied to the plasma membrane of other cell types.

With the use of optical polarization analyses on myelin, Schmidt[5] found a birefringence consistent with the presence of radially oriented molecules, in this case the smectic bilayer of amphipathic molecules wound in sheets around the axone. The same studies implicated elongated mol-

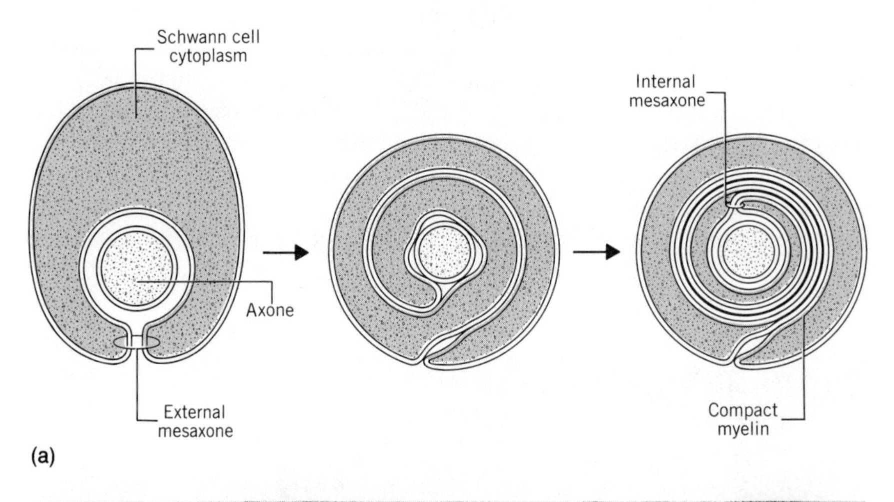

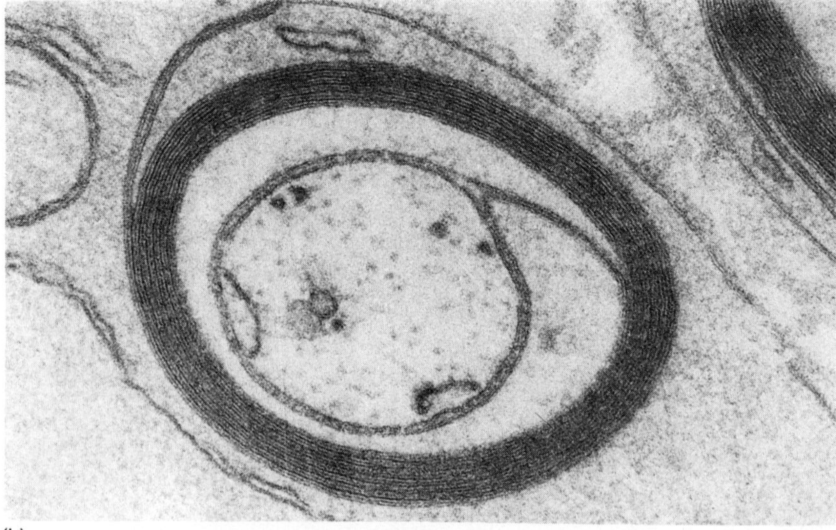

FIGURE 4.3 Relation between the Schwann cell and the myelin sheath of nerve axons. (*a*) A Schwann cell wraps around the axon, first as a loose scroll and later as compact layers wherein both outer (minor bands) and inner (major bands) cell surfaces touch. (*b*) Electron micrographic view of a more advanced stage of sheath formation. (*b*, courtesy Dr. J. D. Robertson.)

ecules of other types oriented in the longitudinal direction of the sheath. These we now interpret to be the proteins of the membrane.

Low angle X-ray diffraction analyses[6,7] supported this view and showed a spacing of 17 to 18 nm bringing into the picture some quantitation of the thickness of the lipid–protein layers.

This degree of spacing was eventually found to be a measure of the distance between inner surfaces of the cell (see Figure 4.3) and therefore represents *two* membrane layers. With electron microscopy, these surfaces are seen as major dense lines. Eventually, electron microscopy revealed an intraperiod line between the dense bands, demarcating a width that reflects the dimensions of a single layer of membrane. The detection of bands with different intensities not only helped define the thickness of the membrane, but was an important clue to a property that we now call membrane asymmetry. The results suggested that the inner and outer surfaces of the membrane differed chemically.

These studies established that the concentric layers seen by electron microscopy in fixed myelin were also present in the fresh material used for the optical and X-ray studies. Therefore, the techniques of electron microscopy were not producing a series of artifacts, an ever-present concern whenever the electron microscope is used.

Robertson's Unit Membrane Concept

Electron microscopy on myelin as well as many other cell types began to reveal a consistent dark–light–dark or railway track pattern to the membrane as illustrated in Figure 4.4. These observations led Robertson

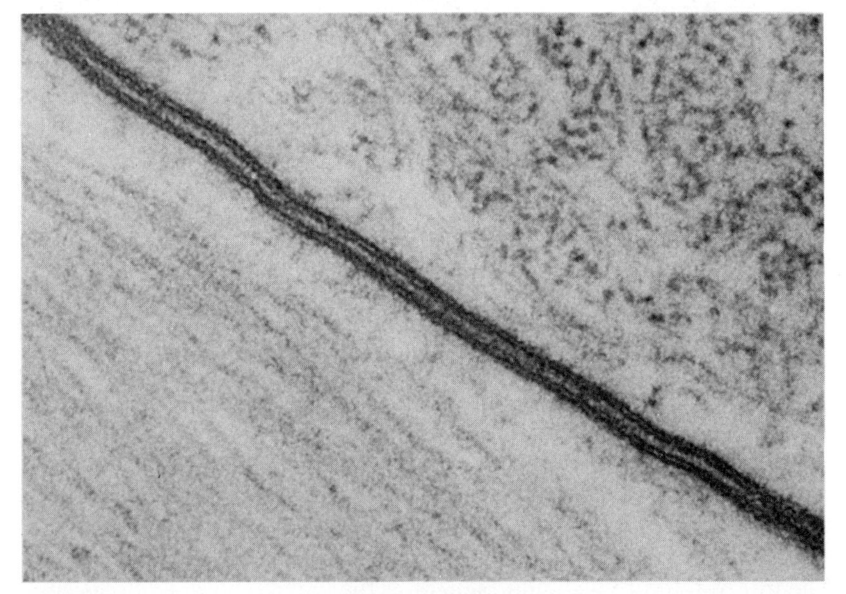

FIGURE 4.4 Electron micrograph of plasma membranes from two adjacent cells. Each membrane has a railway track appearance of dark–light–dark pattern. The dark lines appear to be due to a deposition of osmium in the hydrophilic layers of the membrane. (Courtesy of Dr. Don W. Fawcett.)

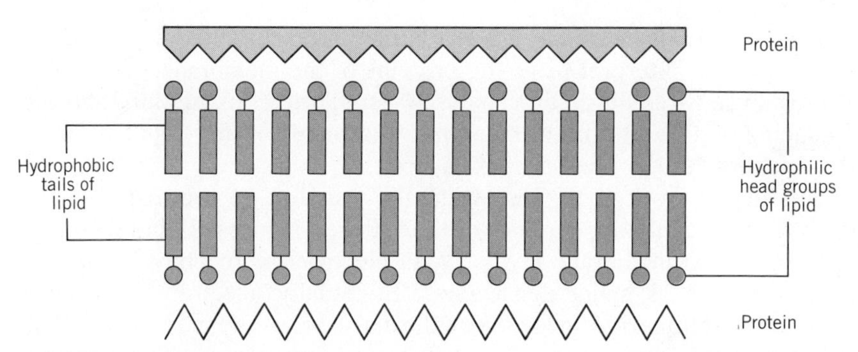

FIGURE 4.5 Schematic diagram of the Robertson model of membrane structure. The lipid layer is defined as bimolecular, and the protein is extended but different on the two faces of the membrane.

to formalize the *unit membrane* concept and to propose in 1959 a model (Figure 4.5) that was quite widely accepted for a number of years.[8] Robertson specified the lipoid layer as a *bimolecular leaflet* in smectic orientation with the polar heads coated with protein in the extended, or β, configuration.The bilayer of lipid is 3.5 nm thick and each protein layer 2.0 nm, making up a 7.5 nm structure that was seen as a common feature of every type of cell examined. Significantly, the model depicted membrane asymmetry, reflecting the fact that the inner and outer surfaces of the Schwann cell show different properties where they are apposed according to both electron microscopy and X-ray diffraction.

One of the major weaknesses of Robertson's model was its failure to address satisfactorily the known permeability and transport properties of the membrane. The model was rigid, much like the electron micrographs from which it was derived. This, of course, was not a unique feature of Robertson's model. All the model building up to that time arose primarily from the results of physical studies on the lipids and proteins and did not address very competently the function of the membrane.

The problem of explaining membrane function in the models was largely a technological one. Theories and models began to proliferate, but scientific data to substantiate them often were lacking. Membrane researchers began to think of the membrane more and more as a dynamic system of interrelated lipid and protein, and the different concepts and pictures that surfaced are too numerous to treat in this chapter.

The Singer–Nicolson Fluid Mosaic Model

Methods I, VI, and VIII

The development of freeze-fracture techniques in electron microscopy and a number of molecular probing procedures, such as circular dichroism and the labeling of membrane proteins, began to throw some new light on the topography of the molecules in the membrane.

The freeze-fracture technique separates the unit membrane into its constituent halves down the middle between the hydrophobic lipid tails. Electron microscopy of the exposed inner faces reveals particles on the order of 8.5 nm in diameter.[9] This is consistent with the presence of globular

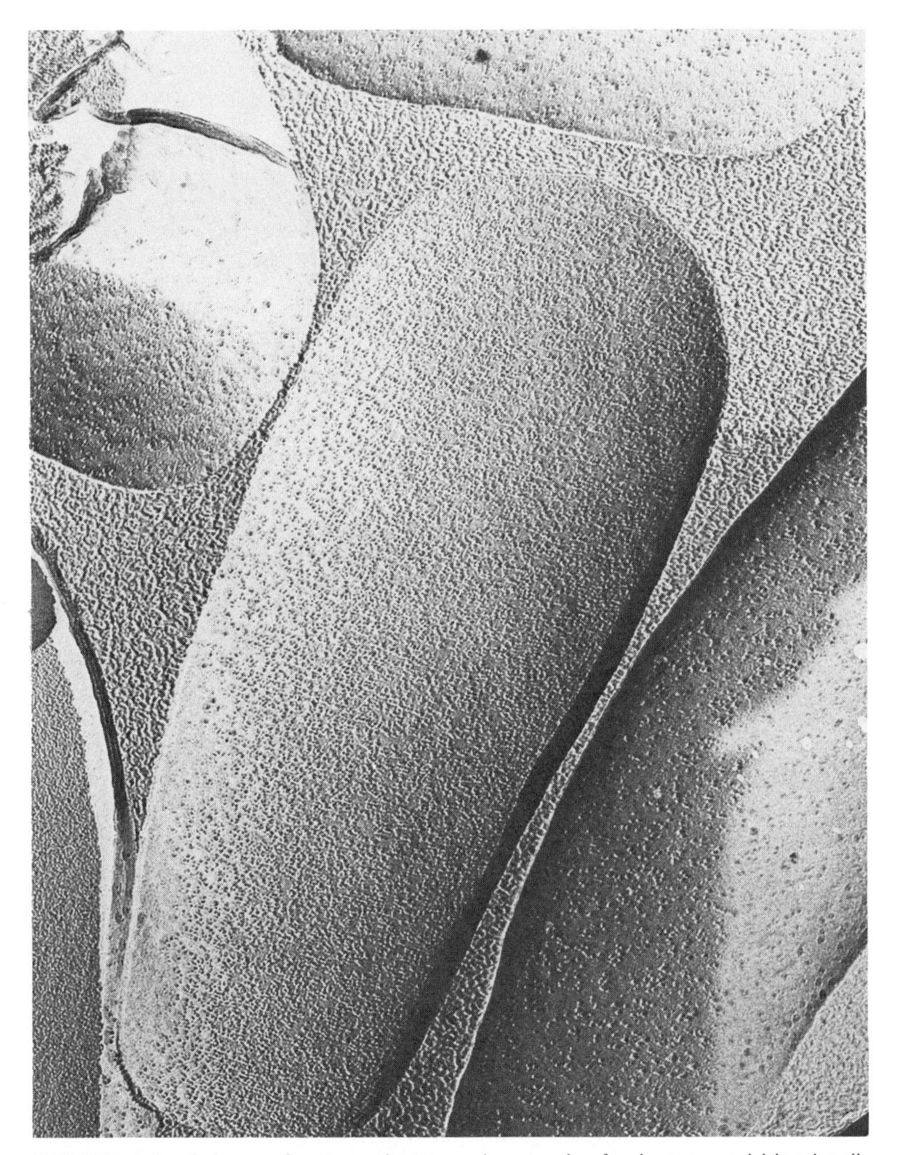

FIGURE 4.6 A freeze-fracture electron micrograph of a human red blood cell. The surface of the fracture contains particles having an approximate diameter of 8.5 nm; these particles are proteins that are embedded in the lipid bilayer in the intact membrane. ×64,000. (Courtesy of Dr. Daniel Branton.)

proteins embedded in the matrix of the membrane. An electron micrograph showing particles of this size in the membranes of erythrocytes is presented in Figure 4.6. It is apparent from this that proteins do not merely coat the surfaces of a homogeneous lipid layer.

Optical measurements on membranes supported the notion that the protein was in globular form rather than the extended conformation. In some membranes, about 40% of the protein length appears to be in the α-helix configuration.

The results of a number of studies along these lines were finally illustrated in the fluid mosaic model as put forth by Singer and Nicolson.[10]

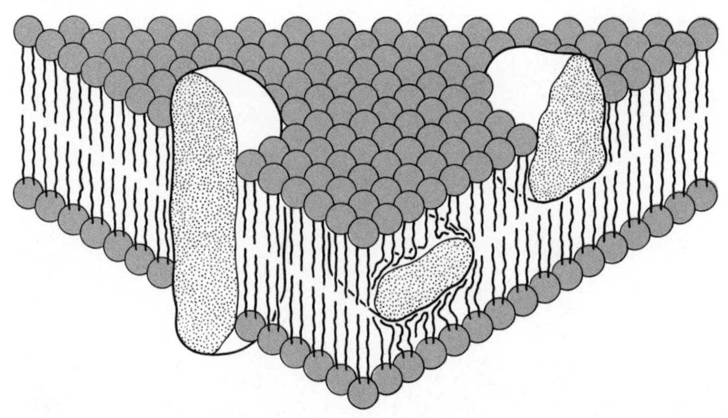

FIGURE 4.7 The fluid mosaic model of membrane structure proposed by Singer and Nicolson. Globular amphipathic proteins float in a fluid bimolecular lipid matrix. Proteins may span the lipid layer completely or may be embedded on one surface or the other or even within the lipid layer. This model can be used to rationalize the flexibility and dynamic properties of membranes as well as many of their physiological functions. (Adapted from a drawing provided by Dr. S. J. Singer.)

This model (Figure 4.7) visualizes proteins as globular and amphipathic, either partially or wholly embedded in the lipid bilayer. This important concept not only is consistent with the known physical and chemical properties of the membrane's molecular constituents, but it is also the first model that explained many of the dynamic features and physiological functions of the membrane, as will become increasingly apparent as this chapter develops.

Briefly, the model portrays proteins as intercalated or floating in a fluid matrix, and by virtue of their mobility and globular form free to carry out the reactions necessary to effect the transport of specific molecules across the membrane. The flexibility seen in the model can quite easily account for many of the dynamic properties of membranes, such as fragmentation, deformation, and growth. The fluid mosaic model is now the most reasonable explanation of plasma membrane architecture.

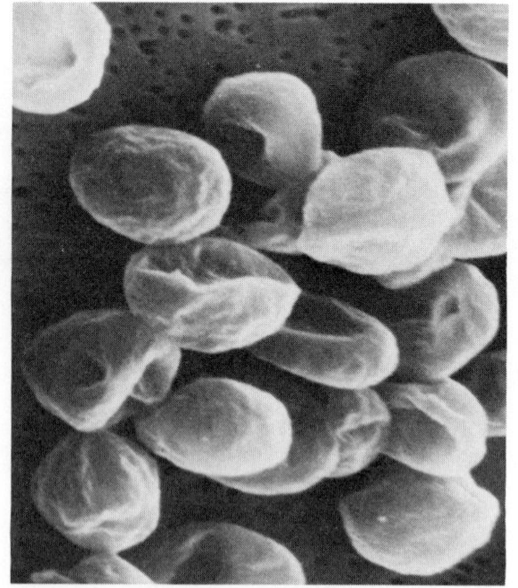

FIGURE 4.8 Electron micrograph of human erythrocyte ghosts at isotonic ionic strength. These are essentially membrane bags from which the content of hemoglobin has been removed. Ghosts are useful model systems for a variety of studies including a source of membranes for chemical analyses and objects for the study of protein and lipid topography in the membrane. (Courtesy of Dr. S. J. Singer.)

4.2 ISOLATION OF PLASMA MEMBRANES

The retrieval of pure plasma membranes from eucaryotic cells has been difficult because of the similarities in properties of the entire membrane network of the cell. For this reason, the mammalian erythrocyte, which is void of a nucleus and other organelles, has been a favored source of boundary membrane material.

Membranes are harvested from erythrocytes quite simply by hypotonic lysis (20 mosM) at pH 7.6 followed by sedimentation of the membranes from the released hemoglobin. The membrane products obtained in this manner are, in effect, evacuated cells, and are referred to as *erythrocyte ghosts*. These ruptured cells can be made to reseal under appropriate conditions, and when in this form they retain many of the physiological properties of intact native erythrocytes. Figure 4.8 shows evacuated human erythrocytes.

In addition to erythrocytes, membranes are most commonly isolated from liver cells and myelin, and to some extent from HeLa cells and ascites tumor cells. Myelin represents a special case, for its membranes have a fairly atypical composition, high in lipid content. Because of this, they can be obtained by flotation techniques using density gradient centrifugation on homogenized nerve tissue.

Method III

Eucaryotic cells other than erythrocytes are homogenized in 0.25 *M* sucrose with Potter Elvejhem or Dounce tissue homogenizers and then subjected to differential and density gradient centrifugation to obtain a membrane fraction. With the best of procedures it is still nearly impossible to prepare pure plasma membranes because the shearing forces of the homogenizers commonly injure about 15% of the lysosomes and 10% of the mitochondria. The fragments of these organelles, as well as pieces of the endoplasmic reticulum and nuclear envelope, invariably end up with the plasma membrane. Membrane preparations obtained in this manner are a morphologically heterogeneous mixture of sheets, fragments, and vesicles. A highly purified preparation is shown in Figure 4.9.

Method II

Certain types of microorganism are good sources of membranes. The cell wall can be enzymatically removed from the yeast *Saccharomyces cerevisiae* with a snail gut enzyme, and the resultant protoplast serves as a source of membrane in a manner similar to that of the mammalian erythrocyte. Since yeast is a eucaryotic microorganism, its plasma membrane bears many similarities to those of animal and plant cells.

The plasma membrane of procaryotic microorganisms is quite different in physical and chemical properties from that of eucaryotes. The mycoplasmas, a group of organisms that do not produce a cell wall, have become a favored model system for the preparation of procaryotic plasma membranes. Our discussion, for the most part, centers on eucaryotic systems with emphasis on erythrocytes.

4.3 THE MOLECULAR ANATOMY OF MEMBRANES

With the refinement of techniques to isolate and purify the plasma membrane, it has become feasible to carry out compositional studies that are reasonably reliable. Probably the most accurate results have been obtained for the human erythrocyte membrane, since this is about the easiest to obtain in pure form. However, the compositions of other plasma membranes are appearing in the literature as methods are being worked out to free them from the rest of the intracellular membrane network.

One of the features of the plasma membrane that becomes apparent from comparative analyses is that it has no unique composition. All plasma membranes have protein, lipid, and generally some carbohydrate, but the ratios of these three constituents show considerable variation among cells of different types, and even in cells of the same type from different species. This immediately suggests that cells with compositionally different plasma membranes have ways of carrying out similar or identical functions, and poses an interesting structure–function question. The erythrocytes of the rat and pig, for example, would be expected to have roughly the same functions, but as we shall see, the plasma membranes are chemically quite different.

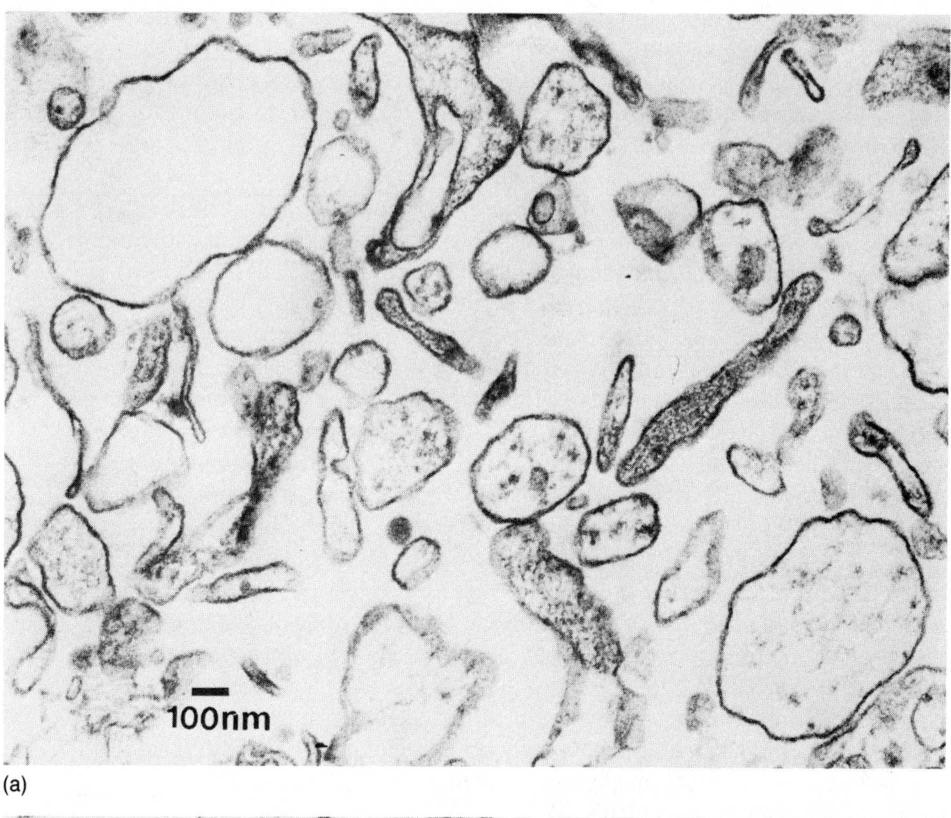

(a)

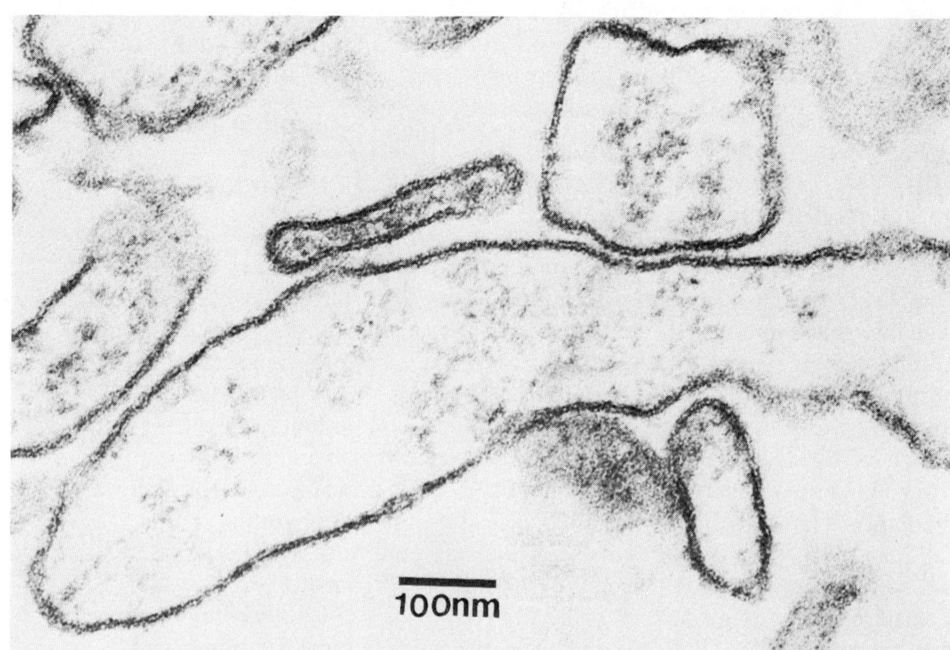

(b)

FIGURE 4.9 Electron micrographs of purified membrane preparations retrieved from lymphoblastoid cells. At lower magnifications (*a*) the preparation is seen to be free from membrane-bound organelles, such as lysosomes and mitochondria, and from ribosomes. At higher magnifications (*b*) the classical trilamellar appearance of the membrane is evident. (Courtesy of Dr. M. J. Crumpton.)

Described in generalities, the plasma membrane contains approximately 40% of its dry weight as lipid and 60% as protein. These two major constituents associate by noncovalent interactions, and under the proper experimental conditions they can be completely dissociated from each other. In addition, carbohydrate is normally present, making up 1 to 10% of the total dry weight of the membrane. In contrast to the noncovalent lipid–protein interactions, carbohydrate is covalently bonded to either the lipid or the protein. Water constitutes about 20% of the total plasma membrane weight in the living cell. It is generally highly ordered and tightly bound to the amphipathic molecules that make up the nonaqueous phase of the membrane.

Lipid Composition

The discovery of polar lipids in membranes and the determination of their amphipathic properties were essential ingredients for the bilayer models of membrane structure. A very high percentage of membrane lipids are polar and thus amenable to placement into bimolecular leaflet structures.

Membrane lipids are extractable with chloroform, ether, and benzene, and with the development of thin layer and gas chromatography, membrane lipid compositions are quite readily determined. Four major classes of lipids are normally present: phospholipids, sphingolipids, glycolipids, and sterols (see margin). It is apparent that these are all amphipathic molecules, possessing both hydrophilic and hydrophobic domains.

Method V

The four classes of membrane lipids can be illustrated by these specific representatives. Colored areas cover the hydrophilic heads.

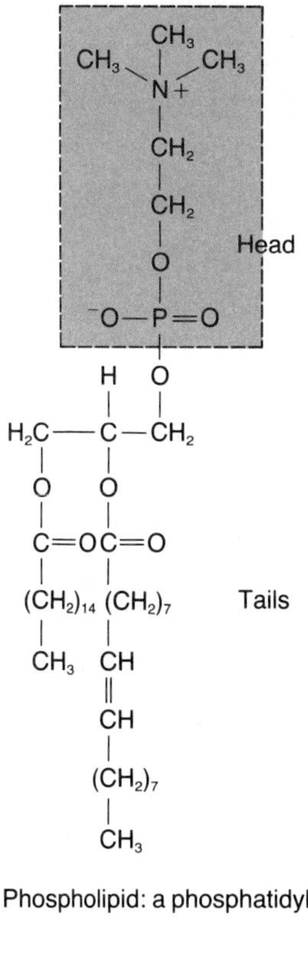

Comparative erythrocyte plasma membrane lipid compositions for four different species are given in Figure 4.10. Cholesterol is the most abundant lipid in all cases and is, in fact, more abundant in the plasma membrane than in the remaining intracellular membranes of eucaryotic cells. The sum of sphingomyelin and phosphatidylcholine makes up a fairly constant percentage of the total lipids present.

Table 4.1 contains a breakdown of the phospholipids present in various membranes of the liver cell. Phosphatidylcholine, phosphatidylethanolamine, and sphingomyelin are again the most prominent phospholipids in the plasma membrane. Table 4.1 has particular value in demonstrating that each of the membranes of the cell is somewhat unique in lipid composition even though similarities are present. Cardiolipin, for example, is not present at significant levels in the plasma membrane, but it is a major constituent of the inner mitochondrial membrane.

Head

Tails

Phospholipid: a phosphatidylcholine

Physical Properties of Lipids

The behavior of amphipathic lipids at water–air interfaces and in solution provides some of the clearest clues as to their likely organization in the membrane. The success of Gorter and Grendel's experiments with erythrocyte lipids and the accuracy of their conclusions is largely due to the formation by polar lipids of a monolayer at air–water interfaces, with their hydrophobic tails interacting and projecting into the air phase (Figure 4.11).

In water, polar lipids, such as phosphatidylcholine, form bilayers and will remain homogeneously in this lamellar form as long as the water

Sphingolipid: sphingomyelin

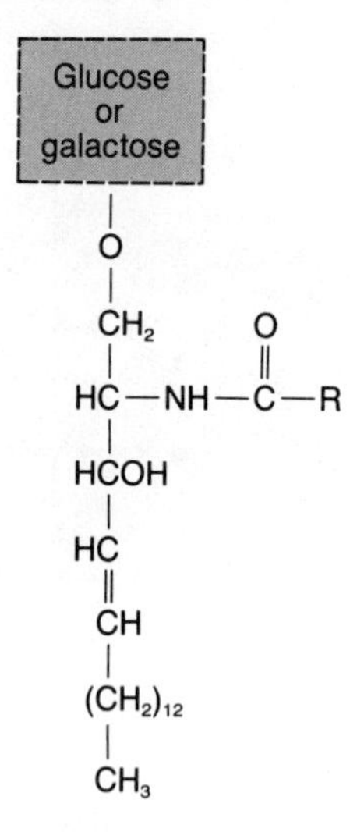

Glycolipid: cerebroside

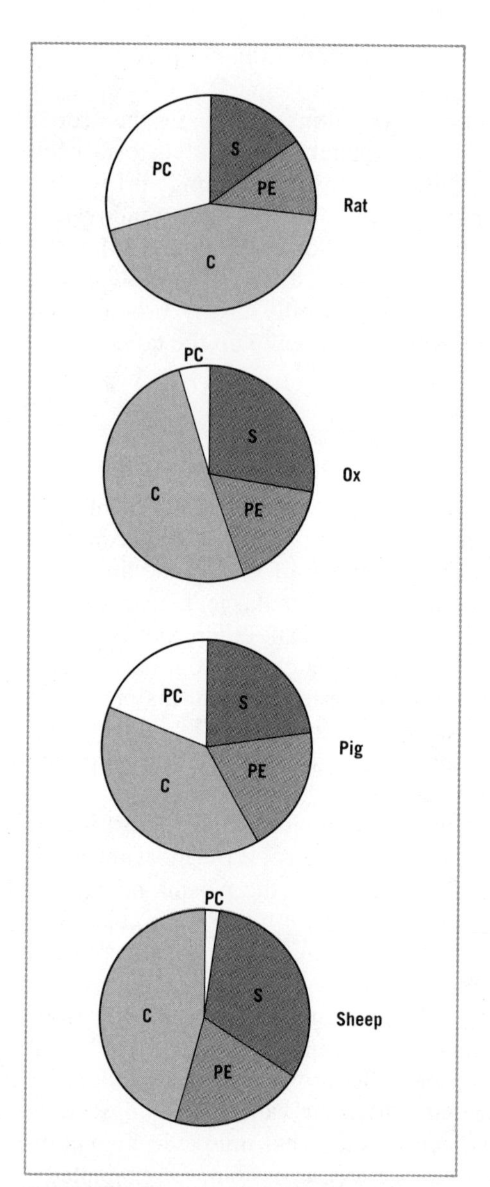

FIGURE 4.10 Comparative lipid compositions in erythrocyte plasma membranes for four different species. Sphingomyelin (S) and phosphatidylcholine (PC) show wide variations in their proportions; however, their sum constitutes a relatively constant proportion of the total lipids. C, cholesterol; PE, phosphatidylethanolamine.

content of the system is maintained between 0 and 40% by weight and the system is kept above its transition temperature (Figure 4.12). As the water content is increased, the aqueous layers increase in width until above the 40% level a two-phase system forms, made up of *multilamellar vesicles* dispersed in the water. These vesicles, called *liposomes* are illustrated in Figure 4.13. Ultrasonication of this two-phase system will

Table 4.1 Phospholipid Composition (mole % of total phospholipid) of Liver Cell Membranes[a]

Phospholipid	Plasma Membrane	Nuclear Membranes	Rough Endoplasmic Reticulum	Golgi Membranes	Mitochondrial Membranes: Inner	Mitochondrial Membranes: Outer	Lysosomal Membranes
Phosphatidylcholine	34.9	61.4	60.9	45.3	45.4	49.7	33.5
Phosphatidylethanolamine	18.5	22.7	18.6	17.0	25.3	23.2	17.9
Phosphatidylinositol	7.3	8.6	8.9	8.7	5.9	12.6	8.9
Phosphatidylserine	9.0	3.6	3.3	4.2	0.9	2.2	8.9
Phosphatidylglycerol	4.8	*	*	*	2.1	2.5	*
Phosphatidic acid	4.4	1.0	1.0	*	0.7	1.3	6.8
Cardiolipin	trace	0	*	*	17.4	3.4	6.8
Lysophosphatidylcholine	3.3	1.5	4.7	5.9	*	*	0
Lysophosphatidylethanolamine	*	0	0	6.3	*	*	*
Sphingomyelin	17.7	3.2	3.7	12.3	2.5	5.0	32.9

[a]Asterisk indicates no value available.

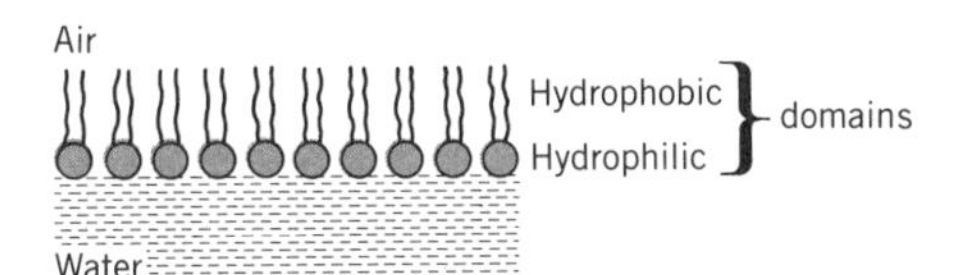

FIGURE 4.11 A monolayer of polar lipids at an air–water interface.

Sterol: cholesterol

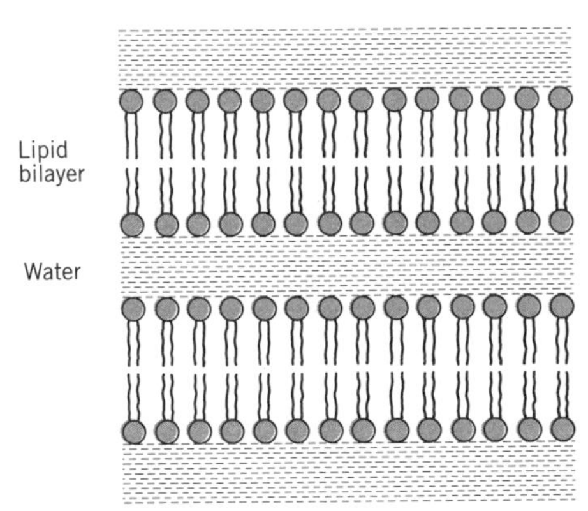

FIGURE 4.12 Schematic depiction of the behavior of polar lipids when mixed in water at a concentration less than 40% water by weight and above its transition temperature. The lipids stack in bilayers with water sandwiched between the hydrophilic surfaces of the lipids.

Cardiolipin, also called diphosphatidylglycerol, is found primarily in bacterial membranes and the inner membrane of mitochondria. It is large and complex.

further disperse it into stable *single bilayer liposomes,* which are sealed and can be used as tiny "model cells" for certain permeability studies. Liposomes exhibit many of the same properties as naturally occurring cellular membranes.

Figures 4.14 and 4.15 illustrate the lamellar and vesicular forms, respectively, of polar lipids in water. There is a remarkable similarity in morphology of these model systems to the myelin sheath and the plasma membranes of cells as seen in a variety of electron micrographs. No cell, of course, has a plasma membrane that is made up of a single type of

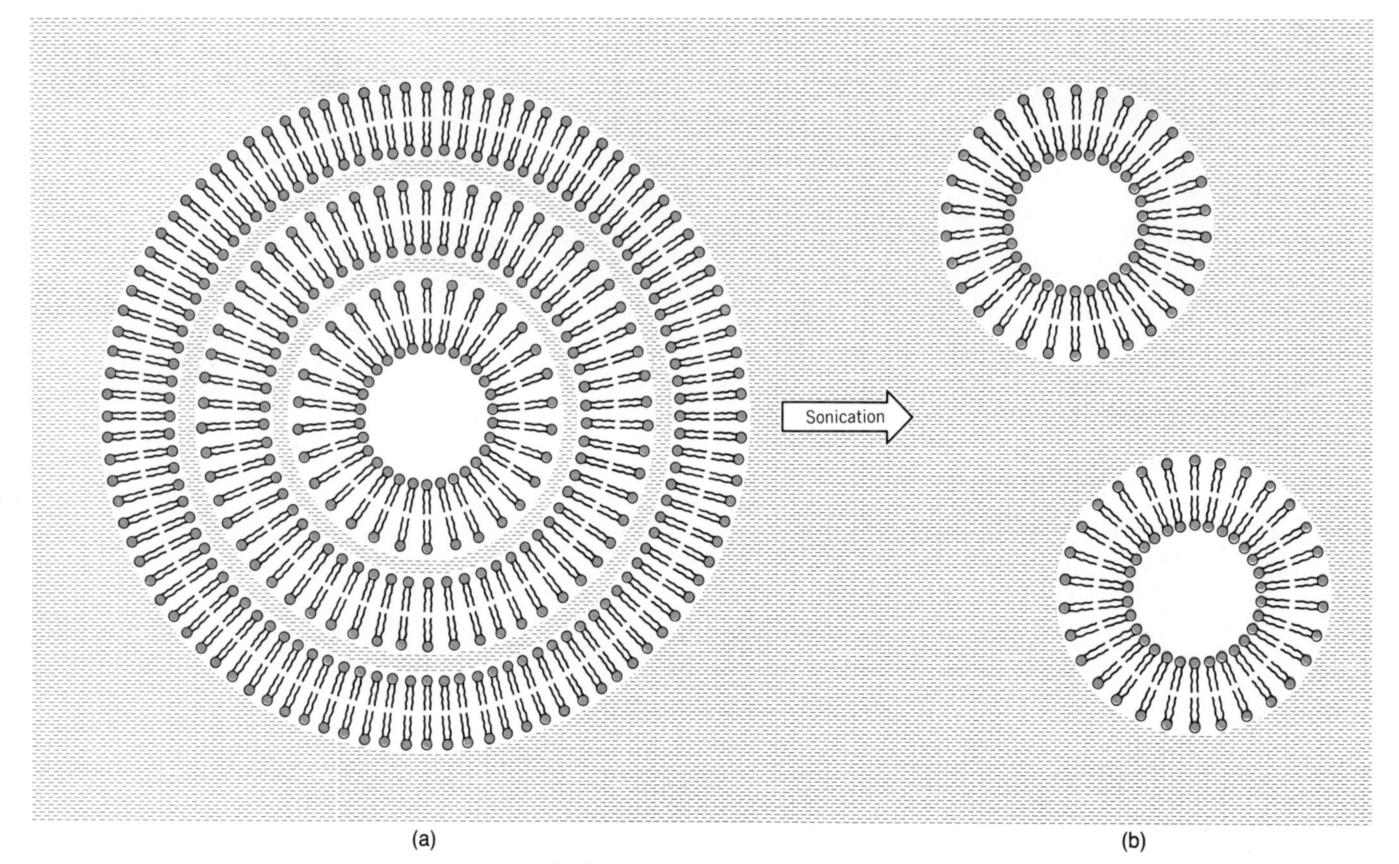

FIGURE 4.13 When polar lipids and water are mixed at water concentrations higher than 40% by weight, multilamellar vesicles called liposomes (*a*) are formed. Sonication of lipid–water systems will disperse the lipids into stable, single bilayer liposomes (*b*).

polar lipid. The closest a natural membrane approaches this is in the myelin sheath, which, as discussed earlier, is a rather atypical membrane. However, these artificial membrane systems have yielded a good deal of information about membrane phenomena, such as the interaction of a variety of metal ions with lipids, the influence of degree of saturation of fatty acids on the liquid–crystal configuration of membranes at physiologic temperatures, the permeability of the membrane to different molecules, and many other physiological properties. Some of these results are discussed in Section 4.5, "Membrane Fluidity."

Protein Composition

Although proteins make up, by weight, a greater portion of the plasma membrane than lipids, less is known of their structures and physiochemical properties because they are present in a much greater variety of forms.

It is now clear from optical studies, freeze-fracture electron microscopy, and other approaches that the predominant and perhaps nearly exclusive form of the protein molecule in the membrane is globular.

Freeze-fractured membranes, such as those shown in Figure 4.16, con-

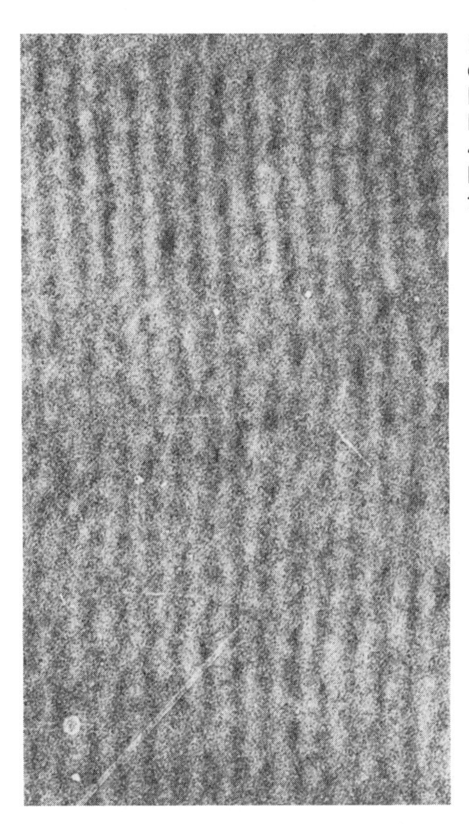

FIGURE 4.14 A polar lipid fixed with osmium tetroxide demonstrating the lamellar form in lipid–water systems. Bands repeat at a distance of about 4 nm, very close in dimensions to the bilayer of lipids in membranes. (Courtesy of Dr. J. D. Robertson.)

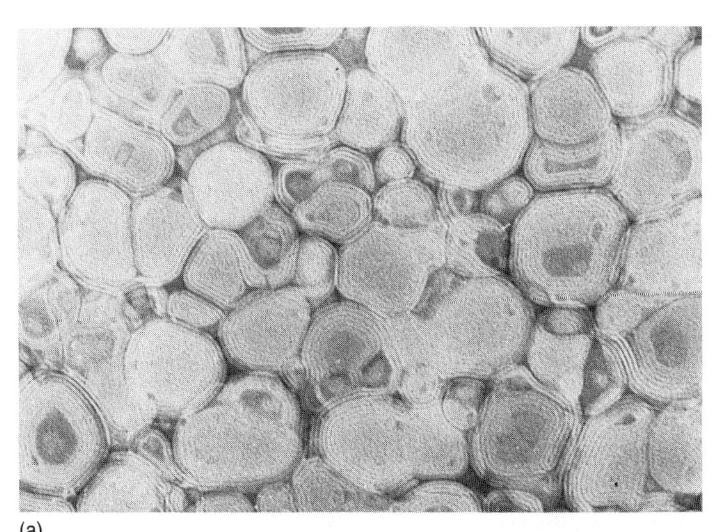

(a)

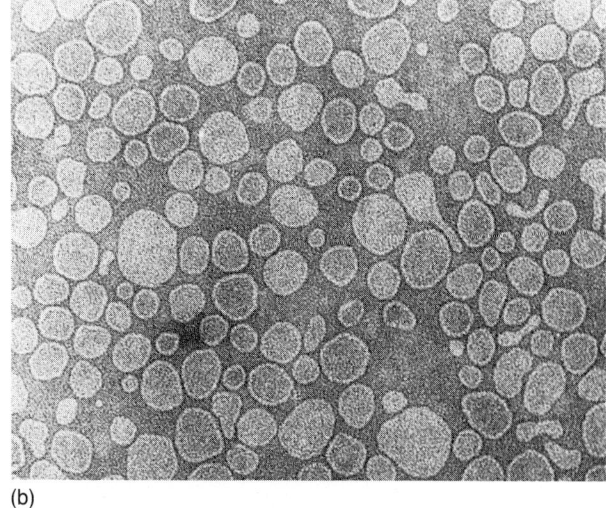

(b)

FIGURE 4.15 Electron micrographs of polar lipids in water that demonstrate the multilamellar vesicle (*a*) and the single bilayer liposome (*b*) forms. The latter are derived by ultrasonic irradiation of the former. (Courtesy of Dr. Alec D. Bangham.)

FIGURE 4.16 Freeze-fracture preparation of an area of the membrane of adjacent Sertoli cells from guinea pig testis. Particles (integral proteins) in this case appear to populate predominantly the P face, which represents the outwardly directed inner half-membrane. The E face is the inwardly directed outer half-membrane. (Courtesy of Dr. Don W. Fawcett.)

tain particles of about 8.5 nm that are well distributed throughout the interior of the plasma membrane. Circular dichroism spectra of the intact erythrocyte membrane indicate about 40% of the proteins are in the α-helix form,[11] dispelling some of the earlier notions of unrolled or extended proteins. Furthermore, it is relatively easy to gain access to certain of the membrane proteins from either the inside or outside surface of the cell, whereas other proteins are inaccessible.

These and other studies have led to a concept of membrane structure that views the topography of membrane proteins as being either *peripheral* or *integral*. The two forms can be distinguished quite readily according to the criteria listed in Table 4.2. It appears that some 20 to 30% of membrane proteins can be considered peripheral. It is obvious, by their easy release from the membrane, that peripheral proteins are only weakly bound and do not interact significantly with the lipid bilayer. Some of the common peripheral proteins are listed in Table 4.3.

In a number of cases, it appears as though peripheral proteins interact with integral proteins that have domains projecting into the aqueous phase. It is often speculated that this type of interaction permits peripheral pro-

Table 4.2 Criteria for Distinguishing Peripheral and Integral Membrane Proteins

Property	Peripheral Protein	Integral Protein
Requirements for dissociation from membrane	Mild treatments sufficient: high ionic strength, metal ion chelating agents	Hydrophobic bond-breaking agents required: detergents, organic solvents, chaotropic agents
Association with lipids when solubilized	Usually soluble free of lipids	Usually associated with lipids when solubilized
Solubility after dissociation from membrane	Soluble and molecularly dispersed in neutral aqueous buffers	Usually insoluble or aggregated in neutral aqueous buffers

Table 4.3 Some Peripheral Proteins and Complexes

Peripheral Protein or Complex	Membrane Localization
Cytochrome *c*	Outer surface of inner mitochondrial membrane
Spectrin	Cytoplasmic surface of erythrocyte
HPr protein	Cytoplasmic surface of bacterial membranes
D-Glyceraldehyde-3-phosphate dehydrogenase	Cytoplasmic surface of erythrocyte membranes
Aldolase	Cytoplasmic surface of erythrocyte membranes
Ribosomes	Cytoplasmic side of rough endoplasmic reticulum
Nectin	Plasma membranes of *Streptococcus faecalis*
Oligomycin-sensitivity-conferring protein (OSCP)	Matrix side of inner mitochondrial membrane
Monoamine oxidase	Outer membrane of mitochondria
Periplasmic binding proteins	Plasma membrane of Gram-negative bacteria

teins to exert some control over integral proteins, and hence over membrane functions.

It has been proposed that there are at least four classes of integral proteins.[12] These are illustrated schematically in Figure 4.17, where the cis face of the membrane is analogous to the surface to which ribosomes are bound on interior membranes.

Type A and C proteins are probably structurally quite similar but are found embedded into different halves of the bilayer.

A well-studied example of a type A protein is cytochrome b_5 of endoplasmic reticulum. This protein is discussed in greater detail in the next chapter because it is not a normal member of the plasma membrane.

A thoroughly studied protein of the erythrocyte membrane, glycophorin, is an example of a type D protein. As we shall see in more detail in the next section, it spans the entire membrane and is made up of a

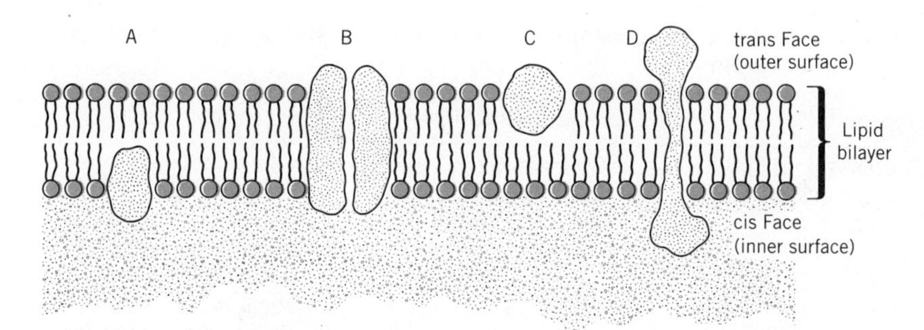

FIGURE 4.17 Four classes of integral membrane proteins. Proteins A and C are only partially embedded in the lipid bilayer, whereas protein D is a transmembrane component. Protein B is a complex of proteins such as are thought to constitute certain transport systems.

hydrophobic section connnecting two hydrophilic ends. Type B proteins are molecular aggregates of a small number of identical or similar subunits. The proteins that have a structure consistent with this are the Na^+, K^+-ATPase and an anion transport protein (band 3) of the erythrocyte. These molecular aggregates are proposed to function in membrane transport mechanisms, a topic discussed in Section 4.6.

Carbohydrates

Almost all plasma membranes, as far as we know, contain carbohydrate in addition to protein and lipid. However, as compared to the protein and lipid components, carbohydrate structures and their functions in the membrane are less clearly defined.

There are two major categories of membrane carbohydrates. One is represented by the complex polysaccharide coating associated with encapsulated bacteria or the cells of connective tissue. This type of carbohydrate is not covalently linked to membrane components but is secreted by the cell and remains only loosely associated with the membrane.

The other category of membrane carbohydrate is made up of material *covalently* linked to the membrane components. In this respect there are two types of linked carbohydrate, molecules bonded to protein (glycoprotein) and those bonded to lipid (glycolipid). Of these two types, the glycolipids are better understood, since their properties are not nearly as diverse as those of the glycoproteins. Carbohydrates bound in this way to membrane components make up the *glycocalyx* of the cell surface.

Glycoproteins

The only membrane glycoprotein on which we have detailed information is glycophorin of human erythrocytes. Studies on this and other glycoproteins have given rise to some generalities.

The carbohydrate moiety of membrane glycoproteins is generally made up of a collection of less than a dozen different kinds of monosaccharide unit. Among these are the simple sugars, D-glucose, D-galactose, D-mannose, L-fucose, L-arabinose, D-xylose, and the amino sugars *N*-acetyl-D-glucosamine, *N*-acetyl-D-galactosamine, and *N*-acetylneuraminic acid (sialic acid). These sugars are normally clustered in oligosaccharide complexes only a few residues long, and it appears that a given glycoprotein may have a number of similar oligosaccharide units that vary only in the degree to which they have been biosynthetically completed. Thus if A-B-C-D-*P* represents an oligosacchride with the monosaccharide unit D attached to the protein *P*, one might expect to see the following variations in oligosaccharide complexes: A-B-C-D-*P*, B-C-D-*P*, C-D-*P*, and so forth.[13]

The monosaccharide most proximal to the protein is linked by a *glycosidic* bond, which may be either an alkali-stable N-glycosidic linkage to the amide nitrogen of asparagine, or an alkali-labile *O*-glycosidic linkage to serine or threonine (see Figure 4.18). In both cases, the proximal monosaccharide is commonly *N*-acetyl-D-galactosamine.

Sialic acid constitutes the *terminal* or most distal residue on most of the oligosaccharide chains of glycophorin, where it imparts a cluster of negative charge on the outer surface of the membrane. It is thought that this is the major mechanism to maintain a net negative charge on the outer cell surface, thus inhibiting erythrocyte contact and aggregation *in vivo*. As erythrocytes age, the sialic acid content decreases, exposing the penultimate sugar galactose, and as this occurs the cells are removed from circulation by the spleen. We thus see a possible role for the carbohydrate portion of the glycoprotein as important in maintaining cellular independence and, by alteration, in marking the termination of the life span of the erythrocyte.

The membrane glycoprotein molecule, and in particular its carbohydrate, may have other physiological functions. Evidence has accumulated showing that glycoproteins bind hormones such as insulin, and in this capacity they may serve as a transmembrane message transmitter for sugar transport. In addition, certain blood group antigens are found on glycoproteins (see Chapter 5) as well as the receptors for influenza virus and lectins. Carbohydrates are also implicated in the important property of cellular adhesion. So the surface carbohydrate of the cell membrane appears to have a multifaceted role vis-à-vis some fundamentally important cell properties.

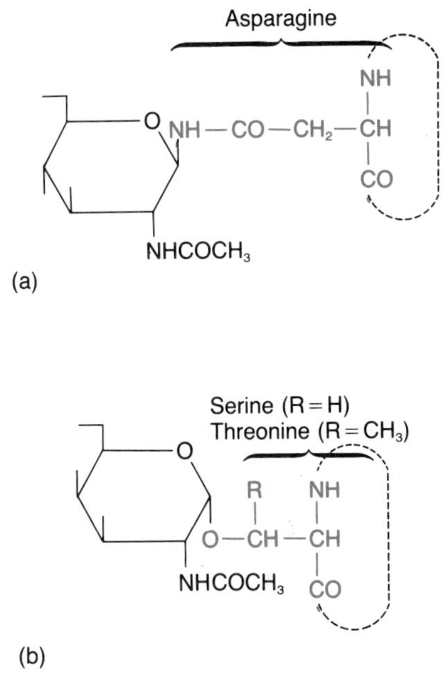

FIGURE 4.18 Glycopeptide bonds that occur between monosaccharide units and proteins. (*a*) An alkali-stable *N*-glycosidic linkage between *N*-acetylglucosamine and asparagine. (*b*) An alkali-labile *O*-glycosidic bond between *N*-acetylgalactosamine and the hydroxyamino acid serine or threonine.

Glycolipids

The glycolipids of the membrane are a collection of amphipathic molecules with large and highly hydrophilic carbohydrate heads. In these molecules, the carbohydrate head is attached by a glycosidic linkage to one of the carbons of the glycerol moiety of the molecule. In this manner, the glycolipids may exist as glycosyl diacylglycerols of mono- or oligosaccharide form (Figure 4.19) or as glycosphingolipids, which may bear highly complex oligosaccharide heads (Figure 4.20). Finally, steryl glycosides constitute a class of glycolipid found mainly in plant membranes. In these molecules, the hydroxyl group of a sterol is bonded by a glycosidic linkage to a monosaccharide.

The Membrane Glycophorin Molecule

One of the most thoroughly studied membrane proteins is glycophorin, an integral protein of the human erythrocyte. Although the erythrocyte

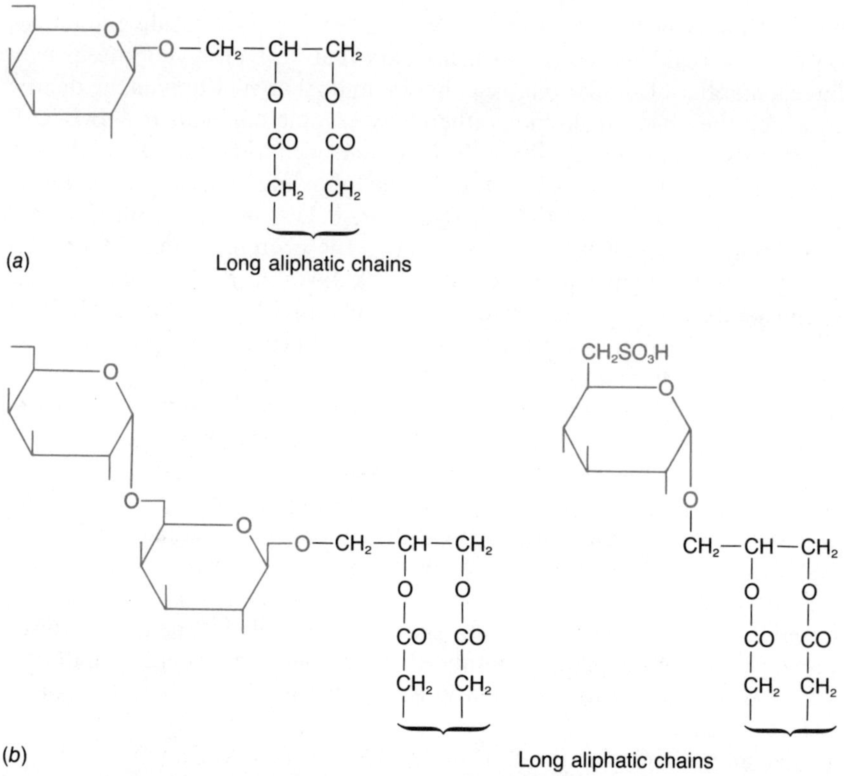

FIGURE 4.19 Membrane glycolipids. (*a*) A monosaccharide diacylglycerol (monogalactosyl diglyceride). (*b*) Oligosaccharide diacylglycerols, α-1,6-galactosyl-β-galactosyl diglyceride (left) and 6-sulfo-α-quinovosyl diglyceride (right). In all cases the hydrophilic domain is rich in alcoholic hydroxyl groups.

is a highly specialized cell, and glycophorin a protein unique to this system, it is quite probable that proteins with similar overall properties exist in other membrane systems. So rather than summarizing the fragments of information available about a variety of integral proteins, we will concentrate on glycophorin.

Figure 4.21 depicts in schematical form the primary structure of the glycophorin molecule. It is a single polypeptide chain of 131 amino acid residues, with a molecular weight of about 55,000 daltons. Approximately 60% of the molecule is carbohydrate, all of which is present as short oligosaccharide chains covalently linked to asparagine, threonine, and serine residues. An interesting, and not incidental, feature of the molecule is that the carbohydrate is confined to the N-terminal 90 to 100 amino acids, giving this half of the molecule strong hydrophilic properties.

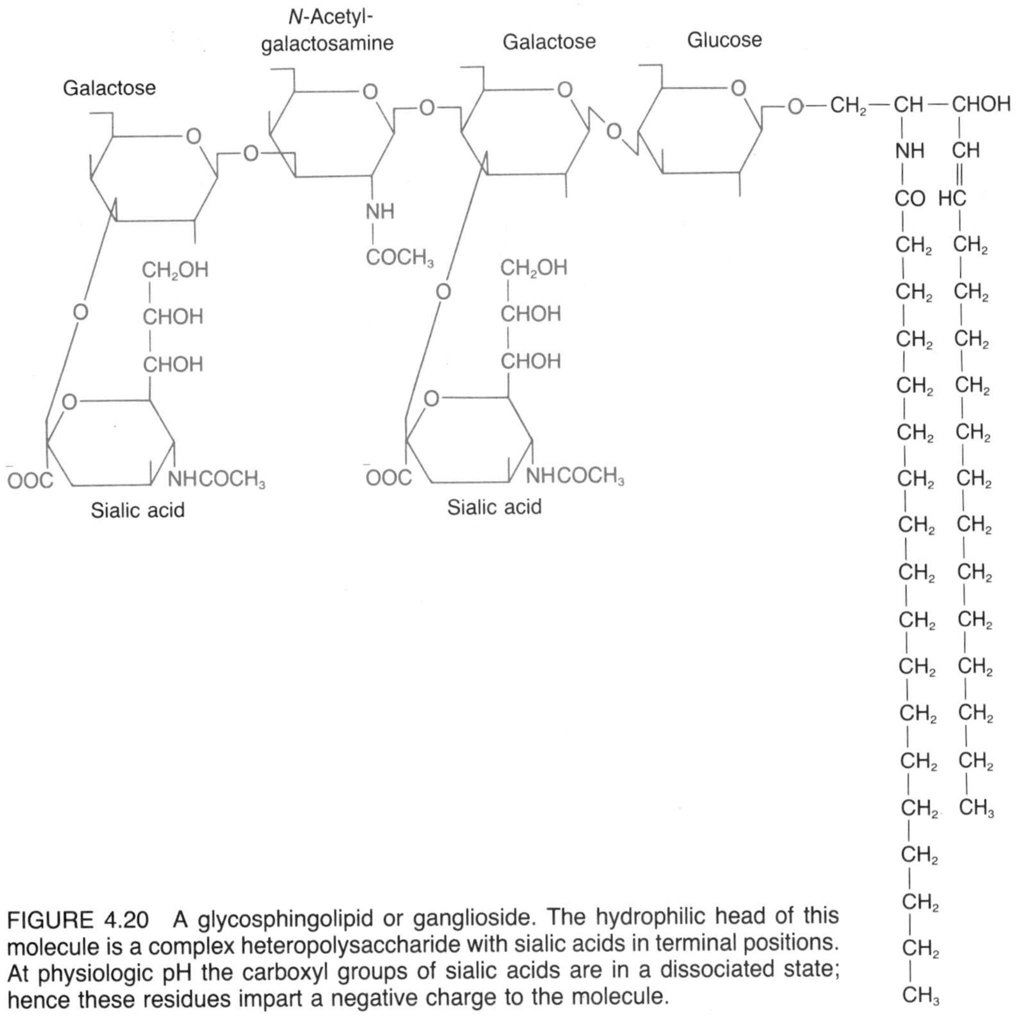

FIGURE 4.20 A glycosphingolipid or ganglioside. The hydrophilic head of this molecule is a complex heteropolysaccharide with sialic acids in terminal positions. At physiologic pH the carboxyl groups of sialic acids are in a dissociated state; hence these residues impart a negative charge to the molecule.

The C-terminal end of glycophorin is also hydrophilic, because there is a high concentration of charged amino acids such as aspartic acid, glutamic acid, and arginine toward this end of the molecule.

These two hydrophilic ends are linked by a hydrophobic region of about 22 amino acids, none of which is ionic at physiological pH.

Glycophorin is therefore a structurally segmented molecule, with alternating hydrophilic–hydrophobic–hydrophilic domains. Based on this information, it would be feasible for the molecule to exhibit type A or C properties, with the middle region embedded in the lipid bilayer in hairpin fashion. But experimental results have shown quite clearly that this is not the case.

In the first place, the carbohydrate portion of the molecule projects from the outer surface of the membrane. Treating intact cells or sealed ghosts with radioiodine and the enzyme lactoperoxidase labels glyco-

Lactoperoxidase is an enzyme of milk that will iodinate tyrosine and histidine residues in proteins when provided with I^- and H_2O_2. When $^{125}I^-$ or $^{131}I^-$ is used, proteins can be easily radioactively labeled.

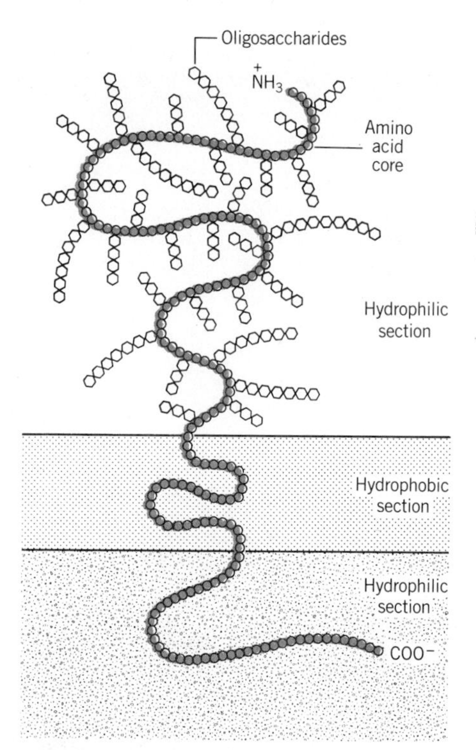

FIGURE 4.21 Representation of membrane glycophorin molecule. It consists of three domains: a hydrophilic N-terminal section of 90 to 100 amino acids to which oligosaccharides are attached, a hydrophobic section of about 22 nonionic amino acids, and a hydrophilic *C*-terminal section rich in charged amino acids.

phorin on the carbohydrate end only. Trypsin, when applied to intact cells, clips off fragments from the N-terminal carbohydrate-containing end only.

Second, if leaky ghosts are labeled, so that the label is presented to both interior and exterior cell surfaces, label appears in the C-terminal hydrophilic section of the molecule as well as the N-terminal end. It is therefore reasonable to conclude that glycophorin spans the entire membrane, in a manner similar to that depicted in Figure 4.22. The physical properties of the molecule seem to be most appropriate for its membrane environment.

4.4 MEMBRANE ASYMMETRY

It was apparent from the early electron microscopic studies carried out on myelin by Robertson that the Schwann cell membrane was not a symmetrical structure. The apposition of inner membrane surfaces produced a major dense line, whereas a line of lighter intensity marked the point at which the outer surfaces came together (see Figure 4.3).

This morphological asymmetry is certainly what one would expect, given our current knowledge of the function of the membrane. We know that materials are not transported with equal facility into and out of cells and that the establishment and maintenance of electrochemical gradients, which are produced by an unequal distribution of ions across the membrane, is a salient feature in the economy of most cells.

In addition to these lines of functional and morphological evidence, there is now a substantial body of chemical and physical information indicating that the plasma membrane is an asymmetric structure. Most of the information we have concerning this property has been gleaned from a study of the erythrocyte membrane, so we are not yet in a position to say that membrane asymmetry is a universal cell property. At present, however, with a limited number of cell types investigated, there is no reason to believe that erythrocyte membranes are unique in this regard.

The study of membrane asymmetry has generally been conducted along three different lines.[14] One approach has been to compare the extent and pattern of labeling of membrane surfaces in intact versus broken cells. Chemical treatment of *intact* cells with nonpenetrating reagents (such as [^{35}S]-diazoniumbenzenesulfonate or [^{35}S]formylmethionylsulfone methylphosphate) results in a labeling of membrane components that differs significantly in pattern and degree from that obtained when *ghosts* or *isolated membranes* are labeled. Alternatively, restricted enzymatic treatment of intact and broken cells (as with proteases, neuraminidase, lipases, and iodination with ^{125}I using lactoperoxidase) shows a differential modification of the membrane components depending on whether one or both surfaces are exposed.

A second approach has been to compare the results of labeling or modifying right-side-out with inside-out membrane vesicles. Closely related to this approach, impenetrable substrates can be added exogenously to these two systems and the relative use of the substrates by membrane-bound enzymes noted.

A third method has employed histochemical labeling of membranes followed by electron microscopy. Techniques have been worked out to stain the Na^+,K^+-dependent ATPase in erythrocytes, and electron-dense reagents, such as ferritin-labeled antibodies and lectins, will interact only with certain membrane components and can be visualized by electron microscopy.

The data that have resulted from these experimental approaches support the notion that *both* lipids and proteins are distributed asymmetrically across the membrane.

Method VII

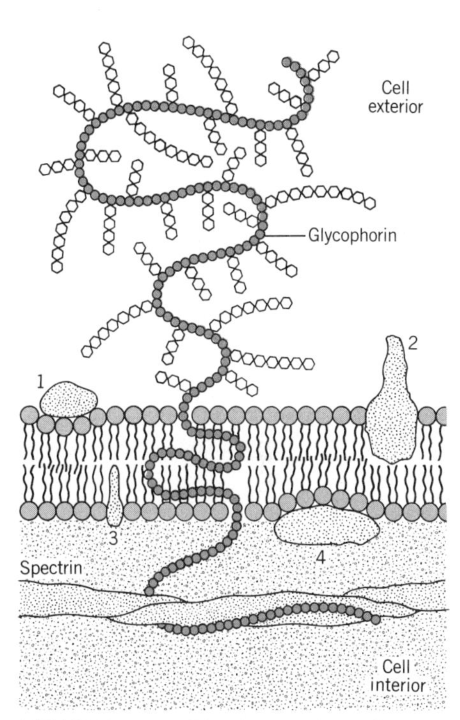

FIGURE 4.22 The interaction of glycophorin and certain other proteins with the lipid bilayer of the membrane. Glycophorin spans the membrane with the lipid bilayer capturing its hydrophobic midsection. The hydrophilic carbohydrate-containing end projects into the cell environment and the opposite hydrophilic tail associates with certain intracellular proteins, such as spectrin. In contrast to glycophorin, proteins 1 and 4 are extrinsic (peripheral) and proteins 2 and 3 are intrinsic (integral), but only partially embedded in the bilayer.

Proteins

The results so far are most detailed for the erythrocyte membrane. Of the nine or so major proteins studied, only two are exposed at the outer cell surface, and, in fact, span the entire membrane. Both are glycoproteins. One is glycophorin, the major erythrocyte protein already discussed, and the other is a high molecular weight (100,000-dalton) protein called component a or 3 by different investigators. The complex oligosaccharide portions of these two proteins are found only on the outer membrane surface.

The other seven proteins do not contain carbohydrate and are found exposed strictly at the inner or cytoplasmic membrane surface.

Other membrane systems that have been studied, such as those of the lymphocyte, show indications of protein asymmetry, and we will see good evidence for structural and functional membrane asymmetry involving proteins in organelles when we discuss mitochondria (Chapter 12).

Lipids

If a protein asymmetry exists in the membrane, one would expect to find an uneven distribution of lipids complementing it simply on the grounds that lipids must fill in the gaps unoccupied by proteins. For the erythrocyte, the evidence is quite compelling that the major phospholipids are unevenly distributed across the membrane.

An enzymatic approach has been most enlightening in this regard[15]. When *intact* human erythrocytes are treated with the phospholipase A_2 of *Naja naja* venum, 70% of the phosphatidylcholine of the membrane is converted to lysolecithin, while sphingomyelin, phosphatidylserine, and phosphatidylethanolamine escape significant degradation. Sphingomyelinase from *Staphylococcus aureus* degrades about 80% of the sphingomyelin of the membrane. Under both these conditions, the cell does not become lysed.

In contrast, phospholipase A_2 treatment of *broken* cells, which gives the enzyme access to both surfaces, results in degradation of all four types of phospholipid.

These observations suggest that phosphatidylcholine and sphingomyelin are preferentially distributed in the outer half of the lipid bilayer and that the other phospholipids are predominantly in the inner half.

One of the important implications of these studies is that the asymmetry is stable, at least over the period the cells were exposed to enzyme degradation. This means the membrane components are not completely free to "flip-flop" from one side of the bilayer to another. Other lines of evidence support this, as the next section reveals.

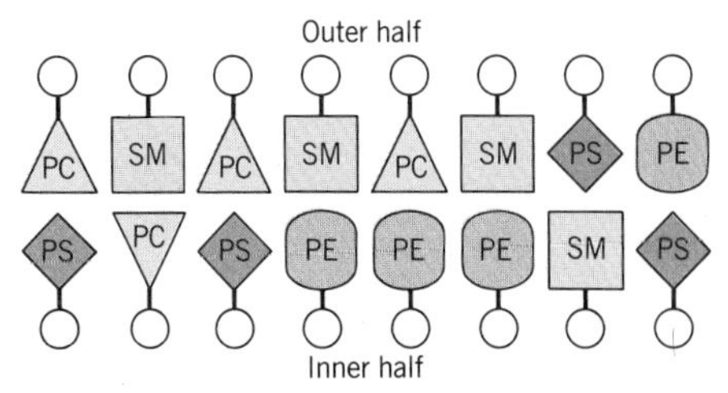

The membrane outer half contains more phosphatidylcholine (PC) and sphingomyelin (SM) than the inner half. The inner half is richer in phosphatidylserine (PS) and phosphatidylethanolamine (PE) than the outer half.

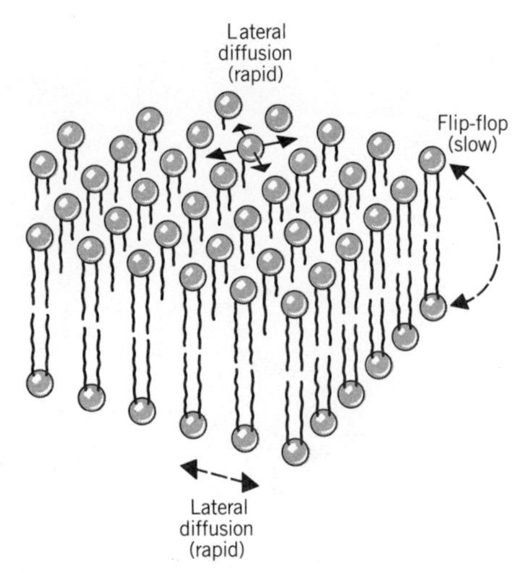

Lateral diffusion and flip-flop mobility of lipids.

4.5 MEMBRANE FLUIDITY

The fluid mosaic model of membranes reflects the view that globular amphipathic proteins are intercalated in a fluid or liquid–crystalline lipid bilayer. The model thereby provides for membrane flexibility and suggests the possibility of both lateral and trans bilayer (flip-flop) mobility of the lipids and proteins that make up the membrane.

Crystalline Gel to Liquid Crystal Transitions

Studies on pure phospholipid systems have provided a means of approaching the question of membrane fluidity. When heated, pure lipid systems undergo a transition from a *crystalline gel* form to a *liquid crystal* form. The transition temperature depends on the length of fatty acid chains and their extent of saturation as well as the degree to which the lipid is hydrated. Another factor, which probably plays an important role in natural membranes, is the amount of cholesterol present. Cholesterol apparently interferes with the formation of the crystalline gel form. Therefore, a membrane containing cholesterol will maintain the liquid crystal form at temperatures lower than normal for a sterol-free membrane. In a similar manner, shorter fatty acid chains or higher degrees of unsaturation promote the liquid crystal state at lower temperatures.

The transition from crystalline gel to liquid crystal is an endothermic process, so it can be followed experimentally with a calorimeter. The effect of the degree of hydration on this phase transition is illustrated in the results of differential scanning calorimetry (DSC) in Figure 4.23. As the water content increases, the transition temperature drops. The effect of cholesterol on this phase transition is shown in Figure 4.24. Cholesterol broadens the transition peak until, at a 1 : 1 ratio, it disappears completely.

It is probably because of the presence of cholesterol that native myelin and erythrocyte ghosts show no phase transitions, hence can exist in the liquid crystal form at physiological temperatures.

The results above are merely examples of numerous studies that have been carried out on chemically defined lipid systems and natural membranes using thermal techniques. Other approaches have utilized X-ray diffraction, infrared spectroscopy, nuclear magnetic resonance (NMR) and electron spin resonance (ESR) spectroscopy. The results have led to a wide acceptance of the concept of membrane fluidity at physiological conditions.

Method VIII

But it is unlikely that all lipid molecules of the bilayer exist in an equally mobile state. There is evidence, for example, that certain of the integral proteins contain tightly bound lipid coats, which are required for optimal enzyme activity. The lipid molecules that constitute this boundary would be severely restricted in mobility. Thus, in any given membrane the amount of lipid in this form would depend on the composition of integral proteins and their tendency to bind lipids.

Lipid Mobility

The concept of lipid fluidity implies a freedom of lateral molecular motion in the plane of the bilayer. Lateral mobility, or lateral diffusion, has been

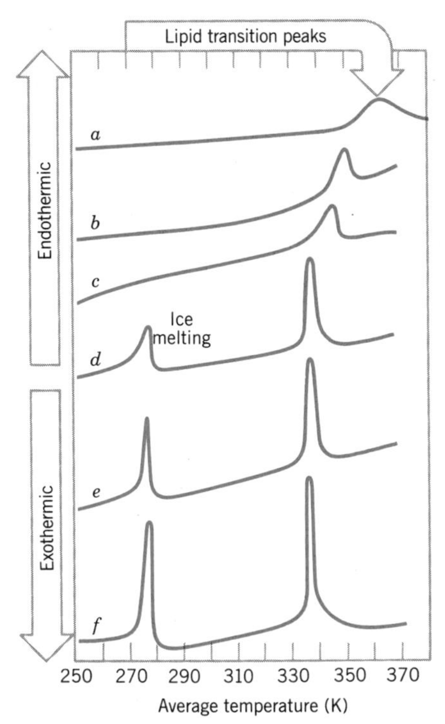

FIGURE 4.23 Demonstration of the effect of state of hydration on the gel to liquid crystalline phase transition by the technique of differential scanning calorimetry. The polar lipid employed is distearoyl phosphatidylcholine. Water contents by weight are represented by curves: *a*, 3%; *b*, 10%; *c*, 20%; *d*, 25%; *e*, 30%; *f*, 40%. At the phase transition of the sample, the extra heat flow is recorded as a peak. The left-hand peak corresponds to the melting of ice in the system. At 20% water content all the water is bound to the phospholipid and does not freeze.

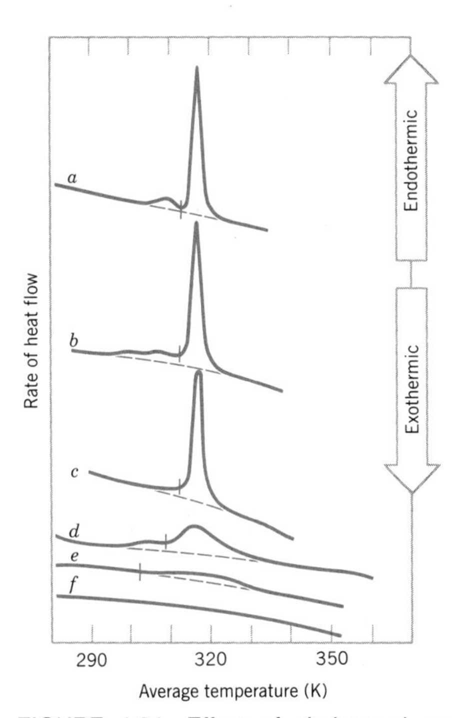

FIGURE 4.24 Effect of cholesterol on phase transitions of polar lipid dipalmitoyl phosphatidylcholine as demonstrated by differential scanning calorimetry. Polar lipid concentration is 50% by weight with increasing concentrations of cholesterol: *a*, 0.0; *b*, 5.0; *c*, 12.5; *d*, 20.0; *e*, 32.0; *f*, 50 mole %.

confirmed by the use of spin label methods and NMR spin-echo techniques. Diffusion coefficients on the order of 1.8×10^{-8} and 0.5×10^{-7} cm^2/sec have been obtained by different investigators.[16] These rapid diffusion rates mean that a given lipid molecule exchanges place with its neighbor at a rate on the order of 10^7 per second. This rate of exchange would be expected to vary somewhat for the different molecular species in a membrane and would not be applicable to the lipid that coats integral proteins.

In contrast to the high degree of lateral motion, lipids apparently are not as free to move from one side of the bilayer to the other. This flip-flop mobility is about 10^{10} times slower than lateral movement. The reasons for this are readily apparent when we consider the nature of the bilayer. Thermodynamically, a much lower free energy state is maintained if the polar head of the lipid can avoid the hydrophobic tail layer of the bilayer. If this type of mobility were not restricted, there would be little possibility of lipid asymmetry in membranes.

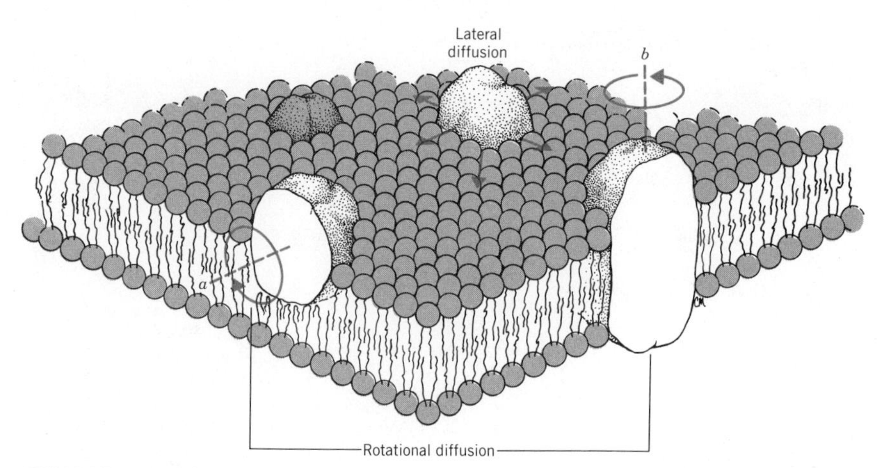

FIGURE 4.25 Mobility of proteins in lipid bilayer may be of three types: lateral diffusion in the plane of the bilayer, rotational diffusion around an axis parallel with the bilayer plane (*a*), and rotational diffusion around an axis perpendicular to the bilayer plane (*b*).

Protein Mobility

The mobility of proteins in the fluid matrix can be thought of as occurring in three different ways: by lateral diffusion, as for the polar lipids, by rotational diffusion about an axis that is perpendicular to the surface of the membrane, and by rotational diffusion about an axis that is parallel to the membrane surface. These types of mobility are illustrated in Figure 4.25.

So far, there is ample evidence for the occurrence of lateral diffusion, limited evidence for type *b* mobility, and little or no direct evidence for type *a* movement.

The lateral diffusion of proteins is generally slower than that of the lower molecular weight phospholipids, as might be expected by the size of the former. Diffusion coefficients of rhodopsin in frog retinal rod membranes of 4×10^{-9} cm^2/sec and of glycophorin in erythrocytes of 3×10^{-12} cm^2/sec have been reported. Both the molecular dimensions of the proteins and the consistency of the fluid bilayer in these two systems probably affect these rates of diffusion.

A dramatic demonstration of the lateral diffusion of proteins was effected by labeling certain integral proteins in human and mouse cells with different fluorescent antibodies and then fusing the two cells with a virus to produce a mouse–human heterokaryon.[17] At first, the different species-specific antigenic markers were localized on their own cell halves, but with a 40 min incubation at 37°C, the integral proteins were completely randomized over the entire heterokaryon surface. This experiment is illustrated in Figure 4.26. Several other cell systems have been studied with similar results.

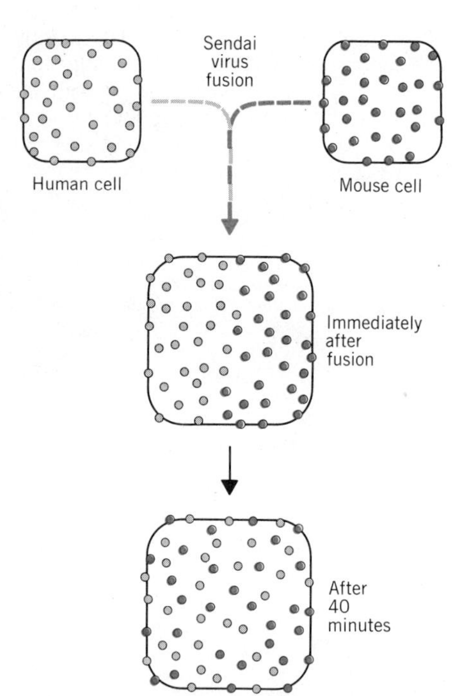

FIGURE 4.26 Demonstration of the lateral diffusion of membrane proteins. Human and mouse cells, each with distinctly labeled proteins, are fused to form a heterokaryon. With time, the distinctive proteins randomize over the surface, having started from discrete regions corresponding to the cells from which they originated.

A quite different approach, yielding similar conclusions, has been to add ferritin-labeled multivalent ligands, such as antibodies or lectins, to their respective receptors on cell surfaces and to note the changes in the membrane that follow. There is, within minutes, a rapid clustering of receptors (integral proteins) into patches or caps. The antibodies or lectins attract and cross-link the mobile integral proteins until a raft is formed on the cell surface.

Rotational diffusion of type *b* has been measured on rhodopsin in the rod outer segment membrane by noting the rate of decay of photodichroism induced by bleaching with a flash of polarized light.[18] Immobilizing the rhodopsin by cross-linking with glutaraldehyde blocks decay. In this experimental setup, a blocking of decay is a measure of rotational restriction.

Constraints on Mobility

It is now quite certain that the biological membrane exists in its native form in a liquid crystal state—ordered but with mobile components. It is also becoming increasingly clear that there is not complete and independent freedom of movement for the various species making up the membrane. As already indicated, mobility of the fraction of the lipid population that is tightly bound to certain of the integral proteins is constrained. There is also, as discussed, the thermodynamic constraint on trans bilayer flip-flop activity.

The finding of ordered structures within the fluid mosaic portion of the membrane has led to the notion that certain of the proteins of the membrane are constrained by protein–protein interactions to form specialized ordered regions that represent from 2 to 20% of the membrane of the system.[19] Examples of these specialized structures are found in the gap junctions of cells, in the synapses of neurons, and in the plaques of halobacteria. Gap junctions are discussed in detail in Chapter 5. The plaques of halobacteria are large planar aggregates of identical proteins that are packed together to function in the photosynthetic activity of the organism.[20] It is obvious that there is restricted mobility of the proteins that are members of a plaque.

Still other forces apparently operate to restrict the mobility of proteins in the membrane. Evidence is beginning to accumulate to suggest that certain peripheral proteins may form a bridgelike latticework between integral proteins, thus restricting their mobility within the plane of the membrane. Spectrin, a peripheral protein of the erythrocyte, appears to bridge and hold in place glycophorin and, possibly, other integral proteins. The intact adult human erythrocyte (which contains the spectrin complex) has a membrane rigidity that does not permit a clustering or capping of integral proteins when the appropriate antibodies or lectins are added. In contrast, when washed erythrocyte ghosts (which lack spectrin) are exposed to the same antibodies or lectins, the integral protein receptors are readily clustered in the plane of the membrane. From this, it would appear that spectrin, by virtue of forming a peripheral scaffolding structure, restricts protein mobility and may thus be responsible for maintaining the rigid biconcave shape that is characteristic of the mature erythrocyte (see Figure 4.27). It has been suggested that similar peripheral proteins, possessing mechanochemical properties, may control cell shape changes in response to stimuli and may thus be governing factors in cell locomotion, endocytosis, cell division, and a number of other cell functions that depend on membrane deformation. Other structures, more deeply embedded in the cell, such as microfilaments and microtubules, may operate in a similar manner to effect shape changes on the cell.

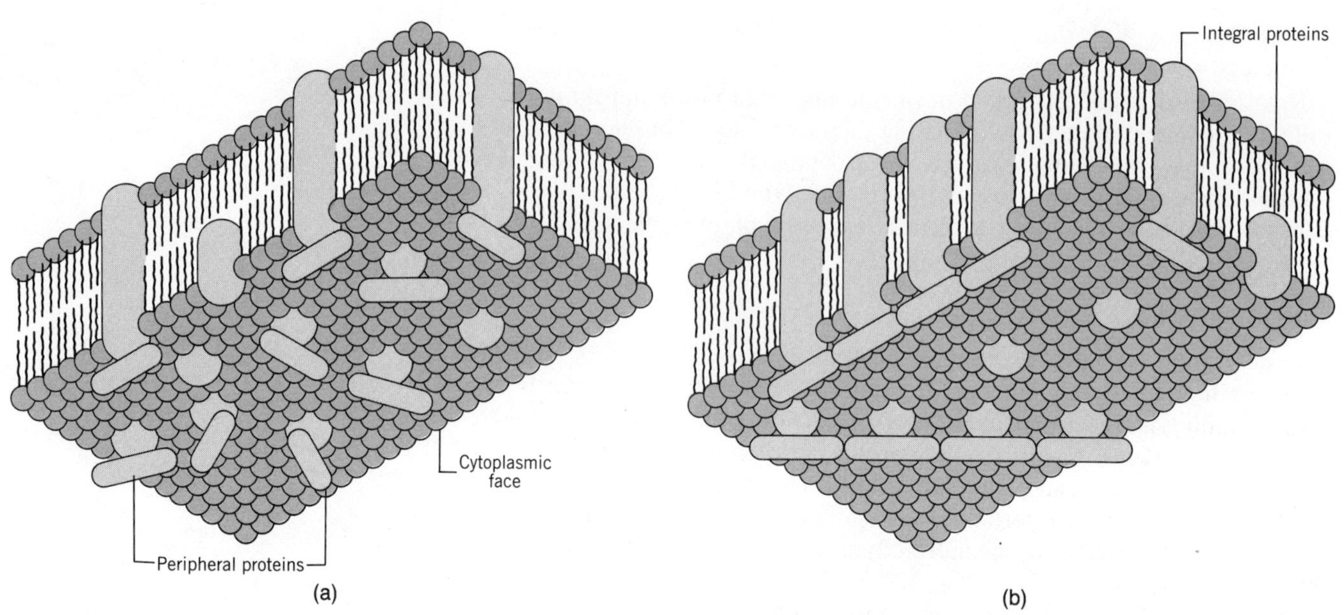

FIGURE 4.27 Schematic depiction of how peripheral proteins may bind to integral proteins at the cytoplasmic face of the membrane and how, depending on whether they are disaggregated (*a*) or aggregated (*b*), these interactions may affect the translational mobility of proteins in the membrane. Spectrin is an example of an erythrocyte protein that may have this role.

4.6 MECHANISMS OF PERMEABILITY AND TRANSPORT

Up to this point, our emphasis has been on the molecular architecture of the membrane and some of its physical properties. We have seen that the concept of membrane structure has evolved from that of an undefined lipoid layer to a protein-intercalated liquid crystal lipid bilayer possessing the physical properties of asymmetry and lateral diffusion of its components. If cells were closed systems, we could pretty much terminate our discussion now with some sense of confidence that the details of membrane structure were essentially complete. And, perhaps most of the essential features are in hand.

But, when we reflect that cells are open-ended systems, constantly in communication with their environment and with other cells, and continually moving some materials in and others out, and that the membrane is the border through which all this activity occurs, then we realize that there is still some work to be done before the mechanisms of membrane function can be pronounced complete.

Transport across the plasma membrane has been an active research field for some time and continues in full swing today. The topic of transport is generally divided to consider what appear to be three different principles of transmembrane movement: simple diffusion, facilitated diffusion, and active transport.

The first two processes describe the transporting of material ''downhill'' from a region of higher concentration to a region of lower concentration. The energy for this movement is thermal molecular motion. Since ATP is not required, diffusion mechanisms are thought of as passive, energy-independent processes. The movement of material across the membrane against an electrochemical gradient, or "uphill," always requires cellular ATP, and is thus designated as an active, energy-dependent process. As we discuss some properties of each of these mechanisms separately, keep in mind that in a living cell they operate simultaneously and in an integrated manner to maintain the necessary balance of biological molecules for the homeostasis of the cell.

Simple Diffusion

Only a limited number of molecular species move across the membrane by simple diffusion, and even for these the membrane poses some hindrance or has a solubility effect on the permeating substances. As discussed earlier, Overton found that a limited number of small molecules, especially hydrophobic species, penetrated plant cells quite readily, whereas larger molecules or ionic species are not transported.

The differentiation of transport between hydrophobic and hydrophilic molecules is generally attributed to a higher solubility of hydrophobic substances in the lipid bilayer. Hence the bilayer poses a lower energy barrier to diffusing molecules that are hydrophobic.

The ability of the cell to discriminate between low molecular weight and high molecular weight hydrophilic materials, permitting the former to diffuse across but not the latter, is often credited to the function of aqueous channels or pores in the membrane. Two types of pore are frequently discussed. One type is an aqueous channel either *through* a protein or *between* clustered integral proteins that extend through the entire bilayer. The other type of pore, referred to as a *statistical pore*, forms randomly through the lipid bilayer as a result of thermal movement of acyl phospholipid chains.

It is not possible to catch a glimpse of the statistical pore with the electron microscope, nor is it certain that spaces exist between integral proteins that could function to allow the diffusion of small molecules across the membrane. If pores exist in the membrane, it is likely that they are not stable structures, since it can be shown by the use of microelectrodes that electrical gradients are maintained between the inside and the outside of the cell. Since this gradient is set up by an unequal distribution of charged species across the membrane, including Na^+ and K^+, the pores must be smaller than their hydration spheres, or must form only transiently.

Facilitated Diffusion

Facilitated diffusion is a carrier-assisted transport mechanism that, like simple diffusion, requires no cellular ATP for its operation. The driving force is a concentration gradient, but the entry into cells is more rapid than would be expected from the principles of Fick's law. According to Fick's law, the rate of movement of a substance is directly proportional to the concentration gradient, which holds for simple diffusion, but in this case saturation kinetics can be demonstrated (i.e., plotting the rate of entry versus concentration gives an asymptotic curve).

Fick's first law of diffusion states that the amount of solute ds diffusing across the area A in a period of time dt is proportional to the concentration gradient dc/dx at that point.

$$\frac{ds}{dt} = -DA\frac{dc}{dx}$$

where D is the diffusion coefficient, defined as the quantity of solute diffusing per second across a surface area of 1.0 cm^2 when there is a concentration gradient of unity. The negative reflects diffusion in the direction of lower concentration.

Facilitated diffusion demonstrates saturation kinetics, as do enzyme reactions. Simple diffusion shows a linear relation between the concentration of solute and its rate of entry into cells.

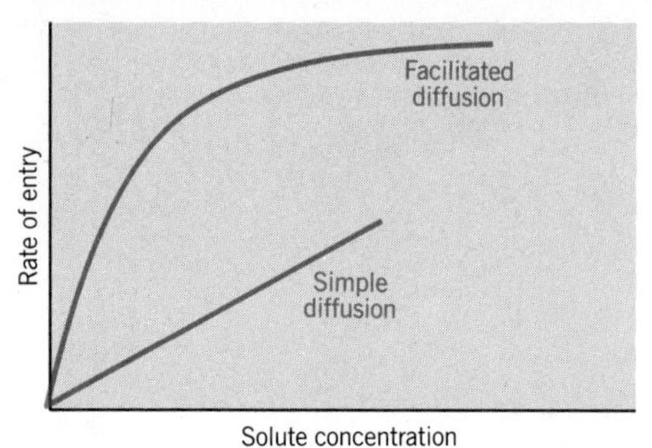

K_m is the Michaelis constant, which in enzyme kinetics is a measure of the affinity of an enzyme for its substrate. It is the amount of substrate required to move the reaction at half maximum velocity. Thus, the lower the K_m, the higher the affinity. The same relation holds for facilitated diffusion.

The kinetic information on transfer in this system simulates enzyme kinetics and suggests that carrier molecules with high specificities facilitate the movement of molecules across the membrane.

Glucose and other hexoses are commonly transported across the erythrocyte membrane by facilitated diffusion. The glucose facilitator demonstrates nicely the specificity of the transport system. D-Glucose has an apparent K_m about 10 times lower than D-galactose and over 1000 times lower than L-glucose. Thus the readily metabolized D-form of glucose competes very effectively for transport into the erythrocyte even if present at lower concentrations than other hexoses.

Because of the kinetic behavior of facilitated diffusion, because enzyme poisons stop this type of transport, and because of the results of other approaches, the carrier molecule is generally thought to be a protein. According to this line of thinking, a given membrane would contain a set of facilitator proteins, each one having a binding site specific for a certain transportable molecule. The carrier protein, after binding to its particular molecule, would have the task of moving the molecule to the other side of the membrane either by rotation, diffusion, a conformational change, or by forming a pore. It is not yet clear which of these, if any, are in fact the means whereby the carrier operates. The models that have been proposed are essentially those that are used to explain the mechanisms of active transport.

Active Transport

The energy-dependent transport of materials across a membrane against an electrochemical gradient takes on several shades of complexity. In its simplest form, a specific molecular species is moved across the membrane in a manner that closely resembles facilitated diffusion, except that ATP is required. In other instances, the movement of a molecular species is coupled to the movement of other species in a variety of different ways or the molecule is chemically modified as it is transported.

A helpful way to classify active transport is to divide it into two major categories: *primary active transport*—transport directly coupled to ATP hydrolysis or electron flow, and *secondary active transport*—transport dependent on membrane potentials or ion gradients (chemiosmotic energy). These two types of active transport are linked in the sense that the mechanisms of primary active transport create the gradients that enable secondary active transport to occur. We will discuss only a few examples that demonstrate these transport principles.

The Na^+, K^+-Translocating ATPase

Most cells maintain a higher K^+ concentration inside than is found in their environments while simultaneously keeping the interior Na^+ concentration lower than that outside the cell. Both K^+ and Na^+ are "pumped" against their concentration gradients, and these pumps operate at the expense of ATP hydrolysis. Thus the influx of K^+ and the efflux of Na^+ are mediated by a membrane enzyme that has ATPase activity. When the pumps are operating they exhibit a high degree of polarity, always pumping Na^+ out and K^+ in. Furthermore, the pumps for the two ions are coupled, apparently requiring simultaneous operation.

ATPase enzymes have been studied in a large number of membrane systems. The enzyme is attached firmly to the plasma membrane and appears to require a certain amount of bound phospholipid for activity. The protein–lipid complex has been isolated from some membranes, and the protein portion of the complex consists of two polypeptide chains, a larger one possessing the catalytic properties of the complex (80,000–100,000 daltons) and a smaller glycoprotein (45,000–60,000 daltons). The larger subunit penetrates the membrane and the small unit, being a sialylglycoprotein, appears to be located toward the outer surface of the bilayer. The purified material has a high molecular weight consistent with a tetrameric form, and can be envisioned as illustrated in Figure 4.28.

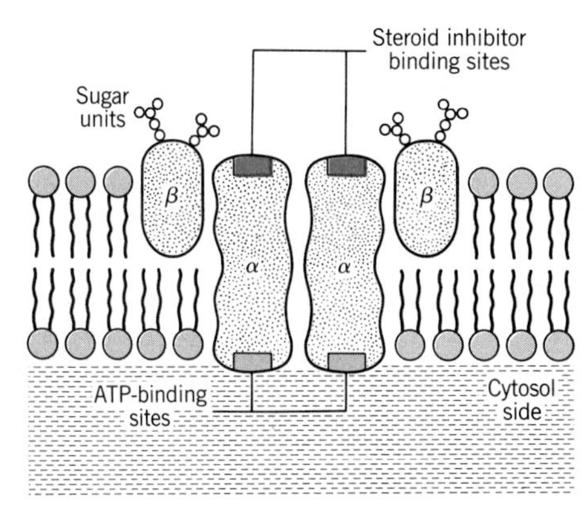

FIGURE 4.28 Schematic diagram of the subunit structure and orientation of the membrane Na^+,K^+-ATPase. It is a transmembrane protein complex that binds ATP on its cytosol side and is subject to steroid inhibitors that may bind to its cell exterior side.

In intact membranes, sodium activates both the pump and the ATPase activity from the *inner* side only, whereas the effect of K^+ occurs from the *outer* side only. Cardiac glycosides, or cardiotonic steroids, which are specific inhibitors of sodium and potassium transport, exert their effects from the outer side only. The evidence is thus quite convincing that the Na^+, K^+-ATPase spans the entire membrane, exposing K^+-sensitive and Na^+-sensitive domains to different sides of the membrane.

It has been confirmed that during transport the ATPase itself becomes transiently phosphorylated and that this phosphorylation depends on the presence of Na^+. Thus, it seems that phosphorylation and Na^+ binding take place on the same side of the membrane, toward the interior of the cell. In contrast, K^+ causes a dephosphorylation of the ATPase. What emerges from these facts is a phosphorylation–dephosphorylation cycle, mediated by sodium and potassium ions, respectively.

The foregoing information is generally integrated into the type of scheme depicted in Figure 4.29. Sodium ion and ATP are first bound to their specific sites on the inner side of the membrane. This is followed by an ATPase reaction in which ATP is hydrolyzed, with the terminal phosphate of ATP covalently attached, through a glutamyl residue, to the enzyme. This causes the enzyme to undergo a conformational change, which has a twofold effect: (1) the Na^+ is translocated to the exterior of the cell where, because of steric changes in the binding site, it is released to the environment and (2) the new conformation uncovers K^+ binding sites to which available K^+ ions on the exterior side are bound. Potassium ion, as we have already noted, stimulates dephosphorylation of the enzyme, which causes the ATPase to revert to its starting conformation and in so doing translocates K^+ across to the inner surface where it is released because of the lower affinity of its binding site in the final conformation.

In recent years, two different models have been proposed to explain the mechanism of coupled Na^+–K^+ transport. One, termed the *rotating carrier* scheme, envisions the ATPase as rotating around an axis that lies in the plane of the membrane (see Figure 4.30). According to this view, the rotation, driven by ATP, causes Na^+ to be picked up on the inner side and K^+ on the outer side, and each is delivered to the opposite side of the membrane. The model has a number of attractive features, but a strong argument has been levied against it on thermodynamic grounds. Rotation would require repeated sweeping of hydrophilic portions of the protein through the hydrophobic region of the lipid bilayer as it tumbles in the membrane. This would be a thermodynamically unfavorable process. Futhermore, it has been found that antibodies or lectins can be attached to transport proteins in intact membranes without altering the

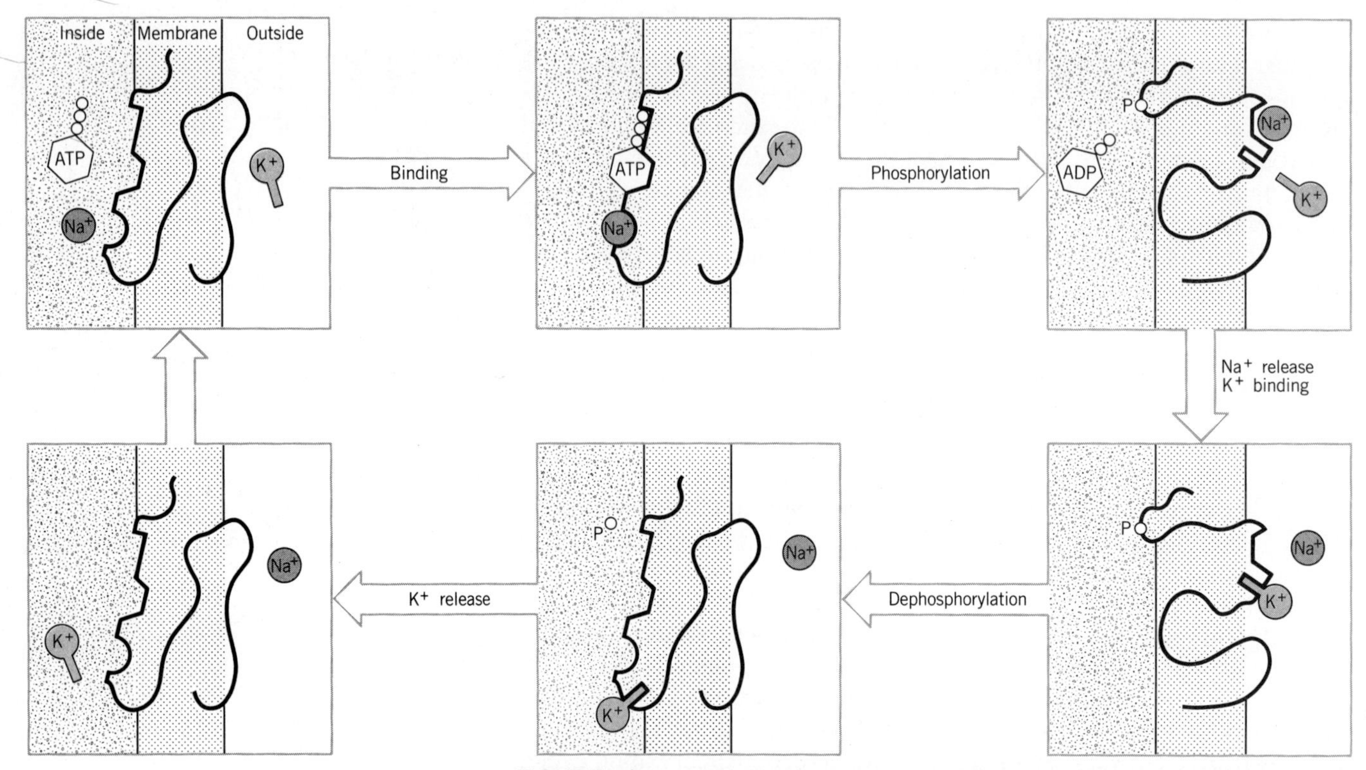

FIGURE 4.29 A schematic summary of the events that take place during the functioning of the Na^+,K^+-ATPase pump. ATP and Na^+ bind to the inner surface of the membrane and after the ATPase is phosphorylated Na^+ is shifted out. K^+ binds, promotes a dephosphorylation, and is shifted in. The shape changes of the ATPase depicted are purely speculative.

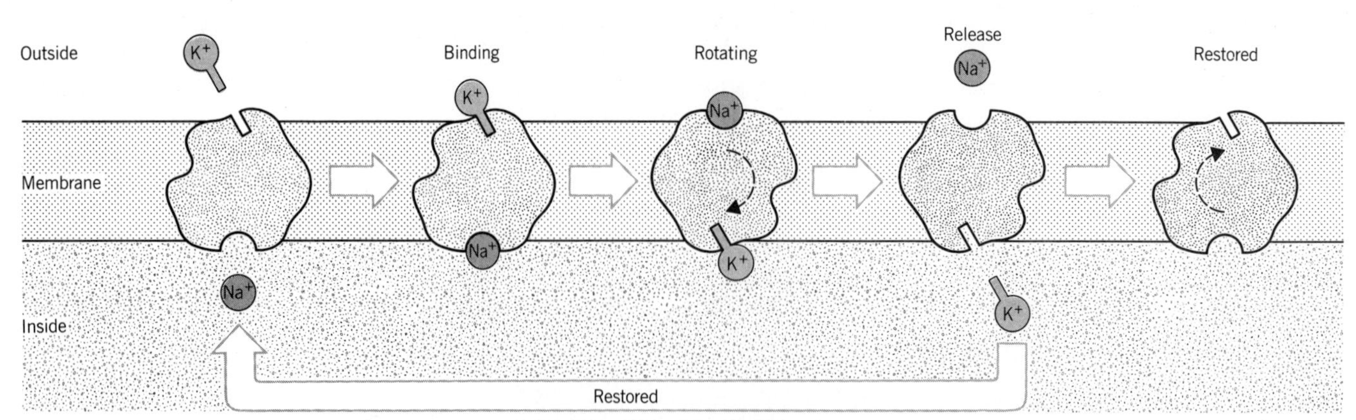

FIGURE 4.30 The rotating carrier scheme of transport. Na^+ and K^+ bind to the carrier on their respective sides of the membrane and are shifted to opposite sides as the carrier rotates 180°. Each is moved against its own gradient with ATP providing the energy for translocation.

ability of these proteins to effect transport. If the transport protein rotates, it would probably have to drag these larger attached molecules through the membrane, a physical feat that is difficult to reconcile with the known properties of the membrane.

The other model, proposed independently by Jardetzky[21] and Singer,[22] views the transporting site as an aggregate of integral proteins or subunits that cluster to form a continuous pore across the entire membrane (see Figure 4.31). An energy-yielding reaction, such as the hydrolysis of ATP, triggers a quaternary rearrangement of the subunits, translocating the binding site from one side to another. To reconcile this model with Na^+–K^+-coupled transport, one would have to imagine an alternating shuttling of Na^+ and K^+ in opposite directions as the quaternary rearrangements take place.

The Na^+, K^+-ATPase system is an example of primary active transport. The result of its operation is for the cell to normally extrude three Na^+ ions and take up two K^+ ions for each ATP molecule hydrolyzed. When the transport system achieves equilibrium, there is generally established a net negative electrical potential inside the cell. Thus although the two models discussed are not electrogenic, (i.e. do not alter the net charge distribution across the membrane), in the real situation the unequal translocation of Na^+ and K^+ operates as an electrogenic pump.

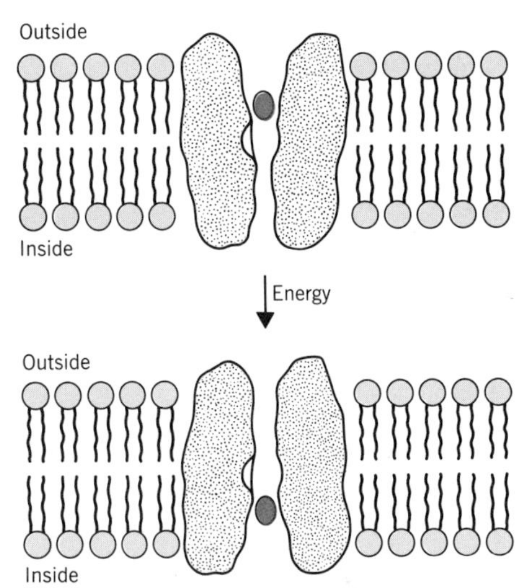

FIGURE 4.31 The protein aggregate rearrangement mechanism of transport. A specific site on the surface of a pore formed by an aggregate binds a solute. An energy-yielding process is then converted into a rearrangement of subunits, which results in translocation of the solute from one side of the membrane to the other.

Sodium-Linked Amino Acid and Sugar Transport

Sugars and amino acids appear to be transported into the animal cell by a type of secondary active transport in which Na^+ is cotransported along with the particular metabolite across the membrane. This mechanism of cell entry is only loosely or indirectly linked to the action of the Na^+, K^+-ATPase. If the latter is inhibited by the specific action of ouabain, a cardiac glycoside, the cotransport carrier will continue to operate as long as a Na^+ gradient exists across the membrane. It is the role of the ATPase to maintain this gradient, thus making this type of mechanism active transport dependent.

A model depicting this mechanism as it would operate in the epithelial cell of the intestine is shown in Figure 4.32. Sodium ion and metabolites, such as certain carbohydrates and amino acids, are cotransported from the lumen across the brush border membranes into the cell. The transport protein E_1 is inhibited by phlorizin. In reconstituted erythrocytes, one glycine molecule is transported together with two Na^+ ions, but in intestinal epithelial cells there is an approximate 1:1 ratio between Na^+ flux and carbohydrate flux. Thus the rules of stoichiometry may vary for different molecular species, and in many cases they are not known. But we do know that the transport is coupled between Na^+ and the metabolite in question. Substitution of Li^+, Mg^{2+}, or choline on the mucosal side abolishes the transport process, so the carrier is highly specific for Na^+.

Phlorizin is found in the root bark of pears, apples, and certain other plants of the rose family. It is used in physiological studies because of its ability to block resorption of glucose by the kidney tubules. It apparently has this effect wherever this type of cotransport mediator operates.

It is obvious from the model that the carrier is only passively driven by the Na^+ gradient. This can be shown experimentally in reconstituted erythrocytes by reversing the gradient, whereupon the glycine influx is converted to a glycine efflux. Once the metabolite has crossed the membrane, it is released on the cytoplasmic side to the interior of the cell.

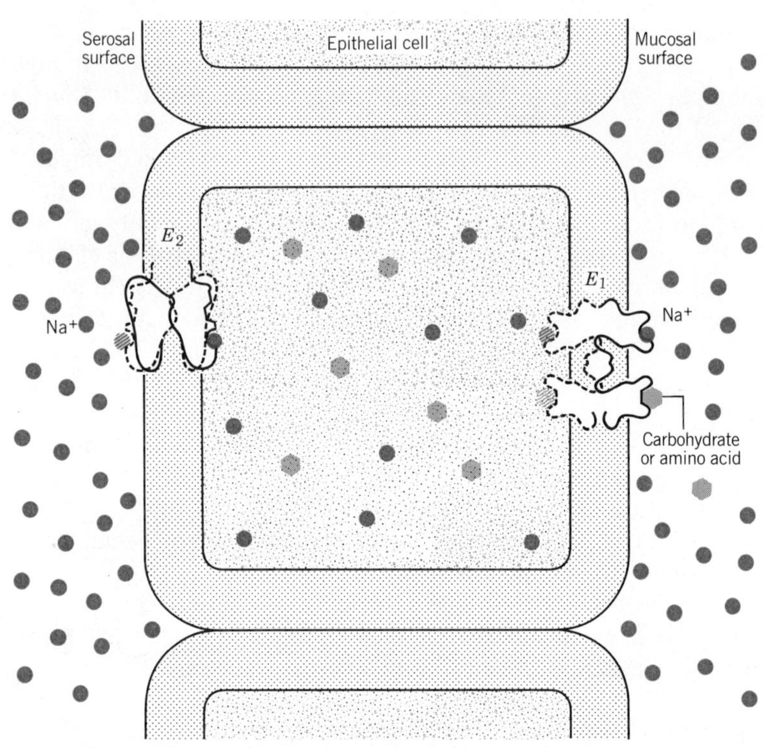

FIGURE 4.32 Model for the coupled transport of Na^+ and certain carbohydrates or amino acids in the intestine. E_1 is the phlorizin-sensitive transport protein and E_2 the ouabain-sensitive sodium pump. Na^+ flows according to its gradient from the extracellular mucosal surface to the cell interior while the metabolite is moved against its concentration gradient. The sodium pump keeps the intracellular Na^+ concentration from building. The complete role of E_2 is shown in Figure 4.29.

Although the mechanism of release is not known, there is evidence that the carrier affinity for the metabolite is regulated by the $Na^+:K^+$ ratios on the two sides of the membrane. The high $Na^+:K^+$ ratio in the lumen promotes a binding of the metabolite to the carrier, whereas the lower ratio on the cytoplasmic side favors a release of the transported molecule. In any event, ATP is not used directly for this process.

The Na^+ gradient that passively drives the transport is maintained by the ouabain-sensitive sodium pump (E_2) operating in the serosal surface membrane of the cell. Blocking the action of this pump will ultimately stop the activity of the cotransporting carrier, since it will no longer continue to withdraw Na^+ from the cell, and the gradient across the mucosal membrane is therefore decreased. It should be noted in this model that even though the flow of Na^+ across the mucosal membrane is according to its concentration gradient, the movement of the metabolites cotransported can be against their concentration gradients. Thus the osmotic energy of the system, the ion gradient, is exchanged for another form, this time a solute gradient.

This is not the only view that has been set forth to describe coupled Na^+–metabolite transport. In some cases it appears that certain amino acids are transported in a manner that is more tightly linked to ATP use.

4.7 MEMBRANE ASSEMBLY

There is hardly any cell structure more important to the immediate health of the cell than the plasma membrane. If it is weakened or injured, the cell loses its ability to maintain gradients, carry out the selective transport of nutrients, and contain the pool of enzymes and organelles essential for homeostasis.

Cell growth, as well as certain other activities such as endocytosis, is marked by membrane dynamics. During these activities the assembly of new membranes is a major cell project. Considering the importance of membrane integrity on the life of a cell, it is essential that membrane growth occur without perturbing the existing membrane to the point of impairing its function or endangering the health of the cell. In other words, new membrane must be added to existing membrane without altering its functions as a barrier and selective transporter.

Another feature of membrane structure puts special demands on mechanisms of membrane assembly. Membranes, as we have discussed, are asymmetric. Proteins are distributed across the membrane unevenly, and carbohydrates are found exclusively on the exterior surface of the plasma membrane. Any change in the orientation of these membrane components would alter membrane function. Thus the membrane must be assembled with precisely the correct molecular topography.

We can imagine several different ways in which membranes may be assembled and still meet these requirements. One would be along the lines of *self-assembly*, which is a well-established mechanism to form such biological particles as ribosomes and viruses. Given the right conditions and the raw materials (proteins and nucleic acids in these cases), mature particles are generated.

Liposomes are also formed by self-assembly mechanisms. Lipids spontaneously take on the structure of a bimolecular leaflet and under appropriate conditions fuse into vesicular conformation. Although liposomes do not contain proteins their ability to form membrane like structures deserves some attention when considering possible mechanisms of membrane assembly.

Natural membranes can be treated with detergents and then dispersed into their constituent protein and lipid components.[23] When the detergent is removed, the membranes reform spontaneously, but the product is significantly different from the native membrane in that it has lost its asymmetry. The proteins are intercalated in their bilayer of lipid, and they are still functional. But they have lost their uneven distribution across the membrane and therefore the membrane as a whole cannot function vectorially. Thus, even if membranes were assembled spontaneously in the cell and fused with the plasma membrane, the added patch of membrane would not be asymmetrical, as is required for its function. Figure 4.33 illustrates the results of self-assembly starting with an intact asymmetric membrane. The final product, although the same in composition as the starting material, is not asymmetric.

A second mechanism of membrane assembly can be imagined wherein the integral proteins are oriented and inserted into the membrane. According to this line of thinking proteins and lipids could be added individually to an existing membrane or whole patches of membrane could

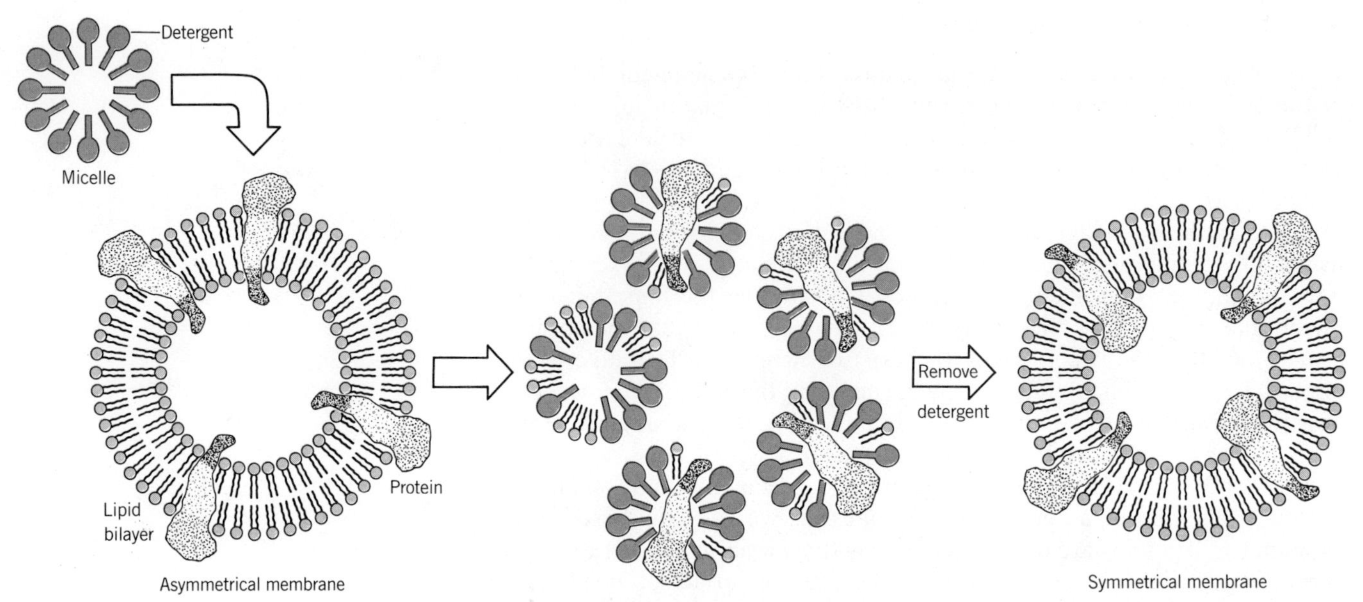

FIGURE 4.33 The effect of self-assembly on membrane asymmetry. An asymmetrical membrane can be disrupted to micellar form by a detergent. Upon removal of the detergent, a new membrane is spontaneously generated but the asymmetry of the starting material is lost.

be added with the proteins and lipids already asymmetrically positioned.

Adding components individually to an existing plasma membrane poses significant thermodynamic problems. Neither component would add easily except to the interior surface. Amphipathic lipids and proteins would have to cross the hydrophobic medium of the bilayer to reach the outer surface, a thermodynamically unfavorable event. We have noted that flip-flop rates for membrane components are extremely low and that the addition of individual components to the membrane would require a kind of flip-flop transmembrane jump.

But in spite of reservations we may have based on thermodynamic considerations, there appear to be instances in which these principles are easily violated. Some of the inner membrane proteins of mitochondria, for example, are synthesized in the cytoplasm and then assembled into the membrane.[24] This suggests that there exists a way of handling the problem of transporting proteins across membranes and inserting them into a hydrophobic layer.

Another apparent violation is the behavior of 5′-nucleotidase during its synthesis. In the early stages of its formation its catalytic site is located on the cytoplasmic face of endoplasmic reticulum and Golgi vesicles. Later it is found on the noncytoplasmic face of secretory granules and eventually on the external (noncytoplasmic) face of the plasma membrane. Somewhere along the line there is a posttranslational transfer of the catalytic site across the lipid bilayer.

We are thus left with a number of possibilities and a number of problems in explaining membrane assembly. Out of these, two general theories have emerged that attempt to deal with the observations and the problems.

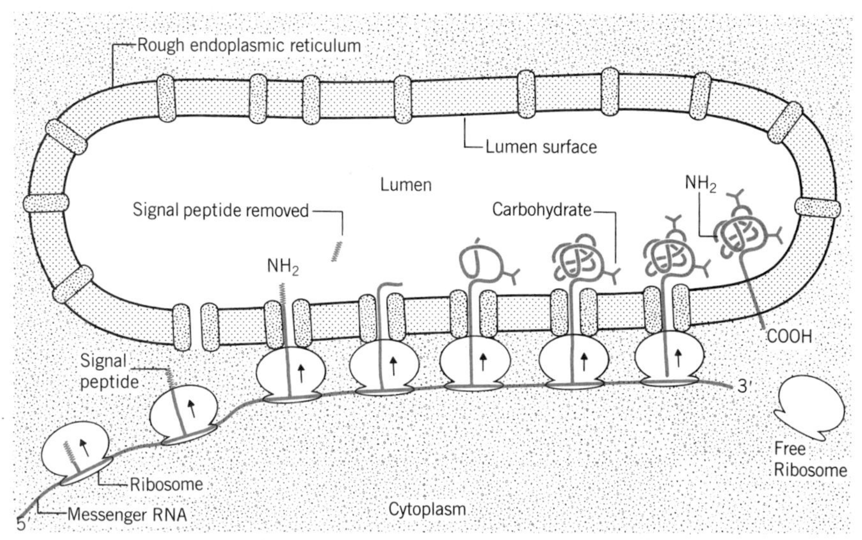

FIGURE 4.34 The signal hypothesis of membrane formation. Proteins, translated from special messenger RNAs that generate an N-terminal signal peptide, are interiorized in the endoplasmic reticulum and are inserted into the membrane facing the lumen. If this membrane is to become plasma membrane, its lumen surface must become the exterior surface of the cell.

The Signal Hypothesis

The *signal hypothesis* was proposed originally to explain the mechanisms of synthesizing and sequestering proteins that are destined for secretion from the cell (see Chapter 10 for more on this topic). It was then extended to include the synthesis of membrane proteins.

Figure 4.34 captures the process as it is envisioned to take place. During the first stage, proteins are added and oriented as they are being synthesized by ribosomes bound to the endoplasmic reticulum.

As proteins are transcribed from the messenger RNA they are fed through a membrane channel into the lumen of the endoplasmic reticulum, where they begin to take on their secondary and tertiary conformations. Carbohydrate is added by enzymes on the interior side of the membrane and oriented with a hydrophilic domain projecting into the ER lumen.

A key part of this hypothesis is the role played by an *N*-terminal hydrophobic stretch of 15 to 30 amino acid residues. This segment, called a *signal peptide,* is commonly observed on membrane and secretory proteins as they are being translated. The signal hypothesis suggests that protein synthesis begins on free ribosomes, with the hydrophobic stretch translated first. As this end is formed, it is attracted to a membrane receptor bringing the ribosome to the surface of the endoplasmic reticulum. This triggers the assembly of a membrane transport system that begins to thread the protein, hydrophobic stretch first, through the bilayer.

The peptide is driven across the bilayer by the force of polypeptide chain elongation, and as long as the ribosome remains tightly bound to the surface, the polypeptide is forced transversely across the membrane.

On the lumen face of the endoplasmic reticulum the hydrophobic signal

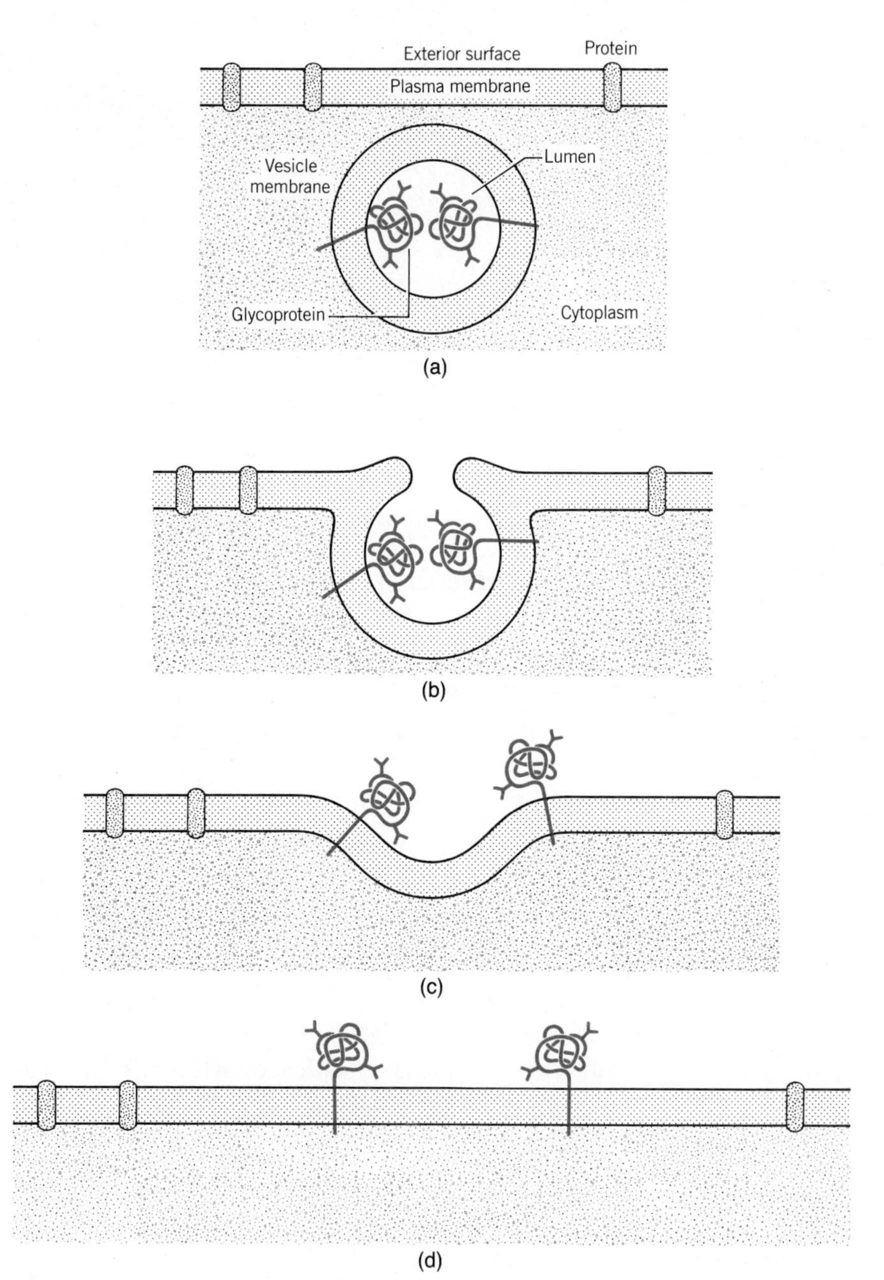

FIGURE 4.35 Fusion of a membrane vesicle, such as one derived from the endoplasmic reticulum, with the plasma membrane. (*a*) Approach. (*b*) Fusion. (*c*) Integration. (*d*) Growth product. This series can be envisioned to occur in a manner that preserves the asymmetry while the membrane is growing. The mechanism suggests also that at no time is the plasma membrane permitted to become perturbed to the point of losing its barrier and functional properties.

peptide is removed by a specific protease. The protein is thus left inserted into the membrane and oriented transversely by virtue of its composition and the thermodynamic constraints imposed by the bilayer.

The second stage is fusion of the resultant vesicle membrane with the plasma membrane. Figure 4.35 shows schematically how a vesicle with

Table 4.4 A Comparison of Two Models for Membrane Assembly

Stage of Synthesis	Signal Hypothesis	Membrane Trigger Hypothesis
Site of initiation	Soluble polysomes	Soluble polysomes
Role of the leader peptide	Recognition by the protein transport channel	To alter the folding pathway
Association of the new protein with membrane	Time: when leader peptide is complete Place: protein transport channel	Time: during *or after* protein synthesis Place: receptor protein or lipid portions of the bilayer
Specific ribosome associations	With the protein transport channel	None
Catalysis for assembly	A specific pore	The effect of the leader peptide on conformation
Driving force for assembly	Polypeptide chain elongation	Protein–protein and protein–lipid associations: self-assembly
Removal of leader peptide	During polypeptide extrusion	During or after polypeptide assembly into bilayer
Segregation of proteins into different cellular membranes	Not clearly addressed	Not clearly addressed
Final orientation	*C*-terminus in, *N*-terminus out	Specified by the primary sequence

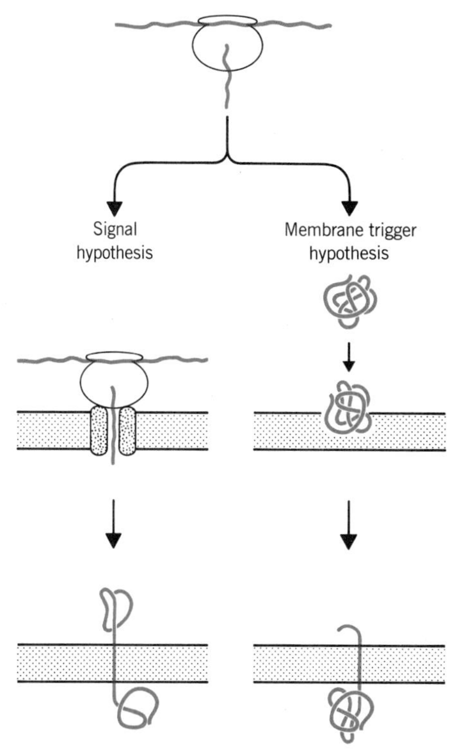

FIGURE 4.36 Membrane trigger hypothesis of membrane growth compared to the signal hypothesis. In the membrane trigger hypothesis, proteins are transcribed on free ribosomes and, because of their composition and conformation, are subsequently self-inserted into the lipid bilayer.

oriented membrane proteins may fuse with the plasma membrane in such a way that the carbohydrate ends up exclusively on the outer surface of the cell.

The Membrane Trigger Hypothesis

The basic features of the *membrane trigger hypothesis* compared to the signal hypothesis of membrane formation are shown schematically in Figure 4.36. The essence of this hypothesis is that proteins destined to become integral membrane proteins are translated on free ribosomes, fold into a conformation that is compatible with solubility in the membrane, and are then inserted.

The driving force of this mechanism is the thermodynamics of protein folding. It is speculated that the leader hydrophobic peptide modifies this pattern of folding, first by directing a conformation that permits the proteins to merge with the membrane, and second, when excised, activating further folding and perhaps protein function.

The two models are compared in the information on Table 4.4.

Choosing Between Membrane and Secretory Proteins

The two hypotheses of membrane assembly, as they have been discussed, do not take into account the necessity for a protein-synthesizing cell to be able to control the destinies of the proteins.

One destiny is simply that of the cytosol. It is not hard to imagine that these proteins are made by ribosomes that are free in the cytosol, and thus the products of synthesis have the same environment as the site of their manufacture.

But it is important to recognize that there are two other important destinies for proteins. One is the membrane, as we have just discussed. A second is the cell exterior, which is the destiny of all secretory proteins. The membrane trigger hypothesis does not address itself to this latter category of proteins, but the signal hypothesis is used to describe the formation of both membrane proteins and secretory proteins (see Chapter 10). Secretory proteins cannot be membrane bound in the same manner that is required of an integral membrane protein. Therefore, how does the cell control the destiny of the particular type of protein being made?

First, according to the signal hypothesis, both membrane and secretory proteins possess an *N*-terminal stretch of amino acids, called a *signal peptide* or an *insertion signal*, which directs the proteins being formed on a ribosome to interact with and insert into the membrane of the endoplasmic reticulum. Protein synthesis then continues, with the product being threaded across the membrane and into the luminal side of the endoplasmic reticulum.

Without some additional control toward the end of synthesis, the product would simply be dumped into the lumen, in which case it would have a secretory fate rather than becoming a membrane protein.

The point of control that operates at this level of synthesis appears to involve another stretch of amino acids closer to the *C*-terminal end of the protein called the *halt-transfer signal*.[25] Several proteins that have been analyzed, including glycophorin, contain a hydrophobic segment of amino acids varying in length from 20 to 30 or so residues, followed by a group of very hydrophilic positively charged amino acids. This polar stretch of the protein may not cross the membrane as easily as the rest of the product, and the hydrophobic section would end up membrane embedded with the *C*-terminal section on the cytosolic surface.

The mechanisms that may be used by a cell to choose between membrane and secretory proteins are summarized in Figure 4.37.

This concept has a degree of versatility that may become important in understanding the relation between synthesis and the eventual topography of the membrane protein. Depending on the positions of the two types of signal in the polypeptide chain, the resulting protein could end up with both *N*- and *C*-terminal ends toward the ER lumen or the cytosol, or there could be imagined a separation of *N*- and *C*-termini toward opposite sides, in either direction. It will be of interest to learn how the actual topograhical positions of polypeptide chains can be made to fit the possibilities of this model.

The Addition of Lipids to Growing Membranes

Lipids are apparently added to growing membranes in a manner different from that of proteins. A clue to how this takes place has come from the study of bacterial membranes as well as membranes of the endoplasmic reticulum. Membrane lipids appear to be synthesized by enzymes that reside within or on membranes that are growing.

Furthermore, these or other proteins may assist the newly formed lipids in finding their final asymmetric distributions by providing channels or

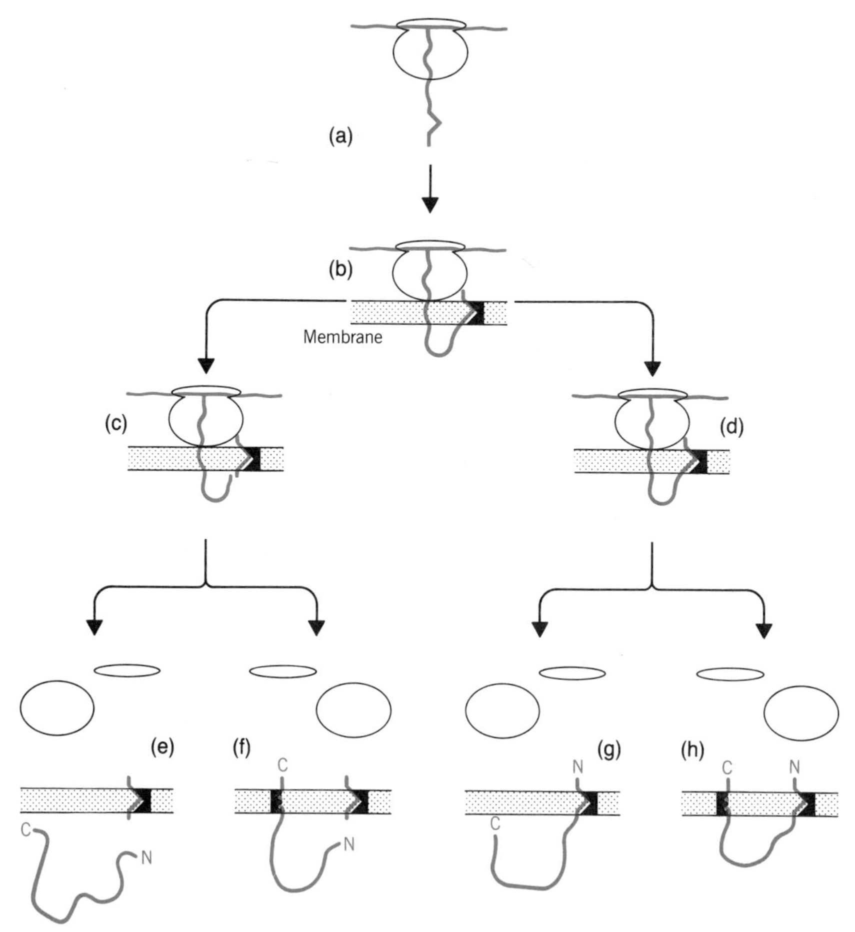

FIGURE 4.37 Models of integral membrane protein insertion. A polypeptide containing an insertion signal emerges from a ribosome (*a*) and directs the attachment of the ribosome to the ER membrane (*b*). The insertion signal may be enzymatically removed (*c*), or it may become a permanent feature of the protein (*d*). In the former case, the protein would become an integral membrane protein only if it contained a halt transfer signal (*f*). In the latter case the protein may end up attached to membrane by either or both ends (*g* and *h*).

promoting flip-flop. Although this is speculative, there is good experimental support in favor of this process.[26]

Thus proteins reside in the membrane in a nonequilibrium state with respect to an asymmetry that is established when they are synthesized and inserted. Lipid asymmetry is a biased equilibrium that occurs during synthesis and depends on the more stable distribution of proteins across the membrane.

Unresolved Problems of Assembly

It is, of course, likely that neither the signal nor the membrane trigger hypothesis as presently developed fits all the situations of membrane assembly completely. It is also likely that ingredients of both provide correct explanations of different types of assembly. Both hypotheses are extremely valuable in that they serve as a framework for testing and discussing the mechanisms of assembly.

But there remain several unanswered questions regarding assembly. One question has to do with the basic problem of translating hydrophobic proteins in an environment (the cytosol) and on a jig (the ribosome) that are primarily hydrophilic. What factors eliminate aggregation and precipitation?

A second question presents a real puzzle. How does a given integral membrane protein end up in the correct membrane? Whereas proteins are synthesized in only two places, on free ribosomes and on ribosomes attached to endoplasmic reticulum, they end up in a variety of different membrane-containing sites. What mechanisms govern the choice of final sites?

Another question relates to the observation that certain proteins appear to cross membranes. How do proteins with polar domains cross the hydrophobic lipid bilayer? Furthermore, why is it that once inserted, this ability to cross is blocked? The final position is stable.

What is ultimately responsible for the transverse asymmetry of integral membrane proteins? Is it only the primary structure of the protein, or may it also be the manner in which the protein is synthesized and the site of the synthesis?

Finally, what controls the lateral interactions and ordering of membrane proteins? Is this random, subject only to the limitations of diffusion rates, or does scaffolding by peripheral membrane proteins play a significant role in lateral organization?

These and other questions that could be raised suggest an interesting and productive future for the work of membranologists. Surely some of these questions are near resolution. But for others, the answers will come with enormous effort and patience.

Summary

Our understanding of membrane structure has evolved over the past 50 years from that of a lipoid halo of undefined thickness to a discrete bimolecular layer of lipid in which proteins float specifically oriented with respect to the two surfaces of the membrane. The plasma membrane is thus asymmetric, with both lipids and proteins demonstrating orientational preference for one or the other side of the membrane. Carbohydrate domains of glycoproteins are oriented exclusively toward the outer surface, whereas the protein itself may project into only one half of the lipid bilayer or may be completely transmembrane in its occupancy.

Both proteins and lipids possess a freedom of movement laterally. In this direction, lipids are restricted mainly if they bind to hydrophobic regions of proteins. Proteins are restricted only if they interact with one another or with peripheral scaffolding proteins. Lipids diffuse laterally much more rapidly than proteins because they are of lower molecular weight.

Both lipids and proteins are restricted from ready flip-flop mobility. Thermodynamically this is an unfavorable event. Nevertheless, these thermodynamic barriers appear to be circumvented by some unknown mechanism during the formation of certain membranes.

Although the plasma membrane is a physical barrier surrounding the cell, it is a selectively permeable barrier. Some materials penetrate the membrane by simple diffusion, although the types of molecule possessing this ability are very limited. Others, such as simple sugars, are facilitated in their transmembrane movement by stereospecific carriers that are a part of the membrane. Neither of these mechanisms is energy dependent.

Active transport is the third type of mechanism whereby materials may cross the plasma membrane. It is energy dependent and with ATP as a source of energy material is moved against a concentration gradient. One form of active transport, such as the form that translocates Na^+ and K^+ across the membrane, is directly linked to the hydrolysis of ATP. In this latter case, a gradient of Na^+ generated by an ATPase provides the driving force for the translocation of certain amino acids and sugars.

When plasma membranes are assembled, they must be put together asymmetrically and without perturbation, which would be deleterious to the cell. Two mechanisms have been proposed. One is an adaptation of the signal hypothesis that was developed to explain the secretory activities of the cell. It proposes that proteins are oriented and inserted into the membrane of the endoplasmic reticulum as they are being synthesized. The endoplasmic reticulum then fuses with existing plasma membrane.

The second hypothesis is called the membrane trigger hypothesis. According to this hypothesis membrane proteins are synthesized on free ribosomes, take on conformations compatible with solubility in the membrane, and spontaneously insert.

Lipids appear to be inserted into membranes by synthetic enzymes that are a part of the membrane. These enzymes may guide them to assume their proper orientation.

References

1. The Probable Origin and Physiological Significance of Cellular Osmotic Properties, E. Overton, *Vierteljahrsschr. Naturforsch. Ges. (Zurich)* 44:88(1899).
2. Membranes in the Plant Cell. I. Morphological Membranes at Protoplasmic Surfaces, Janet Q. Plowe, *Protoplasma* 12:196(1931).
3. On Bimolecular Layers of Lipoids on the Chromocytes of the Blood, E. Gorter and F. Grendel, *J. Exp. Med.* 41:439(1925).
4. A Contribution to the Theory of Permeability of Thin Films, J. F. Danielli and H. Davson, *J. Cell. Physiol.* 5:495(1935).
5. Birefringence and Fine Structure of Nerve Myelin, W. J. Schmidt, *Z. Zellforsch. Mikrosk. Anat.* 23:657(1936).
6. X-Ray Diffraction Studies on Nerve, F. O. Schmitt, R. S. Bear, and G. L. Clark, *Radiology* 25:131(1935).
7. Electron Microscope and Low-angle X-ray Diffraction Studies of the Nerve Myelin Sheath, H. Fernandez-Moran and J. B. Finean, *J. Cell Biol.* 3:725(1957).
8. The Ultrastructure of Cell Membranes and Their Derivatives, J. D. Robertson, *Biochem. Soc. Symp.* 16:3(1959).
9. Fracture Faces of Frozen Membranes, D. Branton, *Proc. Natl. Acad. Sci. U.S.* 55:1048(1966).
10. The Fluid Mosaic Model of the Structure of Cell Membranes, S. J. Singer and G. L. Nicolson, *Science* 175:720(1972).
11. Architecture and Topography of Biologic Membranes, S. J. Singer, in *Cell Membranes*, G. Weissmann and R. Claiborne, eds., HP Publishing Co., New York, 1975.

12. Thermodynamics, the Structure of Integral Membrane Proteins, and Transport, S. J. Singer, *J. Supramol. Struct.* 6:313(1977).
13. Sugars of the Cell Membrane, Saul Roseman, in *Cell Membranes*, G. Weissman and R. Claiborne, eds., HP Publishing Co., New York, 1975.
14. The Molecular Organization of Membranes, S. J. Singer, *Annu. Rev. Biochem.* 43:805(1974).
15. Localization of Red Cell Membrane Constituents, R. F. A. Zwaal, B. Roelofsen, and C. M. Colley, *Biochim. Biophys. Acta* 300:159(1973).
16. Spin-spin Interactions Between Spin Labelled Phospholipids Incorporated into Membranes, P. Devaux, C. J. Scandella, and H. M. McConnell, *J. Magn. Res.* 9:474(1973).
17. The Rapid Intermixing of Cell Surface Antigens After Formation of Mouse-Human Heterokaryons, C. D. Frye and M. Edidin, *J. Cell Sci.* 7:319(1970).
18. Rotational Diffusion of Rhodopsin in the Visual Receptor Membrane, R. A. Cone, *Nature New Biol.* 236:39(1972).
19. On the Fluidity and Asymmetry of Biological Membranes, S. J. Singer, in *Perspectives in Membrane Biology*, S. Estrada, ed., Academic Press, New York, 1974.
20. Rhodopsin-like Protein from the Purple Membrane of *Halobacterium halobium*, D. Oesterhelt and W. Stoeckenius, *Nature New Biol.* 233:149(1971).
21. Simple Allosteric Model for Membrane Pumps, O. Jardetzky, *Nature* 211:969(1966).
22. Molecular Organization of Biological Membranes, S. J. Singer, in *Structure and Function of Biological Membranes*, L. I. Rothfield, ed., Academic Press, New York, 1971.
23. The Assembly of Cell Membranes, H. F. Lodish and J. E. Rothman, *Sci. Am.* 240:48(1979).
24. The Assembly of Proteins into Biological Membranes: The Membrane Trigger Hypothesis, W. Wickner, *Annu. Rev. Biochem.* 48:23(1979).
25. Mechanism for the Incorporation of Proteins in Membranes and Organelles, D. D. Sabatini, G. Kreibich, T. Morimoto, and M. Adesnik, *J. Cell Biol.* 92:1(1982).
26. Rapid Transmembrane Movement of Newly Synthesized Phospholipids during Membrane Assembly, J. E. Rothman and E. P. Kennedy, *Proc. Natl. Acad. Sci. U.S.* 74:1821(1977).

Selected Books and Articles

Books

Biological Membranes, Their Structure and Function, Roger Harrison and George G. Lunt, John Wiley & Sons, New York, 1975.

Cell Membranes, G. Weissman and R. Claiborne, eds., HP Publishing Co., New York, 1975.

Mammalian Cell Membranes, Vols. 2 and 4, G. A. Jamieson and D. M. Robinson, eds., Butterworths, London, 1977.

Molecular Dynamics in Biological Membranes, Milton H. Saier, Jr., and Charles D. Stiles, Springer-Verlag, New York, 1975.

Papers on Biological Membrane Structure. Selected by Daniel Branton and Roderic B. Park, Little, Brown, Boston, 1968.

The Molecular Biology of Cell Membranes, Peter J. Quinn, University Park Press, Baltimore, 1976.

Articles

Mechanisms for the Incorporation of Proteins in Membranes and Organelles (A Review), D. D. Sabatini, G. Kreibich, T. Morimoto, and M. Adesnik, *J. Cell Biol.* 92:1(1982).

Membrane Structure, J. D. Robertson, *J. Cell Biol.* 91:189s(1981).

The Assembly of Cell Membranes, H. F. Lodish and J. E. Rothman, *Sci. Am.* 240:48(1979).

The Fluid Mosaic Model of the Structure of Cell Membranes, S. J. Singer and G. L. Nicolson, *Science* 175:720(1972).

The Molecular Organization of Membranes, S. J. Singer, *Annu. Rev. Biochem.* 43:805(1974).

5

The Cell Surface: Antigens, Receptors, Adhesion, and Communication

Rather than trying to force all biological specificity in the immunological compartment we might have to consider the latter merely as a special case of a more universal biological principle, namely, molecular keylock configuration as a mechanism of selectivity, *whether involving enzymes, genes, growth, differentiation, drug action, immunity, sensory response, or nervous coordination.*

P. WEISS, 1947

. . . complementariness of molecular structure of some sort is responsible for biological specificity in general.

L. PAULING, 1956

The cell surface began to take on a new look for biologists toward the middle of the twentieth century. This new fashion was not simply a fad: it was a change that marked the dawn of a new era in cellular biology on which the sun is far from setting. All indications are that this subspecialty will persist and grow for some time.

It became obvious to scientists that the cell surface could no longer be viewed as an unresponsive, nondescript barrier. Rather, it was showing itself to be a responsive, intricately structured border that participates both in the sending and in the receiving of signals. Cells were observed to be capable of interacting with one another at short distances as well as by means of chemical signals often sent from the farthest reaches of the organism.

There thus emerged the concept of the *cell receptor*, a site on the surface of the cell that can recognize signals or be recognized by signal molecules.

In 1878 J.N. Langley planted one of the historically important seeds that ultimately gave rise to the receptor concept. He was studying saliva production in cats, with particular emphasis on the opposing actions of atropine and pilocarpine on this process.[1] From his observations he concluded that these substances exerted their effects by interacting with some substance in the nerve endings or gland cells. Today, we call these substances receptor sites.

Atropine and pilocarpine are alkaloids that have opposing actions on the parasympathetic system. Atropine, for example, relaxes smooth muscles in various organs, whereas pilocarpine causes muscle contraction.

The first outright definition of the receptor came about 30 years later from Paul Ehrlich.[2] According to his view, cells possessed side chains (receptors) that normally function to bind nutrients to the cell. He suggested that other substances, such as toxins, might also bind to these side chains on target cells because of a highly selective interaction between the toxin and the side chain. The basis of an antitoxin reaction could therefore simply be an excess production of side chains that could be shed by the cell, binding to the toxin some distance away and neutralizing its effect. This *side chain theory*, which arose primarily from studies on toxin–antitoxin interactions, was illustrated in drawings that accompanied a lecture given to the Royal Society in London in 1910 (Figure 5.1). Despite the lack of chemical detail compared to our understanding of receptors today, Ehrlich made a tremendously important contribution to cellular biology by introducing the concept of a *stereospecific* cell surface receptor. Notice that this idea emerged many years before the statements of Weiss and Pauling quoted at the beginning of this chapter.

Gradually, and often meeting great resistance, the receptor theory was extended to areas of biology other than immunology. For example, the aggregation of dispersed sponge cells to form an identifiable organism and the processes of fertilization, embryogenesis, and development were increasingly viewed in the framework of cell adhesion resulting from an interaction between specific receptor sites. The guiding force of hormone action on unique target tissues and the propagation of the nerve impulse in the synapse were also studied more and more with a view to the role of cell receptors. The immune response, which in a sense triggered the notion of stereospecific cell receptors, continues to be studied at the molecular level within the context of surface receptor structure and function.

The emerging use of the electron microscope in the 1950s suggested another possibility for cell communication other than by molecular signals and receptors. For the first time physical linkages between cells involving their adjacent plasma membranes were seen. The discovery of cellular junctions of several identifiable types opened a new area of research and provided significant insight into the mechanisms of direct cell-to-cell communication.

In the material that follows we will discuss a few examples of receptors and junctions that have been most commonly and thoroughly studied. We cannot deal with all receptor types for there appear to be many variations in structure and the amount of detailed information is becoming very extensive. But, before looking at specific systems, it is necessary to review the components of the plasma membrane that affect the architecture of the cell surface.

5.1 THE TOPOGRAPHY OF PLASMA MEMBRANE CARBOHYDRATES

In the preceding chapter, we examined the plasma membrane primarily in cross section. We observed the membrane to be a bimolecular leaflet of lipids intercalated with hydrophobic proteins. We noted in the case of the red blood cell that the glycophorin molecule contains carbohydrate that projects outward from the cell surface. Although glycophorin is a specialized molecule, its property of projecting a hydrophilic carbohydrate arm from the cell surface is a behavior exhibited by many membrane macromolecules in perhaps all cell systems.

This has been an important finding. Carbohydrates of the plasma membrane appear exclusively on the exterior surface of the cell. This, of course, is an important factor in contributing to membrane asymmetry. But beyond this, carbohydrates play a major role in a variety of cell surface phenomena. Furthermore, they are ubiquitous chemical components of membranes, having been reported and analyzed from diverse animal, plant, and bacterial sources.

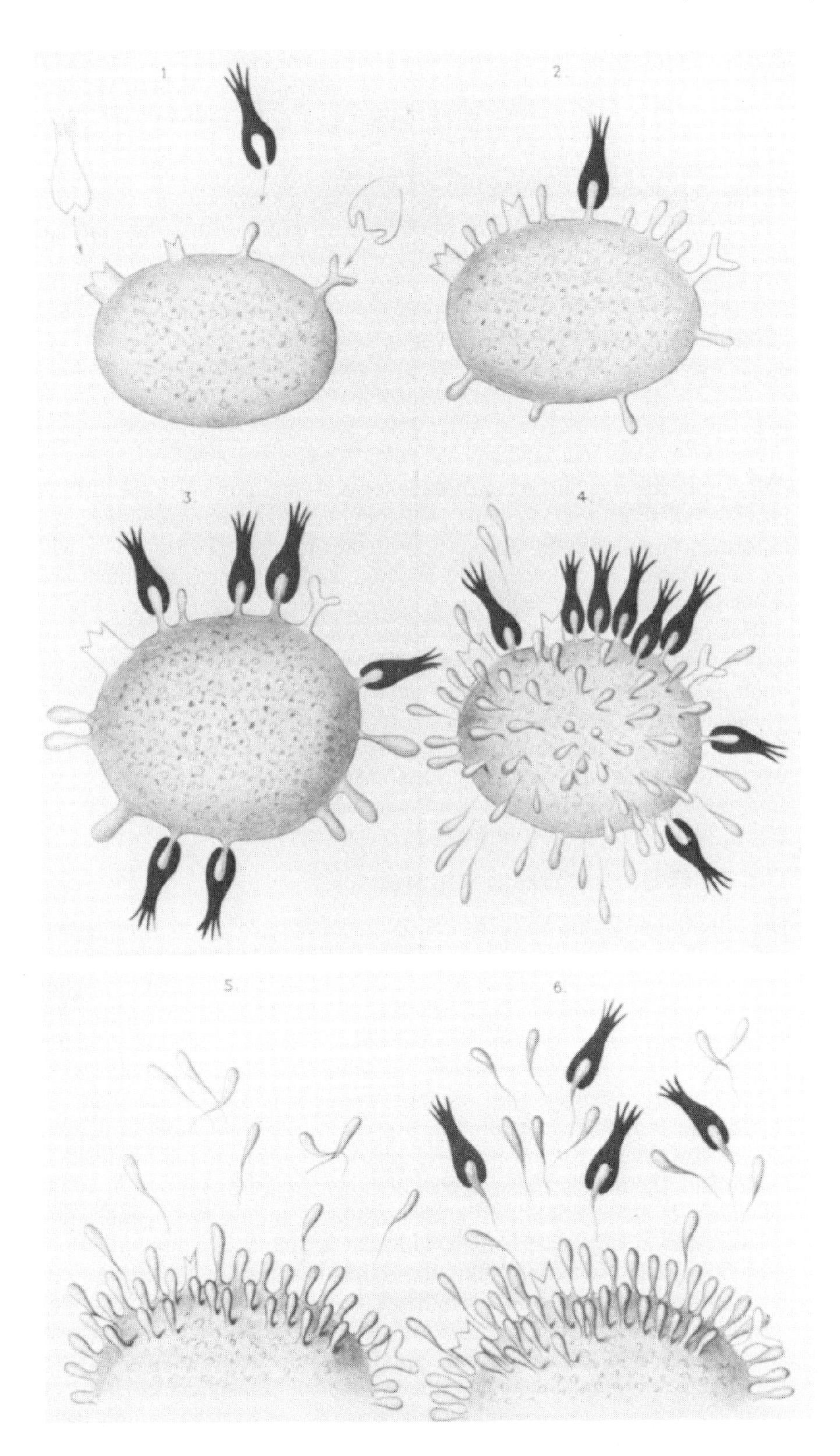

FIGURE 5.1 The side chain theory of Paul Ehrlich as illustrated in 1910. The cell surface was viewed as having side chains (receptors) that stereospecifically fit with certain nutrients or toxins (1). Toxins may bind to sites normally occupied by nutrients (2–4) until the cell becomes physiologically impaired. The basis of an antitoxin response is the production of excess side chains (5), which are released from the cell and neutralize toxins at some distance from the cell surface (6). (Courtesy of the National Library of Medicine.)

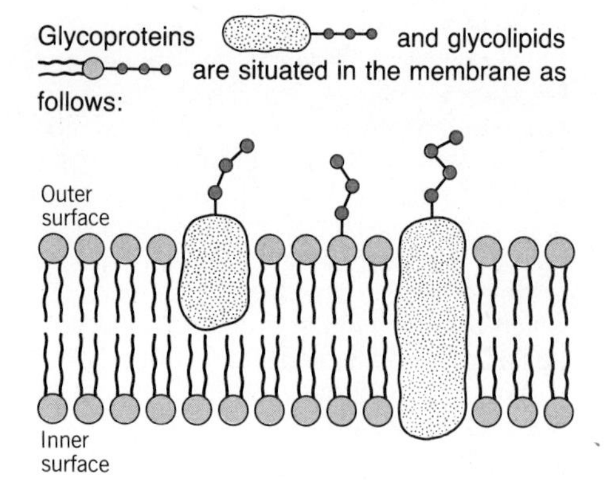

Glycoproteins
and glycolipids
are situated in the membrane as follows:
Outer surface
Inner surface

The carbohydrate of the cell surface does not exist in free form. It is attached either to lipids (glycolipids) or proteins (glycoproteins). The lipid portion of glycolipids is immersed in the bimolecular leaflet region of the membrane, and the protein component of glycoproteins is also anchored in this same hydrophobic region. The carbohydrates of both, however, extend exclusively toward the exterior surface of the plasma membrane.

It would be helpful to keep in mind as we proceed that the lipid and protein components of the membrane are in a fluid state. Therefore, viewed from the surface, these macromolecules can move about in all directions laterally, and may appear to be clustered or dispersed depending on their diffusion properties and their interactions with one another and with the molecules of the environment.

5.2 CELL SURFACE ANTIGENS

Cells are antigenic. This means that when cells of one species are injected into another species, the recipient will recognize the injected cells as being of foreign origin and will produce antibodies to interact specifically with the alien cells. If whole cells are injected, and if they remain intact, the cellular antigens are specific *surface components* of the foreign cells.

Proteins and carbohydrates, or some combination of the two, are the best candidates for surface antigens. Since the composition of the membrane is complex, the number of determinants on a given cell type is enormous.

Several cell surface antigens have been studied in considerable detail. We will consider three of these as representative of those that have been most commonly studied.

The ABO Blood Group Antigens

The ABO blood group antigens are the major antigens present on human erythrocytes. These antigens are under genetic control, so that an individual may possess either the A or B antigen, or both or neither. When neither antigen is present, the individual is called type O. However, individuals with type O erythrocytes possess an antigen called H, which is the structural foundation on which the A and B antigens are built.

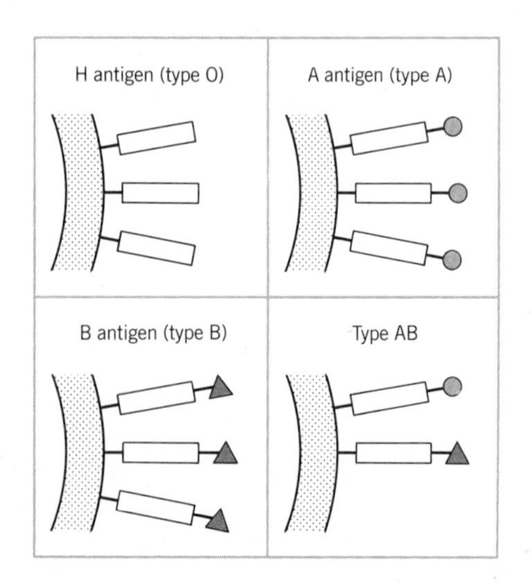

The antigens of the ABO system are glycolipids, with the oligosaccharide portion of the glycolipid responsible for the antigenic properties. Sugars are attached to terminal regions of sphingolipids as outlined in Figure 5.2. The oligosaccharide chains shown are anchored by way of the sphingolipid into the bimolecular lipid leaflet of the plasma membrane.

Two types of oligosaccharide chain exist, giving rise to two subgroups of each antigenic type. The difference is due to the type of linkage between galactose and *N*-acetylgalactosamine at the core of the oligosaccharide.

Erythrocytes that are type O contain fucose linked α-1,2 to galactose, a structure that constitutes the H antigen. From this unit the A and B antigens are enzymatically generated by glycosyl transferases that attach either an *N*-acetylgalactosamine unit (type A) or a galactose unit (type B). The level of genetic control is exerted by the presence or absence of appropriate glycosyl transferases that modify the H antigen, adding either an A or a B type determinant.

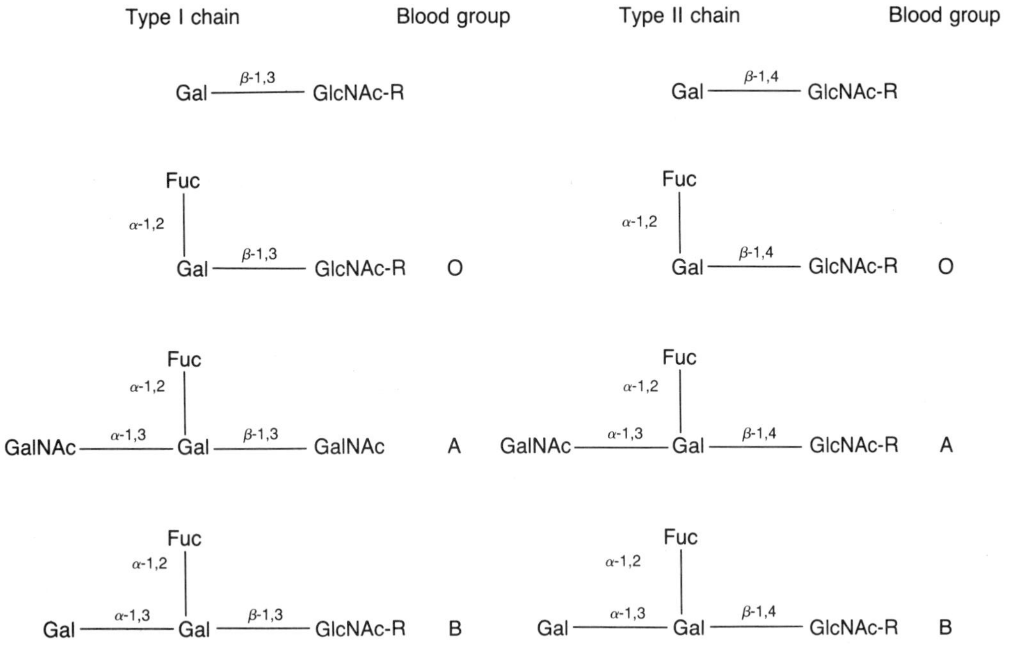

FIGURE 5.2 Oligosaccharide determinants of human ABO blood group antigens. The oligosaccharide antigen is covalently attached to sphingolipids (R), which are immersed in the bimolecular lipid leaflet. There are two types of core chain (I and II) on which saccharide units are built to generate the O(H), A, and B types.

The MN Blood Group Antigens

The second major blood group system in the human is the MN system. This system is also genetically determined. Individuals may be homozygous MM or NN or heterozygous MN.

In this case, the antigenic determinant is a surface glycoprotein, specifically the glycophorin molecule. We discussed this molecule in detail in Chapter 4. Less is known of the chemical structure of these antigenic determinants than is true for the ABO system, but sialic acid, galactose, and *N*-acetylgalactosamine appear to be essential components for this antigenicity (see Figure 5.3). Both M and N determinants, for example, are destroyed when erythrocytes are treated with neuraminidase, an enzyme that removes sialic acid.

Sugars are often abbreviated as in Figures 5.2 and 5.3. The complete terms are:
Fuc = fucose
Gal = galactose
Glc = glucose
GlcNAc = *N*-acetylglucosamine
GalNAc = *N*-acetylgalactosamine
Man = mannose
NeuNAc = *N*-acetylneuraminic acid
A designation such as α-1,2 means that the glycosidic bond extends alpha or down from carbon 1 of one sugar to carbon 2 of the next sugar unit.

Histocompatibility Antigens

Histocompatibility antigens are tissue cell surface proteins that differ from individual to individual. They are recognized as different by the mechanisms of tissue graft rejection. These antigens provide a chemical fingerprint for the tissues of each individual. Given the genetic complexity of the human, there is an enormous number of antigenic combinations possible on tissue surfaces, and this is the basis for the difficulties encountered in getting a good tissue match for skin or organ transplantation surgery.

Some of the histocompatibility antigens are strong antigens and others are weak. In mice, the H-2 system is strongly antigenic and has been most

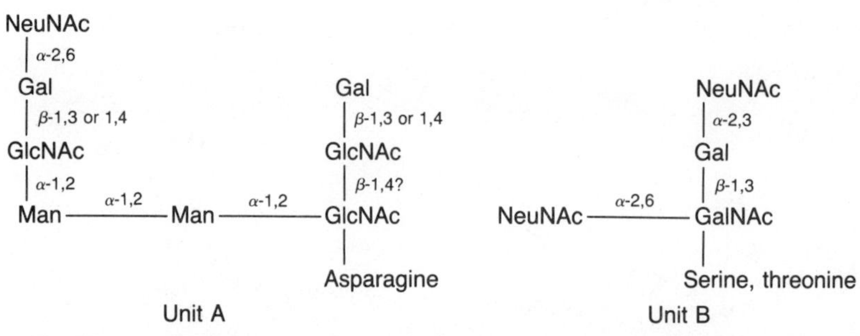

FIGURE 5.3 Oligosaccharide determinants of human MN blood group antigens. In this case, the oligosaccharide is linked to a protein, glycophorin. One type of unit (A) is attached through asparagine and the other (B) through serine or threonine. Variations on these structures exist. They characteristically contain terminal sialic acids (*N*-acetylneuraminic acid).

thoroughly studied. A counterpart of this system in the human is the HLA system. Both these systems have many genetic variants; however, by using inbred mice it has been possible to create relatively homogeneous systems for structural analysis.

The intact H-2 antigen in its native environment of the plasma membrane is made up of two polypeptide chains (see Figure 5.4). One of these chains is "heavy" or long and the other is "light" or short. The heavy and light chains interact, but not by covalent bonds. Heavy chains from two different antigens may interact covalently to generate dimers. At least, when isolated, they are often covalently attached. Carbohydrate is associated only with the heavy chains.

The intact H-2 antigen cannot be separated from the membrane except by the use of solubilizing agents such as detergents. Treating the cell membrane with the enzyme papain cleaves the molecule in such a way that a water-soluble portion called the F_s fragment is released and a smaller piece, the F_m fragment is left in the membrane.

An interesting observation is that the light chain of the H-2 antigen is actually a protein that is manufactured by nearly every cell in the body and is present in a variety of animal species. It is called *β-2-microglobulin*. This protein has been well characterized in terms of both amino acid composition and sequence, and by these techniques has been found to bear a striking resemblance to the constant regions of antibody molecules (Figure 5.5). This finding implies that these proteins are genetically related, perhaps having come from the same ancestral gene.

Partial sequences have been worked out for the heavy chains of both H-2 and HLA antigens. At present, similarities are seen between the two types of heavy chain.

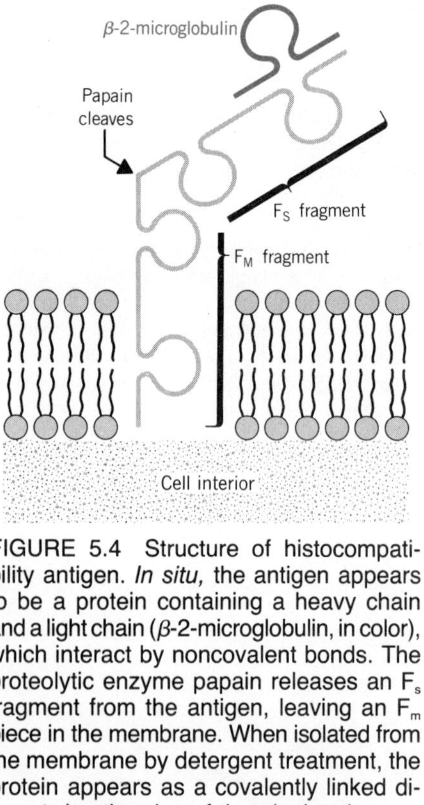

FIGURE 5.4 Structure of histocompatibility antigen. *In situ,* the antigen appears to be a protein containing a heavy chain and a light chain (β-2-microglobulin, in color), which interact by noncovalent bonds. The proteolytic enzyme papain releases an F_s fragment from the antigen, leaving an F_m piece in the membrane. When isolated from the membrane by detergent treatment, the protein appears as a covalently linked dimer, twice the size of that depicted.

The Biological Function of Surface Antigens

The biological significance of surface antigens is largely an enigma. It is obvious that they are useful handles for the researcher and are invaluable as fingerprints in blood typing and tissue matching work. However, as far as is known, the cell does not depend on these antigens for its normal functioning.

Nevertheless, some ideas are beginning to emerge regarding their utility. One postulated role for the H-2 antigen in mice is worthy of mention. It has been proposed that H-2 antigens may function during the mechanism of *immunological surveillance*.[3] According to this notion, abnormalities arising in cell surface structures by mutations or other means are recognized by wandering lymphocytes of the immune system. When a defective cell is found by the lymphocyte, a message is carried back to the lymph nodes where a class of lymphocytes called killer T cells are activated and destroy the abnormal cell.

Two alternative mechanisms that have been suggested to explain the involvement of the H-2 antigen in immunological surveillance are presented in Figure 5.6. The schemes take into account the observation that the target cell *must* have an H-2 antigen to be destroyed. It is thought therefore that the H-2 antigen *plus* an abnormal or modified antigen are recognized by killer lymphocytes. It may be that the normal (H-2) and the abnormal pose together as a hybrid and are then recognized, or that they may attract the lymphocytes without being physically attached to one another. In any event, in the absence of the H-2 antigen, an abnormal cell will avoid destruction and proliferate.

In summary, the chemical natures of several types of surface antigen are now known. Carbohydrates, projecting from the cell surface, are the usual structural elements of these antigens. They are anchored either to protein or to lipid. Except for a proposed role of histocompatibility antigens in eliminating abnormal cells from the organism, the physiological functions of surface antigens remain a mystery. It is likely that some of the surface antigens are merely exposed areas of glycoproteins that have other physiological roles in the cell membrane. But in other cases there may be functions for these surface components that we do not presently comprehend.

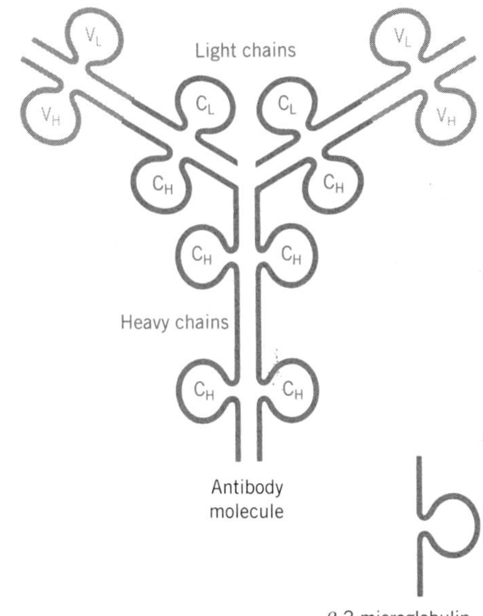

FIGURE 5.5 The light chain component of H-2 and HLA antigens, called β-2-microglobulin, is structurally similar to the constant regions present in antibody molecules (C_L, constant region of light chain; C_H, constant region of heavy chain). The variable regions of antibodies (V_L and V_H) constitute their antigen binding sites.

5.3 THE RECEPTOR CONCEPT

The biology and chemistry of the cell surface receptor has in recent years become a research field of high activity and diversity. It is a complex area of investigation. The reason for complexity becomes readily apparent if the sensory world of the lymphocyte is considered as an example. Figure 5.7 points out the variety of receptors that one would expect to find on the surface of this kind of cell. Each type of receptor must have a unique chemical structure and, when activated, each would drive its own particular biological function of the cell.

Not all cell types have as complex a sensory world as the lymphocyte. However, when one considers the enormous variety of cell types in nature, each with its own complement of receptors, it is possible to appreciate the difficulty of studies in this field and the diversity of the results.

The topological site of the receptor is the exterior surface of the plasma membrane. The role of the receptor, however, generally has an impact beyond the surface of the cell. The receptor is but the first link in a chain of communication that activates a transfer of information across the membrane, normally culminating in a metabolic event in the interior of the cell.

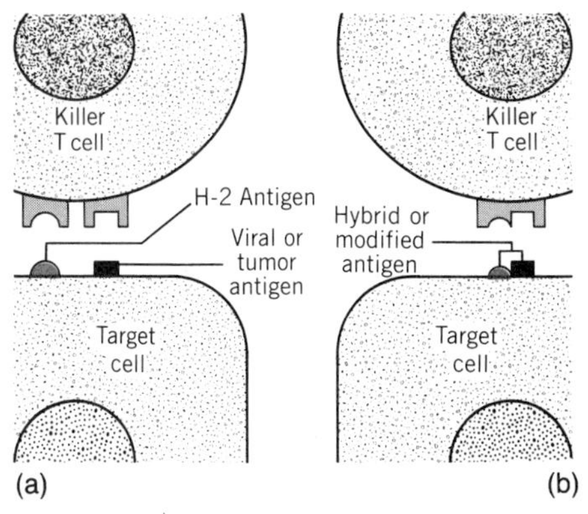

FIGURE 5.6 Alternative models to explain the role of H-2 antigen during immunological surveillance. A cell with a surface abnormality is recognized by a special group of lymphocytes called killer T cells. For a killer T cell to destroy an abnormal cell, it must recognize the abnormal antigen *plus* the H-2 antigen. The antigens may be independent on the surface (*a*), or they may form a hybrid surface structure (*b*).

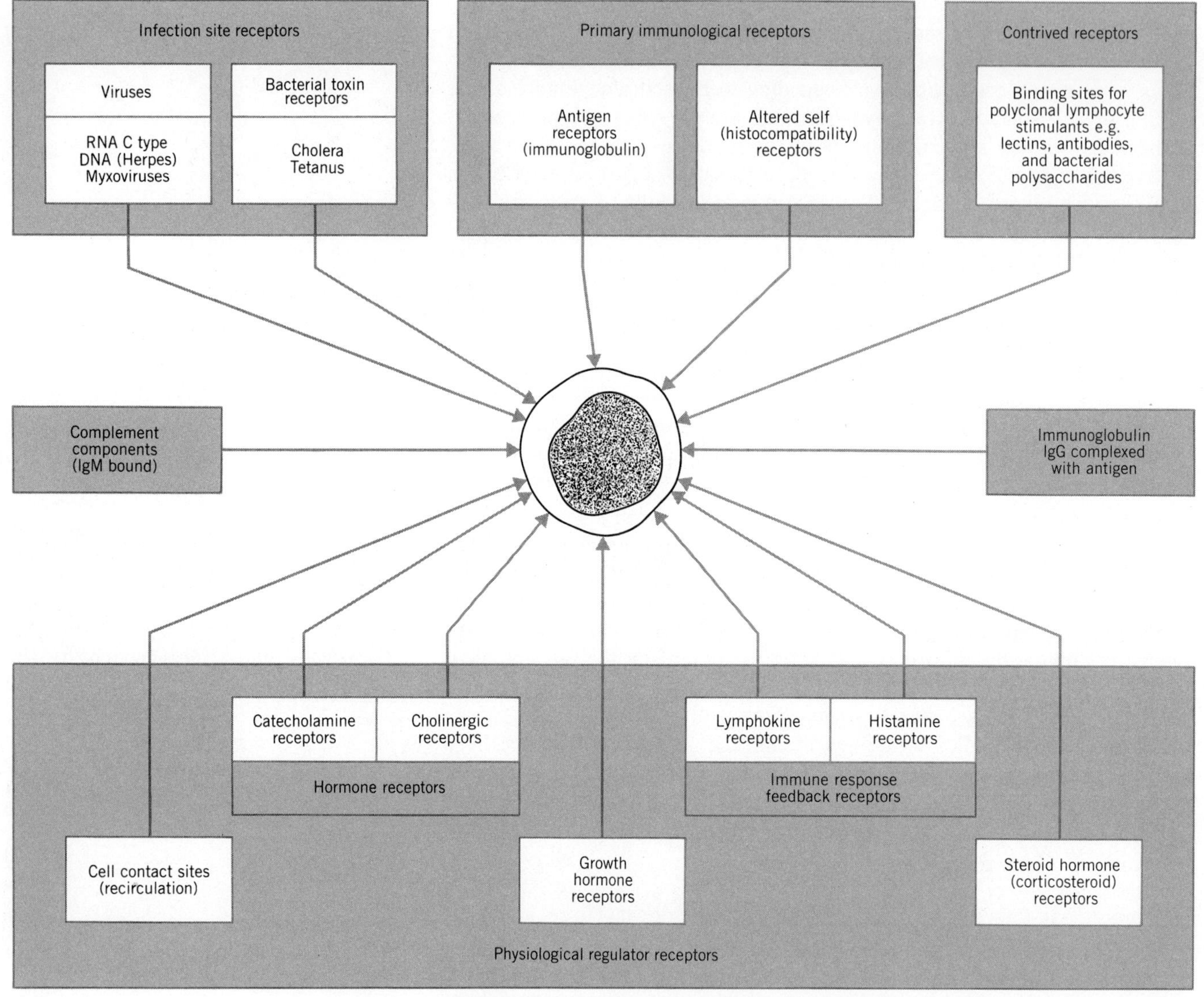

FIGURE 5.7 The sensory world of lymphocytes. Not all lymphocytes have all these receptor sites, but within the lymphocyte family this sensory versatility can be found. Several major categories of receptors are shown.

Receptors: First Links in a Chain of Communication

A model of receptor position and activity is presented in Figure 5.8. In this scheme, the surface receptor has a *cognitive* or *discriminator* role. It recognizes with high specificity the first messenger, which may be an adjacent cell surface, a smaller particle, or a soluble ligand. If the role of the receptor is simply to cause adjacent cells to adhere to one another, then by mere discrimination and binding it has fulfilled its role.

The function of the receptor rarely stops at this point. More generally, it also initiates an internal cellular response. The receptor is thus bifunctional, receiving a signal and passing the information on. It is relatively easy to imagine how a receptor may attract and bind a signal molecule, but it is more difficult to understand how it passes the information on. This may be accomplished either by possessing an inherent *transducer* function or by activating a transducer molecule that resides within the membrane.

The third element is the *effector*, a molecule that operates on the interior side of the plasma membrane.

As we move from the exterior to the interior along the chain of communication across the membrane, we see that the diversity of components is decreasing. Receptors are highly diverse, interacting with a large assortment of messengers. The variety of effectors, however, appears to be much more limited. In many cell types the effector is the enzyme adenylcyclase (for polypeptide hormones and catecholamine neurotransmitters), whereas in other cells Na^+, K^+ -dependent ATPase, guanylcyclase, or iongating molecules serve this role.[4] The transducer probably fits between the discriminator and effector in terms of strucutral diversity.

The last member of the chain, the *second messenger*, probably exists in a very limited number of forms. In several cell systems studied the second messenger has been identified as cyclic AMP.[5] Cyclic AMP (cAMP) generally operates as a positive second messenger, turning on or activating some intracellular process. The mechanism to turn off or deactivate cellular events by way of messengers uses cyclic GMP instead of cAMP. Other second messengers may be operative as message transmitters; these two nucleotides appear to be the most common ones, however.

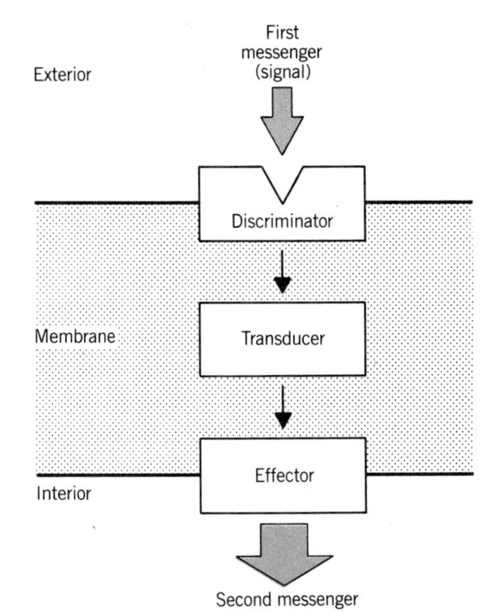

FIGURE 5.8 A model of the chain of communication across the membrane. The receptor discriminates among an enormous variety of messengers, ligands, or adjacent cells. Upon binding with an appropriate complementary material, a transducer is activated which, in turn, passes the message on to the effector, a member of the chain that is situated on the interior side of the membrane. The final member of the chain is a second messenger, a low molecular weight interior molecule.

Receptor Stereospecificity

Receptor structure must be complementary to the first messenger. Stereospecificity between receptor and ligand is required; it is essential if the receptor is to pick out a specific message from a heterogeneous milieu of message bearers.

There is, of course, a precedent for this type of stereospecificity in the complementariness that exists between an enzyme and its substrate and between an antibody and its antigen. We would expect to find the same degree of complementarity between a receptor and its messenger.

The binding of a messenger should also be reversible, as is also true for both the antibody and enzyme systems, so that there can be a discontinuous effect on the cell. Finally, the receptor should be capable of undergoing a conformational change when bound to the ligand, so that the information of binding can be passed into or across the membrane. Conformational changes have been noted for both enzymes and antibodies when they are bound to their respective ligands. It would therefore not be surprising to find a similar effect on the receptor.

The most likely membrane candidate to meet these expectations is a protein, or glycoprotein. Proteins along with their carbohydrate side chains can provide the structural diversity required for stereospecific message reception, and they are flexible enough to change conformation upon binding to a ligand.

The following hormones (first messengers) use cyclic AMP as a second messenger: calcitonin, chorionic gonadotropin, corticotropin, epinephrine, follicle-stimulating hormone, glucagon, luteinizing hormone, lipotropin, melanocyte-stimulating hormone, norepinephrine, parathyroid hormone, thyroid-stimulating hormone, and vasopressin. Adenylcyclase is activated when the hormone binds to a receptor, and the transducer molecule called G protein transmits this information to it.

Cyclic adenosine monophosphate (AMP) and cyclic guanosine monophosphate (GMP), although quite similar structurally, often have exactly opposing effects as second messengers. The effect of cAMP in many cases is to activate protein kinases which, in turn, phosphorylate different proteins, hence modulating their activities.

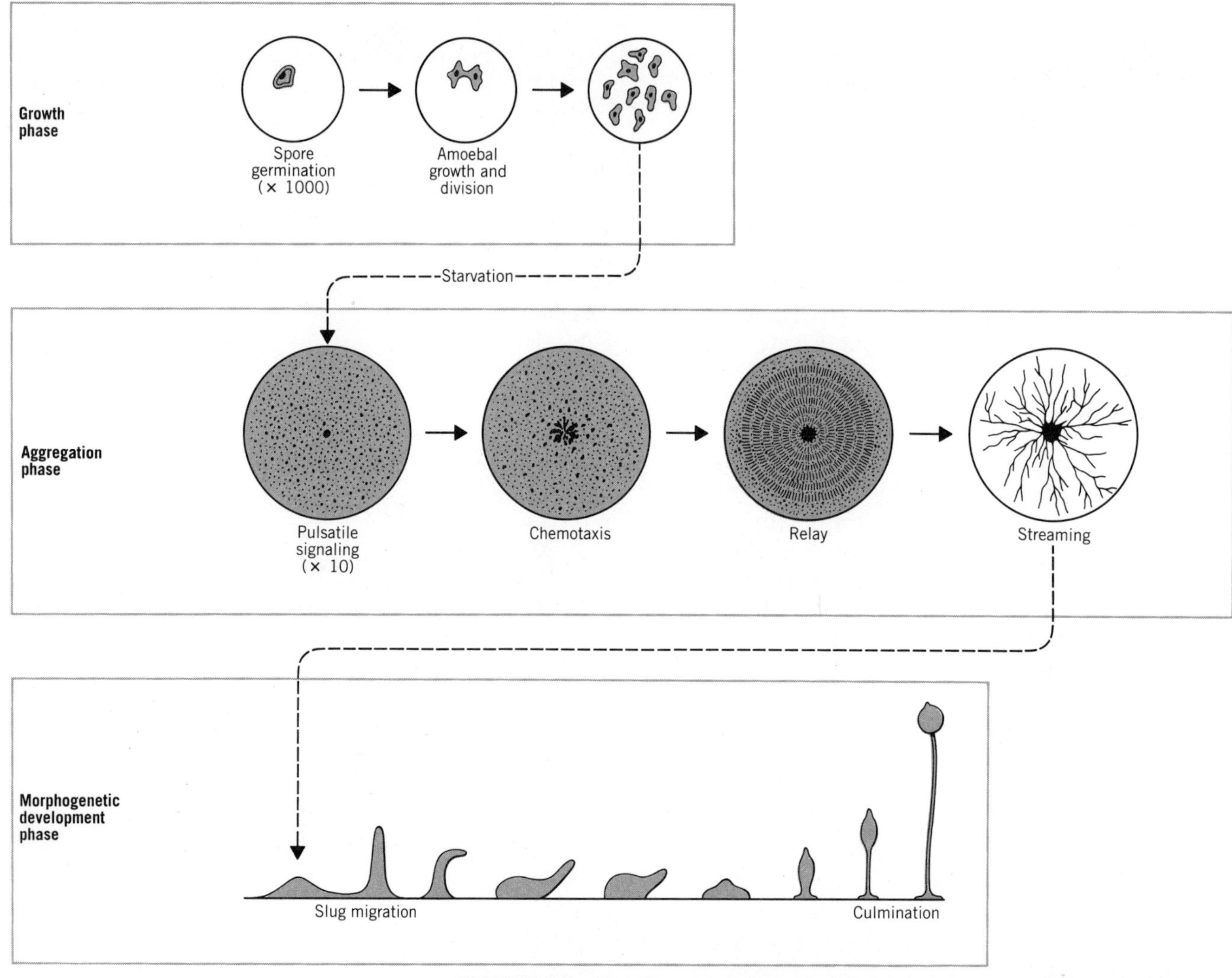

FIGURE 5.9 The life cycle of *Dictyostelium discoideum* showing the unicellular growth phase, a phase of aggregation during which there is a gradual loss of unicellular character, and a stage of multicellular morphogenetic development. These behaviors provide a model system for the study of signal receipt and intercellular adhesiveness.

5.4 RECEPTORS FOR CYCLIC AMP

Dictyostelium discoideum, a cellular slime mold, spends part of its life cycle in unicellular fashion and part in an aggregated multicellular state. As long as sufficient nutrients are available, the unicellular state is maintained. Upon starvation, aggregation is triggered, whereupon the individual amoeboid cells move toward one another, adhere, and form a macroscopic multicellular fruiting body (see Figure 5.9). Because of these behaviors, *Dictyostelium* is a model for the study of two distinct phenomena of the cell surface: the *receipt* of a signal to aggregate and the *adhesiveness* of the cells in the resulting slug.

A few paragraphs back we spoke of cAMP as a second messenger. In

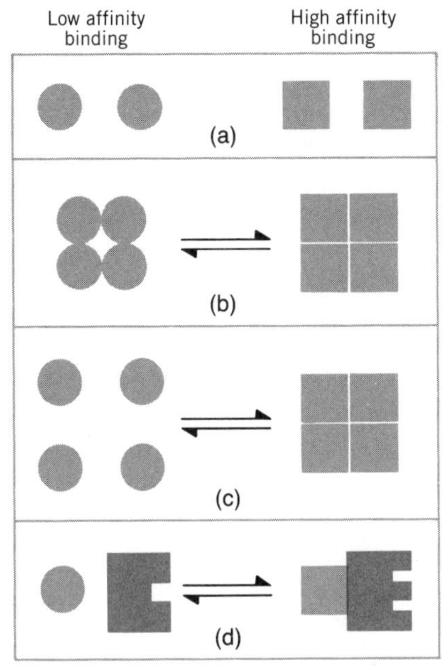

FIGURE 5.11 Models of cAMP receptors. (*a*) Two distinct types of receptor protein with different, fixed affinities. (*b*) Negatively cooperative subunit interactions. (*c*) The affinity of the receptor is affected by association or dissociation of the subunits in the plane of the membrane lipid layer. (*d*) Interaction of receptor with its membrane effector molecule (shown in color) activates the effector and simultaneously converts the receptor to the high affinity form.

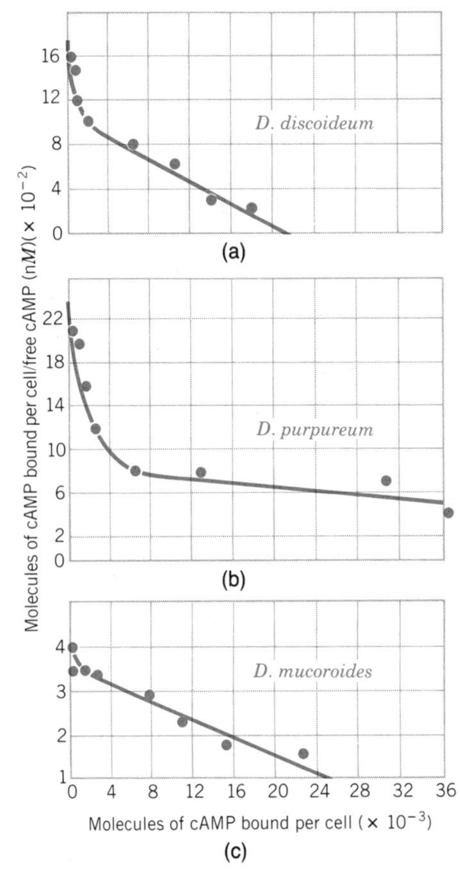

FIGURE 5.10 The binding of cAMP to three species of *Dictyostelium* plotted by the method of Scatchard. The receptor affinities (measured as the equilibrium dissociation constants (K_d) from the slopes of the curves) are seen to vary with the cAMP concentration. (*a*) *D. discoideum*, K_d = 10 to 100 n*M*; (*b*) *D. purpureum*, K_d = 10 to 580 n*M*; (*c*) *D. mucoroides*, K_d < 10 to 95 n*M*.

the case of *Dictyostelium*, cAMP is the *signal* or *first messenger*, which interacts with surface receptors on the organism.

Receipt of the Signal

Upon starvation, pulses of cAMP are released by some of the cells in the population every 5 to 10 min. These pulses generate attracting centers where cAMP is bound to the surfaces of the cells. Two events then immediately follow. The amoebae move toward the attracting center and in turn release a new pulse of cAMP, thus relaying the signal to other cells.

The physicochemical structure of the cAMP receptor has not yet been delineated, but some estimations of the number of receptors per cell have been made. Calculations from equilibrium binding data point to as many as 5 x 10^5 cAMP receptors per cell for *D. discoideum* during aggregation.[6] Scatchard plots of the binding demonstrate a degree of curvilinearity (see Figure 5.10), which can be interpreted in two different ways. Either two or more types of cAMP receptors are present, or the receptors present show variable affinities for cAMP. Figure 5.11 depicts models that have been proposed to fit with the curvilinear Scatchard plots, but the models do not adequately fit some of the kinetic studies related to the dissociation of ligand from cAMP receptors.

Adhesion

The receptor sites involved with cell adhesion have been worked out a little more clearly. Using antibodies or fragments of antibodies containing their binding sites (Fab fragments), investigators have probed the cell surface for sites active in cell adhesion. Two types of contact site have been identified, termed contact sites A (cs-A) and contact sites B (cs-B). There appear to be about 3×10^5 cs-A per cell, and evidence points to cs-A as being a glycoprotein of molecular weight 80,000 to 120,000 daltons.

The evidence is now growing that cellular slime molds produce carbohydrate-binding proteins that promote intercellular adhesion and thus aggregation. These molecules are called *lectins*. One type of lectin, called *discoidin*, has been isolated and purified from *Dictyostelium discoideum*. It has a molecular weight of 100,000 daltons. Other lectins, of similar function but differing physicochemical properties, have been isolated from other species of slime mold. Cell agglutination is thus brought about by a lectin, which acts as a bridge between carbohydrate-containing receptor sites on adjacent cells.

5.5 RECEPTORS OF LYMPHOCYTES

By returning to the subject of lymphocytes, we have completed a cycle in this chapter. Paul Ehrlich's side chain theory concerned a property of cells in the immune system. His work was basic to the current concept of the receptor site.

The lymphocyte is still under intensive investigation as an object for studies on receptor sites. Lymphocytes orginate in bone marrow from hematopoietic stem cells. They subsequently differentiate in primary lymphoid organs. Those that differentiate in the thymus are called T lymphocytes and those that differentiate in the bursa of Fabricius (in birds) or some equivalent tissue in other vertebrates are called B lymphocytes.

Both T and B lymphocytes emerge from primary lymphoid tissues and colonize secondary lymphoid organs, such as lymph nodes and the spleen. In these tissues they are responsive to antigens that they may encounter. B Lymphocytes are activated by antigens to further differentiate into plasma cells which, in turn, produce antibodies. T Lymphocytes are also activated in the presence of antigen and are converted to a form of cell that participates in cell-mediated immunity. The roles of B and T lymphocytes are outlined in Figure 5.12.

In both cases, antigens are recognized by cell surface receptors. This recognition initiates a transmembrane relay of information resulting in cell differentiation and specialized cell functions. We will confine our discussion to the first event: the reception of the signal of antigen presence.

A large number of experimental approaches have now shown that the receptor sites on lymphocytes are immunoglobulins or molecules with immunoglobulinlike properties.

The Nature of B-Lymphocyte Receptors

Late in the differentiation of B lymphocytes, immunoglobulins (Ig) can be detected on the surface of the cell membrane.[7] Two types of immu-

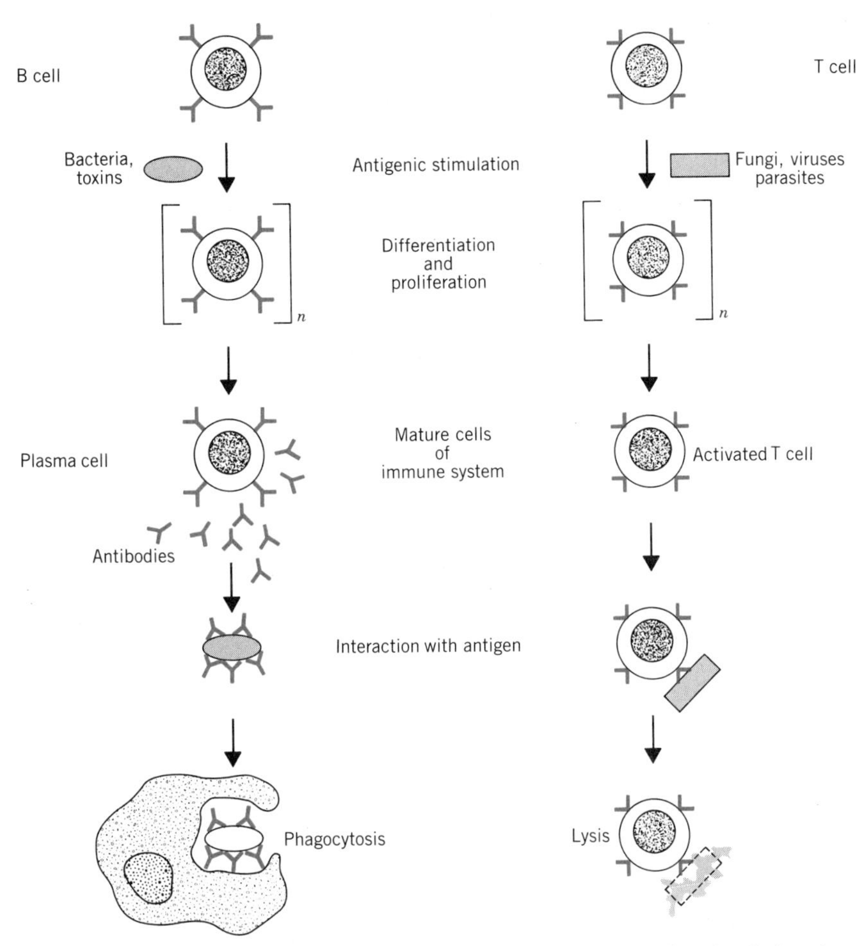

FIGURE 5.12 Roles of B and T lymphocytes. Upon antigenic stimulation, lymphocytes differentiate, interact with their respective antigens, and effect their eventual elimination. A key to this entire process is the initial recognition of the cell surface receptor with its first messenger, the antigen.

noglobulin, IgM and IgD, are found, with IgM appearing first. These are the molecules that act as receptors of antigens with which the cell comes in contact. These immunoglobulins appear to act as integral membrane proteins and are anchored to the membrane by their C-terminal domains.

Immunoglobulins are divided into five classes (IgG, IgM, IgA, IgD, and IgE), based on differences in their molecular structures. IgG is the most abundant and IgM the highest in molecular weight.

The number of receptors per cell is enormous. A single B lymphocyte may have between 10^4 and 10^5 Ig molecules on its surface. However, all the immunoglobulins on a given cell appear to be identical. Thus a given lymphocyte can be activated by only one antigen, and when activated it will differentiate and produce only one type of antibody. This antibody will be similar if not identical to the immunoglobulin that functioned as a receptor on the cell surface.

The Nature of T-Lymphocyte Receptors

The molecular nature of the receptor on the T lymphocyte is more elusive. It does not appear to be a complete immunoglobulin molecule, but rather

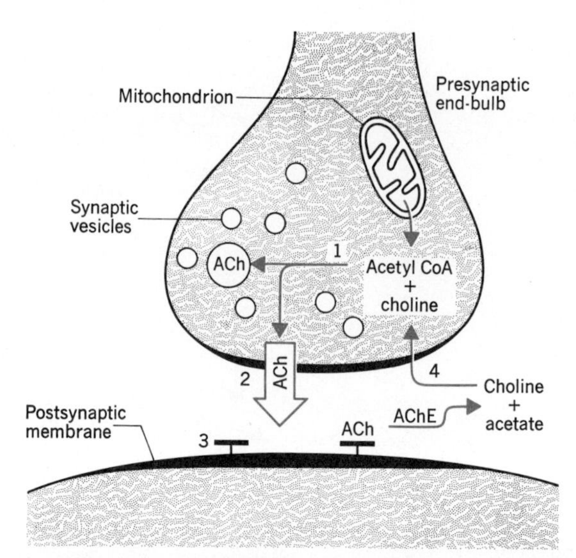

FIGURE 5.13 Diagrammatic representation of the synaptic junction showing chemical transmission of the nerve impulse with acetylcholine (ACh). Acetylcholine is formed in the end plate of the neuron (1), released from synaptic vesicles (2), then migrates the span of the synapse to receptor sites on the postsynaptic membrane (3). Upon hydrolysis of ACh by acetylcholinesterase (AChE), the first messenger is released and its products return to their cell origin (4).

a piece of the immunoglobulin that contains the variable region ordinarily involved with binding to antigen in antigen–antibody reactions.[8]

5.6 RECEPTORS FOR ACETYLCHOLINE

The acetylcholine (ACh) receptor was the first neurotransmitter receptor isolated in pure form. Before discussing the details of this receptor, some background information is provided for the reader who has not studied some basic properties of neurotransmission.

Neurotransmission is an example of short-range communication between cells mediated by chemical messengers. The space between neurons, or between neurons and muscle or gland, is the synaptic junction (Figure 5.13). Acetylcholine is the chemical messenger given off by the neuron sending the signal. The target cell receiving the message contains stereospecific receptors on its surface to receive acetylcholine.

When acetylcholine binds to the receptor molecules on the postsynaptic membrane, the characteristics of the membrane are changed with several possible results. One result is an excitatory response. In this case the postsynaptic membrane becomes locally depolarized from its normal state, which is positive on the inside and negative on the outside. The depolarization is due to a large influx of Na^+ and a smaller efflux of K^+.

One nerve impulse can trigger a localized depolarization by releasing ACh into the synaptic junction, with amounts of ACh on the order 6×10^6 molecules arriving at the postsynaptic membrane to bind to the receptor sites. The number of receptor sites is about 6×10^7, a tenfold excess over messenger molecules.

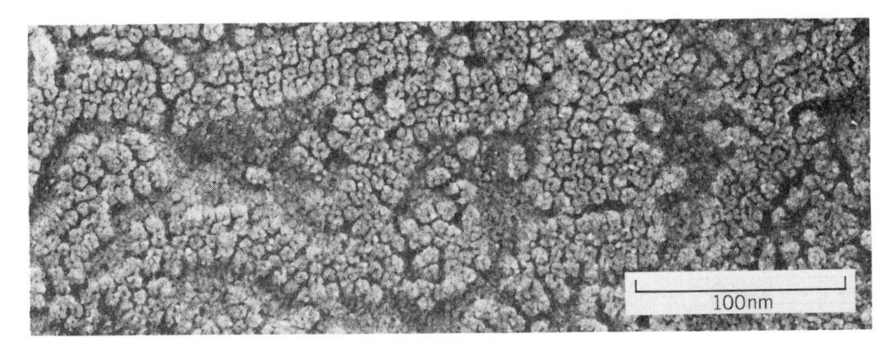

FIGURE 5.14 Electron micrograph of acetylcholine receptors in a postsynaptic membrane. The receptor is apparently a hydrophobic integral membrane protein. (Courtesy of Dr. John Heuser.)

The binding of ACh to the receptor induces a conformational change in the receptor that apparently has an effect on associated macromolecules that function as transmembrane channels. These channels are opened rapidly, permitting the movement Na^+ and K^+ across the membrane. The effect of ACh binding to the receptor is then terminated by high concentrations of acetylcholinesterase in the synapse. This enzyme hydrolyzes ACh and the products no longer bind to the receptor site. The channels then close, stopping the movement of cations across the membrane.

Acetylcholinesterase has a high turnover number of 25,000 sec^{-1}, which permits a rapid repolarization of the postsynaptic membrane.

$$H_3C-\overset{\displaystyle O}{\overset{\|}{C}}-O-CH_2-CH_2-\overset{+}{N}-(CH_3)_3$$

Acetylcholine

$$\xrightarrow[\text{acetylcholinesterase}]{H_2O}$$

$$H_3C-C\begin{matrix}\nearrow O \\ \searrow O^-\end{matrix} + HO-CH_2-CH_2-N-(CH_3)_3 + H^+$$

Acetate Choline

Receptor molecules for ACh have most commonly been isolated from the electric organs of the electric eel or electric ray. The electric organ cell, called an electroplax, is so large that many electrophysiologic studies have been carried out on single cell preparations. Thus, this organ system has been an attractive one for many investigations in addition to serving as a source for the study of receptor molecules.

The chemical nature of the electric organ ACh receptor has been delineated by a number of investigators.[9] It is an acidic glycoprotein with an isoelectric point of pH 4.5 to 4.8. It contains the customary collection of glucose, glucosamine, and *N*-acetylgalactosamine residues seen in many other glycoproteins.

The ACh receptor is thought to be an integral membrane protein because of its hydrophobic properties. It has a high affinity for Ca^{2+}, which accounts for 4 to 5 % of its molecular weight (320,000 daltons). Using sodium dodecyl sulfate (SDS) electrophoresis, the receptor can be separated into subunits that vary in size depending on the treatment to which the receptor has been subjected. One point of view is that there are four 80,000 dalton subunits in the receptor, each subunit containing one ACh binding site.

Method IV

Electron micrographs of negatively stained receptor sites have been prepared,[10] one of which is presented in Figure 5.14. The molecule appears doughnut shaped with an electron-dense core. At present, the significance of this shape is not known although it has the appearance of a channel-containing gate across the membrane.

Even though the ACh receptor has been isolated and studied by way of electron microscopy, its role in changing membrane permeability is still not known. One line of speculation is that the ACh receptor is noncovalently linked by way of Ca^{2+} to another integral membrane component called an *ionic conductance modulator* (ICM). Upon binding to ACh, the receptor undergoes a conformational change resulting in a release of

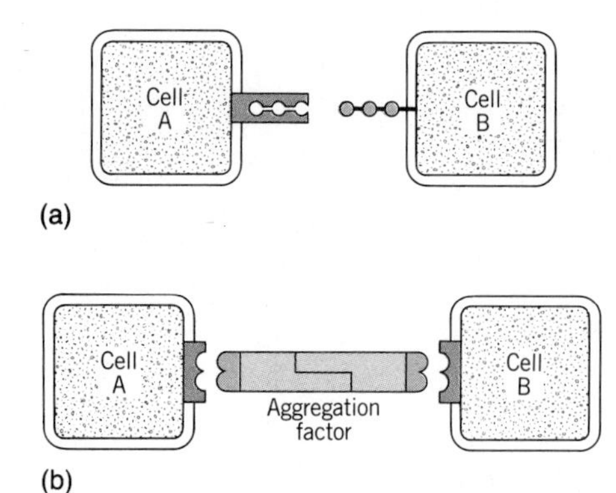

FIGURE 5.15 Models of intercellular adhesion. (*a*) Two cells possess complementary cell surface molecules that interact directly with one another. (*b*) Cells adhere through a mediating aggregation factor, which carries sites that are complementary to cell receptors.

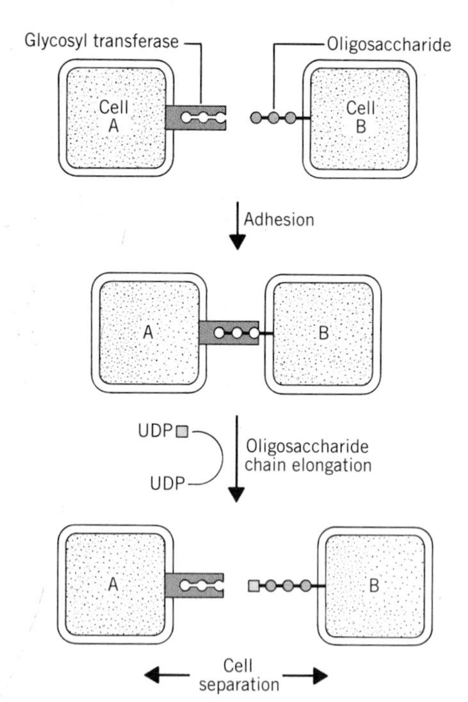

FIGURE 5.16 The Roseman hypothesis of cell–cell interaction. Cell A contains a surface enzyme that binds to its substrate on the surface of adjacent cell B. Modification of the substrate by adding a monosaccharide unit weakens the binding and frees the cells from their interaction. Thus, the hypothesis accounts for a dynamic interaction between cells.

bound Ca^{2+}. This release may in turn induce the ICM to open an ion channel or to shuttle ions across the membrane. Although the ICM concept is gaining credibility, at this point it is not known whether such a modulator is part of the receptor itself or is a separate entity, or whether it even exists as a discrete molecule. It may simply be a function of the overall process.

In any event, despite the speculative nature of the ACh receptor function, the past decade has witnessed some remarkable advances in this area of research.

5.7 DIRECT CELL–CELL INTERACTIONS

An interaction or adhesion between cells is the basis for the structural integrity of multicellular systems. Unicellularity depends on the avoidance of adhesion, but the advantages of being a free spirit cell are somewhat tempered when it is realized that cells in a multicellular system can afford to differentiate and focus their metabolic attention on fewer activities. In short, they have less metabolic responsibilities and can become metabolically specialized.

Cells in multicellular systems that lose their adhesive abilities generally become the instigators of uncontrolled growth, or malignancy. Thus adhesiveness is an important factor in the control of multicellular systems.

The entire process of embryogenesis is highly dependent on specific adherence between cells as they take on the characteristics of tissues and organs. We can see therefore that cell adhesion is not only an important property for a mature tissue, but it is also crucial for the proper development of that tissue.

Two mechanisms of intercellular adhesion are portrayed in Figure 5.15. The simpler of the two models depicts a direct interaction of cell surfaces. For this to occur, apposing cells must carry complementary structures. The second type of intercellular adhesion is facilitated by a bridging molecule or aggregation factor. In this case, the aggregation factor is bifunctional. It forms a link between cells by interacting with sites on the cell surface that are complementary to structures that it possesses.

The Roseman Hypothesis

One way in which cells may interact directly is by virtue of enzyme–substrate binding between adjacent cells. This is the basis of a hypothesis by Roseman[11] (Figure 5.16). Each cell may carry many surface oligosaccharides and surface glycosyl transferase enzymes. Adjacent cells could form multiple noncovalent cross-links by virtue of enzyme–substrate reactions. If the enzymes modify the oligosaccharide chains of the apposing cells, the cells will no longer adhere to one another. Thus, there is in this hypothesis a mechanism to permit cell separation to occur.

The extent to which this mechanism of adhesion operates between cells is not known, but the presence of glycosyl transferase enzymes on the surface of several types of cells is circumstantial evidence that this type of principle may attract cells. Glycosyl transferases have been found on the outer surfaces of chick embryo cells, human blood platelets, mouse fibroblasts, neural retinal cells, and rat intestinal epithelial cells.[12] According to the Roseman hypothesis, cells of these types could adhere to one another by way of glycosyl transferase–oligosaccharide interactions, but until more experimental verification appears, it would not be wise to assume that this is the only mechanism of adhesion operating on these cell surfaces.

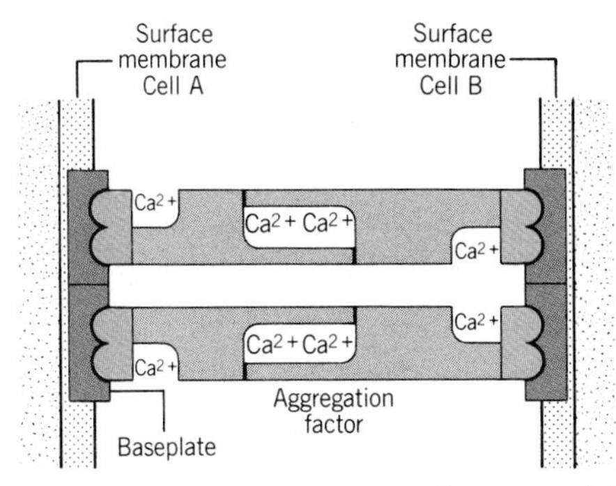

FIGURE 5.17 Aggregation factor model for specific cell–cell interactions in sponges.

Aggregation Factors

Several multicellular systems appear to stay intact with the aid of aggregation factors. One of the most intensively studied of such systems is the sea sponge. As early as 1907, H. V. Wilson described the regeneration of sponges from dispersed cells and noticed that when the cells of two species are mixed there occurs a species-specific reaggregation to form two distinct organisms.[13]

Sponges can be dispersed into single cells mechanically by pressing them through cheese cloth or by removing divalent cations from the medium. In the former case, cells will spontaneously reaggregate to reform the multicellular organism. In the latter case, reaggregation will occur only when the cations are replaced.

Another factor, in addition to divalent cations, is necessary for reaggregation. Such a factor has been purified from the sponge *Microciona parthen*.[14] It is a high molecular weight protein and carbohydrate complex that induces aggregation in the species from which it is isolated, but has no effect among cells of unrelated species of sponges.

Figure 5.17 depicts this type of aggregation. According to this model there are three participants at the site of adhesion. One is the *baseplate* on the cell surface. This is a component, probably glycoprotein, that functions as the receptor site on the membrane. It is probably a peripheral membrane component since it can be released from cells when they are hypotonically shocked (in 0.08 M NaCl).

The second participant is the *aggregation factor*, a high molecular weight proteoglycan molecule. The third is *calcium ions*. Calcium ions stabilize the aggregation factors, two of which interact to form the cross-link. Stereospecific interaction takes place between the baseplate and a portion of the aggregation factor.

Aggregation factors are apparently not restricted to sponges. Factors with similar functions have been isolated from mouse tumor cells that grow in ascites fluid. These are also glycoproteins. In the mouse system, the tumor cells appear to contain baseplates that bind galactose residues present on the aggregation factor.[15]

5.8 JUNCTIONS BETWEEN CELLS

An extreme form of intercellular communication and contact is effected by several types of structure called junctions. These structurally modified regions of plasma membranes bring cells into close proximity, keep them in fixed positions, and in some cases mediate limited forms of communication.

Junctions are common structures among metazoan members of the animal kingdom, including vertebrates and invertebrates. They are found *in vivo* and are also seen *in vitro*, a property that considerably facilitates their study. Several different morphological variants of junctions exist, each having a specific function. We will consider three types of intercellular junction that are representative of the major cell–cell connections.

Gap Junctions: Communicating Junctions

One type of junction is the *gap junction,* a structure that has been the recipient of many other names, among which *nexus* is probably the most common. It was first seen clearly in 1967 by Revel and Karnovsky[16] as a seven-layered or septilaminar structure. The layered appearance, clearly visible in the electron micrographs of Figure 5.18, is due to the close proximity of two unit membranes separated by a gap of 2.0 to 4.0 nm. The width of the entire structure is 15 to 19 nm, or twice that of two unit membranes plus the electron translucent gap.

The gap junction is probably the most common type of junction between cells in animal systems. It has been observed in both excitable and nonexcitable cells, and its ubiquity may include all cell types with only a few exceptions. It has not, for example, been observed between mature skeletal muscle cells and between circulating blood cells.[17]

Method I

Electron microscopy of thin sections has been useful for delineating the thickness of the gap junction and the overall structural integrity of the junction. But it was not until freeze-fracture techniques were used that some of the unique structural properties of this type of junction were understood. An electron micrograph of a replica of a gap junction prepared by this technique is shown in Figure 5.19.

Several important insights have come from the use of freeze-fracturing. In the first place, the gap junction exists as a disc-shaped segregated domain in the membrane. The diameter of the disc may vary considerably among different tissues, with large junctions having diameters near 1 μm. Second, homogeneous 6 to 8 nm particles are embedded in each of the two-unit membranes that contribute to the junction. Furthermore, these particles are clustered in a hexagonal array with a center-to-center spacing of 9 to 10 nm. A central channel of 2.0 to 2.5 nm is present between the subunits that make up the particles. This type of arrangement is more clearly seen in the preparation of isolated gap junctions shown in Figure 5.20. The hexagonal array of subunits making up the 6 to 8 nm particle forms a molecular cylinder in each unit membrane.

Molecular cylinders from adjacent membranes protrude sufficiently far from the unit membrane that they abut, thereby forming an intercellular channel that spans the gap. A model of this physical apposition of cylinders (Figure 5.21) has resulted from the work of Goodenough and co-workers, who have carried out extensive studies correlating the results of electron microscopy with those of X-ray diffraction.[18,19]

Method VIII

Goodenough et al. have named the basic hexagonal unit or particle *connexon.* Two connexon units form the intercellular channel, each 7.5 nm long with a 6.0 nm outside diameter and a 2.0 nm inside diameter. Working with isolated gap junctions these workers have found them to contain only one major protein, which has been termed *connexin.* It has a molecular weight of 18,000 daltons and the shape of a dumbell.

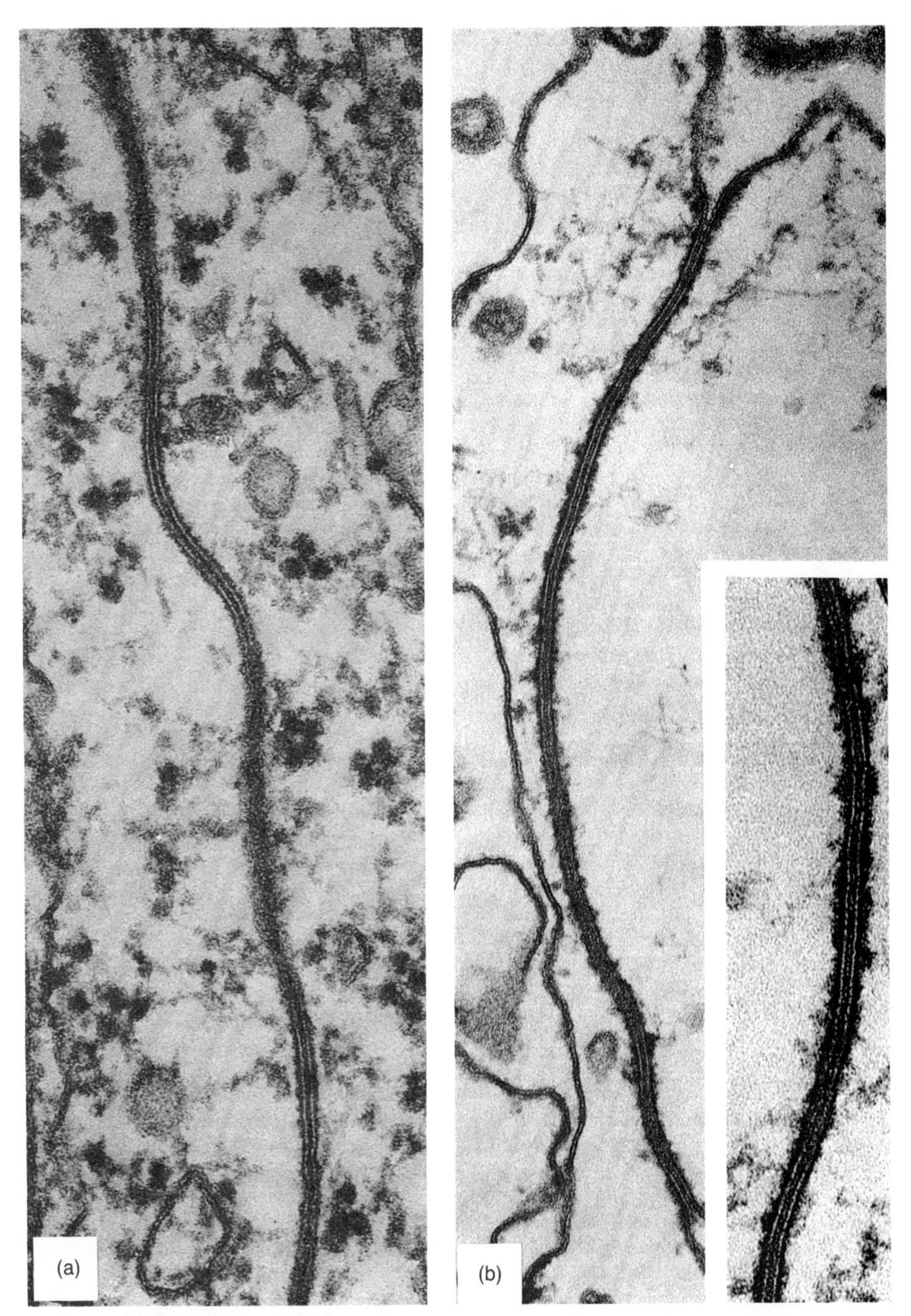

FIGURE 5.18 The septilaminar appearance of gap junctions due to a close apposition of two unit membranes. (*a*) Extensive gap junction between Don hamster fibroblasts in cell culture. Both transverse and slightly oblique planes are shown. (*b*) Gap junction in an enriched plasma membrane subfraction of rat liver. The structure of the gap junction is maintained during isolation (see inset). The 2- to 4-nm electron-translucent gap is apparent in the inset. (Courtesy of Dr. N. B. Gilula.)

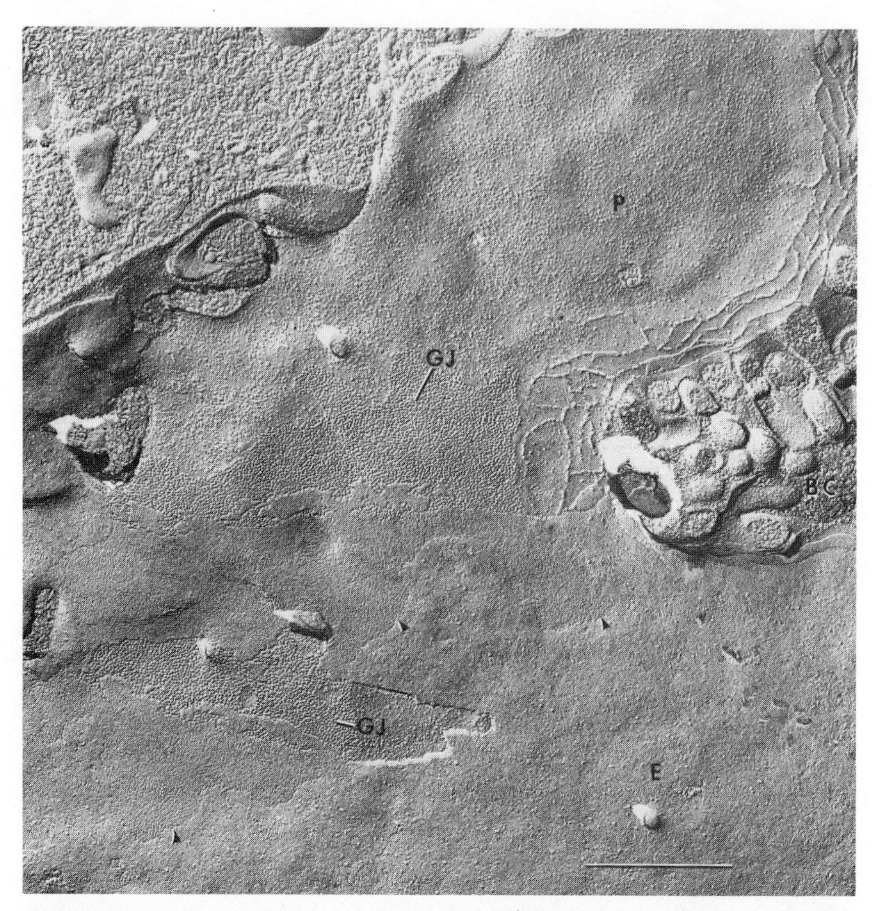

FIGURE 5.19 Freeze-fracture transmission electron micrograph of a murine hepatic bile canaliculus (BC). The gap junctional particles (GJ) are hexagonally packed on the P membrane fracture face. Ordered arrays of pits (arrow heads) can be seen on the E membrane face. ×40,000; bar = 0.5 μm. (Courtesy of Dr. Martin Porvaznik.)

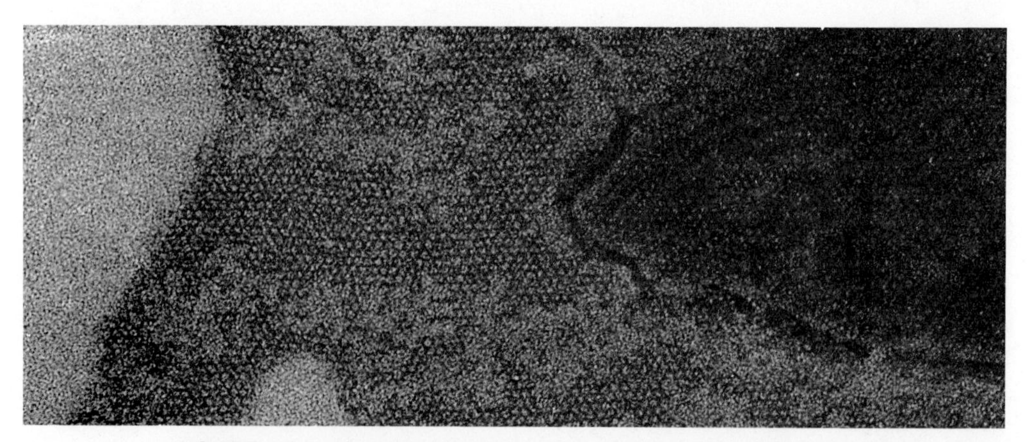

FIGURE 5.20 Negative stained gap junction from isolated rat liver plasma membranes. The polygonal array of particles is apparent. Each particle, in turn, is made up of a hexagonal array of subunits. (Courtesy of Dr. N. B. Gilula.)

The gap junction appears to have two functions. One is cellular adhesion. The means of physical contact that is provided is highly resistant to physical or mechanical stress.

A second function more commonly associated with the gap junction is that of direct intercellular communication. Several experimental approaches have demonstrated that the gap junction serves as a permeability pathway between cells for many different kinds of molecules.

One type of permeability is called *metabolic coupling*. Molecules up to a molecular weight of 1000 daltons pass readily from one cell to another through the junction. This affords a means to distribute ions, sugars, amino acids, nucleotides, vitamins, hormones, and other low molecular weight species among the cells in a given tissue. It is thought that metabolic coupling is used to distribute nutrients to cells during tissue development before there is a functional circulatory system and to provide regulatory molecules, such as cAMP, to cells during tissue differentiation.

A second type of permeability associated with the gap junction is called *ionic* or *electrotonic coupling*. Gap junctions are low resistance electrical pathways between cells: if an electric current is applied within one cell, a voltage shift can be detected in adjacent cells. This type of electrical impulse conduction between cells is nearly instantaneous. Furthermore, it has an advantage over the mechanism of impulse conduction between neurons in that there is no need for a neurotransmitter.

This mode of conductance is found among cells where synchronous action is necessary for the tissue or organ to function properly. Muscle cells of the heart are energized in this way, simplifying the required innervation for the controlled contraction of cardiac muscle. The smooth muscle cells that perform the peristaltic contractions of the intestine also make use of electrotonic coupling provided by gap junctions.

Calcium ions play an important role in controlling the permeability of the gap junction. It is known that high calcium levels within the cytoplasm of the cell uncouple the cell from its normal communication with neighboring cells. The current thinking is that calcium changes the conformation of connexin molecules so that the channel is sealed. The channels will reopen only when the calcium concentration once again drops to normal levels. A speculated reason for this molecular behavior is that when tissue is damaged and cells are ruptured there is an influx of calcium. Internalized calcium immediately seals the channels to adjacent cells, thus protecting them from a loss of vital nutrients and biosynthetic building blocks. As the tissue is repaired, new gap junctions are established and normal intercellular communication is restored.

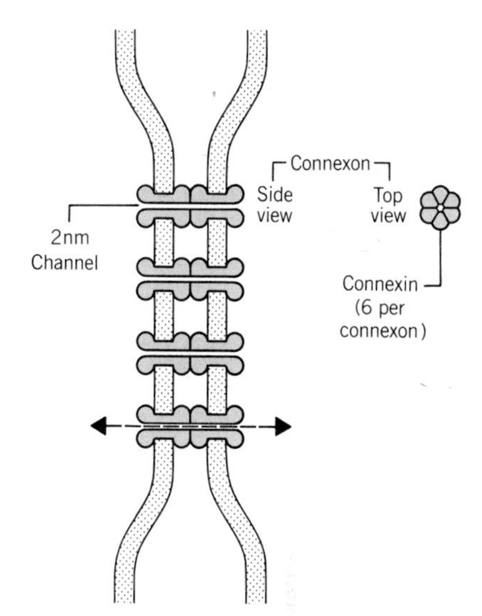

FIGURE 5.21 Model of a gap junction. Pairs of abutting cylinders permit the exchange of nutrients and signal molecules between cells. Each particle is made up of six dumbbell-shaped protein subunits that are integral members of the membrane and span it. The channel in the center of the cylinder is about 2 nm in diameter.

Desmosomes: Adhering Junctions

Desmosomes are areas of contact between cells that have a central role in adhesion. Unlike gap junctions, these structures are not channels of communication. Their overall effect is to provide a means whereby a group of cells can function together as a structural unit.

Desmosomes take on two different structural forms. One, the *spot desmosome*, is a buttonlike point of contact between adjacent cells that anchors the two cell membranes together, much as a spot weld holds sheets of metal together at points.

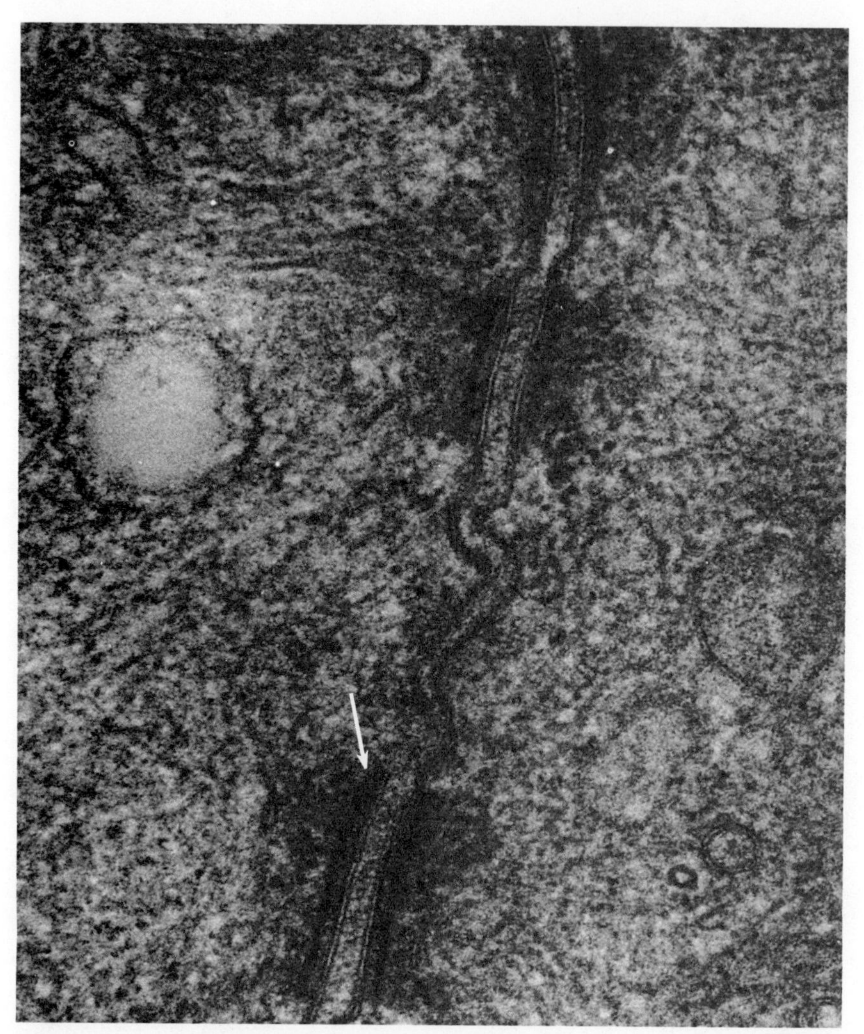

FIGURE 5.22 Spot desmosomes in rat intestinal epithelium. These adhering junctions occur as symmetrical plaques between two adjacent cells. Their characteristic features include a wide (25–35 nm) intercellular space containing dense material, two parallel cell membranes, a dense plaque associated with the cytoplasmic surface (arrows), and cytoplasmic tonofilaments that converge on the dense plaque. (Courtesy of Dr. N. B. Gilula.)

Unlike the spot weld, the membranes are not fused at this point but remain separated as parallel sheets by a distance of about 30 nm. This can be seen easily in the electron micrograph of Figure 5.22 where a thin section has been cut through the spot desmosome.

The cytoplasmic surfaces of the membranes in the desmosome region contain dark staining plaques. These are linked to one another across the intermembrane gap by filaments that extend through the membranes and are anchored to an intermembrane network called the *central stratum*. These *transmembrane linkers* thus directly couple adjacent cells without the membrane surfaces fusing.

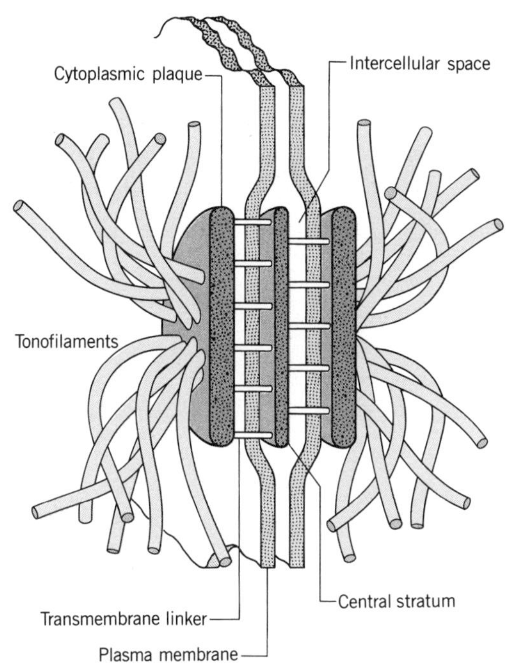

FIGURE 5.23 Model of a spot desmosome. The tonofilaments, 10 nm in diameter, form a tensile network that extends throughout the interior of the cell. They are attached to the plaques of the spot desmosome through poorly defined filamentous structures. Other filaments, called transmembrane linkers, connect the plaques of the spot desmosome across the intercellular space. The junction therefore serves to couple the tonofilament networks of adjacent cells, allowing the dissipation of shearing stresses throughout the tissue.

A model of a spot desmosome interpreting these electron microscopic features is shown in Figure 5.23. Note the parallel configuration of adjacent cell membranes, the cytoplasmic plaques, and the intercellular bridging filaments. The model includes still another feature of the spot desmosome that can also be seen in electron micrographs. This feature is the presence of *tonofilaments*, which comprise a cytoplasmic network of filaments that are anchored to the cell surface at the cytoplasmic plaques.

Bundles of tonofilaments crisscross the cell interior. Since they are also rigidly attached to the plaques, which in turn are linked to plaques in adjacent cells, they provide a structural framework for the cell interior as well as a means of holding adjacent cells together. One can easily picture several cells tightly held together by means of this system of biological cables.

Hemidesmosomes are half-desmosomes that weld the cell to connective tissue, which serves as a foundation under epithelial cells. This prevents the layers of cells from peeling off the underlying matrix.

Tissues that are often subjected to mechanical stress contain an abundance of spot and hemidesmosomes, an example being the cervix of the uterus. During the mechanical stress of parturition, the shear forces to which the tissue is subjected are shared and distributed among the cells by means of the tonofilaments and transmembrane linkers.

The other major form of desmosome, the *belt desmosome*, is present between cells as a band binding adjacent cells together. In cross section, the belt desmosome can be distinguished from the spot desmosome because it lacks the electron-opaque plaque and is associated with filaments that are thinner (7 nm) than the tonofilaments (10 nm). Two types of filament are found associated with the cytoplasmic surfaces in this region. One type runs as bundles parallel to the membrane, and the other extends at right angles out into the cytoplasm. The intermembrane space also contains filaments.

The filaments of the belt desmosome contain actin, a contractile protein. Although the role of this type of desmosome is not as clear as for the spot desmosome, it is speculated that belt desmosomes both serve to anchor adjacent cells together and to contract and close gaps that may form in tissues when individual cells die and are removed.

Tight Junctions: Impermeability Barriers

In contrast to gap junctions and desmosomes, *tight junctions* are points of cell contact where a true fusion of adjacent cell membranes takes place. There is no intercellular space in this type of junction. As described by Farquhar and Palade in 1963,[20] tight junctions may be present in beltlike (zonula occludens) or bandlike (fascia occludens) forms that completely encircle the cell, sealing it in all directions with adjacent cells. Several of these belts may rim the cell, making a network of fusion strands between adjacent cells.

This type of seal between cells normally is positioned near an exposed cell surface, such as is the case for epithelial cells that line the interior of the intestines.

Figure 5.24 is a transmission electron micrograph of a section of tissue from this region. The zonula occludens is found near the lumen. This electron micrograph also illustrates a common feature of contact between epithelial cells, the *junctional complex*. It is made up of the tight junction near the lumen followed by the belt desmosome and the spot desmosome.

A micrograph of a freeze-fractured tight junction (Figure 5.25) depicts very clearly the reticulation of the tight junction, with the ridges and grooves being the lines of contiguous membrane fusion. It is also apparent from this micrograph that the network functions as a *continuous seal* between adjacent cells so that material cannot penetrate between cells in either direction.

In a model illustrating the ultrastructure of the tight junction as it is most commonly envisioned (Figure 5.26), the sealing strand is thought of as a ropelike structure made up of integral membrane proteins contributed by the two adjacent membranes. The cell surfaces are thus fused in lines by a head-to-head interaction between membrane proteins from neighbor cells. In cross section, these lines would appear as points of membrane fusion. The system is further thought to be supported by a network of cytoplasmic filaments that make contact with the integral membrane proteins that participate in the sealing strand.

A conceptual alternative to this model has been more recently proposed in which the sealing strand is viewed as an intramembranous structure

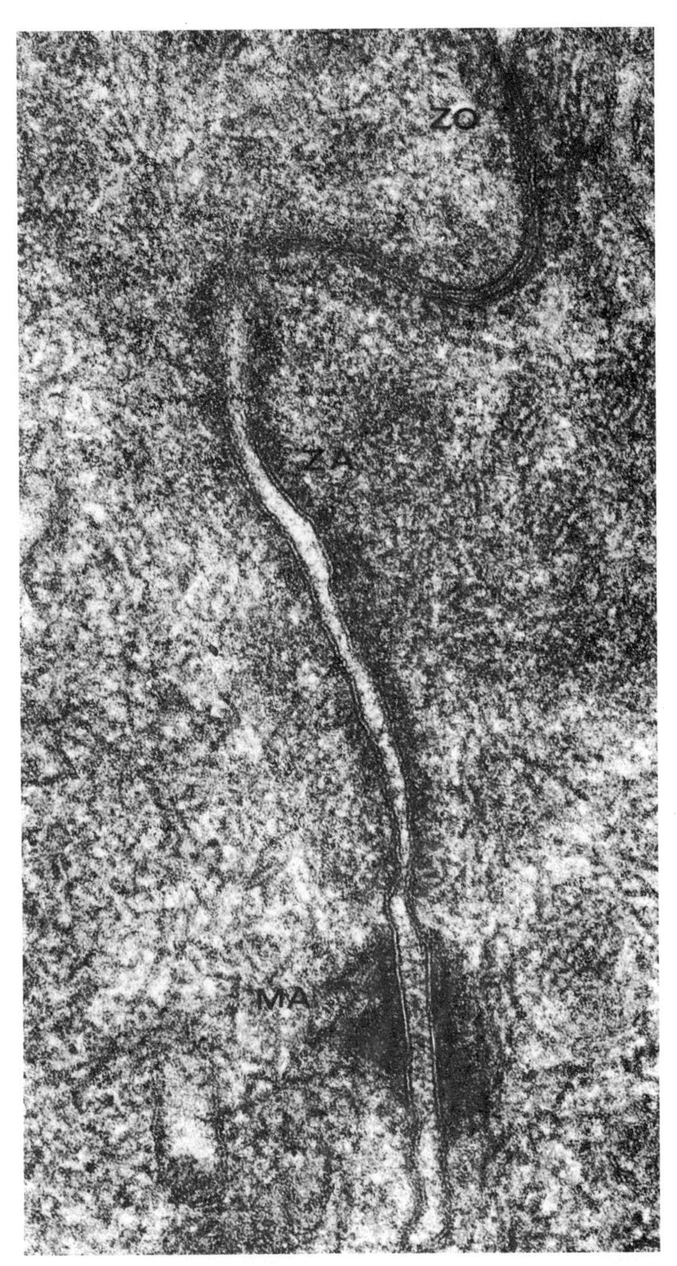

FIGURE 5.24 Electron micrograph of junctional complex between epithelial cells of the rat small intestine. Beginning near the lumen toward the top of the micrograph, membranes fuse to form the tight junction (zonula occludens: ZO); deeper in the belt desmosome (zonula adherens: ZA) and the spot desmosome (macula adherens: MA) are seen. (Courtesy of Dr. N. B. Gilula.)

that is fundamentally an inverted lipid micelle in cylindrical form.[21] Figure 5.27 will help in understanding the thrust of this idea.

This concept is based on an analysis of freeze-fracture electron microscopic studies of tissues prepared with or without fixation before frac-

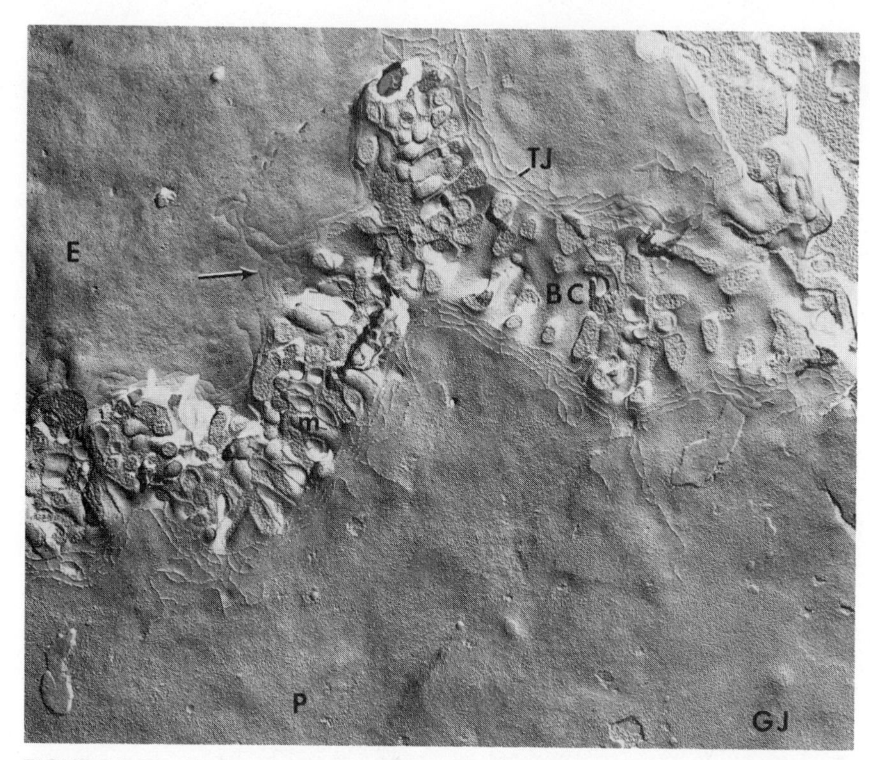

FIGURE 5.25 Freeze-fracture transmission electron micrograph of a murine hepatic bile canaliculus. The elements of the tight junctions (TJ) are apparent as ridges on the P face and as grooves (arrow) on the E face. Many microvilli (m) are observable within the lumen of the canaliculus. The tight junction limits the canaliculus. ×21,400; bar = 1.0 μm. (Courtesy of Dr. Martin Porvaznik.)

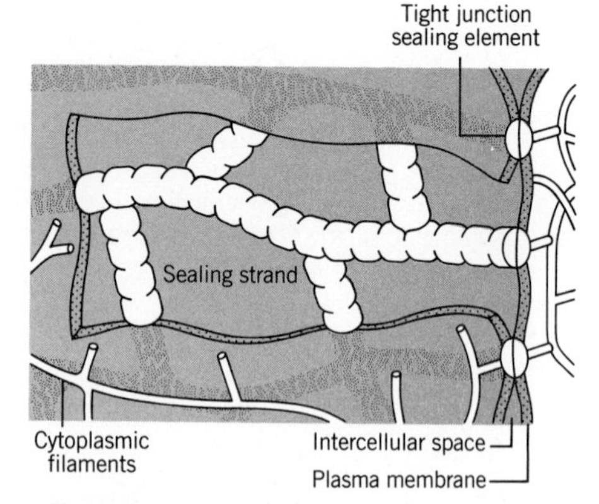

FIGURE 5.26 Model of a tight junction. Adjacent cell membranes are held together by sealing strands, which form lines of attachment. Each sealing strand is composed of rows of particles contributed by adjacent cell membranes. The strands make up a network of adhesion lines that give the tight junction flexibility and the capacity to remain intact under stress. A network of filaments permeates the region near the cytoplasmic surfaces, stabilizing the positions of the sealing strands.

turing. These analyses suggest that when the sealing strand appears as a row of particles, it may be an artifact of fixation or fracturing. The real *in situ* morphology of this strand may be cylindrical, which under certain conditions of tissue preparation gives rise to a continuous linear ridge on one face of the fracture and a continuous linear groove on the matching face.

According to the inverted micelle model, in cross section a true fusion of adjacent cell membranes is proposed, involving primarily their lipids. On surface view, this would take on the appearance of a long cylindrical element lying between the two cells.

Although this model centers on lipids rather than proteins as essential ingredients of the sealing strand, it incorporates the idea that proteins may interact with the cylindrical micelle and thereby impart a stabilizing influence on the structure. As with the preceding model, these proteins could interact with the fibrous cytoskeletal components of the cell to impart additional strength and stability to the cell collections tied together by tight junctions.

The function of the tight junction is at least twofold. In the first place, it prevents leaking of molecules across the epithelium in either direction, from the lumen to the basal regions of the cell, and vice versa. The import of this type of seal is readily apparent if the function of the epithelium is recalled. In the intestine, the epithelial cells selectively transport nutrients to the blood. If no seal were present, the transport could not be selective, and a backflow of already transported material to the lumen could take place.

In the second place, the tight junction forms a physical barrier that limits the movement of other integral membrane proteins laterally within the fluid lipid bilayer. This barrier maintains the functional polarity of the cell. Thus, transport proteins that bring nutrients into the cell are kept at the apical surface while transport proteins that pass the nutrients on to the bloodstream are kept at the basal surface. An intermingling of these proteins would confuse the transport capabilities of the epithelial layer.

One effect, therefore, of the tight junction is to maintain cells, and thus a tissue, in a differentiated state. There is now evidence from studies carried out on fetal rat liver that sealing strands form very early during organ development, first as short rows and later as a network of strands around honeycomb-shaped depressions in the plasma membrane. It would therefore appear that tight junctions establish a polarity in cells very early in their development and maintain this polarity throughout the life of the cell.

In summary, we have discussed three types of intercellular junction, as illustrated schematically in Figure 5.28. The gap junction is an intercellular point of contact functioning in ionic and metabolic coupling. Desmosomes function in intercellular adhesion. The spot desmosome is a localized spot of adhesion between cells and the belt desmosome rims cells with adhesive belts. Tight junctions form a permeability barrier between cells in epithelial tissue. In contrast to the other types of junction, there is a true fusion of membrane surfaces by rows of integral membrane proteins. This blocks the nonspecific passage of molecules between cells and serves to maintain cellular polarity.

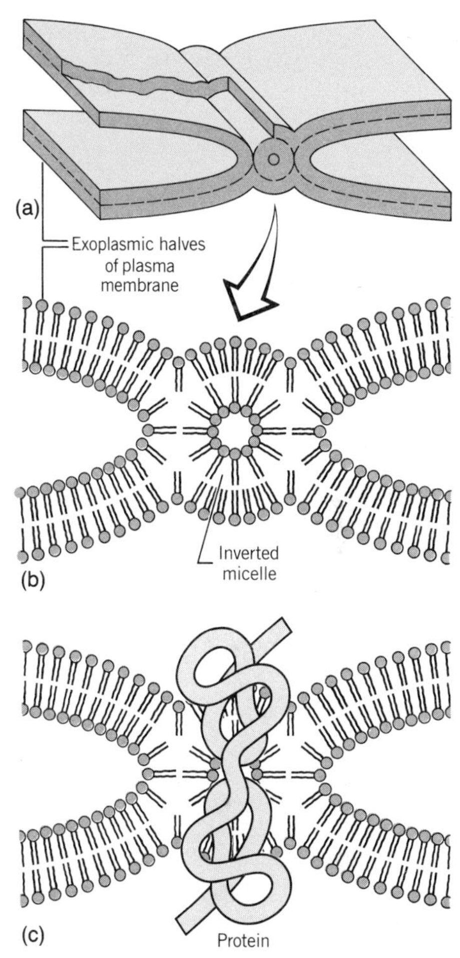

FIGURE 5.27 Inverted lipid micelle model of a tight junction. The exoplasmic halves of the plasma membranes of adjacent cells are fused to form a continuous leaflet. On surface view the inverted lipid micelle appears cylindrical (*a*). The micelle may exist without (*b*) or with stabilizing proteins (*c*).

5.9 CELL–SUBSTRATUM ADHESION

The integrity of a tissue depends not only on adhesion between cells but also on adhesion between a layer of cells and the substratum on which the layer is based. As indicated in Chapter 2, this substratum is made up of collagens and other matrix macromolecules that have roles both in cell adhesion and in cell differentiation.

The adhesion between a layer of cells and its substratum is not a direct one for many cell types. Two components are required, namely, a cell surface receptor and a mediator molecule that bridges the receptor with the matrix molecule. This is apparently the case for cells such as chondrocytes, fibroblasts, myoblasts, hepatocytes, and certain epithelial cells.[22] Several different adhesion factors have been studied that are known to attach cells to collagen. The most extensively studied of these is *fibronectin,* one of the most abundant glycoproteins in matrix material and on the surface of many cells as well as circulating in serum in free form. Blood fibronectin stimulates macrophages to carry out phagocytic activities while other forms of the molecule promote cell–substratum adhesion.

Fibronectin is a high molecular weight protein (see Table 5.1) consisting

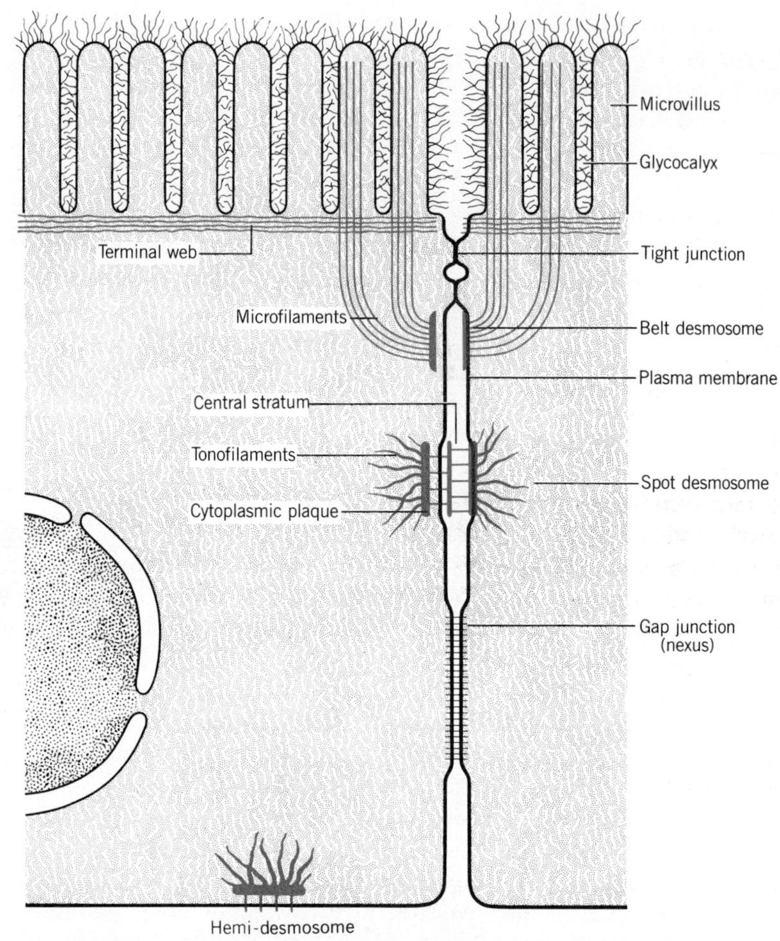

FIGURE 5.28 Distribution of junction types as they would be seen in the epithelium of the small intestine. The tight junction forms a seal between adjacent cells near their lumen surfaces. Below this lies the belt desmosome, girdling the cell membrane with contractile filaments. Deeper into the cell are located spot desmosomes and, at the basal ends, hemidesmosomes. Finally, gap junctions make up patchlike areas of close intercellular contact. Desmosomes are interconnected by crisscrossing bundles of tonofilaments. In sum, adjacent cells are coupled mechanically (spot desmosomes) and metabolically (gap junctions) while the digestive fluids in the intestinal lumen are kept away from tissue fluids (tight junctions).

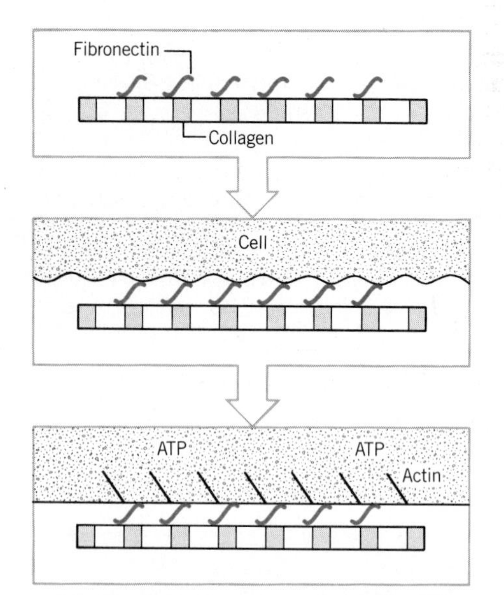

FIGURE 5.29 Schematic model of fibronectin-mediated cell adhesion to collagen *in vitro*. Fibronectin binds to a specific sequence of amino acids in collagen. Then, in the presence of Ca^{2+} or Mg^{2+}, cells bind to the fibronectin–collagen complex. In the presence of ATP and with the aid of contractile filaments, the cells spread.

of two similar or identical polypeptide chains covalently linked by disulfide bonds. It is a glycoprotein, containing about 6% carbohydrate that is attached to the protein by way of asparagine residues.

One role of fibronectin is that of promoting adhesion between a cell and its collagenous substratum. This appears to involve three steps, outlined schematically in Figure 5.29. First, fibronectin binds to collagen, a binding that appears to involve a specific binding domain on fibronectin and a particular amino acid sequence in collagen. Next, the cell binds to the fibronectin–collagen complex by virtue of receptor sites on the cell

Table 5.1 Characteristics of Adhesion Factors

Characteristic	Fibronectin	Chondronectin	Laminin
Molecular weight	450,000 ↓ Reduction 220,000	180,000 ↓ Reduction 80,000	800,000 ↓ Reduction 200,000, 400,000
Activity in attachment	1–5 μg	5–50 ng	1–5 μg
Serum amount	200 μg/ml	Below 20 μg/ml	Below 1 μg/ml
Tissue location	Fibrous connective tissue, some basement membranes	Cartilage, vitreous body	Basement membrane
Heparin binding	Yes	Yes	Yes
Heat lability	57°C	52°C	49°C

surface. Finally, *in vitro,* the cell will spread out on the surface, a movement that requires ATP production on the part of the cell.

Components of the matrix other than just fibronectin may be essential for this type of adhesion, as evidenced by the observation that fibronectin binds to proteoglycans, which contain hyaluronic acid and heparan sulfate.

The least well characterized component of this adhesive activity is the cell membrane surface receptor. It is thought in this case to be a ganglioside, with the adhesive activity residing in the carbohydrate portion of the glycolipid. A model depicting how these components may cooperate in effecting cell–substratum adhesion is presented in Figure 5.30.

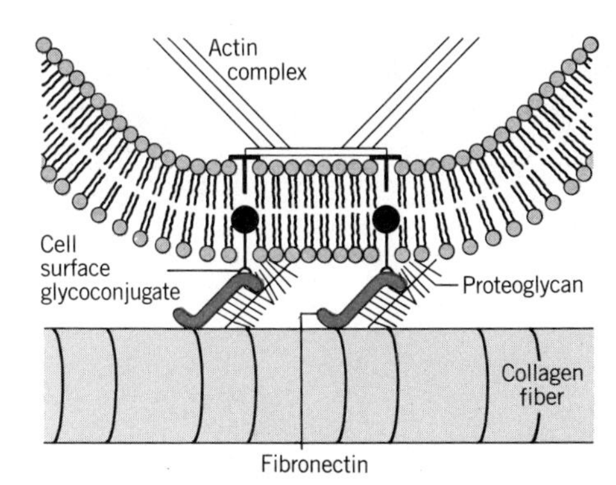

FIGURE 5.30 Schematic model of an adhesion site. Collagen fibers contain specific sites to which the fibronectin molecules bind. Fibronectin contains a region that recognizes collagen and another that recognizes the cell surface. Proteoglycans are also present to stabilize the complex.

Two other adhesion factors are included in Table 5.1. One is chondronectin, a factor found in cartilage instead of fibronectin. It appears to have a function similar to that of fibronectin but with a specificity for chondrocytes.

The attachment factor between epithelial cells and type IV collagen is *laminin,* a high molecular weight glycoprotein found in basement membranes. Once again, the function of laminin appears to parallel that of fibronectin, but its specificity is for a different cell type.

The importance of the matrix and a proper adhesion of cells to the matrix can hardly be overemphasized. Tumorigenic and metastatic cells have properties that weaken or obviate this adhesive interaction. In some cases less collagen and fibronectin are synthesized by the cells, and thus the basis for a strong interaction with the substratum is reduced.

5.10 DISEASES DUE TO FAULTY CELLULAR COMMUNICATION

The discovery of mediators and cell receptors, and of cellular communication in general, has paved the way for a new approach in dealing with a number of diseases. Several diseases are now understood to be due to a problem in cell communication. In some cases the level of mediator molecules is the problem, either overstimulating or understimulating target tissues. In other cases the receptors malfunction so that the interior of the cell receives the wrong information and responds abnormally.

Among the diseases that are now attributed to errors in communication are cholera, hyperthyroidism (Graves' disease), myasthenia gravis, certain types of diabetes, and neurotransmitter diseases such as Parkinson's

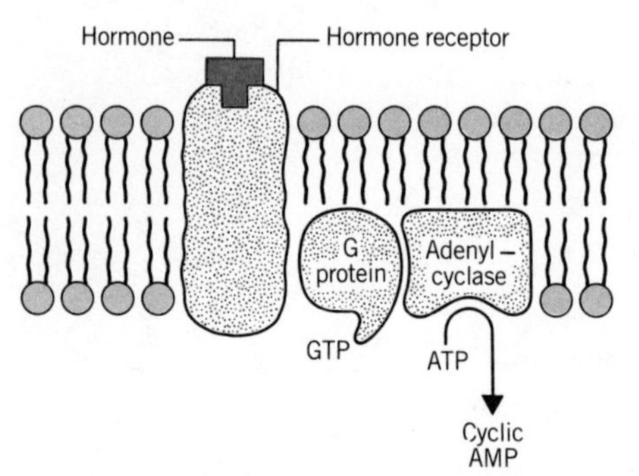

FIGURE 5.31 Flow of information between hormone activation of the receptor and formation of cAMP in the cell interior. The G protein plays a crucial role in this chain of communication.

disease and Huntington's chorea.[23] Let us briefly examine just two diseases, cholera and diabetes.

Cholera

Cholera is brought about by ingestion of the bacterium *Vibrio cholerae,* which is picked up from contaminated water or food supplies. Clinically, the disease is recognized by violent purging, vomiting, burning thirst, muscular cramps, suppression of urine, and rapid collapse. It is a terrifying endemic infectious disease in India and China with a mortality of 70% or more. It is also an old disease, known at the time of Alexander the Great, having originated near Calcutta and on the delta of the Ganges River. Now, nearly 24 centuries later, there has emerged a molecular understanding of this scourge.

One of the hallmarks of cholera is a loss of fluid, up to 10 to 20 liters per day. The production of fluid in the intestine is greater than the amount that can be reabsorbed, so death is caused by rapid and extensive dehydration.

While growing in the lumen of the small bowel, *Vibrio* produces a toxic material that is the source of the problem. This material is a protein consisting of three polypeptides, A_1, A_2, and B. A_1 and A_2 are covalently linked to form a dimer designated subunit A. The A subunit is noncovalently associated with a variable number of B chains.

The binding of cholera toxin to the epithelial cell membrane takes place by interaction of the B subunit with a ganglioside in the membrane. The A_1 chain, brought into position on the cell surface by this binding, activates adenylcyclase causing the formation of cAMP within the cell. Increased levels of cAMP trigger an efflux of chloride and bicarbonate ions from the epithelial cells into the intestinal lumen. This loss of ions is followed by an efflux of water in an attempt to maintain the osmotic balance within the cell. The high level of adenylcyclase activity, fanned by the bound toxin, brings about the copious loss of water.

The normal flow in information in the activation of adenylcyclase is indicated schematically in Figure 5.31. Under ordinary circumstances, the entry of food into the stomach causes a mediator to be sent to these intestinal cell receptors, to establish the proper alkaline fluid environment in the small intestine for the digestive process. The amount of fluid produced by the normal mediator is readily reabsorbed farther along in the intestinal tract after digestion is completed.

A key participant in this entire process is a protein called G protein, which functions as a transducer. This protein binds guanosine triphosphate (GTP) when the receptor binds the signal. The GTP–G protein complex then activates adenylcyclase and cAMP is formed. The G protein also possesses a GTPase activity, so with time the bound GTP is hydrolyzed and the entire system thereby deactivated.

The A_1 subunit of the cholera toxin has its impact at the level of the G protein. It catalyzes a reaction that places an ADP-ribose unit on to the G protein, which in effect locks the G protein in its activated state. The GTPase activity of the protein cannot deactivate the system. Thus, the adenylcyclase enzyme is also locked into an activated state, accounting for the uncontrolled efflux of ions and water into the intestinal lumen.

G protein — cholera toxin →
Nicotinamide
NAD⁺: Adenine – Ribose – (P) – (P) – Ribose- Nicotinamide

G protein -Ribose-(P)-(P)-Ribose-Adenine

This is covalently linked to an arginine side chain, locking the G protein in an activated state.

The molecular level of sabotage in this chain of communication is not the receptor but the transducer. The receptor is apparently circumvented by a toxin that gains direct access to the G protein. The effector molecule, adenylcyclase, is not structurally modified by the toxin but is simply continuously activated.

Diabetes

Diabetes mellitus occurs in several forms. Basically, there is either underproduction of insulin or defective functioning of the receptors to which insulin binds. Under normal circumstances, insulin activates receptors in the cell membranes of muscle and fat cells. The receptors, in turn, facilitate the entry of glucose into the cell by forming an insulin–receptor complex that moves to the membrane of the cell nucleus. There the insulin exerts its effects on biosynthesis and in regulating the metabolism of carbohydrates.

In certain types of diabetes there are reduced numbers of receptor on the cells of several types of tissues. Even though in these cases the output of insulin is normal, the decreased availability of receptors affects the transportability of glucose into the cell.

In other types of diabetes there is a decreased affinity of the receptor for insulin. This is true in a rare form of genetic disease known as ataxia telangiectasia.

In still other cases there is insufficient insulin produced as a result of malfunctioning of the β cells of the pancreas. In most of these cases the receptors are not affected; therefore, the administration of insulin to the individual is sufficient to correct the metabolic problem associated with this type of diabetes.

Summary

The surface of the plasma membrane is a mosaic of proteins and lipids that along with attached carbohydrates, impart antigenic properties to the cell and function as sites for the receipt of a variety of signals.

The ABO blood group antigens are glycolipids, the MN blood group antigens are glycoproteins, and the histocompatibility antigens are proteins that consist of heavy and light chains. The natures of the antigenic determinants in the ABO systems have been delineated as subtle differences in carbohydrate structure. The chemical determinants for the MN system and for the histocompatibility antigens have not been determined with as high a degree of molecular resolution.

Although the biological function of antigens is generally not known, histocompatibility antigens are speculated to have a role in the phenomenon of immunological surveillance.

The cell receptor is the first component in a chain of communication that generally operates to transfer information across the plasma membrane to activate or inhibit a cell process. The chain of command includes, in addition to the receptor, a transducer and an effector. The effector brings about the formation of a second messenger that functions directly within the cell. Cells have many different kinds of receptors but only a few effectors. The kinds of first messenger are great, but the kinds of second messenger very limited.

Receptors have been characterized in part in B- and T-type lymphocytes and in postsynaptic membranes of the synapse. In the former two they are immunoglobulin-type molecules, in the latter an aggregate of proteins with a doughnut-shaped quaternary structure.

Gap junctions, desmosomes, and tight junctions represent modified areas of the plasma membrane that function in cell adhesion and in cell communication. The chief role of the gap junction is to effect direct cell–cell communication by permitting the passage of low molecular weight molecules between cells. Desmosomes function primarily to keep cells together. Tight junctions block the movement of substances between cells, such as from a tissue lumen to the nonlumen surfaces, and maintain the functional polarity of the cell by keeping transport proteins on their respective apical or basal surfaces.

The molecular bases of some diseases have been traced to that of faulty communication mechanisms. Cholera is an infectious disease that falls into this category and certain kinds of diabetes do as well. In cholera, adenylcyclase is continuously stimulated to make cAMP which, in turn, brings about a loss of salt and water into the intestinal lumen. In certain types of diabetes the number of receptors or the affinity of receptors for insulin is altered.

References

1. On the Physiology of Salivary Secretions, J. N. Langley, *J. Physiol. London* 1:339(1878).
2. *The Collected Papers of Paul Ehrlich,* vol. 2, Pergamon Press, Oxford, 1957.
3. A Certain Symmetry: Histocompatibility Antigens Compared with Immunocyte Receptors, F. M. Burnet, *Nature* 226:123(1979).
4. Cell Surface Receptors, a Biological Perspective, M. F. Greaves, in *Receptors and Recognition,* P. Cuatrecasas and M. F. Greaves, eds., John Wiley & Sons, New York, 1976.
5. *Cyclic AMP,* G. A. Robinson, R. W. Butcher, and E. W. Sutherland, Academic Press, New York, 1971.
6. Short-term Binding and Hydrolysis of Cyclic 3′:5′-Adenosine Monophosphate by Aggregating *Dictyostelium* Cells, D. Malchow and G. Gerisch, *Proc. Nat. Acad. Sci. U.S.* 71:2423(1974).
7. B-Lymphocyte Receptors and Lymphocyte Activities, G. J. V. Nossal, in *International Cell Biology,* B. R. Brinkley and K. R. Porter, eds., Rockefeller University Press, New York, 1977.
8. T-Lymphocyte Receptors and Cell Interactions in the Immune System, D. H. Katz, in *International Cell Biology,* B. R. Brinkley and K. R. Porter, eds., Rockefeller University Press, New York, 1977.
9. Acetylcholine Receptors, M. E. Eldefrawi and A. T. Eldefrawi, in *Receptors and Recognition,* 4 Series A, P. Cuatrecasas and M. F. Greaves, eds., Chapman and Hall, London, 1977.
10. Molecular and Functional Properties of the Acetylcholine Receptor, M. E. Eldefrawi, A. T. Eldefrawi, and A. E. Shamov, *Ann. N.Y. Acad. Sci.* 264:183(1975).
11. The Synthesis of Complex Carbohydrates by Multiglycosyl Transferase Systems and Their Potential Role in Intercellular Adhesion, S. Rosemen, *Chem. Phys. Lipids* 5:270(1970).
12. The Complex Carbohydrates of Mammalian Cell Surfaces and Their Biological Roles, R. C. Hughes, in *Essays in Biochemistry,* vol. 11, P. N. Campbell and W. N. Aldridge, eds., Academic Press, New York, 1975.

13. On the Phenomena of Coalescence and Regeneration in Sponges, H. V. Wilson, *J. Exp. Zool.* 5:245(1907).
14. Characterization of Sponge Aggregation Factor, A Unique Proteoglycan Complex, P. Henkart, S. Humphreys, and T. Humphreys, *Biochemistry* 12:3045(1974).
15. Utilization of L-Glutamine in Intercellular Adhesion: Ascites Tumour and Embryonic Cells, S. B. Oppenheimer, *Exp. Cell Res.* 77:175(1973).
16. Hexagonal Array of Subunits in Intercellular Junctions of the Mouse Heart and Liver, J. P. Revel and M. J. Karnovsky, *J. Cell Biol.* 33:C7(1967).
17. Junctions Between Cells, N. B. Gilula, in *Cell Communication,* R. P. Cox, ed., John Wiley & Sons, New York, 1974.
18. Gap Junction Structures I. Correlated Electron Microscopy and X-ray Diffraction, D. L. D. Caspar, D. A. Goodenough, L. Makowaki, and W. C. Phillips, *J. Cell Biol.* 74:605(1977).
19. Gap Junction Structures. II. Analysis of the X-ray Diffraction Data, L. Makowski, D. L. D. Caspar, W. C. Phillips, and D. A. Goodenough, *J. Cell Biol.* 74:629(1977).
20. Junctional Complexes in Various Epithelia, M. G. Farquhar and G. E. Palade, *J. Cell Biol.* 17:375(1963).
21. On Tight Junction Structure, P. Pinto da Silva and B. Kachar, *Cell* 28:441(1982).
22. Role of Collagenous Matrices in the Adhesion and Growth of Cells, H. K. Kleinman, R. J. Kleve, and G. R. Martin, *J. Cell Biol.* 88:473(1981).
23. Diseases Caused by Impaired Communication Among Cells, E. Rubenstein, *Sci. Am.* 242:102(1980).

Selected Books and Articles

Books

Cell-Cell Recognition, A. Curtis, ed., Cambridge University Press, Cambridge, MA, 1978.

Cell Communication, R. P. Cox, ed., John Wiley & Sons, New York, 1974.

Cell Surface Receptors, G. Nicolson, M. A. Raftery, M. Rodbell, and C. F. Fox, eds., Alan R. Liss, Inc., New York, 1976.

Cell Surface Reviews, vol. 3, Dynamic Aspects of Cell Surface Organization, G. Poste and G. L. Nicolson, eds., North-Holland, Amsterdam, 1977.

Immunology, L. E. Hood, I. L. Weissman, and W. B. Wood, Benjamin/Cummings Publishing Co., Menlo Park, CA, 1978.

Membrane Receptors of Lymphocytes, M. Seligmann, J. L. Preud'homme, and F. M. Komilsky, eds., North-Holland, Amsterdam, 1975.

Receptors and Recognition, P. Cuatrecasas and M. F. Greaves, eds., John Wiley & Sons, New York, 1976.

Articles

Acetylcholine Receptors, M. E. Eldefrawi and A. T. Eldefrawi, in *Receptors and Recognition,* P. Cuatrecasas and M. F. Greaves, eds., Chapman and Hall, London, 1977.

Cell Adhesion, J. G. Edwards, in *Mammalian Cell Membranes,* vol. 4, G. A. Jamieson and D. M. Robinson, eds., Butterworths, London, 1977.

Chemical Signals of Social Amoebae, John Tyler Bonner, *Sci. Am.* 248:114(1983).

Junctions Between Living Cells, L. A. Staehelin and B. E. Hull, *Sci. Am.* 238:140(1978).

On Tight-Junction Structure, P. Pinto da Silva and B. Kachar, *Cell* 28:441(1982).

The Antigenic Nature of Mammalian Cell Surfaces, J. Weiss and J. V. Klavins, in *Mammalian Cell Membranes,* vol. 4, G. A. Jamieson and D. M. Robinson, eds., Butterworths, London, 1977.

The Erythrocyte: Topomolecular Anatomy of MN-Glycoprotein, J. P. Segrest, in *Mammalian Cell Membranes,* vol. 3, G. A. Jamieson and D. M. Robinson, eds., Butterworths, London, 1977.

The Structure and Function of Histocompatibility Antigens, Bruce A. Cunningham, *Sci. Am.* 237:96(1977).

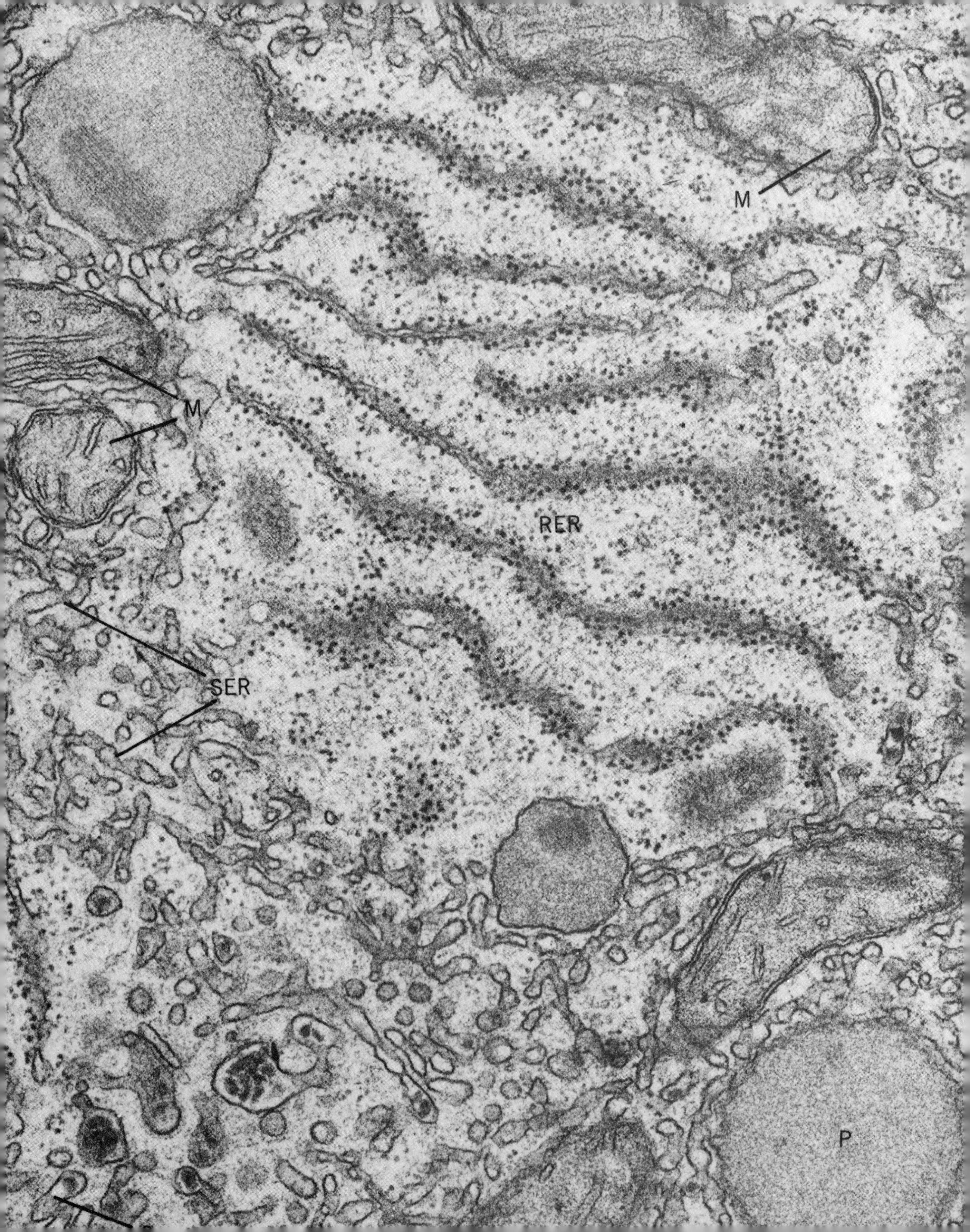
M
M
RER
SER
P

SECTION 3

INTERIOR MEMBRANES: SITES OF SPECIALIZED BIOSYNTHESIS AND OXIDATION

The interior of the eucaryotic cell is permeated by a number of membrane systems possessing a variety of activities and enclosing several different organelles. Chapters 6 and 7 focus on two interior membranes that appear to be related biogenetically, the endoplasmic reticulum and microbodies. Although the endoplasmic reticulum may bind ribosomes and thus have a function in protein synthesis, we are more concerned with the membrane itself, which mediates hydroxylation reactions and is deeply involved in the pathways of sterol biosynthesis. Microbodies possess a group of oxidative enzymes that degrade several different kinds of substrate and function in specialized metabolic pathways.

6

Endoplasmic Reticulum

You can fool some of the enzymes most of the time, you can fool most of the enzymes some of the time, but you cannot fool all of the enzymes all of the time. We can fool many enzymes particularly after mistreating them thus permitting us to study them under unusual circumstances. But can we expect them to behave normally when we poison them with cacodylate buffer or put them to sleep with veronal buffer? Perhaps in the end the enzyme is fooling us.

EFRAIM RACKER, 1976

We now take for granted that the cell is more than just a nucleus floating about in a sea of amorphous, unresolvable cytoplasm. We know that the cell is intercalated and compartmentalized with membrane structures and that no molecule is free to diffuse from one end to the other without weaving through an intricate obstacle course.

But it was not always so. Cell biologists using ordinary light microscopy were unable to see the membrane network that we now call the *endoplasmic reticulum*. In fact, it was not clearly seen until the early 1950s when electron microscopy was gaining widespread use for the study of cell ultrastructure. Several years later, the resolution of the electron microscope was improved to the point that the endoplasmic membranes were seen to be of various forms and dimensions.

The endoplasmic reticulum is not a static, easily defined organelle of the cell. It is rather one morphological component of a greater system of dynamic intracellular membranes, which includes all the membranous organelles that exist within the boundaries of the plasma membrane. This network, sometimes referred to as the *cytocavitary network,* includes the mitochondria, lysosomes, the Golgi complex, microbodies, and the nuclear envelope, all of which are membrane-limited cell components (see Figure 6.1). Together, they form a network within the cell separating it into two compartments, a cytoplasmic and an intracavitary compartment. This means that the membranes that make up the network have one side facing the cytosol and the other facing the lumen of the network. It is important to keep this in mind when we discuss the properties and characteristics of the membranes.

It is generally thought that all the members of the cytocavitary network are *continuous in time,* that is, that there is communication or exchange between certain pairs or neighbors in the network, so that viewed as a whole all members are interacting either directly or indirectly. However, the extent to which the cytocavitary network is *continuous in space* is probably fairly limited. At various times direct continuity has been observed between the endoplasmic reticulum and the plasma membrane or the outer membrane of the nuclcar envelope, but the degree to which all the members of the network are simultaneously interconnected is not known.

Although the endoplasmic reticulum is a part of a greater intracellular membrane system, we shall deal with it as though it were a separate entity

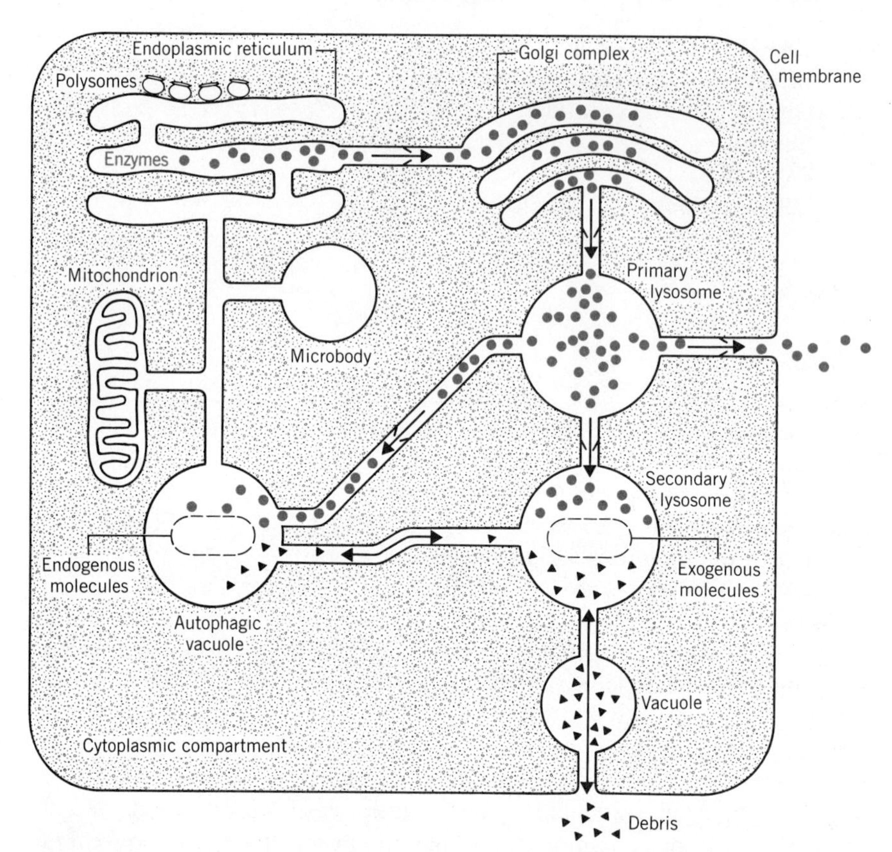

FIGURE 6.1 A conceptualization of the cytocavitary network. This membrane system includes the endoplasmic reticulum and several other membrane-bound organelles. The network divides the cell into two compartments, one continuous with the cell environment (intracavitary compartment) and the other separated from the cell exterior (cytoplasmic compartment).

so that the subject is manageable. It should be borne in mind, however, that both its structure and its function are related to and dependent on other members of the cytocavitary network. As other members are discussed in subsequent chapters, the material of this chapter will serve as a useful foundation.

6.1 GENERAL FEATURES OF THE ENDOPLASMIC RETICULUM

Were it not for the presence of a dynamic membrane system, the interior of the eucaryotic cell would be most difficult to reach. Substrates would not be able to diffuse to their enzymes as rapidly as needed, and waste products and essential building blocks would accumulate at useless or toxic concentrations. The heart of the cell would be little better than a stagnant cesspool. This is not a significant problem for the procaryotic cell in which the distances are short enough that diffusion can meet the metabolic demands. But distance is a major consideration in the eucaryotic cell.

The cytocavitary network in general, and the endoplasmic reticulum in

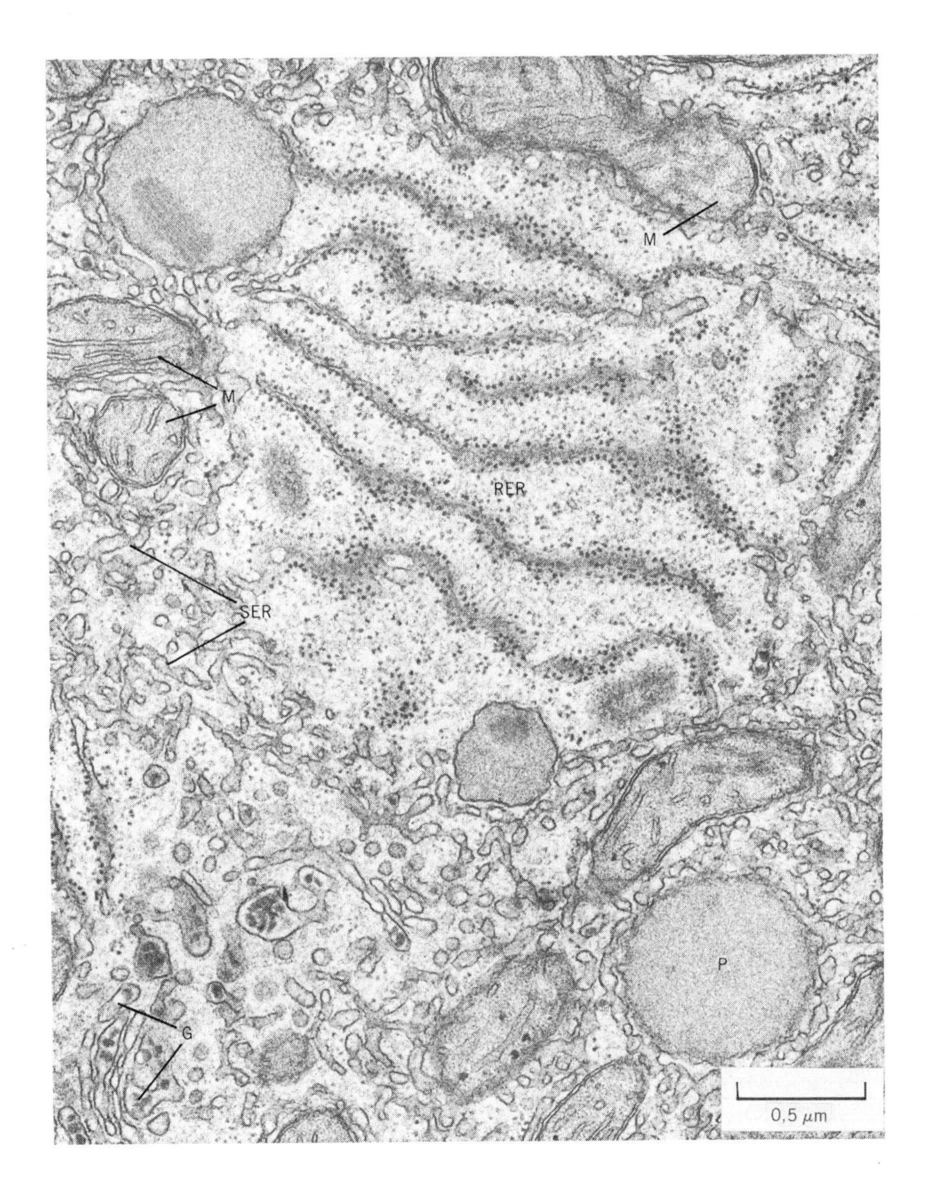

FIGURE 6.2 An area of hepatocytic cytoplasm illustrating morphological differences in endoplasmic reticulum. In between the mitochondria (M) and the peroxisome (P) there reside both rough (RER) and smooth (SER) endoplasmic reticula. The membranes of the Golgi complex (G) are also smooth. ×34,400 (Courtesy of Dr. Robert Bolender.)

particular, prevent stagnation from occurring in that they form an intracellular circulatory system and a widely dispersed network of enzymes for catabolic and anabolic activities. Vital substrates are able to reach the cell interior rapidly by membrane fusions and movements, and materials that are synthesized and assembled within the depths of the cell can be rapidly transported to its surface.

Microscopy of the Endoplasmic Reticulum

Some aspects of the functional diversity of the endoplasmic reticulum are also apparent as morphological differences, as illustrated in the electron micrograph of Figure 6.2. Here one can view in a small region of a single hepatocyte the two major forms of membranes making up the endoplasmic reticulum. Membranes that have a rough appearance, because they are

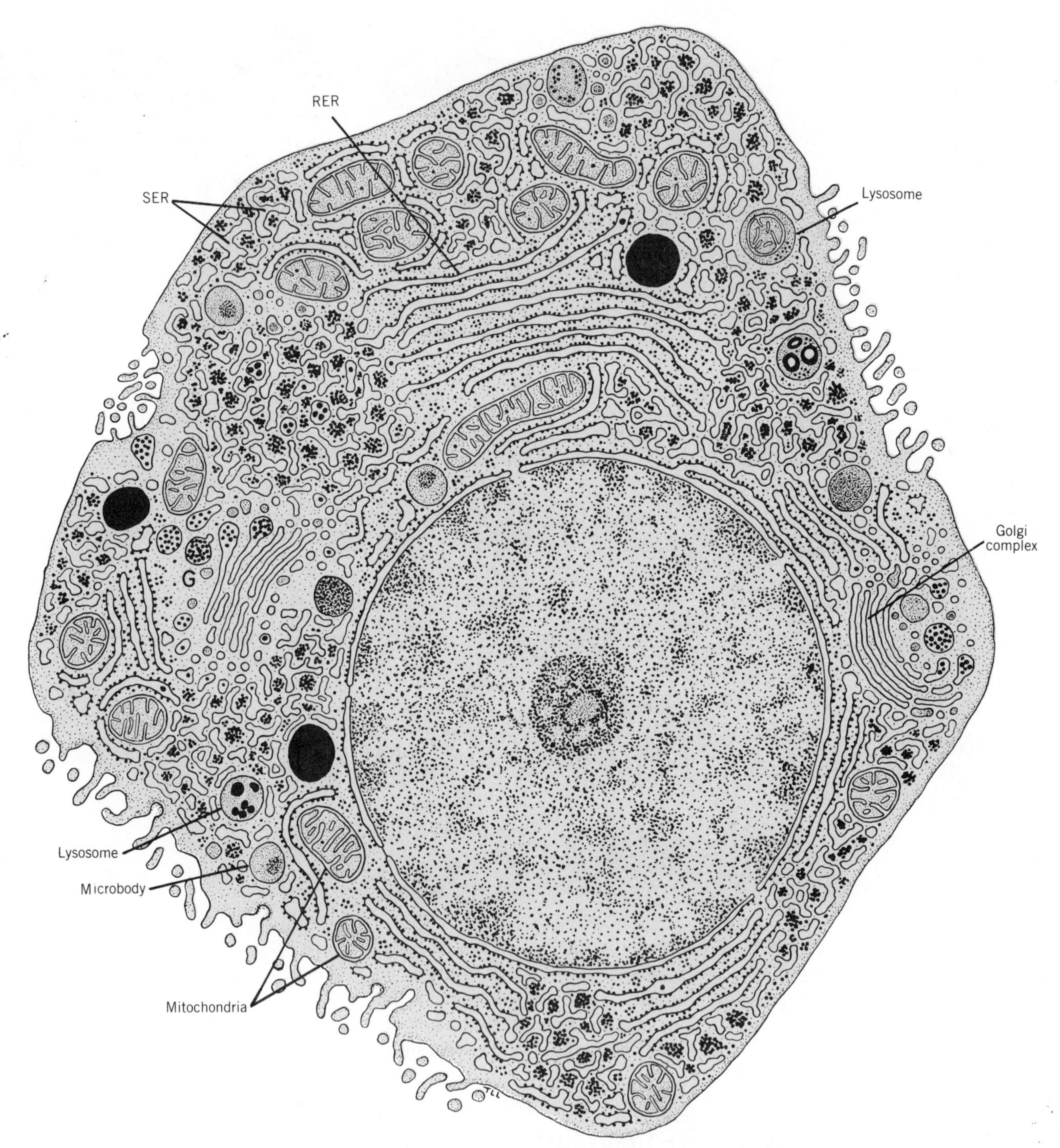

FIGURE 6.3 An idealized drawing of the hepatocyte showing both RER and SER and the variety of organelles typical of this versatile metabolic factory.

coated with ribosome particles, are called *rough endoplasmic reticulum* (RER) and membranes lacking these particles are termed *smooth endoplasmic reticulum* (SER).

Rough and Smooth Endoplasmic Reticula

The RER is the site of protein synthesis in the cell, in particular, the synthesis of secretory proteins and protein constituents of the endoplasmic reticulum itself, whereas the SER mediates a variety of other synthetic reactions as well as chemical modifications of low molecular weight materials.

The presence of both RER and SER in the hepatocyte is reflective of the multifaceted role played by the liver in metabolism. Virtually all the reactions of intermediary metabolism take place in the liver, and some of the reactions are divided up between the two types of membrane. For a better overall picture of the hepatocyte, it might be useful to compare the electron micrograph of Figure 6.2 with the idealized drawing of Figure 6.3. Note in particular the presence of both RER and SER.

Cells of different types contain different ratios of rough and smooth endoplasmic reticulum. Cells that are heavily involved in the synthesis of secretory proteins, such as the pancreatic acinar or plasma cells, have a predominance of RER (see Figure 6.4). Their main task is to assemble and export proteins, and their endoplasmic reticulum is exceptionally well geared up for this. Cells that are primarily concerned with steroid production, such as those making up the tissue of the adrenal cortex, contain endoplasmic reticulum that is predominantly in the smooth form. The synthesis of cholesterol takes place in these membranes as well as some of the reactions that modify the steroids to form progesterone and deoxycorticosterone. Other cells with abundant SER are retinal pigment cells, sebaceous cells, the interstitial cells of the testis, and the corpus luteum (Figure 6.5).

Lamellar, Vesicular, and Tubular Forms of Endoplasmic Reticulum

Besides existing as RER and SER, the endoplasmic reticulum is commonly present in three different physical forms in the cell. The RER is most frequently seen in the *lamellar* form. This form is not simply a stack of membrane sheets, but rather collections of flattened membrane sacks. One feature of the lamellar form that reveals this is the asymmetric attachment of ribosomes to these membranes which enclose spaces called *cisternae*. The cisternae are components of one of the two major compartments of the cell, being in a sense an outside compartment within the cell. This is most easily envisaged by picturing the results of continuity of the endoplasmic reticulum with the plasma membrane. In this type of continuity, the environment around the cell would be continuous with the cisternae.

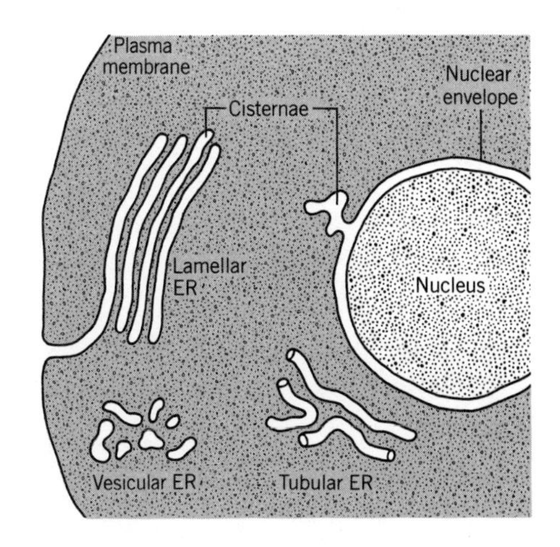

A second form of endoplasmic reticulum is *vesicular*. This form is most common among SER membranes and can be readily seen in electron micrographs of many types of cells bearing SER. Membranes have natural tendencies to form closed vesicles *in vitro*, and their appearance as vesicles in the cell probably reflects the *in vivo* counterpart of this property.

The third form of endoplasmic reticulum, *tubular*, probably best dem-

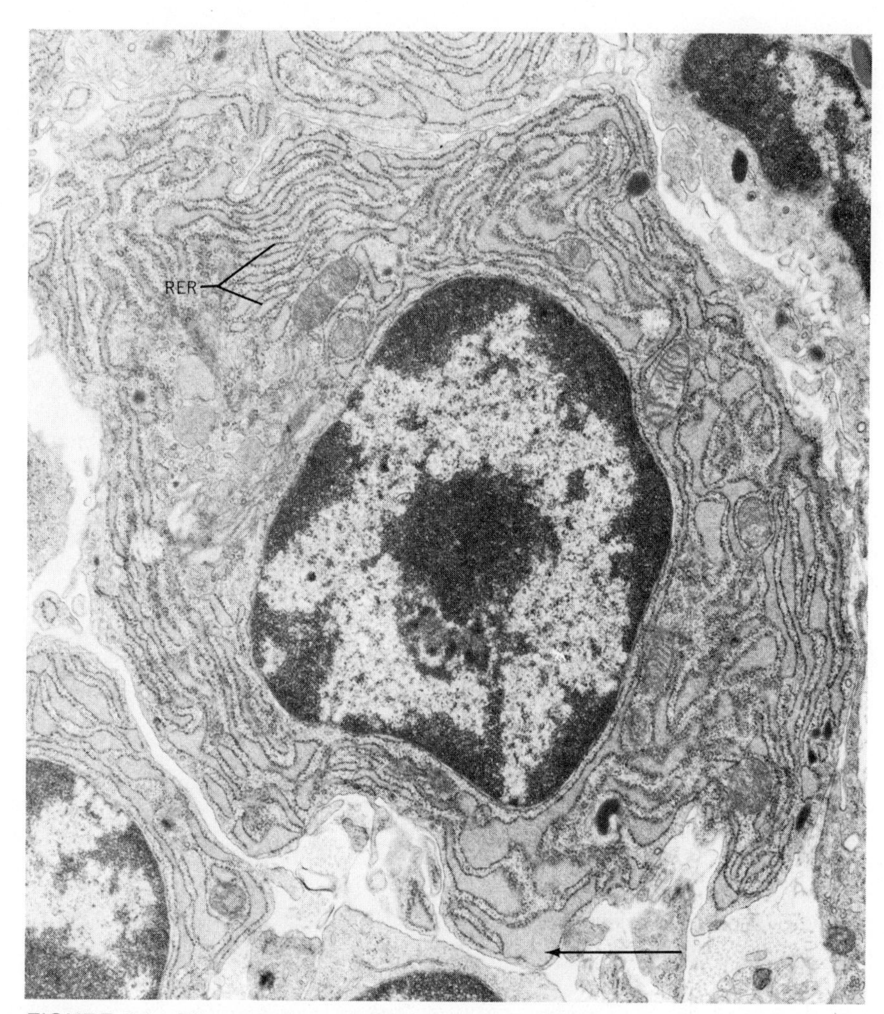

FIGURE 6.4 Electron micrograph of a plasma cell. Plasma cells are geared up for the production of antibodies and therefore contain endoplasmic reticulum predominantly in the rough form. The protein that is produced is stored temporarily in the cisternae of the endoplasmic reticulum (arrow) before it is secreted from the cell into the bloodstream. (Courtesy of Dr. Stanley Erlandsen.)

onstrates the dynamic nature of the endoplasmic reticulum. It is primarily a form of the SER, and may be associated with membrane movements, fission, and fusion within the cytocavitary network.

Isolation: The Microsomal Fraction

Methods II and III

The membranes of the endoplasmic reticulum can be isolated by subjecting homogenized tissues to differential centrifugation. They are spun down as a pellet after the mitochondria have been removed. Electron microscopy of ER preparations obtained in this manner reveal that the membranes form closed vesicles of either a rough or a smooth form as illustrated in the electron micrograph of Figure 6.6. These membranous entities were coined "microsomes" by Claude[1] in 1940, and the relation-

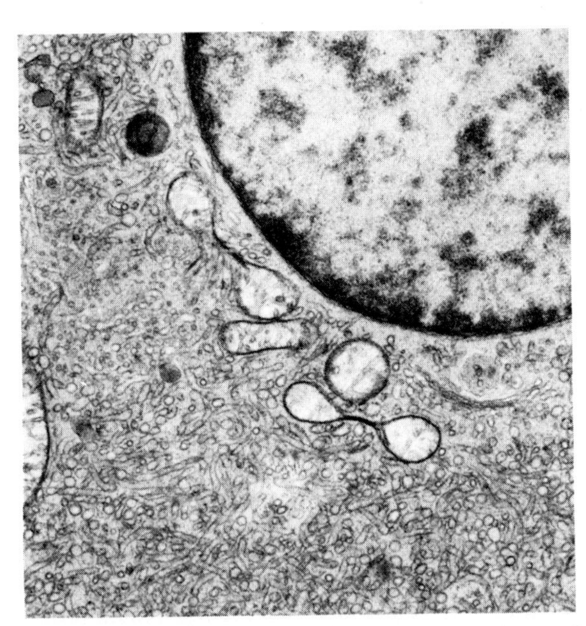

FIGURE 6.5 The interstitial cell of the testis. The endoplasmic reticulum appears exclusively in the smooth form. ER membranes make up an interconnected network of vesicles and tubules that permeate the cytoplasm. (Courtesy of Dr. Don W. Fawcett.)

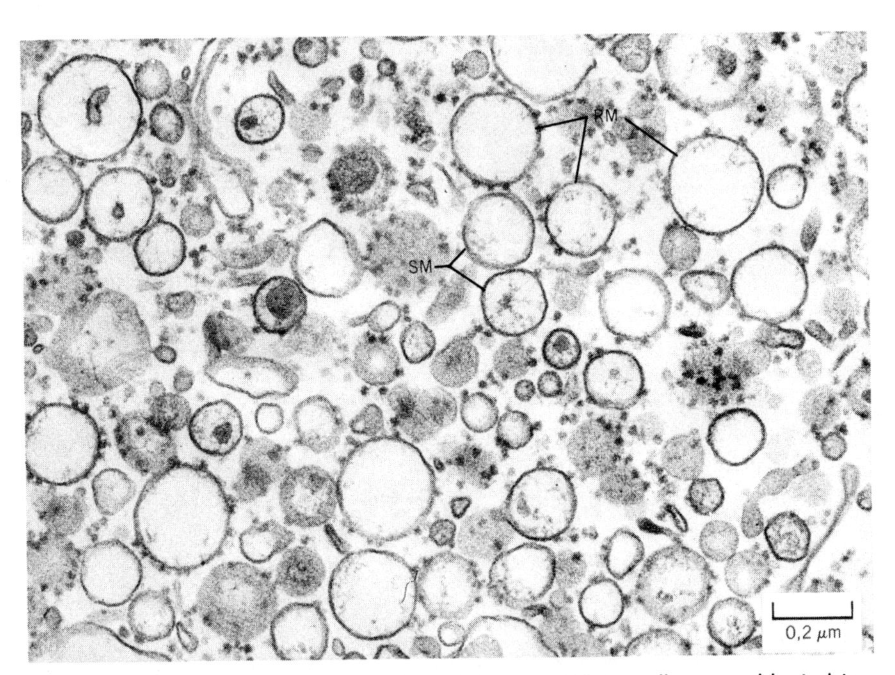

FIGURE 6.6 The microsomal fraction of the cell. When cells are subjected to homogenization, the endoplasmic reticulum is fragmented and appears after centrifugation as a collection of vesicles. Those membranes derive from RER are coated with ribosomes (RM) and those from SER have smooth surfaces (SM). Thus microsomes are artifacts never seen within a cell, only *in vitro*. (Courtesy of Dr. Robert Bolender.)

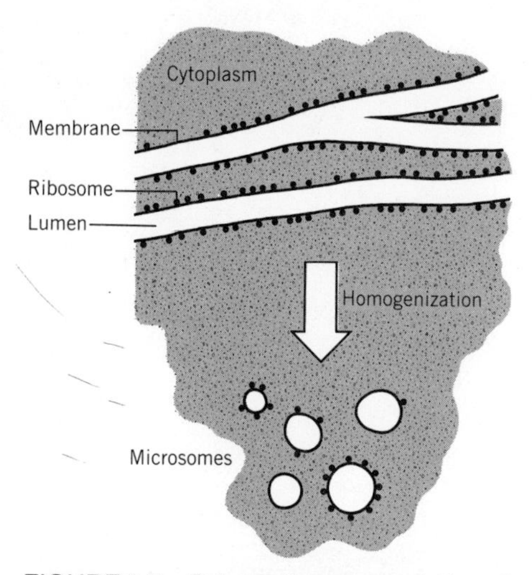

FIGURE 6.7 Schematic representation of what happens to the membranes of the endoplasmic reticulum when subjected to homogenization. Microsomes are vesicular pieces of the original endoplasmic reticulum.

ship between microsomes and the elements of the endoplasmic reticulum in the intact cell was established by Palade and Siekevitz in 1956.[2]

It is apparent from Figure 6.6 that the endoplasmic reticulum is severely disrupted from its native form during homogenization and centrifugation and is finally retrieved as a collection of membrane vesicles. The formation of vesicles rather than sheets or tubules represents a movement of the isolated membranes to their lowest free energy state.

The vesicles derived from portions of the RER appear with ribosomes coating their outer surfaces, giving them a studded appearance. This appearance and a number of other experimental results have led to the conclusion that essentially all microsomal vesicles have the same inside–outside orientation such that the outside surface corresponds to the cytoplasmic surface in the intact cell (Figure 6.7). As we shall see later, this property has been most useful for studying the topography of the molecules in the ER membrane.

It should be apparent that microsomes are not true cell organelles but are, in reality, artifacts of the isolation procedure. The term is therefore a functional term. We use "microsomes" when referring to *isolated* endoplasmic reticulum and confine the use of the term "endoplasmic reticulum" to the membrane structures as they would exist in the intact cell.

6.2 STRUCTURE AND COMPOSITION OF THE ENDOPLASMIC RETICULUM

Physical Structure

The general physical features of the membranes of the endoplasmic reticulum seem to fit the fluid mosaic model of membrane structure. They do appear to differ, however, from the plasma membrane in some of their finer details.

The results of electron microscopy of thin sections through the endoplasmic reticulum suggest that the transverse dimensions of its membranes are less than those of the plasma membrane, on the order of 5 nm in osmium-fixed material compared to 8 to 10 nm for the plasma membrane. Figure 6.8 illustrates this in schematic form. This difference is presumably a function of composition, the endoplasmic reticulum differing from the plasma membrane by having a higher ratio of protein to lipid and a lower concentration of cholesterol. More protein gives the endoplasmic reticulum greater structural stability than plasma membranes, perhaps by immobilizing a greater proportion of the amphipathic lipid molecules. The ER membrane is thus less fluid in its properties.

A trilaminar–electron-dense–trilaminar sequence can be observed by electron microscopy of ER.

This may be due to micellar bridges.

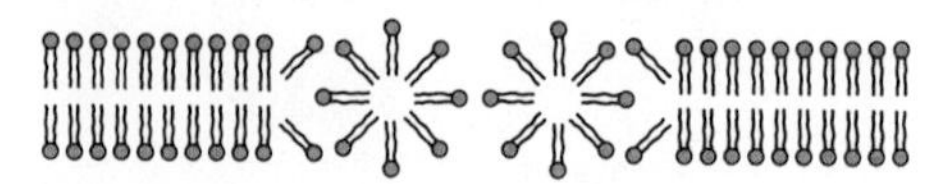

Electron microscopy has also revealed that the endoplasmic reticulum may not exist solely in the trilaminar form. There is evidence for the appearance of electron-dense stretches in the membrane of one layer only, suggesting the presence of lipoprotein micelles that bridge trilaminar regions. The significance of this is not clear, but there is speculation that these micellar regions may be dynamic points in the system that facilitate membrane fission and fusion activities.

Table 6.1 Transverse Distribution of Various Enzymes of the Endoplasmic Reticulum

Enzyme	Surface Localization
Cytochrome b_5	Cytoplasm
NADH-cytochrome b_5 reductase	Cytoplasm
NADPH-cytochrome *c* reductase	Cytoplasm
Cytochrome P-450 (most abundant)	Cytoplasm
	Lumen
ATPase	Cytoplasm
5′-Nucleotidase	Cytoplasm
Nucleoside pyrophosphatase	Cytoplasm
GDP-mannosyl transferase	Cytoplasm
Nucleoside diphosphatase	Lumen
Glucose-6-phosphatase (enzyme marker)	Lumen
Acetanilide-hydrolyzing esterase	Lumen
β-Glucuronidase	Lumen

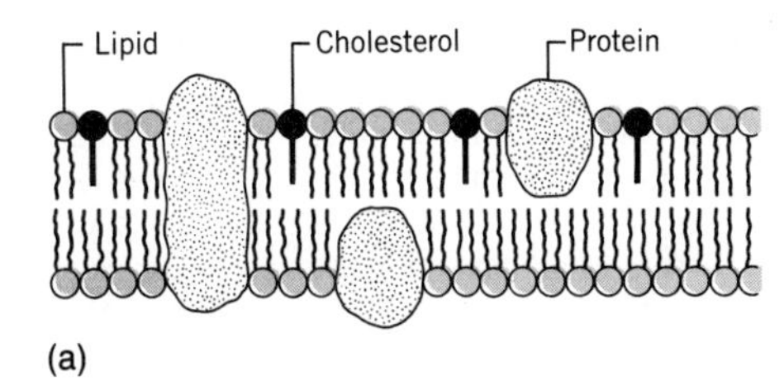

(a)

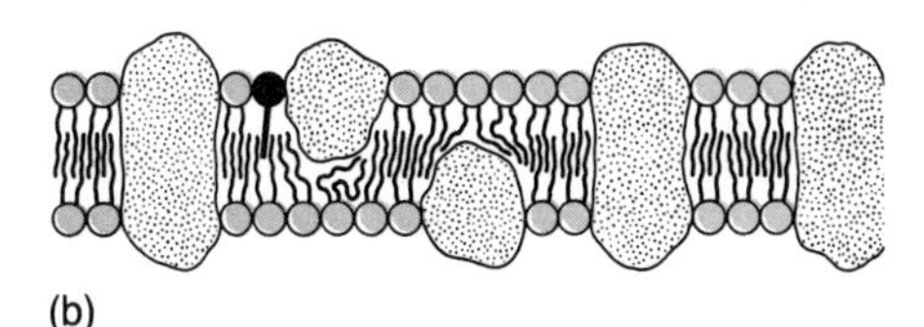

(b)

FIGURE 6.8 Schematic representation of the plasma membrane (*a*) and endoplasmic reticulum (*b*). ER membranes are thinner, more stable, and less fluid than plasma membranes.

Chemical Composition

In rat liver, the membranes of microsomes are 60 to 70% protein and 30 to 40% phospholipid by weight. At least 33 polypeptides of different physical and chemical properties have been observed in these membranes, and the distribution of microsomal phospholipids is approximately as follows: phosphatidylcholine (55%), phosphatidylethanolamine (20–25%), phosphatidylserine (5–10%), phosphatidylinositol (5–10%), and sphingomyelin (4–7%).[3]

These values vary a good deal in the literature, but in general it is fair to say that ER membranes contain more proteins than plasma membranes and that they are richer in phosphatidylcholine and poorer in sphingomyelin. Table 4.1 (see p. 121) contains comparative information on lipid compositions for the membranes of the cytocavitary network.

Enzyme Constituents

The endoplasmic reticulum is the cellular site of a large number of enzymes. Probably no class of enzymes is as difficult to study as membrane enzymes, especially if they are integral membrane proteins. They are normally hydrophobic and are isolated and purified only with the aid of detergents. It is thus not always clear that the behavior of the enzymes *in vitro* is a faithful reproduction of their activities *in situ*. It is in this sense that membrane enzymes are prone to fool researchers. Nevertheless, several of these enzymes have been studied in considerable detail, and some of the commonly observed ones are listed in Table 6.1.

Glucose-6-phosphatase is generally made use of as the enzyme marker for ER membranes. We will discuss its role when we deal with carbohydrate metabolism in the endoplasmic reticulum.

The most abundant enzyme is cytochrome *P*-450. It occurs at levels as high as 10% of microsomal protein. This cytochrome is one member of an endoplasmic reticulum electron transport chain that includes NADPH-

Nicotinamide adenine dinucleotide (NAD) and nicotinamide adenine dinucleotide phosphate (NADP) are coenzymes or electron carriers that exist in oxidized (NAD^+ and $NADP^+$) and reduced (NADH and NADPH) forms, the latter state occurring when the coenzymes are carrying electrons.

cytochrome *P*-450 reductase and phosphatidylcholine. This complex is involved in hydroxylation reactions that we discuss in more detail in Section 6.4. Another important membrane enzyme that participates in certain aspects of this electron transport is cytochrome b_5. It also has a reductase, an NADH-containing enzyme.

There are many more ER enzymes known that we cannot discuss within the scope of this text. Many of them have biosynthetic functions, as in the synthesis of the ether-linked glycerolipids, a complex group of membrane lipids. Others have catabolic roles and are concerned with the breakdown of membrane lipids. Together, these two classes of enzymes regulate membrane lipid homeostasis and turnover.

Topography of Enzymes in the Lateral Plane

Microsomal vesicles are heterogeneous with respect to size, density, and surface charge, which permits their subfractionation into groups according to these properties. The smaller the vesicles, the greater the degree of heterogeneity. This is not especially surprising, since one would predict that smaller vesicles would have room for fewer enzymes, to the point where statistically one would expect to find vesicles deficient in certain of the enzyme components even if the enzymes were randomly distributed in the membrane. If, on the other hand, there existed a marked degree of lateral association of certain groups of enzymes, one may be able to observe this as a limiting effect on vesicular heterogeneity.

The evidence so far indicates that quantitative but not qualitative differences exist in the enzyme makeup of vesicles. Thus, for example, all vesicles appear to contain glucose-6-phosphatase, cytochrome b_5, and NADPH-cytochrome *P*-450 reductase.[3] However, essentially all the enzymes that have been studied for distribution characteristics using the subfractionation approach have shown heterogeneous distributions. Although all enzymes are present in each vesicle, different vesicles may contain the enzymes in different ratios.

The significance of this is not yet certain, but there are several possible explanations. The heterogeneity may reflect a restricted mobility of the proteins in the bimolecular lipid layer or may reveal a degree of structural interaction essential for their enzymatic activities. Alternatively, the heterogeneity may be due to variations in properties of the endoplasmic reticulum in a single cell or in different cells in a single tissue or organ, rather than heterogeneity within a continuous span of ER membrane.

Topography of Enzymes in the Transverse Plane

Table 6.1 includes the topographical localization of various ER enzymes. The results are somewhat clearer than for the studies on lateral distributions. All the enzymes listed, except for cytochrome *P*-450, are asymmetrically distributed across the membrane, preferring either the luminal (cisternal) or the cytoplasmic side of the membrane. Cytochrome *P*-450 appears to be a transmembrane protein. These positions of the enzymes within the membrane are undoubtedly related to their catalytic functions.

Detailed studies on cytochrome b_5 have provided some remarkable insight with respect to the topographical properties of this enzyme. Cytochrome b_5 is an integral membrane protein (molecular weight 11,000 daltons) that is skewed toward the cytoplasmic surface of the endoplasmic reticulum. The results of studies in a number of laboratories have led to the conclusion that the enzyme has a hydrophilic head containing its catalytically active site and a hydrophobic tail that is not catalytically active. The hydrophobic tail is thought to be dissolved in the lipid bilayer of the membrane, thus serving as an anchoring point, and the catalytic head is free to associate with the more hydrophilic surface regions of the membrane where it interacts with appropriate substrate molecules.

There is also evidence that NADH-cytochrome b_5 reductase, the enzyme that supplies electrons to cytochrome b_5, is situated topographically in the membrane in a similar way. The two enzymes may therefore interact physically in carrying out their catalytic functions.

Cytochrome b_5 has topographical properties such that its catalytic head is exposed to the cytoplasmic surface.

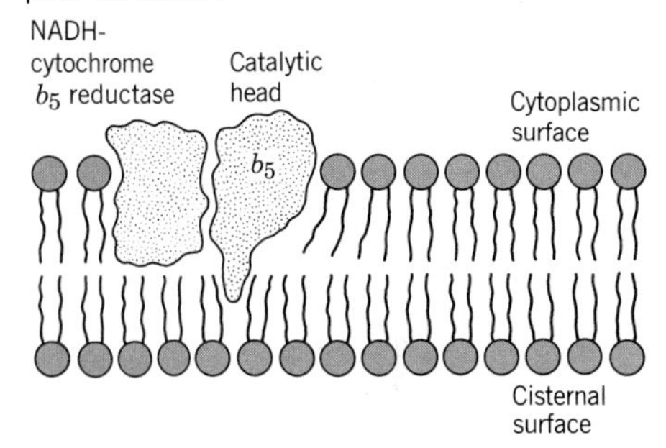

The topographical positions of these two enzymes in the membrane of the endoplasmic reticulum is reminiscent of glycophorin in the plasma membrane, as discussed in Chapter 5. Although this type of interaction of integral membrane proteins with both the hydrophobic and hydrophilic portions of the membrane may be common, it apparently is not the rule. Certain other membrane proteins display less amphipathic properties, and may be more, or completely, immersed in the hydrophobic region of the membrane.

6.3 RELATION BETWEEN ROUGH AND SMOOTH ENDOPLASMIC RETICULA

As already indicated, there is a functional difference between the smooth and rough endoplasmic reticula. The SER is involved in the synthesis and metabolism of certain small molecules and cellular detoxification mechanisms, whereas the RER is engaged in the synthesis and transport of proteins. In the latter case the endoplasmic reticulum serves as a physical support for polysomes in addition to possessing enzymatic functions itself.

Rough Endoplasmic Reticulum Gives Rise to Smooth Endoplasmic Reticulum

The relation between the smooth and the rough forms of the endoplasmic reticulum is a research problem that has long been pursued. The two morphologically different forms are contiguous in the cell, at least intermittently, and one form, the SER, is probably derived from the other. As a case in point, phenobarbital stimulates the proliferation of ER membranes. This appears to occur, however, by a two-stage process. First, there is a proliferation of the RER, and later the SER proliferates. Since it is the function of the SER to deal with the phenobarbital by detoxification reactions, it must be first formed by synthesis of the appropriate detoxifying enzymes, which are incorporated into membranes by bound ribosomes. Hence the appearance of RER first, followed by increases in the SER.

Thus, the RER and the SER are functionally related but apparently not identical. A good number of studies have demonstrated that rough and

FIGURE 6.9 Freeze-fracture image of polysome imprints in RER of *Micrasterias*. The 11 nm particles are regularly spaced in a position peripheral to the imprints of the polysome spirals. ×76,300 (Courtesy of Dr. Thomas H. Giddings, Jr.)

smooth microsomes, which are separable by density gradient techniques, have many membrane proteins in common. Nevertheless, many investigators have reported qualitative differences between rough microsomes and smooth microsomes, and there is now good evidence for the presence of at least two membrane proteins that are characteristic of rough microsomes and appear to function as binding sites for ribosomes.[4] These proteins, called *ribophorins*, have molecular weights of 63,000 and 65,000 daltons. It is speculated that they may not only function to bind ribosomes to the membrane, but may form a scaffolding network throughout the membrane interconnecting ribosome binding sites. This filamentous network may confer on RER its characteristic lamellar form, which tends to be more rigid and stable than the vesicular and tubular forms of the SER.

Ribosome Binding Sites on the Endoplasmic Reticulum

Protein synthesis is carried out both on free ribosomes and on ribosomes bound to the endoplasmic reticulum. Free ribosomes produce the soluble proteins of the cytosol, whereas bound ribosomes translate secretory proteins and integral membrane proteins. In animal cells ribosomes commonly line up on a messenger RNA forming polysomes, which may also be free or bound to endoplasmic reticulum when producing proteins.

When ribosomes are bound to endoplasmic reticulum, they adhere through their large subunits. It is generally accepted that ribophorins I and II, the two integral transmembrane proteins, are the proteins to which the ribosomes bind and thereby are attached to the membrane.

Method I

Using freeze-fracture techniques 11-nm particles have been observed in ER membranes that appear to be the structural equivalents of ribosome binding sites.[5] It is thought that each particle is made up of one molecule each of ribophorins I and II plus a nascent polypeptide chain.

Figure 6.9 is an electron micrograph of a polysome imprint on RER membranes. In general, there is one 11-nm particle for each ribosome imprint in a polysome and this particle is located peripherally to the imprint. It would therefore appear as though the ribosome large subunit is anchored to the integral membrane particle not directly in the center of the subunit but rather to one side.

The ribosome is apparently bound to the endoplasmic reticulum by its edge.

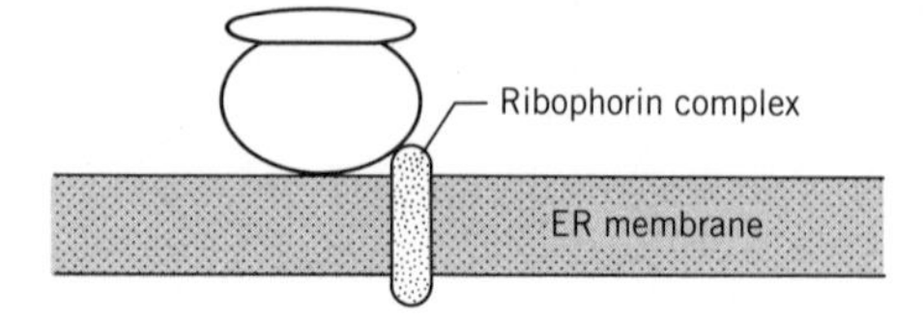

The relation between ribophorins and the particles viewed by electron microscopy is not certain at the moment, but it is believed that they are one and the same. Thus we may be viewing the protein complex responsible for binding ribosomes to RER.

SER and RER are quite certainly functionally and structurally related, yet different in certain of their fundamental properties. It is anticipated that research being presently conducted will shed a great deal of light on these differences and similarities during the next few years.

Although it would be appropriate to discuss protein synthesis in this chapter, we will deal with the ultrastructure and molecular architecture of this process in the chapters on the secretory pathway (Chapter 10) and the ribosome (Chapter 17).

6.4 THE HYDROXYLATION SYSTEM OF THE ENDOPLASMIC RETICULUM

The ability of the membranes of the endoplasmic reticulum to hydroxylate substrates provides for the cell both an anabolic and a protective function.

Table 6.2 Some Hydroxylation Reactions Catalyzed by Cytochrome P-450

Aromatic hydroxylation

$$C_6H_5\text{—}NHCOCH_3 \xrightarrow{[OH]} HO\text{—}C_6H_4\text{—}NHCOCH_3$$

Aliphatic hydroxylation

$$R\text{—}CH_3 \xrightarrow{[OH]} R\text{—}CH_2\text{—}OH + H^+$$

***N*-Dealkylation**

$$R\text{—}NH\text{—}CH_3 \xrightarrow{[OH]} [R\text{—}NH\text{—}CH_2OH] \longrightarrow RNH_2 + CH_2O \text{ (formaldehyde)}$$

***O*-Dealkylation**

$$R\text{—}O\text{—}CH_3 \xrightarrow{[OH]} [R\text{—}O\text{—}CH_2OH] \longrightarrow ROH + CH_2O \text{ (formaldehyde)}$$

Deamination

$$R\text{—}CH(NH_2)\text{—}CH_3 \xrightarrow{[OH]} [R\text{—}C(OH)(NH_2)\text{—}CH_3] \longrightarrow R\text{—}CO\text{—}CH_3 + NH_3$$

Sulfoxidation

$$R\text{—}S\text{—}R' \xrightarrow{[OH]} [R\text{—}SOH\text{—}R']^+ \longrightarrow R\text{—}SO\text{—}R' + H^+$$

***N*-Oxidation**

$$(CH_3)_3N \xrightarrow{[OH]} [(CH_3)_3NOH]^+ \longrightarrow (CH_3)_3NO + H^+$$

Anabolically, the endoplasmic reticulum is involved in the synthesis of cholesterol, steroid hormones, and bile acids. In these reactions, the ER often collaborates with mitochondria, which contain certain key enzymes in the metabolic pathways.

Protectively, the endoplasmic reticulum chemically modifies xenobiotics, toxic materials of both endogenous and exogenous origin, making them in general more hydrophilic, hence more readily excreted. Among these materials are drugs, insecticides, anesthetics, petroleum products, and carcinogens. The endoplasmic reticulum is thus the chief detoxification site in the cell.

Table 6.2 contains some of the hydroxylation reactions that are mediated by the enzymes of endoplasmic reticulum. These are all catalyzed by cytochrome *P*-450.

The Mixed-Function Oxidase System

The hydroxylation reactions carried out by the enzymes of the endoplasmic reticulum are ones in which molecular oxygen is divided between a substrate and water with the aid of an electron-donating carrier:

$$NADPH + H^+ + RH + O{:}O \rightarrow NADP^+ + ROH + H_2O$$

The enzymes catalyzing these reactions are referred to as *mixed-function oxidases* or as *monooxygenases*.

At the heart of the reaction is an electron transport chain of the en-

The term "mixed-function oxidase" indicates that both a substrate and a coenzyme (NADPH) are oxidized. The term "monooxygenase" reflects the transfer to a substrate of one atom of an oxygen molecule to effect a hydroxylation.

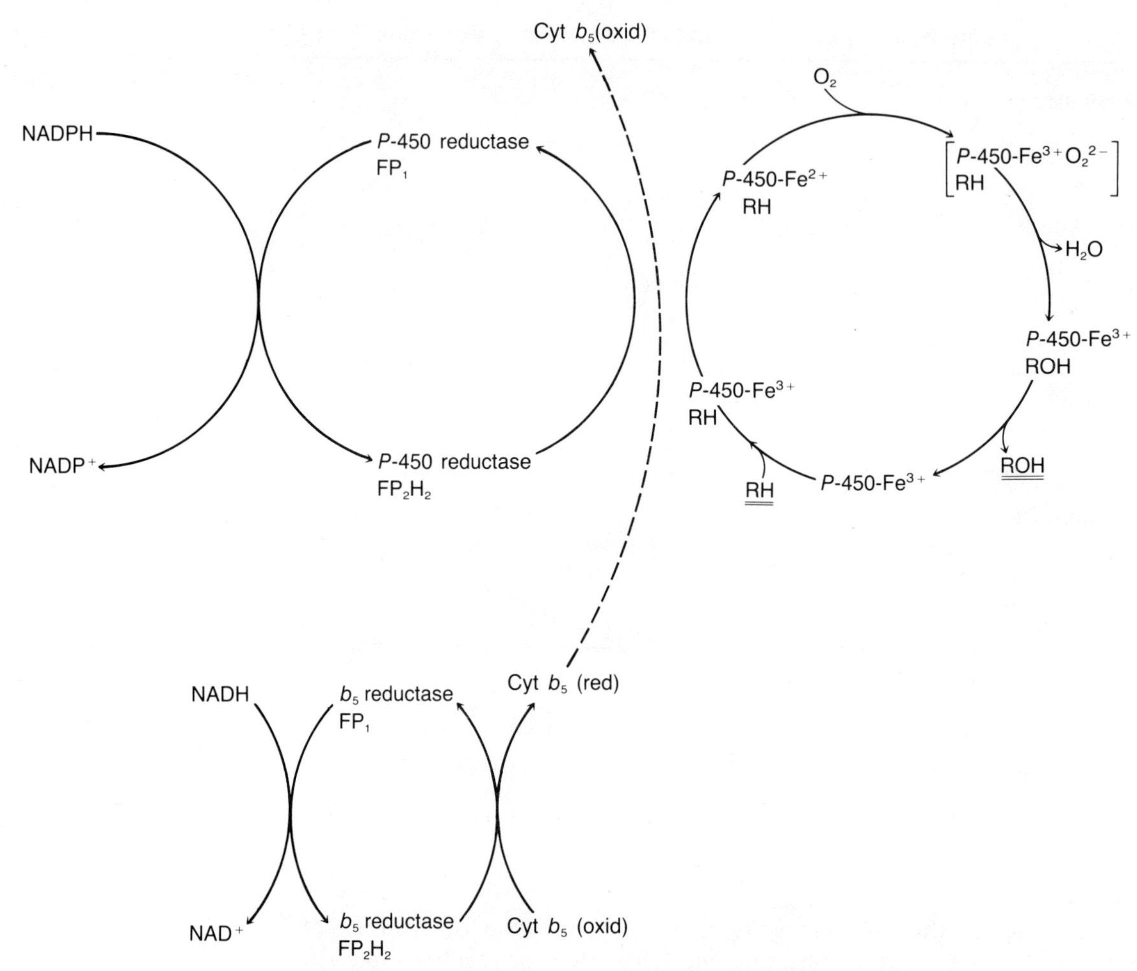

FIGURE 6.10 Main features of the reactions of the electron transport chain in the endoplasmic reticulum. Cytochrome *P*-450 with its iron in the Fe^{3+} state binds to substrate (RH) and the complex is reduced to Fe^{2+}. This reduction step is mediated by the flavoprotein NADPH-cytochrome *P*-450 reductase. After reaching the reduced state, *P*-450 binds atmospheric oxygen, half of it being used to reoxidize the iron and half of it appearing in the oxidized substrate.

doplasmic reticulum that consists of at least two proteins: cytochrome *P*-450 and NADPH-cytochrome *P*-450 reductase. The two enzymes, together with the electron donor NADPH, are proposed to function in the concerted manner depicted in Figure 6.10. A second electron transport chain consisting of cytochrome b_5 and NADH-cytochrome b_5 reductase is also shown to be involved in this cycle, although it may operate somewhat peripherally to the main reaction sequence. The two electron transport chains can be separated and can function independently *in vitro*.

Based on the stoichiometric relationships of the members of these electron transport chains and their relative accessibilities to the membrane surfaces, the cluster model of the monooxygenases has been proposed as shown in Figure 6.11. In this model, the cytochrome *P*-450 reductase is depicted as a core around which several *P*-450s are clustered. This complex, in turn, interacts with the b_5 reductase and its cytochromes. This model may not adequately take into account the amphipathic nature of the individual members and the steric problems of putting together so many large molecules, but it does help us to envision how the system may operate in a concerted manner.

As complex as this model may appear, the *in vivo* situation is probably even more complicated. The actual ratios of cytochromes to reductases, for both electron transport chains, may be as high as 20 to 30. Furthermore, it is known that cytochrome *P*-450 exists in multiple forms in the endoplasmic reticulum and that the different forms show variations in their catalytic abilities.[6]

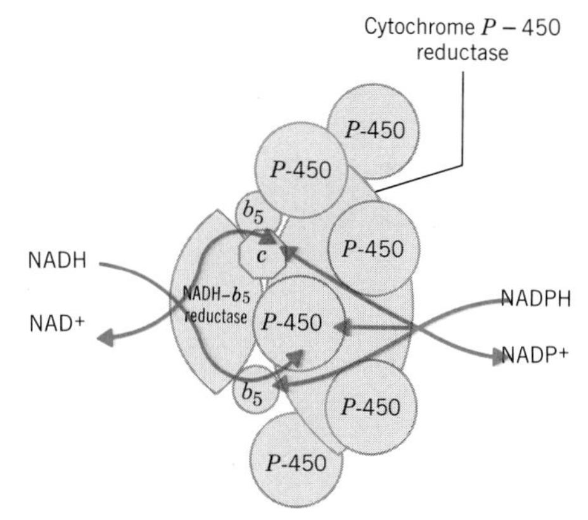

FIGURE 6.11 The cluster model of monooxygenases. Several cytochrome *P*-450s cluster around a *P*-450 reductase, which interacts with the b_5 reductase and its cytochromes. This side view cross section of one complex in the membrane reflects no consideration for membrane asymmetry.

Induction of Enzymes of the Endoplasmic Reticulum

A wide variety of drugs, when administered to animals, will bring about the proliferation of the ER membranes and/or enzyme activities of the liver cell. For example, compounds like 3-methylcholanthrene, β-naphthoflavone, 2,3,7,8-tetrachlorodibenzo-*p*-dioxin, and phenobarbital are commonly used to induce the enzymes of the endoplasmic reticulum.

When phenobarbital is used as an inducer, enhanced enzyme activities are observed first in the RER, and after about 6 hr following administration of the drug the activity increases in the SER.[7] It thus appears that the drug induces the synthesis of new enzymes, which are observable first as they are being translated on ribosomes and later when they are incorporated into new smooth membranes. Experimental verification for the synthesis of new enzymes is found in the fact that actinomycin D inhibits the enhancement partially and puromycin abolishes it.[8]

Phenobarbital, a depressant of the central nervous system, induces enzyme activity in the endoplasmic reticulum.

Drug induction of the endoplasmic reticulum stimulates several biosynthetic processes. The biosynthesis of cholesterol is stimulated as well as the production of phospholipid. One would expect the formation of the latter to be prerequisite to membrane proliferation, and indeed phospholipid synthesis occurs early, preceding the synthesis of new protein. Components of the monooxygenase system are also enhanced, in particular cytochrome *P*-450 and NADPH-cytochrome *P*-450 reductase. Cytochrome *P*-450 may exhibit as much as a two- to fivefold increase upon induction.

Actinomycin D binds to DNA and inhibits transcription. Thus its effects would not be immediate, since RNA and enzymes previously formed would continue their template and catalytic activities. Puromycin inhibits protein synthesis by causing the release of proteins from ribosomes before their synthesis is complete. Therefore the effect is much more rapid.

Membranes that are proliferated in this manner persist in the cell for several days and are then gradually removed, apparently by the formation of autophagosomes (see Chapter 9, "The Lysosome").

An interesting result of ER membrane proliferation is that it does not bring about a wholesale enhancement of all the ER enzymes. For example, phenobarbital markedly increases the ER content of cytochrome *P*-450 and its reductase, but cytochrome b_5 is enhanced only slightly, and enzymes such as glucose-6-phosphatase, ATPase, and NADH-cytochrome b_5 reductase actually show a reduction of specific activity. Thus, inducers appear to stimulate principally their own metabolism and sometimes the metabolism of a limited variety of compounds. To accommodate this, the membranes proliferate and become enriched only in certain of the drug-metabolizing enzymes.

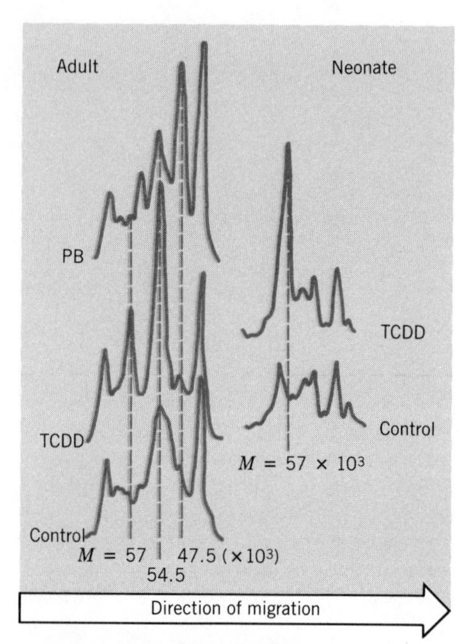

FIGURE 6.12 Densitometric tracings of Coomassie-stained gels after electrophoresis of purified cytochrome *P*-450 from rabbit liver microsomes. Three prominent peptides are elevated in preparations induced by phenobarbital (PB) and 2,3,7,8-tetrachlorodibenzo-*p*-dioxin (TCDD) as compared to controls in the adult. In the neonate the highest molecular weight peptide is induced by TCDD. *M*, molecular weight.

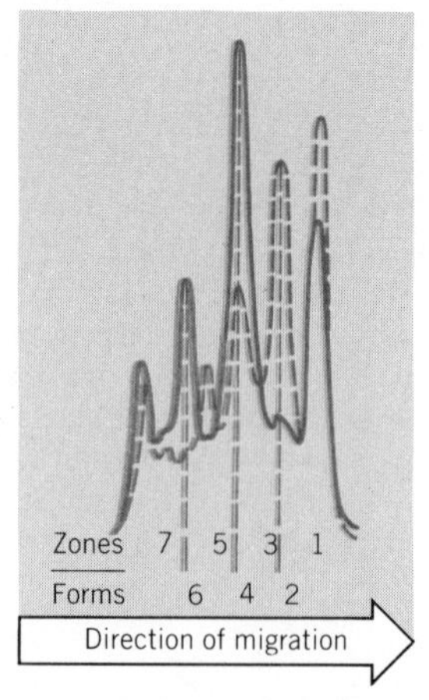

FIGURE 6.13 Electrophoretic designations of multiple forms of rabbit cytochrome *P*-450. The densitometric tracings represent a molecular weight range corresponding to about 45,000 to 60,000 daltons for adult rabbit PB microsomes (dashed curves) and TCDD microsomes (solid curves). Forms 2 and 4 have been subjected to detailed compositional and structural studies.

Properties of Cytochrome *P*-450

In situ, cytochrome *P*-450 appears to be a transmembrane integral protein. It is the chief enzyme of the endoplasmic reticulum in liver, occurring at levels as high as 10% of the microsomal protein in noninduced systems. This concentration is enhanced upon drug induction to levels two- to fivefold over controls, boosting the concentration of *P*-450 up to 25 to 30% of the total protein. This has made possible the isolation and purification of cytochrome *P*-450 to apparent homogeneity and in sufficient quantities for detailed physicochemical studies.

Using polyacrylamide gel electrophoresis in the presence of sodium dodecyl sulfate, the molecular weights of cytochrome *P*-450 from several species have been determined. In rabbit, rat, and mouse liver, the molecular weights range from 45,000 to 60,000 daltons, and the proteins appear to exist as single polypeptide chains.

Under appropriate conditions of electrophoresis, purified cytochrome *P*-450 resolves into multiple forms that vary in intensity according to the age of the animal and the type of inducer it received.[6] This is illustrated nicely in the densitometric tracings of Figure 6.12. Three prominent peptides having molecular weights of 47,500, 54,500, and 57,000 daltons are apparent. The lowest molecular weight peptide is induced by phenobarbital (PB), and 2,3,7,8-tetrachlorodibenzo-*p*-dioxin (TCDD) induces the middle and high molecular weight peptides. The first two of these three, called forms 2 and 4, respectively (Figure 6.13), have been purified from rabbit to an extent that has permitted detailed compositional and structural studies. Their amino acid compositions are similar but not identical. They have different *C*-terminal amino acids (arginine for form 2 and lysine for form 4), they vary in tryptophan content, and form 4 contains 60 more residues than form 2.

The picture that is emerging regarding the form and function of cytochrome *P*-450 is complex and difficult to focus. A number of important questions await answers. What is the *in vivo* relationship between the *P*-450 and b_5 electron transport chains and what is its physiological significance? What is the significance of multiple forms of cytochrome *P*-450 and what mechanism controls the specific induction of the various forms? The answers to these two questions alone will require years of research, demanding the development and use of sophisticated hardware and experimental approaches. But it will undoubtedly take place, and the results will fill a void that presently exists regarding an important function of the endoplasmic reticulum.

6.5 STEROL METABOLISM

The endoplasmic reticulum contains several of the key enzymes that catalyze the synthesis of cholesterol. Cholesterol, in turn, serves as a point from which two important metabolic pathways diverge: the formation of steroid hormones and the biosynthesis of bile acids. We will consider some of the salient features of these three connected pathways.

Cholesterol Biosynthesis

The synthesis of cholesterol from acetate has been the object of an enormous amount of research, beginning back in the 1940s and continuing

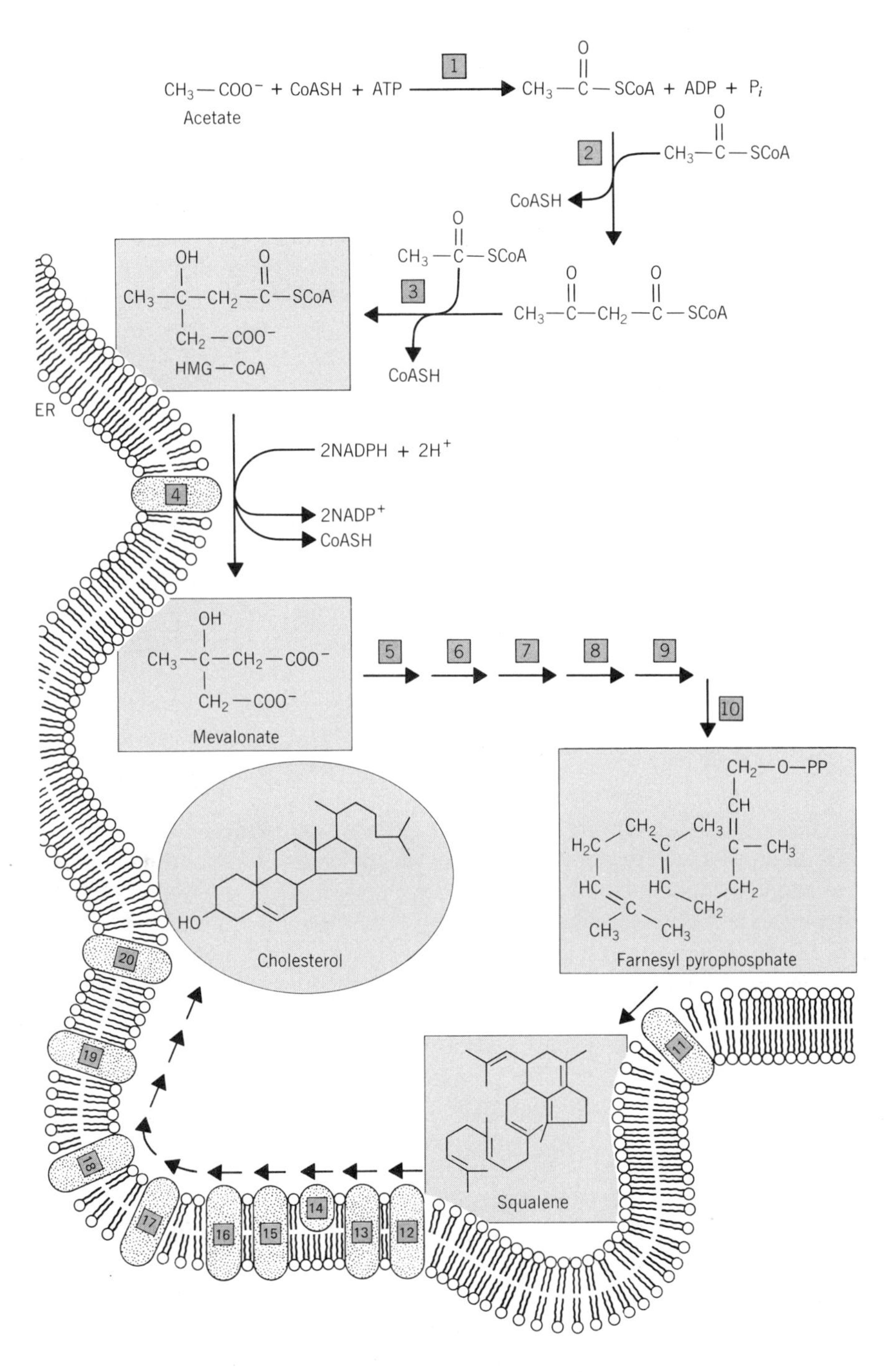

FIGURE 6.14 Steps in the synthesis of cholesterol showing the location in the cell of the various enzymes. The two membrane-bound enzymes that have been purified and studied in detail are HMG-CoA reductase (4) and squalene synthetase (11). Enzymes 1 to 3 and 5 to 10 are soluble and relatively easy to isolate and characterize, but enzymes 12 to 20 are uncooperative molecules. They are hydrophobic and buried in the lipid environment of the endoplasmic reticulum.

today. An outline of some of the key reactions is presented in Figure 6.14, which differentiates between reactions that are catalyzed by *soluble* enzymes and those reactions carried out by enzymes that are *bound to the membranes* of the endoplasmic reticulum.

β-Hydroxy-β-methylglutaryl-coenzyme A (HMG-CoA) is formed in the cytosol from acetate with the aid of three different enzymes. All three of these enzymes are also located in mitochondria, but it is believed that their role in mitochondria is in ketogenesis, and not in cholesterogenesis.[9]

Fat broken down by the reactions of β-oxidation in mitochondria results in an abundance of acetyl coenzyme A. If this production is greater than can be handled by the citric acid cycle, "ketone bodies" will form: 3-hydroxybutyrate, acetoacetate, and acetone.

Two membrane enzymes that participate in cholesterol biosyntheses have been isolated and studied in detail. They are HMG-CoA reductase and squalene synthetase.

The enzyme that converts HMG-CoA into mevalonic acid, HMG-CoA reductase, is the first microsomal enzyme involved in cholesterol biosynthesis, and it operates as the rate-limiting point of regulation in the pathway. HMG-CoA reductase is one of only two membrane enzymes of cholesterol biosynthesis that have been purified. Elaborate techniques using detergents, freeze-thawing, snake venom, and lyophilization have been worked out to solubilize the enzyme out of its membrane environment. Given this traumatic treatment, it is not clear that the isolated enzyme retains all its native physicochemical and enzymatic properties. The enzyme appears to be a trimer of identical subunits, each of about 65,000 daltons molecular weight.

The activity of HMG-CoA reductase is subject to unusually diverse regulatory controls, and being rate limiting in the formation of cholesterol, so is cholesterol biosynthesis. The rat liver enzyme shows a five- to tenfold diurnal variation that seems not to be clearly related to light–dark cycles or feeding periods. However, these two factors do have an influence on the circadian-type rhythm of reductase activity.

The reductase activity is also stimulated nutritionally, by a high fat diet, and hormonally, by insulin and thyroid hormone. The hormonal control is probably quite complex, since glucagon and glucocorticoids block an increase in the activity of the reductase.

The enzyme is also subject to feedback repression. Cholesterol, the product of the pathway, inhibits its own synthesis at the level of the reductase. There is also evidence that bile acids exert control on the pathway at the same level.

It is apparent from the considerations above that HMG-CoA reductase is subject to widely different control mechanisms. A unified mechanism explaining this control has not yet been presented. It is obvious that the reaction is a crucial one, and the membrane-bound nature of the enzyme is likely an important factor in its mechanisms of reaction and control.

The next six steps in the pathway, converting mevalonic acid to farnesyl pyrophosphate, are catalyzed by soluble cytosol enzymes. The product is then the substrate of the membrane-bound enzyme, squalene synthetase. This enzyme is the other membrane-bound cholesterogenic enzyme that has been purified to apparent homogeneity and studied in detail (from yeast, not liver). The monomeric form of the enzyme has a molecular weight of 450,000 daltons. It catalyzes the condensation of two molecules of farnesyl pyrophosphate to presqualene pyrophosphate. The reduction of this substance to squalene with the aid of NADPH can apparently be carried out only with a polymeric form of the enzyme that has a molecular weight of several million daltons.

The remaining reactions in the biosynthesis of cholesterol are mediated by membrane-bound enzymes of the endoplasmic reticulum. This stretch can be viewed as having two major components: the oxidative cyclization of squalene to lanosterol, and the oxidative demethylation, double-bond removal, and double-bond transfer in the conversion of lanosterol to cholesterol (Figure 6.15). Squalene epoxidase, a mixed-function oxidase using NADPH and molecular oxygen, apparently is not a cytochrome *P*-450 system, even though the type of reaction it catalyzes appears to be one in which *P*-450 could participate.

The conversion of lanosterol to cholesterol takes place as a series of

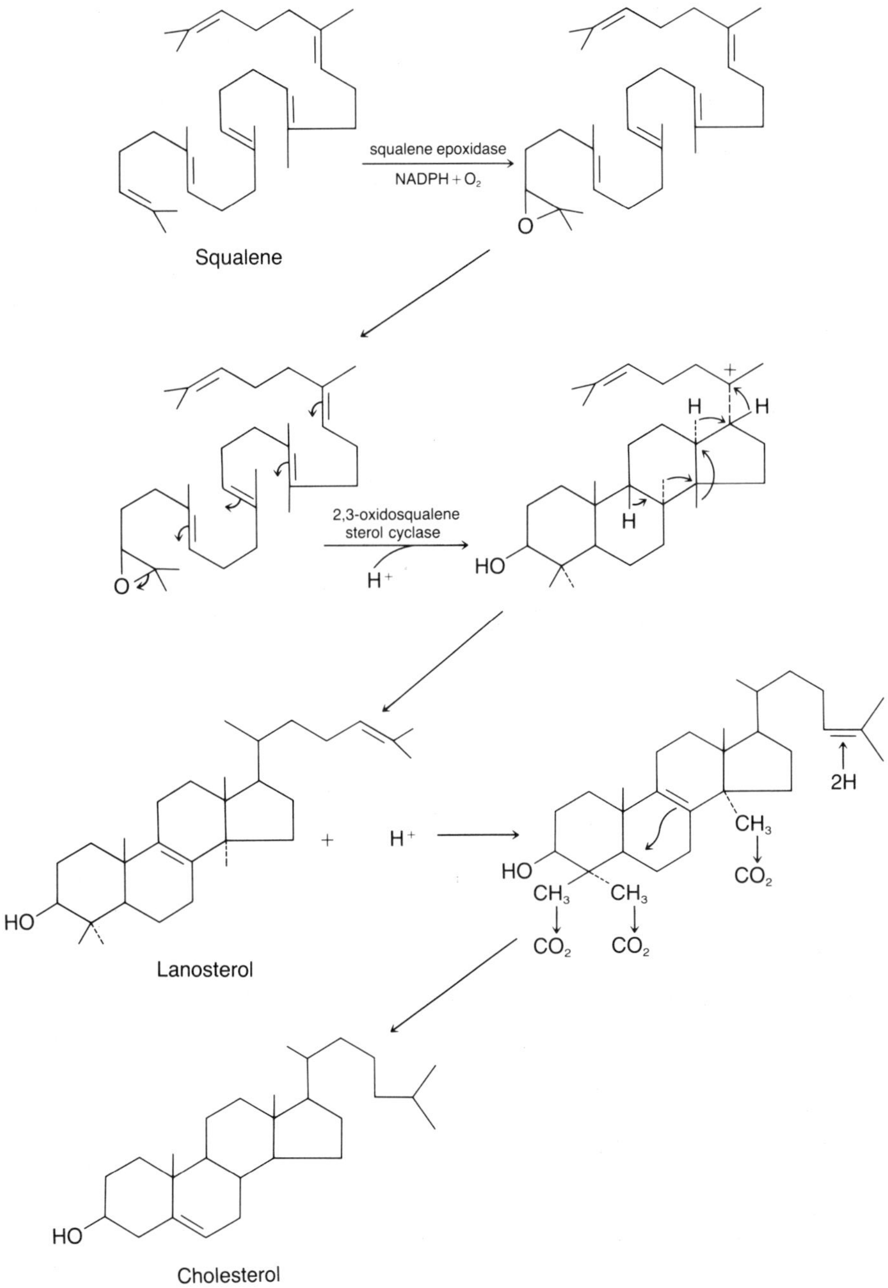

FIGURE 6.15 The conversion of squalene to cholesterol. Lanosterol is formed by an oxidative cyclization of squalene involving a mixed-function oxidase without using the cytochrome *P*-450 system. Cholesterol is then generated by reductions, oxidations, and double-bond shifts. These reactions are all mediated by enzymes that are a part of the endoplasmic reticulum.

three processes: reduction of the Δ^{24} double bond, oxidative removal of three methyl groups, and a shift of the double bond from the Δ^8 position in lanosterol to the Δ^5 position in cholesterol. Because of the difficulty in working with the enzymes involved in these steps, the reactions from

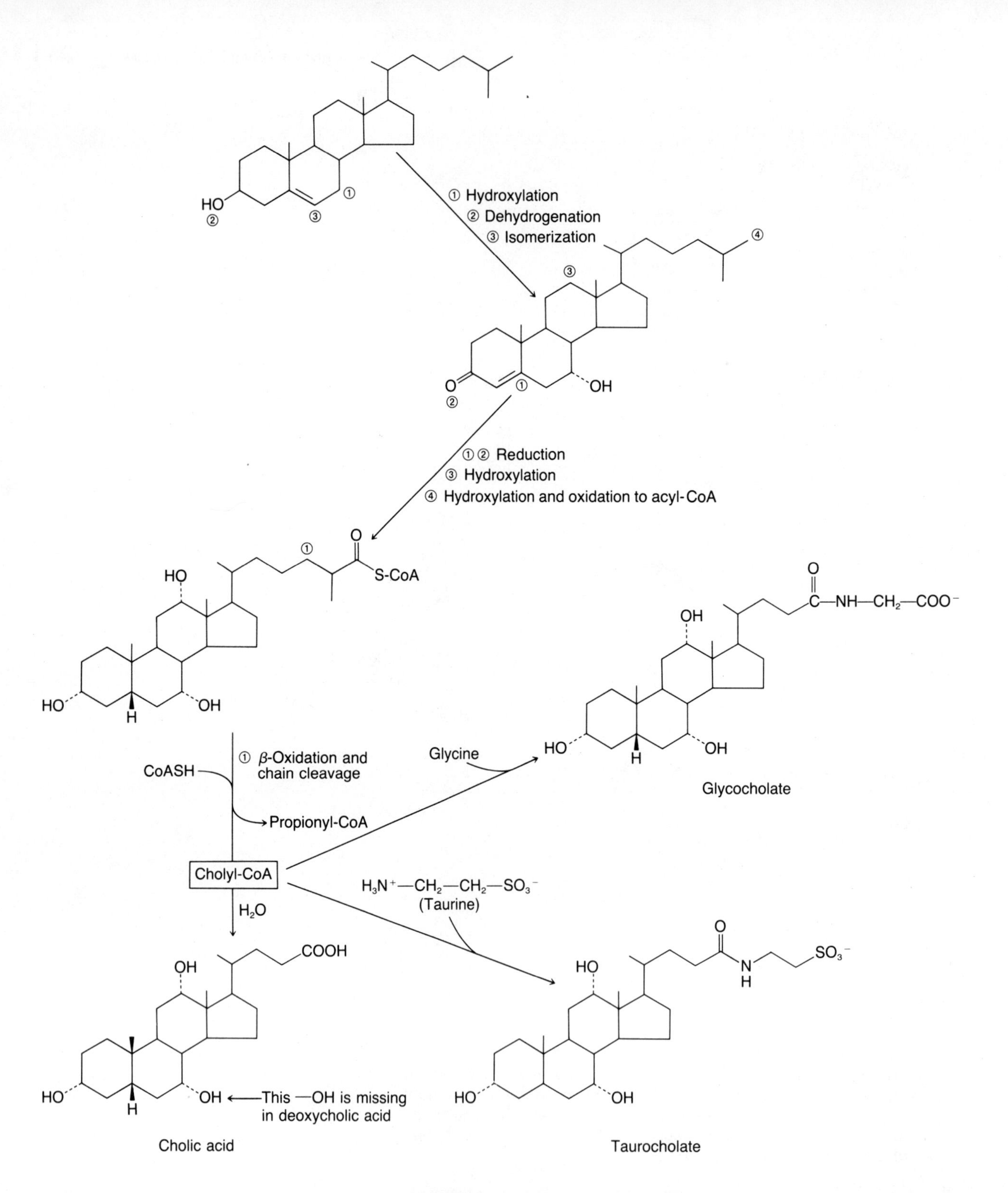

FIGURE 6.16 Biosynthesis of bile acids from cholesterol. Only a few of the enzymatic steps are shown. Cholesterol 7α-hydroxylase mediates the hydroxylation in position 1 of the first structure. Cholyl-CoA is an active intermediate in the synthesis of most bile salts. Glycocholate is the major bile salt. One major difference between cholesterol and the bile salts is that the latter are more hydrophilic, enabling them to function as very effective detergents in the intestine.

lanosterol to cholesterol have not been delineated with certainty. The oxidative steps catalyzing the conversion of methyl to carboxyl appear to be carried out by one mixed-function oxidase that requires molecular oxygen and a reduced pyridine nucleotide, but again, cytochrome *P*-450 does not seem to be involved in these reactions.

Bile Acid Biosynthesis

The biosynthesis of the bile acids represents a very complex pattern of enzymes and products (Figure 6.16). Most of the enzymes coordinating these pathways are microsomal. They include monooxygenases, dehydrogenases, isomerases, and reductases.

The one enzyme of this group that has been studied most intensively is cholesterol 7α-hydroxylase. This enzyme catalyzes the first reaction in the sequence and, as was true for HMG-CoA reductase in cholesterol synthesis, is the rate-limiting enzyme for bile acid synthesis. Although it has not been possible to solubilize cholesterol 7α-hydroxylase and purify it, a good deal has been learned concerning its enzymatic properties. As for HMG-CoA reductase, it is subject to a type of diurnal regulation that can be influenced by hormones.

The hydroxylases that convert 7α-hydroxycholesterol to bile acids require NADPH and molecular oxygen, and in this case depend on cytochrome *P*-450 and NADPH-cytochrome *P*-450 reductase.

Steroid Hormone Biosynthesis

Steroid hormones are synthesized in the cells making up the tissues of the adrenal cortex, the ovaries, the testes, and the placenta. The synthesis of the glucocorticoids and the mineralocorticoids, as well as traces of the sex hormones, takes place in the adrenal gland. The sex glands produce only the sex hormones, the male sex glands producing minor portions of the female sex hormones, and vice versa. We will focus on steroid synthesis as it takes place in the adrenal gland. The reactions are outlined in Figure 6.17.

Glucocorticoids promote gluconeogenesis in the liver with the accumulation of glycogen. An example is cortisol. Mineralocorticoids, such as aldosterone, regulate water and electrolyte metabolism. Aldosterone promotes the resorption of sodium ions in the kidney tubules. Glucocorticoids also possess weak mineralocorticoid activity.

In contrast to the synthesis of cholesterol and the bile acids, which is facilitated on or near the endoplasmic reticulum, the steroid hormones are produced in a coordinated manner by enzymes located partly in the mitochondria and partly in the endoplasmic reticulum. Because of this, key metabolites must shuttle back and forth across the mitochondrial membrane to complete the biosynthetic pathways.

Pregnenolone is the chief precursor to all the steroid hormones. It is derived from cholesterol by an oxidative cleavage of the side chain within the mitochondrion. This reaction, the first in the pathway to steroid hormone biosynthesis, is the rate-limiting reaction. It demarcates the third point of control in sterol metabolism and illustrates the cellular principle of assigning the point of control in a reaction sequence to the first reaction unique to that sequence.

Figure 6.18 presents schematically the coordinated roles of endoplasmic reticulum and mitochondrial enzymes in the biosynthesis of steroid hormones. Note the role of the endoplasmic reticulum in converting progesterone, the sex hormone intermediate, into 11-deoxycorticosterone and 11-deoxycortisol.

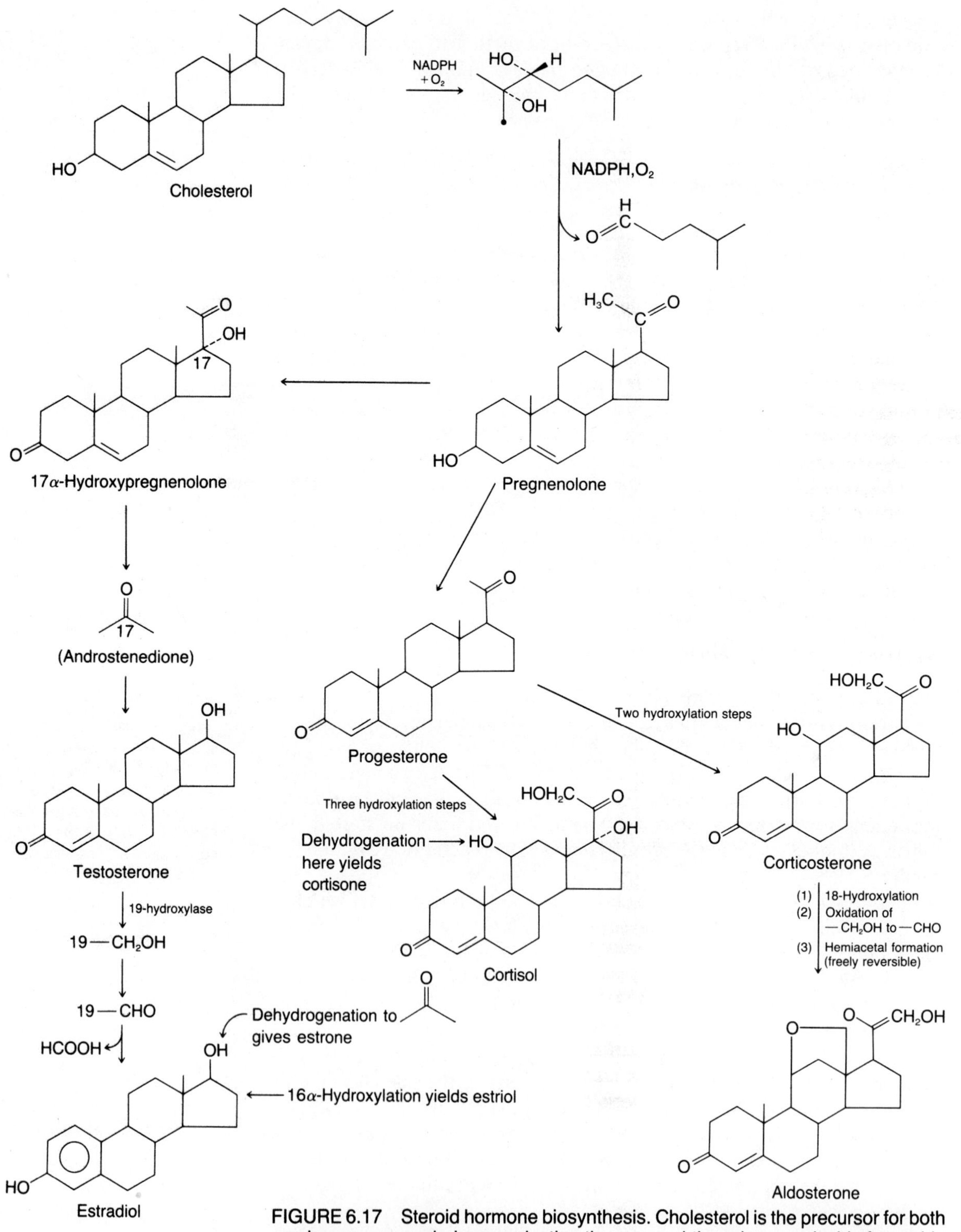

HO

Cholesterol

NADPH $+ O_2$

HO H

OH

$NADPH, O_2$

H

$O{=}C$

H_3C $C{=}O$

O

OH

17

O

17α-Hydroxypregnenolone

HO

Pregnenolone

O

17

(Androstenedione)

O

OH

O

Progesterone

Two hydroxylation steps

HOH_2C O

HO

O

Corticosterone

O

Testosterone

Three hydroxylation steps

HOH_2C O

OH

Dehydrogenation ⟶ HO
here yields
cortisone

19-hydroxylase

19—CH_2OH

O

Cortisol

(1) 18-Hydroxylation
(2) Oxidation of —CH_2OH to —CHO
(3) Hemiacetal formation (freely reversible)

19—CHO

HCOOH

O

Dehydrogenation to
gives estrone

OH

⟵ 16α-Hydroxylation yields estriol

O CH_2OH

O

HO

Estradiol

O

Aldosterone

FIGURE 6.17 Steroid hormone biosynthesis. Cholesterol is the precursor for both sex hormones, made in reproductive tissues, and the adrenocorticoids formed in the adrenal gland. Note that oxidations and hydroxylations are the themes for these transformations.

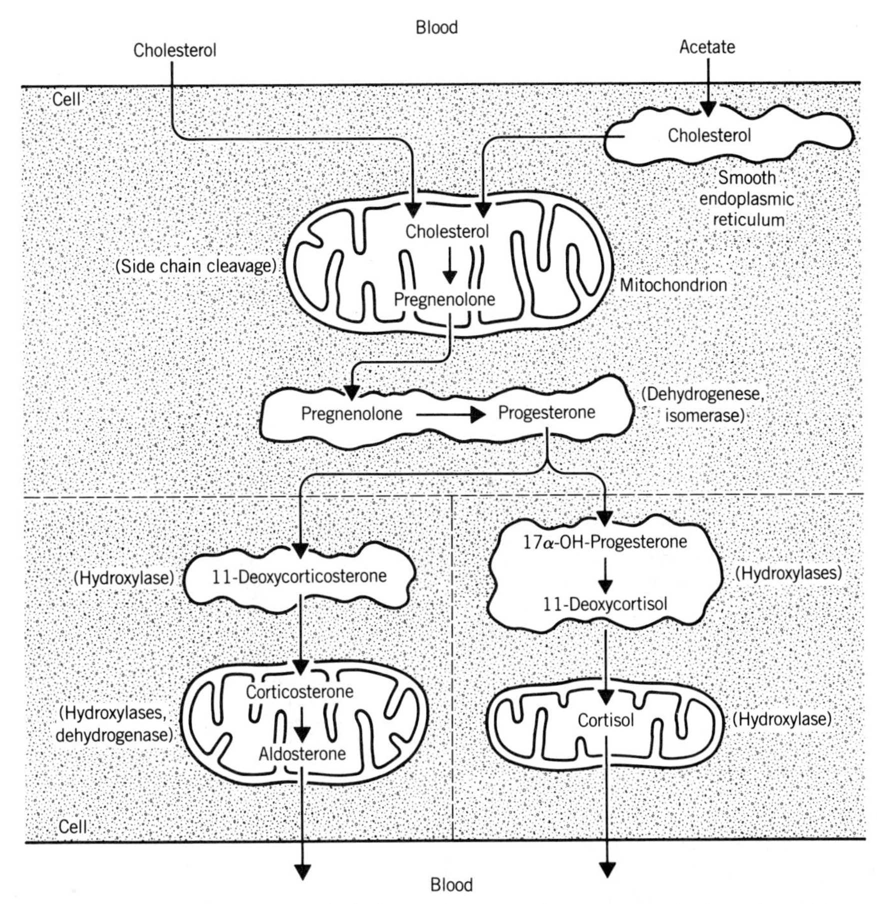

FIGURE 6.18 Conversion of cholesterol to steroid hormones in the adrenal cortex. Organelles with single line boundaries represent SER, those with double lines mitochondria. The intermediates of the biosynthetic pathways are shuttled back and forth between organelles to complete the appropriate transformations. Whereas all cells of the adrenal cortex carry out the pathway to progesterone, the formation of cortisol occurs in the inner cortex, and corticosterone and aldosterone are made in the glomerulosa (regions demarcated by dashed lines).

The final hydroxylation reactions that affix hydroxyl groups at the 11β positions of these two products are catalyzed by mixed-function oxidases operating within mitochondria. These depend on a cytochrome *P*-450 electron transport chain similar in many respects but not identical to its *P*-450 counterpart in the endoplasmic reticulum. The mitochondrial reactions are thus necessary to complete the formation of the three major adrenal steroid hormones: corticosterone, aldosterone, and cortisol.

6.6 CARBOHYDRATE METABOLISM

It is now generally thought that the endoplasmic reticulum plays a role in carbohydrate metabolism, but the evidence for its direct involvement is not as clear and conclusive as we would like. We will briefly discuss two systems in which the endoplasmic reticulum may have a crucial role.

(1) Glucose-6-P + H_2O→glucose + P_i

(2) PP_i + H_2O→$2P_i$

(3) PP_i + glucose→glucose-6-P + P_i

(4) Carbamyl-P + glucose → glucose-6-P + NH_3 + CO_2

FIGURE 6.19 Reactions of glucose-6-phosphatase. This enzyme of the endoplasmic reticulum is usually thought of as responsible for maintaining glucose levels in the blood (reaction 1). However, it is a very versatile enzyme, capable of catalyzing the hydrolysis of inorganic pyrophosphate (2) and the synthesis of glucose-6-P (3 and 4). It is an enzyme that requires further study before we may understand its multifaceted role.

Blood Glucose Homeostasis

Glucose-6-phosphatase is an enzyme that appears to be an integral ER enzyme or at least is tightly bound to ER membranes. It is normally thought of as a gluconeogenic phosphohydrolase that catalyzes the release of free glucose from its phosphorylated form in liver, thus operating to maintain homeostatic levels of glucose in the blood for the maintenance of red blood cells and nerve tissue. In this capacity it may function in conjunction with glycogen breakdown.

In mammals, the enzyme is absent before birth, as might be anticipated because the embryo receives a constant supply of glucose via the placenta from the maternal circulation. Immediately after birth there is a marked increase in glucose-6-phosphatase activity. In animal life that is begun in an egg, enzyme is already present in the egg, probably because there is no exogenous supply of glucose.

Thus there is circumstantial evidence favoring a role for glucose-6-phosphatase in blood glucose homeostasis. However, a rather puzzling property of the enzyme is its multifunctionality. It can catalyze a number of reactions including the synthesis of glucose-6-phosphate at rates that are comparable to or exceed its hydrolysis[10] (Figure 6.19). The mere fact that the enzyme is versatile does not negate its postulated function in blood glucose homeostasis, but it does suggest that the fullness of its physiological impact may not yet be realized.

The location of glucose-6-phosphatase is not limited to the liver, although its activity in both synthetic and hydrolytic forms has been observed in the liver of more than 20 species of mammals, birds, amphibians, and invertebrates.[10] But it has also been commonly observed in kidney tissue and has been monitored in epithelial cells of small intestine, and in lower but significant amounts in organs such as the brain, pancreas, adrenal glands, and testes. It likely appears in almost all tissues at some level with the endoplasmic reticulum its primary site.

Glycoprotein Synthesis

Most of the proteins that are destined for transport to the exterior of the cell are glycoproteins. These are complex proteins to which one or more carbohydrate units are attached.

The endoplasmic reticulum is the starting point for materials that are transmitted to the cell exterior by way of the secretory pathway (see Chapter 10 on secretion). The role of the endoplasmic reticulum is to sequester secretory proteins into its cisternae and begin their initial modification by attaching carbohydrate residues most proximal to the polypeptide backbone as summarized in Figure 6.20.

Proteins that are segregated into the lumen of the endoplasmic reticulum are modified to the glycoprotein form at several sites.[11] Sugars that are immediately adjacent to the polypeptide chain (namely, *N*-acetylglucosamine and mannose) are added either while the polypeptide is still attached to polysomes or soon after its release. Other sugars, more distal to the polypeptide chain, are added in the SER as the glycoprotein is moved through the secretory pathway. Finally, carbohydrates are added as the proteins are moved through the Golgi complex.

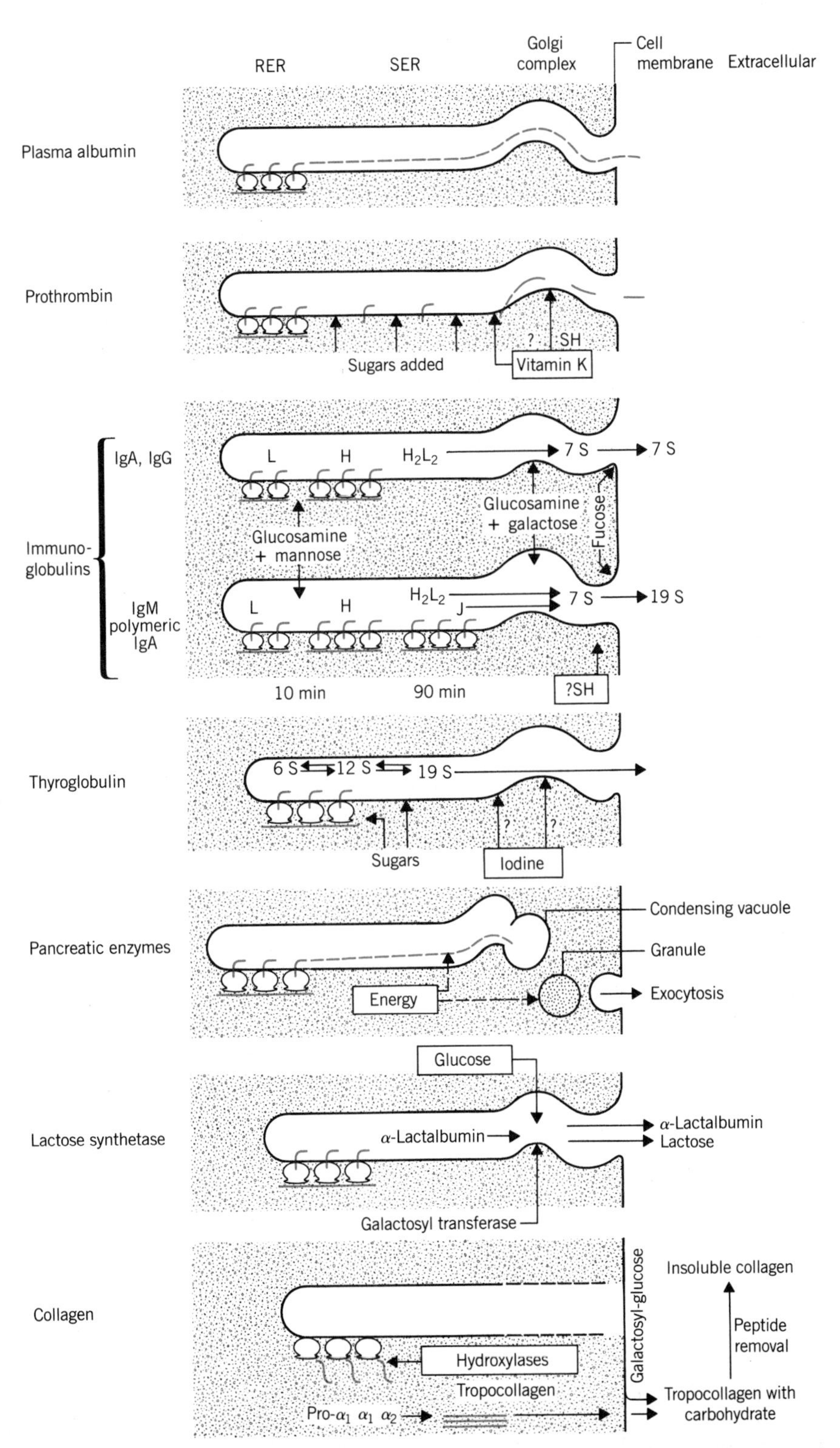

FIGURE 6.20 A summary of the role of endoplasmic reticulum in a variety of secretory mechanisms. As a rule, the SER adds sugar residues to proteins that have been produced by RER. The Golgi complex further modifies these products before they are moved to the cell exterior. A more detailed treatment of the pancreatic enzymes and the role of the ER in their formation in the secretory pathway is given in chapter 10.

Summary

The endoplasmic reticulum is a system of membranes that penetrate all regions of the eucaryotic cell between the plasma membrane and the nuclear envelope. Rough endoplasmic reticulum has ribosomes attached to the membranes, whereas smooth endoplasmic reticulum is ribosome free. RER appears to be the biogenetic precursor of SER. Endoplasmic reticulum may be present in a cell in lamellar, vesicular, and tubular forms.

The endoplasmic reticulum can be isolated from other cell constituents by homogenization and sedimentation techniques. But the conditions necessary to achieve this fragment the membranes severely and convert them to vesicles called microsomes. The microsomal fraction is not made up of true *in situ* organelles but rather of modified membrane fragments.

Membranes of the endoplasmic reticulum are thinner than plasma membranes and contain a higher ratio of protein to lipid. They contain many proteins in common with the plasma membrane but also harbor a number of enzymes that are predominantly associated with the endoplasmic reticulum. The marker enzyme for ER is glucose-6-phosphatase.

Except for cytochrome *P*-450 and the ribophorins, most of the membrane proteins studied are distributed asymmetrically across the membrane, preferring either the cytoplasmic or the luminal surface. The exceptions are transmembrane proteins.

The enzymes of the endoplasmic reticulum mediate hydroxylation reactions and as such have important roles in detoxification mechanisms and in the synthesis of cholesterol, steroid hormones, and bile acids. The hydroxylations are carried out by a mixed-function oxidase system that uses an electron transport system consisting of cytochrome *P*-450 and NADPH-cytochrome *P*-450 reductase. Another electron transport system employing cytochrome b_5 and NADH-cytochrome b_5 reductase may also participate. The cytochrome *P*-450 system is readily induced by phenobarbital.

Cholesterol is synthesized by a complex pathway catalyzed by enzymes, some of which are soluble and some microsomal. The microsomal enzymes are key to the pathway in that they serve as regulatory points for biosynthesis.

Cholesterol is the progenitor of a group of steroid compounds called bile acids, as well as the steroid sex hormones and steroid hormones of the adrenal gland.

The endoplasmic reticulum, by virtue of its enzymes, appears to have a role in blood glucose homeostasis and in the early stages of glycosylation of glycoproteins.

References

1. Particulate Components of Normal and Tumor Cells, A. Claude, *Science* 91:77(1940).
2. Liver Microsomes: An Integrated Morphological and Biochemical Study, G. E. Palade and P. Siekevitz, *J. Biophys. Biochem. Cytol.* 2:171(1956).

3. Enzyme Topology of Intracellular Membranes, J. W. DePierre and L. Ernster, *Annu. Rev. Biochem.* 46:201(1977).
4. Proteins of Rough Microsomal Membranes Related to Ribosome Binding. I. Identification of Ribophorins I and II, Membrane Proteins Characteristic of Rough Microsomes, G. Kreibach, B. Ulrich, and D. D. Sabatini, *J. Cell Biol.* 77:464(1978).
5. Ribosome Binding Sites Visualized on Freeze-fractured Membranes of the Rough Endoplasmic Reticulum, T. H. Giddings and L. Andrew Staehelin, *J. Cell Biol.* 85:147(1980).
6. Multiple Forms of Cytochrome P-450: Criteria and Significance, E. F. Johnson, *Rev. Biochem. Toxicol.* 1:1(1979).
7. Drug-Induced Changes in the Liver Endoplasmic Reticulum: Association with Drug-Metabolizing Enzymes, H. Remmer and H. J. Merker, *Science* 142:1657(1963).
8. Phenobarbital-induced Synthesis of the Microsomal Drug-metabolizing Enzyme System and Its Relationship to the Proliferation of Endoplasmic Membranes, S. Orrenius, J. L. E. Ericsson, and L. Ernster, *J. Cell Biol.* 25:627(1965).
9. Membrane-bound Enzymes of Sterol Metabolism, R. E. Dugan and J. W. Porter, in *The Enzymes of Biological Membranes,* vol. 2, A. Martonosi, ed., Plenum Press, New York, 1976.
10. Glucose-6-phosphatase, R. C. Nordlie and R. A. Jorgenson, in *The Enzymes of Biological Membranes,* vol. 2, A. Martonosi, ed., Plenum Press, New York, 1976.
11. Role of Endoplasmic Reticulum and Golgi Apparatus in the Biosynthesis of Plasma Glycoproteins, J. Molnar, in *The Enzymes of Biological Membranes,* vol. 2, A. Martonosi, ed., Plenum Press, New York, 1976.

Selected Books and Articles

Books

The Structural Basis of Membrane Function, Y. Hatefi and L. Djavadi-Ohaniance, eds., Academic Press, New York, 1976.

Articles

Enzyme Topology of Intracellular Membranes, B. F. Trump, in *Cell Membranes,* G. Weissmann and R. Claiborne, eds., HP Publishing Co., New York, 1975.

Hepatic Cytochrome P-450, A. J. Paine, in *Essays in Biochemistry,* vol. 17, P. N. Campbell and R. D. Marshall, eds., Academic Press, London, 1981.

Membrane-bound Enzymes of Sterol Metabolism, R. E. Dugan and J. W. Porter, in *The Enzymes of Biological Membranes,* vol. 2, A. Martonosi, ed., Plenum Press, New York, 1976.

Membranes of the Endoplasmic Reticulum and the Secretory System and Their Role in Plasma Membrane Regulation, J. J. Geuze, M. F. Kramer, and J. C. H. de Man, in *Mammalian Cell Membranes,* vol. 2, G. A. Jamieson and D. M. Robinson, eds., Butterworths, London, 1977.

Ribosome Binding Sites Visualized on Freeze-Fractured Membranes of the Rough Endoplasmic Reticulum, T. H. Giddings and L. A. Staehelin, *J. Cell Biol.* 85:147(1980).

Role of Endoplasmic Reticulum and Golgi Apparatus in the Biosynthesis of Plasma Glycoproteins, J. Molnar, in *The Enzymes of Biological Membranes,* vol. 2, A. Martonosi, ed., Plenum Press, New York, 1976.

The Endoplasmic Reticulum, M. J. Chrispeels, in *The Biochemistry of Plants,* vol. 1, N. E. Tolbert, ed., Academic Press, New York, 1980.

The Role of the Endoplasmic Reticulum in the Biosynthesis and Transport of Macromolecules in Plant Cells, M. J. Chrispeels, in *International Cell Biology,* B. R. Brinkley and K. R. Porter, eds., Rockefeller University Press, New York, 1977.

7

Microbodies: Peroxisomes and Glyoxysomes

It is almost textbook knowledge that peroxisomes arise as budding outgrowths of the endoplasmic reticulum.

CHRISTIAN DE DUVE, 1973

During the last seven years, we have formulated and tested a number of specific versions of this theory in order to see whether any of its possible biochemical implications could be verified . . . the results have been uniformly negative.

CHRISTIAN DE DUVE, 1974

Not everything that is widely accepted in the scientific world and broadly disseminated in textbooks is solidly undergirded with experimental verification. As a matter of fact, most theories hang by a pretty thin thread for a long time before they earn their rights to firm and comfortable niches. Usually they become textbook knowledge before they reach that stage, and once they appear on those sacred pages they can be killed about as easily as the proverbial cat. According to de Duve, the theory of peroxisome biogenesis had not quite made the grade by the mid-1970s, so it is our task to review its present status as we move along in the chapter. But before considering this problem, let us turn to some of the basics.

Since the mid-1950s electron microscopists have observed small structures or bodies in cells that on morphological grounds have been aptly termed *microbodies*. These structures are spherical or oblate in form and possess diameters that range between 0.2 and 1.5 μm. They are thus about the same size as mitochondria and lysosomes.

Peroxisomes can be assayed by their ability to effect the following reaction.

$$H^{14}COOH \xrightarrow[\;\to\, H_2O]{H_2O_2\,\to} {}^{14}CO_2 + H_2O$$

Two major classes of microbodies have been identified and studied. Those that contain catalases and oxidases constitute one class. They are called *peroxisomes*.[1] They were found originally in mammalian systems and were named for their peroxidative release of $^{14}CO_2$ from radioactive HCOOH. The microbodies that contain part or all of the enzymes of the glyoxylate cycle, in addition to the catalases and oxidases, comprise the other class. They are usually called *glyoxysomes*.[2] These microbodies are generally found in the endosperm of seeds, where they have a specific role in germination.

Endosperm is a storage site for lipids that are oxidatively degraded to provide energy for germination.

Peroxisomes and glyoxysomes appear to be both structurally and functionally related and perhaps differ only quantitatively in terms of their enzyme contents. Thus we use the term "microbody" when referring to them in a general sense, and resort to "peroxisomes" and "glyoxysomes" when a class distinction is necessary.

7.1 THE ULTRASTRUCTURE AND COMPOSITION OF MICROBODIES

Structure and Distribution according to Electron Microscopy

Microbodies are most easily distinguished from other cell organelles by their content of catalase. This enzyme can be visualized with the electron microscope when cells are treated with the stain 3,3′-diaminobenzidine (DAB). The product is electron opaque and appears as dark regions in the cell where catalase is present.

DAB has the following structure.

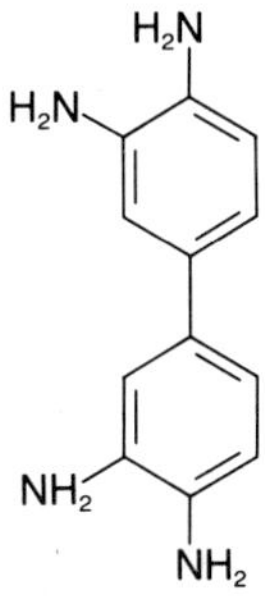

In the presence of H_2O_2 and OsO_4, catalase will oxidatively polymerize DAB (by way of indamine and phenazine polymers) to an electron-opaque material called osmium black.

Electron micrographs illustrating *in situ* microbodies in animal and plant tissues (Figures 7.1 and 7.2) reveal that the organelles are bounded by a single membrane, and the interior, or *matrix,* is amorphous or granular. In certain tissues, the microbody matrix contains a crystalline nucleoid structure. This is generally a crystalline form of urate oxidase, one of the enzymes of the matrix. A large crystal of this type is illustrated in the electron micrograph of Figure 7.3.

In animal tissue, microbodies are distributed quite diffusely throughout the cells, but are commonly in the vicinity of the endoplasmic reticulum. In plant cells microbodies are frequently associated with chloroplasts (see Figures 7.2 and 7.3). This ultrastructural affiliation is thought to reflect a metabolic relationship between the two organelles in providing for the

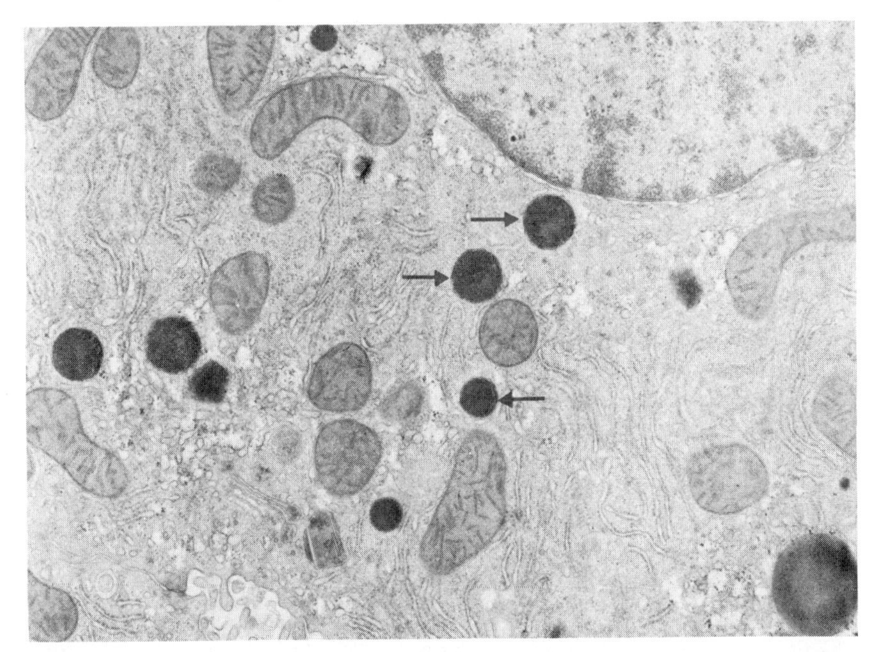

FIGURE 7.1 Microbodies as they appear in normal rat liver tissue. Peroxisomes are detected by staining with 3,3′-diaminobenzidine. They appear significantly darker than other organelles (see arrows). ×11,300 (Courtesy of Dr. Paul B. Lazarow.)

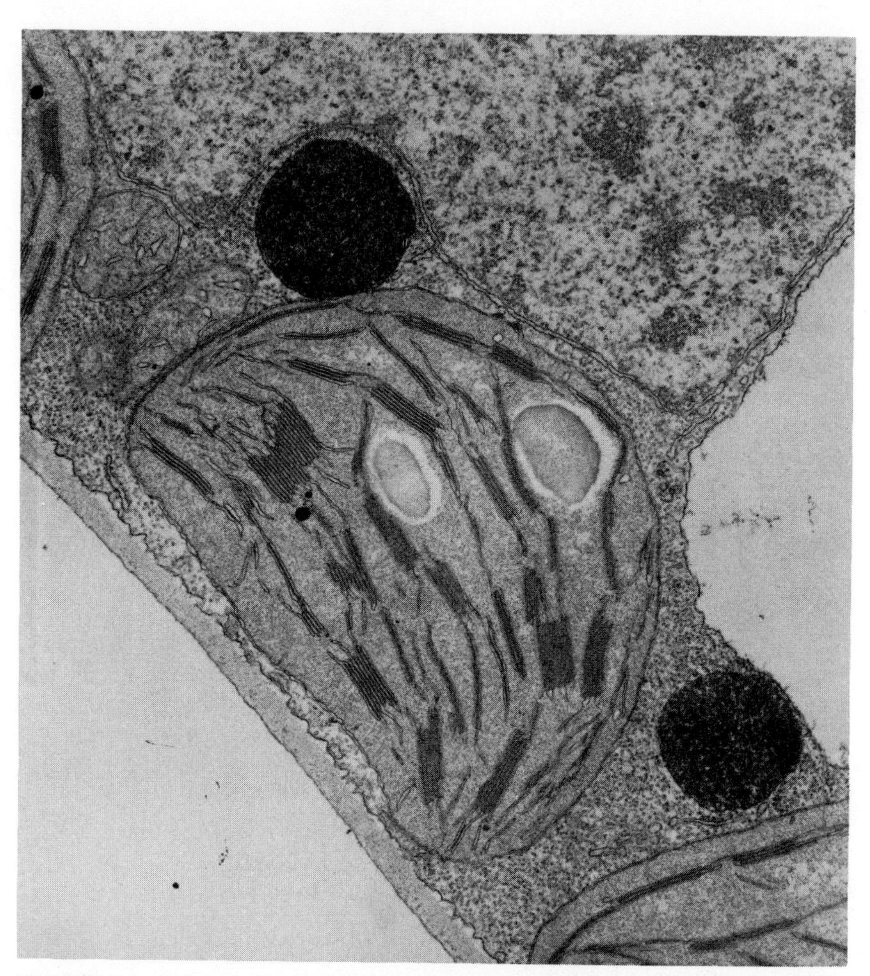

FIGURE 7.2 Tobacco leaf mesophyll cell stained with DAB to reveal the catalase in the dark-appearing microbodies. In these cells, the microbodies take up positions adjacent to chloroplasts in conjunction with which they carry out certain metabolic pathways. (Courtesy of Drs. S. E. Frederick and E. H. Newcomb.)

reactions of the glycolate pathway. We will turn to more detail on this topic later.

Isolation

Microbodies have been observed via DAB staining in a wide variety of mammalian tissues and have been isolated from liver, kidney, intestine, and brain. They have also been observed in the tissues of many non-mammalian metazoans and have been isolated from amphibians, birds, and plants. In addition, they have been identified in and isolated from certain protozoans and from yeast. They thus appear to be quite ubiquitous and may be a vital organelle in all eucaryotic cells.

Procedures to isolate microbodies generally follow a three-step pattern. First, tissues are ground carefully, a step that invariably causes rupture and destruction of a significant fraction of the microbody population. This

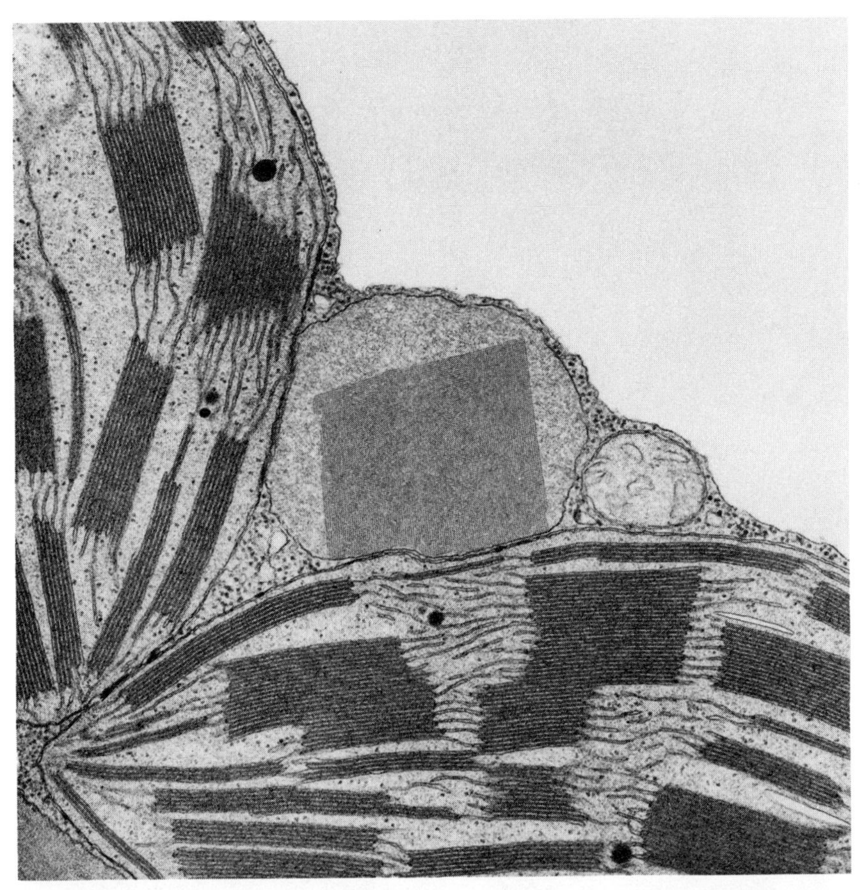

FIGURE 7.3 A microbody in a tobacco leaf cell. This one contains a large regular structure that is a crystalline form of the enzyme urate oxidase. Note once again the close proximity of the microbody to chloroplasts. (Courtesy of Drs. S. E. Frederick and E. H. Newcomb.)

is a special problem in isolating microbodies from plants where the grinding technique must be harsh enough to disrupt the cell wall, yet gentle enough to preserve the fragile organelles. Yields from 1 to 10% are not uncommon and up to about 50% maximum.

Method III

The second step is generally a differential centrifugation to obtain a fraction of the cell homogenate that is enriched in microbody organelles.

Third, and finally, the enriched fraction is subjected to isopycnic ultracentrifugation on discontinuous or continuous sucrose density gradients. In gradients, microbodies equilibrate at a density of 1.20 to 1.25 g/cm^3 depending somewhat on the tissue source. This is sufficiently far from other cell organelles to permit the preparation of reasonably pure specimens.[3] Figures 7.4 and 7.5 are electron micrographs of isolated microbodies. Lysosomes, along with other cell debris, can be seen in these preparations. Generally preparations of this sort contain isolated "cores" as well as intact microbodies. The tissue fractionation is rough enough to strip the membranes from some of the microbodies, leaving a crystallinelike core, which is taken through the procedure and ends up with the intact microbodies.

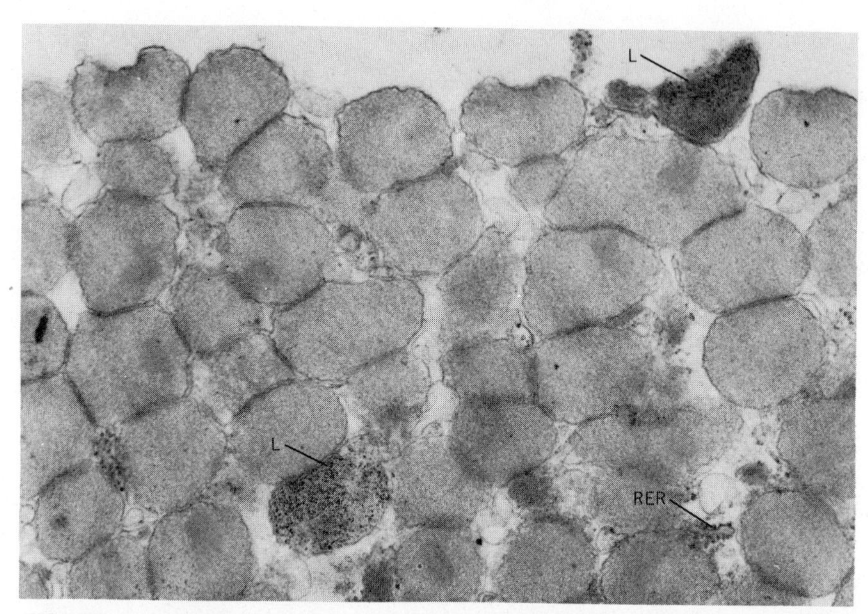

FIGURE 7.4 Section through a peroxisome-rich fraction. Electron-dense cores are barely visible in most of the organelles. Although this preparation is indeed "peroxisome rich," lysosomes (L) do appear in this fraction as well as low levels of RER. ×21,300 (Courtesy of Dr. Pierre Baudhuin.)

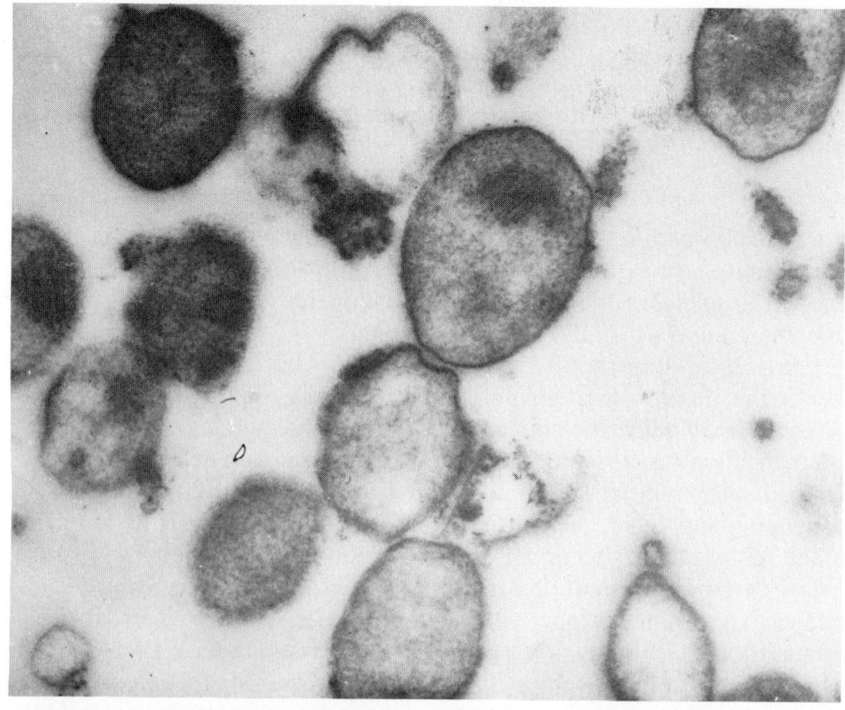

FIGURE 7.5 Microbodies at higher magnification. The organelle contains a single membrane boundary and has an interior that may be finely granular or dense with a crystallinelike character. ×56,000 (Courtesy of Dr. Pierre Baudhuin.)

Chemical Composition and Permeability of Microbody Membranes

The membrane that limits the microbody has the typical trilamellar characteristics of other membranes according to electron microscopy, but is thinner (6–8 nm) than the plasma membrane. In this regard it resembles endoplasmic reticulum and the outer membrane of mitochondria. The intact organelle is quite stable osmotically but can be ruptured in pyrophosphate solutions. One volume of an organelle preparation added to one volume of 0.01 *M* pyrophosphate will cause rupture of the membrane for reasons that are not clearly understood. Nevertheless, this has become a standard way of breaking open microbodies.

Inorganic pyrophosphate is

$$\begin{array}{c} O^- \\ | \\ O^-\!-\!P\!=\!O \\ | \\ O \\ | \\ O^-\!-\!P\!=\!O \\ | \\ O^- \end{array}$$

But even with broken preparations it is difficult to separate the membranes from the enzymes of the matrix. This is partly because the enzymes often reside as a core that has a density (1.17 g/cm^3) quite similar to that of the separated membranes (1.19–1.21 g/cm^3). In other instances the enzymes adhere to the membranes. These difficulties, combined with the low yields of microbodies that are obtainable with conventional techniques, have limited the level of sophistication with which compositional analyses can be approached.

Enzymes

For the reasons just cited, an analysis of the enzyme content of the microbody membrane is in its infancy. So far, two enzymes, which are also integral proteins of the endoplasmic reticulum, have been consistently associated with the microbody membrane: cytochrome b_5 and NADH-cytochrome b_5 reductase. Because of the relatively large amounts of endoplasmic reticulum in the cell compared to microbody membranes, it is not yet absolutely certain that these enzymes are not contaminants.

Several other enzymes are membrane associated, at least in glyoxysomes. They may be peripheral proteins rather than integral proteins of the membrane, since they are easily extracted from the membrane with 0.05 to 0.15 *M* KCl. Among these are the citrate and malate synthetases, malate and 3-hydroxyacyl-CoA dehydrogenases, and crotonase. The rather strong association of these enzymes with glyoxysomal membranes has been used to advantage in studies on microbody biogenesis (see Section 7.3).

Lipids

The lipid composition of the microbody membrane simulates that of microsomes, as can be seen in Table 7.1. Peroxisome and microsome membranes from rat liver are experimentally indistinguishable, but they differ significantly from mitochondrial membranes in their lower content of cardiolipin. However, since cardiolipin is found almost exclusively in the *inner* mitochondrial membrane, this comparison is not a valid discriminator between the *outer* mitochondrial membrane and the membranes of the peroxisome.

Table 7.1 also shows that glyoxysome membranes from castor bean endosperm differ more markedly in composition from the endoplasmic reticulum of the same tissue as compared to the two membranes in rat liver. Glyoxysome membranes have a lower level of phosphatidylinositol and possibly phosphatidylserine and a higher level of an unidentified lipid

Table 7.1 Phospholipid Composition of Microbodies as a Percentage of the Total

	Rat Liver			Castor Bean Endosperm		
Phospholipid	Peroxisomes	Mitochondria	Microsomes	Glyoxysomes	Mitochondria	Microsomes
Phosphatidylcholine	55.1	44.5	49.8	49.0	36.9	50.0
Phosphatidylethanolamine	16.0	28.1	18.8	31.4	30.9	26.6
Phosphatidylinositol[a]	19.7	7.1	19.7	6.1	14.3	18.9
Phosphatidylserine	7.4	1.9	8.5	0.0	4.1	1.8
Cardiolipin	1.6	18.4	3.1	2.4	13.7	2.7
Unidentified				11.4	0.2	0.0

[a]Combined with sphingomyelin.

that about makes up for the difference. It is too early to know for certain, but the lipid differences between rat liver and endosperm microbody membranes may be related to the different roles these organelles play in the two types of tissue.

Membrane Permeability

In many respects, the permeability of microbodies to a variety of molecules is quite similar to that of microsomes. This, of course, is not surprising in view of their similar compositions. The microbody membrane appears to be freely permeable to a number of substances that are natural substrates of some of the enzymes within, such as amino acids, α-hydroxy acids, and uric acid. Sucrose also diffuses easily across the membrane. This relatively nonrestrictive property of the microbody explains why it is so osmotically stable.

By contrast, the pyridine nucleotides, NADH and NADPH, are not permeative to the microbody membrane. This raises an important question, since these coenzymes are used in the organelle as electron acceptors for certain oxidative enzymes. If these coenzymes cannot cross the membrane, how are they regenerated to the oxidized form that permits them to continue to function with the oxidases?

It now appears that microbodies handle this problem by using substrate-linked hydrogen transport systems to shuttle reducing equivalents across the membrane. In this way, reduced NADH, for example, can be reoxidized without crossing the membrane by shuttling out its electrons on substrates that are freely permeative to the membrane. This shuttle mechanism is similar to that which operates across the mitochondrial membrane.

One type of shuttle that is thought to operate is a malate–oxaloacetate (aspartate) shuttle in plant peroxisomes. Malate has the ability to diffuse freely across the membrane. The malate dehydrogenase in the matrix removes electrons from NADH, transferring them to oxaloacetate and producing malate (Figure 7.6). Malate crosses the microbody membrane into the cytosol, where a different malate dehydrogenase (an isoenzyme) converts the malate back to oxaloacetate by electron removal. Since oxaloacetate does not have complete freedom in crossing back into the matrix, it is aminated to aspartic acid, which then moves into the microbody. The removal of the amino group from aspartic acid, producing

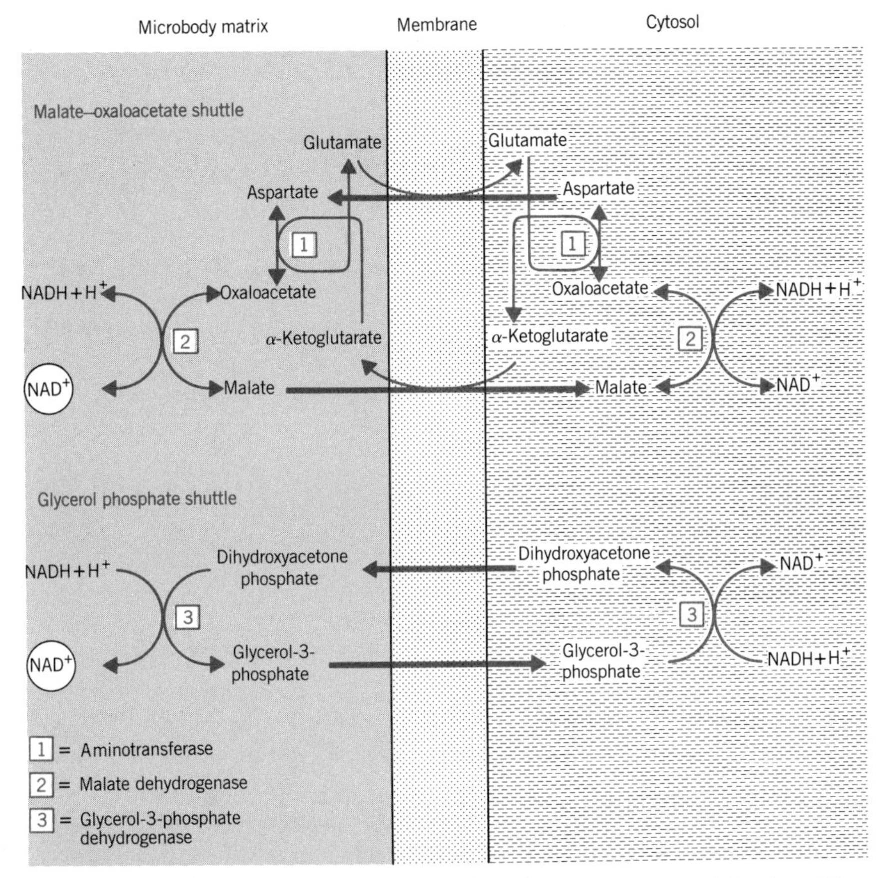

FIGURE 7.6 The proposed shuttles that function to generate oxidized pyridine nucleotides within microbodies. The malate–oxaloacetate shuttle is thought to operate in plant peroxisomes, whereas the glycerol phosphate shuttle is characteristic of liver peroxisomes.

oxaloacetate, then completes the cycle. For the shuttle mechanism to function properly, glutamate and α-ketoglutarate are also exchanged across the membrane.

The peroxisomes of liver or kidney appear to contain an NAD-linked glycerol phosphate shuttle rather than a malate shuttle. This is also illustrated in Figure 7.6. The net effect is to regenerate oxidized NAD^+ *within* the microbody without necessitating the movement of the coenzyme out of the organelle.

Enzyme Content of the Matrix

Table 7.2 lists some of the most common microbody enzymes. Not all these enzymes are found together in a given tissue, or even a given organism. There is rather considerable tissue and species variability with respect to the microbody content of enzymes.

The enzymes that appear to be common denominators among most or all tissues are the catalases and a variety of hydrogen peroxide producing flavin oxidases. In addition to these, plant leaf peroxisomes contain the

Table 7.2 Microbody Enzymes[a]

Catalase[b]
NAD-Malate dehydrogenase[b]
Hydroxypyruvate reductase
Serine–glyoxylate aminotransferase
Glycolate oxidase
Glutamate–oxaloacetate aminotransferase[b]
Glutamate–glyoxylate aminotransferase
NADP-Isocitrate dehydrogenase[b]
Urate oxidase
Xanthine dehydrogenase
NAD–Glycerol phosphate dehydrogenase
Malate synthetase
Citrate synthetase
Aconitase
Isocitrate lyase
Fatty acyl-CoA synthetase
β-Oxidation enzymes
Thiolase
Uricase
Allantoinase
Amino acid oxidase

[a]Listed in approximate decreasing order of their relative activities in microbodies on their respective substrates.
[b]Indicates enzymes that are common to both peroxisomes and glyoxysomes.

enzymes of the glycolate pathway, and plant endosperm glyoxysomes contain the enzymes for the glyoxylate cycle and for the β-oxidation of fatty acids.

In general, it can be stated that the enzymes of microbodies are a specialized group of oxidation enzymes that act on a fairly limited number of substrates in a degradative manner. The enzymes of mammalian tissues do not fit into a coherent metabolic pattern at present, but the plant microbody enzymes appear to have roles that make more metabolic sense.

Microbodies possess several enzymes for which there are counterparts in other regions of the cell.[4] For example, there are distinctive isoenzymes of malate dehydrogenase in microbodies, mitochondria, and the cytosol. Thus identical reactions can be carried out in different cell compartments, but there is no mixing of the enzymes from different regions of the cell.

Another example is citrate synthetase, which appears in two isoenzyme forms in castor bean endosperm and corn scutella. One form is associated with the glyoxysomes in these tissues and the other with mitochondria.

A more complex distribution of isoenzymes exists for asparate–α-ketoglutarate transaminase. In spinach four distinct isoenzymes are separable by starch gel electrophoresis. One each is present in the cytosol, chloroplasts, mitochondria, and leaf peroxisomes. In cucumber as many as six isoenzymes are separable. Two of these are glyoxysomal.

The presence of isoenzymes in different parts of the cell enables the cell to compartmentalize various metabolic pathways, a factor that may help in the regulation of metabolism. We have already seen the role of isoenzymes in shuttles, and we encounter more examples of their importance for the proper functioning of microbodies later in the chapter.

7.2 THE FUNCTION OF MICROBODIES

Substrate Oxidations in Mammalian Systems

The oxidative reactions occurring in the peroxisomes of mammalian tissues are mediated by flavin oxidases, which use molecular oxygen as electron acceptor, converting it to hydrogen peroxide (H_2O_2). The H_2O_2 thus formed, normally toxic to the cell, remains compartmentalized within the peroxisome where it is acted on by catalase and converted into H_2O and O_2. The reactions can be summarized as follows.

$$O_2 \xrightarrow[RH_2 \curvearrowright R]{\text{oxidase}} H_2O_2 \xrightarrow{\text{catalase}} H_2O + \tfrac{1}{2}O_2$$

$$H_2O_2 \xrightarrow[R_{red} \curvearrowright R_{ox}]{\text{peroxidase}} H_2O \text{ (an } \textit{in vitro} \text{ reaction)}$$

A specific example of this generalized type of reaction is the fate of D-amino acids as they enter the peroxisome. D-Amino acids are oxidatively deaminated to the corresponding α-keto acids with an FAD-oxidase:

$$\text{D-amino acid} + H_2O + \text{E-FAD} \rightarrow \alpha\text{-keto acid} + NH_3 + \text{E-FADH}_2$$

Flavin adenine dinucleotide (FAD) is a coenzyme that functions as an electron carrier. When carrying electrons it is written in its reduced form, $FADH_2$. An FAD-oxidase is an enzyme that mediates oxidation–reduction reactions by employing its FAD coenzyme to carry electrons.

The reduced coenzyme resulting from this oxidation is regenerated to oxidized form by reaction with molecular oxygen. Hydrogen peroxide is formed.

$$\text{E-FADH}_2 + O_2 \rightarrow \text{E-FAD} + H_2O_2$$

The hydrogen peroxide is immediately broken down by catalase.

$$H_2O_2 \xrightarrow{\text{catalase}} H_2O + \tfrac{1}{2}O_2$$

Flavin-containing enzymes are certainly not unique to microbodies. They are also participants during electron transport in mitochondria. But their catalytic activity in microbodies is fundamentally different from the reactions they mediate in mitochondria. In microbodies their electrons are transmitted *directly* to molecular oxygen rather than to another type of acceptor, such as coenzyme Q or nonheme iron. A direct transfer generates H_2O_2; hence the requirement for catalase in the vicinity to defuse the system from toxic effects.

The presence of D-amino acid oxidases in mammalian tissue has been an enigma ever since the discovery of these enzymes. It makes little metabolic sense because mammals do not normally contain D-amino acids, nor are these substances brought in as nutrients. Since the cell wall of bacteria is one of the few places in nature where D-amino acids exist (see Chapter 3), a presumed role for liver and kidney D-amino acid oxidase is to initiate the degradation of D-amino acids that may arise from breakdown and absorption of the peptidoglycan material of intestinal bacteria.

β-Oxidation of Long Chain Acyl-CoA's in Mammals

In recent years some newer roles have been found for the peroxisomes of mammalian systems which may change some of our thinking concerning the chief cellular site of fatty acid oxidation. For some time the

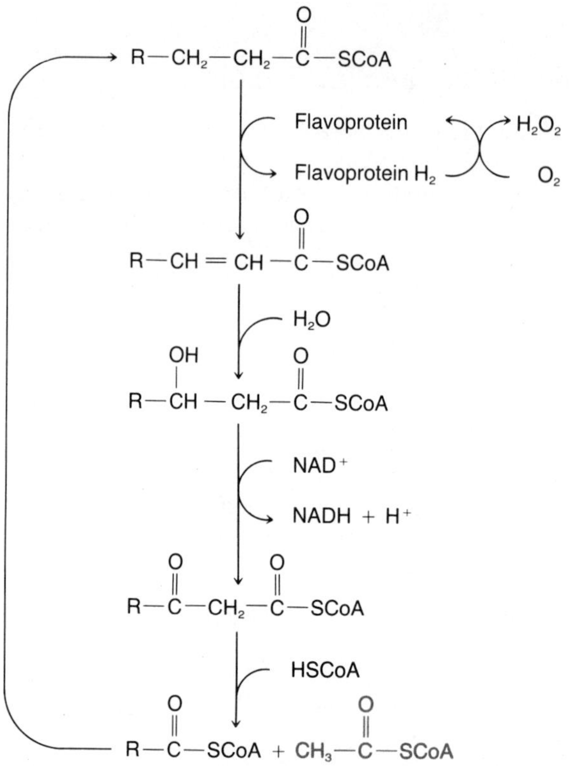

$$R-CH_2-CH_2-\overset{O}{\overset{\|}{C}}-SCoA$$

$\downarrow$ Flavoprotein → Flavoprotein H_2 (Flavoprotein H_2 + O_2 → Flavoprotein + H_2O_2)

$$R-CH=CH-\overset{O}{\overset{\|}{C}}-SCoA$$

$\downarrow$ H_2O

$$R-\underset{}{\overset{OH}{\overset{|}{C}H}}-CH_2-\overset{O}{\overset{\|}{C}}-SCoA$$

$\downarrow$ NAD^+ → $NADH + H^+$

$$R-\overset{O}{\overset{\|}{C}}-CH_2-\overset{O}{\overset{\|}{C}}-SCoA$$

$\downarrow$ HSCoA

$$R-\overset{O}{\overset{\|}{C}}-SCoA + CH_3-\overset{O}{\overset{\|}{C}}-SCoA$$

FIGURE 7.7 Scheme of β-oxidation of fatty acyl-CoA's for glyoxysomes of germinating castor bean endosperm as proposed by Cooper and Beevers. Apparently the same pathway operates in mammalian peroxisomes.

Carnitine acyltransferase operates to shuttle $R-\overset{O}{\overset{\|}{C}}-$ groups across the membranes by transferring acyl groups from the sulfur of acyl-CoA to the hydroxyl of carnitine.

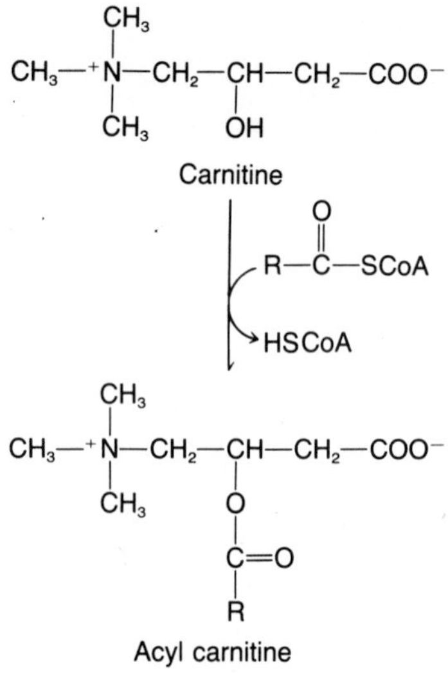

$$CH_3-\overset{+}{N}(CH_3)_2-CH_2-\underset{OH}{\underset{|}{C}H}-CH_2-COO^-$$

Carnitine

$\downarrow$ $R-\overset{O}{\overset{\|}{C}}-SCoA$ → HSCoA

$$CH_3-\overset{+}{N}(CH_3)_2-CH_2-\underset{O-\overset{}{C}(=O)-R}{\underset{|}{C}H}-CH_2-COO^-$$

Acyl carnitine

The acyl carnitine then diffuses across the membrane. The reverse reaction occurs on the other side of the membrane, regenerating acyl-CoA's.

dogma has been that neutral lipids (triacylglycerols), which are stored in the cytosol, are hydrolyzed by lipases to free fatty acids. These, in turn, are transported by carriers into mitochondria where they are oxidatively degraded to acetyl-CoA's. This has been thought to occur exclusively in mitochondria.

It is now known that rat liver peroxisomes are capable of oxidizing palmitoyl-CoA to acetyl-CoA, using molecular oxygen and NAD as electron acceptors.[5] The reaction pathway is outlined in Figure 7.7. This is, of course, consistent with the finding of carnitine acyltransferases in peroxisomes to shuttle substrates and products back and forth across the membrane. Presumably the acetyl-CoA formed by this process would be transported eventually to the mitochondria where it could enter into the citric acid cycle. If, alternatively, it remained in the cytosol, it could be reconverted to fatty acids and ultimately to neutral fats. In any event, it has been proposed that the β-oxidation of fatty acids in mammalian peroxisomes may be a major pathway in the overall economy of the cell.

This pathway of oxidation is very similar to the one that occurs in mitochondria *with one very important exception*. In mitochondria, the flavin dehydrogenase donates its electrons to the respiratory chain. It does not react with molecular oxygen. In microbodies, the dehydrogenase

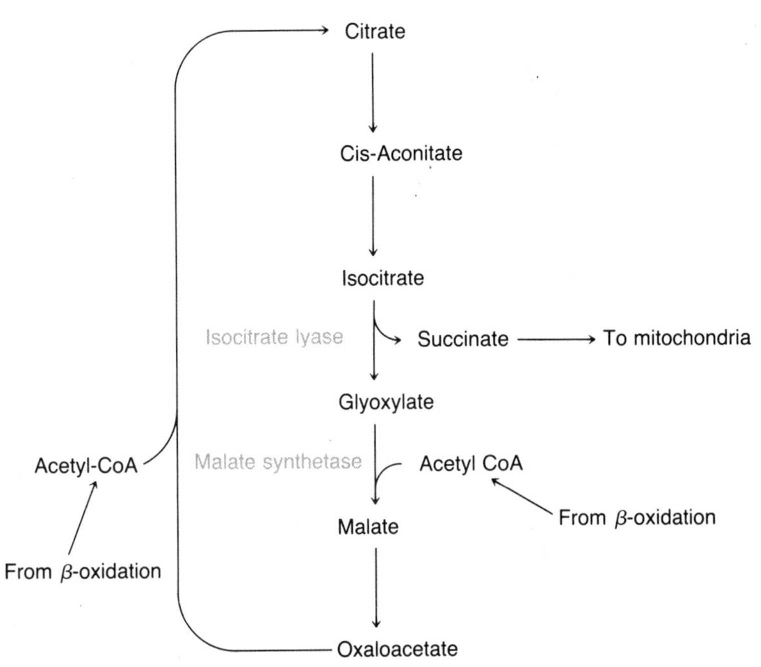

FIGURE 7.8 The glyoxylate cycle. Succinate is formed from acetyl-CoA, which, in turn, is generated by the reactions of β-oxidation. This cycle is similar to the citric acid cycle except that it bypasses the reactions that yield CO_2. Thus there can be a net synthesis of carbohydrate from fat.

reacts directly with O_2, and in so doing generates H_2O_2. Mitochondria contain no catalase and therefore cannot deal with the formation of toxic hydrogen peroxide. For microbodies it is not a problem.

β-Oxidation of Fatty Acids in Plant Endosperm

The enzymes necessary for the β-oxidation of fatty acids in microbodies were first found in the glyoxysomes of plant endosperm by Cooper and Beevers.[6] The pathway of β-oxidation is apparently the same in mammalian peroxisomes and plant glyoxysomes.

Plant endosperm is a fat storage tissue. It provides the nutrients that are necessary to sustain the process of seed germination. The energy source for germination is carbohydrate rather than lipid, so a conversion is made from the latter to the former with the aid of glyoxysomal enzymes. The product of fatty acid oxidation, acetyl-CoA, is then apparently used right in the glyoxysome to form C_4 acids by way of the glyoxylate cycle.

The Glyoxylate Cycle in Plants

The reactions in the *glyoxylate cycle* are illustrated in Figure 7.8. In germinating seeds, this cycle operates completely within the glyoxysome. In other organisms, such as yeast and *Tetrahymena,* the cycle is shared between glyoxysomes and mitochondria.

The glyoxylate cycle is considered to be a modified form of the citric acid cycle in that it bypasses the steps in the latter that bring about the evolution of carbon dioxide (CO_2). Thus the reactions permit the use of fatty acids (in the form of acetyl-CoA) as the sole source of carbon to bring about the *net synthesis* of carbohydrate from fatty acids. For every two molecules of acetyl-CoA fed into the cycle, one molecule of succinate is formed, which may then be used as a precursor for gluconeogenesis in mitochondria. There succinate is converted to malate and eventually, in the cytosol, to phosphoenol pyruvate in the synthesis of new glucose.

Starting with succinate, part of gluconeogenesis takes place in mitochondria and the rest in the cytosol.

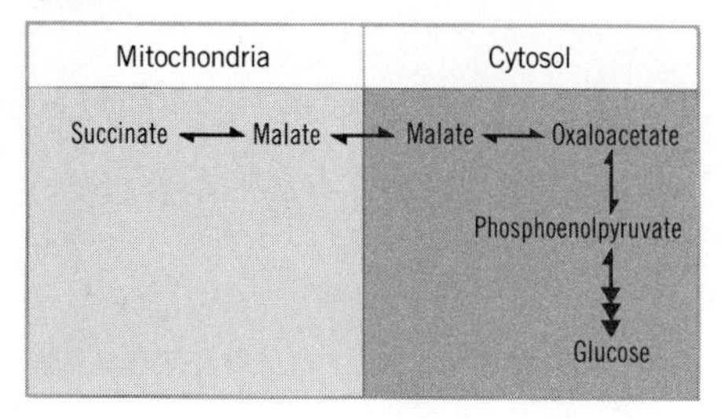

Higher animals cannot carry out the net synthesis of glucose from fatty acids because they lack the two essential glyoxysomal enzymes *isocitrate lyase* and *malate synthetase*. Instead, the two carbon atoms of acetyl-CoA are brought into the citric acid cycle and are ultimately lost as CO_2.

The Glycolate Pathway

The *glycolate pathway* (Figure 7.9) is a sequence of reactions that takes place in leaf peroxisomes in conjunction with the carbon cycle of chloroplasts. This pathway is a shared pathway, with chloroplasts, peroxisomes, mitochondria, and the cytosol all making contributions. The pathway brings about the conversion of nonphosphorylated compounds to glycine, serine, and C_1 compounds and in this capacity serves as the generator of precursors for nucleic acid biosynthesis.

The glycolate pathway can be viewed as beginning in the chloroplast where phosphoglycolate, glycolate, and phosphoglycerate are formed photosynthetically. The chloroplast possesses phosphatases, which remove phosphate from the two different phosphate-containing substrates, generating glycolate.

COO^-
H—C—OH
$CH_2OPO_3^{2-}$
Phosphoglycerate

COO^-
$CH_2OPO_3^{2-}$
Phosphoglycolate

COO^-
CH_2OH
Glycolate

COO^-
CH=O
Glyoxylate

COO^-
COO^-
Oxalate

Glycolate then leaves the chloroplast and is transported to the peroxisome by a postulated glycolate–glyoxylate shuttle. Within the peroxisome, glycolate is oxidized via a hydrogen peroxide producing oxidase to glyoxylate, which may either remain to be animated to serine or return via the shuttle to the chloroplast. The return of glyoxylate completes a small side cycle of the glycolate pathway. It is thought that this side cycle may be a mechanism to consume excess photosynthetic reducing power (NADPH) within the chloroplast, thus serving as a means of regulating net photosynthesis.

Notice that NADPH is reoxidized within chloroplasts by this mechanism without the production of H_2O_2, an event that could not be tolerated because chloroplasts contain no catalase. Instead, H_2O_2 is generated in the peroxisome, where there is ample catalase.

Glycine, arising from glyoxylate, undergoes interconversion reactions within the mitochondrion to serine, a part of the cycle that is not clearly understood. Serine is transported back into the peroxisome, where it is deaminated to oxalate and reduced to glycerate, then transported back into the chloroplast, where it is phosphorylated to phosphoglycerate. Thus the cycle is completed. Notice that part of the glycolate pathway is unidirectional and part reversible. Thus serine can be produced more directly from phosphoglycerate than from phosphoglycolate.

The cycle may accomplish either of the following. It may produce one mole of CO_2 and one of serine (or glycerate) from two moles of phosphoglycolate, or one mole of serine or one of glycine plus "C_1" from one

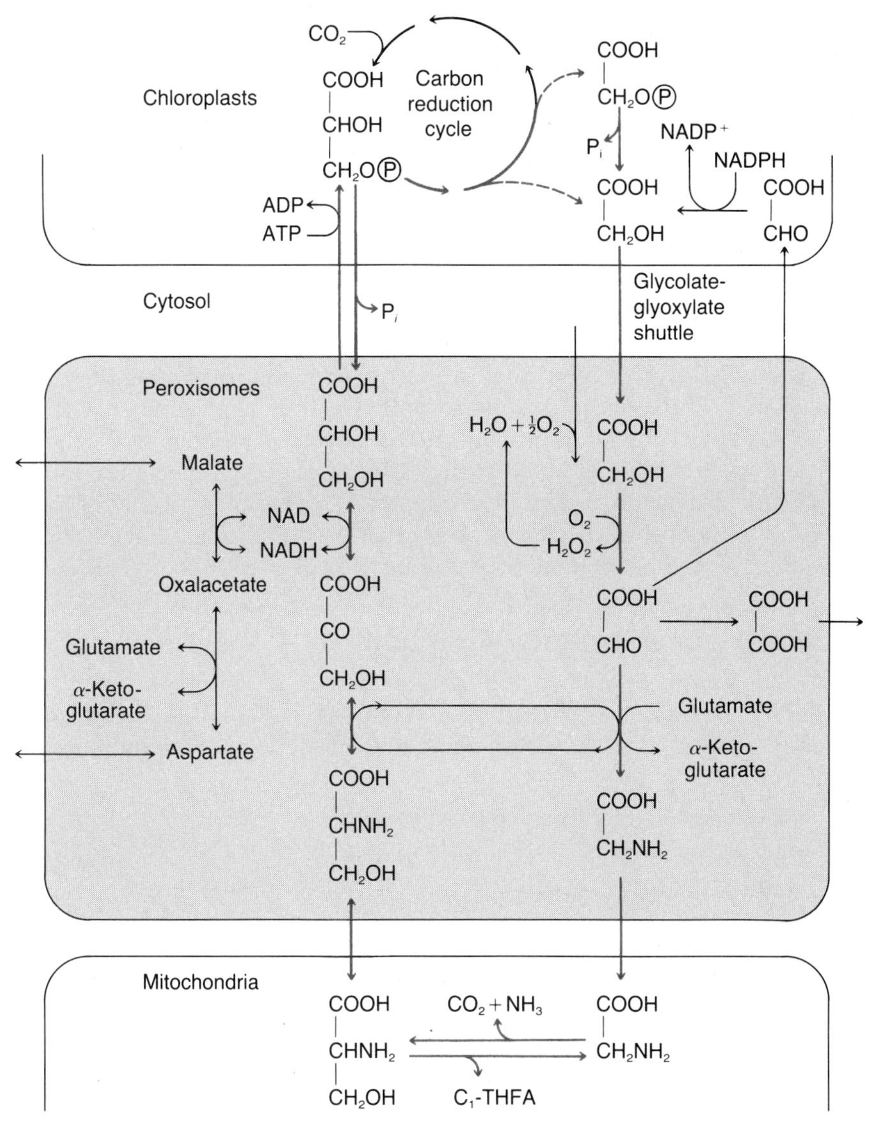

FIGURE 7.9 Reactions of the glycolate pathway. Four different cellular components are required to complete the cycle: chloroplasts, peroxisomes, mitochondria, and the cytosol.

3-phosphoglycerate. This metabolic scheme appears to be a very important one within the plant cell, since as much as half of the carbon fixed appears to go through it.

Reactions of glycolate also appear to be at the heart of the phenomenon called *photorespiration*. Photorespiration is a light-stimulated production of CO_2 that is different from the generation of CO_2 by mitochondria in the dark. It takes place in peroxisomes, where glycolate is oxidized to glyoxylate and then further to CO_2 and formate. The suggested reactions are as follows.

$$CH_2OH\text{-}COOH + O_2 \rightarrow CHO\text{-}COOH + H_2O_2$$
$$CHO\text{-}COOH + H_2O_2 \rightarrow HCOOH + CO_2 + H_2O$$

Photorespiration is driven by atmospheric conditions in which the O_2 tension is high and the CO_2 tension low. Apparently O_2 competes with CO_2 for the enzyme ribulose diphosphate carboxylase, which normally is the key enzyme in CO_2 fixation. When O_2 is used by the enzyme, an unstable intermediate is formed that breaks down into 3-phosphoglycerate and phosphoglycolate. The phosphoglycolate that arises by this reaction then adds to the glycolate concentration by removal of its phosphate group, and therefore more glycolate is available for additional oxidation and CO_2 release.

It is apparent that photorespiration is a detrimental process for the plant cell. It ties up some of the CO_2-fixing enzyme and returns a portion of the fixed CO_2 to the atmosphere. The rate of photorespiration may approach 50% of the net rate of photosynthesis, thus significantly reducing the efficiency of the latter process. It is a particular problem in C_3 plants that are more readily affected by low CO_2 tensions; C_4 plants are much more efficient in this regard (see Chapter 13). One of the goals of agriculture is to develop species of plants that carry out more efficient photosynthesis because of a decreased ability to carry out photorespiration.

Summary of Microbody Functions

Microbodies appear to have different functions in different organisms and in different tissues. In spite of these variations, certain of their functions overlap. Taken together, we summarize the functions of microbodies as follows:

1. Protection of cells from high oxygen toxicity.
2. Compartmentalization of purine and pyrimidine catabolism and D-amino acid destruction.
3. Gluconeogenic conversion of fats to sugars.
4. The formation of glycine and serine, which are the precursors for all C_1 biosynthesis.
5. Regulation of growth in plants by consumption of excess reducing power.

7.3 THE BIOGENESIS OF MICROBODIES

As suggested at the beginning of this chapter, microbodies are quite commonly thought to arise by an outgrowth and budding off of the endoplasmic reticulum. This point of view, however, is by no means universally accepted. There is a rather persuasive school of thought that the replication of microbodies is independent of the endoplasmic reticulum, with no convincing proof of membrane merger between these two systems. To complicate the biogenesis story, peroxisomes and glyoxysomes are generally studied independently by different research groups, and at the moment the data suggest that these two families of microbodies may arise in the cell by different mechanisms.

Rat Liver Peroxisomes

In trying to draw a conclusion regarding peroxisome biogenesis, several different experimental tacts and their results must be rationalized. Basically, these approaches and their outcomes can be summarized in the following categories.[7]

The Biochemical Homogeneity of Peroxisomes

Although peroxisomes may vary slightly in size, morphology, and density, all appear to contain approximately the same distribution of enzymes. Furthermore, there does not seem to be any difference in the size or enzyme contents of old as compared to young peroxisomes.

These results could be interpreted to mean that peroxisomes exist as a separate cellular entity and that there is never observed a "transitional peroxisome," one that is being generated from other cell components. Alternatively, it could be argued that when the point is reached for a budding off of the endoplasmic reticulum to produce peroxisomes, the endoplasmic reticulum involved and its contents have differentiated so much that the immediate product is no longer biochemically distinguishable from mature organelles.

Results of Electron Microscopy

The results of electron microscopy have been interpreted in ways that are often diametrically opposed. On the one hand, several research groups have reported sightings of physical continuity between peroxisome membranes and smooth membranes of the endoplasmic reticulum, such as that depicted in Figure 7.10. Furthermore, peroxisomes are often seen to lie adjacent to membranes of the endoplasmic reticulum, especially in cells that have been treated with hypolipidemic drugs, which stimulate the proliferation of these organelles.

However, peroxisomes have been seen by many researchers to contain extensions that may take on the form of tails or dumbbell shapes. These tails, rather than being interpreted as continuities with the endoplasmic reticulum, have been envisioned as replicative forms of peroxisomes because they stain positively for catalase throughout their interiors (see Figure 7.10).

Other morphological studies have suggested that membrane continuity between peroxisomes and the endoplasmic reticulum is not real, only apparent. When apparent, careful serial sectioning or tilting of the electron microscopic section to bring the membranes sharply into focus has shown that there is not true continuity between these two membrane systems.

Thus, depending on the interpretation of electron micrographs, either school of biogenesis can be supported. Either there is continuity between the ER membranes and peroxisomes or there is not. But in the case of rat liver cells, it is difficult to be certain.

Biochemical Differences between Peroxisomes and Endoplasmic Reticulum

Electron microscopy combined with cytochemistry has brought to light some striking differences between the endoplasmic reticulum and peroxisomes that have been interpreted as a reflection of no biogenetic relationship between the two.[8]

The approach used is based on the employment of glucose-6-phosphatase as a marker enzyme for endoplasmic reticulum and catalase activity

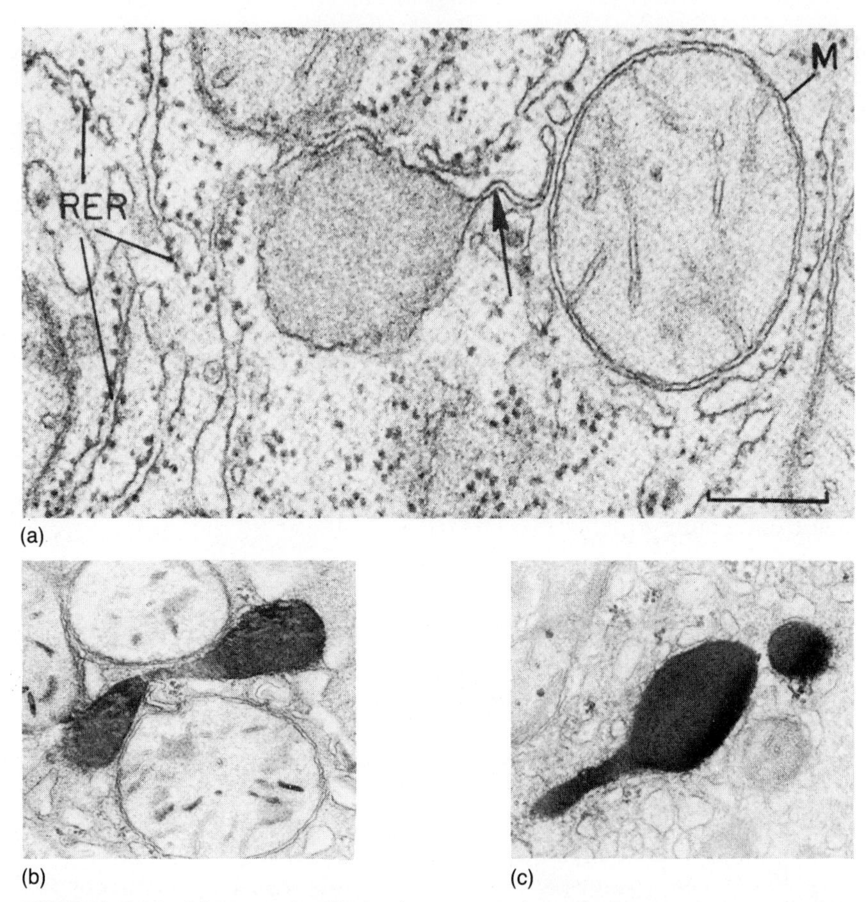

FIGURE 7.10 Micrographs illustrating morphological diversity among peroxisomes. (*a*) A peroxisome tail (arrow) appears to be a piece of smooth membrane, perhaps SER. (*b*) and (*c*) These extensions of the organelle stain for catalase throughout, as though all parts constitute a single organelle. (Electron micrograph (*a*) courtesy of Dr. George E. Palade; (*b*), (*c*) courtesy of Dr. Paul B. Lazarow.)

as a marker for the peroxisome matrix. Glucose-6-phosphatase is situated in the ER membranes in such a way that it releases its products into the cisternae of the endoplasmic reticulum.

The rationale for this approach is illustrated in Figure 7.11. If ER and peroxisome membranes are fused and thus their interiors connected, enzymatic products released in one organelle should be detected in the adjacent one simply by diffusion. Thus, phosphate generated in endoplasmic reticulum when glucose-6-phosphatase releases it from its substrate should diffuse into any attached peroxisome chambers and be deposited there as a lead precipitate. The same process, but in the reverse direction, would occur if catalase were provided with its substrate.

The results of these cytochemical approaches have been quite clear. First, as can be seen in Figure 7.12, peroxisomes are the only cellular organelles that are positive for DAB cytochemistry, even when ER membranes are in close proximity to the peroxisomes.

Second, as revealed in Figure 7.13, glucose-6-phosphatase activity is strong in the endoplasmic reticulum but completely absent from the peroxisomes.

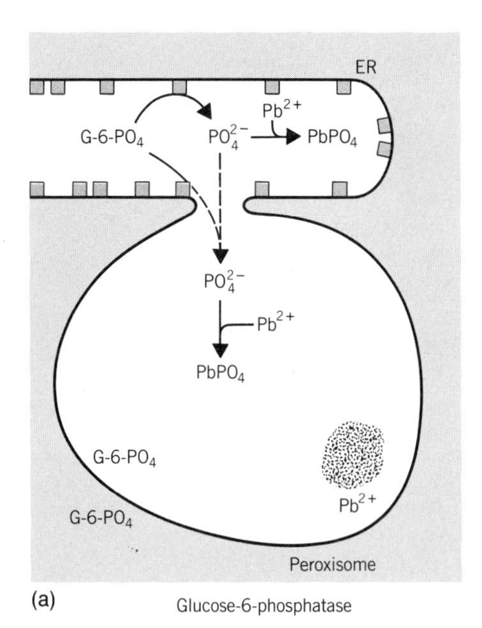

(a) Glucose-6-phosphatase

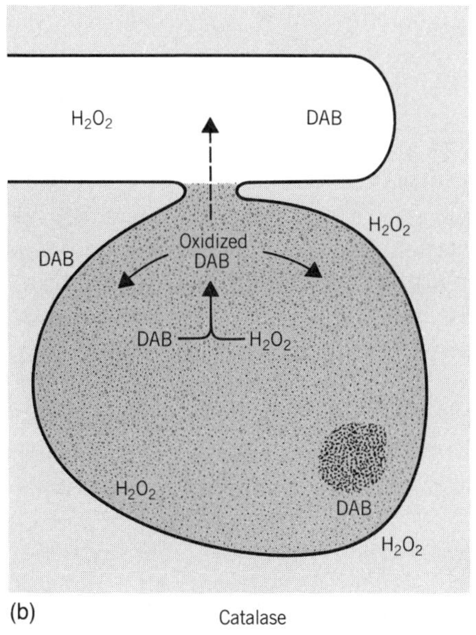

(b) Catalase

FIGURE 7.11 Expected diffusion pathways of products when endoplasmic reticulum and peroxisomes are connected. In the endoplasmic reticulum (*a*) glucose-6-phosphatase would release PO_4^{2-} from its substrate, the phosphate would diffuse into the peroxisome lumen and there be precipitated by lead. In the reverse direction, catalase in peroxisomes (*b*) would oxidize DAB, some of which would be expected to diffuse into the ER lumen. (Drawing courtesy of Dr. Paul B. Lazarow.)

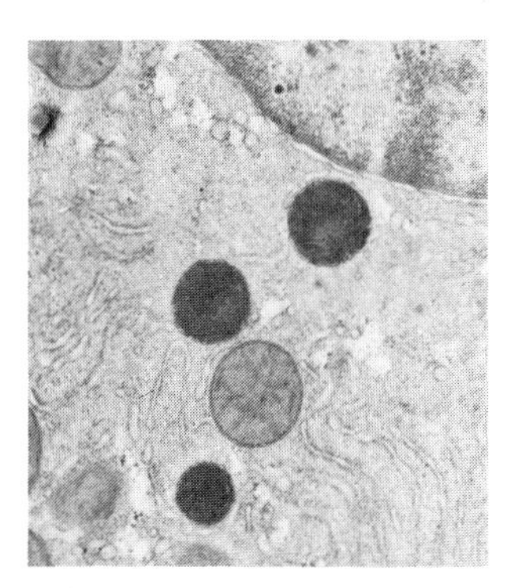

FIGURE 7.12 Exclusive staining of peroxisomes with DAB. Even though the peroxisomes are physically close to membranes of the endoplasmic reticulum and to mitochondria, only peroxisomes are catalase positive. (Courtesy of Dr. Paul B. Lazarow.)

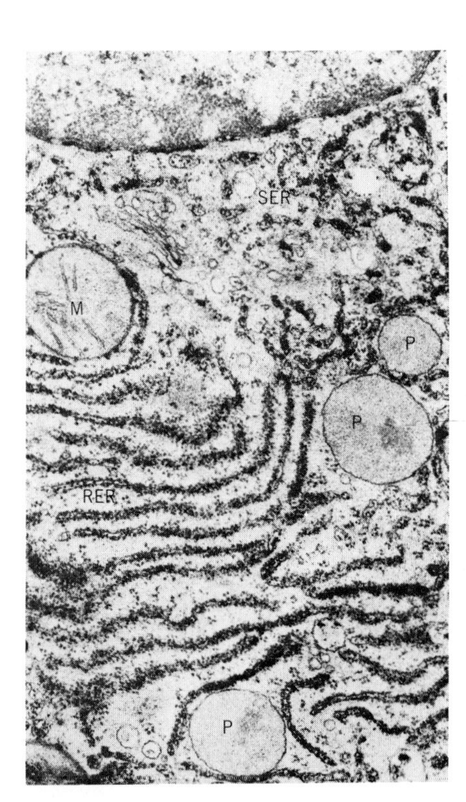

FIGURE 7.13 Exclusive staining of endoplasmic reticulum for glucose-6-phosphatase. Black patches, indicating the presence of the enzyme, are associated with SER and RER and with the nuclear envelope, but not with peroxisomes (P) and mitochondria (M). ×23,200. (Courtesy of Dr. Paul B. Lazarow.)

In a recent study of polypeptide and phospholipid compositions in highly purified organelle membranes, Paul Lazarow and his co-workers reported that peroxisome and ER membranes are quite dissimilar.[9] The peroxisomal membrane was found to contain three major polypeptides (21,700, 67,700, and 69,700 daltons), according to SDS–polyacrylamide electrophoresis, none of which was present in the endoplasmic reticulum. Conversely, most of the proteins possessed by the endoplasmic reticulum were absent from peroxisome membranes. The same studies showed the phospholipid composition of peroxisome membranes to be qualitatively similar to that present in the endoplasmic reticulum, but the ratio of phospholipid to protein is lower in peroxisomes.

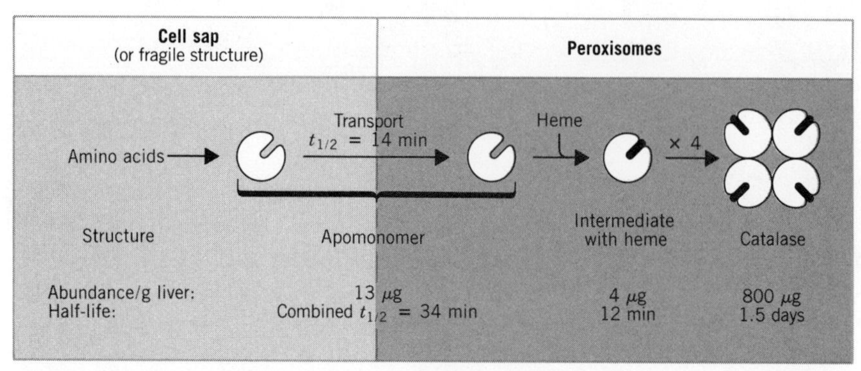

FIGURE 7.14 Proposed topology of catalase biosynthesis. Catalase is formed as the apomonomer on ribosomes, is released into the cytosol, and only later is incorporated into the peroxisomes. A small particulate cell component (designated fragile structure) appears to be involved in the process, but its structure and function are uncertain.

It is not surprising that these results have been interpreted to mean that rat liver peroxisomes are not connected to the endoplasmic reticulum but rather exist as separate compartments in the cell.

Studies on Biogenesis

Our conclusions thus far have been based on observations of membrane or organelle composition and morphologies that are not necessarily related to biogenesis. To get more directly at the events of biogenesis, Lazarow and de Duve have studied the appearance and movement of catalase in the cell as an indicator of peroxisomal birth.[10,11] The thought behind this approach is that catalase or its precursor should be seen in the endoplasmic reticulum before it appears in peroxisomes if the latter are indeed generated from the former.

An apomonomer is a protein subunit that does not possess a particular prosthetic group, in this case the heme group.

The results of these studies are consistent with the interpretation that catalase is formed as the apomonomer on ribosomes and is released mainly into the cytosol rather than the cisternae of the endoplasmic reticulum. Once inside the organelle, the monomer protein acquires the heme prosthetic group and then aggregates into its final tetrameric configuration. These results are summarized in schematic form in Figure 7.14.

These results lend credence to the contention that peroxisomes are not generated from endoplasmic reticulum because at no time is there evidence of catalase passing from the ER cisternae to peroxisomes. One must, however, entertain the possibility that catalase has its own peculiar route to peroxisomes that may be unrelated to the biogenesis of the organelle per se. Therefore, the results, although quite convincing, do leave open a crack for other possibilities.

Models of Peroxisome Biogenesis

Because of the difficulties involved in studying peroxisome biogenesis and the varied interpretations of the results, several models of peroxisome biogenesis have arisen over the years.[7] Briefly, they can be summarized as follows.

Budding from the Endoplasmic Reticulum Budding was an early model of peroxisome biogenesis in which it was assumed that proteins that were to become the contents of peroxisomes were discharged vectorially into the cisternae of the endoplasmic reticulum by attached ribosomes followed by a budding of appropriate ER sections to give rise to peroxisome progeny. This is the "textbook" model that is still widely upheld, despite a lack of clear and direct experimental support.

Synthesis on Free Polysomes with Subsequent Uptake This model suggests that endoplasmic reticulum may have no role in peroxisome genesis. Proteins, according to this view, are made on polysomes that are not attached to the endoplasmic reticulum. Once formed, they are freed in the cytosol and taken up by preexisting peroxisomes. The uptake of high molecular weight proteins has been one of the major speculative stumbling blocks to this model, since it is generally assumed that material this large is not carried by transport mechanisms across a membrane against a concentration gradient. However, in completely unrelated studies it has been shown that mitochondrial and chloroplast proteins enter these organelles posttranslationally, apparently by transmembrane movements. Thus there is precedence for this in other systems that lends credibility to this model.

Combined Budding and Uptake This model is a hybrid of the preceding two. It suggests that peroxisome membranes are formed by budding from the endoplasmic reticulum but that the matrix enzymes enter into the organelles by posttranslational uptake. Although this model fits nicely with some of the data, it does not fit well with other results, such as those showing a lack of glucose-6-phosphatase in peroxisome membranes.

The Peroxisome Reticulum Hypothesis A relative newcomer to the line of models has been proposed by Paul Lazarow.[7] He and his coworkers suggest that peroxisomes are interconnected with one another but not with the endoplasmic reticulum. They are thus free to undergo fission and fusion movements, giving rise to the often-sighted tailed and dumbbell-shaped organelles. Since they would be a part of a dynamic interacting system, the peroxisome would evidence biochemical homogeneity among the entire population of organelles.

This model suggests that peroxisome biogenesis occurs by fission or a budding from the system of membranes making up this specialized reticulum in the cell. Thus, the peroxisome reticulum can be thought of as structurally discontinuous at any one instant, but it can be viewed with time as a functionally continuous membrane system.

Glyoxysomes

Endosperm material has also served as a source for studies on microbody biogenesis. Perhaps the most progress and the most clear-cut results have been derived from the study of glyoxysomes.

In dry seeds, glyoxysomal enzymes are present at near zero levels.[4] In fatty seedling tissues, however, the numbers of glyoxysomes increase during the time wherein the lipid of the endosperm is being used for plant growth. When this reserve fat is depleted, the glyoxysomes also disappear.

Clearly in this situation glyoxysomes are being formed, and it is a reasonable object for the study of mechanisms of biogenesis.

One approach in these studies has been to follow the incorporation of [^{14}C]choline into the membranes of endoplasmic reticulum and glyoxysomes. When this is done, endoplasmic reticulum is labeled first and glyoxysomes secondarily. Furthermore as the radioactivity declines in the endoplasmic reticulum, it increases in glyoxysomes. This has generally been interpreted to mean that phosphatidylcholine is synthesized in ER membrane followed by a conversion of ER membrane to glyoxysome membrane. Another possible interpretation is that phospholipids are transferred individually from endoplasmic reticulum to glyoxysomal membranes. This is considered to be unlikely, because such a mechanism would require glyoxysome precursors into which the phospholipids would be incorporated. At present there is no evidence for such precursor structures in endosperm.

Another finding that links glyoxysomes biogenetically to the endoplasmic reticulum is that ER contains the variety of enzymes necessary for the synthesis and interconversion of all the phospholipids found ultimately in glyoxysome membranes, especially the major lipids phosphatidylcholine, phosphatidylinositol, and phosphatidylethanolamine. Glyoxysomes do not possess these biosynthetic enzymes and are therefore dependent on endoplasmic reticulum for their production.

As discussed earlier, certain of the glyoxysomal enzymes, in particular malate synthetase and citrate synthetase, are tightly bound to glyoxysomal membranes. Both enzymatic activities appear in endoplasmic reticulum before they appear in glyoxysomes, and the percentage of total activity rises in glyoxysomes while it declines in ER. Additionally, studies using antibodies have demonstrated that the malate synthetase that is formed in ER is identical to that which appears in glyoxysomes and that there is a quantitative recovery of the ER enzyme in the glyoxysomes.

These studies have strengthened the idea that microbodies arise from a budding of endoplasmic reticulum and have further confirmed this mechanism as textbook knowledge.

But the picture is obviously not yet clear. If microbody biogenesis occurs as a budding off of endoplasmic reticulum, an important question to answer is how the particular enzymes that are characteristic of the microbody are herded together and separated from other cisternal enzymes. If, on the other hand, microbodies are formed by a selective intake of enzymes from the cytosol, researchers must answer questions concerning the transport of proteins like catalase (60,000 daltons) across the membrane against a concentration gradient and the selection of a certain limited group of proteins from the cytosol.

What is clear, however, is that microbodies lack their own genetic machinery and, unlike mitochondria and chloroplasts, cannot maintain even a semiautonomous existence. They are completely dependent on other parts of the cell for their biogenesis.

In fatty seedling tissues, glyoxysomes disappear as the lipid is depleted. In some plants, such as the castor bean, the disappearance is due to an autophagic activity of the cell. Glyoxysomes are fused with digestive organelles and then broken down. In other systems different mechanisms may operate to decrease the glyoxysomal numbers.

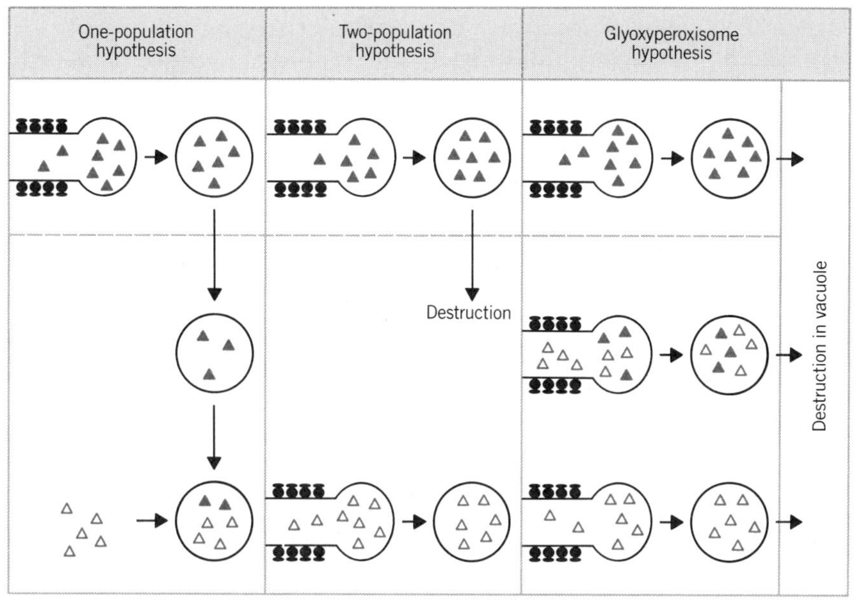

FIGURE 7.15 Models proposed to explain the transition of glyoxysomes into leaf peroxisomes in cotyledons. Phase I, above the dashed line, shows glyoxysome formation. Phase II, below the line, indicates modes of transition.

Microbody Transitions

The reserve fat in many seeds, which is drawn on during germination and the early stages of growth, is found in cotyledons. These emerge from the ground and for a brief period become functional leaves.

During this transient phase of development, glyoxysomes, which function in mobilizing fat, decline in numbers, and leaf peroxisomes appear. The decrease in glyoxysomes appears to be quite tightly linked to the increase in peroxisomes.

This cotyledon transition phenomenon has attracted a great deal of attention because it raises some obvious questions. Are glyoxysomes destroyed and peroxisomes formed *dé novo*? Are glyoxysomes converted into peroxisomes? Or is there yet another mechanism to explain the glyoxysome-to-peroxisome transition in cotyledons?

These questions have not been settled to the satisfaction of microbody researchers. Instead there have arisen three schools of thought represented by the three models in Figure 7.15. Each of these schools of thought is based on experimental work that is consistent with or supportive of the particular model. But in many instances more than one interpretation may be placed on the results to fit more than one model.

The One-Population Hypothesis

One school of thought suggests that a single population of microbodies in converted from glyoxysomal to peroxisomal in function by a change in enzyme content. This hypothesis assumes that the microbody membrane persists during transition. Furthermore, the enzymes that are characteristic of glyoxysomes, such as enzymes of the glyoxylate cycle and β-oxidation, are lost or destroyed. Enzymes characteristic of peroxisomes, such as those of the glycolate pathway, must be added to the transitional organelle.

The Two-Population Hypothesis

The second school of thought adheres to the notion that glyoxysomes, along with their complement of enzymes, are destroyed. Peroxisomes are then formed from the endoplasmic reticulum by a mechanism similar to that by which glyoxysomes are generated, except that a different population of enzymes is incorporated into the organelles. Thus the cotyledon sees two completely different populations of microbodies. One is not the precursor of the other.

The Glyoxyperoxisome Hypothesis

The third school of thought pictures a dynamic state of continual microbody production and destruction in the cotyledon. The basis of transition from glyoxysome to peroxisome is due, in this case, to a transition in the enzyme complement with time produced by endoplasmic reticulum. According to this hypothesis, there is an intervening stage wherein a microbody of mixed glyoxysome–peroxisome enzyme content is present, the glyoxyperoxisome.

Summary

Two classes of microbodies differing somewhat in their enzyme populations are known. One, the peroxisome, is found in both plant and animal systems; the other, the glyoxysome, is a plant microbody. Both classes contain catalases and oxidases, and the glyoxysome contains in addition enzymes of the glyoxylate cycle.

Microbodies are small membrane-bound organelles (0.2–1.5 μm in diameter) that are generally found in close proximity to mitochondria and chloroplasts. They are quite difficult to isolate in high yields, and their membranes are especially hard to separate from their enzyme cores. The enzymes they contain are noted for their oxidative abilities on a number of different substrates.

The enzymes of microbodies cooperate in a two-step process. The first is to use molecular oxygen to oxidize a given substrate, with the generation of hydrogen peroxide. The second is to degrade the H_2O_2 to water and oxygen to protect the cell from its toxic effects.

In animals, microbodies carry out substrate oxidations, such as on D-amino acids, as well as the β-oxidation of long chain acyl-CoA's. The latter is also conducted in plants, along with the reactions of the glyoxylate cycle and the glycolate pathway. The glyoxylate cycle is an important pathway for the use of fatty acids as a source of carbon, such as occurs during seed germination; the glycolate pathway feeds one-carbon biosynthesis and may serve as a regulator of photosynthesis.

The biogenesis of microbodies may be by a budding off of endoplasmic reticulum or by a mechanism of gathering together microbody enzymes that is not yet understood. In the fatty tissues of cotyledons, which serve briefly as functional leaves, there is a transition of microbodies from glyoxysomal to peroxisomal. The mechanism underlying this transition is highly controversial.

The subject remains unsettled but still of great interest. A solution to this problem of transition is certain to cast some light on the problem of microbody biogenesis in general. Considering the rate of progress in this field over the past decade, it should not be long before a satisfying answer is in.

References

1. Peroxisomes (Microbodies and Related Particles), C. de Duve and P. Baudhuin, *Physiol. Rev.* 46:323(1966).
2. Glyoxysomes of Castor Bean Endosperm and Their Relation to Gluconeogenesis, H. Beevers, *Ann. N.Y. Acad. Sci.* 168:313(1969).
3. Microbody Membranes, N. E. Tolbert and R. P. Donaldson, in *Mammalian Cell Membranes,* vol. 2, G. A. Jamieson and D. M. Robinson, eds., Butterworths, London, 1977.
4. Microbodies in Higher Plants, H. Beevers, *Annu. Rev. Plant Physiol.* 30:159(1979).
5. A Fatty Acyl-CoA Oxidizing System in Rat Liver Peroxisomes; Enhancement by Clofibrate, a Hypolipidemic Drug, P. B. Lazarow and C. de Duve, *Proc. Natl. Acad. Sci. U.S.* 73:2043(1976).
6. Mitochondria and Glyoxysomes from Castor Bean Endosperm, T. G. Cooper and H. Beevers, *J. Biol. Chem.* 244:3507(1969).
7. Biogenesis of Peroxisomes and the Peroxisome Reticulum Hypothesis, P. B. Lazarow, H. Shio, and M. Robbi, in *Biological Chemistry of Organelle Formation,* T. Bücher, W. Sebald, and H. Weiss, eds., Springer-Verlag, Berlin, 1980.
8. Relationship Between Peroxisomes and Endoplasmic Reticulum Investigated by Combined Catalase and Glucose-6-phosphatase Cytochemistry, H. Shio and P. B. Lazarow, *J. Histochem. Cytochem.* 29:1263(1981).
9. Polypeptide and Phospholipid Composition of the Membrane of Rat Liver Peroxisomes: Comparison with Endoplasmic Reticulum and Mitochondrial Membranes, U. Fujiki, S. Fowler, H. Shio, A. L. Hubbard, and P. B. Lazarow, *J. Cell Biol.* 93:103(1982).
10. The Synthesis and Turnover of Rat Liver Peroxisomes. IV. Biochemical Pathway of Catalase Synthesis, P. B. Lazarow and C. de Duve, *J. Cell Biol.* 59:491(1973).
11. The Synthesis and Turnover of Rat Liver Peroxisomes. V. Intracellular Pathway of Catalase Synthesis, P. B. Lazarow and C. de Duve, *J. Cell Biol.* 59:507(1973).

Selected Articles

Biochemical Studies on the Occurrence, Biogenesis and Life History of Mammalian Peroxisomes, C. de Duve, *J. Histochem Cytochem.* 56:507(1973).

Biogenesis of Peroxisomes and the Peroxisome Reticulum Hypothesis, P. B. Lazarow, H. Shio, and M. Robbi, in *Biological Chemistry of Organelle Formation,* T. Bücher, W. Sebald, and H. Weiss, eds., Springer-Verlag, Berlin, 1980.

Microbodies in Higher Plants, H. Beevers, *Annu. Rev. Plant Physiol.* 30:159(1979).

Microbodies in the Living Cell, C. de Duve, *Sci. Am.* 248:74(1983).

Microbodies—Peroxisomes and Glyoxysomes, N. E. Tolbert, in *The Biochemistry of Plants,* vol. 1, N. E. Tolbert, ed., Academic Press, New York, 1980.

Microbody Membranes, N. E. Tolbert and R. P. Donaldson, in *Mammalian Cell Membranes,* vol. 2, G. A. Jamieson and D. M. Robinson, eds., Butterworths, London, 1977.

Role of Peroxisomes in the Hepatic β-Oxidation of Long Chain Fatty Acyl-CoA's, P. B. Lazarow, *J. Cell Biol.* 75:203a(1977).

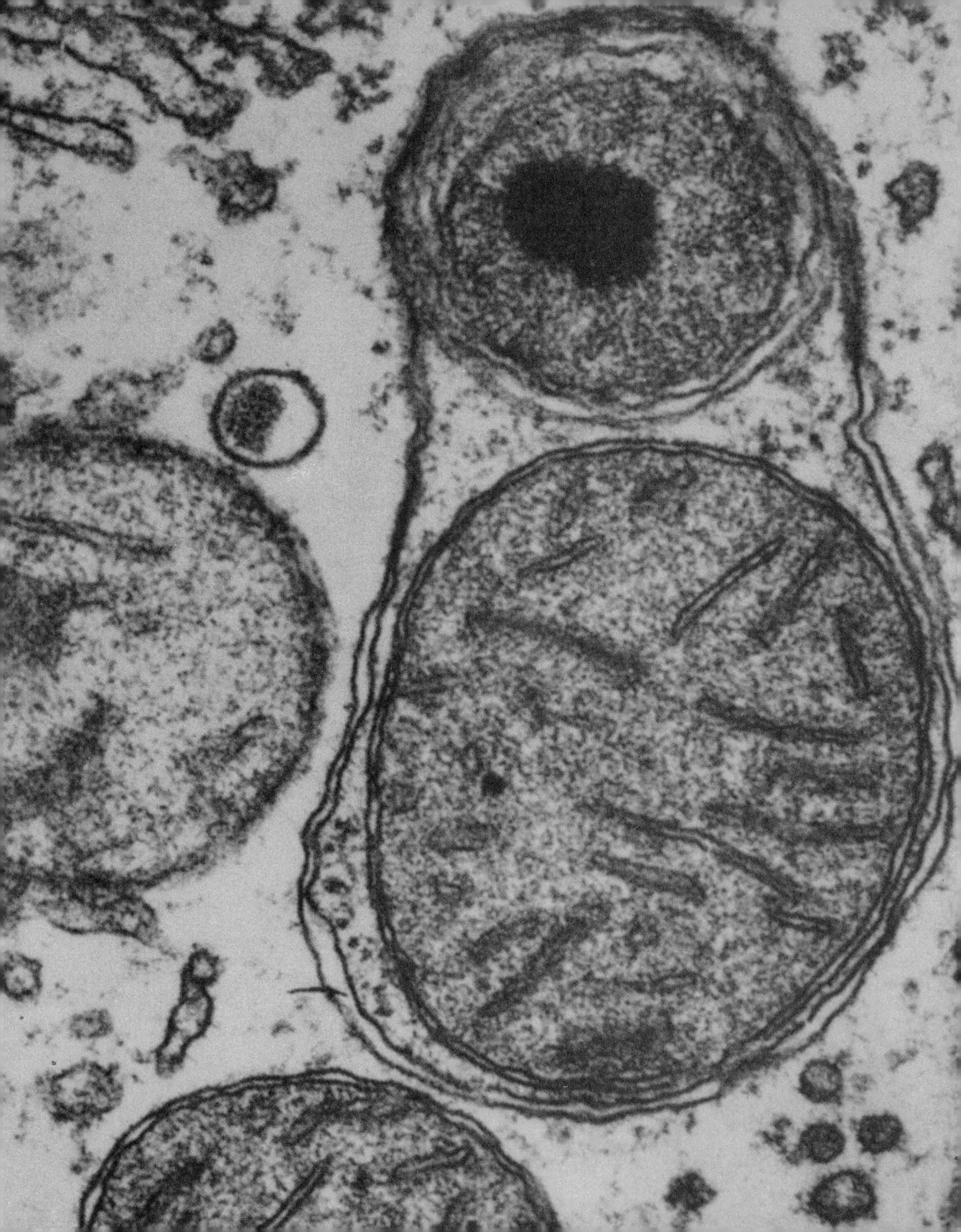

SECTION 4

BULK TRANSPORT AND INTRACELLULAR DIGESTION

The selective transport of molecules across the plasma membrane discussed in Section 2 is one means of bringing nutrients into a cell. It is a highly discriminatory mechanism that operates with little noticeable morphological impact on the cell or its plasma membrane.

Another means of transport used by most cells is bulk transport, or endocytosis. Although somewhat selective, it is less discriminatory than selective transport and is generally triggered by a population of molecules with certain chemical characteristics or by a large "foreign" particle. Bulk transport makes a tremendous morphological impact on the membrane as it invaginates, encircles a large volume of the environment, and sweeps it in.

Material brought into the cell by endocytosis is thus bounded by a membrane. This vacuole fuses with a lysosome, which releases its hydrolytic enzymes to degrade the material to its constituent subunits.

If the products of this hydrolysis are small enough, they are released from this digestive vacuole to the cytosol and used as nutrients. If the enzymes of the lysosome cannot thoroughly degrade the material, it may accumulate in the cell, forming residual bodies, or it may after a time be ejected from the cell by exocytosis.

Since bulk transport is a precursor process for intracellular digestion, we deal with these processes as a logical unit in the two chapters that follow.

8

Endocytosis

Phagocytosis is a phenomenon of considerable complexity. When it is exhibited by leucocytes, these cells are in the first place affected by various substances which possess an attraction for them. They proceed towards these substances by means of their amoeboid movements and then englobe them. Intracellular digestion may afterwards occur. Here then we have phenomena of sensibility, contraction, ingestion, and production of digestive fluids.

É. METCHNIKOFF, 1891

The phagocyte theory presupposes extraordinary powers on the part of the protoplasm of leucocytes, to which are attributed sensations, thoughts and actions, in fact a kind of psychical activity.

C. FRÄNKEL

The first quotation at the beginning of this chapter is a statement of the theory of phagocytosis as developed and championed by the Russian-born zoologist, Élie Metchnikoff. The second quotation, written at about the same time, indicates something less than a complete embrace of his ideas. But Metchnikoff was not easily thwarted by those who were reluctant to except the results of his methodical comparative approach to the study of immunity. This was fortunate for a great deal of biological science, but in particular for an understanding of the cellular defense mechanisms in host resistance.

Metchnikoff's early studies on phagocytosis were quite simple and straightforward. In fact, a pivotal study took place right at home, not in the laboratory of some great institute, as told in these famous autobiographical lines.

"One day when the whole family had gone to a circus to see some extraordinary performing apes, I remained alone with my microscope, observing the life in the mobile cells of a transparent star-fish larva, when a new thought suddenly flashed across my brain. It struck me that similar cells might serve in the defense of the organism against intruders. Feeling that there was in this something of surpassing interest, I felt so excited that I began striding up and down the room and even went to the seashore in order to collect my thoughts.

I said to myself that, if my supposition was true, a splinter introduced into the body of a star-fish larva, devoid of blood-vessels or of a nervous system, should soon be surrounded by mobile cells as is to be observed in a man who runs a splinter into his finger. This was no sooner said than done.

There was a small garden to our dwelling, in which we had a few days previously organized a "Christmas tree" for the children on a little tangerine tree; I fetched from it a few rose thorns and introduced them at once under the skin of some beautiful star-fish larvae as transparent as water.

I was too excited to sleep that night in the expectation of the result of my experiment, and very early the next morning I ascertained that it had fully succeeded.

That experiment formed the basis of the phagocyte theory, to the development of which I devoted the next twenty-five years of my life."[1]

There were, of course, forerunners to Metchnikoff's line of thinking. Darwin's published work on the theory of natural selection influenced his approach to science profoundly, and Ernst Haeckel had earlier, in 1859, observed the ability of white blood cells to take up foreign matter in sites of inflammation. In addition, Robert Koch had reported bacteria in white blood cells in wound infections in 1878. This was an important observation for the development of the theory of inflammation, but Koch thought of it in terms of a mechanism of bacterial multiplication rather than destruction, a point squarely opposed by Metchnikoff.

Metchnikoff had considerable support on his side in fighting the phagocyte battle. Pasteur provided a place for him to work at the newly founded Institut Pasteur in 1888, and he was encouraged by the great Rudolph Virchow, a proponent of the theory of cellular pathology. C. Claus, director of the Zoological Institute in Vienna, invited Metchnikoff to publish his results in his journal. In fact, the Viennese group helped him select the term *phagocyte* to properly describe the cell that he was studying.

Apparently unknown to Metchnikoff, but very important for the outcome of the struggle, was the work of George M. Sternberg in the United States. In 1881, at a meeting of the American Association for the Advancement of Science in Cincinnati, Sternberg voiced an opinion that was nearly identical to Metchnikoff's theory of phagocytosis in inflammation. Since Sternberg's work actually predated Metchnikoff's by about a decade, it is only proper to acknowledge his contributions as pioneering efforts in this field. In any event, the last decade or two before the turn of the century was a time ripe for a major advance in the understanding of the cellular role in defense.

Certainly part of the problem in convincing the scientific world of the validity of phagocytosis as a defense mechanism in inflammation was that it was only a part of the story of immunity. At the time of Metchnikoff's work, humoral agents were being found for bacteria and toxins, and even the humoral factor complement was being described. These findings attracted a very strong following of adherents to the humoral theory of immunity. This, by itself, was as incomplete as the cellular basis of immunity proclaimed by Metchnikoff. The complete story demanded a merger of ideas. Eventually, this came about because of the relentless determination of scores of researchers who insisted that these controversial matters be tested and tried in the objective experimental setting.

"Humor" means a liquid or moisture in Latin. Humoral agents, in this context, are proteins such as antibodies, which are soluble in body fluids. Phagocytes are viewed as *particulate* or *cellular* blood components and are not a part of humoral defense mechanisms.

This story is too long and complicated to repeat here. But at least it is safe to report that the phagocyte storm is over. Phagocytosis is well documented as a mechanism operative in resistance and immunity as well as in a number of other cellular processes.

If phagocytes have history books, they would do well to note the events of 1908 in special detail. In that year Metchnikoff shared the Nobel prize in physiology and medicine with Paul Ehrlich, who was one of the principal advocates of the humoral theory of immunity (see Chapter 5 on receptors). It was thus astutely recognized that each theory by itself was inadequate to explain the complex mechanism of host resistance in higher organisms but that both were valid and essential components in the machinery of immunity. The theory of phagocytosis was now formally recognized.

8.1 THE NOMENCLATURE OF ENDOCYTOSIS

So far we have been using the term *phagocytosis* exclusively to describe the cellular engulfment of material. In modern parlance this term has a more restricted meaning, namely, *the engulfment of particulate material.* The term *pinocytosis* (cellular drinking), coined by W.H. Lewis in 1931,[2] designates *the engulfment of soluble materials,* such as proteins or other smaller molecules in solution. A scaled-down version of pinocytosis in which only very tiny amounts of liquid are taken up is referred to as *micropinocytosis*. The general property of cellular engulfment, whether of particulate material or of substances in soluble form, is called *endocytosis*.

Figure 8.1 is an attempt to depict the rather arbitrary distinctions that have been made between the three endocytic processes. It is relatively easy to distinguish the engulfment of a large particle such as a bacterium or latex bead from that of a solution of low molecular weight proteins,

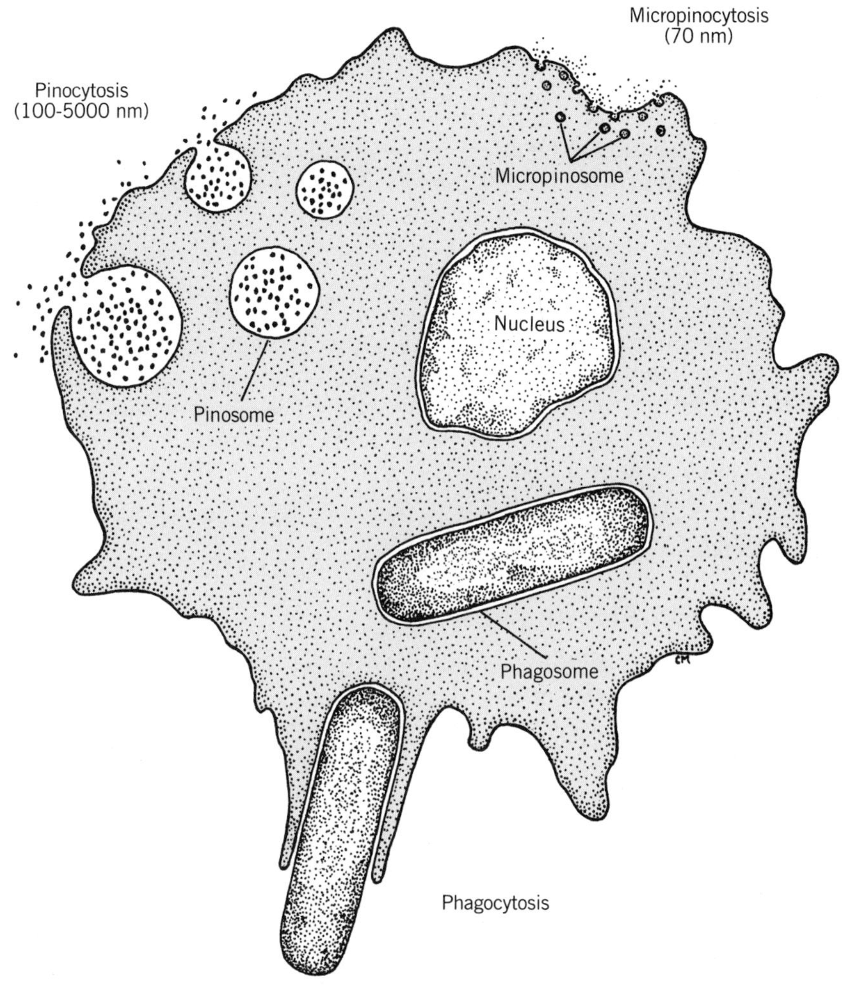

FIGURE 8.1 The types of endocytosis. Particles are engulfed by phagocytosis. The internalized particle is called a phagosome. Relatively large volumes of soluble materials are brought in by pinocytosis, and smaller volumes enter by micropinocytosis. The internalized products of the latter two processes are pinosomes and micropinosomes.

but a reasonable demarcation between high molecular weight protein aggregates and submicroscopic particles becomes essentially impossible.

The distinction between pinocytosis and micropinocytosis is even more arbitrary, and their definitions may vary somewhat according to the research group studying these processes. One manner of distinction is that with light microscopy, pinocytosis is observable and micropinocytosis is submicroscopic. With the electron microscope, however, both processes are observable.

Some workers have placed size restrictions on the vesicles that form, to assist in classifying them as resulting from micropinocytosis. Thus vesicles that are about 70 nm in diameter are *micropinosomes,* products of micropinocytosis. Larger vesicles, ranging from 100 to 5000 nm in diameter are *pinosomes,* derived by the process of pinocytosis. Since micropinosomes may fuse to form larger vesicles, a clear distinction based on vesicle size is not always possible.

At present there is not uniform agreement among researchers regarding the validity of viewing phagocytosis and pinocytosis as different processes. There are those who stress the similarities if not identities of these processes, pointing to the results of morphological and energy requirement studies. Others, noting the results of similar studies, are sure that there are some fundamental differences in these mechanisms of bulk transport. We take neither side, but deal instead with the three terms separately; this makes an organized presentation a bit easier. Keep in mind, however, that at the very least, phagocytosis, pinocytosis, and micropinocytosis are closely related cellular activities. Functionally, they are processes that effect transport of materials across the plasma membrane in bulk form. Structurally, there may be some subtle differences between processes, related perhaps to the kinds of materials transported.

8.2 PHAGOCYTOSIS

The Common Phagocytic Cells

Phagocytosis has been observed in a wide variety of cell and tissue types. Many of the lower forms of life, such as the amoebae, ciliates, and phytoflagellates, use phagocytosis as a common form of feeding.

Amoebae, ciliates, and phytoflagellates are protozoans—unicellular organisms. All animals except the protozoans are multicellular. They are metazoans.

In lower metazoans, (e.g., sponges, coelenterates, and turbellarians), nutrition often depends on this process too. Phagocytic cells are found distributed in many types of tissue in higher metazoans. In mammals, phagocytes circulate in the blood and lymphatic systems and are distributed in fixed positions in many organ and tissue sources.

The phagocytic cell that through the years has been the object of close scientific scrutiny is the *mononuclear phagocyte*, the self-same cell to which Metchnikoff eventually gave most of his attention.

The Mononuclear Phagocyte System

The mononuclear phagocyte, or *macrophage* as it is commonly called, is ubiquitously distributed in mammalian tissues, as can be seen by its anatomical sites listed in Table 8.1. There is hardly a tissue that escapes its pervading presence.

The liver houses more macrophages than any other single organ, where these cells keep the blood clear of foreign materials that may have entered the circulatory system at any point prior to its location. It is evident, however, that this function is not restricted to liver macrophages. Macrophages line the sinuses of the spleen, lymph nodes, and bone marrow and are embedded in a number of other tissues, where they carry out phagocytosis on any particle that may be harmful to the organism, such as bacteria, fungi, protozoa, effete (worn-out) erythrocytes, and other foreign particles.

To gain a better appreciation for the role played by macrophages in higher organisms, it is well to underscore the tissue system of which they are a part. In 1924 Aschoff[3] developed the concept of the *reticuloendo-*

Table 8.1 The Distribution of Macrophages in Mammals

Anatomical Site	Localization of Cells
Liver	Kupffer cells lining hepatic sinusoids; connective tissues of portal tracts
Spleen	Lining venous sinuses and enmeshed in Billroth cords of red pulp; scattered among lymphocytes of Malpighian follicles in white pulp; in marginal zones
Lymph nodes	Lining subcapsular and medullary sinuses; scattered in medullary pulp and in lymphoid follicles of cortex
Bone marrow	Lining venous sinuses of red marrow; scattered monocytes and macrophages in extrasinusoidal tissues
Thymus	Scattered throughout cortex and medulla; within Hassall's corpuscles in some species
Lung	Within interstitial tissue of alveolar wall; in alveolar spaces
Central nervous system	Microglia
Serous cavities	Peritoneal and pleural fluids; "milk spots" of peritoneum and pleura
Adrenals	Lining sinusoids of cortex (particularly zona reticularis); scattered in medulla
Joints	Type A or M cells of synovial lining; within synovial fluid
Subcutaneous tissue, alimentary tract, pituitary, testis, ovary, endometrium, kidney	Connective tissues generally, lining vascular channels in pituitary, testis, ovary and decidua
Blood	Monocytes and some macrophages
All sites	Inflammatory exudates

thelial system (RES), a terminology in wide usage today. This system is made up of reticular cells of the spleen and lymph nodes along with the reticuloendothelial cells of lymph and blood sinuses. When viewed in its broadest sense, this includes histiocytes, splenocytes, and monocytes, and the low activity endothelial cells and fibrocytes (Table 8.2). The cells in this system differ in a number of characteristics, such as morphology, arrangement, and their rate of particle uptake. They are, however, all phagocytic cells.

Reticular cells are often highly branched, and because of this can form a netlike barrier. They may have the following appearance:

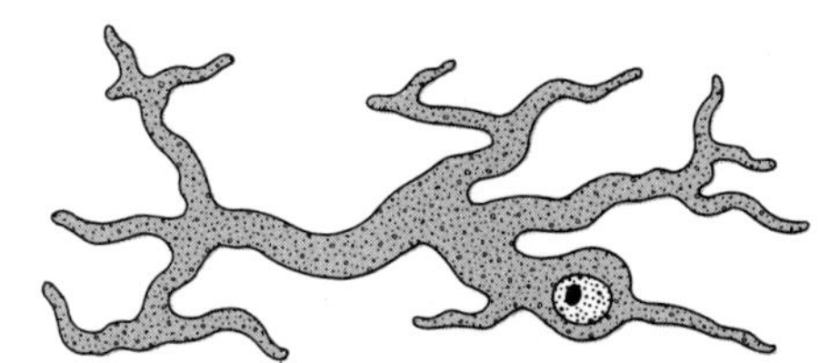

The endothelium is the layer of cells that lines cavities of the body, such as within the heart, and the blood and lymph vessels.

In recent years the concept of the RES has been criticized because of its inclusion of such diverse cell types, and a newer term has evolved that provides a better framework for the accumulated knowledge concerning the phagocytic mononuclear cells. The concept of the *mononuclear phagocyte system* (MPS) was developed primarily by van Furth and ratified by a committee at the Conference on Mononuclear Phagocytes in Leiden, the Netherlands, in 1969.[4] This system, depicted in Table 8.3, is made up of cells that are related by morphology, function, and origin. The phagocytes of the MPS are characterized by avid phagocytosis and pinocytosis and by firm adherence to glass surfaces. These are the macrophages that have been so extensively studied.

Table 8.2 Components of the Reticuloendothelial System

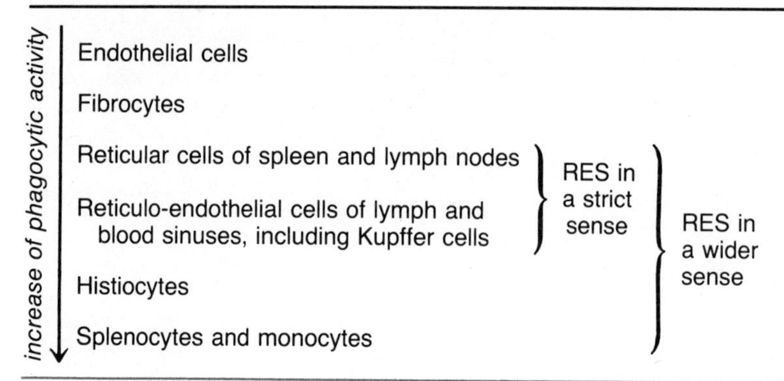

increase of phagocytic activity ↓		
Endothelial cells		
Fibrocytes		
Reticular cells of spleen and lymph nodes	RES in a strict sense	RES in a wider sense
Reticulo-endothelial cells of lymph and blood sinuses, including Kupffer cells	RES in a strict sense	RES in a wider sense
Histiocytes		RES in a wider sense
Splenocytes and monocytes		RES in a wider sense

Note. Cells are listed in order of increasing phagocytic activity. The system may be defined in a strict sense, a wider sense, or in its broadest sense. It is apparent from the list that the cells in this system differ in many of their properties.

Table 8.3 Components of the Mononuclear Phagocyte System

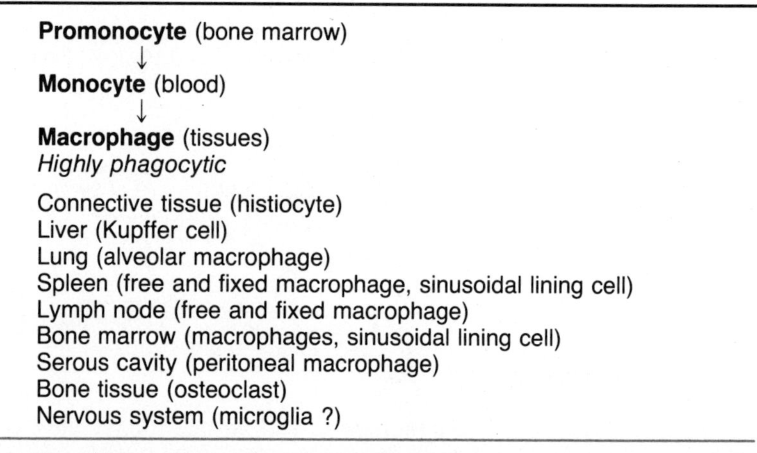

Promonocyte (bone marrow)
↓
Monocyte (blood)
↓
Macrophage (tissues)
Highly phagocytic

Connective tissue (histiocyte)
Liver (Kupffer cell)
Lung (alveolar macrophage)
Spleen (free and fixed macrophage, sinusoidal lining cell)
Lymph node (free and fixed macrophage)
Bone marrow (macrophages, sinusoidal lining cell)
Serous cavity (peritoneal macrophage)
Bone tissue (osteoclast)
Nervous system (microglia ?)

Note. The cells making up this system arise as promonocytes in bone marrow. Monocytes appear in the circulation on their way to tissue sites. The cells of the MPS are thus related by their common origin as well as in structure and function.

Figure 8.2 is an idealized drawing of the macrophage. It is a cell rich in lysosomes and engulfed particles with an outer membrane that is highly indurated and flexible. An electron micrograph better depicting this latter property is shown in Figure 8.3.

Another type of phagocytic white blood cell in mammals that has received considerable research attention is the *polymorphonuclear leukocyte* (PMN). The PMNs, together with the mononuclear phagocytes, are called *professional phagocytes* in honor of their full-time phagocytic careers.[5] All other phagocytic cells are referred to as *facultative* or *nonprofessional phagocytes,* suggesting something less than their primary raison d'être. Our discussions continue with emphasis on the professional types.

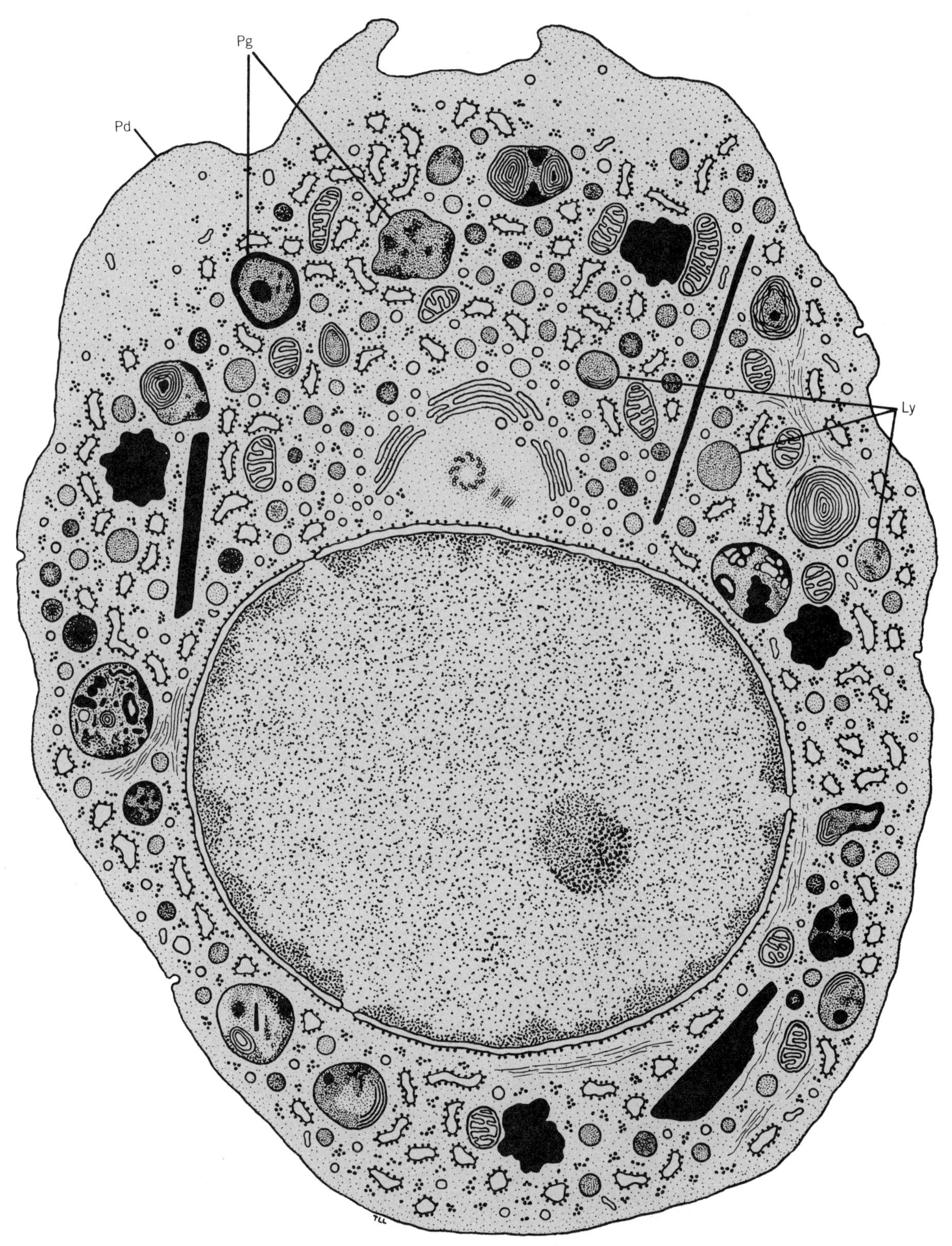

FIGURE 8.2 Idealized drawing of a macrophage. The cell is rich in lysosomes (Ly) and engulfed materials within phagosomes (Pg). The pseudopodia (Pd) are generally devoid of organelles.

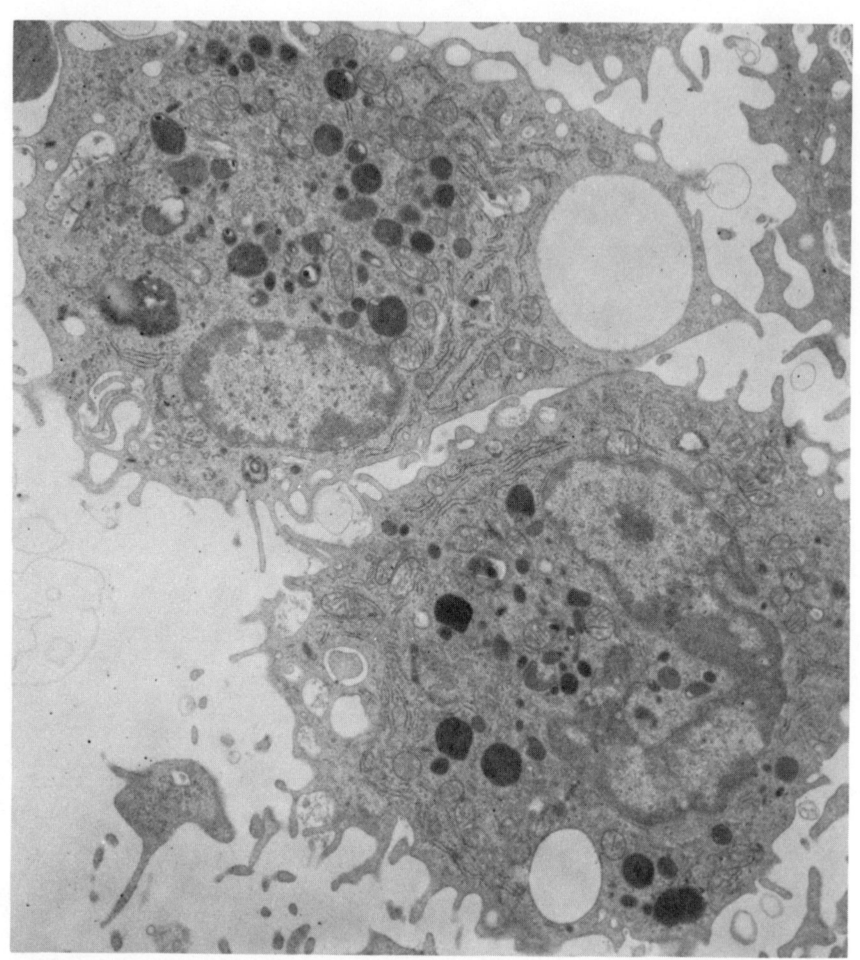

FIGURE 8.3 Peritoneal macrophages showing numerous cytoplasmic extensions and prominent dense bodies or lysosomes. Some of the latter are small and homogeneous and others large and heterogeneous. Large lipid vacuoles are obvious. Lipid is used to stimulate these cells. ×8,200. (Courtesy of Dr. Ian Carr.)

The Stages of Phagocytosis

Phagocytosis is a continuum of cellular activity that for purposes of discussion is generally viewed as occurring in stages. One way to divide the activities is as follows: the signal, pursuit, surface recognition, and engulfment.[6] Let us examine these stages separately in some detail, recognizing that they most certainly overlap under normal circumstances.

The Signal

The first stage of phagocytosis can be thought of as a sudden "awareness" on the part of the cell that foreign, engulfable material is in its vicinity. In other words, some type of signal or first messenger starts the whole process. The origin of this initial event has been very hard to pin down experimentally, but it is generally thought that something must reach the phagocyte surface to trigger the events that follow.

The placement of fixed macrophages at strategic thoroughfares in the organism makes it plausible to assume that the initial signal results from an interaction between the foreign material and the macrophage itself. This may result in the release of a signal factor that relays information of the encounter to other macrophages. Theoretically, the factor could arise from the macrophage itself, or from the tissue in which it is located. Different phagocytic cells may be signaled by different mechanisms.

These are largely speculations that are awaiting some good experimental validation. This stage of phagocytosis is difficult to study because it is the beginning of a continuum and apparently depends on the earliest possible interactions between the phagocytic cell and the signal. In addition, an *in vivo* signal system may be very hard to reproduce and study *in vitro*.

Pursuit

Once the signal has been given, macrophages exhibit the cellular curiosity of moving toward the scene of action. This aspect of phagocytosis, properly called *chemotaxis,* has received some careful and illuminating study, and has been nicely reviewed by Wilkinson.[7]

Chemotaxis is a common phenomenon in the biological world. It is the movement of an organism (or in this case a cell) in response to a chemical gradient. Chemotaxis may be either negative (moving down the gradient) or positive (moving up). In the case of macrophages, it is positive.

For the chemotactic phenomenon of directional migration to occur, there must be a gradient of chemotactic substances. In the absence of a gradient, or in regions of high chemotactic activity, cells tend to migrate randomly.

Cells must therefore possess a mechanism for detecting and responding to a gradient of molecules. Probably the most widely accepted hypothesis in this regard is that multiple receptor sites are scattered over the surfaces of mobile cells. When a gradient exists across the cell, the receptor sites on the side of the cell proximal to the center of origin of the attracting substance will bear a heavier load of attractant species than the sites on the distal side. The detection of this gradient across the cell is then translated into a motor event to cause migration upstream in the gradient.

A number of substances, including polysaccharides, polypeptides, and proteins, have been shown to be effective chemotactic mediators *in vitro*.[8] Certain of these "cytotaxins," such as the split products of complement, appear to act directly on the macrophage surface; others, called "cytotaxigens," act indirectly by generating or unmasking the direct-acting cytotaxins. Two categories of chemotactic substances have therefore been established reflecting different modes of action. Some of the more common cytotaxins and cytotaxigens are listed in Table 8.4.

Complement is a family of serum proteins that bind to antigen–antibody complexes and trigger the lysis of foreign cells.

Cytotaxins have been obtained from a number of exogenous and endogenous sources. A majority of leukocyte cytotaxins are polypeptides and proteins. These materials all must be hydrosoluble and diffusible. Apparently no polysaccharide, lipid, or nucleic acid has been found to meet the criteria of cytotaxins.

Most of the cytotaxigens act on the serum complement system to produce cytotactic fragments; however they are not limited to operating by way of complement. Cytotaxigens do not have to be hydrosoluble and diffusable. In fact, they are often particulate. But many others are soluble biological molecules, including proteins, polysaccharides, lipids, and even low molecular weight species such as guanosine. The molecular rationale

Table 8.4 **The Classification of Chemotactic Substances According to Their Mode of Action**

Cytotaxins	Agents with direct chemotactic effect on cells *(a) Endogenous:* Present in normal and antigen–antibody (Ag/Ab) activated sera; split products of complement components C_3 and C_5; in Ag/Ab activated plasma; in exudate fluids; in postgranular supernatants from neutrophils *(b) Exogenous:* Bacterial culture filtrates; casein; peptone Cytotaxins can be further classified by their cell specificity.
Cytotaxigens	Agents without direct chemotactic effect on cells; they induce formation of cytotaxins or unmask them *(a) Endogenous:* Granules from macrophages, neutrophils, liver, and heart; postgranular supernatant from macrophages, neutrophils, and liver; plasmin, trypsin; C_3 convertase, C_3 inactivator complex *(b) Exogenous:* Ag/Ab complexes, bacteria, endotoxin, aggregated γ-globulin, guanosine, glycogen, tuberculin, zymosan Some cytotaxigens require incubation with unheated normal serum, others with plasma, and still others with complement components to generate chemotactic mediators for either neutrophils or macrophages.

for this heterogeneous population of molecules possessing similar activities is not at all clear.

The picture that emerges is a very complex one, probably involving humoral blood factors as well as the macrophage itself. Furthermore, it is possible that a gradient of several *different* cytotaxins attracts the *same line* of macrophages or that *different* cytotaxins influence the migration of functionally *different* macrophages at different sites in the organism. It is of particular interest in light of the chapter on lysosomes that follows, that intracellular agents of leukocytes, such as lysosomes, when incubated with serum or plasma, attract other leukocytes. Therefore, either released or breakdown products of macrophages may serve to accelerate other cells to the site of attack.

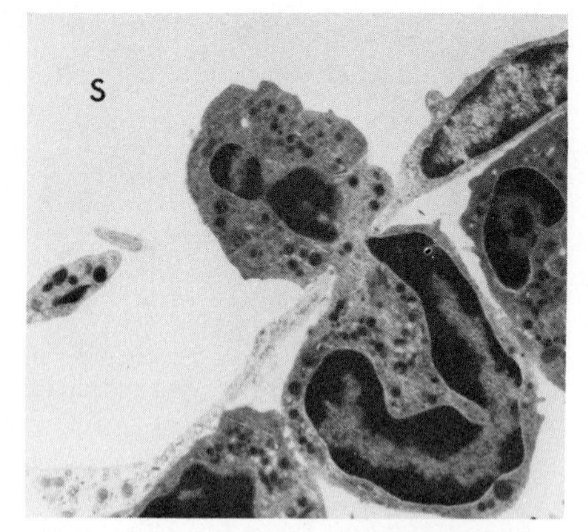

FIGURE 8.4 A leukocyte passing through an endothelial cell of a marrow vascular sinus. The cell exits from the marrow into the sinus (S) by squeezing through an attenuated region in the cytoplasm of the endothelial cell. The challenge for the egressing cell is about as great as for a human to slip through a knothole in a board fence. Fortunately, the leukocyte is better suited for the task. (Courtesy of Dr. Marshall A. Lichtman.)

Macrophages in pursuit emigrate through the endothelial lining of blood vessels and other tissue spaces to the site of foreign particles and there continue their phagocytic activities. The egression of leukocytes between or through endothelial cells is a remarkable event of cellular contortions. In one example of this (Figure 8.4), a leukocyte passes through an endothelial cell as it leaves the marrow and moves into a vascular sinus. When neutrophils exit the circulation, their exit is intercellular rather than through an attenuated region of the cytoplasm of a cell.

At the site of inflammation, the increased concentration of phagocytic cells engulfing foreign material, effecting the release of tissue substances, triggering the blood clotting mechanism, and dying and disintegrating probably all affect the chemical gradients that entice more phagocytic cells to the region. It will take some time before the contributions of all these factors are sorted out and accounted for experimentally.

Surface Recognition

At the site of foreign particle intrusion, physical contact is made with the macrophage, and engulfment proceeds. This particular stage of phagocytosis is governed by an interaction between the foreign particle and receptor sites on the surface of the phagocytic cell. Recognition is a crucial stage in phagocytosis, for it is in this interaction that the cell discriminates between foreign and autochthonous (host) material.

It is generally agreed by workers in this field that recognition of a foreign particle and specific attachment to it are regulated by a number of forces. Electrostatic forces are thought to be important initially in the interactive process because they operate over greater distances than other forces. However, forces such as van der Waals forces and hydrophilic–hydrophobic interactions probably contribute to recognition at shorter distances.

Studies on the interaction of peritoneal macrophages with glutaraldehyde-treated sheep red blood cells *in vitro* have shown that the charge density on the cell surfaces markedly influences phagocytosis.[9] When cells are combined in a low ionic strength medium, which produces an increased negative cell surface charge, there is a concomitant reduction in phagocytic activity. Treatment with protamine sulfate, which reduces the surface charge of both cell types, enhances phagocytic activity. An analysis of the data reveals that the phagocytic activity of macrophages decreases linearly with an increase in the product of the negative charge density on the surfaces of both treated red cells and macrophages.

Protamines are highly basic proteins of low molecular weight (~ 5000 daltons) and rich in arginine residues. At physiological pH they thus possess a net positive charge and interact with the negatively charged components on the cell surface.

The role played by electrostatic interactions on phagocytosis once the phagocytic cell has made contact with the foreign particle is not fully understood, but the data cited above imply that these forces govern some type of recognition during the initial phases of cell–particle interaction.

Some researchers have centered their attention on the effect of other forces on phagocytosis. For example, an intriguing series of studies was carried out by van Oss and associates, who demonstrated that there is a good correlation between the degree of hydrophobicity of bacteria and their susceptibility to phagocytosis.[6] Bacteria with highly hydrophilic capsules, often characteristic of the most virulent forms, resist phagocytosis. These workers propose that phagocytosis takes place primarily because of differences in the surface free energies of phagocyte and intruding particle.

The experimental approach used was a fairly simple and old technique. The contact angle, θ, between a surface and a tangent to a liquid drop at the point the drop meets the surface was measured to within $<1°$ with a rotatable goniometer (Figure 8.5). For these experiments, the surface consists of prepared monolayers of various bacteria, neutrophils, or macrophages on which there is applied a drop of saline solution. It is important to understand this experimental setup for the discussion that follows.

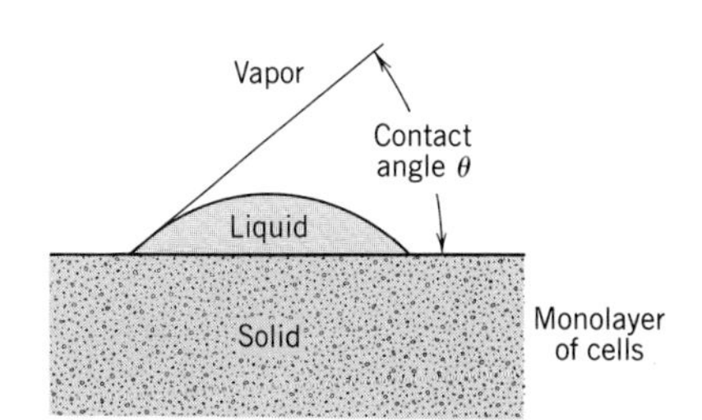

FIGURE 8.5 The contact angle between a liquid droplet and a solid surface. In practice, a monolayer of bacteria or phagocytic cells constitutes the solid surface. The more hydrophobic the monolayer, the greater the contact angle.

Table 8.5, summarizing the results, demonstrates a reasonably direct correlation between contact angle and susceptibility to phagocytosis. A high contact angle in this case means that the bacterial monolayer is highly hydrophobic. Conversely, a low contact angle reflects a greater wetting action of the droplet due to spreading on a more hydrophilic surface. Table 8.5 indicates that the organisms that have contact angles *lower* than

Table 8.5 Degree of Phagocytosis and Contact Angles of a Number of Bacteria

	Organism	Contact Angle ± < 1°[a]	Average Number of Bacteria Phagocytized per Neutrophil[a]
↑ Hydrophobic	*Mycobacterium butyricum*	70.0°	6.4 ± 1.1
	Brucella abortus	27.0°	1.8 ± 0.3
	Neisseria gonorrhoeae	26.7°	1.9 ± 0.3
	Listeria monocytogenes (rough form)	26.5°	2.1 ± 0.4
	Staphylococcus epidermidis	24.5°	2.5 ± 0.4
	Escherichia coli, type 07	23.0°	2.2 ± 0.4
	Enterobacter aerogenes	21.5°	2.0 ± 0.3
	Streptococcus pyogenes	21.3°	2.1 ± 0.3
	Salmonella typhimurium	20.2°	1.4 ± 0.2
	Streptococcus faecium	20.0°	2.1 ± 0.3
	Salmonella arizonae	19.0°	1.6 ± 0.3
	Staphylococcus aureus	18.7°	1.6 ± 0.3
	Haemophilus influenzae (rough form)	18.6°	1.6 ± 0.3
	Shigella flexneri	18.1°	1.0 ± 0.2
	HUMAN NEUTROPHILS	(17.5°–18.5°)[b]	
Hydrophilic ↓	*Haemophilus influenzae* (group B)	17.6°	1.1 ± 0.1
	Streptococcus pneumoniae, type unknown	17.3°	0.8 ± 0.2
	Escherichia coli, type 0111	17.2°	0.6 ± 0.2
	Streptococcus pneumoniae, type 1	17.0°	0.5 ± 0.1
	Klebsiella pneumoniae	17.0°	0.4 ± 0.1
	Salmonella typhimurium	17.0°	0.3 ± 0.1
	Staphylococcus aureus, strain Smith	16.5°	0.2 ± 0.1

[a] ± The standard error.
[b] The contact angles of peripheral human neutrophils from different donors tend to vary between these values.

Table 8.6 The Influence of the Capsule of *Staphylococcus aureus*, Strain Smith on Its Phagocytosis and on Its Contact Angle

Organism	Contact Angle, ± < 1°	Average Number of Bacteria Phagocytized per Neutrophil
Staphylococcus aureus, strain Smith	16.5°	0.2 ± 0.1
Staphylococcus aureus, strain Smith, decapsulated	26.0°	2.1 ± 0.3

neutrophils (17.5°–18.5°), hence are *more hydrophilic*, are quite resistant to phagocytosis.

The influence of the capsule in resisting phagocytosis, and therefore in enhancing the virulence of the bacterium in an organism, is shown in Table 8.6. The capsule both decreases the contact angle and lowers the rate at which the bacterium is phagocytized.

Using an entirely different experimental approach, Stiffel and co-workers also demonstrated the sparing effect of the capsule on phagocytic digestion.[10] They studied the kinetics of phagocytosis of bacteria in mice by injecting them with ^{131}I-labeled bacteria and noting the rate of blood

In this experiment, when radioactivity disappears from the blood, the labeled bacteria have also disappeared. The only manner in which they can disappear is by being engulfed by phagocytic cells and thereby removed from the circulation.

clearance of radioactivity, a measure of phagocytosis. Three strains of *Salmonella typhi* were used, differing in their surface properties. It was found that *S. typhi* T_{415}, an encapsulated organism, persisted in the blood at high levels for prolonged periods of time compared to the rough, non-encapsulated strain R_2 (Figure 8.6). These *in vivo* results are completely in accord with those of van Oss on interfacial energies.

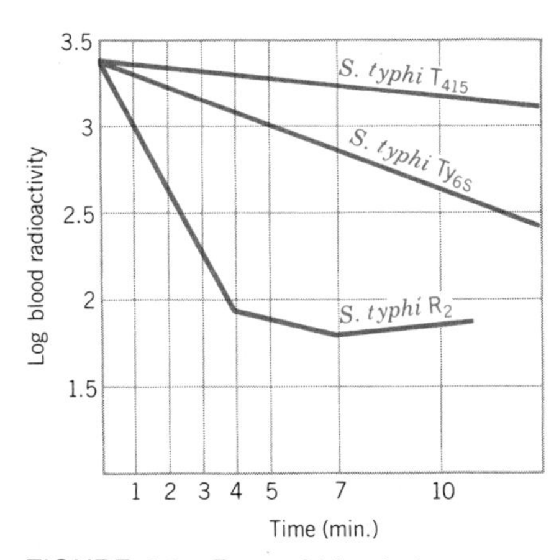

FIGURE 8.6 Rate of blood clearance of three strains of labeled *Salmonella typhi* injected intravenously into mice. Strain T_{415} persists for a long time and is therefore not phagocytized. Strain R_2 is rapidly removed from the blood by engulfment. *S. typhi* T_{415}, encapsulated with O antigen; *S. typhi* R_2, no capsule or O antigen; *S. typhi* Ty $_{6S}$, strain with Vi antigen, no O or H antigen.

Although it is clear from both approaches that phagocytosis is not an on–off, either–or mechanism, it is equally clear that the surface chemistries of the interacting cells do play an important role in recognition and engulfment.

Modifying the surfaces of either the phagocytic cell or the target particle has a pronounced effect on phagocytosis (Table 8.7). In general, any substance that increases the hydrophobicity of the particle with respect to that of the phagocytic cell, or increases the hydrophilicity of the phagocyte with respect to the particle, enhances phagocytosis.

Of special interest in this regard is the effect on phagocytosis of serum factors, such as complement and specific antibodies. For some time it has been observed that certain antibodies, termed *opsonins,* which interact with and coat bacteria, also enhance phagocytosis. Thus a normally virulent and phagocyte-resistant strain of bacteria can be forced to succumb to engulfment within the organism if it is first coated with specific antibodies that have been formed against its outer surface antigens. Complement, a family of serum proteins that bind to antigen–antibody complexes, also promotes phagocytosis.

Antibodies are proteins made up of four polypeptide chains that can be cleaved by proteolytic enzymes. One such enzyme, papain, produces three major fragments: two hydrophilic pieces termed Fab and one hydrophobic piece called Fc.

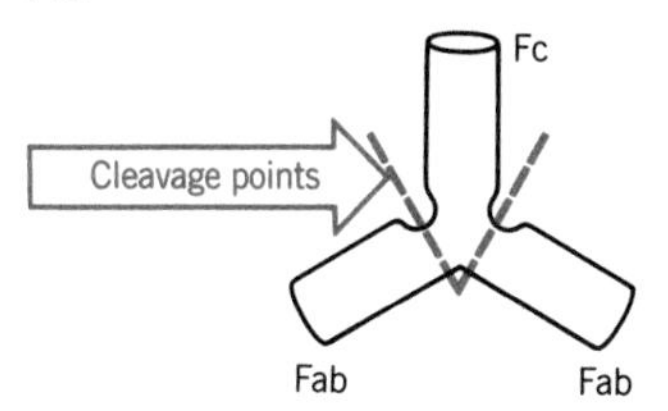

It appears from the studies of van Oss that opsonins exert their effects by coating relatively hydrophilic particles with a more hydrophobic layer, thereby changing the interfacial energy between bacterium and phagocyte surface and promoting engulfment. The structure of the antibody molecule with its two rather hydrophilic combining sites (Fab pieces) and protruding hydrophobic tail (Fc piece) is uniquely designed to effect this type of surface recognition and modification (Figure 8.7).

One of the primary functions of macrophages in higher organisms is to remove effete red blood cells from the circulation. Evidence is now accumulating that red blood cells lose their surface sialic acids as they age, a process that also changes their surface chemistry to one of a more hydrophobic nature.

Erythrocytes artificially aged in saline for 24 hr at 37°C or treated at pH 5 show increased contact angles (Table 8.8). Cells that are modified in this way are phagocytized by neutrophils, whereas untreated cells are not. These results indirectly suggest that increased hydrophobicity may be a factor in enhanced erythrophagocytosis, which is a major mechanism for the constant removal of effete erythrocytes in the living organism. The red blood cell, in this case, has not been converted to a foreign particle on aging, but it has become susceptible to the fate of other particles in the organism that are more hydrophobic than the phagocytic leukocytes.

Although the recognition of particles by phagocytes may involve interfacial energies, this may not represent the complete story of the process. It does appear, however, that the interaction of surfaces before engulfment depends to a significant extent on these surface properties. What actually ensues *in vivo* is probably more complicated, and the recognition of a particle as being foreign may depend also on whether serum antibodies and other factors are bound to it.

Table 8.7 Schematic Presentation of Six Modes of Action of Various Substances

Mode of Action	Substances	Action on Contact Angles of Phagocytes	Action on Contact Angles of Microorganisms	Action on Phagocytosis
1. Substances that make only the surfaces of phagocytes more hydrophilic enhance phagocytosis	Levamisole, heparin, hydrophobic particles or droplets, Hageman factor, MIF	↘	—	↗
2. Substances that make only the surfaces of phagocytes more hydrophobic depress phagocytosis	Lectins	↗	—	↘
3. Substances that make only the surfaces of microorganisms more hydrophobic enhance phagocytosis	Antibodies and complement, α_2HS, glycoprotein	—	↗	↗
4. Substances that make only the surfaces of microorganisms more hydrophilic depress phagocytosis	Penicillin, chloromycetin, polymyxin, bacitracin, α_1 acid glycoprotein	—	↘	↘
5. Substances that act on the surfaces of both phagocytes and microorganisms generally tend to affect both in the same way; these substances lower the contact angles of both phagocytes and microorganisms so that they approach one another in value, which results in a depression of phagocytosis	Surfactants, ampicillin	↘	↘	↘
6. A few substances may simultaneously cause the surfaces of phagocytes to become more hydrophilic and those of microorganisms to become more hydrophobic, which results in a strong enhancement of phagocytosis	Gentamycin	↘	↗	↗↗

Table 8.8a The Effect of Saline-Aging on the Contact Angle of Human Erythrocytes

Erythrocytes	Contact Angle
Untreated	15.0° ± 0.3°
Incubated for 24 hr at 37°C in saline	23.0° ± 0.2°

Table 8.8b The Effect of pH 5 for 30 min at 37°C on the Contact Angle of Human Erythrocytes and their Phagocytosis by Human Neutrophils

Erythrocytes	Contact Angle	Phagocytosis[a]
Untreated	15.0° ± 0.3°	0
Treated with acetate buffer, pH 5 at 37°C	23.0° ± 0.3°	0.8 ± 0.9

[a]Mean number of erythrocytes phagocytized per neutrophil ± standard error.

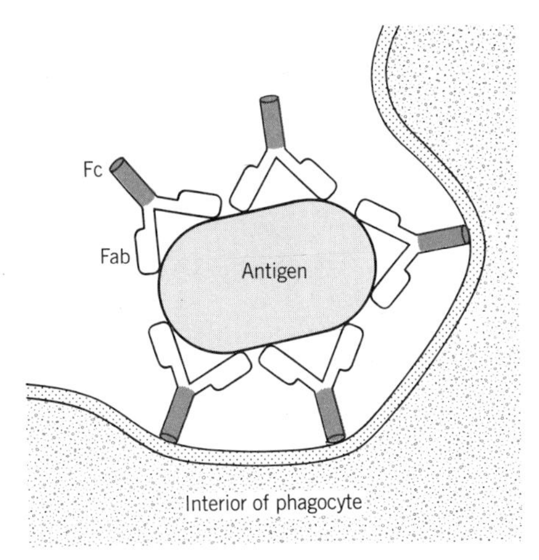

FIGURE 8.7 Schematic version of an antigen to which are attached antibodies by their hydrophilic Fab ends. The hydrophobic Fc pieces form an outer surface that promotes phagocytosis.

Engulfment

Once the phagocyte has made contact with a particle and has identified it as foreign, there is set in motion an irreversible series of membrane invaginations culminating in the deposition of the particle on the inside of the plasma membrane encased in an outside-in membrane barrier. This process, as studied in macrophages and neutrophils, depends on movements within the cytoplasm in its early stages.

The formation of pseudopodia appears to be a general mechanism of cell surface movement that operates during normal motility as well as during phagocytosis. This morphological change is accompanied by changes in the consistency in the cytoplasm just beneath the membrane in the protruded region. Referred to as "sol–gel" transformations, the relatively organelle-free region of the pseudopod undergoes physical changes that appear to reflect polymerization and depolymerization reactions of contractile proteins within the cell (see Section 8 for a more detailed discussion of this phenomenon).

The phagocyte extends its flaplike processes of cytoplasm over the surface of the object until the membrane surfaces meet distally to completely encase the particle (Figure 8.8). When the particle is less than 2 μm in diameter (which would be true for *Escherichia coli* in Figure 8.8); a single phagocyte simply wraps around it. Particles larger than 10 μm are often cornered by several macrophages, which then intrude small cytoplasmic projections into the foreign substance to break it up and eventually surround it. Particles of sizes intermediate between these two extremes are subject to a mix of intrusion and ingestion.

The space between the bacterial surface and the phagocyte membrane remains quite minimal and constant during engulfment, as can be seen from this transmission electron micrograph. As the process of engulfment proceeds, the membrane of the phagocyte forms a continuous collar that conforms to the shape of the ingested particle. The result is that no extraneous material is brought into the phagosome along with the particle.

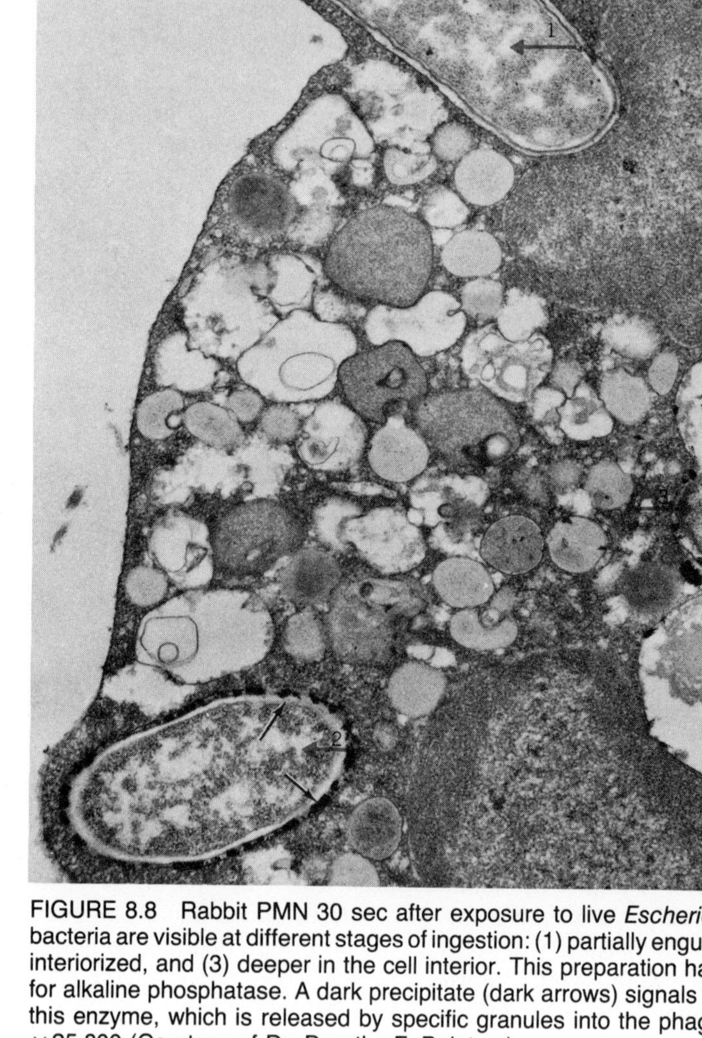

FIGURE 8.8 Rabbit PMN 30 sec after exposure to live *Escherichia coli.* Three bacteria are visible at different stages of ingestion: (1) partially engulfed, (2) recently interiorized, and (3) deeper in the cell interior. This preparation has been reacted for alkaline phosphatase. A dark precipitate (dark arrows) signals the presence of this enzyme, which is released by specific granules into the phagocytic vacuole. ×25,800 (Courtesy of Dr. Dorothy F. Bainton.)

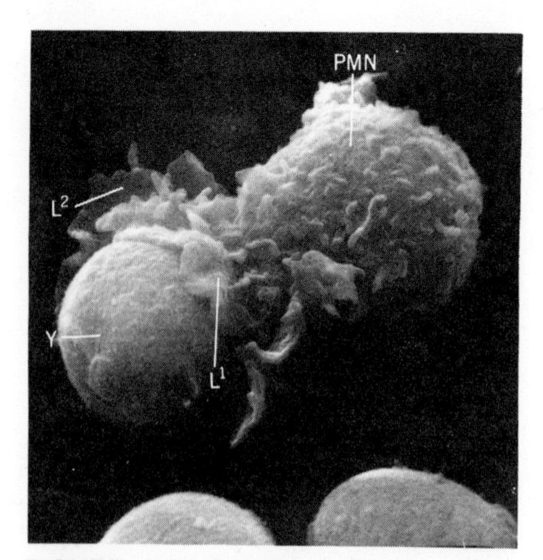

FIGURE 8.9 Scanning electron micrograph of a PMN engulfing a yeast cell (Y). Two layers of lamellipodia lie near the edge of the advancing ruffle. One is closely adherent to the yeast cell surface (L^1) and the other at some distance from the yeast (L^2). ×6,400 (Courtesy of Dr. Dorothy F. Bainton.)

Figure 8.9, a scanning electron micrograph, illustrates the wrapping of membrane-bound cytoplasmic extensions around the particle being engulfed. In this case the victim is a yeast cell. After contact with the yeast, lamellipodia emerge from regions of cytoplasm lacking cellular organelles and begin wrapping around the particle. Often waves of lamellipodia advance and progressively engulf the particle. These scanning views are helpful in presenting a clearer picture of the spatial relations between cytoplasmic extensions and the particle being engulfed.

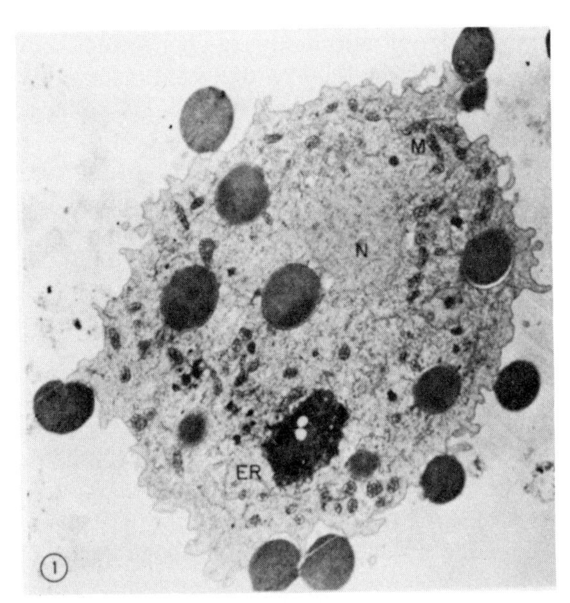

FIGURE 8.10 Cross section of *Acanthamoeba* phagocytosing 1.90-μm latex beads. Because of their large size, these beads are taken in individually. $\times$8100. (Courtesy of Dr. Edward D. Korn.)

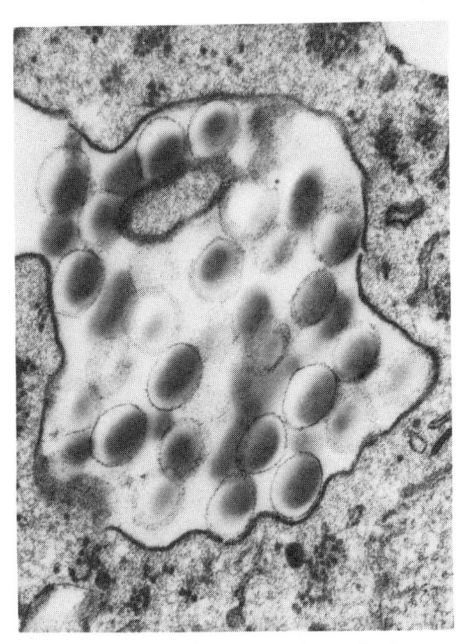

FIGURE 8.11 Cross section of phagocytic vesicle being formed around 0.264-μm latex beads. Since the beads are small, they are brought in as a group. Extraneous material may be brought into the cell when this occurs. $\times$33,600. (Courtesy of Dr. Edward D. Korn.)

Studies on the phagocytosis of latex beads have been useful in defining the size parameters of engulfment. Latex beads are relatively hydrophobic ($\theta > 60°$) and are readily taken up by most phagocytic cells. Korn and Weisman[11], studying the uptake of latex beads by *Acanthamoeba,* found that beads 1.305, 1.90, and 2.68 μm in diameter are engulfed individually, whereas smaller beads (0.557, 0.264, 0.126, and 0.088 μm) are first accumulated at the surface and then taken in as a group. These workers propose that phagocytic vesicles have an optimum size for smaller beads, and until there is enough material to reach this critical volume, phagocytosis will not occur. Figures 8.10 and 8.11 are electron micrographs illustrating the engulfment of latex particles. When a collection of beads is brought in together, it is apparently impossible for the cell to exclude all extraneous material. Under these conditions smaller, soluble substances are brought in as baggage between particles.

The process of engulfment is considered to be complete when the extended cytoplasmic projections undergo membrane fusion and a phagosome detached from the surface membrane of the cell is formed. The final event of engulfment is shown in Figure 8.12. Although not apparent in this cross-sectional view, the opening between the phagocytic vacuole and the extracellular space decreases to a small circular channel just before membrane fusion. The process of membrane fusion is therefore restricted to a very limited site on the cell surface.

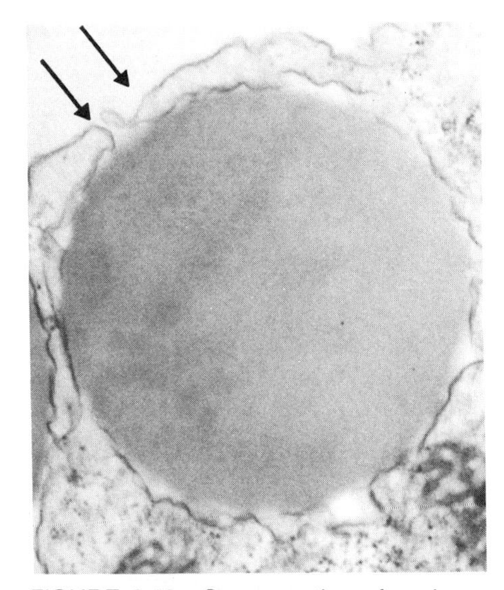

FIGURE 8.12 Cross section of a phagocytic vesicle just before interiorizing a 2.68-μm latex bead. The membrane surfaces that have enwrapped the particle are about ready to fuse (arrows). $\times$22,000. (Courtesy of Dr. Edward D. Korn.)

The Fate of Phagosomes

Under ordinary circumstances phagosomes migrate away from the periphery of the cell and for awhile may retain their original size and shape. A phagosome containing a collection of latex particles is shown in Figure

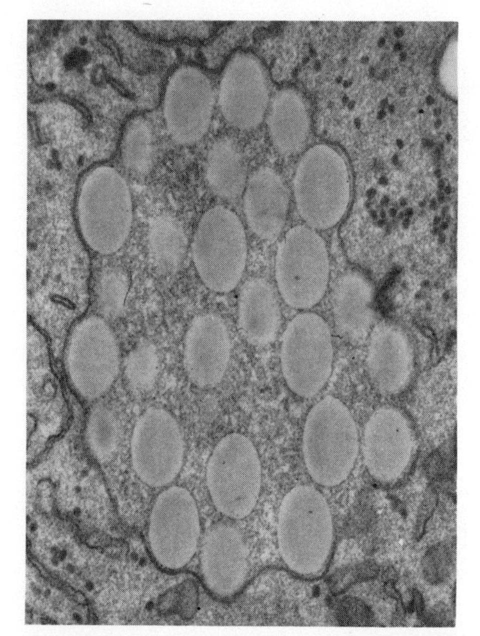

FIGURE 8.13 A phagosome containing a collection of 0.264-μm latex beads. The phagosome is completely encased by its own membrane, which has now separated from the plasma membrane. The presence of extraneous material in this phagosome is readily apparent. (Courtesy of Dr. Edward D. Korn.)

8.13. After a time phagosomes undergo morphological changes and gain increased electron transparency as the ingested particles are dismantled. Before dismantling, the phagosome must combine with a lysosome to form a digestive vacuole.

En route to its point of fusion with a lysosome, a given phagosome may coalesce with neighboring phagosomes or may even fragment, giving rise to multiple phagosomes. The fate of the digestive vacuole is taken up in the chapter on lysosomes.

8.3 PINOCYTOSIS

Pinocytosis is the bulk uptake of soluble molecules. It bears many features in common with phagocytosis. We consider pinocytosis as taking place at two different morphological levels. At one level, uptake of the liquid may occur by way of the formation of "large" vesicles (100 nm or larger). At the other level, uptake commences with the formation of "small" vesicles (~70 nm in diameter). We refer to the general phenomenon of cellular drinking as "pinocytosis" and use the term "micropinocytosis" when only small vesicles are formed.

The Stages of Pinocytosis

Pinocytosis is viewed as a distinctly two-stage process: a reversible adsorption of dispersed molecules onto the cell surface and an irreversible flow and synthesis of membranes resulting in the incorporation of the extracellular molecules into vesicles on the interior side of the plasma membrane.

The adsorption of solute molecules onto the surface of the cell is mediated by specific receptor sites, which concentrate the extracellular material and activate the membrane invagination mechanisms. The term *adsorptive pinocytosis* is applied to this event to distinguish it from the entrance of nonspecific substances swept in within the fluid content of the endocytic vesicle. The latter process is called *fluid phase pinocytosis*. Obviously, fluid phase pinocytosis does not concentrate extracellular molecules and operates in a manner detached from the molecular and membrane mechanisms of endocytosis.

Inducers of Pinocytosis

A protein that has an isoelectric point of pH 5 will bear a net negative charge above that pH. At pH 5 it possesses a net zero charge. Lower than pH 5 it is positive.

$pH < 5$ (+++−)

$pH = 5$ (++−−)

$pH > 5$ (+−−−)

Pinocytosis is induced by biological molecules of many different types. In macrophages, anionic substances are generally more effective than cationic ones. Thus acidic mucosubstances, DNA and RNA, fatty acids, and proteins with isoelectric points of pH 5 or less are effective inducers.

In contrast, pinocytosis in amoebae is stimulated by cationic materials.[12] Proteins are found to be inducers when they are in a solution that is on the acidic side of the isoelectric point, thus possessing a net positive charge.

These findings do not necessarily mean that the fundamental mechanisms of pinocytosis in different cell systems are significantly different. It may simply mean that receptor sites differ in composition and require oppositely charged inducers to set in motion the invagination processes.

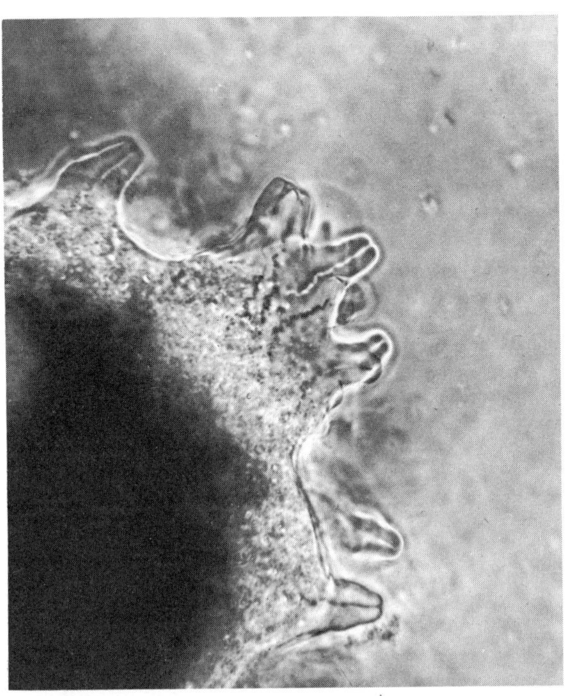

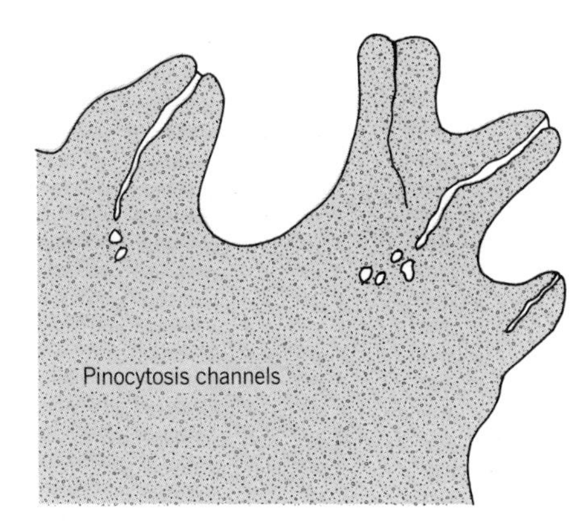

FIGURE 8.14 Photomicrograph and drawing of pinocytosis in an amoeba. Pinosomes bleb off from membrane-invaginated channels. (Micrograph courtesy of Dr. Richard A. Boolootian, Science Software Systems.)

Micrographic Examples of Pinocytosis

It is now thought that virtually every kind of eucaryotic cell is capable of carrying out some level of pinocytosis. In some it is more prominent than in others. For example, pinocytosis is especially prominent in certain protozoans and in capillary endothelial cells.

A light micrograph of pinocytosis in an amoeba is shown in Figure 8.14. Pinosomes form from the ends of channels deeply embedded into the cell interior. Once this process has been triggered in amoebae it continues for about a half-hour until an apparent state of cellular exhaustion is reached. Another cycle can be started only after a 2- to 3-hr rest period.

Micropinocytosis is readily apparent in the endothelial cells that make up the capillary channels in the circulatory systems of higher organisms. Figure 8.15 shows micropinocytic vesicles within these thin endothelial cells. These vesicles are open toward both the capillary lumen and the extracapillary space. It is thought that the formation of micropinocytic vesicles in these cells is a shuttle mechanism providing nutrients to meet the high energy demands of the adjacent muscle tissue.

The micropinosomes in these cells are smooth on their interior surfaces, reflecting the smooth exterior surface of the endothelial cells. In cells of other types (e.g., erythroblasts), the micropinosomes have coated or fuzzy interior surfaces (Figure 8.16). This results when certain receptor sites cluster on the surface of the cell and subsequently become concentrated on the inner surface of the pinocytic vesicle. In the case of erythroblasts the receptor sites appear to attract ferritin, an iron-containing protein.

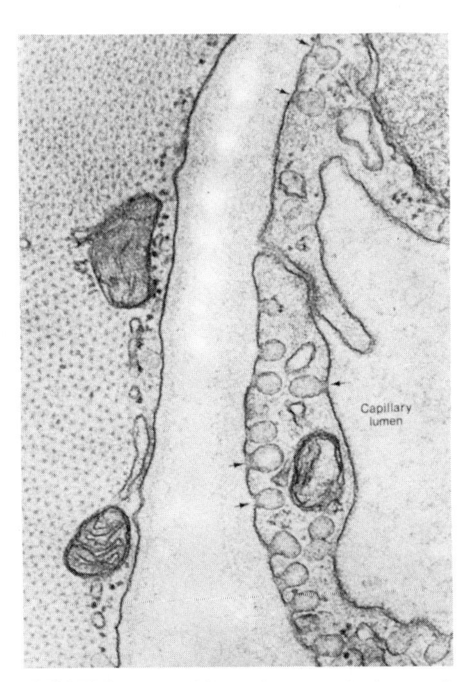

FIGURE 8.15 Micropinocytosis in capillary endothelium from mammalian cardiac muscle. Since these vesicles are seen to be open toward both cell surfaces, they are thought to function as shuttles. (Courtesy of Dr. Don W. Fawcett.)

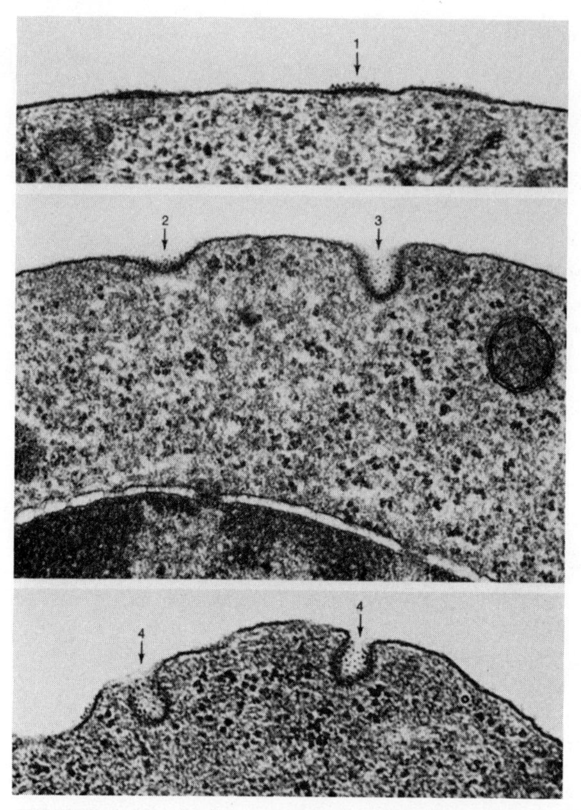

FIGURE 8.16 Surface areas of erythroblasts from guinea pig bone marrow. Particles of ferritin adhere to a region with an inner coat and a fuzzy external layer (1). These regions begin to depress (2) and eventually form deep pits (3) until the pit neck narrows (4) and ultimately fuses. (Courtesy of Dr. Don W. Fawcett.)

The ferritin is thus brought into the cell and serves as a donor of iron for hemoglobin synthesis.

8.4 THE METABOLIC REQUIREMENTS OF ENDOCYTOSIS

Endocytosis, viewed in its entirety, requires energy. But not all stages of the process are energy requiring.

Some energy is expended by any phagocytic cell that must move toward a site of activity. This, however, is preliminary to engulfment and is not a factor that we consider as a part of the energy requirements for endocytosis. We will explore the energy requirements for two separate stages of endocytosis: adsorption of particles or soluble molecules to the cell surface and ingestion or engulfment of the adsorbed material.

Temperature Effects

Phagocytosis by macrophages and PMNs operates by an on–off temperature-dependent mechanism.[13] Below a certain critical temperature threshold

Table 8.9 Metabolic Requirements for Phagocytosis and Pinocytosis in Cultivated Mouse Macrophages

Inhibitor	Concentration	Inhibition (%)	
		Phagocytosis	Pinocytosis
Nitrogen-CO_2	—	<10	84
Cyanide	10^{-4} *M*	<10	87
Antimycin A	5×10^{-7} *M*	<10	92
Oligomycin	2 μg/ml	<10	90
2,4-Dinitrophenol	5×10^{-6} *M*	<10	87
2-Desoxyglucose	10^{-2} *M*	>80	90
Iodoacetate	10^{-4} *M*	>80	60
Fluoride	10^{-3} *M*	>80	60
p-Fluorophenylalanine	10^{-3} *M*	<10	85
Puromycin	0.1 μg/ml	<10	90
Cycloheximide	0.2 μg/ml	<10	92
Actinomycin D	0.01 μg/ml	<10	81

(18–21°C), particles, such as sheep red blood cells, are bound to the cell surfaces but are not ingested. When the temperature is raised above the threshold, the particles are engulfed.

We can interpret these results in two different ways. One is that adsorption is the result of random molecular collisions requiring no energy beyond thermal molecular movement, whereas membrane invagination requires a supply of ATP that is generated in sufficient amounts only at higher temperatures. The other interpretation is that the temperature effect is primarily on the fluid–crystal phase transitions of the bimolecular lipid leaflet of the plasma membrane. At higher temperatures the membrane would be in a more fluid state and more amenable to shape changes and engulfment. It is possible, of course, that both interpretations have validity with respect to the mechanism of endocytosis.

Pinocytosis does not exhibit the critical threshold levels that are true for phagocytosis. For example, the rate of solute uptake by pinocytosis in mouse fibroblasts[13] is directly proportional to temperature in a range of 2 to 38°C.

The differences observed in these two instances may be more a function of the volume of ingested material than of differences in membrane properties or energy requirements.

Effect of Metabolic Inhibitors

In a study published by Cohn[14] the effect of various metabolic inhibitors on pinocytosis in mouse macrophages was determined under *in vitro* conditions. The results (Table 8.9) show that pinocytosis is strongly inhibited by agents that block electron transport (cyanide) and oxidative phosphorylation (2,4-dinitrophenol), as well as protein synthesis (puromycin). In contrast, phagocytosis is most strongly inhibited by compounds that interfere with glycolysis. Similar results have been reported by a number of research groups. It is therefore quite widely assumed that pinocytosis and phagocytosis are energy-consuming events but that the ATP providers differ for the two types of endocytosis.

The energy for pinocytosis can be derived from either glycolysis or oxidative metabolism, whereas the energy for phagocytosis comes mainly from glycolysis or any ATP-generating reaction not dependent on the electron transport chain. The results support the notion held by many investigators that there is a distinct difference between the metabolic mechanisms of the two endocytic processes.

Mazur and Williamson,[15] also studying the effect of metabolic inhibitors and pharmacologic agents on phagocytosis in macrophages, found that phagocytosis was inhibited by specific inhibitors of glycolysis as well as by agents that interact with contractile proteins.

The cellular logic of the results above is not yet entirely obvious. There is evidence that the cellular *level* of ATP is not the only factor that assures energy for endocytosis. In some systems studied, reduced levels of cellular ATP are not followed by reduced endocytic activities. Thus it is not clear that ATP is the *immediate* energy source for endocytosis.

ATP may function indirectly, for example, for the generation of new membrane components consumed during endocytosis or for the operation of contractile proteins.

Surface membrane consumption is high during endocytosis. Studies on the amoeba *Chaos chaos*[16] have revealed that under optimal conditions, on a diet of *Paramecium aurelia,* more than 100 phagosomes are formed in a single 24-hr cycle of growth. About 10% of the plasma membrane is consumed in the formation of each phagosome. Thus during a 24-hr period of feeding, more than 10 times the amount of plasma membrane normally contained by the organism is consumed. There are obvious ATP requirements for replacing the lost membrane by biosynthetic mechanisms.

The other likely candidate requiring ATP is the contractile apparatus of the cell. Actin and myosin proteins within endocytic cells require ATP for contraction and are being increasingly implicated in the membrane movements of endocytosis.

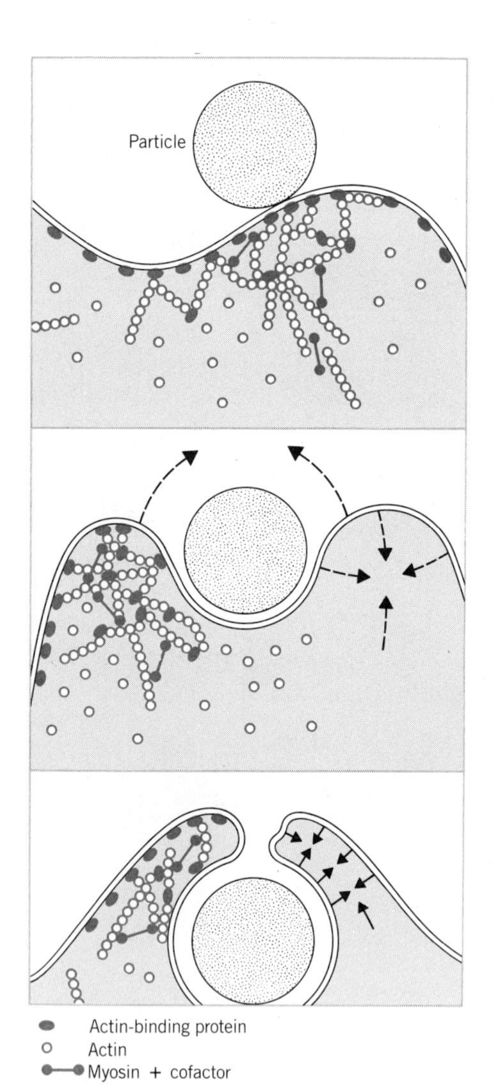

FIGURE 8.17 Hypothetical mechanism implicating contractile proteins during engulfment. Upon contact of a particle with the phagocytic cell surface, an actin-binding protein underlying the membrane is altered or activated to promote polymerization of actin filaments. Myosin contracts the actin filaments, bringing new membrane sites in contact with the particle, thus propagating the process.

8.5 RECEPTOR SITES AND CONTRACTILE PROTEINS

The presence and properties of contractile proteins in cells and their relationships to microtubules, microfilaments, and cell movement constitute a comparatively new research field that is being actively pursued. Both migration of the endocytic vesicles and the membrane movements of endocytosis are being increasingly thought of as contractile protein-related phenomena.

Actin, myosin, and a high molecular weight actin-binding protein have been extracted from macrophages and subjected to assembly reactions.[17] Cytochalasin B, a fungal metabolite that inhibits phagocytosis by macrophages, also inhibits the assembly and gelation of actin in macrophage extracts. It is speculated that the contact of a particle with the macrophage plasma membrane in turn induces the actin-binding protein to promote the assembly and gelation of actin in the cell periphery. Myosin may then further compress the gel to form pseudopods and trigger the engulfment process (Figure 8.17). Presumably, these contractile proteins also act in promoting the migration of the endocytic vesicles toward the center of the cell, or they deliver the vesicles to microtubules, which ultimately direct the intracellular movements. Colchicine, a drug that depolymerizes microtubules, blocks this migration within the interior of the cell.[12]

The contractile proteins of the cell interior are plausible recipients of a signal generated by receptors on the cell surface. Studies carried out by Schekman and Singer on mature erythrocytes of neonatal humans[18] have supported the hypothesis that a clustering of receptors on the cell surface is prerequisite to endocytosis. In these studies, concanavalin A (Con A), a lectin purified from jack bean meal, binds to specific receptors on the cell surface. These receptors possess a lateral mobility, as would be expected for the fluid properties of the membrane, and cluster under the influence of the multivalent ligand Con A until a critical accumulation is reached. Endocytosis is then triggered. These workers suggest that endocytosis in this sytem is a locally activated intracellular event signaled by the formation of clusters on the cell surface.

Lectins are proteins that are readily extracted from plant seeds and certain other sources. They have the inherent ability to agglutinate red blood cells. Con A combines specifically with α-glucosyl and α-mannosyl residues that project from the cell surface.

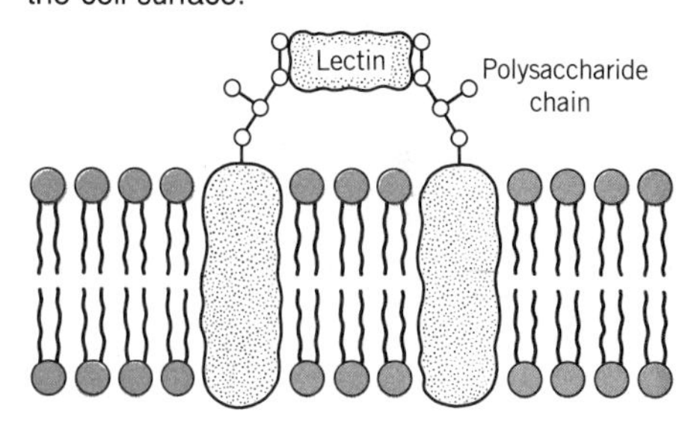

Erythrocytes contain spectrin, which is a family of actin- and myosinlike proteins. These are known to be localized toward the cytoplasmic surface of the membrane, where they are viewed as being bound to integral proteins at the cytoplasmic face (see Figure 4.27). The clustering phenomenon mediated on the outside surface and the cross-linking of contractile proteins occurring on the inner surface are concerted in some unknown manner to effect the ultimate shape of the membrane. Sheetz and Singer propose that the polymerization of the spectrin complex forces an expansion of the inner surface area relative to the outer of the lipid bilayer, thus causing an invagination.[19]

A generalized hypothesis for endocytosis occurring when specific proteins are brought into the cell via receptors is presented in Figure 8.18. Binding of the proteins to the receptors is followed by clustering and membrane invagination. As we have discussed previously, energy is required only for the last stage in this process. Clustering, although energy independent, is promoted at 37°C, whereas binding of the proteins to receptor sites can take place at 4°C.

8.6 THE FATE OF PLASMA MEMBRANE DURING ENDOCYTOSIS

During endocytosis high amounts of plasma membrane are interiorized, significantly reducing the surface area of the cell with each endocytic event. In the case of *Chaos chaos* mentioned earlier, the formation of only 10 phagosomes would be sufficient to consume all the surface plasma membrane of the cell.

Clearly there must be mechanisms either to recycle ingested plasma membrane or to replace it biosynthetically at high rates. This matter has intrigued researchers for some time, but until recently it has been a problem quite resistant to successful experimental techniques.

One can imagine three possible fates for the plasma membrane that is interiorized. One would be a recycling of the membrane, essentially unmodified, by a return mechanism similar to the pathway of entrance into the cell. A second fate would be partial breakdown of the membrane with a recycling of fragments. A third fate, the most drastic of the three, would be complete breakdown of the membrane to its lipid and protein com-

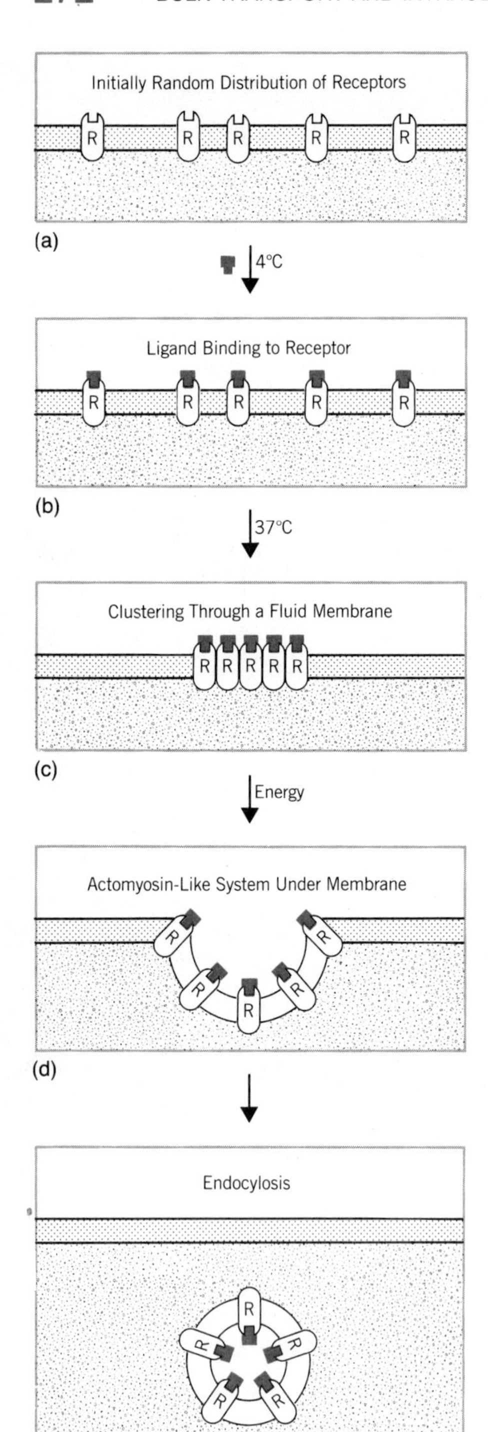

FIGURE 8.18 Generalized hypothesis for the mechanism of uptake of specific proteins by cells. Randomly positioned receptors (R) on the plasma membrane (*a*) bind specific proteins (*b*) which, at 37°C, leads to a clustering of receptors with bound protein (*c*). An energy-requiring process, possibly involving contractile proteins, brings about membrane invagination (*d*) and the formation of a pinosome (*e*).

ponents. These could be reused in assembling a new plasma membrane, or they could become precursors for the biosynthesis of new lipids and proteins for new membrane.

Recent work being carried out by Christian de Duve and his colleagues[20] has provided evidence that in cultured rat fibroblasts plasma membranes are *incompletely* degraded in lysosomes and membrane *fragments* are then returned to the plasma membrane.

The experimental approach used by these investigators was to react fibroblasts with labeled antibodies (IgG) to plasma membranes and follow the distribution of the label within the cell with time.

Their results showed that lysosomes were the principal and possibly sole sites of antibody storage and processing. All labeled IgG molecules were subject to at least partial degradation, and many were digested completely. But an important find was that the IgG fragments that were resistant to breakdown (in particular the Fab fragment of the molecule) were returned to the surface of the cell as plasma membrane material. This is taken to mean that membrane patches that were brought into the cell and fused with lysosomes are returned to the cell surface and refused with plasma membranes.

Although the evidence favors a recycling of plasma membranes interiorized by endocytosis, the precise pathway is essentially unknown. It is presumed that material destined for recycling is pinched off from the lysosomal surface and returned as a closed vesicle, which fuses with the plasma membrane just as a secretory vesicle would during the process of exocytosis.

Much remains to be worked out before recycling is understood. If it is a major mechanism of plasma membrane regeneration, one would suspect that it may not operate independently of the normal mechanisms of membrane assembly, which involve the endoplasmic reticulum and the Golgi complex (see Chapter 4). It is, of course, possible that other mechanisms are used to restore plasma membranes, and the complete picture may be a mixture of biosynthesis and recycling.

Summary

Endocytosis is a mechanism of bulk transport across the plasma membrane carried out by virtually every type of eucaryotic cell. Materials may be brought in as particles (phagocytosis) or as soluble substances in relatively large (pinocytosis) or small (micropinocytosis) volumes.

Phagocytosis has been studied most extensively on professional phagocytes, such as macrophages and polymorphonuclear leukocytes, which are cellular agents active in inflammatory reactions and resistance mechanisms in higher organisms. The stages of phagocytosis include the initial signal, pursuit, recognition, and final engulfment.

Pinocytosis is frequently studied in capillary endothelial cells and in erythroblasts. Two distinct stages mark the process: adsorption of molecules onto the cell surface and ingestion of the adsorbed material.

The interaction of the surface of endocytic cells with material destined to be transported in depends on the surface chemistry of the endocytic cell and the chemical properties of the particle surface or the soluble substance ingested. In some cases specific receptors are needed on the surface of the endocytic cell. Adsorption is energy independent and will take place at low temperatures (4°C). Engulfment is energy and temperature dependent and is blocked by metabolic inhibitors.

Actin and myosin contractile proteins beneath the plasma membrane appear to be receptors of surface signals and effectors of the membrane movements during endocytosis. The movements of these proteins require energy, as does the replacement of membrane consumed during invagination.

The fate of endocytic vesicles formed during engulfment is generally fusion with lysosomes to form digestive vacuoles. The material brought into the cell is thus dismantled and used as a source of energy.

Plasma membranes that are interiorized by endocytosis appear to be recycled by a shuttle mechanism that returns fragments of membrane from lysosomes to the cell surface.

References

1. *Life of Élie Metchnikoff,* O. Metchnikoff, pp. 116–117, Constable, London, 1921.
2. Pinocytosis, W. H. Lewis, *Bull. Johns Hopkins Hosp.* 49:17(1931).
3. Das Reticulo-Endotheliale System, L. Aschoff, *Ergeb, Inn. Med. Kinderheilk.* 26:1(1924).
4. *Mononuclear Phagocytes,* R. van Furth, ed., Blackwell Scientific Publications, Oxford, 1970.
5. Phagocytosis: The Engulfment Stage, M. Rabinovitch, *Semin. Hematol.* 5:134(1968).
6. Phagocytic Engulfment and Cell Adhesiveness, C. van Oss, C. F. Gillman, and A. W. Neumann, Marcel Dekker, New York, 1975.
7. *Chemotaxis and Inflammation,* P. C. Wilkinson, Churchill Livingstone, Edinburgh, 1974.
8. Chemotaxis of Mononuclear and Polymorphonuclear Phagocytes, E. Sorkin, J. F. Borel, and V. J. Stecher, in *Mononuclear Phagocytes,* R. van Furth, ed., Blackwell Scientific Publications, Oxford, 1970.
9. Studies on the Mechanism of Phagocytosis. I. Effect of Electric Surface Charge on Phagocytic Activity of Macrophages for Fixed Red Cells, H. Nagura, J. Asai, and K. Kojima, *Cell Struct. Funct.* 2:21(1977).
10. Kinetics of the Phagocytic Function of Reticulo-Endothelial Macrophages *in Vivo,* C. Stiffel, D. Mouton, and G. Biozzi, in *Mononuclear Phagocytes,* R. van Furth, ed., Blackwell Scientific Publication, Oxford, 1970.
11. Phagocytosis of Latex Beads by *Acanthamoeba.* II. Electron Microscopic Study of the Initial Events, E. D. Korn and R. A. Weisman, *J. Cell Biol.* 34:219(1967).
12. The Role of Lysosomes in Protein Catabolism, A. H. Gordon, in *Lysosomes in Biology and Pathology,* vol. 3, J. T. Dingle, ed., North Holland, Amsterdam, 1973.
13. Endocytosis, S. C. Silverman, M. Steinman, and Z. A. Cohn, *Annu. Rev. Biochem.* 46:669(1977).
14. Endocytosis and Intracellular Digestion, Z. A. Cohn, in *Mononuclear Phagocytes,* R. van Furth, ed., Blackwell Scientific Publications, Oxford, 1970.

15. Macrophage Deformability and Phagocytosis, M. T. Mazur and J. R. Williamson, *J. Cell Biol.* 75:185(1977).
16. A Study of Phagocytosis in the Amoeba *Chaos chaos,* R. G. Christiansen and J. M. Marshall, *J. Cell Biol.* 25:443(1965).
17. Interaction of Actin, Myosin, and an Actin-Binding Protein of Rabbit Pulmonary Macrophages. III. Effects of Cytochalasin B, J. H. Hartwig and T. P. Stossel, *J. Cell Biol.* 71:295(1976).
18. Clustering and Endocytosis of Membrane Receptors Can Be Induced in Mature Erythrocytes of Neonatal but not Adult Humans, R. Schekman and S. J. Singer, *Proc. Nat. Acad. Sci. U.S.* 73:4075(1976).
19. On the Mechanism of ATP-Induced Shape Changes in Human Erythrocyte Membranes. I. The Role of the Spectrin Complex, M. P. Sheetz and S. J. Singer, *J. Cell Biol.* 73:638(1977).
20. Fate of Plasma Membrane during Endocytosis. III. Evidence for Incomplete Breakdown of Immunoglobulins in Lysosomes of Cultured Fibroblasts, Y.-J. Schneider, C. de Duve, and A. Trouet, *J. Cell Biol.* 88:380(1981).

Selected Books and Articles

Books

Chemotaxis and Inflammation, Peter C. Wilkinson, Churchill Livingstone, Edinburgh, 1974.

Mononuclear Phagocytes, Ralph van Furth, ed., Blackwell Scientific Publications, Oxford, 1970.

Phagocytic Engulfment and Cell Adhesiveness. Carel J. van Oss, Cetewayo F. Gillman, and A. Wilhelm Neumann, Marcel Dekker, New York, 1975.

The Macrophage: A Review of Ultrastructure and Function, Ian Carr, Academic Press, New York, 1973.

The White Cell, Martin J. Cline, Harvard University Press, Cambridge, MA, 1975.

Articles

Endocytosis, S. C. Silverman, R. M. Steinman, and Z. A. Cohn, *Annu. Rev. Biochem.* 46:669(1977).

Endocytosis, Thomas P. Stossel, in *Receptors and Recognition,* P. Cuatrecasas and M. F. Greaves, eds., John Wiley & Sons, New York, 1977.

9

Lysosomes

As scientists we do not simply read the book of nature. We write it. Even the physicist has had to admit a subjective aspect to his discipline. How much more so then the biologist, who deals with a reality of such elusive complexity that only deliberate simplification can cloak it with the appearance of intelligibility. Nevertheless, that is the way our science progresses. But we must accept our concepts for what they are, provisional approximations that are as much fictions of our minds as they are faithful depictions of facts.
So it is with the lysosome . . .

C. DE DUVE, 1969

The thoughts at the beginning of this chapter, written by Christian de Duve after nearly two decades of tracking and stalking the lysosome, reflect both the historical mystique of this elusive cell organelle and the efforts of researchers to write the chapter in the book of nature concerning its existence. In contrast to most other cell structures, which were seen long before their functions were determined, the lysosome emerged first as a concept and only some time later as a distinct and identifiable organelle in the cell. The manner in which this happened is a fascinating episode in the history and development of cell biology.

Some threads of the early history of the lysosome can be picked up with the work of Élie Metchnikoff on phagocytosis in the late 1800s[1] and of Paul Ehrlich on leukocyte granules near the turn of the century.[2] An understanding of phagocytosis as a cellular digestive process and the implication that leukocyte granules were involved in this process were essential to the development of the lysosome concept. But in spite of these early beginnings, it was not until the 1950s that the picture began to take on a cleaner focus under the persisting prodding of Christian de Duve and his colleagues in Belgium and Alex Novikoff and his research group in the United States. De Duve's approach to the problem was a biochemical one, whereas Novikoff's was morphological and cytochemical, thereby giving the lysosome the distinction of being the first cellular ultrastructure to be subjected to a powerful dual scrutiny now common in biological research.

Some of the early discoveries came by accident. De Duve marks December 16, 1949, as the date of serendipity. He and his research team were studying the effect of insulin on isolated rat liver tissue and had taken their work in the direction of purifying glucose-6-phosphatase by differential centrifugation of gently homogenized liver. This was a new approach for them. Assay for the enzyme acid phosphatase, operating on the substrate β-glycerophosphate, was routinely carried out as a control along with the search for glucose-6-phosphatase activity in the separated fractions. The glucose-6-phosphatase activity was rapidly and clearly identified with the microsomal fraction, but the amount and distribution of acid phosphatase activity presented a puzzle to the researchers on December 16. The activity of the enzyme in the homogenates was only

10% of what they had been observing in water extracts of the tissue in their previous work, and the activities of the fractions summed up amounted to twice the activity in the whole homogenate. Either the experiment was a fluke or something was misbehaving.

Fortunately, the results were not passed off as error and the experiment was not abandoned. The samples were refrigerated and 5 days later assayed again. This time the acid phosphatase activity was at "normal" levels and a high activity peak was noted in the mitochondrial fraction (Figure 9.1). Whatever had masked the activity in the earlier assays was no longer operating. It was further found that the acid phosphatase activity in the *aged* mitochondrial fraction was converted to a *nonsedimentable* form, suggesting that the original activity was bound to a heavy structure that no longer held the enzyme. This "latency" phenomenon, an expression of activity upon aging, was consistent with the idea that the acid phosphatase activity of the cell was membrane bound and released only after deterioration of membranes.

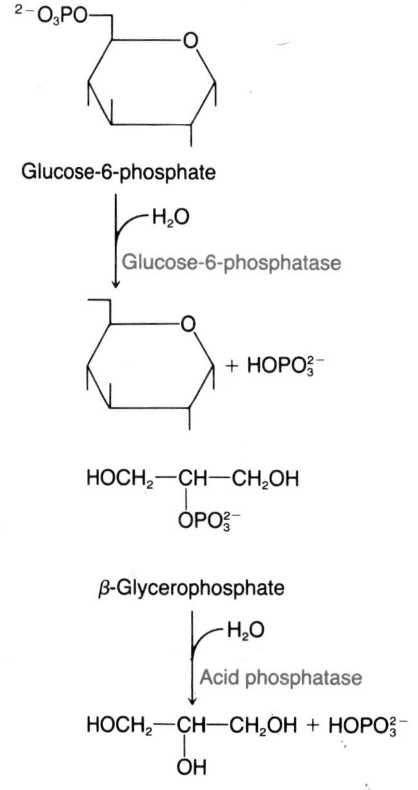

A model of latency is depicted in Figure 9.2. A number of treatments generally considered to be destructive of membranes were found to trigger this type of enzyme release.

The lysosome was not yet identified as a distinct particle, but the concept was beginning to grow and take on detail. Acid phosphatase activity was associated primarily with the mitochondrial fraction, and indeed it was thought that the activity was contained within mitochondria. It was not until centrifugation techniques were worked out to separate the acid phosphatase activity from the main body of the mitochondrial fractions that the search for a distinct particle began in earnest.

De Duve began to collaborate with Novikoff, who was independently studying a number of enzymes and their distributions in varous fractions of cell homogenates. Novikoff supplemented his enzyme studies with an exhaustive microscopic screening of the fractions and came up with the observation that acid phosphatase activity was not clearly associated with a certain fraction but was intermediate in its location. This was, of course, exactly what de Duve had found. Once the two research groups had began to pool their materials, their techniques, and their thinking, it did not take

For these studies, extreme caution was exercised so that the cell could be fractionated with *minimum injury* to the organelles.

The two enzyme activities being observed were

$^{2-}O_3PO$–

Glucose-6-phosphate

$\downarrow$ H_2O Glucose-6-phosphatase

$+ HOPO_3^{2-}$

$HOCH_2—CH(OPO_3^{2-})—CH_2OH$

β-Glycerophosphate

$\downarrow$ H_2O Acid phosphatase

$HOCH_2—CH(OH)—CH_2OH + HOPO_3^{2-}$

Fraction	Phosphatase acide en μg P/20 min.	
	16/12/49	21/12/49
Homogénéisat	10	89
Noyaux	2	10
Mitochondries	7	46
Microsomes	6	10
Decantat	6	9

FIGURE 9.1 Acid phosphatase activity of fresh and aged fractions separated from rat liver homogenates. These results suggest that the enzyme activity is released with time, as would happen if membrane-bound vesicles released their contents upon deterioration. (This copy of an old slide was kindly provided by Dr. Pierre Baudhuin.)

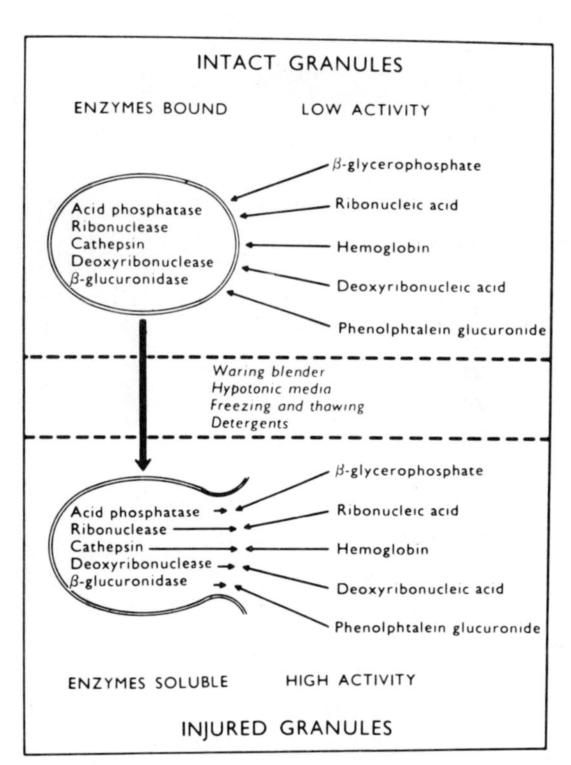

FIGURE 9.2 A model of latency. Treatments that disrupt membranes free the enzymes for ready detection. (Copy of a slide provided by Dr. Pierre Baudhuin.)

long for the first electron micrograph to appear, showing the presence of dark-staining bodies along with mitochondria in a "lysosome-rich" fraction of homogenized rat liver. Figure 9.3 is one of the early electron micrographs resulting from the collaborative studies. The sighting of distinct organelles was exhilarating, to say the least, for both groups of investigators.

Operating in part on tips from other laboratories, de Duve and his workers began to look for other enzymes that might show distribution patterns similar to that of acid phosphatase. Of a number of enzymes studied, four others resembled acid phosphatase in their locations in cell fractions, and, significantly, demonstrated the latency property too. These were *ribonuclease, deoxyribonuclease, cathepsin D,* and *β-glucuronidase*. It was of immediate interest that these five enzymes not only appeared to be compartmented together, but were all *hydrolases* that operated optimally at *acidic pH* values, but on entirely different substrates. This result was taken to be something other than coincidence. It suggested a common but nonspecific lytic role for members of this cellular compartment and led de Duve in 1955 to christen the organelle "lysosome," meaning lytic body.

The biological role of the lysosome in the intact cell was initially difficult to surmise. What could such a destructive particle be doing in the normal life of a healthy cell? Its most obvious role was thought to be associated with cell death and dissolution, popularizing the term "suicide bag" and temporarily confining its assumed role to this self-destructive function.

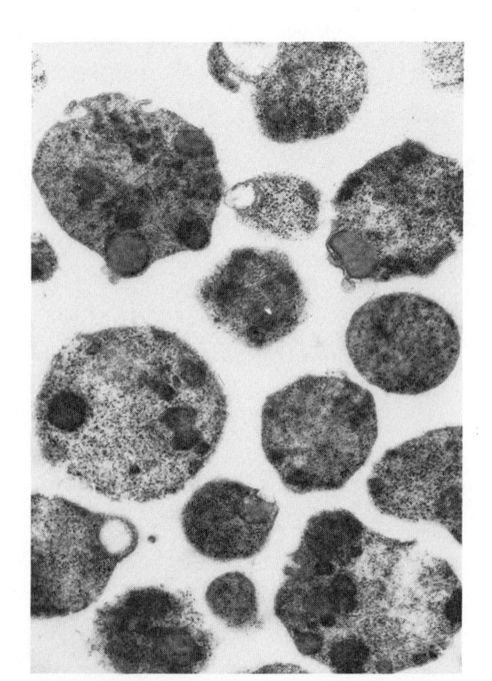

FIGURE 9.3 An electron micrograph of a lysosome-rich fraction from rat liver. ×30,900. (Courtesy of Dr. Pierre Baudhuin.)

It was at this point that Werner Straus entered the picture with his work on kidney tubule cells.[3] He found that the droplets of tubule cells, taken to be sites of storage and breakdown of reabsorbed proteins, were biochemically similar to lysosomes with respect to acid phosphatase and protease activity. This was an important observation, for it bridged the gap between the cell process of endocytosis and the activity of lysosomal digestion and gave birth to the thinking that intracellular digestion may be a process common to the *normal* life of many cell types, and that lysosomes may be the agents of this process.

Since the mid-1950s, the lysosome has been an intense object of study in many laboratories. Its occurrence, structure, and composition in various tissue types have been studied by most of the biological, chemical, and physical techniques that are available to the present-day researcher. Its roles in cell life and cell death, and in normal and abnormal processes have been implicated, studied, and confirmed or denied. But in spite of all that is known about the organelle, there is still something about the lysosome that maintains it a shrouded "reality of such elusive complexity."

9.1 A WORKING DEFINITION OF THE LYSOSOME

As more laboratories began to take on the lysosome as an object of inquiry, it became apparent that work could be well coordinated only if there was a consensus regarding the major properties of this new cellular organelle. The acid phosphatase activity of a certain type of mitochondrial fraction, as a definition, was too closely tied to techniques of centrifugation worked out on special tissues in one or two laboratories.

Eventually there emerged a set of criteria by which a lysosome could be tested and identified. It has been quite widely adopted. According to these criteria, any cellular particle identified as a lysosome should (1) be bounded by a limiting membrane, (2) contain two or more acid hydrolases, and (3) demonstrate the property of enzyme latency when treated in a way that adversely affects membrane structure. These criteria have been particularly useful in the study of lysosomes in animal tissues, but they have presented some problems in pinpointing lysosomes in plant tissues. In plants the hydrolytic compartments may have somewhat different properties and functions. We will discuss this in a later section of this chapter.

Nomenclature and the Functional Forms

Before going further into a detailed discussion of lysosomes, it is appropriate to look briefly at their physiological role in the cell. Lysosomal nomenclature, as it has developed, is closely linked with their physiological activities. Figure 9.4 schematically summarizes the different physiological states of lysosomes.

The lysosome, as initially formed by the cell and before it has participated in any cellular event, is called the *primary lysosome*. Other synonyms that are frequently used are *pure, true, virginal,* or *original* lysosome. This form is one physiological state of the lysosome.

Lysosomes that are engaged in digestive activity in the cell are termed *secondary lysosomes*. These lysosomes may have two distinct functions. One is the digestion of extracellular material taken in by endocytosis. A

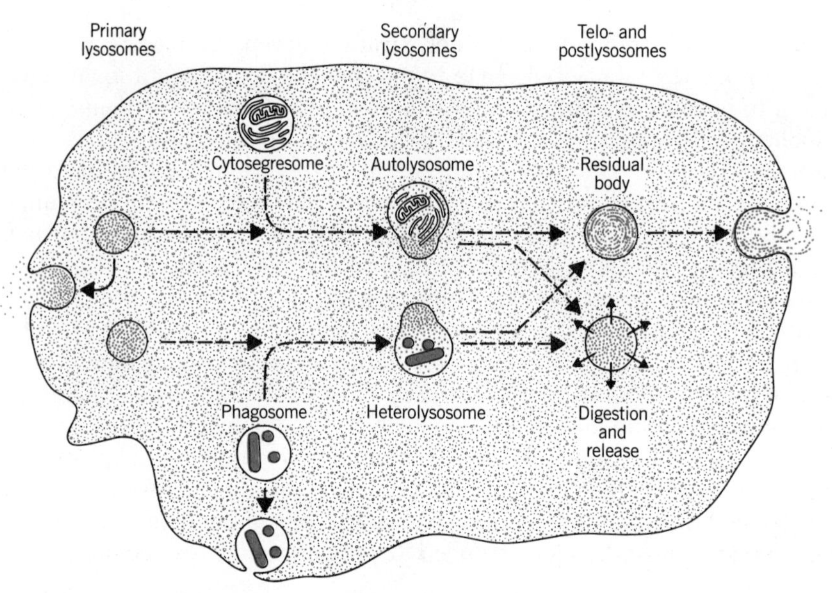

FIGURE 9.4 Schematic summary of lysosome nomenclature as related to the various physiological states of the organelles.

second function is the digestion of some of the cell's own intracellular material. These two digestive forms are called *heterolysosomes* or *digestive vacuoles,* and *autolysosomes* or *autophagic vacuoles,* respectively. The autolysosome arises by a fusion of the primary lysosome with a *cytosegresome*.

Late or aged varieties of these two types of lysosome are referred to as *telolysosomes*. They may degenerate and become permanent static residues in the cell, in which case they are called *postlysosomes* or *residual bodies*. On occasion, the material within a telolysosome may be released to the exterior environment. This process has been called cellular defecation.

Some of these functional states of lysosomes can be viewed in the electron micrographs of Figure 9.5.

It is important to keep these different forms in mind as we discuss ultrastructure and composition, for the particular appearance and content of a lysosome may vary with its physiological state.

Formation and Fate of Lysosomes

Any theory that deals with the formation of the primary lysosome in the cell must take into account the origins of its two main components: the limiting membrane and its enzymatic contents. It is fairly easy to speculate how the hydrolytic enzymes are packaged in a membrane sack, but it has proved to be difficult to test the ideas.

Two main schools of thought on the subject have emerged. One parallels the mechanism proposed for the formation of secretory granules such as those that arise in the pancreas. According to this line of reasoning the hydrolytic proteins are formed on the ribosomes of the rough endoplasmic reticulum (RER) from which they are discharged into the channel system of the endoplasmic reticulum and delivered to the Golgi complex. As the

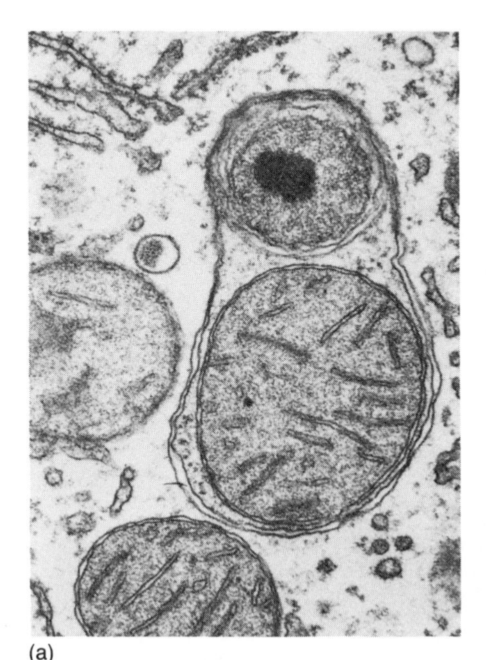

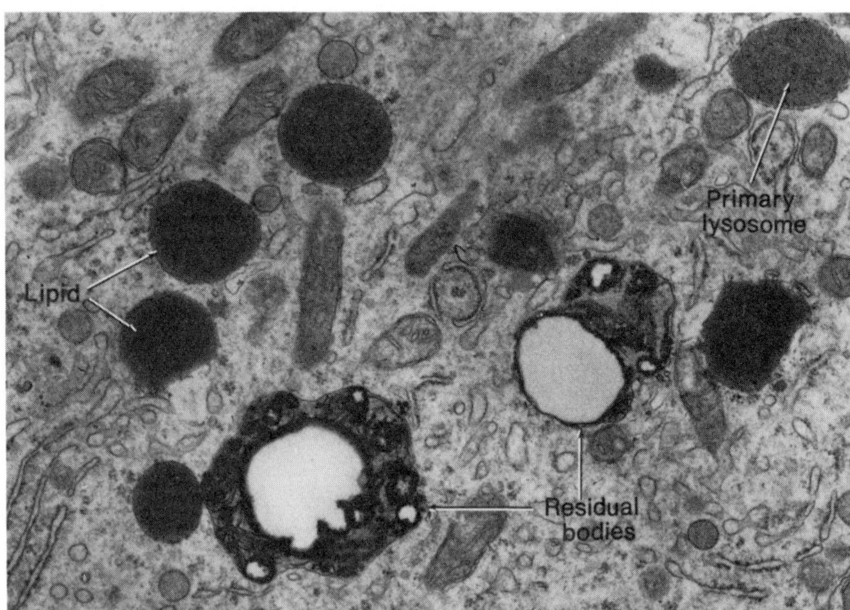

(a) (b)

FIGURE 9.5 Different physiological forms of lysosomes. (*a*) An autophagic vacuole containing a mitochondrion and a microbody. (*b*) Residual bodies, resulting from heterophagy in the Sertoli cells of the testis. These cells degrade the residual cytoplasm left behind by spermatozoa after their release. (Micrographs courtesy of Dr. Daniel Friend, a, and Dr. Don W. Fawcett, b.)

proteins are worked through the Golgi from convex to concave faces, they are processed and packaged with membrane wrappings, emerging as primary lysosomes.

The second school of thought, developed by Alex Novikoff, starts at the same place, the RER. However, in this case the enzymes are thought to bypass the Golgi complex and to be packaged by a budding off of a specialized smooth endoplasmic reticulum (SER) in a region adjacent to the maturing or concave face of the Golgi. This region, and the theory itself, are referred to as the GERL (*G*olgi-associated *e*ndoplasmic *r*eticulum giving rise to *l*ysosomes).

These two ideas of lysosome formation are schematized in Figure 9.6. Although these schemes are helpful in providing a framework to think about the genesis of lysosomes, they do not adequately answer some very tough questions. How, for example, does the cell sort out only hydrolytic enzymes and package them free from the other enzymes of the cytosol? Or, what are the driving forces of enzyme transport and what controls their direction of movement in the cell? These and other questions will undoubtedly be clarified as work continues on these intriguing and complex problems.

The fate of the primary lysosome, once formed, is to become a secondary lysosome. This takes place by fusion with a phagosome or with a vacuole formed by sequestering some of the internal material of the cell into a compartment. Depending on the type of cell or its physiological state, the digestive or autophagic vacuole may in turn have one of three

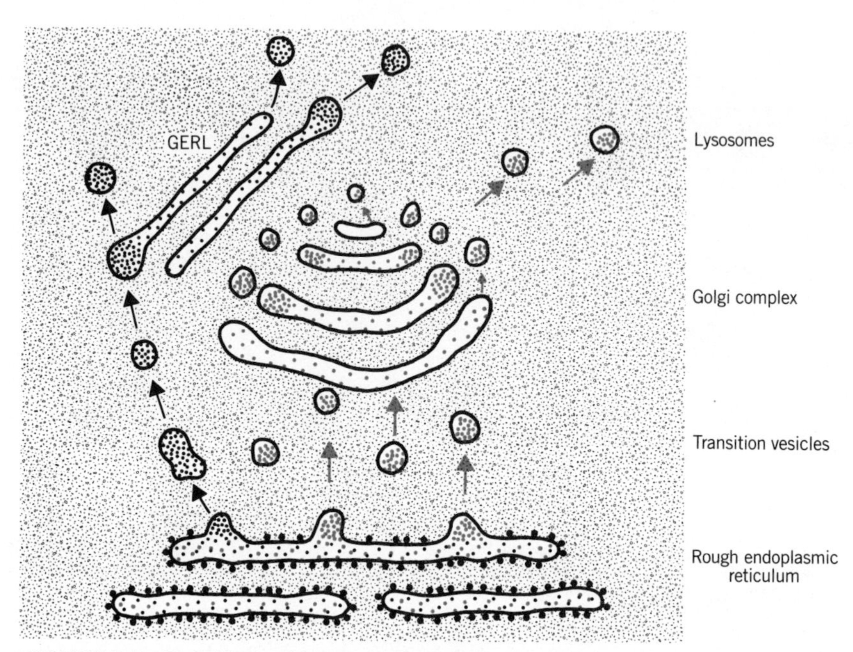

FIGURE 9.6 Two ideas of lysosome formation. According to one school of thought, the route involves the Golgi complex directly (color). The second school of thought envisions lysosomal packaging as involving the GERL, not the Golgi.

fates: (1) it may empty its contents to the outside of the cell—reverse endocytosis (exocytosis) or cellular defecation, (2) it may be left with a nondegradable residue and become a residual body devoid of hydrolases, remaining indefinitely in the cell, or (3) it may completely hydrolyze its contents to diffusible, low molecular weight materials and stand ready for a new cycle of activity or, if depleted of enzymes, be completely recycled by the cell. The factors that control the fate of a particular lysosome are largely unknown.

The reactions that are used to reveal acid phosphatase activity are as follows.

$$CH_2OH-CH(-O-P(=O)(OH)-O^-)-CH_2OH$$

$$\xrightarrow[\text{pH 5, 37°C}]{\text{Acid phosphatase}}$$

$$CH_2OH-CHOH-CH_2OH + HO-P(=O)(OH)-O^-$$

$$\xrightarrow{Pb^+}$$

$$Pb_3(PO_4)_2$$

(Electron microscopy)

$$\xrightarrow{(NH_4)_2S}$$

$$\downarrow \underline{\underline{PbS}} \downarrow$$

(Light microscopy)

9.2 OCCURRENCE AND CHARACTERISTICS IN DIFFERENT CELL TYPES

Methods of Detection

Historically, lysosomes were detected and identified mainly by their possession of acid phosphatase activity. This has become the principal marker enzyme for lysosomes. The acid phosphatase reaction is one of a number of enzymatic histochemical techniques that has been developed for their identification and study.

The histochemical procedure used most widely is that developed by Gomori[4] in which a slice of tissue is incubated with the substrate β-glycerophosphate in the presence of lead nitrate at pH 5. At sites where the phosphatase enzyme is located, inorganic phosphate is released from the substrate and is immediately trapped in insoluble form as lead phosphate (as shown in the margin). The high electron density of the lead is directly useful for electron microscopy or, alternatively, by further rinsing the tissue in dilute ammonium sulfide a black lead sulfide precipitate forms that can be visualized in the light microscope.

Many modifications of this procedure are in use today.[5] One that produces brightly colored or fluorescent products is the azo dye method, in which the hydrolysis products of organic esters are coupled to aromatic amines to form insoluble azo dyes, highly visible with light microscopy. This technique is less useful than the Gomori procedure for electron microscopy.

The finding that lysosomes possess a veritable host of hydrolytic enzymes stimulated the development of other enzyme assays to reveal these particles. All these approaches are based on the same principle of presenting a specific substrate to enzyme-containing tissue followed by a reaction that pins down the product and visualizes it where it is generated in the cell. In this manner the lysosomal enzymes β-glucuronidase, aryl sulfatase, *N*-acetyl-β-glucosaminidase, non-specific esterases, and amino and dipeptidyl aminopeptidases are commonly assayed for in a variety of tissues.

The location of lysosomes in cells can also be pinpointed by nonenzymatic histochemical techniques. For example, lysosomes demonstrate metachromasia with toluidine blue and give a positive periodic acid Schiff reaction. They concentrate the vital anionic dyes of the trypan blue group (Niagara and Evans blue) and even possess some autofluorescent properties. These procedures are especially useful when a tissue system under study is quite thoroughly defined in terms of the presence and location of lysosomes. However, they must be used in a guarded manner when exploring new systems, since a positive reaction is not sufficient to verify the presence of a lysosome according to the threefold criteria of identification discussed earlier.

A whole new area of histochemistry has been developed because of the diverse properties of the lysosome. In this sense, the lysosome has made a fine contribution to an important area of cytological technology.

Occurrence in Animal Cells

Rats have also donated without obvious reservation to the development of this field. The major share of the early work carried out on lysosomes probed two principal tissue systems, rat liver and rat kidney. This was due in part to the discovery of lysosomes in rat liver tissue and to the particular properties of these tissues that lend them to thin sectioning, homogenization, and differential centrifugation. Figure 9.7 is an electron micrograph of liver tissue displaying lysosomes.

Electron micrographs show a remarkable amount of ultrastructural detail compared to what can be seen with the light microscope. Nevertheless, light microscopy combined with histochemical techniques is an important tool in surveying the cellular and tissue distribution of lysosomes. A close examination of micrographs of lysosomes often shows them to be close to the Golgi region of the cell and prevalent in tissues that are working absorptive areas of the organ surface.

In addition to liver and kidney tissues, lysosomes have been detected and studied in many different tissues. Some of the more common tissue sources are listed in Table 9.1. It now appears that all mammalian cell types contain lysosomes except mature red blood cells.

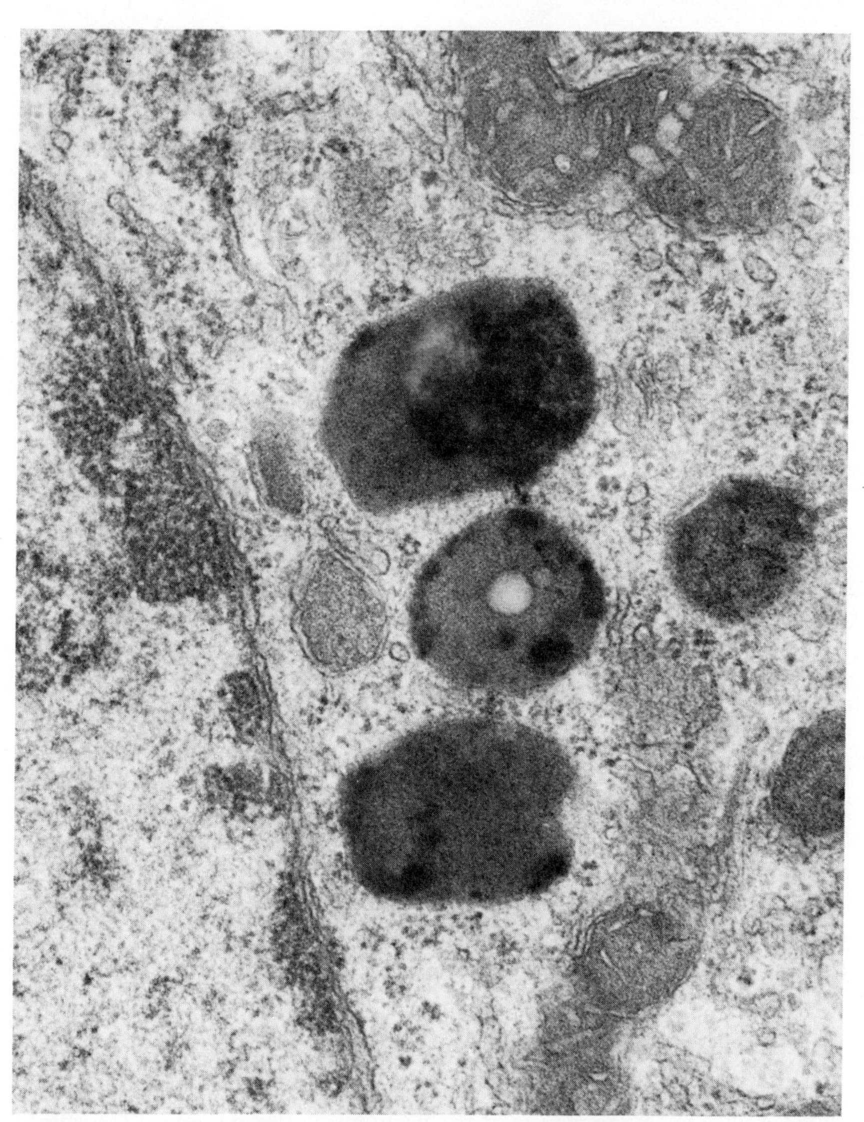

FIGURE 9.7 Lysosomes in rat liver tissue. Primary lysosomes tend to be homogeneous and spherical; secondary lysosomes contain granular elements and may have variable shapes.

In many cases, work on lysosomes in a particular tissue type has been hampered by the reluctance of the tissue to yield its lysosomes to the techniques of isolation without damage to the organelle. Lysosomes are especially difficult to secure from tissues that are fibrous or hard, like skin, cartilage, and muscle. But in each of these tissues lysosomes have been detected by electron microscopy or other means, and in most cases specialized techniques have been worked out to permit the isolation of lysosomes.

Occurrence in Plant Cells

Plants contain several hydrolases, but they are not always as neatly compartmentalized as they are in animal cells. Many of these hydrolases are

Table 9.1 Common Tissue Sources of Lysosomes

Brain and nerve tissues
Bone
Cartilage
Cells of lungs
Intestine
Leukocytes (especially macrophages and lymphocytes)
Liver
Muscle
Phagocytic cells of bone marrow
Prostate
Reproductive tracts (both male and female)
Spermatozoa
Spleen
Thyroid

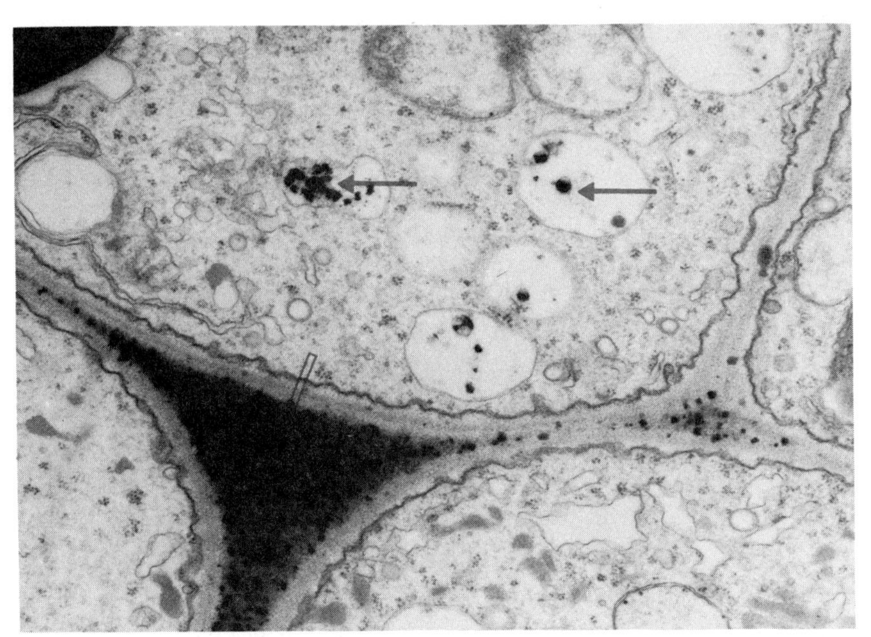

FIGURE 9.8 Electron micrograph of cultured carrot cells: acid phosphatase activity is found in the middle lamella of the wall (open arrow) and in small vesicular bodies within vacuoles (solid arrows). ×24,000. (Courtesy of Dr. Walter Halperin.)

found bound to and functioning within the vicinity of the cell wall and are not necessarily contained in membrane-bound vacuoles at these sites. But the presence of acid phosphatase has been documented in cell vacuoles of several kinds of plants.[6] Electron micrographs showing cell wall and vacuole sites of acid phosphatase activity are presented in Figure 9.8. The available evidence suggests that the plant vacuole is an organelle analogous to the animal cell lysosome, but perhaps not identical in structure and function.

The isolation of vacuoles from plant cells is a difficult task because of their fragility and because of the tough and fibrous structure of the plant cell. The conditions of homogenization that are sufficiently abrasive to break up the cell wall are usually destructive for the vacuole. A productive approach to the problem of isolation has been to strip off the cell wall with a snail gut enzyme and then gently lyse the spheroplasts by osmotic shock. This has worked for the lower plants, such as yeast, but has not solved the problem of liberating intact vacuoles from higher plants. Therefore most of the information that we have on the lysosomelike organelles of plants has come from the study of yeast cells. Two further complicating factors in the study of plant lysosomes are that enzymes other than hydrolases appear to be associated with the vacuoles, and the vacuoles often contain certain hydrolases that are found in other parts of the cell as well. It is evident that much more work will have to be done in this area before the hydrolytic system in plants is thoroughly understood.

Occurrence in Eucaryotic Protists

Acid hydrolase activity has been found in diverse types of protists such as protozoa, slime molds, fungi, and algae. In protozoa, the acid phosphatase activity has been noted in food vacuoles following endocytosis. This has been studied, for example, in *Tetrahymena pyriformis* where the lysosome appears to function both in intracellular digestion and in secretion of enzymes into the extracellular environment. Although lysosomes have been observed by cytochemical techniques in slime molds, fungi, and algae, their roles in these systems have not yet been very clearly delineated.

9.3 ISOLATION AND PURIFICATION OF LYSOSOMES

Fragility and Heterogeneity

Method II

As suggested in the preceding section, lysosomes are fragile organelles often susceptible to damage by the abrasive techniques of cell disruption and fractionation. Because of this, extreme caution must be used to ensure that the organelles survive the process of tissue homogenization and are further protected from osmotic shock and lysis once they are released from the tissues. Both the homogenizers of Potter and Elvehjem and of Dounce can be used to break cells without excessive destruction of lysosomes, but the use of these instruments is as much an art as it is a predictable science. Sucrose, at a concentration of 0.25 *M*, is commonly used as an osmotic protectant during and after the homogenization process. Often buffers and other substances such as EDTA, a chelating agent, are added to enhance the preservation of the particles and to inactivate the enzymes.

Ethylenediamine tetraacetic acid (EDTA).

$$\begin{array}{ll} {}^{-}OOC{-}CH_2 & CH_2{-}COO^{-} \\ \quad\quad N{-}CH_2{-}CH_2{-}N & \\ {}^{-}OOC{-}CH_2 & CH_2{-}COO^{-} \end{array}$$

EDTA is widely used to bind unwanted metal ions that may interfere with biochemical reactions.

The success of isolation has always been frustrated because lysosomes are in reality a heterogeneous population of closely related morphological and biochemical entities. Although the exact reasons are obscure, several factors may contribute to this heterogeneity. One is the nature of the tissue source. The cells of a given liver or kidney, for example, cannot be considered to be a homogeneous population. Liver, a case in point, is composed primarily of two distinct cell types, hepatocytes or parenchymal cells (90–95% of the liver weight) and reticuloendothelial or Kupffer cells (5–10% of the liver weight).[7] The heterogeneity of isolated lysosomes may reflect the heterogeneity of their cellular origins. But even when homogeneous cell populations, such as eucaryotic microorganisms, rabbit heterophil leukocytes, or Ehrlich ascites tumor cells, have been used as source materials, isolated lysosomes show biochemical and physical heterogeneity.

Another factor that may account for heterogeneity is differences in age among the lysosomes in a given cell. As we saw in Section 9.1, lysosomes may be caught in various stages of their normal cell functions as well as in different stages of their origin and development to the level of primary lysosome. Since in a given cell we see a continuum of functional stages, it is not surprising to note this reflected in the isolated organelles as well.

The pleomorphism of lysosomes is apparent in the electron micrographs that have been presented thus far in the chapter. Particles ranging in size from 0.25 to 0.8 μm are commonly seen. An interesting question that has been pursued by a number of workers is whether this morphological heterogeneity is paralleled by biochemical heterogeneity. Are the acid hydrolases grouped in different combinations in different lysosomes?

In several tissue systems studied, the distribution of several acid hydrolases monitored together in partially fractionated lysosome peparations has shown different patterns. This means that all lysosomes do not contain the same complement of enzymes. In fact, the lysosome population of liver has repeatedly been partially resolved into two populations: light lysosomes rich in acid phosphatase and β-glucuronidase and heavy lysosomes rich in acid nucleases and acid cathepsin. Thus, although the evidence for biochemical heterogeneity is compelling, it is not clear how this property may be related to morphological heterogeneity. In the case of liver, the two populations of lysosomes may simply arise from the two major cell types, parenchymal and Kupffer cells. Within each cell type there may be additional, but more subtle, enzymatic heterogeneity. This sticky problem probably will not be resolved until better techniques are developed to fractionate and purify lysosomes.

In any event, it is difficult to apply either morphological or biochemical criteria to assess the purity of lysosome preparations because the populations that have been studied so far seem to violate the restrictions that these criteria would impose on them.

Still another factor that makes lysosomes difficult to isolate is their particular physical properties, which resemble quite closely those of certain other cell organelles. This was evident during the early experiments carried out by de Duve and co-workers[8] using differential centrifugation on rat liver homogenates. The hydrolase activity of the cell was found to some degree in the nucleus and soluble fractions, as well as in the heavy mitochondrial, light mitochondrial, and microsomal fractions (see Table 9.2) These results are a further indication of lysosome heterogeneity with respect to size and density.

Differential Centrifugation

Method III

The use of differential centrifugation alone to isolate pure preparations of lysosomes has been futile for the reasons discussed above. The basis of this problem was considerably clarified by Beaufay and co-workers[9]

Table 9.2 Distribution of Reference Enzymes in Cell Fractions[a] after Fractionation of the Liver by Differential Centrifugation in 0.25 M Sucrose

	Recovery (%)	Percentage of Recovered Amount					Relative Specific Activity[b]				
		N	M	L	P	S	N	M	L	P	S
Nitrogen	98.9	13.4	16.5	7.5	24.7	37.9	1.00	1.00	1.00	1.00	1.00
Enzymes											
Acid phosphatase	101.8	3.5	23.8	40.0	19.7	13.0	0.26	1.44	5.33	0.80	0.34
Acid deoxyribonuclease	95.3	5.6	35.3	32.7	6.6	19.8	0.42	2.14	4.36	0.27	0.52
Cytochrome oxidase	88.9	11.4	64.8	19.6	4.2	—	0.85	3.93	2.61	0.17	—
Urate oxidase	94.9	8.6	13.5	49.8	21.3	6.8	0.64	0.82	6.64	0.86	0.18
Glucose-6-phosphatase	92.8	7.3	2.9	7.5	79.4	2.9	0.54	0.18	1.00	3.21	0.08

[a]N, nucleus; M, heavy mitochondrial fractions; L, light mitochondrial fractions; P, microsomal fractions; S, soluble fractions.
[b]Relative specific activity = percentage of enzyme in the fraction/percentage of nitrogen (or protein) in the fraction.

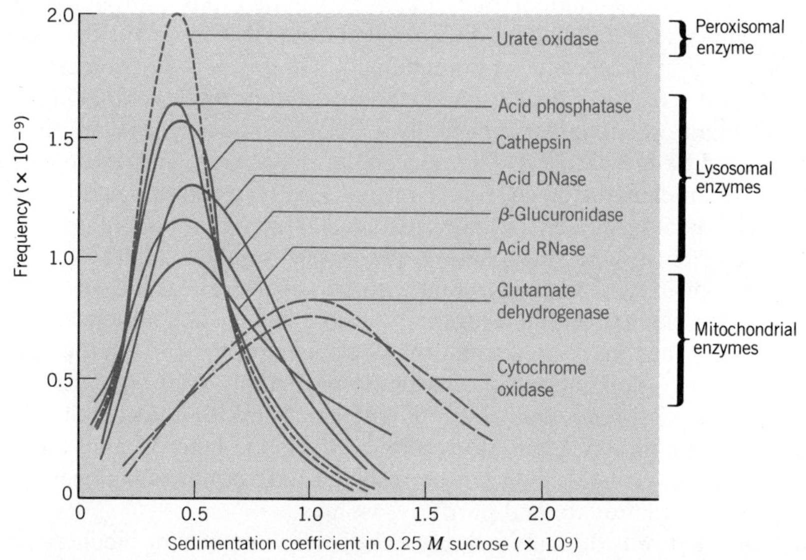

FIGURE 9.9 Frequency distribution curves of the sedimentation coefficients of various enzymes. Some enzymes are associated with mitochondria, some with lysosomes, and some with peroxisomes.

who studied the frequency distribution curves of sedimentation coefficients for mitochondria, lysosomes, and peroxisomes of rat liver (Figure 9.9). It is clear from their results that the distribution of mitochondrial enzymes (glutamate dehydrogenase and cytochrome oxidase) overlaps the curves of lysosomal enzymes (acid phosphatase). Furthermore, the distribution curve of urate oxidase, characteristic of the cellular organelle peroxisome, is essentially coincident with that of the lysosomal enzymes. These observations underscore the problem of separating these three organelles—lysosomes, peroxisomes, and mitochondria—strictly by differential centrifugation. Nevertheless, differential centrifugation can be used to obtain enriched fractions of lysosomes, which can be used for a variety of studies or as a starting material for further purification.

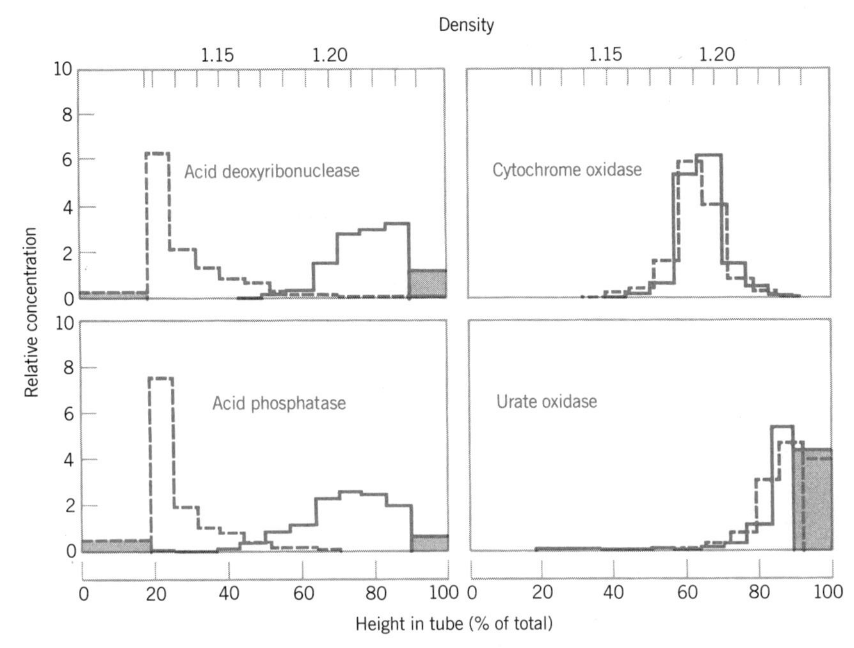

FIGURE 9.10 Distribution histograms of marker enzymes after density equilibration of fractions isolated from a normal rat (solid lines) and a rat injected with Triton WR-1339 (dashed lines). The colored areas represent the percentages of enzyme activity floating or packed in the pellet.

Isopycnic Centrifugation

Isopycnic centrifugation makes use of a gradient that allows particles to come to rest at a position in the gradient corresponding to their respective densities. Sucrose-H_2O is commonly used for this, with the particles layered in 0.25 *M* sucrose on top of the gradient. In the sucrose gradient system, the median equilibrium density of lysosomes (1.22) lies very close to those of mitochondria (1.19) and peroxisomes (1.23–1.25).[10] Therefore contamination of the lysosome fraction with the other two is inevitable.

An ingenious trick has been developed to overcome all these problems of separation based on density similarities. Animals that are injected with the detergent Triton WR-1339 accumulate it in liver lysosomes.[11] These "loaded" lysosomes, or *tritosomes,* have densities near 1.12, sufficiently less than either mitochondria or peroxisomes to permit a fairly clean separation from these normal contaminants. The techniques that have been developed using Triton loading are quite laborious, but they hold more promise than most of the other procedures.

The effect of Triton in shifting the density of lysosomes out of the mitochondrial and peroxisome region can be seen in Figure 9.10, where marker enzymes are assayed in the gradients. Lysosomes isolated in this way are distended (Figure 9.11) and may be detrimentally modified for certain kinds of studies, an unfortunate side effect of the procedure.

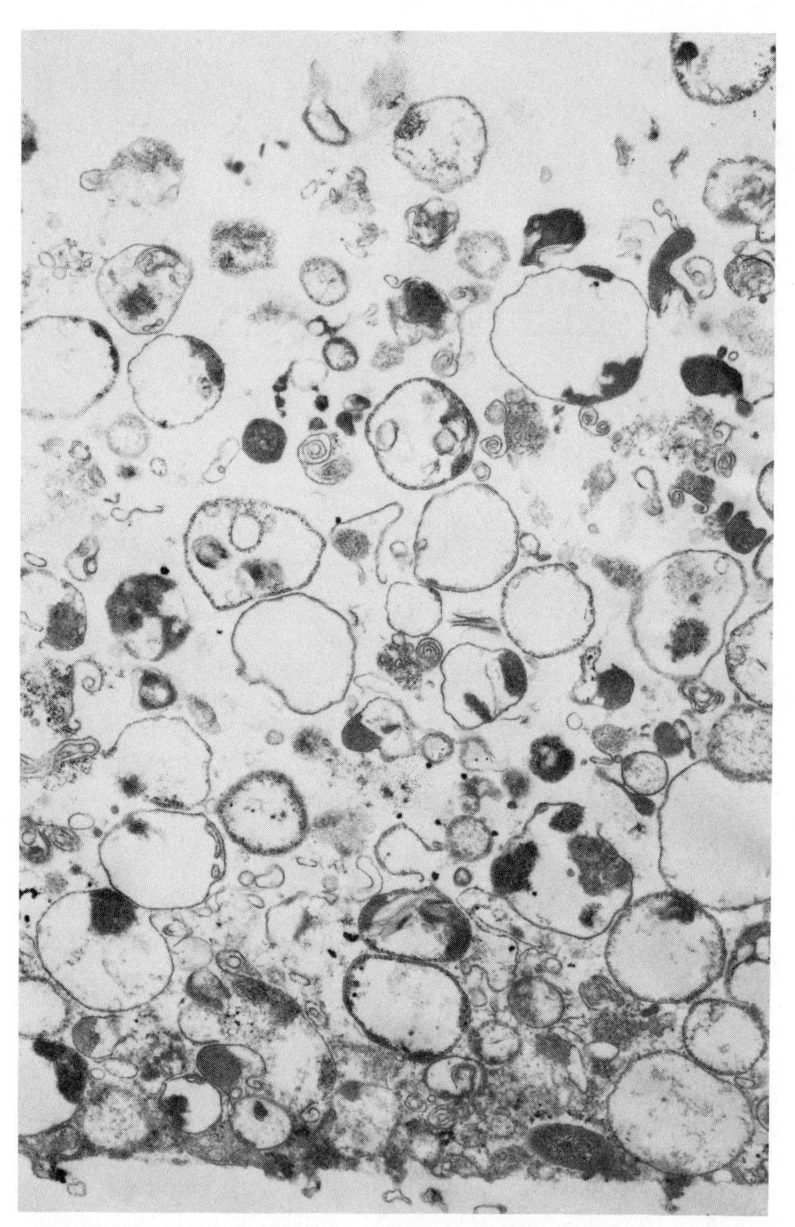

FIGURE 9.11 Section through a lysosome fraction isolated from animals injected with Triton WR-1339. The swollen lysosomes appear to contain remnants of their original matrix. This preparation contains amorphous material and membrane fragments, either from damaged lysosomes or other contaminant parts of the cell. ×18,000. (Courtesy of Dr. H. Beaufay.)

Carrier-Free Continuous Electrophoresis

Method IV

Using an entirely different approach, Stahn and his colleagues[12] have subjected crude lysosomal preparations, separated by differential centrifugation, to electrophoresis in a carrier-free electrophoretic apparatus. Lysosomes move toward the anode at pH 7.4 more rapidly than mitochondria, peroxisomes, or microsomes, allowing a fair degree of purification. Although the preparations are not yet pure, this procedure does have an advantage over others in that it is not necessary to load the lysosomes with damaging detergents.

In summary, both biochemical and morphological studies have shown lysosomes to be heterogeneous. Furthermore, different lysosomes from the same cell population appear to contain different sets of enzymes. Since lysosomes are similar in density to both mitochondria and peroxisomes, it is nearly impossible to purify them by conventional differential or isopycnic centrifugation. Modified lysosomes can be isolated, however, in a fairly high state of purity by Triton loading, but some of their native properties may be irreversibly altered in the process. Carrier-free continuous electrophoresis proves to be a promising technique for the isolation of unmodified lysosomes.

9.4 THE ULTRASTRUCTURE AND COMPOSITION OF LYSOSOMES

The Membrane

The problem of characterizing the lysosomal membrane rides hard on the problem of isolating the organelle in a pure form. The most highly purified lysosomes are obtained by Triton or Dextran 500 loading, but it is not certain that these techniques leave the membrane in its native configuration. Membranes have been prepared from "purified" fractions of lysosomes by sonication, osmotic shock, or by successive freezing and thawing followed by centrifugation at high speeds. Since no specific marker enzyme for the lysosomal membrane has been discovered, it is difficult to monitor the degree to which these membranes are separated from the other membrane components of the cell.

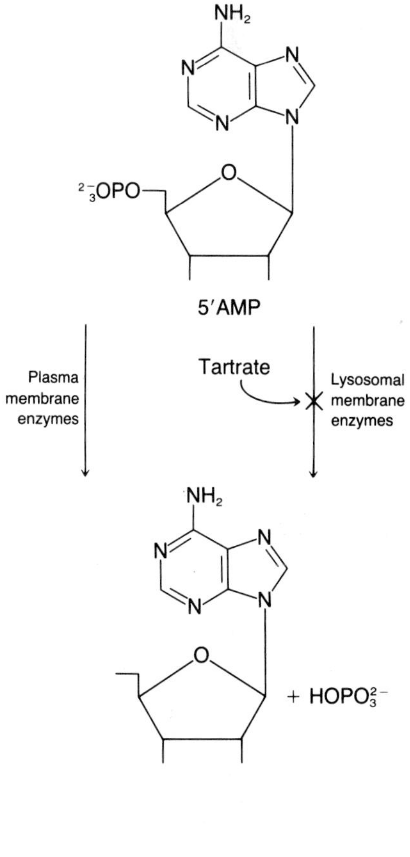

A number of studies have been carried out to determine the enzymatic properties of the lysosomal membrane and to compare these with the plasma membrane. Although the results are equivocal, there appear to be some differences between the two membranes. Lysosomes have an enzyme that acts upon 5′-AMP at alkaline pH and is inhibited by L(+)-tartrate. The analogous plasma membrane enzyme is unaffected by tartrate. Lysosomes also possess an acid phosphodiesterase that has properties that differ from the comparable alkaline enzyme of the plasma membrane. Other studies have suggested that the marker enzymes for plasma membranes either are absent or are present at low concentrations in lysosomal membranes.

Once again, it is important to keep in mind the dynamic nature of the lysosome and the effect that the physiological form may have on membrane properties. Tritosomes in particular would be expected to have membranes with plasma membrane markers, since they have arisen by endocytosis and fusion of the two membrane systems. At the same time, we must also entertain the possibility that plasma membrane enzymes may be inactivated once they are incorporated into the lysosome. Thus, even though present, they may resist enzymatic characterization.

Comparative amino acid analyses of the proteins of plasma membranes and tritosome (and dextranosome) membranes show no unusual compositional differences, but on disc electrophoresis the protein profiles for the two types of membrane are completely different.[13] Studies on their lipid compositions show an array of phospholipids, but of special interest is the presence of cholesterol and sphingomyelin in tritosome membranes; these two lipids are found in plasma membranes but are relatively rare in other cytomembranes. Lysosomal membranes appear to be richer than plasma membranes in neutral carbohydrates, hexosamines, and *N*-acetylneuraminic acid.

The ultrastructure of the lysosomal membrane as examined by electron microscopy reveals nothing unusual. It appears to have the unit membrane structure with a width of 9 nm, similar to the plasma membrane but thicker than that of mitochondria.

Certainly one of the outstanding physical characteristics of lysosomal membranes is their ability to fuse with other selective membranes of the cell. Most commonly, fusion occurs between primary lysosomes and phagosomes during intracellular digestion and between lysosomes and the plasma membrane during cellular secretion. In both cases, the fusion is between lysosomal membranes and plasma membranes or a derivative of plasma membranes.

Retinol (vitamin A_1).

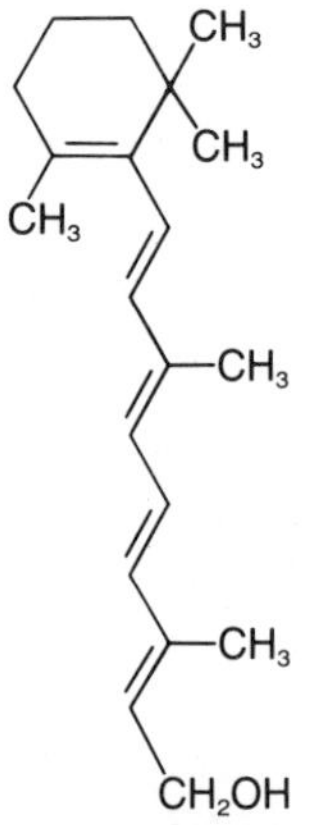

CH_3 CH_3 CH_3 CH_3 CH_3 CH_2OH

Retinol, one of the "fat-soluble" vitamins, is readily immersed in the bimolecular leaflet. It occurs in abundance in the membranes of the light receptor cells of the eyes.

J.A. Lucy, who has long been a proponent of the micellar hypothesis of membrane structure, has addressed the question of membrane fusion as being favored when the membrane is perturbed to take on a lipid micellar form.[14] Two adjacent membranes possessing this form could fuse by interdigitation of the micelles (Figure 9.12).

It is proposed that lysosomal membranes may have a high proportion of their lipids in the micellar configuration because of the presence of lysolecithin, a surface-active molecule. Conversely, membranes that are in the bimolecular leaflet configuration would not be inclined to fuse. Thus any agent that increases the proportion of micelles in a membrane would facilitate membrane fusion. Retinol (vitamin A alcohol) is a surface active agent that destabilizes lysosomes at relatively high concentrations (5–30 μg/ml) *in vitro,* causing a release of enzymes. In cells, the vitamin enhances fusion of lysosomal membranes with the plasma membrane. Lucy attributes this retinol effect to an increase in the amount of micellar configuration in these membranes.

Just the opposite effect is produced by hydrocortisone and related steroids. These compounds stabilize lysosomes and inhibit enzyme release and membrane fusion as well. Although the experimental data are still somewhat wanting, it is believed by some workers that the steroids exert their stabilizing effects by promoting the bimolecular leaflet configuration of the lysosomal membrane, and thus membrane fusion and enzyme leaking are inhibited.

The use of cortisone and related steroids for the treatment of joint inflammation has its basis in the stabilizing effect of these drugs on membranes, in particular lysosomal membranes.

The Contents

The number of enzymes associated with the lysosome has increased from the original 5 to somewhat over 50. Table 9.3 lists these enzymes. It is

Table 9.3 Enzymes of Lysosomes[a]

Oxidoreductases Acting on Hydrogen Peroxide as Acceptor	
Peroxidase	
Hydrolases Acting on Ester Bonds	
Arylesterase	Phosphoprotein phosphatase
Triacylglycerol lipase	Phosphodiesterase I
Phospholipase A_2	Deoxyribonuclease II
Cholesterol esterase	Sphingomyelin phosphodiesterase
Phospholipase A_1	Ribonuclease II
Acid phosphatase	Arylsulfatase A and B
Phosphatidate phosphatase	Chondroitinsulfatase
Hydrolases Acting on Glycosyl Compounds	
Lysozyme	β-*N*-Acetylglucosaminidase
Neuraminidase	β-Glucuronidase
α-Glucosidase	Hyaluronoglucosidase
β-Glucosidase	α-*N*-Acetylgalactosaminidase
α-Galactosidase	α-*N*-Acetylglucosaminidase
β-Galactosidase	α-L-Fucosidase
α-Mannosidase	L-Iduronidase
β-Mannosidase	
Hydrolases Acting on Peptide Bonds	
Glutamate carboxypeptidase	Lysosomal elastase
Lysosomal carboxypeptidase A	Cathepsin G
Lysosomal carboxypeptidase B	Neutral proteinase
Lysosomal carboxypeptidase C	Plasminogen activator
Lysosomal dipeptidase	Cathepsin B
Dipeptidylpeptidase I	Cathepsin D
Dipeptidylpeptidase II	Cathepsin E
Kininogenin	Lysosomal collagenase
Acrosin	Renin
Hydrolases Acting on Carbon–Nitrogen Bonds Other Than Peptide Bonds	
Acylsphingosine deacylase	Amino acid naphthylamidase
Aspartylglucosylaminase	Benzoylarginine naphthylamidase
Hydrolases Acting on Acid Anhydrides	
Inorganic pyrophosphatase	
Hydrolases Acting on Phosphorus–Nitrogen Bonds	
Phosphoamidase	
Hydrolases Acting on Sulfur–Nitrogen Bonds	
Heparin sulfamidase	

[a]Enzymes are listed in order according to enzyme commission number.

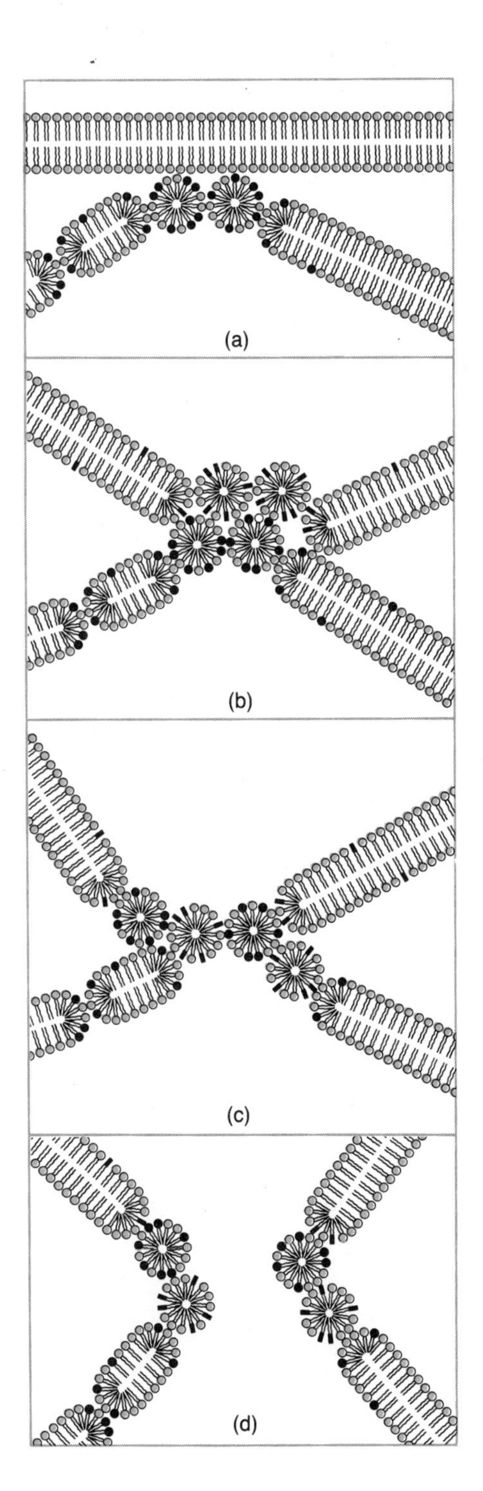

FIGURE 9.12 The proposed role of lipid micelles in membrane fusion. (*a*) Two membranes approach, one having a stable lipid bilayer construction and the other containing lipid micelles. Fusion is unlikely as long as one membrane is predominantly in bilayer form. (*b*) Some molecule (▬) that perturbs the stable bilayer triggers micelle generation so that micelles of the two membranes may interdigitate (*c*), reorganize, and separate, creating newly joined membranes (*d*).

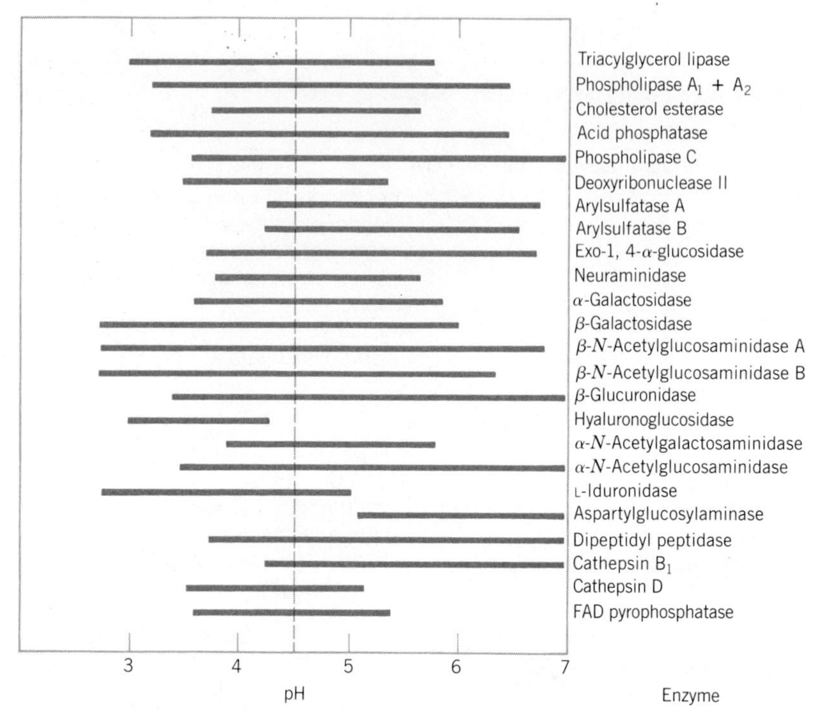

FIGURE 9.13 Activity pH ranges of lysosomal acid hydrolases. Ranges are those giving at least 25% of maximum activity of hydrolases from bovine, human, porcine, or rat liver.

well to reiterate that probably no single lysosome, or even cell, contains all these enzymes. Table 9.3 is a composite of enzymes that have been measured in the lysosomes from a variety of tissues. Some of these enzymes, such as the cathepsins, ribonuclease, and acid phosphatase, have been characterized in considerable detail, whereas others are only poorly characterized.

Some generalities emerge from the studies that have been conducted. All the enzymes are *hydrolases,* and as such cleave their substrates by the common mechanism of water insertion.

$$\text{S—S}' + \text{H—O—H} \rightleftharpoons \text{S—H} + \text{S}'\text{—O—H}$$

In general acid catalysis involving the carboxyl group, the catalytic proton is provided by a carboxyl group in the reactive center to a leaving group.

Enzyme —C(O—H)(=O)   O(R)(R¹)

↓

Enzyme —C(O⁻)(=O)   HO(R)(R¹)

The acid hydrolases appear to operate at pH values where this type of carboxyl ionization could take place.

Ribonuclease is technically a transferase, transferring a phosphate group from the 5′ position of one nucleotide to the 2′ position of another forming a 2′,3′-cyclic phosphate diester. This is followed, however, by a slow hydrolysis.

Another feature held in common by the lysosomal enzymes is their *acid pH optima.* This is clearly illustrated in Figure 9.13, which plots the activity pH ranges of 24 acid hydrolases. The pH-rate profiles of several of the enzymes suggest the involvement of one or more carboxyl groups at their catalytic centers. This implies that general acid catalysis may be a common mechanism of reaction among these acid hydrolases.

Several of the lysosomal enzymes also possess carbohydrate. This is known to be true for acid phosphatase, deoxyribonuclease II, hyaluronidase, β-acetylglucosaminidase, β-glucuronidase, cathepsin C, and cathepsin D. This list will likely get longer as other enzymes are prepared in large enough quantities and in sufficient purity for meaningful chemical analyses.

Finally, lysosomal enzymes share the property of resistance to autolysis. This is obviously an important characteristic for these enzymes and probably is reflective of their peculiar chemical and physical properties.

It is beyond the scope of this text to comment on each of the hydrolytic lysosomal enzymes, but let us single out a couple of the better characterized ones and examine some of their properties.

The presence of acid phosphatase in a cellular organelle is part and parcel of the definition of a lysosome. As discussed earlier, acid phosphatase is normally detected by a release of inorganic phosphate from an appropriate organic phosphate substrate. The enzyme is reactive with several different phosphate-containing substrates.

Substrates of acid phosphatase (↓, hydrolysis point).

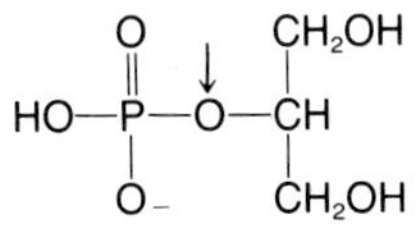

Glycero-2-phosphate

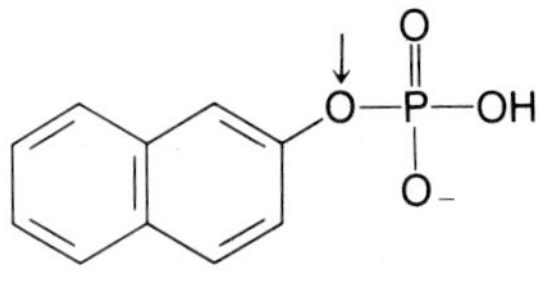

2-Naphthyl phosphate

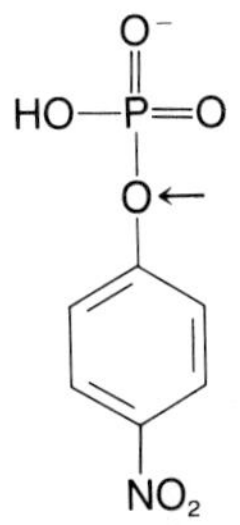

p-Nitrophenyl phosphate

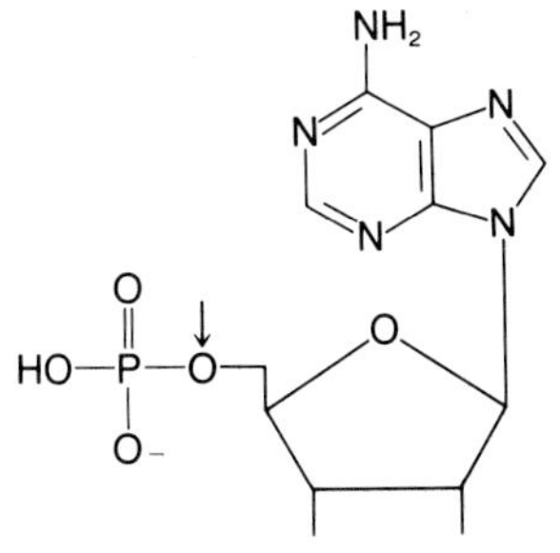

Adenosine 5′-monophosphate

The pH optimum for the enzyme has been found to vary with the tissue source, with values obtained that range from pH 5.4 for rat liver, to 3.0–4.8 for bovine spleen, to 4.0–4.4 for the human placenta enzyme.

The molecular weight of acid phosphatase isolated from rat liver is 100,000 to 120,000 daltons and that of the human placenta enzyme 105,000. The enzyme is inactivated by Hg^{2+}, *p*-chloromercuribenzoate, and other heavy metals, indicating a sulfhydryl group in its catalytic center. Sedimentation constants near 5.6 S have been obtained, and isoelectric points ranging from 4.5 to 7.0 for enzymes from different sources.

The literature shows a great deal of variation with respect to the physical properties of acid phosphatase. The values above are representative of literature values and should not be considered to be comprehensive.

The acid phosphatase of prostatic tissue has been studied in considerable depth. The enzyme from human tissue has a molecular weight of 109,000 daltons and a sedimentation constant of 5.7 S. As is true for a number of the lysosomal enzymes, prostatic acid phosphatase appears in multiple forms when subjected to starch gel electrophoresis, a property that is tentatively attributed to differing amounts of covalently bound sialic acid. In spite of the information that is available on this particular enzyme, it is not yet clear that all the acid phosphatase of the prostate is of lysosomal origin. A good deal of the enzyme is contained in secretory vesicles that may or may not be of lysosomal origin.

The cathepsins of the lysosome are peptidases or proteinases that are acidic hydrolases. Cathepsin D, the major acid proteinase of tissue, has been associated with the lysosome by many research groups. The enzyme has low activity against low molecular weight synthetic substrates but is highly reactive against hemoglobin, with which it is commonly assayed. The pH optimum for the enzyme is 3.0–3.5 and it is unaffected by thiol and serine inhibitors. Estimates of molecular weights for the enzyme from various sources range from 35,000 to 60,000 daltons.

The specificity of cathepsin D resembles that of pepsin quite closely, as can be seen from its action on the oxidized B chain of insulin shown in Figure 9.14: obviously the enzyme can degrade proteins very efficiently, especially those rich in aromatic and hydrophobic aliphatic amino acids.

The acidic pH optimum of lysosomal enzymes implies that the pH of the interior of the lysosome must be lower than that of the surrounding cytosol to promote efficient enzyme activity. A number of different ap-

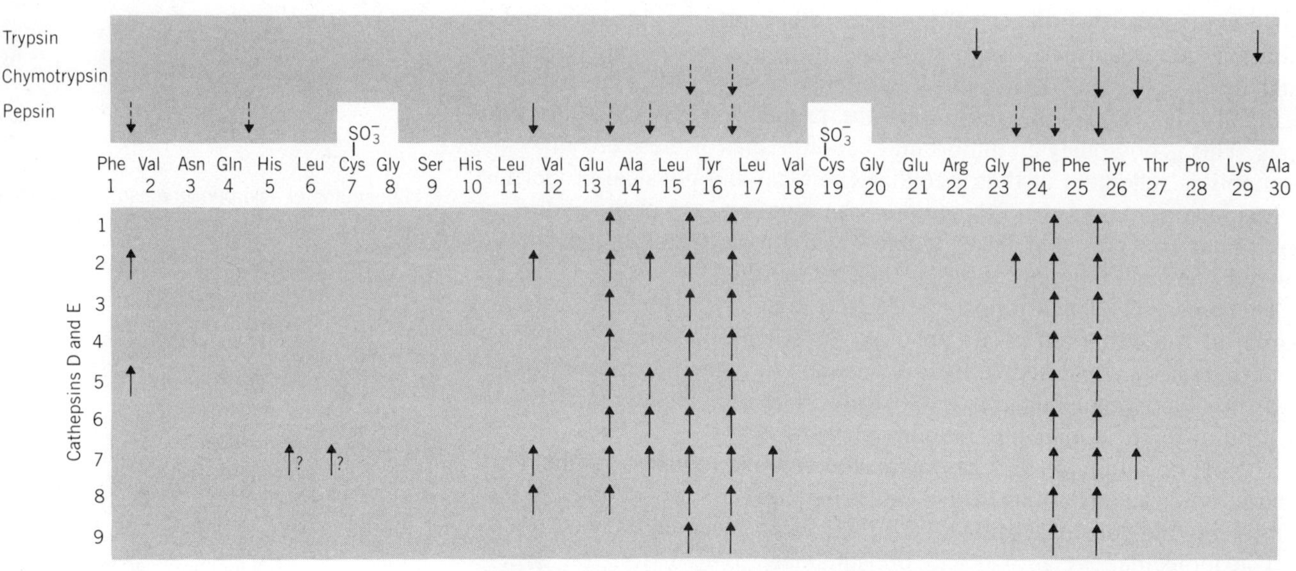

FIGURE 9.14 Action of several cathepsins on the oxidized B chain of insulin compared with the specificities of three other proteolytic enzymes (on top). The cathepsins most closely resemble pepsin, which is an acid proteinase of the stomach. Cathepsin preparations from nine different tissue sources are shown. Preparations 1–7 are cathepsin D.

proaches have been used to estimate the intralysosomal pH, and all point to a significantly acidic value. This has been found true for the digestive vacuole of blood leukocytes[15] which is initially neutral after endocytosis and then drops rapidly to pH 4.0 within 7 to 14 min after fusion with the primary lysosome. If this is a universal phenomenon, we may anticipate a common presence of some mechanism to effect a pH gradient across the lysosomal membrane, such as an H^+-pumping ATPase.

A brief reflection on the enzyme content of the lysosome makes it obvious that this cellular organelle possesses all the enzymes necessary to degrade most of the complex molecules of cells and tissues. All the evidence we have points to a strictly catabolic role for the lysosome, a role that we shall shortly see is indispensable to normal cellular processes.

9.5 THE DIGESTIVE ROLE OF THE LYSOSOME

The physiological role of the lysosome in the cell is now understood to be multifaceted, but each facet is related in some fashion to digestion, most of which is intracellular. Intracellular digestion of *exogenous* cellular material is termed *heterophagy*. As well as being a common form of feeding in lower animals, such as protozoans, it is now believed to be a process carried out by cells of virtually all types, possibly including plants as well. Intracellular digestion of *endogenous* material is called *autophagy*. This process is also apparently common to most organisms, whether protozoan or metazoan, and may be an important normal process in cell turnover or may be stimulated by abnormal factors such as stress or starvation.

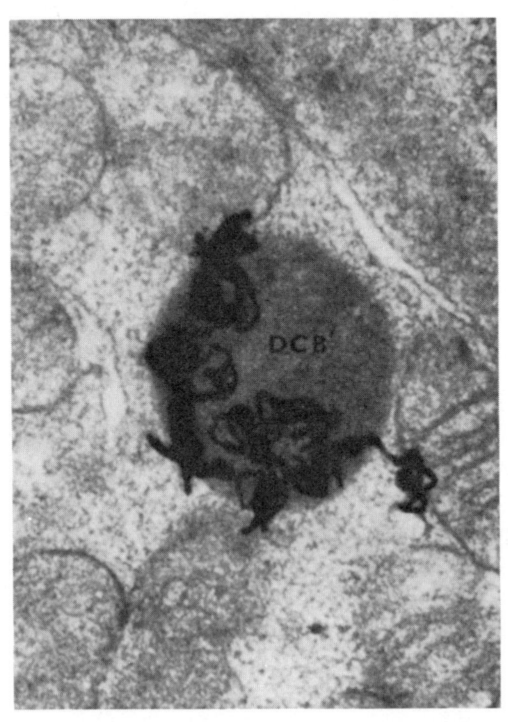

FIGURE 9.15 Electron microscope autoradiograph of a section of proximal tubule of rat kidney demonstrating heterophagy. The tubule was labeled with [^{125}I]-albumin and incubated for acid phosphatase. The radioactivity, evidenced as dense dots and lines, is confined to the acid phosphatase organelles, lysosomes. DCB′ refers to dense cytoplasmic body, an early designation for this organelle. (Electron micrograph by Dr. A. B. Maunsbach.)

Heterophagy

Endocytosis is, of course, prerequisite for the intracellular digestion of high molecular weight exogenous materials. The evidence is now substantial that food vacuoles, or phagosomes, resulting from endocytosis fuse with primary lysosomes to produce the digestive pot, or secondary lysosome.

This was demonstrated in elegant fashion by Straus,[16] who injected horseradish peroxidase into rats and observed the appearance of the enzyme in the cells of the convoluted tubules of the kidney. Benzidine, which forms a *blue* chromophore with peroxidase, was used as a histochemical marker for the enzyme. The enzyme was seen to be taken up by pinocytosis. A double marker technique was then worked out, labeling lysosomes with a histochemical azo dye method and peroxidase-containing pinosomes with the benzidine. Fusion of the pinosomes with the lysosomes resulted in *purple* vacuoles, because of a mixing of the two chromophoric labels.

Straus's experiment can be schematically illustrated this way.

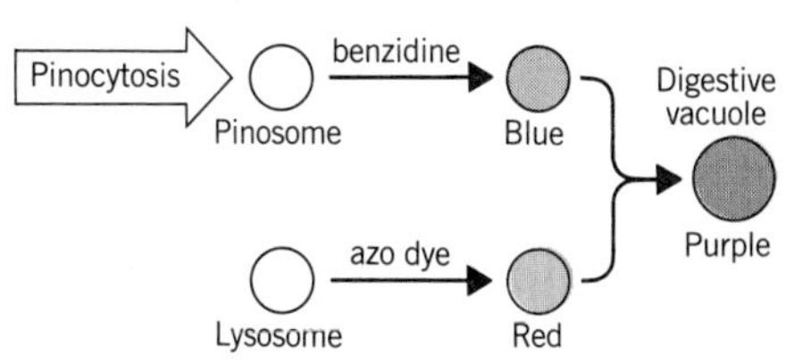

Other approaches, using ^{125}I-labeled rat serum albumin and autoradiography at the electron microscope level[17] have confirmed the work done with light microscopy as illustrated in the electron micrograph of Figure 9.15.

Method VI

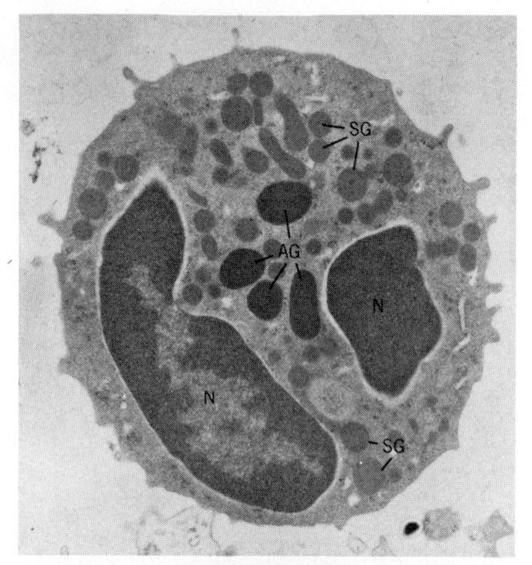

FIGURE 9.16 Neutrophil from rabbit bone marrow. This spherical cell with multilobated nucleus (N) contains two major types of granule in its cytoplasm. The larger dense granules, about 800 nm in diameter, are azurophil granules (AG). The smaller, less dense organelles are specific granules (SG). The azurophil granules are a type of primary lysosome. ×10,000. (Courtesy of Dr. Dorothy F. Bainton.)

Table 9.4 Individual Content of Rabbit PMN Granules

Content	Azurophil	Specific
Myeloperoxidase (pH 4.5)	+	−
Acid hydrolases (pH 4–5)	+	−
Acid mucosubstances	+	−
Cationic antibacterial proteins	+	−
Phagocytin (pH 5.6)	+	+
Lysozyme (pH 7.5)	+	+
Lactoferrin (stable pH 7.4–3.5)	−	+
Collagenase (pH 7.5)	−	+
Alkaline phosphatase (pH 9.4)	−	+

Blood phagocytes are cells that capitalize on the process of phagocytosis (heterophagy) and in so doing provide an effective barrier against the invasion of the organism by foreign agents. There are four types of phagocyte: polymorphonuclear neutrophilic leukocytes (PMNs) or neutrophils, eosinophils, basophils, and monocytes. All these function in host resistance to disease organisms. Although all four types circulate in the blood, the neutrophils and monocytes have the ability to leave the bloodstream and wander throughout the tissue, purging the tissue of encroaching foreign material by phagocytosis. In addition, organs such as the spleen, liver, bone marrow, and the lymph system are lined and infiltrated by phagocytic cells called "fixed" macrophages, which function throughout the body as a system called the reticuloendothelial system (more detail on this can be found in Chapter 8). All circulating body fluids are screened for foreign agents by this effective and efficient dragnet.

Let us look briefly at just one of these important phagocytic cell types, the neutrophils. Neutrophils, which comprise 65% of the total white blood cell population in the human, are distinguishable from other leukocytes by their cytoplasmic granules. They possess two types of granule, which can be seen readily in electron micrographs (Figure 9.16). The lighter staining granules, called *specific granules,* do not appear to be lysosomal because they are devoid of the acid hydrolases (see Table 9.4). However, they do contain hydrolases, but these enzymes have pH optima that are nearer neutrality. The deeper staining graules, called *azurophil granules,* are a type of primary lysosome, containing in addition to the acid hydrolases the enzyme myeloperoxidase.

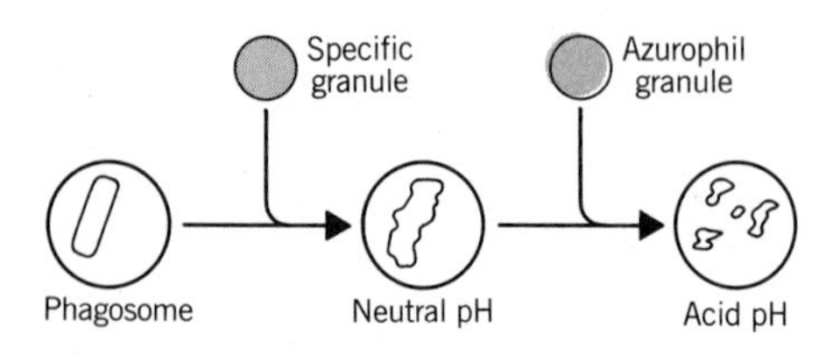

Armed with these two types of enzyme compartment, a very fascinating series of events takes place during the phagocytosis of bacteria. There is first an intake of bacteria by endocytosis, as shown dramatically in Figure 9.17. The phagosomes that result combine *first* with the *specific granules* of the cell, producing a digestive vacuole that remains at *neutral* pH and permits lysozyme to begin degrading the peptidoglycan of the bacterial cell wall. The digestive vacuole *then* combines with the acid hydrolase containing *azurophil granule.* Within a few minutes the pH drops to 4.0, and the acid hydrolases and myeloperoxidase, a potent antibacterial agent, further degrade and destroy the foreign cell.[18] This compartmentalization of enzymes and the sequential fusion process permit the enzymes to work independently on the bacteria at the pH optima where they are most effective. This could not take place by a one-step fusion with a single type of hydrolytic vacuole.

Autophagy

Autophagy is now recognized as a common cellular event. The main line of evidence for this is electron microscopic observations of organelle remnants in vacuoles containing acid phosphatase. A typical electron micrograph showing this was presented in Figure 9.5a.

The origin of this digestive vacuole, variously called autophagic vacuole or autolysosome, is not as clearly delineated as for the digestive vacuole that arises from phagocytosis. Four proposed schemes for the formation of its precursor, the cytosegresome, are shown in Figure 9.18.

The interjection of lysosomal enzymes into the cytosegresome occurs subsequently by fusion with a primary lysosome. Alternatively, if the membranes making up the cytosegresome are derived from preexisting membranes of the endoplasmic reticulum or Golgi complex, the lysosomal activity of the resulting particle may derive from the cellular membranes of which it is made.

Autophagic vacuoles are thought to be responsible for several types of cellular activity,[19] including cell turnover, cellular remodeling and tissue metamorphosis, physiological survival during adverse nutritional conditions, a response to cell injury, and accidental endocytic trapping. It is not possible to discuss all these in detail, but a few features should be underscored.

First, in regard to cell turnover, it is well established that the materials of the cell, both the organelles and the soluble macromolecules, are characterized by specific turnover rates.[20] Liver mitochondria, for example, have half-lives of approximately 10 days, and the estimated half-life of parenchymal liver cells is on the order of 150 days. Organelles other than mitochondria are subject to turnover, and it is likely that all the components of the cell are forced into some kind of controlled disassembly and recycling. There is little doubt that lysosomes are intimately involved in these processes.

A second area of autophagic activity is the type of activity that occurs during tissue development or remodeling. Studies that have been conducted on metamorphosis in insect and amphibian systems suggest that tissue regression is directly related to increased hydrolase activities in these tissues. In particular, the later stages of tail regression in *Xenopus,* a type of frog, is accompanied by extensive scavenging by lysosome-rich macrophages.

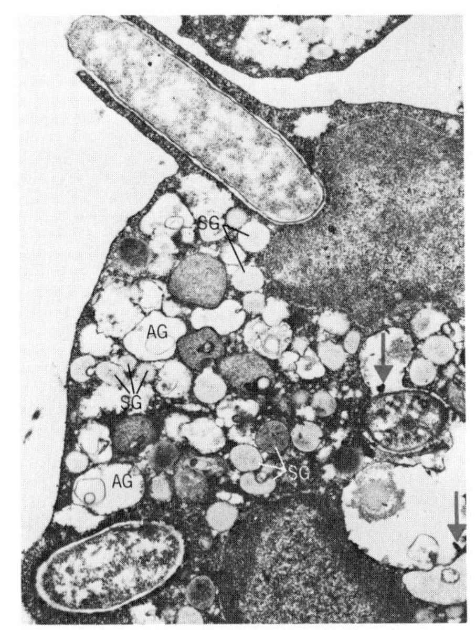

FIGURE 9.17 Phagocytic cell showing engulfed bacteria and specific (SG) and azurophil (AG) granules. The phagosome in the lower left contains alkaline phosphatase (arrows), indicating that it has fused with SGs. Some of this is also apparent in the digestive vacuole on the right. Alkaline phosphatase is extracted from unreacted SGs in this procedure so they do not appear dark. ×13,600. (Courtesy of Dr. Dorothy F. Bainton.)

The Products of Digestion and Their Fates

Reflecting on the properties of the enzymes contained in the lysosome, it is evident that the organelle has the potential for breaking down most biological macromolecules to their elemental units. In this regard, it is

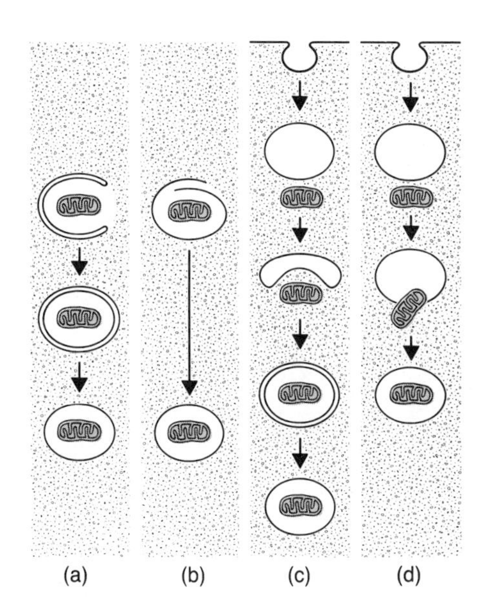

FIGURE 9.18 Schemes proposed to explain the formation of cytosegresomes. (*a*) A membrane-bound cisterna encircles a mitochondrion; the internal membrane degenerates. (*b*) An individual membrane wraps around a mitochondrion, enclosing it completely. (*c*) An endocytic vesicle encircles a mitochondrion. (*d*) A mitochondrion enters an endocytic vesicle through a hole or weak point. Each of these would become an autolysosome (Figure 9.4) by fusion with a primary lysosome.

Table 9.5 **Molecules That Are Subject (+) and Not Subject (○) to Hydrolysis by Lysosomal Enzymes in Macrophages (column 1) and Those Products Retained (+) and Not Retained (○) by Lysosomes (column 2)**

Compound	Susceptibility to Macrophage Hydrolases	Retention in Macrophage Secondary Lysosomes
Monosaccharides C_5 C_7	—	○
Amino Acids	—	○
Disaccharides		
Maltose	+	○
Lactose	+	○
Sucrose	○	+
Cellobiose	○	+
Polysaccharides		
Ficoll	○	+
Dextran	○	+
Inulin	○	+
Dextran-SO_4	○	+
Peptides		
$(\text{L-Ala})_2$	+	○
$(\text{L-Ala})_3$	+	○
$(\text{L-Glu})_2$	+	○
D-Leu–L-Tyr	+	○
$(\text{D-Ala})_2$	○	○
$(\text{D-Ala})_3$	○	+
$(\text{D-Glu})_2$	○	+

proper to ask whether molecules larger than the ultimate products of enzymatic hydrolysis will be released by lysosomes for metabolism in the cytosol.

To study this problem, Cohn and his associates[21] devised techniques to evaluate the permeability of the lysosomal membrane to a variety of molecular species. Table 9.5 tabulates the results for a number of carbohydrates and peptides. The di- and polysaccharides for which there are no lysosomal hydrolytic enzymes (sucrose, cellobiose, Ficoll, etc.) are retained in secondary lysosomes after engulfment by the cell, whereas sugars with a molecular weight below 200 daltons are released.

Peptides composed of D-amino acids (L-amino acid peptides being susceptible to hydrolysis by lysosomal enzymes) were also examined for their retention properties and, as Table 9.5 indicates, the tripeptide of alanine is retained whereas its dipeptide is not. The dipeptide of glutamic acid, larger than the homologous alanyl peptide, is also retained. Thus the cutoff point of lysosomal retention is somewheres between the di- and tripeptide levels.

By using radioactively tagged macromolecules, such as albumin and hemoglobin, the extent of degradation by hydrolases in the lysosome has been monitored. Iodinated albumin, for example, is engulfed and segregated in the lysosomes, then degraded to the extent of 80 to 95% in 20 hr as revealed by a release of low molecular weight radioactivity into the extracellular medium. Hemoglobin labeled with [^{3}H]-leucine is degraded so completely that only the tritiated leucine is released as a radioactive product.

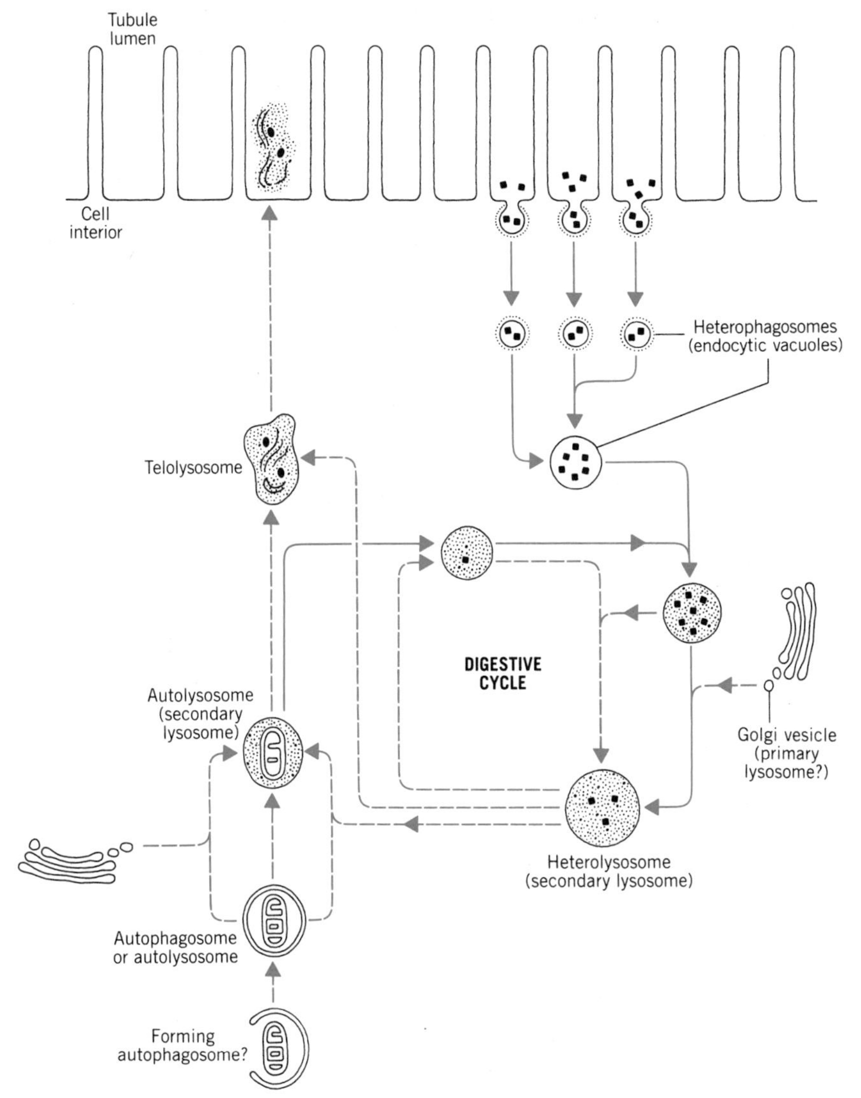

FIGURE 9.19 Relationships between autophagy (*left*) and heterophagy (*right*) as they would take place in kidney proximal tubule cells. During heterophagy, protein is taken into endocytic vacuoles, fuses with primary lysosomes, and undergoes hydrolysis. During autophagy, cytosegresomes form, become autolysosomes by fusion with lysosomes, and enter into the pathways. Solid arrows represent pathways that are strongly supported by experimental work; broken arrows show pathways for which there is less experimental support.

The evidence is thus quite strong that biological macromolecules for which there are lysosomal hydrolytic enzymes are degraded to the point where their products permeate the lysosomal membrane. Presumably if the products are metabolizable, they are then made available to the catabolic or anabolic machinery of the cytosol.

The relationships between heterophagy and autophagy, in schematic and summary form, are shown in Figure 9.19.

Accumulation or Discharge

Occasionally the end product of intracellular digestion is an undigestible residue. These residual bodies may remain in the cell for the rest of its life span and may adversely influence the normal processes of the cell and be responsible in part for cell aging. An example of this type of accumulation is that which occurs in the aging human liver, heart, and nerve tissues, where lipofuscin is stored. In other cases, materials are accumulated in the lysosomes because of an inherited enzyme defect. We will discuss the latter phenomenon in more detail in Section 9.8.

Lipofuscin, also called age pigment, apparently consists of lipid–protein complexes that precipitate. For some reason it cannot be degraded by lysosomal enzymes.

Other cells, such as protozoa and a limited number of animal cells, discharge the residues of the lysosome to the cell exterior. Kidney cells, for example, have the ability to discharge a myelinated substance to the lumen of the proximal tubule by exocytosis (see Figure 9.19).

9.6 THE ROLE OF LYSOSOMES IN SECRETION

Animal Cells

Lysosomes carry out some very important activities related to events that take place outside the cell. To investigate this property, a very large area of research is opening up that concerns itself with the role of lysosomes in hormone release in the pituitary and thyroid glands. In these systems lysosomal enzymes appear to operate *within* the cell to assist in the preparation of material destined to function *outside* the cell.

Thyroxine.

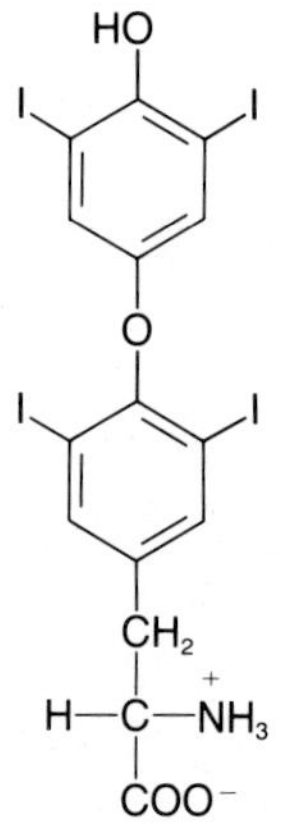

Triiodothyronine is identical except that one I is missing.

The events that take place in the thyroid gland serve as a good example.[22] The thyroid hormones thyroxine and triiodothyronine are covalently linked to a protein, thyroglobulin, and held in such form in the follicles of the gland. However, the form of hormone that appears in the blood when the thyroid is stimulated by the pituitary gland hormone TSH is no longer tied to a protein carrier. The sequence of events that takes place between storage and release involves the lysosome as follows (Figure 9.20). TSH stimulates pinocytosis on the lumen side of the epithelial cell of the follicle. The thyroglobulin droplets merge with primary lysosomes in which the thyroid hormone is excised from its carrier protein. The hormone is then released from the cell to the blood capillary either by diffusion or selective transport. The fate of the thyroglobulin protein core is unknown.

Lysosomes are also known to play a secretive role in the turnover of bone.[23] Two types of bone cell are particularly important in this turnover: osteoblasts, which form the calcified component of bone, and osteoclasts, which effect breakdown of the calcified bone matrix.

Figure 9.21 summarizes the proposed involvement of osteoclasts and lysosomal enzymes in this process. This scheme traces the process back to the point of cell stimulation by parathyroid hormone (PTH), a stimulation that may increase both glycolysis and the synthesis of lysosomal enzymes. The former process, producing lactate, drops the pH in the bone matrix next to the osteoclast, and the latter process prepares more lysosomes for digestion of the matrix.

Thus geared up, the cell has a twofold task to accomplish, which it apparently executes by a two-phase process. The task consists of solubilizing the mineral content of the bone matrix and degrading the organic content (mainly collagen, mucopolysaccharides, and mucoproteins). This is done by the two-phase process of secreting H^+ and releasing lysosomal enzymes to solubilize and fragment the matrix and engulfing organic fragments followed by intracellular digestion. One of the unresolved problems associated with this scheme is the uncertainty of the presence of collagenase in osteoclast lysosomes. There is some thought that two separate mechanisms may be superimposed in the process, namely, an exocytosis of lysosomal enzymes and a secretion of collagenase from the cell independent of the lysosomes.

Plant Cells

As discussed earlier, the elaboration of the presence, structure, and properties of lysosomes in plant cells needs further work. In plants, lysosomes may be one component of a larger system of hydrolases that carries out a variety of functions.

Plant cells are generally unable to engulf large particles, presumably because of the restrictions imposed on the cell by the cell wall. The secretion of hydrolases to carry out extracellular digestion therefore becomes an important plant cell process. Hydrolases are commonly secreted by fungi, enabling the organism to degrade and grow on macromolecules it cannot transport into the cell. Higher plants also secrete hydrolases, a

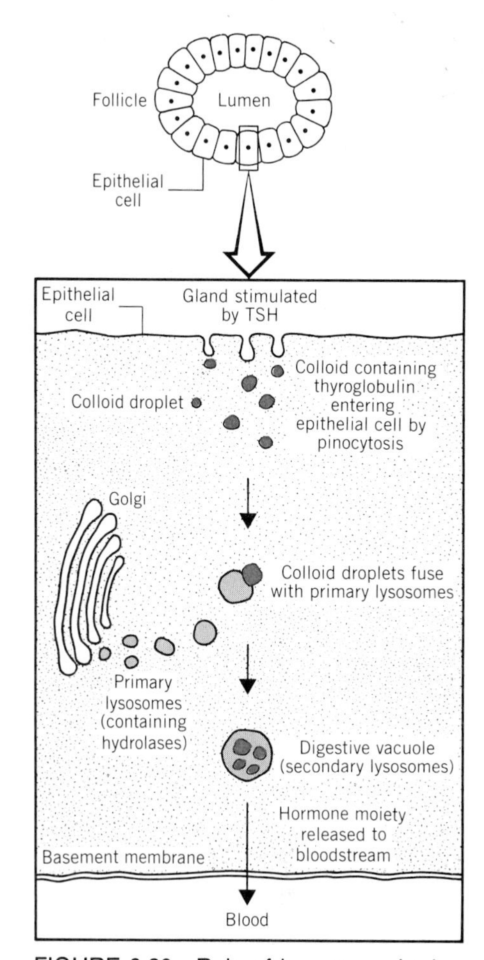

FIGURE 9.20 Role of lysosomes in thyroid gland epithelial cells.

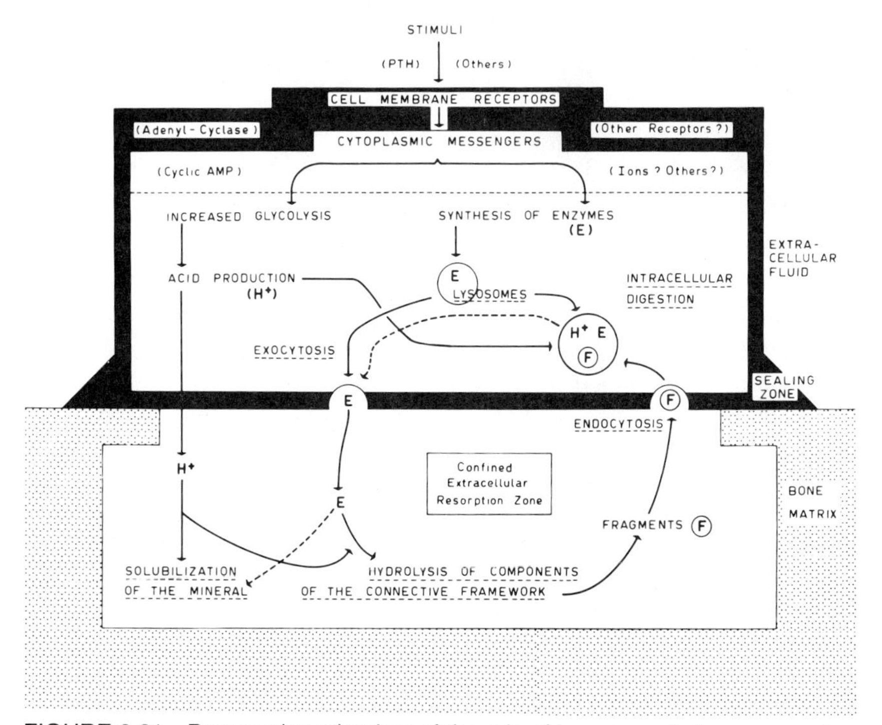

FIGURE 9.21 Proposed mechanism of the role of lysosomes in bone resorption carried out by osteoclasts. (Courtesy of Dr. Gilbert Vaes.)

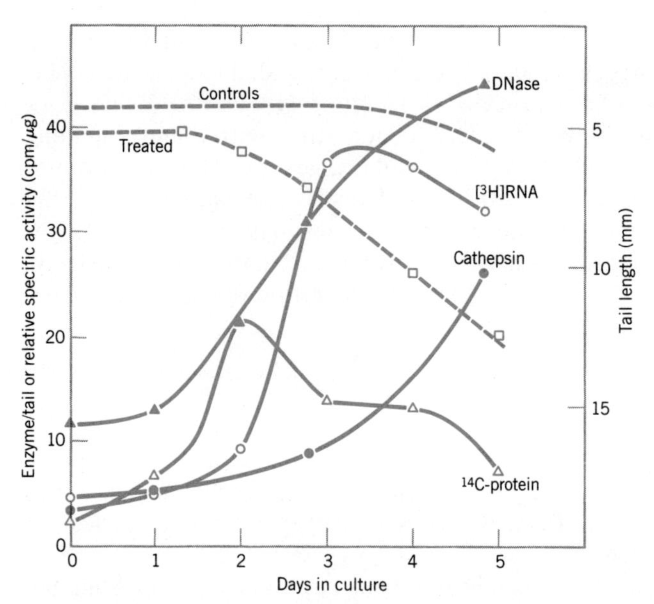

FIGURE 9.22 Changes in activity of hydrolases during tadpole tail regression. Tail length of controls decreases only slightly over the experiment period, whereas tails treated with thyroid hormone decrease rapidly after a couple of days. RNA and the lysosomal hydrolases (represented by cathepsin) increase dramatically during regression.

notable example being the insectivorous pitcher plants, which produce a proteinase-containing liquid in which the victims are trapped and digested. The lysosomal role in these secretive processes has not been thoroughly delineated, but it seems reasonable to assume that there is a relation between lysosomes and these activities in plants.

9.7 THE ROLE OF LYSOSOMES IN DEVELOPMENT AND TISSUE DYNAMICS

Metamorphosis

The role of lysosomes vis-à-vis regression of the tail tissue in frogs has already been suggested as a type of autophagy. This process has been quite extensively studied in frogs because it is possible to maintain the tail tissue in organ culture and to stimulate regression by applying thyroid hormone. Tissue regression is accompanied by an increase in lysosomal and hydrolase activity and an enhanced macrophage activity.

Thyroid hormone stimulates the synthesis of RNA and protein in cultured cells at about the same time that increased hydrolytic activity is noticed. This is taken to mean that mRNA is first produced, after which there is synthesis of hydrolytic enzymes that participate in tissue regression. Some of the activity changes of hydrolases during tail tissue regression are summarized in Figure 9.22.

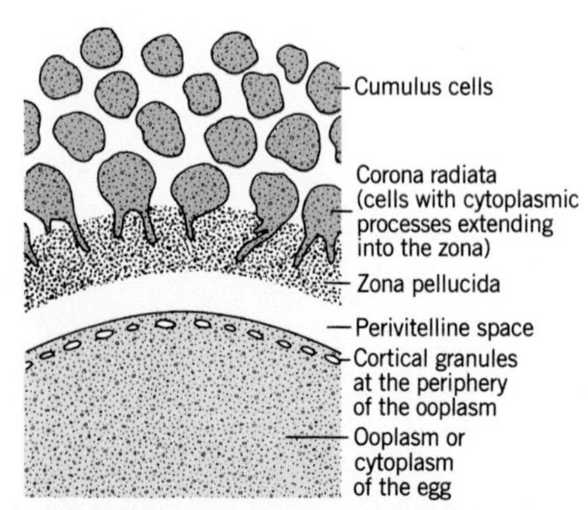

FIGURE 9.23 Section through a mammalian egg showing the protective maze of cells that must be penetrated by sperm before fertilization can take place.

It is not precisely clear whether increased lysosomal enzyme activities *cause* regression or simply *accompany* it as a mopping-up operation of dead cells. If lysosomes are directly involved, their hydrolytic activities must be carefully controlled to permit only the appropriate rate and amount of tissue regression. It is proposed that in tadpoles the hormone prolactin may have a regulatory feedback effect to balance the stimulation by TSH.

In insects, metamorphosis is stimulated by ecdysone, a hormone that brings about a regression of a number of larval tissues such as the fat bodies and salivary glands. The literature contains numerous electron micrographs of regressing tissue showing membrane- and particle-containing secondary lysosomes. Juvenile hormone appears to balance the effect of ecdysone in controlling metamorphosis in insects.

Fertilization in Mammalian Systems

The union of a sperm with an egg in mammals takes place only after the sperm has successfully penetrated the extracellular protective maze of the egg. The structure of this maze, beginning at the plasma membrane and working out, is first a perivitelline space followed by the zona pellucida and two amorphous layers of cells (Figure 9.23). The cumulus cells are embedded in a matrix of hyaluronic acid, and the zona pellucida contains the polysaccharide chondroitin sulfate as well as proteinaceous material. Combined, these materials make up a barrier that must be loosened or weakened before the membranes of the sperm and egg can fuse.

The sperm is uniquely designed to effect penetration because it possesses a hydrolytic warhead, the *acrosome*. This caplike structure fits snuggly over the anterior half of the sperm nucleus (Figure 9.24). The acrosome resembles lysosomes of other cells in the following ways: it derives from the Golgi complex during spermatogenesis, it stains orange-red with acridine orange, and it stains cytochemically for typical lysosomal enzymes, such as acid phosphatase, esterase, and arylsulfatase.[24] In addition to the acid hydrolases, the acrosome contains some enzymes active at alkaline pH.

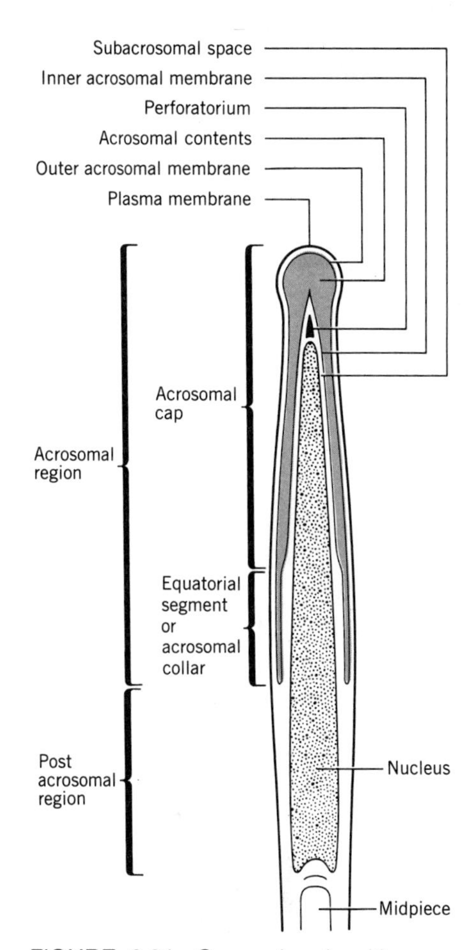

FIGURE 9.24 Sperm head, with acrosome fitting over the anterior portion of nucleus. The acrosome is a type of lysosome and contains hydrolytic enzymes necessary for penetrating the protective shell around the egg.

Only sperm that are activated by incubation in the female tract are able to carry out fertilization. This activation, called *capacitation*, is characterized in its final stages by a fusion of the *outer* acrosomal membrane with the adjacent plasma membrane. The fused system becomes vesicular and porous, and the enzymatic content of the acrosome is released (Figure 9.25). This "acrosomal reaction" takes place in the vicinity of the egg just before penetration by the sperm.

The acrosome of most domesticated and laboratory animals contains a proteolytic trypsinlike enzyme, called *acrosin*, which has been quite thoroughly investigated. The application of cytochemical and subcellular fractionation techniques has suggested that the enzyme is bound to the *inner* acrosomal membrane near the nucleus. The enzyme has been purified and in this form has the ability to solubilize the zona pellucida. Although the evidence is only circumstantial, the binding of acrosin to the inner acrosomal membrane makes a lot of sense, for it assures a supply of enzyme to break down the zona pellucida *after* the acrosome reaction has released most of the other enzymes and has used them to penetrate the outer layers of the egg.

Hyaluronidase, another acrosomal enzyme, has been studied in a manner similar to that of acrosin. It too is bound to the inner acrosomal membrane, but not as tightly or completely as acrosin. Thus, hyaluron-

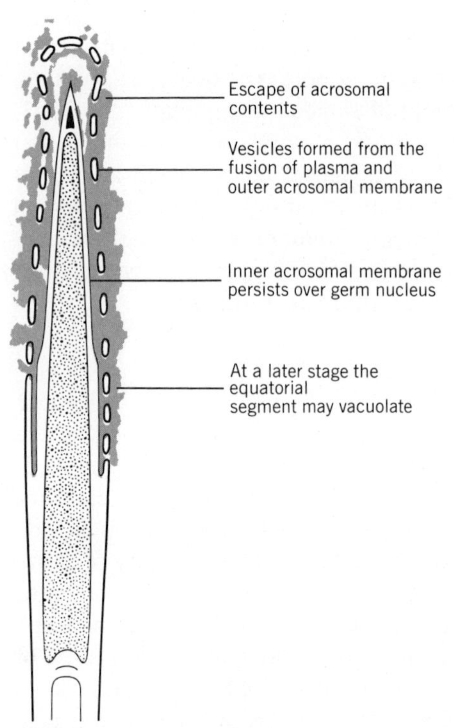

FIGURE 9.25 Changes in the sperm head during the acrosome reaction. The plasma membrane and the outer acrosome membrane fuse, then become porous, and the hydrolytic enzymes of the acrosome are lost to the environment. This occurs just before fertilization.

idase is released earlier than acrosin and, at least in some species, this may be an important factor in aiding the sperm in penetrating between cumulus cells.

Uterus and Mammary Gland Involution

The massive tissue changes that take place in the uterus during the normal menstrual and estrus cycles and in the uterus and mammary glands during reproduction are rapidly conducted.[25] Although the changes are obviously complex, they do appear to be ultimately under hormonal control and to use the hydrolytic activities of lysosomes during involution.

During the estrus and menstrual cycles, macrophages and phagocytic cells reach a peak in their activities in the endometrium during the time of maximal tissue breakdown. During the postpartum involution of the uterus, one of the most striking examples of rapid tissue reorganization known, there is an elevation of lysosomal hydrolase activity due to an increased population of phagocytic cells that infiltrate the various layers of the uterus. When the mammary gland involutes after enforced cessation of lactation, there is also an increase of acid hydrolase concentration with concomitant autophagic and heterophagic activities.

Although these lysosome-associated events have been studied in the uterus and mammary gland, the questions involving the mechanism of hormonal control of these processes are largely unanswered. Both the uterus and the mammary gland afford good model systems for the study of the coordination of hormone and hydrolase activity.

Germination of Seeds

The role of lysosomes in plant development has not been unequivocally worked out, but there is enough evidence implicating hydrolytic enzyme activities during the germination of seeds to make them good candidates as agents of plant tissue development.

One of the most intensively studied plant systems in this regard is the endosperm in cereal seeds, such as barley. Figure 9.26 summarizes the sequence of events involving hydrolases that is thought to take place as the endosperm reserves are mobilized for use by the growing plant.

This is another example of the activity of hydrolases being hormonally controlled. Gibberellic acid, a plant growth hormone, is released by the embryo to the aleurone layer where, in turn, the hydrolases are released to the endosperm. Gibberellin apparently operates by derepressing appropriate genes in the aleurone cells, which then begin to crank out new hydrolytic proteins. Although there is ultrastructural evidence available showing lysosomelike structures in the cotyledons of germinating pea, tobacco, and cotton seeds, convincing proof of their involvement in germination is lacking.

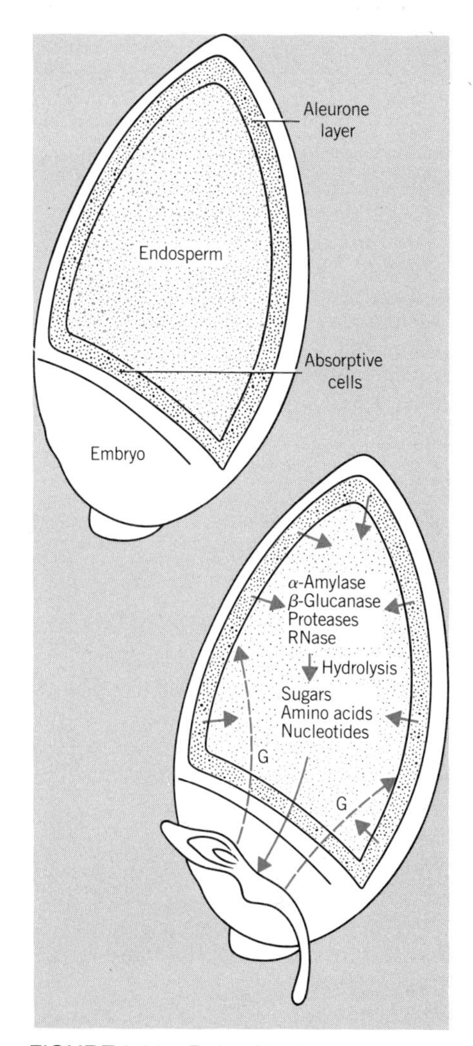

FIGURE 9.26 Role of hydrolytic enzymes in seed germination. (*a*) Ultrastructure of seed before germination. (*b*) During germination, gibberellins (G) pass from the embryo to the aleurone layer, where the *de novo* synthesis of hydrolytic enzymes is induced. These enzymes break down the macromolecular stored reserves and the low molecular weight products are transported to the embryo, where they function as nutrients.

9.8 THE LYSOSOME AND DISEASE

The central role that lysosomes play in a number of cell processes is well established. Thus it should also be apparent that a malfunction of lysosomes, in terms of either hyper-or hypoactivity, could be seriously detrimental to the cell and to the entire organism.

There are now several diseases that can be understood in the context of lysosome malfunction. Some of these are inborn lysosomal diseases, characterized by a genetic defect in the lysosome, and others are induced by exogenous environmental materials.

Inborn Lysosomal Diseases

Table 9.6 lists several diseases that are apparently related to genetic abnormalities of lysosomes. All these are so called "deposition" diseases, characterized by the accumulation of a particular complex macromolecule in the cells of different tissues. The sequence of events that is proposed to account for this is as follows.

Mutation of a gene →
Alteration of one enzyme →
Disturbance in the degradation of one or several macromolecules →
Clinical manifestation

Studies on Pompe's disease, a glycogen storage disorder, have revealed a complete absence in the lysosomes of acid α-glucosidase, which degrades glycogen and oligosaccharides to glucose. Lysosomes that take up

Table 9.6 Summary of Lysosomal Disorders

Disorder	Enzyme Deficiency	Metabolite Primarily Affected
Mucopolysaccharidoses		
Hurler and Scheie syndromes	α-L-Iduronidase	Dermatan sulfate, heparan sulfate
Hunter syndrome	Iduronate sulfatase	Dermatan sulfate, heparan sulfate
Sanfilippo syndrome		
A subtype	Heparan *N*-sulfatase	Heparan sulfate
B subtype	*N*-Acetyl-α-glucosaminidase	Heparan sulfate
Maroteaux–Lamy syndrome	*N*-Acetylgalactosamine sulfatase (arylsulfatase B)	Dermatan sulfate
β-Glucuronidase deficiency	β-Glucuronidase	Dermatan sulfate, heparan sulfate
Morquio syndrome	Uncertain	Keratan sulfate
Sphingolipidoses		
GM_1 gangliosidosis	β-Galactosidase	GM_1 ganglioside, fragments from glycoproteins
Krabbe's disease	β-Galactosidase	Galactosylceramide
Lactosylceramidosis	β-Galactosidase	Lactosylceramide
Tay–Sachs disease	Hexosaminidase A	GM_2 ganglioside
Sandhoff's disease	Hexosaminidases A and B	GM_2 ganglioside, globoside
Gaucher's disease	β-Glucosidase	Glucosylceramide
Fabry's disease	α-Galactosidase	Trihexosylceramide
Metachromatic leukodystrophy	Arylsulfatase A	Sulfatide
Niemann–Pick disease	Sphingomyelinase	Sphingomyelin
Farber's disease	Ceramidase	Ceramide
Disorders of Glycoprotein Metabolism		
Fucosidosis	α-L-Fucosidase	Fragments from glycoproteins, glycolipids
Mannosidosis	α-Mannosidase	Fragments from glycoproteins
Aspartylglycosaminuria	Amidase	Aspartyl-2-deoxy-2-acetamidoglucosylamine
Other Disorders with Single Enzyme Defect		
Pompe's disease	α-Glucosidase	Glycogen
Wolman's disease	Acid lipase	Cholesterol esters, triglyceride
Acid phosphatase deficiency	Acid phosphatase	Phosphate esters
Multiple Enzyme Deficiencies		
Multiple sulfatase deficiency	Sulfatases (arylsulfatase A, B, C: steroid sulfatases; iduronate sulfatase; heparan *N*-sulfatase)	Sulfatide, steroid sulfate, mucopolysaccharide
I cell disease and pseudo-Hurler polydystrophy	Almost all lysosomal enzymes deficient in cultured fibroblasts; present extracellularly	Mucopolysaccharide and glycolipids
Disorders of Unknown Etiology		
Cystinosis	Accumulation of cystine in lysosomes	Cystine
Mucolipidoses I, IV	Ultrastructural evidence of lysosomal storage	Unknown

glycogen by endocytosis in liver and muscle cells are unable to break it down, and it accumulates in large amounts. This eventually results in breakage of the lysosomes and a release of the lysosomal enzymes to the cell. The effect of this is most adversely noticed in muscle, where the released cathepsins digest muscle fibers and the patient succumbs from muscle destruction.

Each of the other inborn lysosomal diseases has its own specific but related mechanisms.

At this point, one can imagine how advantage can be taken of the endocytic abilities of cells to "treat" cells with stored materials. If the appropriate enzyme can be introduced into the cell, stored materials may be degraded and the genetic problem reduced in intensity.

Uric acid, (structure: OH, N, N, OH, HO, N, N) is a normal metabolite arising from the degradation of purines. In approximately 3 out of 1000 persons there is a metabolic defect that is clinically expressed as gout. Either uric acid is overproduced (because purines are overproduced as a result of enhanced biosynthetic activity) or the kidneys malfunction and do not excrete uric acid properly.

To this end, several workers have prepared liposomes, lipid vesicles containing the missing enzyme, and have administered them to laboratory animals.[26] Theoretically, the replacement enzyme should be taken up by the cell and the stored substrate degraded *in situ* as illustrated schematically in Figure 9.27. Some of the results are quite encouraging, but there are still many obstacles to be overcome before enzyme replacement therapy is a reality.

One of the oldest and most prominent inborn errors of metabolism, gout, is now becoming understood at the cellular and molecular level.[27] The identification of uric acid in the urine and tophi of patients with gouty arthritis dates back to the late eighteenth century, but it is only recently that the detrimental effect of monosodium urate (MSU) crystals in joints has been recognized as a membrane phenomenon, in particular involving lysosomal membranes.

Tophi are chalky deposits that accumulate in tissues in the vicinity of joints.

The sequence of events that is now thought to take place in acute gouty inflammation is summarized in Figure 9.28. In persons who are predisposed to crystal formation, phagocytic white cells engulf MSU crystals and incorporate them into the phagosome of the cell, which merges, in normal fashion, with primary lysosomes. The hydrolytic enzymes of the lysosome strip the crystal of any coated plasma proteins, exposing a highly concentrated surface of hydroxyl groups, which are now free to hydrogen bond with the polar portions of the amphipathic lipids of the membrane. The result is a weakening of the bimolecular leaflet structure of the lysosomal membrane followed by its rupture. The hydrolytic enzymes of the lysosome are thereby dumped into the cytosol, eventually destroying the leukocyte and releasing to the environment of the cell toxic cell materials and hydrolytic enzymes. This sequence of events could be repeated by another cell picking up the same crystal and succumbing to the same fate. The cellular materials released are responsible for the inflammation and tissue destruction typical of arthritic gout.

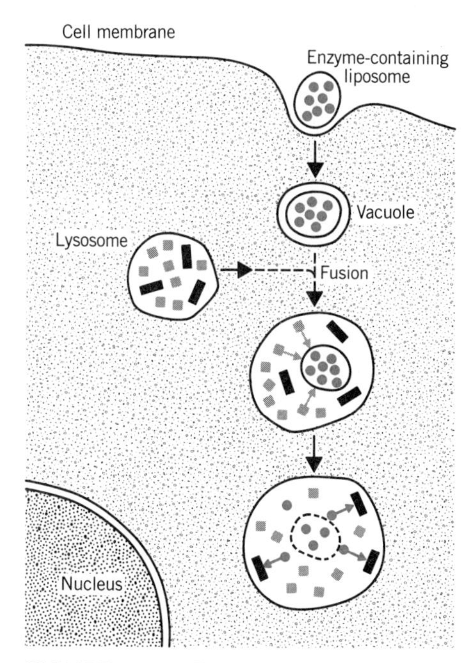

FIGURE 9.27 The principles of enzyme replacement therapy. An appropriate enzyme is housed in a liposome, enters the cell by endocytosis, and fuses with the malfunctioning lysosome to provide the necessary enzyme for degradation.

Environmentally Induced Diseases

Several diseases have been studied in which the detrimental effect on the lysosome has come from outside the organism. Such is the case for silicosis and asbestosis. Silicon and asbestos particles, when inhaled, are taken up by the alveolar macrophages in which they combine with primary lysosomes to form digestive vacuoles. In the case of silicosis, silicon dioxide particles have an adverse effect on the lysosomal membrane, causing it to disrupt and empty its contents to the interior of the cell and eventually the surrounding tissue. This process activates fibroblasts to initiate the deposition of fibrous tissue at the site of these events, resulting in an overall impairment of lung function. This process is illustrated in Figure 9.29.

Alveoli are the small saclike structures in the lungs from which oxygen is transported into the blood.

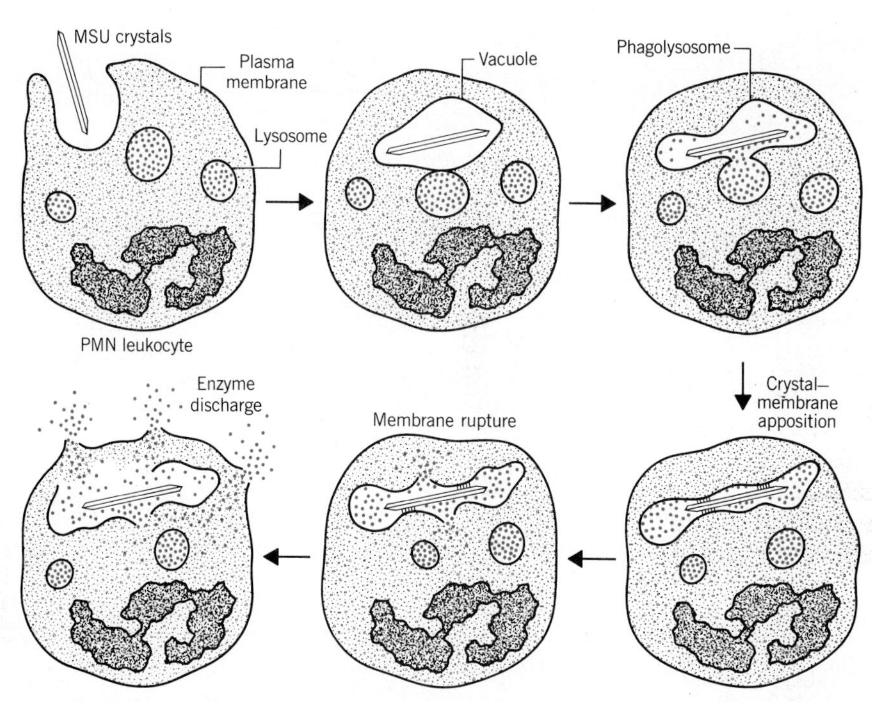

FIGURE 9.28 Sequence of events in acute gout. Crystals of monosodium urate (MSU) are taken into cells, weaken lysosomal membranes, and trigger a release of hydrolytic enzymes into the cell.

Lysosomotropic Chemotherapy

Theoretically, it should be possible to introduce at will a variety of materials specifically into the lysosome by the mechanisms of endocytosis. This is how, for example, Triton WR-1339 is loaded into lysosomes. In this case the detergent is nondigestible and is retained by the lysosome. Drugs that are effective chemotherapeutic agents but unable to penetrate the plasma membrane might be brought into cells this way, and neoplastic cells, usually having a higher endocytic activity than normal cells, could be target cells for certain inhibitory drugs. This is the concept behind a newly emerging line of treatment called lysosomotropic chemotherapy, that is, the administration of a protected drug that is engulfed and fused with the lysosomes of actively endocytic cells. Once exposed to the hydrolases of the lysosomes, the protective coating or carrier is enzymatically removed and the drug is released from the lysosome to the cytosol to carry out its function in the cell. Figure 9.30 illustrates the principle of lysosomotropic chemotherapy.

A neoplasm is a new or abnormal growth, such as a tumor.

Applying this principle, some remarkable success has been achieved by Trouet and his co-workers on leukemic mice and terminally ill human cancer patients.[28]

In mice, the powerful but toxic antitumor agent adriamycin has been used. The drug was complexed with DNA, which served as an endocytosis-stimulating carrier and, after absorption, was readily degraded by lysosomal enzymes. Mice administered the free drug succumbed rapidly from toxicity, but mice treated with the drug–DNA complex survived much longer, and 60% of them recovered completely.

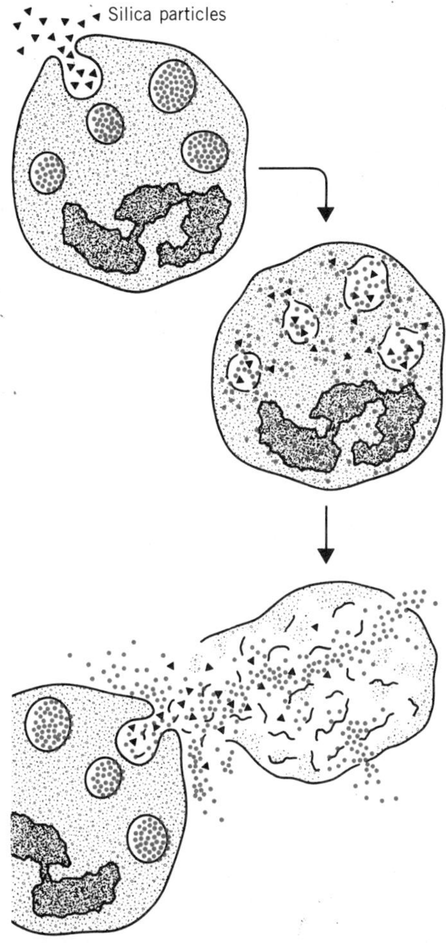

FIGURE 9.29 Sequence of events in silicosis. Silicon dioxide particles are engulfed by phagocytic lung cells, within which they weaken lysosomal membranes. Hydrolytic enzymes are released, and cell destruction occurs.

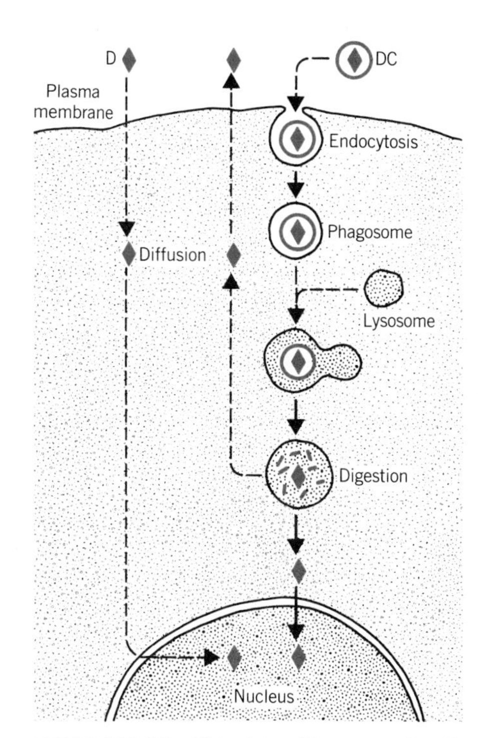

FIGURE 9.30 Principle of lysosomotropic chemotherapy. Drugs (D) that either cannot penetrate the plasma membrane or penetrate only slowly can be brought into the cell interior when combined with a carrier (DC) that is brought in by endocytosis. Once in, the carrier is stripped off by lysosomal enzymes and the drug made immediately available to the cell interior.

Terminally ill human subjects with acute myelocytic leukemia were treated in a similar manner with the drug daunorubicin complexed with DNA. Of 15 patients so treated, 9 demonstrated a complete remission and 4 had partial remissions.

Myelocytes are bone marrow cells. They normally develop into leukocytes of the blood. In certain forms of leukemia, myelocytes appear in the blood.

Apparently the lysosomotropic drugs are less toxic than normally administered free drugs because they target on the cells, mostly tumor cells, that have a higher than normal mitotic rate and elevated endocytic activities. These results are striking and exciting, and may well open the door to a most effective and inventive mode of cancer chemotherapy that holds some real promise of success.

Summary

Lysosomes are membrane-bound, hydrolytic, enzyme-containing organelles that are ubiquitous in eucaryotic cells. They exist in several physiological forms. The lysosome as initially formed by the cell is called a primary lysosome. After fusion with an endocytic vacuole it is called a secondary lysosome. Telolysosomes are aged varieties that may become residual bodies or may release their contents by cellular defecation.

Lysosomes are formed in the region of the Golgi complex by a packaging of enzymes produced in the rough endoplasmic reticulum. Their fate is normally fusion with endocytic vacuoles or with vacuoles containing intracellular material, or fusion with the plasma membrane to expel their contents to the cell exterior.

The enzymatic activities of lysosomes often form the basis for their detection. In particular, acid phosphatase activity can be quite specifically associated with lysosomes by way of histochemical techniques.

Lysosomes are fragile, hence difficult to isolate. They are also heterogeneous, even within a given cell population. Techniques have been worked out to isolate relatively pure populations of these organelles by differential and isopycnic centrifugation, often combined with detergent loading of the lysosomes before centrifugation.

The membrane of the lysosome has the typical unit membrane appearance and contains certain components in common with the plasma membrane. In other respects it appears to have its own unique properties. Somewhat more than 50 hydrolytic enzymes have been found in lysosomes from different sources.

Lysosomes have important roles in the digestion of extracellular (heterophagy) and intracellular (autophagy) material. They thus play important roles in resistance mechanisms in higher organisms, in obtaining nutrients in lower forms, and in cell turnover. The enzymes of lysosomes also have functions in certain types of cell secretion, such as in the secretion of thyroid hormone from the thyroid gland and in the turnover of bone.

In development, lysosomes are implicated in fertilization and in the tissue dynamics associated with metamorphosis. They are also involved with the large-scale tissue changes during the involution of the uterus and mammary glands. They appear to have a role in the germination of seeds.

A number of diseases are associated with lysosomal malfunction. Some of these are inborn deposition diseases and others are triggered by elements in the environment.

Certain approaches to therapy appear to be feasible by taking advantage of the endocytic activity of cells to introduce drugs into cells that would be released when combined with the lysosomes.

References

1. Leçons sur la Pathologie Comparée de l'Inflammation, É. Metchnikoff, Masson et Cie., Paris, 1892.
2. *Collected Papers,* vol. 1, P. Ehrlich, Pergamon Press, London, 1956.
3. Concentration of Acid Phosphatase, Ribonuclease, Desoxyribonuclease, β-Glucuronidase, and Cathepsin in "Droplets" Isolated from the Kidney Cells of Normal Rats, W. Straus, *J Biophys. Biochem. Cytol.* 2:513(1956).
4. *Microscopic Histochemistry,* G. Gomori, University of Chicago Press, Chicago, pp. 189 and 198, 1952.
5. Histochemistry, F. Beck, J. B. Lloyd, and C. A. Squier, in *Lysosomes—A Laboratory Handbook,* J. T. Dingle, ed., North Holland, Amsterdam, 1972.
6. *The Lytic Compartment of Plant Cells,* P. Matile, Springer-Verlag, New York, 1975.

7. The Heterogeneity of Lysosomes, M. Davies, in *Lysosomes in Biology and Pathology,* vol. 4, J. T. Dingle and R. T. Dean, eds., North Holland, Amsterdam, 1975.
8. Tissue Fractionation Studies. 6. Intracellular Distribution Patterns of Enzymes in Rat-Liver Tissue, C. de Duve, B. C. Pressman, R. Gianetto, R. Wattiaux, and F. Appelmans, *Biochem. J.* 60:604(1955).
9. Tissue Fractionation Studies. 13. Analysis of Mitochondrial Fractions from Rat Liver by Density Gradient Centrifuging, H. Beaufay, D. S. Bendall, P. Bauduin, R. Wattiaux, and C. de Duve, *Biochem. J.* 73:628(1959).
10. Methods for the Isolation of Lysosomes, H. Beaufay, in *Lysosomes—A Laboratory Handbook,* J. T. Dingle, ed., North Holland, Amsterdam, 1972.
11. Influence of the Injection of Triton WR-1339 on the Properties of Rat-Liver Lysosomes, R. Wattiaux, M. Wibo, and P. Baudhuin, in *Ciba Foundation Symposium on Lysosomes,* A. V. S. de Reucke and M. P. Cameron, eds., Churchill, pp. 176–200, London, 1963.
12. A New Method for the Preparation of Rat Liver Lysosomes, R. Stahn, K. P. Maier, and K. Hannig, *J. Cell Biol.* 46:576(1970).
13. Isolation and Chemical Composition of the Lysosomal and the Plasma Membrane of the Rat Liver Cell, R. Henning, H. D. Kaulen, and W. Stoffel, *Hoppe-Seylers Z. Physiol. Chem.* 351:1191(1970).
14. Lysosomal Membranes, J. A. Lucy, in *Lysosomes in Biology and Pathology,* vol. 2, J. T. Dingle and H. B. Fell, eds., North Holland, Amsterdam, 1969.
15. Primary Lysosomes of Blood Leukocytes, D. F. Bainton, B.A. Nichols, and M. G. Farquhar, in *Lysosomes in Biology and Pathology,* vol. 5, J. T. Dingle and R. T. Dean, eds., North Holland, Amsterdam, 1976.
16. Lysosomes, Phagosomes and Related Particles, W. Straus, in *Enzyme Cytology,* D. B. Roodyn, ed., pp. 239–319, Academic Press, New York, 1967.
17. Absorption of ^{125}I-Labeled Homologous Albumin by Rat Kidney Proximal Tubule Cells, A. B. Maunsbach, *J. Ultrastruct. Res.* 15:197(1966).
18. Temporal Changes in pH Within the Phagocytic Vacuole of the Polymorphonuclear Neutrophilic Leukocyte, M. S. Jensen and D. F. Bainton, *J. Cell Biol.* 56:379(1973).
19. Mechanisms of Cellular Autophagy, J. L. E. Ericsson, in *Lysosomes in Biology and Pathology,* vol. 2, J. T. Dingle and H. B. Fell, eds., North Holland, Amsterdam, 1969.
20. Lysosomes and Intracellular Protein Turnover, H. L. Segal, in *Lysosomes in Biology and Pathology,* vol. 4, J. T. Dingle and R. T. Dean, eds., North Holland, Amsterdam, 1975.
21. Endocytosis and Intracellular Digestion, Z. A. Cohn, in *Monouclear Phagocytes,* R. van Furth, ed., Blackwell Scientific Publications, Oxford, 1970.
22. Secretion of Thyroid Hormone, S. H. Wollman, in *Lysosomes in Biology and Pathology,* vol. 2, J. T. Dingle and H. B. Fell, eds., North Holland, Amsterdam, 1969.
23. Lysosomes and the Cellular Physiology of Bone Resorption, G. Vaes, in *Lysosomes in Biology and Pathology,* vol. 1, J. T. Dingle and H. B. Fell, eds., North Holland, Amsterdam, 1969.
24. Lysosomal Enzymes in Mammalian Spermatozoa, D. B. Morton, in *Lysosomes in Biology and Pathology,* vol. 5, J. T. Dingle and R. T. Dean, eds., North Holland, Amsterdam, 1976.
25. The Physiology of the Uterus and Mammary Gland, J. F. Woessner, in *Lysosomes in Biology and Pathology,* vol. 1, J. T. Dingle and H. B. Fell, eds., North Holland, Amsterdam, 1969.
26. *Enzyme Therapy in Lysosomal Storage Diseases,* J. M. Tager, G. J. M. Hooghwinkel, and W. T. Daems, eds., North Holland, Amsterdam, 1974.
27. The Molecular Basis of Acute Gout, G. Weissmann, in *Cell Membranes,* G. Weissman and R. Claiborne, eds., HP Publishing Co., New York, 1975.

28. Lysosomotropic Cancer Chemotherapy, A. Trouet, D. Deprex-de Campeneere, A. Zenebergh, and G. Sokal, in *Activation of Macrophages,* W. H. Wagner and H. Hahn, eds., Elsevier, New York, 1974.

Selected Books and Articles

Books

Enzyme Therapy in Lysosomal Storage Diseases, J.M. Tager, G. J. M. Hooghwinkel, and W. T. Daems, eds., North Holland, Amsterdam, 1974.

Lysosomes–A Laboratory Handbook, J. T. Dingle, ed. North Holland, Amsterdam, 1972.

Lysosomes and Cell Function, Dennis Pitt, Longman, New York, 1975.

Lysosomes in Biology and Pathology, vol. 1, J. T. Dingle and Honor B. Fell, eds., North Holland, Amsterdam, 1969.

Lysosomes in Biology and Pathology, vol. 2, J. T. Dingle and Honor B. Fell, eds., North Holland, Amsterdam, 1969.

Lysosomes in Biology and Pathology, vol. 3, J. T. Dingle and Honor B. Fell, eds., North Holland, Amsterdam, 1973.

Lysosomes in Biology and Pathology, vol. 4, J. T. Dingle and R. T. Dean, eds., North Holland, Amsterdam, 1975.

Lysosomes in Biology and Pathology, vol. 5, J. T. Dingle and R. T. Dean, eds., North Holland, Asterdam, 1976.

The Lytic Compartment of Plant Cells, P. Matile, in *Cell Biology Monographs—Continuation of Protoplasmatologia,* vol. 1, Springer-Verlag, New York, 1975.

Articles

Inherited Disorders of Lysosomal Metabolism, E. F. Neufeld, T. W. Lim, and L.J. Shapiro, *Annu. Rev. Biochem.* 44:357(1975).

The Molecular Basis of Acute Gout, Gerald Weissman, in *Cell Membranes,* G. Weissmann and R. Claiborne, eds., HP Publishing Co., New York, 1975.

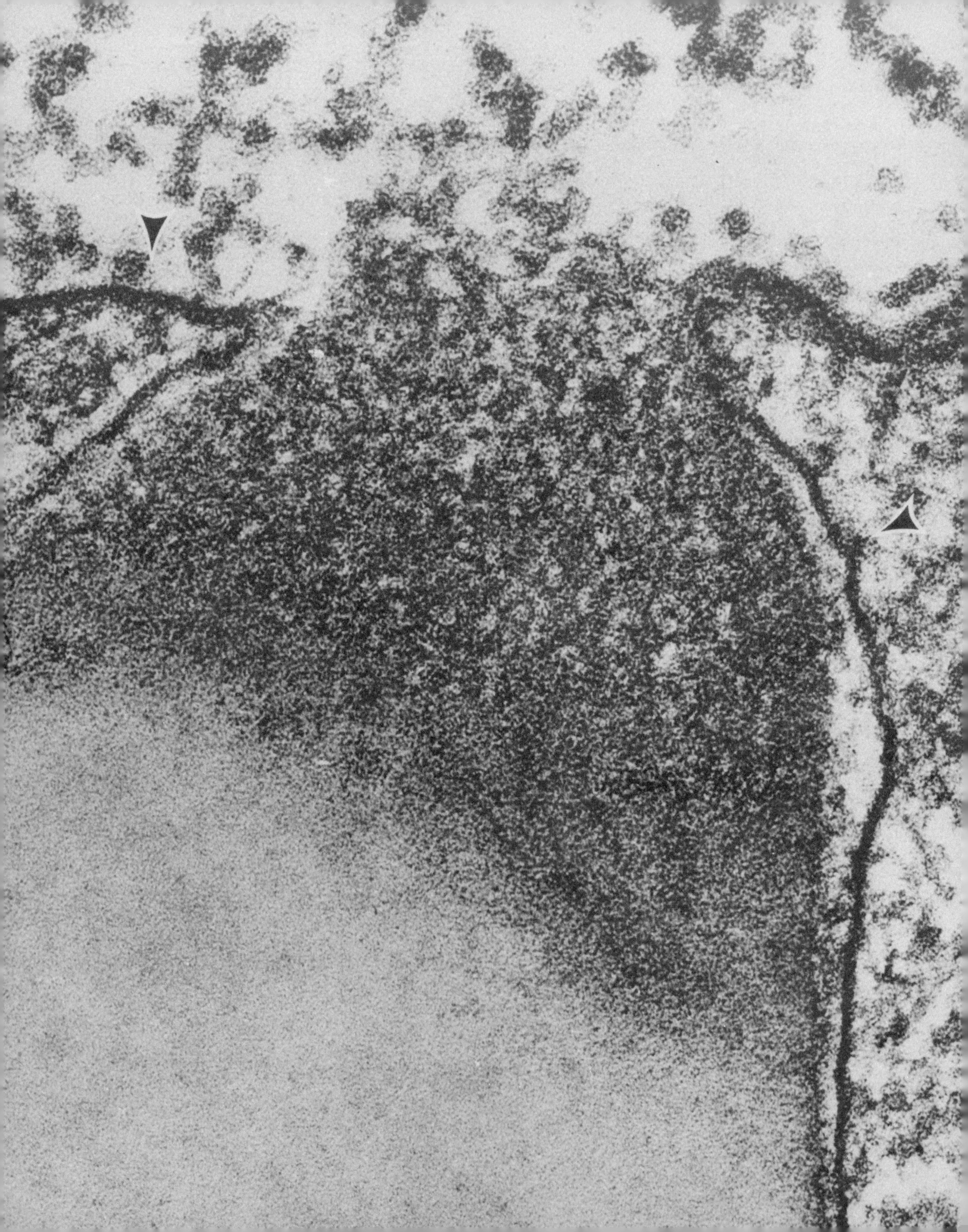

SECTION 5

MODIFICATION AND EXPORT

Cells are open-ended systems. Some materials enter, as by selective or bulk transport, and others exit. Between entry and exit there takes place a phenomenal amount of cellular activity. Certain materials that enter serve strictly as energy sources, and their fate is complete oxidation to carbon dioxide and water. Other compounds are only partially degraded before they are snatched from a catabolic fate and used in the manufacture of something structural or enzymatic needed by the cell. Still others escape significant breakdown.

In addition to making macromolecules that are essential for the replication and survival of the cell, most cells make materials that are secreted and find some role to play outside the cell, sometimes at great distances from the cell itself. These secretory cells synthesize, transport, and expel their products by a common mechansim that is called the secretory pathway. It is a controlled way of handling substances within the cell that are destined for export.

As materials move along the secretory pathway in the cell they are generally chemically modified. The Golgi complex is the major organelle involved in modifying these materials as well as in packaging them for export. In the two chapters that follow we first take a look at the complete secretory pathway and then focus on the Golgi complex, a key element in the pathway.

It is an interesting intracellular journey.

10

The Secretory Pathway

But perhaps the most important factor in this selection [the choice of the pancreatic exocrine cell for the study of secretion] was the appeal of the amazing organization of the pancreatic acinar cell, whose cytoplasm is packed with stacked endoplasmic reticulum cisternae studded with ribosomes. Its pictures had for me the effect of the song of a mermaid: irresistible and half transparent. Its meaning seemed to be buried only under a few years of work, and reasonable working hypotheses were already suggested by the structural organization itself.

GEORGE PALADE, December 12, 1974

Nature has graciously provided researchers with a model system for the study of almost every cellular process. Each cell or tissue or organ seems to have some outstanding feature that makes it a prototype for all the rest. The pancreas became the archetypal mammalian secretory organ and has been exhaustively studied to unravel the secretory mechanisms of cells.

As Palade suggested, its very appearance is that of a miniaturized factory: long rows of parallel assembly lines stuffing products into packages that accumulate in piles near the shipping door (Figure 10.1). For Palade, and many others who worked with him or followed in his steps, the exocrine cell of the acinar regions of pancreatic tissue was the obvious model system for the study of the secretory process.

It has been a very productive system. The results gleaned from its study have been so meaningful that they attracted for George Palade the Nobel prize for physiology and medicine in 1974, shared with Albert Claude and Christian de Duve. It was during the reception of this international recognition that Palade made the statement quoted above, alluding to the structural amenabilities of the pancreatic exocrine cell for research. Without this level of cellular cooperation, the work could never have been accomplished. It would therefore seem most appropriate, before we get into the heart of the chapter, to acknowledge the contribution made by the pancreatic cell itself to our understanding of its function. It has been a most helpful prototype and deserves at least a small portion of the prize.

The pancreatic acinus is a group of cells, one layer thick, clustered around a lumen into which they secrete digestive enzymes. Collectively, many such acini provide the exocrine function of the gland.

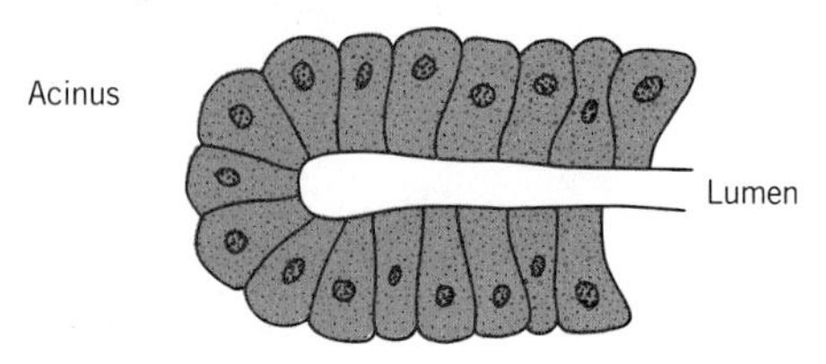

10.1 THE TYPES OF SECRETORY CELL

For a long time scientists viewed the role of cells in secretion in a rather limited sense. It was restricted to only a few cell types, namely, those that are prodigious and obvious secreters. These cells often display a specialized function and are structurally highly differentiated.

Among these prolific secreters are three major types: producers of digestive enzymes, such as the pancreatic exocrine cell and the cells of the parotid gland, deliverers of hormones, exemplified by such endocrine cells as those of the adrenal medulla and the pituitary gland, and providers of neurotransmitters to the synapse, the neurons.

Most cells that secrete large amounts of material are *intermittent* secreters. They supply copious quantities of secretion products to the cell environment in response to hormonal or neural stimulation. Other secretory cells are *continuous* secreters, delivering smaller amounts of material, but in a more constant fashion.

As more cells have received detailed study for their ability to secrete, it has become apparent that secretion is much more the rule than the exception in the overall behavior of cells. All plant eucaryotic cells, for example, can be thought of as secretory cells, since they must produce internally and secrete externally the polysaccharides and proteins that make up their cell walls. Even more fundamentally, however, all cell types, of both plant and animal origin, probably produce lysosomes. These are packaged by a mechanism similar to that of pancreatic secretory granules and can be viewed as secretory products whose fate is extracellular in the sense of being fused with phagosomes or pinosomes. Some

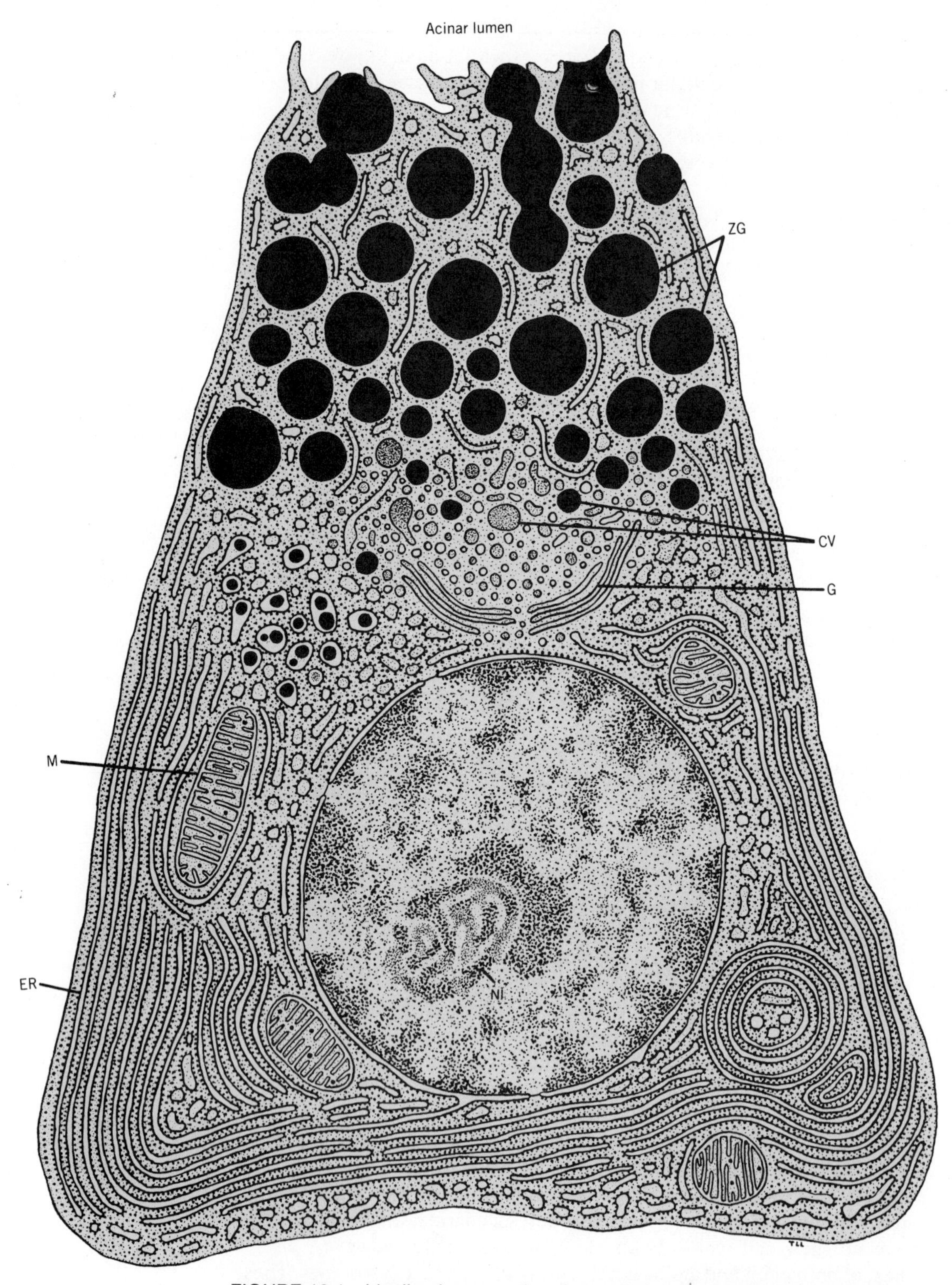

FIGURE 10.1 Idealized pancreatic acinar cell showing stacked endoplasmic reticulum (ER), mitochondria (M), the Golgi complex (G), condensing vacuoles (CV), and zymogen granules (ZG). The apical end of the cell is at the top, beyond which lies the acinar lumen.

lysosomes secrete their contents directly into the extracellular environment, as is true for osteoclasts, and thus these cells are truly secretory in the same manner as are pancreatic cells.

Stretching this line of thinking a bit farther, the glycocalyx material attached to the outer surface of the plasma membrane is a secretory product common to most if not all cells. Applying the foregoing logic, it may be stated that all cells are secretory cells, probably without exception.

In any event, if secretion is viewed in this broadened sense, allowing for lysosomes to pose as secretory granules and for the glycocalyx to be viewed as a secretory product, it becomes clear that the mechanisms of secretion have a universal importance in cell economy and, with high probability, operate by way of a common chemical and physical denominator.

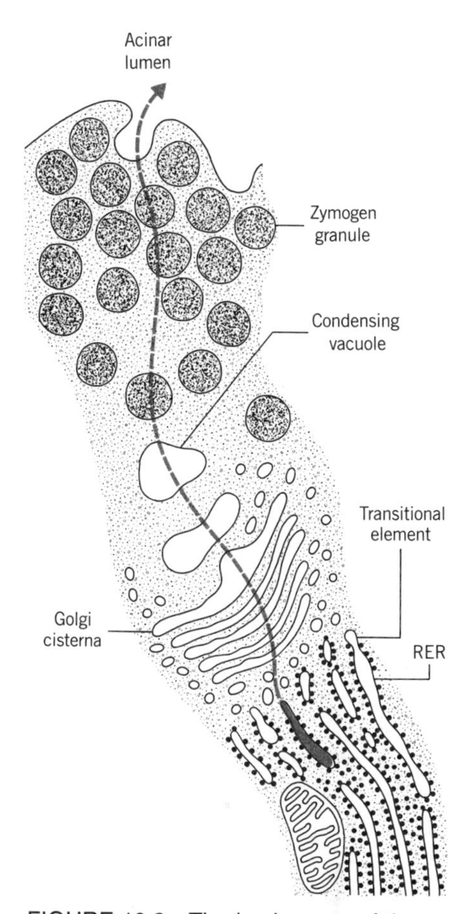

FIGURE 10.2 The basic route of the secretory pathway begins in the rough endoplasmic reticulum (RER), moves through or among the Golgi membranes, and includes stages in condensing vacuoles and zymogen (storage) granules before terminating in the acinar lumen.

10.2 THE SECRETORY PATHWAY

Although all cells may possess secretory activity, it would have been poor scientific judgment to choose to study one that is only weakly active. Palade's choice of the pancreatic cell was wise, for this cell possesses obvious secretory activity and has associated subcellular structures that are readily studied. Perhaps fortuitously, the pancreatic cell revealed in its activity a "complete" story of secretion that is demonstrable only in part in many other cells. It is from the study of this cell that the concept of the *secretory pathway* was worked out in terms of the subcellular compartments and organelles of which it is constituted. Therefore a good deal of the discussion that follows centers on this cell.

The basic route of the secretory pathway is illustrated in Figure 10.2. It originates in the rough endoplasmic reticulum (RER) where the secretory proteins are synthesized. In pancreatic cells, which are highly polarized both morphologically and functionally, the RER is generally distributed predominantly toward the basal half of the cell. From its origin in the RER, the secretory pathway takes the products through or to the vicinity of the Golgi complex and eventually concentrates and stores them in the secretory or storage granules. These are congregated in a region that extends from the Golgi complex to the apical plasma membrane at the luminal surface of the cell. Finally, the secretory pathway ends in the lumen itself, into which the products are discharged by the mechanism of exocytosis.

Palade has recognized and delineated six successive stages in the secretory pathway,[1] which we use as a framework for our discussion of the details of the process. We also use the pancreatic cell as a model for secretion, to develop a general and complete story to which variations can be added later. The six stages of secretion are schematically outlined in Figure 10.3.

10.3 THE KEY STAGES OF SECRETION

Stage I: Synthesis

Proteins that are destined to become secretory products are synthesized by the normal mechanisms of protein synthesis on polysomes that are

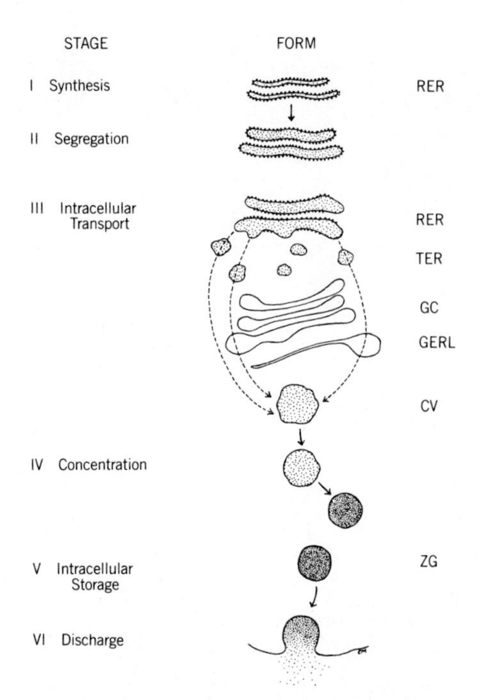

FIGURE 10.3 Summary of the six stages evident in the "complete" secretory pathway. For details of each stage and abbreviations, see text.

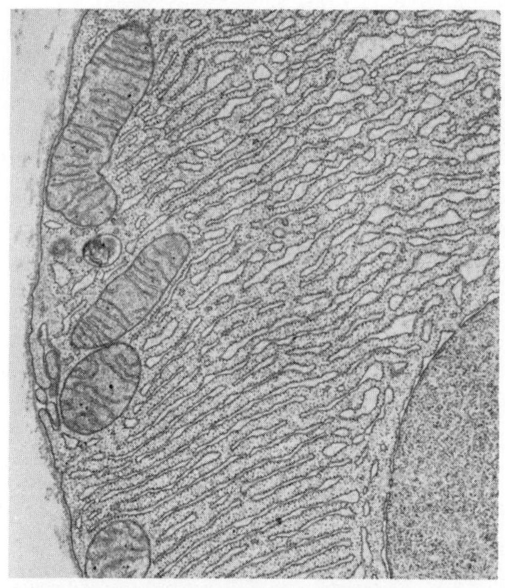
(a)

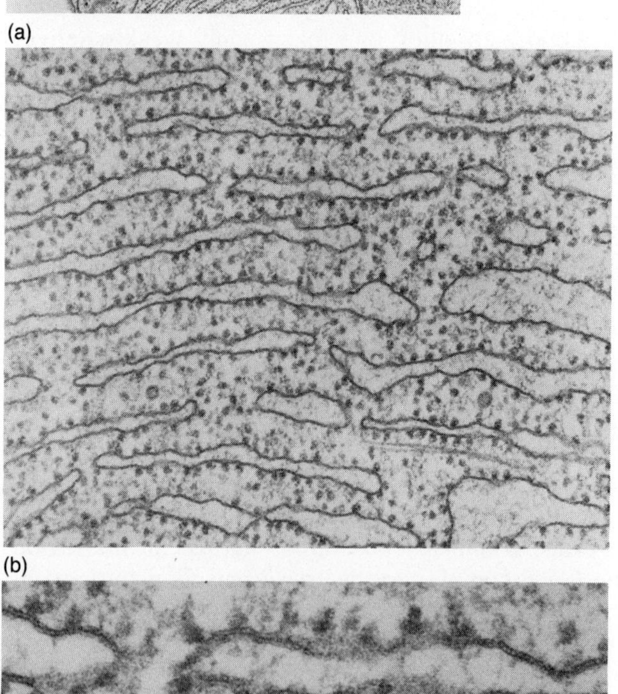
(b)

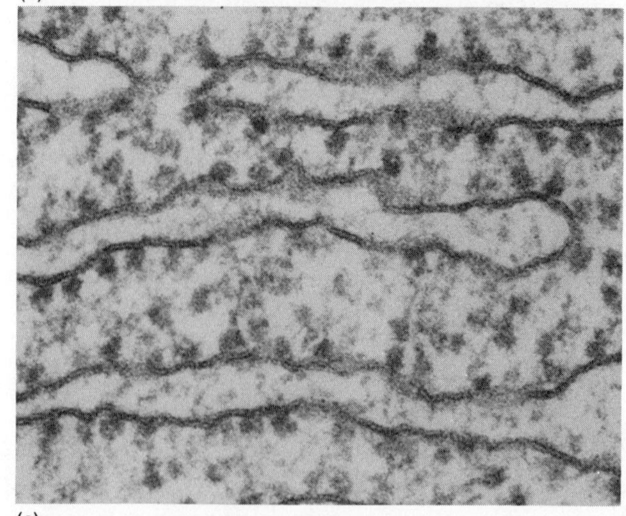
(c)

FIGURE 10.4 Pancreatic exocrine cells shown at increasing magnifications. (*a*) the lamellar form of rough endoplasmic reticulum is apparent in the region between the nucleus and the plasma membrane. (*b*) Ribosomes are evident as occupying only one surface of the ER membranes, providing for cisternae with smooth surfaces. (*c*) The ER membranes have the typical railroad track appearance. Ribosomes are seen to be attached directly to these membranes. (Courtesy of Dr. George E. Palade.)

Table 10.1 Relative Cytoplasmic Volumes and Membrane Surface Areas of Secretory Compartments in Resting Guinea Pig Pancreatic Exocrine Cells

Compartment	Relative Cytoplasmic Volume (%)[a]	Membrane Surface Area (μm^2/cell)
Rough endoplasmic reticulum	~20	~8,000
Golgi complex	~8	~1,300
Condensing vacuoles	~2	~150
Secretory granules	~20	~900
Apical plasma membrane		~30
Basolateral plasma membrane		~600

[a]Percentage relative to cytoplasmic volume exclusive of the nucleus.

attached to the membranes of the endoplasmic reticulum. The pancreatic exocrine cell is exceptionally well tooled up for the synthesis of large amounts of protein because it contains an enormous surface area of endoplasmic reticulum. This is nicely illustrated in the electron micrographs of Figure 10.4, which show increasing magnifications of the rough endoplasmic reticulum.

The RER takes up only about 20% of the cytoplasmic volume of the cell, but it contains nearly 77% of the internal membrane surface area (see Table 10.1). This design is probably essential for a cell that produces digestive enzymes, for even though the demands on the cell are sporadic, when they are made they are heavy.

The pancreatic cell synthesizes a mixture of proteins simultaneously, and apparently transports them together through the secretory pathway.[2] Typically, it may produce amylase, ribonuclease, chymotrypsinogen, trypsinogen, lipase, procarboxypeptidase A and B, proelastase, and certain other enzymes. All these proteins appear together in the same storage granule even though they possess a wide range of molecular weights (6,000–70,000 daltons), isoelectric points (pH 3–11), and chemical compositions (some are glycoproteins, others not).

Using sodium dodecyl sulfate (SDS)–gel electrophoresis to examine the products transported and discharged from pancreatic cells, Palade's research group made the following observations.

1. *The only proteins synthesized at an appreciable rate are the secretory proteins.*
2. *No extensive proteolysis of these proteins occurs between the time of their synthesis and discharge.*
3. *All of the major proteins originate in the rough endoplasmic reticulum and pass by way of the Golgi complex, its associated condensing vacuoles, and zymogen granules, prior to discharge.*
4. *A quantitatively identical protein mixture is discharged when lobules are stimulated by any of a variety of stimulants.*
5. *When carbamylcholine is the stimulant the kinetics of discharge of all the proteins are similar.*
6. *The kinetics of intracellular transport of all the proteins are similar.*[2]

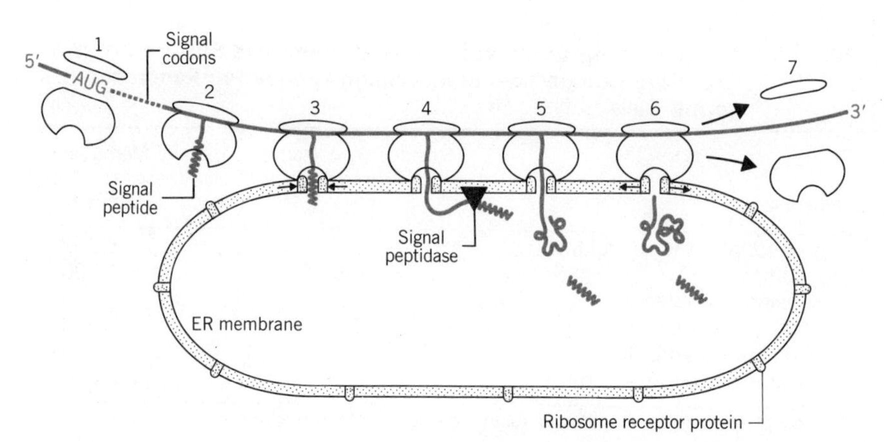

FIGURE 10.5 Schematic illustration of the signal hypothesis. mRNAs containing signal codons cause the formation of signal peptides, which are attracted to special receptor proteins on the ER membrane. The receptor proteins form a channel through which the emerging protein moves into the lumen of the endoplasmic reticulum. Once in, the signal peptide section is excised from the protein by a signal peptidase. The protein is eventually completed and sequestered in the lumen.

Although these conclusions carry us beyond just a consideration of synthesis, they point out quite convincingly that different enzymes must be synthesized simultaneously if they are to be transported in this parallel fashion.

Stage II: Segregation

Proteins that are synthesized on polysomes attached to the endoplasmic reticulum are next found in the cisternal spaces of this membrane network. Although it has been understood for some time that proteins are segregated in the cisternal spaces by some type of vectorial transport from the ribosomes, the process that enables this is only recently becoming clear.

For a cell to successfully segregate secretory proteins from the rest of the cell milieu, it must (1) be able to discriminate between secretory and other cellular proteins and (2) transfer secretory proteins vectorially across the membrane of the endoplasmic reticulum.

The Signal Hypothesis

A hypothesis put forth by Blobel,[3] termed the *signal hypothesis*, addresses these requirements and provides a way of explaining how segregation might be accomplished. Illustrated in Figure 10.5, the signal hypothesis describes protein segregation as dependent on two important physical parameters of the system: the ribosome–membrane junction and a specific sequence of nucleotides in mRNA.

Beginning with mRNA, it is proposed that all secretory proteins are coded for by mRNAs that contain a specific sequence of codons, called *signal codons*, on the 3′ side of the AUG initiation codon. Using free ribosomes, the signal codons are translated into a unique *signal peptide* at or near the *N*-terminal end of the secretory protein being made. The proposed function of the signal peptide, which begins to emerge from the large subunit of the ribosome, is to attract specific receptor proteins in the ER membrane that move laterally in the fluid bilayer and aggregate around the signal peptide. When in juxtaposition with the ribosome, the receptor proteins bind to multiple sites on the ribosome, sweeping away lipid molecules and anchoring the ribosome to the membrane. The resulting interaction between ribosome and membrane proteins creates a pore, or channel, through which the growing polypeptide chain passes unimpeded by the bimolecular leaflet.

The 3′ side of a codon, or a single nucleotide, is by convention toward the end of the nucleic acid that is terminated by a 3′-OH group not linked to another nucleotide, as follows.

The next step in the hypothetical scheme is the enzymatic removal of the signal peptide from the rest of the polypeptide chain by a *signal peptidase*, located on the cisternal surface of the membrane or within the cisternal space. Removal of this unique peptide fragment permits the polypeptide to begin to fold and take on its native secondary and tertiary structure. The change in conformation of the protein blocks any possibility of return to the extracisternal space because of the general impermeability of ER membranes to molecules greater than 2 nm in diameter. Upon chain termination, the secretory protein is released, completely segregated in the cisternal lumen, and the ribosome is displaced from the membrane by a *detachment factor*, a substance only recently isolated. The ER ribosome receptor proteins are then free to diffuse laterally throughout the ER membrane, eliminating the channel and assuring a secure compartmentalization of the secretory proteins.

The signal hypothesis is based on experimental evidence that has emerged along three separate lines. Briefly, they are summarized as follows.

1. Certain secretory proteins produced in the absence of microsomal membrane *in vitro* (and thus the absence of the proposed signal peptidase) retain an *N*-terminal "signal sequence" of 16 to 23 amino acids not found in the normally formed secretory product.
2. Combining mRNA (for the light chain of *immunoglobulin*, a secretory protein) with rabbit *reticulocyte* ribosomes and dog *pancreas* microsomes results in the production of segregated immunoglobulin. A similar system using mRNA for rabbit globin (a nonsecretory protein) results in no segregation. Thus the signal factor is mRNA associated. Furthermore, the signal peptide of the secretory protein can be recognized by a heterologous membrane, resulting in the attachment of ribosomes and segregation. In addition, the signal peptide can be removed by the heterologous dog microsomal signal peptidase once segregated.
3. Chain completion in rough microsomes yields only "processed" chains (i.e., chains from which the signal peptide has been excised). Detached polysomes, prepared from rough microsomes after detergent solubilization of the membrane (and presumably destruction of the signal peptidase), yield some "unprocessed" precursor too. Therefore, in rough microsomes, some nascent chains still contain the signal sequence, suggesting the time of excision to be a cotranslational event.

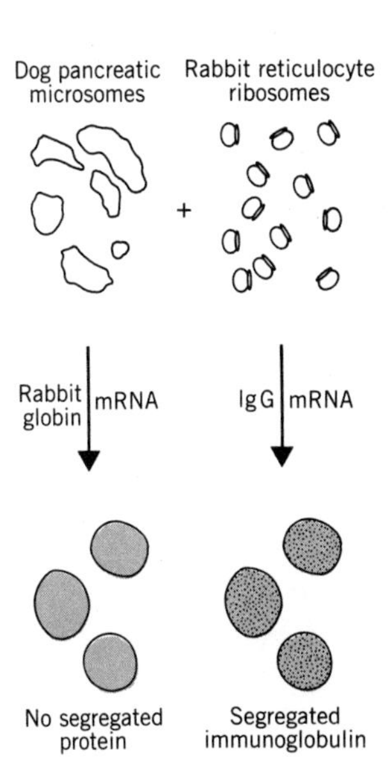

Implications of the Signal Hypothesis

Stepping back and looking at these data and the postulated signal hypothesis, we see a remarkable series of events unfolding during segregation. Molecules of mRNA that code for secretory proteins possess special signal codons to assure that ultimately these proteins will be properly segregated from other proteins for transport. The whole process depends directly on the formation of a unique signal peptide, which plays a crucial role in directing the proteins to the ER cisternae and is then dispensed with by enzymatic excision.

These studies also suggest that the signal codon and the ribosome receptor proteins have been highly conserved during evolution, thus permitting the heterologous systems studied to effect segregation.

It is apparent from considerations of the signal hypothesis that for proteins to be segregated for transport, they must be vectorially translocated across the membrane of the endoplasmic reticulum from cytosol to lumen side. The *synthesis* of proteins actually takes place in the cytosol for that is where ribosomes, mRNA, and all the enzymatic machinery for translation are found. But the products of synthesis in the case of secretory proteins are never found in the cytosol. The *location* of these proteins is always the ER lumen.

The molecular basis of translocation is currently a very active research endeavor. Although the molecular detail cannot yet be explained, several features of this process that are taking form appear to be at the heart of translocation mechanisms.

One feature is the presence of a pair of integral membrane glycoproteins referred to as *ribophorins* I and II (see Chapter 6 for more information on these proteins).[4] These proteins, characteristic of RER and not SER, appear to be the receptor sites to which ribosomes bind after the signal peptide has been formed.

A second feature is the discovery of an integral membrane protein that appears to be required for translocation.[5] When rough microsomes are treated with elastase, a proteolytic enzyme, and high salt (0.5 M KCl), a peptide fragment having a molecular weight of 60,000 daltons is released from the cytosol surface of the membrane. Without this peptide in place, nascent peptides are not translocated vectorially across the membrane. Translocating activity can be restored by recombining the peptide fragment with the membrane preparation from which it was released. Neither proteolysis alone nor high salt alone will release the fragment from the membrane.

It therefore appears as though the peptide fragment is a cytosolic domain of an integral membrane protein that exhibits strong electrostatic interaction with other membrane components. The evidence suggests that it may have an important role in translocating secretory proteins once the ribosome is attached to the surface of the endoplasmic reticulum. The properties of the entire protein and how it relates to the ribosome are not yet known. Nor has it been proposed how the protein may function as a translocator.

Stage III: Intracellular Transport

The third stage of the secretory pathway describes a long and complex jump from the rough endoplasmic reticulum on one side of the Golgi complex to condensing vacuoles on the other side. The subcellular region in which this takes place is shown in the electron micrograph of Figure

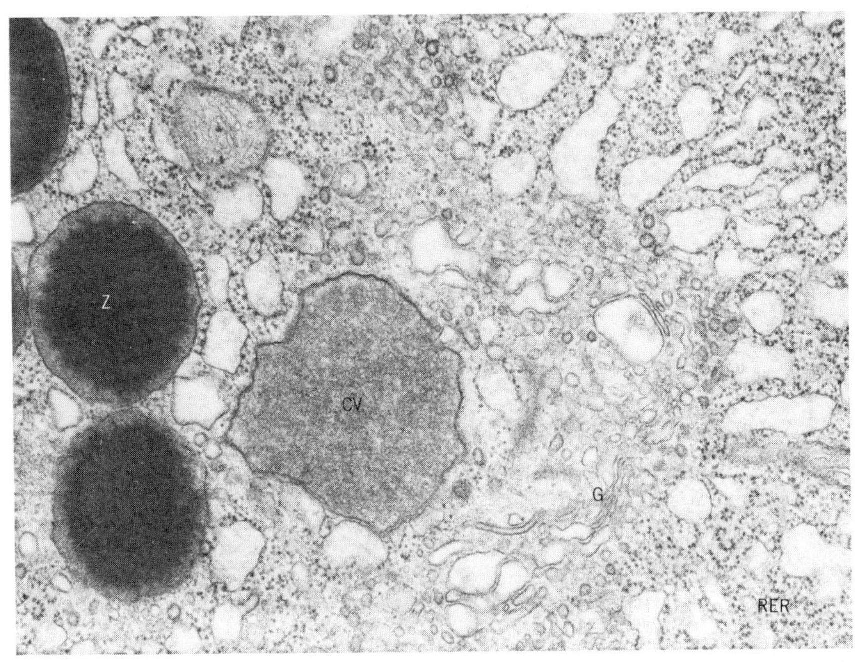

FIGURE 10.6 The subcellular region in which stage III (intracellular transport) takes place. The stage includes a jump from the RER in the upper and lower right, through the Golgi complex (G) to the condensing vacuoles (CV). The storage product is the zymogen granule (Z). (Courtesy of Dr. J. D. Jamieson.)

10.6. The details of this stretch of pathway are still somewhat enigmatic at both the membrane and molecular levels. But autoradiography and electron microscopy are being used to slowly sort out some of the facts.

The General Pattern of Transport

Autoradiographs have established very nicely the general pattern of flow of secretory proteins through this part of the pathway.[6] Figure 10.7 presents results that are typically obtained when pancreatic tissue is given a pulse of radioactive leucine, L-[4,5-^{3}H] leucine, followed by a chase for a prolonged period with cold leucine. The label appears first in the RER (stages I and II), then within 7 min it moves into the transitional elements of the endoplasmic reticulum and the Golgi complex (stage III), and by 37 min appears in the condensing vacuoles on the distal side of the Golgi complex (late stage III and stage IV). By 80 min the radioactivity appears in zymogen granules (stage V) and in the lumen of the pancreas (stage VI).

Method VI

A "pulse" of radioactive leucine is a short exposure of the tissue to radioactive leucine. The radioactive leucine is then taken away and replaced by nonradioactive leucine—the "chase." The pulse–chase procedure permits the investigator to follow the *movement* of material in the cell. It is similar to studying the current of a river by throwing in a bucket of paint.

The same sequence of events is graphed in Figure 10.8, where it is evident that as the radioactivity declines in the RER it mounts successively in the Golgi complex, the condensing vacuoles, and the zymogen granules. These results, although most helpful in a general sense in indicating the *direction of flow* in the secretory pathway, provide no information about the behavior of individual proteins or the mechanism of protein transport.

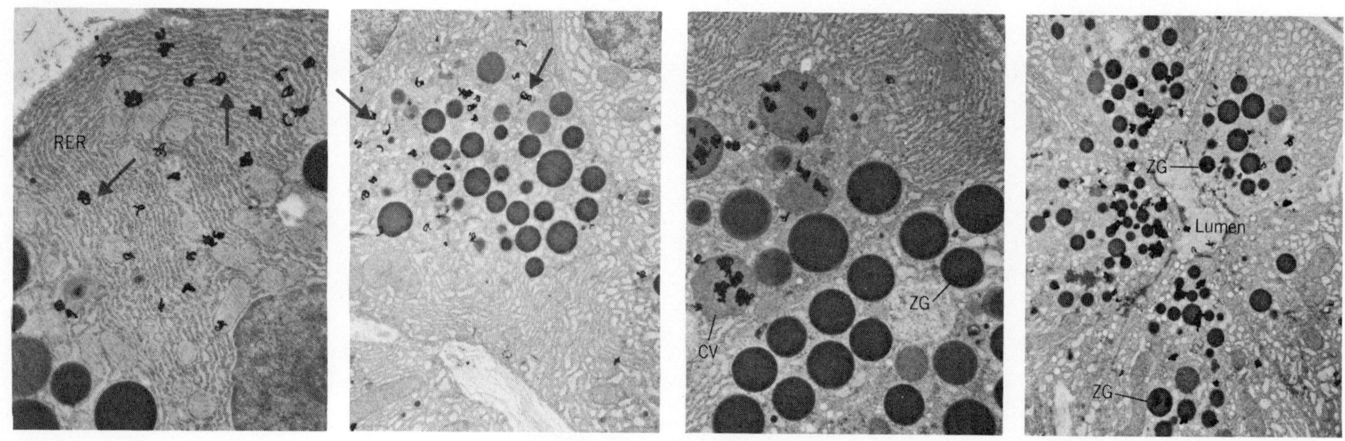

FIGURE 10.7 Pulse-chase experiment in tracking proteins. Radioactive leucine is added to pancreatic tissue, and the course of radioactivity in the cell is monitored by autoradiographic techniques (see arrows). After receiving a pulse with radioactive label, the tissue is exposed to nonradioactive leucine, permitting the route of radioactivity to be followed with time, in this case 3, 7, 37, and 80 min after the pulse. (Courtesy of Drs. J. D. Jamieson and G. E. Palade.)

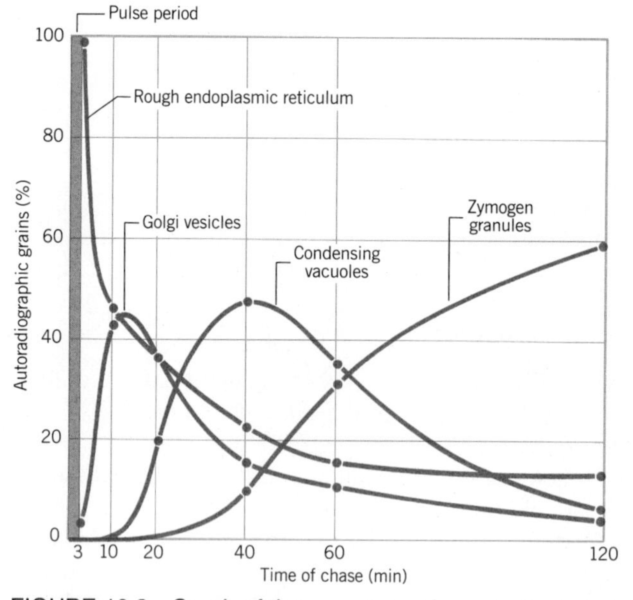

FIGURE 10.8 Graph of the sequence of events illustrated in Figure 10.7. These curves are based on counts of autoradiographic grains at successive intervals after the pulse. As radioactivity leaves the RER, it appears in sequence in the Golgi complex, condensing vacuoles, and zymogen granules.

Stage III, intracellular transport, can be viewed for our purpose as occurring in two substages. The first is that of transport between the RER and the Golgi complex, and the second is the movement of proteins from the Golgi region to the condensing vacuoles.

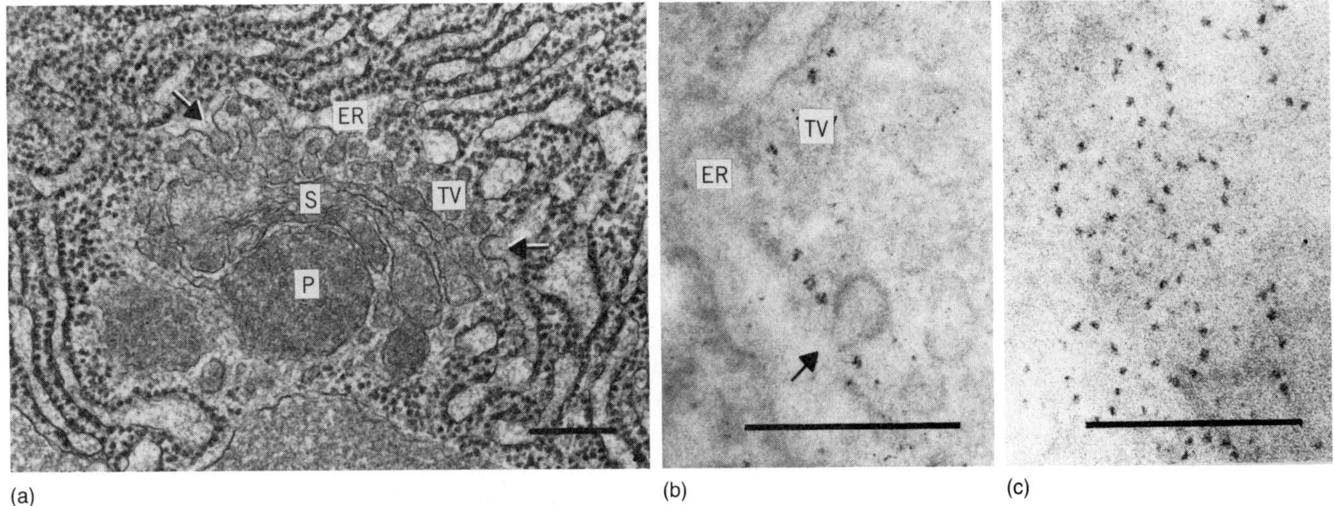

FIGURE 10.9 Formation of transition vesicles and bead rings. (*a*) Transition endoplasmic reticulum (arrows) gives rise to transition vesicles (TV) and then to electron-dense saccules (S) and protein granules (P). ×55,000. (*b*) Beads are associated with the neck (arrow) of a forming transition vesicle. ×140,000. (*c*) The beads are arranged in rings when viewed toward the ER surface. ×140,000. Bar 0.2 μm. (Courtesy of Dr. David A. Brodie.)

Transport between the Rough Endoplasmic Reticulum and the Golgi Complex

The mechanism whereby proteins are transported between the RER and the Golgi region is a matter of some debate. There are two main schools of thought. One is that the two regions are directly connected by membrane tubules that permit diffusion of products directly from the RER to the Golgi. This would imply that this part of the secretory pathway is *structurally continuous*. The other view is that the two regions are connected only intermittently by membrane vesicles, which shuttle the products from one point to another. In this view, the pathway, though functionally continuous, is *structurally discontinuous*.

Transport possibilities between RER and Golgi.

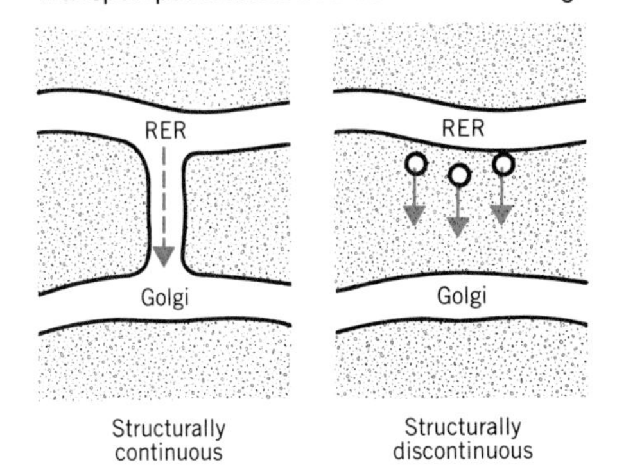

Figure 10.9 contains electron microscopic evidence that "transition elements," or transition endoplasmic reticulum (TER) vesicles, bleb off the smooth portions of the RER near the forming face of the Golgi complex. It is envisioned that each vesicle moves with its content of RER product to the small peripheral vesicles of the Golgi complex. But it is not yet clear whether the TER vesicles fuse with Golgi membranes directly or whether the Golgi is skirted and plays only an indirect role in processing transition vesicles or in escorting them to the site of condensation. Either or both mechanisms may operate, and there may be specific cellular variations on this.

The shuttle mechanism between the RER and the Golgi complex, whether continuous or discontinuous, is an energy-requiring step in the secretory pathway. Inhibitors of ATP production abruptly stop intracellular transport in pancreatic exocrine cells,[7] which is taken as evidence that there is the functional equivalent of a lock-gate or transportation valve at this point in the pathway. The location of this mechanism is not yet known, but it apparently lies somewhere at the level of the transition vesicles: perhaps at the points of TER formation or their fusion with condensing vacuoles, or in their propulsion from one site to another. Inhibiting protein synthesis with cycloheximide does not stop transport, so the driving force is apparently not due to a gradient of newly forming protein.

Some recently reported studies on transport between the endoplasmic reticulum and the Golgi complex suggest that the formation of transition vesicles may depend on 10 to 12 nm particles that appear as a ring of beads at the site of transition vesicle formation.[8] This has been studied most intensively in fat body tissue from the larvae of *Calpodes ethlius* when it is actively manufacturing and secreting hemolymph proteins.

An important result from studies on this system is the finding that the ring of beads arrangement is ATP dependent. When ATP formation is blocked by inhibitors of oxidative phosphorylation (antimycin A, cyanide, anoxia) transition vesicles do not form and the ring of beads collapses.

Figure 10.9 illustrates the formation of transition vesicles in this tissue system and the association of beads with the neck of the transition vesicle that is taking form. On surface view, the beads appear as rings.

When an inhibitor of oxidative phosphorylation is present, the spacings between beads and between the beads and the membrane appear to be unchanged, however the ring arrangement disappears and a mass of beads is generally observed instead.

The molecular import of these observations is not yet clear, but it does appear that the integrity of the ring of beads is necessary for transition vesicle formation. At this point one could speculate that the ring structure selects a membrane region that is appropriate in composition to become part of the forming face of the Golgi complex. By binding to this membrane in ring form, a type of fission is triggered to generate the transition vesicle. The ring structure, hence transition vesicle formation, is in turn dependent on ATP. Thus, the ring of beads may be the morphological correlate of the energy lock-gate in this region of the secretory pathway.

Thiamine pyrophosphate (TPP) is a coenzyme.

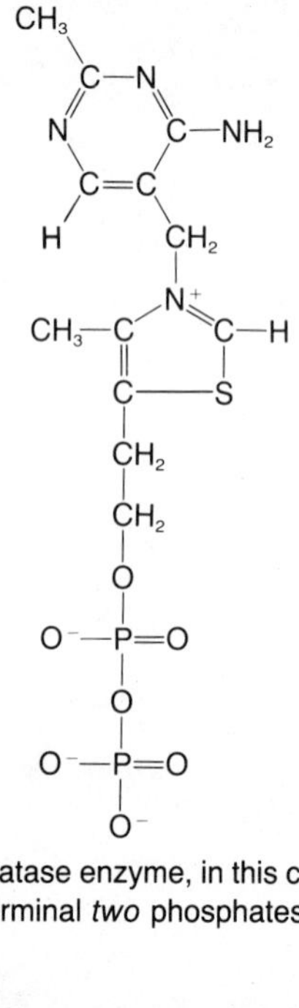

CH_3
C—N
N C—NH_2
C=C
H CH_2
N^+
CH_3—C C—H
C—S
CH_2
CH_2
O
O^-—P=O
O
O^-—P=O
O^-

A pyrophosphatase enzyme, in this case, would remove the terminal *two* phosphates.

Transport Between the Golgi Complex and Condensing Vacuoles

The second substage of intracellular transport, the movement of material from the Golgi to the condensing vacuole, may involve a specialized membrane region on the mature face of the Golgi complex mentioned in Chapter 10, referred to as the GERL (Golgi-associated endoplasmic reticulum involved in the formation of lysosomes) by Novikoff and coworkers.[9]

The Golgi and the GERL can be differentiated from one another by cytochemical techniques. The Golgi complex exhibits thiamine pyrophosphatase (TPPase) activity, whereas the GERL is positive for the enzyme acid phosphatase (AcPase). By using these cytochemical distinguishers, Novikoff and his research group found that condensing vacuoles in mammalian pancreatic exocrine cells appear to be expanded portions of the cisternae of the GERL.[10] Moreover, structural continuities were seen between the RER and the condensing vacuoles but not between the Golgi and the GERL. These results suggest that condensing vacuoles arise either directly from the RER or with the aid of the GERL and cast some shadow of doubt on the direct role of the Golgi complex at this stage.

Figures 10.10 and 10.11 demonstrate the mutually exclusive AcPase

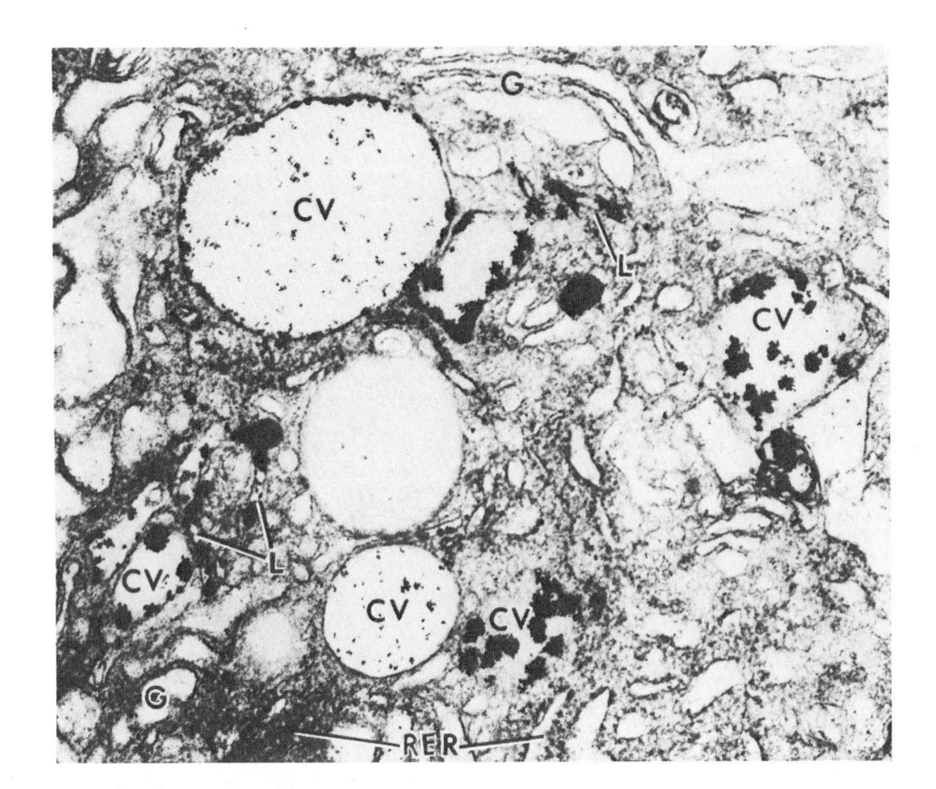

FIGURE 10.10 Acid phosphatase activity is demonstrated in these pancreatic exocrine cells by components of the GERL, condensing vacuoles (CV), and the rigid lamellae (L). No such activity is seen in the Golgi complex (G) or in the RER. (Courtesy of Dr. A. B. Novikoff.)

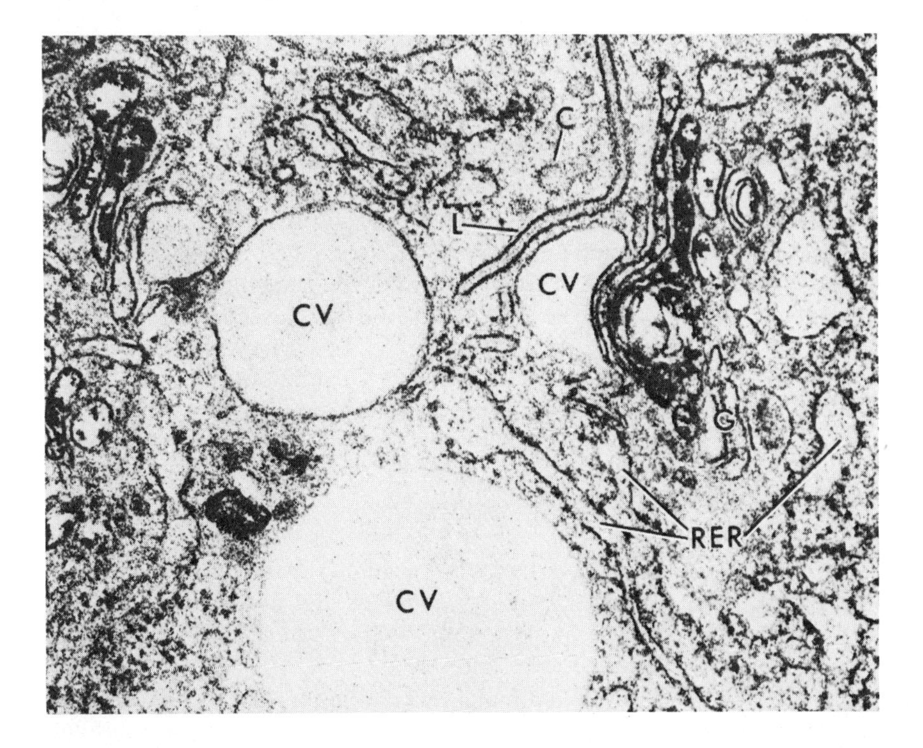

FIGURE 10.11 These portions of pancreatic exocrine cells show that thiamine pyrophosphatase activity is exclusively associated with elements of the Golgi complex. The GERL is void of this activity, as are the RER and the condensing vacuoles. (Courtesy of Dr. A. B. Novikoff.)

and TPPase activities, respectively, of the GERL and the Golgi complex. It is clear from these micrographs that *both* the condensing vacuoles and the rigid lamellae of the GERL are positive for AcPase, whereas the Golgi shows no such activity. The Golgi complex, on the other hand, demon-

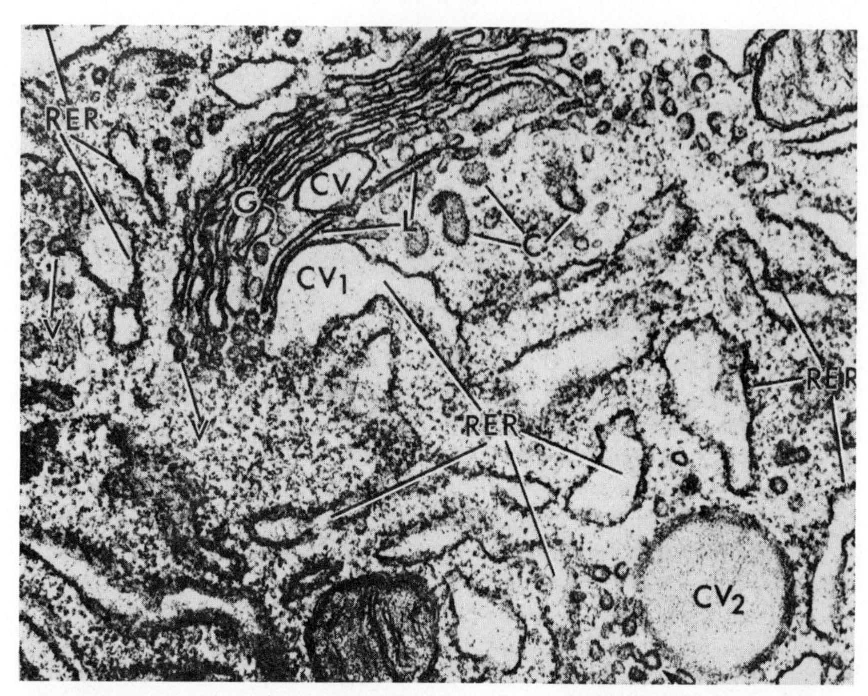

FIGURE 10.12 A portion of a pancreatic cell of a rabbit showing the positional relationships between RER, the Golgi complex (G), transition vesicles (V), condensing vacuoles (CV), coated vesicles (C), and the rigid lamellae (L) of the GERL. This region of the cell is in the heart of the secretory pathway, a complex and dynamic part of the cell. (Courtesy of Dr. A. B. Novikoff.)

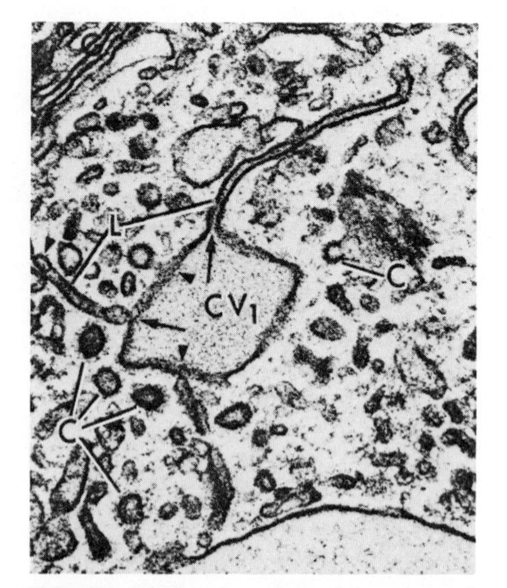

FIGURE 10.13 A portion of a pancreatic cell of a guinea pig. A continuity of a rigid lamella of the GERL and a condensing vacuole is strikingly revealed in this micrograph (upper arrow). Because of sectioning, such continuity is not always obvious, as indicated by the lower arrow and arrowheads. A coated vesicle (C) appears to be attached to a condensing vacuole. (Courtesy of Dr. A. B. Novikoff.)

strates exclusive TPPase activity. These results establish, cytochemically, a closer link between GERL and condensing vacuoles than between Golgi and condensing vacuoles.

The electron micrograph of Figure 10.12 shows the positional relationships between the RER, the Golgi complex, the transition vesicles, the condensing vacuoles, and the GERL. The GERL is seen on the mature face of the Golgi and is separated from it by transition vesicles and condensing vacuoles. Note that the RER is present at both faces of the Golgi and at one point is continous with a condensing vacuole.

A direct connection between a rigid lamella of the GERL and a condensing vacuole is apparent in Figure 10.13. This is a striking illustration in support of the notion that condensing vacuoles form directly from the GERL.

All this appears to make a very provocative picture out of what one would hope to be a simple secretory pathway. But at least for the pancreatic exocrine cell, the secretory pathway does not seem to involve the membranes of the Golgi complex directly.

In other secretory cells, the Golgi may be more directly involved in transport. For example, studies on the acinar cells of the rat exorbital lacrimal gland[11] have yielded information suggesting that the Golgi participates in the *transport* of the secretory enzyme peroxidase and that the GERL is concerned with the *formation* of the secretory granule. In this system, peroxidase can be followed cytochemically by its reaction product

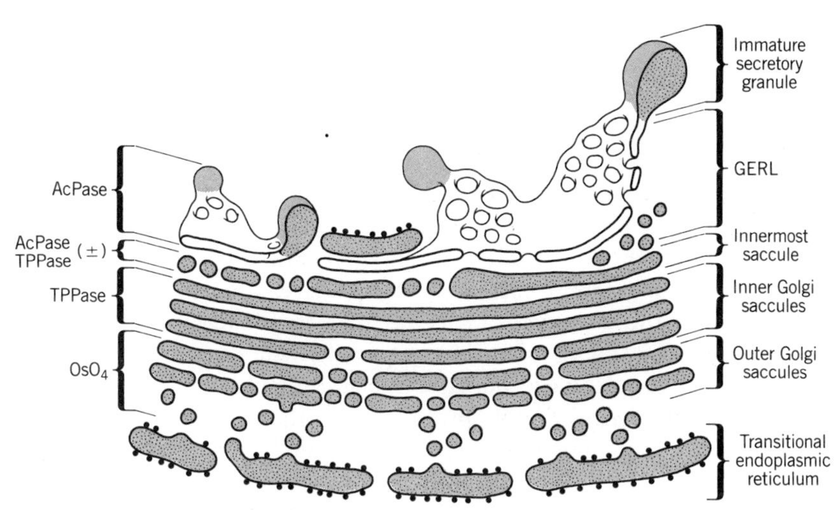

FIGURE 10.14 Morphological and cytochemical features of the Golgi complex and GERL as they would represent the acinar cells of rat exorbital lacrimal glands. The GERL is depicted as fenestrated sheets or tubular regions from which secretory granules are formed. Secretory proteins (colored regions) are found in RER, Golgi saccules, and immature granules, but not GERL. GERL and immature granules contain AcPase activity; TPPase activity is found in the inner Golgi saccules.

with diaminobenzidene. It is found in the cisternae of the nuclear envelope, the RER, the Golgi membranes, and the secretory granules. GERL is essentially free of this enzyme activity. There therefore appears to be a fundamental difference between the lacrimal gland cells and pancreatic cells with respect to the involvement of the GERL in secretion.

The exorbital lacrimal gland is a tear-producing gland located near the eye.

Even though the GERL possesses no peroxidase activity, multiple tubular connections are seen between it and the immature granules that form at the mature face of the Golgi complex. Thus the GERL does not appear to be directly associated with the secretory protein but may assist in the formation of the secretory granule. This process is diagrammed schematically in Figure 10.14.

The end product of this substage of intracellular transport is a condensing vacuole. It emerges as an irregular, relatively electron-transparent vacuole that undergoes a morphological transformation to that of a sphere as it is acted on by the forces that effect concentration.

Stage IV: Concentration

As the condensing vacuole is converted into a mature secretory granule, the secretory products are condensed considerably and the form of the vesicle is changed from irregular and somewhat electron transparent to spherical and electron dense when fixed with osmium tetroxide (Figure 10.6). The extent to which the products are concentrated during this stage is not known for sure, but it is estimated that between the endoplasmic reticulum and the mature granule, which would in effect constitute nearly the whole of the secretory pathway, secretory proteins undergo about a 20- to 25-fold concentration. It is likely that a major part of this effort is carried out in the condensing vacuole.

It is now clear that the mechanism of concentrating products in the vacuole depends primarily neither on protein synthesis nor on an energy source,[12] dispelling speculations that intracellular protein gradients forced concentration or that membrane-associated ion pumps were responsible for the efflux of water from the structure. Rather, a novel and *energy-independent mechanism* is presently being postulated for this stage.

It is based on the finding that condensing vacuoles contain large polyanions that belong to a sulfated peptidoglycan class of macromolecules.[13] The Golgi complex carries out sulfation reactions and, in general, is involved with the synthesis of carbohydrate-rich macromolecules. Apparently one of its major contributions to the secretory pathway is to synthesize and supply these negatively charged macromolecules to the secretory vesicles. The secretory proteins, in contrast to this carbohydrate polyanion, are predominantly basic. It is thus envisioned that an electrostatic interaction is set up between these two systems in the condensing vacuole, causing the formation of huge protein–polyanion complexes. These would lower the osmotic pressure of the vacuole and trigger the release of water by the nonenergy-requiring mechanism of osmosis.

The energy-independent mechanism of concentration in secretory granules would work as follows.

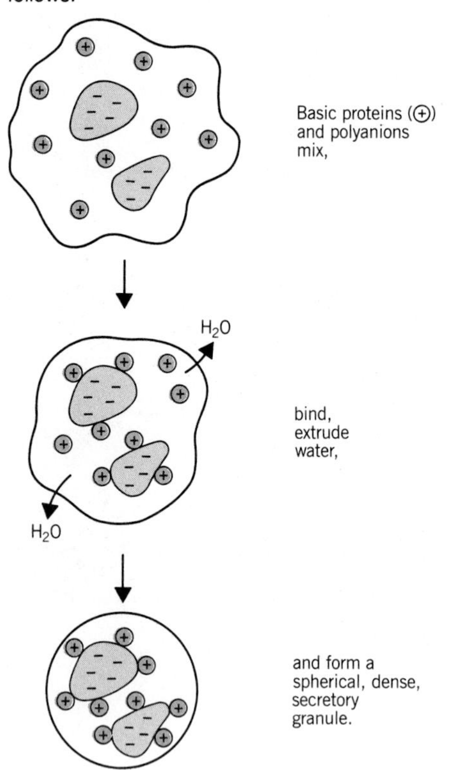
Basic proteins (⊕) and polyanions mix,
H_2O
bind, extrude water,
H_2O
and form a spherical, dense, secretory granule.

This postulate is most intriguing and attractive. Undoubtedly we will shortly know whether any solid experimental evidence exists to support it and, if so, the extent to which it may be a generalized mechanism of condensation for secretory products.

The membranes of the secretory vesicle undergo a change in chemical composition and physical properties, presumably at the level of the Golgi complex, which begins to play an important role in the maturation and ultimate stability of the secretory granule. The transition is from endoplasmic reticulum-like to plasma membrane-like, reflected by an increase in cholesterol and sphingomyelin and a decrease in unsaturated fatty acids. This change is viewed as converting the membrane from one of high permeability to one of low permeability, a property that is important if the contents are to be securely stored and sheltered from the cytosol. As we shall see, however, the chemical composition of the secretory granule membranes is distinct from that of the plasma membrane; thus the change that these membranes have undergone falls short of a complete transformation to that of the cell boundary membrane.

The point in the secretory pathway where condensation takes place probably varies somewhat from one cell type to another. In the guinea pig pancreatic exocrine cell, condensation occurs in the mature face region of the Golgi, but in other species and in other glandular cell types the Golgi seems more directly involved, especially with its outer two or three mature saccules.

Stage V: Intracellular Storage

The mature secretory granule in the pancreatic exocrine cell is called the *zymogen granule*. As is true for the pancreatic cell itself, the zymogen granule is the archetypal secretory granule for a variety of studies.

The individuals making up the population of zymogen granules in a given tissue or cell are not uniform in size. They often vary in the approximate range of 250 to 1500 nm in diameter. These variations are easily seen in the electron micrographs of this chapter. It is tempting to speculate that this morphological heterogeneity results from a discriminatory packaging of different secretory proteins, but the evidence suggests that this is quite certainly not the case, at least for the pancreas.

Immunocytochemical results have revealed that all the secretion granules of all the cells examined contain all the proteins secreted. In other words, each individual granule contains a mixed bag of molecular species in the same proportion as when first synthesized and segregated. In fact, this finding is probably the strongest indicator that a mixed population of secretory proteins is synthesized in the endoplasmic reticulum in the first place. So even though the cell appears to have a remarkable mechanism to discriminate between secretory and nonsecretory proteins, it does not differentiate between the individual members of the secretory class, despite the existence of marked differences in their physicochemical properties. This is an outstanding example of cellular wisdom, for there is no need in this case to sort out the different enzymes delivered to the pancreatic duct. Instead, the level of discrimination is placed where it is most efficient and effective, at the level of the endoplasmic reticulum.

Secretory granules are accumulated apically from the Golgi complex, arriving there either by the process of simple diffusion or with the assistance of microtubules. At the present time, the evidence favors the former, since it is generally understood that the process requires no energy, and newly formed granules distribute randomly among the preexisting zymogen granule population (as determined by labeling and autoradiography).[14] Quite recently, some investigators[15] have reported microtubules bound to secretory granules and their membranes in pituitary cells, but it is too early to discern the significance of this finding in the cell. It may be that microtubules either passively channel the migration of granules toward the lumen or actively propel them in that direction. Definitive results are not yet in.

Stage VI: Discharge

The final step in the secretory pathway, termed discharge or exocytosis, is the object of quite intensive investigation. It has been obvious for a long time that the ultimate fate of the secretory contents of the granule is the apical lumen of the cell. This is shown clearly in the electron micrograph of Figure 10.15. Recently this process has been scrutinized more closely, both biochemically and morphologically.

Biochemically, exocytosis is dependent on a supply of energy. This step is thus the second and final energy lock-gate in the secretory pathway. In addition, the process must have either hormonal or neural stimulation of the appropriate cell membrane receptors. The term *stimulation–secretion coupling* is used to describe the sequence of events linking cell stimulation to discharge. This sequence includes the involvement of the second messenger Ca^{2+} and/or a cyclic nucleotide generated in the process of stimulation. We shall look at this in greater detail in the next section. The precise level at which energy is required is not clear yet, but there are several reasonable candidates, including granule movement, membrane fusion and fission interactions, and the generation of the requisite cyclic nucleotides.

Morphologically, the membrane of the secretory granule fuses with the

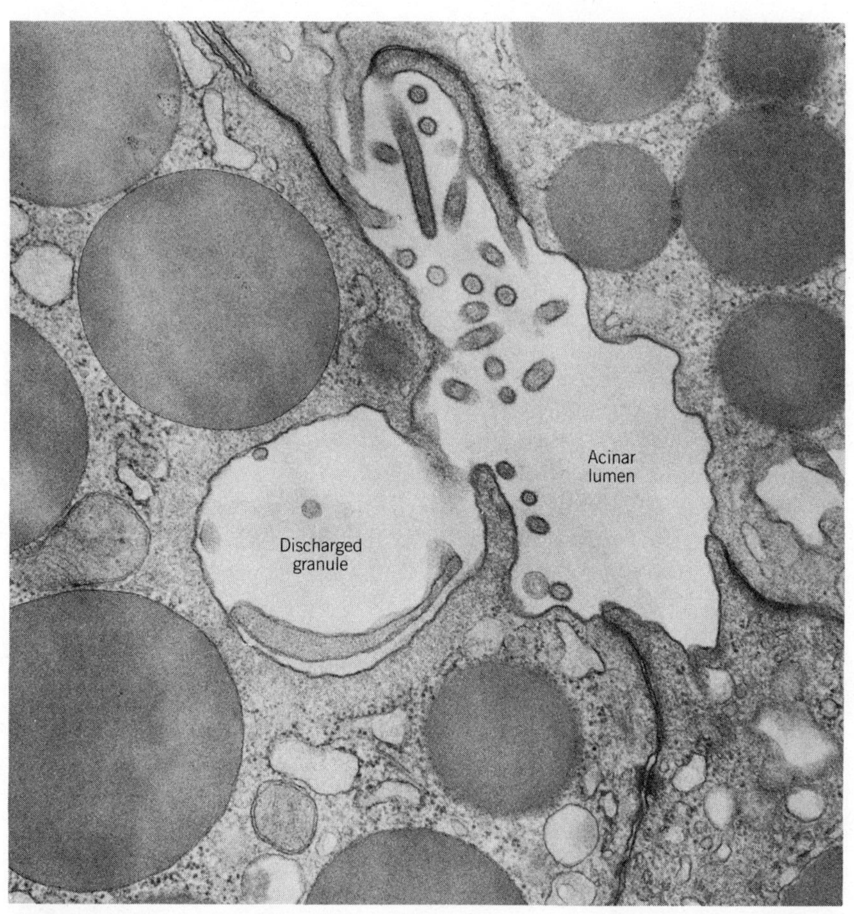

FIGURE 10.15 Discharge of a zymogen granule into the acinar lumen. Several other zymogen granules are in line for a similar fate. (Courtesy of Dr. J. D. Jamieson.)

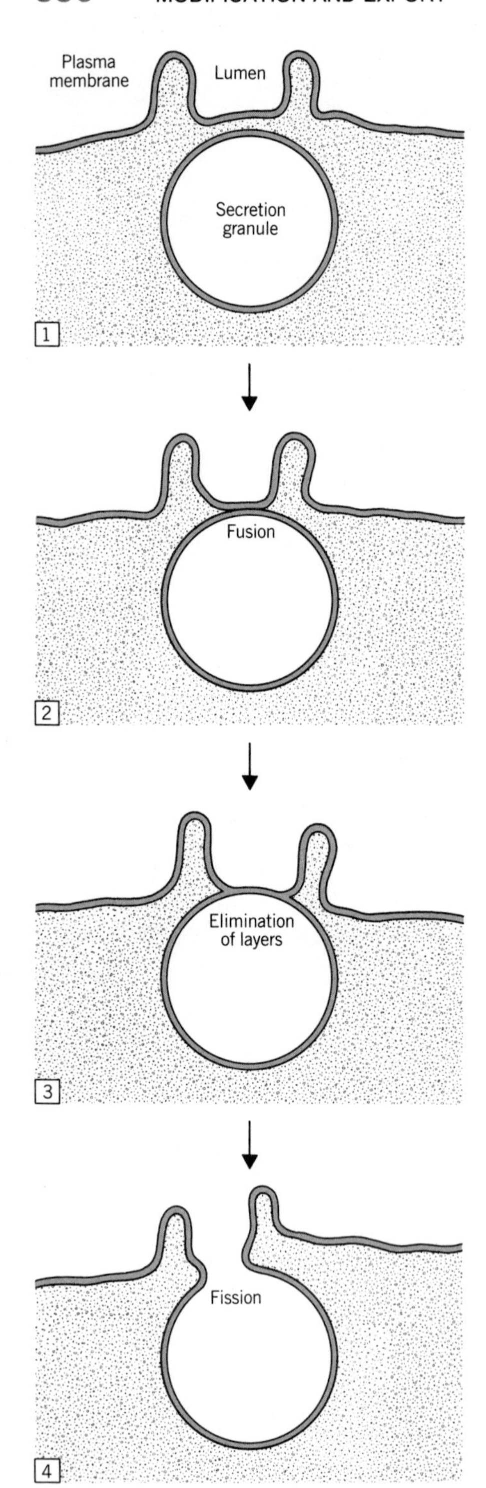

FIGURE 10.16 Interaction of the membranes of the secretory granule and the plasma membrane during discharge. Each membrane is depicted by two lines, representing an inner and outer half of the membrane. Note that at stage 3 a hybrid membrane exists.

plasma membrane, followed by fission of the membranes in a manner that delivers the secretory material to the lumen without ever exposing it to the interior of the cell. This might be thought to occur at any point in the cell where a secretory granule encounters a plasma membrane, except that there is known to exist a highly regimented recognition mechanism that permits fusion only at the apical end of the pancreatic cell.

A nondiscriminant release of enzymes to the extracellular environment at other points would, of course, be disastrous for the tissue. To avoid any possibility of intercellular leakage, adjacent pancreatic cells form a barricade in the vicinity of the lumen (see Figure 10.15). This structure, termed the *tight junction*, was described in Chapter 5. It prevents luminal enzymes from working back between the cells and also prevents other materials from moving in the luminal direction. Thus it is clear that membrane fusion must occur in a specified area of the cell and in a highly controlled manner.

Those who have studied the mechanism of exocytosis have divided the sequence of events during membrane fusion into four stages:[16] (1) close approximation and contact, (2) induction, (3) fusion proper, and (4) stabilization.

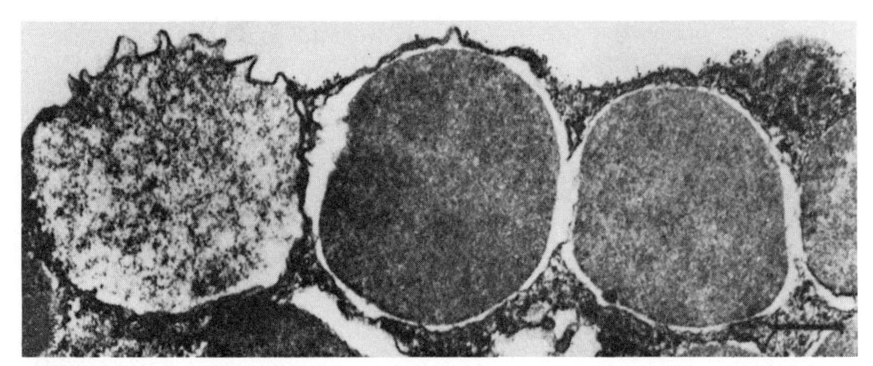

FIGURE 10.17 Changes that take place as granules approach the plasma membrane (with increasing approach from right to left). Cytosol is apparent between the granule and plasma membranes in the approach zone on the right. In the middle the cytosol is being squeezed out as membranes move together. Finally, the membranes appear to fuse completely, eliminating cytosol from the approach zone. (Courtesy of Dr. D. Lawson.)

In step 1, which is prerequisite to membrane fusion, the recognition mechanisms must operate to control the specificity of the fusion reaction. Step 2, induction, has to do with the establishment of an appropriate physicochemical environment in the region of fusion, to encourage the ensuing molecular rearrangements. Fusion proper, step 3, deals with the actual intermembrane molecular linkages that are the result of molecular rearrangements, and stabilization, step 4, refers to the return of a physicochemical equilibrium in the region of fusion to a state of normality.

Mechanisms of Membrane Fusion

A theory concerning the mechanism of membrane fusion, or fusion proper, is beginning to emerge that in turn is viewed as another multistep process. Although the work on which this discussion is based arises from studies on some specialized cells (*Phytophthora palmivora* zoospores and rat peritoneal mast cells),[16,17] it is likely that the mechanisms proposed have some general significance.

Mast cells are participants in the immune reaction. Immunoglobulin E binds to mast cells. Interaction of the appropriate antigen with the bound IgE triggers the mast cell to release the contents of its intracellular vacuoles, which contain histamine and heparin, a sulfated polysaccharide.

For the points that follow, it is important to keep in mind the lipid bilayer structure of the membrane and to think of the bilayer as consisting of an inner and an outer half-surface. The initial point of contact between the membranes of the secretory granule and the plasma membrane involves the most closely apposed layers—the inner half of the plasma membrane and the outer half of the granule membrane. Illustrated in schematic form in Figure 10.16, the secretory granule moves toward the lumen plasma membrane and fuses with it; then fission occurs. This process can also be viewed in the electron micrograph of Figure 10.17, where fusion is taking place in mast cells. As the membrane surfaces approach one another, the cytosol is dispersed or squeezed out, as is noted in these pictures by decreased amounts of labeled protein in these regions.

When membrane contact is made, the two apposing half-layers fuse, triggering a molecular rearrangement that results in the formation of a single bilayer shared by the granule and the plasma membrane. The single bilayer is very clearly shown in the micrograph of Figure 10.18.

On surface view, the formation of the common bilayer membrane takes

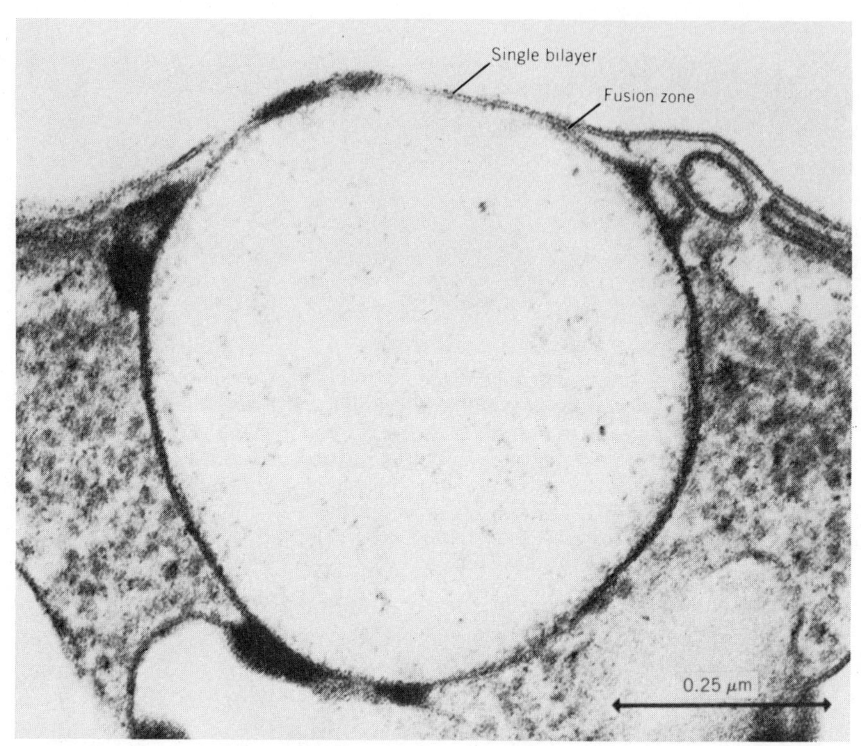

FIGURE 10.18 Highly magnified section showing complete fusion of plasma membrane and granule membrane to form a single bilayer. At the edge of the single bilayer there is a fusion zone where the outer dense line of the granule membrane and the inner dense line of the plasma membrane merge and disappear. ×120,400. (Courtesy of Dr. P. Pinto da Silva.)

Inner and outer half-surfaces of plasma membrane and secretory granule membrane and their fusion.

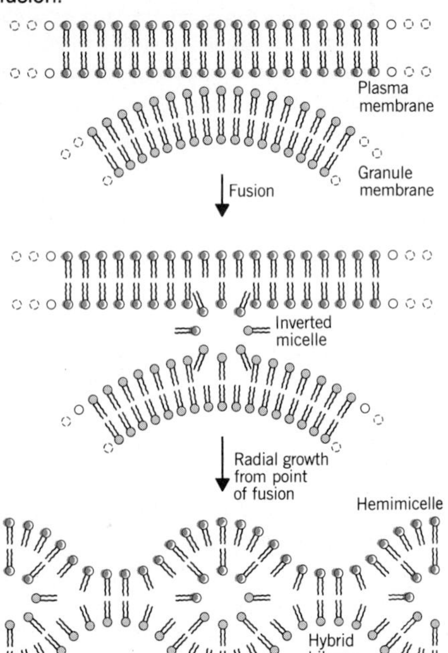

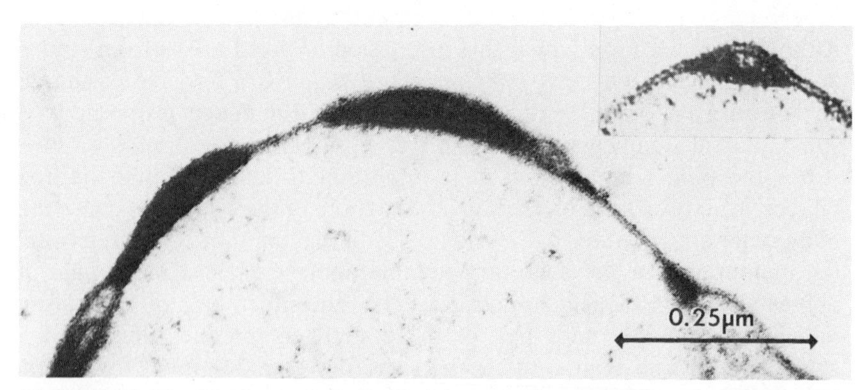

FIGURE 10.19 Electron micrograph demonstrating formation of trapped pools and tunnels of cytosol between the approaching granule and plasma membranes. This would occur whenever multiple contacts were made between the two membranes rather than fusion emanating from a single contact point. ×125,000. Inset, ×211,200. (Courtesy of Dr. P. Pinto da Silva.)

place radially from the point of contact, as in the ripple effect of dropping a stone in still water. If multiple contact points form between the apposing membrane surfaces, trapped pools and tunnels containing small amounts of cytosol are formed. These are evident in the micrograph of Figure 10.19.

The formation of the new lipid bilayer leaves a residual membrane micelle at the point of initial disruption. The bilayer grows radially as the diameter of an interacting ring of lipids, called the *toroid hemimicelle*, expands. This growth is illustrated schematically in Figure 10.20. This hemimicelle sweeps before it integral membrane proteins that according to freeze-fracture studies on these systems, are almost nonexistent in regions of the newly formed bilayer. In Figures 10.21 and 10.22, membrane particles are absent at sites of exocytosis.

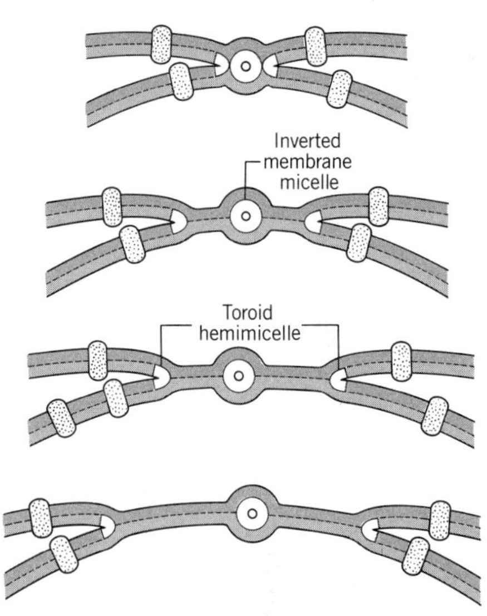

FIGURE 10.20 Steps during membrane fusion leading to formation of diaphragm bilayer. The hydrophobic junction between membrane halves is indicated by a dashed line. The postulated inverted membrane micelle (circular) is where fusion begins, and the toroid hemimicelle ring (wedges) sweeps back from that point. Integral membrane proteins, especially those spanning the lipid bilayer, are swept back and cannot penetrate the new diaphragm bilayer.

Implications of the Concept of Membrane Fusion

A little reflection concerning the properties of the new membrane that has formed will bring into focus three important facts. One, the new bilayer is a hybrid of plasma membrane and granule lipids and is thus a compositionally unique patch on the cell surface. Two, the bilayer exists without the stabilizing effect of integral membrane proteins. Three, the osmotic properties of the granule are changed as its surroundings are switched from granule–cytosol to granule–cell exterior. These physical changes are probably sufficient to destabilize the bilayer to the point of

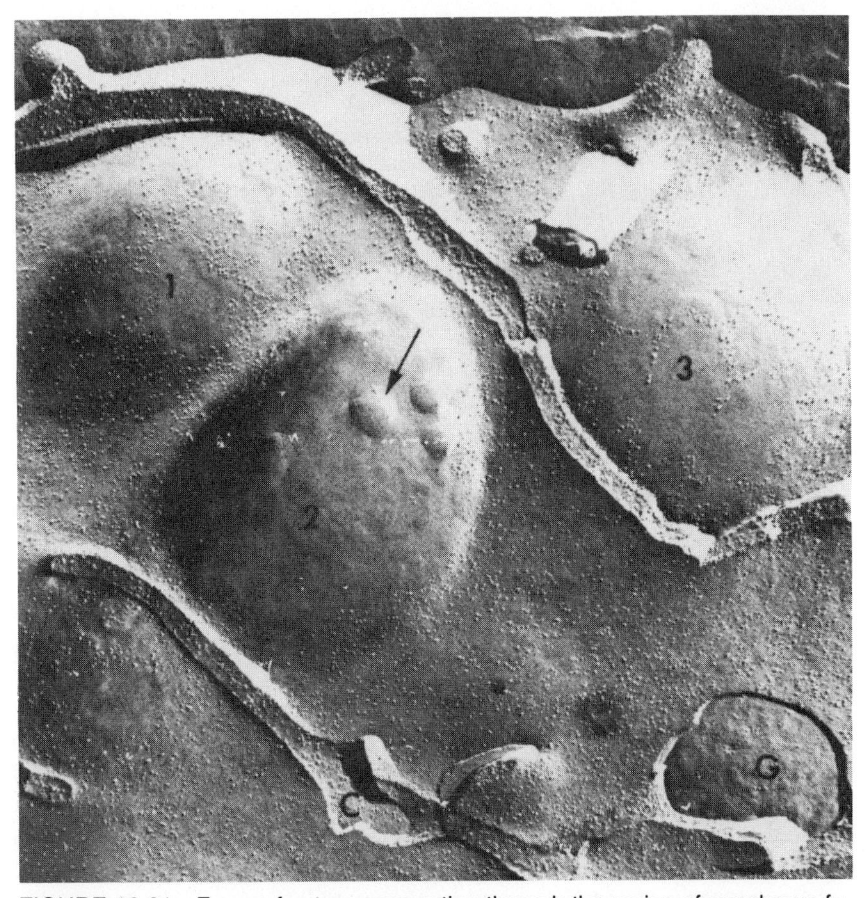

FIGURE 10.21 Freeze-fracture preparation through the region of membrane fusion (which splits open along the dashed line of Figure 10.20). Integral membrane proteins, appearing as particles, are randomly distributed throughout the plasma membrane, but are absent in regions 1, 2, and 3, where the membranes fuse. C, cytoplasm; G, granules. (Courtesy of Dr. D. Lawson.)

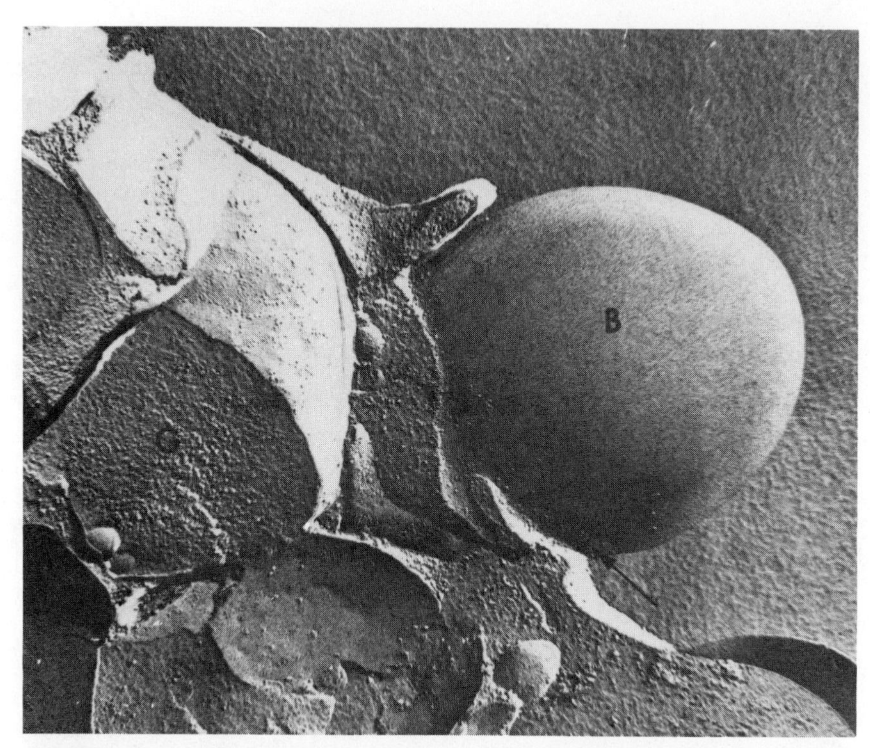

FIGURE 10.22 Blebs, bulged regions of diaphragm membrane, often are observed during exocytosis. Particles are absent from these regions but are accumulated (arrow) near the base of the bleb (B). (Courtesy of Dr. D. Lawson.)

rupture, whereupon the contents of the granules are released into the lumen.

So far, our discussion has assumed that the secretory granule is only passively involved in discharging its contents from the cell. It moves by simple diffusion or is propelled by contractile filaments and locates, by chance, receptor membrane proteins (or recognition sites) on the plasma membrane before membrane fusion. The process is probably not this uncontrolled, for it is well documented, at least in cells that secrete intermittently, that discharge is a triggered phenomenon. Secretory granules may lie in the proximity of the plasma membrane for prolonged periods without imminent membrane fusion. The following question therefore arises: Is one or are both membranes activated to initiate the fusion process and, in particular, what is the role of the secretory granule membrane vis-à-vis discharge?

In this regard, some evidence has been uncovered that ATP, when added to an *in vitro* system, stimulates the formation of pseudopodia on isolated secretory granules in rat parotid cells.[18] This behavior may well be an indication of what happens in the intact cell, for when slices containing intact cells are incubated with secretion inducers (isoproterenol or butyryl cAMP), pseudopodia directed toward the acinar lumen of the tissue are observed on the granules. Figure 10.23 illustrates these results.

It is therefore tempting to think that tips of pseudopodia arise from the secretory granule as a result of hormonal or neural stimulation and initiate the first localized contact with the plasma membrane. This starts an irreversible process of membrane fusion and discharge.

If the membrane systems that undergo fusion possessed like charges, the apposition and contact of rather large surface areas would not be expected to take place because of electrostatic repulsion. The pseudopodia could, theoretically, probe microareas of the plasma membrane where, because of local surface properties, contact would be encouraged.

It is beginning to appear that the discharge process described above may not be a universal rule among secretory cells. Studies reported recently on horseshoe crab amoebocytes using rapid freezing and freeze-substitution techniques instead of normal fixatives suggest some variations from the results of the cell systems described before.[19]

Horseshoe crab amoebocytes are triggered by endotoxin to degranulate by exocytosis within 30 sec of endotoxin application. Rapid-freeze techniques allow for observation of discharge during these rapidly occurring events.

The first stage observed in this system is an *invagination* of the plasma membrane to form a pedestallike structure that makes contact with the surface of the storage granule. A particularly interesting observation in this system is that filamentous structures appear to connect the plasma membrane with the surface membrane of the secretory granule. These *may* be contractile filaments that pull the membranes together when the cell is triggered to discharge. Figure 10.24 depicts this hypothetical sequence of events.

The next stage in this sequence is the formation of several punctate pentalaminar junctions at the point of membrane apposition. One of these rearranges into a small pore, which quickly widens to permit discharge of the granule contents to the cell exterior. Figure 10.25 is an electron micrograph showing granule discharge at a stage comparable to stage *d* in Figure 10.24.

The main differences between these observations and those reported for mast and other cell types have to do with the early events of discharge. Membrane apposition may be different, and the mechanism of initial pore formation seems to vary. In the amoebocyte there is no observed sweeping aside or clearing of membrane particles from the region of membrane apposition, and an extensive area of pentalaminar junction is not seen. Furthermore, the unilaminar or single bilayer diaphragm seen in other systems is not noted in amoebocytes. Therefore, pore opening is not as clearly related to simple lipid destabilization, since integral proteins may not be moved out from the region of membrane apposition.

At the moment it is not clear whether the differences observed are related to different experimental approaches, different cell systems, or truly different mechanisms of discharge. The results do show, however, that the mechanisms of discharge must be constantly reexamined in the light of new data. As is true in most explorations of this type, certain elements of all systems studied are probably common to the mechanism of discharge and may be used some day to explain a universal principle of exocytosis. We appear to be very close to that point, with some room yet for additional work.

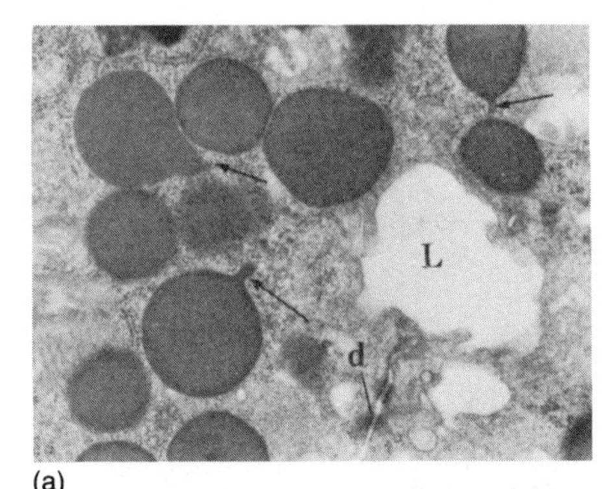

(a)

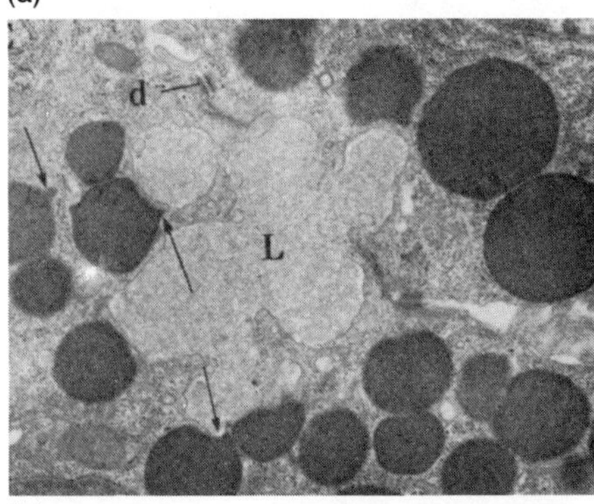

(b)

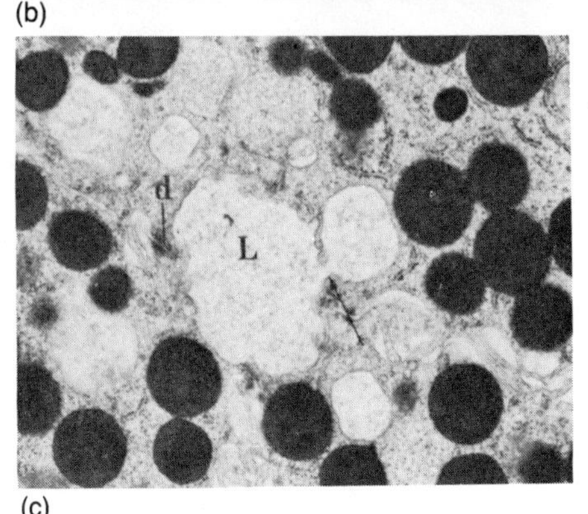

(c)

FIGURE 10.23 Pseudopodia formation by secretory granules within intact cells of rat parotid slices. Pseudopodia form when slices are incubated with secretion inducers isoproterenol (*a*) and butyrl cAMP (*b*). The pseudopods (arrows) appear to be oriented toward the lumen (L). In (*c*) an empty ghost of a secretory granule is connected to the lumen by a narrow connecting tube (arrow) that may have been a former pseudopod. d, desmosome. (Courtesy of Dr. Z. Selinger.)

10.4 VARIATIONS IN THE SECRETORY PATHWAY

Before discussing variations in the secretory pathway, it is important to point out once again that there are basically two kinds of secretory cell.

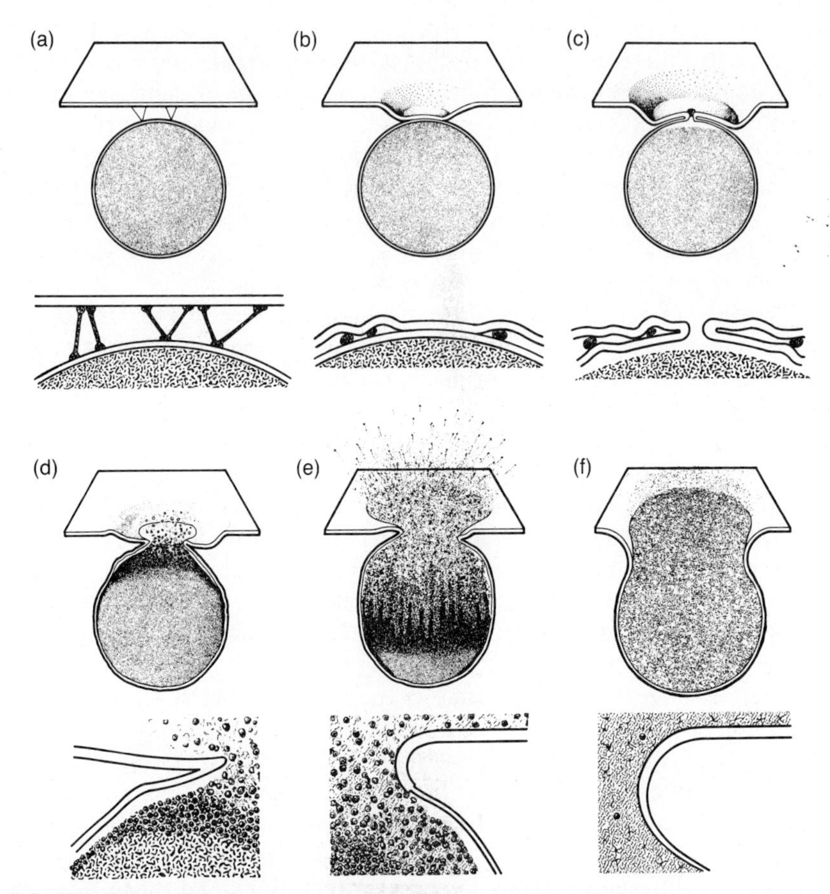

FIGURE 10.24 Hypothetical sequence of events taking place when amoebocytes from the horseshoe crab secrete their granule contents into the surrounding blood. (*a*) Unstimulated cells. Morphological changes after stimulation with endotoxin as follows: (*b*) and (*c*) for 1 to 5 sec, (*d*) for 5 to 10 sec, (*e*) and (*f*) for 10 to 15 sec. A small pore is formed, which rapidly widens to permit discharge. (Courtesy of Dr. Richard Ornberg.)

One type, exemplified by the pancreatic exocrine cell, secretes intermittently, normally exuding large amounts of material in a short time. The salivary parotid gland cell discussed above, and any secretory cell that produces storage granules, can be placed in this category.

In the other type of cell, secretion is a much more continuous process. The plasma cell, which produces and secretes immunoglobulins, is an example of this cell type. The major distinction between these types involves storage of the secretory material. In general, intermittent secreters store secretion products and continuous secreters do not.

An outline of variations in the secretory pathway as put forth by Palade is presented in Figure 10.26. The six stages that we have discussed are outlined, with stage VI broken down into two possibilities: discharge at the cell surface and discharge into endocytic vacuoles. Variations occur mainly with respect to stages IV and V. These two stages are omitted in

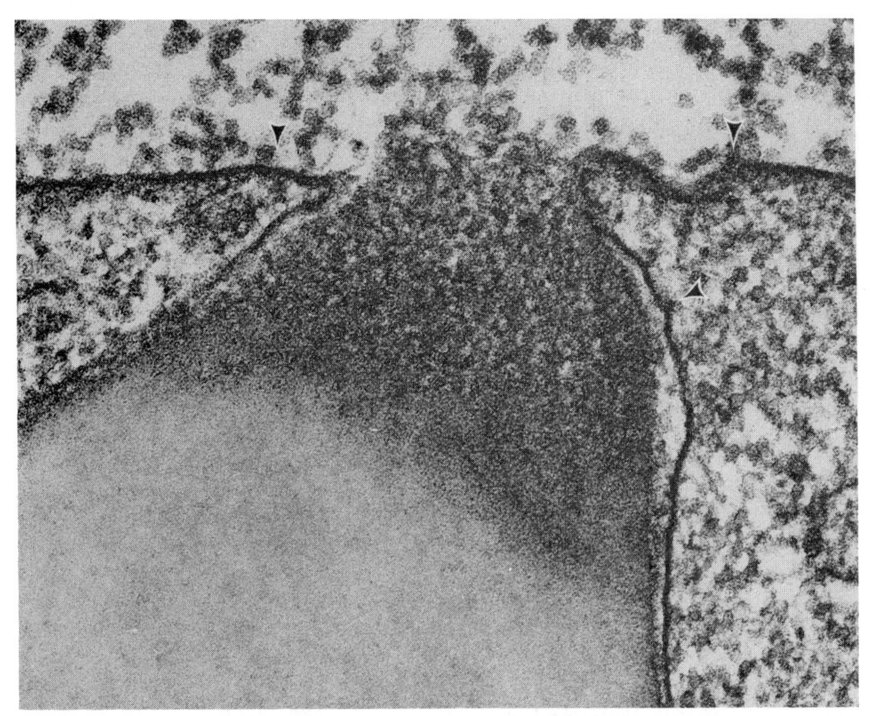

FIGURE 10.25 Secretory granule discharging its contents 5 sec after application of endotoxin. Shoulders of the original pedestal are still apparent (between arrowheads at top). The pore neck at left consists of a sharp corner that may represent the original joint between the membranes of the granule and the cell surface. ×34,500. (Courtesy of Dr. Richard Ornberg.)

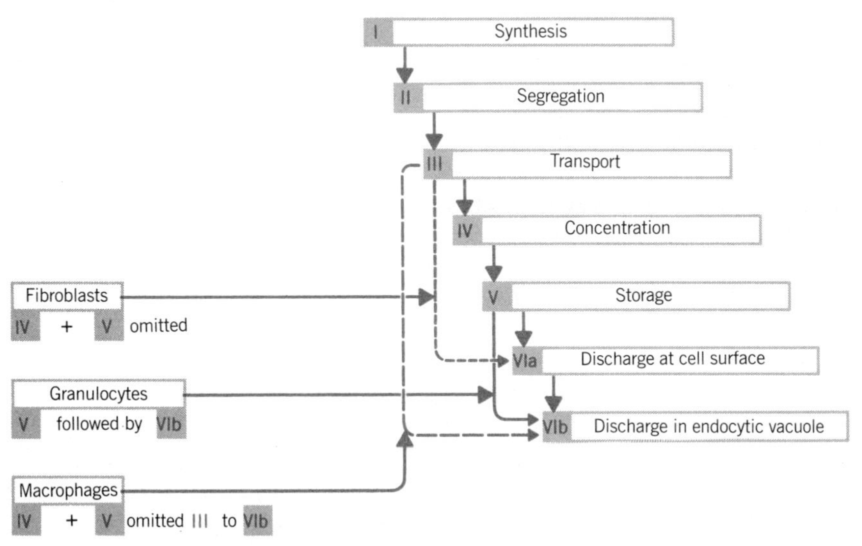

FIGURE 10.26 Variations in the secretory pathway. The variations occur mainly with respect to stages IV (concentration) and V (storage). Continuous secreters often omit these stages.

cells such as fibroblasts, chondrocytes, and plasma cells. These cells discharge their secretory materials more or less continuously as they are formed, and often do so along the entire plasma membrane rather than at a particular cell surface, as is true for the pancreatic cell. It is possible, however, that these continuous secreters have equivalents of secretion granules that are less apparent than in cells that store the material and secrete it in large amounts intermittently.

Granulocytes, such as polymorphonuclear neutrophils and eosinophils, form "secretion-type" granules that are temporarily stored and then released into endocytic vacuoles following endocytosis. The contents of these granules are largely hydrolytic enzymes. These are the lysosomes and specific granules common to these particular phagocytic cells (see Chapter 9). Macrophages, another category of phagocytic cell, seem to put less effort into concentrating and storing the acid hydrolases, but ultimately they secrete these enzymes into endocytic vacuoles.

Lysosomes are used in some glandular cells of the anterior pituitary to degrade secretory proteins that may be produced in excess. This is a mechanism of control, termed *crinophagy*. It is yet another variation of the secretory pathway in which the protein granules are discharged into lysosomes, whereupon they are enzymatically dismantled.

The formation of cyclic adenosine monophosphate (cAMP).

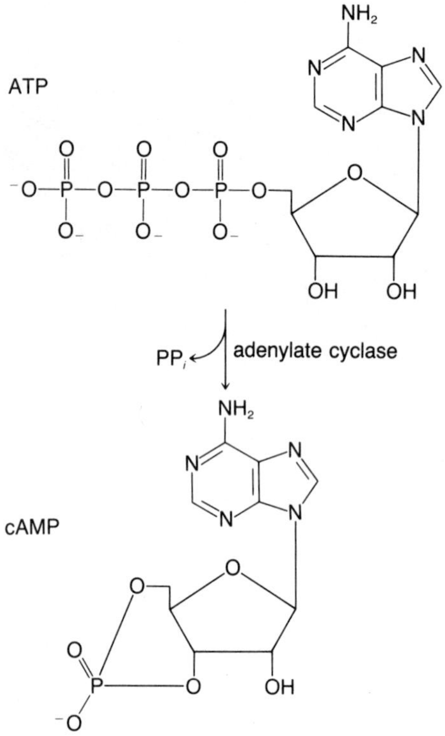

ATP

adenylate cyclase

cAMP

10.5 STIMULUS–SECRETION COUPLING

The discharge of secretory products in intermittent secreters, and perhaps in continuously secreting cells as well, is triggered by a stimulus that is either hormonal or neural. The term *stimulus–secretion coupling* denotes these linked phenomena.[20]

The general idea of this effect is as follows. Secretory cells are fitted with surface receptor sites that bind specific activators. In this respect there is high specificity between the activator and the receptor site. For example, thyroid-stimulating hormone (TSH) finds a specific receptor site on thyroid cells, vasopressin on the cells of the adrenal medulla, and so on. The receptor protein, present on the outer surface of the membrane, activates adenylate cyclase on the inner side of the membrane by way of an intermediate transducer molecule. Adenylate cyclase then converts ATP into its 3′,5′-cyclic product, cAMP. Figure 10.27 illustrates this activation process.

cAMP is viewed as a versatile cell messenger, transmitting the stimulus of the activator to specific enzymatic reactions within the cell. In many different cell systems it activates a family of molecules called protein kinases by releasing an inhibitory subunit from the inactive form. The active protein kinase is a phosphorylating enzyme that with the contribution of phosphate from ATP, carries out a variety of cellular phosphorylation reactions.

cGMP often has exactly the countereffect of cAMP. In addition, regulation by these second messengers frequently depends on Ca^{2+}. There is thus a small family of low molecular weight materials that play crucial roles in the regulation of a number of cellular activities. Given this general information, let us now return to the pancreatic cell to see the role of second messengers in the process of secretion.

In the pancreatic exocrine cell, secretion requires ATP and calcium ion. Pancreozymin, a hormone of the duodenal mucosa, triggers the process. Based on a careful study of the relation of cyclic nucleotides and Ca^{2+} to enzyme secretion in the pancreas, the following theory has been proposed to explain the events that take place after stimulation of the cell.[21] Protein kinases, which have been activated via cAMP, phosphorylate the serine and threonine residues of proteins in the membranes of zymogen granules. This causes a change in the properties of the zymogen membrane, presumably to destabilize it or to prepare it for easier interaction with the plasma membrane. cAMP is also thought to mobilize Ca^{2+} from intracellular pools. Available Ca^{2+} can then act as a catalyst for the hydrolysis of ATP by a Ca^{2+}, Mg^{2+}-ATPase. The hydrolysis of ATP provides the energy for fusion of the plasma membrane with the modified granule membrane, effecting exocytosis.

The foregoing scheme is highly conjectural and is certainly painted in very broad strokes. The mechanism of stimulus–secretion coupling is simply another example in which a long-observed phenomenon cannot yet be explained in molecular detail. Future work in this area will certainly generate some general properties of exocytosis that will make sense at the molecular level and will likely uncover some variations among different cell types.

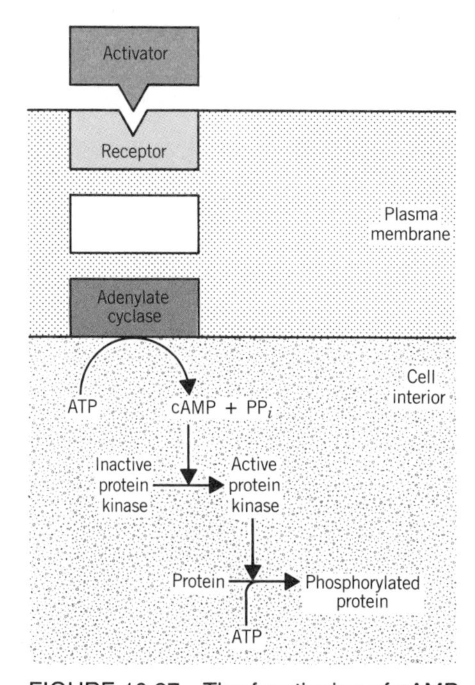

FIGURE 10.27 The functioning of cAMP as a messenger in a number of different systems. cAMP removes from a family of protein kinases an inhibitory subunit that activates them to phosphorylate various enzymes or other proteins. When phosphorylated, they may be either activated or inactivated, and thus control is effected.

10.6 MEMBRANE REGULATION

In all secretory cells, huge amounts of proteins are synthesized, transported by way of the secretory pathway, and after a very transient or extended storage period, secreted. Exocrine secretion from the pancreas accounts for more than 90% of its total protein synthesis, and at any one time the products represent more than 30% of the total protein of the gland.

According to estimates made for the pancreas of the guinea pig, each day an amount of membrane equaling the area of the *entire plasma membrane* is incorporated into the *luminal portion* of the cell, an area less than 10% of the total plasma membrane.[22] Considering the mechanism of exocytosis just discussed, the events of membrane fusion would have at least a twofold effect on the luminal plasma membrane: its surface area would increase dramatically during exocytosis and its composition would change as a result of the different chemical properties of the two merging membranes. It is obvious that these changes cannot take place in a secretory cell. Rather, the cell must have a mechanism to deal with the piling up of redundant membrane.

In this regard, several questions are posed for which researchers are presently seeking answers. What mechanism maintains the homeostatic composition of the plasma membrane? Are membranes recycled as fragments, or as soluble lipid and protein molecules? Are only the newly added patches resulting from exocytosis recycled, or is the recycling mechanism more random?

It is not now possible to answer any of these questions in terms of a general rule of membrane regulation. But a number of observations are beginning to point to some interesting mechanisms that may have general utility in secretory cells.

It has been well documented that the membranes making up the components of the secretory pathway differ quite markedly in their lipid and protein compositions as well as in terms of enzyme activities. Figure 10.28

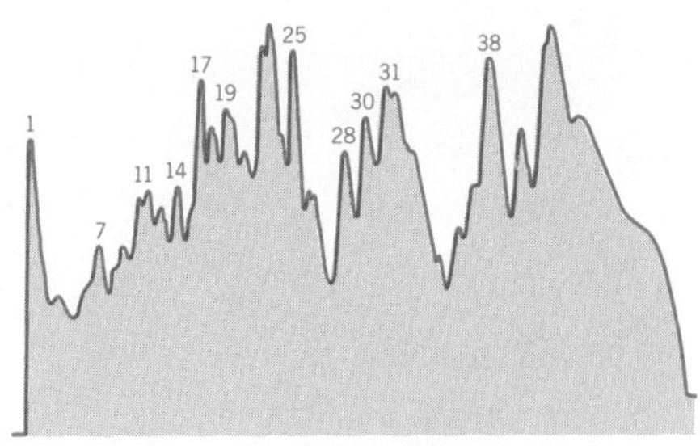

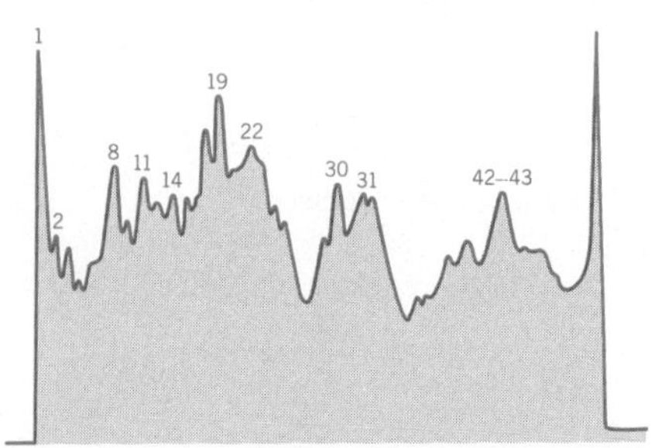

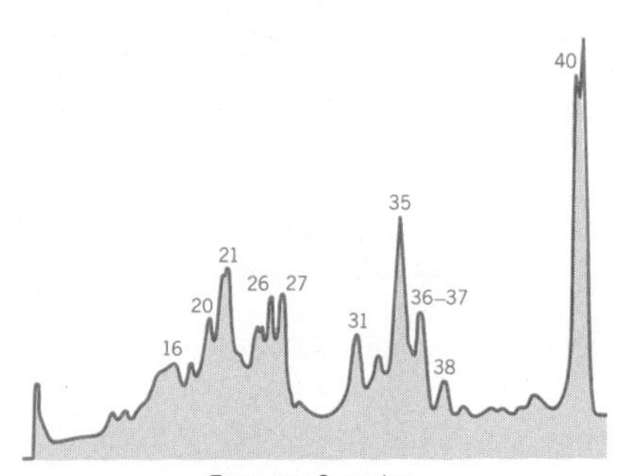

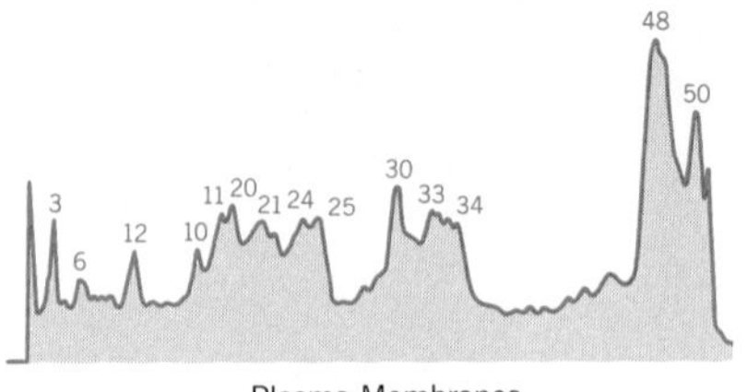

FIGURE 10.28 Densitometric tracings of electrophoretograms prepared on SDS–polyacrylamide gels. The direction of migration is from left to right. These protein profiles, with components numbered to help in comparing them, differ markedly in the different membrane preparations.

Table 10.2a Lipid Composition of Subcellular Fractions Isolated from Guinea Pig Pancreas

Lipid	Rough Microsomes	Smooth Microsomes	Zymogen Granules	Plasma Membrane
PLP: protein	0.32	0.57	0.46	0.44
Sphingomyelin[a]	3.40	14.20	23.50	19.20
Lecithin + lysolecithin[a]	47.80	39.40	30.50	32.00
Phosphatidyl-ethanolamine[a]	35.80	36.50	31.50	34.40
Phosphatidylserine + Phosphatidylinositol[a]	5.60	4.70	5.00	4.60
Total cholesterol:PLP[b]	0.12	0.47	0.55	0.51

[a]Percent total phospholipid (PLP).
[b]Molar ratio.

Table 10.2b Distribution of Enzyme Activities in Subcellular Fractions Isolated from Guinea Pig Pancreas

Enzymes	Rough Microsomes	Smooth Microsomes	Zymogen Granules	Plasma Membrane
NADH-cytochrome *c* reductase (rotenone insensitive)[a]	181.0	130.0	32.8	47.2
NADPH-cytochrome *c* reductase[a]	27.7	17.8	0	3.5
TPPase[b]	0	1.49	0	0
Galactose transferase[c]	1.3	22.8	0.8	1.1
5′-Nucleotidase[b]	0	13.2	4.2	18.3
β-leucylnaphthylamidase[b]	1.0	41.2	6.7	44.3
Mg^{2+} ATPase[b]	0	11.0	38.0	46.0

[a]Nanomoles of cytochrome *c* reduced at 25°C/(min)(mg) phospholipid.
[b]Micromoles of P_i or naphthylamine released at 37°C/(60 min)(mg) phospholipid.
[c]Nanomoles of galactose transferred at 37°C/(60 min)(mg) phospholipid.

and Table 10.2 summarize the results of some detailed studies carried out on membrane compositions. The membranes of the zymogen granule especially have unique protein compositions. This is true not only for the pancreas, but for the secretory granules of parotid and adrenal medulla as well. These compositional differences argue indirectly that a unidirectional membrane flow cannot take place without grave consequences for the target plasma membrane, and they support the notion of a mandatory recycling mechanism.

Labeling experiments have revealed the important finding that the average half-life of secretory proteins is on the order of hours, and the half-life of the membranes in the secretory pathway is on the order of days. This implies that membranes are conserved and suggests that the membrane products of exocytosis may not be completely dismantled to the molecular level. Apparently the physical movement of membranes is great, but the turnover is low. In other words, membrane biogenesis and secretory protein transport are independently controlled on different time scales.

When pancreatic cells are stimulated with pilocarpine to secrete, they show increased endocytic activity.[23] This is especially obvious at the luminal boundary, where deep caveolae form and generate multivesicular bodies which move back into the deeper recesses of the cell.

These bodies appear not only to be membrane bound themselves but to contain membrane fragments in their interiors, perhaps budding off from their own boundaries. They fuse with lysosomes, suggesting that at least some of the membrane that is taken in by this mechanism is catabolized by lysosomal hydrolases. When this occurs, the membrane patches must be broken down to their molecular building blocks, or at least substantially modified.

Endocytic activity in stimulated cells is not confined to the apical surface of the cells. It has also been seen in the lateral walls of the cell, where tubular or sheetlike infoldings form. The vesicles that form in this case also fuse with lysosomes. Finally, several hours after stimulation, there is an increase in coated vesicles in the cell, apparently produced by endocytic withdrawal of the plasma membrane.

Based on these observations, a three-phase mechanism has been proposed to describe membrane regulation.[23] Illustrated in Figure 10.29, it consists of one major and two minor modes. The major phase, which occurs first, takes place on the apical plasma membrane (I) where deep caveolae form as a result of exocytoses of secretory granules that approach the membrane in tandem. The second and more minor phase (II) involves the lateral membranes. It is believed that redundant membranes may pile up on the lateral surfaces as a result of flow from the apical oversupply of membranes. The fate of the vesicles formed by both of these processes is breakdown in lysosomes.

Phase III is the formation of small coated vesicles at the apical plasma membrane. This occurs later than the other phases and is thought of as a fine-tuning control mechanism, since it does not involve the movement of large quantities of membrane.

There are obviously some large gaps to be filled in the pattern before we can feel comfortable about our understanding of membrane regulation. It is still not clear, for example, if membrane is returned as fragments or as individual molecules, or as some combination of the two. Nor is it known at what level in the secretory pathway reentry is made.

Other models have been proposed that emphasize a bidirectional shuttle mechanism between each component membrane system of the secretory pathway by nonrandom fusion–fission reactions.[22] It is virtually certain that a well-controlled mechanism exists, but for the present its details await delineation by future research.

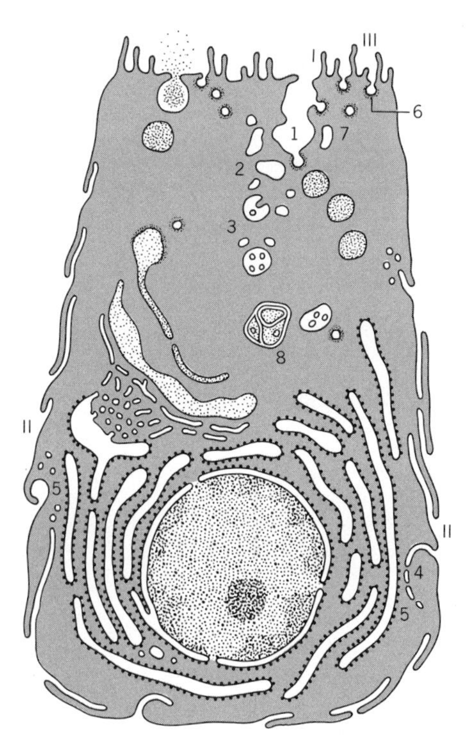

FIGURE 10.29 Three modes of redundant membrane withdrawal in a pancreatic acinar cell. In phase I, caveolae form (1), giving rise to endocytic vesicles (2) and multivesicular bodies (3). In phase II, lateral infoldings occur (4), which generate tubules and vesicles (5). In phase III, coated pits form in the apical membrane (6) and become coated interior vesicles (7). At least some of these endocytic structures are thought to be degraded by lysosomes (8).

10.7 POSTTRANSLATIONAL PROCESSING OF SECRETORY PROTEINS

Proteins that are discharged from a secretory cell probably never have the same structure they had when entering into the secretory pathway. In transit, they are subjected to proteolytic attack and to glycosylations and disulfide bond formation, all of which alter their primary, secondary, and tertiary structures and prepare them for their ultimate biological roles.

The Golgi complex is the principal site for glycosylation reactions and is dealt with more specifically in Chapter 11. In this chapter we emphasize posttranslational proteolytic cleavages.

Protein	−25 … −20	Hydrophobic domain (−15 … −10)	Flexible domain (−5 … −1 +1)
Proparathyroid hormone (bovine)	Met-Met-Ser-Ala-Lys-Asp-Met-Val-	Lys-Val-Met-Ile-Val-Met-Leu-Ala-Ile-Cys-Phe-	Leu-Ala-Arg-Ser-Asp-Gly-Lys-
Proinsulin (rat)	I. (Met)Ala-Leu-Trp-Met-Arg-Phe-	Leu-Pro-Leu-Leu-Ala-Leu-Leu-Val-Leu-Trp-Glu-	Pro-Lys-Pro-Ala-Gln-Ala-Phe-
	II. Ile	Phe Phe Ile	
Growth hormone (rat)	Met-Ala-Ala-Asp-Ser-Gln-Thr-Pro-Trp-	Leu-Leu-Thr-Phe-Ser-Leu-Leu-Cys-Leu-Leu-Trp-	Pro-Gln-Glu-Ala-Gly-Ala-Leu-
Prolactin (rat)	Met-Asn-Ser-Gln-Val-Ser-Ala-Arg-Lys-Ala-Gly-Thr-	Leu-Leu-Leu-Leu-Met-Met-Ser-Asn-Leu-Leu-Phe-	Cys-Gln-Asn-Val-Gln-Thr-Leu-
Lysozyme (chicken)	Met-	Arg-Ser-Leu-Leu-Ile-Leu-Val-Leu-Cys-Phe-Leu-	Pro-Leu-Ala-Ala-Leu-Gly-Lys-
Ovomucoid (chicken)	(Met)Ala-Met-Ala-Gly-Val-Phe-	Val-Leu-Phe-Ser-Phe-Val-Leu-Cys-Gly-Phe-Leu-	Pro-Asp-Ala-Ala-Phe-Gly-Ala-
Conalbumin (chicken)	Met-Lys-	Leu-Ile-Leu-Cys-Thr-Val-Leu-Ser-Leu-Gly-Ile-	Ala-Ala-Val-Cys-Phe-Ala-Ala-
Trypsinogen (dog)	I.	Ala-Lys-Leu-Phe-Leu-Phe-Leu-Ala-Leu-Leu-	Leu-Ala-Tyr-Val-Ala-Phe-Val-
	II.	Phe Pro Phe	
Myeloma L-chain			
(Mouse MOPC-41)	Met-Asp-Met-Arg-Ala-	Pro-Ala-Gln-Ile-Phe-Gly-Phe-Leu-Leu-Leu-Leu-	Phe-Pro-Gly-Thr-Arg-Cys-Gln-
(Mouse MOPC-321)	Met-Glu-Thr-	Asp-Thr-Leu-Leu-Leu-Trp-Val-Leu-Leu-Leu-Trp-	Val-Pro-Gly-Ser-Thr-Gly-Gln-
(Mouse MOPC-104E)	Met-Ala-Trp-	Ile-Ser-Leu-Ile-Leu-Ser-*-Leu-Leu-Ala-Leu-	Ser-Ser-Gly-Ala-Ile-Ser-Gln-
(Mouse MOPC-315)	Met-Ala-Trp-	Thr-Ser-Leu-Ile-Leu-Ser-*-Leu-Leu-Ala-Leu-	Cys-Ser-Gly-Ala-Ser-Ser-Gln-
V_{λ}L-chain gene (mouse)	Met-Ala-	Trp-Thr-Ser-Leu-Ile-Leu-Ser-Leu-Leu-Ala-Leu-	Cys-Ser-Gly-Ala-Ser-Ser-Gln-
Proalbumin (bovine)	Met-	Lys-Trp-Val-Thr-Phe-Leu-Leu-Leu-Leu-Phe-Ile-	Ser-Gly-Ser-Ala-Phe-Ser-Lys-
Promellitin (honeybee)	Met-Lys-Phe-Leu-	Val-*-Val-Ala-Leu-Val-Phe-Met-Val-Val-Tyr-	Ile-*-Tyr-Ile-Tyr-Ala-Ala-
Lipoprotein (*Escherichia coli*)	Met-Lys-Ala-	Thr-Lys-Leu-Val-Leu-Gly-Ala-Val-Ile-Leu-Gly-	Ser-Thr-Leu-Leu-Ala-Gly-Cys-
Penicillinase (*E. coli* plasmid pBR322)	Met-Ser-Ile-Gln-His-Phe-	Arg-Val-Ala-Leu-Ile-Pro-Phe-Phe-Ala-Ala-Phe-	Cys-Leu-Pro-Val-Phe-Ala-His-
Ovalbumin (chicken)		Met-Gly-Ser-Ile-Gly-Ala-Ala-Ser-Met-	Glu-Phe-Cys-Phe-Asp-Val-↑Phe- (Cleavage site)

FIGURE 10.30 Amino acid sequences of various presecretory peptide extensions. A region rich in hydrophobic amino acids is found between positions −7 and −17, and between this region and the cleavage site small neutral amino acids predominate.

Proteolytic cleavage of secretory proteins takes place at more than one point in the secretory pathway. Early cleavages, occurring in the endoplasmic reticulum, are carried out on a category of proteins called *presecretory proteins*. Cleavage further along the pathway, mainly in the vicinity of the Golgi complex, is conducted on the so-called *prosecretory proteins* or *proproteins*.[24]

Presecretory Proteins

Ovalbumin is an exception to the category of presecretory proteins. It is formed without a signal peptide. But it does have a hydrophobic sequence near its *N*-terminus that may *function* as a signal peptide.

The category of presecretory proteins includes most of the secreted proteins known, with only rare exceptions. These are the proteins that contain the *N*-terminal signal peptide, a peptide extension that is presumed to be essential for the segregation stage in the secretory pathway. As indicated earlier in this chapter, this sequence is removed early during segregation and may be as much a cotranslational event as a posttranslational cleavage. Because of this early cleavage, presecretory proteins are present in very low concentrations in cells. Most have apparently lost their *N*-terminal signal peptide during or immediately after the nascent protein was released into the lumen of the endoplasmic reticulum.

Amino acids with hydrophobic and bulky side chains in the signal peptide are leucine, phenylalanine, isoleucine, and tryptophan. Amino acids with smaller, neutral side chains are glycine, alanine, valine, and serine.

Studies on the primary structures of the signal peptide stretch of a number of secretory proteins have revealed some common features that have led to some interesting proposals. Most of the signal peptides contain a region rich in hydrophobic amino acids with bulky side chains between positions −7 and −17 (see Figure 10.30). Between this hydrophobic domain and the cleavage site is a stretch of amino acids with smaller neutral side chains. The latter domain is thought to be quite flexible, with a high potential to form a β conformation.

These features have led D. F. Steiner and his co-workers to propose a *loop mechanism* to set the stage for segregation and early cleavage. This

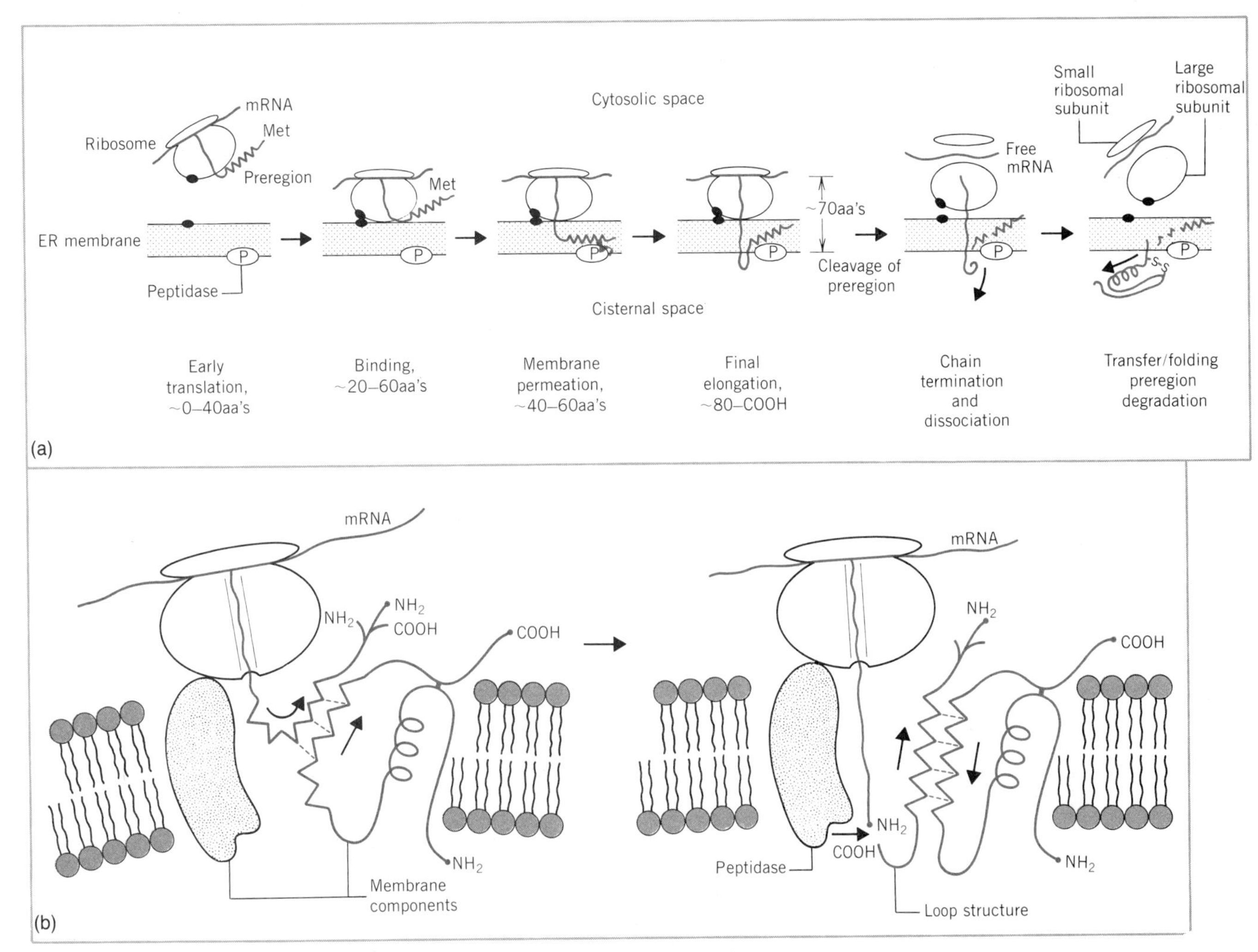

FIGURE 10.31 (*a*) The loop mechanism proposed for the segregation of presecretory proteins into the cisternae of rough endoplasmic reticulum. (*b*) Details of interaction of β strands of integral membrane protein and presecretory protein being generated. The signal peptide, in β conformation, immerses in the membrane, and by interacting with a preexisting membrane protein, supplies the necessary grip and leverage to drive the rest of the protein into the lumen.

mechanism is summarized in Figure 10.31. According to this proposal, the signal peptide does not pass through the bilayer but partitions in it in such a way that a loop is formed with the turn in the loop near the cisternal surface of the membrane. The hydrophobic domain may then form a β-pleated sheet with an integral membrane protein, such as the ribophorins or the translocator protein discussed earlier.

This postulated interaction could accomplish two things. It could control the orientation of the signal peptide sequence with respect to the signal peptidase so that the signal sequence is efficiently and properly excised. Second, it could provide the necessary leverage to drive the forming protein into the cisternal space during the early stages of its translation.

The signal peptide, which is released initially by the signal peptidase,

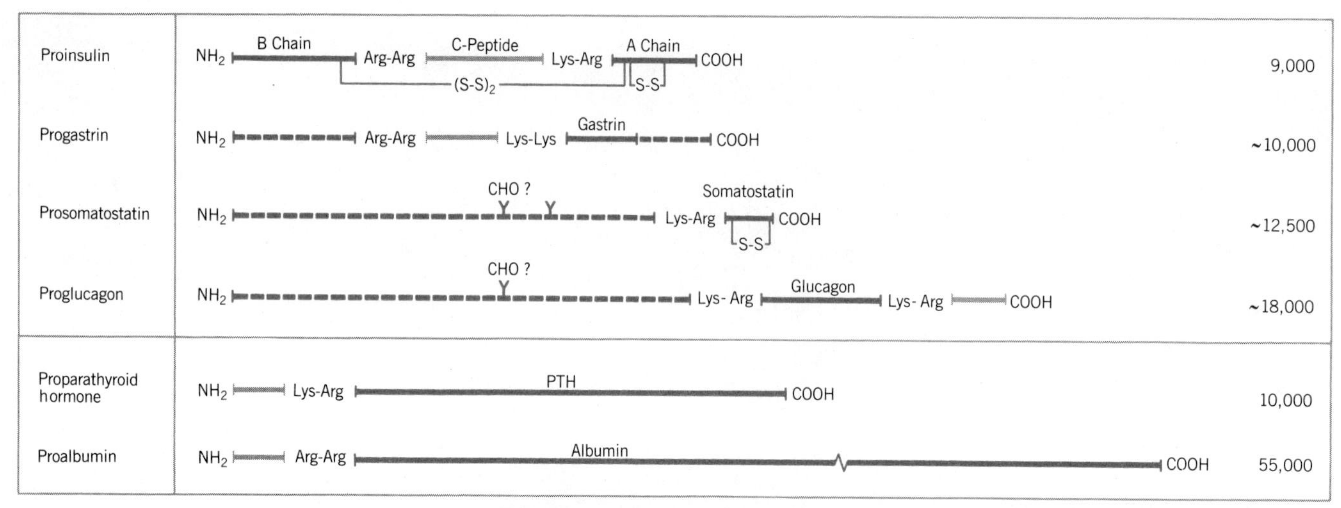

FIGURE 10.32 Structures of some proproteins. Paired basic residues (as Lys–Arg) are found at the cleavage site, where "useless" regions (gray) are excised from biologically active or "useful" regions (solid colored lines). Unsequenced regions are demarcated by dashed lines.

is subsequently degraded completely by other peptidases. Thus the pre-*N*-terminal sequence that characterizes presecretory proteins never makes an appearance in the secretory granule.

Prosecretory Proteins

Many secretory proteins are subjected to proteolysis at later stages in the secretory pathway. Two categories of such proteins are recognized. One encompasses a large number of secretory proteins including the small polypeptide hormones, such as proinsulin or proparathyroid hormone. The other group includes a diverse collection of cellular and organelle components (chloroplasts and mitochondria) as well as animal virus capsule proteins.

The polypeptide hormones have received the most attention. Figure 10.32 contains some structural information that has been accumulated for several proproteins, including proinsulin. In general it is of interest to note that a pair of basic residues (Lys–Arg) is commonly found at the site of proteolytic cleavage, apparently an important processing point. The presence of long stretches of "useless" polypeptide, often separating "useful" regions in the proproteins, has also been of considerable research interest. It is speculated that a secretory protein must be of some minimal length (65–70 residues) to enable it to span the distance from its point of chain elongation on the ribosome to the cisternal side of the membrane. Thus the function of these useless regions is to get the nascent protein properly segregated in the lumen. These regions are, after that operation, dispensable, and are removed by proteolysis later in the secretory pathway.

For a number of proproteins that have been studied, including proinsulin and proparathyroid hormone, the proteolytic conversion to mature form (insulin and parathyroid hormone) takes place in the region of the Golgi complex. It is not certain exactly where in the secretory pathway this conversion takes place. It may be within the complex itself, or it may occur within the condensing vacuoles.

Some studies have suggested that lysosomal enzymes are cosegregated with proproteins into newly forming secretion granules in the Golgi complex, thereby providing the proteolytic activities necessary for conversion. In other cases the proteolytic enzymes appear to be added to immature granules by a vesicle fusion process. The conversion of proalbumin to albumin occurs in mature secretory vesicles, an exception to the general rule for other proproteins.

In spite of the information that is accumulating to further our understanding of the events that take place along the secretory pathway, there are still a number of puzzles to be solved. Why, for example, do some secretory proteins have a prosecretory structure (proalbumin, proinsulin, and proparathyroid hormone) while a large number of others (growth hormone, various pancreatic digestive enzymes) do not? What functions may the "useless" sequences have other than assuring a minimum length to the product? Is limited proteolysis necessary to unveil an activity not present in the proprotein? Answers to these and other related questions will surely be forthcoming within the next few years, and they will probably both clarify and intensify the complexity of the secretory pathway.

Summary

All cells appear to have secretory functions of one type or another. However, cells that secrete digestive enzymes, hormones, or neurotransmitters are more commonly thought of as secretory cells.

The mechanism of secretion is viewed as involving a six-stage pathway including the following components in order as they participate in secretion: synthesis, segregation, intracellular transport, concentration, intracellular storage, and discharge.

The overall pathway contains a mechanism to ensure that only proteins that are bound for secretion are segregated and incorporated into the pathway. The basis of segregating proteins resides in their content of an *N*-terminal signal sequence of amino acids, which directs the ribosome to the ER membrane and initiates vectorial transport into the lumen. Once segregated, secretory proteins continue through the pathway sheltered from the rest of the cell by membrane boundaries.

Variations on the six-stage theme occur in different cell types. The pancreatic exocrine cell exhibits all six stages, but cells like plasma cells tend to omit stages concerned with concentration and storage.

The secretion of materials from a secretory cell is linked to a stimulus that is either hormonal or neural. In the pancreatic cell cAMP and Ca^{2+} play important roles as mediators between the hormonal stimulus and the discharge of secretory material.

During discharge, an excess of redundant membrane accumulates on the discharging surface by incorporation of the secretory granule membrane into the plasma membrane. This redundancy is regulated by mechanisms of endocytosis that return the membranes or their components to the interior of the cell.

During transport through the secretory pathway, proteins are processed by proteolytic enzymes and modified chemically in other ways. Proteolysis takes place at two levels. First, presecretory proteins are shortened by excision of the signal peptide and, finally, proteases remove "useless" regions from prosecretory proteins in the vicinity of the Golgi complex.

References

1. Intracellular Aspects of the Process of Protein Synthesis, G. Palade, *Science* 189:347(1975).
2. Parallelism in the Processing of Pancreatic Proteins, A. M. Tartakoff, L. J. Greene, J. D. Jamieson, and G. E. Palade, in *Advances in Cytopharmacology*, vol. 2, B. Ceccarelli, F. Clementi, and J. Meldolesi, eds., Raven Press, New York, 1974.
3. Synthesis and Segregation of Secretory Proteins: The Signal Hypothesis, G. Blobel, in *International Cell Biology*, B. R. Brinkley and K. R. Porter, eds., Rockefeller University Press, New York, 1977.
4. Proteins of Rough Microsomal Membranes Related to Ribosome Binding. I. Identification of Ribophorins I and II, Membrane Proteins Characteristic of Rough Microsomes, G. Kneibach, B. L. Ulrich, and D. D. Sabatini, *J. Cell Biol.* 77:464(1978).
5. Identification and Characterization of a Membrane Component Essential for the Translocation of Nascent Proteins across the Membrane of the Endoplasmic Reticulum, D. I. Meyer and B. Dobberstein, *J. Cell Biol.* 87:503(1980).
6. Membranes and Secretion, J. D. Jamieson, in *Cell Membranes*, G. Weissmann and R. Claiborne, eds., HP Publishing Co., New York, 1975.
7. Intracellular Transport of Secretory Proteins in the Pancreatic Exocrine Cell. IV. Metabolic Requirements, J. D. Jamieson and G. E. Palade, *J. Cell Biol.* 39:589(1968).
8. Bead Rings at the Endoplasmic Reticulum–Golgi Complex Boundary: Morphological Changes Accompanying Inhibition of Intracellular Transport of Secretory Proteins in Arthropod Fat Body Tissue, D. A. Brodie, *J. Cell Biol.* 90:92(1981).
9. Golgi Apparatus, GERL, and Lysosomes of Neurons in Rat Dorsal Root Ganglia, Studied by Thick Section and Thin Section Cytochemistry, P. M. Novikoff, A. B. Novikoff, N. Quintana, and J. J. Hauw, *J. Cell Biol.* 50:859(1971).
10. Studies of the Secretory Process in the Mammalian Exocrine Pancreas. I. The Condensing Vacuoles, A. B. Novikoff, M. Mori, N. Quintana, and A. Yam, *J. Cell Biol.* 75:148(1977).
11. Relationship between the Golgi Apparatus, GERL, and Secretory Granules in Acinar Cells of the Rat Exorbital Lacrimal Gland, A. R. Hand and C. Oliver, *J. Cell Biol.* 74:399(1977).
12. Condensing Vacuole Conversion and Zymogen Granule Discharge in Pancreatic Exocrine Cells: Metabolic Studies, J. D. Jamieson and G. E. Palade, *J. Cell Biol.* 48:503(1971).
13. Sulfated Compounds in the Secretion and Zymogen Granule Content of the Guinea Pig Pancreas, H. Reggio and G. E. Palade, *J. Cell Biol.* 70:360a(1976).
14. Intracellular Transport of Secretory Proteins in the Pancreatic Exocrine Cell. II. Transport of Condensing Vacuoles and Zymogen Granules, J. D. Jamieson and G. E. Palade, *J. Cell Biol.* 34:597(1967).
15. Binding of Microtubules to Pituitary Secretory Granules and Secretory Granule Membranes, P. Sherline, Y.-C. Lee, and L. S. Jacobs, *J. Cell Biol.* 72:380(1977).
16. Membrane Fusion During Secretion. A Hypothesis Based on Electron Microscope Observation of *Phytophthora palmivora* Zoospores During Encystment, P. Pinto da Silva and M. L. Nogueira, *J. Cell Biol.* 73:161(1977).
17. Molecular Events During Membrane Fusion. A Study of Exocytosis in Rat

Peritoneal Mast Cells, D. Lawson, M. D. Raff, B. Gomperts, C. Fewtrell, and N. B. Gilula, *J. Cell Biol.* 72:242(1977).
18. Modification of the Secretory Granule During Secretion in the Rat Parotid Gland, Z. Selinger, Y. Sharoni, and M. Schromm, in *Advances in Cytopharmacology*, vol. 2, B. Ceccarelli, F. Clementi, and J. Meldolesi, eds., Raven Press, New York, 1974.
19. Beginning of Exocytosis Captured by Rapid-freezing of *Limulus* Amoebocytes, R. L. Ornberg and T. S. Reese, *J. Cell Biol.* 90:40(1981).
20. Stimulus–Secretion Coupling: The Concept and Clues from Chromaffin and Other Cells, W. W. Douglas, *Br. J. Pharmacol.* 34:451(1968).
21. Molecular Basis of Enzyme Secretion by the Exocrine Pancreas, J. Christophe, P. Robberecht, M. Deschodt-Lanckman, M. Lambert, M. Van Leemput-Coutrez, and J. Camus, In *Advances in Cytopharmacology*, vol. 2, B. Ceccarelli, F. Clementi, and J. Meldolesi, eds., Raven Press, New York, 1974.
22. Secretory Mechanisms in Pancreatic Acinar Cells. Role of the Cytoplasmic Membranes, J. Meldolesi, in *Advances in Cytopharmacology*, vol. 2, B. Ceccarelli, F. Clementi, and J. Meldolesi, eds., Raven Press, New York, 1974.
23. Redundant Cell-Membrane Regulation in the Exocrine Pancreas Cells after Pilocarpine Stimulation of the Secretion, M. F. Kramer and J. J. Geuze, in *Advances in Cytopharmacology*, vol. 2, B. Ceccarelli, F. Clementi, and J. Meldolesi, eds., Raven Press, New York, 1974.
24. Proteolytic Cleavage in the Posttranslational Processing of Proteins, D. F. Steiner, P. S. Quinn, C. Patzelt, S. J. Chan, J. Marsh, and H. S. Tager, in *Cell Biology: A Comprehensive Treatise*, vol. 4, D. M. Prescott and L. Goldstein, eds., Academic Press, New York, 1980.

Selected Books and Articles

Books

Advances in Cytopharmacology, vol. 2, *Cytopharmacology of Secretion*, B. Ceccarelli, F. Clementi, and J. Meldolesi, eds., Raven Press, New York, 1974.

Articles

Beginning of Exocytosis Captured by Rapid-freezing of *Limulus* Amoebocytes, R. L. Ornberg and T. S. Reese, *J. Cell Biol.* 90:40(1981).
Intracellular Aspects of the Process of Protein Synthesis, G. Palade, *Science* 189:347(1975).
Membranes and Secretion, J. D. Jamieson, in *Cell Membranes*, G. Weissmann and R. Claiborne, eds., HP Publishing Co. New York, 1975.
Protein Secretion and Transport, C. M. Redman and D. Banerjee, in *Cell Biology: A Comprehensive Treatise*, D. M. Prescott and L. Goldstein, eds., Academic Press, New York, 1980.

11

The Golgi Complex

Taking advantage of the method, found by me, of the black staining of the elements of the brain, staining obtained by the prolonged immersion of the pieces, previously hardened with potassium or ammonium bichromate, in a 0.50 or 1.0% solution of silver nitrate, I happened to discover some facts concerning the structure of the cerebral gray matter that I believe merit immediate communication.

CAMILLO GOLGI, 1873

This paragraph introduced a preliminary communication published by Camillo Golgi in 1873.[1] It is a modest understatement of a major scientific breakthrough. The silver chromate method, termed *la reazione nera,* which Golgi discovered and developed, opened a new field of scientific inquiry called neuromorphology. Aided by the silver chromate technique, scores of scientists working over a period of many years were able arrive at a basic understanding of the neuronal architectonics of the main centers of the higher vertebrate nervous system. Indeed, the basic Golgi procedure, modified in various ways and even adapted to electron microscopy, is still used today. It is probably one of the most useful neurohistological techniques available for studying the neuron network.

The "some facts" reported by Golgi have so increased in number that it has become impossible for our gray matter to sort out all the facts available concerning itself. If Golgi only had known how students of this subject just a century after his time would long for fewer facts to deal with, he might not have been so anxious to pioneer a new field. Be that as it may, we have a subject to discuss in this chapter because of a perceptive researcher who caught a glimpse of a consistent cell property and pursued it until it was taken up for study by many others, who together formulated and verified a major cell concept, *the Golgi complex.*

It was while applying *la reazione nera* to the Purkinje cells of a barn owl that Golgi repeatedly observed an internal reticular network of material with a high affinity for the stain. He reported the sighting of this *apparato reticolare interno* (internal reticular apparatus) in a scientific journal in 1898,[2] and in so doing precipitated a lengthy and heated controversy regarding the form and function, and even the existence, of this cellular organelle.

The Golgi controversy, as it was called, lasted in one form or another up until the 1960s. Some researchers believed that there was no hard evidence to support the existence of a Golgi substance, and others believed that whatever was being observed appeared in too many different forms in cells to justify the attachment of Golgi's name to the material.

Nevertheless, Golgi continued to visualize this network in cells from disparate sources and recorded his findings with drawings such as those in Figure 11.1. His work shows remarkable clarity and detail. We could hardly do better today, using the finest of light microscopes with the best photographic systems.

A solid supporter of Golgi during the years of controversy was Santiago Ramón y Cajal. The Spanish histologist refined Golgi's method of staining and became a pioneer student of the nervous system. But he and Golgi clashed on an interpretation of the structure of the nervous system; Golgi viewed it as a continuous network and Cajal pictured it as composed of individual cells that interacted with one another. Nevertheless, Cajal verified Golgi's finding of a special internal cell complex and observed its morphology and behavior under a variety of metabolic states.

In spite of the data that began to accumulate, many scientists resisted conversion to the Golgi point of view. A substantial population of researchers remained convinced that the material was artifact right up until the era of the electron microscope, a half to three quarters of a century after the discovery of the structure. Now, of course, the electron micro-

The first electron micrographs of biological materials appeared as early as 1934. However, commercial electron microscopes were not made until 1939 in Germany and 1941 in America, and the main breakthroughs in preparing and studying biological materials had to await the early 1950s.

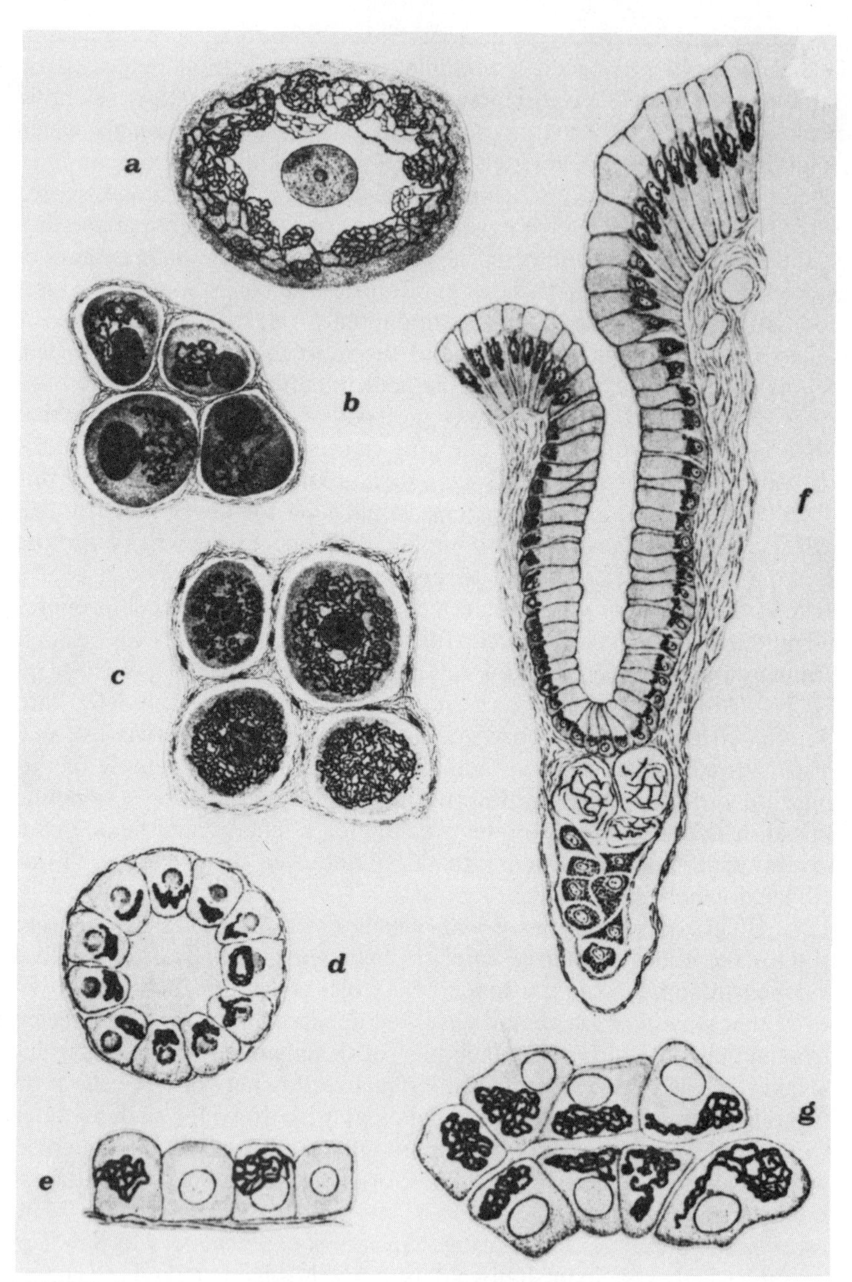

FIGURE 11.1 Some of Golgi's original drawings of the inner reticular apparatus in spinal ganglion nerve cells of (*a*) horse, (*b*) cattle fetus, (*c*) rabbit, and the complex as it stains in (*d*) renal tubule, (*e*) thyroid cells (*f*) mucous gastric cells, and (*g*) hepatic cells.

scope has amply vindicated Golgi's light microscopic observations, and many researchers have devoted their careers to ferreting out the ultrastructure and function of the Golgi complex.

In 1906 Camillo Golgi and Santiago Ramón y Cajal were jointly awarded the Nobel prize. In retrospect, we can now see that this award was prophetic of the ultimate significance of their work, as much as it was a recognition of the import of the early studies. Now hundreds have joined the work and thousands of papers have been published. In the material that follows in this chapter, we can barely put a serious scratch on the surface.

11.1 THE STRUCTURE AND COMPOSITION OF THE GOLGI COMPLEX

The cellular organelle named after Golgi has taken on a number of auxiliary names over the years, including "apparatus," "body," "complex," "vesicle," and "material." A very common designation is "apparatus." But this term seems more applicable to a piece of farm machinery or a satellite communications system than to a delicate and highly versatile membrane region in the cell. For reasons that are no better than these subjective connotations, except that it is truly a complex structure, we will consistently use the name *Golgi complex*.

Morphology

The ordinary light microscope is nearly useless in defining the ultrastructure of the Golgi complex except to reveal the important fact that it is extremely pleomorphic. In some cell types it appears compact and limited, in others spread out and reticular (netlike). It may occupy positions that range from circumnuclear to central, to apical or to dispersed, and may occur in numbers that range from one to hundreds per cell. It is no wonder that the complex catalyzed a controversy.

Method I

Phase contrast microscopy on *living cells* reveals distinguishable Golgi regions. This technique was useful in providing evidence in support of the Golgi concept because regions in live cells could hardly be artifacts of fixation or staining. Similarly, polarization microscopy picks up a positive birefringence in these regions. Beyond that, light microscopy is of little help.

The electron microscope has firmly substantiated the existence of the Golgi complex as a bona fide region of specialized membranes. But it has also complicated the picture somewhat by increasing the number of observable forms that the structure can assume in different cells. Figure 11.2 and 11.3 depict two commonly seen forms, one loosely dispersed in the cell and the other more compact. A common feature is seen, however, even in these two highly divergent examples: a stack of flattened membrane sacks. This stack represents a core unit of the complex in most cell types on which is built a number of peripheral structural variations.

Nomenclature of the Complex

The idealized drawing of the Golgi complex presented in Figure 11.4 is a composite of features possessed by the organelle in a variety of cells. Each flattened sack is called a *cisterna, saccule,* or *lamella*. Each saccule tends to be disc shaped and made up of a smooth 7.5-nm unit membrane within which there is a 15-nm-wide space, the *lumen*. A stack of saccules is sometimes referred to as a *dictyosome*. This term is often applied to the entire Golgi complex in plant cells, where this layered structure is often more evident. The individual saccules in a dictyosome are spaced by a 20-nm gap. A complex array of associated vesicles and anastomosing

To anastomose means to communicate with one another.

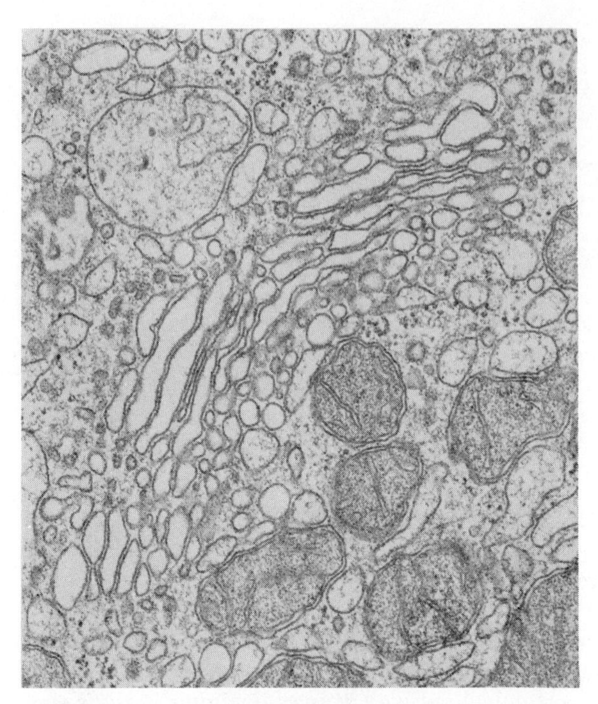

FIGURE 11.2 Electron micrograph of loosely organized Golgi complex in rat epididymis tissue. Several mitochondria are seen in the lower right quarter, with the Golgi membranes spread out diagonally from lower left to upper right. ×32,200. (Courtesy of Dr. Daniel Friend.)

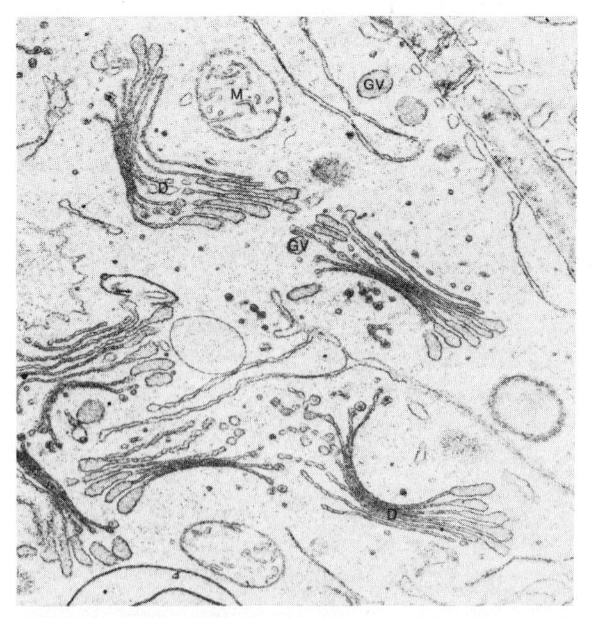

FIGURE 11.3 Electron micrograph of tightly organized Golgi complexes in the root cap cells of corn. Layered complexes of this sort, sometimes referred to as dictyosomes (D), are more common in plants than in animals. In this case the complex appears to be generating membrane-bound vesicles (GV). (Micrograph by Dr. W. Gordon Whaley.)

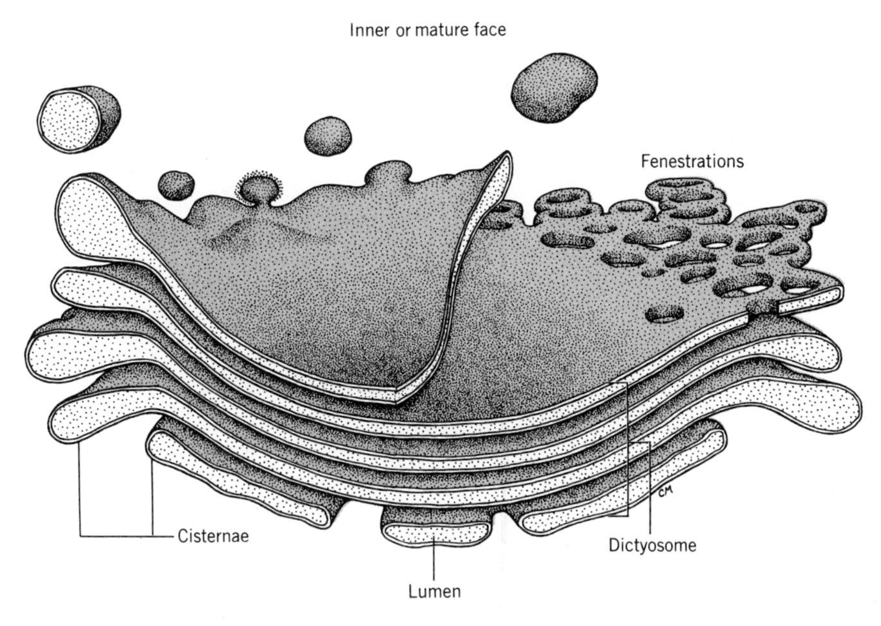

FIGURE 11.4 Idealized version of the Golgi complex.

tubules surrounds the dictyosome and radiates from it. These peripheral structures show up as a severe membrane discontinuity in the vicinity of the dictyosome (note Figure 11.3).

Studies on freeze-etched preparations of cells and thin sections made parallel to the saccule face have shown that this region of apparent discontinuity is actually fenestrated (lacelike) in structure. Thus there is overall structural continuity in a saccule, but it may be disrupted by a number of openings. Fenestrations may penetrate across an entire saccule; however, this generally occurs only in saccules that are at one of the outer faces of the stack.

Method I

Structural and Functional Polarity of the Golgi Complex

The Golgi complex frequently exhibits a morphological polarity that has the appearance of stacked cups or saucers. The curvature of the dictyosome in this form may be quite severe or fairly subtle, depending on the species and cell type. These morphological traits are directly related to a functional polarity of the organelle. Several examples of this will be given as the chapter develops.

In secretory cells, the Golgi complex constitutes a central functional link in the *secretory chain* or *pathway*. The linking elements of this chain are the endoplasmic reticulum, the Golgi complex, the secretory granules, and the plasma membrane. Figure 11.5 illustrates the position of the Golgi complex and its polar orientation in this pathway. During secretion there is a flow of material in the direction indicated by the vertical arrow. The convex face of the Golgi complex, which is oriented toward the rough endoplasmic reticulum is called the *outer* face or *forming* face. The concave face, called the *inner* face or *mature* face, is oriented toward the

The secretory pathway has the following links.

Endoplasmic reticulum
↓
Golgi complex
↓
Secretory granules
↓
Plasma membrane
↓
Cell exterior

See Chapter 10 for more on this pathway.

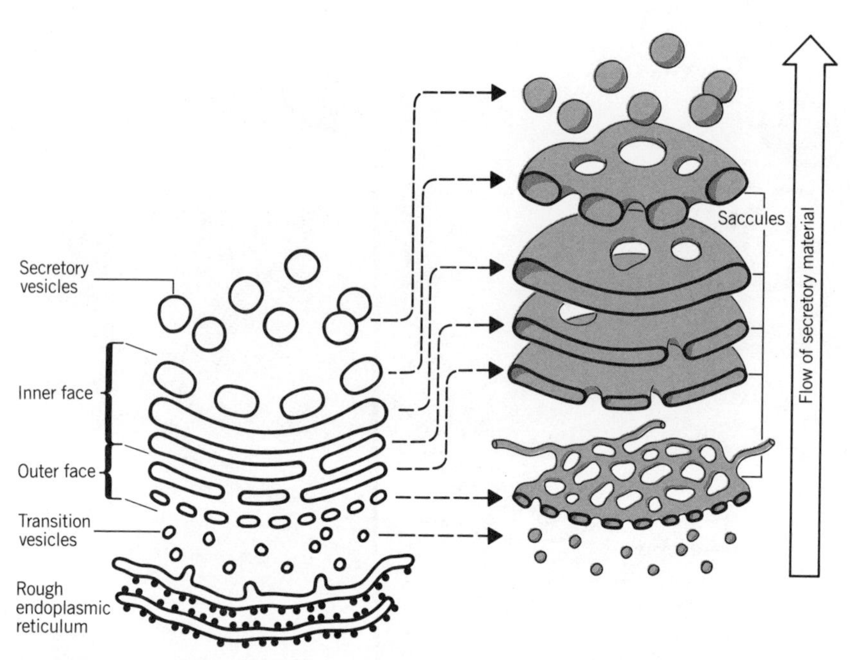

FIGURE 11.5 The position and orientation of the Golgi complex in the secretory pathway. *Left:* The components as they might be seen by electron microscopy of thin sections; *right*: a three-dimensional reconstruction.

cell surface. Secretory vesicles appear to arise from the mature face. Small membrane-limited vesicles called *transition vesicles* are frequently seen between the rough endoplasmic reticulum and the forming face of the Golgi complex.

During secretion, material formed in the endoplasmic reticulum moves through the Golgi complex from forming to mature face, then to the secretory granule in the apical region of the cell, and finally to the plasma membrane where it is expelled from the cell. The function of the Golgi complex in this pathway and its associated polarity is discussed in more detail later in the chapter. But it is important to keep this pathway and the cellular dynamics it requires in mind as we continue to discuss morphology and composition.

Cytochemical Properties

Once the Golgi complex had been validated from a morphological point of view, a number of investigators made attempts to distinguish it chemically from the rest of the cell. The fact that the complex has a greater affinity for *la reazione nera* than the rest of the cell is immediate cytochemical evidence that there is some degree of compositional individuality about the organelle.

Osmium tetroxide (OsO_4) adheres well to lipids, especially phospholipids and unsaturated fats.

Other cytochemical techniques have substantiated this individuality and, in addition, have revealed some internal differentiation *within* the complex. Osmium, for example, selectively impregnates the outer face of the complex. In this capacity it is a useful stain to distinguish the complex from the rest of the internal membranes of the cell as well as to pinpoint the outer face (see Figure 11.6).

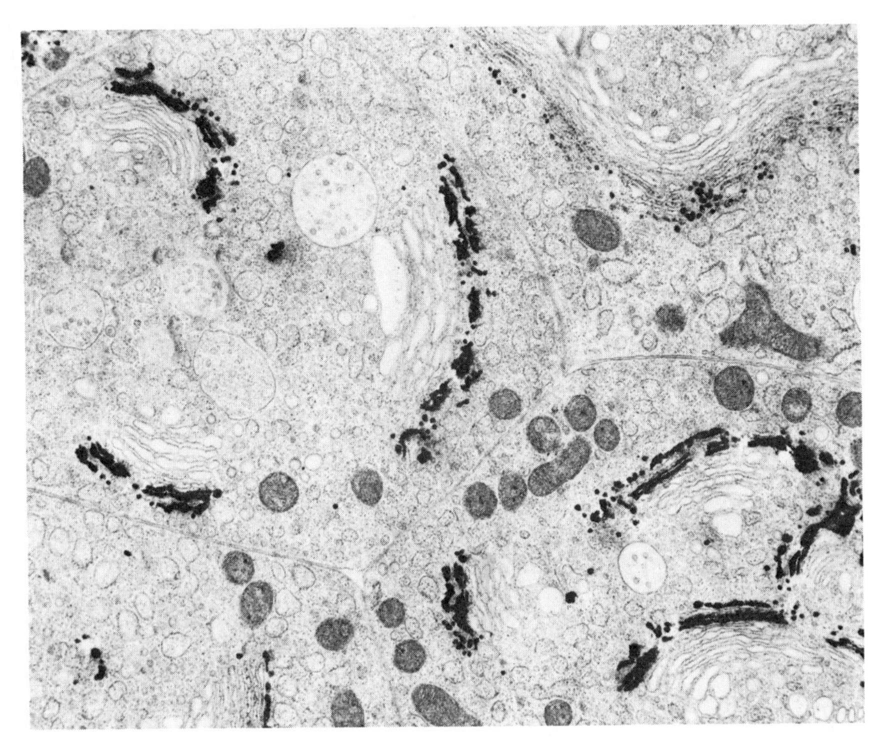

FIGURE 11.6 Mouse epididymis impregnated with reduced osmium tetroxide. The outer face of each Golgi complex contains electron-dense material; the inner face is electron transparent. (Courtesy of Dr. Daniel S. Friend.)

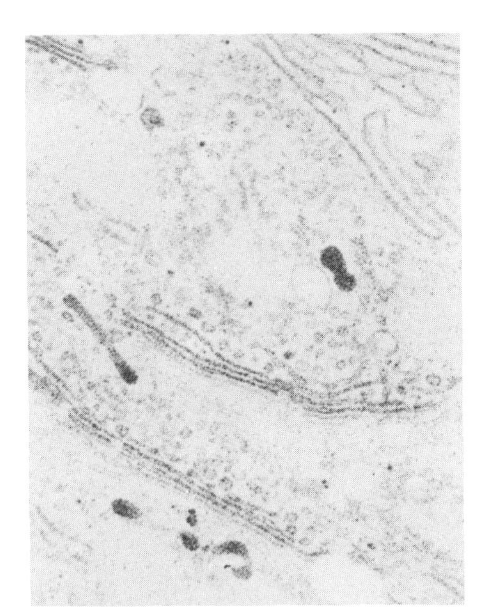

FIGURE 11.7 Electron micrograph demonstrating carbohydrate in saccules of the Golgi complex. In contrast to the results in Figure 11.6, the *inner* face contains the highest level of electron-dense material. (Courtesy of Dr. C. P. Leblond.)

The inner face can be selectively stained by cytochemical reactions for polysaccharides either by unmasking 1,2-glycol bonds or by detecting acidic carboxyl or sulfate functional groups with appropriate stains. When these techniques are used, the Golgi complex appears to contain polysaccharide in the lumen of the saccules. This is shown quite clearly in Figure 11.7. A gradient can also be seen *across the complex,* with the highest concentration of polysaccharide at the mature face. This has been observed in a number of different cell types that are involved in secretory activities. The gradient, as we shall see, is reflective of the function of the complex. Phosphotungstic acid also selectively stains the mature face saccules.

1,2-Glycol bonds in carbohydrates can be "unmasked" by oxidation with periodic acid. Aldehyde groups will result that are reactive with Schiff's reagent.

Certain enzymes, such as the transferases, can be localized cytochemically in the Golgi. Some of these enzymes also show a gradient of distribution from one face to the other. But it is not always clear whether the enzyme activities are confined to the Golgi membranes or occur inside the saccule lumen.

Phosphotungstic acid ($H_3PO_4 \cdot 12WO_3 \cdot 24H_2O$) is an anionic stain with a special affinity for polysaccharides and proteins.

Isolation

To probe questions of enzyme activity and others more thoroughly, it is necessary to isolate the organelle from other cellular debris, a task that was not accomplished with good success until the mid-1960s. Differential centrifugation combined with density gradient techniques were the most useful tools of separation. A flow chart for a typical procedure is presented

Method III

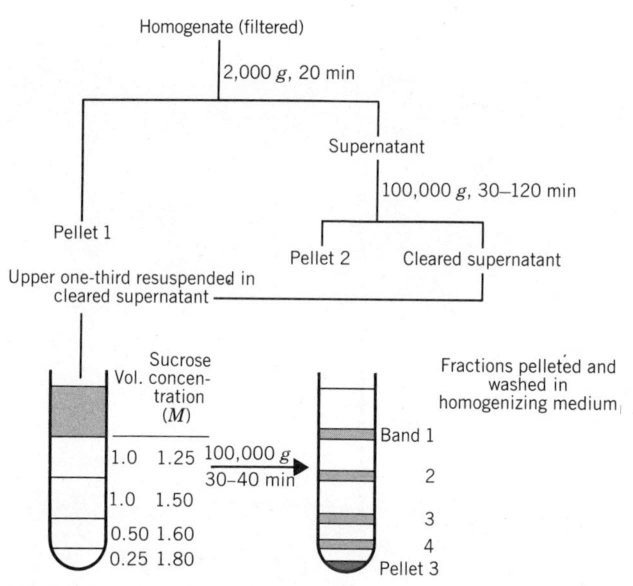

FIGURE 11.8 Flow chart outlining the steps in the isolation of Golgi material. The Golgi-rich fraction separates from other material of the homogenate as band 1.

in Figure 11.8. The Golgi-rich fraction ends up as the band of lowest density in a sucrose gradient after centrifugation for 30 to 40 min at 100,000*g* (see band 1).

Plants proved to be the best source of the Golgi complex in the early studies, at least partly because of their easily recognized dictyosomes which can be tracked in the enriched fractions with electron microscopy. Working with plant systems, Morré and his associates[3] developed a procedure that has been applied successfully to other types of tissue as well, such as liver. This procedure, adapted to rat liver, is the one illustrated in Figure 11.8. At best, these preparations of Golgi should be viewed as *enriched* Golgi fractions, since it is not clear on either morphological or chemical grounds that they constitute ''pure' cellular organelles.

An example of the degree of purity obtainable by cell fractionation techniques is shown in the electron micrographs of Figure 11.9. This series presents four membrane fractions: rough endoplasmic reticulum, smooth endoplasmic reticulum, the Golgi fraction, and plasma membranes. It is apparent that each fraction has distinctive morphological characteristics. However, spillover between fractions is also noticeable, a property of the isolation procedure that makes it essentially impossible to obtain pure organelles.

The Golgi complex fraction has the expected parallel saccules and tubular and spherical vesicles, but it also has other cell contaminants, such as fragments of the endoplasmic reticulum. This type of spillover appears because homogenization, even if gentle, breaks up the membranes of the endoplasmic reticulum into fragments and fused vesicles that morphologically resemble the components of the Golgi complex. They cannot be separated from one another on the basis of density differences.

The presence of intact dictyosomes in isolated fractions is possible only if the homogenization procedure is *extremely* gentle. Standard homoge-

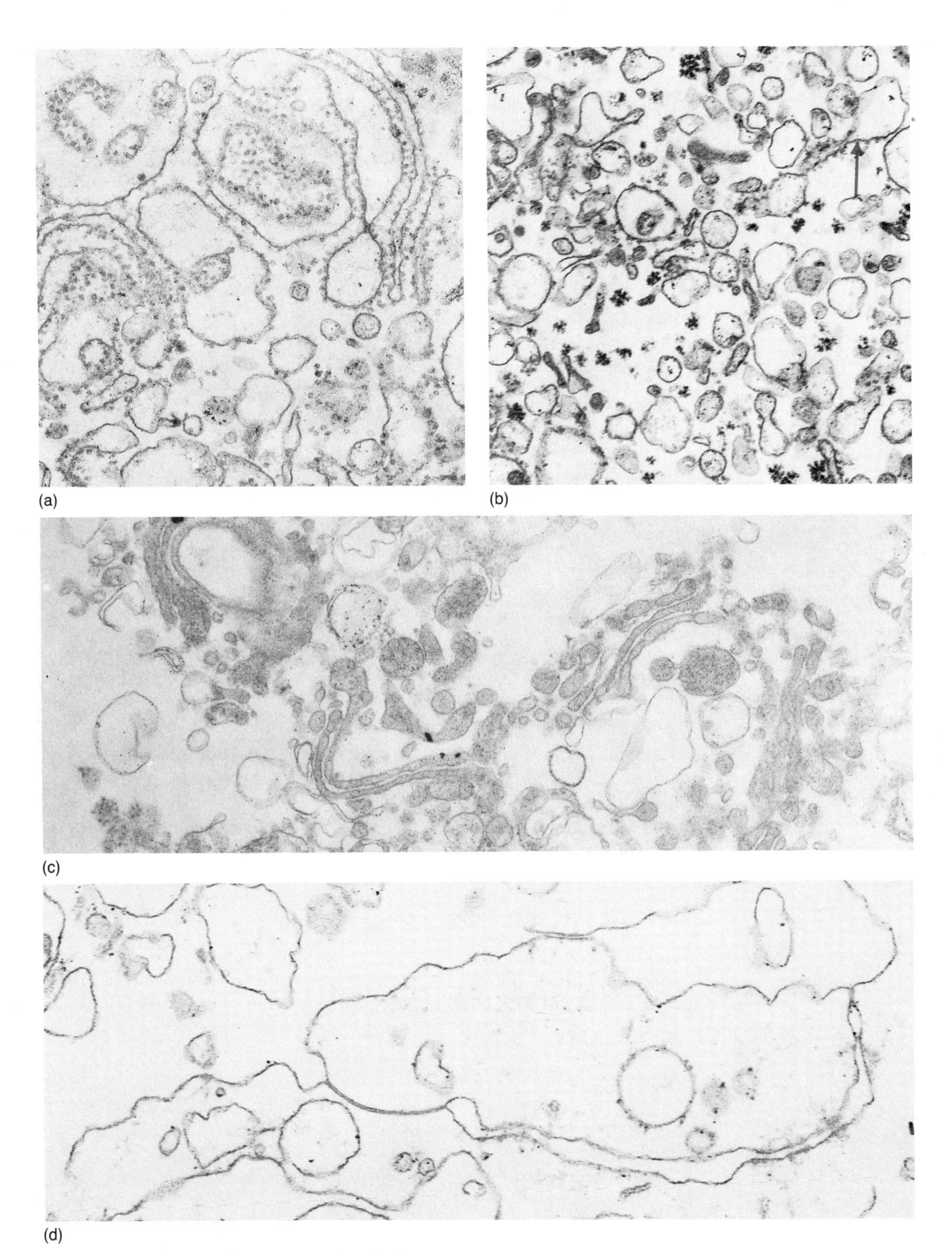

FIGURE 11.9 Membrane fractions showing their distinct morphological traits. (*a*) Rough endoplasmic reticulum. Ribosomes are attached to membrane surfaces. (*b*) Smooth endoplasmic reticulum. Thin (~6 nm) membranes lack ribosomes. In some places RER and SER are joined (arrow). (*c*) Golgi complex enriched fraction. (*d*) Plasma membrane fraction, consisting of thick (~10 nm) membranes in large vesicles or sheets. (Courtesy of Dr. D. James Morré.)

Table 11.1 Estimates of Fraction Purity of Isolated Cell Fractions from Rat Liver Based on Analyses of Marker Enzyme Activities

Enzyme Activity[a]	MITO	ER	GC	PM
Succinate dehydrogenase	25	0.2	0.2	1.3
Monoamine oxidase	33	—	0.5	—
Glucose-6-phosphatase	—	34.0	3.0	1.3
Galactosyl transferase	—	0.003	0.22	Trace
5′-Nucleotidase (AMP)	—	0.1	2.5	77.5
Contamination (%)[b]				
Mitochondria		1	1	5
Endoplasmic reticulum		—	9	4
Golgi complex		1	—	0
Plasma membrane		1	3	—
Total		3	13	9
Fraction purity (%)		97	87	91

[a]Units are micromoles of product formed or substrate consumed per hour per milligram of protein. MITO, mitochondria; ER, endoplasmic reticulum; GC, Golgi complex; PM, plasma membrane.

[b]In calculating percentage contamination, each of the reference fractions is assumed to be of absolute purity. The percentages contamination of Golgi complex by endoplasmic reticulum and plasma membrane are maximum values, since some endoplasmic reticulum and plasma membrane activities may be indigenous to the fraction.

The effect of homogenization on dictyosome structure.

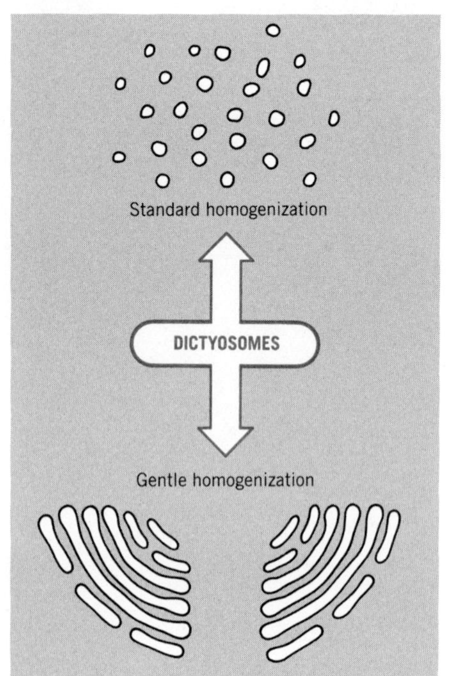

nization techniques are severe enough to completely disrupt the membranes of the dictyosomes. This results in a Golgi fraction that contains small membranous vesicles that are very hard to distinguish from smooth microsomes.

In spite of these difficulties, a remarkable degree of purification can be achieved. The fractions illustrated in the micrographs are approximately 90% pure according to morphological and biochemical criteria. Table 11.1 lists the results of examining these fractions for various enzyme markers. According to these biochemical criteria, the Golgi complex is 87% pure.

Enzymes of the Golgi Complex

Enzymatic analyses of isolated Golgi complexes have turned up a heterogeneous array of enzymes, many of which can be found in other parts of the cell as well. Table 11.2 gives enzymes found in the Golgi complexes of animal cells. Although the transferases seem to be predominant, enzymes of several other kinds are present in the tissues studied.

Many investigators have tried to find a marker enzyme that is unique to the Golgi complex so that the organelle can be identified and its purity assessed by means more sensitive and discriminatory than simply membrane morphology.

One approach to monitoring purity is to follow a decrease in certain enzyme activities normally associated with membranes other than the Golgi as the Golgi fraction is enriched. For example, a decrease of 5′-nucleotidase activity during fractionation would mean that the material isolated was being separated from contaminating amounts of plasma membranes. This approach has its pitfalls, since there can be a loss of enzyme activity during purification without a parallel loss of enzymes. The enzymes may simply be denatured or inactivated. The measurement of en-

Table 11.2 Enzymes of the Golgi Complex of Animal Cells[a]

Glycosyltransferases: Glycoprotein Biosynthesis
Sialyltransferase
UDP-Galactose: *N*-acetylglucosamine galactosyltransferase
Glycoprotein β-galactosyltransferase
UDP-*N*-Acetylglucosamine-glycoprotein *N*-acetylglucosaminyltransferase
Sulfo- and Glycosyltransferases: Glycolipid Biosynthesis
Galactocerebroside sulfotransferase
CMP-NANA: lactosylceramide sialyltransferase
CMP-NANA: GM_1 sialyltransferase
CMP-NANA: GM_3 sialyltransferase
UDP-Galactose: GM_2 galactosyltransferase
UDP-GalNAc: GM_3 *N*-acetylglucosaminyltransferase
Oxidoreductases
NADH-cytochrome *c* reductase
NADPH-cytochrome reductase
Phosphatases
5′-Nucleotidase
Adenosine triphosphatase
Thiamine pyrophosphatase
Kinases
Casein phosphokinase
Mannosidases
α-Mannosidase
Transferases: Phospholipid Synthesis
Lysolecithin acyltransferase
Glycerolphosphate phosphatidyltransferase
Phospholipases
Phospholipase A_1
Phospholipase A_2

[a]Enzymes listed in order according to Enzyme Commission number.

zyme activity alone would therefore lead to the wrong conclusions regarding the purity of the fractions.

Table 11.3 lists for six enzymes relative distributions in three types of cell membrane: rough microsomes, the Golgi fraction, and plasma membranes. In this case, the distributions are based on enzyme activities. The Golgi fraction has an intermediate position with respect to the activities of several of the enzymes. It is most clearly distinguished from the plasma membrane by its lower 5′-nucleotidase activity and from the endoplasmic reticulum by its lower glucose-6-phosphatase activity.

It is beginning to appear that the glycosyltransferases are associated primarily with the Golgi complex. These enzymes catalyze the transfer of sugars from UDP carriers to appropriate protein acceptor molecules. A number of workers have found about half the total glycosyltransferase activity of the cell in Golgi-enriched fractions, while at the same time the endoplasmic reticulum and plasma membrane fractions appear to be devoid of such activity. Indeed, it appears as though glycosyltransferase activity fulfills all the criteria associated with a true marker enzyme for the Golgi complex, at least in rat liver, where this property has been most vigorously pursued.

Glycosyltransferases catalyze the following reaction, using glucose as an example.

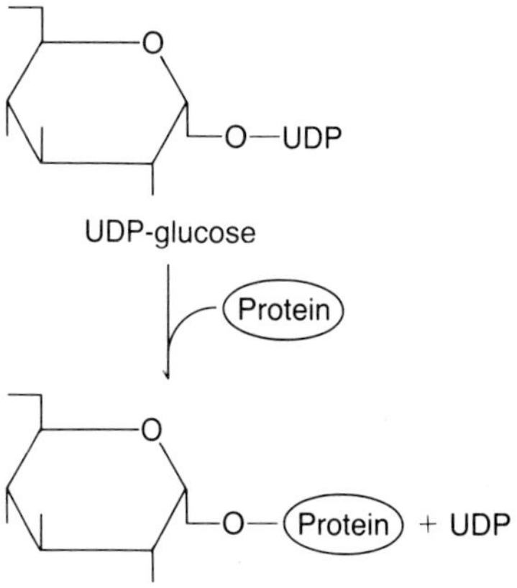

Glycosyltransferases are the marker enzymes of the Golgi complex.

Table 11.3 Enzyme Profiles of Rough Microsomes, Golgi Fraction, and Plasma Membranes of Bovine Liver[a]

Enzyme	Rough Microsomes	Golgi-rich Fraction	Plasma Membrane
TPPase	0.17	0.25	0.20
Acid phosphatase	0.026	0.022	0.016
5′-Nucleotidase	0.029	0.16	0.81
Glucose-6-phosphatase	0.30	0.023	0.048
NADH-Cytochrome *c* reductase	2.3	1.4	0.01
NADPH-Cytochrome *c* reductase	0.089	0.094	0.055

[a]All phosphatases are expressed as micromoles of P_i released per minute per milligram of protein. Cytochrome *c* reductases are expressed as micromoles of cytochrome *c* reduced per minute per milligram of protein.

Table 11.4 UDP-Galactose:*N*-Acetylglucosamine Galactosyltransferase Activity in Cell Fractions Isolated from the Liver of Ethanol-Treated Rats

Fraction	Protein (mg/g liver)[a]	Specific Activity[b]	Recovery (%)
Homogenate	195	1.7	100
Nuclear/mitochondrial fraction	97	1.1	32
Microsomal fraction 1[c] (initial)	26	9.0	70.7
Golgi fractions	1.5	170	71.0
Microsomal fraction 2[d] (residual)	20	0	0

[a]Wet weight.
[b]Nanomoles of galactose transferred per hour per milligram of protein.
[c]Microsomal fraction before the removal of Golgi fractions by flotation.
[d]Microsomal fraction left after the removal of Golgi fractions.

Table 11.4 shows the high level of transferase activity typical of the Golgi fraction compared to other fractions of rat liver cells. The transferase studied in this case is the one that catalyzes the transfer of galactose from UDP-galactose to *N*-acetylglucosamine. It appears that this transferase, as well as several others that have been studied, is located in the membranes of the Golgi, not in the lumen of the saccules.

The presence of specific transferase activity will undoubtedly continue to be one of the most powerful techniques for discriminating between the Golgi complex and other cellular membranes.

Sphingomyelin

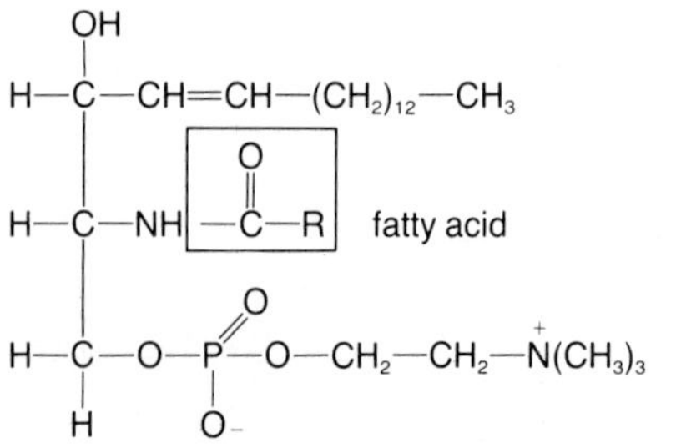

OH
H—C—CH=CH—$(CH_2)_{12}$—CH_3
O
H—C—NH —C—R fatty acid
O
H—C—O—P—O—CH_2—CH_2—$\overset{+}{N}(CH_3)_3$
H O⁻

Lipid Contents

The membranes of the Golgi complex also differ somewhat from other cell membranes in their lipid contents. Biochemical studies on phospholipid and fatty acid compositions of various cell fractions have shown the Golgi to be intermediate between the endoplasmic reticulum (or nuclear envelope) and the plasma membrane. This is illustrated nicely in Table 11.5.

Note in particular the content of sphingomyelin and phosphatidylcho-

Table 11.5 Lipid Composition of a Golgi-Rich Cell Fraction and Comparison with Other Cell Fractions from Rat Liver

	Percentage of Total Fraction		
Compound	plasmic Reticulum	Golgi Complex	Plasma Membrane
Total phospholipids[a]	84.9	53.9	61.9
Sphingomyelin[b]	3.7	12.3	18.9
Phosphatidylcholine	60.9	45.3	39.9
Phosphatidylserine	3.3	4.2	3.5
Phosphatidylinositol	8.9	8.7	7.5
Phosphatidylethanolamine	18.6	17.0	17.8
Lysophosphatidylcholine	4.7	5.9	6.7
Lysophosphatidylethanolamine	[c]	6.3	5.7
Total neutral lipids[d]	15.1	46.1	38.1
Cholesterol	24.6	16.5	34.5
Free fatty acids	40.6	38.9	35.1
Triglycerides	24.7	35.1	22.4
Cholesterol esters	10.1	9.6	8.0

[a]Mean value of three determinations, phospholipid = phosphorus value × 25.
[b]Individual phospholipids expressed as percentage of total lipid phosphorus. Values are mean values.
[c]Not detected.
[d]Mean value of three to four determinations. Individual neutral lipids expressed as percentage of total neutral lipid fraction.

line, which places the Golgi between the values obtained for the endoplasmic reticulum and the plasma membrane. Although these results can be viewed as evidence for the chemical distinctiveness of the Golgi complex, these data have been used more commonly as biochemical evidence for membrane flow in the cell. We discuss this topic in more detail later in the chapter.

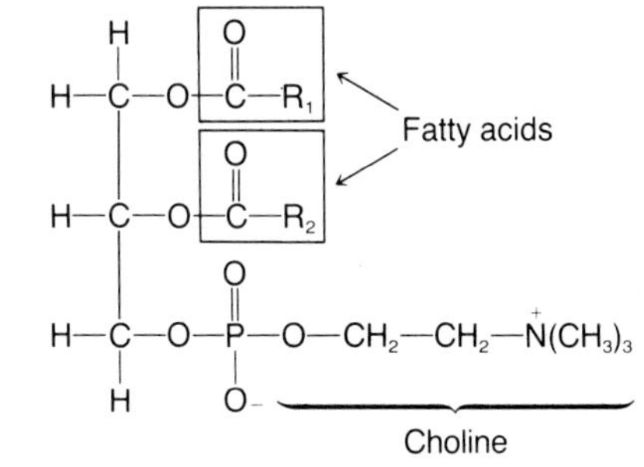

In summary, the Golgi complex has both distinctive and intermediate features compared to other organelles of the cell. It has a distinctive morphology, can be selectively stained, and contains certain enzymes that are not present at significant levels in other parts of the cell. These properties set it apart as a distinguishable organelle. It is intermediate between the endoplasmic reticulum and the plasma membrane in its content of lipid and certain other enzyme markers. In this respect it seems to be an organelle in transition between the other two systems.

Phosphatidylcholine

Fatty acids

$H-C-O-P-O-CH_2-CH_2-\overset{+}{N}(CH_3)_3$

Choline

11.2 THE FUNCTION OF THE GOLGI COMPLEX

The Golgi complex plays an indispensable role in a number of cell processes during which its general pattern of function is to act as a modifying center. This is especially evident in the secretory pathway, where the role of the Golgi has been most intensely studied. It is becoming apparent, however, that this pleomorphic organelle may perform several seemingly unrelated but basic cell functions.

In the material that follows we shall consider a few of the functions that have been the focus of a good deal of study.

Table 11.6 The Biosynthesis of Animal Glycoproteins

Ultimate fate of Glycoprotein	Organ of Origin	Glycoprotein Product	Carbohydrate–Protein Linkage Type[a]
Endocrine secretion	Liver	Plasma glycoproteins	Asn-GlcNAc
	Thyroid	Thyroglobulin	Asn-GlcNAc
	Plasma cells	IgG	Asn-GlcNAc
Exocrine secretion	Bovine pancreas	Ribonuclease B; DNase	Asn-GlcNAc
	Hen oviduct	Ovalbumin	Asn-GlcNAc
Exocrine secretion	Submaxillary gland	Submaxillary mucin	Ser(Thr)-GalNAc
	Mucous glands	ABO–Lewis blood group glycoproteins	Ser(Thr)-GalNAc
Intercellular ground substance (a "secretion")	Connective tissue cells	Collagen; basement membrane	Hyl-Gal
		Chondroitin sulfate	Ser-Xyl
Cellular membranes	All cells	Membrane-bound glycoproteins	Asn-GlcNAc Ser(Thr)-GalNAc Others?

[a]Asn, asparagine; Ser, serine; Thr, threonine; Hyl, hydroxylysine; GlcNAc, *N*-acetyl-D-glucosamine; GalNAc, *N*-acetyl-D-galactosamine; Gal, D-galactose; Xyl, D-xylose.

Common carbohydrates of glycoproteins and glycolipids.

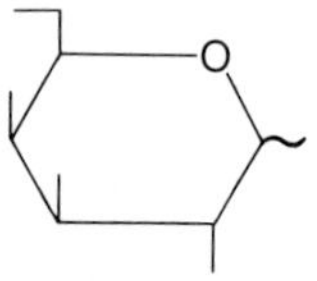

Galactose

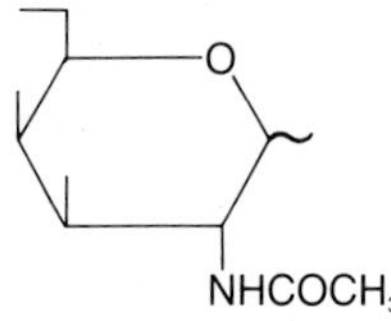

$NHCOCH_3$

N-Acetyl-D-galactosamine

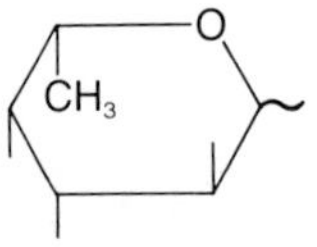

CH_3

Fucose

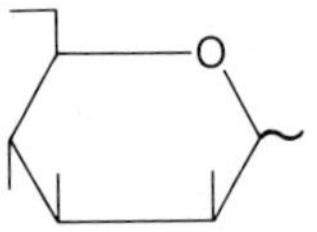

Mannose

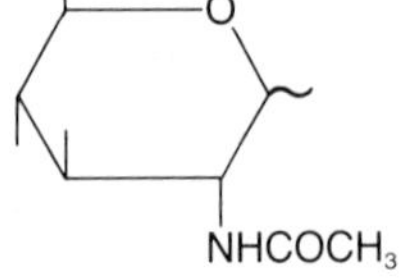

$NHCOCH_3$

N-Acetyl-D-glucosamine

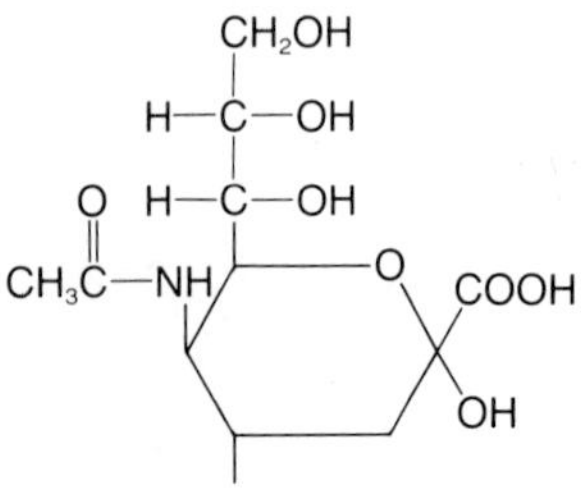

CH_2OH, H—C—OH, H—C—OH, $CH_3C(=O)$—NH, COOH, OH

N-Acetylneuraminic acid

The Assembly of Carbohydrate-Rich Macromolecules in Animals

A large number of experiments have demonstrated that the Golgi complex functions in the biosynthesis of glycoproteins and glycolipids. Glycoproteins are those proteins that contain covalently linked carbohydrates, commonly D-galactose, D-mannose, L-fucose, *N*-acetyl-D-glucosamine, *N*-acetyl-D-galactosamine, and *N*-acetylneuraminic acid (sialic acid). These monosaccharide units are linked in oligosaccharide chains that are normally quite short. One form of linkage is a covalent *N*-acetylglucosamine–asparagine bridge.[4] In other cases the carbohydrate–protein bridge is between serine, threonine, hydroxylysine, and certain other sugars.

Several types of glycoprotein, their tissue sources, and the nature of their respective carbohydrate–protein bridges are listed in Table 11.6. It is of interest that the *types* of carbohydrate–protein linkage are few, even though the glycoproteins vary widely with respect to structures and organ sources. This suggests a priori that the glycoproteins may be assembled by common mechanisms.

Glycolipids are complex molecules often associated with membranes and cell surfaces. They are made up of lipid cores to which various carbohydrates are covalently linked.

Figure 11.10 illustrates some typical carbohydrate–protein and carbohydrate–lipid linkages in membrane glycoproteins and glycolipids. The role of the Golgi in forming these membrane ingredients is to condense monosaccharide units onto the growing oligosaccharide chains through

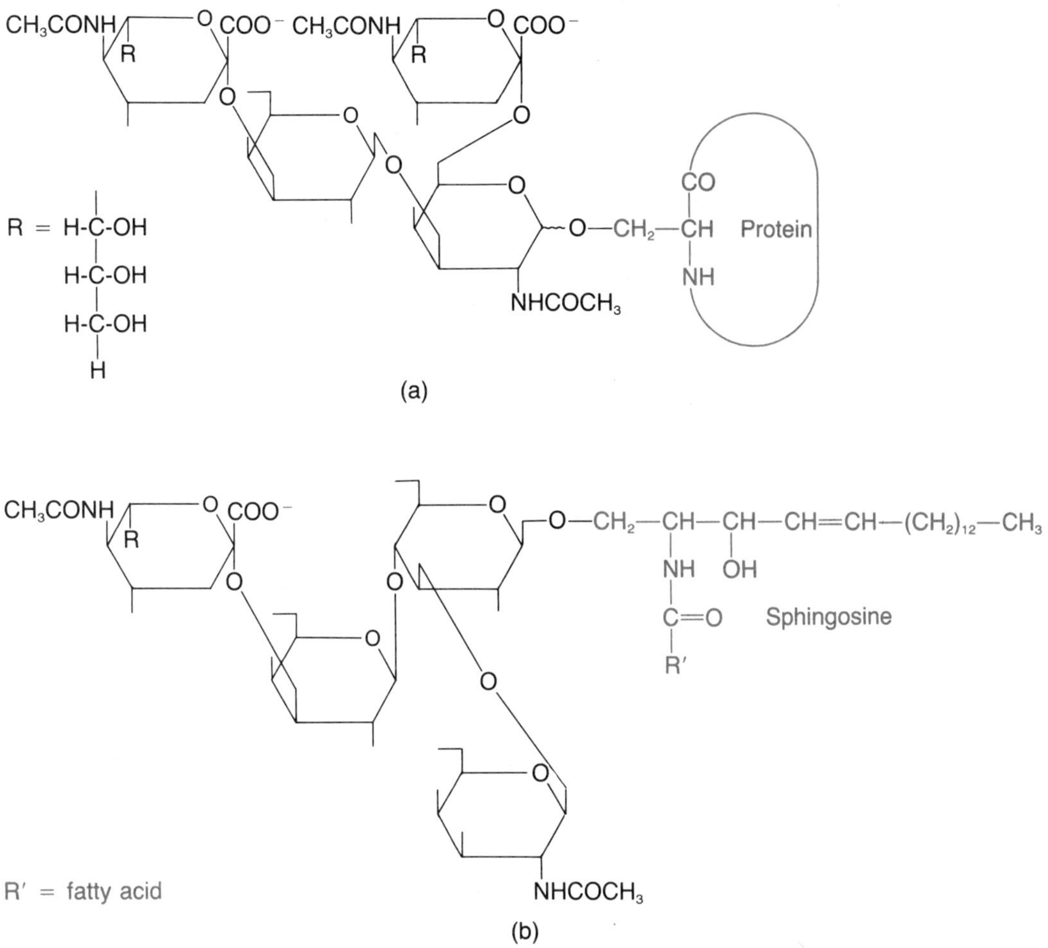

CH_3CONH O COO^- CH_3CONH O COO^-
R R
O O
O O
O CO
R = H-C-OH O—CH_2—CH Protein
H-C-OH NH
H-C-OH $NHCOCH_3$
H

(a)

CH_3CONH O COO^- O —O—CH_2—CH—CH—CH=CH—$(CH_2)_{12}$—CH_3
R
O O NH OH
O C=O Sphingosine
O R′
O

R′ = fatty acid $NHCOCH_3$

(b)

FIGURE 11.10 Typical carbohydrate–protein and carbohydrate–lipid linkages found in membrane molecules. (*a*) A glycoprotein containing a serine-carbohydrate linkage. (*b*) A ganglioside, made up of carbohydrate linked to sphingosine.

the action of the glycosyltransferases on appropriate protein and lipid substrates.

We now turn to a few examples of cell events in which the Golgi complex participates in the synthesis of carbohydrate-rich molecules.

Secretions of the Thyroid Gland

One role of the Golgi complex in glycoprotein biosynthesis is seen quite clearly in thyroid follicle cells. These cells are engaged in the synthesis of thyroglobulin, a glycoprotein that is secreted by the cell into the follicle, where it is iodinated and stored. Figure 11.11 shows a cross section of the thyroid gland. The follicles are the large intercellular spaces, and the epithelial cells surrounding the follicles are those concerned with the synthesis of thyroglobulin.

Thyroglobulin is the protein that carries thyroxine, the hormone.

Thyroglobulin contains two types of polysaccharide chain. One has only the disaccharide of *N*-acetylglucosamine attached to mannose. The second type is more complex, containing in addition to the sugars above,

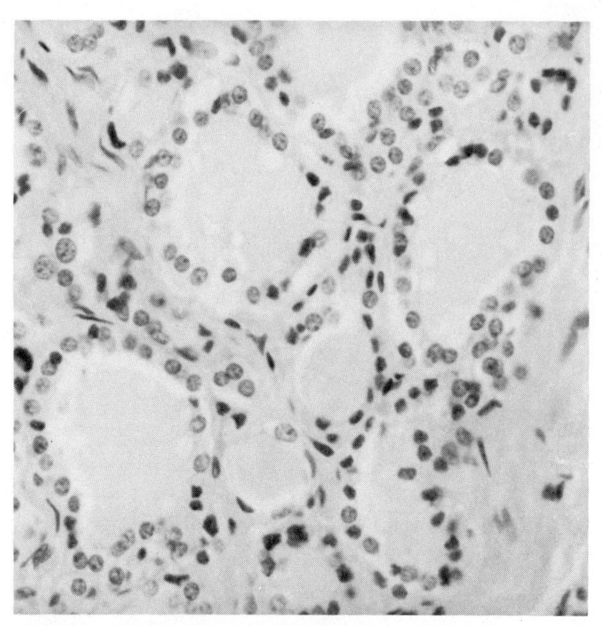

FIGURE 11.11 Cross section of the thyroid gland showing epithelial cells surrounding thyroglobulin-containing follicles.

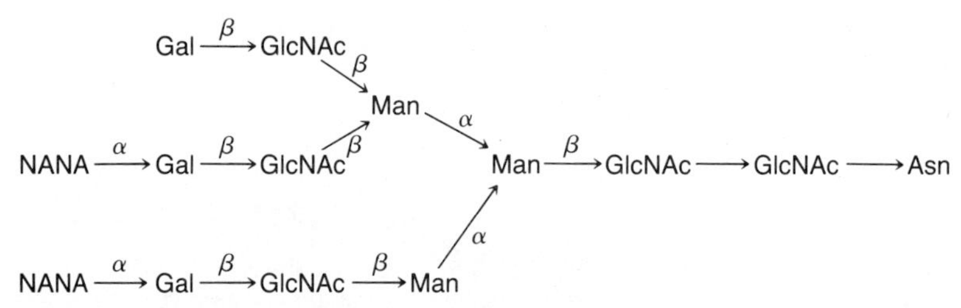

FIGURE 11.12 One type of polysaccharide side chain possessed by thyroglobulin. Asn, asparagine; GlcNAc, *N*-acetylglucosamine; Man, mannose; Gal, galactose; NANA, sialic acid.

galactose and fucose or sialic acid. The more complex of these two structures is shown in Figure 11.12.

Method VI

Autoradiography has been used to monitor the incorporation with time of [^{3}H]leucine, [^{3}H]mannose, and [^{3}H]galactose into thyroglobulin.[5] When this is done, [^{3}H]leucine and [^{3}H]mannose appear first in the region of the rough endoplasmic reticulum (5–15 min after incorporation), and after 1 to 2 hr are transferred to the Golgi complex and eventually into the follicle. This movement is blocked by puromycin, an inhibitor of protein synthesis. These results are consistent with the expected synthesis of the polypeptide chains of thyroglobulin (the [^{3}H]leucine marker) on the ribosomes of the rough endoplasmic reticulum and suggest, in addition, that carbohydrates that have positions *close to the protein* on the oligosaccharide chain (the [^{3}H]mannose marker) are added by transferases operating in the channels of the rough endoplasmic reticulum.

When puromycin is present, the growing peptide is released before it is completed. Puromycin is attached covalently to the carboxyl end of the released peptide.

In contrast to this, [^{3}H]galactose is observed first in the Golgi complex region (within 10 min) and is moved from that point toward the follicle.

This movement is not blocked by puromycin. We may interpret these results to mean that once a protein has been made (and thus no longer subject to blockage by puromycin), it picks up carbohydrate units in the Golgi complex that are more distally positioned in the oligosaccharide than is mannose. [^{3}H]Fucose, when followed in a similar manner, is also taken up first in the Golgi complex.[6] Fucose, remember, is a terminal carbohydrate that may appear in the place of sialic acid.

Studies on the incorporation of *N*-acetylglucosamine into thyroglobulin have shown that it is added both in the endoplasmic reticulum and in the Golgi complex. This agrees with the known structure of the oligosaccharide because one glucosamine is in a proximal position and the other more distally located in the chain (see Figure 11.12).

Based on tracer studies like these, the sequence of events during the formation of thyroglobulin can be summarized according to Figure 11.13. Thyroglobulin is transported vectorially in the cell, commencing in the rough endoplasmic reticulum, where the polypeptide chain is formed and the most proximal monosaccharides are attached. The incomplete glycoprotein continues through the Golgi complex, where the monosaccharides, which are nearer the nonreducing termini of the oligosaccharides, are added. Secretory vesicles containing the completed glycoprotein are formed apically from the Golgi complex and are moved toward the lumen of the follicle, where the last step, iodination, takes place. The iodinated glycoprotein is then stored in the follicle until the epithelial cell is stimulated by TSH from the pituitary gland to reabsorb thyroglobulin, excise the iodinated thyroid hormone, and release the hormone to the blood, where it is carried to appropriate target tissues.

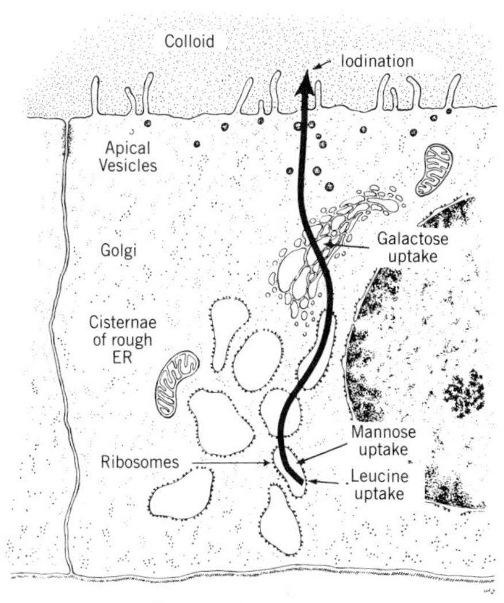

FIGURE 11.13 Sequential assembly of thyroglobulin. Mannose, a proximal monosaccharide on the polysaccharide side chain, is attached in the vicinity of the rough endoplasmic reticulum. Galactose, more distally located in the side chain, is added in the Golgi complex. (Courtesy of Dr. C. P. Leblond.)

TSH, or thyrotrophin, is a hormone given off by the pituitary gland; its target tissue is the thyroid gland.

Thyroid cells are thus involved in a cycle of activities that employs several organelles. Synthesis takes place in the endoplasmic reticulum, carbohydrate incorporation and packaging in the Golgi complex, transport and exocytosis on the apical plasma membrane, endocytosis of thyroglobulin at the same surface, release of the thyroid hormone in lysosomes (see Chapter 9, Section 6), and finally transport of the thyroid hormone across the basal membrane to the capillary lumen.

Mucigen Formation in Intestinal Goblet Cells

The goblet cells of the colon secrete large amounts of glycoprotein material into the lumen of the intestine. This material, referred to as *mucigen,* is high in carbohydrate, and because of its slippery physical properties has a lubricating function in assisting the passage of foods through the gastrointestinal tract. With a light microscope, the goblet cell can be spotted easily because of its hollowed-out appearance, due either to the content of mucigen taking up the bulk of the apical end of the cell or to the empty space left on evacuation of the mucigen granule into the lumen of the intestine. Figure 11.14 illustrates the appearance of the goblet cell with light microscopy.

One of the features of the goblet cell that makes it especially attractive for studies on the Golgi complex is that it is a highly polarized cell with well-defined basal and apical ends. The Golgi complex is normally located apically from the nucleus and is oriented in a cuplike manner, opened toward the secretory end of the cell. Three-dimensional and cross-sectional views of the goblet cell (Figure 11.15) illustrate the obvious polarity

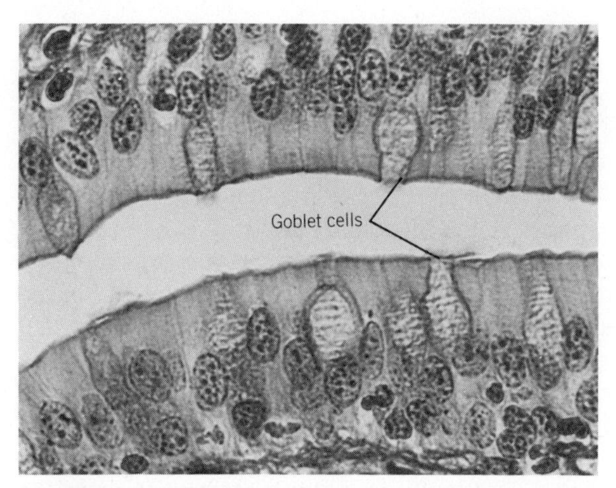

FIGURE 11.14 Photomicrograph of intestinal goblet cells. The enlarged region between the nucleus and the intestinal lumen contains mucigen. (Carolina Biological Supply Co.)

of the cell and the severely curved morphology of the Golgi.

According to a series of experiments roughly parallel to those described for the thyroid follicle cell, [^{3}H]glucose and [^{3}H]galactose, when administered to young rats and followed autoradiographically in intestinal tissue, are found first in the Golgi complex region of the goblet cell nearest the nucleus (5 min). Somewhat later (20 min), the label appears in the distal portions of the Golgi complex and in nearby mucigen granules, and finally (40 min) in mucigen granules alone. Similarly, sulfate is taken up in the Golgi complex before it appears in the polysaccharide of the completed mucigen granule.

Active sulfate is 3′-phosphoadenosine 5′-phosphosulfate.

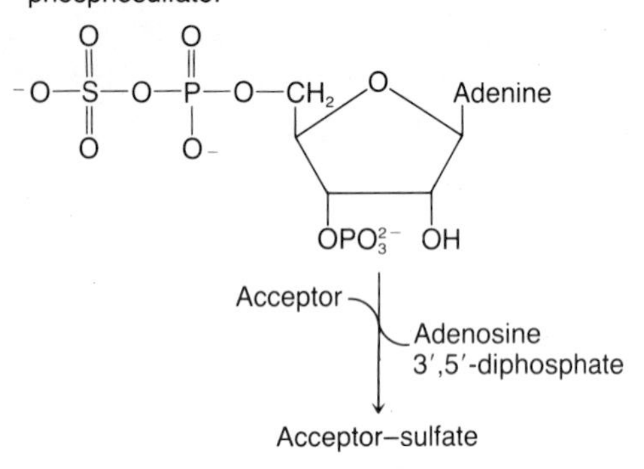

The interpretation of these events is shown schematically in Figure 11.16. Protein synthesized in the rough endoplasmic reticulum is channeled through the Golgi complex. In the Golgi it is derivatized with complex carbohydrates before moving on to the apical end of the cell.

The uptake of ^{35}S in the Golgi complex of the goblet cell probably reflects a general mechanism of sulfation reactions. This pattern of incorporation has been found in such widely divergent cell types as fibroblasts, ovarian follicular and interstitial cells, keratocytes, mast cells, myelinated and unmyelinated Schwann cells, myelocytes, and reticular cells lining the sinusoids, ameloblasts, osteoblasts, and chondrocytes.[7] Apparently, sulfate, activated by ATP, is transferred to the appropriate acceptor molecules by the sulfotransferases, which are known to be localized in the Golgi complex (see Table 11.2).

Glycoproteins of the Plasma Membrane

The assembly of the plasma membrane was discussed at length in Chapter 4. Now we simply underscore the role of the Golgi complex in this process.

Studies on the columnar cells of the intestinal epithelium suggest that plasma membrane glycoproteins and the secretory glycoproteins thyroglobulin and mucigen are formed along the same route.[8] Labeled protein precursors are found associated first with the rough endoplasmic reticulum and [^{3}H]galactose and [^{3}H]fucose are incorporated initially in the Golgi

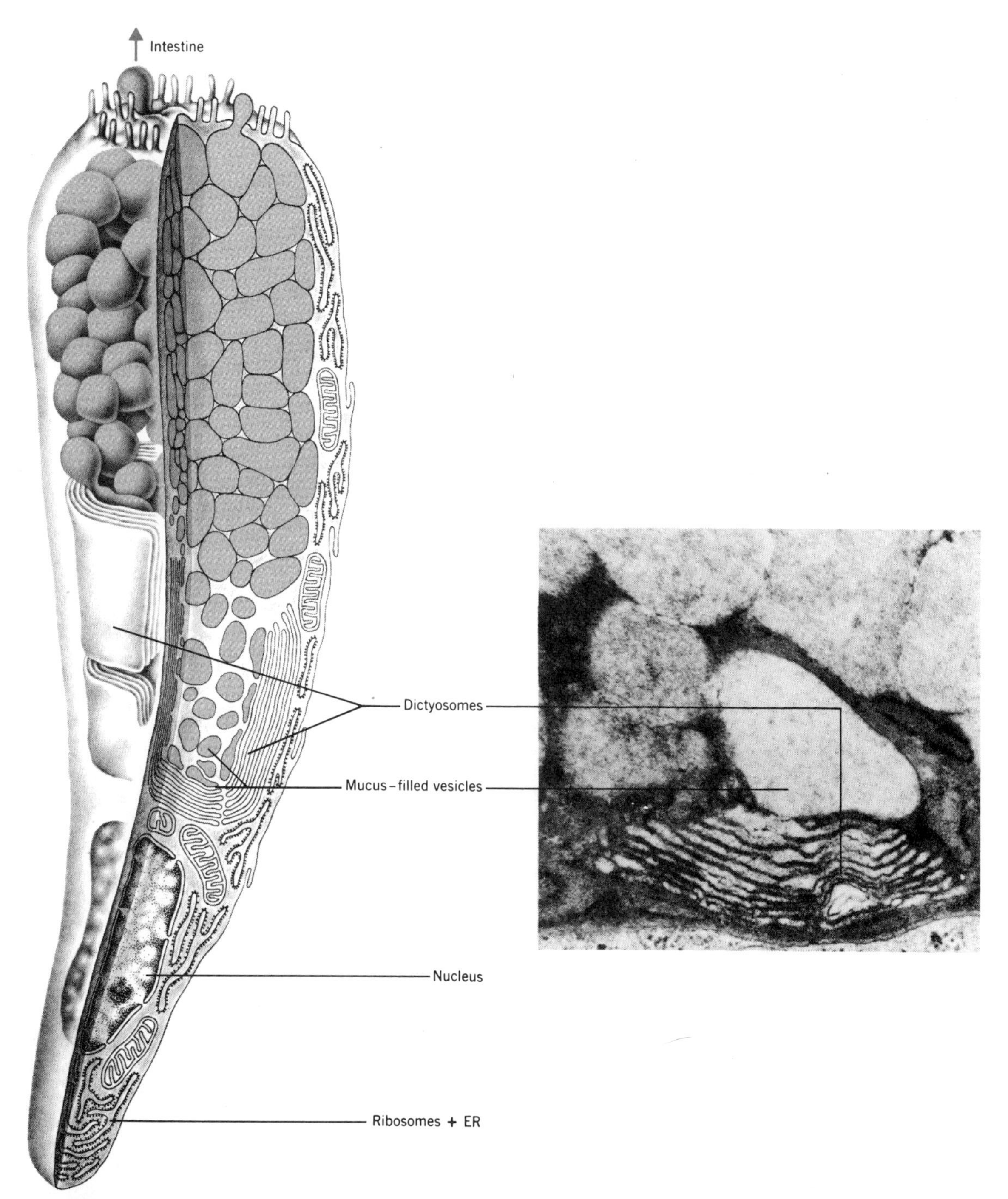

FIGURE 11.15 The goblet cell in three dimensions and in section. The cell is highly polarized and contains a severely curved Golgi complex. (Micrograph courtesy of Dr. E. G. Pollock.)

region. Then vesicles are derived from the mature face of the Golgi complex that transport the labeled glycoproteins to the apex of the cell.

There is a fundamental difference between this mechanism of glycoprotein processing and that of the secretory proteins. Since membrane

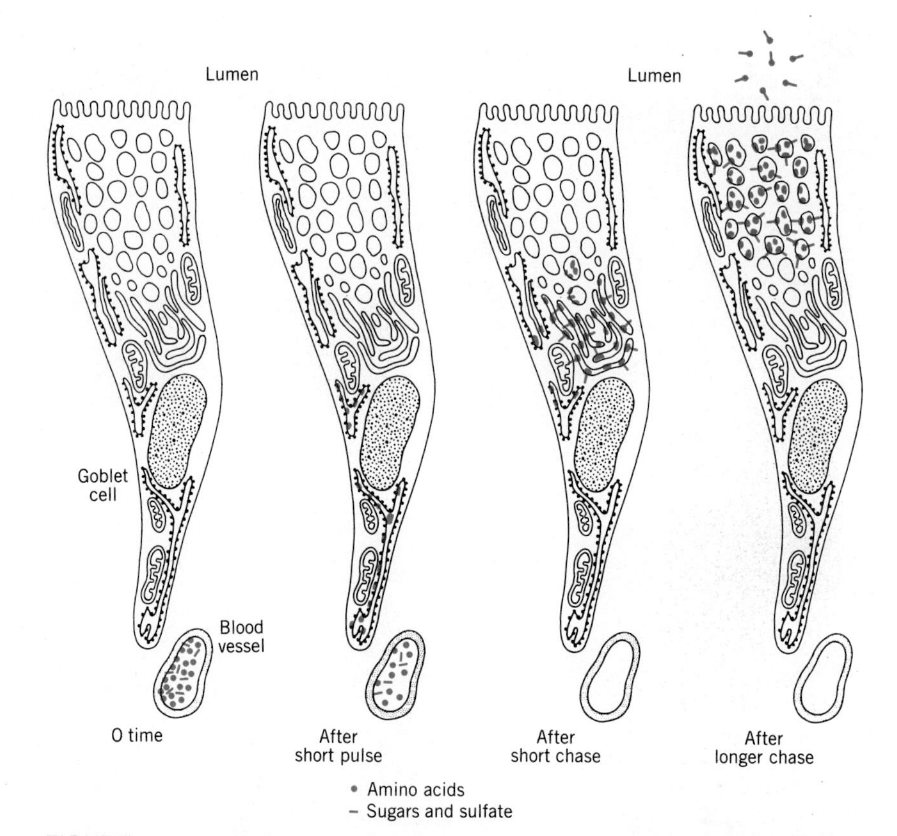

FIGURE 11.16 Movement of radioactive precursors in the goblet cell with time. Amino acids (solid spheres) are incorporated into protein in the rough endoplasmic reticulum and move the entire length of the cell to the lumen. Carbohydrates and sulfate (solid rods) are added in the Golgi complex to complete the synthesis of the mucigen.

glycoproteins are not destined to be secreted from the cell but rather must be retained by the plasma membrane, they apparently are never released into the saccule lumen of the Golgi complex. This is illustrated in Figure 11.17. Secretory vesicles containing *integral membrane glycoproteins* bleb off of the Golgi complex and fuse with the plasma membrane by the mechanism of exocytosis. Instead of dumping their contents into a lumen, the integral membrane proteins end up with their carbohydrate-rich ends on the exterior surface of the plasma membrane. Hence the glycocalyx is formed.

Mechanism of Glycosylation

The proposed mechanism of glycosylation of proteins in the rough endoplasmic reticulum and the Golgi complex is summarized in Figure 11.18. Glycosyltransferases are viewed as integral membrane proteins that may initiate glycosylation reactions while the proteins are still being made off the ribosomes. Proximal carbohydrates are added in the rough endoplasmic reticulum and more distal carbohydrates by glycosyltransferases in the Golgi complex membranes. The resulting glycoprotein either is released into the lumen of the Golgi saccules or is retained in the membranes (if it is destined for the plasma membrane). Nonglycoproteins are

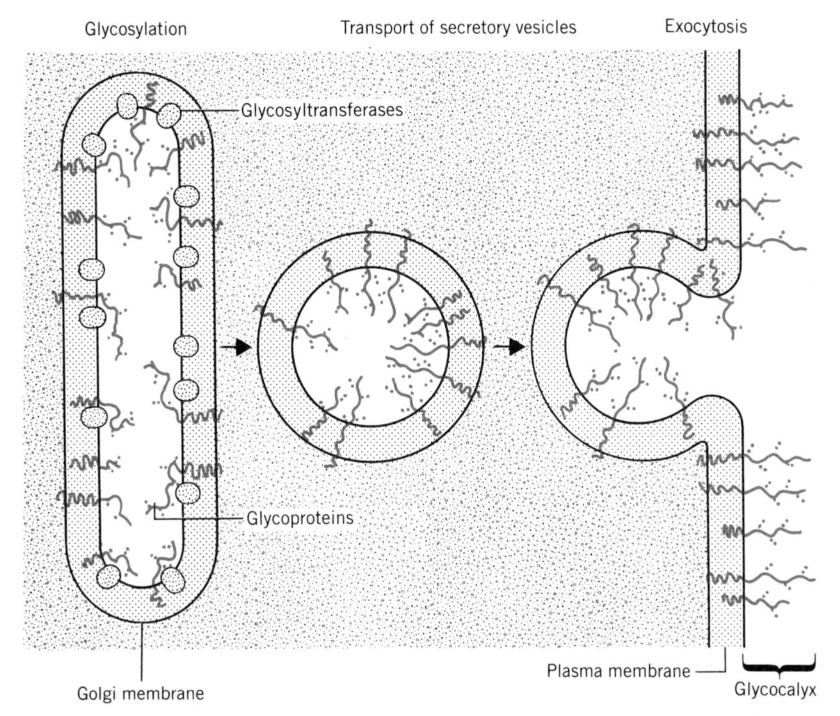

FIGURE 11.17 Mechanism for the incorporation of glycoproteins into the plasma membrane. According to this scheme, membrane glycoproteins are never found free from membranes in the cell.

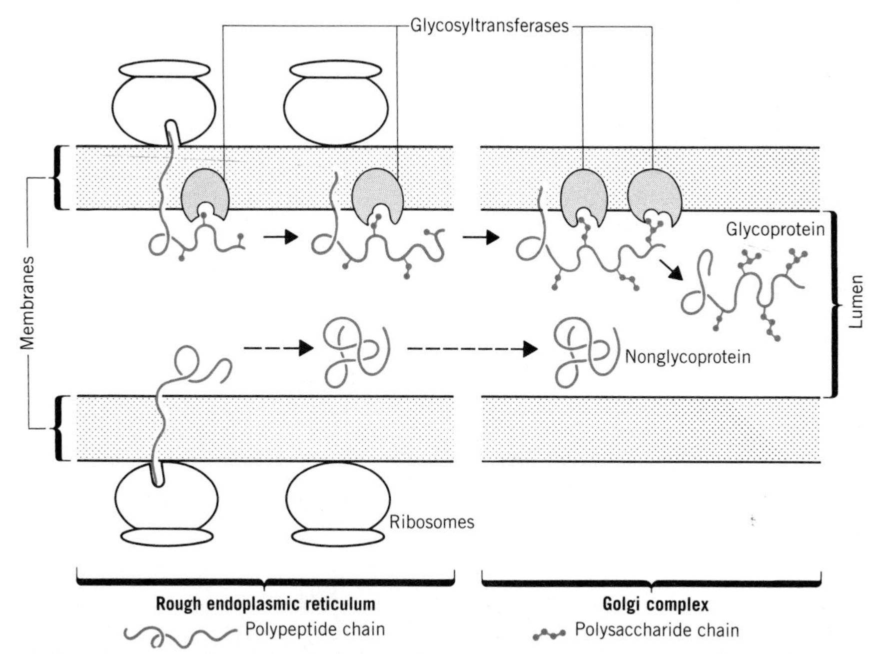

FIGURE 11.18 Successive steps in the glycosylation of proteins. Glycosyltransferases are integral membrane proteins found both in rough endoplasmic reticulum and in Golgi complex membranes.

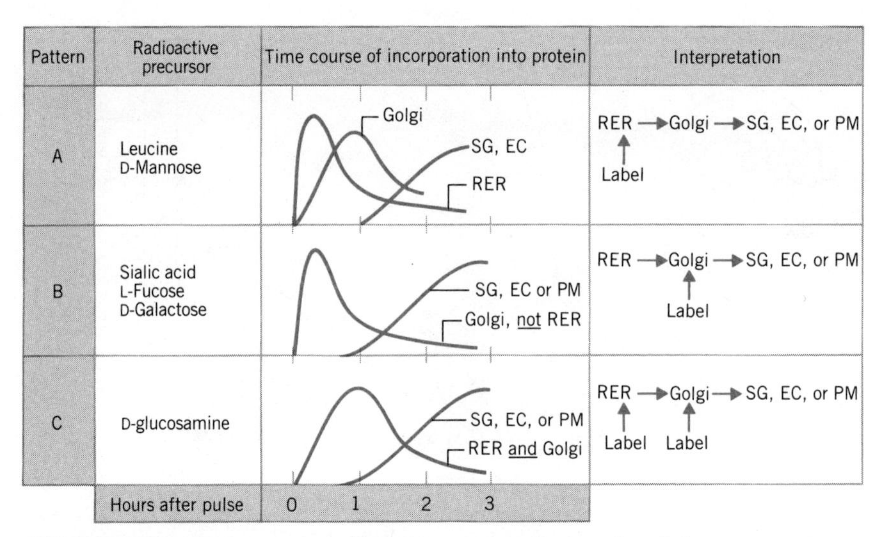

FIGURE 11.19 Incorporation of label into organelle proteins. SG, secretory granule; EC, extracellular space; PM, plasma membrane; RER, rough endoplasmic reticulum.

not subject to the action of the glycosyltransferases and move into the lumen of the endoplasmic reticulum as soon as they are formed.

Summary of Glycoprotein Formation

Figure 11.19 summarizes the work of several research groups obtained from a variety of tissues (liver, thyroid, duodenal columnar cells, kidney tubule cells, mucous glands). For these results, the incorporation of the radioactive precursor has been followed either by electron microscopic autoradiography or by the determination of radioactivity in isolated subcellular fractions.

The data fall into three distinct patterns. Pattern A shows an uptake of radioactive leucine and mannose by the rough endoplasmic reticulum, followed by a movement of this material through the Golgi complex and eventually to the secretory granule or plasma membrane. Apparently mannose and leucine show the same pattern of incorporation because mannose is a proximal member of the oligosaccharide chain in these systems and is covalently linked to the protein while still in the rough endoplasmic reticulum.

Pattern B is typical of the sugars sialic acid, fucose, and galactose. These are taken up initially in the Golgi complex and move from that point in the secretory chain to secretory granules or other destinations such as the plasma membrane. These particular sugars are positioned in the oligosaccharide prosthetic group more distally than mannose, toward the nonreducing terminus of the chain.

Finally, radioactive glucosamine follows pattern C incorporation. It appears in both the rough endoplasmic reticulum and the Golgi complex at roughly the same time, then moves on to the secretory granules. This is commensurate with the known structure of the oligosaccharide, since it has both a proximal and a distal glucosamine moiety.

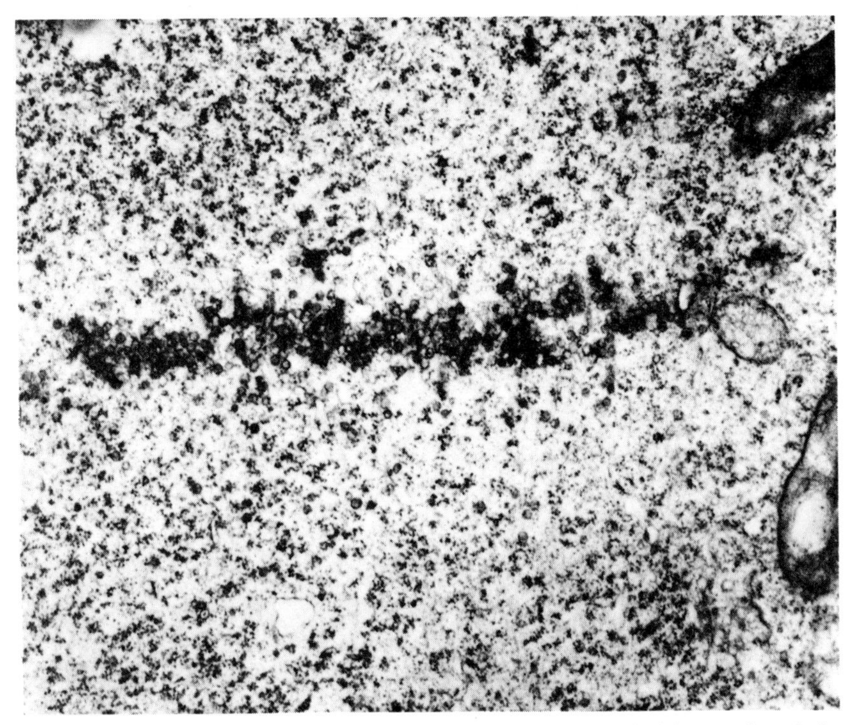

FIGURE 11.20 The fusion of vesicles derived from the Golgi complex during telophase in *Zea mays* root apex cells. These appear to provide the materials to form the new cell wall. (Micrograph by Dr. W. G. Whaley.)

The Formation of Plant Cell Walls

In plants, the Golgi complex appears to play a dominant role in the formation of the wall materials during cell replication and growth. The polysaccharide of the newly forming matrix is accumulated in the Golgi complex before it is transferred into the new wall.

In a dividing plant cell, a cell plate is generated between the two daughter nuclei by way of membrane vesicles that condense, beginning in the center of the cell and moving out toward the plasma membrane. The evidence is good that the vesicles and their contents of pectins and hemicelluloses arise from the Golgi complex and fuse to form the cell plate.[9] The fusion of Golgi-derived vesicles is shown in Figure 11.20.

In addition to the initial events in wall building, the Golgi complex has a continuing role in the laying down of secondary wall materials at later stages of wall development. The electron micrograph of Figure 11.21 can be interpreted as illustrating this function. Here prominent dictyosomes are evident, with Golgi vesicles blebbing off from the main stacks and moving to the plasma membrane, where they fuse and empty their contents onto the cell wall.

The studies on plant systems carried out so far certainly have not implicated the Golgi complex as the only cellular organelle that is directly involved with the formation of the cell wall. But the Golgi clearly does have a role in directing the implantation of at least some of the cell wall materials into their new sites.

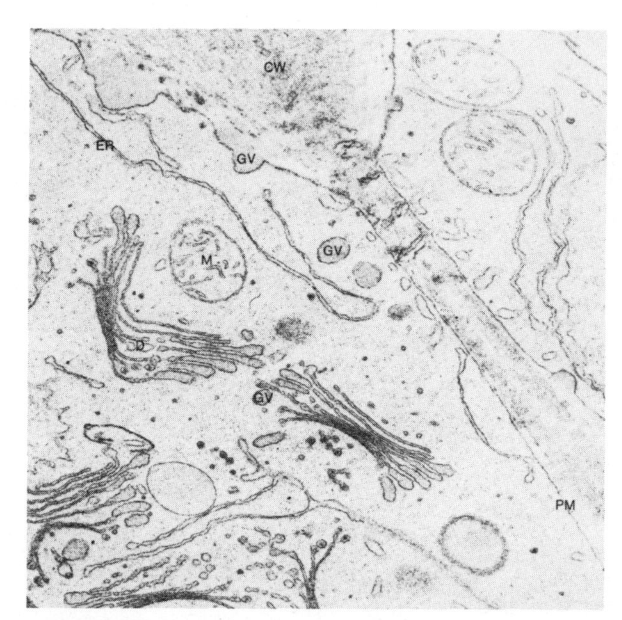

FIGURE 11.21 Electron micrograph of root cap cells of corn suggesting a functional realtionship between the Golgi complex and the cell wall (CW). Vesicles (GV) appear to bleb off the dictyosomes (D) and move to the plasma membrane (PM) where their contents are added to the growing cell wall. (Micrograph by Dr. W. G. Whaley.)

The Formation of Lysosomes and Acrosomes

The two major theories concerning the formation of lysosomes were discussed in Chapter 9. We shall not reiterate them here but rather emphasize that a large number of experimental results are consistent with the idea that primary lysosomes emerge from the maturing face of the Golgi complex, where they are packaged with membrane boundaries. The sighting of various hydrolases in vesicles formed in the Golgi region has been an important factor in connecting Golgi complex activity with the formation of lysosomes.

One of the most dramatic examples of this connection is the role of the Golgi in the formation of the acrosome, a lysosomelike organelle in the sperm of a large number of animals. In mammals the acrosome forms a cap overlaying the nucleus on the anterior end of the sperm (Figure 11.22). The hydrolytic role of this organelle is discussed at more length in Chapter 9.

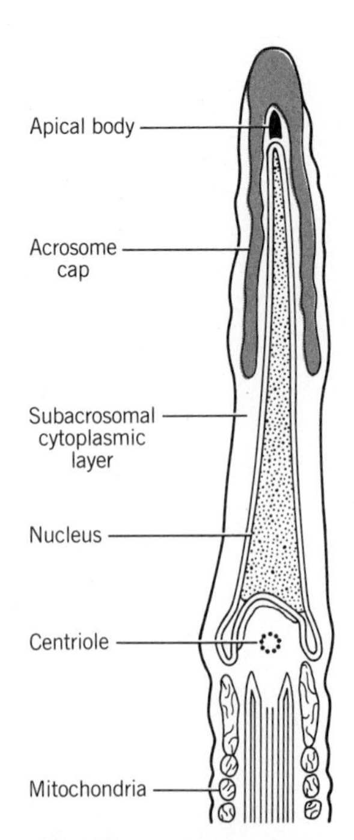

FIGURE 11.22 Idealized version of the mammalian spermatozoan. The acrosome cap at the anterior end of the nucleus is a lysosomelike organelle.

Patterns of protein transfer from the rough endoplasmic reticulum to the Golgi complex with the formation of glycoproteins in the complex have been demonstrated in maturing spermatocytes just as was discussed for the thyroid and goblet cells.[10] Figure 11.23 presents the Golgi complex as packaging and transporting hydrolytic enzymes as well as contributing membrane to the forming acrosome. The acrosomic granule grows with the deposited material until it fills the head cap with the hydrolytic enzymes. This function therefore represents an instance in which the Golgi complex plays an important role in cellular differentiation.

Going hand in hand with the role of the Golgi complex in spermatogenesis is the formation of protective substances around the developing

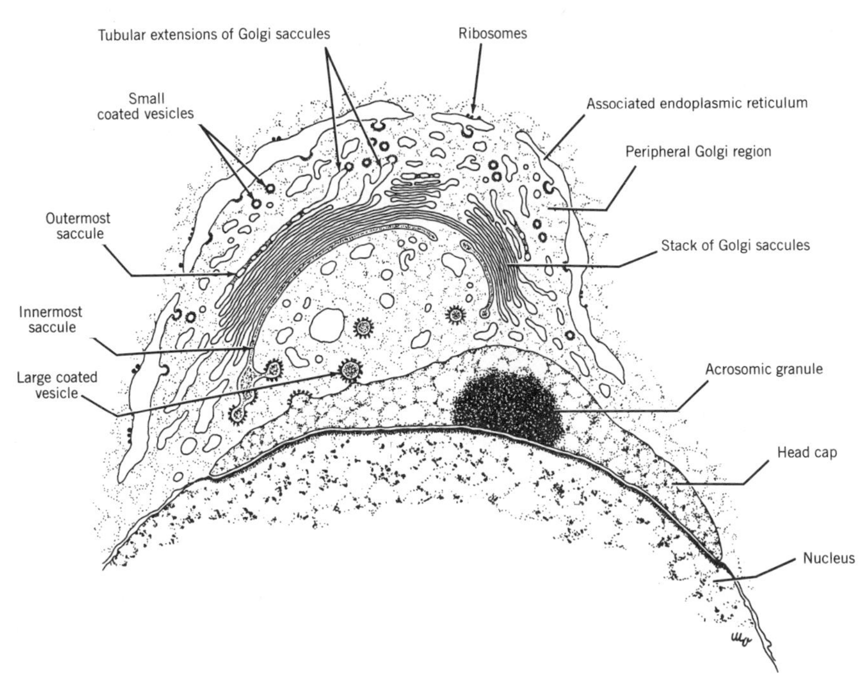

FIGURE 11.23 The relation between the Golgi complex and the acrosomic granule in an early-stage spermatid in the rat. Coated vesicles are viewed as moving from the mature face of the Golgi to the membrane of the acrosome. (Courtesy of Dr. C. P. Leblond.)

oocyte. In fact, the presence of a protective coating around the oocyte mandates the function of the acrosome in the sperm. There is evidence that the Golgi complex of the oocyte elaborates the materials that are eventually secreted to coat the mature cell.

We therefore see in the developing sperm and egg an example of synergism at the level of cellular organelles. The Golgi complex of the oocyte produces a material that has a chemical composition that can be decoded only by enzymes elaborated by the Golgi complex of the sperm. The Golgi organelles thus function to provide protection as well as to make available a means of permitting a species-specific interaction.

11.3 INTRACELLULAR MEMBRANE DIFFERENTIATION AND FLOW

It is becoming increasingly clear as a result of studies such as those reported by Morré and his associates[11] that the Golgi complex is not a physically static member of the secretory chain. In contrast, it is probably an organelle in transition, being endoplasmic reticulumlike on its forming face and plasma membranelike on its mature face. In some systems there appears to be an actual flow of membrane.

Endoplasmic reticulum → Golgi → Plasma membrane

and at the same time as this movement occurs, the membranes differentiate, changing in their composition from ER-like to plasma membranelike.

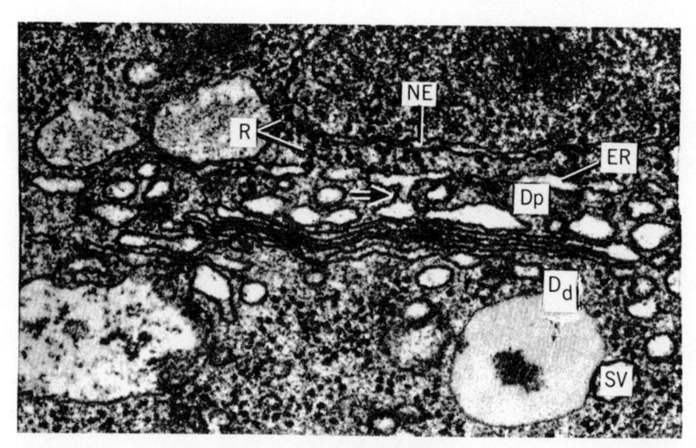

FIGURE 11.24 Transverse section of a dictyosome of a hypha of the fungus *Pythium middletonii.* The cisterna at the proximal pole (D_p) or forming face of the dictyosome is near the nuclear envelope (NE) and adjacent to the endoplasmic reticulum (ER) where ribosomes (R) can be seen. A tubular connection can be seen between an ER cisterna and the first dictyosome cisterna (arrow). Note especially the increase in thickness of the membranes from ER-like at the forming face to thicker and more like plasma membranes at the distal pole (D_d) or mature face where a secretory vesicle (SV) is found. (Courtesy of Dr. D. James Morré.)

Dimensional Differences between Endomembranes

Methods I and VIII

Generating elements: RER and outer nuclear envelope. Transition elements: SER, Golgi, and early secretory vesicles. End products: lysosomes, secretory granules, and plasma membrane.

Morphological studies on internal membranes have revealed dimensional differences among the various endomembranes of the cell. According to both electron microscopy and low angle X-ray diffraction analysis, the outer nuclear envelope and rough endoplasmic reticulum membranes are thin (5–6 nm thick), the plasma membrane, secretory product, and lysosomal membranes thick ($\sim$ 10 nm), and the membranes of the smooth endoplasmic reticulum, Golgi complex, and primary and secondary secretory vesicles intermediate (6–9 nm thick). These three membrane systems have been termed *generating elements, end products,* and *transition elements,* respectively. Electron microscopic evidence for this is presented in Figure 11.24 and a schematic illustration of the results of low angle X-ray analysis in Figure 11.25.

Biochemical Differences between Endomembranes

If membrane differentiation is truly taking place along the secretory pathway, it ought to be accompanied by discernible biochemical changes in membrane markers. Apparently this is the case. Differences in phospholipid composition between the endomembranes were presented in Table 11.5. Golgi membranes occupy an intermediate position. Similarly, the Golgi membranes are intermediate with respect to their content of cholesterol, cerebroside, ganglioside, sialic acid, and total glycoprotein.

Enzymatic studies on rat liver show that plasma membrane markers, such as 5′-nucleotidase, Mg^{2+}-ATPase, and a UDP-galactose hydrolase, appear in increasing concentrations along the secretory chain: endoplasmic reticulum $<$ Golgi complex $<$ plasma membrane. Enzyme markers of the endoplasmic reticulum, such as nucleoside diphosphatase, ap-

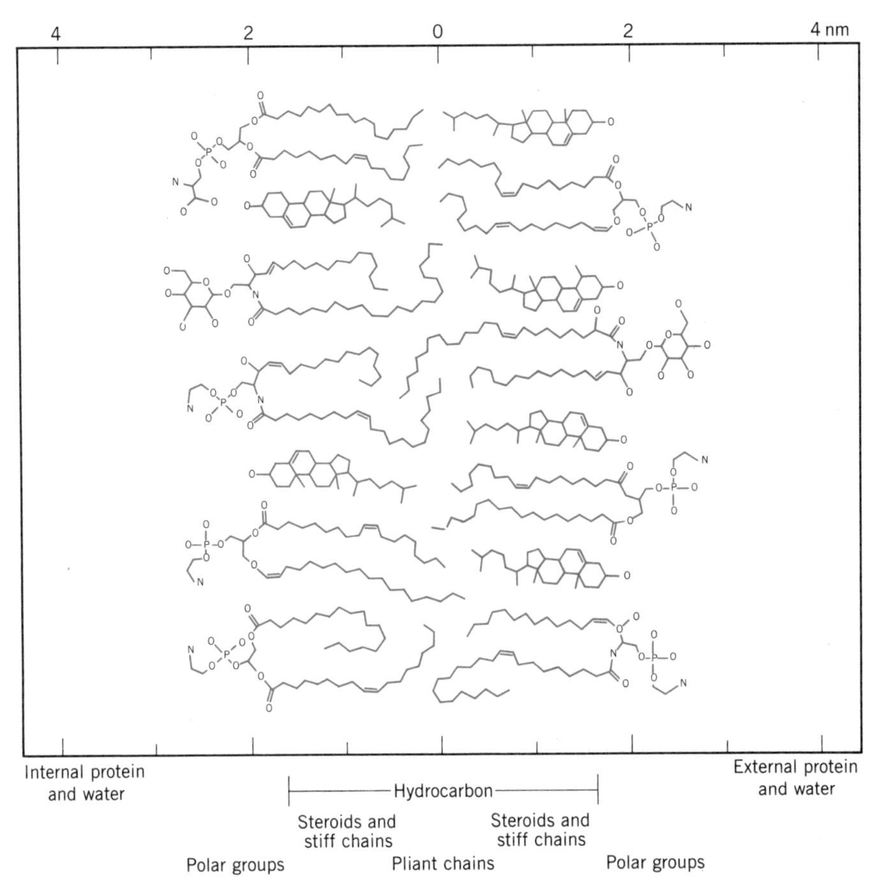

Fraction	Distance between phospholipid polar groups in bilayer (nm)
Endoplasmic reticulum	4.0
Golgi complex	4.5
Plasma membrane	5.0

FIGURE 11.25 Schematic illustration of membrane structure as interpreted from low angle X-ray analysis. Values are from rat liver fractions.

pear in the reverse intensity (ER > GC > PM). Thiamine pyrophosphatase and glycosyl transferase, on the other hand, remain concentrated in the Golgi complex. This observation would seem to suggest that these enzymes operate within the Golgi cisternae, perhaps on the transitional membranes, but are not an integral part of the membranes themselves.

Thus, there are both structural and biochemical grounds for the view that the Golgi complex is a transforming element in the differentiation of the endomembrane systems of the cell.

Movement of Membranes through the Golgi Complex

Membrane differentiation of the type discussed above assumes that there is an accompanying physical movement of the membrane elements from the endoplasmic reticulum to the plasma membrane by way of the Golgi membranes. The evidence for this occurring is also both structural and biochemical.

Figures 11.26 and 11.27 are electron micrographs demonstrating membrane activities that may facilitate membrane transfer. In Figure 11.26, tiny primary vesicles, usually referred to as transition vesicles, can be seen between the endoplasmic reticulum and the forming face of the Golgi

Relative concentrations of different membrane markers.

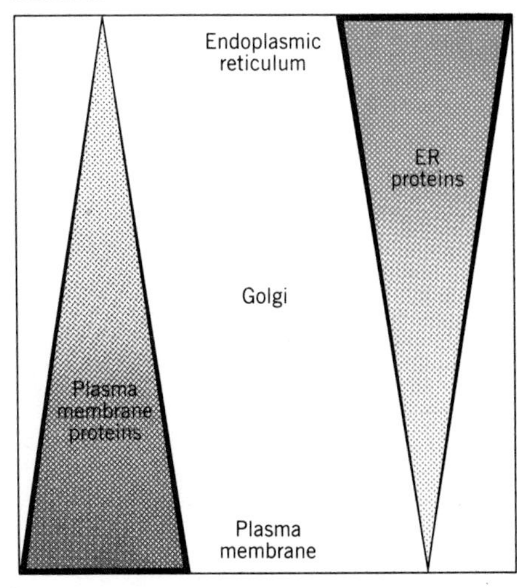

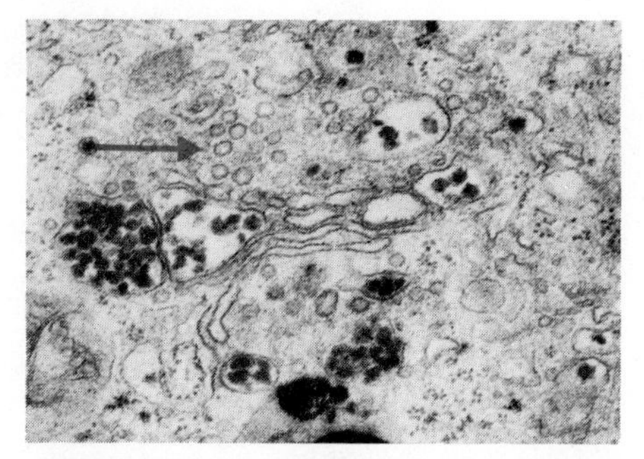

FIGURE 11.26 Golgi complex of rat liver showing primary vesicles (arrow) that are presumed to arise from the endoplasmic reticulum (above vesicles) and move toward and fuse at the forming face (below vesicles) of the Golgi. (Courtesy of Dr. D. James Morré.)

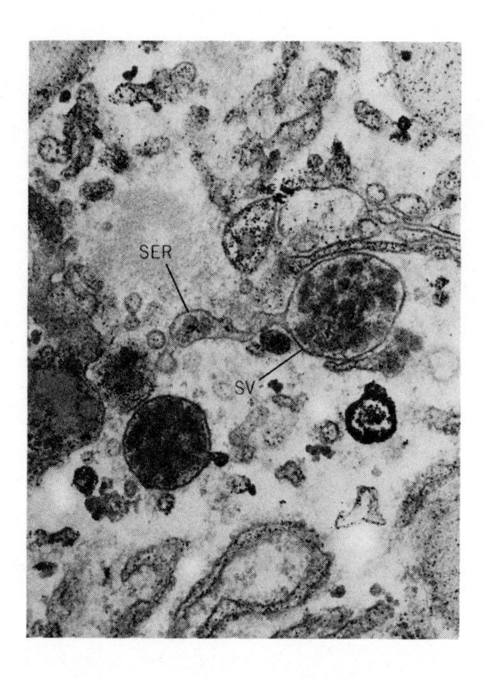

FIGURE 11.27 Region of membrane continuity between smooth endoplasmic reticulum (SER) and secretory vesicles (SV) of the Golgi complex. In this case lipoprotein particles appear to be transported from SER to the secretory vesicles. Tubule interiors were stained for arylsulfatases to increase their visibility. (Courtesy of Dr. D. James Morré.)

Radioactive arginine.

$$\overset{+}{N}H_3{-}CH{-}COO^-$$
$$(CH_2)_3$$
$$NH$$
$$\boxed{C}{=}\overset{+}{N}H_2$$
$$NH_2$$

complex. In Figure 11.27 there appear to be direct membrane connections between the smooth endoplasmic reticulum, the Golgi membranes, and secretory vesicles. Such electron micrographs, though static pictures of membrane systems, are believed to capture a dynamic vectorial flow of membrane material in the cell.

Labeling studies have supported the microscopic work. When [^{14}C]guanido-L-arginine is used to pulse label rats, the label appears in the membranes of liver fractions in the following time sequence: endoplasmic reticulum (and nuclear envelope), Golgi complex, and plasma membrane. This is an indirect demonstration consistent with the idea that there is a physical transfer of membrane proteins from one endomembrane component to the next.

For membrane flow to be more thoroughly established experimentally, it is necessary to show that the *same protein* appears in the different endomembranes. For example, the sequential flow of labeled arginine could be the result of metabolic transformations between endomembrane compartments, not a true physical transfer. Careful studies on the NADH-cytochrome *c* oxidoreductase complex, a family of integral membrane proteins, have helped settle this issue to some extent.[11] The NADH oxidoreductases of endoplasmic reticulum and Golgi complex are indistinguishable when compared by polyacrylamide gel electrophoresis, enzyme kinetics (K_m), sensitivity to specific inhibitors, stimulation by Triton X-100 and sodium deoxycholate, and by cytochemical techniques. By itself this statement does not verify membrane flow, but it strongly supports such a phenomenon. Taken with the results of labeling studies, the picture becomes quite convincing.

The experimental results of a number of approaches leave open the possibility that some membrane flow in the cell may bypass the Golgi complex, going directly from the smooth endoplasmic reticulum to the plasma membrane. The kinetics of labeling studies are consistent with this occurring to an extent that may equal the pathway including the Golgi complex.

Figure 11.28 summarizes the sequence of events during membrane flow. According to this diagram, the fenestrated saccule on the forming face arises when transition vesicles coming from the endoplasmic reticulum fuse. This saccule is moved into the interior of the stack as more synthesis takes place on the forming face. Eventually the saccule reaches the maturing face, where it fragments and gives rise to secretory vesicles. During the movement of the saccule from forming to maturing face, the membrane of the saccule is modified to become more plasma membranelike, while at the same time the material transported in the lumen is glycosylated.

Excellent research has been done on the Golgi complex. Our understanding of the structure and function of this complex organelle has expanded considerably over the last decade. But some intriguing questions remain unanswered. What, for example, are the driving forces in membrane flow? What causes the formation of transition vesicles and their coalescence on the forming face? By what molecular mechanisms do the membranes differentiate as they move from the endoplasmic reticulum to the plasma membrane? When is the Golgi used and when is it not in the secretory pathway?

The answers will surely come, but they will require time, patience, and persistence. Perhaps some who are students of cell biology will have the opportunity to get in on the act and provide vital missing links in the secretory chain.

Summary

The Golgi complex is a system of membrane sacks arranged in stacked form and surrounded by vesicles and tubules. It is a polar organelle, consisting of a forming face on one side of the stack and a mature face on the other. The organelle is oriented with its forming face close to the endoplasmic reticulum and its mature face toward the cell surface, where secretory materials are exported.

The Golgi complex can be selectively stained in the cell with reagents that have an affinity for glycoproteins and oligosaccharides. It can be isolated by differential and density gradient centrifugation techniques.

A variety of enzymes are found in the Golgi complex, but the glycosyltransferases are the best marker enzymes. The lipid content of the Golgi is intermediate between that of the endoplasmic reticulum and the plasma membrane.

The Golgi complex is responsible for the attachment of certain of the carbohydrates to growing oligosaccharide chains during the biosynthesis of glycoproteins. Some of the more commonly studied proteins are thyroglobulin, mucigen, and the plasma membrane glycoproteins. The com-

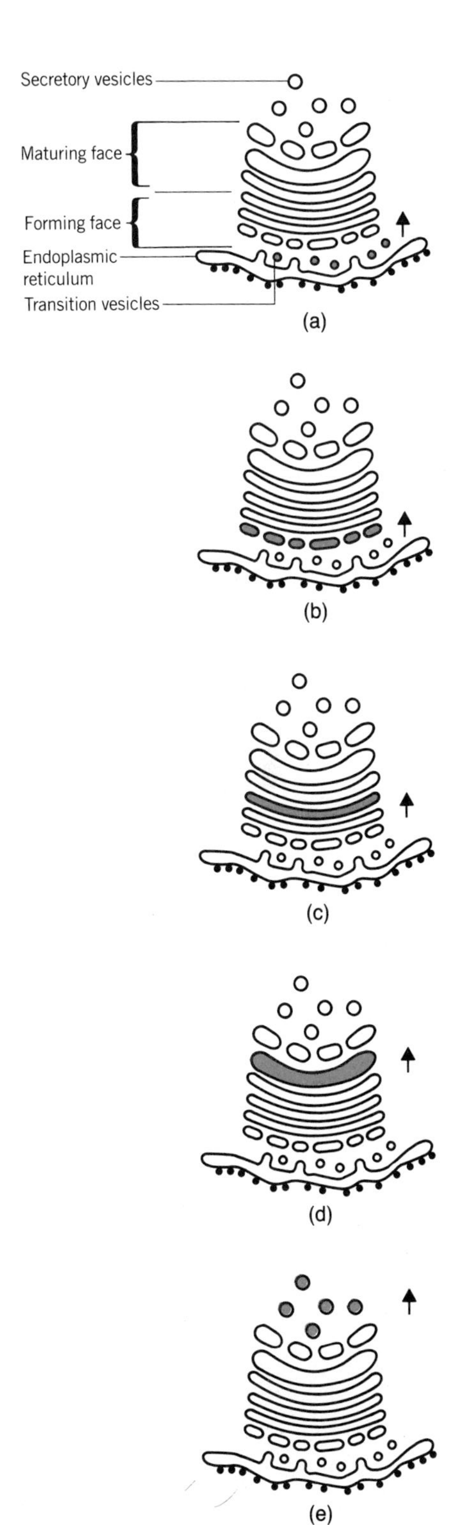

FIGURE 11.28 The Golgi complex as it participates dynamically in membrane flow. The arrow indicates the direction of movement of membranes from the rough endoplasmic reticulum to secretory vesicles. Transition vesicles bud off the endoplasmic reticulum (*a*), fuse with the Golgi membranes (*b*), and then move through the Golgi complex (*c* and *d*) before they exit as secretory vesicles (*e*).

plex also functions in the formation and transport of polysaccharide that is incorporated into the material of the plant cell wall.

The Golgi is involved in the formation of lysosomes and the lysosome-like structure of the sperm called the acrosome.

The Golgi complex is an important link in the secretory pathway in the cell. In this capacity it functions as an intermediary between the endoplasmic reticulum and the secretory granule or plasma membrane. There is evidence that the Golgi not only modifies proteins that pass through it but also modifies itself during the vectorial flow of membrane material in the cell.

References

1. Sulle Sostanza Grigia del Cervello, C. Golgi, *Gaz. Med. Ital.* (1873).
2. Sur la Structure des Cellules Nerveuses des Ganglions Spinaux, C. Golgi, *Arch. Ital. Biol.* 30:278(1898).
3. Isolation of a Golgi Apparatus-rich Fraction from Rat Liver. I. Method and Morphology, D. J. Morré, R. L. Hamilton, H. H. Mollenhauer, R. W. Mahley, W. P. Cunningham, R. D. Cheetham, and V. S. Lequire, *J. Cell Biol.* 44:484(1970).
4. *The Golgi Apparatus,* W. G. Whaley, Springer-Verlag, New York, 1975.
5. Radioautographic Visualization of the Incorporation of Galactose-^{3}H and Mannose-^{3}H by Rat Thyroids *in Vitro* in Relation to the Stages of Thyroglobulin Synthesis, P. Whur, A. Herscovics, and C. P. Leblond, *J. Cell Biol.* 43:289(1969).
6. Radioautographic Study of *in Vivo* and *in Vitro* Incorporation of Fucose-^{3}H into Thyroglobulin by Rat Thyroid Follicular Cells, A. Haddad, M. D. Smith, A. Herscovics, N. J. Nadler, and C. P. Leblond, *J. Cell Biol.* 49:856(1971).
7. The Role of the Golgi Complex in Sulfate Metabolism, R. W. Young, *J. Cell Biol.* 57:175(1973).
8. Transport of Glycoprotein from Golgi Apparatus to Cell Surface by Means of Carrier Vesicles, as Shown by Radioautography of Mouse Colonic Epithelium after Injection of ^{3}H-Fucose, J. Michaels and C. P. Leblond, *J. Micros. Biol. Cell.,* 25:243(1976).
9. Patterns of Incorporation of ^{3}H-Galactose by Cells of *Zea mays* Root Tips, M. Dauwalder and W. G. Whaley, *J. Cell Sci.* 14:11(1974).
10. Changes in the Golgi Apparatus During Spermiogenesis in the Rat, R. F. Susi, C. P. Leblond, and Y. Clermont, *Am. J. Anat.* 130:251(1971).
11. Membrane Flow and Differentiation: Origin of Golgi Apparatus Membranes from Endoplasmic Reticulum, D. J. Morré, T. W. Keenan, and C. M. Huang, *Advances in Cytopharmacology,* vol. 2, B. Ceccarelli, F. Clementi, and J. Meldolesi, eds., Raven Press, New York, 1974.

Selected Books and Articles

Books

Advances in Cytopharmacology, vol. 2, *Cytopharmacology of Secretion,* B. Ceccarelli, F. Clementi, and J. Meldolesi, eds., Raven Press, New York, 1974.

Golgi Centennial Symposium: Perspectives in Neurobiology, M. Santini, ed., Raven Press, New York, 1975.

The Golgi Apparatus, G. M. W. Cook, Oxford University Press, London, 1975.

The Golgi Apparatus, W. G. Whaley, Springer-Verlag, New York, 1975.

Articles

Membranes of the Golgi Apparatus, P. Favard, in *Mammalian Cell Membranes,* vol. 2, G. A. Jamieson and D. M. Robinson, eds., Butterworths, London, 1977.

Role of the Golgi Apparatus in Terminal Glycosylation, C. P. Leblond and G. Bennett, in *International Cell Biology 1976–1977,* B. R. Brinkley and Keith R. Porter, eds., Rockefeller University Press, New York, 1977.

The Golgi Apparatus: Form and Function, G. W. M. Cook, in *Lysosomes in Biology and Pathology,* J. T. Dingle, ed., North Holland, Amsterdam, 1973.

The Golgi Apparatus, H. H. Mollenhauer and D. J. Morré, in *The Biochemistry of Plants,* vol. 1, N. E. Tolbert, ed., Academic Press, New York, 1980.

The Golgi Apparatus, M. Neutra and C. P. Leblond, *Sci. Am.* 220:100(1969).

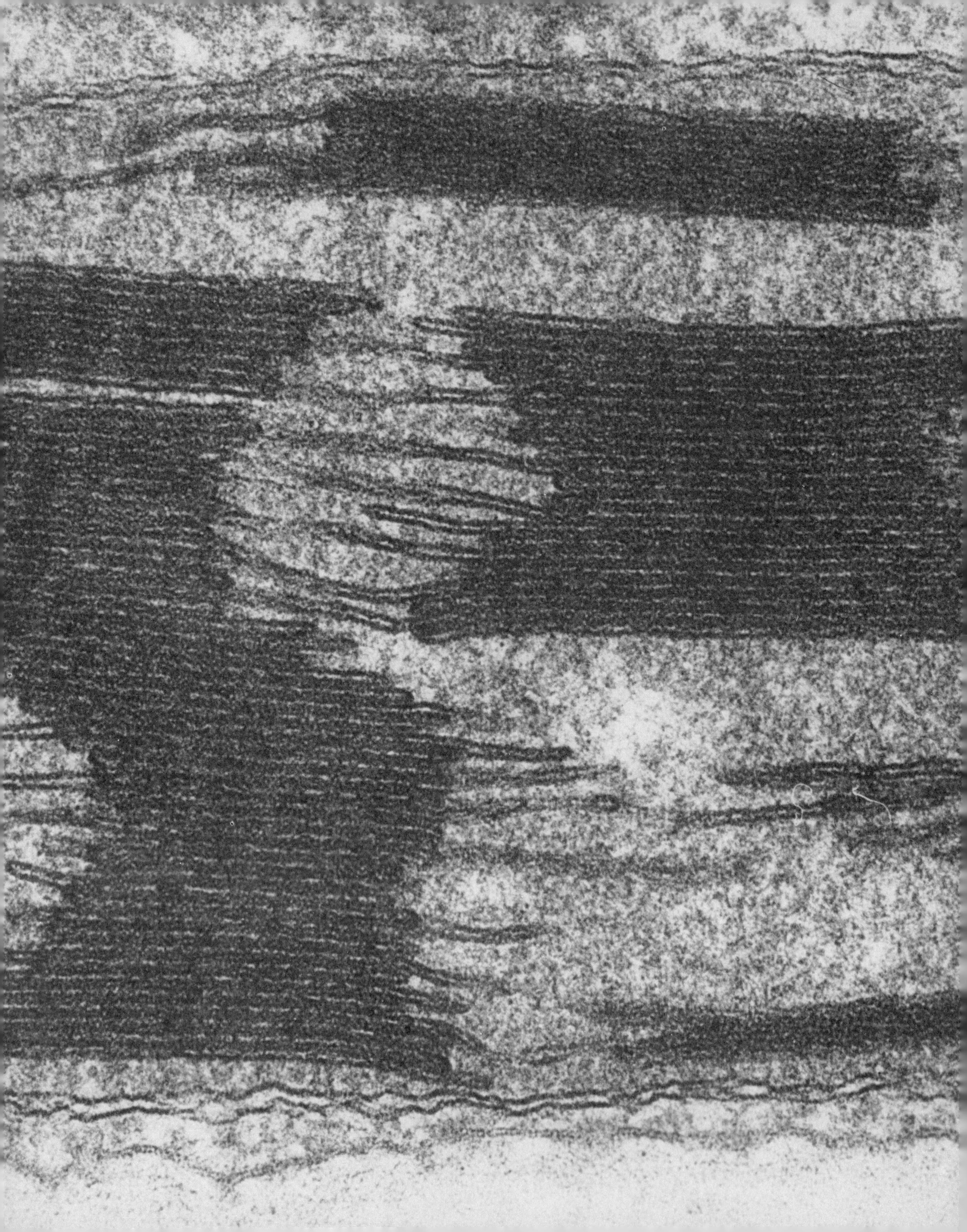

SECTION 6

ENERGY TRANSDUCTION

Cells use many different sources as energy for all the activities in which they engage. For some the source is chemical compounds, for others it may be radiant energy. But whatever the source, it must be transduced into a form that is useful to the cell, namely, ATP.

The main energy-transducing organelles in animal cells are the mitochondria. In plants this function is assumed by chloroplasts. Both types of organelle are of interest not only for their transducing capabilities but also for their structural and functional similarities, their metabolic roles that lead up to and follow the transducing reactions, and their semiautonomous life-styles. These and other properties of these eucaryotic organelles form the topic of the next two chapters.

12

Mitochondria

Current research on the mitochondrion may very aptly be pictured as a busy scene of construction, where the rising structure is still hidden by scaffolding. Here artisans of many guilds ply their traditional skills, each group intent on its specific task. There is a great sound of hammers and the shouts of workmen are often strident. To formulate an impression of this new edifice, we must tour the site of construction, peer behind the scaffolding, gather reports from the architects and artisans, and then retreat out of earshot to reflect on the purpose and direction of this effort.

ALBERT L. LEHNINGER, 1964

By the time students read this chapter, the scene of construction referred to by Albert Lehninger above will find only a finishing crew carefully putting on the final touches—polishing the brass railings and doorknobs and rearranging the furnishings in ever so subtle ways. The scaffolding will be down and the structure will loom starkly, with perhaps a shrub here and there beginning to conceal some of its more striking features. And, sadly and unfortunately, much of the excitement and anticipation generated during the construction will be quieted. Many will walk by with hardly a sidelong glance, for the structure will have the familiarity of an old shoe.

The conceptual construction of the mitochondrion has lent itself, in an unusual way, to the participation of a diverse menagerie of artisans. Important contributions have been made at various stages in the building by microscopists, biochemists, enzymologists, physiologists, geneticists, developmental biologists, physical chemists, and many others. This collection of workers with strikingly different interests and skills has constituted a powerful workforce for the project, but has generated a good deal of confusion for the average sidewalk engineer gazing with curiosity on the growing edifice. Different workers periodically have emerged from their work with plans and projections that appeared to be unrelated or in direct conflict, or perhaps more pertinent to projects other than the one under construction. So, just as was true for the plasma membrane, the molecular perspectives of the mitochondrion have been subject to change with time. It was not until the scaffolding was nearly dismantled that the final form could be viewed with confidence.

Although most of the important structural and functional features of the mitochondrion are now worked out, not everything is completed. Some of the carpets do not match, and there is still not complete agreement about the trim. It will probably be up to some of you to complete the project.

12.1 THE HISTORICAL CONTEXT

Various types of granule and filament were observed in and teased out of tissues in the late 1800s, but systematic studies on the cellular inclusions that eventually were named mitochondria were not carried out until 1890, when Richard Altmann developed and employed a stain that had useful specificity for this organelle. He speculated that these "bioblasts" were autonomous elementary living particles that made a genetic and metabolic impact on the cell. He was, in many respects, very much on target.

During its history, the mitochondrion has been the bearer of a bewildering bevy of names. The name *mitochondrion* was applied to the organelle by Benda around the turn of the century but was not widely adopted until a good number of years later. About the same time, Michaelis used the supravital stain Janus green to demonstrate that mitochondria were oxidation–reduction sites in the cell, but it was not until 1912 that Kingsbury suggested that the oxidation reactions mediated by these organelles were *normal* cellular processes. Warburg, who is thought of as the father of respirometry, found in 1913 that respiration, or oxygen consumption, was also a property of mitochondria, but this observation did not really attract its deserved attention until about three decades later.

The cytological approach to studying mitochondria began to merge significantly with the biochemical approach around the early 1930s when the first serious attempts were made to isolate large quantities of mitochondria by differential centrifugation techniques. Before this, biochemical studies made tremendous strides in ferreting out the activities and relationships of the respiratory enzymes, but as was generally the case for biochemists of that time, the implications of these reactions vis-à-vis cell organelles was largely ignored. Nevertheless, research groups under the leadership of investigators such as Warburg, Keilin, Szent-Györgyi, Krebs, and Lehninger pieced together many of the fundamental pathways of oxidative degradation reactions, and Lohmann discovered ATP in muscle in 1931. Warburg linked the phenomenon of ATP formation to the oxidation of glyceraldehyde phosphate, Meyerhof showed the formation of ATP from phosphopyruvate, and finally, Kalckar related oxidative phosphorylation to respiration. Thus the stage was set to relate these important biochemical events to cell ultrastructure.

The convergence of these two approaches took place in an exemplary manner at the Rockefeller Institute in the late 1940s and early 1950s. Here a group of cytologists refined the technique of differential centrifugation for isolation (Claude) and developed the use of osmium tetroxide as a fixative (Palade) and thin-sectioning as a way to better visualize these membrane bound organelles (Porter). At the same time, a group of biochemists and enzymologists (Hogeboom, Hotchkiss, Schneider) found that the enzymatic properties of the isolated organelles could account for the same catalytic activities present in intact cells. Their work also revealed the cellular practice of compartmentalizing different metabolic activities, housing the tricarboxylic acid and respiratory enzymes in the mitochondrion, and limiting the glycolytic enzymes to the extramitochondrial cytosol.

Methods I and III

The work since the early 1950s has been focused on the molecular architecture of the mitochondrial membranes, the mechanisms of transport and oxidative phosphorylation, and mitochondrial biogenesis. The progress has been remarkable and the findings most significant for our understanding of cellular respiration. In the pages that follow we systematically examine some of what is known of mitochondrial structure, function, and biogenesis.

12.2 GENERAL STRUCTURAL CHARACTERISTICS OF MITOCHONDRIA

Gross Morphological Properties and Dynamics *in Situ*

Altmann's systematic study of cellular inclusions before the turn of the century was, of course, limited to the use of staining and light microscopy. Nevertheless, he was able to document the occurrence of mitochondria in a wide variety of tissues and published his observations with a beauty and clarity typified by the lithographs in Figure 12.1. It is obvious even from these drawings that mitochondria appear in different forms and numbers in different tissues.

Size and Shape

Mitochondria are about the same size as average bacillus bacteria. To students who have observed bacteria with the ordinary light microscope, the limitations of this tool for the study of mitochondria will be readily apparent. Liver mitochondria are generally somewhat elongated, being about 0.5 to 1.0 μm wide and about 3 μm long. These dimensions are fairly typical for mitochondria of the type that are free in the cytoplasm, such as is true for kidney and pancreas, as well as liver tissue. In other tissues where the topographical freedoms of mitochondria are more restricted, a greater variety of forms and sizes is found.

Location

Within the cell, mitochondria may be situated randomly, as is the case for liver, or they may exhibit some type of ultrastructural association. Perhaps the most common example of the latter is the ordered arrangement of mitochondria between the muscle fibers in striated muscle (Figure 12.2). Another example of an ultrastructural association is the position of the mitochondrion in the flagellar region of spermatozoa (Figure 12.3).

In general, mitochondria are found nearest to sites in the cell where the energy requirements are the heaviest. In the examples above, both striated muscle and flagella depend on a ready supply of ATP for their function. The proximity of mitochondria to these structures assures an abundant source of energy for contraction.

Number

The number of mitochondria per cell varies widely with cell type, ranging from zero to hundreds of thousands. The colorless algae *Leucothrix* and *Vitreoscilla* have no mitochondria, certain spermatozoans and flagellates like *Chromulina* contain only one per cell, liver has an average of about 800 per cell, and some sea urchin ova and the giant amoeba *Chaos chaos* contain up to 500,000 mitochondria. In some cases, there appears to be a direct relation between the number of mitochondria per cell and the metabolic demands on the cell. In other cases, however, the ultrastructure of the mitochondria, not the number, is a reflection of metabolic demands.

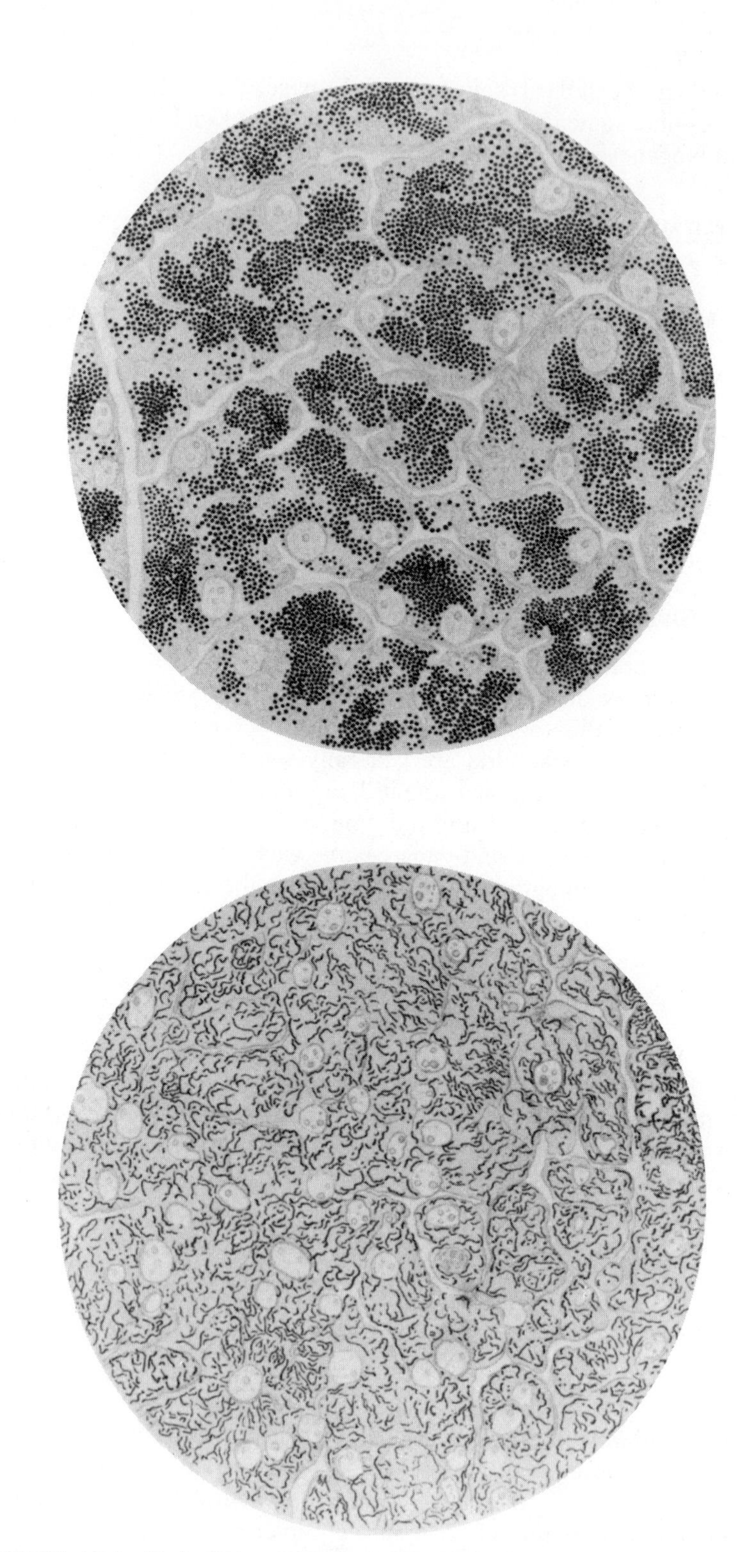

FIGURE 12.1 Early lithographs depicting mitochondria in (top) frog liver and (bottom) mouse pancreas. Note the remarkable detail and clarity obtained from light microscopy before the turn of the century.

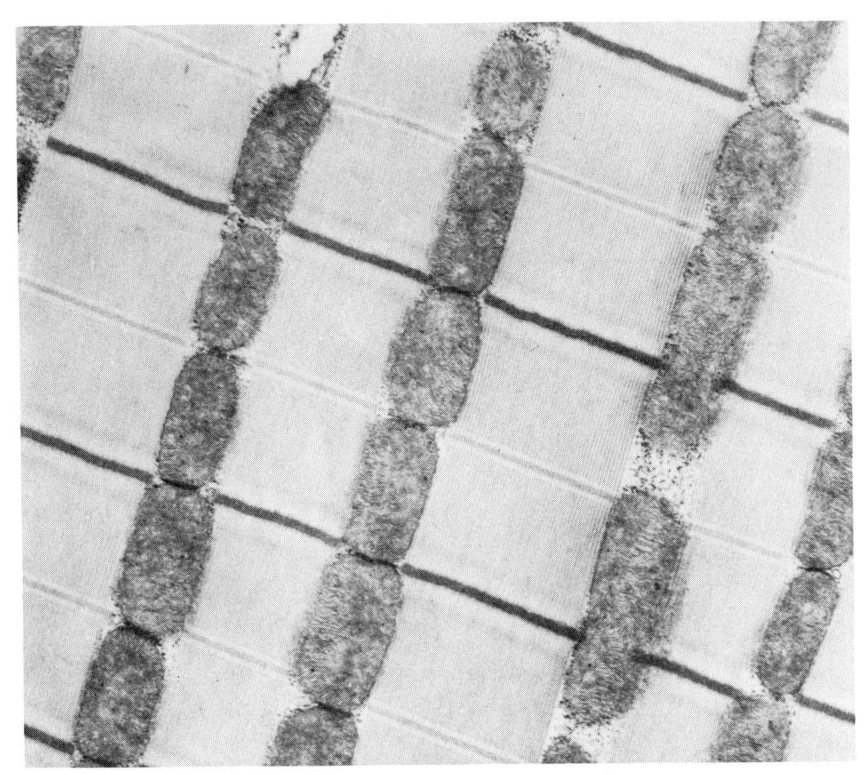

FIGURE 12.2 Ordered arrangement of mitochondria in striated muscle, in this case flight muscle of the wasp *Polistes*. This arrangement assures an even dispersal of mitochondria throughout the tissue and places them where energy needs are highest. (Courtesy of Dr. David S. Smith.)

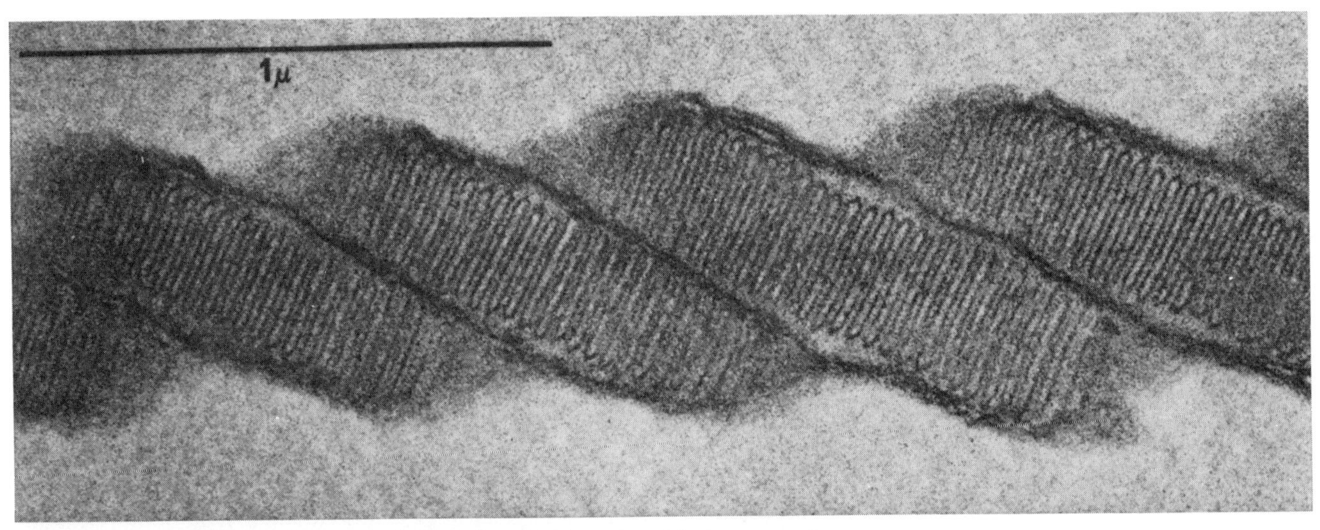

FIGURE 12.3 Mitochondria wrapped tightly in a highly ordered fasion around a sperm tail of a snail. (Courtesy of Dr. Winston A. Anderson.)

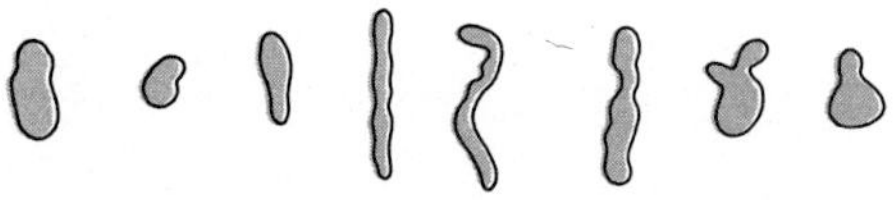

FIGURE 12.4 Typical shape changes in a single mitochondrion with time. This organelle can undergo an endless array of contortions, all of which may be evident within a minute.

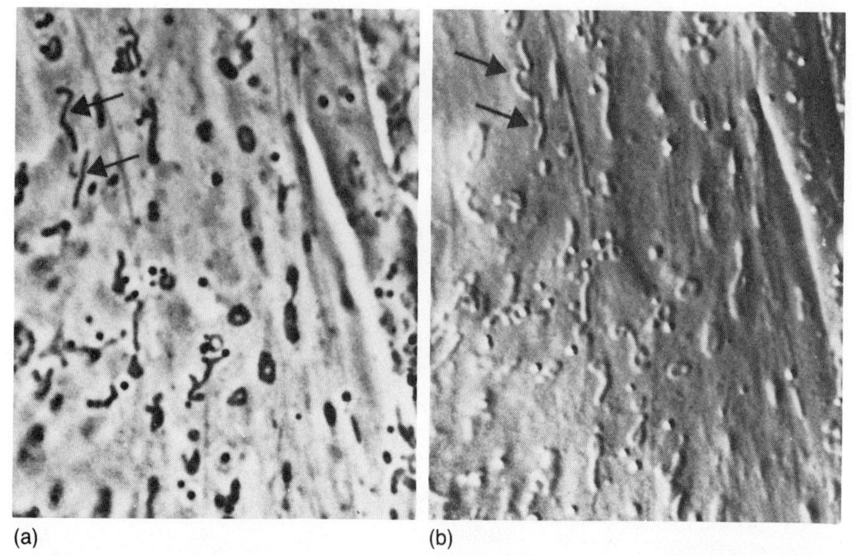

(a) (b)

FIGURE 12.5 The dynamic properties of mitochondria as viewed by phase contrast (*a*) and Nomarski (*b*) optics. These two micrographs are of the same field, observed only a few minutes apart. The arrows point to two mitochondria that have undergone both shape and position changes. (Courtesy of Dr. R. L. Goldman.)

Plasticity

Method I

Phase contrast microscopy and Nomarski optics have been helpful in demonstrating the plasticity of mitochondria *in situ*. Cinematographic studies using these approaches have shown that mitochondria spin and contort through an endless variety of conformations. An example of shape changes with time for a single mitochondrion is presented in Figure 12.4; the micrographs in Figure 12.5 contain a distribution of forms.

In striated muscle and other cells wherein mitochondria are not free in the cytosol, there is less structural plasticity. There is, perhaps, some cellular logic to this. The plasticity and movement of mitochondria in cells assures a wide distribution of ATP throughout the cell in places it is needed. In the liver cell this would be wherever synthetic reactions are taking place in the cytosol. In muscle, where the cells have a more specialized contractile function, the energy requirements are more intense but localized within the cell where the contractile filaments are operating.

Properties of Isolated Mitochondria

Rat liver is the most common tissue source used for the isolation of mitochondria. This tissue is quite easily homogenized and the mitochondria, as already pointed out, are relatively free in the cytosol. Another very common source of mitochondria is beef heart. This tissue has been used because of its obvious capacity to generate ATP for muscle contraction.

Methods II and III

Figure 12.6 diagrams the general procedure used in isolation. Following homogenization in 0.25 *M* sucrose or a comparable isosmotic medium, the disrupted cell contents are subjected to differential and density gradient centrifugation in a manner that permits separation of the mitochondria from the other cell components. Some damage to mitochondria is

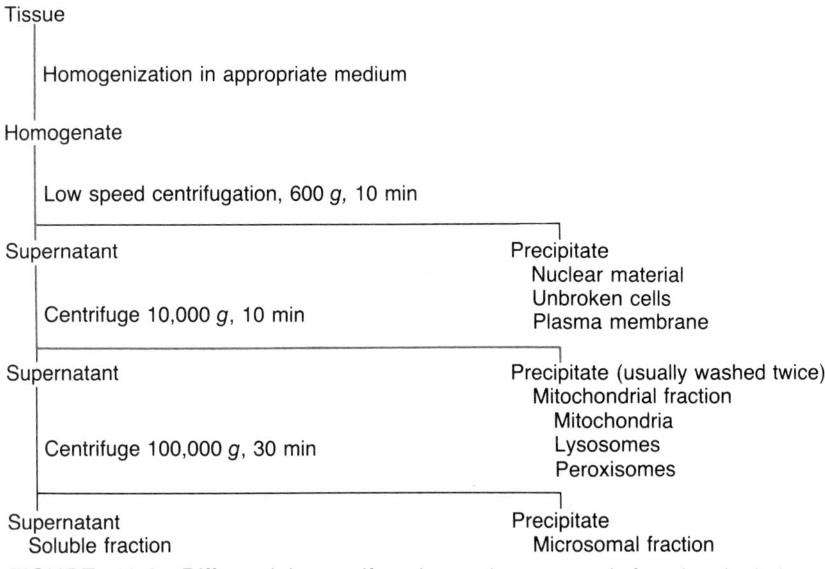

FIGURE 12.6 Differential centrifugation scheme used for the isolation of mitochondria.

inevitable with this procedure, and a complete separation of mitochondria from lysosomes has been a special problem. Any preparation of isolated mitochondria has the potential of containing fragments of every other membrane component of the cell. This is a problem that is not unique to the isolation of mitochondria, for it is also encountered when attempting to purify any cellular ultrastructure. But the existence of this contamination forces the investigator to approach with caution the results of enzymatic activities or trace analyses of mitochondrial preparations.

The structures of properly isolated mitochondria appear to reflect their morphology *in situ*. However, factors other than isolation can introduce structural variations. For example, the treatment used in preparation for visualization in the electron microscope can modify their appearance. In addition, they will vary in form according to the tissue source and their metabolic state before isolation.

12.3 ULTRASTRUCTURE OF MITOCHONDRIA

Terminology

Innumerable electron micrographs of mitochondria, both *in situ* and when isolated, have amply confirmed the general features of this double membrane-containing organelle. They are exceptionally well revealed in the electron micrograph of Figure 12.7.

The border of the mitochondrion is rimmed by a barrier of two membranes, each of which has the unit membrane characteristics of other cellular membranes. The two membranes appear discontinuous in Figure 12.7, and for the most part they are. However, the inner membrane is continuous with the crisscrossed membranes of the interior. These inner membranes, running in perpendicular and obtuse angles from the mito-

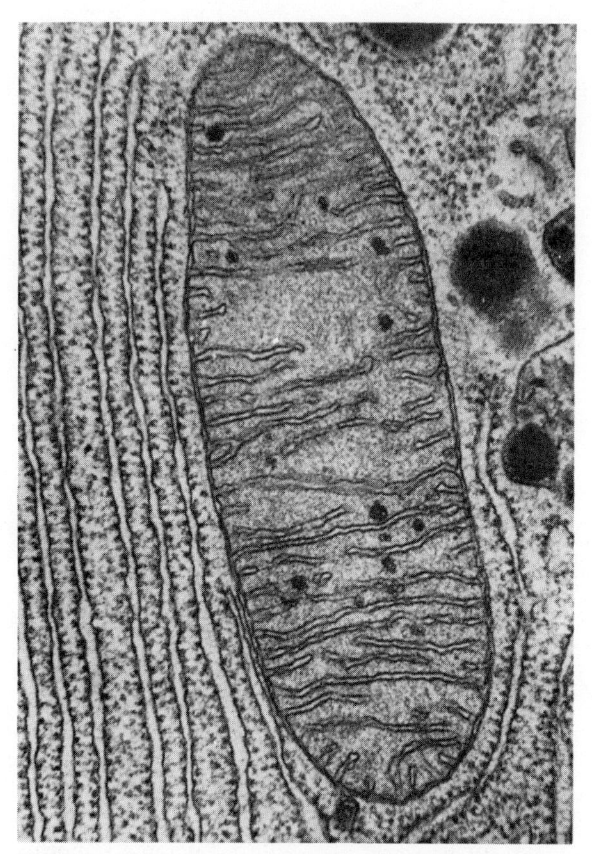

FIGURE 12.7 Electron micrograph of bat pancreas showing a mitochondrion as double membrane-bound organellea. The inner membrane is infolded to form the cristae. (Courtesy of Dr. Keith Porter.)

chondrial border, are infolded and flattened portions of the inner membrane. They have been given the name *cristae*.

Figure 12.8 depicts the fundamental features of the mitochondrion shown in Figure 12.7. This three-dimensional array indicates the outer covering membrane and the infolded inner membrane. There is evidence that the inner and outer membranes are attached to each other in certain types of mitochondrion, but the attachment is rather loose, since it is possible to separate the two membranes without breaking covalent bonds.

Because of this double membrane structure, mitochondria possess two discontinuous spaces, the *intermembrane space* and the *matrix,* or innermost compartment. The volume of the intermembrane space is quite small because of the close apposition of the inner and outer membranes (note Figure 12.7), but it has access to a large surface area of inner membrane due to the infolding of the latter.

The matrix may appear smooth and amorphous at low magnifications, but at higher magnifying powers some particulate materials are generally seen. These range all the way from the dense and easily resolved *matrix granule* to *ribosomes, polyribosomes,* and filaments of *DNA* (see Figure 12.9). The structures and functions of these are discussed later in the chapter.

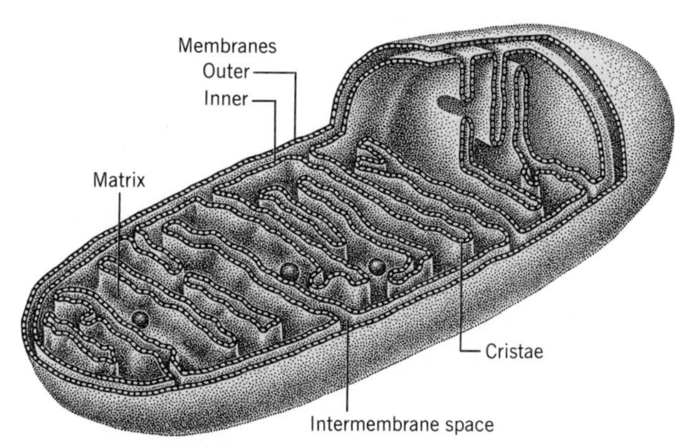

FIGURE 12.8 Fundamental features of the mitochondrion, emphasizing its membranes.

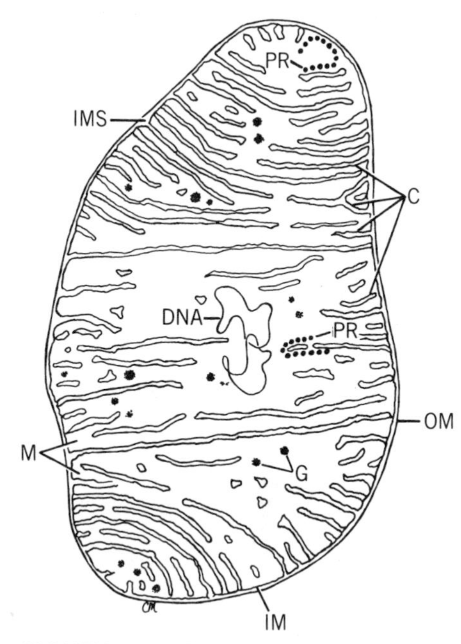

FIGURE 12.9 Drawing of thick section of mitochondrion showing typical internal structures. OM, outer membrane; IM, inner membrane; IMS, intermembrane space; C, cristae; G, granule; PR, polyribosomes; M, matrix.

Variations in Cristae

The most variable morphological structures of mitochondria are the cristae. Within a given cell type, they are generally quite uniform and characteristic of the cell, but an impressive array of different forms exists in different cell types.

The majority of mitochondria have cristae that are either *lamellar* or *tubular*. The lamellar form, in which the cristae are relatively parallel or stacked, is probably the most common and is exemplified by the pancreas mitochondrion shown in Figure 12.7 and the kidney mitochondria of Figure 12.10. In kidney, the cristae are generally quite numerous and look a bit like stacked coins. These cristae are apparently continuous with the inner bordering membrane through narrow necks or stems.

Another variation of the basic lamellar structure is seen in the mitochondria of insect flight muscle (Figure 12.11). Note the fenestrated appearance of these cristae.

An important functional implication of the cristae is emerging as we compare their conformations in pancreas, kidney, and insect muscle. In each case, by packing and by modifying the structure of the cristae, the area of the inner membrane is significantly affected. Thus, by regulating the surface area of the inner membrane, the cell receives the appropriate complement of respiratory enzymes to meet its metabolic demands in terms of ATP production. This will become more apparent as we examine the molecular structure of the inner membrane.

In other mitochondria, the cristae are predominantly in a tubular form. This is true, for example, for most protozoa and for a variety of mammalian steroid-producing cells. Figure 12.12 shows a type of tubular occurrence in which the tubules appear to be predominantly transversely oriented in the matrix. In some mitochondria, the tubules are extremely ordered in arrangement, such as is the case for the amoeba *Chaos chaos*. In these organelles there is not only a regularity in tubule morphology, but an interdigitation of linear zigzag tubules with a reticuloform system of tubes (see Figure 12.13).

In tissues such as the adrenal gland, from which steroids are secreted, the cristae often are more vesicular in nature and in many cases appear

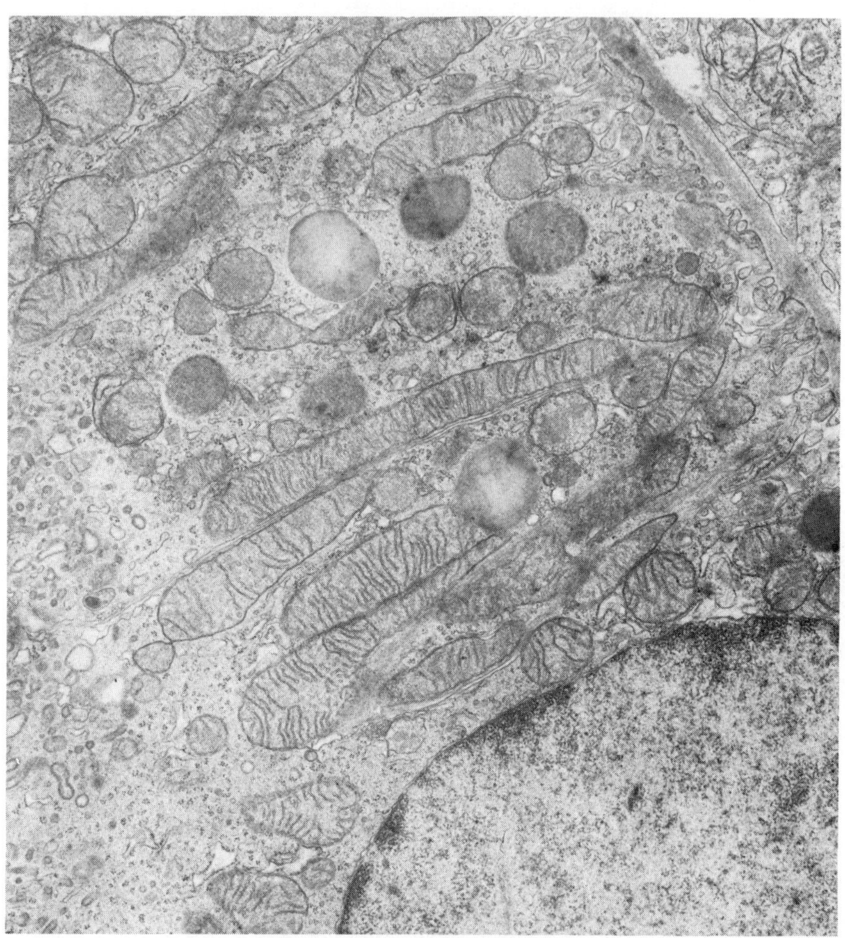

FIGURE 12.10 Electron micrograph of thin section of rat kidney tubule cell. Cristae have a "stacked coin" appearance but are not independent of the inner membrane.

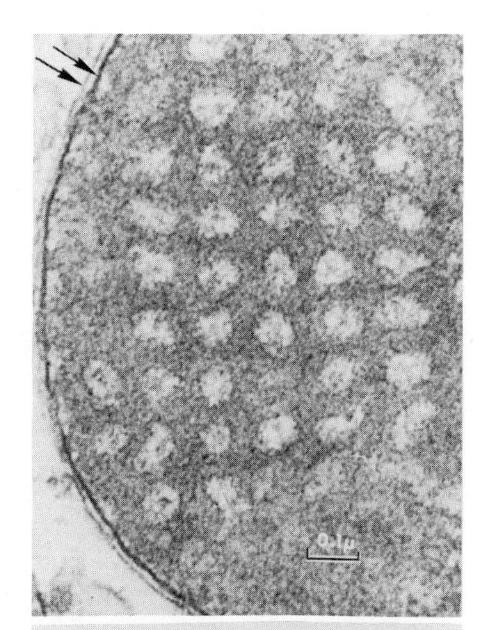

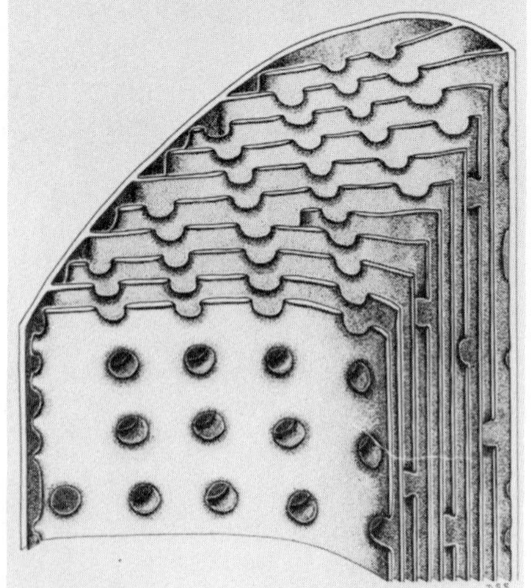

FIGURE 12.11 (*Top*) Mitochondrion of housefly (*Musca*) flight muscle, sectioned parallel with one of the fenestrated cristae. Inner and outer membranes are marked by arrows. (*Bottom*) A model depicting the interior of such a mitochondrion with fenestrated cristae. (Courtesy of Dr. David S. Smith.)

to be randomly distributed throughout the matrix (Figure 12.14). In other instances they may be packed in a more regular array.

These are only a few examples of the almost endless variety of cristal morphologies that are seen in mitochondria from different sources. The precise reason for the existence of the different forms is far from clear, but the general effect of the cristae generating an increased surface area of the inner membrane seems to be an important product of the morphological variations.

A simple calculation of a model liver cell mitochondrion may serve to emphasize this effect.[1] Assuming the liver cell to be a sphere of 30 μm diameter with 1000 cylindrical mitochondria (3.5 $\times$ 10μm), its surface area would be about 3000 μm^2 ($4\pi r^2$), and the combined outer surface area of the mitochondria about 13,000 μm^2. If one further assumes that each mitochondrion has 10 cristae, it can be calculated that the total cristal area of a liver cell is on the order of 16,000 μm^2, or about 10 μm^2 per mitochondrion. This should be sufficient to demonstrate the effectiveness of this mechanism to provide large amounts of surface area in a single

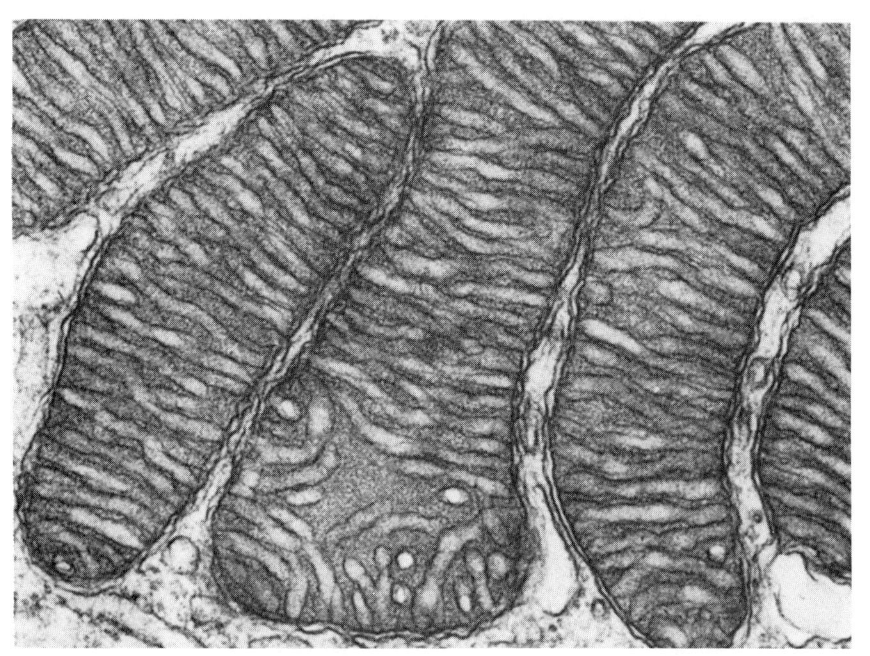

FIGURE 12.12 Tubular arrangement of cristae in mitochondria from the zona fasciculata of the hamster adrenal cortex. Although predominantly transverse, they appear to be random and intertwined in certain regions. (Courtesy of Dr. Don W. Fawcett.)

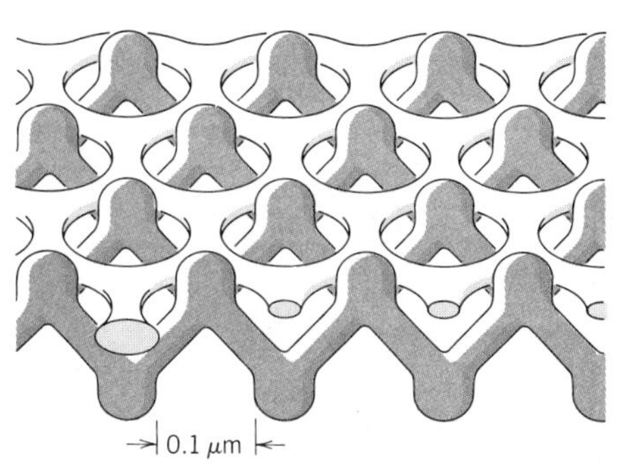

FIGURE 12.13 Drawing depicting the form of cristae in the mitochondria of the amoeba *Chaos chaos*. The tubular cristae with a regular wave form interdigitate with those having a reticuliform morphology.

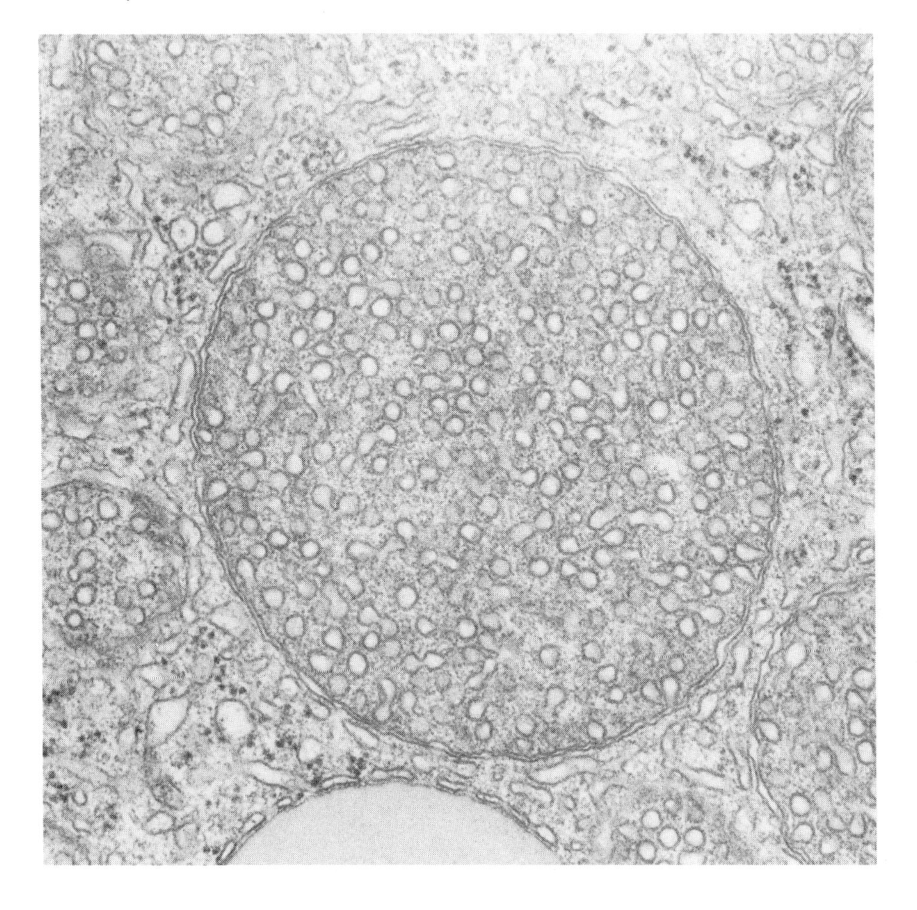

FIGURE 12.14 Electron micrograph of tubulovesicular cristae of adrenal cortical cell mitochondrion. (Courtesy of Dr. Daniel S. Friend.)

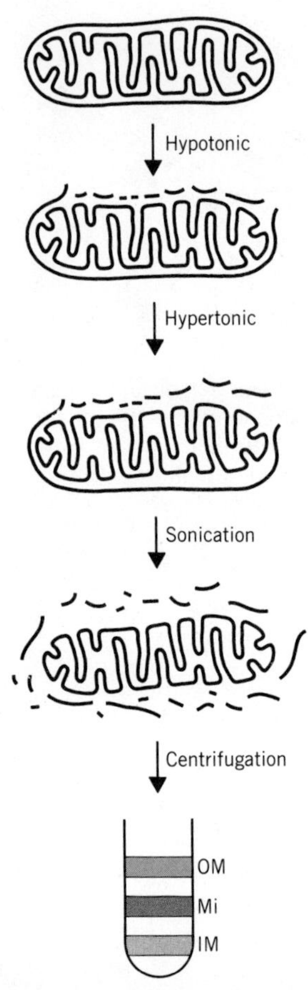

FIGURE 12.15 Steps used to separate outer from inner membranes of mitochondria. OM, outer membrane; IM, inner membrane; Mi, intact mitochondria.

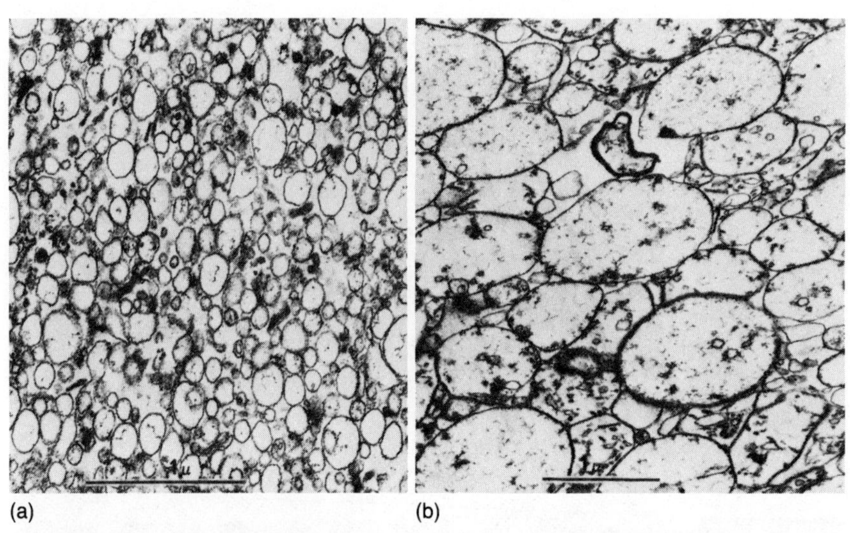

FIGURE 12.16 Mitochondrial outer (*a*) and inner (*b*) membranes obtained from rat liver by the technique outlined in Figure 12.15. Since the surface area of the inner membrane is larger, it forms larger membrane vesicles. (Courtesy of Dr. Lars Ernster.)

cell, and one can imagine how this must multiply in magnitude in mitochondria containing tubular and other modified forms of cristae.

Subfractionation of Mitochondria

In 1966 a very significant breakthrough in mitochondrial research helped to define the structures and roles of the outer and inner membranes. Parsons and his co-workers reported a procedure whereby the two membranes of mammalian liver mitochondria could be separated from each other.[2] Very shortly after this report, other researchers worked out techniques that could be used to separate the double membrane system in the mitochondria of many different cell sources.

The procedures that have been developed commonly involve four stages (see Figure 12.15). First, isolated mitochondria are placed in a *hypotonic* medium in which the outer membrane ruptures. The inner compartment remains intact under these conditions. Second, the medium is enriched with sucrose to a *hypertonic* state and ATP is added. This causes a contraction of the inner membrane-bound matrix. Third, mild *sonication* serves to shake the ruptured outer membrane from the inner compartment at points where they may be attached or nonspecifically adhering. Finally, the mixture is subjected to sucrose density gradient *centrifugation*. Centrifugation separates the mixture into three fractions, the lightest containing outer membranes, the heaviest inner membranes, and the intermediate fraction intact mitochondria.

Figure 12.16 is an electron micrograph illustrating the type of membrane preparations obtained by applying this procedure to rat liver mitochondria. Based on morphological differences only, the two fractions are distinguishable primarily in that the inner compartments form larger vesicles because of their larger surface areas. Some damage results to the inner membrane during this procedure. Outer membranes are present in assorted fragments and as resealed vesicles of varying dimensions.

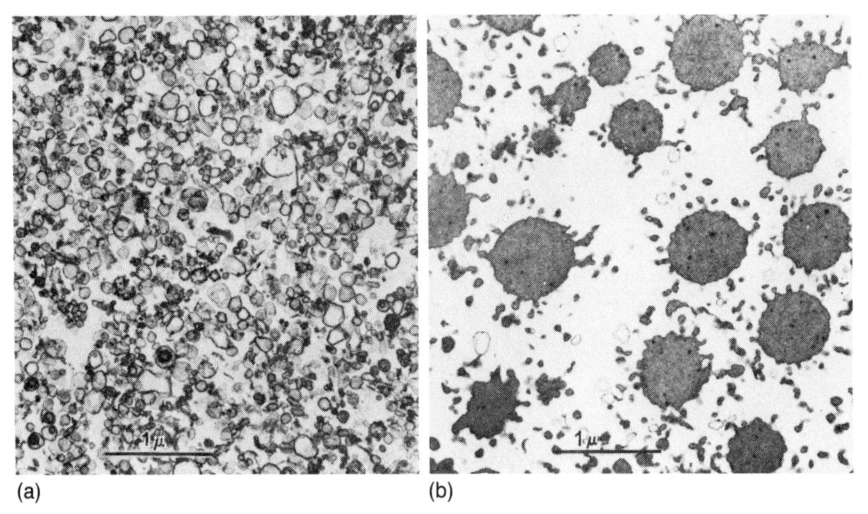

(a) (b)

FIGURE 12.17 Mitochondrial outer (*a*) and inner (*b*) membranes obtained by treatment with digitonin followed by differential centrifugation. (Courtesy of Dr. John W. Greenawalt.)

A modified separation procedure uses digitonin to selectively dismantle the outer membrane with little injury to the inner membrane. This, followed by differential and/or density gradient centrifugation, gives intact inner compartments in which the cristae appear to assume everted configurations (see Figure 12.17).

Chemical Composition of Membranes

The ability to separate the inner and outer membranes has permitted careful chemical analysis and comparison of the two systems.

Considering first the intact mitochondrion, water is quantitatively the predominant component. It is found in all parts of the mitochondrion except in the smectic lipid bilayers of the membranes and the interiors of the macromolecules. In addition to its chemical role in enzymatic reactions, it serves as the physical medium through which metabolites diffuse among the enzyme systems, many of which are anchored, and enter and exit the mitochondrion.

The major component of the dry mass of mitochondria is protein. The precise percentage occupied by protein is related to the amount of inner membrane present, for much of the protein, both enzymatic and structural, is found in the inner membrane. In some mitochondria, the inner membrane may contain as much as 60% of the total protein of the organelle. Based on enzyme distributions in rat liver mitochondria, it has been shown that the inner membrane contains 21.3% of the total mitochondrial protein and the outer membrane 4%.[3] According to these calculations, 67% of the protein is present in the matrix, and the rest is found in the intermembrane space.

The proteins of mitochondria can be classified broadly in two forms: soluble and insoluble. The soluble proteins are largely the enzymes of the matrix and certain of the peripheral or extrinsic membrane proteins. Insoluble proteins generally make up an integral part of the membranes. Some of these are structural and some are enzymatic proteins. For mi-

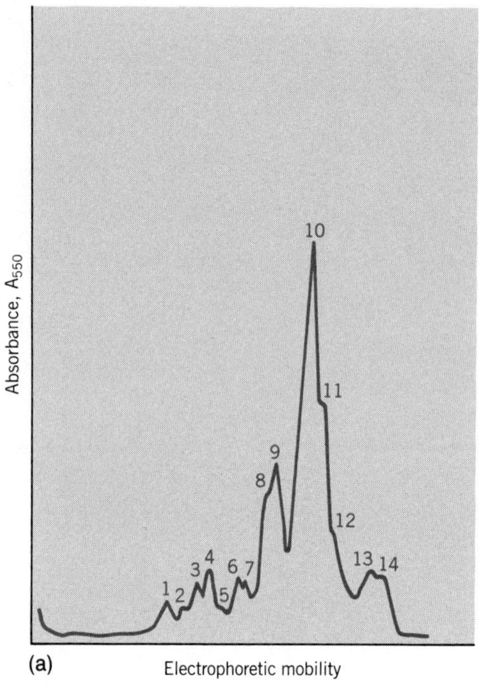

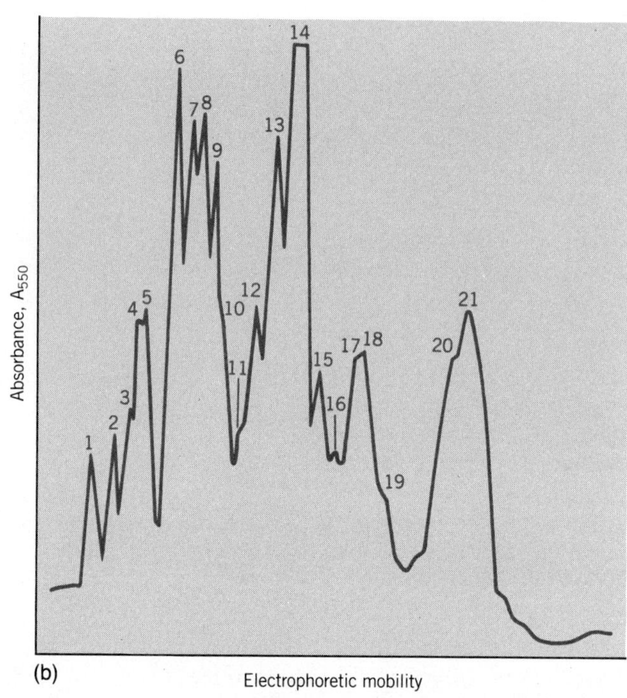

FIGURE 12.18 Densitometric traces of mitochondrial membrane proteins following their separation by polyacrylamide gel electrophoresis (PAGE). (*a*) Outer membrane proteins in 5% PAGE; molecular weights range from 220,000 (peak 1) to 12,000 daltons (peak 14). (*b*) Inner membrane proteins in 10% PAGE with 1% SDS; molecular weights range from 90,000 (peak 1) to 10,000 daltons (peak 21).

tochondria isolated from liver, approximately 50 to 70% of the protein is soluble,[3] whereas in ox heart only about 15% is easily extracted and considered soluble.[4] These values probably reflect the relative amounts of cristae that project into the matrix compartment.

The insolubility of mitochondrial proteins has made the elucidation of molecular structure a most difficult task. In many cases this property has permitted the isolation of only relatively large protein—lipid complexes that retain portions of the enzymatic activities of the intact membranes. When these complexes are further disrupted with detergents, the molecular components rapidly lose their activities.

Figure 12.18 compares the complexities of the outer and inner membranes by way of the electrophoretic patterns of their proteins. About 14 different proteins are discernible in the outer membrane and 21 in the inner membrane. As we shall see shortly, there are many more enzymatic activities associated with the inner membrane, in agreement with these electrophoretic results.

Cardiolipin (diphosphatidylglycerol) is found at significant levels only in bacterial membranes and in the *inner* membrane of mitochondria.

The lipid composition of mitochondria varies considerably with their source, but in all cases phospholipid is the predominant form, generally accounting for over three quarters of the total (Table 12.1). Phosphatidylcholine and phosphatidylethanolamine generally occur in the highest amounts (Table 12.2), but a significant level of cardiolipin is found, and cholesterol is present at low levels. The abundance of cardiolipin and the dearth of cholesterol set the mitochondrion apart, compositionally, from the other membranes of the cell.

Table 12.1 Lipid Content of Mitochondria Isolated from Various Animal Organs (mg/g protein)

Organ	Total Lipid	Neutral Lipid	Cholesterol	Phospholipid
Human heart	400	—	—	335
Ox heart	320	18	4	283
Ox kidney	240	17	11	190
Ox liver	180	16	4	145
Guinea pig liver	—	—	2	159

Table 12.2 Phospholipid Composition of Mitochondria Isolated from Various Sources (% total phospholipid)

Source	Phosphatidyl-choline	Phosphatidyl-ethanolamine	Cardiolipin	Phosphatidyl-inositol (and others)
Human heart	43	34	18	5
Pig heart	36	25	13	23
Ox heart	41	37	19	3
Ox kidney	40	38	19	4
Ox liver	43	35	17	5
Guinea pig liver	40	28	22.5	7
Blowfly flight muscle	13.4	67	16.8	—
Housefly flight muscle	8.0	58.5	24.6	8.9
ELD ascites tumor cells	48.9	27.3	11.3	12.5
Yeast (wild type)	32.5	22.0	15.6	29.9
Crown-gall	44	30	10	16
Cauliflower	42	30	12	16
Mung bean	44	29	11	16

Table 12.3 Phospholipid Composition of Inner and Outer Membranes of Liver Mitochondria (% of total phospholipid in each fraction)

	Guinea Pig			Rat			
Phospholipid	Whole	Inner Membrane	Outer Membrane	Whole	Inner Membrane	Outer Membrane	"Soluble"
Phosphatidylcholine	40	44.5	55.2	47	41	49	64
Phosphatidylethanolamine	28	25.3	27.7	26	35	31	28
Cardiolipin	22.5	21.5	3.2	14	21	3	8
Phosphatidylinositol } Phosphatidylserine	7 / —	4.2 / —	13.5 / — (combined)	13	2	17	—

The phospholipid distributions in the outer and inner membrane are shown in Table 12.3. The most noticeable difference between the two membranes is the higher levels of cardiolipin in the inner membrane than in the outer. Distribution studies on cholesterol have shown it to be associated primarily with the outer membrane. The different distributions of lipids noted undoubtedly have some structural and functional significance, not yet clearly understood in detail. In general, the *outer* membrane

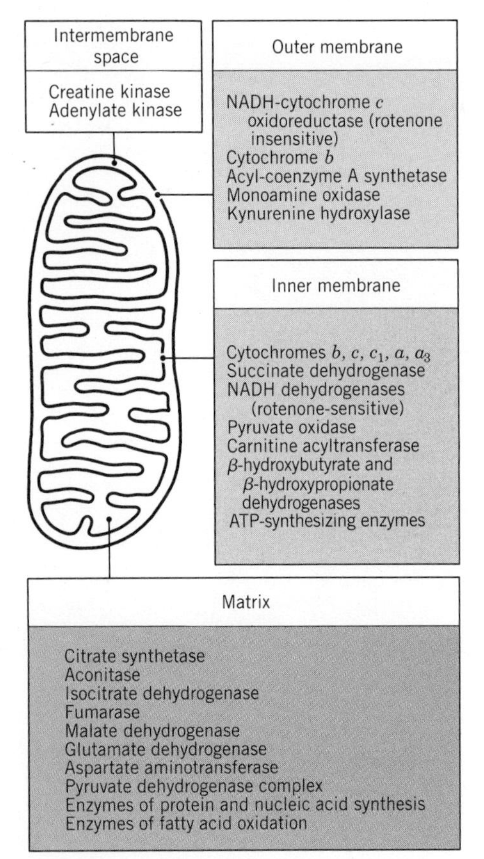

FIGURE 12.19 Compartmentation of some of the major enzymes of the mitochondrion.

Monoamine oxidase (MAO) is a flavin enzyme that serves in the catabolism of the neurotransmitters dopamine and norepinephrine. In so doing it oxidatively removes amino groups.

$$R{-}CH_2{-}NH_3^+ \xrightarrow[H_2O]{MAO} R{-}CH{=}O + NH_4^+$$

resembles the other intracellular membranes more than the inner membrane.

It is known that the electron transport system depends on phospholipids for its operation[5] and the interactions of coenzyme Q and the other redox molecules with their neighbors or carriers appear also to be lipid dependent.[6]

A number of different small organic molecules, in addition to the lipids cited above, are membrane associated in mitochondria. Several of these are redox molecules that participate in electron transport. The ubiquinones (coenzymes Q), flavins (FMN and FAD), and the pyridine nucleotides (NAD^+) are normally membrane bound, associated most often with the inner membrane.

Enzymatic Compartmentation

About 120 enzymes have been identified as mitochondrion associated. Approximately 37% of these are oxidoreductases, 10% are ligases, and less than 9% hydrolases.[6] Of these, only about 30 have been characterized in significant biochemical and biophysical terms.

A schematic of the compartmentation of some of the major enzymes is presented in Figure 12.19. Of the several enzymes found in the outer membrane, monoamine oxidase is the most thoroughly characterized. It contains 1 mole of covalently bound flavin, 2 moles of sialic acid, and 8 to 12 moles of hexosamine. It has a molecular weight of 115,000 daltons.[7] This protein has become the marker enzyme for outer membrane material.

The inner membrane, even though more complex than the outer, is considerably better characterized. Succinate dehydrogenase, a thoroughly studied protein, is the marker enzyme for the inner membrane. The enzymes of electron transport and oxidative phosphorylation are exclusively associated with the inner membrane. Most of the proteins of the inner membrane are separable into one of five complexes by applying subfractionation procedures to isolated inner membranes, as discussed in the next section.

The matrix contains a collection of enzymes that are mediators of the reactions of the tricarboxylic acid (TCA) cycle and are concerned with protein and nucleic acid synthesis. All the enzymes of the TCA cycle are free within the matrix except for succinate dehydrogenase, which as just noted is a component of the inner membrane. Thus, for pyruvate to be completely oxidized by way of these enzymes to carbon dioxide and water in the matrix, the metabolite succinate must make contact with the inner membrane before it is oxidized to fumarate (Figure 12.20).

Subfractionation of the Inner Membrane

The inner membrane can be disrupted by chemical and enzymatic treatment and separated into four lipid–protein complexes, each of which is capable of carrying out a fraction of what is accomplished by the complete electron transport chain. In addition, the procedure can be used to isolate the ATP synthetase system.

By studying each of these complexes in detail, attempts have been made, particularly by Green and co-workers, to reconstruct the intact electron transport system as it exists in the natural state.[8] Let us, before discussing the electron transport chain and the related event of oxidative phosphorylation, consider the physicochemical and enzymatic properties

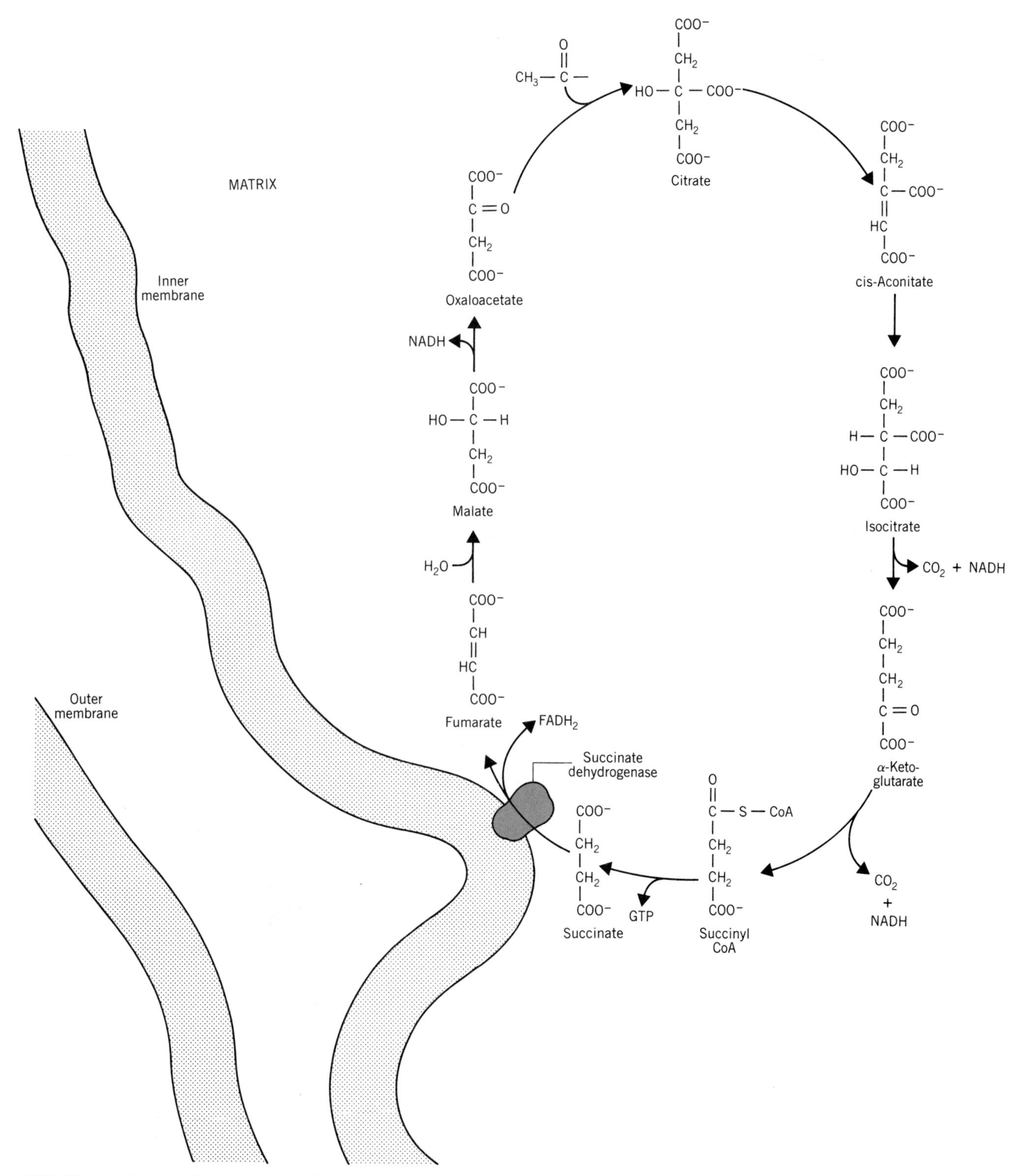

FIGURE 12.20 The tricarboxylic acid cycle operates completely with matrix enzymes except for one step, the conversion of succinate to fumarate. This enzyme, succinate dehydrogenase, is an integral protein of the inner membrane. Because of this, it is used as a convenient marker enzyme for inner membrane material.

Table 12.4 Properties of the Electron Transport Complexes I to IV from Ox Heart

Complex	Enzyme Activity	Components	Phospholipid (%)	Particle Weight ($\times 10^{-4}$ daltons)
Complex I	NADH-coenzyme Q oxidoreductase	Nonheme iron protein, structural protein, flavin	22	117
Complex II	Succinic–coenzyme Q oxidoreductase	Nonheme iron protein, cytochrome *b*, flavin	20	20
Complex III	Reduced coenzyme Q-cytochrome *c* oxidoreductase	Nonheme iron protein, cytochrome *b*, cytochrome c_1, core protein, fifth small protein	20–30	30
Complex IV	Cytochrome *c* oxidase	Cytochromes *a*, a_3, copper protein	35	26–36

of each of these complexes. These are briefly summarized in Table 12.4.

Complex I: NADH Coenzyme Q Oxidoreductase

Complex I electron transfer activity:
NADH (NADPH) ⟶ Q

The major component of complex I is the enzyme NADH dehydrogenase, an FMN flavoprotein of molecular weight about 70,000 daltons.[7] A densitometric trace of the complex after polyacrylamide gel electrophoresis (Figure 12.21*a*) shows that several other protein components are present in the complex. However, except for the presence of NADH dehydrogenase activity and nonheme iron centers, these materials have not been well characterized. The molecular weight of the entire complex is about 550,000 daltons, and a considerable amount of lipid is present, much of it tightly bound to the protein.

The enzymatic ability of the complex is to transfer electrons from the coenzyme NADH to coenzyme Q. NADH is, of course, generated as a result of three different reactions of the TCA cycle (Figure 12.20). So the role of complex I appears to be the first step in handling the electrons produced in the mitochondrial matrix.

Complex II: Succinic Coenzyme Q Oxidoreductase

Complex II electron transfer activity:
Succinate ⟶ Q

Smaller than complex I, complex II has a total molecular weight of about 220,000 daltons. The enzyme succinate dehydrogenase is the major protein component, made up of two polypeptides, having molecular weights of 70,000 and 23,000 daltons, respectively. The former carries a flavin prosthetic group (FAD) and the other nonheme iron. The flavin is covalently linked to the protein through its 8α position to a histidine residue.

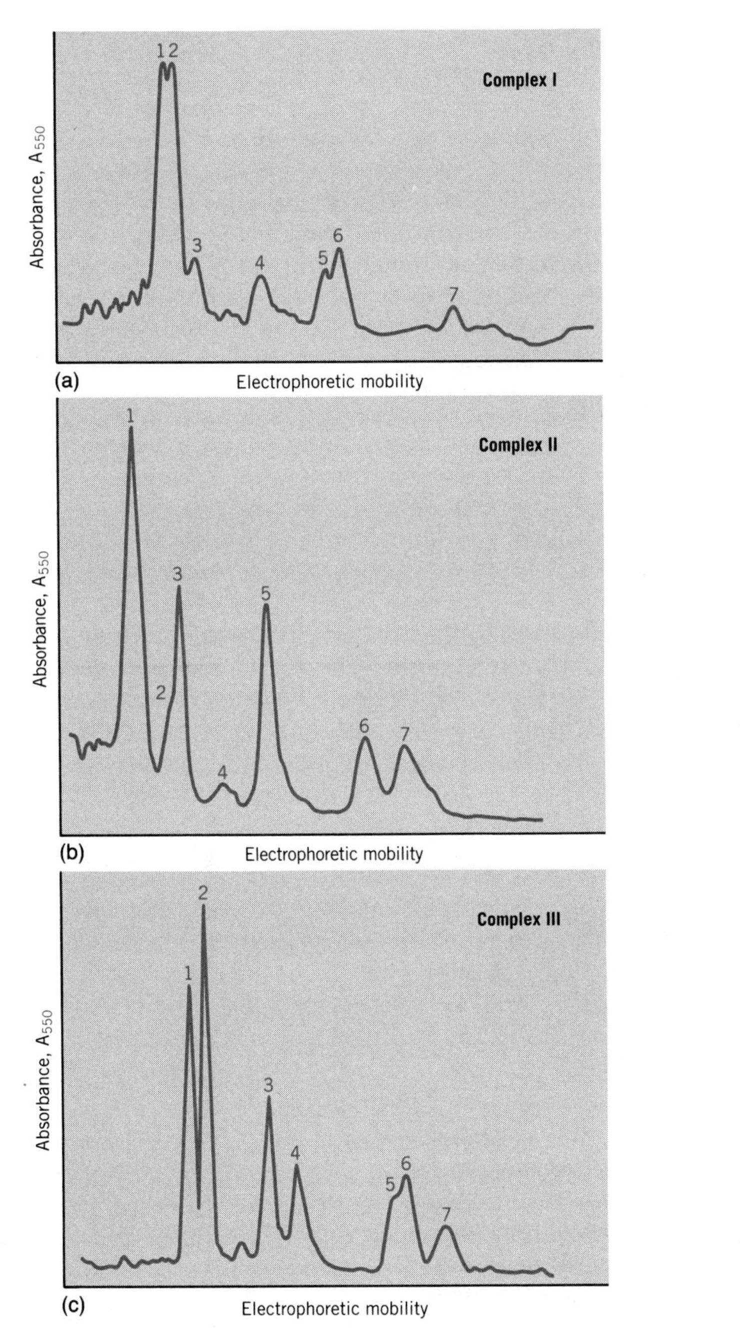

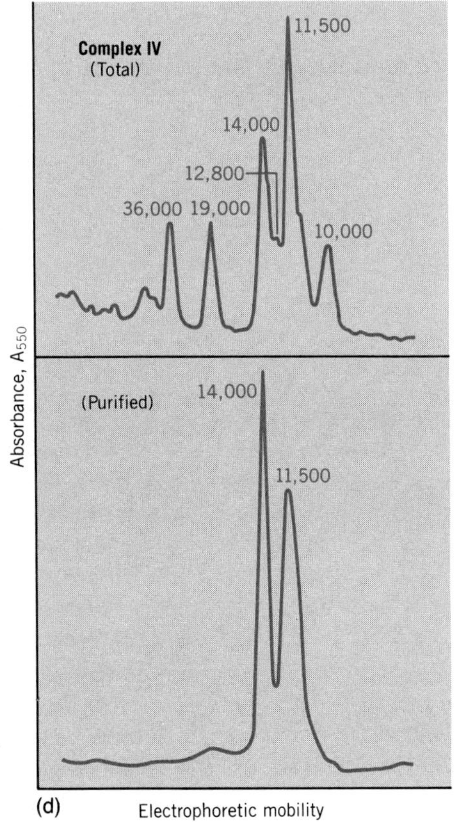

FIGURE 12.21 Densitometric traces of inner membrane enzyme complexes after separation in a 10% polyacrylamide gel.

The densitometric tracing of the results of electrophoresis on this complex (Figure 12.21*b*) indicates that other components are present in the complex, some in fairly significant amounts. Spectrophotometric studies on the complex show the presence of a cytochrome with *b*-like properties, and other studies reveal another nonheme iron protein.

Complex II has the ability to oxidize succinate to fumarate and transfer the extracted electrons to oxidized coenzyme Q. It thus appears to be a second feeder line to coenzyme Q, along with complex I. Since succinate is a normal product of the TCA cycle, the succinate-to-fumarate step is unique in the cycle in that it is the only step mediated by a membrane-bound enzyme.

Complex III: Reduced Coenzyme Q Cytochrome c Oxidoreductase

Complex III electron transfer activity:
$QH_2 \longrightarrow$ Cytochrome *c*

Complex III has been studied in great detail, and many of the physicochemical properties of its components are known. Its total particle weight is around 210,000 daltons, very close to that of complex II. The complex contains cytochrome c_1, two *b* cytochromes called b_k and b_t, one or more nonheme proteins, and a core protein.

Cytochrome c_1 is a heme glycoprotein made up of two polypeptides (29,000 and 15,000 daltons). Cytochrome b_k is apparently the original cytochrome *b* studied by Keilin. The two *b* cytochromes have molecular weights around 30,000 and 50,000 daltons. One nonheme iron protein that has been purified has a molecular weight around 26,000 daltons, and the core protein is about 50,000 daltons.

The densitometric tracing of the electrophoretic pattern of complex III (Figure 12.21*c*) illustrates the complexity of this particle.

Complex III uses reduced coenzyme Q as a substrate and passes its electrons to cytochrome *c*. The latter is an extremely well characterized protein, the only member of the electron transport chain that is not tightly bound to the inner membrane. Complex III, by virtue of its electron-transferring ability, functions as an electron lead from coenzyme Q, where the pathways of complexes I and II converge.

Complex IV: Cytochrome c Oxidase

Complex IV electron transfer activity:
Cytochrome *c* $\longrightarrow O_2$

Complex IV, another particle of 210,000 daltons, is made up of seven different proteins (in yeast—there are six in beef heart), which range in molecular weight from about 10,000 to 36,000 daltons. Two hemes are associated with the complex, but they are chemically distinguishable. One reacts with carbon monoxide and is identified with cytochrome a_3, and the other, associated with cytochrome *a,* does not form a similar complex. Two copper moieties are also characteristic of the complex. The heme and copper are associated with polypeptides that have molecular weights of 11,500 and 14,000 daltons.

A densitometric tracing of the components of complex IV is presented in Figure 12.21*d,* along with a purified preparation of the heme- and copper-associated proteins.

Complex IV functions as the terminal oxidase system in electron transport, removing electrons from reduced cytochrome *c* and transferring them to oxygen where, with protons available from the environment, water is formed.

Figure 12.22 is a schematic representation of the coordinated activities of these four complexes in carrying out the whole of electron transport.

ATP Synthetase

A fifth molecular aggregate that is separable from the inner membrane is a complex characterized by its ability to hydrolyze ATP when in isolated form. The reaction of the complex *in situ* is actually the reverse of this, ATP synthesis.

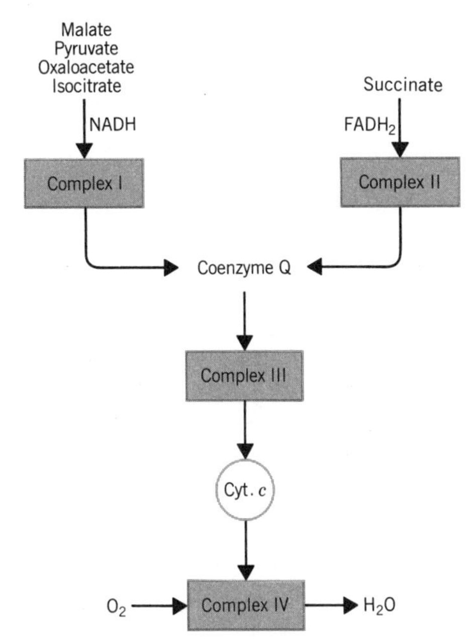

FIGURE 12.22 Schematic representation of coordination of activities by the inner membrane complexes to effect complete electron transport from a substrate to oxygen.

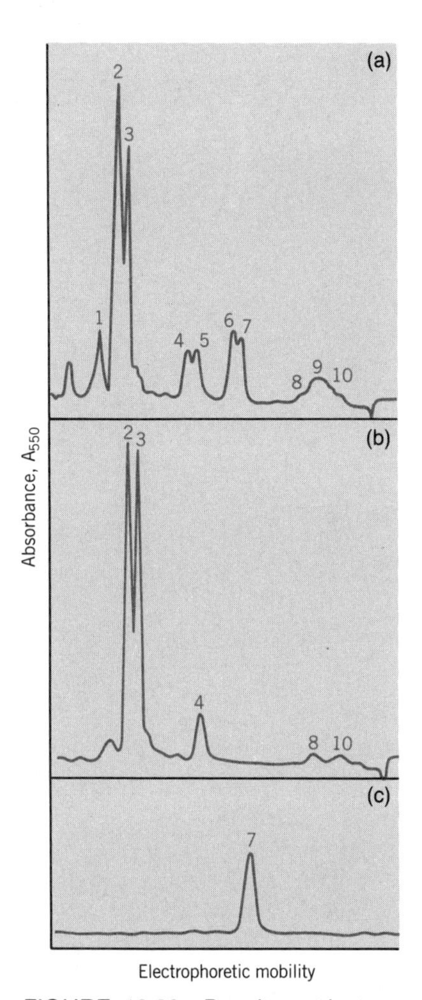

FIGURE 12.23 Densitometric traces of different fractions of ATP synthetase (complex V) after separation in 10% PAGE. (*a*) Total complex. (*b*) F_1 ATPase (extrinsic proteins). (*c*) Oligomycin-sensitivity-conferring protein (OSCP): one of the five intrinsic proteins).

ATP synthetase is a complex of 10 polypeptides, of which half are intrinsic membrane proteins and half are extrinsic, projecting toward the matrix side of the inner membrane. The extrinsic proteins constitute a subaggregate of the entire complex, often referred to as F_1 ATPase, where the reversible ATPase activity is located. Each of the five extrinsic proteins has been purified and subjected to detailed chemical analysis, including determination of amino acid composition.

The five intrinsic proteins, a water-insoluble aggregate called F_0, contain receptor sites for oligomycin and dicyclohexylcarbodiimide (DCCD), specific inhibitors of ATP production. A 19,000 dalton member of this complex, thought to be responsible for oligomycin sensitivity, is called OSCP, an acronym for oligomycin-sensitivity-conferring protein. Another, 10,000 daltons in molecular weight and extremely hydrophobic, is thought to be the precise site of DCCD binding. Densitometric tracings of the total complex, F_1, and OSCP appear in Figure 12.23.

The ATP synthetase complex is now known to be structurally related to the stalked particles or knobs that were seen in electron micrographs of inner membranes several years before the molecular details of the complex were revealed. These knoblike projections are visible in the electron micrograph of Figure 12.24. They are readily seen by negatively staining inner membrane preparations, but are generally not observed in sectioned specimens.

Inner membranes that are severely disrupted are generally observed as a collection of membrane vesicles with the knobs projecting outward, an observation that for some time was interpreted as meaning that the knobs were projecting into the intermembrane space in intact mitochondria, away from the matrix. It is now known that these rough vesicles are formed by a sealing of fragments of the cristae after they are torn apart by the disrupting processes, as shown in Figure 12.25.

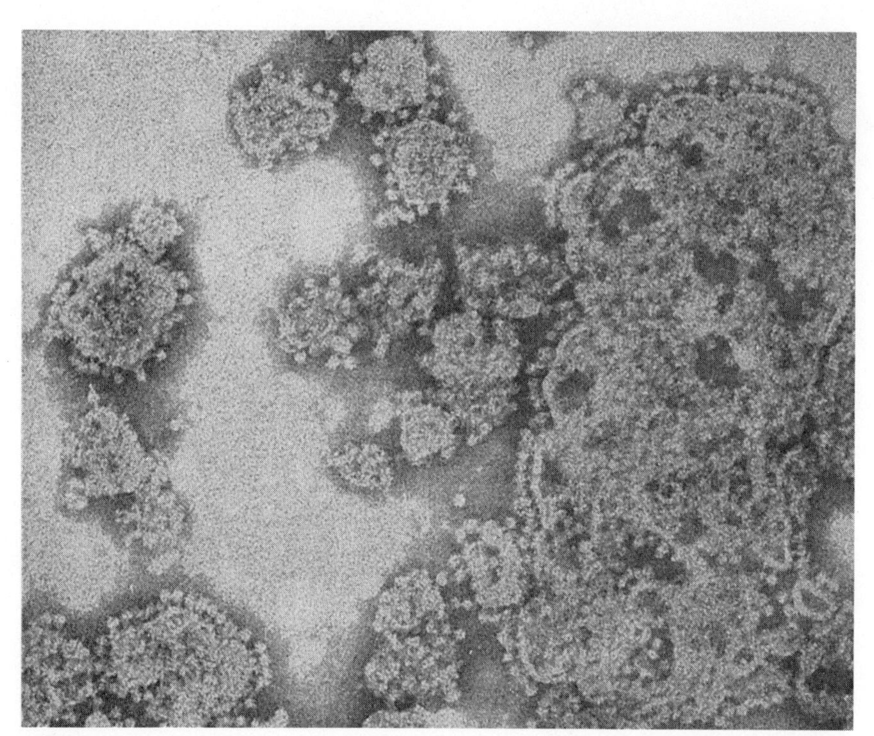

FIGURE 12.24 Submitochondrial particles showing the ATP synthetase complex, which exists as a knob attached to a stalk fixed to the inner membrane. This structure projects into the mitochondrial matrix. ×163,000. (Courtesy of Dr. Efraim Racker.)

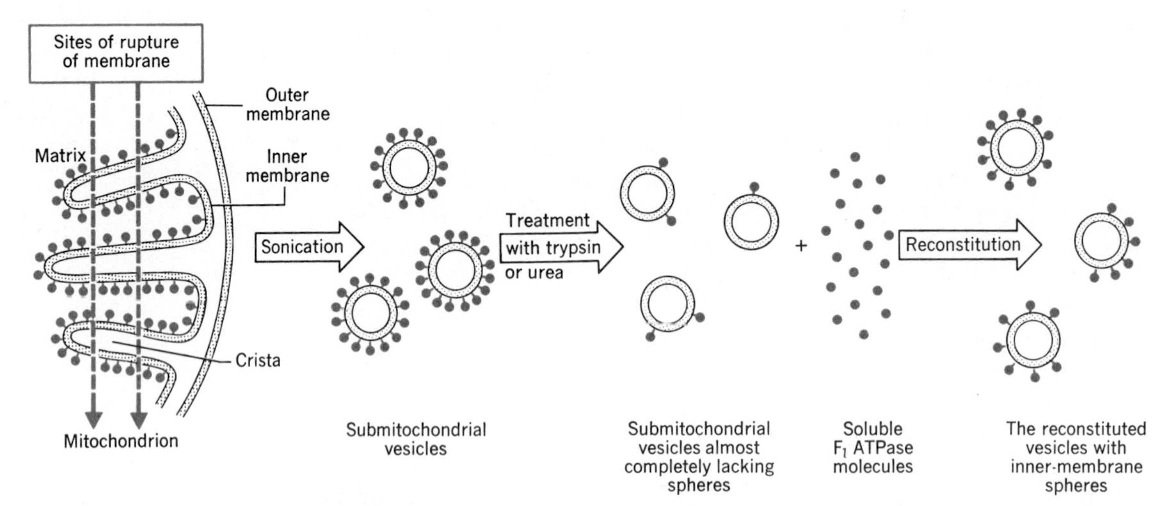

FIGURE 12.25 The effect of disrupting the inner mitochondrial membrane and the manner in which submitochondrial vesicles can be partially disassembled and reconstituted.

The abilities of researchers to bring about the formation of these inside-out knobbed vesicles, to monitor them in the electron microscope, and to test them enzymatically have provided ways to determine whether the projecting structures indeed have the ATPase activity of the inner membrane. The F_1 particles, 85 nm spheres, can be removed from the vesicles by treatment with urea,[9] and the smooth membranes that result lack ATPase activity (see Figure 12.26). Under appropriate experimental conditions, the F_1 particles can be reattached to the membrane vesicles, effecting a regain of ATPase activity. Although it is possible that urea may have a damaging effect on an intrinsic membrane ATPase, or that the F_1 particle must be physically attached to the membrane to activate such an enzyme, it is now generally accepted that the morphologically observed knobs are the sites of ATPase activity, and thus ATP synthesis in the intact mitochondrion.

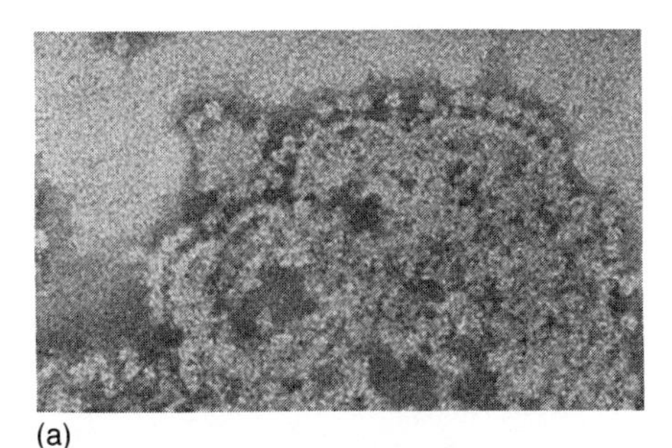

(a)

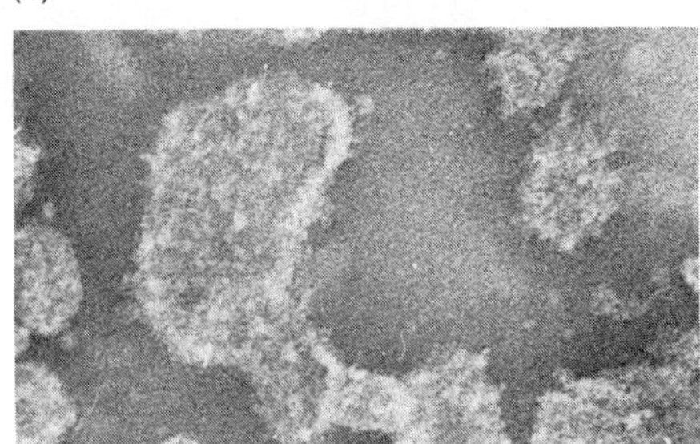

(b)

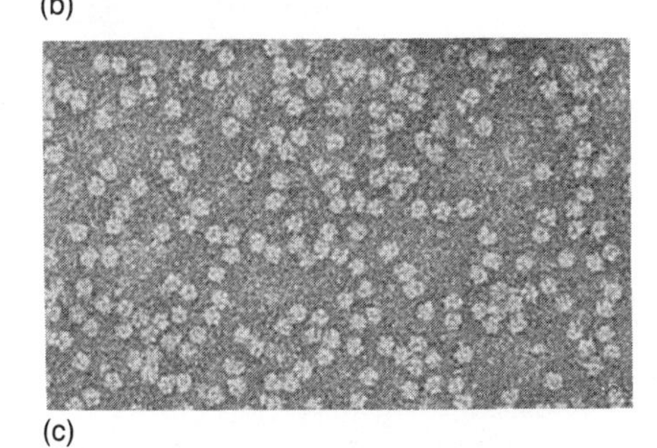

(c)

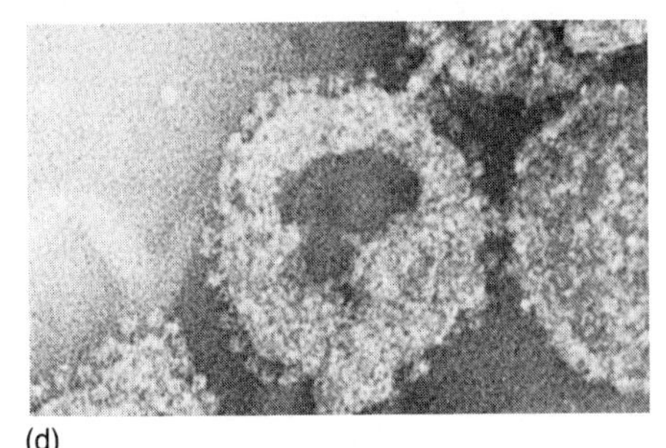

(d)

FIGURE 12.26 Series of electron micrographs demonstrating the classic experiment of Efraim Racker that showed ATPase activity to be associated with F_1 particles. (*a*) Submitochondrial particles (inner membrane pieces) contain ATPase activity and F_1 particles. (*b*) Treatment with urea removes F_1 particles and ATPase activity. (*c*) Isolated F_1 particles. (*d*) Inner membranes with reattached particles. The ATPase activity is restored. (Courtesy of Dr. Efraim Racker.)

Stoichiometry and Organization of Complexes in the Membrane

There appear to be quite tightly controlled stoichiometric ratios of the five complexes in the inner mitochondrial membrane. For example, for every 2 molecules of cytochrome *c*, there are 1 of complex III, 2 of complex IV, and 16 of coenzyme Q.[10] At present, there is not agreement on the amounts of complex I and complex II relative to that of complex III, but they appear to be equal to or somewhat less than the amount of complex III. The ratio of ATP synthetase to complex III is 1, thus establishing a stoichiometry suggesting that there is only one site for ATP synthesis for each complete electron transport chain.

By studying intact mitochondrial inner membranes and inside-out vesicles of inner membranes, it has been possible to show an asymmetric distribution of components across the membrane. Antibodies have been prepared against specific membrane molecules. These have been combined with native inner membranes and with subinner membrane vesicles, noting the ability of the antibodies to inactivate the various membrane enzymes. The technique used (Figure 12.27) is powerful because different sides of the inner membranes are protected from antibodies in the two membrane preparations.

The F_1 complex, as we have already noted, is clearly on the matrix side of the membrane according to this technique. Cytochrome *c*, on the other hand, is accessible only from the outer surface of the membrane. Cytochrome oxidase is a transmembrane protein, surfacing on both sides of the membrane. Other members of the electron transport chain are thought to be oriented according to the schematic summary in Figure 12.28. An asymmetric distribution of the members of the electron transport chain is depicted, a feature that becomes important when we consider the mechanism whereby the energy of electron flow is transformed so that it can be used for ATP synthesis.

12.4 MECHANISMS OF ENERGY TRANSFORMATION

Before discussing the current views of energy transformation, it may be helpful to step back and take a brief look at the relationship of the material presented up to this point to intermediary metabolism. Metabolic products resulting from the extramitochondrial reactions of glycolysis enter the mitochondrion and are acted on by the matrix enzymes of the TCA cycle. One enzyme of this cycle, succinate dehydrogenase, is an intrinsic inner

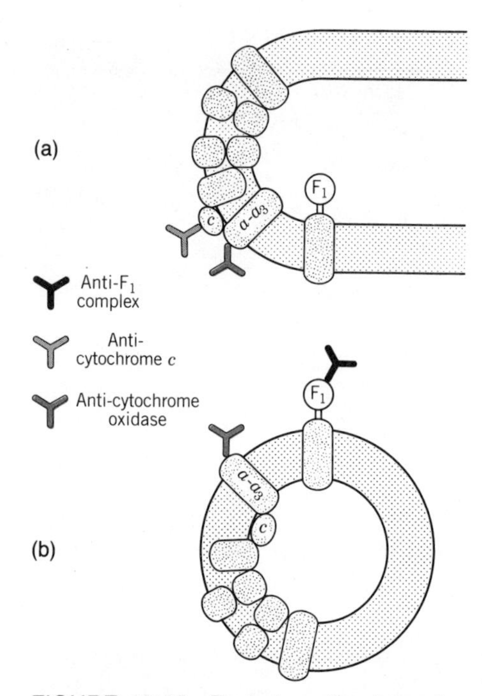

FIGURE 12.27 Technique for determining the topography of enzymes in the membrane by the use of antibodies. Specific antibodies will bind to and inactivate only enzymes that are exposed at a given surface. An enzyme that can be inactivated by providing antibodies to both surfaces of a membrane is a transmembrane protein. (*a*) Intact inner membrane. (*b*) Inside-out inner membrane vesicle.

membrane protein. The oxidoreductases of the cycle sequester electrons from the substrates, loading them on to the NADH and $FADH_2$ electron carriers. The electrons then flow through a system of membrane enzymes involving complexes I, II, III, and IV, which are known to have a particular spatial organization in the membrane.

The energy released as the electrons flow is captured in part by the phosphorylation of ADP to ATP, mediated by the ATP synthetase complex. This electron flow is *coupled* to phosphorylation so that for each pair of electrons that enters by way of complex I, three ATPs are generated per mole of oxygen consumed, giving a P/O ratio of 3. Electrons entering through succinate dehydrogenase, complex II, result in a P/O ratio of only 2. A variety of chemical agents, most notably dinitrophenol, can *uncouple* the normally tightly related events of electron transport and oxidative phosphorylation. Uncouplers permit continued electron transport but inhibit phosphorylation.

One of the most puzzling and active areas of biochemical research has been the elucidation of the mechanisms whereby the energy of electron flow is coupled to the synthesis of ATP. Because of the structural and enzymatic knowledge that has accumulated during the past few years, we are at a historic point, that is, the mechanism of energy transformation is becoming clear.

The Steps of Oxidative Phosphorylation

Oxidative phosphorylation can be viewed as a three-step process.[11] Step 1 involves the *transformation* of the energy of oxidation to a different form, which may be represented by the creation of an activated or conformationally changed intermediate, or by a directional flow of chemical agents. Step 2 is a *transmission* of energy to a coupling device, which may be thought of as a transmission of an activated or conformational state to another molecule, or a flow of gradient-producing ions. In step 3, energy is again *transformed,* this time in the formation of ATP.

Historically, three main theories have been set forth to explain these steps. Although only one of these, the chemiosmotic theory, is becoming widely accepted and trusted, the other two are very briefly discussed. Any such theory must be able to answer two overriding questions:

1. How is the energy of electron flow (the respiratory chain) transformed and transmitted?
2. How is the energy finally used (by ATP synthetase)?

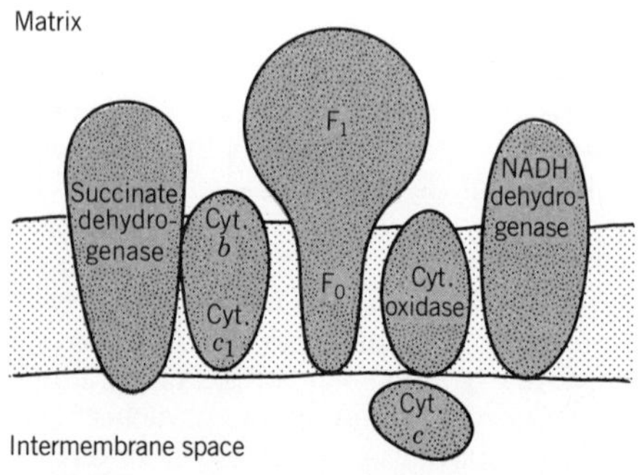

FIGURE 12.28 Illustration of the orientation of several inner membrane proteins of the mitochondrion.

The Chemical Hypothesis

The chemical hypothesis, first propagated by E. C. Slater in 1953, proposes that step 1 of oxidative phosphorylation gives rise to a high energy intermediate molecule, presumably an activated protein or similar molecule (see Table 12.5). This activated molecule, represented as $A \sim X$, transmits its energy to another species, $X \sim Y$, which is an intermediate coupler, operating between the respiratory chain and the site of ATP synthesis. The final step takes place when $X \sim Y$ undergoes phosphorylation to $X \sim P$ and then transphosphorylation to ADP, generating ATP.

Table 12.5 Three Steps and Three Hypotheses of Oxidative Phosphorylation

	Coupling Mechanism		ATP-Pump Mechanism
Hypotheses	Step 1: Energy Transformation	Step 2: Energy Transmission	Step 3: Energy Transformation
Chemical	$AH_2 + B + X \rightleftarrows A \sim X + BH_2$	$A \sim X + Y \rightleftarrows X \sim Y + A$	$X \sim Y + P_i + ADP \rightleftarrows X + Y + ATP + H_2O$
Conformational	$AH_2 + B \rightleftarrows A^* + BH_2$	$A^* \rightleftarrows X^*$	$X^* + P_i + ADP \rightleftarrows X + ATP + H_2O$
Chemiosmotic	$AH_2 + B \rightleftarrows A + B_{red} + 2H^+_{M\text{-side}}$	Proton flux	$P_i + 2H^+_{M\text{-side}} + ADP \rightleftarrows ATP + H_2O + 2H^+_{C\text{-side}}$

This theory is largely modeled after the mechanism of substrate phosphorylation that takes place in glycolysis, where an energy-rich intermediate is formed prior to the phosphorylation of ADP.

The chemical hypothesis was the original hypothesis of oxidative phosphorylation, and served as a framework for huge amounts of research. But the experimental results generally do not favor its main thrust, that of possessing an activated intermediate coupler, and most researchers, including Slater, have come to favor the chemiosmotic theory.

In glycolysis the energy-rich intermediate is 1,3-diphosphoglycerate (1,3-diPG)

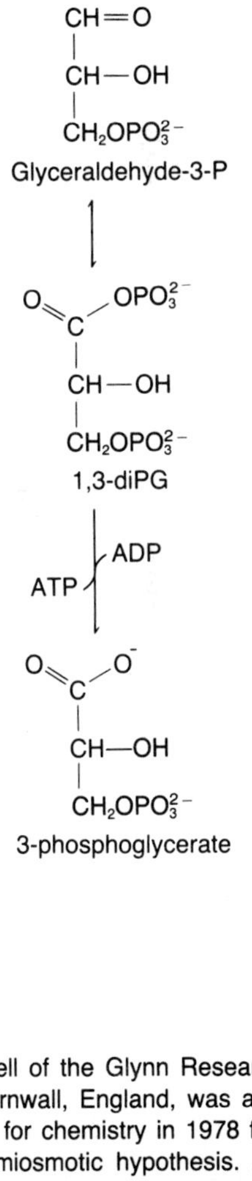

CH=O
|
CH—OH
|
$CH_2OPO_3^{2-}$
Glyceraldehyde-3-P

↓

$O{=}C{-}OPO_3^{2-}$
|
CH—OH
|
$CH_2OPO_3^{2-}$
1,3-diPG

ADP → ATP

$O{=}C{-}O^-$
|
CH—OH
|
$CH_2OPO_3^{2-}$
3-phosphoglycerate

The Conformational Hypothesis

The conformational hypothesis, proposed by Paul Boyer in 1965,[12] suggests that step 1 results in a conformational change in a respiratory component $A \rightharpoonup A^*$, which is then transmitted to another component $X \rightharpoonup X^*$, which brings about phosphorylation of ADP and release of ATP from the ATP synthetase. As tantalizing a mechanism as this has been to explain the relation between electron transport and oxidative phoshorylation, there has not been appropriate experimental verification of the theory. Boyer recently modified his theory to include the basic features of the chemiosmotic proposal.

The Chemiosmotic Hypothesis

The chemiosmotic hypothesis was formulated by Peter Mitchell before a good deal of our present detailed knowledge of the mitochondrial inner membrane was available. In fact, it was devised to explain transport mechanisms in bacterial membranes. Nevertheless, its basic features are entirely consistent with what we now know of the molecular organization of the mitochondrial inner membrane.

The hypothesis is very uncomplicated in principle: the proteins of the respiratory chain cause a flux of protons across the membrane, generating an electrochemical proton gradient that consists of both a difference in hydrogen ion concentration (ΔpH) and a difference in electric potential. This electrochemical gradient is the driving force, or protonmotive force, to cause a reversal of proton flow, providing the energy for the operation of ATP synthetase. The fundamental features of this hypothesis are illustrated schematically in Figure 12.29.

Peter Mitchell of the Glynn Research Laboratories in Cornwall, England, was awarded the Nobel prize for chemistry in 1978 for his work on the chemiosmotic hypothesis. As he indicated in his acceptance speech in Stockholm, the hypothesis has been so thoroughly embraced by the scientific community that it has been given the status of theory.

Given this scheme, we can now address the two major questions posed previously. First, how is the energy of electron flow converted into an electrochemical gradient? For the present, this can be answered only speculatively, drawing on the information available about the molecular

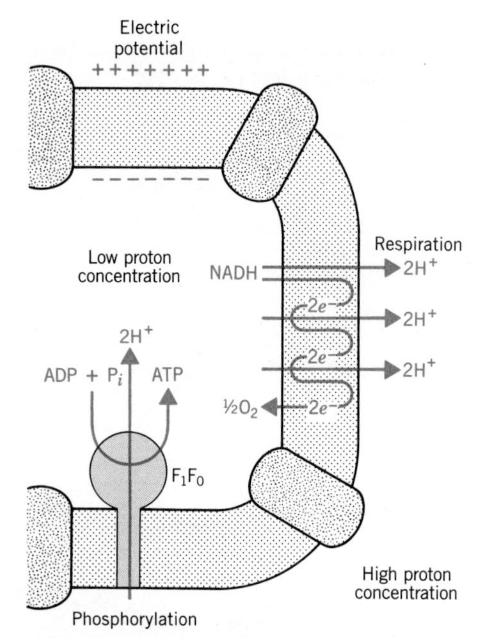

FIGURE 12.29 Basic features of the chemiosmotic hypothesis. Electron transport from reduced coenzyme to molecular oxygen results in a flux of hydrogen ions to the outside of the inner membrane. This electrochemical gradient drives the phosphorylation of ADP to ATP.

architecture of the inner membrane. From this scheme, we see that for every pair of electrons removed from NADH and eventually transferred to oxygen, six protons are translocated across the membrane from matrix to outside. We can now take into consideration the topography of the membrane enzymes to explain the proton flux.

Figure 12.30 illustrates in detail how this may occur. The coenzyme FMN, which is a reversible redox system attached to an enzyme of complex I, picks up a pair of electrons and one proton from NADH, sequesters another proton from the aqueous environment, and becomes fully reduced: $FMNH_2$. The reduced coenzyme then releases two protons to the outside of the membrane, presumably by channeling them through the protein by a conformational change, and passes the pair of electrons on to the next member of the respiratory chain, an iron–sulfur protein. This returns the flavin coenzyme to its oxidized state, whereupon it can repeat the translocation process. The electrons picked up by the iron–sulfur proteins are transferred, either in pairs or singly, to coenzyme Q. Coenzyme Q is a reversible redox molecule that can exist in three oxidation states, a fully oxidized or quinone form, a semiquinone, and a hydroquinone.

Mitchell has proposed a ''Q cycle'' to explain the next series of events in electron transport. In this cycle, coenzyme Q operates in close cooperation with cytochrome *b*, a member of complex III. It is envisioned that two Q's each pick up an electron from the iron–sulfur proteins and a proton from the matrix, converting to the semiquinone form. Cytochrome *b* then provides each semiquinone with an additional electron, enabling each to retrieve an additional H^+ from the matrix. The two Q's, now in hydroquinone form, migrate across the membrane to the outside, where two protons are initially released to the outside aqueous environment and two electrons are given up to the next member of the respiratory chain, cytochrome c_1. The two Q's, returned now to semiquinone form, revert to the fully oxidized quinone state by each giving up an additional proton to the outside and the remaining electron to cytochrome *b*. This completes the cycle.

Coenzyme Q is the name given to a family of molecules also called ubiquinones because of their ubiquitous occurrence. The electron-transferring heart of the molecule is the quinone ring.

CH_3O CH_3 CH_3O CH_3 H n

Fully oxidized ubiquinone

H

OH OH H O OH

Semiquinone ring Hydroquinone ring

The most common form of Q in mammals contains 10 isoprene units (n = 10) and is called Q_{10}.

Thus far, we can see a net movement of six hydrogen ions across the membrane, but the original pair of electrons that entered into the electron transport chain has not completed its journey. They were deposited by coenzyme Q to cytochrome c_1, an enzyme exposed to the exterior surface, which transfers them by way of oxidation and reduction reactions of its heme group to the extrinsic water-soluble protein cytochrome *c*. Cytochrome *c* is a protein link between complex III and complex IV. The electrons are finally shuttled once more across the membrane by way of the heme-containing cytochromes *a* and a_3, whereupon they are donated to oxygen and, combined with two available protons from the matrix environment, water is formed.

Viewing the results of this electron journey, we can see that a pair of electrons, originating in the matrix, has been returned there, but not before making three round trips across the inner membrane. For each loop made, two protons have been translocated to the outside of the membrane. The theory assumes that the positions of the respiratory proteins are fixed fairly tightly and that the spatial relationships among the proteins are not a matter of random positioning. This is consistent with the known asymmetry of the membrane and the unlikelihood of ''flip-flop'' mechanisms operating across the membrane. On the other hand, the theory does pro-

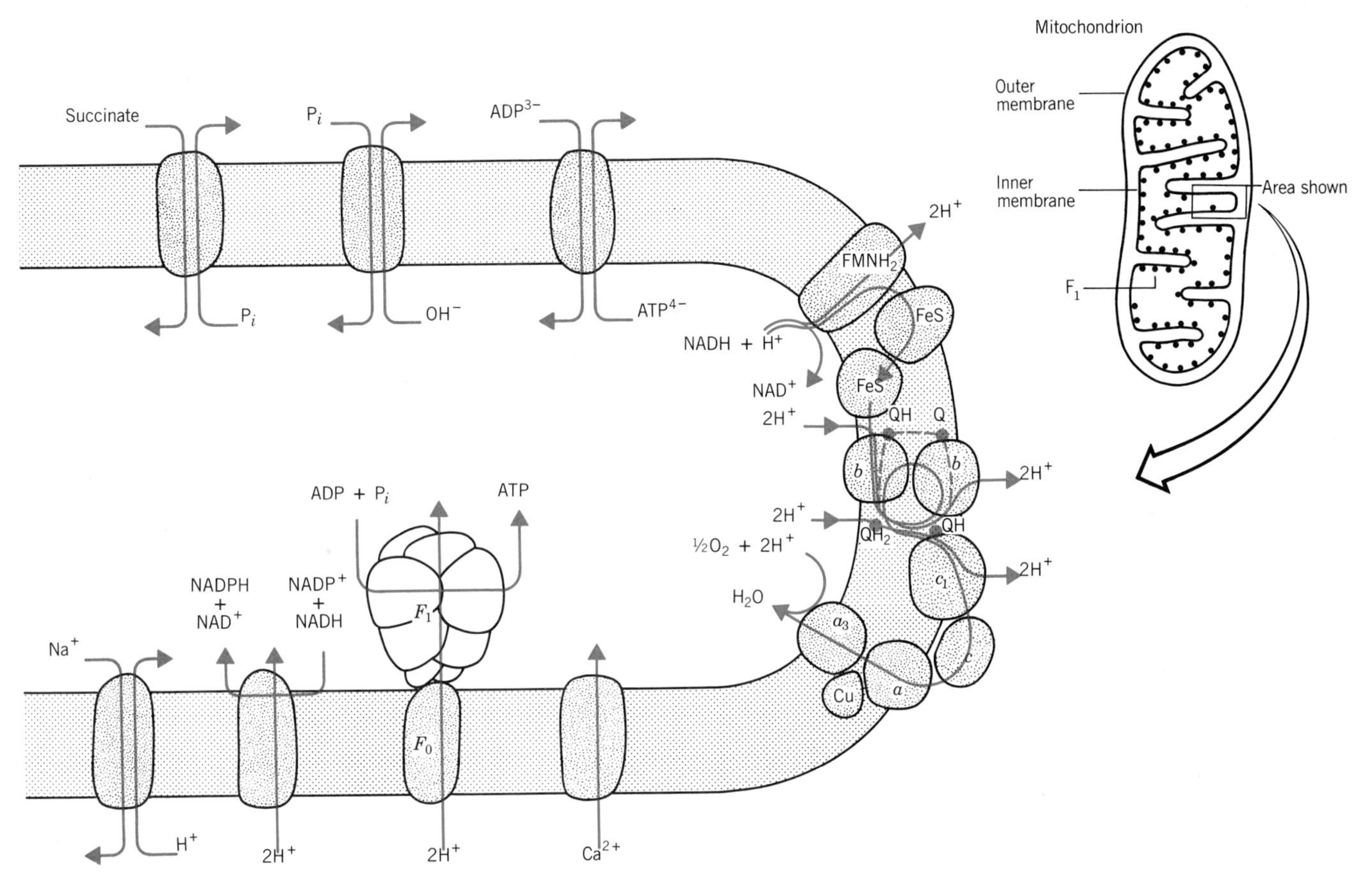

FIGURE 12.30 Schematic illustration of the inner mitochondrial membrane showing the participants in electron transport, oxidative phosphorylation, and several other carriers. The details of electron flow are enumerated in the text.

pose that coenzyme Q is highly mobile and moves across the membrane, translocating four of the six protons that are finally moved across. This is not unreasonable, for coenzyme Q is a lipid-soluble molecule and is present at a high ratio compared to the other members of the respiratory chain. If indeed it has this shuttle role, it would probably take a group of such carriers to keep the cycle operating.

More important than the particular route traversed by the electrons is the transmembrane electrochemical gradient that has resulted from electron transport. It is this gradient that is proposed to drive the ATP synthetase reaction.

Now let us face the second major question. How is the energized state of the membrane used to synthesize ATP? This is the most controversial facet of the mechanism. But it does seem to have some precedent. At least two other unrelated systems that have been studied in considerable detail can generate ATP. These are the Ca^{2+} pump of the sarcoplasmic reticulum and the Na^+, K^+ pump of the plasma membrane. These, of course, normally operate as ATP-driven ion pumps, but they can also operate in reverse, using ion gradients to generate ATP.

The proton-translocating ATPase of the mitochondrial inner membrane is much more complicated in structure than the pumps of the other two

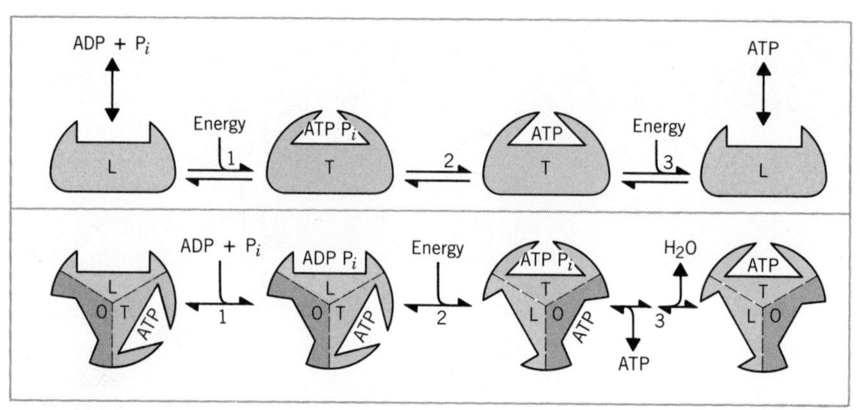

FIGURE 12.31 Models of ATP synthesis invoking conformational changes related to energy-dependent binding of ligands.

systems. Whereas the Ca^{2+} pump and the Na^+, K^+ pump are made up of only 1 or 2 polypeptide chains, the F_1 ATPase contains 10 protein subunits, half of which are intrinsic membrane proteins and half projected into the matrix as a morphological knob.

Step 3, the mechanism of ATP generation, is presently viewed as taking place by one of two different hypothetical schemes, called the *indirect* and *direct* mechanisms of ATP generation. The terminology refers to whether the protonmotive force is used directly or indirectly by the ATPase.

Indirect Mechanisms of ATP Formation

Two indirect mechanisms have been considered to explain ATP generation. One invokes the involvement of a *phosphoenzyme intermediate* and the other *conformational coupling* reactions.

An important component of the indirect mechanisms is that the F_1 ATPase undergoes a conformational change. In one view, the conformational change may be induced by Mg^{2+}, which in turn permits the formation of a high energy phosphoenzyme intermediate. Protons, driven by the electrochemical gradient, displace the Mg^{2+}, whereupon ADP is phosphorylated and the enzyme conformation is returned to its original state. Since a phosphoenzyme intermediate has never been found experimentally, this view seems quite untenable.

In the other view of the indirect mechanism, a conformational change is induced by protons after ADP and P_i have bound to the F_1 complex. In this mechanism the driving force for ATP production—and very crucially, its release—is a conformational change that is produced by a proton flux.

The manner in which the electrochemical gradient drives ATP synthesis by way of conformational changes is not known with any degree of certainty. But one line of thinking that appears to be emerging and gaining considerable acceptance is that the protonmotive force affects the binding of ADP, P_i, and ATP to the F_1 ATPase. Two schemes that deal with the energy-dependent binding and the effect of this on F_1 conformation are presented in Figure 12.31.

In the first scheme, F_1 is depicted as interconvertable between two forms. One, termed L, is catalytically inactive and binds ligands loosely. The other, designated T, binds ligands tightly and is catalytically active. The mechanism is dependent on energy in steps 1 and 3.

In the second scheme the F_1 ATPase consists of three interacting catalytic subunits. Energization in step 2 results in these subunits undergoing interconversion between three forms. Two of these forms, termed L and T, function as the L and T forms in the first scheme. The third form, O, marks a conformation with a low affinity for ligands and catalytic inactivity. It is thought of as an open form.

These models, although gaining credibility and experimental support, simply reflect developing understanding of ATP synthesis. They are most certain to change as time goes on.

A Direct Mechanism of ATP Formation

According to the direct scheme, the electrochemical gradient drives protons through a channel in the membrane leading to the F_1 ATPase. In the ATPase, the protons interact directly with a bound phosphate ion, extracting an oxygen to form water, and thereby establish a favorable situation for a nucleophilic attack on the phosphorus by an adjacently bound ADP. The completed ATP then dissociates from the F_1 complex.

There is now a substantial body of evidence that the F_1 ATPase undergoes conformational changes during the generation of ATP.[13] Although directly supportive of the indirect mechanism, the finding of conformational changes does not exclude direct mechanisms, for conformational changes would be expected to accompany a binding of ADP and P_i and release of ATP even if the process were basically driven by protons.

Proton Circuitry and Coupling between Electron Transport and Phosphorylation

In Mitchell's view of the chemiosmotic hypothesis, electron flow must be coupled with a transmembrane flux of protons. This flux operates independently of ATP generation, that is, it is entirely possible to build up an electrochemical gradient without the occurrence of phosphorylation. Protons that are translocated, according to this scheme, are free to move about outside the membrane, but eventually are driven back across through the proton channel formed by the F_0 portion of the F_0–F_1 complex. Thus there is a proton circuit between the reactions of the respiration chain and the ATP synthetase resembling that depicted in Figure 12.32*a*. Among other things, this view would suggest that a stoichiometric relationship between the components effecting the energy-delivering reactions and ATP generation is not necessarily mandatory.

Other investigators, namely, Williams and Boyer, feel that the evidence supports a tighter link between the electron-transferring and ATP-producing systems. There is, for example, a solid stoichiometric relationship between the components of the two systems, suggesting that they operate in a concerted manner as a single complex, or at least as tightly linked complexes. Furthermore, the membrane potentials in mitochondria seem to be lower than one would expect if this were the primary driving force of ATP synthesis.[13]

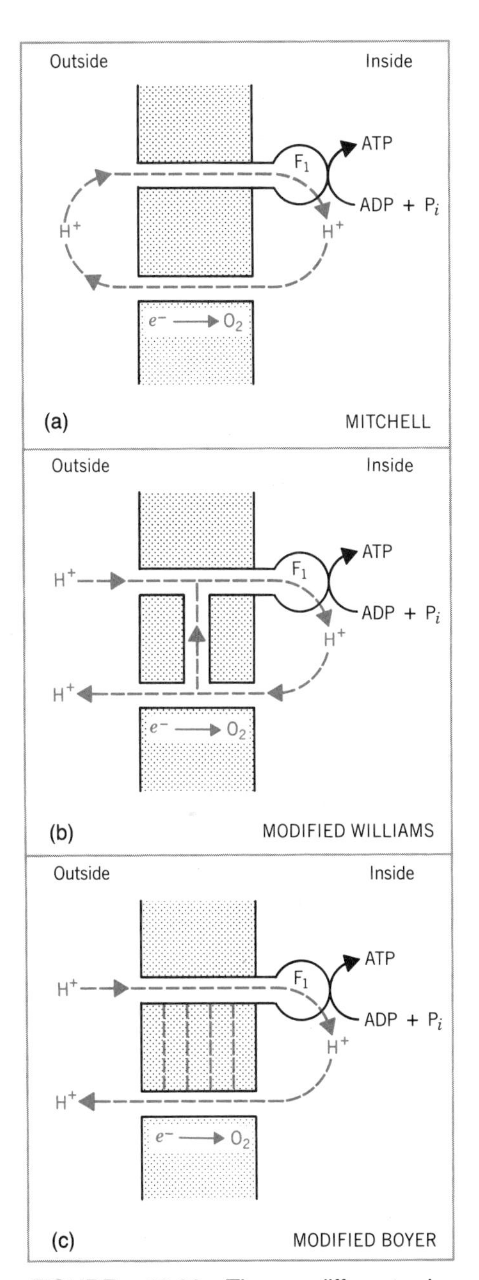

FIGURE 12.32 Three different hypotheses of coupling between electron transfer, proton translocation, and phosphorylation.

Williams has therefore proposed (Figure 12.32*b*) that the protons translocated by the electron transport chain are not extruded outside the membrane but are channeled *within* the membrane, thereby developing a driving force that does not depend solely on an electrochemical gradient across the entire membrane. Boyer has taken a different approach (Figure 12.32*c*). He envisages electron transfer as bringing about conformational changes in the electron transfer proteins that, because of their proximity to the F_0–F_1 ATPase complex, induce conformational changes in these latter proteins. Thus, both a gradient and a tightly linked system of conformational changes can drive the ATP synthetase.

Uncoupling of Phosphorylation from Electron Transport

The uncoupling of respiration from phosphorylation is a phenomenon that has been known since 1948, when W. Loomis and F. Lipmann observed that dinitrophenol has this effect. Since then a number of other molecules have demonstrated this ability. The molecular basis for this uncoupling behavior has been shrouded in enigma until only quite recently.

It now appears that uncouplers act by dissipating the electrochemical gradient, that is, by removing the driving force for ATP generation. Electron transport is unaffected by this action. It should be kept in mind that the electrochemical gradient has two components, a pH increment and a charge increment. Certain compounds, such as Valinomycin, dissipate only the membrane potential, whereas others, such as Nigericin, neutralize the pH gradient. Each compound by itself will not completely inhibit phosphorylation, but when together they completely block ATP generation. Proton ionophores, acting singly, can abolish both increments because they transport protons across the membrane.

The effectiveness of molecules in acting as uncouplers is related to their ability to act as ionophores—molecules that carry ions across membranes. Valinomycin carries potassium ions (K^+) and Nigericin exchanges H^+ for K^+, thereby eliminating the pH gradient. These compounds are lipid-soluble agents that are incorporated into the lipid bilayer of the membrane and, acting from that position, induce the reverse flow of materials that have contributed to the electrochemical gradient.

Inner Membrane Metabolite Transport Systems

The inner mitochondrial membrane is impermeable to a wide variety of small molecules, including simple sugars such as glucose or sucrose, anions such as Cl^- and Br^-, and cations such as K^+ and Na^+. In addition, it is impermeable to many coenzymes, such as NAD^+ and $NADP^+$ in both oxidized and reduced forms, to nucleotides, such as AMP, CDP, CTP, GDP, and GTP, and to coenzyme A and acyl-CoA's.

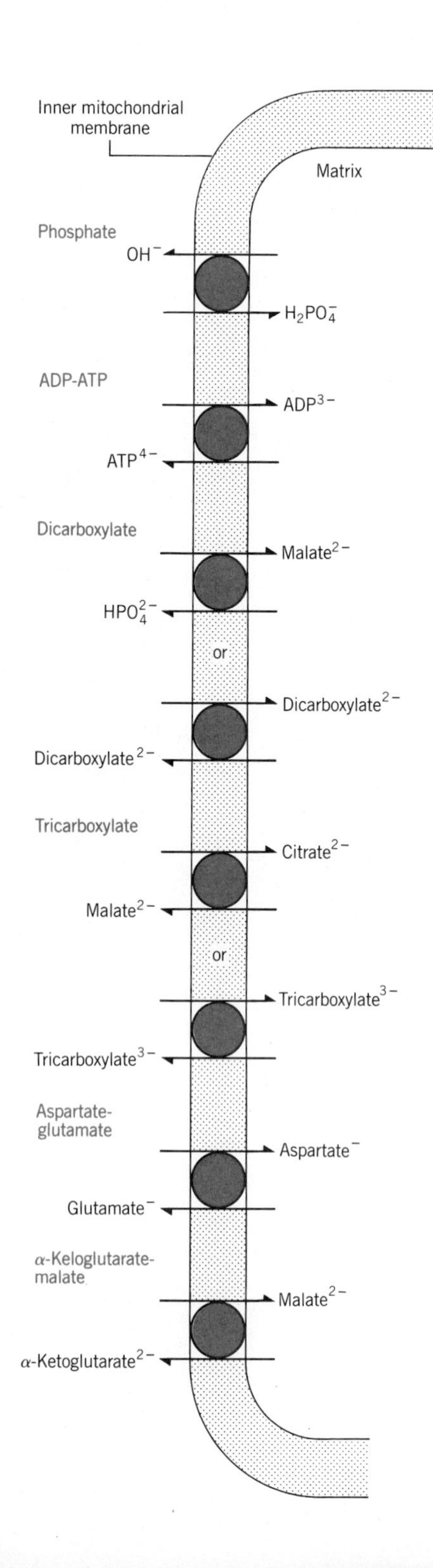

FIGURE 12.33 Some of the important transport systems of the inner mitochondrial membrane. They can function in either direction depending on concentration gradients. The dicarboxylate carrier can exchange malate, succinate, and fumarate with each other or with phosphate. The tricarboxylate carrier can exchange citrate and isocitrate with each other or a tricarboxylate with a dicarboxylate.

In contrast, several specific nucleotides and metabolites are allowed passage. These include ADP and ATP, inorganic phosphate, pyruvate, citrate, succinate, α-ketoglutarate, malate, glutamate, and aspartate. Transport of these materials is mediated by membrane transport systems that operate with an extremely high specificity.

Figure 12.33 summarizes some of the important transport systems of the inner membrane. Of special note are the ADP–ATP and phosphate carriers, which provide the constant supply of ingredients for phosphorylation reactions by the ATP synthetase and channel the product, ATP, out into the cytosol for energy-requiring reactions. The carriers illustrated are termed *passive carriers,* in that under normal circumstances they operate in response to concentration gradients, transporting materials *down* the gradient. However, these carriers can operate in reverse when coupled to electron transport, driving metabolites *up* their particular gradients. In this manner, for example, phosphate can be accumulated within the mitochondrion as a result of the electrochemical gradient established by electron transport (see Figure 12.34). The operation of the phosphate carrier in this way leads, in a coupled way, to enhanced activity of other carriers as illustrated in the diagram.

One of the important coupling actions of the phosphate carrier with other carriers is the influence it has on the accumulation of Ca^{2+} by mitochondria. Mitochondria can accumulate large amounts of Ca^{2+} against its concentration gradient by linking the ion's transport with the cotransport of equivalent amounts of $H_2PO_4^-$. When Ca^{2+} is accumulated in this manner, no ATP is formed because the electrochemical gradient is utilized fully for the transport of Ca^{2+}. Calcium ion and phosphate deposits are readily seen in electron micrographs of mitochondria. They are believed to function in biological calcification processes.

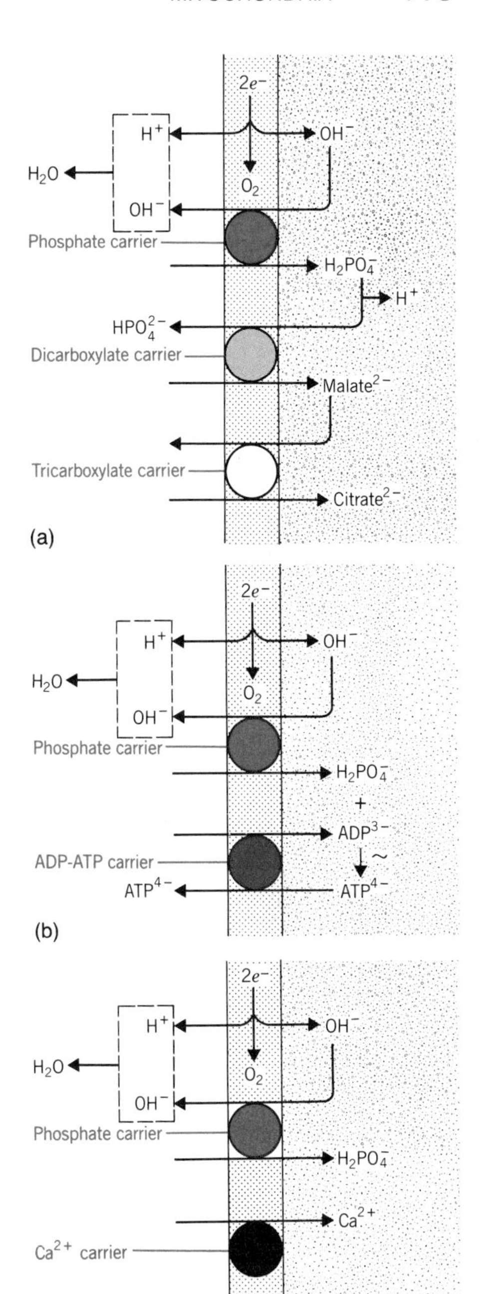

FIGURE 12.34 The coupling of the activity of the phosphate carrier to the operation of other carriers. Thus the phosphate carrier plays a crucial role between electron transport and transmembrane transport. (*a*) Inward transport of dicarboxylates and tricarboxylates coupled to electron transport. (*b*) Entry of ADP^{3-} and exit of ATP^{4-} via the ADP-ATP carrier. (*c*) Inward transport of Ca^{2+}; other cations, such as Mn^{2+}, Fe^{2+}, and K^+, may also enter in response to the negative inside potential.

12.5 GENETICS AND BIOGENESIS OF MITOCHONDRIA

Structure and Function of Mitochondrial DNA

Mitochondrial DNA can be readily isolated and has been the object of intensive characterization studies.[14] The DNA exists primarily as circular strands in multiple identical copies within the mitochondrial matrix, apparently attached at points to the inner membrane. Molecular weights range between 10×10^6 daltons in animal mitochondria and 70×10^6 daltons in higher plants, a fairly narrow range (see Table 12.6).

A single mitochondrion generally has between two and six copies of DNA, allowing the number of mitochondrial DNAs (mtDNAs) per cell to reach magnitudes of 10^8 or better, depending, of course, on the number of mitochondria in a given cell type. The results of DNA–DNA renaturation and restriction endonuclease studies strongly support the contention that all the mtDNAs in a single organism are identical.

A typical electron micrograph of isolated mtDNA is presented in Figure 12.35. It is commonly observed to exist in both single circular and catenated forms. In the latter configuration, two or more strands are linked together like links in a chain.

The role of mtDNA in the mitochondrion is similar to the role of nuclear DNA in the nucleus of the eucaryotic cell, that of producing rRNA, tRNA,

Table 12.6 Size and Structure of Mitochondrial DNAs

Species	Structure	Molecular Weight (daltons × 10^{-6})
Animals (from flatworm to man)	Circular	9–12
Higher plants	Circular	70
Fungi		
Baker's yeast (*Saccharomyces*)	Circular	49
Kluyveromyces	Circular	22
Protozoa		
Acanthamoeba	Circular	27
Malarial parasite (*Plasmodium*)	Circular	18
Paramecium	Linear	27
Tetrahymena	Linear	30–36
Kinetoplastidae	Circle network	2000–20,000
Trypanosoma brucei	Minicircle	0.6
	Maxicircle	13
Crithidia luciliae	Minicircle	1.5
	Maxicircle	22

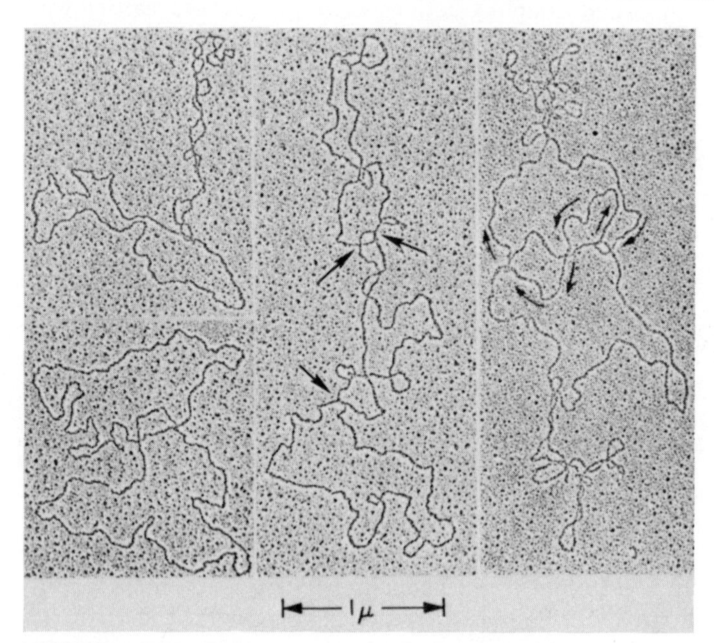

FIGURE 12.35 Electron micrographs of shadowed preparations of catenated DNA from mitochondria. (*Lower left*) A dimer with two relaxed circular DNAs. (*Upper left*) A dimer of a relaxed and a twisted circle. (*Middle*) A trimer of three relaxed circles. (*Right*) A tetramer of two relaxed and two twisted circles. (From L. Pikó, D. Blair, A. Tyler & J. Vinograd.)

and mRNA, the latter of which is translated into protein. Although the *roles* are similar, the *products* are not. The nuclear and mitochondrial genetic apparatus appear to yield no products in common. In spite of this, or perhaps because of this, the mitochondrial genetic system, as we shall

Table 12.7 **Biosynthesis of Major Mitochondrial Enzyme Complexes in Yeast**

		Number of Subunits	
Enzyme Complex	Total	Made on Cell-Sap Ribosomes	Made on Mitochondrial Ribosomes
Cytochrome oxidase	7	4	3
Cytochrome bc_1 complex	7	6	1
ATPase (oligomycin-sensitive)	9	5	4
Large ribosomal subunit	30	30	0
Small ribosomal subunit	22	21	1

Table 12.8 **Mitochondrial Components in Yeast Petite Mutants with Grossly Altered mtDNA**

Absent

A functional respiratory chain; functional cytochromes *b* and aa_3
A functional energy transfer system
A functional protein-synthesizing system; ribosomes

Present

Outer membrane
Inner membrane (altered)
Parts of respiratory chain: cytochromes *c* and c_1, some subunits of cytochrome aa_3
Parts of the energy transfer system: F_1-ATPase
Permeability barrier for H^+ and K^+; translocators for adenine nucleotides, phosphate, dicarboxylic acids, and succinate
The Krebs cycle
Parts of the system for protein and nucleic acid synthesis: DNA and RNA polymerases, ribosomal proteins, elongation factors

see, is highly dependent on the nuclear system, and the cell itself is dependent on the functions of the mitochondrion.

Yeast mtDNA contains one gene for each of the rRNAs and at least 20 4S RNA genes. It is likely that yeast mtDNA can specify all the tRNAs necessary for reading 61 condons in mRNA. In addition, the DNA contains the information to specify a number of protein subunits of important mitochondrial enzyme complexes. Some of these are listed in Table 12.7. Of special importance in this regard is the composition of the complexes: they are made of products of *both* mitochondrial and cytosol protein synthesis.

A significant amount of information about the role of mtDNA has been gained from studies on petite mutants. Petite mutants are deletion mutants in which large regions, between 50 and 100%, of their DNA are missing. Since yeast can live without a functioning mitochondrion, it is possible to observe the extent to which a grossly altered mtDNA structure influences the assembly and function of mitochondrial components. Table 12.8 summarizes results of studies along these lines. Since the mtDNA in these is essentially nonfunctional, the mitochondrial proteins that are present must be specified by nuclear DNA and synthesized on cytosol ribosomes.

Transcription and Translation in Mitochondria

When it comes to the mechanism of transcription and translation within mitochondria, there is an absolute dependency on the nuclear genetic machinery. Certain of the required materials are produced by mitochondria, independently of the nucleus. These include ribosomal, transfer, and messenger RNA. However, certain other obligatory proteins are specified by the nucleus, such as ribosomal proteins, RNA and DNA polymerases, aminoacyl tRNA synthetase, and protein synthesis factors.[15]

Brief reflection will reveal a very interesting phenomenon: mtDNA cannot be expressed without assistance from the nucleus. In fact, it cannot even be replicated without such assistance. This is the level at which the molecular dependency of the mitochondrion on the nuclear genetic machinery is observed.

Another intriguing facet of this dependent relationship is the difference between the proteins that are imported into the mitochondrion from cytosolic sites of synthesis and the molecules that are involved in nucleocytoplasmic transcription and translation. In general, the former have higher affinities for mitochondrial substrates. Thus the two genetic systems, although carrying out identical functions (i.e., transcription and translation), use no components in common to execute these functions.

The semiautonomous properties of the mitochondrion in this regard are most clearly illustrated by the manner in which its ribosomes are synthesized. The RNA of the mitochondrial ribosome is transcribed from mtDNA, whereas ribosomal proteins are transcribed from nuclear DNA, translated on cytoplasmic ribosomes, and transported into mitochondria for assembly of the final nucleoprotein particle (Figure 12.36).

Mitochondrial ribosomes are similar to procaryotic and eucaryotic ribosomes in certain of their basic properties (see Chapter 17). For example, they are made up of two subunits and translate mRNA using the typical aminoacyl tRNAs and protein initiation and elongation factors. However, they differ very significantly from *both* ribosome systems in their fine structure and even show considerable species variations. Their sedimentation coefficients range from 55 to 80 S, and variations are seen in their buoyant densities and guanine plus cytosine (GC) contents. Some of these variations are shown in Table 12.9.

The Assembly of Mitochondria

The process of mitochondriogenesis consists of two phases.[16] The first is the gross replication or *growth* of the mitochondrial membranes and reproduction of individuals. The second phase consists of the *differentiation* of the organelle into an active compartment of oxidative phosphorylation. These two phases are under independent control so that the mitochondrion cannot operate as a self-replicating entity.

The mitochondria of *Neurospora crassa*[17] and HeLa cells[18] have been observed to grow by the addition of new components to old structures, with subsequent division of one mitochondrion into two. Figure 12.37 depicts mitochondrial fission schematically. After elongating, one or more centrally located cristae form a partition by growing across the matrix and fusing with the opposite inner membrane. This separates the matrix into two compartments. The outer membrane then invaginates at the

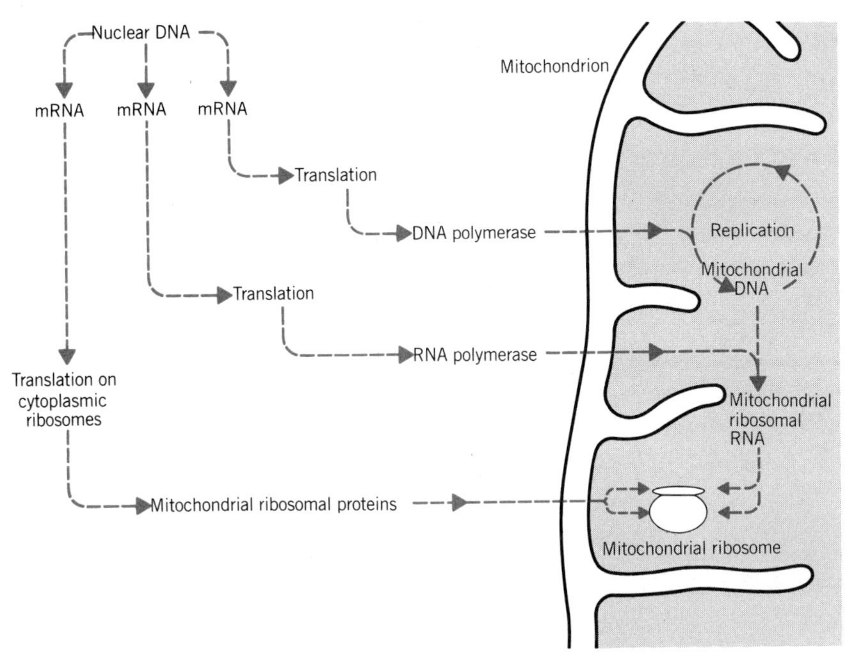

FIGURE 12.36 Schematic summary of the functional relationship between nuclear and mitochondrial DNA.

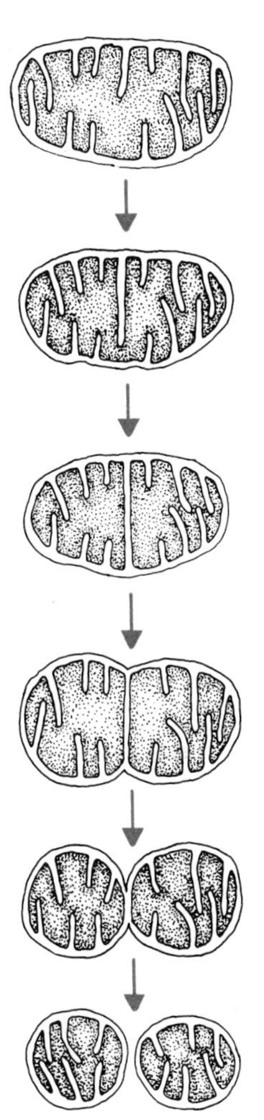

FIGURE 12.37 Mitochondrial fission by partition formation.

Table 12.9 Properties of Mitochondrial Ribosomes

Subclass	Kingdom	Example	Sedimentation Coefficient (S)	Buoyant Density (g/cm^3)	Molecular Weight of RNA ($\times 10^{-6}$ daltons)	RNA Base Composition (% GC)
1	Animals	Vertebrates	55–60	1.44	0.35, 0.54	40–45
2		Insects	60–71	—	0.3, 0.5	19–32
3	Protists	*Euglena*	71	1.61	0.56, 0.93	27
4		*Tetrahymena*	80	1.46	0.47, 0.90	29
5	Fungi	*Candida*	72	1.48	0.71, 1.21	34
6		*Neurospora*	73–80	1.52	0.72, 1.28	38
7		*Saccharomyces*	72–80	1.64	0.70, 1.30	30
8	Plants	Maize	77	1.56	0.76, 1.25	—

partition plane, constricting until there is membrane fusion between the two inner membrane walls. Thus two separable daughter mitochondria are formed. Figure 12.38 is an electron micrograph illustrating mitochondrial fission.

The entire process of growth, represented by the increased membrane surface areas and fission, appears to be under the control of the nucleus of the cell. A number of experimental approaches have shown that this phase operates independently from the phase of differentiation. For example, in the presence of chloramphenicol, mitochondria continue to proliferate in HeLa cells, but they are functionally defective. Hence, mitochondrial protein synthesis does not appear to be a prerequisite for the gross formation and division of mitochondria, nor for the assembly of the framework of the inner membrane.

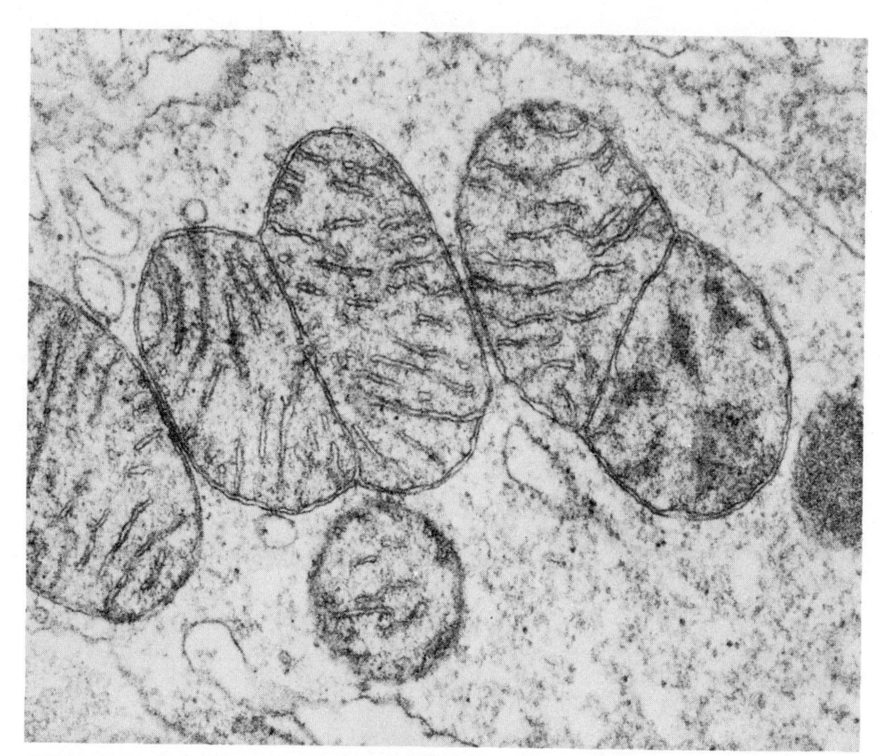

FIGURE 12.38 Electron micrograph showing mitochondrial fission. This stage of division corresponds to the fourth stage of Figure 12.37. (Courtesy of Dr. William J. Larsen.)

The differentiation of the mitochondrion into a properly functioning organelle depends on the mitochondrial genome to the extent that it operates in concert with the nuclear genome directing the assembly of the respiratory enzymes. It is at this level of operation that we see most clearly the interdependency of the two genomes of the cell. Neither genome functioning independently can differentiate the mitochondrion.

The cooperation of the two genomes has been clarified considerably by studies on the molecular assembly of cytochrome oxidase.[19] This cytochrome, as studied in *Saccharomyces cerevisiae,* is made up of seven polypeptide subunits for a combined molecular weight of 139,000 daltons. Three of the polypeptides are coded by mtDNA and assembled on mitochondrial ribosomes. They are very hydrophobic and high in molecular weight (23,000–40,000 daltons). The remaining four subunits are coded by nuclear DNA and made on cytoplasmic ribosomes. These are hydrophilic polypeptides of lower molecular weight (4500–14,000 daltons).

The topographical relationships of these seven subunits making up the intact enzyme have been studied by using various radioactive surface probes and antibodies directed against specific subunits. Again, intact mitochondrial membranes and the inverted submitochondrial vesicles were used to establish the accessibility of cytochrome oxidase subunits to the matrix and intermembrane space sides of the inner membrane. The resulting model, depicting the manner in which the enzyme is thought to span the membrane, is presented in Figure 12.39. The presumed juxtaposition of cytochrome *c* to the cytochrome oxidase complex of subunits is also shown.

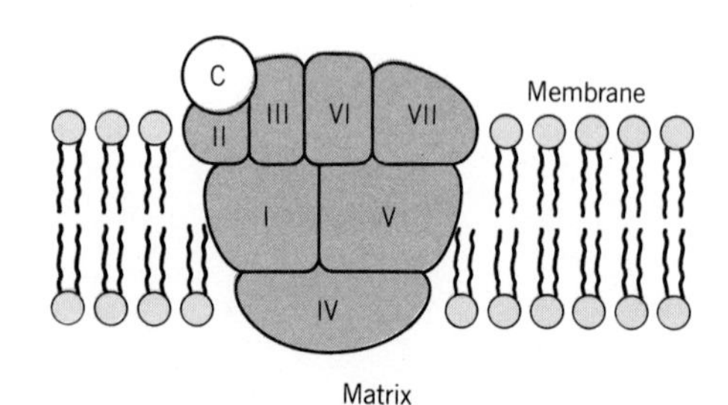

FIGURE 12.39 The topographical relationships of the seven subunits of cytochrome oxidase in the inner membrane. The numbering of the subunits corresponds to that employed for the yeast enzyme.

An interesting conclusion that has emerged from a variety of studies on this system is that one role of the mitochondrially made subunits is to anchor the cytoplasmic subunits to the inner membrane. In petite mutants, where the three mitochondrial subunits are no longer made, the cytoplasmic subunits are present but no longer firmly associated with the inner membrane.

The cytochrome oxidase heme prosthetic group also has a role in the assembly of the functional enzyme. In mutants unable to synthesize heme in the absence of added δ-aminolevulinic acid, some of the subunits were missing, or at least undetectable as normal subunits, and others no longer bound together as in wild type organisms. Apparently the heme affects subunit associations and/or conformations that are mandatory for the assurance of a proper assembly.

In summary, the picture of mitochondriogenesis that has emerged is one of an independently controlled two-phase process. The formation of the basic mitochondrial framework, including both outer and inner membranes, is under nuclear control. The differentiation of the organelle into a site of oxidative phosphorylation is under cooperative control by both the nuclear and mitochondrial genome. The replication of mtDNA and its expression depends on the nuclear genome because mtDNA and RNA polymerases are coded by the nuclear genome and made on cytoplasmic ribosomes. Thus the replication and transcription of the mitochondrial genome depend on the nucleus, but mitochondrial protein synthesis is independent of the nucleus.

Why Two Separate Cellular Genomes?

An intriguing philosophical question involves the reason for the existence in the cell of a separate mitochondrial genome. In one sense, it is a very extravagant system, for it takes about 100 proteins to provide for DNA replication and transcription and for protein synthesis. By all experimental appearances, the mitochondrion does have its own such genetic system. The genetic cost to support a minimal genetic system is high. The amount of DNA required to code for 100 proteins is about 50×10^6 daltons, or five times the amount present in animal mtDNA. From these facts alone, it is obvious that the mitochondrion operates at considerable genetic expense to the cell.

In spite of these apparent disadvantages, the mitochondrion has persisted during cellular evolution and has become an integral and mandatory part of most eucaryotic cells. As P. Borst has pointed out, there must be some explanation for this.[14]

Generally, the question is answered along one of three lines. In the first place, mtDNA is thought to provide a useful form of amplification of the nuclear genome. By this means the cell could contain multiple gene copies for certain proteins that are needed in large amounts. This view is weakened by the observation that certain of the mitochondrial proteins, such as cytochrome oxidase, are coded by both nuclear and mitochondrial genomes. If this mechanism is to be invoked, why are not all the genes coding for cytochrome oxidase equally amplified and present together in the mitochondrion?

Another view is that it is necessary to have a separate protein–synthesizing system to manufacture hydrophobic proteins that cannot be transported across the membrane but must be assembled near their site of operation. This view is somewhat more attractive than the first, since there is some experimental support for it. So far, all the proteins known to be synthesized in mitochondria are hydrophobic. But until we know more about the overall topography of the proteins of the inner membrane, this view will be quite shaky.

The third line of argument is based on the belief that mitochondria are primitive bacterialike endosymbionts that have lost most of their genomes to the nucleus. According to this view, the loss was interrupted at a point, leaving what we see today as the mitochondrial genome. The cell has become dependent on certain of these mitochondrial functions and can no longer tolerate a loss of the mitochondrial genome. In the absence of satisfying experimental support, this point of view is not particularly attractive either.

It may well be, as Borst has stated, that

none of these explanations provides compelling arguments for the existence of a mitochondrial genetic system in nearly all eucaryotic cells. It is possible, therefore, that we are still on the wrong track and that separate organelle genes provide advantages to the cell that remain to be discovered.

For those who aspire to enter the world of cellular research some day, it should be a relief to know that all the answers are not in. Furthermore, when the basics are complete, the furnishings can always be redone and rearranged.

Summary

Mitochondria are organelles bounded by two membranes that are separated by a space, the intermembrane space. The inner membrane is highly invaginated by cristae that project into the matrix of the organelle. The matrix contains the soluble enzymes of the tricarboxylic acid cycle.

The inner and outer membrane can be separated and studied independently. They differ in composition, enzymatic properties, and permeabilities. The marker enzymes for the outer and inner membranes are monoamine oxidase and succinate dehydrogenase, respectively.

The inner membrane contains all the enzymes necessary to carry out electron transport from reduced coenzymes to molecular oxygen as well as the coupled process of phosphorylation. These processes have been studied both by breaking up the membrane into complexes that can conduct portions of the pathways and by examining the topology of enzymes with respect to the two surfaces of the membrane. These approaches have resulted in membrane models depicting an asymmetric cross-sectional arrangement of enzymes with spatial clustering in the lateral directions.

The chemiosmotic hypothesis of energy transformation is generally embraced as the most reasonable explanation of the driving force for oxidative phosphorylation. According to this hypothesis, as electrons are removed from reduced coenzymes in the matrix, they make three loops across the membrane before returning to the matrix side. Coincident with each loop made is an efflux of H^+ to the cytosol side of the membrane. This generates an electrochemical gradient that when reversed through the ATP synthetase, is the driving force for ATP production.

Electron transport and phosphorylation are coupled in the sense that the former creates a gradient to drive the latter. Agents that uncouple the two processes do so by destroying the electrochemical gradient.

Mitochondria contain DNA, which directs the formation of mitochondrial rRNAs and tRNAs and a number of mitochondrial protein subunits of enzyme complexes. However, ribosomal proteins, RNA and DNA polymerases, and certain other enzymes necessary for translation are specified by nuclear DNA. Thus the two genomes must operate in concert to produce structurally and functionally complete mitochondria.

References

1. *The Mitochondrion, Molecular Basis of Structure and Function,* A. L. Lehninger, W. A. Benjamin, New York, 1964.
2. Characteristics of Isolated and Purified Preparations of the Outer and Inner Membranes of Mitochondria, D. F. Parsons, G. R. Williams, and B. Chance, *Ann. N.Y. Acad. Sci.* 137:643(1966).
3. Enzymatic Properties of the Inner and Outer Membranes of Rat Liver Mitochondria, C. A. Schnaitman and J. W. Greenawalt, *J. Cell Biol.* 38:158(1968).
4. Observations on the Fragmentation of Isolated Flight-Muscle Mitochondria from *Calliphora erythrocephala* (Diptera), G. D. Greville, E. A. Mann, and D. S. Smith, *Proc. R. Soc. London, Ser. B* 161:403(1965).
5. Removal and Binding of Polar Lipids in Mitochondria and Other Membrane Systems, S. Fleischer and B. Fleischer, in *Methods in Enzymology,* vol. 10, R. W. Estabrook and M. E. Pullman, eds., pp. 406–433, Academic Press, New York, 1967.
6. *The Structure of Mitochondria,* E. A. Munn, Academic Press, New York, 1974.
7. The Structure of Mitochondrial Membranes, R. A. Capaldi, in *Mammalian Cell Membranes,* vol. 2, G. A. Jamieson and D. M. Robinson, eds., Butterworths, London, 1977.
8. Formation of Membranes by Repeating Units, D. E. Green, D. W. Allmann, E. Bachmann, H. Baum, K. Kopaczyk, E. F. Korman, S. Lipman, D. H. McLennan, D. G. McConnell, J. F. Perdue, J. S. Rieske, and A. Tzagoloff, *Arch. Biochem. Biophys.* 119:312(1967).
9. Partial Resolution of the Enzymes Catalyzing Oxidative Phosphorylation. XIII. Structure and Function of Submitochondrial Particles Completely Resolved with Respect to Coupling Factor, E. Racker and L. L. Horstman, *J. Biol. Chem.* 242:2547(1967).
10. *Biochemistry of Mitochondria,* E. C. Slater, Z. Kaniuga, and L. Wojtczak, eds., Academic Press/Polish Scientific Publishers, p. 1, London/Warsaw, 1967.
11. Mechanisms of Energy Transformation, E. Racker, *Annu. Rev. Biochem.* 46:1006(1977).
12. Carboxyl Activation as a Possible Common Reaction in Substrate-level and Oxidative Phosphorylation and in Muscle Contraction, P. O. Boyer, in *Oxidases and Related Redox Systems,* vol. 2, T. E. King, H. S. Mason, and M. Morrison, eds., John Wiley & Sons, New York, 1965.
13. Mechanism of Oxidative Phosphorylation, E. C. Slater, *Annu. Rev. Biochem.* 46:1015(1977).
14. Structure and Function of Mitochondrial DNA, P. Borst, in *International Cell Biology,* B. R. Brinkley and K. R. Porter, eds., Rockefeller University Press, New York, 1977.
15. Transcription and Translation in Mitochondria, T. W. O'Brien, in *International Cell Biology,* B. R. Brinkley and K. R. Porter, eds., Rockefeller University Press, New York, 1977.

16. The Biogenesis of Mitochondria in HeLa Cells: A Molecular and Cellular Study, G. Attardi, P. Costantino, J. England, D. Lynch, W. Murphy, D. Ojala, J. Posakony, and B. Stonie, in *Genetics and Biogenesis of Mitochondria and Chloroplasts,* C. W. Birky, P. S. Perlman, and T. J. Byers, eds., Ohio State University Press, Columbus, 1975.
17. Genesis of Mitochondria in *Neurospora crassa,* D. J. L. Luck, *Proc. Natl. Acad. Sci. U.S.* 49:233(1963).
18. Expression of the Mitochondrial Genome in HeLa Cells. XV. Effect of Inhibition of Mitochondrial Protein Synthesis on Mitochondrial Formation, B. Storrie and G. Attardi, *J. Cell. Biol.* 56:819(1973).
19. The Assembly of Mitochondria, J. Saltzgaber, F. Cabral, W. Buchmeier, C. Kohler, T. Frey, and G. Schatz, in *International Cell Biology,* B. R. Brinkley and K. R. Porter, eds., Rockefeller University Press, New York, 1977.

Selected Books and Articles

Books

Genetics and Biogenesis of Mitochondria and Chloroplasts, C. William Birky, Jr., Philip S. Perlman, and Thomas J. Byers, eds., Ohio State University Press, Columbus, 1975.

Mitochondria, A. Tzagoloff, Plenum Press, New York, 1982.

Mitochondria: Structure, Biogenesis and Transducing Functions, H. Tedeschi, Springer-Verlag, New York, 1976.

The Biogenesis of Mitochondria—Transcriptional, Translational and Genetic Aspects, A. M. Kroon and C. Saccone, eds., Academic Press, New York, 1974.

The Enzymes of Biological Membranes, vol. 4, *Electron Transport Systems and Receptors,* A. Martonosi, ed., Plenum Press, New York, 1976.

The Genetic Function of Mitochondrial DNA, C. Saccone and A. M. Kroon, eds., North Holland, Amsterdam, 1976.

The Structure of Mitochondria, E. A. Munn, Academic Press, New York, 1974.

Articles

How Cells Make ATP, Peter C. Hinkle and Richard E. MaCarty, *Sci. Am.* 238:104(1978).

Inner Mitochondrial Membranes: Basic and Applied Aspects, Efraim Racker, in *Cell Membranes,* G. Weissmann and R. Claiborne, eds., HP Publishing Co., New York, 1975.

Oxidative Phosphorylation and Photophosphorylation, P. D. Boyer, B. Chance, L. Ernster, P. Mitchell, E. Racker, and E. S. Slater, *Annu. Rev. Biochem.* 46:955(1977).

The Structure of Mitochondrial Membranes, R. A. Capaldi, in *Mammalian Cell Membranes,* vol. 2, G. A. Jamieson and D. M. Robinson, eds., Butterworths, London, 1977.

13

Chloroplasts

That the vegetable creation should restore the air which is spoiled by the animal part of it looks like a rational system.

BENJAMIN FRANKLIN, 1773

The idea that plants restore "spoiled" air, and the enormous scientific implications of that process, did not dawn on a lone researcher or research group, or even on a single research era. It was too much to be worked out except by a long and persistent historical surge of investigation that began in the 1700s and continues today.

One wonders why it took so long for plants to receive the credit they deserve for their maintenance of a balanced ecosystem by virtue of the reactions of photosynthesis. But the culprit was probably the age-old one of technological capabilities lagging behind the inquiries of the mind. The experimental exploration of photosynthesis had to await the development of new tools to measure gas production and consumption, to monitor reactions spectrophotometrically, to separate small quantities of biological molecules, and to trace the fate of radioactive carbons. Prehistoric observers may indeed have wondered why green plants did not do well in the back of the cave, but they lacked the devices to measure the effect of monochromatic light at a given wavelength on the evolution of oxygen and the production of reduced pyridine nucleotides. Even if the devices had been available, there was no place to plug in the equipment. Neither were such aids accessible to a long train of people who dressed better and walked straighter and put their thoughts down on paper. We do indeed live in fortunate times, at least for pursuing an interest in photosynthesis.

Joseph Priestley is often credited with discovering some of the early clues to photosynthesis in 1771, when he showed that a sprig of mint in a closed container could restore the air "spoiled" by a mouse or a burning candle. Although the experiment did not do much to distinguish between a candle and a mouse, it did expose one of the great principles of nature, namely, that matter is exchanged between the photosynthetic and nonphotosynthetic kingdoms. Thus Priestley concluded that plants produce oxygen, reversing the effects of animals, and Benjamin Franklin placed his benediction on the concept. This type of teamwork may have started a new tradition, for even today it is important for an area of research to receive a political blessing so that it can secure the social acceptance and funding necessary to move ahead.

Jan Ingenhouse is credited with observing that the evolution of oxygen requires two important factors, light and the green portion of the plant. Robert Mayer enlarged the story by showing that the assimilation of carbon also requires light, and Julius von Sachs demonstrated that starch was produced in a light-dependent manner in chloroplasts. This work, with contributions by many others, led in the mid-1800s to the general equation:

$$CO_2 + H_2O \xrightarrow{\text{light}} O_2 + \text{organic matter}$$

This reaction, as written, suggests a photolytic split of CO_2, which was thought for some time to be the case. In reality, the oxygen released during photosynthesis comes from water alone.

This is still an accurate representation of what happens in the overall process of photosynthesis.

In the late 1800s, Theodor Engelmann showed with remarkable technical ability that oxygen was evolved only from the parts of the chloroplast that were illuminated. He did this indirectly, by noting the movement of oxygen-requiring bacteria toward illuminated portions of the large ribbonlike chloroplast of *Spirogyra*.

The biochemical approach to the study of photosynthesis was born in the 1930s. The work of Robert Hill and Cornelius Van Niel began to suggest that CO_2 consumption and O_2 evolution were completely separate processes and that both oxygen and reducing equivalents were derived from water alone. Using the labile isotope ^{11}C, Samuel Ruben, Martin Kamen, and William Hassid demonstrated that photosynthesis was not simply a photoreduction of CO_2, as the general equation was often interpreted.

Method VI

In 1940 the isotope ^{14}C was discovered by Ruben and Kamen. This provided the technology for a remarkable series of experiments launched by Melvin Calvin. By exposing photosynthesizing systems to $^{14}CO_2$ and separating and monitoring the products via chromatography, Calvin delineated the entire pathway of carbon movement from carbon dioxide to carbohydrate in less than a decade. Efraim Racker showed that this cycle of reactions would take place in the absence of light, provided the appropriate reducing power and phosphate energy were available. Subsequently, the light production of reduced coenzyme was shown by Anthony Pietro and Helge Lang, and the basics of photosynthetic phosphorylation were pinned down by Daniel Arnon, Mary Allen, and Frederick Whatley.

At about this time in the historical development of photosynthetic research, the work began to merge and be correlated with studies on mitochondria. It soon became obvious that the structures and actions of the two organelles were related, although essentially reversed from each other. This broadened the base of research efforts and attracted the contributions of many other researchers, too numerous to name.

The global enormity of the reactions of photosynthesis is too great to comprehend. It has been quite properly stated that "the reduction of carbon dioxide by green plants is the largest single chemical process on earth."[1] About 10^{11} tons of CO_2 are fixed each year by the reactions of photosynthesis; this is nearly two orders of magnitude greater than the amount of fossil fuel that is consumed yearly in a process that, of course, reverses the efforts of photosynthesis. Photosynthesis has generated a net global deposit of fossil fuel reserve that is estimated to be around 10^{13} tons.

In view of the magnitude of this process, one would think that research on plant biochemistry, and in particular photosynthesis, would receive top funding priorities as the world awaits a better understanding of the mechanism of this phenomenon. But such has not been the case. Most politicians apparently are not vegetarians, and society in general is more concerned about the fearful problems of infectious diseases, heart attacks, and cancer than about comprehending the earth's greatest chemical reaction and perhaps making use of it to better feed the world's people. So work on chloroplasts has moved more slowly than on mitochondria, their

ATP-producing counterparts in animal cells. Mitochondria are, both symbolically and literally, closer to the human heart.

A large fraction of the 10^{11} tons of material handled by the reactions of photosynthesis each year moves across the chloroplast membranes twice—going in in one form and out in another. Since virtually all the organic nutrients that we consume have come about by this process, it follows that we have all been through the chloroplast in one form or another. Although the prophet Isaiah was not referring to photosynthesis when he said ''All flesh is grass,'' he was hitting on a molecular truth that has moved into the light of understanding only within the past few decades.

13.1 THE OCCURRENCE OF PHOTOSYNTHESIS

We normally associate photosynthesis with the higher green plants, but they are responsible for less than half the photosynthesis that takes place on the earth's surface. A wide variety of other eucaryotes participate in the process, some unicellular and some multicellular. Among these are the brown, green, and red algae, and the marginally plantlike organisms euglenoids, dinoflagellates, and diatoms.

Important contributions to the whole of global photosynthesis are also made by procaryotes. Among these are the blue-green algae (cyanobacteria) and two groups of photosynthetic bacteria, the green sulfur and the purple groups. Some of these are considered to be descendants of primitive photosynthetic cells.

The bulk of photosynthesis is carried out by microscopic organisms making up the phytoplankton of the ocean. They thus initiate great food chains that culminate in sharks and whales in the ocean and spread by way of ocean food consumers to humans.

The mechanism of photosynthesis as carried out by bacteria has been an active arena of research, and we will refer to it where helpful and appropriate. For the most part, however, we will deal with photosynthesis as it is known to take place in the chloroplasts of eucaryotic cells.

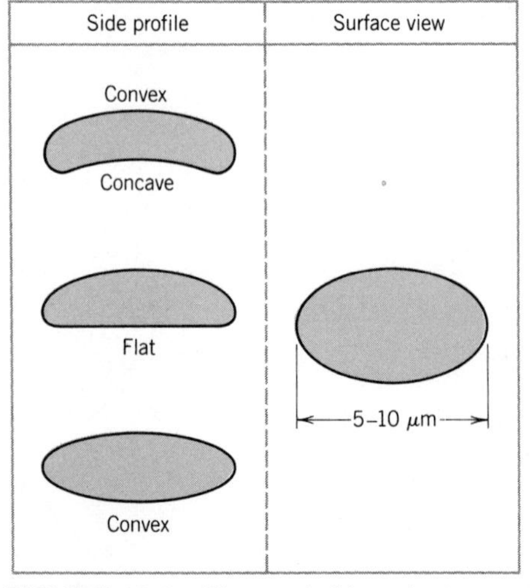

FIGURE 13.1 Classical chloroplast morphologies.

13.2 THE STRUCTURE AND ULTRASTRUCTURE OF CHLOROPLASTS

Isolation

Certainly one of the most common sources from which chloroplasts are isolated is spinach leaves. Leaves are gently homogenized in NaCl (0.35 *M*) or sucrose (0.3–0.4 *M*) near neutrality (pH 8), and cell wall debris and nuclei are removed by filtration and centrifugation. The cleared homogenate is then centrifuged at 1000*g* to bring down intact chloroplasts. Chloroplasts isolated in this manner maintain their normal *in situ* functions of photosynthesis.

Abundance, Size, and Shape

The cells of most higher plants generally contain between 50 and 200 chloroplasts. These normally have the classical lens shape when viewed

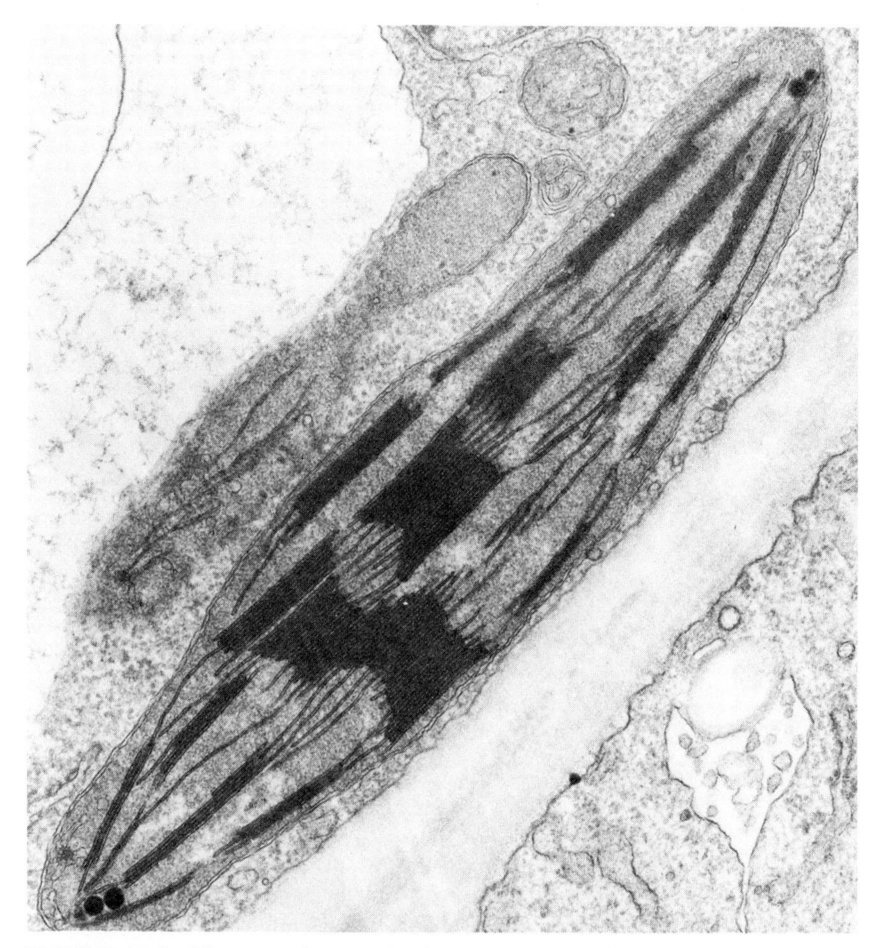

FIGURE 13.2 Electron micrograph of a thin section through a chloroplast. The double membrane system that forms a boundary around the organelle is apparent, and several stacks of internal membranes are present. (Courtesy of Dr. Harry T. Horner.)

in side profile with one surface convex and the other concave, flat, or convex (see Figure 13.1). The long axis of these chloroplasts frequently measures 5 to 10 μm. Viewed from the top, these organelles look ellipsoidal.

In lower plants, and especially in some of the microorganisms, highly divergent forms are seen and often fewer chloroplasts are present per cell. The much-studied *Euglena gracilis,* frequently used for chloroplast research, contains about 10 chloroplasts per cell, and *Chlamydomonas* only one cup-shaped organelle. *Spirogyra* represents a morphological extreme with respect to both size and shape, possessing a ribbonlike chloroplast that spans the length of the cell.

Compartments and Nomenclature

Ultrathin sections of chloroplasts in side profile reveal an ultrastructure that is similar in many respects to that of the mitochondrion. Figure 13.2, a typical electron micrograph, illustrates these features, and the model in

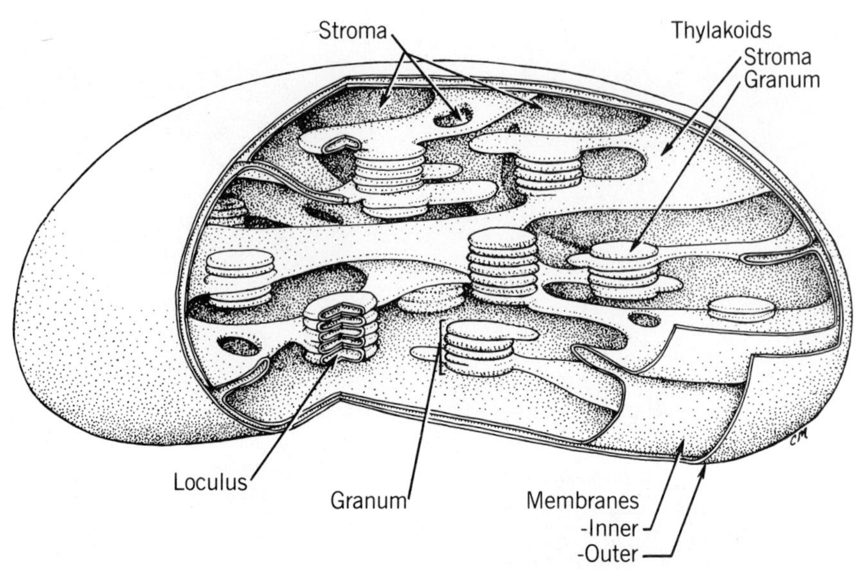

FIGURE 13.3 Model of the chloroplast emphasizing the terminology of its components.

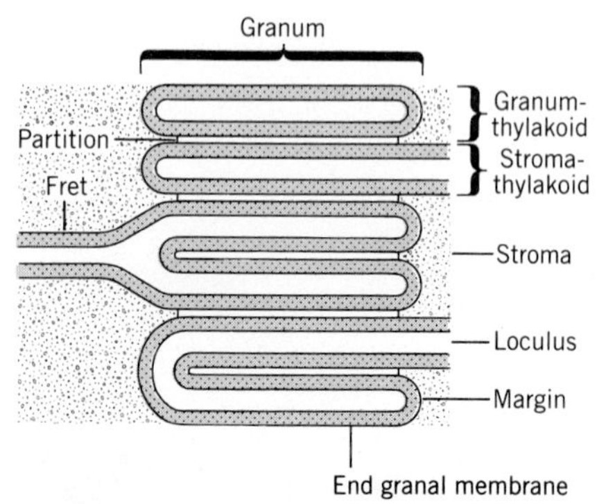

FIGURE 13.4 Model showing details of the internal membrane structure of a granum.

Figure 13.3 contains the terminology used to describe the various components of the organelle.

Fundamentally, the chloroplast is bounded by a double membrane system. The *outer membrane* is separated from the *inner membrane* by an *intermembrane space,* and the inner membrane is connected in a limited manner with a complex of membranes transversing the interior. In many cases the inner membranes appear to be independent of the two outer membranes. Thus, the organelle is a three-membrane system.

The most common form of interior membrane is a flattened sack called a *thylakoid*. Several of these may form a stack, called a *granum,* the individuals of which are then referred to as *granum thylakoids*. When thylakoids reach out through the *stroma,* the individuals are termed *stroma thylakoids*. Thylakoids that cross an intergrana region form structures called *frets*. The interior of the thylakoid is called the *loculus*. A model depicting the details of the granum and an electron micrograph of a highly magnified region of grana are shown in Figures 13.4 and 13.5, respectively.

The membranes of the chloroplasts demarcate two and possibly three separate compartments: the intermembrane space, the stroma, and the loculus. Although the loculus may be transiently continuous with the intermembrane space, it appears to carry out reactions not common to the latter. Most of the current experimental evidence favors the notion that the loculus is a compartment separate from the intermembrane space.

It is now generally accepted that the light-dependent reactions of photosynthesis are carried out in the thylakoids, whereas the dark reactions of CO_2 assimilation take place in the stroma. There is thus a compartmentalized division of labor within the chloroplast. This is reflected in the composition of the membranes and in their development during chloroplast biosynthesis.

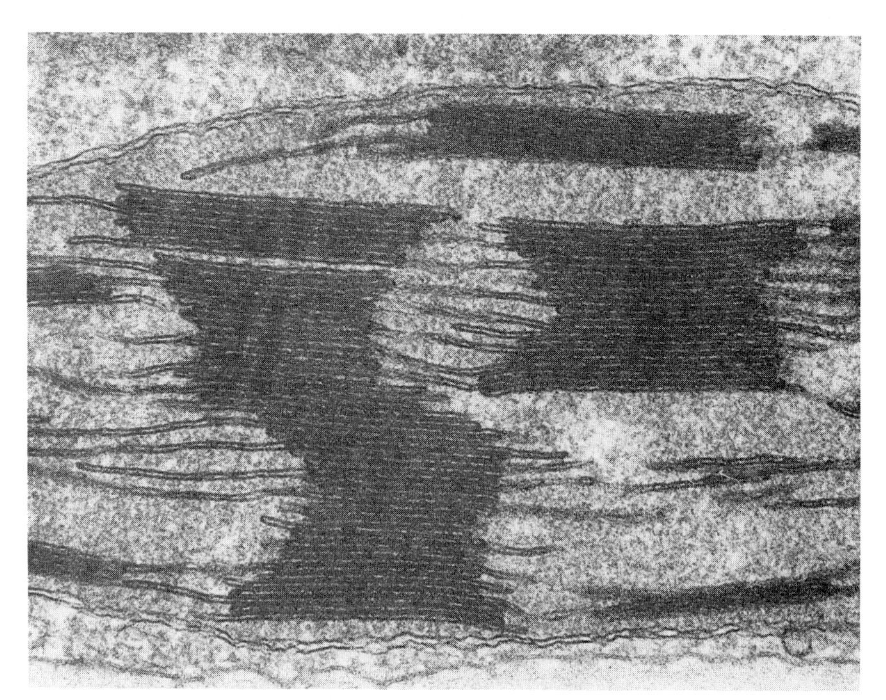

FIGURE 13.5 Highly magnified thin section through several grana showing features illustrated in FIGURE 13.4 (Courtesy of Dr. Harry T. Horner.)

Outer Membrane

The outer membrane of higher plant chloroplasts is separated from the inner membrane by an electron-translucent space of about 10 nm. This membrane is nonspecifically permeable to a wide variety of compounds of low molecular weight including nucleotides, inorganic phosphate, phosphate derivatives, carboxylic acids, and sucrose.[2] Thus the intermembrane space has free access to most of the nutrient molecules of the cytosol. Viewed in terms of their permeability characteristics, the outer membranes of chloroplasts and mitochondria apparently serve similar functions.

Inner Membrane

The inner membrane operates as the functional barrier between the cytosol and the stroma of the chloroplast. It is impermeable to sucrose, sorbitol, and a variety of anions (e.g., di- and tricarboxylates, phosphate, and compounds like nucleotides and sugar phosphates).

Although impermeable to many types of compounds, the inner membrane is permeable to carbon dioxide and certain monocarboxylic acids (e.g., acetic acid, glyceric acid, glycolic acid). It is somewhat less permeable to amino acids.

Like the mitochondrial inner membrane, the chloroplast inner membrane contains specific carriers to translocate the important metabolic agents of photosynthesis: phosphate, phosphoglycerate, dihydroxyacetone phosphate, dicarboxylates, and ATP.

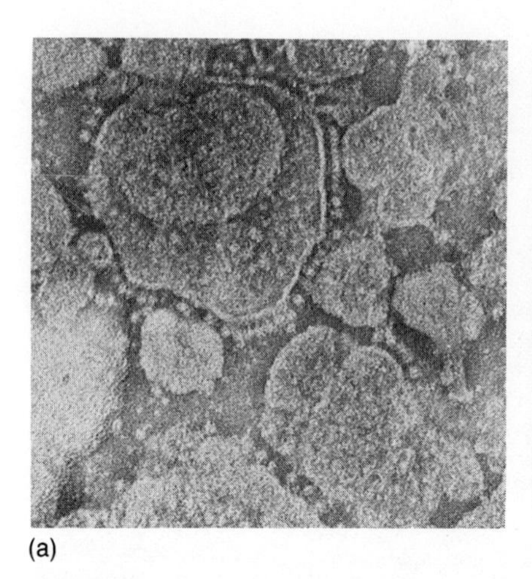

(a)

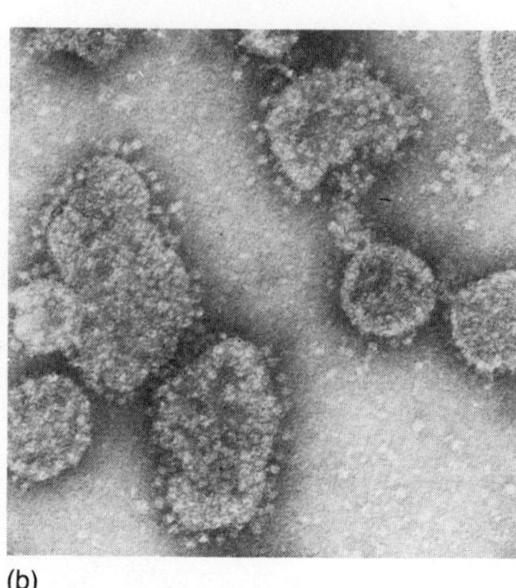

(b)

FIGURE 13.6 Electron micrograph showing the presence of knoblike structures protruding from the membranes of chloroplasts (*a*) and mitochondria (*b*). These knobs contain the ATPase activity of isolated membranes and the ATP synthesizing ability of the intact organelles. (Courtesy of Dr. John N. Telford.)

The two major photoreceptor molecules are chlorophyll *a* and *b*:

— CH_3 in chlorophyll *a*
— CH=O in chlorophyll *b*

$$R' = -CH_2-CH=C(CH_3)-CH-(CH_2-CH_2-CH(CH_3)-CH_2)_3H$$

They differ only by a —CH_3 in *a* replaced by a —CH=O in *b*. Chlorophylls are magnesium porphyrin molecules rich in alternating single and double bonds. This enables them to be exceptional absorption sinks of light in the visible spectrum where solar output is maximal. The absorbing abilities of the chlorophylls (extinction coefficients) are among the highest known for organic compounds. The two chlorophylls have slightly different absorption spectra such that they complement each other. The isoprene unit side chain confers hydrophobicity to that portion of the molecule, an important property for its membrane location.

Internal Membrane System

The internal membranes that pervade the stroma form an extremely complex network. Electron microscopy of ultrathin sections suggests that the lamellar membranes of the grana and the various interconnecting stretches of membrane comprise a discontinuous system of membrane pieces and fragments. This scattered fragment point of view has evolved over the years, and some now accept the notion that the entire internal membrane system is made up of a single folded and perforated sheet that encloses a continuous space that is separate from the stroma. This is very difficult to picture three dimensionally, and experimental confirmation is extremely hard to realize.

Whatever the precise physical form of these membranes, it is quite clear that the thylakoid membrane contains a complete complement of enzymes to carry out the light-dependent reactions of photosynthesis. This is the organelle location of chlorophyll, electron carriers, and the factors that couple electron transport to phosphorylation.

Electron micrographs of the internal membrane system have shown the presence of protruding knoblike structures, comparable to those found attached to the inner membranes of the cristae of mitochondria (see Figure 13.6). In both cases, the knobs project into enzymatically comparable compartments, the stroma and the matrix.

The Stroma and Its Inclusions

The stroma houses the enzymes necessary to carry out the assimilation of CO_2 into carbohydrate. The roles of these enzymes are discussed later in the chapter. Several kinds of particle are also often present in the stroma. The most obvious one is the starch grain, which may be up to 2 μm long. With osmium tetroxide fixation, small round particles termed

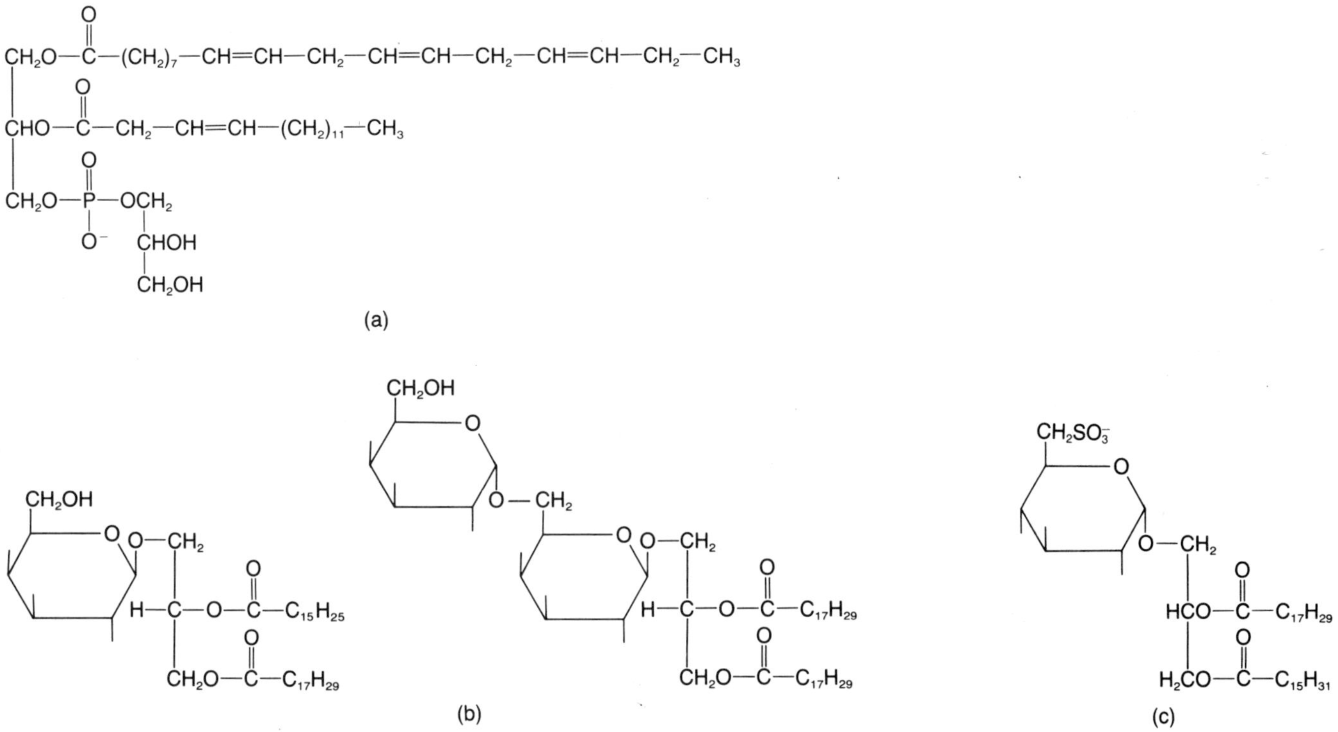

FIGURE 13.7 Structures of some thylakoid membrane lipids. (*a*) The major phosphatidylglycerol of spinach chloroplasts. (*b*) Galactolipids of green plants. (*c*) A plant sulfolipid.

plastoglobulin are seen lying freely in the stroma. They apparently are sites of lipid storage, particularly that of plastoquinone and tocophorylquinones.

The stroma also contains ribosomes and DNA fibrils.

13.3 THE CHEMICAL COMPOSITION OF CHLOROPLAST MEMBRANES

We know a good deal more about the composition and properties of the internal membrane system than the outer membranes. Since most of the chemical activities of the chloroplast membranes take place in the interior, most of the research attention has been focused there.

Thylakoids can be isolated from intact chloroplasts by osmotic shocking techniques and have been subjected to careful chemical analyses. Thylakoid membranes are about 50% lipid by weight, with only about 10% of this amount phospholipid. The lipids that appear to be unique to chloroplasts, galactolipids and sulfolipids, constitute 40 and 4%, respectively, of the total lipids present. In addition, a number of different specialized lipid molecules are present, including chlorophylls, carotenoids, and plastoquinones. Chlorophyll is a major lipid component, comprising about 20% of the total lipid of the thylakoid membrane.

The structures of some of the thylakoid membrane lipids are presented in Figure 13.7.

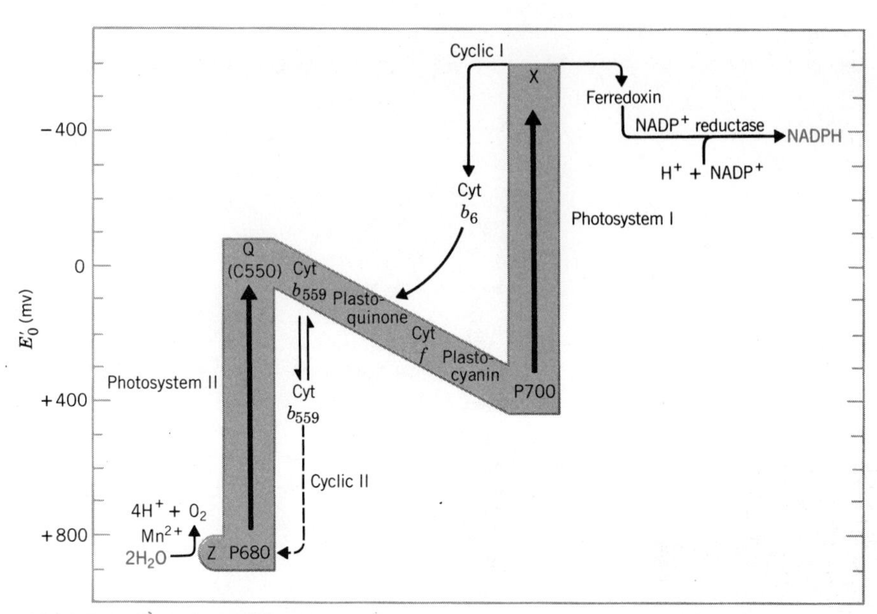

FIGURE 13.8 The Z scheme demonstrating the flow of electrons during photosynthesis from water to the reduced coenzyme NADPH.

13.4 THE PATHWAYS OF PHOTOPHOSPHORYLATION

Oxidoreductases are enzymes that generally employ coenzymes as electron carriers, either to remove electrons from a substrate (oxidation) or to add electrons to the substrate (reduction). The two pyridine coenzymes are nicotinamide adenine dinucleotide (NAD) and nicotinamide adenine dinucleotide phosphate (NADP).

Oxidoreductases that mediate catabolic reactions generally employ NAD; anabolic reducing enzymes use NADP. Since the products of photosynthesis are needed for growth (anabolism, reduction reactions), NADPH is formed. It is essential for the reductive assimilation of CO_2 into organic form.

The light-induced reactions of photosynthesis give rise to the formation of two products: ATP and the reduced coenzyme NADPH. The formation of ATP in this manner is referred to as *photophosphorylation,* in contrast to the reactions of oxidative phosphorylation that take place in mitochondria. The reduced coenzyme generated by this process is often called *reducing power.* Both these products are essential for the reactions of CO_2 assimilation.

The light-induced reactions, which we refer to collectively as *the light reaction,* take place in the thylakoids. They constitute a complex series of reactions in which the flow of electrons is coupled to ATP synthesis in a manner that strikingly parallels the mechanisms we discussed for oxidative phosphorylation in mitochondria. Two different photochemical systems participate in the overall scheme of the light reaction. Before discussing these individually, we will take a look at the general features of the pathway.

General Features of Electron Flow

The flow of electrons in the light reaction is commonly illustrated by the Z scheme (see Figure 13.8), in which the flow of electrons originates in water and terminates in the reduced coenzyme NADPH. Between these termini there are two points at which pigment electrons are boosted by light to higher energy levels; from these points they move down the redox scale through either an electron transport chain involving cytochromes or a nonheme iron protein.

It can be seen from this pathway that water is oxidized, oxygen is evolved, and a substrate is reduced. These relations are expressed by a simple equation.

$$H_2O + A \xrightarrow[\text{chloroplasts}]{\text{light}} AH_2 + \tfrac{1}{2}O_2$$

This is called the *Hill reaction* after Robert Hill, who first observed the process in isolated chloroplasts.

Cyclic and Noncyclic Electron Transport

The pathway of electrons as we have described it so far represents a *noncyclic* flow from water to coenzyme. This noncyclic flow is the major form of electron movement in photosynthesis. However, electron flow is not restricted to a noncyclic pattern. Two cyclic patterns have also been observed.[3] The predominant one involves a cyclic return of an excited electron to P700 by way of cytochrome b_6. A second cycle, which may operate only under certain special circumstances, utilizes cytochrome b_{559} to return an excited electron to P680. (The reaction centers P700 and P680 are so named because they are pigments that absorb strongly at 700 and 680 nm, respectively.) When the term "cyclic flow" is used, it normally refers only to the former cycle, designated cyclic I in Figure 13.8.

The Photochemical Systems

The Z scheme as it is here depicted can be divided into two separate but linked light-harvesting systems designated photosystem I (PS I) and photosystem II (PS II). There is now a great deal of experimental evidence to support the existence and operation of two separate photoreactive centers in chloroplasts. Some of the early work, however, which serves as a basis for our present understanding of photosystems, was carried out by Emerson as he observed the effect of light of different wavelengths on the quantum yield of photosynthesis.

The Red Drop and Enhancement Phenomena

In the red drop and enhancement experiments, such photosynthetic activities as O_2 generation, CO_2 fixation, and NADP reduction were measured as a function of wavelength. The results, when plotted, are referred to as *action spectra,* since they reflect the amount of action taking place at different wavelengths.

Using these kinds of experimental approach, Emerson made several important observations that helped to clarify the roles of various pigments in photosynthesis. One discovery was that termed the *red drop* (Figure 13.9).

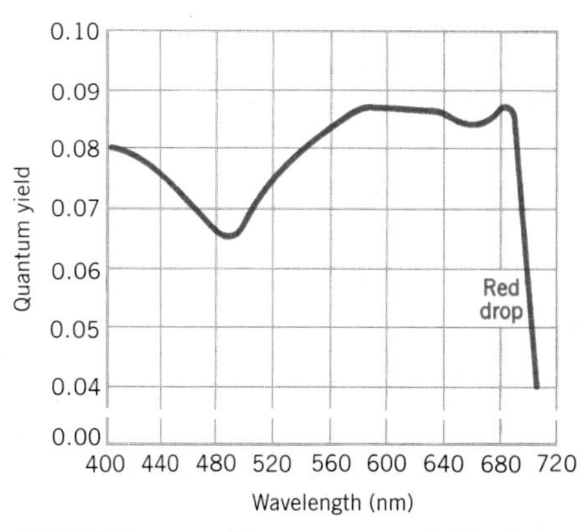

FIGURE 13.9 The quantum yield of photosynthesis as a function of the wavelength of light. The precipitous drop near 700 nm, when only chlorophyll *a* is absorbing light, is the red drop.

Figure 13.9 shows that the quantum yield from photosynthesis drops off sharply as the wavelength that is used to activate the process approaches 700 nm. This is not due to an inability of pigments to absorb at these longer wavelengths. Chlorophyll *a* continues to absorb light; in fact it is the only absorbing pigment at this wavelength, but its efficiency in photosynthesis drops markedly. At shorter wavelengths, accessory pigments absorb as well.

A second important observation made by Emerson was the phenomenon of *enhancement*. When red light (long wavelength and poor chlo-

Accessory pigments play important roles in photosynthesis by absorbing and transmitting energy to chlorophyll *a*, the heart of the photosynthetic apparatus. Different organisms may have different accessory pigments that affect their appearance:

Green algae	chlorophyll *b*
Brown algae	carotenoids
Red or blue-green algae	phycobilins

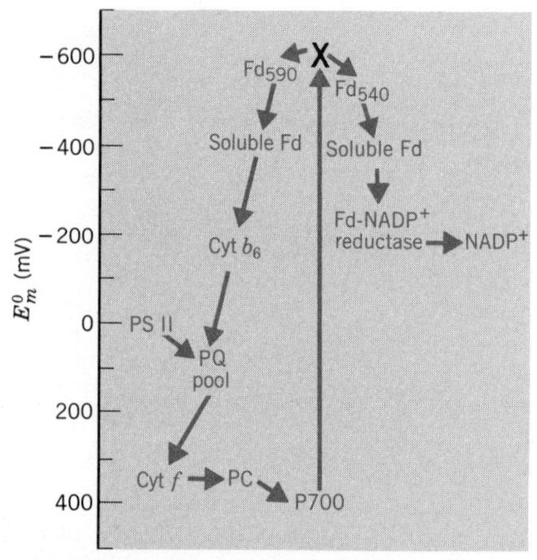

FIGURE 13.10 The components of photosystem I positioned on a potential scale with the hydrogen electrode at pH 7.0 having E^0_m = 0.

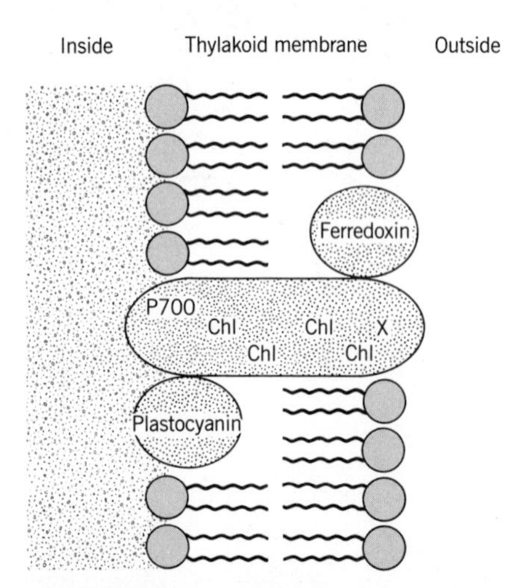

FIGURE 13.11 Model of reaction center of photosystem I showing a proposed cross-membrane jump of electrons from P700 to component X. Chl, chlorophyll.

rophyll *a* efficiency) is supplemented with light of a shorter wavelength, which is absorbed by accessory pigments, the rate of photosynthesis that results is greater than the sum of the rates when the two wavelengths are used separately. The rate of photosynthesis is thus enhanced by the use of both wavelengths simultaneously.

By varying the wavelength of the supplementary light beam, it was noticed that the action spectrum that resulted corresponded to the absorption spectrum of the accessory pigment. Thus, as long as energy can be absorbed by the accessory pigment, it can be transmitted to chlorophyll *a* to enhance its activity.

Another observation important to a consideration of photochemical systems was that the beams of light did not have to be used simultaneously to evoke enhancement. Alternately flashing beams at intervals of 0.6 to 1.6 sec separation were able to achieve enhancement.

The results obtained from studies on action spectra have been interpreted to mean that two photochemical systems exist and that they are separated by chemical reactions that can continue in the absence of a light flash of particular wavelength. This should become more apparent as we look into the reactions of the photosystems.

Photosystem I Photoreactions

We begin our discussion of photosystems with PS I, since it was the first of the two to be researched and is the best understood.

The components of PS I are summarized in Figure 13.10. The system is activated when light impinges on a *photosynthetic* unit, a concept introduced by Emerson and Arnold[4] and subsequently supported by several lines of experimental research. A photosynthetic unit is presently understood to be a collection of photoreactive molecules; an average unit contains about 300 chlorophyll molecules.

When light is absorbed by a chlorophyll molecule, it is boosted to an S_1, or excited singlet, state.[5] This excited state is then shifted among the chlorophylls of the unit by a mechanism that is not yet understood until the excitation reaches reaction center P700, which appears to be a dimer of chlorophyll *a*. P700 serves as the electron lead from the unit.

Once P700 is activated, there is a rapid one-electron transfer from P700 to a primary electron acceptor, designated X in Figure 13.10. The nature of this acceptor is controversial. For some time it has been thought that a membrane-bound ferredoxin is the primary acceptor, but now there is evidence from electron paramagnetic resonance (EPR) studies that another type of molecule, perhaps a quinone, may act as primary acceptor.[6]

It has been proposed that the electron jump from P700 to X is a cross-membrane jump (see Figure 13.11). This would separate P700 from X by a distance of 4.0 to 4.5 nm, the thickness of the membrane. In this view, certain chlorophyll molecules are assumed to form an overlapping π-system bridge or tunnel across the membrane.

The photooxidized form of P700 is reduced back to its original state by receiving an electron from plastocyanin (PC). Both plastocyanin and cytochrome *f* are topographically closely associated with P700, and it appears that they operate in tandem with single electron transfers, deriving their electrons from PS II.

The electron captured by the primary acceptor X is transferred down its redox scale to ferredoxin. Two such ferredoxins have now been resolved, one with a redox potential of −590 V, Fd_{590}, and one of −540 V, Fd_{540}. The pathway from X to $NADP^+$ is not clearly worked out yet, but it definitely involves ferredoxin-$NADP^+$ reductase, which appears to accumulate 2 electron equivalents before reducing the coenzyme.

Cyclic electron flow from P700 back to P700 makes use of factor X and cytochrome b_6. The electron is returned to a plastoquinone pool that operates normally in the electron transport chain between PS II and P700. It is thought that the relation between the extent of cyclic and noncyclic electron flows is governed by the redox state of the chloroplast.[7] When NADPH occurs in excess, the flow through Fd_{540} is blocked and thus channeled through Fd_{590} along the cyclic pathway.

A model proposed for the functioning of PS I in the thylakoid membrane is presented in Figure 13.12.

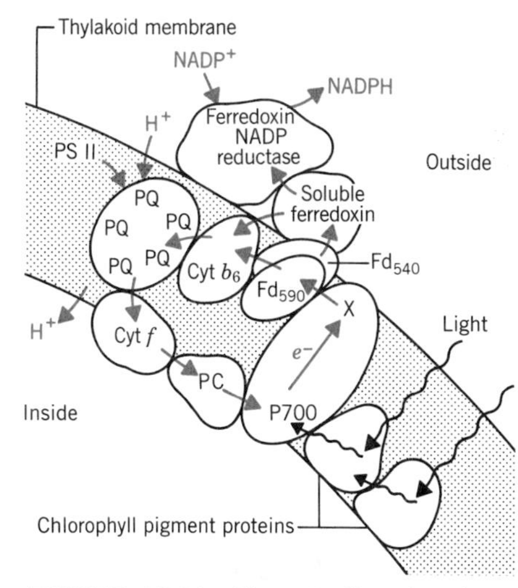

FIGURE 13.12 Electron flow in photosystem I as it is proposed to take place in the thylakoid membrane. Both cyclic and noncyclic pathways are shown. Electrons enter this system from PS II through the plastoquinone pool (PQ).

Photosystem II Photoreactions

Our knowledge of the reactions of PS II is less complete than for those of PS I. Nevertheless, the overall scheme of PS II appears to have the same basic features as PS I.

The components of PS II are presented in Figure 13.13. Let us begin with the primary donor of the excited electron, P680. This reaction center is comparable to P700, the reaction center of PS I. Light energy excites an electron in the photoreactive pigment molecule, which is apparently a dimer of chlorophyll *a* as is the case for P700.[8]

The acceptor of the electron is designated Q because it can be observed to quench the fluorescence of chlorophyll *a* in PS II. The identity of this acceptor is not certain, but there is evidence that it is a plastoquinone. Another component, designated C550, undergoes photoreduction with about the same kinetic properties as Q. At present, there is evidence both pro and con that Q and C550 may be the same. If they are not the same, they are certainly topographically very close together.

After the electron is received by the electron acceptor Q, it moves down its potential scale by way of cytochrome b_{559} and the plastoquinone pool on to P700. Plastoquinone refers to a variety of benzoquinone derivatives that occur in varying amounts in chloroplasts.

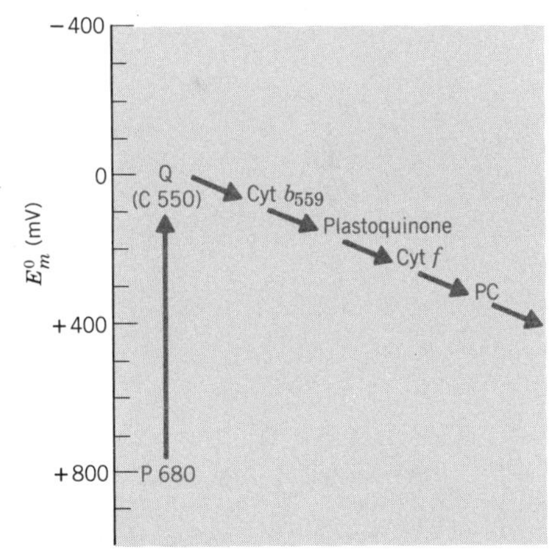

FIGURE 13.13 The components of photosystem II placed on a scale to reflect their relative redox potentials.

The photochemical center P680 is reduced by an electron that is sequestered from water. The result is the evolution of oxygen.

Even now, more than 200 years since Priestley's observations of the evolution of oxygen, the specific mechanism by which water is photooxidized remains a puzzle. The equation that describes the process is simply:

$$2H_2O \rightarrow 4e^- + 4H^+ + O_2$$

A kinetic model proposed by Kok et al.[9] to explain oxygen evolution is presented in Figure 13.14. According to this model the system (S) involved in O_2 evolution undergoes four successive light-induced oxidation steps:

$$S_0 \xrightarrow{hv} S_1 \xrightarrow{hv} S_2 \xrightarrow{hv} S_3 \xrightarrow{hv} S_4$$

Each step creates an additional charge, which is stored until the S_4 state is reached. Upon reaching this oxidation level, two molecules of water

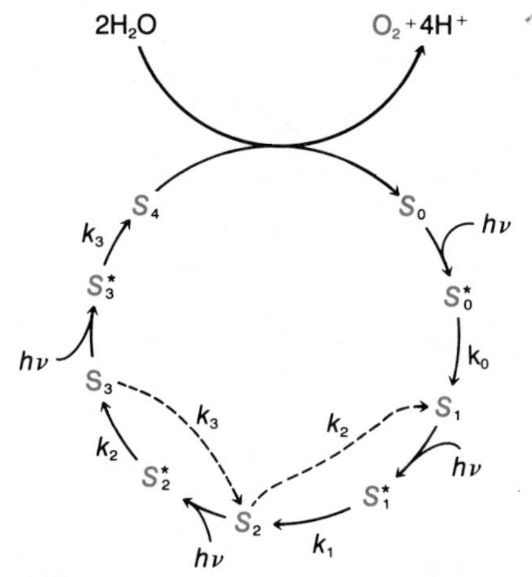

FIGURE 13.14 Kinetic model of oxygen evolution proposed by Kok. Subscripts (n = 0–4) represent the oxidation state of the O_2 system. Phototransitions are shown as $S_n \rightarrow S_n^*$ and dark relaxations are shown as $S_n^* \rightarrow S_{n+1}$. Dashed arrows represent deactivation reactions. Oxygen is evolved during the $S_4 \rightarrow S_0$ transition.

are oxidized to O_2 in a concerted manner with the reduction of S_4 to S_0:

$$S_4 + 2H_2O \rightarrow S_0 + O_2 + 4H^+$$

The chemical identity of the charge collector (S states) is unknown, however, there is some indirect evidence that suggests it may be thylakoid manganese.[10]

The Coupling of Electron Transport to Phosphorylation

So far we have seen that electron transport produces two products: oxygen and NADPH. Oxygen is for the most part a useless product, since the bulk of it is given off by the plant to the environment (some of it may be used by mitochondria for oxidative phosphorylation), whereas reduced coenzyme is an obligatory reducing agent in the biosynthesis of sugars. But another important product of photosynthesis, generated indirectly from electron transport, is ATP. ATP as produced in this manner comes about by a process called *photophosphorylation*.

The energy that drives photophosphorylation is provided by a mechanism similar to that employed for oxidative phosphorylation, namely, a chemiosmotic process. The basics are outlined in Figure 13.15.

Light impinges on the thylakoid membrane activating the two photochemical systems to transport electrons from water to $NADP^+$. The result is that an electrochemical gradient is established across the membrane, setting up the driving force for phosphorylation.

Although what is proposed here is similar to that which occurs in the mitochondrial inner membrane, there are some important differences. The electrons are transported from *inside* the thylakoid to *outside* via three passes across the membrane, whereas in mitochondria the electrons make three round trips returning to the inside, their point of origin. The flux of protons across the thylakoid membrane is inward, whereas it is outward in mitochondria. Finally, the CF_1 knob of the CF_1–F_0 ATP synthetase is oriented outward rather than inward as is the case for mitochondria.

The Topography of Enzymes in the Membrane

A vectorial transport of electrons outward and protons inward requires that the components of the photochemical systems have topographical constraints regarding their sites in the membrane. A variety of studies using antibodies, nonpenetrating agents, and other approaches have verified that this is indeed the case. CF_1, for example, has been demonstrated unambiguously to project into the stroma, hence outward. The electron acceptors of both photosystems I and II are also on the outside of the thylakoid membrane, while the electron donor of PS I is on the inner surface. Topographical studies of this sort have confirmed that $NADP^+$ is reduced on the outer surface and ATP is synthesized in the stroma as well.

Figure 13.16 illustrates the presumed topographical relationships between components of the electron transport system and the orientation of the CF_1–F_0 ATP synthetase in the membrane.

The import of vectorial electron and proton transport is that it results in an electrochemical gradient to drive the phosphorylation of ADP. Thus electron transport and photophosphorylation are coupled in chloroplasts, just as is true for mitochondria.

The model in Figure 13.16 depicts both PS I and PS II as operating in close spatial proximity within the thylakoid membrane linked by a plastoquinone shuttle system. This fits well with the overall scheme of electron transport and proton flux across the membrane, but it may not represent the *in vivo* picture.

A number of studies have revealed that there exists an extreme lateral heterogeneity in the distributions of PS I and PS II related to inner membrane structures.[11] Figure 13.17 points out that the membranes that make up a granum may differ depending on whether they are exposed to the stroma or appressed with other membranes in the stack.

As a result of working out techniques whereby stroma and grana thylakoids can be separated and studied independently, it is becoming clear that the two photochemical systems are not distributed equally among these membranes. Photosystem I is predominantly localized in stroma thylakoids, whereas PS II is restricted mainly to grana membranes. This finding is depicted in Figure 13.18. The photosystems are not the only membrane constituents that are segregated in this way. Both the ATPase coupling factor and ferredoxin-NADP$^+$ reductase are located exclusively in exposed thylakoid membranes. Thus, the final reactions of electron transport where reduced coenzyme is formed and ATP is generated take place on membrane surfaces maximally exposed to the stroma.

The lateral separation of photosystems poses some challenging questions regarding the mechanism whereby they may be linked to carry out the reactions of noncyclic electron flow. It is being speculated, for a number of reasons, that plastoquinone is the most likely candidate to serve as a mobile electron carrier between the two photosystems.

One reason has to do with the relative abundance of plastoquinone molecules compared to other members of the electron transport chain. There are 10 to 15 times as many plastoquinones as there are P680s or P700s. Another reason relates to the unusual degree of fluidity of thylakoid membranes due to their high content of unsaturated lipids. It is thought

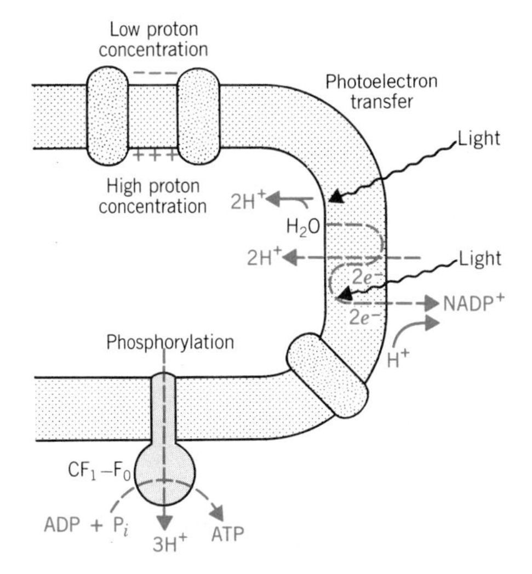

FIGURE 13.15 The basics of the chemiosmotic process compared in the two energy-producing organelles, chloroplasts (a) and mitochondria (b). In terms of electrochemical gradients and sites of ATP formation, one organelle is an inside-out version of the other.

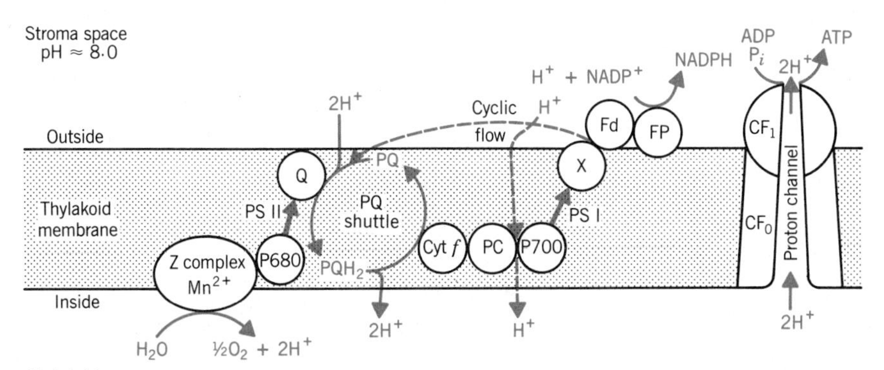

FIGURE 13.16 Diagram showing photosynthetic electron flow in the thylakoid membrane coupled to the formation of ATP. (Courtesy of Dr. Richard G. Jensen.)

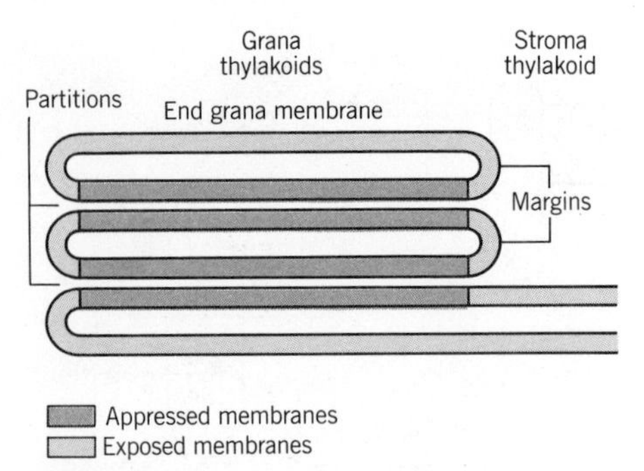

FIGURE 13.17 Drawing of granum membranes suggesting that membranes exposed to the stroma may be different from those not in contact with the stroma.

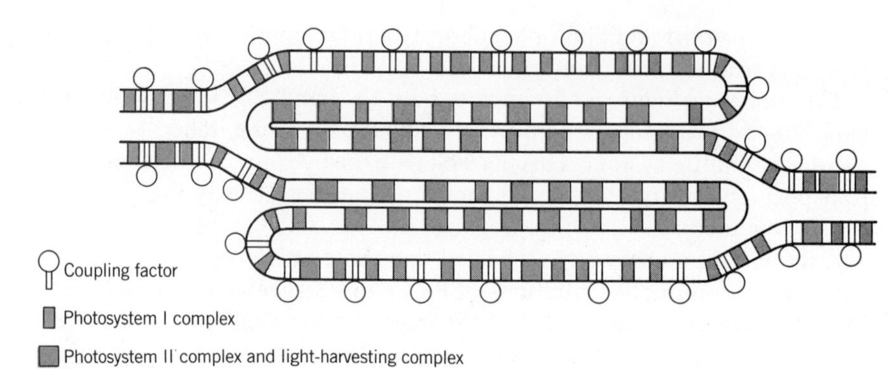

FIGURE 13.18 Proposed localization of photosystem I and photosystem II complexes and their associated light-harvesting chlorophyll *a*/*b*–protein complexes in grana membranes.

that plastoquinone, especially in its reduced form, should have a very rapid rate of lateral diffusion in this fluid medium which is well within the rate-limiting step of photosynthesis.

It therefore appears as though the challenge to explaining electron transport provided by lateral photosystem heterogeneity may not be insurmountable if experimental conditions can be worked out to make a careful analysis of the role and behavior of plastoquinone in the membrane. This problem is certain to spark a surge of research interest in this important organelle, and the results, as they are unraveled, will be of great interest to the scientific community.

The Structure of the ATPase Complex

The head or knob of the CF_1–F_0 ATPase complex is hydrophilic and can be removed from the membrane with EDTA at low ionic strength. Relatively pure CF_1 particles are obtainable this way and have been subjected to chemical analysis.

Method IV

Gel electrophoresis of CF_1 results in resolution of its proteins into five bands designated α, β, γ, δ, and ϵ in decreasing molecular weights from 59,000 to 13,000 daltons.[12] Agreement has not been reached on the relative compositions of the proteins in the CF_1 particle, but a stoichiometry of $\alpha_2\beta_2\gamma\delta_2$ is viewed as consistent with recent evidence.

The F_0 portion of the ATPase is hydrophobic and buried in the thylakoid membrane. It is much more difficult to study because of this. The F_0 piece appears to contain about six major polypeptides ranging in molecular weight from 42,000 to 11,000 daltons.

The Mechanism of ATP Formation

The mechanism by which the CF_1–F_0 complex brings about photophosphorylation is quite controversial. However, it is clear that it is driven by a proton gradient of sufficient magnitude that the ΔpH across the membrane is on the order of 3 units. The number of protons translocated per ATP formed (H^+/P ratio) has been the object of considerable study. It is generally agreed that a H^+/P ratio of 3 holds for most systems studied.

In this respect the ATPase of chloroplasts differs from that of mitochondria, where a ratio of 2 is found.

Most of the theories of phosphorylation that have been proposed for mitochondria (see Chapter 12) have also been applied to the chloroplast. Although it is tempting to discuss the various ideas that have been proposed to rationalize the existing evidence, such a discussion may not be very enlightening. For now we will resign ourselves to a recently drawn conclusion in a review article.

. . ., although a wealth of information has been gathered on the events during ATP formation by the ATPase complexes in energy-transducing membranes, the molecular details of this process and how energy is used to drive it remain uncertain. We must await more structural and chemical information on the ATPase complex in order to unravel the mechanism of energy transduction.[12]

The interested reader is encouraged to pursue the appropriate review articles listed at the end of this chapter.

13.5 THE METABOLISM OF CARBON IN THE CHLOROPLAST

The light reactions of photosynthesis generate ATP and reduced coenzyme which have a variety of metabolic functions in the cell. But their most immediate use in the cell economy is to provide the energy for carbon–carbon bond formation and reducing power as carbon dioxide is fixed into the form of carbohydrate.

Thus CO_2 fixation requires light only indirectly—to supply the ATP and NADPH. As long as these are available, the dark reaction of CO_2 reduction can take place independent of the light reaction.

The immediate products of CO_2 fixation are three or four carbon compounds, which are converted to hexoses and/or amino acid precursors. Depending on the physiological condition of the plant, hexoses may be used for the biosynthesis of starch, which may be stored in the chloroplast as a starch granule. As much as half of the CO_2 fixed may be converted to starch.

Plants possess two major pathways of CO_2 fixation that are designated according to the number of carbons in the primary fixation products. Plants that produce three-carbon compounds utilize the C_3 pathway of CO_2 fixation. Those that generate four-carbon compounds make use of the C_4 pathway.

The C_3 Pathway of Carbon Dioxide Fixation

Carbon dioxide is fixed in C_3 plants by entering into a cyclic series of reactions referred to as the *photosynthetic carbon reduction cycle* or, more simply, as the *Calvin cycle*. The reactions of this cycle are presented in Figure 13.19. Overall it has a very complex appearance, but it can be viewed more simply as consisting of three phases. Phase 1 begins with the incorporation of CO_2 into a triose by carboxylating ribulose-1,5-diphosphate and splitting it into two molecules of 3-phosphoglycerate. This phase is mediated by an enzyme, ribulose diphosphate-carboxylase/ox-

Phase 1 reactions of the Calvin cycle.

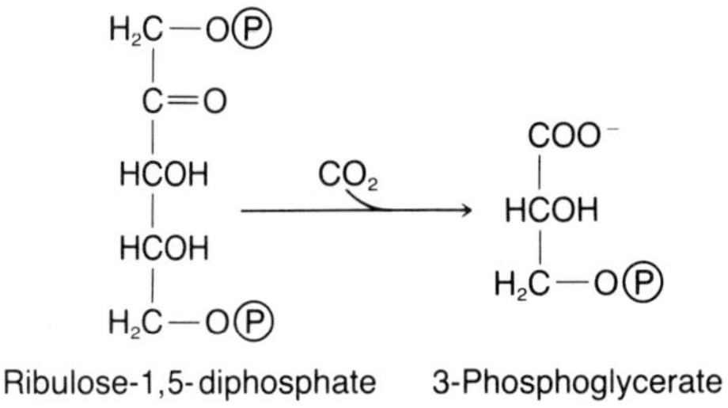

$H_2C-O(P)$ / $C=O$ / $HCOH$ / $HCOH$ / $H_2C-O(P)$ $\xrightarrow{CO_2}$ COO^- / $HCOH$ / $H_2C-O(P)$

Ribulose-1,5-diphosphate 3-Phosphoglycerate

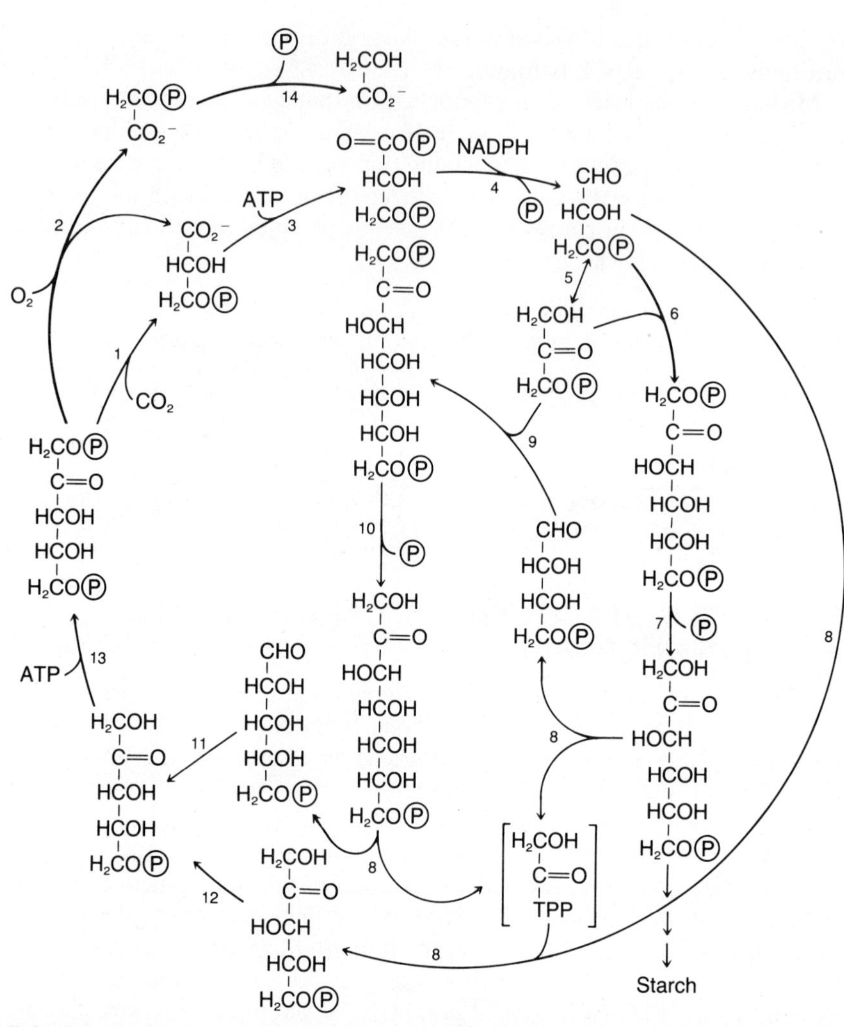

FIGURE 13.19 Photosynthetic carbon reduction or Calvin cycle. The enzymes corresponding to the numbers as follows are: 1, 2, ribulose II diphosphate carboxylase/oxygenase; 3, 3-phosphoglycerate kinase; 4, NADP-glyceraldehyde-3-phosphate dehydrogenase; 5, triosephosphate isomerase; 6, 9, aldolase; 7, fructose diphosphatase; 8, transketolase; 10, sedoheptulose diphosphate-1-phosphatase; 11, ribose-5-phosphate isomerase; 12, ribulose-5-phosphate epimerase; 13, ribulose-5-phosphate kinase; 14, phosphoglycolate phosphatase.

Phase 2 reactions of the Calvin cycle.

3-Phosphoglycerate

ATP

ADP

NADPH + H^+

NADP$^+$

Glyceraldehyde-3-phosphate

ygenase, which has an important role in addition to this first reaction in CO_2 fixation. It may use the same substrate, ribulose-1,5-diphosphate, and convert it to phosphoglycolate and 3-phosphoglycerate. The latter enters the Calvin cycle, but the former is dephosphorylated to glycolate, in which form it leaves the chloroplast. Glycolate may diffuse into peroxisomes, where it may serve as a substrate for the glycolate pathway or it may be oxidized by the reactions of photorespiration (see Chapter 7).

The second phase of the cycle is a two-step conversion of 3-phosphoglycerate to glyceraldehyde-3-phosphate. These two enzymatic reactions demonstrate vividly the manner in which light reactions are coupled to dark reactions. The reaction mediated by 3-phosphoglycerate kinase requires ATP to phosphorylate the substrate 3-phosphoglycerate, whereas NADP-glyceraldehyde-3-phosphate dehydrogenase utilizes NADPH to reduce its substrate to the aldehyde level. Hence the primary products of the light reaction are used immediately in phase 2 of the Calvin cycle.

Phase 3 consists of a series of interconversions that ultimately regenerate pentose monophosphates to complete the cycle and hexose monophosphates for the biosynthesis of high molecular weight polysaccharides.

By examining the cycle it can be seen that for every CO_2 reduced to the level of carbohydrate, three ATPs are required (one in step 13, two in step 3) and two NADPH coenzymes are employed (in step 4).

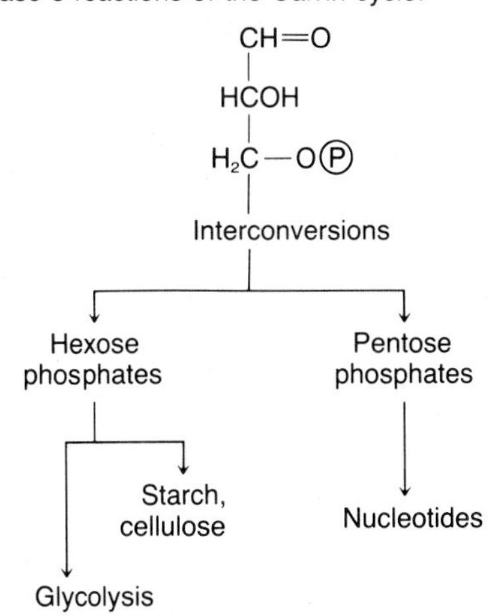

Phase 3 reactions of the Calvin cycle.

The C_4 Dicarboxylic Acid Pathway

Carbon dioxide fixation in the tropical grasses, including the familiar corn, sorghum, and sugarcane plants, is carried out by a variation or elaboration of the Calvin cycle. The primary fixation products are C_4 dicarboxylic acids: malate, aspartate, and oxaloacetate.

The three major variations of C_4 metabolism are outlined in Figure 13.20. First, notice that there are two categories of chloroplast-containing cells: mesophyll cells and bundle sheath cells (Figure 13.21). The complete photosynthetic carbon reduction (PCR) or Calvin cycle operates only in the bundle sheath cells. In addition to chloroplasts, cytoplasm and even mitochondria contribute to C_4 metabolism.

To begin, CO_2 is brought into the cytoplasm of mesophyll cells, where by action of phosphoenolpyruvate carboxylase a C_4 acid, oxaloacetate, is formed. Oxaloacetate may be either reduced to malate in the chloroplast or transaminated to aspartate in the cytoplasm, giving rise to two products that leave mesophyll cells and move into bundle sheath cells.

The three types of bundle sheath cell depicted in Figure 3.20 represent three types of C_4 plant. A given species would have only one type of bundle sheath cell. Thus oxaloacetate is reduced to malate only in species of the NADP-ME type.

Continuing with the NADP-ME-type species, malate enters bundle sheath chloroplasts, where it is decarboxylated, forming CO_2 and pyruvate. Pyruvate may diffuse back to mesophyll cells to complete a metabolic cycle, whereas the CO_2 is brought into the Calvin cycle. Variations on this scheme are apparent in the other two types of bundle sheath chloroplast.

The import of this rather complicated scheme is that CO_2 is concentrated in bundle sheath chloroplasts at a level higher than can be achieved by simple diffusion from the atmosphere. Phosphoenolpyruvate carboxylase has a very high affinity for CO_2 compared to ribulose diphosphate carboxylase (in the Calvin cycle) and therefore serves as a more efficient lead in draining CO_2 from the atmosphere.

Increasing the concentration of CO_2 in the thick-walled bundle sheath cells has a twofold effect. First, a higher level of CO_2 favors increased CO_2 fixation, thus increasing the effectiveness of ribulose diphosphate carboxylase in this reaction. Second, CO_2 at higher levels competes more effectively with O_2 for ribulose diphosphate carboxylase (see Figure 13.19), to favor CO_2 fixation and hold down the reactions of photorespiration. Furthermore, if some CO_2 should be formed by photorespiration, the high affinity phosphoenolpyruvate carboxylase will refix it at the level of oxaloacetate.

It is thus apparent that C_4 plants fix CO_2 more efficiently than do C_3

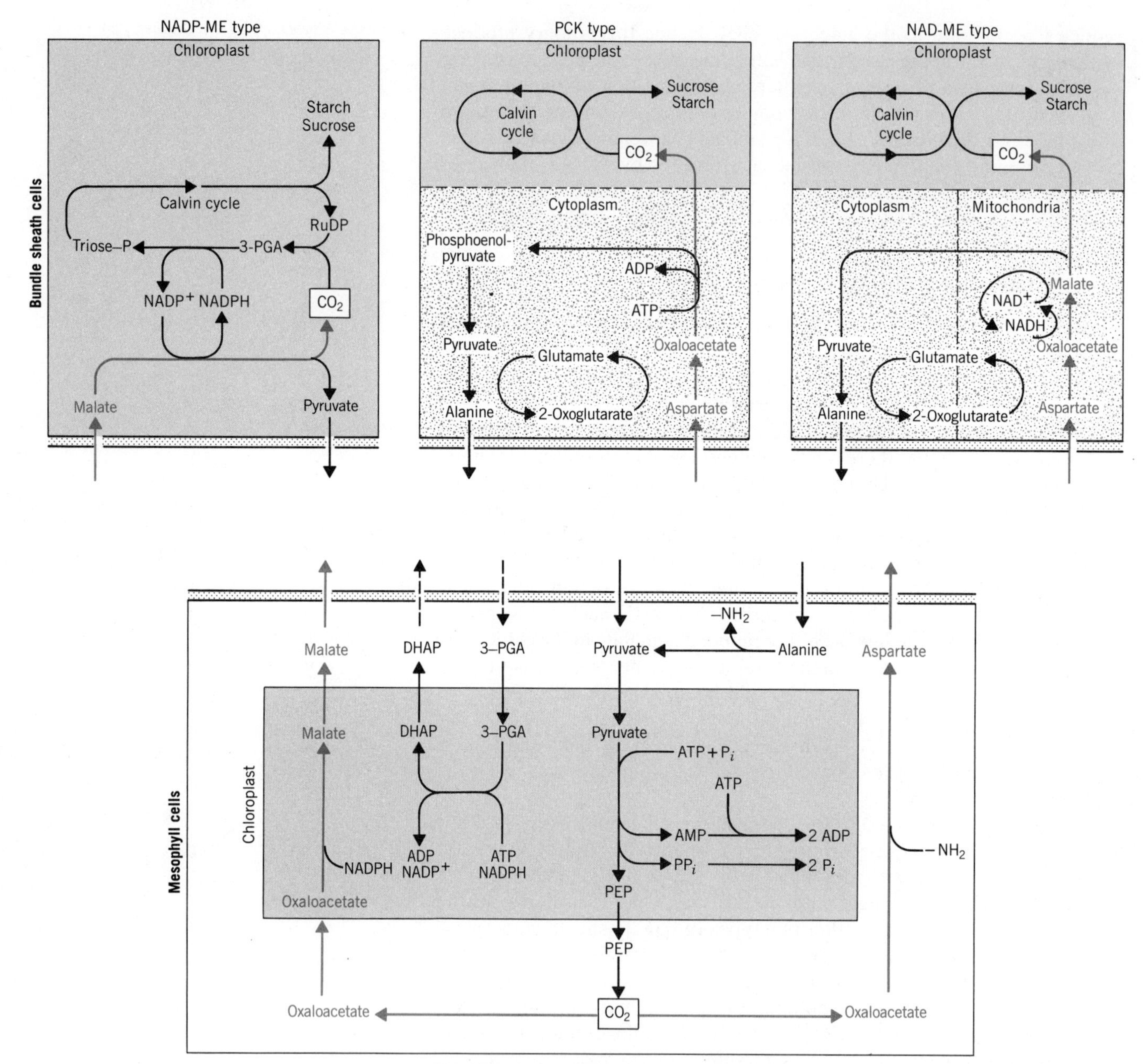

FIGURE 13.20 Models of C_4 metabolism in mesophyll and bundle sheath cells. Colored arrows delineate enzymatic pathways by which CO_2 is concentrated in bundle sheath cells. The three variations of C_4 metabolism are designated according to the names of the main enzymes that release CO_2, namely, NADP malate enzyme (NADP-ME type), phosphoenolpyruvate carboxykinase (PCK type), and NAD malate enzyme (NAD-ME type).

plants because of the high affinity for CO_2 and the low sensitivity to oxygen-induced photorespiration of the former. These plants are able to produce high yields of organic matter and are better adapted to arid, hot regions where photorespiration could result in a serious loss of CO_2.

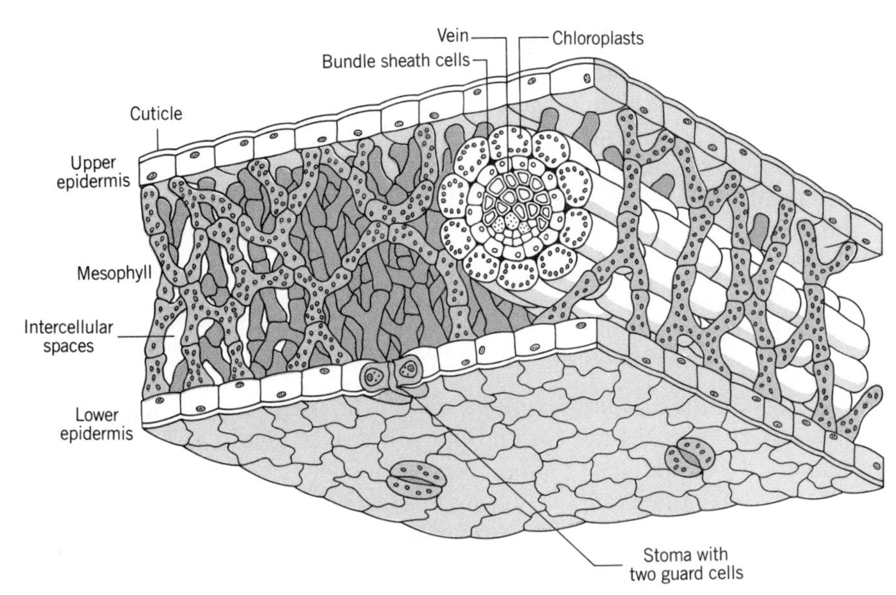

FIGURE 13.21 Mesophyll cells are distributed between the epidermal cell layers that constitute the surfaces of a leaf. They are generally columnar (palisade parenchyma) or of irregular (spongy parenchyma) shape.

Bundle sheath cells form one or more layers around the veins, which contain the xylem and phloem. Although they may resemble mesophyll cells morphologically, they have the special task of guarding the xylem–phloem tracts and can carry out the complete sequence of the Calvin cycle.

13.6 THE STRUCTURE AND PROPERTIES OF THE CHLOROPLAST GENOME

Chloroplasts possess a degree of autonomy within the cell that is in many ways similar to that of mitochondria. They do contain in the stroma a DNA that is unique to the organelle. With this genome a number of chloroplast-specific proteins are made using ribosomes that are also located in the stroma. Like mitochondria, chloroplasts replicate and thereby demonstrate a measure of reproductive autonomy.

The Physical Properties of Chloroplast DNA

The entire chloroplast genome resides within a single circular chloroplast DNA (ctDNA) molecule. However, the DNA is generally present in multiple copies with as many as 20 to 60 ctDNAs per chloroplast.

Depending on the species of organism, molecular weights of ctDNA commonly range from 85 to 140 $\times$ 10^6 daltons. The contour length is frequently around 45 μm, but may range from about 40 to 60 μm, depending on species. An electron micrograph of a single ctDNA of average contour length is presented in Figure 13.22.

Isolated ctDNA typically exists in a variety of forms and conformations. Some are unicircular, as in Figure 13.22, and others appear as interlocked dimers. Two types of dimer are found: circular dimers, which are formed by recombination between two monomers, and catenated dimers, in which two monomers interlock like links in a chain. Circular dimers may con-

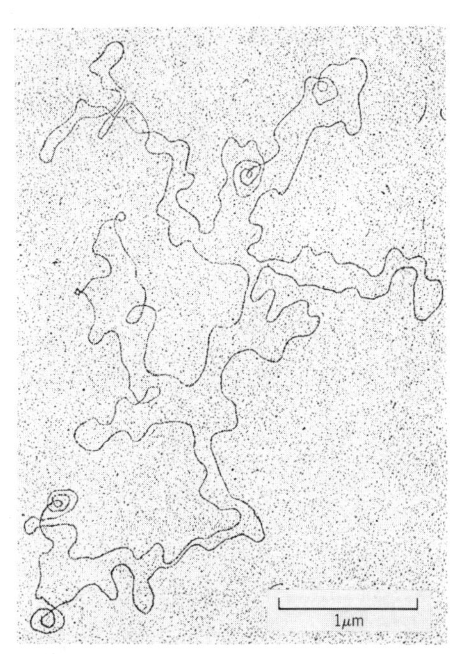

FIGURE 13.22 Electron micrograph of an open circular DNA molecule isolated from tobacco chloroplasts. (Courtesy of Dr. R. Herrmann.)

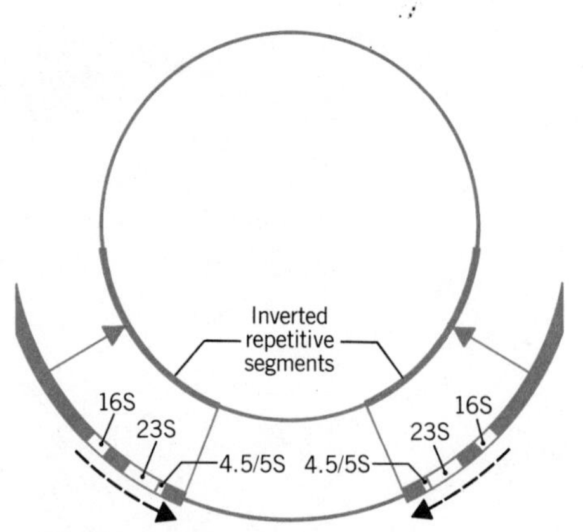

FIGURE 13.23 Circular physical map of chloroplast DNA from *Spinacia oleracea.* Thick arc regions represent the inverted repetitive segments. The expanded segments show the locations of the two sets of genes for RNA. The arrows indicate the polarity of rDNA transcription, which proceeds from the 16S genes toward the 23S genes.

stitute up to 10% of ctDNA and catenated dimers about 2.5%. The monomers often appear as relaxed (open) circular duplexes *in vitro,* but *in situ* the closed (supercoiled) form is predominant.

The Ultrastructure and Gene Content of Chloroplast DNA

Chloroplast DNA contains long repetitive sequences making up 20 to 30% of the contour length of the monomer. Shorter repetitive sequences that are inverted are also found in ctDNA, a rather unusual property for a genome.

The repetitive and nonrepetitive regions are organized in segments in all ctDNAs (see Figure 13.23). The genome is thus divided into four regions, two of which contain repetitive sequences that are inverted with respect to each other and two that are made up of nonrepetitive sequences. Physical maps of this sort have revealed that rRNA genes appear in the order 16, 23, and 5S RNAs, similar to that found in *E. coli.* Transcription is from 16 to 23S (note the figure). These two genomes are separated by approximately 2100 base pairs (in *Zea mays*). A 4.5S RNA, characteristic of chloroplast ribosomes, is coded for by a gene in the vicinity of the 5S RNA gene.

Genes coding for chloroplast tRNAs are scattered over the genome and are found both in the inverted repeat regions and in nonrepetitive regions. It is not presently known if ctDNA contains all the genes necessary to code for tRNAs for all amino acids, but as techniques are being refined it is beginning to look as though this may be the case. More than 20 tRNA genes have been identified.

Although ctDNA probably codes for a number of proteins, only two protein genes have been identified and positioned in the genome. One is the gene for the large subunit of ribulose-diphosphate carboxylase (also called fraction-1 protein). The other is a gene for a thylakoid peptide of unknown function, having a molecular weight of 32,000 to 35,000 daltons. Both genes are located in the large single copy region with one gene copy for each of the two proteins made.

Although gene positions for other proteins are not yet known, isolated intact chloroplasts synthesize about 90 soluble and insoluble peptides that contribute to the structure and function of the organelle. This obviously represents an area in which a good deal of work remains.

The Division of Labor between Nuclear and Chloroplast Genomes

Chloroplast DNA is large enough to code for more than 150 proteins of 50,000 daltons molecular weight each. This, coincidently, is approximately the number of different kinds of protein in the organelle, counting both structural proteins and enzymes necessary for photosynthesis and the synthesis of carbohydrates, lipids, and proteins.

But in fact, chloroplasts do not by themselves code for the synthesis of all these proteins. Rather, it is now widely accepted that *both replication*

Table 13.1 **Identified Chloroplast Gene Products**

Ribosomal RNAs	23S, 16S, 5S, 4.5S
Transfer RNAs	25–30 species
Messenger RNAs	For large subunit of ribulose diphosphate carboxylase/oxygenase, for P-32,000
Proteins	Ribosomal proteins: 2–5 species
	Soluble proteins (*in vitro* labeled ~ 90 bands)
	Large subunit of ribulose diphosphate carboxylase/oxygenase
	Cytochrome 552
	Elongation factor G; elongation factor Tu (?)
	Fatty acid synthetase (?)
	Thylakoid-bound proteins (*in vitro* labeled > 15 bands)
	Protein P-32
	Cytochrome *f*
	Three subunits of the CF_1 coupling factor (ATPase)
	Certain subunits of PS I and PS II (reaction center, water splitting)

and differentiation are controlled partly by the nuclear genome and partly by ctDNA.

The number of constituents unequivocally coded for by ctDNA and subsequently synthesized within the organelle continues to grow. A list of products presently known to be chloroplast gene products is presented in Table 13.1. Although certainly incomplete, the table suggests that the chloroplast is far from autonomous in its biosynthetic abilities.

The results from several laboratories have revealed that many of the stroma and thylakoid membrane proteins are encoded completely from nuclear DNA and generated on cytoplasmic ribosomes. For example, the small subunit of ribulose diphosphate carboxylase, other enzymes of the Calvin cycle, nucleic acid polymerases, and aminoacyl-tRNA synthetases all appear to be incorporated into chloroplasts but made in the cytoplasm under direction from the nucleus.

According to the multisubunit completion principle, the origin of a given protein, such as ribulose diphosphate carboxylase, from two different cellular sites and genomes represents a mechanism of regulation for the synthesis of organelle proteins, hence organelle function. These regulatory mechanisms have not been worked out, but it is apparent that the chloroplast depends on the nuclear genome for the operation of the Calvin cycle and photosynthetic phosphorylation. The plant cell, in turn, depends on chloroplast products, both as a source of energy and as building blocks for biosynthesis.

There is no question whether these coordinated activities between genomes are regulated to provide optimal levels of energy and substrates for the proper growth and reproduction of the cell. Herein lies another area of research opportunity for young scientists who wish to bring an additional measure of understanding to a basic biological process.

13.7 CHLOROPLAST REPLICATION AND DIFFERENTIATION

Chloroplasts arise from preexisting chloroplasts during the life cycle of higher plants and are passed on to progeny cells during cell division. The precise manner in which chloroplasts divide is not known, but electron microscopic evidence suggests that division is executed by a type of binary fission similar to that which occurs in bacteria and mitochondria. A constriction develops near the center of the plastid, and two progeny chloroplasts result from a pinching together of membranes in this region.

Chloroplast division appears to be generally asynchronous in a given plant tissue or cell. However, a number of environmental factors are known to influence replication and differentiation. Therefore, peaks of replication may be observed when the environmental conditions are optimal and synchrony or near synchrony can be forced by manipulating environmental factors.

Factors Influencing Replication and Differentiation

Of the several environmental factors that affect plastid development, light is the most important. But several others have important influences, including temperature, nutrition, hormones, nucleic acid analogues, levels of DNA (both nuclear and plastid), and water stress.[13,14] We examine the effect of several of these on division and differentiation.

Since a number of terms are associated with chloroplasts at different stages of their development, these are dealt with first.

The term *proplastid* is used to designate the small colorless or pale green plastid in meristematic cells of the shoot or root. They are undifferentiated, being bounded by a double membrane but having a homogeneous matrix with few if any vesicles. An older term for these was leukoplast.

In the dark, plastids called *etioplasts* form that are characterized by one or more crystalline prolamellar bodies. These plastids are also bounded by a double membrane. In their interiors double membrane lamellae extend from the prolamellar bodies and eventually become thylakoids.

With these two basic forms in mind, plus the intact mature chloroplast, let us consider the effect of certain factors on their development.

Light

Light per se is not generally a requirement for chloroplast division: when spinach is cultured under diurnal conditions, for example, chloroplasts increase in size and number during the light period, while in the dark they increase in number but decrease in size. This result also suggests that whereas light may not be prerequisite for division, it is important for maturation.

On the other hand, when grown in the light, chloroplast numbers and size appear to be related to the quantity (intensity × time) of light to which they are exposed. Thus the effect of light on division is not simple.

Somewhat more clear is the effect of light on differentiation. Two types of plant must be considered here, however.

In one type of plant light is required for the production of chlorophyll.

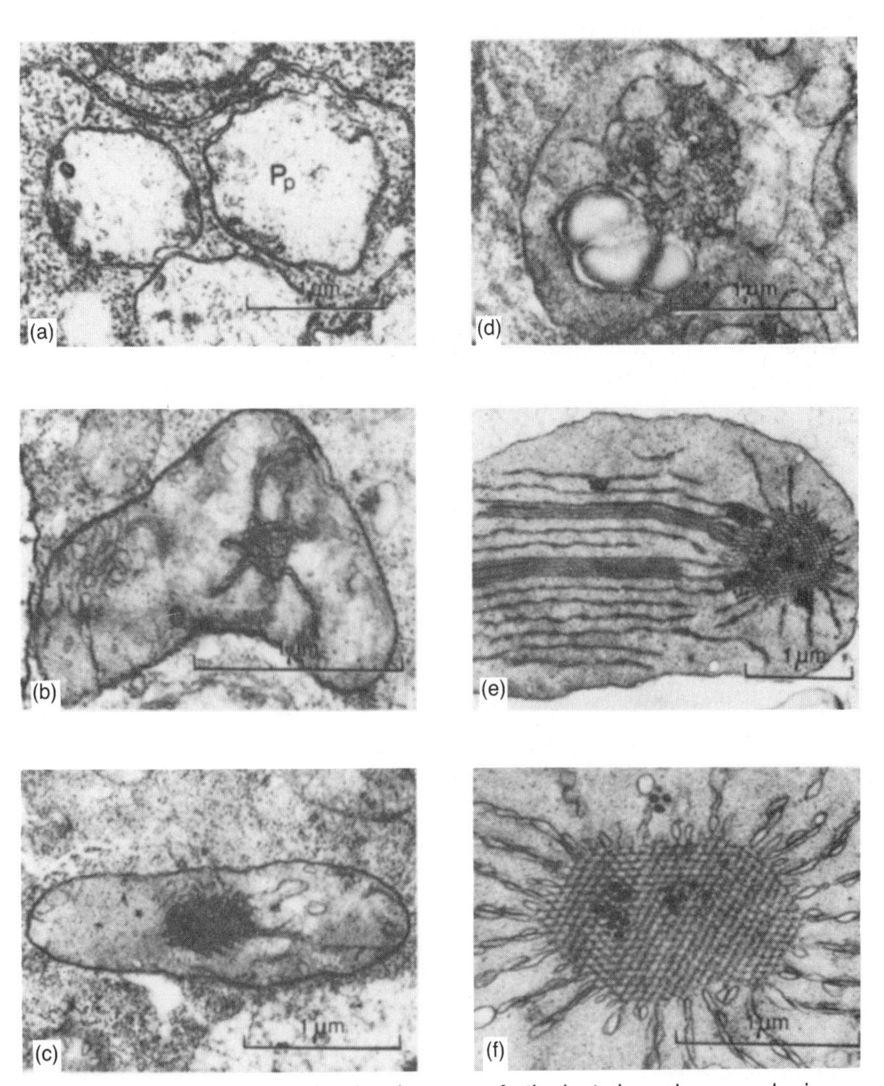

FIGURE 13.24 Stages in the development of etioplasts in embryos and primary leaves of *Hordeum vulgare* in the dark at six ages: (*a*) 0 hr, (*b*) 48 hr, (*c*) 72 hr, (*d*) 96 hr, (*e*) 120 hr, (*f*) 168 hr, (*a*) is a proplastid (P_p); (*e*) and (*f*) are fully developed etioplasts. (Courtesy of Dr. B. Sprey.)

In the absence of light, only etioplasts will form. These are undifferentiated and contain within the prolamellar body protochlorophyll, which is a precursor of chlorophyll. Stages in the development of etioplasts are depicted in Figure 13.24. Etioplasts differentiate into fully developed chloroplasts in the light and therefore appear to possess the necessary ingredients to produce the mature plastid. The prolamellar body stores membranes in the form of tubular, branched, and interconnected units, which have the appearance of a crystalline lattice. There is an equilibrium between this form of membranes in the prolamellar body and the form in thylakoids that is affected by the intensity and quality of light.

This type of light-induced differentiation affects a number of chloroplast ingredients. Changes occur related to protein and lipid biosynthesis, pigments other than chlorophyll, and certain enzymes. The impacts on these ingredients do not appear to be direct, however. The immediate effects appear to be on the phytochrome system. Red light having a wavelength of approximately 660 nm is most effective in promoting certain chloroplast transformations, and this effect is more plausibly explained as an effect on phytochrome.

The phytochrome-governed effect relates to enzyme synthesis and activities and even to RNA content.

Blue light is also necessary for chloroplast maturation. Certain enzymes (e.g., glyceraldehyde-3-P:NADP oxidoreductase) are formed only after irradiation with blue light. The blue light action spectrum (250–500 nm) is similar to the absorption spectra for various carotenoids and flavoproteins. This may be interpreted to mean that they are the immediate recipients of blue light and thus initiate the blue light effect in the chloroplast.

In a second type of plant, such as seedlings of some gymnosperms (e.g., *Pinus*), chlorophyll is synthesized in the absence of light. The chloroplasts in these plants appear to be fully differentiated, complete with thylakoids.

Temperature

The division of chloroplasts is temperature sensitive. At low temperatures (12°C) chloroplasts become larger than at higher temperatures (24–35°C), but their rate of division is lower.

The temperature dependence of biosynthesis for a variety of chloroplast constituents, including chlorophyll and rRNA, generally shows a maximum that corresponds to the optimal growth temperature for the plant. Thus below and above a certain temperature range that varies with species, there is a falling off of biosynthetic activity which, in turn, affects chloroplast development.

The temperature effect on chloroplast replication and development can be reasonably attributed to a normal temperature effect on enzymes, in this case especially the enzymes concerned with the biosynthesis of structural components and the molecular machinery of photosynthesis.

Chemical Factors

Several kinds of chemical are known to affect chloroplast differentiation. These can be grouped as follows: metabolites (e.g., precursors for chloroplast components or energy providers), growth regulators, metabolic inhibitors, herbicides, and mineral nutrients. Let us examine some representative chemical factors whose effects are best understood.

Growth Regulators. There are several growth regulators, or plant hormones, that influence chloroplasts when provided exogenously. Certain of these are also found in chloroplasts. They may thus exert their effect on chloroplasts either directly, from a site within the organelle, or more indirectly, by regulating a cytoplasmic activity that is essential for chloroplast replication or differentiation.

Cytokinins are a family of chemical agents that exert an influence by preserving chlorophyll during periods of senescence and promoting chlorophyll formation in dark-grown leaves that are subsequently exposed to light. The cytokinins are often viewed as antiaging factors that restore or maintain a variety of biosynthetic enzyme activities but do not stimulate biosynthetic pathways concerned with differentiation to operate in excess of normal.

Chloroplast components other than chlorophyll are affected by cytokinins. Plastoquinone, α-tocopherol, and carotenoids are influenced such that the syntheses of some are promoted and others deterred.

Certain photosynthetic enzyme activities, such as NADP-glyceraldehyde-3-phosphate dehydrogenase and ribulose diphosphate carboxylase, do appear to be increased by cytokinins. Ultrastructural differentiation is also promoted, such as grana and thylakoid formation.

The cytokinin effect is quite obviously a complex one. Basically, it influences enzyme behavior, but at the moment it is not clear at which level this takes place. It may be at the level of transcription or translation, where the resultant effect would be on enzyme amount, or it may be an effect directly on the mechanism of enzyme action.

Gibberellic acids cause a paling effect in plants that appears to be correlated with a decrease in chlorophyll. Other pigments show increases or decreases, the overall effect being very difficult to interpret. In some cases, the gibberellic acids appear to prevent chlorophyll breakdown.

Abscisic acid, a regulator found within chloroplasts, affects their differentiation by decreasing levels of chlorophyll, carotenoids, and vitamin K_1. One thought is that abscisic acid acts at the level of the gene to depress a normal light-induced activator of genes. The result is thus a decrease in pigment and thylakoid formation. In general, abscisic acid accelerates senescence, which is characterized by a breakdown of chlorophyll.

Indole acetic acid, an auxin, also has a multiple effect on plants. It appears to elevate the number of reaction centers in the chloroplasts as evidenced by an increase in concentration of chlorophyll P700 and Hill activity rates. In certain cotyledons an enhanced production of chlorophylls, carotenoids, and plastoquinone by the regulator is coupled to an increase in thylakoid function. Indole acetic acid may also promote starch formation.

Mineral Nutrients. Plant macronutrients (e.g., N, P, K, Ca, S, Mg) have a profound effect on plant homeostasis because they are components of vital structural and photosynthetic parts of the chloroplast.

A nitrogen deficiency, for example, decreases chlorophyll and carotenoid formation. This inhibition at the biosynthetic level also influences the structure of the chloroplasts, reducing them to smaller organelles. A calcium deficiency causes chloroplast lamellae to break down, and a magnesium deficiency affects both ultrastructure and rates of photosynthesis.

Micronutrients (e.g., Fe, Mn, Cu, B, Mo, Zn) play important roles in chloroplasts that become especially evident when these elements are deficient. Iron, for example, influences two biosynthetic steps in chlorophyll production, and is therefore essential for proper chloroplast differentiation. Manganese is associated with PS II activity, in which capacity it governs the extraction of electrons from water and the evolution of oxygen.

Copper, a component of plastocyanin, is essential for electron flow during photophosphorylation.

Other micronutrients have their own unique effects on photosynthesis and chloroplast development. A deficiency of a given mineral may promote an increase or a decrease of a given chloroplast component. Decreases occur when biosynthesis or activity is held in check by a lack of the vital ingredient. Increases may take place, however, because lack of a mineral may shift the equilibrium that exists among the various metabolic pathways such that the biosynthesis of a particular component actually may be favored during a deficiency.

Thus the study of deficiencies versus their effect at the molecular and ultrastructural levels becomes a very complex matter. The effect may take place at several different levels, such as at the gene or the active site, and the result may be seen on concentration, activity, and ultrastructure.

Summary

Although the overall process of photosynthesis can be expressed by a simple equation involving CO_2, H_2O, O_2, and organic matter, the molecular pathways by which this comes about are extremely complex.

The process takes place within chloroplasts in eucaryotes, organelles that possess three membrane systems separated by an intermembrane space and a stroma. The most interior of these membranes make up the thylakoids, flattened membrane sacks, which form stacks called grana.

The light reactions of photosynthesis—the generation of ATP, NADPH, and O_2—are enacted in the thylakoids. The dark reactions—the reduction of CO_2—are carried out in the stroma.

The pathway of electron flow is illustrated by a Z scheme with electrons emanating from water and ending in the reduced coenzyme NADPH. Electrons may follow both a cyclic and a noncyclic pattern of flow.

The electron pathway is divided into two photochemical systems, designated photosystem I and photosystem II. Each of these has a unique photosynthetic unit out of which electrons are boosted to participate in the pathway.

The flow of electrons occurs from inside the thylakoid to outside, moving across the membrane via the electron carriers that make up the Z scheme. Concomitant with electron flow is a flux of protons across the thylakoid membrane opposite from the direction of electron flow. The result is an electrochemical gradient that provides the protonmotive force for ATP generation.

With the ATP and NADPH provided by the light reaction, CO_2 is reduced in the stroma. One pathway of reduction, called the C_3 pathway, generates three-carbon products; the other, the C_4 pathway, produces four-carbon dicarboxylic acids.

Because of their higher affinity for CO_2 and their lower sensitivity to oxygen-induced photorespiration, C_4 plants fix CO_2 more efficiently than do C_3 plants.

Chloroplasts contain DNA molecules in multiple copies. The genome contains the information for transcribing ribosomal RNAs, transfer RNAs, and information for approximately 90 proteins that contribute to the structure and function of the organelle.

Although chloroplasts possess a degree of autonomy because of their own genomes, replication and differentiation are controlled partly by the nuclear genome and partly by chloroplast DNA.

Several environmental factors, including light, temperature, growth regulators, and minerals, also make contributions to chloroplast replication and differentiation.

References

1. *Photosynthesis and Related Processes,* E. I. Rabinowitch, Wiley-Interscience, New York, 1945.
2. Metabolite Transport in Intact Spinach Chloroplasts, H. W. Heldt, in *The Intact Chloroplast, Topics in Photosynthesis,* vol. 1, J. Barber, ed., Elsevier, New York, 1976.
3. The Two Photosystems and Their Interactions, W. P. Williams, in *Primary Processes of Photosynthesis, Topics in Photosynthesis,* vol. 2, J. Barber, ed., Elsevier, New York, 1977.
4. A Separation of the Reactions in Photosynthesis by Means of Intermittent Light, R. Emerson and W. Arnold, *J. Gen. Physiol.* 15:391(1932).
5. Photosystem I Photoreactions, J. R. Bolton, in *Primary Processes of Photosynthesis, Topics in Photosynthesis,* vol. 2, J. Barber, ed., Elsevier, New York, 1977.
6. Electron Spin Resonance Spectrum of Species "X" Which May Function as the Primary Electron Acceptor in Photosystem I of Green Plant Photosynthesis, A. R. McIntosh and J. R. Bolton, *Biochim. Biophys. Acta* 430:555(1976).
7. Regulation of Ferredoxin-Catalyzed Photosynthetic Phosphorylations, D. I. Arnon and R. K. Chain, *Proc. Natl. Acad. Sci. U.S.* 72:4961(1975).
8. Primary and Associated Reactions of System II, J. Amesz and L. N. M. Duysens, in *Primary Processes of Photosynthesis, Topics in Photosynthesis,* vol. 2, J. Barber, ed., Elsevier, New York, 1977.
9. Cooperation of Charges in Photosynthetic O_2 Evolution. I. A Linear Four Step Mechanism, B. Kok, B. Forbush, and M. McGloin, *Photochem. Photobiol.* 11:457(1970).
10. Mechanisms of Oxygen Evolution, R. Radmer and G. Cheniae, in *Primary Processes of Photosynthesis, Topics in Photosynthesis,* vol. 2, J. Barber, ed., Elsevier, New York, 1977.
11. Consequences of Spatial Separation of Photosystems 1 and 2 in Thylakoid Membranes of Higher Plant Chloroplasts, J. M. Anderson, *FEBS Lett.* 124:1(1981).
12. Energy Transduction in Chloroplasts: Structure and Function of the ATPase Complex, N. Shavit, *Annu. Rev. Biochem.* 49:111(1980).
13. Factors in Chloroplast Differentiation, C. Sundquist, L. O. Björn and H. I. Virgin, in *Chloroplasts,* J. Reinert, ed., Springer-Verlag, New York, 1980.
14. Plastid Replication and Development in the Life Cycle of Higher Plants, J. V. Possingham, *Annu. Rev. Plant Physiol.* 31:113(1980).

Selected Books and Articles

Books

Biochemistry of Photosynthesis, R. P. F. Gregory, John Wiley & Sons Ltd., London, 1977.

Chloroplast Metabolism, B. Halliwell, Clarendon Press, Oxford, 1981.

Chloroplasts, J. Reinert, ed., Springer-Verlag, New York, 1980.

Primary Processes of Photosynthesis, Topics in Photosynthesis, vol. 2, J. Barber, ed., Elsevier, New York, 1977.

The Intact Chloroplast, Topics in Photosynthesis, vol. 1, J. Barber, ed., Elsevier, New York, 1976.

Articles

Biochemistry of the Chloroplast, R. G. Jensen, in *The Biochemistry of Plants,* vol. 1, N. E. Tolbert, ed., Academic Press, New York, 1980.

Energy Transduction in Chloroplasts: Structure and Function of the ATPase Complex, N. Shavit, *Annu. Rev. Biochem.* 49:111(1980).

Enzyme Topology of Intracellular Membranes, J. W. DePierre and L. Ernster, *Annu. Rev. Biochem.* 46:201(1977).

How Cells Make ATP, P. C. Hinkle and R. E. McCarty, *Sci. Am.* 238:104(1978).

The C_4 Pathway of Photosynthesis: Mechanism and Function, M. D. Hatch, in *CO_2 Metabolism and Plant Productivity,* R. J. Burris and C. C. Black, eds., University Park Press, Baltimore, 1976.

The Structure of Chloroplast DNA, J. R. Bedbrook and R. Kolodner, *Annu. Rev. Plant Physiol.* 30:593(1979).

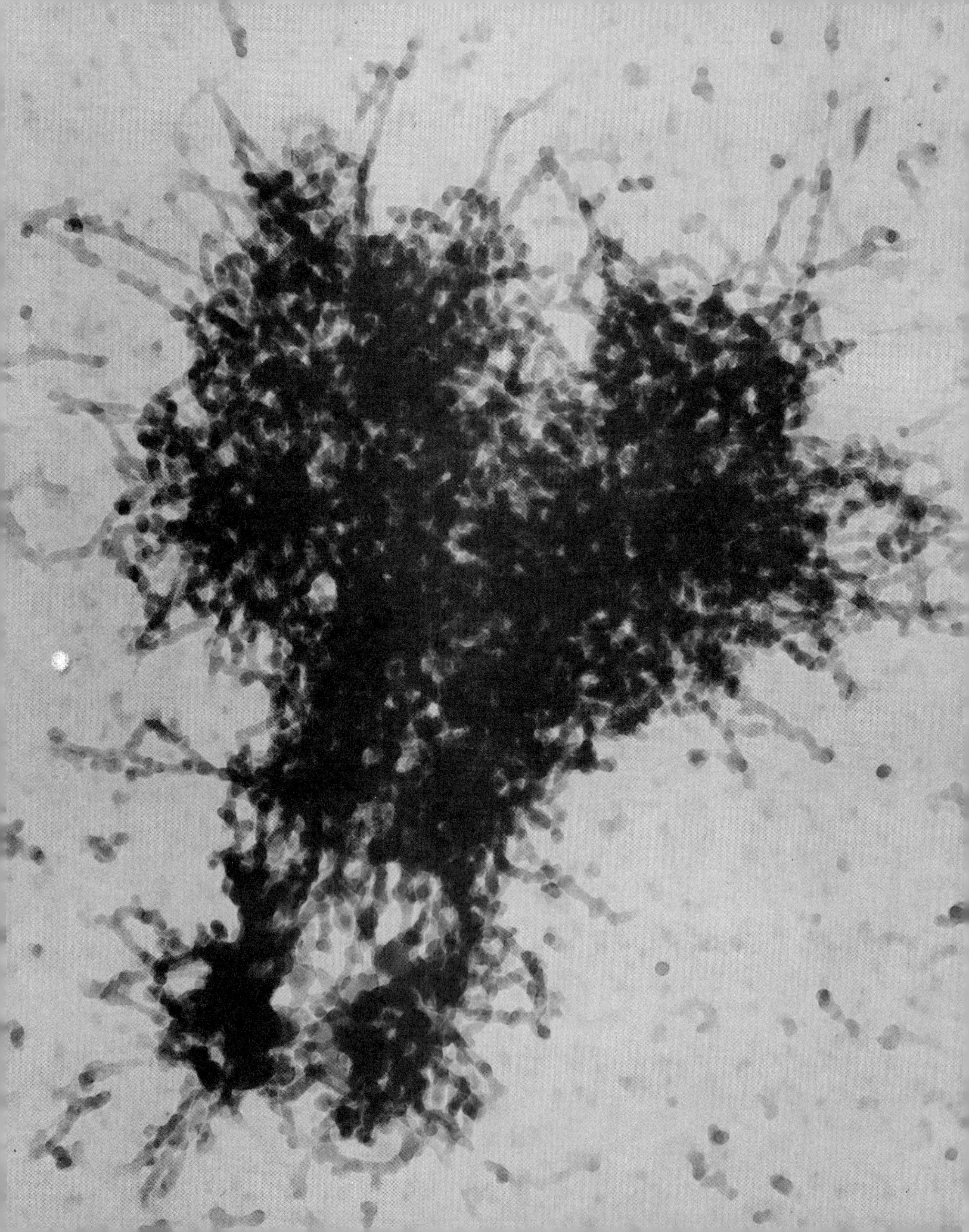

SECTION 7

COMPARTMENTALIZATION AND INTRACELLULAR FLOW OF INFORMATION

The information center of the procaryotic cell resides in a region of the cell called the nucleoid on a naked circular DNA molecule. The flow of information is from DNA (transcription) to ribosomes (translation) to proteins, which have structural and enzymatic roles. The information center is not sheltered from the rest of the cell, and the flow is relatively unhindered by intracellular structures.

Viruses possess even simpler information centers than do procaryotic cells, but they may be either linear or circular, and composed of either DNA or RNA. Because of their relative simplicity, they have yielded invaluable information regarding genome structure and function that has not been so readily retrieved from more complex organisms.

The information center in the eucaryotic cell is housed in a membrane-bound compartment on nucleoprotein complexes called chromosomes. The flow of information is the same as for procaryotes, but in this case there is a nuclear envelope to impede or influence the flow.

The nuclear interior consists, in addition to chromosomes, of a complex maze of fibrils and particles. Certain of these particles contain high molecular weight precursors of messenger RNA. Others are ribosome precursors, assembled in the nucleolus. The fibrils appear to have roles in directing the biosynthetic events taking place in the nucleus as well as the flow of information toward exit points from the nucleus.

In both procaryotic and eucaryotic systems ribosomes are the sites at which the genetic information is handled and made into something useful to the cell.

In the four chapters that follow we will view cellular information, its form and flow, and the complex devices used by the cell to handle both form and flow.

14

The Nuclear Envelope

The cell nucleus, central and commanding, is essential for the biosynthetic events that characterize cell type and cell function; it is a vault of genetic information encoding the past history and future prospects of the cell, an organelle submerged and deceptively serene in its sea of turbulent cytoplasm, a firm and purposeful guide, a barometer exquisitely sensitive to the changing demands of the organism and its environment. This is our subject—to be examined in terms of its ultrastructure, composition, and function.

VINCENT ALLFREY, 1968

We now turn our attention to the "central and commanding" organelle of the cell, which serves more than any other feature to distinguish the eucaryotic from the procaryotic cell. This is the *true nucleus,* as opposed to the *nuclear region* of the procaryotic cell. In this chapter we will focus on that which delimits the nucleus, forming its borders, namely, the nuclear envelope.

Every student having even a casual acquaintance with a light microscope has looked at eucaryotic cells and has observed two or three outstanding characteristics: cytoplasm, a nucleus, and perhaps nucleoli. The very existence of nuclei as distinct compartments suggests that they must contain something not possessed by the rest of the cell. It does not take a venturesome mind to propose that *something* must gather together and contain the elements of the nucleus, or they would be dispersed throughout the cytoplasm or clumped as an irregular mass rather than so clearly separated from the rest of the cell interior. But with a light microscope alone it is a tricky matter to provide experimental support for the structure of a nuclear boundary.

Nevertheless, the existence of a membrane surrounding the nucleus was suspected as early as the late 1800s. Around the turn of the century, experiments were carried out on cells and on isolated nuclei that showed the nuclei to behave as osmometers, a behavior that is consistent with a limiting border of membrane type.

Between 1920 and the 1950s nuclei were observed by means of phase contrast microscopy, which confirmed the conclusions drawn from the more indirect osmotic studies. Also, during this period, ideas emerged suggesting that the nuclear membrane was in fact a membrane envelope, that is, a double membrane wrapped around the nucleus with a space between membrane layers.

But as with most other cell organelles, an insight into the ultrastructure of the envelope had to await the electron microscope and the 1950s and 1960s. Once the electron microscope had been used successfully in probing the nuclear surface, not only was the existence of the envelope verified, but the view that it was a static sack holding the chromosomes had to be revised. Pores were seen to pocket the surface of the nucleus, and continuities were noted between the outer membrane of the envelope and the membranes of the endoplasmic reticulum.

The structure was increasingly viewed as a dynamic gateway between nucleus and cytoplasm, compartmentalizing the cell on the one hand and mediating transport on the other.

If we reflect briefly on some of the traffic between the nucleus and the cytoplasm, it is clear that the membrane system around the nucleus must be able to respond to the passage of a variety of molecular species in a manner different from that characteristic of the plasma membrane.

Proteins are made on ribosomes *outside* the nucleus, but some of these proteins are needed *inside* the nucleus for DNA replication, for transcription, and for the formation of preribosomal particles. They must move in.

RNAs of messenger, transfer, and ribosomal types are made *within* the nucleus, but all these have ultimate destinies and functions *outside* the nucleus. They unite in the complex act of protein synthesis in the cytoplasm. They must all move out.

The nuclear envelope is a busy checkpoint for materials that are passed from their point of origin to their place of function.

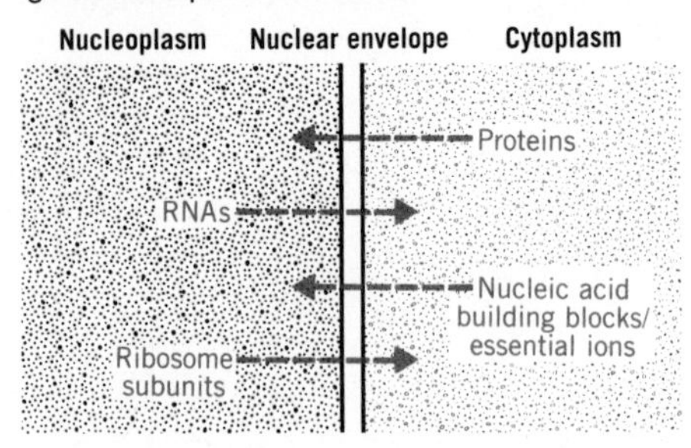

In some cases the traffic is back and forth, as for proteins made in the cytoplasm that become ribosome subunits in the nucleoplasm but end up functioning in the cytoplasm.

In addition, smaller molecules, which are the building blocks of the polymers formed in the nucleus, must be able to move to sites in the cell where they are needed. They are certainly required in the nucleus during cell replication and growth, and they must get there by way of the cytoplasm.

But this requirement for porosity must be reconciled with the osmotic properties of the nucleus and the fact that electrical potentials exist across the nuclear envelope. These characteristics are consistent with a high degree of restriction in the flow of small molecules between nucleus and cytoplasm.

It is now over 80 years since the existence of the nuclear envelope was first surmised. Much has been learned about its structure and function, but there is room for exploration in both areas. There is no longer any doubt about the membrane nature of the envelope, but its role as a dynamic intersection between nucleus and cytoplasm is still somewhat mystical.

14.1 GENERAL MORPHOLOGY AND NOMENCLATURE OF THE NUCLEAR ENVELOPE

The nuclear envelope is a double membrane sheath that defines the outer limits of the nucleus. The two membranes are roughly parallel, separated by a space except in limited areas where the membranes join and become a part of the pore complex. Since a considerable amount of nomenclature has developed pertinent to the nuclear envelope, we will begin our discussion by defining terms.

Figure 14.1 shows a cross section through a typical nuclear envelope consisting of an *inner* and an *outer* membrane, the latter occasionally continuous with the endoplasmic reticulum. The gap between the membranes, which is discontinuous from the rest of the cell, is called the *perinuclear space*. The only exception to its discontinuity occurs when there is a fusion of outer membrane with the cisternae of the endoplasmic reticulum.

The outer and inner membranes fuse in localized regions, forming small circular channels across the nuclear envelope. These openings, along with associated electron-dense materials in this region, are referred to as the *pore complex*.

The pore complex has several components, the presence of which depends to some extent on the species and type of cell and the manner in which it is prepared for electron microscopy. A cylindrical nonmembranous structure called the *annulus* rims the inside of the pore. Inside this lies a *central granule*. *Fibers* can be seen in some preparations extending from the central granule and the material of the annulus perpendicular to the plane of the envelope. In addition to these particulate structures, amorphous material appears to form a diaphragm over the pore from rim to rim.

FIGURE 14.1 Electron micrograph of thin section of nuclear envelope in rat liver. The inner (IM) and outer (OM) membranes are separated by the perinuclear space. A pore is apparent where these two membranes fuse.

Ribosomes may adhere to the outer nuclear envelope membrane, while a continuous, fibrous layer of electron-dense material coats the nuclear side of the inner membrane. This layer has been given the names *fibrous lamina, zonula nucleum limitans, internal dense lamella,* and most recently, *nuclear cortex*. We shall use the last name because it is the newest

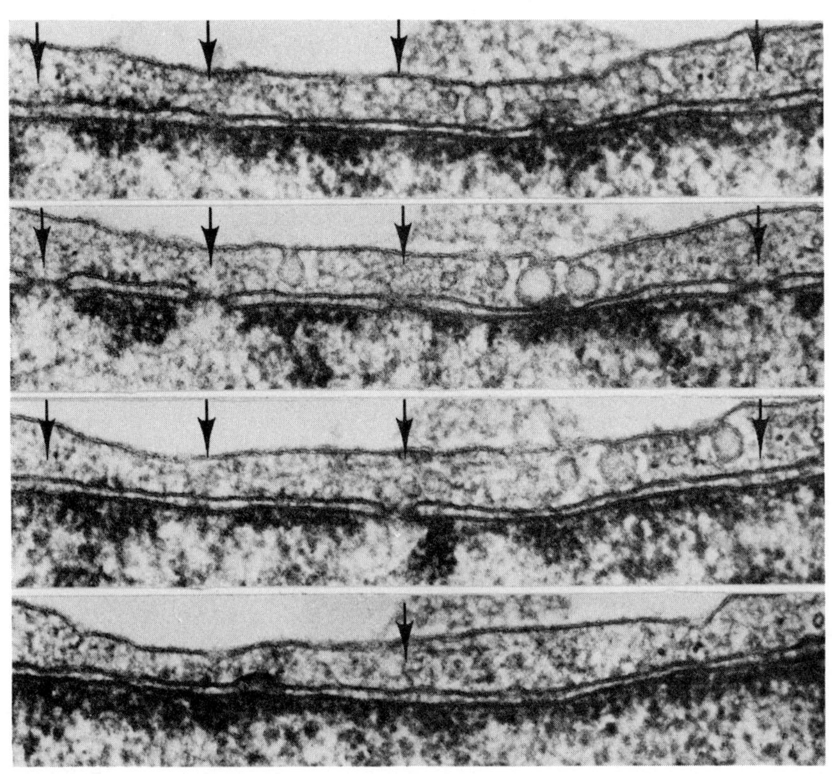

FIGURE 14.2 Serial cross sections through a nuclear envelope. By moving vertically along a line of arrows, one can see that the appearance of a pore depends on where the cut is made. The trilamellar characteristic of each membrane is clear in this preparation. ×100,000. (Courtesy of Dr. Gerd G. Maul.)

and is most clearly descriptive of the position of this layer in the ultrastructure of the nucleus.

"Cortex" is the Latin word for bark, rind, or shell. The nuclear cortex is a shell or rind for the nucleus that fits immediately *inside* the nuclear envelope. The term suggests the nucleus could exist without a limiting membrane. Such, in fact, is the case.

14.2 THE ULTRASTRUCTURE OF THE ENVELOPE

The Nuclear Envelope in Cross Section

Transmission electron micrographs of thin sections cut perpendicular to the surface of the envelope have consistently revealed the membranes to be 7 to 8 nm wide, with the same trilamellar appearance of most other membranes (see Figure 14.2). The outer membrane is a dynamic structure, fusing at points with the endoplasmic reticulum. Figure 14.3 shows membrane continuity between the nuclear envelope and the endoplasmic reticulum, along with additional ultrastructure typical of the envelope. In regions where ribosomes are attached, the outer membrane resembles rough endoplasmic reticulum. In other regions it has the appearance of smooth endoplasmic reticulum.

Both the outer and inner membranes have attachments to the rest of the cell interior. In certain cells filaments approximately 10 nm thick extend from the envelope outer surface into the cytosol, sometimes appearing to anchor their far ends to other organelles or to the plasma membrane.

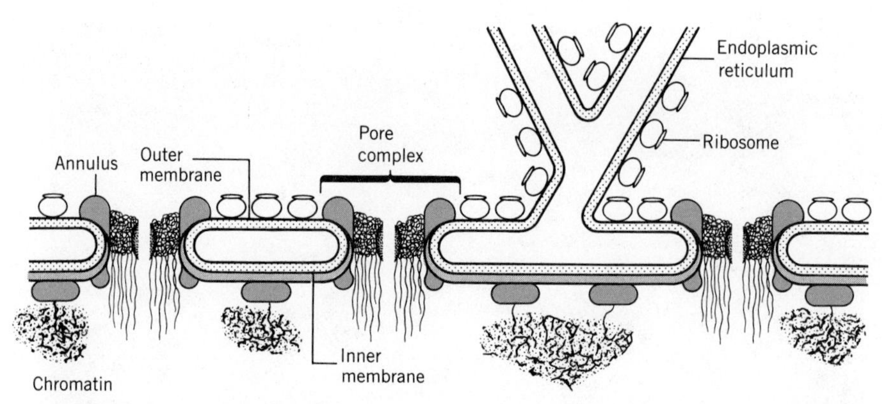

FIGURE 14.3 Ultrastructural details of the nuclear envelope and its immediate environment.

The inner surface of the envelope is also often coated with filaments and fibrous structures, some of which may extend to the interior of the nucleus, and others may affix to chromatin material.

The nucleus therefore does not appear to be a free-floating element in the cell. It may well be held in position by a web of filaments extending from its surfaces throughout the cell interior. The membranes themselves, at least from the point of view of ultrastructure, appear to be ordinary membranes except for regions where they are interrupted by pores.

The Perinuclear Space

The perinuclear space is a fluid-filled compartment between the inner and outer membranes that ranges in width from 10 to 70 nm. Although the spacing of the membranes is irregular, a gap width of about 20 nm is common.

Even though the perinuclear space generally appears to be inert and free of ultrastructure, a number of exceptions have been observed. There have been scattered reports of fibers, crystalline deposits, lipid droplets, and other electron-dense materials spanning the gap. But the exact biochemical compositions of these inclusions and their significance between the envelope membranes remain elusive.

The Nuclear Pore Complex

Perhaps the most intriguing part of the nuclear envelope is the pore. From the time it was first observed in 1950 by A. Callan and S. Tomlin[1] until the present it has been intensively investigated. This ubiquitous structure occurs in all eucaryotic cells of both plant and animal sources. The pore is outlined by a circular fusion of inner and outer nuclear membranes, but it exists in association with the nonmembrane structures that together form the pore complex.

Diameter, Density, and Distribution of Pores

Pore dimensions are generally quite consistent in a given tissue and among similar cell types. But dimensions have appeared in the literature ranging from 40 to more than 100 nm (see Table 14.1).

Table 14.1 **Some Estimates of Pore Diameter and Density for Various Mammalian Nuclear Envelopes**

Tissue	Method	Pore Diameter (nm)[a]	Pore Density (pores/μm^2)
Rat liver	Whole tissue fixed (glutaraldehyde and osmium) and embedded	68 ± 6	16.3 ± 1.5
Rat liver	Whole tissue fixed and freeze-etched	88 ± 8	14.1 ± 2.3
Rat liver	Isolated nuclei fixed and freeze-etched	78 ± 6	24.9 ± 3.0
Rat liver	Isolated nuclear envelopes, fixed and freeze-etched	89 ± 7	24.3 ± 7.5
Rat liver	Isolated envelopes, fixed and negatively stained	67 ± 2	35.8 ± 4.3
Rat liver	Isolated nuclei, glycerinated and freeze-etched	80–88	
Rat liver	Whole tissue fixed (osmium) and embedded	40	
Rat pancreas	Whole tissue fixed (osmium) and embedded	40	
Rat thyroid	Whole tissue fixed (osmium) and embedded	50	
Monkey kidney cell cultures	Whole tissue fixed (osmium) and embedded	80–82	18–23
Monkey kidney cell cultures	Whole tissue fixed (permanganate) and embedded	99–110	16–20
Monkey kidney cell cultures	Whole tissue fixed (glutaraldehyde), postosmicated and embedded	75–80	16–23
Mouse 3T3 cultured cells	Whole tissue fixed (glutaraldehyde), postosmicated and embedded	68–72	19–24
HeLa cells	Whole tissue fixed and embedded	65	13
HeLa cells	Isolated envelopes fixed and negatively stained	68 ± 3	46 ± 8
Rat brain	Isolated envelopes fixed and negatively stained	68–80	
Human lymphocytes	Whole cells glycerinated and freeze-etched		3–4
Human melanoma	Whole cells glycerinated and freeze-etched	98–110	

[a]Values given are to the nearest whole number.

Table 14.2 Pore Densities and Pores per Nucleus in Seven Cell Types

Cell Type	Pore Density (pores/μm^2)	Pores per Nucleus
Human lymphocyte	3.2	405
Leopard frog embryo	5.6	1729
Human embryonic lung	8.5	2788
African green monkey kidney	8.6	4277
Newt heart	7.6	12,707
Xenopus laevis oocyte	51	37.7×10^6
Triturus alpestris oocyte	50	57×10^6

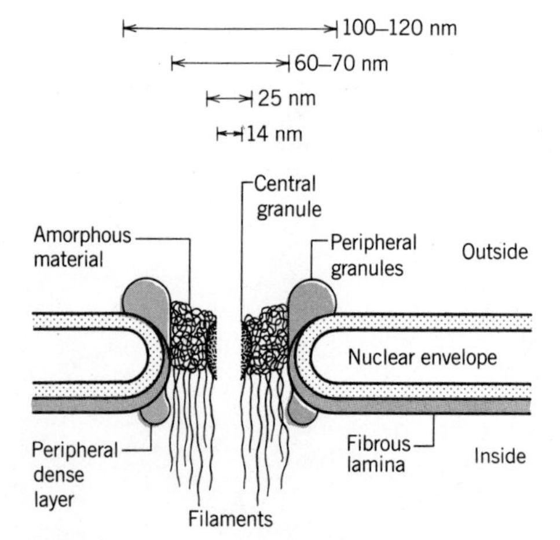

FIGURE 14.4 Schematic representation of the nuclear pore complex.

There are probably several reasons for these variations that have little to do with the actual diameter of the pore. Often the method of tissue preparation, the plane through which the pore is sliced, and the inclusion or exclusion of pore complex material influence the measurements.

It can be seen quite readily from Figure 14.4 that pore dimensions depend on the site of measurement. A compromise diameter that is commonly used is 80 nm.

The number of pores per unit area of the nuclear envelope varies with cell type and with the physiological state of the cell. There appears to be a general relationship between pore density and the degree to which a nuclear envelope is transporting RNA from the nucleus. The density is low in slowly metabolizing cells and in other cells during relatively inactive phases of their cell cycles. In the new postmitotic nucleus, when RNA transport is high and protein synthesis rapid, the density of pores is high.

Pore densities as low as ~3 pores/μm^2 are seen in nucleated red blood cells and lymphocytes. These cells are highly differentiated but metabolically inactive, and they are nonproliferating cells. The majority of proliferating cells have pore densities between 7 and 12 pores/μm^2. Among cells of a third type, differentiated but highly active, pore densities are often 15 to 20 pores/μm^2. Liver, kidney, and brain cells fall into this category. Still higher pore densities are found in specialized cells, such as salivary gland cells (~40 pores/μm^2) and the oocytes from *Xenopus laevis* (~50/μm^2).

The total pore number per nucleus varies greatly, ranging from 100 to 5×10^7. Table 14.2 lists some representative values found in widely differing tissues. These results suggest that the total pore number is related to pore density, but such a generality does not hold in all cases. Scheer[2] has shown in a careful study that during the growth of the oocyte of *Xenopus laevis* from a diameter of 300 to 1200 μm, the surface area of the nucleus increases from about 200,000 to 800,000 μm^2, while the pore density remains nearly constant, around 50 to 60 pores/μm^2. The total number of pores increases from 10.1×10^6 to 37.7×10^6, but they maintain a nearly constant diameter of 75 nm. The percentage of nuclear surface occupied by the pores in this study remained near 25%.

The distribution of pores over the surface of the nuclear envelope is generally not random in somatic cell types. Pore arrangements in other cell types range from rows to clusters to hexagonal packing orders. Examples of these are illustrated in the electron micrographs of Figure 14.5.

In any event, whether the pores are random or nonrandom, they are

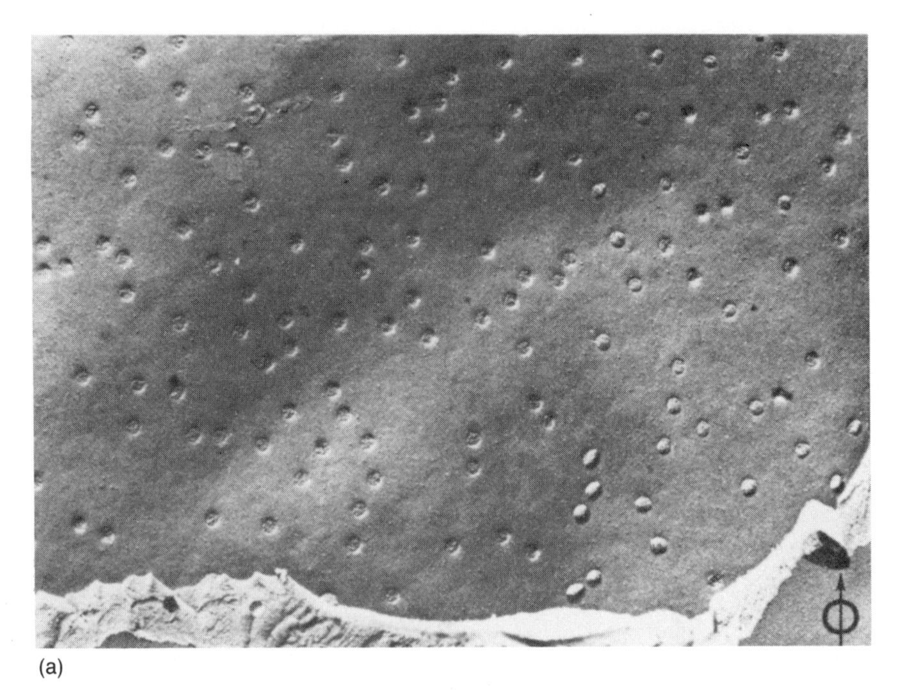

(a)

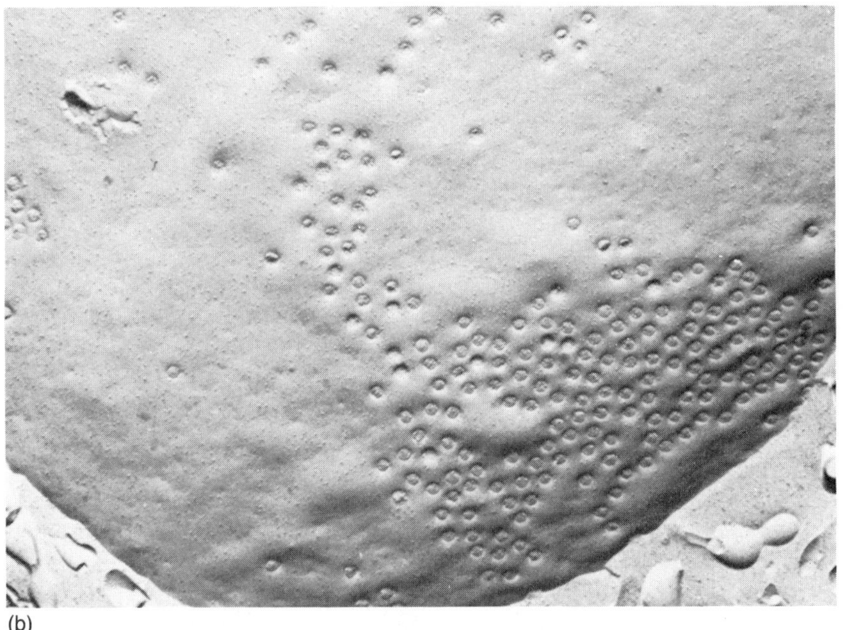

(b)

FIGURE 14.5 Examples of different pore arrangements in nuclear envelopes. (*a*) Pores are rather evenly or randomly distributed, quite typical of somatic cells. (*b*) A clustered pore arrangement, exhibiting regions of high pore density as well as extensive pore-free areas. This is typical of the spermatocyte nucleus. (Courtesy of Dr. Don W. Fawcett.)

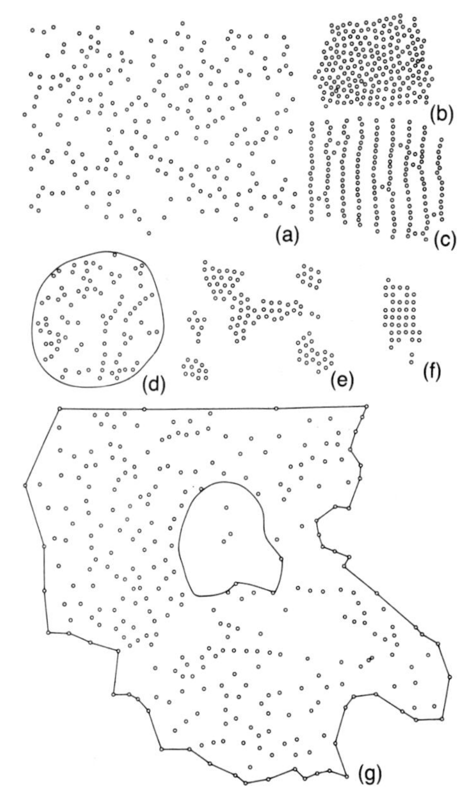

FIGURE 14.6 Graphic representation of various pore arrangements. (*a*) Random. (*b*) Close packing as in *Xenopus laevis* oocytes (see Table 14.2). (*c*) Parallel rows as in *Equisetum* spores. (*d*) Gently curving rows, in rat liver. (*e*) Hexagonal arrangement, from Malpighian tubule of leaf hopper. (*f*) Orthogonal arrangement, as in *Rana* oocytes. (*g*) Pore distribution over nucleolar areas in cultured cell line. Pore frequency is reduced over the nucleolus. (Courtesy of Dr. Gerd G. Maul.)

probably not free to move laterally in the plane of the nuclear envelope because of the surface interactions of the envelope with fibrils on both inner and outer sides. Some commonly observed pore arrangements are shown schematically in Figure 14.6.

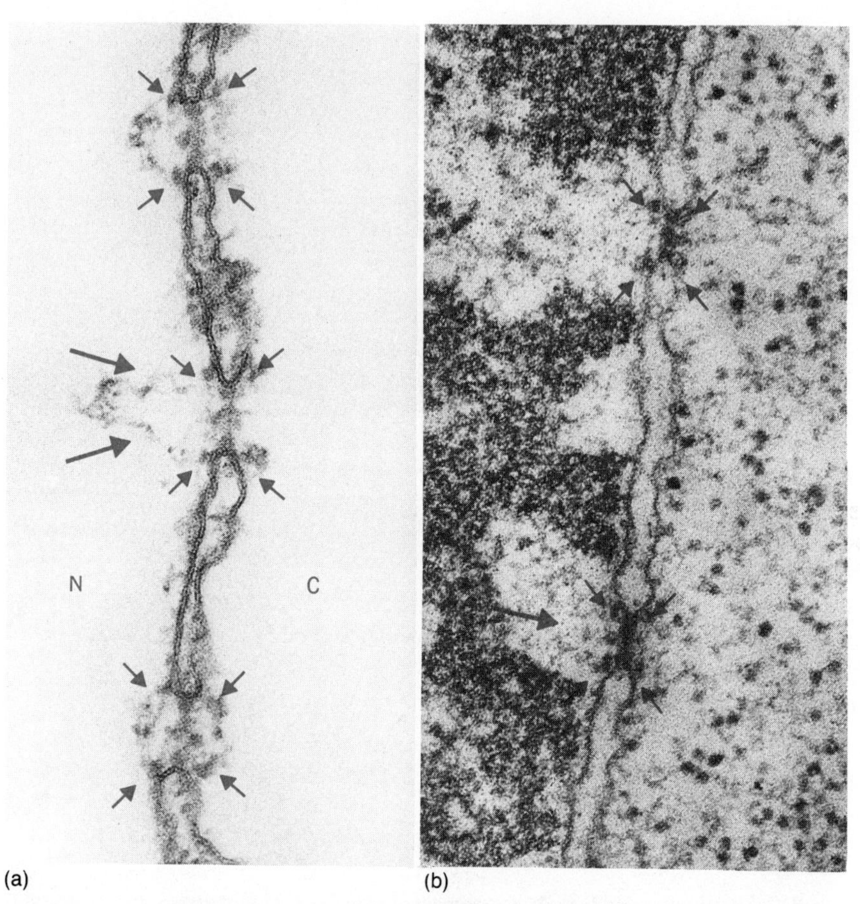

FIGURE 14.7 Highly magnified, transversely sectioned nuclear envelope showing details of nuclear pore complex on nucleoplasmic (N) and cytoplasmic (C) sides. (*a*) Maturing *Xenopus laevis* oocytes. ×114,000. (*b*) Onion root meristem. ×93,000. Small arrows point to annular granules on pore margins. Larger arrows denote nucleoplasmic fibrils, which terminate in the vicinity of the central granule, a densely staining particle in the pore center. (Courtesy of Dr. W. W. Franke.)

The Annulus

The morphological and ultrastructural details of the annulus have been most difficult to unravel experimentally. In cross section, the annulus is an electron-dense material that rims the pore and projects from both the nucleoplasmic and cytoplasmic surfaces of the nuclear envelope as a diffuse and particulate cylinder. This is evident in Figure 14.7.

On surface view, using high resolution, the annulus appears as a ring of subunits (see Figure 14.8). In newt oocytes, eight subunits are clearly seen to rim the pore, each with a diameter of about 17 nm.[3] In light of the resistance of these subunits to scatter by electron microscopic techniques, it is thought that the annulus is a closed self-assembly structure that interacts with the nuclear membrane during the final stages of its formation.

In addition to subunits, the annulus has a matrix that varies in consistency from amorphous to fibrillar.

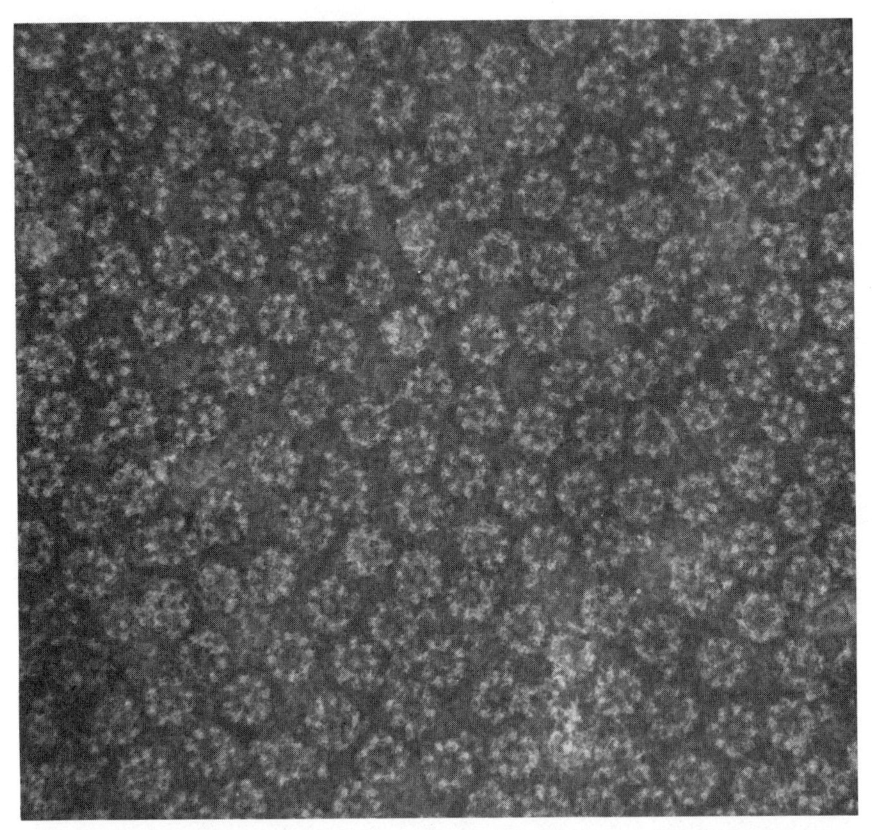

FIGURE 14.8 Negative stain of nuclear envelope of oocyte of *Taricha granulosa.* The annuli appear as rings of eight subunits each. ×60,000. (Courtesy of Dr. Alexander C. Fabergé.)

The independence of the annular material from the nuclear envelope membrane has been shown by some remarkable experiments in which nuclei were treated with the nonionic detergent Triton X-100.[4] This detergent solubilizes the nuclear envelope, leaving an intact nucleus devoid of inner and outer membranes.

The surface of such a nucleus still consists of annular material as a ring of subunits, as shown in Figure 14.9. The results have established quite clearly that the annular portion of the nuclear pore complex is anchored to the interior of the nucleus, specifically to the nuclear cortex. They have also shown in dramatic form that *the nucleus can exist intact without a surrounding membrane*. When the nuclear cortex is solubilized by treatment with Triton and deoxycholate, the subunit ring structure disappears.

A very insightful study of the outer surface of the nuclear envelope of oocytes from *Xenopus laevis* has been published by Schatten and Thoman.[5] These cells are 1.25 to 1.4 mm in diameter and have nuclei that are large enough to be isolated from individual cells and split open manually with forceps. The outer and inner surfaces can be exposed by manipulation, and even the surfaces of the perinuclear gap can be exposed and examined.

Figure 14.10 is a scanning electron micrograph of the outer surface of these nuclei. The surface is characterized by a pore density of about 58

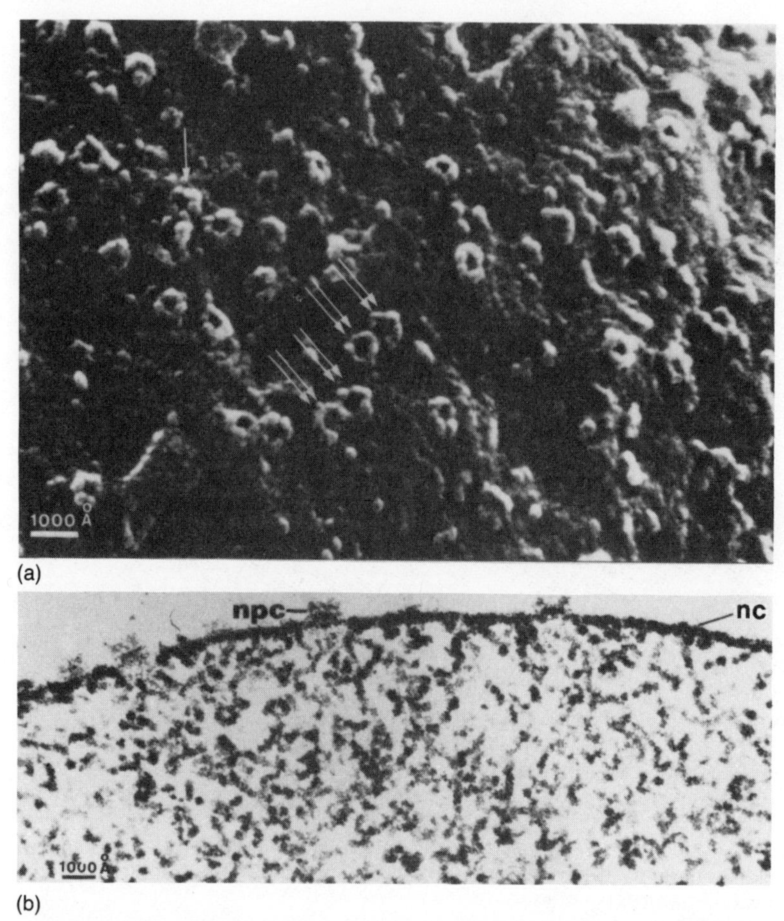

FIGURE 14.9 Surface detail of Triton-treated nucleus (*a*). Nuclear pore complexes are prominent and in some eight subunits can be discerned (single arrow). In others, there is a distinct cleavage between some subunits but not between all. A few nuclear pore complexes seem to share a common subunit (double arrows). Infrequent pores not surrounded by annuli are also seen. ×66,400. The transmission electron micrograph (*b*) shows nuclear pore complexes interspersed on the nuclear cortex, and the granular characteristics of the cortex. npc, nuclear pore complex; nc, nuclear cortex. ×48,000. (Courtesy of Dr. Terence Martin.)

pore complexes per square micrometer. Each pore complex has an outer diameter of 90 nm, an inner diameter of 25 nm, and a wall thickness of about 30 nm. The wall is made up of 30-nm particles, arranged in tetrameric fashion around the pore opening. In this system pore complex triplets were often seen in which the three complexes shared common walls.

Studying the same *Xenopus* oocyte system, but with transmission electron microscopy, Unwin and Milligan have delineated four discrete constituents of the pore complex.[6] These are most readily seen when the complexes are released from the nuclear envelope by Triton X-100 treatment as shown in Figure 14.11.

The four constituents are *rings,* having an inside diameter of 90 nm and an outer diameter of 120 nm, the same as the pore complex, *plugs,* particles of diameter near 35 nm at the centers of the pore complexes, *smaller*

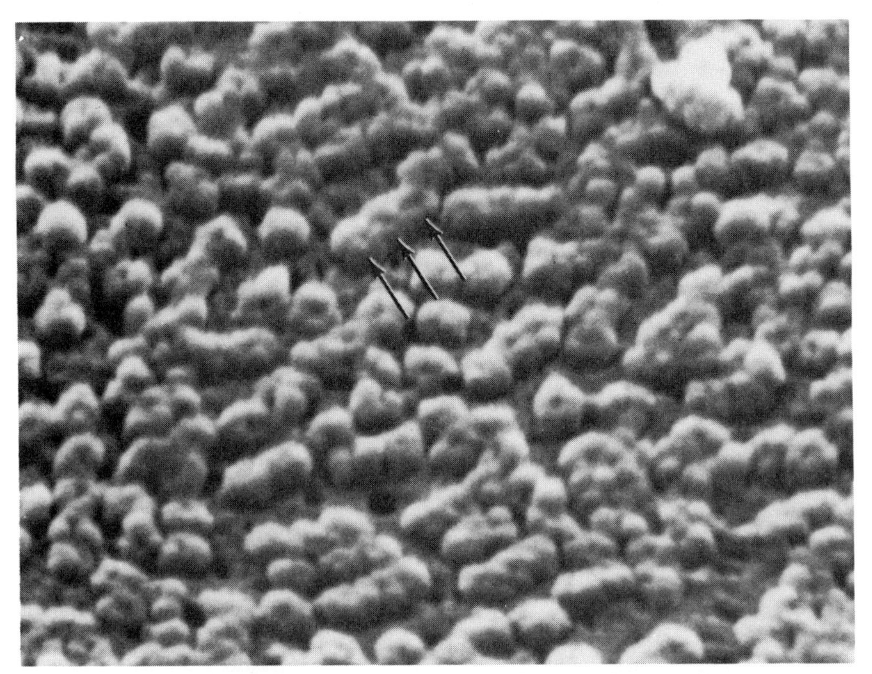

FIGURE 14.10 Scanning electron micrograph of the outer surface of the nuclear envelope. Pore complexes project as tightly packed doughnuts from the surface. Triplets sharing common walls (arrows) are often seen. ×65,000. (Courtesy of Dr. Gerald Schatten.)

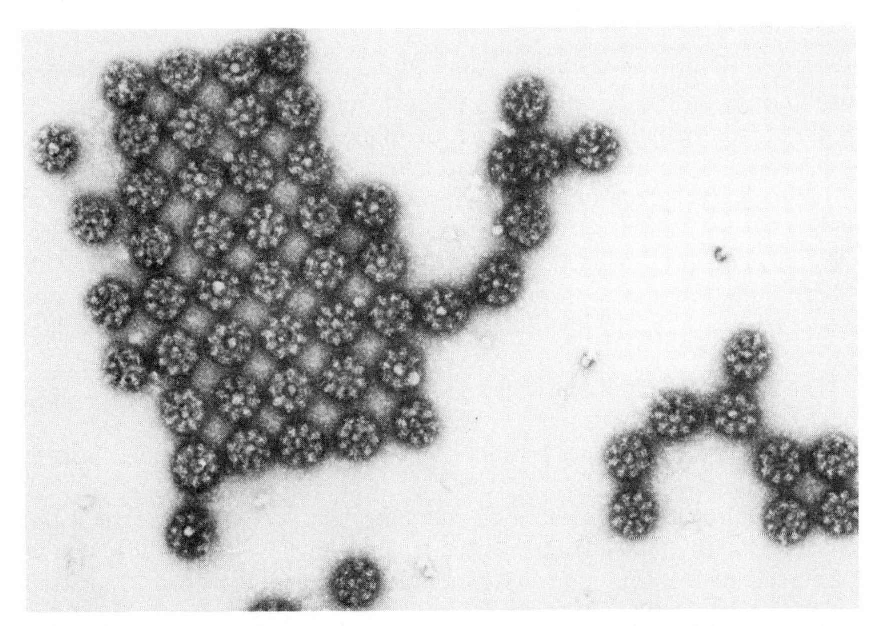

FIGURE 14.11 These detached pore complexes were released from a nuclear envelope onto an electron microscope grid by immersion in a low salt medium containing 0.1% Triton X-100. Uranyl acetate stain. ×52,000. (Courtesy of Dr. P. N. T. Unwin.)

(a)

(b)

FIGURE 14.12 Projection maps of pore complexes. Resolution is about 9 nm. Broken lines indicate the position of the membrane border in the intact systems. (*a*) Detached pore complex such as those in Figure 14.11. Broad and sharp peaks of density (shaded) are consistently seen at radii of 45 and 55 nm, respectively. (*b*) Pore complex attached to the nuclear envelope, containing particles not present in images of isolated pore complexes. (Courtesy of Dr. P. N. T. Unwin.)

particles, of a diameter near 22 nm, arranged around the circumference of the rings on the cytoplasmic side of the envelope, and *spokes,* matter that extends radially outward from the plugs toward the rings.

The ring constituent of the pore complex is made up of globular subunits arranged in octagonal symmetry when viewed from the surface and mirror symmetry when viewed from the side. The octagonal symmetry is clearly evident in the projection map of pore complexes presented in Figure 14.12. When detached and well-preserved pore complexes are studied, 16- and 24-fold harmonics are apparent as well as the basic 8-fold harmonic evident in this figure.

The particles that are attached to the cytoplasmic surfaces of the rings are thought to be ribosomes or ribosome subunits, since they are very easily detached from these positions by manipulations of the ionic strength of the medium.

The relation between the eight subunits that surround pores and the tetrameric structures depicted earlier is not yet clear. There is speculation that each of the four units is dimeric, more visible with transmission electron microscopy, or that there are two sets of four particles staggered. It may be that the limitations of scanning and transmission electron microscopy differ sufficiently that slightly different results can be obtained when viewing the same thing. In any event, these differences do not appear to be serious, and it is likely that they do not represent actual discrepancies.

The Nuclear Cortex

Generally, when we think of the nuclear envelope, we picture a view of its surface from the outside. The inner surface is harder to get at, but it turns out to be more interesting and certainly more complex.

The inner surface is plastered with a material that looks quite nondescript until it is examined with high resolution scanning electron microscopy. Instead of being amorphous, the nuclear cortex is a highly structured fibrous layer.

The scanning electron micrograph of Figure 14.13 gives a striking view of the nuclear cortex from the inside of the nucleus. The cortex is a 300 nm thick matrix of fibers arranged in funnellike whorls that narrow in toward the inner nuclear membrane. The fibers are proteinaceous and of about the same diameter and appearance as actin polymers (see Chapter 18). Chains of particles are often associated with or inserted into the channels.

An interesting observation is that the ratio of pores to funnels is not 1. The density of intranuclear channels is $3.2/\mu m^2$, whereas that of pore complexes is much higher ($\sim 55/\mu m^2$). This has led investigators to speculate that the fibrous whorls might feed the pore openings by one of the mechanisms included in Figure 14.14.

Schatten and Thoman developed a model putting together the results of their studies (Figure 14.15). The outer surface of the nuclear envelope is shown packed by pore complexes that extend about 20 nm from the surface of the outer membrane. These investigators believe the annulus to be made up of two sets of tetramers that separate from each other when the nuclear envelope is split down the perinuclear gap. Both inner surfaces of the perinuclear gap harbor pore complexes that project about the same distance, 20 nm, from the membranes.

The pore complexes project very little from the inner surface of the inner membrane. When isolated nuclear envelopes are treated with 0.6 *M* potassium iodide, the fibers associated with the inner surface can be removed to permit an unobstructed look at the inner membrane surface. The frequency of pores is the same as for that observed on the outer surface, but the annular rims protrude less.

The nuclear cortex is the fibrous region next to the inner membrane. Rather than obstructing the pore, the cortex appears to function as a funnel to direct exiting materials toward the pore channel. Granules and chains of granules, often seen in the intranuclear channel, are thought to be ribonucleoprotein particles on their way from the cell nucleus to the cytoplasm. These could be ribosome precursors or high molecular weight pre-RNA complexed with protein.

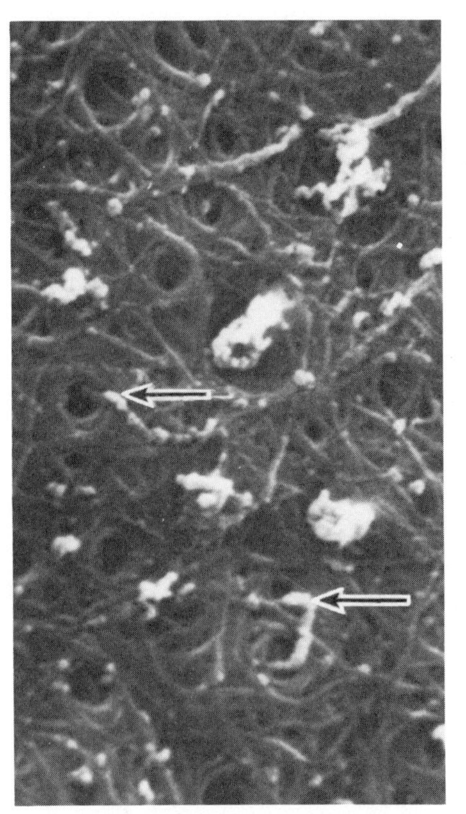

FIGURE 14.13 Scanning electron micrograph of the nuclear cortex viewed from the nucleoplasmic side of the envelope. This structure can be observed only after a very rapid fixation. Chains of particles can be seen inserted into the funnels (arrows). $\times$13,500. (Courtesy of Dr. Gerald Schatten.)

Models of the Pore Complex

Over the years the concept of the nuclear envelope pore has evolved a good deal, and it does not appear to have reached its terminus yet. Many different models have been proposed to illustrate the ultrastructure of the pore, but each one seems to be specialized for the particular type of tissue or cell studied and therefore not applicable for nuclear envelopes in a general sense.

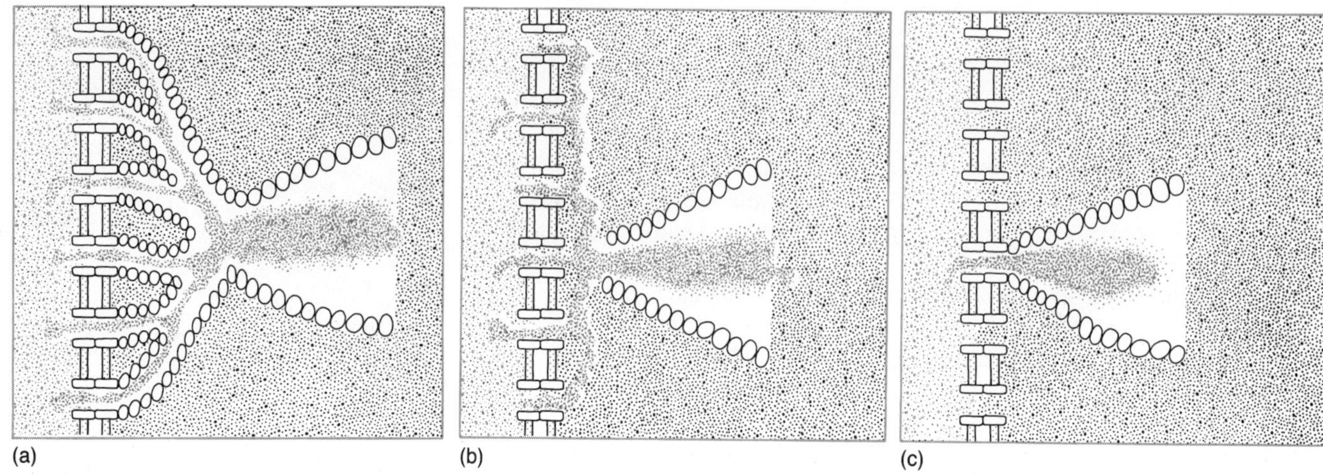

FIGURE 14.14 Models of relation between funnels and pores. Three possibilities are proposed. (*a*) A funnel might bifurcate and provide channels to several pores. (*b*) There might be no direct attachment between funnels and pores. (*c*) A funnel might attach to a given pore and, after some time, break and reattach to another pore. (Courtesy of Dr. Gerald Schatten.)

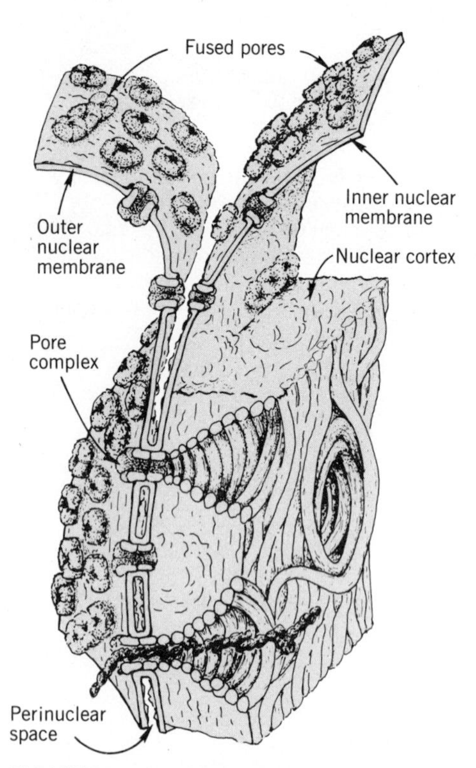

FIGURE 14.15 Model of the nuclear surface complex as proposed by Schatten and Thoman, representing a compilation of data obtained by study of the four membrane faces and the nuclear cortex. (Courtesy of Dr. Gerald Shatten.)

A rather popular view of the nuclear pore is depicted in Figure 14.16. This model takes into account the octagonal properties of the pores of certain systems, and it illustrates the granular and fibrous nature of the pore region. Some adjustments of this model would be necessary to fit some of the observations that have been made, such as those which led to the model of the nuclear surface complex in Figure 14.15.

Based on the symmetry studies discussed earlier, Unwin and Milligan have published the model shown in Figure 14.17. These side and surface views of the nuclear pore complex show the spatial relationships of the four constituents. The spokes (S) radiate out from the central plug (C), where they interact with the rings (R) at the periphery of the complex. Particles (P), which are located on the cytoplasmic surface of the complex, are anchored to the cytoplasmic ring.

Thus at this point in nuclear envelope research, full understanding of the structure of the pore may be near. We do not know the extent to which pores vary from cell to cell and which features are common among all cells. But more exact information along these lines will certainly be appearing in the literature shortly.

14.3 THE BIOCHEMISTRY OF THE NUCLEAR ENVELOPE

Chemical and enzymatic analyses of the nuclear envelope are now possible because techniques have been worked out to isolate nuclei in high purity and to separate the envelope from the nuclear matrix. An analysis of the envelope gives some clues to its biogenetic and functional relationships with the rest of the membranous elements of the cell.

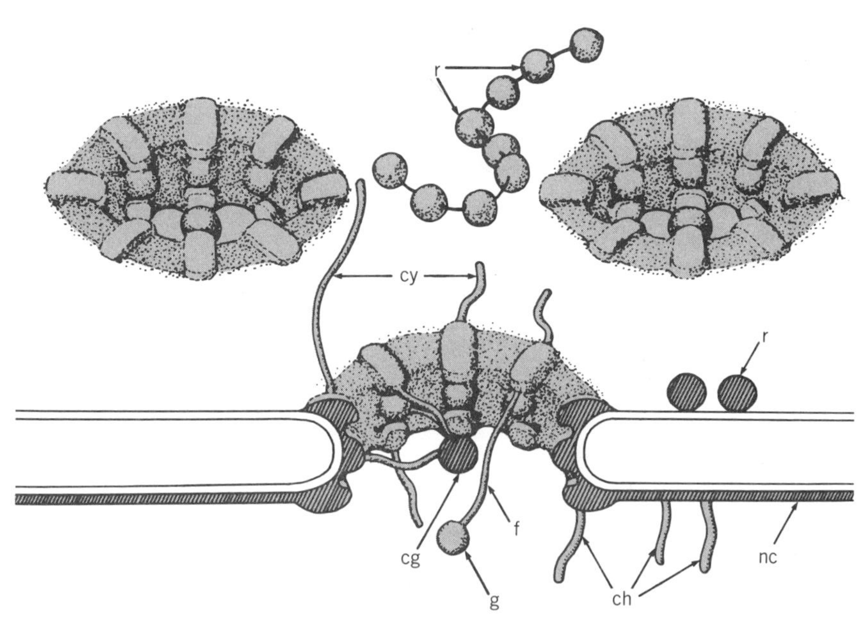

FIGURE 14.16 The nuclear pore complex. Fibrils (f) are associated with annular granules and with other granules (g) above or below the plane of the complex. A central granule (cg) is positioned in the center of the complex. Other fibers project into the cytoplasm (cy) or into the nucleus. The nuclear cortex (nc), a coating of the inner membrane seen especially in animal cells, may be connected to chromatin (ch) material. Ribosomes (r) may adhere to the outer membrane on its cytoplasmic side. (Courtesy of Dr E. G. Jordan.)

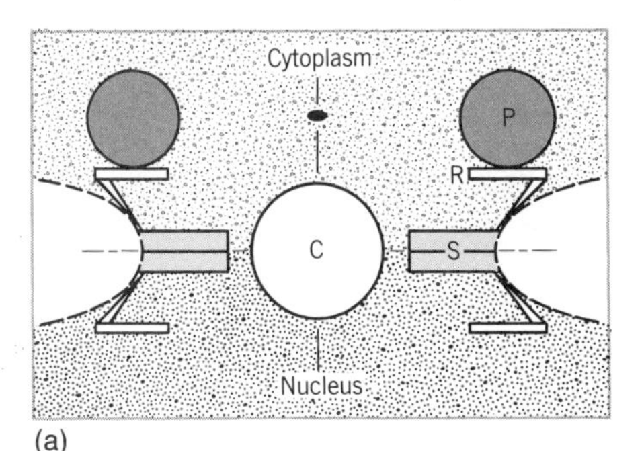

(a)

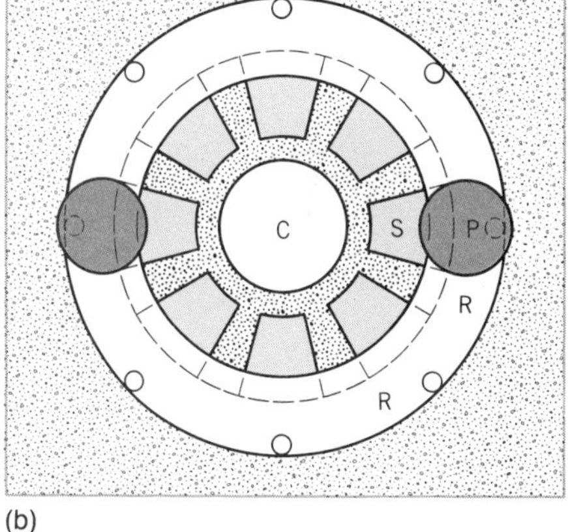

(b)

FIGURE 14.17 Central cross section (*a*) and projection down the octad axis (*b*) of the nuclear pore complex, based on symmetry studies. C, central plug; S, spokes; R, rings; P, particles resembling ribosomes. The broken lines represent the nuclear envelope membranes. Particles are present only when pore complexes are membrane attached. (Courtesy of Dr. P. N. T. Unwin.)

Methods II and III

Isolation of Envelopes

Many different approaches have been devised for the isolation of nuclear envelopes. Some of these are listed in Table 14.3. Nuclei are first of all separated from the rest of the cell. This is normally accomplished by disrupting tissue in homogenizers wherein the clearance is such that nuclei are not broken but the plasma membrane and endoplasmic reticulum are severely disrupted. Nuclei can then be harvested by differential centrifugation.

They are then lysed by sonication and their envelopes separated on density gradient centrifugation. Alternatively, DNase digestion followed by extraction with salts releases envelopes which, again, can be banded on sucrose or cesium chloride gradients. Other modifications to these protocols are used in various laboratories.

In any case, there is always concern that the envelope preparations are not 100% pure. Low levels of enzymatic activity of certain biological compounds may represent contamination either from the cytoplasm or the nucleoplasm.

Chemical Composition

Relative concentrations of protein, phospholipid, RNA, and DNA in nuclear envelopes of liver and thymus isolated by a variety of technqiues

Table 14.3 Composition of Some Nuclear Envelope Preparations (% weight of sum)

Tissue	Method for Isolating Nuclear Envelopes[a]	Density (g/cm^3)	Protein	Phospholipid	RNA	DNA
Rat liver	Low salt lysis,	1.16	46.5	49.5	3.8	0.4
	sonication	1.19	60	34.5	4.2	1.6
Rat liver	Sonication	1.27	64	23	5	8
Rat liver	Sonication and K	1.16–1.18	64.5	32.1	3.4	0
	citrate extraction	1.18–1.20	67.4	26.1	6.6	0
Rat liver	Sonication and high salt (KCl) extraction	1.18–1.21	77.4	16.5	3.9	2.2
Mouse liver	Sonication and high salt (KCl) extraction	1.18–1.21	76	18.2	3.4	2.3
Rat liver	Sonication and high salt (KCl) extraction	1.21–1.23	67.1	23.5	6.1	3.4
Rat liver	$MgCl_2$ extraction	1.18	73	23	3	0.6
Rat liver	Short DNase digestions		65.7	26.7	3.6	3.9
Bovine liver	Long DNase digestion and $MgCl_2$ extraction	$\leq$1.23	70.4	22.7	5.8	1.1
Calf thymus	Long DNase digestion	$\leq$1.13	71.1	25.1	3.8	0
	and NaCl extraction	1.19–1.21	88.9	9.7	1.4	0
Calf thymus	Sonication and high-salt (KCl) extraction	1.19	71.6	16.5	4.3	7.7
Rat thymus	Sonication and high-salt (KCl) extraction	1.19	*ca.* 78	*ca.* 15	*ca.* 4	*ca.* 3

[a]In all cases the nuclei were isolated in sucrose solutions.

are given in Table 14.3. Most preparations of envelope material have densities in the vicinity of 1.18 g/cm^3. The predominant constituent of the envelope is protein, making up 65 to 75% of the material. Lipid makes up most of the remainder, with low levels of nucleic acids generally present.

Neutral lipids, although not included in Table 14.3, are also present at levels ranging from 15 to 35% of the total lipids. Of these, the predominant lipid is cholesterol.

The presence of DNA and RNA is probably due to contamination of nuclear membranes by closely associated chromatin material. But since the problems of isolation are very great, the extent to which these amounts represent contamination is not completely clear. It is apparent from Table 14.3 that materials prepared in different ways contain different amounts of nucleic acids, suggesting contamination to be a factor in some preparations. But the levels of phospholipid also vary, pointing to difficulty in obtaining envelopes of uniform purity for compositional studies.

Proteins

Most preparations of nuclear envelopes contain about 20 different proteins as distinguished by gel electrophoresis. They range in molecular weight from 16,000 to 160,000 daltons.

Many of the proteins appear to be the same, at least electrophoretically, as those that appear in microsomes. But about a half-dozen envelope proteins are not reflected in electrophoretic patterns of microsome proteins. They thus appear to be unique to the envelope.

Lipids

The lipids of the nuclear envelope are present in relative concentrations quite similar to that seen in the endoplasmic reticulum. However, some differences have been observed, indicating that the nuclear envelope is more than the mere continuation of the endoplasmic reticulum around the chromosomes.

For example, in some systems the nuclear envelope contains lower concentrations of unsaturated fatty acids in lecithin and phosphatidylethanolamine but higher levels of cholesterol and triglycerides as compared to microsomes. These compositional characteristics suggest that the envelope may be more stable than the membranes of the endoplasmic reticulum.

The comparative compositions of envelope and other membranes raise questions concerning the biogenetic relation between the intracellular membrane systems and the role differentiation may play in distinguishing them. These questions pose some challenging research problems, but they are presently unanswered.

Enzyme Activities

The nuclear envelope contains a variety of enzyme activites that in general bear similarities to those observed in the endoplasmic reticulum. Glucose-6-phosphatase, an enzyme characteristic of the endoplasmic reticulum, is often found in preparations of nuclear envelopes. Cytochemically, the enzyme is found associated with both the inner and outer membranes.

The envelope also possesses a number of enzymes with electron transport activities similar to those of microsomes. Among these are NADH-cytochrome *c* reductase, NADH-cytochrome b_5 reductase, and NADPH-cytochrome *c* reductase. The properties of these enzymes were discussed in Chapter 6, dealing with the endoplasmic reticulum. The enzymes appear to have similar properties in the nuclear envelope.

Cytochrome *P*-450, which is characteristic of endoplasmic reticulum and acts as an electron acceptor from NADPH-cytochrome *c* reductase, is also present in nuclear envelope membranes. But as is true for the enzymes with electron transport activities, the concentration of *P*-450 is lower in envelope membranes than in membranes of the endoplasmic reticulum.

This general finding of envelope enzymes similar to those in endoplasmic reticulum but of lower concentration has led some investigators to speculate that perhaps they are found only on the outer membrane, the one that shows physical continuity with the endoplasmic reticulum.

In addition to the enzymes cited above, several others have been found to be envelope associated. Exhaustive lists are available of hydrolytic enzymes, transferases, oxidoreductases, and others.[7] The significance of all these activities in the nuclear envelope remains to be elucidated. But, in general, the activities reflect a similarity between the nuclear envelope and the endoplasmic reticulum, with some distinctive features for each set of membranes.

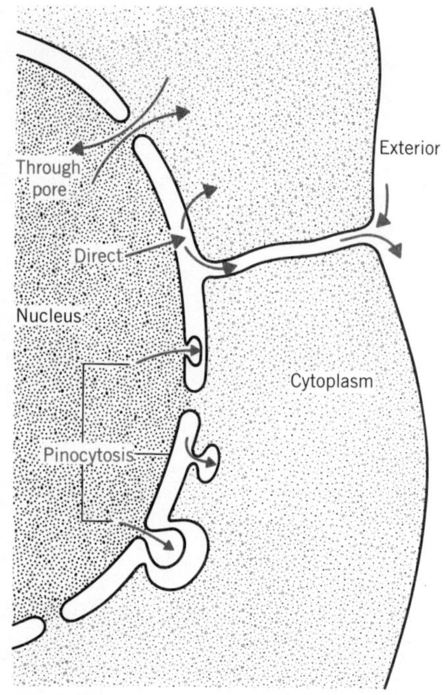

FIGURE 14.18 Summary of ways in which transport may be envisioned to occur across the nuclear envelope.

14.4 THE ROLE OF THE NUCLEAR ENVELOPE IN NUCLEOCYTOPLASMIC INTERACTIONS

On the surface the nuclear envelope appears to separate the cell interior into two major compartments: cytoplasm and nucleoplasm. Only a few decades ago this, indeed, was perceived as the structure of the eucaryotic cell.

But as we have seen, the cell interior is much more complex than this, and so is the nuclear envelope. Furthermore, not only is its structure complex, but its function is almost contradictory. On the one hand it is a barrier, but on the other it is the means of transport between compartments.

Because of its structure we could speculate over several ways in which the envelope is involved in transport. The pore complex with its channellike characteristics may be one means of transport directly from the nucleoplasm to the cytoplasm, or vice versa. Another means could be transport across the inner membrane either directly or by pinocytosis. This would bring material into the perinuclear space from the nucleus, and from the perinuclear space to the cisternae of the endoplasmic reticulum and even potentially to the cell exterior. In this sense the nucleoplasm has a direct line of communication with the extracellular environment, and vice versa. Finally, pinocytosis of the outer nuclear membrane or of the entire envelope could bring vesicles into the cytoplasm. Once again, the process could be thought of as taking place in the opposite direction.

Figure 14.18 summarizes schematically the various ways by which transport could take place. Even though the nuclear envelope may potentially participate in any or all of these ways, it is likely that certain of the mechanisms have extremely limited usage, whereas others represent the major modes of transport.

Permeability to Inorganic Ions and Small Organic Molecules

The envelope is a physical barrier in the cell, and one would therefore expect it to hinder the free movement of molecules to and from the cytoplasm. However, several lines of investigation have shown that water, ions, and small organic molecules such as glycerol and sucrose cross the envelope so rapidly that their rates of movement cannot be readily measured.[8] One estimate of the delay imposed on diffusing molecules by the envelope is 35 msec, or 70% of the time required to move 15 μm from cytoplasm to nucleus. This is not considered to be a significant delay. Diffusion itself may be as rate limiting in movement as physical impedance by the nuclear envelope. Investigators generally believe that the kinetics of movement from nucleus to cytoplasm are consistent with a channel diameter in the pore complex of 9 nm.

Electrophysiological studies on cells of several kinds, such as brain and *Chironomus* salivary glands, have indicated potential differences between nucleoplasm and cytoplasm. For some time it was thought that this meant the nuclear envelope could not be freely crossed by ions, and that an ion gradient existed across the membrane demonstrated by the potential difference between inner and outer spaces. But it is currently believed that this is better explained by recognizing that the nucleus has polyanionic macromolecules that preferentially attract and fix cations in the nucleus interior. These fixed cations would not be free to migrate across the membrane and would contribute to a charge gradient across the envelope.

It thus appears that the nuclear envelope permits free passage of molecules the size of those that are the nucleoside phosphate building blocks of nucleic acids as well as substrates that may be needed to feed metabolic pathways within the nucleus.

At physiological pH, nucleic acids possess a high net negative charge. Although RNA does leave the nucleus, DNA is fixed in the nucleoplasm. Thus by electrostatic interaction with nucleic acids, cations may be fixed and immobilized in the nucleus.

Permeability to Proteins and Macromolecular Complexes

It is obvious that many different proteins must gain access to the nucleus to carry out the enzymatic functions of biosynthesis and to serve as structural and regulatory agents. Most of those proteins concerned have molecular weights ranging between 20,000 and 90,000 daltons. If one assumes a globular form for the proteins, their molecular radii would be in the range of 2 to 4 nm. They should therefore be small enough to penetrate the pore complex channels, and indeed there is ample experimental evidence of the ready movement across the envelope of proteins of similar size (e.g., DNase, RNase, trypsin).

The nuclear envelope begins to impose restrictions on movement for particles with diameters larger than 9 nm. Colloidal gold particles with diameters of 2.5 to 5.5 nm move into the nuclei of amoebae within minutes. Ferritin, having a diameter of 9.5 nm, takes several hours, and particles of diameter 15 nm are excluded. In the amphibian oocyte ferritin is excluded. Thus there are species differences in envelope permeability, but most show a relatively unhindered passage for proteins required in the nucleus for DNA replication and transcription.

Besides proteins moving into the nucleus, a variety of products must move out. This includes the families of RNA as well as the ribonucleoprotein particles that are the precursors of RNA and of ribosomes.

Ribosomes do not mature in the nucleus, but they leave the nucleus as particles about 23 nm in diameter. Messenger RNA, before transport to the cytoplasm, combines with proteins to form macromolecular complexes 40 to 50 nm in diameter. Both these ribonucleoprotein families are too large to penetrate the 9 nm pore channel without some modification of either the pore or the particle.

Electron microscopic studies of particle movement between nucleus and cytoplasm have caught what appear to be ribonucleoprotein particles moving through the pore complex of the nuclear envelope. This phenomenon is illustrated in Figure 14.19. As the large granules move through the pore, they undergo a configurational change, narrowing and lengthening to traverse the tight channel of the pore. The factors that drive the change in shape and the extent to which the pore complex is involved, either actively or passively, are not known.

14.5 THE FATE OF THE NUCLEAR ENVELOPE DURING THE CELL CYCLE

During the cell cycle the nuclear envelope pulls a disappearing act par excellence. One instant it is there, seemingly pulled tightly around thick-

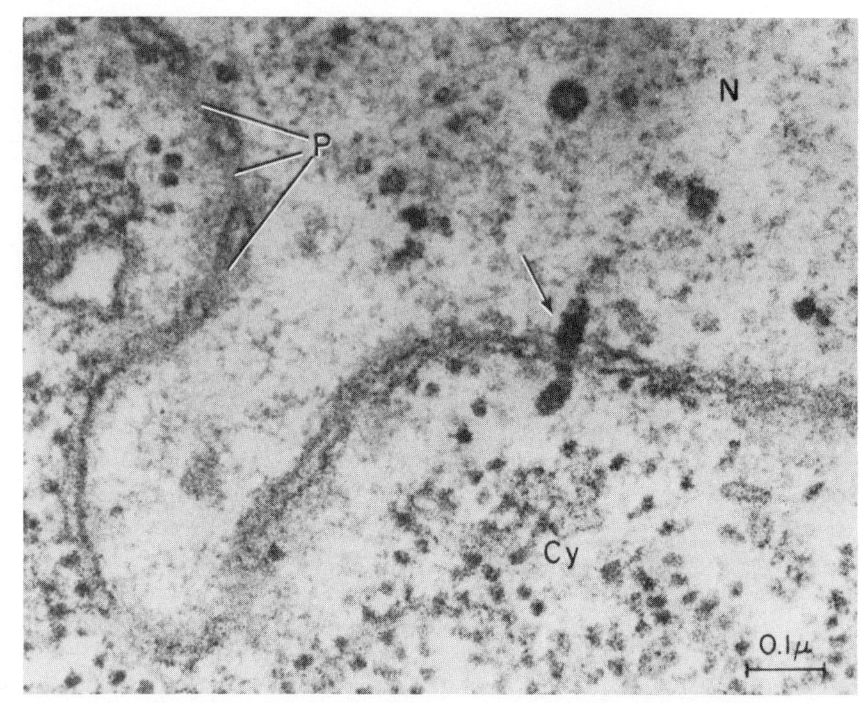

FIGURE 14.19 Penetration of nuclear pore by an electron-dense particle (arrow) in the salivary gland cell. The particles are thought to be transit forms of granules from Balbiani rings. This may be a ribonucleoprotein particle containing mRNA moving from its site of manufacture in the nucleus (N) to its site of function in the cytoplasm (Cy). Several nuclear pores (P) not involved in transporting particles are also apparent. (Courtesy of Dr. Hewson Swift.)

ening prophase chromosomes. The next instant it has disappeared, and the chromosomes appear to be much freer to move about and to become oriented on the metaphase plate. In late anaphase, the reappearance of the envelope is almost as sudden and mysterious as its disappearance.

Several questions concerning this phenomenon have haunted investigators for years. What triggers the breakdown of the nuclear envelope? Does the envelope break down partially or completely, and if completely is the membrane dismantled to its molecular components? Are the dismantled fragments or molecules salvaged and used for the production of new envelope, or is the envelope synthesized entirely *de novo*?

Not all these questions can yet be answered with certainty, but several of them are nearing resolution.

The cell cycle contains the following phases:

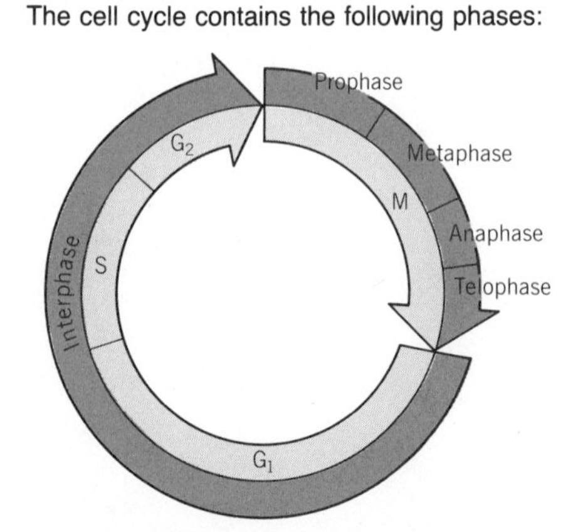

DNA replication takes place during the S phase, necessitating an increase in nuclear size, hence nuclear envelope area.

The Association of Chromatin with the Interphase Nuclear Envelope

During the S phase of the cell cycle DNA is replicated, demanding an increase in volume for the nucleus. Between G_1 and the end of interphase the surface of the nuclear envelope increases by about 60%. Thus it is during this period of the cell cycle that envelope membranes are synthesized to accommodate the increase in nuclear contents.

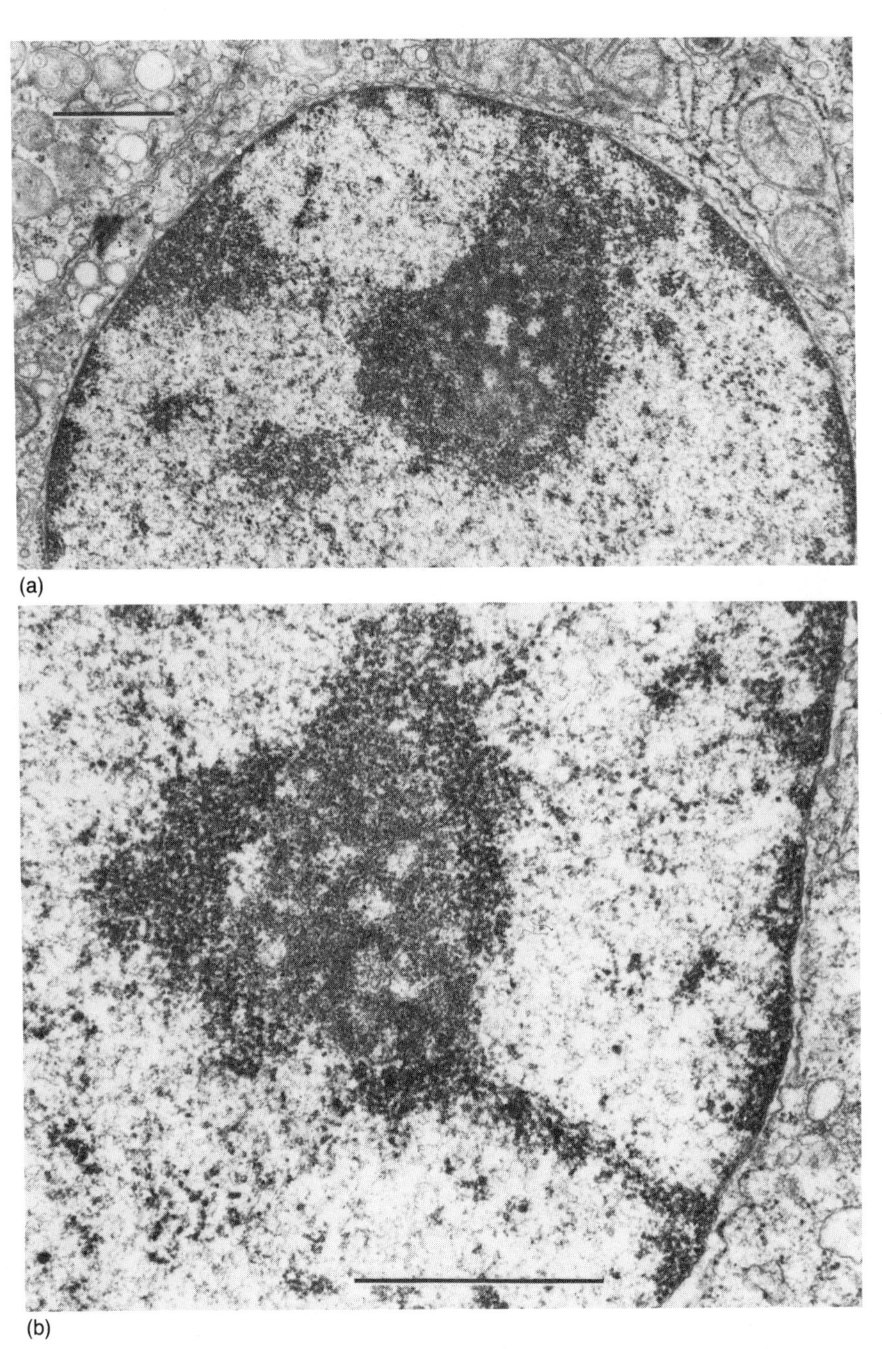

FIGURE 14.20 Cross section through periphery of nuclei showing a high degree of association between condensed chromatin and the nuclear envelope. This chromatin is continuous with the heterochromatin surrounding the nucleoli by way of bridges of chromatin. Pores do not appear to be obstructed by associated chromatin. (*a*) ×16,000. (*b*) ×33,000. Bars, 1 μm. (Courtesy of Dr. W. W. Franke.)

During interphase chromatin has a high degree of association with the inner surface of the nuclear envelope. This is illustrated in the electron micrograph of Figure 14.20. The interaction of chromatin with the envelope is interrupted only at the pore complexes, which appear to be left open as nucleoplasmic channels.

The association of chromatin with the envelope membrane is very stable, making it difficult to isolate either one completely free from the other. Although the nature of the binding is unclear, the DNA–membrane association is resistant to high salts, hydrogen bond disrupting agents such as urea and detergents, and shearing forces and sonication.

The centromere is the part of the chromosome to which the spindle fiber becomes attached. The telomeres are the arms. In *Allium* the two parts would be skewed across the nucleus:

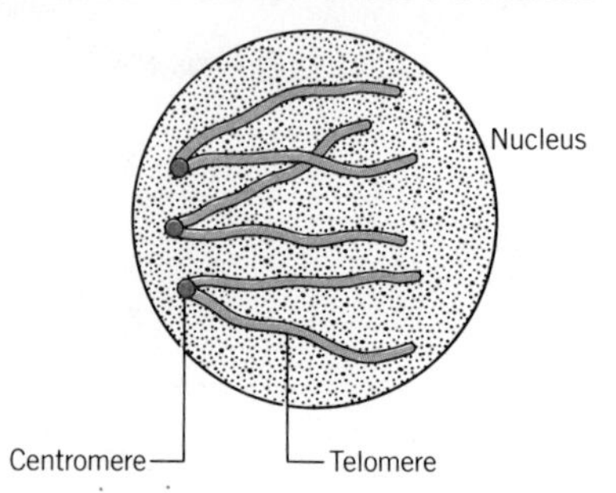

The interaction between chromosomes and membrane is not random. Depending on the particular species being observed, the binding appears to show centromeric or telomeric preferences that are reproducible and predictable. Recent studies on *Allium* have shown that the centromeric regions of interphase chromosomes are grouped to one side of the nucleus and their telomeric regions are scattered in an arc on the other side. A similar situation appears to exist in the nuclei of dipteran salivary glands.

Not only do specified regions of the chromosomes bind to the nuclear envelope, but there appear to be a limited number of specific sites on the inner surface where binding may take place.

The impact of this specific and reproducible association between chromatin and nuclear envelope membrane is being studied, and several speculations suggest the future course of research in this area.

One speculation is that the binding may serve to keep the chromosomes in an ordered arrangement during the cell cycle. Chromosome condensation appears to be initated at these sites. Also, the formation of the synaptinemal complexes in meiotic prophase occurs so that the ends of the pair of chromosomes are attached to the nuclear envelope. Thus the envelope may aid the pairing activity of chromosomes during meiosis.

Other possibilities have to do with DNA replication. The interaction may be a regulatory mechanism to initiate DNA replication or to control its rate. RNA synthesis may also be regulated to some degree by this interaction.

Whatever the role of the chromatin–envelope interaction, the stability of this association suggests that the nuclear envelope may never completely disintegrate during the cell cycle. It may break up and remain associated with chromatin in fragments, which in turn may be the seed bed for the formation of new envelope.

Prophase and Telophase Envelope Dynamics

Toward the end of mitotic or meiotic prophase, the nuclear envelope breaks down and loses its characteristic morphology. Vesicles containing the perinuclear space and bilamellar fragments form, lose their pores, and become morphologically identical with smooth endoplasmic reticulum.

The degree of envelope disintegration and its dispersion throughout the cell varies somewhat with cell type. In some cells the breakdown is essentially complete, so it is difficult to tell whether fragments are retrieved in reforming a new envelope. In other cells large sections of the envelope remain attached to the chromosomes throughout the remainder of the cell cycle and apparently serve as starter fragments on which the new envelope is built.

Before the breakdown of the envelope, changes are taking place inside and outside the nucleus that may have a bearing on its disintegration. On the inside, chromatin is condensing next to the inner envelope membrane and the chromosomes are taking form. On the outside, microtubules are radiating out of the centrioles toward the chromosomes. The nuclear envelope appears to weaken and break, first of all between the chromosomes and the centrioles, where the microtubules are reaching toward its surface. It is therefore believed that the nuclear envelope may be weakened in part by the probing forces of the microtubules against the nuclear surface.

Other cytoplasmic factors are also probably involved in envelope breakdown. For example, mitosis can be deferred by certain chemicals that block transcription and translation. Cells that are thus deferred can be reversed to continue mitosis by fusing them with normal cells in G_2 phase. This indicates that cytoplasmic factors, such as proteins, may well be needed to trigger a breakdown of the nuclear envelope. During telophase the nuclear envelope reforms. The source of membrane for this probably depends on the extent to which the old nuclear envelope was destroyed.

In late anaphase vesicles quite indistinguishable from the endoplasmic reticulum condense around the chromosomes and coalesce into bilamellar sheets that coat the chromosomes. The sheets enlarge and fuse together until there is a continuous bilamellar sheet surrounding the entire nucleus. Finally, pores begin to reappear in the sheets.

In some cases, such as in sea urchin embryos, an envelope forms around each individual chromosome before fusion takes place to form a single nuclear envelope. In other cases the chromosomes are not individually wrapped with membrane.

It is apparent that there is no universal pattern of nuclear envelope breakdown and reformation. Breakdown may range from partial to essentially complete, with membrane pieces of varying size remaining attached to the chromosomes throughout the cell cycle.

Reformation of the nuclear envelope may be partially *de novo* or may originate from the endoplasmic reticulum and from surviving fragments of old nuclear envelope.

Summary

The nucleus is surrounded by a double membrane layer interrupted by pores. The outer membrane often contains attached ribosomes and at times is structurally continuous with the endoplasmic reticulum. A gap called the perinuclear space separates the two membranes.

Pores frequent the nuclear envelope to a degree that is related to the physiological activities of the cell. An increased pore density is associated with a high rate of traffic between nucleus and cytoplasm. Pores are found both randomly and in ordered arrays on the nuclear surface.

The pore complex consists of fibrous and particulate structures that rim the region of fusion between inner and outer membranes. The morphological unit, called the annulus, is made up of two sets of four to eight protein subunits, each set projecting away from the envelope toward either cytoplasm or nucleoplasm. The two sets appear to abut in the plane of the envelope.

The inner surface of the envelope is plastered by a fibrous coating called the nuclear cortex. The fibers are arranged in whorls in such a way that they funnel material from the nucleoplasm toward the pore complexes.

The nuclear envelope resembles the endoplasmic reticulum in its composition and possession of enzyme activities. However, the enzyme activities appear to be lower in the envelope than in the endoplasmic reticulum.

The envelope is highly permeable to inorganic ions and small organic molecules and in general permits passage of material up to a molecular weight of those proteins needed inside for nucleic acid replication and transcription. Materials with diameters greater than 9 nm are excluded.

During the cell cycle the nuclear envelope disappears and reappears. The extent to which it is disassembled seems to vary with species and cell type.

References

1. Experimental Studies on Amphibian Oocyte Nuclei. I. Investigation of the Structure of the Nuclear Membrane by Means of the Electron Microscope, A. G. Callan and S. G. Tomlin, *Proc. R. Soc. London Ser. B* 137:367(1950).
2. Nuclear Pore Flow Rate of Ribosomal RNA and Chain Growth Rate of Its Precursor during Oogenesis of *Xenopus laevis*, U. Scheer, *Dev. Biol.* 30:13(1973).
3. The Nuclear Pore Complex: Its Free Existence and an Hypothesis as to Its Origin, A. C. Fabergé, *Cell Tissue Res.* 151:403(1974).
4. Characterization of the Nuclear Envelope, Pore Complexes, and Dense Lamina of Mouse Liver Nuclei by High Resolution Scanning Electron Microscopy, R. H. Kirschner, M. Rusli, and T. E. Martin, *J. Cell Biol.* 72:118(1977).
5. Nuclear Surface Complex as Observed with the High Resolution Scanning Electron Microscope, G. Schatten and M. Thoman, *J. Cell Biol.* 77:517(1978).
6. A Large Particle Associated with the Perimeter of the Nuclear Pore Complex, P. N. T. Unwin and R. A. Milligan, *J. Cell Biol.* 93:63(1982).
7. An Enzyme Profile of the Nuclear Envelope, I. B. Zbarsky, in *Int. Rev. Cytology*, vol. 54, G. H. Bourne, J. F. Danielli, and K. W. Jeon, eds., Academic Press, New York, 1978.
8. The Movement of Material Between Nucleus and Cytoplasm, P. L. Paine and S. B. Horowitz, in *Cell Biology, A Comprehensive Treatise*, vol. 4, D. M. Prescott and L. Goldstein, eds., Academic Press, New York, 1980.

Selected Articles

Characterization of the Nuclear Envelope, Pore Complexes, and Dense Lamina of Mouse Liver Nuclei by High Resolution Scanning Electron Microscopy, R. H. Kirschner, M. Rusli, and Terence E. Martin, *J. Cell Biol.* 72:118(1977).

Chemical and Biochemical Properties of the Nuclear Envelope, C. B. Kasper, in *The Cell Nucleus*, vol. 1, H. Busch, ed., Academic Press, New York, 1974.

Dynamic Properties of the Nuclear Matrix, Ronald Berezney, in *The Cell Nucleus*, vol. 7, H. Busch, ed., Academic Press, New York, 1979.

Structures and Functions of the Nuclear Envelope, W. W. Franke and U. Scheer, in *The Cell Nucleus*, vol. 1, H. Busch, ed., Academic Press, New York, 1974.

The Movement of Material Between Nucleus and Cytoplasm, Philip L. Paine and Samuel B. Horowitz, in *Cell Biology, A Comprehensive Treatise*, vol. 4, D. M. Prescott and L. Goldstein, eds., Academic Press, New York, 1980.

The Nuclear Envelope in Mammalian Cells, D. J. Fry, in *Mammalian Cell Membranes*, vol. 2, G. A. Jamieson and D. M. Robinson, eds., Butterworths, London, 1977.

15

The Genetic Material

I feel that chromosome workers will ignore . . . advances in molecular biology at their peril. It is not enough, in order to understand the Book of Nature, to turn over the pages looking at the pictures. Painful though it may be, it will also be necessary to learn to read the text. Only with the assistance of molecular biology will this be possible.

F. H. C. CRICK, 1977

At this point one is in effect sitting in a darkened theater, having seen the first act and the interplay of the main characters: the DNA and the histones. The full cast has not yet appeared; no one knows quite how the plot will unfold.

ROGER D. KORNBERG, 1981
AARON KLUG

When Charles Darwin died in 1882, a mere century ago, the genetic material on which his theory depended had not been correctly identified. Although DNA was discovered during Darwin's lifetime and chromosomes were observed as well, the idea of putting together chromosomes and nucleic acids with heritable traits and genetic change had not yet emerged. It was not until the early part of the twentieth century that the chromosome was implicated as the bearer of heredity.

Karl Nägli observed rodlike chromosomes in the nuclei of plant cells in 1842 and was probably the first to report their sighting. This was well before Ernst Abbe and Otto Schott developed the Jena glass that made possible lenses that virtually eliminated spherical and chromatic aberrations (1866). So Nägli was working under the handicap of primitive lens systems.

Thirty years after Nägli's report, in 1872, E. Russow made the first serious attempts to describe chromosomes. Only one year later A. Schneider published a most significant paper dealing with the relation between chromosomes and the stages of cell division. The thought was thus beginning to form that chromosomes had some role in cell replication, but the nature of this role was unclear.

In 1879 Walther Flemming introduced the term *chromatin* to describe the threadlike material of the nucleus that became intensely colored after staining. Although he could not see chromosomes clearly because of the limitations of microscopes of his time, he drew detailed pictures of their activities during mitosis that are strikingly similar to what one is able to observe today with the compound light microscope. Some of his drawings are reproduced in Figure 15.1.

In 1902 Walter S. Sutton reported some important observations that gave him the distinction of being credited as the originator of the theory of the chromosomal basis for heredity. He observed that the chromosome pair in synapsis is made up of one maternal and one paternal member. He believed that chromosomes, acting in this way, may be the physical basis for the Mendelian laws of heredity.

A highly significant contribution in establishing a cytological basis for the laws of heredity was made by Thomas Morgan and Hermann Muller

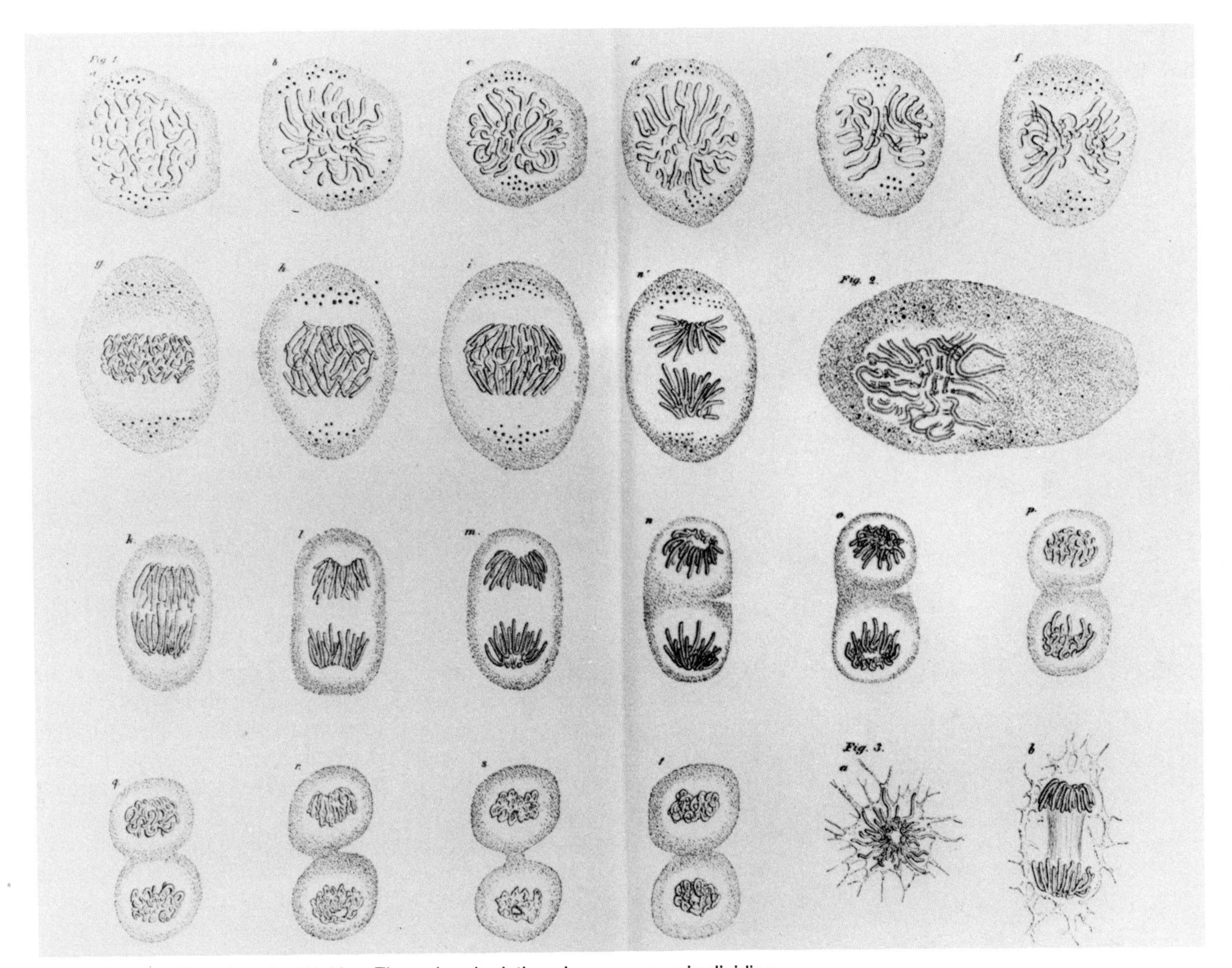

FIGURE 15.1 Drawings by Walther Flemming depicting chromosomes in dividing salamander larvae cells. Figure 1 shows a series of stages through mitosis. Figure 2 suggests that chromosomes are doubled when visible and Figure 3 depicts a mitotic spindle apparatus.

in the early 1900s. Working with *Drosophila* chromosomes, they located 2000 genetic factors on the four fruitfly chromosomes by 1922. So it was that chromosomes, still of unknown chemical nature, began to have an impact on the ideas of heredity.

A signficantly different but parallel approach to heredity was begun in 1869 by Friedrich Miescher. Working in the laboratory of Felix Hoppe-Seyler in Tübingen, Miescher discovered DNA. He termed the material *nuclein*, which he isolated from pus cells that he washed from bandages discarded by a nearby clinic.

Richard Altmann was able to obtain nuclein free of protein in 1889. It was he who first suggested the use of the term *nucleic acid* to describe the phosphorus-containing nuclein studied initially by Miescher.

Around the turn of the century Albrecht Kossel began a very intensive chemical investigation of nucleic acids. He was the first to recognize carbohydrate as a component of these materials. Using rather harsh techniques, he was also able to determine the elemental composition for several of the nucleic acid bases, and depending on some brilliant synthetic work of Emil Fischer, advanced the understanding of purine and pyrimidine structures. But it was not until the 1950s that the internucleotide bond was established (by Todd). The chemical structure of nucleic acid was thus slowly taking form, but its role in heredity was not known.

The first conclusive experimental support for nucleic acid as the genetic material came in 1944 in a paper published by O. T. Avery, C.M. MacLeod, and M. McCarty.[1] These investigators isolated the material that transformed a type of pneumococcus devoid of a capsule into an encapsulated form. The conclusion to their paper sums up the results.

The data obtained by chemical, enzymatic and serological analyses together with the results of performing studies by electrophoresis, ultracentrifugation and ultraviolet spectroscopy indicate that, within the limits of the methods, the active fraction contains no demonstrable protein, unbound lipid or serologically reactive polysaccharide and consists principally, if not solely, of a highly polymerized, viscous form of desoxyribonucleic acid. . . . The evidence presented supports the belief that a nucleic acid of the desoxyribose type is the fundamental unit of the transforming principle of Pneumococcus Type III.

Once it was accepted that nucleic acid was the genetic material of the cell, an enormous amount of effort in scores of laboratories was brought to bear on questions of nucleic acid form and function. Some workers chose to study eucaryotic systems, where the genetic material is more visible than in other simpler systems. But in eucaryotes, both the amount and complexity of genetic material is great.

Others focused on bacterial genomes, which offered the advantages of being readily subject to a variety of laboratory manipulations for growth and testing as well as existing in a "naked" form, free of the proteins present in eucaryotic genetic material. The procaryotic genome, although much simpler than its eucaryotic counterpart, is still a giant macromolecule and a vigorous challenge for investigators.

Still other researchers investigated the viral genome, which may be only one twentieth the size of the procaryotic genome. It is from this group that some of the most dramatic advancements have been made at the level of molecular resolution. But although viral genomes are simpler than others, they are in a sense incomplete and therefore come with their own set of disadvantages. They are acellular genomes that cannot function independently of a cell.

Nevertheless, much of the information gained from studies on viral genomes has been found to be applicable to the most complex procaryotic and eucaryotic genetic systems. Thus even though viruses are only subcellular particles, they have served as prototypes in studies of genetic ultrastructure and therefore deserve consideration in this chapter.

15.1 ACELLULAR GENOMES: VIRAL NUCLEIC ACIDS

Studies on virus genomes have given rise to a number of surprising findings. One is that genetic information is not restricted to the form of DNA;

RNA has genomic potentials. A second is that single-stranded DNA may serve as a genome as well as double-stranded DNA.

Viruses contain either DNA or RNA, never both, which suggests without further investigation that genomes can be either DNA or RNA. Some of the DNA genomes are double stranded, as in procaryotic and eucaryotic cells, and others are single stranded. Most of the RNA genomes are single stranded, but some are double stranded, and the replicative form of the RNA genome is double-stranded RNA. A final twist to virus genomes is that some are linear and others circular.

Virus genomes may be
Double-stranded DNA
Single-stranded DNA
Single-stranded RNA
Double-stranded RNA
Circular O
or linear ———

Hence a number of possibilities exist with respect to the structure of the nucleic acid that serves as genome material in viruses. Some of the more common viruses with descriptions of their genomes are listed in Table 15.1.

The Genetic Material of ϕX174

In 1959 Robert Sinsheimer made two important observations on the DNA of ϕX174, a small polyhedral virus.[2] First, the ratios of A/T and G/C were not one as is mandated by the rules of base pairing in the double helix. Second, the amino groups of the purine and pyrimidine rings reacted readily with formaldehyde, a property that is generally seen only with RNA or heat-denatured DNA.

These characteristics led Sinsheimer to conclude that the ϕX174 genome is a *single DNA strand*, a conclusion that has been amply verified.

Another important finding of Sinsheimer's research group, published in 1962, was that ϕX174 DNA is circular.[3,4] Exonucleases were unable to use the intact DNA as a substrate because of a lack of free ends. In support of circularity, endonuclease activity, when conducted under conditions that permitted only one hit per DNA, resulted in the release of only one fragment, not two. Thus, the endonuclease was merely opening a ring, not cleaving a linear genome.

The next step in confirming the circularity of ϕX174 DNA was to attempt to see these macromolecules with the electron microscope. Success in sighting the circular forms was reported in 1963 and 1964, first on the double-stranded replicative form of the genome,[5] then on the single-stranded form present within the intact virion.[6]

One hit per DNA will give one of two results, depending on whether the DNA is circular or linear.

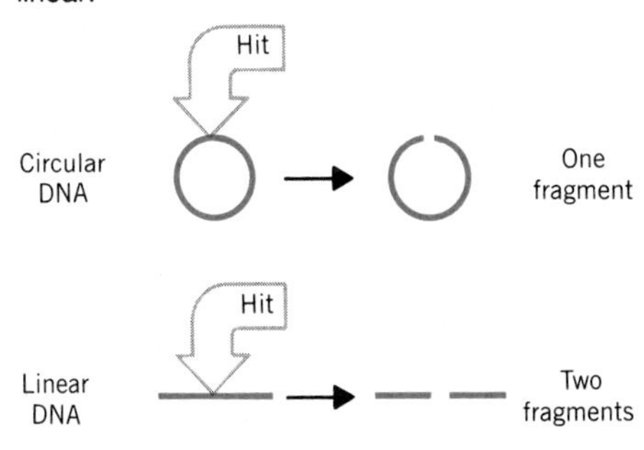

Electron micrographs showing this circularity are presented in Figure 15.2. The contour lengths of the double- and single-stranded forms were reported to be 1.64 ± 0.11 and 1.77 ± 0.13μm, respectively, in the two original articles.

Continued work on this simple virus after 1964 led to some remarkable insights regarding the molecular biology of DNA. In 1967 the *in vitro* synthesis of the double-stranded replicative form of ϕX174 was accomplished,[7] as demonstrated in the schematic representation of Figure 15.3. This success opened a new era of nucleic acid replication work and ushered into routine use the enzymes of DNA replication: DNA polymerase and DNA ligase.

In 1977, nearly 20 years after the single-strandedness of ϕX174 DNA was reported, its complete nucleotide sequence was published by Frederick Sanger et al.[8] The techniques of determining DNA primary structure developed by Sanger, enabling the success of the ϕX174 work, was of such consequence that he was awarded the Nobel prize for his accom-

Table 15.1 The Size and Structure of the Genomes of a Few Representative Viruses and Bacteria

Organism	Description	Molecular Weight ($\times 10^{-6}$ daltons)	Approximate Number of Genes[a]
Plant Viruses			
Potato spindle tuber virus	Linear single-stranded RNA	0.025–0.11	<1
Turnip yellow mosaic virus	Linear single-stranded RNA	2.0	7
Tobacco mosaic virus	Linear single-stranded RNA	2.0	7
Small RNA Viruses			
R17 coliphage	Linear single-stranded RNA	1.3	4 (3)
Polio virus	Linear single-stranded RNA	2.6	9
Small DNA Viruses			
ϕX174 coliphage	Circular single-stranded DNA	1.7	6 (8)
fd Coliphage	Circular single-stranded DNA	2.1	7
Polyoma virus	Circular double-stranded DNA	3.0	5
SV40 virus	Circular double-stranded DNA	3.2	6
Large RNA Viruses			
Influenza virus	Linear single-stranded RNA (readily fragments)	3.6	12
Vesicular stomatitis virus	Linear single-stranded RNA	3.6–4.5	15
Sindbis virus	Linear single-stranded RNA	4.6	14
Mouse leukemia virus (Rauscher)	Linear single-stranded RNA (can be dissociated into fragments)	11.2	37
Reovirus	Linear double-stranded RNA (readily fragments)	15	25
Large DNA Viruses			
PM2 *Pseudomonas* phage	Circular double-stranded DNA	6	10
P4 coliphage	Linear double-stranded DNA	6.7	10
Adenovirus type 12	Linear double-stranded DNA	21–22	38
T7 coliphage	Linear double-stranded DNA	25.2	42
T5 coliphage	Linear double-stranded DNA	75	125
Herpes simplex	Linear double-stranded DNA	96	160
Insect granulosis virus	Circular double-stranded DNA	100	170
T4 coliphage	Linear double-stranded DNA	110	180
Shope fibroma virus	Linear double-stranded DNA	153	250
Bacteria			
Mycoplasma arthritidus	Circular double-stranded DNA	440	730
Haemophilus influenzae	Circular double-stranded DNA	1660	2800
Bacillus subtilis	Circular double-stranded DNA	2400	4000
Escherichia coli	Circular double-stranded DNA	2500	4200
Salmonella typhimurium	Double-stranded DNA	3000	5000
Serratia marcescens	Double-stranded DNA	3600	6000

[a]Estimated by assuming that an "average" gene contains 1000 nucleotides (or nucleotide pairs). Figures in parentheses are more reliable estimates from genetical data.

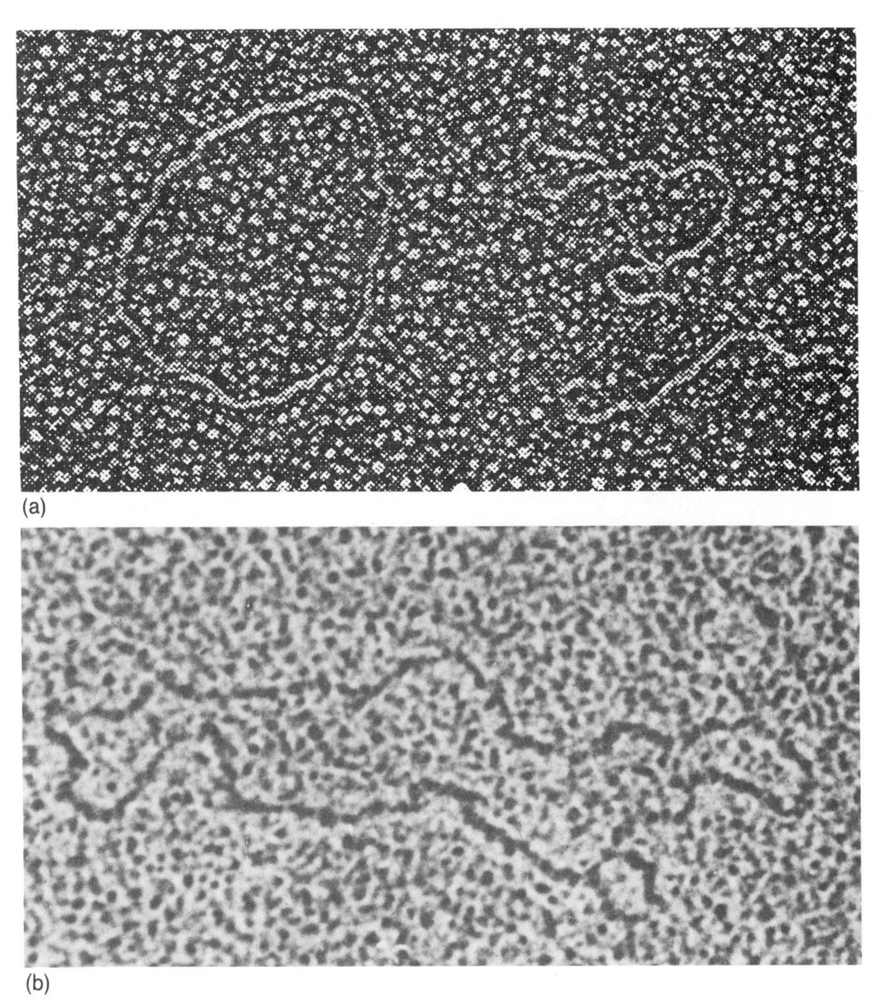

FIGURE 15.2 Electron micrograph of the double-stranded replicative (*a*) and the single-stranded nonreplicative (*b*) forms of ϕX174 viral DNA. Both are circular. These micrographs represent historical milestones in the sighting of viral DNA and in confirming their circular structures. (Micrographs courtesy of Drs. Robert L. Sinsheimer, a, and David Freifelder, b.)

plishments in 1980. The molecular detail of this molecule has added greatly to our understanding of how genes are situated within the genome.

The DNA of ϕX174 contains 5375 nucleotides making up nine genes in the circle. A surprising finding was that certain of these genes overlap. For example, the start nucleotide sequence, which in every case is ATG, is merged with the end sequence in several cases. Thus the start of gene D is found within the end signal (TGA) of gene C: ATGA.

An even more surprising and extremely important find was that two genes, B and E, are contained *within* the two genes A and D, respectively (see Figure 15.4). Indeed, in one region of the DNA the nucleotide sequence serves three different functions: ribosome recognition for a start for gene J, coding for amino acids in genes D and E, and the termination of gene D (see Figure 15.5).

Genes F, G, and H are separated from the rest of the genome by

ØX174 is transcribed by first making a complementary antiparallel strand and then using this new strand as a template to make mRNA. Thus a sequence such as –A–T–G– has a complementary sequence –T–A–C– which, when transcribed, produces –A–U–G–, the start codon for protein synthesis.

5′ –A–T–G– 3′ the original strand
3′ –T–A–C– 5′ the complementary strand
5′ –A–U–G– 3′ the start codon in mRNA

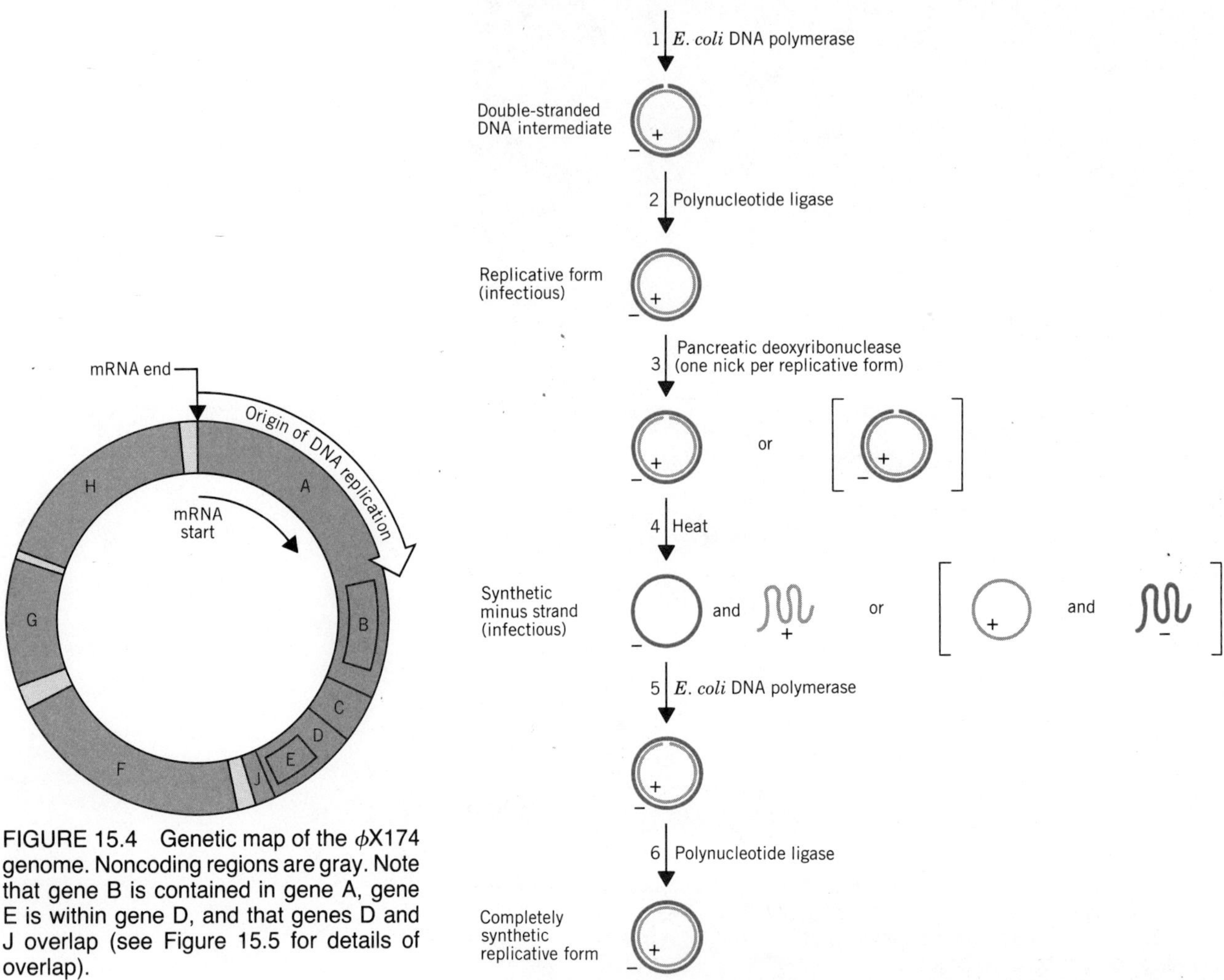

FIGURE 15.4 Genetic map of the ϕX174 genome. Noncoding regions are gray. Note that gene B is contained in gene A, gene E is within gene D, and that genes D and J overlap (see Figure 15.5 for details of overlap).

FIGURE 15.3 Enzymatic synthesis of infectious phage ϕX174. Template DNA containing tritiated thymidine was used to make a complementary(−) strand containing ^{32}P and 5-bromouracil instead of thymine (step 1). Given the different radioactivities of the two strands and their different densities (due to 5-bromouracil), all forms are separable after step 4 before the use of DNA polymerase again. By selecting the (−) strand and making a strand complementary to it, a complete synthetic replicative form is produced.

FIGURE 15.5 Multiple use of a nucleotide sequence in ϕX174 DNA. Messenger RNA, which is produced from a DNA strand complementary to that shown here, contains the information indicated. Note that where the D and J genes overlap there is a frame shift in reading the nucleotide sequence.

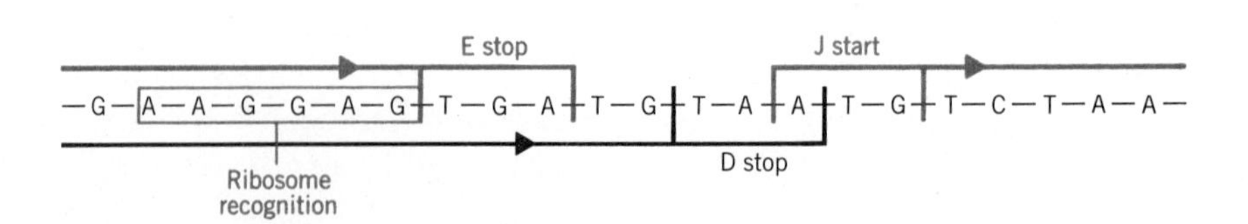

noncoding spaces. Knowledge of the nucleotide sequence has provided the insight that three of these regions contain sequences that permit self-complementary base pairing, as illustrated in Figure 15.6. These regions are thought to function as regulatory regions where RNA polymerase binds and initiates transcription.

Although it is beyond the scope of this text to expand on the ramifications of these studies on ϕX174 DNA, it is easy to see that the molecular detail of this viral genome has answered many questions regarding the relation of structure to function in genetic material. In particular, it has shown that genes are not necessarily situated in a linear array, carefully spaced by nongene regions. It has also suggested the possibility that intrastrand base pairing may occur, providing means by which transcriptases may recognize and bind to specific sites on the template nucleic acid. Above all, determining the complete primary structure of ϕX174 has disrobed the gene, robbing it of some of its mystery and intrigue. It is, after all, only a string of nucleotides. In that sense, a gene is a gene is a gene.

Other Viral Genomes

ϕX174 is not the only virus that possesses a single-stranded DNA genome. But many viruses that contain DNA possess the double-stranded, or duplex, form. Some of these are circular, such as is true for SV40, and others linear, like the bacteriophages in the T series and the lysogenic virus λ.

Therefore, from the point of view of ultrastructure, these DNA genomes exist in a variety of forms. Some of these are summarized in Figure 15.7.

The genomes of RNA viruses also exist in a variety of forms, as summarized in Figure 15.8. These RNAs come in sizes that range from just 3569 nucleotides in the small bacteriophage MS2 to about four times this size in paramyxoviruses. The MS2 virus has a single-stranded RNA genome that has been completely sequenced by Walter Fiers and his colleagues. In contrast to ϕX174, the MS2 genome is linear and contains only three genes that are separated by noncoding regions.

Supercoiled DNA

In 1965 another very significant breakthrough concerning the ultrastructure of genetic material took place. In the laboratory of Jerome Vinograd of the California Institute of Technology, it was found that polyoma virus is easily renatured after heat denaturation and that the individual strands of the duplex molecule do not separate from each other. It was concluded that the two strands were covalently closed and intertwined with each other.

Sedimentation studies on polyoma DNA revealed three sedimenting components. One of these, viewed by means of electron microscopy, was linear, or a broken-open form of the genome. The other two components were circular, but differed in that one was a twisted loop and the other a relaxed loop (see Figure 15.9).

It was soon realized that the difference between the latter two forms

FIGURE 15.6 The sequence of bases in the non-coding regions between genes of ϕX174 is such that base-pairing can take place to form hairpin loops. These may function in signaling the RNA polymerase to recognize start and stop points.

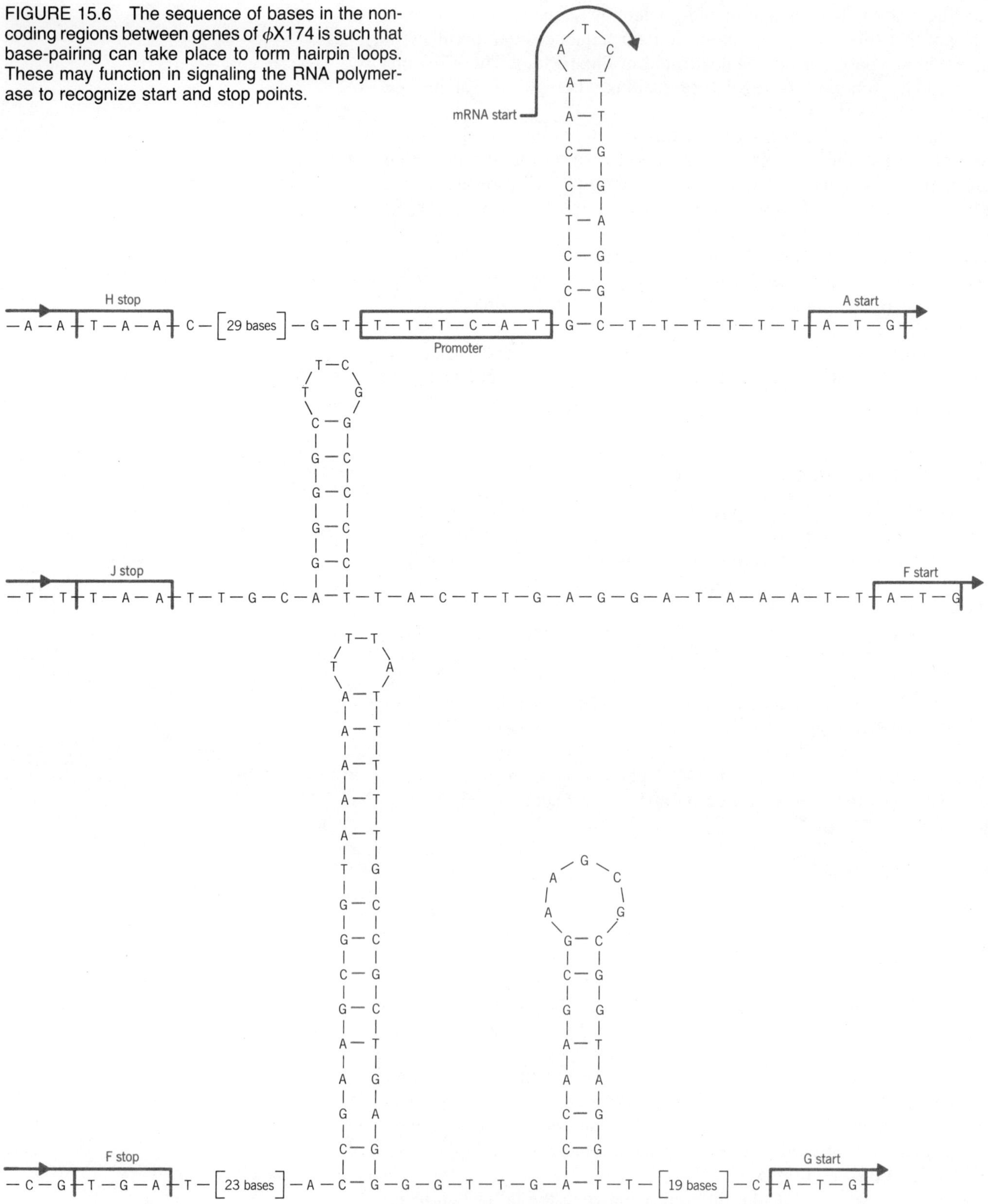

mRNA start

Loop: A T C
A — T
A — T
C — G
C — G
T — A
C — G
C — G

H stop | Promoter | A start

— A — A ┼ T — A — A ┼ C — [29 bases] — G — T ┼ T — T — T — C — A — T ┼ G — C — T — T — T — T — T — T ┼ A — T — G ┼

Loop: T — C, T, G
C — G
G — C
G — C
G — C
G — C

J stop | F start

— T — T ┼ T — A — A ┼ T — T — G — C — A — T — T — A — C — T — T — G — A — G — G — A — T — A — A — A — T — T ┼ A — T — G

Loop: T — T, T, A
A — T
A — T
A — T
A — T
A — T
T G
G — C
G — C
C — G
G — C
A — T
A G
G A
C — G

Loop: A, G, C, A, G
G — C
C — G
G G
A — T
A A
C — G
C — G

F stop | G start

— C — G ┼ T — G — A ┼ T — [23 bases] — A — C — G — G — G — T — T — G — A — T — T — [19 bases] — C ┼ A — T — G ┼

Types of viral DNA molecules	Examples
Linear single strand	Parvoviruses
Circular single strand	ϕX174 and other bacteriophages
Linear duplex	T7; many phages and animal viruses
Linear duplex single chain breaks	T5
Duplex with closed ends	Vaccinia
 Closed circular duplex with and without supercoils	Papovaviruses, bacteriophage PM2, and cauliflower mosaic virus (forms without supercoils may still have a protein bound)

FIGURE 15.7 Forms of DNA molecules found in a variety of viruses.

Types of RNA	Examples
(+) Linear, single strand, infectious, "positive" (+) strand	Most plant viruses, RNA bacteriophages, picornaviruses, togaviruses
(−) Linear, single strand, noninfectious, "negative" (−) strand	Rhabdovirus, paramyxovirus
(+) Segmented, positive strands	Brome mosaic virus (covirus of plants)
(−) Segmented, negative strands	Myxoviruses
RNAs Double strand, segmented	Reoviruses, wound-tumor virus of plants, cytoplasmic polyhedrosis virus of insects

FIGURE 15.8 Forms of RNA in different viruses.

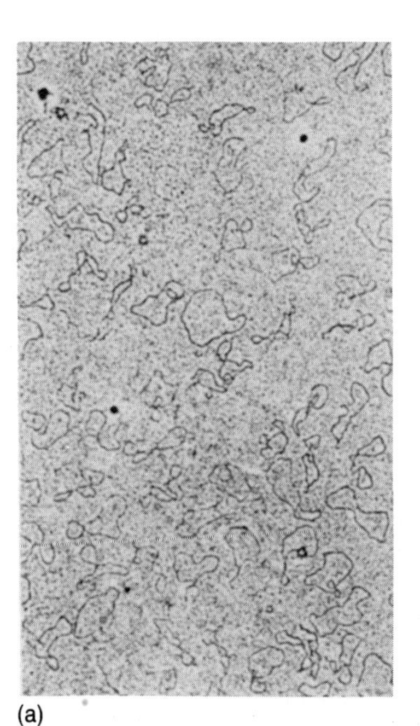
(a)

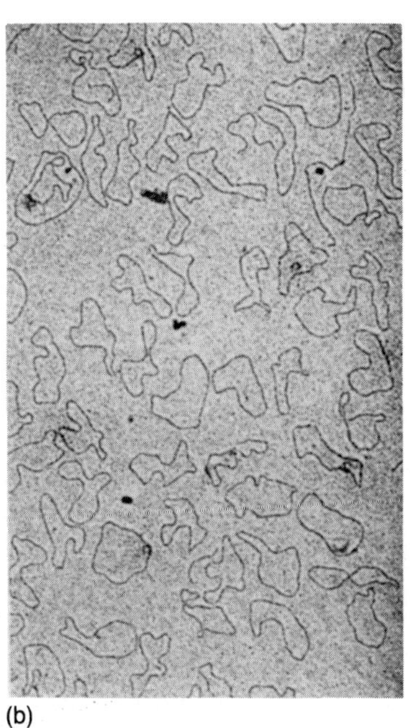
(b)

FIGURE 15.9 Electron micrographs of polyoma DNA. The supercoiled or twisted loops (*a*) are converted to relaxed open loops (*b*) by treatment with a DNase that nicks one of the strands and relieves the winding. (Courtesy of Dr. Philip Laipis.)

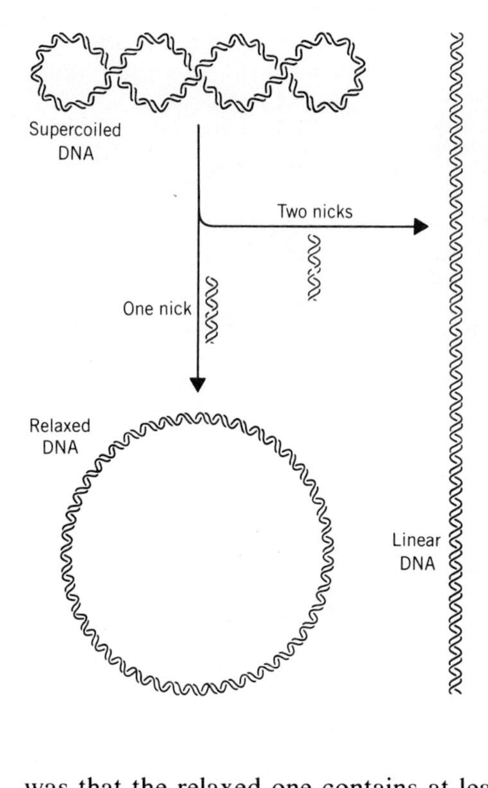

FIGURE 15.10 Representation of three forms of polyoma DNA. A supercoiled conformation, resulting from underwinding during synthesis, can be relaxed to a simple circular form if one strand of duplex DNA is severed. If both strands are severed sufficiently close together, a linear molecule results.

was that the relaxed one contains at least one nick, or break, in one of the strands. This relaxes the winding of the normal intertwined duplex and the twisting or supercoiling disappears. Vinograd and his associates proposed that the circular form of the duplex is slightly *underwound* compared to the linear or nicked form.[9] This could happen during biosynthesis as a result of chain closing before base pairing is completed. The strain of this underwound form is relieved by supercoiling, which makes the structure more dense, hence separable from the relaxed form by centrifugation. Figure 15.10 illustrates how supercoiling and nicking relieve the stress of underwinding.

The topology of supercoiling can perhaps be understood best as follows. A covalent relaxed circle will form if two ends of a linear DNA duplex molecule are attached. If, however, before the ends are joined one strand is rotated with respect to the other, a strain will result that is relieved by forming a figure 8. Since it is necessary to rotate one strand 360° with respect to the other before joining, one twist in the opposite direction will relieve the strain. Thus supercoiling is a response to a duplex twist.

Supercoiling is now recognized to be a common property of DNAs from widely varying sources, ranging from viruses to bacteria to eucaryotic cells. But it is especially common among the smaller DNAs, including the small DNA elements that are used in genetic engineering.

Moreover, it is beginning to appear that the insertion of a small piece of DNA into a larger one, which is the basis for the recombination of episomes into normal host genomes, requires that the integrating DNA be supercoiled.

What forces in nature would twist the double helix causing supercoiling, and for what purpose would this be done? Twisting is apparently not a type of faulty replication but rather a normal enzyme activity. A twisting enzyme called *DNA gyrase*, which has been discovered in *E. coli*, can use a relaxed covalently closed circular DNA as substrate and convert it to a twisted form.[10] The role of supercoiling in nature is presently the subject of intensive investigation.

The model in Figure 15.11 suggests a mechanism of DNA supercoiling by DNA gyrase.

15.2 PROCARYOTIC GENOMES: BACTERIAL DNA

The procaryotic cell contains a DNA that is huge compared to that present in viruses. Table 15.1 listed a few representative bacterial DNAs in terms of their molecular weights and approximate gene numbers. These values should be compared to similar values for viruses, which generally are on the order of 2–100 × 10^6 daltons molecular weight and contain from only a few to about 200 genes.

Because of its size, the bacterial genome has been extremely difficult to observe intact. John Cairns was the first investigator to visualize intact *E. coli* DNA by autoradiographic techniques.[11] He estimated its length to be several hundred micrometers and its molecular weight to be 10^9 daltons or more.

As a result of putting better techniques into practice, the genome of *E. coli* is now measured at 1.4 mm in its extended state. *In vivo*, it is circular and naked: the DNA is a single duplex molecule without free ends and without associated proteins.

Figure 15.12 depicts a bacterial genome that has been released from the cell as a completely intact circular macromolecule. In this case the nucleoid has been spread so that a regular array of loops emanates from a higher density core region. The loops appear to have a limited size and to number around 100 per genome. The loop, visualized here, is apparently related to the *domain*, which we discuss shortly.

Within the bacterial cell the 1400 μm length of DNA is packed into a nucleoid body approximately 1 μm in diameter, occupying about 25% of the volume of the cell. The degree of entanglement would appear to be enormous, but the helical duplex can be replicated and the resultant strands separated from one another every 17 min under ideal growth conditions, apparently without a hitch.

Supercoiling may solve some of the entanglement problems, for procaryotic DNA is also supercoiled, apparently to about the same degree as has been observed for the simpler closed circular species of DNA in viruses. Some workers have suggested that the *E. coli* genome is supercoiled into about 50 loops, all of which emanate from a central core held in shape by RNA.

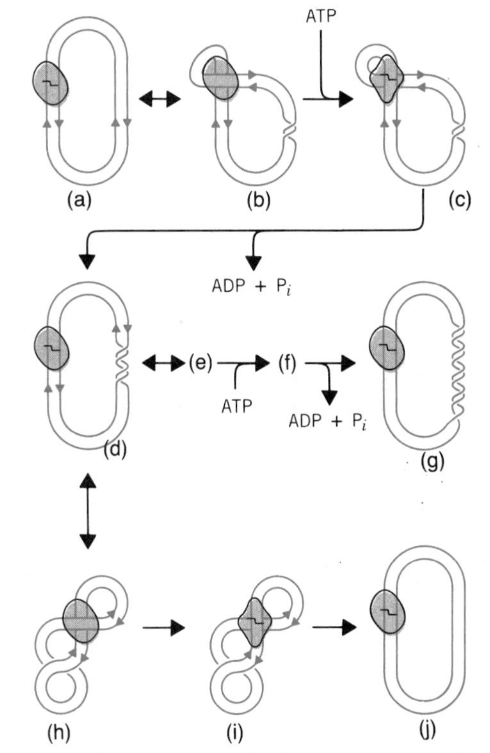

FIGURE 15.11 Model for mechanism of DNA supercoiling by DNA gyrase. The enzyme, by binding to a certain site on DNA (*a*), induces a portion of the DNA to wrap around it (*b*). After ATP binds, the upper double helix is transiently broken in both strands and is moved by the enzyme past the lower double helix; then the strands are rejoined without being rotated relative to one another (*c*). ATP is then hydrolyzed and the translocated segment released. The cycle can be repeated, each time adding another loop to the molecule (*d–g*). The same enzyme can effect relaxation of the supercoiled structure (*h–j*). (Courtesy of Dr. Martin Gellert.)

The Gross Organization of Procaryotic Genomes *in Situ*

The nucleoid, although not surrounded by a membrane, is a well-defined region of the procaryotic cell, as illustrated in Figure 15.13.

When cells are grown in rich media and the rate of synthesis is high, nucleoid surfaces are convoluted or fringed, whereas the nucleoids of inactive cells are more confined and dense. One interpretation of this observation is that a dense nucleoid is a view of DNA in a highly condensed and inactive state. When ample nutrients are available and syn-

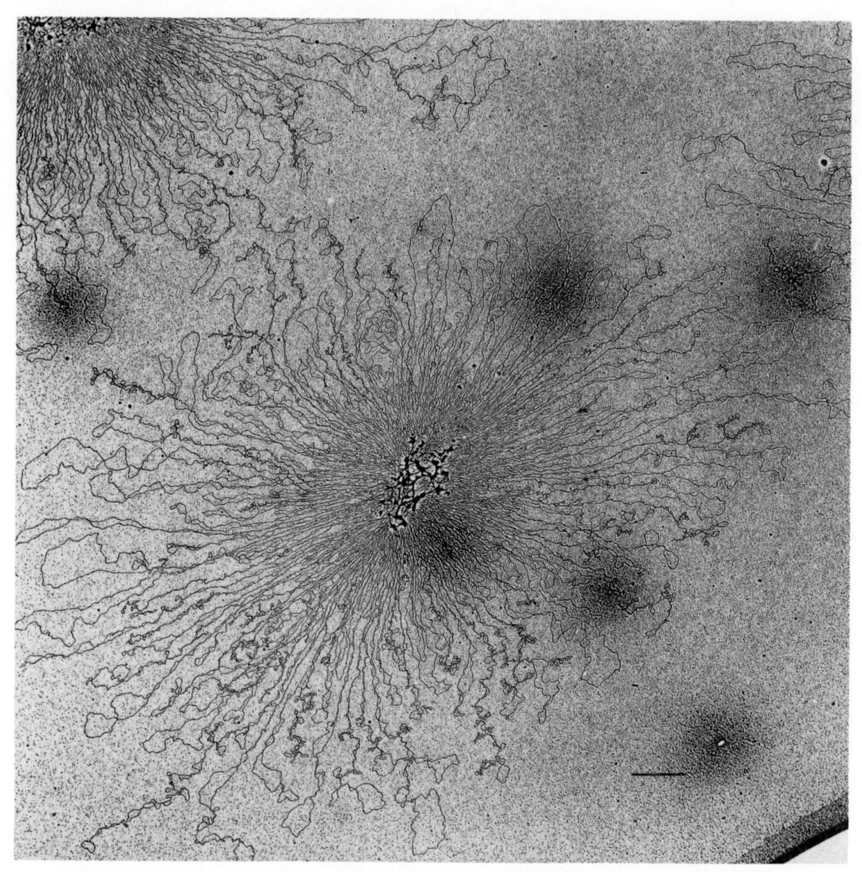

FIGURE 15.12 Electron micrograph of a bacterial genome that has been spread to reveal its circular, macromolecular nature. (Courtesy of Dr. Ruth Kavenoff.)

thesis high, DNA loops out of the nucleoid to provide a template for transcription by RNA polymerase enzymes.

Properties of Isolated Procaryotic Genomes

In the 1970s techniques were developed to isolate condensed bacterial nucleoids and study them *in vitro*. A good deal of caution must be used in their isolation because nucleoids are very unstable and must be protected or stabilized by the use of controlled levels of cationic counterions. In addition, nucleoid preparations commonly contain some amount of membrane envelope to which they are anchored in the intact cell.

Figure 15.14 shows isolated nucleoids as observed by scanning electron microscopy. The structure as viewed here appears to be doubletlike. It is thought that this doublet is actually two nucleoids that are bridged, having originated from multiple nucleoid-containing cells.

The molecular composition of nucleoids, summarized in Table 15.2, shows that they contain significant amounts of RNA as well as some

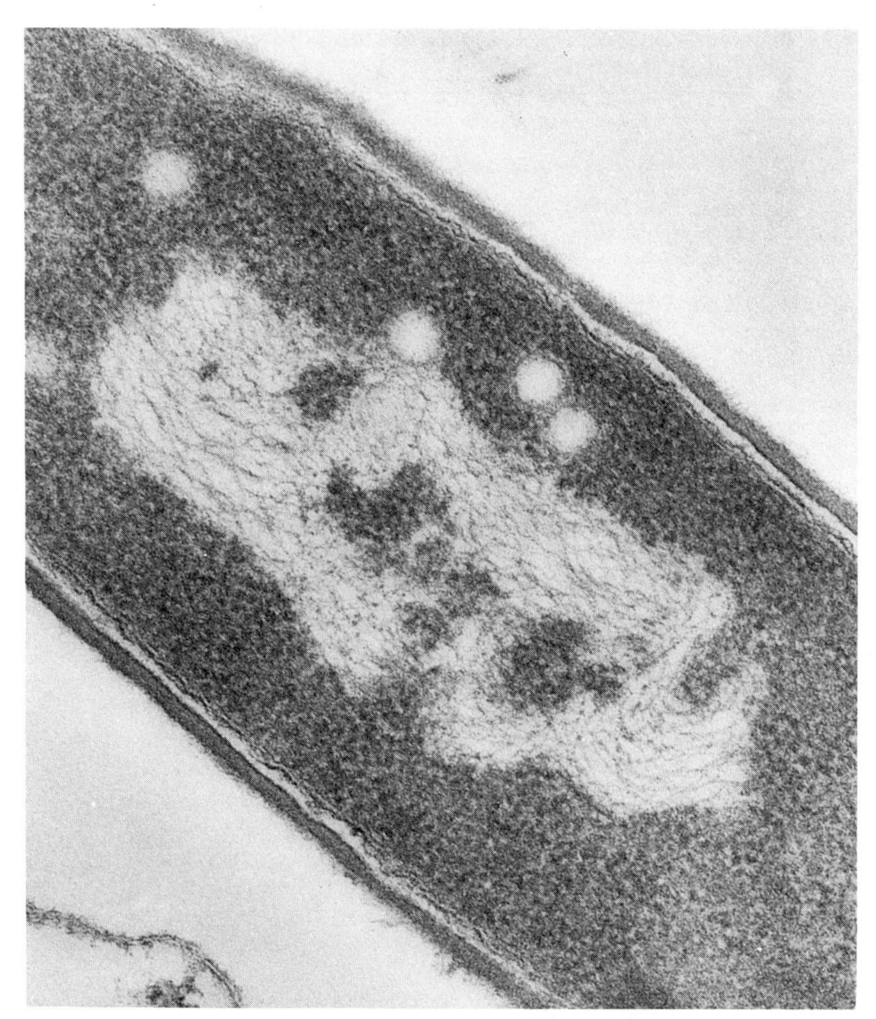

FIGURE 15.13 Electron micrograph of a thin section through a nucleoid of a procaryotic cell. Ribosomes surround the lighter nucleoid region, which contains fibers of DNA. (Courtesy of Dr. Dwight Anderson.)

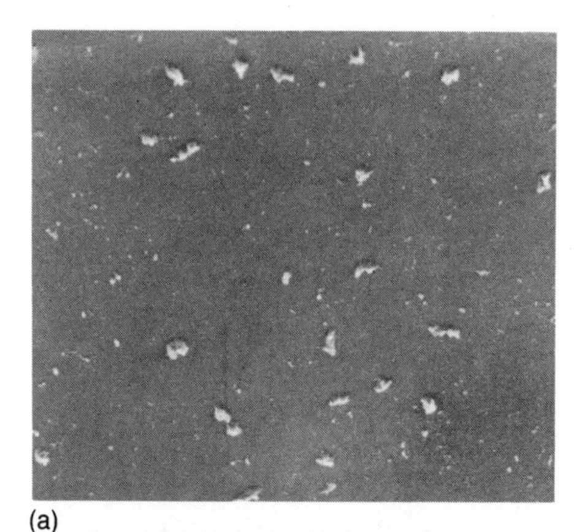

(a)

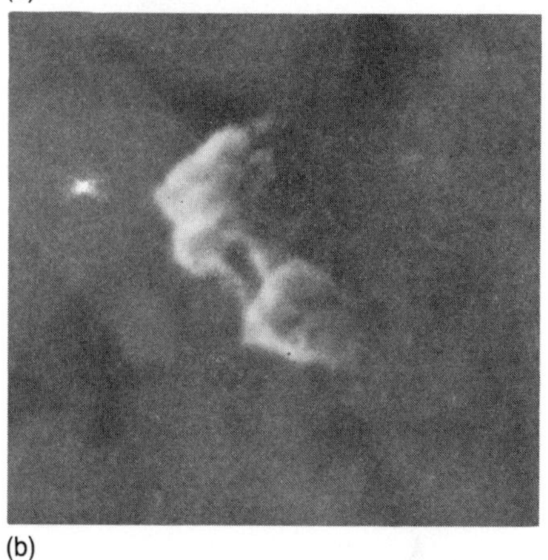

(b)

FIGURE 15.14 Isolated bacterial nucleoids observed by scanning electron microscopy. (*a*). A field showing many nucleoids. (*b*) One selected nucleoid showing the dumbbell appearance that may result upon nucleoid replication. (Courtesy of Dr. David E. Pettijohn.)

protein, particularly RNA polymerase, in addition to the major DNA component. The presence of this triad suggests that isolated nucleoids have been caught in a state in which transcription is taking place. Thus, the DNA of the nucleoid (perhaps in extended surface regions) is being transcribed by RNA polymerase, and nascent RNA products are still a part of the complex.

The DNA in isolated bacterial nucleoids is supercoiled to an extent similar to that of viral nucleic acids. Both nucleoid proteins and RNA appear to stabilize the supercoiled state.[12] Evidence in support of this comes from studies where nucleoid proteins are dissociated or denatured or nucleoids are subjected to RNase treatment. In both cases, the DNA loses its supercoiled state, and when RNA hydrolysis is complete, the DNA is completely unfolded.

Figure 15.15 models the supercoiled state of isolated nucleoid. An im-

Table 15.2 Properties of Nucleoids Isolated from *Escherichia coli* (30-min generation time)

Property	Membrane-free Nucleoid	Membrane-associated Nucleoid[a]
DNA content (weight fraction)[b]	0.6	0.4
RNA content (weight fraction)[b]	0.3	0.15–0.35
Protein content (weight fraction)[b]	0.05–0.1	~0.4
Lipid content (fraction of total labeled lipid)[c]	<1%	~20%
Sedimentation rate (weight average)[d]	1900S°	5500S°
Buoyant density in CsCl	1.69 ± 0.02 g/cm^3	1.46 ± 0.02 g/cm^3
DNA mass per singlet nucleoid	$9 \pm 1.0 \times 10^{-9}$ μg	—
DNA mass per doublet nucleoid	$15 \pm 2 \times 10^{-9}$ μg	—
Genome equivalents per singlet	2.2	—
Genome equivalents per doublet	3.6	—
Fraction of total cellular membrane bound to nucleoids	<0.01	0.2
Bound proteins		
RNA polymerase	Yes	Yes
Envelope proteins	Little or none	Yes

[a]Membrane-associated nucleoids can have variable amounts of membrane; therefore, these properties may vary.
[b]Weight fraction means fraction of total nucleoid dry weight.
[c]Lipids labeled with [^{14}C]oleic acid.
[d]Corrected for rotor speed effect indicated by S°.

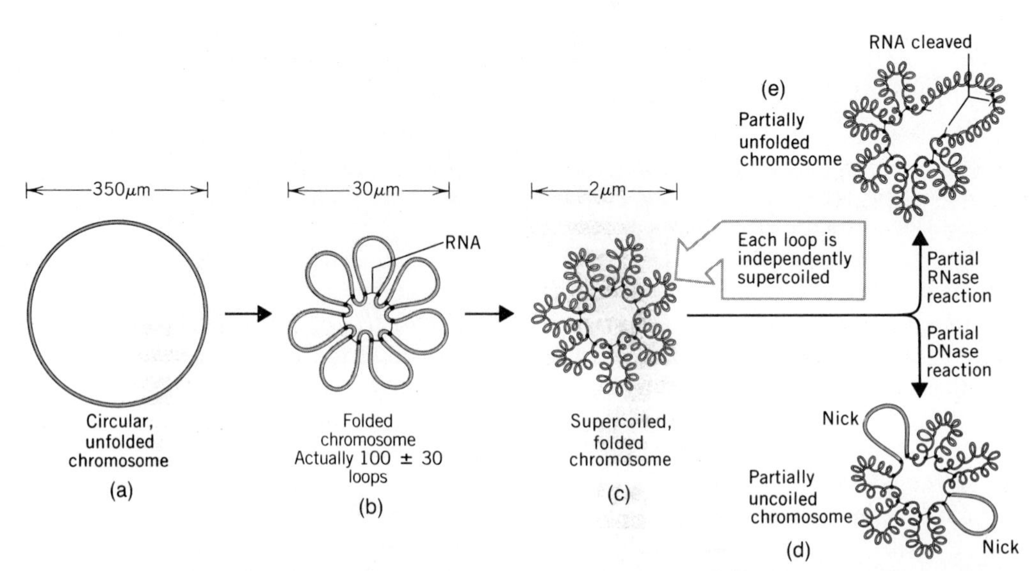

FIGURE 15.15 Model of DNA organization in an isolated bacterial chromosome. (*a*) The nonreplicating, circular unfolded DNA. (*b*) DNA containing folds due to interaction with RNA. Seven domains are shown, but the actual number is 100 ± 30. (*c*) DNA folded and supercoiled. (*d*) Chromosome with two single strand DNA breaks in different domains. The supercoiling in each domain is relaxed. (*e*) Partially unfolded DNA resulting from hydrolysis of some stabilizing RNA. (Courtesy of Dr. David E. Pettijohn.)

portant feature of this model is the role of RNA. RNA is viewed as restraining the circular DNA molecule by bringing together sites that are ordinarily far apart on the unfolded strand. This results in a loop or *domain*. Each domain is supercoiled, having on the average 1 superhelical turn per 200 base pairs.

When the DNA in a given domain is nicked, that domain will unfold and lose its supercoiling, while the remainder of the nucleoid DNA is unaffected. When gamma irradiation is used to produce nicks, about 100 $\pm$ 30 nicks are required to completely unfold the DNA molecules. Thus it is assumed that there is a similar number of domains per nucleoid genome.

Although this model accounts for experimental results obtained by working with nucleoids *in vitro*, it is not clear that the *in vivo* setting is precisely the same. An important inference from these studies is that the nucleoid *in vivo* is probably stabilized both by proteins and by RNA to give to the genome a degree of order that may be necessary for replication, for the separation of daughter genomes, and for the appropriate transcriptional activity of the genome.

15.3 EUCARYOTIC GENOMES: CHROMOSOMES

The genome of the eucaryotic cell is sequestered in the nucleus as part of a nucleoprotein complex referred to as *chromatin*. In most cells, the genome is present in units of chromatin, each unit a condensed form of chromatin called a *chromosome*.

The Cell Cycle and Chromosome Morphology

The physical form of chromosomes varies according to the cell cycle, as summarized in Figure 15.16. During interphase, chromosomes are least readily observed. They are extended or uncoiled, and if it is necessary to rely on light microscopy, one concludes that they have "disappeared" from the nucleus. However electron microscopy confirms that the chromosomes have not disappeared, but are still present as fibrillar masses generally associated with the nuclear envelope as illustrated in Figure 15.17.

During prophase chromosomes coil and shorten until they finally reach thymus, liver, or any other desired source. Nuclei are then lysed with ily observed with the light microscope, whose resolution is sufficient to permit viewing the chromatids of the newly replicated chromosome.

At metaphase each chromosome has replicated and is present as a pair of chromatids. The chromatids in a given pair move to separate daughter cells during anaphase.

Electron microscopic studies on these relatively simple-looking organelles have eliminated any hope of simplicity in chromosome structure. Each chromatid is an intricate maze of fibers that seems to make as much structural sense as a backlash on a fishing reel.

The chromatids move apart during anaphase to separate locations in the dividing cell. Upon gathering in the new location, telophase is reached. A nuclear envelope is formed around the group of chromosomes and they begin to uncoil and associate with the membrane material of the envelope. Once again interphase is reached, during which time DNA replication takes place.

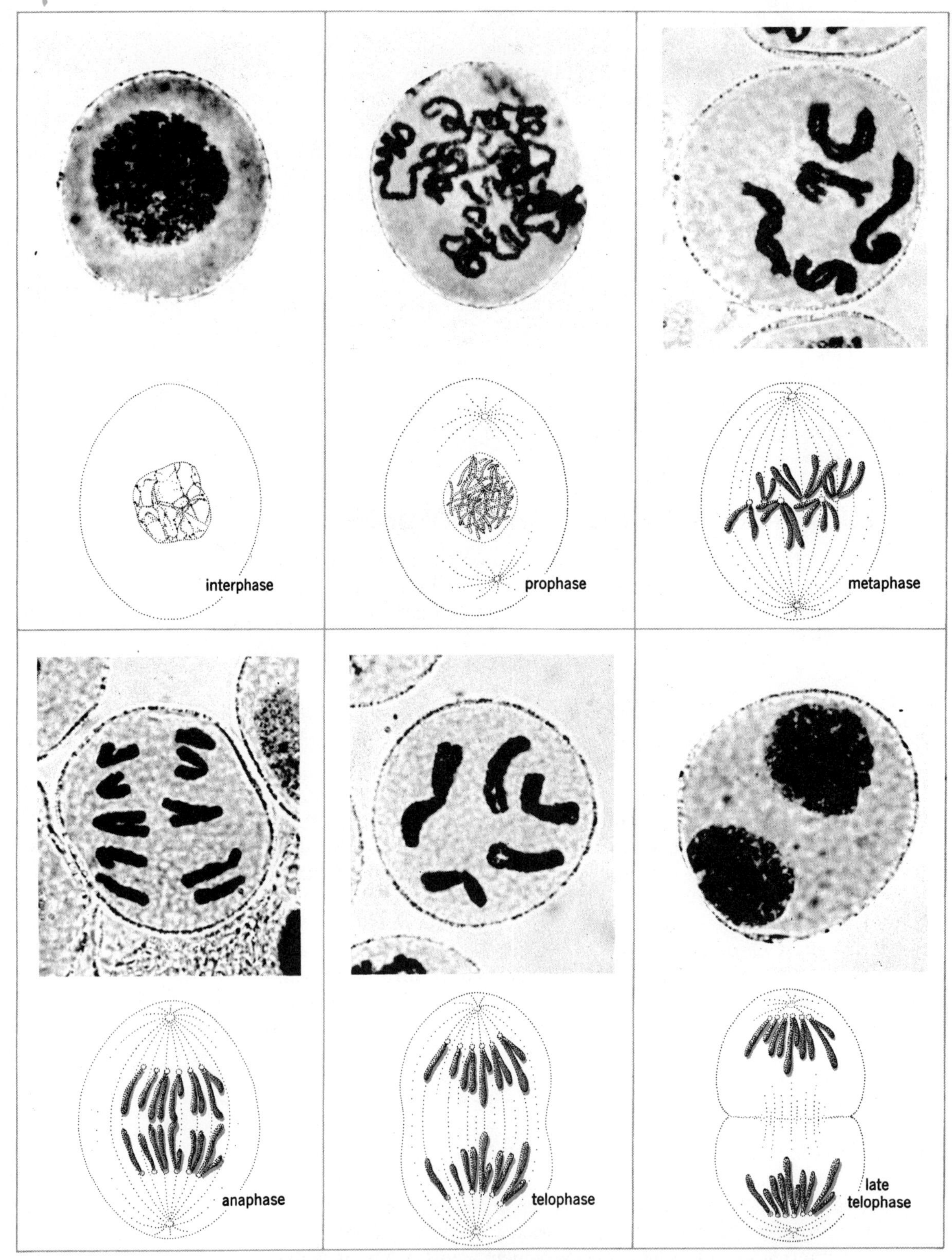

FIGURE 15.16 Phases of the cell cycle showing how chromosomes vary in their morphology during mitosis. (Micrographs by Dr. A. H. Sparrow.)

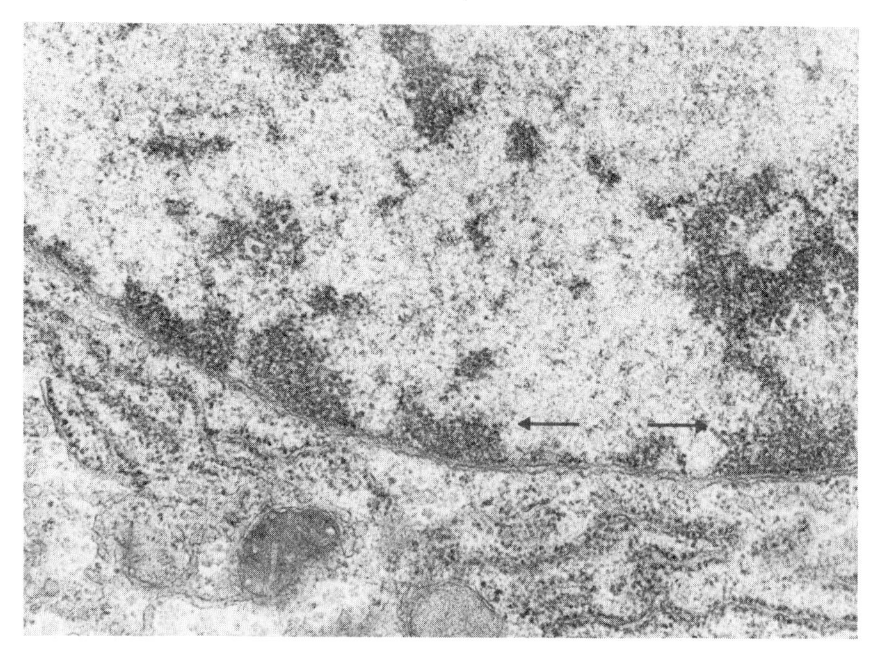

FIGURE 15.17 Thin section of rat liver cell through nuclear envelope. Arrows point to fibrillar masses of chromatin showing their characteristic association with the inner membrane of the envelope.

The Composition of Chromatin

Several methods have been developed over the years to prepare chromatin. A relatively simple approach is to first prepare purified nuclei from thymus, liver, or any other desired source. Nuclei are then lysed with detergent and the chromatin pelleted at 20,000 rpm (about 50,000*g*) in a preparative centrifuge.

Method I

Chromatin, such as that isolated from rat liver, contains DNA, RNA, and protein. Two types of protein are present, the *histones* and the *nonhistone proteins*.

Arginine and lysine at physiological pH are cationic (basic) and can interact electrostatically with anionic nucleic acids.

$$\cdots NH-\underset{\substack{| \\ (CH_2)_4 \\ | \\ \overset{+}{N}H_3}}{CH}-CO-NH-\underset{\substack{| \\ (CH_2)_3 \\ | \\ NH \\ | \\ H_2N-C=\overset{+}{N}H_2}}{CH}-CO\cdots$$

Lysine Arginine

Rat liver chromatin has been used as a model for chromatin. It possesses a histone to DNA ratio near 1:1, a nonhistone protein to DNA ratio of 0.6:1, and an RNA/DNA ratio of 0.1:1.

Histones

Histones are very basic proteins, basic because they are enriched in the amino acids arginine and lysine to a level of about 24 mole percent.

Histones are released from chromatin by increasing the salt concentration of their environment from 0.15 to 2 M, suggesting that the interaction between these proteins and the acidic nucleic acid is electrostatic. Five classes of histones are released. They have been given the designations listed in Table 15.3.

All five classes are quite low in molecular weight, a property that has made them amenable to intensive chemical studies. In fact, the primary structure is now known for each class.

Table 15.3 Characterization of the Histones[a]

Class	Fraction	(Lys + Arg)/(Glu + Asp)	Lys/Arg Ratio	Total Residues	Molecular Weight (daltons)	*N*-Terminal	*C*-Terminal
Very lysine-rich	H1 (I, f1, KAP)	7.11	22.0	~215	~21,500	Ac-Ser	Lys
Lysine-rich	H2A (IIb1, f2a2, ALK)	2.89	1.17	129	14,004	Ac-Ser	Lys
	H2B (IIb2, f2b, KSA)	2.80	2.50	125	13,774	Pro	Lys
Arginine-rich	H3 (III, f3, ARK)	3.44	0.72	135	15,324	Ala	Ala
	H4 (IV, f2a1, GRK)	3.57	0.79	102	11,282	Ac-Ser	Gly

[a]All data for histones of calf thymus.

Histone H4

10 20
Ac-Ser–Gly–Arg-Gly–Lys-Gly-Gly-Lys-Gly-Leu-Gly-Lys-Gly-Gly–Ala-Lys-Arg-His–Arg-Lys-

30 40
Val–Leu-Arg-Asp-Asn-Ile–Gln-Gly–Ile–Thr-Lys-Pro–Ala-Ile–Arg-Arg-Leu-Ala–Arg-Arg-

50 60
Gly–Gly-Val-Lys–Arg-Ile–Ser-Gly-Leu-Ile–Tyr-Glu-Glu-Thr-Arg-Gly-Val-Leu-Lys-[Val/Ile]-

70 80
Phe-Leu-Glu–Asn-Val–Ile–Arg-Asp-Ala–Val–Thr-Tyr-[Thr/Ser]-Glu-His-Ala-[Lys/Arg]-Arg-Lys-Thr-

90 100
Val–Thr–Ala-Met-Asp-Val-Val-Tyr–Ala–Leu-Lys-Arg-Gln-Gly-Arg-Thr-Leu-Tyr-Gly-Phe-

Gly–Gly-COOH

FIGURE 15.18 Amino acid sequence of calf histone H4. Complete sequences are also available for H4 from rat, pig, and pea, and all the substitutions for these histones are shown in boxes. At present, there are no known insertions or deletions, making for a highly conserved structure.

One of the significant revelations that has come from chemical studies is that the primary structures of histones have been highly conserved during evolutionary history. For example, histone H4 of calf and of garden pea contains only two amino acid differences in a protein of 102 residues (see Figure 15.18).[13] These organisms are estimated to have an evolutionary history of at least 600 million years, during which time they diverged structurally. This conservation of structure suggests that over the eras, histones have had a very similar and crucial role in maintaining the structural and functional integrity of chromatin.

Nonhistone Proteins

After histones have been released from chromatin, the remaining complex can be treated with sodium dodecyl sulfate and the nonhistone chromosomal proteins separated from DNA.

In contrast to the modest population of histones in chromatin, several hundred nonhistone proteins have been resolved by polyacrylamide gel electrophoresis. The particular collection of nonhistone proteins and their relative abundance varies with the tissue or cell source.

Approximately half the mass of the nonhistone chromosomal proteins is structural proteins. About a dozen or so of the major ones appear to be electrophoretically similar in a wide variety of organisms. One of the major proteins is a 45,000 dalton molecular weight actin, a type of contractile protein (see Chapter 18). In addition, all chromatins appear to possess α- and β-tubulins having molecular weights of 50,000 and 55,000 daltons; these are the proteins of microtubules. A higher molecular weight protein of 225,000 daltons also present is myosin. Although for some time these contractile proteins were thought to be contaminants, it is now believed that they are vital ingredients of the chromosome, functioning during chromosome condensation and in the movements of chromosomes during mitosis and meiosis.

Many of the remaining 50% of nonhistone chromosomal proteins have enzymatic activity. In this category are found RNA polymerase, serine protease, and acetyl transferase activities. Approximately 2 to 4% of the nonhistone chromosomal proteins bind specifically to the homologous DNA from which they were released.

Thus, it appears that half the nonhistone proteins include all the enzymes and factors that are involved in DNA replication, in transcription, and in the regulation of transcription. These proteins are not as highly conserved among organisms, although they must carry out similar enzymatic activities. Apparently they are not as important as the histones in maintaining chromosomal integrity.

The Fibrous Nature of Eucaryotic Chromatin

Fiber Diameter

During the 1960s chromatin was investigated very intensively by electron microscopists. Preparations of metaphase chromosomes repeatedly revealed chromatin to be fibrous, as illustrated in Figure 15.19.

The measured diameter of the fiber ranges from 3 to 25 nm, depending on the source, the manner in which the material is prepared for electron microscopy, and the stage of cell cycle during which the chromosome is observed. A fiber diameter of very close to 20 nm is commonly found for human chromatin at interphase and metaphase, and we use this dimension as a reference model.

Fiber Mass and Volume

The mass and volume of chromatin fibers may vary somewhat from preparation to preparation and among species and cell types. But as an example of what has been encountered, observations and calculations on the lymphocyte nucleus give a mass of 45×10^{-12} g dry mass of chromatin per nucleus.[14] For these fibers, a mean mass of 5.95×10^{-16} g per micrometer of fiber $\pm$ 29% was determined, which in this instance gives a total chromatin fiber length per nucleus of 7.6 cm.

Considering the fiber to be a 20-nm cylinder 7.6 cm long, a volume of 23.9×10^{-12} cm^3 can be calculated. This finally means that the chromatin

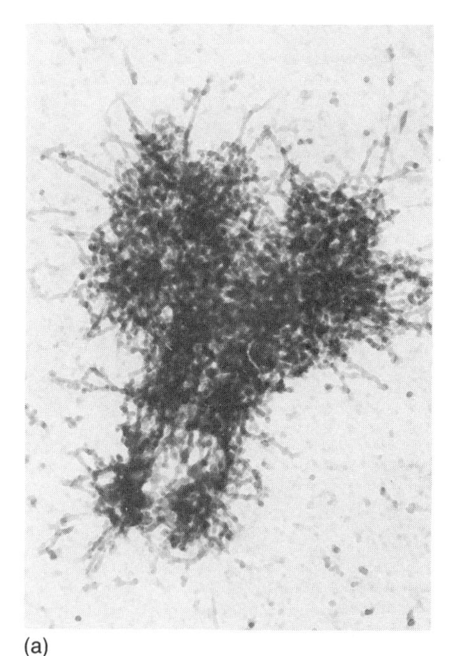
(a)

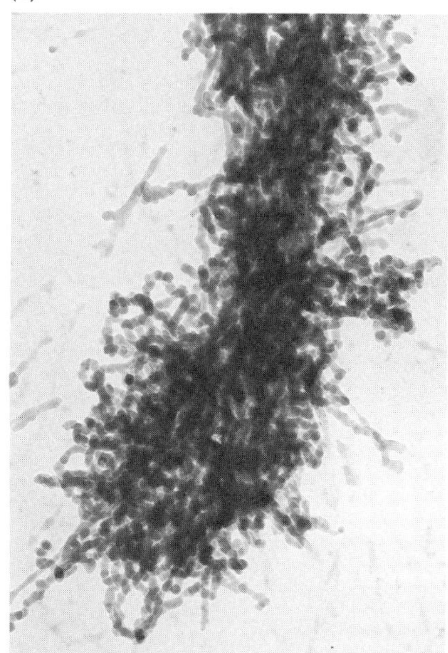
(b)

FIGURE 15.19 Metaphase chromosomes showing 20nm fibers. (*a*) A fully condensed chromosome containing two chromatids. Toward its periphery the fiber loops can be seen. ×20,600. (*b*) Higher magnification of region of chromosome, showing the fiber to be smooth in certain stretches and knobbed or irregular in other regions. ×34,300. (Courtesy of Dr. Gunter F. Bahr.)

fibers take up 20% of the volume of an average 120×10^{-12} cm^3 nucleus.

The exact numbers used above are not so crucial, since they would certainly vary with the investigator, the techniques, and the system studied. But the results do give us a helpful picture of the space available to the chromatin in the nucleus.

Fiber Configuration and Arrangement

The 20-nm fiber is not a smooth cylinder but is irregular, with bumps and knobs interrupting smooth runs. Stretching the chromosome reduces the fiber diameter in regions down to a lower limit near 3 nm, beyond which further stretching will shear the fiber.

A question that has intrigued investigators over the years is the relation of the 20-nm fiber to the DNA of chromatin. The question has two components. How is the DNA packed within the fiber, and does each chromatid possess one continuous fiber with a single DNA molecule?

A chromomere is a dense region on a chromatid where the chromatin fiber is tightly packed.

One approach to answering these questions has been to calculate the total length of DNA and the total length of chromatin fiber in a human diploid nucleus.[14] According to these calculations a DNA/chromatin fiber ratio of 220:7.6 cm or 29:1 is obtained. Therefore each length of fiber contains about 29 lengths of DNA helix.

It is obvious that the DNA helix, having a diameter of 2 nm, is not fully stretched in the chromatin fiber, but must be present as a coiled structure. Recent studies on the fiber indicate that this is clearly the case. We discuss this coiling a bit later in the chapter.

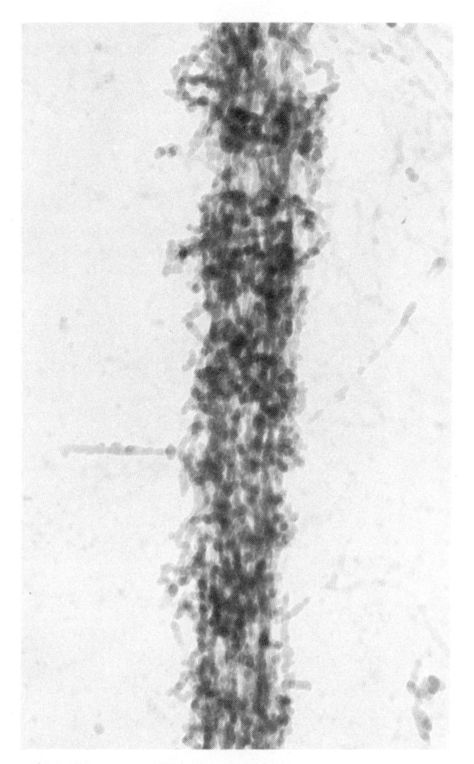

FIGURE 15.20 Electron micrograph demonstrating the presence of several longitudinal runs of the chromosome fiber present between the more highly condensed chromomeres. A chromomere appears to be a local accumulation of looped fibers. ×31,300. (Courtesy of Dr. Gunter F. Bahr.)

Many attempts have been made to describe the manner in which the chromatin fiber is folded within the chromatid. The models have attempted to account for chromomeres and for an average of 8 to 15 longitudinal runs in the interchromomeric region (see Figure 15.20).

One model, the folded-fiber model developed originally by Dupraw in 1966,[15] has received particular attention. Figure 15.21 is a revised schematic of this model. Here the principle of folding is that the fiber begins to form a chromomere by folding, then moves on to another chromomere, partially forming it, and continues by moving to new chromomeric regions and returning to old ones. By weaving back and forth along the length of the chromatid, the runs of interchromomeric regions are laid down.

The model assumes that a single chromatin fiber makes up an entire chromatid and is furthermore consistent with one point of view that the DNA in a chromatid is a single DNA helix. This point of view is referred to as the *unineme* (or *mononeme*) hypothesis.

Although widely accepted, the unineme hypothesis is not universally adopted among investigators in this field. Evidence from studies of replication, recombination, and sister chromatid and DNase kinetics can most easily be understood in terms of the unineme hypothesis, but research based on electron microscopy has suggested that the human mitotic chromatid is at the very least a *bineme*, or two-stranded, structure.[16]

Furthermore, the question of whether eucaryotic DNA is linear or circular has not been completely put to rest by unequivocal experimental results. If the DNA is circular, one must assume that either the chromatin fiber is also circular with a DNA/fiber ratio of one, or that two strands of helix must appear together in cross section of a linear chromatin fiber.

We must therefore await additional experimental results to be absolutely certain of fiber ultrastructure.

Chromatin Subunits

Until the early 1970s the chromatin fiber was viewed as discussed above—a somewhat irregular but nevertheless basically homogeneous 20-nm fiber in which DNA was coiled and associated with histones spread evenly along the DNA. Then two independent lines of investigation brought out pieces of information that have revolutionized our understanding of chromosome ultrastructure perhaps more than any other research result since their discovery.

One approach used isolated nuclei that were subjected to autodigestion by endogenous deoxyribonuclease.[17] The DNA isolated from this system was made up of fragments corresponding in length to about 200 base pairs or multiples of this. The immediate implication of this result was that the DNA helix is accessible to endonuclease attack at regular intervals and that any protecting histone protein is not, therefore, uniformly distributed along the strand.

Continued nuclease digestion shortens the piece down to about 140 base pairs, whereupon a nuclease-resistent DNA–histone core remains.

The second approach used was electron microscopy of chromatin prepared at low salt concentrations.[18] Chromatin fibers took on the appearance of a string of beads. The beads, or chromatin subunits, had a regular diameter of 10 nm and the strand connecting the subunits was a regular length with a diameter of 2 nm (see Figure 15.22). The term *nucleosome* was applied to the subunit in 1975 and it has been widely adopted.[19]

The picture beginning to take form was that the chromatin fiber consists of nucleosomes aggregated in some fashion to form a cylinder of the appropriate diameter.

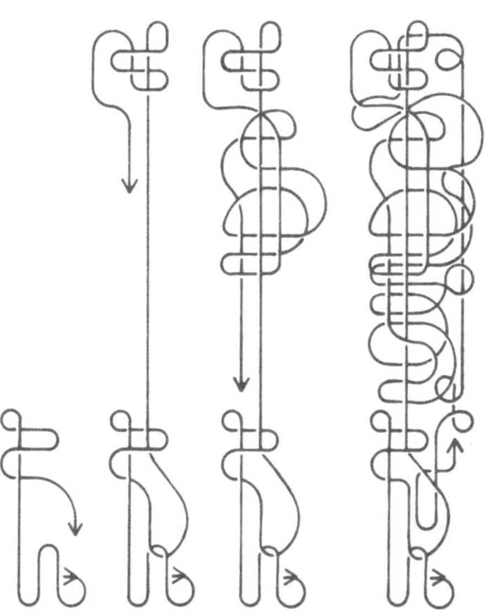

FIGURE 15.21 A model of fiber folding as it is envisioned to occur in a chromatid. Arrow indicates the transcriptional polarity of the fiber. (Courtesy of Dr. Gunter F. Bahr.)

The Core Particle

The nucleosome consists of a core particle plus a stretch of spacer DNA containing the H1 histone. The composition of the core particle as derived from a number of studies is now known to be 146 base pairs of DNA plus an octamer of histones. The eight histones are represented as four pairs each of H2A, H2B, H3, and H4. Histone H1 is not present in the core particle but is rather associated with the strand of spacer DNA that extends between the nucleosomes. The 200 base pair fragment that is released from chromatin during the early stages of endonuclease digestion does contain histone H1, but continued digestion removes the DNA to which it is bound.

In 1974 Roger Kornberg[20] proposed a model of the core particle (Figure 15.23) based on several lines of evidence, including the endonuclease and electron microscopic studies already discussed.

Other pieces of evidence are as follows. When chromatin is subjected to gradually increasing concentrations of salt, the histones are released in pairs, suggesting that pair formation may be an important property of the nucleosome structure. Furthermore, the pairing interactions are not random, but conform to a set of preferences.[21] The pairs H2A and H2B show strong interactions, as do H2B and H4; pairs H3 and H4 interact

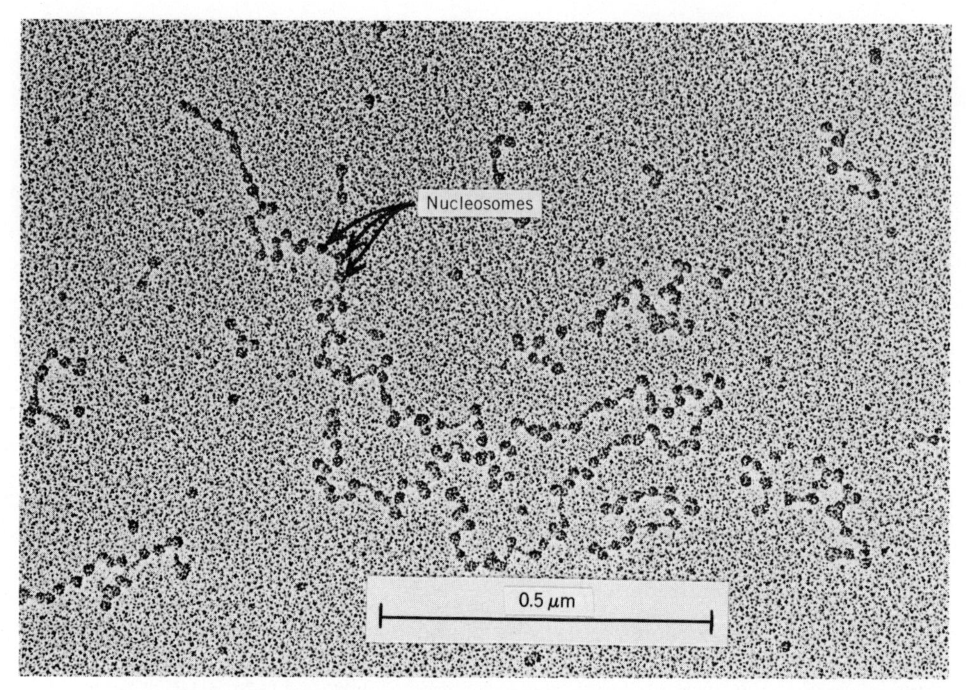

FIGURE 15.22 A darkfield electron micrograph of the edge of a chicken erythrocyte nucleus. The chromatin is well spread and appears as a string of beads. The beads, called nucleosomes, are held together by a strand of DNA. (Courtesy of Dr. Ada L. Olins and Dr. Donald E. Olins.)

to form a tetramer. The association between H2A and H3 is weak, and that between H2A and H4 and H2B and H3 even weaker. These interactions are illustrated in schematic form in Figure 15.24.

When chromatin is exposed to bifunctional protein cross-linkers, a stable octamer of histones results, again supporting the notion that the nucleosome core contains eight histones in aggregated form. Finally, quantitative measurements of histones present in chromatin give amounts of one H1 to two each of the other four histones.

The core particle.

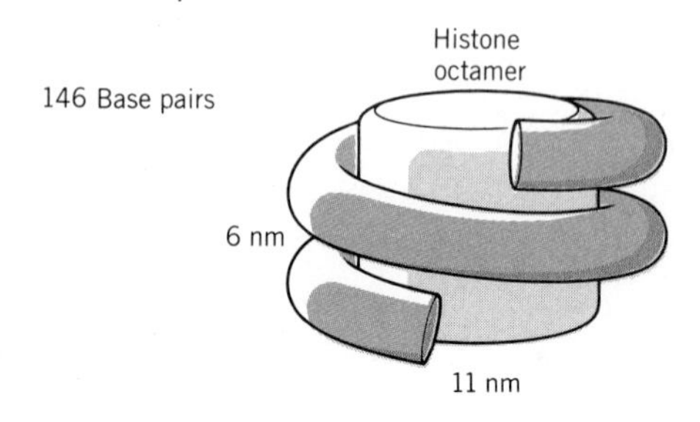

The conformation of the DNA with respect to the core particles has been extensively studied, but the exact relation between the DNA and the histones remains to be clarified. The core particle is roughly cylindrical, with a diameter of 11 nm and a height of 6 nm. The DNA is wrapped around the outside of this core, encircling the core two times.

When nucleosome cores are extensively digested with endonuclease, DNA fragments 10 base pairs long or multiples of 10 are released (see Figure 15.25). This is currently attributed to a fairly tight fit between DNA and the histone octamer. Ten base pairs are required for one complete turn of the DNA helix. Therefore cleavage occurs where the helix is properly oriented for enzymatic binding and hydrolysis. Not all workers agree on this interpretation, however. Some view the DNA as kinked at regular intervals, rendering the helix more susceptible to cleavage at specific points, which results in the production of fragments of uniform size.

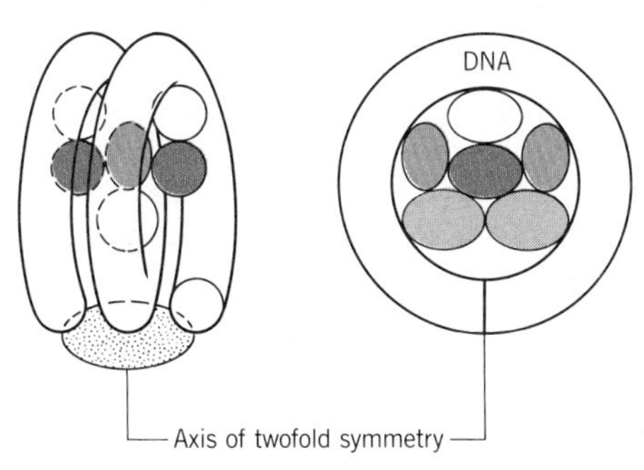

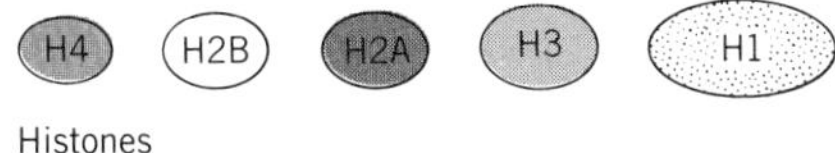

FIGURE 15.23 Schematic diagram of a nucleosome revised from that originally proposed by Kornberg. The DNA doubly encircles an octamer of histones consisting of two molecules each of 2A, 2B, 3, and 4. Histone 1 is not a member of the inner circle, and is technically not a part of the core particle.

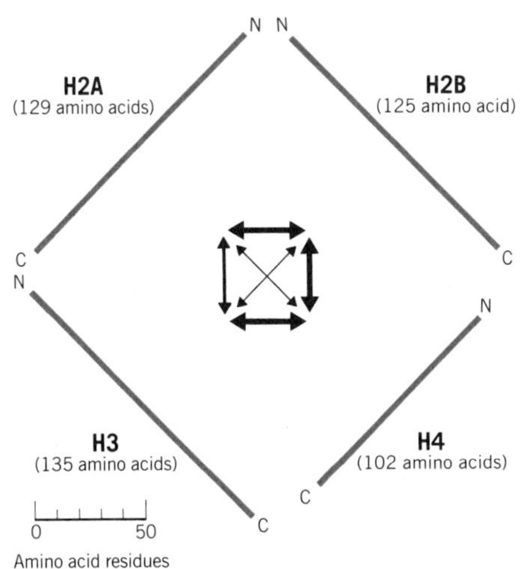

FIGURE 15.24 Schematic representation of interactions between the four histones of the nucleosome core particle, using calf thymus histones as an example. The polypeptide backbones are represented by thick colored lines drawn to the scale indicated. Their *N*- and *C*-terminal ends are shown. The relative thicknesses of the double-headed arrows in the center indicate the order of interaction strengths between histones.

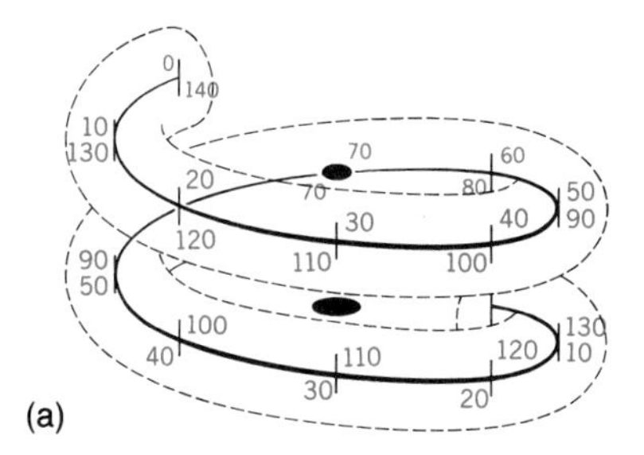

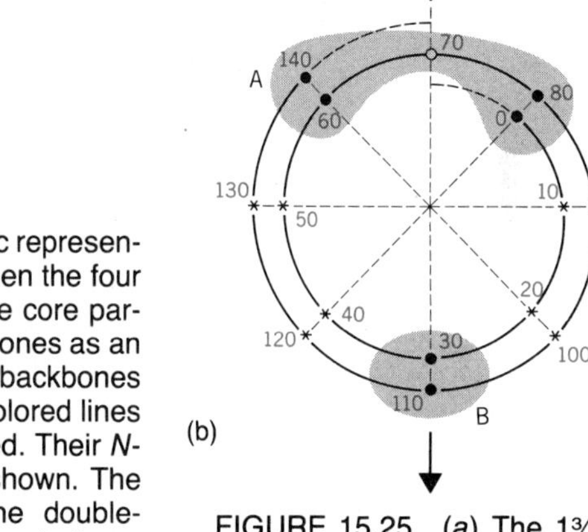

FIGURE 15.25 (*a*) The 1¾ turns of the DNA proposed for the 140 base pair nucleosome core, drawn roughly to scale. The top set of numbers gives distances in bases of the DNase I cutting sites from the 5′ end of one strand and the bottom set refers to the other strand, related to the first by the dyad axis shown. (*b*) the 1¾ turns of one strand of DNA show how the supercoiling brings sites 80 bases apart close together and groups the sites of low or medium cutting frequency by DNase I into two diametrically opposite areas A and B (shaded). The arrow indicates the dyad. Numbers are bases from the 5′ end. DNase I cutting; *, high frequency; ●, low frequency; ○, medium frequency.

A number of physical approaches to histone structure, including circular dichroism, laser Raman, and infrared spectroscopy, have shown that these proteins are about 50% α helical with little or no β-pleated sheet. The total length of α helix is sufficient to traverse the core particle from top to bottom once for every turn of the DNA duplex. With these facts in mind, the particle could be imagined to be constructed like a wooden barrel with α helices simulating staves and the DNA serving as hoops.

The Nucleosome

The next order of complexity of chromatin subunit is the nucleosome, which includes the core particle plus the "spacer" or "linker" DNA and the H1 histone. This is the unit that contains 200 base pairs of DNA.

Several studies have now suggested that H1 may play a most interesting role with respect to the spacer DNA. When nucleosomes are digested with endonuclease starting with the 200 base pair form, an intermediate of 160 base pairs containing H1 is commonly observed as a relatively

The nucleosome.

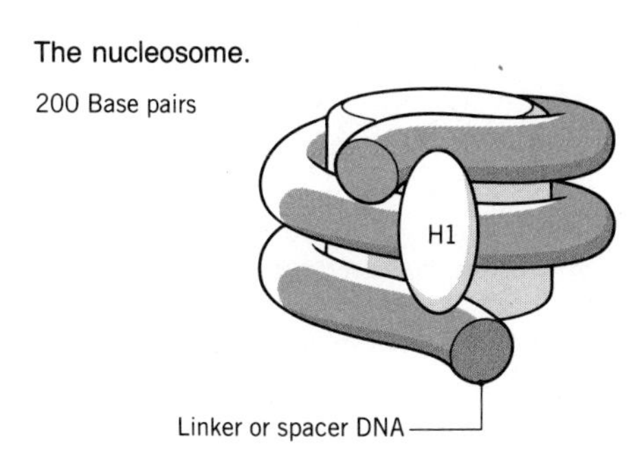

stable intermediate. The H1 histone appears to be bound near the octamer at a point where the entering and exiting strands of DNA are found, occupying about 20 base pairs of the two strands.

The picture we get from these observations is similar to that of a Western boa where the pendant (histone H1) keeps the two strands (incoming and outgoing DNA) snuggly around the neck (histone octamer) of an individual. Although this picture may be scientifically debasing, it is easier to envision than some afforded by other descriptions.

Even though the average length of DNA in the nucleosome is around 200 base pairs, there is considerable variation in the length of spacer region among different species, different tissues within a single species, and perhaps even within the same cell. Spacer lengths vary from zero base pairs in lower eucaryotes up to about 80 base pairs in sea urchin sperm, with 40 representing a commonly observed length in other systems.

The significance of irregular spacing among or within species is not at all certain. It has been proposed that the size and configuration of the H1 histone may have an impact on spacer length or that the phenomenon is in some way related to gene activity.

The H1 histone is the least stringently conserved histone protein. It contains 210 to 220 amino acids and may be represented by a variety of forms even within a single tissue. This variation probably reflects the fact that a closely related family of H1s may meet the criteria for binding to DNA strands outside the core particle, whereas the physical constraints for functioning within the particle are much more severe.

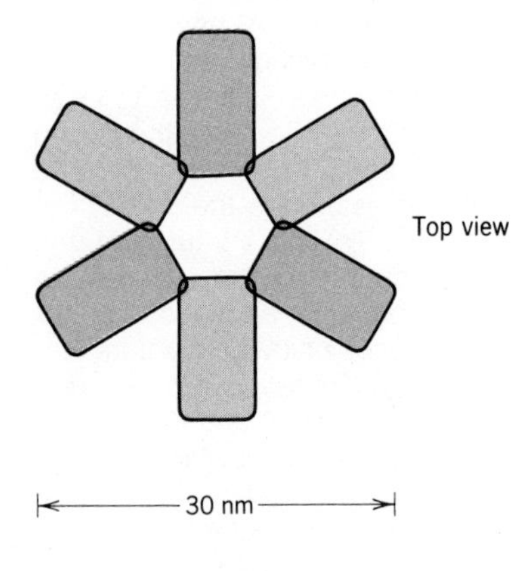

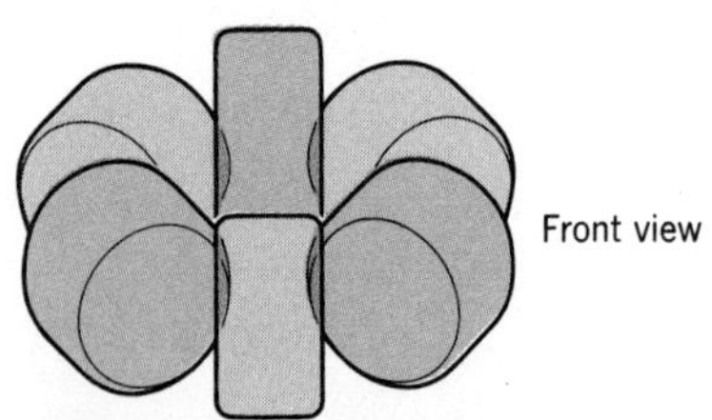

FIGURE 15.26 A model depicting one way in which nucleosomes may be arranged in the 30-nm chromatin solenoid. Six nucleosomes are shown arranged radially. Each series may form a complete ring or a segment of the helix in the solenoid.

Nucleosome Packing

It is clear, given the diameter of the nucleosome, that a 20 to 30 nm chromatin fiber is not made up of a linear array of nucleosomes packed end to end. Some degree of coiling must account for its width.

It is now generally accepted that the chromatin fiber has a solenoid-type ultratructure, with the run of the coil made up of linear nucleosome units. Adjacent nucleosomes, if thought of as cylinders, could be envisioned packed in the solenoid in several different ways. The 10-nm fiber is thought to consist of a series of nucleosomes packed edge to edge with their flat faces nearly parallel to the axis of the fiber. However, the spatial orientation between adjacent nucleosomes in the 30-nm solenoid may be quite different. One view of their orientation is depicted in Figure 15.26. Viewed along the axis of the solenoid (top view) the nucleosomes are in a radial arrangement with their edges forming the outside surface and the inner core of the solenoid. Adjacent nucleosome faces lie at 60° angles to each other.

Viewed from the side, the fiber consists of six or seven nucleosomes per turn. Each turn may form a complete ring, beginning and ending on the same plane, or the ring may be a spiral with each nucleosome raised an increment along the axis of the solenoid. It is not known with certainty whether H1 is inside or outside the solenoid at this level of organization.

This type of model has the advantage of considerable flexibility. A larger diameter chromatin fiber can be explained in terms of more nucleosomes per turn or a longer spacer DNA or some combination of both.

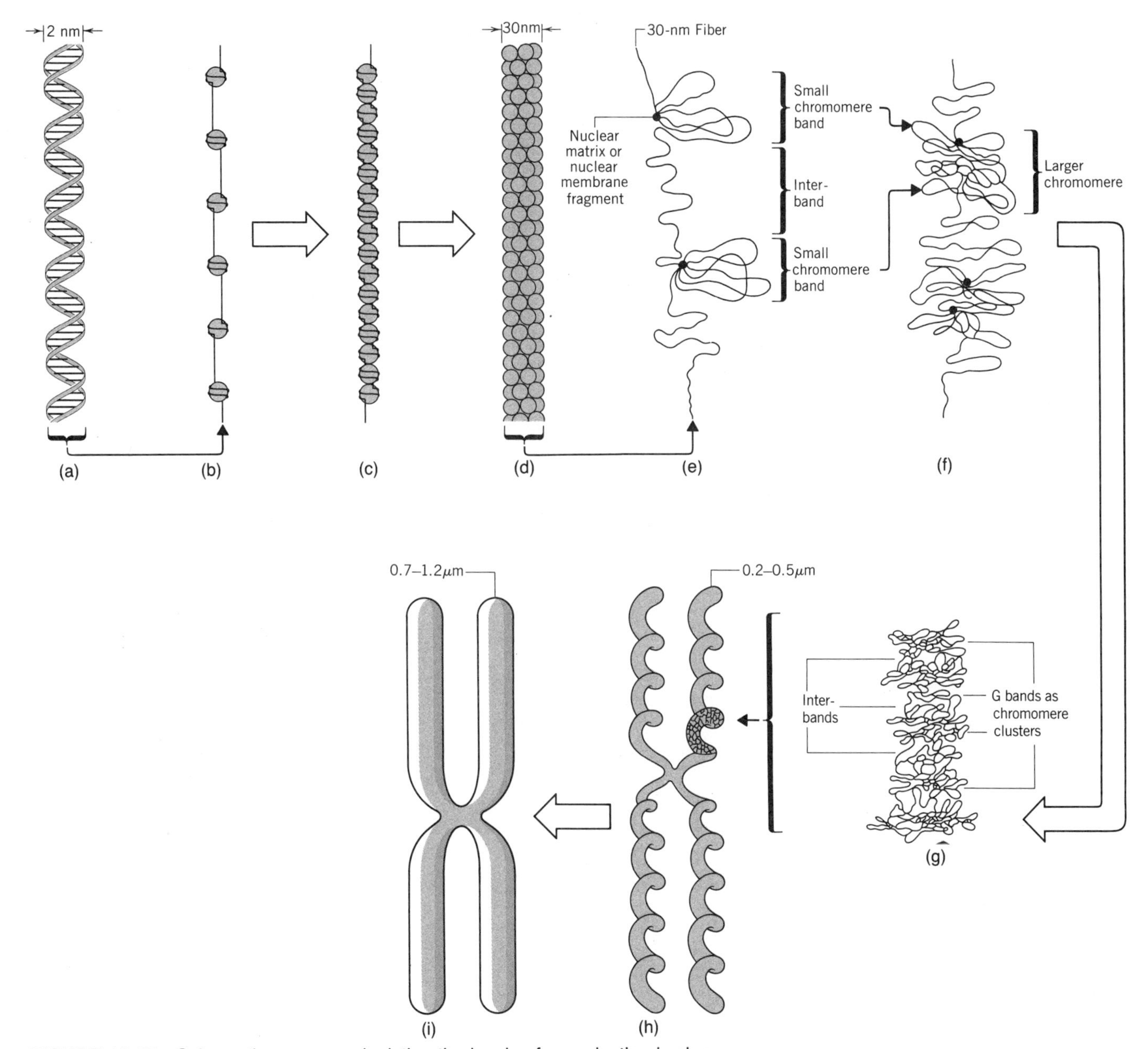

FIGURE 15.27 Schematic summary depicting the levels of organization in chromosomes. (*a*) DNA. (*b*) Extended nucleosomes. (*c*) 10-nm fiber. (*d*) Condensed nucleosomes, 30-nm fiber. (*e*) Chromomeres and interband chromatin. (*f*) Clustering of chromomeres. (*g*) Chromosome bands. (*h*) Spiralized chromosome. (*i*) Compact unbanded chromosome.

Higher Order Packaging in Mitotic Chromosomes: The Supersolenoid

Before moving on, let us reflect briefly on where we are. Refer to Figure 15.27 to assist in picturing the levels of organization in chromosomes.

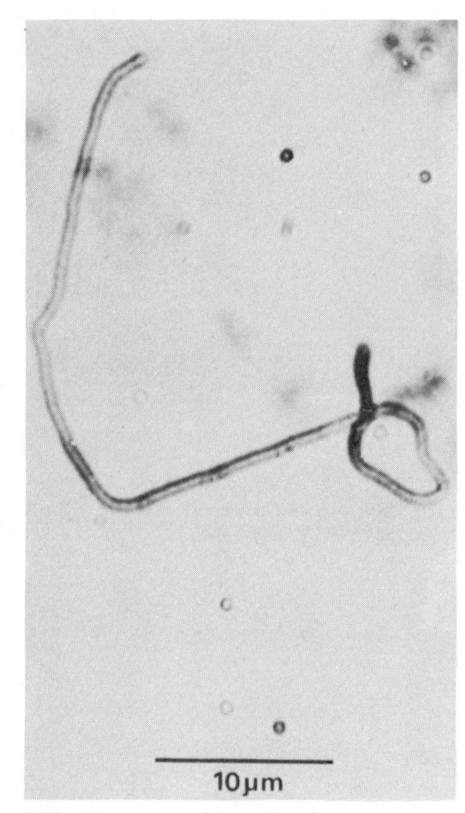

FIGURE 15.28 Extended unit fiber from mitotic human fetal fibroblast. Photographed at a point above the focal plane to show the doubleness suggesting a tubular structure. (Courtesy of Drs A. Leth Bak and J. Zeuthen.)

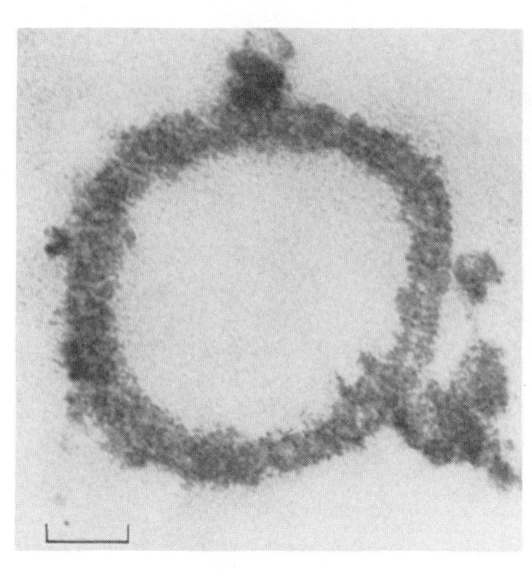

FIGURE 15.29 Electron micrograph of a 60 to 80-nm cross section of unit fiber. A tubular structure with a diameter (400-nm) similar to that seen in light microscopy is observed. The results suggest a helical structure formed by coiling a filament of about 30 nm diameter that corresponds to the solenoid nucleofilament. Bar = 0.1 μm. (Courtesy of Drs. A. Leth Bak and J. Zeuthen.)

The genetic material in chromosomes viewed at highest power of resolution is a strand of duplex DNA. The DNA strand is wound around histone octamers with nearly two turns of DNA required per octamer. These nucleohistone octamers, or nucleosome cores, are spaced on the DNA strand so that microscopically they have the appearance of beads on a string. In the presence of histone H1, the spacer histone, the beads may be somewhat drawn together, providing a smoother appearance. This is the 10-nm fiber.

One type of helical coiling model is based on studies carried out on chromosomes prepared from human fetal fibroblasts.[22] Mitotic chromosomes, when incubated for a short period of time in a special chromosome-isolation buffer, repeatedly take on the form of an extended tubular unit fiber of 0.40 μm. Figure 15.28 shows these unit fibers as they appear with light microscopy, and Figure 15.29 illustrates the cross-sectional appearance of the fibers in the electron microscope. A model of this unit fiber, termed the *supersolenoid*, is presented in Figure 15.30.

Dissecting the supersolenoid, we see it to be a structure containing a hierarchy of helices. Starting with the 10-nm string of nucleosome fiber, it is coiled to form a 30-nm solenoid. The 30-nm solenoid is supercoiled in a very shallow helix of 30 nm pitch with a mean diameter of 400 nm. Each turn of the supersolenoid helix contains 150,000 base pairs of DNA, which gives a packing ratio (length ÷ 30 nm pitch) for DNA of about 1400. This agrees well with direct estimates that have been made for the required packing ratio in metaphase chromosomes.

In this model, there are therefore three hierarchies of helices. One helix is in the nucleosome, a second in the solenoid, and a third in the supersolenoid.

Finally, the diameter of wet, unfixed human chromatids is estimated at about 1.4 μm. Thus, to account for this final dimension, the supersolenoid must be further contracted by a factor of about 5. This additional contraction may be either helical or by a folding mechanism.

Chromosome Scaffolding

Using chromosomes from which histones were removed by competing with the polyanions dextran sulfate and heparin, Laemmli and co-workers have uncovered a proteinaceous scaffolding structure in the core of the chromosome.[23] A fundamental and reproducible observation is that DNA strands in histone-depleted chromosomes cluster around and are anchored to a central protein scaffolding that still retains the skeletal structure of the two chromatids. This is illustrated dramatically in the electron micrograph of Figure 15.31.

A careful analysis of favorable micrographs has revealed that the DNA strand runs as a continuous loop from a point of exiting the scaffold to an adjacent point of entry (Figure 15.32). The majority of loops are about 15 to 30 μm long, representing 45,000 to 90,000 base pairs.

The same workers have successfully isolated intact scaffolds depleted of DNA. A remarkable finding is that the scaffold structure is of the same

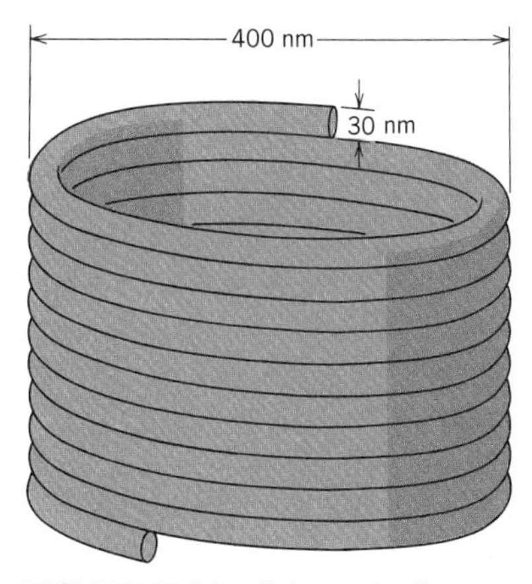

FIGURE 15.30 Schematic diagram of the supersolenoid of the unit fiber. The wall thickness of 30 nm as determined from electron microscopy, would correspond to the diameter of a solenoid made up of about six nucleosomes per turn.

FIGURE 15.31 Electron micrograph of a histone-depleted chromosome from HeLa cells. The central protein scaffold demarcates the skeletal structure of the two chromatids. A complex maze of DNA emanates from the region of the scaffold. (Courtesy of Dr. U. K. Laemmli.)

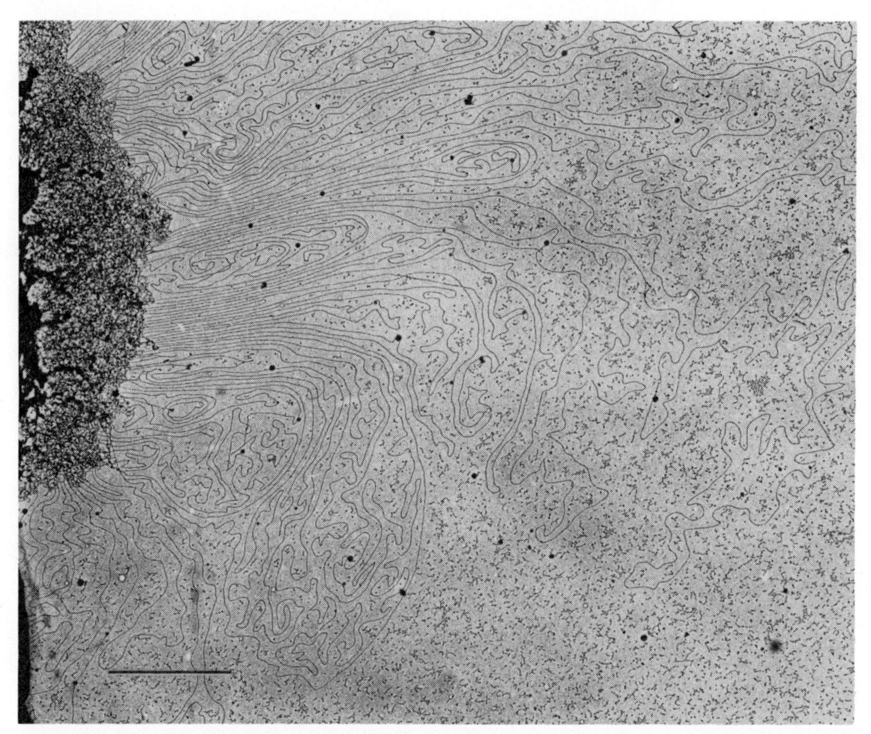

FIGURE 15.32 A histone-depleted metaphase chromosome spread with cytochrome *c*, stained with uranyl acetate, and rotary shadowed with platinum. This remarkable micrograph shows that DNA loops out from the scaffold, exiting and entering the scaffold at about the same place. Bar, 1 μm. (Courtesy of Dr. U. K. Laemmli.)

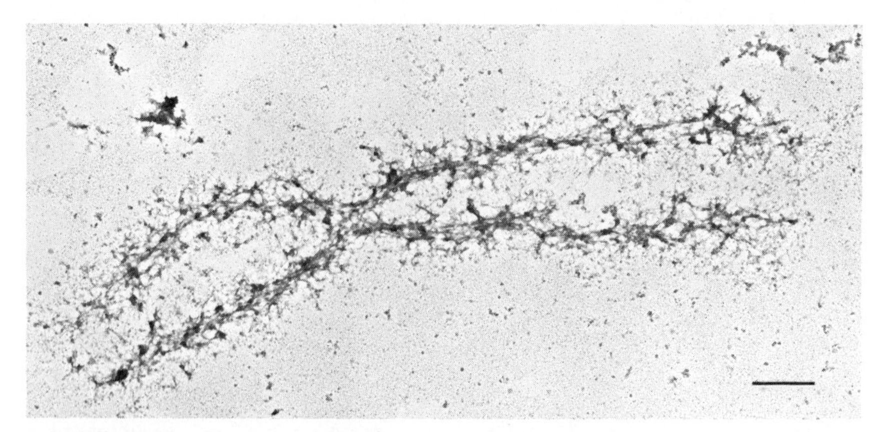

FIGURE 15.33 Electron micrograph of a DNA-depleted chromosome scaffold. Chromosomes are digested with micrococcal nuclease, diluted into 2 *M* NaCl to dissociate the histones and spread with cytochrome *c* for microscopy. Bar, 1 μm. (Courtesy of Dr. U. K. Laemmli.)

length as the original intact chromosome and has very much the same morphology as the chromosome from which it is isolated (Figure 15.33). The material of the scaffold is nonhistone protein. It is resistent to RNase but is disrupted by proteases, urea, and the detergent sodium dodecyl sulfate.

A scaffold model of the chrmosome has been proposed to illustrate these experimental results. In this model, (Figure 15.34), the scaffold is comprised of two nonhistone protein backbones, which are joined at the centromere. The model proposes that DNA is organized by the scaffold proteins alone, and not by proteins that cross-link DNA strands either within a loop or between loops.

The scaffold does not appear to be made up of a single type of protein. Electrophoresis of "purified" scaffolds in SDS–polyacrylamide gels produces about 30 protein bands, but it is not yet clear that these bands arise from the structural scaffolding proteins alone. There may be some degree of contamination.

Within the cell, the DNA loop is thought to exist in the solenoid and supersolenoid conformations discussed earlier. Assuming a cylindrical scaffold of 0.4 μm diameter and a length of 10 μm, calculations have been made illustrating that 8000 fiber loops of 25 nm diameter could have room to anchor along the scaffold axis if closely packed. This is sufficient to account for the amount of DNA in a chromatid of this size.

Other workers, using different extraction techniques, have arrived at protein scaffolds that resemble those obtained by Laemmli. Wray and his colleagues, for example, have derived scaffolds from Chinese hamster chromosomes that are very similar in appearance to those obtained from human HeLa cells.[24]

With the finding of nonhistone scaffolding systems in the metaphase chromosome, there naturally follows the question of their location and role during the rest of the cell cycle. The answer is not yet in final form, but researchers are tempted to speculate that these proteins may have contractile properties and may therefore operate to condense chromosomes during prophase, thereby keeping some order and organization to the chromosome throughout the cell cycle.

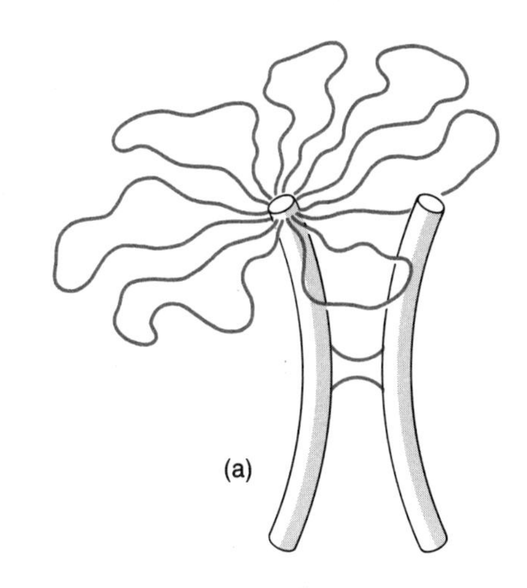

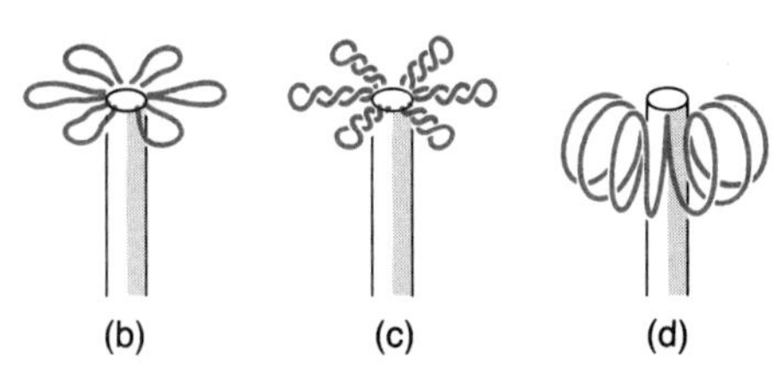

FIGURE 15.34 The scaffold model of chromosome structure. Nonhistone proteins joined at the centromere form a stable scaffold from which DNA loops out (*a*). The DNA in the loops is condensed by interaction with histones and may take on a variety of conformations (*b–d*). (Courtesy of Dr. U. K. Laemmli.)

Giant Chromosomes

Nature has provided investigators with chromosomes of unusual structure that have been especially helpful in answering certain questions about structure and function. Of special value in this regard are the lampbrush chromosomes of oocytes and the polytene chromosomes of the salivary glands of many dipteran species.

Lampbrush Chromosomes: Partially Extended–Partially Condensed

Lampbrush chromosomes, so named because of their similarity in appearance to the nineteenth century lampbrush, were first observed in salamander oocytes in the late 1800s. This type of test tube brush chromosome structure is now recognized to exist commonly in developing oocytes in the ovaries of most animals.

These chromosomes are bivalents that persist through a prolonged diplotene phase of first meiotic prophase. It is during this stage of oocyte development that mRNA is synthesized for use during the early postfertilization stages of development. The lampbrush appearance is illustrated in the micrograph of Figure 15.35.

The term "bivalent" refers to a condition of two homologous chromosomes (or chromatids) found together. In the diplotene stage in meiosis, chromatids appear to exchange regions along their length, giving rise to cross-shaped figures.

Ultrastructure. The ultrastructure of the lampbrush chromosome is explained schematically in Figure 15.36. A lampbrush chromosome consists

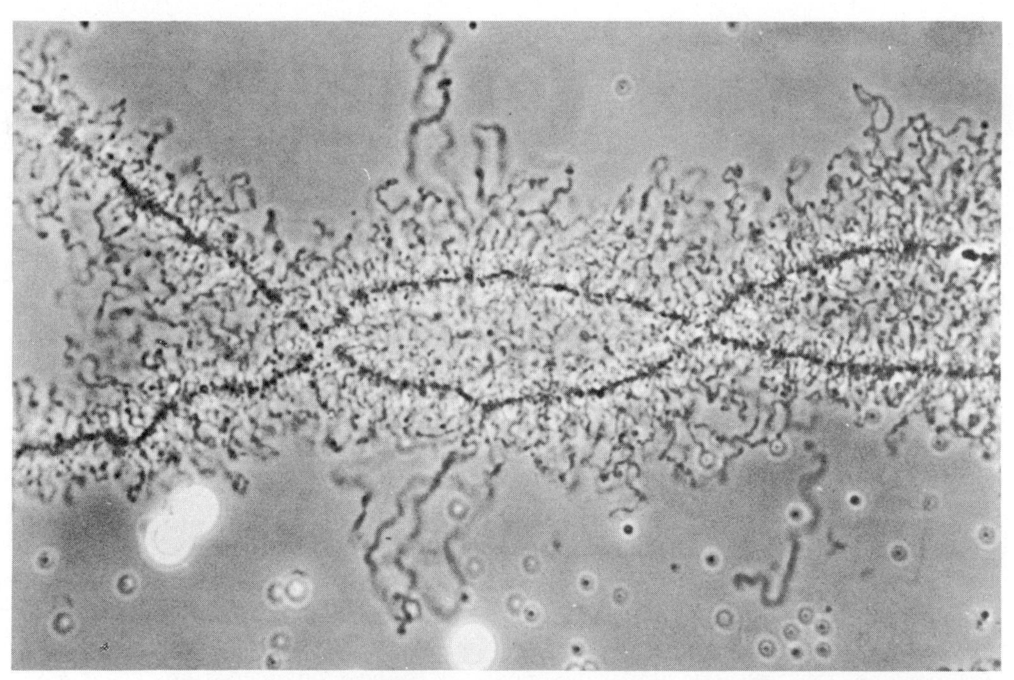

FIGURE 15.35 Phase contrast micrograph of lampbursh chromosomes in the oocyte of the newt, *Triturus viridescens*. Homologous chromosomes of the bivalent are joined at the two chiasmata. At this stage, each chromosome (each half of the bivalent) consists of two chromatids, not apparent in this micrograph. Lateral loops make up the bristles of the brush. (Courtesy of Dr. J. G. Gall.)

of alternating regions of high density granules and low density loops. The high density regions are called *chromomeres*. They are about 0.25 to 2.0 μm in diameter and are spaced about 2 μm from center to center along the chromatid axis.

Two loops, or a multiple of 2, extend from each chromomere laterally about 5 to 50 μm from the main axis. There is a pattern of loop appearance and disappearance during oocyte development, with a peak of loop occurrence at first meiotic prophase. Then, as the oocyte approaches maturity, the loops disappear and the chromosomes revert to a contracted or condensed morphology that is without loops. Chromomeres are thus regions that remain condensed, while the loops are an extended form of the chromatid.

When lampbrush chromosomes are pulled, the chromomere separates transversely at a point and a double bridge is formed. Each component of the bridge is actually one loop that has been stretched to bridge the separated chromomeric regions. Each chromosome is therefore viewed as consisting of two adjacent and identical chromatids. The chromomere is a condensed (coiled or supercoiled) region from which identical loops extend.

Transcriptional Activity. Three questions can be posed regarding the lampbrush chromosome. What is the significance of this strange mor-

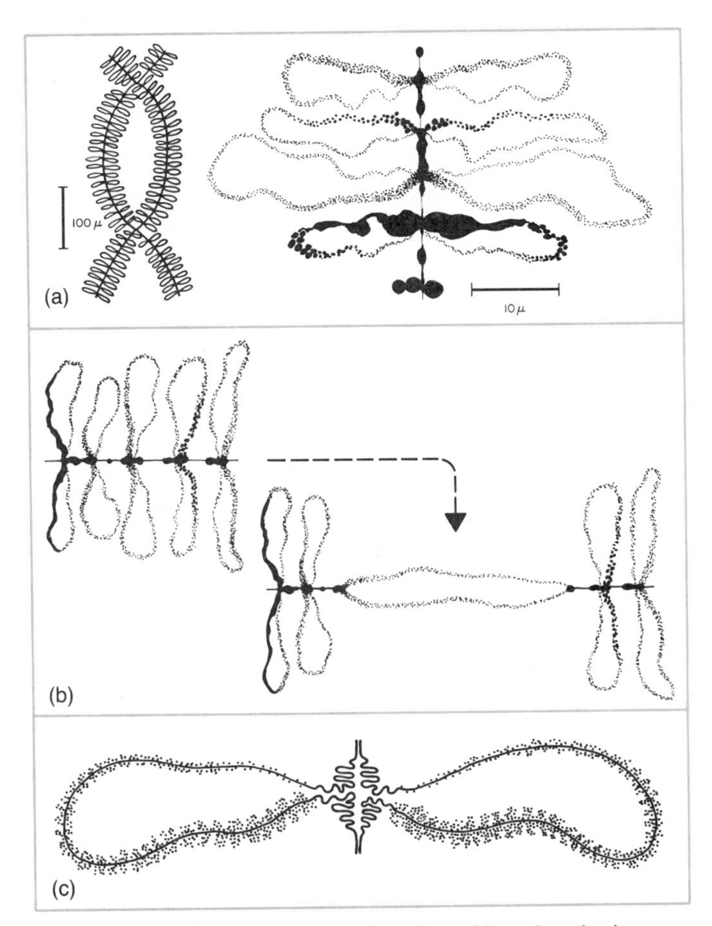

FIGURE 15.36 Schematic explanation of lampbrush chromosome structure. (*a*) Lampbrush chromosomes (*left*) when seen in higher magnification (*right*) have morphologically identical loops that emanate from a central axis. The central axis is made up of two chromatids. When stretched (*b*), the loops and central axis are seen to be continuous. A single DNA molecule is continuous along the axis and through a given loop (*c*). Along the DNA molecule a gradient of nascent RNA is found.

phology, what function is it performing, and how does this relate to chromosome ultrastructure as we have discussed it?

First, let us consider the events that appear to be taking place on the loop. A careful micrographic analysis of the loop reveals that its width steadily changes from thin to thick, beginning at a point near a chromomere and moving around the loop to an adjacent point on the other side of the chromomere (see Figure 15.36). An electron micrograph of a section along a loop is shown in Figure 15.37. The core thread of the loop is the chromosome fiber. The fine strands of increasing length are RNA molecules being formed from the complementary DNA template. Although this feature is not evident in every case, each strand is anchored to the chromosome fiber by an RNA polymerase, which reads the template.

These observations tell us something about the relation of morphology

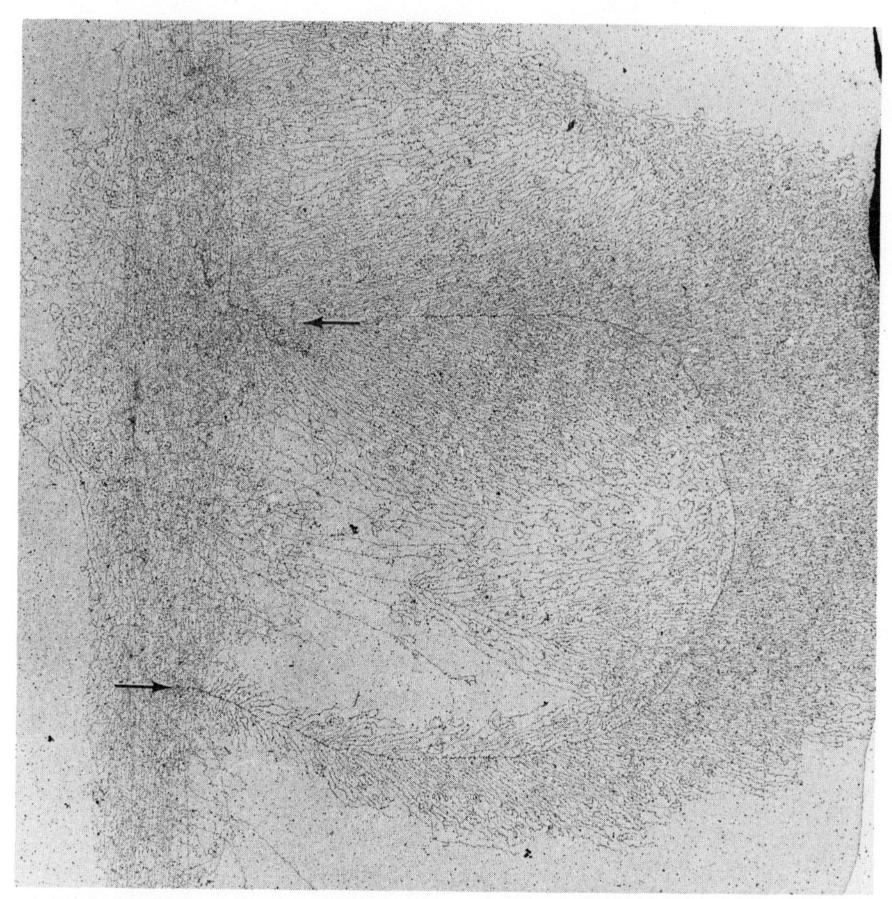

FIGURE 15.37 Electron micrograph of a chromomere showing one of its extended loops. This loop, inserted on both sides of the chromomere (arrows) is made up of closely spaced granules on a DNA strand from which lateral fibrils branch out increasing in length along the loop axis. (Courtesy of Dr. Ulrich Scheer.)

to function, namely, that transcription requires extended and not condensed chromosome fibers.

The loop itself may be viewed in two different ways. It may be static, unchanging in length and constantly exposing the same stretch of chromosome fiber. Or, it may be dynamic, with new loop material spinning out of one side of a chromomere and returning to a condensed state on the other side. According to this view, the *spinning out and retraction hypothesis,*[25] a chromomere would be completely transcribed from end to end by spinning out a transient loop. At the moment, researchers are apparently not certain which of these views is correct.

The precise structure and function of the chromomere is not clear. It certainly is an aggregate of deoxyribonucleoprotein, but its fine structure is not known. The amount of DNA housed by one chromomere is large—as much as is contained by the entire *E. coli* genome. It is not known, however, whether the DNA is structural or regulatory (having some genetic function) or both, or whether in fact it is transcribed.

In light of the model for chromatin that we discussed earlier, the chromomere would appear to be a region of the chromosome containing solenoid supercoiling. The loop, on the other hand, is probably a linear strand of nucleosomes.

A question that quite naturally arises regards the relation between the linear length of the loop and the length of RNA product. Is the loop one transcriptional unit?

An obvious approach to answering this question is to measure the length of the primary transcript product and relate its length to loop length. But the problem is not this simply solved. It is known from a variety of studies that functional mRNA that is translated for ribosomal proteins is derived from a high molecular weight precursor by trimming and splicing mechanisms (see Chapter 16). It is likely, although not certain, that similar processing mechanisms may operate on all precursors to functional mRNA of all types, both ribosomal and nonribosomal RNA.

Most precursor molecules are not a homogeneous lot of RNAs but rather arise as a high molecular weight collection of molecules referred to as heterogeneous nuclear RNA (hnRNA). It is not clear whether hnRNA is the primary transcript product or the first stable form of transcript product that can be isolated and studied. It may therefore be already processed to some degree, representing something less in length than the primary product.

Chromosomes swell and disperse in low salt and slightly alkaline conditions. In this state, they can be centrifuged onto electron microscopic grids and appear finally in a "spread" condition. Using this approach, it has been concluded that *one or more* transcriptional units may be contained by a single loop.[26]

Figure 15.37 shows a spread loop region including a chromomere. The chromomere consists of a parallel array of deoxyribonucleoprotein (DNP) axes free of lateral ribonucleoprotein (RNP) fibrils. The DNP axis of the loop, however, is laced with RNP fibrils increasing in lengths from the thin to the thick insertion sites. The loop axis appears to be transcribed entirely and continuously from end to end; the length of the transcriptional unit in this case is 15 μm.

In other instances a given loop will contain more than one transcriptional unit. The morphology of the loop that forms the basis of this conclusion is when more than one thin-to-thick fibril gradient is seen along its axis.

Three different gradient arrangements have been observed. One type is a thin-to-thick gradient tandemly repeated with the same polarity along the loop axis. A second type is noted when the polarity of repeated transcriptional units is reversed, so that different regions of the loop have opposite polarities. A third type is seen when the polarity changes more than once within a loop. The electron micrographs of Figures 15.38, 15.39, and 15.40 show tandemly arranged transcriptional units with the same, opposite, and alternating polarities, respectively. As many as five transcriptional units have been seen on a single loop. They may be each of similar length, or the length may vary.

Fibril gradients have been likened to Christmas trees (the old-fashioned kind) in appearance. Different types of gradients may be seen.

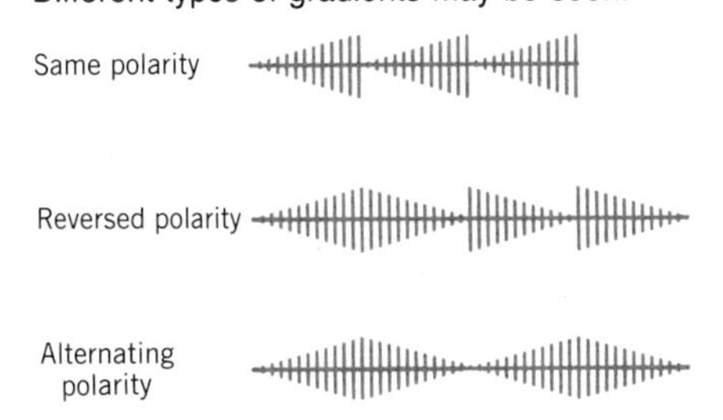

These results have some implications for transcription. For one thing there may be more than one set of initiation and termination points for

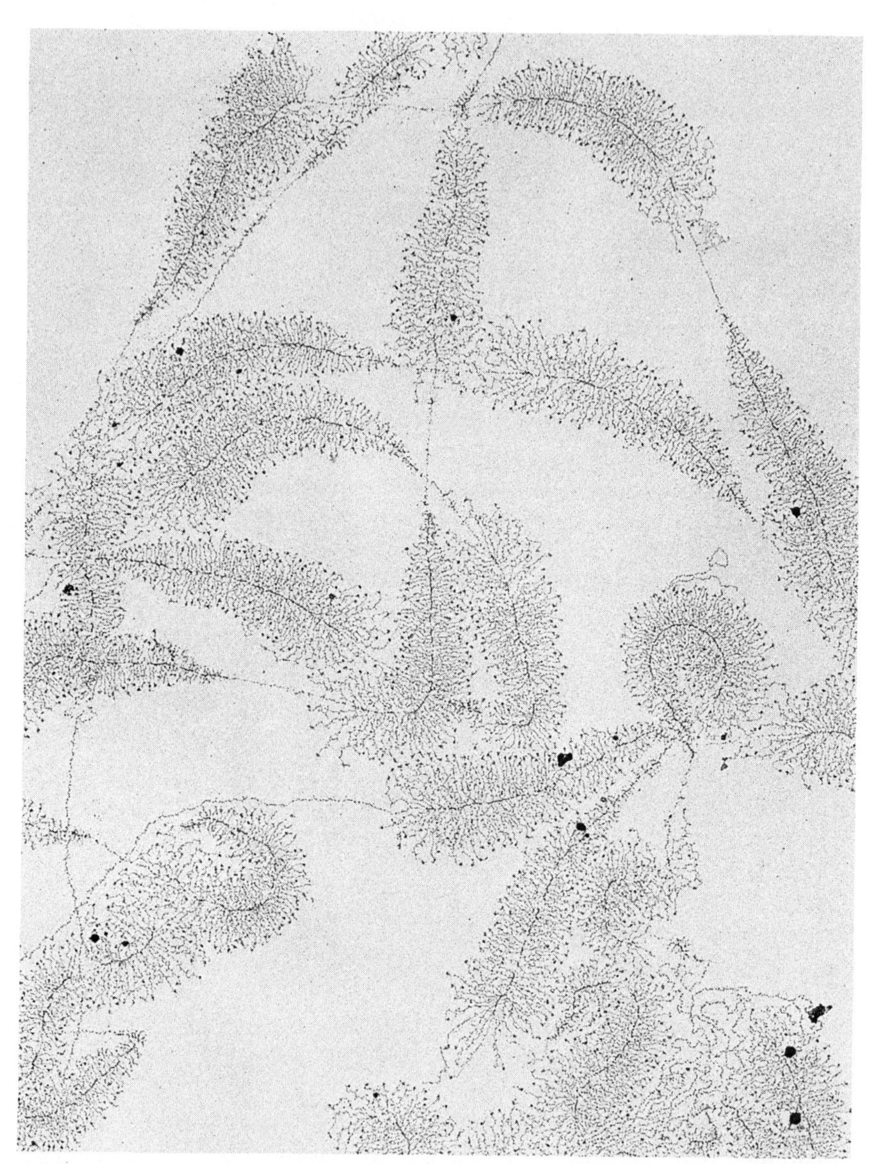

FIGURE 15.38 Electron micrograph showing tandemly arranged transcriptional units with the same polarity. By following the main strand it can be seen that all the Christmas trees face in the same direction. In this case, we are viewing the transcription of rRNA in the nucleolus of the spotted newt *Triturus*. ×17,700 (Courtesy of Dr. Oscar L. Miller, Jr.)

If the DNA strand is always read 5′ → 3′ during transcription, one strand is selected and read to produce same polarity gradients and both strands are read in different regions of the duplex to produce reversed polarity gradients.

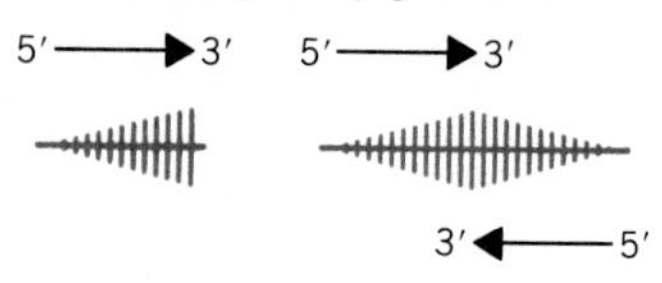

RNA polymerase in a single loop. A loop in this state is therefore polygenic, containing either repeated genes or different genes.

A more puzzling implication emerges from the observation of transcriptional units having opposite or alternating polarities within a loop. The implication is that the DNA axis is transcribed in opposite directions. One must assume, in light of what is known about the directionality of transcription, that the individual DNA strand (half a duplex) is not being read in opposite directions. This being the likely case, it means that *different* DNA strands are being transcribed, both in a 5′→3′ direction but

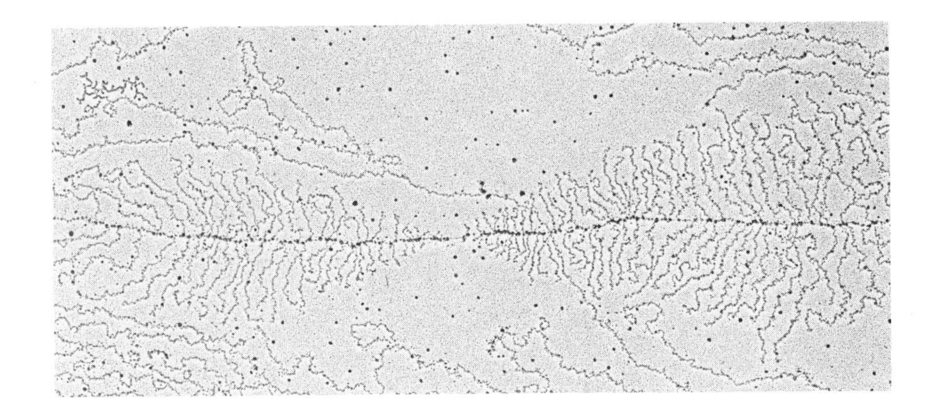

FIGURE 15.39 Two transcriptional units of opposite polarity. All units extending to the left may have the same polarity as the unit on the left and those to the right may maintain the opposite polarity. Thus the change in polarity may be only at one point in the chromosome. These are units from a lampbrush chromosome of an oocyte of *Pleurodeles waltlii*. ×18,500. (Courtesy of Dr. Ulrich Scheer.)

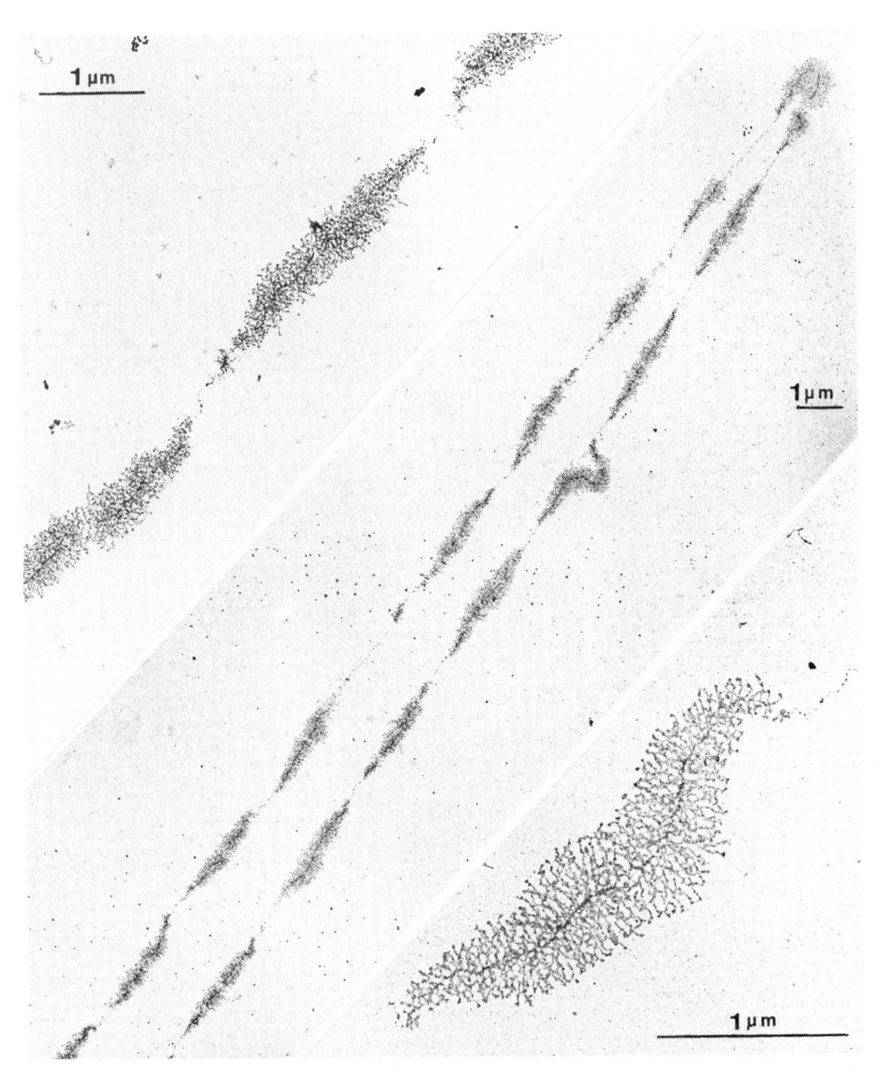

FIGURE 15.40 Alternating transcriptional units of rRNA genes in *Acetabularia exigua*. This is an example of strict alternation; all units alternate in polarity from their respective adjacent units. The units are positioned in such a manner that spacer regions are present between Christmas tree tips and not between bases. (Courtesy of Dr. Sigrid Berger.)

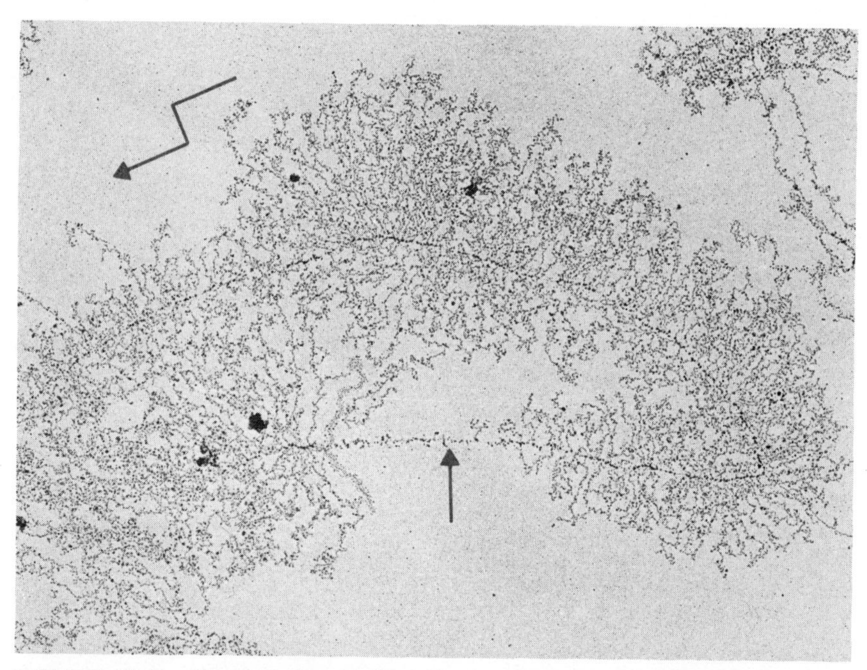

FIGURE 15.41 Transcriptional units with a morphology suggesting trimming of the nascent ribonucleoprotein fibrils. An abrupt reduction in fibril length is shown by the step arrow. This is quite different from the morphology of the normal spacer region, indicated by the straight arrow. These are lampbrush chromosomes from oocytes of *Xenopus laevis*. (Courtesy of Dr. Ulrich Scheer.)

with an antiparallel result in forming the RNA fibril gradient. The molecular basis for strand selection, hence directionality in transcription, is yet to be understood.

A study of electron micrographs that have been published on spread chromatin showing "Christmas tree" patterns leads one to the ready conclusion that although the lengths of lateral RNP fibrils increase gradually, they do not increase linearly as a function of the distance between the initation site and the position of the transcriptional complex on the DNA axis. The primary transcription complex is shorter than the DNA transcribed.

This is indicative of condensation and packing of the RNA as it is being transcribed into ribonucleoprotein particles (discussed further in Chapter 16). This feature of transcription appears to be a general one and negates efforts to measure the length of the transcriptional unit by simply measuring the length of the transcription product.

Another transcription activity further obviates a simple relationship between these lengths. In some cases the RNP fibrils are sheared to a uniform length once they have reached a designated size. This effect is seen in Figure 15.41. The significance of this is not known for certain, but this cleavage appears to be an early step in the trimming to which RNA is subjected before it is brought to its ultimate functional size.

During midoogenesis when RNA production is high, the lateral loops of lampbrush chromosomes are fully extended and densely packed with transcriptional complexes. When oocytes begin to mature, the loops retract and the density of transcriptional complexes lessens. Thus the morphology of the lampbrush chromosome with its lateral loops permits a visualization of the molecular process of transcription.

The Fate of the Nucleosome during Transcription Can DNA be transcribed when wound around core nucleosomes?

This difficult question, which is related to the morphology and function of the loop, touches on the fate of the nucleosome during transcription. Most of the experimental evidence, however, shows that histones are not released from chromatin during transcription. But the nucleosome is apparently not retained either in the form we see it during metaphase. It is believed, therefore, that nucleosomes change their conformation during transcription to permit DNA to unwind. Alternatively, if the DNA does not unwind completely, the change in nucleosome conformation may open up the DNA to better access for RNA polymerase.

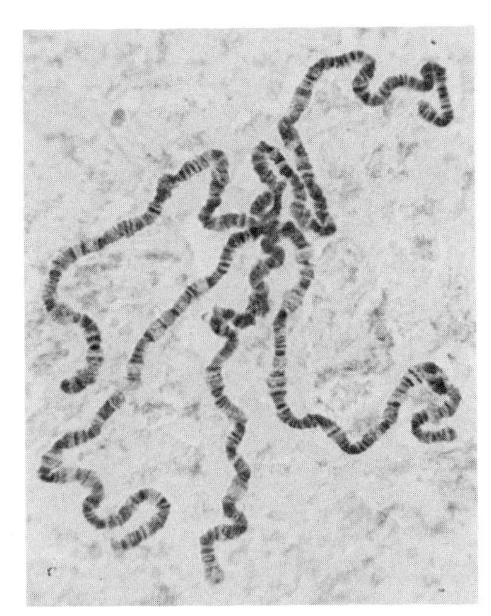

FIGURE 15.42 Giant polytene chromosomes from the salivary gland of *Drosophila melanogaster*. The arms of the chromosomes are held together in a region called the chromocenter. A large number of bands and several puffs can be seen in these chromosomes. (Courtesy of Dr. B. P. Kaufmann).

Polytene Chromosomes: Special Interphase Chromosomes

Polytene chromosomes are found in certain cells of dipteran larvae. The most commonly studied source is salivary gland tissue of *Drosophila* and *Chironomus*.

These giant chromosomes are generated during early larval development as a result of cessation of mitotic divisions in these tissues without a concomitant halt of DNA replication. DNA continues to replicate as many as 10 times, yielding 2^{10} copies of DNA fibers, which remain together and exactly aligned.

The result is a giant chromosome that is easily visualized with light microscopy (see Figure 15.42). These chromosomes have been extremely valuable in cytogenetic studies because they portray readily observed properties that for the most part are beyond microscopic resolution in other chromosomes. Furthermore, these chromosomes are interphase chromosomes. They therefore provide insight into chromosome structure during this phase of the cell cycle.

Banding and Genes. One of the characteristics of interphase chromosomes is banding. Bands of varying widths are separated by lighter staining interband regions. Band numbers and positions are predictable, with as many as 6900 bands reported present on the four polytene chromosomes of *Drosophila*.

The dark-staining bands are chromomeres, regions in which the DNA fiber is supercoiled in a manner such as that depicted in Figure 15.43. Adjacent fibers are aligned laterally by chromomeres, thus giving visible banding in the light microscope.

It was once thought that each band represented a single gene; however improved methods of quantitating DNA have shown that an average band contains about 30,000 base pairs, or 20 times the amount of DNA required to code for an average protein. But 50 years after the one band–one gene condition was proposed, the situation is still not clarified. At least two obvious interpretations can be advanced to explain the high level of DNA in a band. One is that a given band is polygenic, as appears to be the case for certain loops in lampbrush chromosomes. According to this interpretation, a gene may be repeated until it consumes the length of the chromomere; or several different genes may reside in a given band.

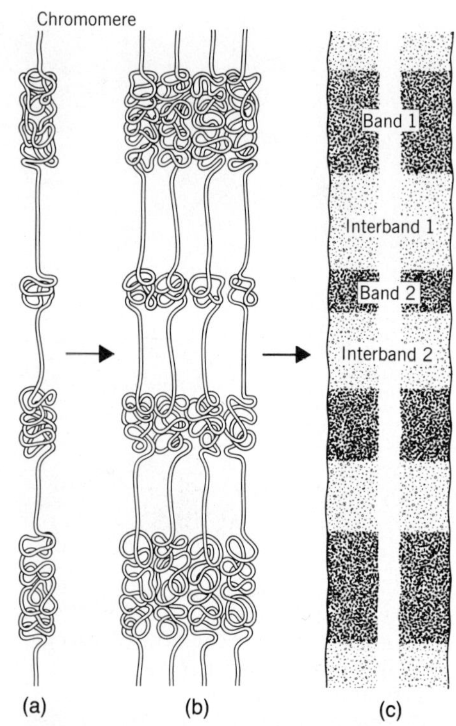

FIGURE 15.43 Diagram to indicate a mechanism of chromosome folding that could account for banding patterns in giant chromosomes. A single chromosome (*a*) aligns laterally with others (*b*), and somatic pairing between two polytene homologues (*c*) gives rise to a single giant structure with bands.

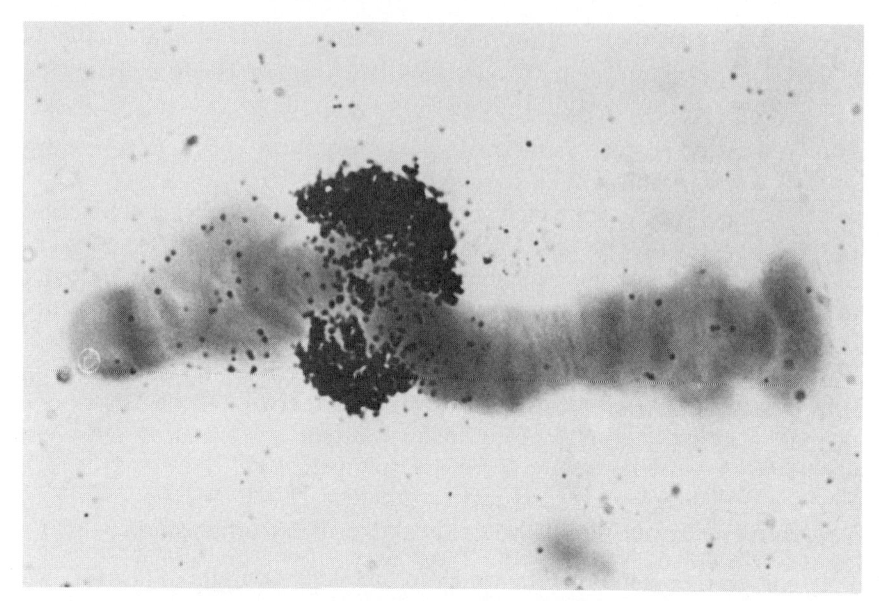

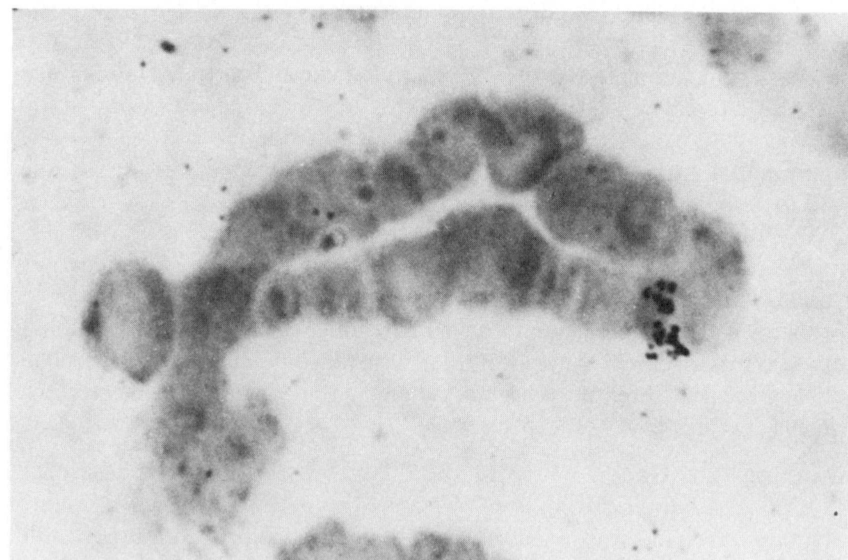

FIGURE 15.44 Giant salivary gland chromosome IV with BR 2 (Balbiani ring) gene in the transcriptionally active state (*top*) and the same chromosome from the Malpighian tubule in the inactive state (*bottom*). The puff activity was detected by labeling BR 2 RNA and hybridizing it with the chromosomes. (Courtesy of Dr. Bertil Daneholt.)

Another interpretation is that the chromomere contains long stretches of DNA that are not transcribed. With this point of view, the one band–one gene hypothesis could hold, provided a sufficient stretch of baggage DNA is present in the band.

Both notions appear to be testable, but until more experimental data are in hand it is tenuous to choose sides.

Puffing and Gene Expression. In the early 1950s it was noticed that polytene chromosomes undergo morphological changes that can be cor-

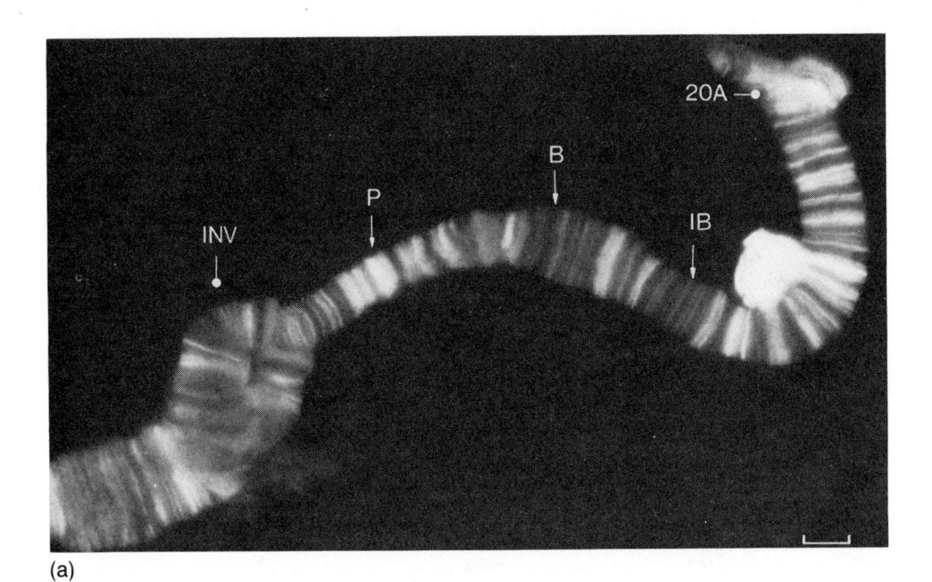

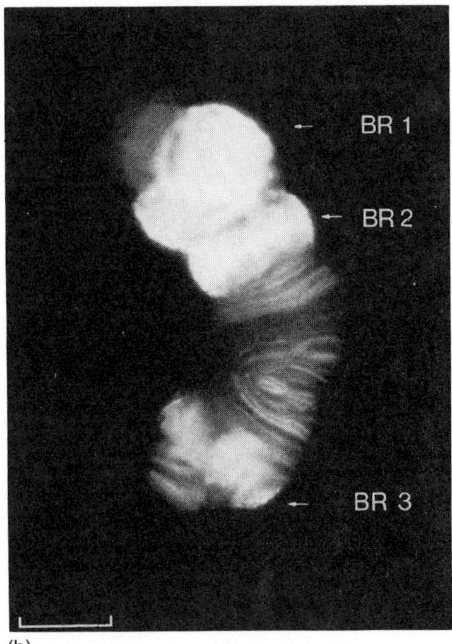

FIGURE 15.45 (*a*) Distribution of RNA polymerase II detected by indirect immunofluorescence in native isolated salivary gland chromosome I of the midge *Chironomus tentans*. Puffs (P) of ordinary size and interbands (IB) fluoresce; bands (B) are devoid of immunofluorescence (free of enzyme). An inversion (INV) and puff I (20A) are marked. (*b*) Native salivary gland chromosome IV of *tentans* after immunofluorescence staining. Whenever the Balbiani ring loci are stimulated to form giant puffs (BR 1, BR 2, BR 3), an enormous quantity of RNA polymerase II is found in them, and vice versa. Bar = 7 μm. (Courtesy of Dr. Heinz Sass.)

related with gene expression.[27] In particular, chromosomes undergo a phenomenon called puffing in which a densely staining chromomere is converted to a dispersed, lightly staining "puff." A given chromosome will contain a pattern of puffs that is a reflection of gene expression in the cell.

A number of observations have given rise to the conclusion that puffing is related to gene expression.[28] Different cell types have differences in their puff patterns, as would be expected where different genes are responsible for dictating the cell type. Temporal changes in a given cell type can also be correlated with changes in puff patterns. Thus, during larval development when different features of the organism are taking form with time, some genes are being turned on and others off. The presence of certain cellular products, such as salivary proteins, is related to the presence of certain puffs. There is thus a puff–gene product correlation that can be made.

Method VI

The evidence cited above is circumstantial, but more direct chemical data have supported these observations completely. Radioactivity in the form of [^{3}H]uridine can be incorporated into polytene nuclei, and when this is done there is a localization of radioactivity (new RNA) to puff regions. This would indicate that the puff is the site of mRNA formation or the morphological unit of gene expression in these chromosomes.

Two types of puff are commonly seen on polytene chromosomes. The major type is a small puff, which causes only a moderate disintegration of the chromosome region in which it forms. A minor type of puff, but more extreme, has been designated the Balbiani ring. Such puffs may have diameters three times the diameter of nonpuffed regions of the chromosome. Figures 15.44 and 15.45 illustrate puffs on salivary gland chromosomes as visualized by the application of radioactive and immunofluorescent techniques, respectively.

Method VII

In addition to the incorporation of radioactivity as observed by autoradiograms, radioactive RNA can be extracted from salivary glands and this, in turn, can be used in hybridization studies to determine whether a particular gene product will bind to specified regions on nonlabeled chromosomes.

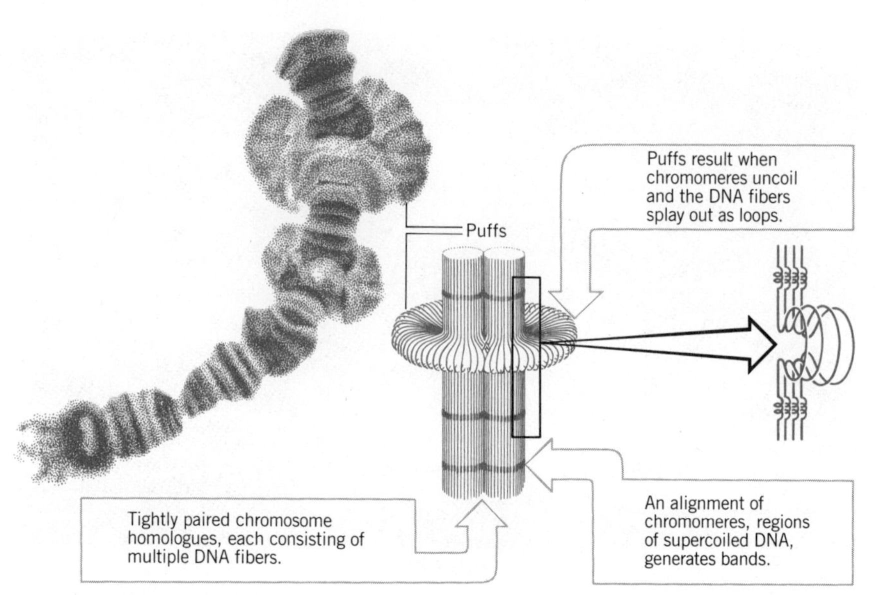

FIGURE 15.46 Schematic interpretation of a puff region in a polytene chromosome. What appears to be a single thick chromosome is in reality two homologues, paired gene for gene.

As a result of the observations and experimental approaches above, it is certain that puffs represent uncoiled regions of the chromosome fiber where active transcription is taking place. It should therefore be obvious that the salivary gland chromosomes provide a unique system on which molecular events (transcription) can be viewed by simple morphological changes. An interpretation of the puff is given in Figure 15.46.

A fascinating orchestrated activity of genes during larval development has been studied by observing puffing patterns with time.[29] The puffing patterns are observed to take form in a highly coordinated and regular manner.

Figure 15.47 summarizes in schematic fashion the induction and regression of puffs during a particular phase of development in *Drosophila melanogaster*. The puffs appear and disappear predictably. In other words, the puffing pattern is a fingerprint of a physiological process. No other system will so easily and readily show the switching on and off of genes.

Changes in puffing patterns are known to occur in response to various stimuli, including developmental, hormonal, metabolic, and nutritional.

Ecdysone is a hormone produced by the prothorax gland located in the thorax. It initiates molting in dipterans, as from the larva into the pupa and the pupa into the adult.

By reading the puffing patterns, it is possible to study the effect of a wide range of phenomena on gene expression. In this manner, ecdysone has been studied. When ecdysone is injected into larvae, it activates a pattern of puffing that corresponds to the pattern seen during molting. Ecdysone, therefore, is the hormonal trigger that initiates gene expression and the onset of this phase of development.

So it is that dipteran chromosomes, which were popular during the early years of study on heredity and subsequently largely abandoned in favor of the simpler viral and procaryotic genomes, are once again gaining favor as objects for the study of gene expression and development. It is obvious that these chromosomes, which so blatantly carry their pedigrees on their backs, provide researchers with powerful levers for gaining access to the molecular basis of developmental phenomena.

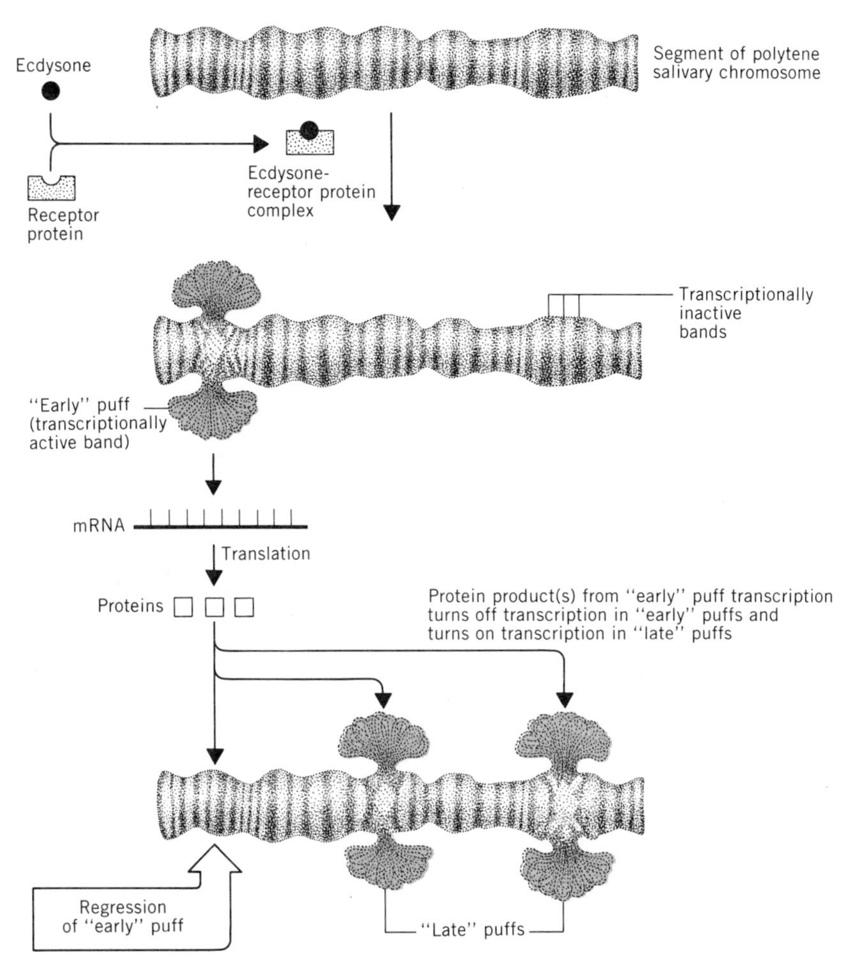

FIGURE 15.47 Schematic illustration of sequences of chromosome puffing in *Drosophila* larvae induced by the hormone ecdysone. After injection of ecdysone, a specific set of bands puffs. After a few hours this set ceases puffing and new bands begin to puff. After about 10 hr some 100 to 125 bands have formed puffs, with the late puffs requiring protein products derived from the early puffing activity.

Summary

The genetic material of viruses has been intensely studied and, as a result, is better understood at the molecular level than any other type of genome. Primary structures are now known for several viral genomes and the positions of genes within genomes have been determined. Genes are found to be spaced, or overlapped, or even coincidently placed along the DNA duplex. Small genomes are often supercoiled to relieve the stress created by underwinding of the duplex.

Procaryotic genomes are circular duplex DNA molecules that are organized and stabilized by attachments to proteins and RNA. The DNA extends out from the organizing center in domains that are supercoiled.

Eucaryotic genomes are complex structures that are made up of DNA, histones, and nonhistone proteins arranged in several orders of helices.

The DNA strands exist in a duplex helix. They are wound around a core of histone octamers to make up the nucleosome core. When histone H1 is present, the complete nucleosome is formed consisting of 200 base pairs of DNA and 9 histone proteins. When nucleosomes are spaced, the appearance of the genome is that of beads on a string. When nucleosomes are pulled together, a 10-nm fiber is formed.

Nucleosomes are arranged in helical fashion to generate a solenoid that has a diameter of 20 to 30 nm. In mitotic chromosomes the solenoid is packed into another helix, the supersolenoid, with a pitch of 30 nm and a diameter of 400 nm. The supersolenoid is condensed further to produce the final shape and dimensions of the mitotic chromosome.

Chromatin fibers are organized by attachment to a protein scaffold that retains the gross morphology of a chromatid pair even after the nucleic acids have been depleted from the structure.

Lampbrush chromosomes are eucaryotic chromosomes in a state that is partially condensed and partially extended. Condensed regions, called chromomeres, unwind to form the loops on which transcription takes place. The loop is thus an extended form of the genome (the $\leq$ 10 nm fiber) and the chromomere is a solenoid or supersolenoid structure.

Polytene chromosomes from dipteran species are banded. When genes are expressed, bands, or chromomeres, puff out by an uncoiling of the elementary chromatin fiber. Transcription takes place in puff regions.

During development in dipteran larvae, the salivary gland chromosomes exhibit a pattern of puffing that is initiated and controlled by hormones. Puffing permits a visualization of gene expression and an opportunity to correlate gene activities with a variety of developmental phenomena.

References

1. Studies on the Chemical Nature of the Substance Inducing Transformation of Pneumococcal Types. I. Induction of Transformation by a Desoxyribonucleic Acid Fraction Isolated from Pneumococcus Type III, O. T. Avery, C. M. MacLeod, and M. McCarty, *J. Exp. Med.* 79:137(1944).
2. A Single-Stranded Deoxyribonucleid Acid from Bacteriophage ϕX174, R.L. Sinsheimer, *J. Mol. Biol.* 1:43(1959).
3. The Structure of the DNA of Bacteriophage ϕX174. I. The Action of Exopolynucleotidases, W. Fiers and R. Sinsheimer, *J. Mol. Biol.* 5:408(1962).
4. The Structure of the DNA of Bacteriophage ϕX174. III. Ultracentrifugal Evidence for a Ring Structure, W. Fiers and R. L. Sinsheimer, *J. Mol. Biol.* 5:424(1962).
5. Electron Microscopy of the Replicative Form of the DNA of the Bacteriophage ϕX174, A. K. Kleinschmidt, A. Burton, and R. L. Sinsheimer, *Science* 142:961(1963).
6. Electron Microscopy of Single-Stranded DNA: Circularity of DNA of Bacteriophage ϕX174, D. Freifelder, A. K. Kleinschmidt, and R. L. Sinsheimer, *Science* 146:254(1964).
7. Enzymatic Synthesis of DNA. XXIV. Synthesis of Infectious Phage ϕX174 DNA, M. Goulian, A. Kornberg, and R.L. Sinsheimer, *Proc. Natl. Acad. Sci. U.S.* 58:2321(1967).
8. Nucleotide Sequence of Bacteriophage ϕX174, F. Sanger, G. M. Air, B. G. Barrell, N. L. Brown, A. R. Coulson, J. C. Fiddes, C. A. Hutchison III, P. M. Slocombe, and M. Smith, *Nature* 265:687(1977).

9. The Twisted Circular Form of Polyoma Viral DNA, J. Vinograd, J. Lebowitz, R. Radloff, R. Watson, and P. Laipis, *Proc. Natl. Acad. Sci. U.S.* 53:1104(1965).
10. DNA Gyrase Action Involves the Introduction of Transient Double-Strand Breaks into DNA, K. Mizuuchi, L. M. Fisher, M. H. O'Dea, and M. Gellert, *Proc. Natl. Acad. Sci. U.S.* 77:1847(1980).
11. A Minimum Estimate for the Length of the DNA of *Escherichia coli* Obtained by Autoradiography, J. Cairns, *J. Mol. Biol.* 4:407(1962).
12. Chemical, Physical, and Genetic Structure of Prokaryotic Chromosomes, D. E. Pettijohn and J. O. Carlson, in *Cell Biology, A Comprehensive Treatise,* vol. 2, D. M. Prescott and L. Goldstein, eds., Academic Press, New York, 1979.
13. Calf and Pea Histone. IV, R. DeLange, D. Fambrough, E. Smith, and J. Bonner, *J. Biol. Chem.* 244:319(1969).
14. Chromosomes and Chromatin Structure, G. F. Bahr, in *Molecular Structure of Human Chromosomes,* J. J. Yunis, ed., Academic Press, New York, 1977.
15. Evidence for a "Folded-fiber" Organization in Human Chromosomes, E. J. DuPraw, *Nature* 209:577(1966).
16. Chromosome Structure: Evidence Against the Unineme Hypothesis, J. H. Ford, in *The Eukaryote Chromosome,* W. J. Peacock and R. D. Brock, eds., Australian National University Press, Canberra, 1975.
17. Chromatin Substructure: The Digestion of Chromatin DNA at Regularly Spaced Sites by a Nuclear Deoxyribonuclease, D. Hewish and L. Burgoyne, *Biochem. Biophys. Res. Commun.* 52:504(1973).
18. Spherical Chromatin Units (nu Bodies), A. Olins and D. Olins, *Science* 183:330(1974).
19. Electron Microscopic and Biochemical Evidence the Chromatin Structure Is a Repeating Unit, P. Oudet, M. Gross-Belland, and P. Chambon, *Cell* 4:281(1975).
20. Chromatin Structure: A Repeating Unit of Histones and DNA, R. D. Kornberg, *Science* 184:868(1974).
21. Histones, I. Isenberg, *Annu. Rev. Biochem.* 48:159(1979).
22. Higher-order Structure of Mitotic Chromosomes, A. Leth Bak and J. Zeuthen, *Cold Spring Harbor Symp. Quant. Biol.* 42:367(1978).
23. Metaphase Chromosome Structure: The Role of Nonhistone Proteins, U. K. Laemmli, S. M. Cheng, K. W. Adolph, J. R. Paulson, J. A. Brown, and W. R. Baumbach, *Cold Spring Harbor Symp. Quant. Biol.* 42:351(1978).
24. Metaphase Chromosome Architecture, W. Wray, M. Mace, Y. Daskal, and E. Stubblefield, *Cold Spring Harbor Symp. Quant. Biol.* 42:361(1978).
25. ^{3}H-Uridine Incorporation in Lampbrush Chromosomes, J. G. Gall and H. G. Callan, *Proc. Natl. Acad. Sci. U.S.* 48:562(1962).
26. Organization of Transcriptionally Active Chromatin in Lampbrush Chromosome Loops, U. Scheer, H. Spring, and M. F. Trendelenberg, in *The Cell Nucleus,* vol. VII, H. Busch, ed., Academic Press, New York, 1979.
27. Chromosomenkonstanz und spezifische Modifikationen der Chromosomenstruktur in der Entwicklung und Organdifferenzierung von *Chironomus tentans,* W. Beerman, *Chromosome (Berlin)* 5:139(1952).
28. Transcription in Giant Chromosome Puffs, C. Pelling, in *Developmental Studies on Giant Chromosomes,* W. Beerman, ed., Springer-Verlag, New York, 1972.
29. Puffing Patterns in *Drosophila melanogaster* and Related Species, M. Ashburner, in *Developmental Studies on Giant Chromosomes,* W. Beermann, ed., Springer-Verlag, New York, 1972.

Selected Books and Articles

Books

Chromatin and Chromosome Structure, H. J. Li and R. A. Eckhardt, eds., Academic Press, New York, 1977.

Cold Spring Harbor Symposium on Quantitative Biology, vol. 43, Cold Spring Harbor Laboratory, Cold Spring Harbor, NY, 1979.

The DNA Molecule, Structure and Properties, Original Papers, Analyses, and Problems, D. Freifelder, W. H. Freeman & Co., San Francisco, 1974.

The Eukaryote Chromosome, W. J. Peacock and R. D. Brock, eds., Australian National University Press, Canberra, 1975.

Articles

Chromosome Structure and Levels of Chromosome Organization, H. Ris and J. Korenberg, *Cell Biology, A Comprehensive Treatise,* vol. 2, D.M. Prescott and L. Goldstein, eds., Academic Press, New York, 1979.

Nucleosome Structure, James D. McGhee and Gary Felsenfeld, *Annu. Rev. Biochem.* 49:1115(1980).

Nucleosomes: Composition and Substructure, Randolph L. Rill, in *Molecular Genetics,* Part III, *Chromosome Structure,* J.H. Taylor, ed., Academic Press, New York, 1979.

Supercoiled DNA, W. R. Bauer, F. H. C. Crick, and J. H. White, *Sci. Am.* 243:118(1980).

The Nucleosome, R. D. Kornberg and A. Klug, *Sci. Am.* 244:52(1981).

The Nucleotide Sequence of a Viral DNA, J. C. Fiddes, *Sci. Am.* 237:55(1977).

16

The Nuclear Interior

Any attempt made at the present time to describe the role of ribonucleoprotein (RNP) complexes in the processing of nuclear RNA, the transport of mRNA to the cytoplasm, and the regulation of translation of that mRNA on polyribosomes must begin with a clear statement of our ignorance of the molecular details of these processes. . . . The models illustrated . . . should therefore only be taken as ways of thinking about the structure and physiology of ribonucleoprotein complexes rather than as documented facts, or even generally recognized hypotheses.

TERENCE E. MARTIN
JAMES M. PULLMAN
MICHAEL D. McMULLEN, 1980

The foregoing comments introduced an excellent review article dealing with the structure and function of ribonucleoprotein particles.[1] The somewhat apologetic tone reflects the attitude of investigators who, after working long and hard for years, are acutely aware that there is yet a great deal of form and function within the nucleus that lies beyond our understanding. But the comments should not be interpreted to suggest that little has been accomplished from studies of the nucleus in the past decade or that nothing of interest has been uncovered regarding the nuclear interior.

Quite the opposite is, in fact, the case. But before enjoying the present state of the nuclear arts, let us look briefly at the pathway that brought us to this point.

Early light microscopists were able to distinguish the nucleus from the rest of the cell, but the nucleolus was not recognized as a nuclear organelle until 1774. Credit is given to Fontana for the first reported sighting of the nucleolus. This must have created a flurry of nuclear investigation, for in 1898, more than 100 years later, a 300-page report was published summarizing a variety of nucleolar morphologies as observed in many different species.[2] Perhaps this sounds like something less than a flurry, but scientific advances often moved with great deliberation in the eighteenth and nineteenth centuries. Investigators did not work under the publish-or-perish cloud of today but appeared to relish the fruits of their labors in isolation for years without bothering to write them up. So it would be safe to assume that much more was studied about nucleoli during that period than was reported.

In the late nineteenth century basic dyes were employed with light microscopy, a technique that on the one hand enabled investigators to study nucleolar morphology and number and on the other hand introduced the possibility that nucleoli were artifacts of staining. Thus the results obtained for a period of time were subject to controversy. But the phase contrast microscope put the matter to rest, for with this instrument nucleoli were observed in living cells as well as in isolated nuclei.

Studies carried out on nucleoli near the turn of the century were largely overshadowed by a great interest at that time in chromosome structure and function. Light microscopy was experiencing a heyday, enabling investigators to propose morphological explanations for the questions of heredity. However, a by-product of these studies was the observation that specific regions of certain chromosomes were associated with nucleoli and appeared to control, or at least influence, the disappearance and reorganization of the nucleolus during the cell cycle. These specific regions were reported by E. Heitz in 1933 and were referred to as SAT regions because it was thought that they lacked DNA (i.e., they were supposedly sine acido thymonucleinico).[3]

At about the same time, B. McClintock, while studying secondary chromosomal constrictions in maize, noticed that mutants that lacked the constrictions also failed to form nucleoli.[4] She subsequently called the secondary chromosomal constrictions *nucleolus organizer regions*, or NORs, a label that is still used today.

Primary constrictions are sites of kinetochores. Secondary constrictions are nucleolus organizer regions (NORs), sites associated with nucleoli.

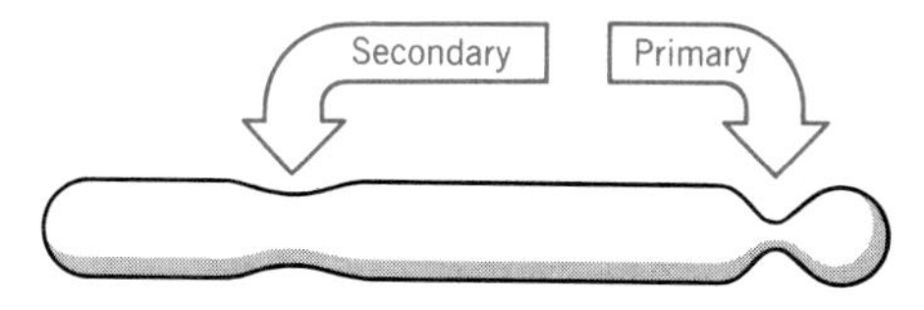

The use of tritium-labeled isotopes grew in the 1950s, and with the concomitant development of autoradiography, yet another approach to studying the nucleus was employed. These techniques caused another flurry of activity, this time in the more modern sense.

It was quickly established that RNA precursors were rapidly incorporated into nuclei, with the highest specific activity of the RNA showing up in the nucleolus. Furthermore, temporal studies revealed that newly formed RNA undergoes a unidirectional flow to the cytoplasm. These results attracted speculations that the nucleolus is a center of the synthesis of RNA, which ultimately has a role in the cytoplasm. But it was awhile before it was possible to sort out with certainty the *kind* of RNA being generated and the *form* it took while in transit to the cytoplasm.

In the 1950s the electron microscope was applied to nuclear studies with mixed blessings. Some things were definitely clearer. It was clear that the nucleolus was not limited by a membrane and that it was indeed associated with chromatin. But with electron microscopy all the elements of the nuclear interior, including the nucleoli, were found to be complex mazes of particles and fibrils, some of which were chromatin and others not.

In recent years the experimental approach to the nucleus has shifted more toward the biochemical, or at least a combination of biochemical and morphological, with heavy emphasis on the former. This combination has made it possible to define a variety of ribonucleoprotein particles and fibrils that function as intermediates in the generation of ribosomes and mRNA.

At this point the picture has become very complex, and we must admit our ignorance of molecular explanations. It is also precisely at this point that the story becomes very exciting, for it is certain that the answers to a number of questions about the nuclear interior are very close at hand. With some appreciation of the past, let us move into the present.

16.1 THE NUCLEAR MATRIX

Chapter 14 mentioned that the nuclear envelope can be solubilized with Triton X-100, leaving the pore complexes in place, still attached to the nuclear cortex. This observation and others like it have substantially changed a long-held view that the nuclear envelope serves as a bag to retain the chromosomes and maintain them in a compartment separate from the rest of the cell.

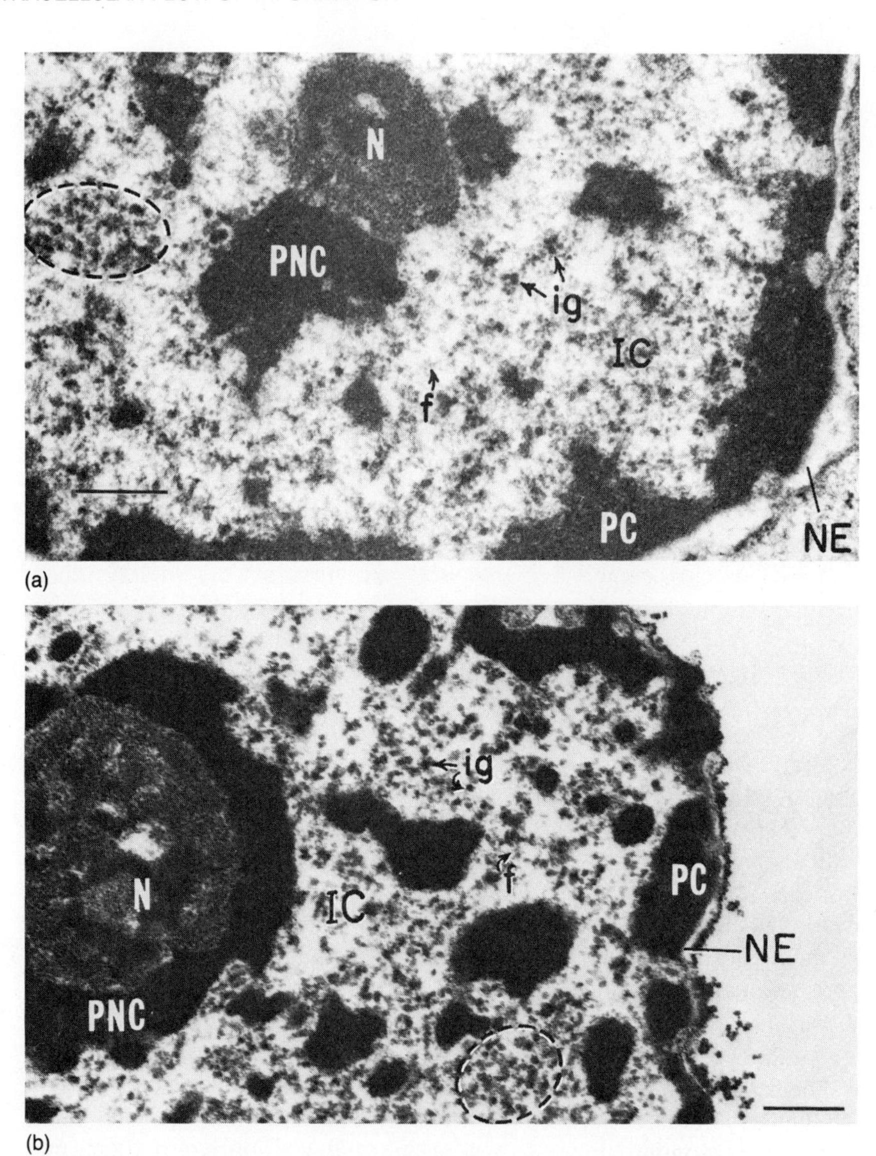

FIGURE 16.1 Electron microscope sections of rat liver nuclei *in situ* (*a*) and in the isolated nuclear fraction (*b*). The most prominent features are NE, nuclear envelope; N, nucleolus; PNC, perinucleolar condensed chromatin; PC, peripheral condensed chromatin; IC, interchromatinic area. The IC contains particles 15 to 25 nm in diameter termed interchromatinic granules (ig) and fibrous material (f). A broken line in (*a*) surrounds a cluster of ig. Bar = 0.4 μm. (Courtesy of Dr. Ronald Berezney.)

It is now clear that the physical integrity of the nucleus is not dependent on its envelope. Nor is it dependent on the chromatin of the nucleus. As a result of studies carried out within the past decade, it is becoming evident that the form and function of the nucleus is related to a supramolecular matrix structure in the nuclear interior.

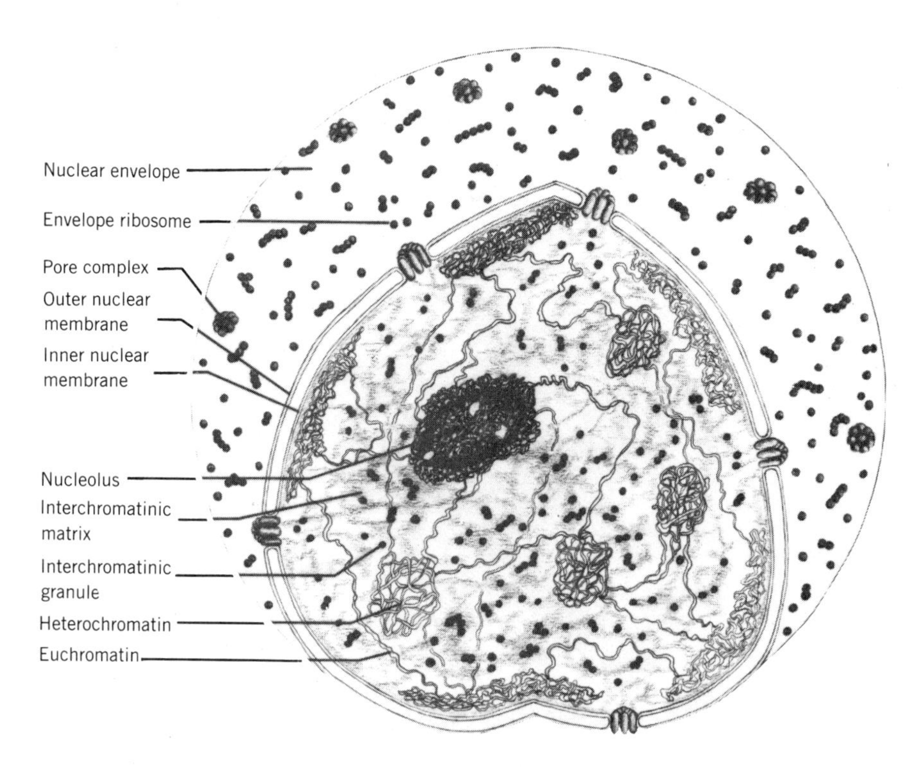

FIGURE 16.2 Schematic model of a eucaryotic nucleus illustrating the components apparent in Figure 16.1. Note especially the four distinct nuclear regions discussed in the text. (Courtesy of Dr. Ronald Berezney.)

Four Distinct Nuclear Regions

Electron microscopy of intact cells from tissues like liver has enabled researchers to define four distinct regions of the nucleus. These are illustrated in Figure 16.1.

The first region is the *nuclear envelope*, which was discussed in Chapter 14. The second is the *nucleolus*, or, if present in multiples, the *nucleoli*. The third region is one of condensed chromatin called *heterochromatin*, present typically in patches near the nuclear envelope, scattered about the interior, or surrounding the nucleoli. The fourth region, which lies between all the rest, is a region of granular and fibrous material. Its main components are extended or diffuse chromatin, called *euchromatin*, as well as a structural network of nonchromatin material. The fourth element of the nucleus has been coined the *interchromatinic matrix*. These regions are illustrated in schematic form in Figure 16.2.

It would be a mistake at this point to begin to think of the matrix as a static structure with the features that have been captured by electron microscopy fixed in position. We know from the varied activities engaged in by the nucleus and its internal elements that the matrix must certainly be a dynamic structure. In fact, the matrix probably does not simply respond to the dynamics of the nucleus but may direct and organize some of these activities.

Table 16.1 Steps in the Isolation of the Nuclear Matrix

			Total Nuclear Material Extracted (Cumulative)[a]			
Step	Consecutive Extractions	Resultant Nuclear Sphere	Protein (%)	DNA (%)	RNA (%)	Phospholipid (%)
0	Isolated nuclei	Intact nuclei	0	0	0	0
1	DNase I (5 μg DNase/10^8 nuclei/ml)	DNase-treated nuclei	—	—	—	—
2	Low magnesium treatment (0.2 m*M* $MgCl_2$)	Nuclear matrix I	52.0 ± 2.23	75.8 ± 2.78	19.7 ± 1.83	2.5 ± 1.13
3	High salt treatment (2*M* NaCl)	Nuclear matrix II	83.7 ± 2.17	98.6 ± 3.2	66.0 ± 1.64	6.4 ± 1.65
4	Detergent treatment (1% Triton X-100)	Nuclear matrix III	90.0 ± 1.26	98.7 ± 1.26	71.0 ± 1.68	97.8 ± 1.51
5	Nuclease treatment (DNase–RNase)	Nuclear matrix IV (nuclear protein matrix)	90.2 ± 1.55	>99.9	98.0 ± 1.12	98.0 ± 1.47

[a]Percentage extracted determined relative to isolated nuclei centrifuged in 0.25*M* sucrose TM buffer an equivalent number of times as each nuclear sphere fraction. SEM (±) for five separate preparations.

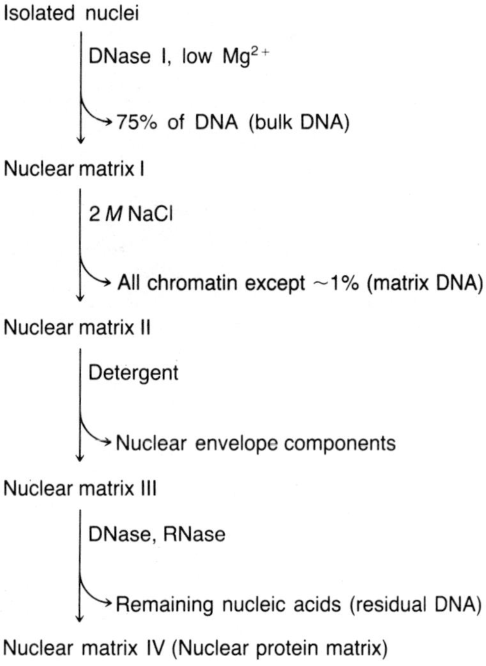

Isolated nuclei
↓ DNase I, low Mg^{2+}
→ 75% of DNA (bulk DNA)
Nuclear matrix I
↓ 2 *M* NaCl
→ All chromatin except ~1% (matrix DNA)
Nuclear matrix II
↓ Detergent
→ Nuclear envelope components
Nuclear matrix III
↓ DNase, RNase
→ Remaining nucleic acids (residual DNA)
Nuclear matrix IV (Nuclear protein matrix)

Isolation of Nuclear Matrix Material

Perhaps the clearest way to arrive at an understanding of matrix structure is to follow the steps taken to isolate matrix material.[5] Table 16.1 lists consecutive steps used in obtaining nuclear matrix material from isolated nuclei.

Isolated nuclei are first predigested with low levels of DNase I followed by an extraction with a low magnesium buffer, which releases about 75% of the DNA. This process changes the morphology of the nuclear interior so that the patches of heterochromatin are lost. This is apparent in the electron micrographs of Figure 16.3. The material that remains at this stage is referred to as nuclear matrix I.

The next step is an extraction with 2 *M* NaCl, leaving a nuclear sphere called nuclear matrix II. This step removes all the remaining chromatin material (98.6 ± 3.2%) except for about 1% that remains tightly bound to the interchromatinic matrix. The nucleus at this stage of the procedure still contains its nuclear envelope.

The third step removes lipid components of the nuclear envelope by extraction with Triton X-100. This results in a material called nuclear matrix III, which emerges from the treatment having lost 99% of the DNA, 98% of the phospholipid, 90% of the protein, and 71% of the RNA of the original intact isolated nucleus.

The electron micrograph of Figure 16.4 illustrates the morphological characteristics of the matrix at this stage. Three main components persist: remnants of the nuclear envelope, interchromatinic matrix, and residual nucleoli.

Treatment of nuclear matrix III with DNase removes the remaining DNA fibrils, and combined treatment with DNase and RNase releases essentially all the nucleic acid material from the matrix. That which remains has about 10% of the protein that was present in the original nuclei. This is called nuclear matrix IV or the nuclear protein matrix.

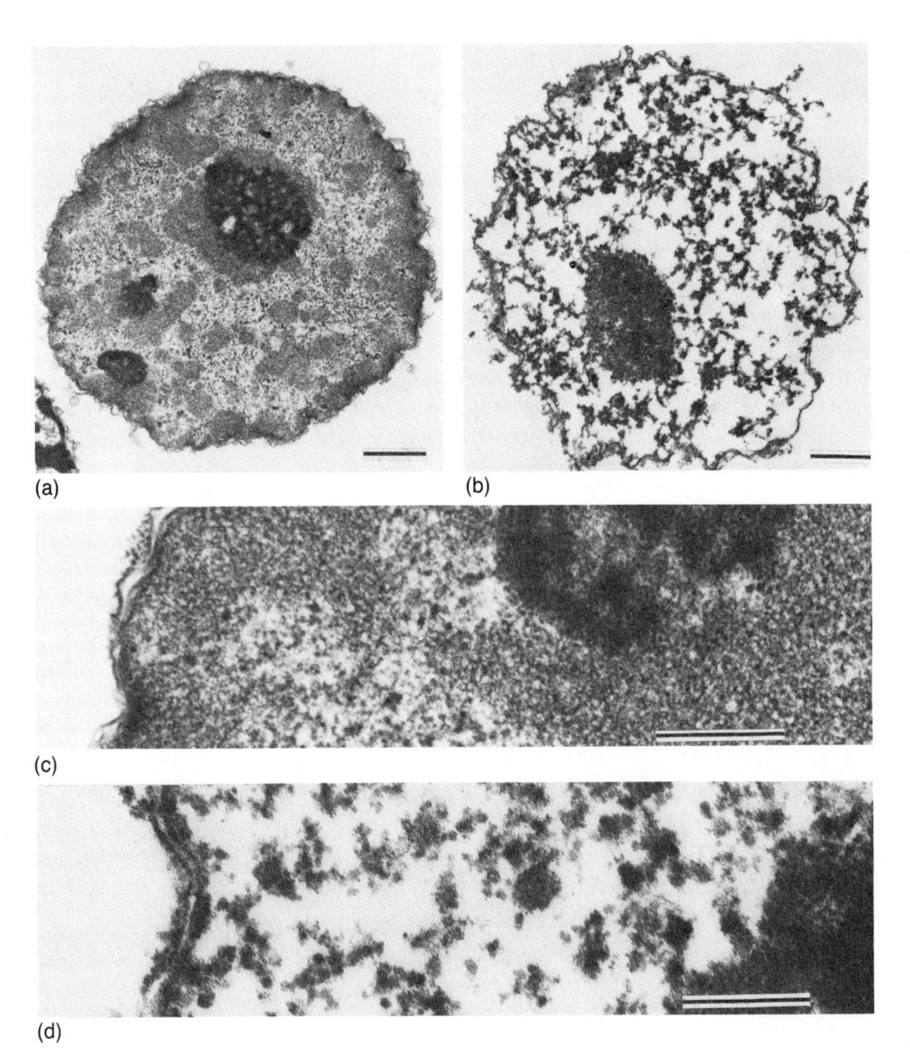

FIGURE 16.3 Comparison of untreated nuclei (*a* and *c*) and nuclear matrix I (*b* and *d*). Condensed chromatin, characteristic of the untreated nucleus, is absent in the preparations of nuclear matrix. Bars in a and b = 1 μm; in c and d = 0.4 μm (Courtesy of Dr. Ronald Berezney.)

In the electron microscope (see Figure 16.5), this material appears as the skeleton of the nucleus. It is surrounded by a residual envelope layer. On the inside is the residual nucleolus, which is continuous with an extensive granular and fibrous matrix structure. High magnifications of this material reveal a close structural tie between the remaining envelope material and the matrix. Furthermore, the residual nucleolus, seen in high magnification (Figure 16.6), appears to be continuous with the internal matrix.

Chemical Components of the Nuclear Protein Matrix

The major chemical component of the internal matrix is protein, as we have seen from the results tabulated in Table 16.1. Depending on the type of cell or tissue from which matrices are prepared, protein accounts for

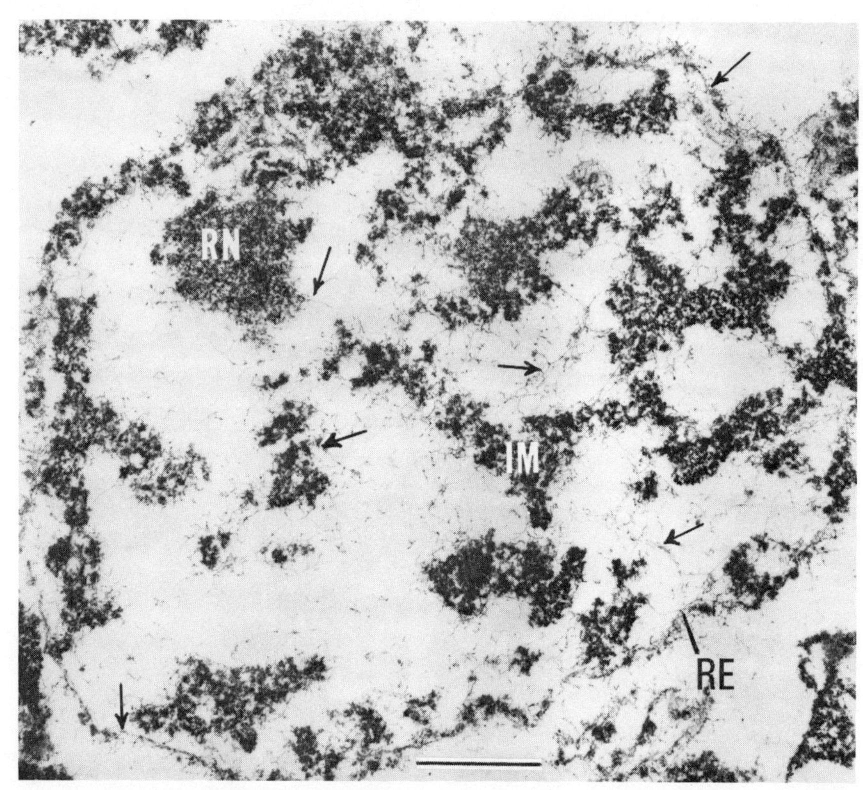

FIGURE 16.4 Electron micrograph of nuclear matrix III. DNA fibrils are apparent throughout the matrix structure (arrows). RE, residual nuclear envelop; IM, internal matrix; RN, residual nucleolus. Bar = 1 μm. (Courtesy of Dr. Ronald Berezney.)

more than 90% of the material, with low levels of RNA, DNA, and phospholipid making up the remainder.

The matrix proteins are largely acidic, possessing a ratio of acidic to basic amino acids of 1.46 to 1.55 in one series of studies.[5] These proteins resolve on SDS–acrylamide gels into three major bands with molecular weights that range from 60,000 to 70,000 daltons. Electrophoretic profiles of the proteins in whole isolated nuclei and in various stages of nuclear matrix isolation are presented in Figure 16.7. These profiles demonstrate a dramatic increase in purity of matrix proteins as compared to the very heterogeneous group of proteins present in untreated nuclei; at the same time, they indicate the presence of several other proteins, both higher and lower in molecular weight than the major matrix proteins.

Method IV

All three of these major proteins are present in the residual nuclear membranes as well as in residual nucleoli. However, their relative concentrations differ somewhat in the different nuclear regions. Furthermore, chromatin contains significant amounts of the major matrix proteins. It thus appears that the superstructure of the nucleus, including residual nucleoli and membranes, consists of a few rather low molecular weight proteins.

One of the unusual properties of nuclear protein matrices is their ability

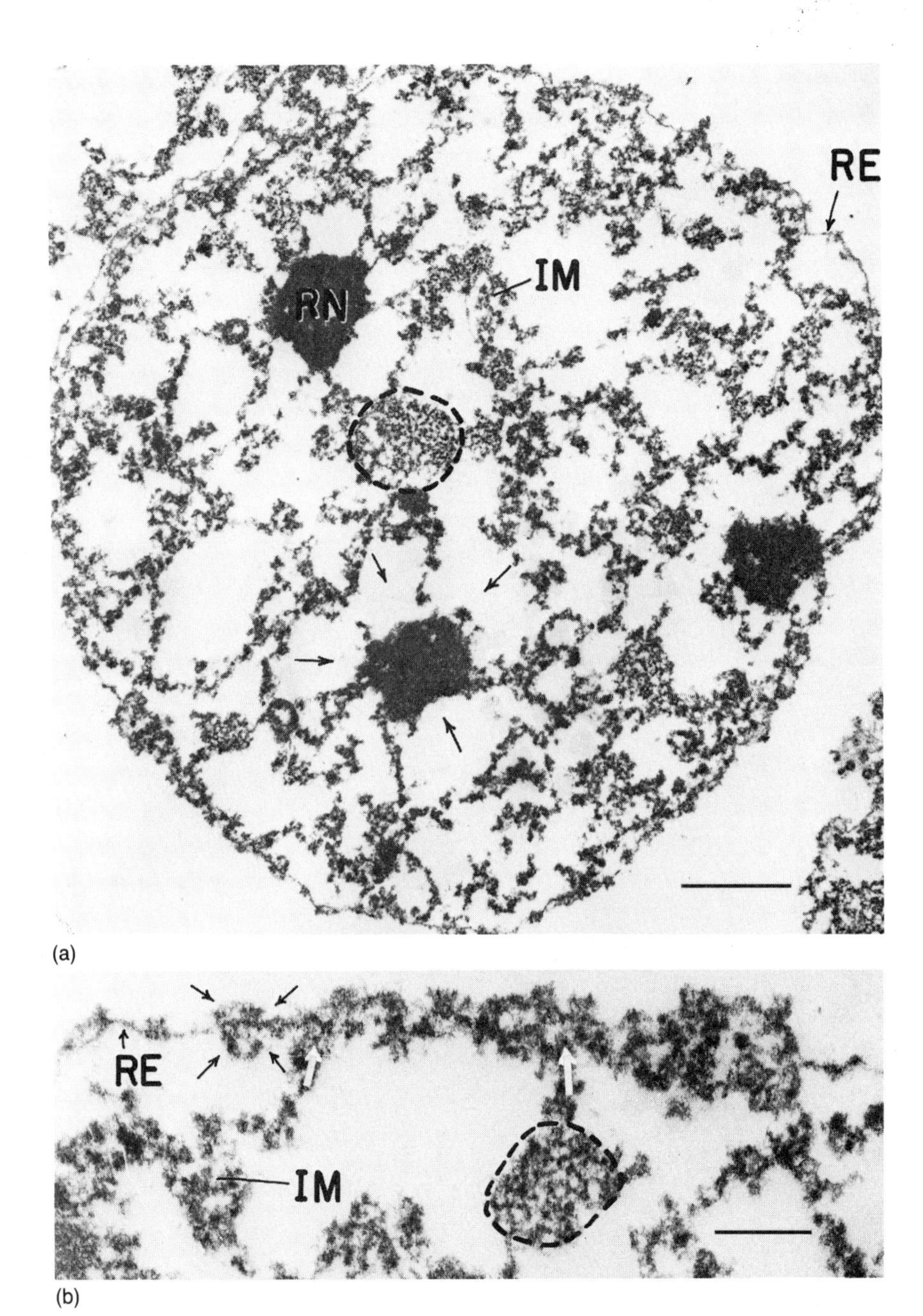

FIGURE 16.5 Sections through the nuclear protein matrix. (*a*) The structure of the entire matrix. RN, residual nucleolus; IM, internal matrix; RE, residual nuclear envelope. Empty spaces surround the residual nucleoli. (*b*) Higher magnification shows a close association of the internal matrix with the residual nuclear envelope (white arrows). Black arrows denote a residual nuclear pore complex. Bar in a = 1 μm; in b = 0.4 μm. (Courtesy of Dr. Ronald Berezney.)

to expand and contract as a function of Ca^{2+} and Mg^{2+} concentration. The mechanism of this dynamic property is not yet understood, but it does not appear to be due to an actin–myosin type of contraction.

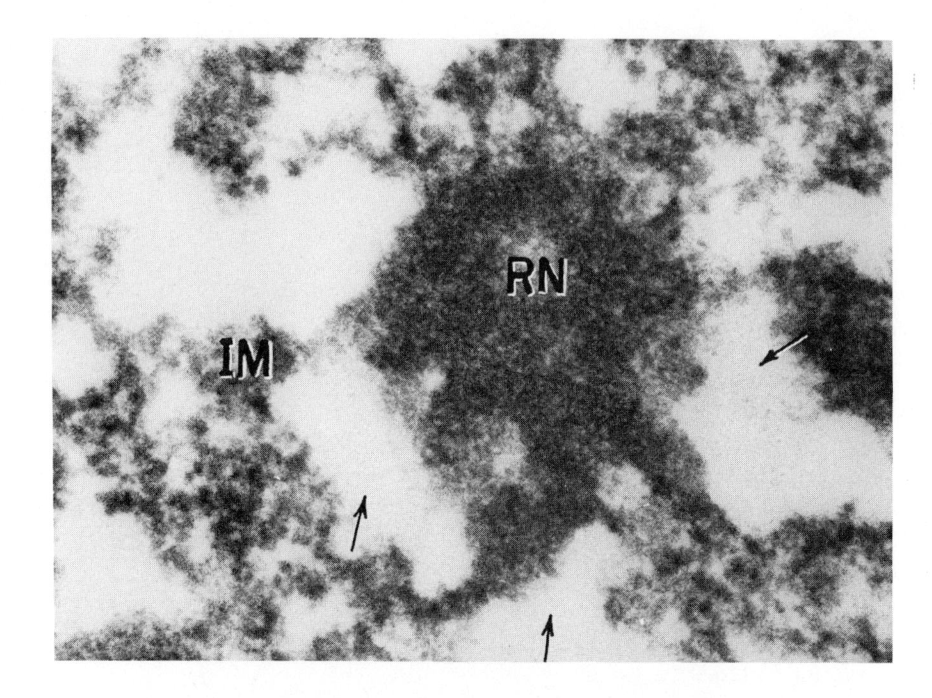

FIGURE 16.6 High magnification of the nuclear protein matrix in the vicinity of the residual nucleolus (RN). There appears to be continuity between the RN and the internal matrix (IM). Empty spaces surrounding the RN (arrows) were probably occupied by perinucleolar condensed chromatin in untreated nuclei. (Courtesy of Dr. Ronald Berezney.)

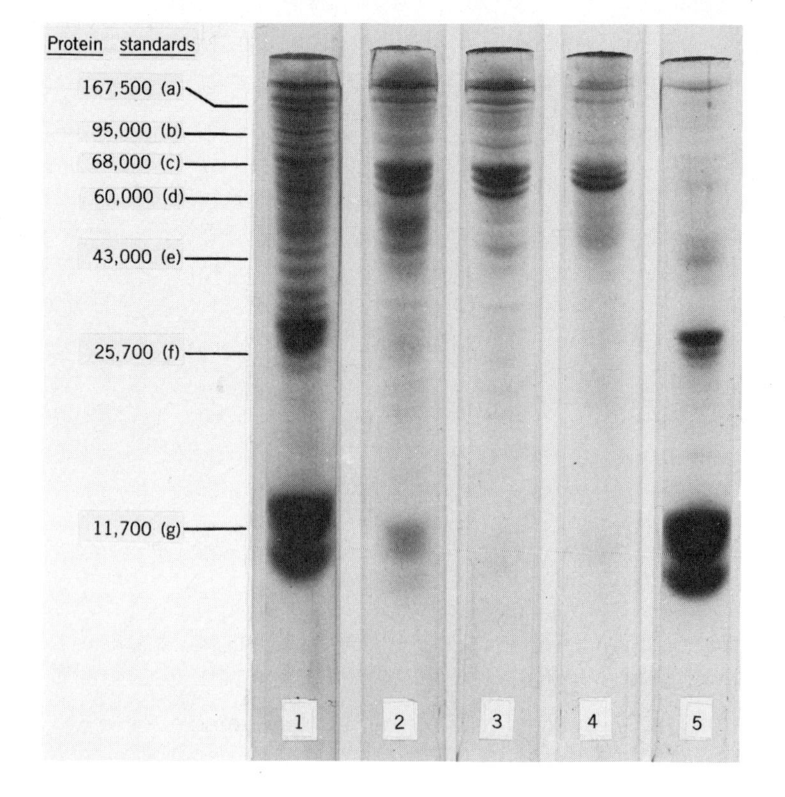

FIGURE 16.7 SDS–PAGE profiles of nuclear matrix proteins. 1, Untreated isolated nuclei. 2, Nuclear matrix II. 3, Nuclear matrix III. 4, Nuclear protein matrix. 5, Isolated chromatin. (*a–g*) standard proteins of known molecular weight (daltons). Three major proteins can be seen in the nuclear protein matrix, run 4. (Courtesy of Dr. Ronald Berezney.)

Matrix Function

The function of the nuclear matrix superstructure is the object of considerable research effort and the subject of several lines of speculation. Basically, the matrix is proposed to have a role in three major nuclear events: replication, transcription, and posttranscriptional processing and transport phenomena. It is not possible to develop any of these proposed functions in depth because they are still in their infancy. But future research will likely be concentrated along these three lines; thus some very brief comments are in order.

Replication

The replication of DNA in the interphase nucleus is commonly thought to occur in the vicinity of the nuclear envelope, where chromatin is often seen localized and condensed. But serious research students of the nuclear matrix suggest that the matrix may be the fundamental site of DNA replication.

As was apparent in Table 16.1, which outlined the steps during matrix isolation, about 75% of nuclear DNA is extracted with 0.2 m*M* $MgCl_2$ (termed bulk DNA), and the remaining 25% is released only after additional treatment of the matrix (matrix DNA). High salt treatment solubilizes most of the matrix DNA while 1% of the original DNA (termed residual DNA) remains tightly adsorbed to nuclear matrix III. This DNA is visible as fibrils in electron micrographs of matrix prepared up to this stage (see Figure 16.4).

Thus, based on morphological and chemical analyses, it is fair to conclude that DNA is indeed matrix associated.

Method VI

Radioactive labeling of DNA during its synthesis followed by a time course study of its appearance in nuclear fractionation has supported the proposed role of the matrix during replication. The results of this type of approach are summarized in Figure 16.8. It is apparent that after supplying the system with [^{3}H]thymidine, the specific activity is highest initially in matrix DNA and lowest in bulk DNA. In fact, during the first few minutes after the radioactive pulse, the residual DNA possesses the highest specific activity. With time the radioactivity associated with matrix DNA decreases, while simultaneously there is an increase in label associated with bulk DNA. Within the matrix DNA fraction itself there is also a shift of relative amounts of label from residual to high salt-soluble form.

These temporal labeling studies thus point to a flow of DNA during synthesis that is suggested as follows.

Residual matrix DNA → High salt-soluble DNA → Bulk DNA

These results lend support to the notion that the nuclear matrix is the meshwork throughout the nucleus where DNA replication first takes place. Replication appears to be initiated where the DNA is most tightly bound to the matrix. Since the matrix extends to the nuclear envelope, this does not contradict the tenet that chromatin is replicated in that region of the nucleus as well. But it may be more closely associated with the matrix than with the nuclear envelope in peripheral nuclear regions.

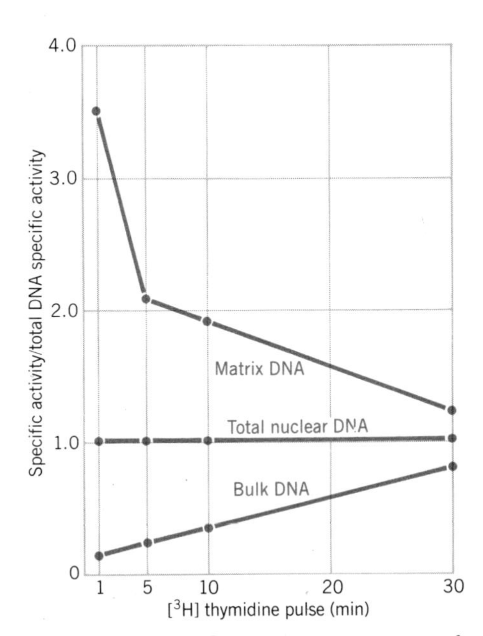

FIGURE 16.8 Graphical summary of temporal labeling studies on DNA in the nucleus. The relative specific activities were determined in a 20-hr regenerating liver after a single *in vivo* injection of [^{3}H]thymidine into the hepatic portal vein.

Transcription

During the isolation of nuclear matrix, a residual amount of RNA is also found associated with the matrix protein. During pulse labeling experiments the residual RNA demonstrates the most rapid uptake of radioactivity as compared to the other RNA of the nucleus, with the highest specific activity associated with high molecular weight (>45S) RNA. It would therefore appear that the immediate product of transcription, a high molecular weight form, is matrix associated. When completed, it may adhere less tightly to the matrix and present itself for processing and transport.

Posttranscriptional Processing and Transport

Since the matrix is a continuous nuclear superstructure extending from nuclear interior to the envelope and pore complexes, it is a likely candidate as a structural pathway for the transport of RNA from its site of transcription to the nuclear pore destinations, where it is released into the cytosol. It is known from a number of studies (see later in this chapter) that high molecular weight RNA is processed to smaller pieces before being exported from the nucleus.

Although the picture is not yet well focused, the current thinking is that the matrix may coordinate and regulate the dynamic intranuclear events that take place from the time of transcription until release from the nucleus, including the mechanisms of transport and posttranscriptional processing.

It will be interesting to observe how this uncertain picture is clarified by research in the next few years.

16.2 THE NUCLEOLUS

In Figure 16.6 we saw electron microscopic evidence suggesting that a residual nucleolus is continuous with the internal matrix of the nucleus. Unfortunately, the physical and chemical relationships between the matrix superstructure and the nucleolus are not clearly understood, partly because the matrix and the nucleolus have generally been studied independently. Keeping in mind that there may be a structural continuity between the two, we discuss the nucleolus as a separate organelle, anticipating that research in the future will clarify its relationship to the rest of the nuclear interior.

The Gross Structure of the Nucleolus

Method I

The nucleolus is seen readily by light microscopy on stained cells or by using phase contrast microscopy on living cells or isolated nuclei. Typically, the diploid eucaryotic nucleus possesses one large or two smaller nucleoli, but there are many variations on this depending on cell type and the physiological state of the cell during the cell cycle. Light micrographs of cells with clearly apparent nucleoli are presented in Figure 16.9.

Nucleoli are situated in the nucleus at specific sites along certain chromosomes, which are responsible for their generation. These sites are termed secondary constrictions or, as noted previously, nucleolus organizer regions (NORs). The number of nucleoli thus appearing in a nucleus is related to the number of chromosomes bearing these organizing regions.

During the cell cycle nucleolar morphologies change. They are most prominent during interphase and become progressively dispersed until at metaphase, when chromosomes are highly condensed, they are no longer apparent. They normally disappear during prophase, reappear during telophase, and are most irregular in form during interphase.

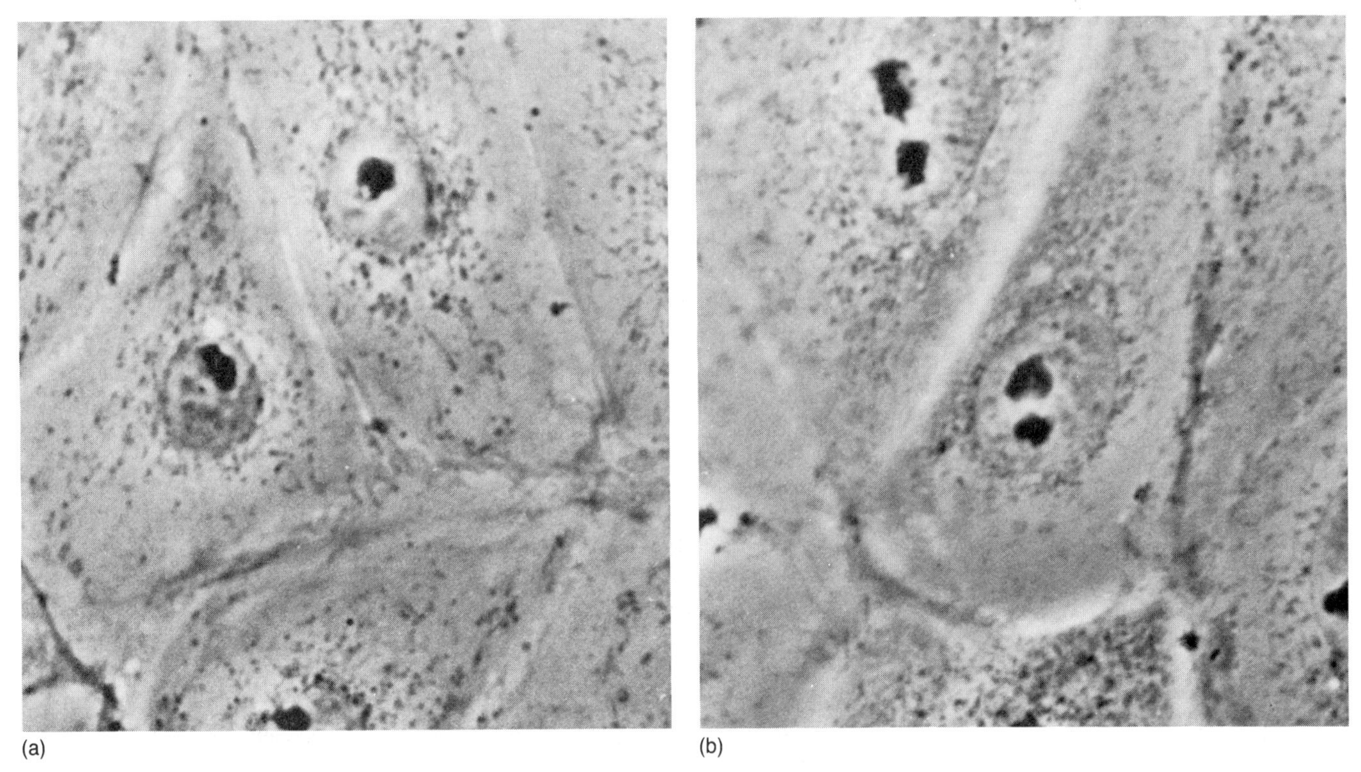

FIGURE 16.9 Light micrographs of rat kangaroo cells. The male contains one nucleolus per cell (*a*) whereas the female possesses two (*b*). (Courtesy of Dr. Michael W. Berns.)

Nucleolar Ultrastructure

Electron microscopy has resolved the general irregularities of the nucleolus into two identifiable and characteristic components. One of these components is granular and the other fibrillar.

Granular and Fibrillar Components

The granular and fibrillar components of the nucleolus are illustrated in Figure 16.10. Although it may at first seem to require a trained eye to distinguish the two components, a little study of the micrograph will reveal the finer texture of the fibrillar as compared to the granular regions.

In general, the granular component is a collection of 15 nm granules surrounded by packed fibrils that are about 5 nm in diameter. Both regions are embedded in a matrix of amorphous material that may harbor other fibers and granules as well.

Nucleolar-Associated Chromatin

Chromatin fibrils surround the nucleolus and penetrate it to varying extents. Referred to as perinucleolar and intranucleolar chromatin, respectively, these fibrils are thought to be the DNA that makes up the NORs of the chromosomes.

A close analysis of nucleolar ultrastructure suggests that chromatin

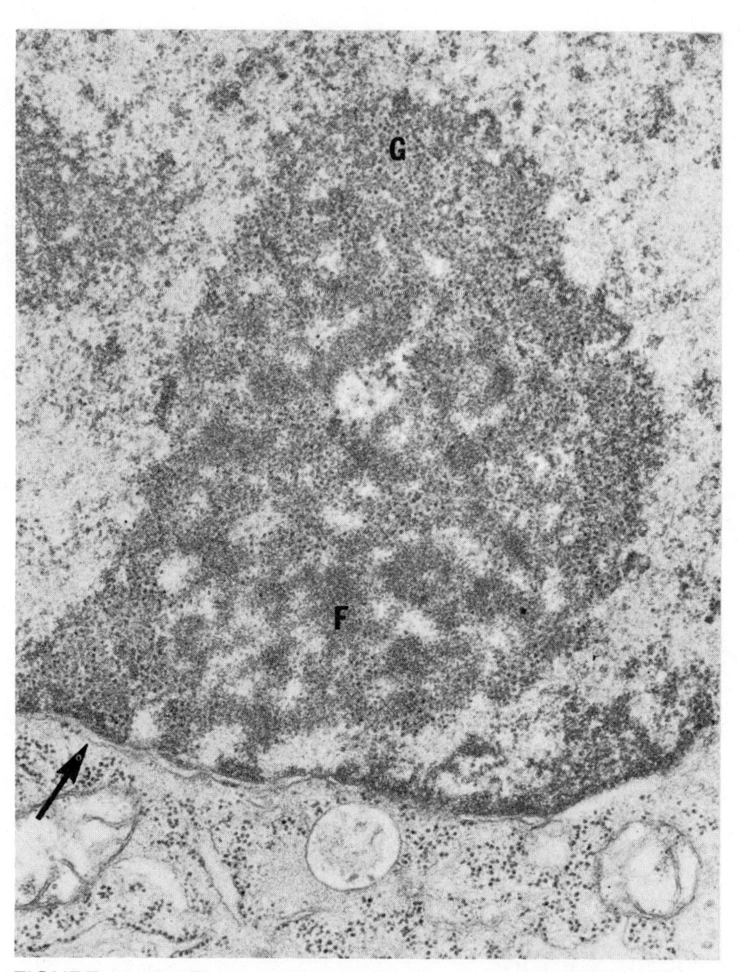

FIGURE 16.10 Electron micrograph showing nucleolus in a rat hepatocyte. The nucleolus appears to be mounted on the nuclear envelope (arrow). Within the nucleolus granular (G) and fibrillar (F) regions can be seen. (Courtesy of Dr. Harris Busch.)

fibrils and the fibrillar component of the nucleolus are in close structural proximity, while the granular component makes up the remainder of the nucleolus. Thus, when viewing electron micrographs, we can think of the NORs as meandering through the nucleolus in regions where the fibrillar component is apparent. Figure 16.11 illustrates this relationship most dramatically.

Configurations of the Nucleolus Organizer

Depending on the physiological state of the cell, the NOR may exist in several different configurations or states of condensation.[6] These states are related to the function of the nucleolus and influence the observed morphology of the organelle.

Fully Dispersed. This state of the NOR is its least condensed or most extended state. When fully dispersed, the genes of the NOR are being transcribed into RNA products. Nucleoli in this state are shown in the

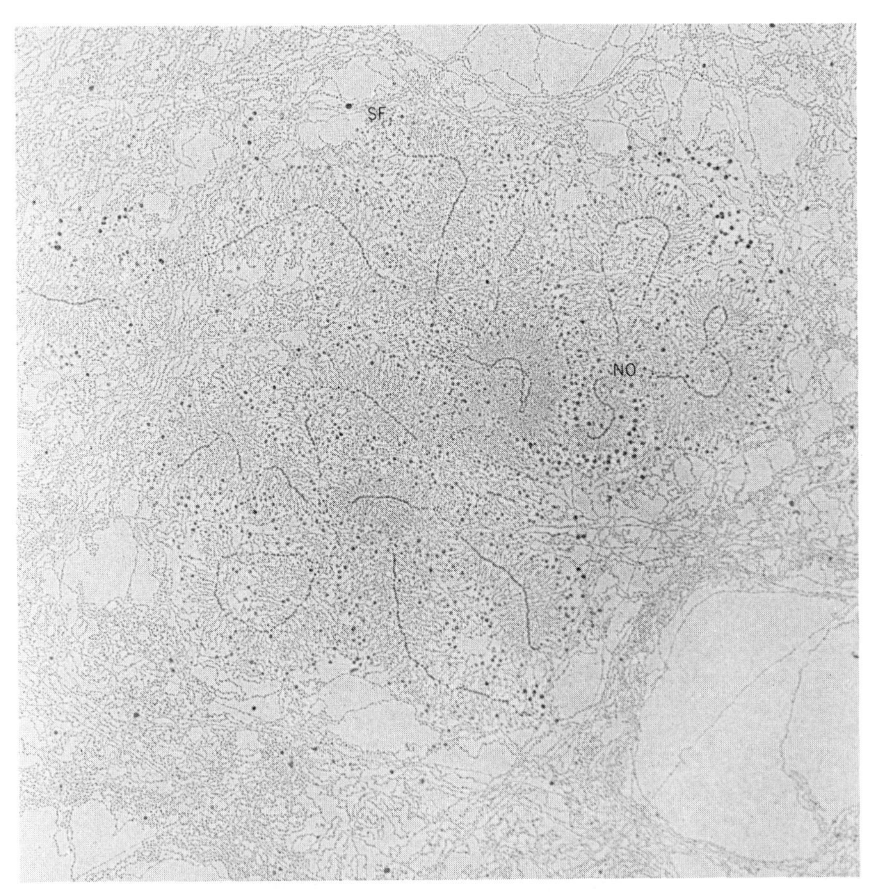

FIGURE 16.11 Electron micrograph of spread nucleolus organizer region. A DNA fiber meanders through the region. RNA strands are splayed out from the polymerase enzymes clustered on the DNA. Granules are present on the ends of the RNA strands. (Courtesy of Dr. Steven L. McKnight and Dr. Oscar L. Miller, Jr.)

electron micrographs of Figures 16.10 and 16.11. Fully dispersed chromatin has filaments about 2 nm in diameter.

Fully Condensed. In contrast to the dispersed NOR, the fully condensed state is a transcriptionally inactive state. Morphologically, the NOR in this state appears as 20 nm fibrils that are indistinguishable from those of other condensed chromatin regions.

Intermediate State of Condensation. An intermediate state of condensation, somewhere between the two extremes above, is the most common condition of the NOR. In this state the chromatin fibrils measure about 10 nm in diameter. These are about twice the diameter of the fibrils that make up the fibrillar component of the nucleolus.

Three Major Morphological Types

It is generally recognized that nucleoli are found as three major morphological types based on the distribution of granular elements.[7] One type is homogeneous, due to a uniform distribution of granules. A second is heterogeneous or mottled in appearance, due to clustering of granules. A

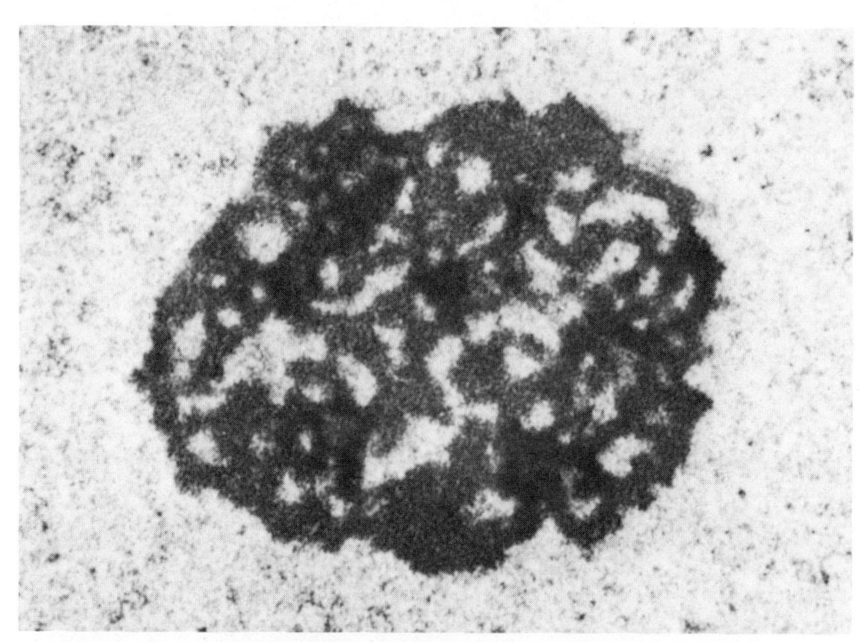

FIGURE 16.12 A nucleolus from a spermatogonium of opossum testis containing an irregular three-dimensional network of clustered granules called the nucleolonema. (Courtesy of Dr. Don W. Fawcett.)

third is ring shaped, due to granules clustered around the periphery of the nucleolus.

The electron micrograph of Figure 16.12 contains a nucleolus with a mottled appearance. The regions of clustered or segregated granules are called *nucleolonema*.

The Function of the Nucleolus

Before citing experimental evidence that has led to our present understanding of nucleolar function, we state its function briefly. The nucleolus is the cellular location for the formation and accumulation of ribosomal precursors. Other sites in the cell contribute to the manufacture of ribosome components, but it is in the nucleolus that these molecular components are first assembled as particles that will eventually become the seats of protein synthesis in the cytosol. It is therefore proper to think of the nucleolus not as an organelle within which events are taking place but rather as a depot of products being formed and accumulated before their transport to the cytosol.

Table 16.2 lists experimental evidence from many lines of research that implicates the nucleolus as the nuclear site of ribosomal RNA production. A careful study of the table will provide important background information for the discussion that follows.

Three events characterize the function of the nucleolus: the transcription of the genes that code for ribosomal RNA, the processing of the preribosomal RNA molecule, and the assembly of ribosome subunits. Let us examine these briefly in the order in which they take place in the nucleolus.

Table 16.2 Evidence That the Nucleolus Is the Site of Synthesis of Ribosomal RNA

1. The nucleolus contains rapidly synthesized 28S RNA, which is very similar in composition to ribosomal 28S RNA.
2. If the nucleolar RNA is labeled with a pulse of radioactive precursor, the label is initially found in 45S RNA and then is transferred from 45S to 35S to 28S RNA; *in vivo* treatment of cells with actinomycin D results in a rapid loss of 45S RNA initially, 35S later, and 28S finally; the general pathway is 45S RNA → 35S RNA → 28S RNA; since in some nucleolar preparations 85S and 60S RNA are labeled more rapidly than is 45S RNA, these may be oligomers of nucleolar 45S RNA and/or 28S RNA.
3. Isolated nucleoli contain a DNA-dependent RNA polymerase capable of biosynthesis of RNA.
4. Inhibition of biosynthesis of ribosomal RNA results from ultraviolet–microbeam irradiation of nucleoli.
5. Synthesis of ribosomes or ribosomal RNA does not occur in mutants that do not contain nucleolus organizers.
6. There is selective hybridization of 28S ribosomal RNA with nucleolus organizers and/or nucleolar DNA.
7. Hybridization of ribosomal RNA occurs with DNA of *Drosophila melanogaster* in direct proportion to the number of nucleolus organizers present.
8. Hybridization of nucleolar 28S RNA with nucleolar DNA is inhibited 85% by ribosomal 28S RNA and not by ribosomal 18S RNA; ribosomal 18S and 28S RNA inhibit hybridization of nucleolar 45S RNA by 15 and 40%, respectively.

Transcription of Ribosomal RNA Genes

The nucleolus organizer region of a chromosome, viewed in molecular dimensions, is a stretch of multiple copies of genes that code for ribosomal RNA (rRNA). The redundancy of these genes is especially high in plants, where as many as 31,900 copies have been determined in a tetraploid hyacinth.[6]

Method VI

When rat liver cells are pulse labeled with tritiated uridine, [^{3}H]UdR, and followed by autoradiography with the electron microscope, label appears first in the transition regions between intranucleolar chromatin and the adjacent fibrillar component of the nucleolus.[8] Within 5 min the label appears in the fibrillar zones and from there becomes associated with the granular regions and is spread throughout the nucleolus. Figure 16.13 demonstrates this sequential labeling of nucleolar components.

The fibrillar centers of the nucleolus are thought to be sites of the NORs where high molecular weight 45S pre-rRNA is synthesized.

Because of a technique developed by Miller and Beatty for viewing genetic transcription,[9] the activity of these multiple gene copies has been demonstrated in dramatic fashion. Figure 16.14 contains electron micrographs of dispersed and spread fibrillar components with highly active transcriptional complexes.

Several features of these spread preparations are apparent. Each rRNA gene being transcribed takes on the appearance of a Christmas tree, with fibrils splayed out from the DNA like branches from a trunk. Multiple copies of the genes in transcription are a striking feature, with each one separated from adjacent genes by spacer sequences.

Figure 16.15 illustrates transcriptional complexes in more structural

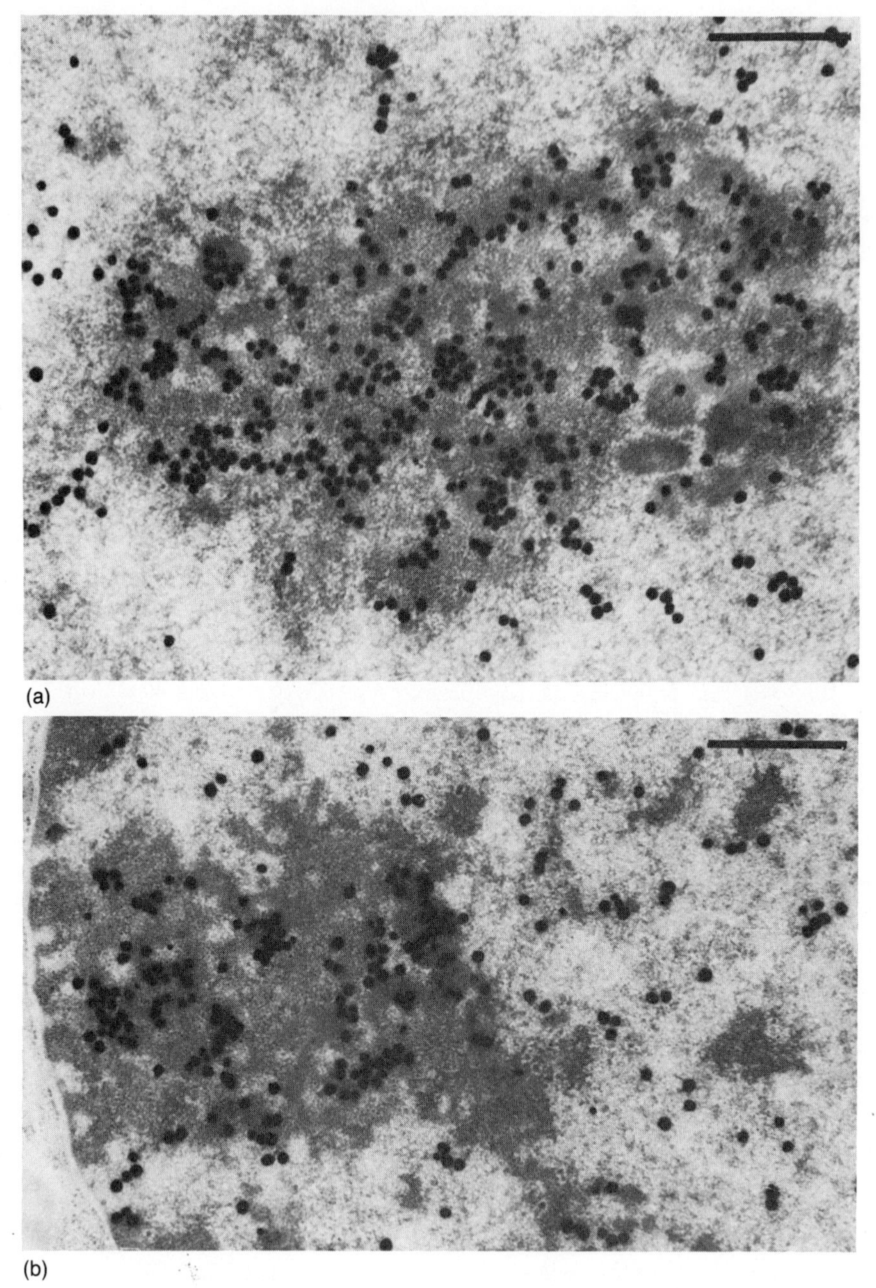

FIGURE 16.13 Demonstration of sequential labeling of the nucleolar components. (*a*) After 5 min incubation with [^{3}H]UdR, radioactivity is found primarily over the fibrillar component. (*b*) After 30 min, the radioactivity is distributed throughout the entire nucleolus. Bar = 1 μm. (Courtesy of Dr. S. Fakan.)

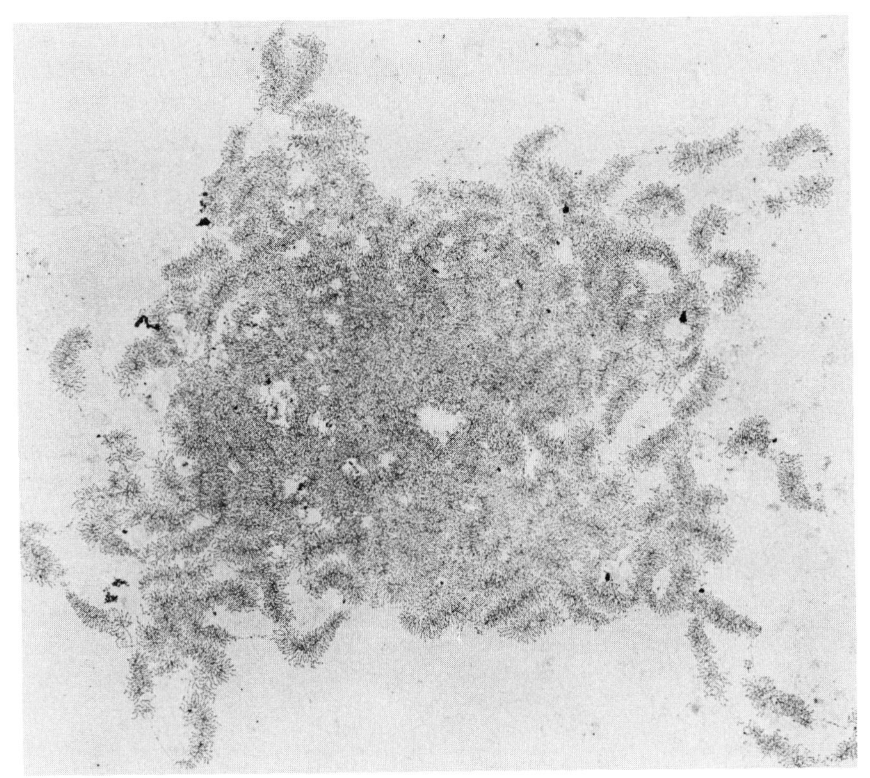

FIGURE 16.14 Transcriptional complexes isolated from *Acetabularia.* This is the fibrillar component of the nucleolus that has been dispersed and spread for electron microscopy. Although the complex is still extremely labyrinthine, fibrils are clumped together in units and extend from a single fiber axis. (Courtesy of Dr. Michael F. Trendelenburg.)

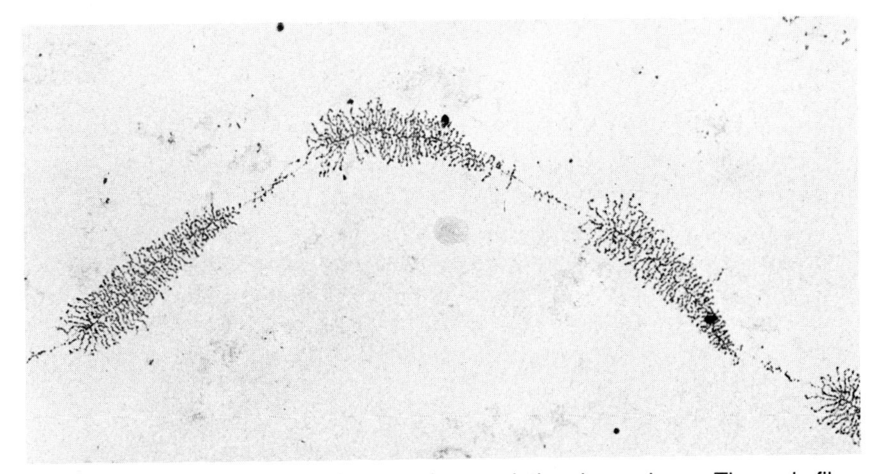

FIGURE 16.15 A well-spread group of transcriptional complexes. The main fiber axis is a DNA strand to which polymerase enzymes are bound. The fibrils that extend laterally are folded nascent chains of precursor ribosomal RNA. (Courtesy of Dr. Michael F. Trendelenburg.)

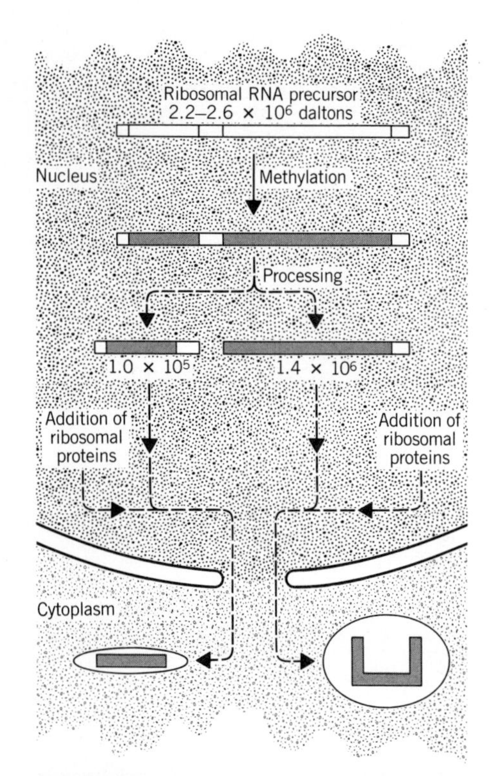

FIGURE 16.16 Ribosomal RNA processing. The initial product of synthesis is a high molecular weight 45S RNA that is methylated, cleaved, and ultimately reduced in size to an 18S and a 28S unit. These are complexed with proteins before exiting from the nucleus in the form of large and small ribosome subunits.

Eucaryotic ribosomes contain two subunits. The larger contains three rRNA molecules with sedimentation coefficients of 28, 5.8, and 5S. The smaller subunit contains a single 18S rRNA. The 45S rRNA particle houses one each of 28, 18, and 5.8S RNA. The 5S RNA is synthesized independently.

detail. The axis of the complex is a strand of DNA to which are attached RNA polymerases. From each RNA polymerase there extends a nascent chain of pre-rRNA. The shorter chains are toward the point of transcription initiation, and the longer chains are the products of reading longer and longer stretches of rRNA gene codons. As the nascent RNA chains grow longer, a terminal granule appears, giving the Christmas tree some peripheral trim.

Electron micrographs like this one of spread nucleolar components have made visible the molecular event of transcription in almost unparalleled dramatic fashion.

An interesting observation on transcriptionally active nucleolar chromatin is that it is devoid of nucleosomes, which are the fundamental chromatin units in nonnucleolar chromatin.[10] The significance of this is not certain, but it does appear that nucleolar DNA is fully extended or at least in a configuration different from that of most eucaryotic chromatin.

Processing of Preribosomal RNA

The RNA product generated by the transcription of rDNA is a 45S species that has a molecular weight on the order of 2.2–2.6 $\times$ 10^6 daltons. Numerous experimental approaches (see Table 16.2) have verified that this species, produced initially in the fibrillar component, is the same material that appears later in the granular regions. However, it is subjected to processing in transit from fibrillar to granular regions.

Figure 16.16 shows the sequence of events leading to the fully processed or mature rRNA. Basically, the precursor 45S rRNA is methylated, cleaved, and reduced in size to two smaller 18 and 28S units (rRNA processing is discussed at greater length in Chapter 17).

Methylation appears to take place primarily on the 45S precursor rRNA, either while it is still being formed or immediately thereafter. When radioactive methionine is used as the donor of the methyl group, the radioactivity appears first in the 45S material and later in the lower molecular weight processed products.

The principal site of methylation is the 2′-hydroxyl group of ribose, with lesser amounts of *N*-methylated bases also resulting.

After methylation, the 45S rRNA is cleaved by nucleases at several points, releasing the two major 28 and 18S ribosomal RNA. The excess non-rRNA fragments excised from the 45S precursor molecule are rapidly degraded. Their role in the overall process of rRNA generation has been the subject of speculations. One is that they may play a role in regulating transcription. Another, considered more likely, is that they may be essential for maintaining a correct rRNA conformation required for the methylation reaction or for complexing with proteins during the early stages of ribosome subunit formation.

Assembly of Preribosomal Subunits

The fully mature ribosome with its two subunits and four types of RNA is not assembled in the nucleus. In fact, it makes its first appearance as a complete particle in the cytosol during translation while complexed with mRNA.

But preribosomal subunits are formed in the nucleus, with the early stages of assembly taking place in the nucleolus, perhaps even before the rRNAs are completely transcribed and processed. The granular elements making up a significant portion of the nucleolus are ribonucleoprotein particles, the precursors of ribosome subunits. These are presumably the terminal granules that appear first of all on nascent RNA chains during transcription.

It is generally believed that preribosomal subunits migrate to the cytoplasm by way of the nuclear envelope pore, but the evidence that this is the case is not direct. Some final processing of the subunit may take place at the envelope, and a conformational change may be necessary before the subunit can be shuttled across to the cytosol.

The assembly of the mature ribosome is taken up in Chapter 17.

16.3 THE MORPHOLOGY OF EXTRANUCLEOLAR RIBONUCLEOPROTEIN COMPLEXES

Complexes of ribonucleic acid and protein, referred to as RNP, are now known to occupy the extranucleolar regions of the nucleus. Although some of these complexes have arisen from the nucleolus, and are thus preribosomal subunits, others have different fates and functions.

Figure 16.17 shows a nucleolus in a rat liver cell nucleus, along with a variety of granules that extend throughout the nucleus beyond the zone occupied by the nucleolus. Table 16.3 contains a brief description of the granule types depicted in the electron micrograph.

Perichromatin Fibrils

It is generally agreed that transcription does not take place on condensed chromatin. But many different experimental approaches have shown that the preferred site of transcription is in the peripheral regions of condensed chromatin. This is the region containing the morphological elements called perichromatin fibrils (PF). Note the location of these in Figure 16.17.

As chromatin decondenses and as extranucleolar RNA synthesis is stepped up, the density of PF increases. These and other observations have led to the conclusion that perichromatin fibrils are the first morphological elements that represent newly synthesized extranucleolar RNA.[8]

Perichromatin fibrils are strands of newly synthesized RNA that have been transcribed outside the nucleolus. These are early forms of mRNA.

When isolated rat liver cells are labeled with [^{3}H]UdR, the label appears initially in regions of PF. When chased, the label is seen to spread throughout the nucleoplasm, presumably throughout the interchromatin spaces. This movement of label upon chasing is illustrated in Figure 16.18.

Along with the movement of PF from sites of transcription toward the inner nuclear regions, there is a decrease in mean size of the RNA component of the fibril. This is interpreted to mean that the RNA is processed in transit, not necessarily to its final form, but at least the initial steps in processing are enacted during these events soon after transcription.

Perichromatin Granules

Granules that average about 35 to 50 nm in diameter are commonly observed in extranucleolar regions of the nucleus. It is commonly assumed that these are ribonucleoproteins and that they may be of two distinct types.

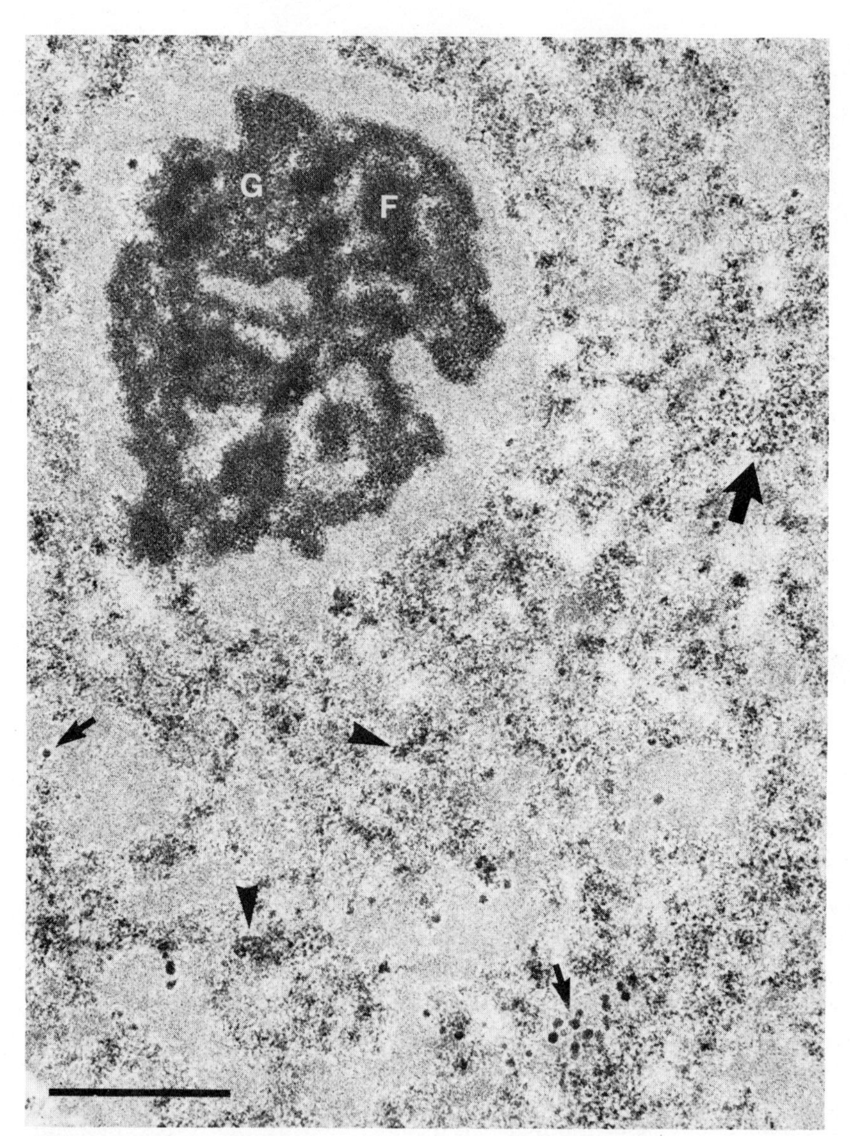

FIGURE 16.17 A general view of a rat liver nucleus showing the most prominent ribonucleoprotein particles and fibrils. PF, perichromatin fibrils (arrowheads); PG, perichromatin granules (small arrows); IG, interchromatin granules (large arrows); F, fibrillar and G, granular nucleolar components. Bar = 1 μm. (Courtesy of Dr. S. Fakan.)

Extranucleolar granules, called perichromatin granules, may have two different origins. Some may arise from the nucleolus (and thus contain rRNA), and others may come from extranucleolar DNA (and thus contain mRNA).

One type of perichromatin granule (PG) is thought to arise from the granular element of the nucleolus, thus containing rRNA complexed with protein. The other type of PG is believed to be generated from perichromatin fibrils as a result of processing and condensing with proteins.

Perichromatin granules appear to be storage forms of RNA produced from a variety of transcription sites. Although it has been difficult to prove experimentally that the extranucleolar PG does not contain rRNA, it is generally thought that pre-mRNA is the main product stored in these granules.

Table 16.3 Extranucleolar Fibril and Granule Types

Type	Dimensions (nm)	Location	Composition	Function
Perichromatin fibril (PF)	3–5, and up to 20	Peripheral region of condensed chromatin	Ribonucleoprotein	Newly synthesized extranucleolar RNA
Perichromatin granule (PG)	35–50	Peripheral region of condensed chromatin, plus border of nucleolus-associated chromatin; interchromatin space	Twisted 3 nm fibrils; thought to contain RNA	Store of pre-mRNA
Interchromatin granule (IG)	20–25	Random clusters	Limited amount of RNA	Not certain, may be part of interchromatinic matrix

If this is indeed the case, this form of RNA must move to the cytoplasm to participate in translation. But intact granules have never been seen to pass through the nuclear pore. They have, however, been observed near pores and are thought to uncoil and pass across the nuclear envelope as a bundle of filaments.

High resolution electron microscopy has revealed that perichromatin granules are actually a mass of filaments. This can be seen in the highly magnified preparations of PG shown in Figure 16.19. The filaments are about 3 nm thick, a diameter that would allow for RNA complexed with protein.

Interchromatin Granules

A final type of granule distributed randomly in clusters throughout the nucleoplasm is the interchromatin granule (IG). The IG are smaller (20–50 nm) than the PG and are interconnected by fibrils to form a network.

Clusters of IG and their interconnecting filaments are evident in Figure 16.20.

In contrast to perichromatin granules, these particles seem to possess only a limited amount of RNA. Their structure and function are for the present an enigma.

Some researchers have proposed that IG are a part of a network extending throughout the nucleus. It may be that these elements are a part of the interchromatinic matrix that we discussed at the beginning of this chapter.

16.4 HETEROGENEOUS NUCLEAR RIBONUCLEOPROTEIN COMPLEXES

In eucaryotic cells, messenger RNA is not formed as free mRNA in the nucleus, but is generated as a high molecular weight heterogeneous collection of molecules bound to proteins. The RNA is referred to as het-

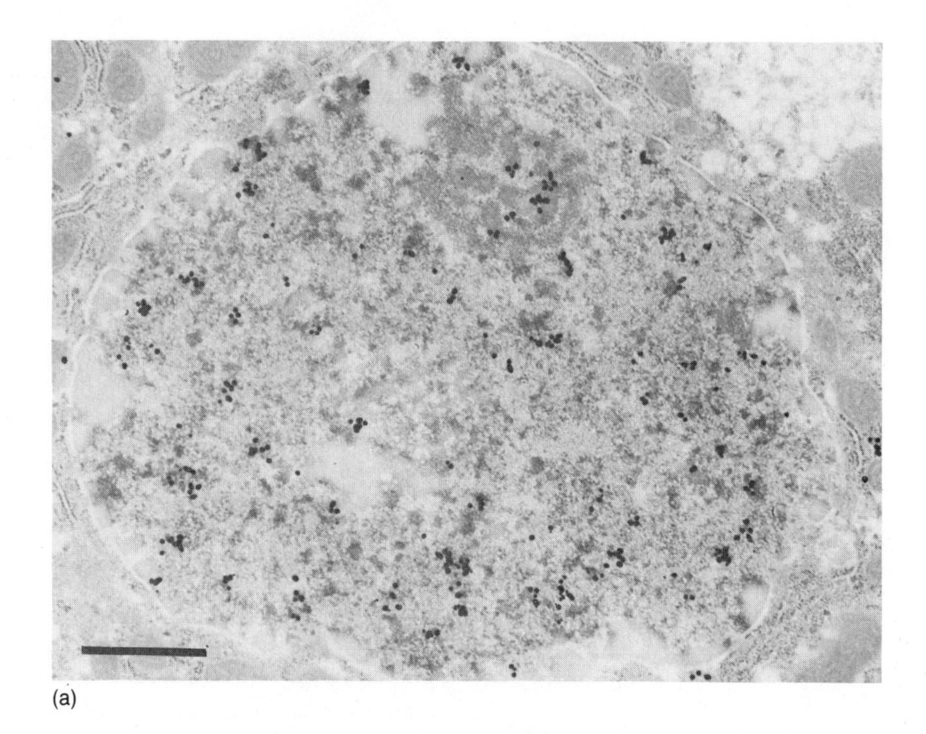

(a)

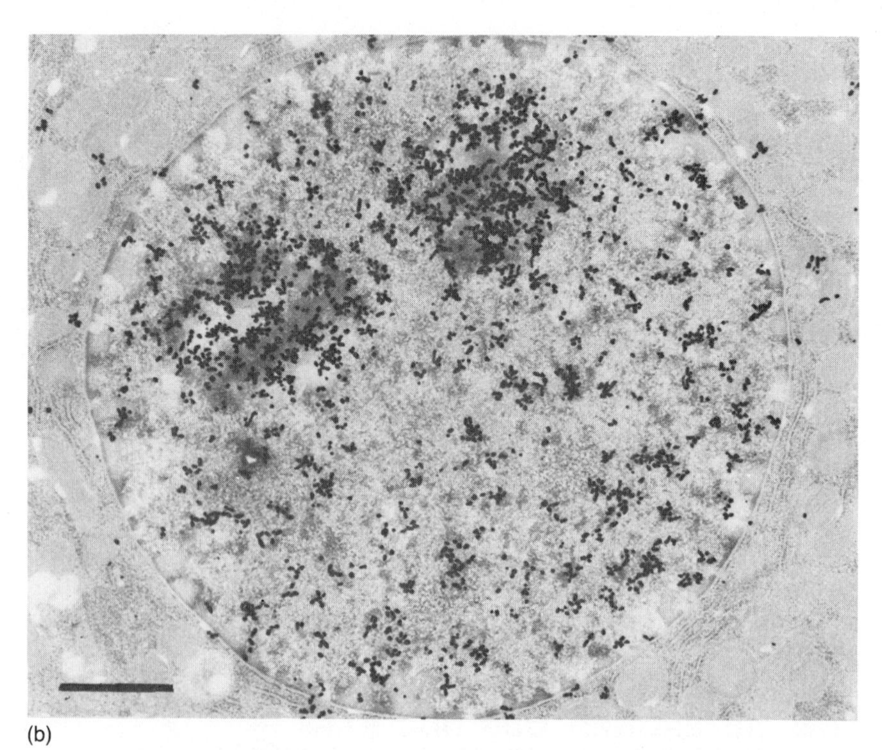

(b)

FIGURE 16.18 (*a*) Isolated rat liver cell labeled for 2 min with tritiated deoxyribouridine, [^{3}H]UdR. Radioactivity is mainly associated with PF in the proximity of condensed chromatin. (*b*) After a 5-min label with [^{3}H]UdR followed by 2 hr of cold UdR chase. Extranucleolar radioactivity, often associated with PF, is now distributed throughout the nucleoplasm. Bars = 1 μm. (Courtesy of Dr. S. Fakan.)

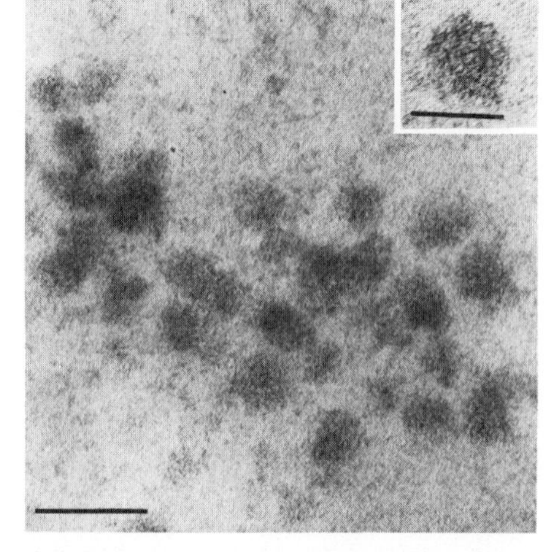

FIGURE 16.19 A group of perichromatin granules (PG) in the nucleoplasm of a rat liver cell. Bar = 0.1 μm. *Inset*: the irregularly coiled filamentous detail of a PG. Bar = 0.05 μm. (Courtesy of Dr. S Fakan.)

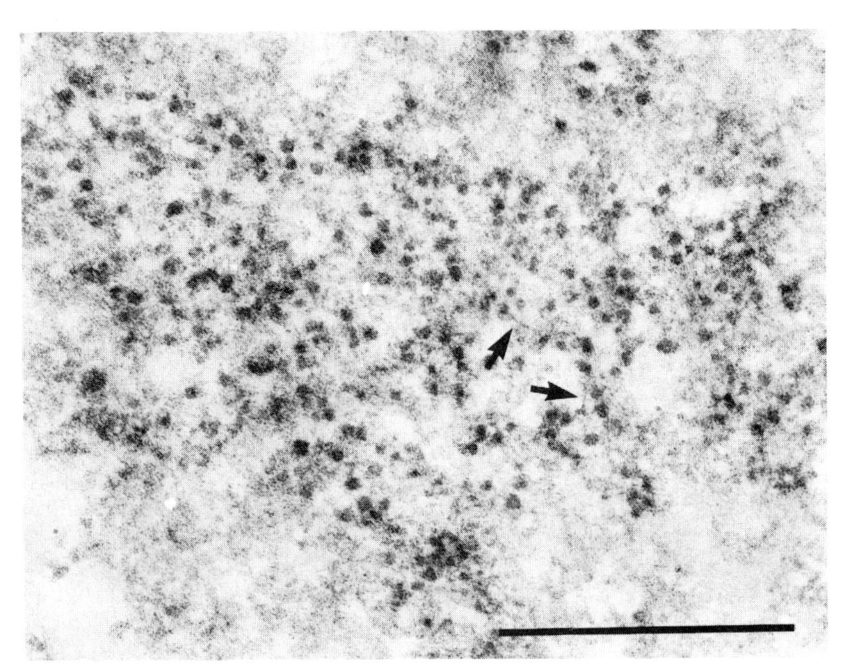

FIGURE 16.20 Clusters of interchromatin granules (IG) and their connecting filaments (arrows) in a rat liver cell. Bar = 0.5 μm. (Courtesy of Dr. S. Fakan.)

erogeneous nuclear RNA (hnRNA) or as premessenger RNA (pre-mRNA). The complex containing both RNA and protein is given the designation heterogeneous nuclear ribonucleoprotein (hnRNP). These complexes, discovered by Georgiev and co-workers[11] in 1965, are apparently the same as the perichromatin granules discussed in the preceding section.

Both hnRNA and pre-mRNA are appropriate designations for the form of RNA generated by transcription in extranucleolar regions of the nucleus. The first, hnRNA, reflects the high molecular weight and heterogeneity of the RNA. The second suggests its ultimate function and fate, that of becoming free messenger RNA to be translated into protein on ribosomes in the cytoplasm.

Composition and Properties of Heterogeneous Nuclear Ribonucleoprotein

Several procedures have been worked out for the isolation of hnRNP. Nuclei are first isolated from cells and the hnRNP purified by one of two general techniques. One is an isotonic extraction procedure by which it is possible to obtain high yields of RNP in the form of a 30S particle, a substructure of hnRNP. This procedure leaves the nucleus intact. A low level of nuclease activity appears to be necessary to completely free the RNP from internal nuclear structures.

Method II

The second procedure is to disperse the nuclear contents by sonication or cavitation induced by high pressure. This procedure permits a very rapid release of RNP.

DNA and RNA appear to have similar higher order morphologies within the nucleus.

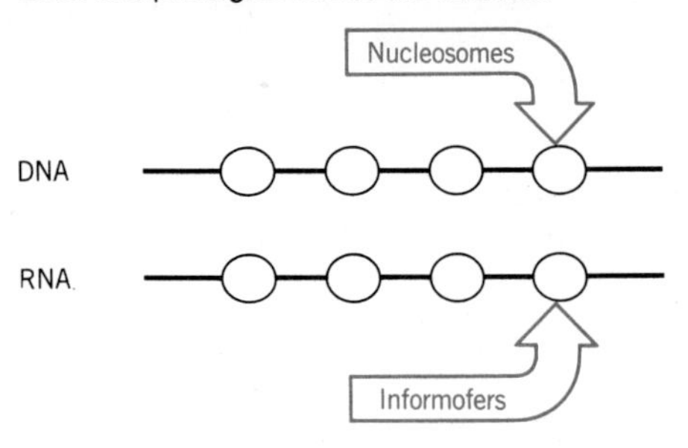

Purified hnRNP consists of protein and RNA and little else. Depending on how measurements are conducted, the particles contain between 75 and 90% protein. Conversion of the hnRNP to the 30S substructure does not appear to alter the ratio of protein to RNA.

Mild nuclease treatment of isolated hnRNP material converts it to a homogeneously sedimenting 30S unit. This unit contains about 700 to 1000 nucleotides of RNA and approximately 10^6 daltons of protein. The large complexes that yield these 30S substructures appear as a chain of monomers in the electron microscope. This observation, along with other experimental approaches, has led researchers to conclude that the physical form of hnRNP is a simple polymer of "beads on a string" like that noted for chromatin. In this case the string is RNA and the beads are identical protein particles called *informofers*.

Characteristics of Ribonucleoprotein RNA

The large hnRNP complexes contain an RNA that is also large and heterogeneous. The precise physical properties of this hnRNA are very difficult to establish because isolation results in partial breakdown and nicking of the RNA. It is now clear, however, that the primary transcript product is much larger than the mature mRNA that is derived from it.

According to the results of hybridization studies, only 5 to 10% of hnRNP-RNA is homologous to cytoplasmic mRNA, but all mRNA sequences appear to be represented in the 30S RNP particles.[1] These results suggest that only a fraction of the RNA in hnRNA ends up as mature mRNA. The remaining 90% or so must have other nonmessenger functions.

The large hnRNP complexes contain in addition to the 30S substructures a 15S nucleoprotein containing about 200 nucleotides in length. It is characterized as having a long poly A sequence of nucleotides. A poly A sequence, as we shall see shortly, is present on the 3′ end of mature mRNA.

Characteristics of Ribonucleoprotein Proteins

The number of proteins associated with RNA as RNP and their molecular weight distributions are unsettled research problems. Early studies suggested a single type of protein present in multiple copies in the 30S particle, but recent work has revealed the RNP to be somewhat more complex. It is generally agreed, however, that the 30S subcomplex contains a small number, perhaps less than a dozen, mildly basic proteins (pI's 8–9) that are relatively conserved among higher eucaryotes. There is a common finding of high proportions of glycine, low levels of cysteine, and the presence of the unusual amino acid *N,N*-dimethylarginine. These proteins have molecular weights in the vicinity of 34,000 to 40,000 daltons.

Some studies have yielded results indicating that large hnRNP complexes contain as many as 50 proteins. Thus there may be a variety of proteins associated with the RNA in spaces between the 30S subcomplexes.

The 15S polyadenylated (poly A) nucleoprotein subcomplex contains five major proteins ranging in molecular weight from 58,000 to 140,000 daltons. All these proteins can be found in unmodified hnRNP material.

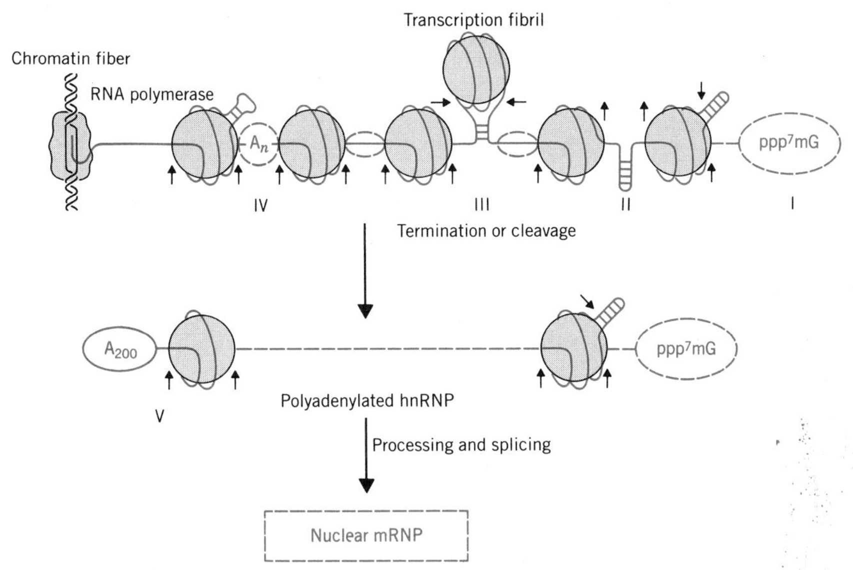

FIGURE 16.21 A model of hnRNP. The major hnRNP proteins fold the RNA chain to yield 30S RNP substructures. There are numerous RNase-sensitive sites (small arrows) between and perhaps within the 30S units. The 5′-capping group (I) and the double strand (II and III) and oligo A (IV) regions are not tightly bound to the 30S unit but may be associated with different proteins (broken lines). The 3′-terminal poly A_{200} is associated with a distinct class of polypeptides.

A Model for Heterogeneous Nuclear Ribonucleoprotein

Even though the physical and chemical details of the primary transcription products have not been formulated, it is helpful to view them in terms of a speculative model. One such model is presented in Figure 16.21.

The model explains a number of features of hnRNP as determined experimentally. As transcription is effected by RNA polymerase, the nascent RNA combines with proteins generating a ribonucleoprotein fibril. The binding of protein to the RNA takes place in such a manner that RNase-sensitive stretches are left between the RNP subcomplexes. Cleavage of the transcription fibril (hnRNP) with nucleases releases the 30S RNP subcomplexes.

A number of RNA stretches are not tightly adherent to the proteins of the 30S subcomplexes, and some of these runs are in hairpin loop conformation. These are removed by RNase treatment along with an oligo A region, which may be associated with its own set of proteins.

The two ends of the hnRNP are unique among RNA molecules. The 5′ end has a "cap" of unusual composition wherein the terminal ribose is linked 5′→5′ with a triphosphate bridge to 7-methyl guanosine (see Figure 16.22). The 3′ end, which is the last to be transcribed, contains a poly A stretch ranging in length from 100 to 200 nucleotides. This stretch is associated with its own distinctive set of proteins, none of which are major proteins in the 30S subcomplex. Both caps are posttranscriptional additions to the hnRNP.

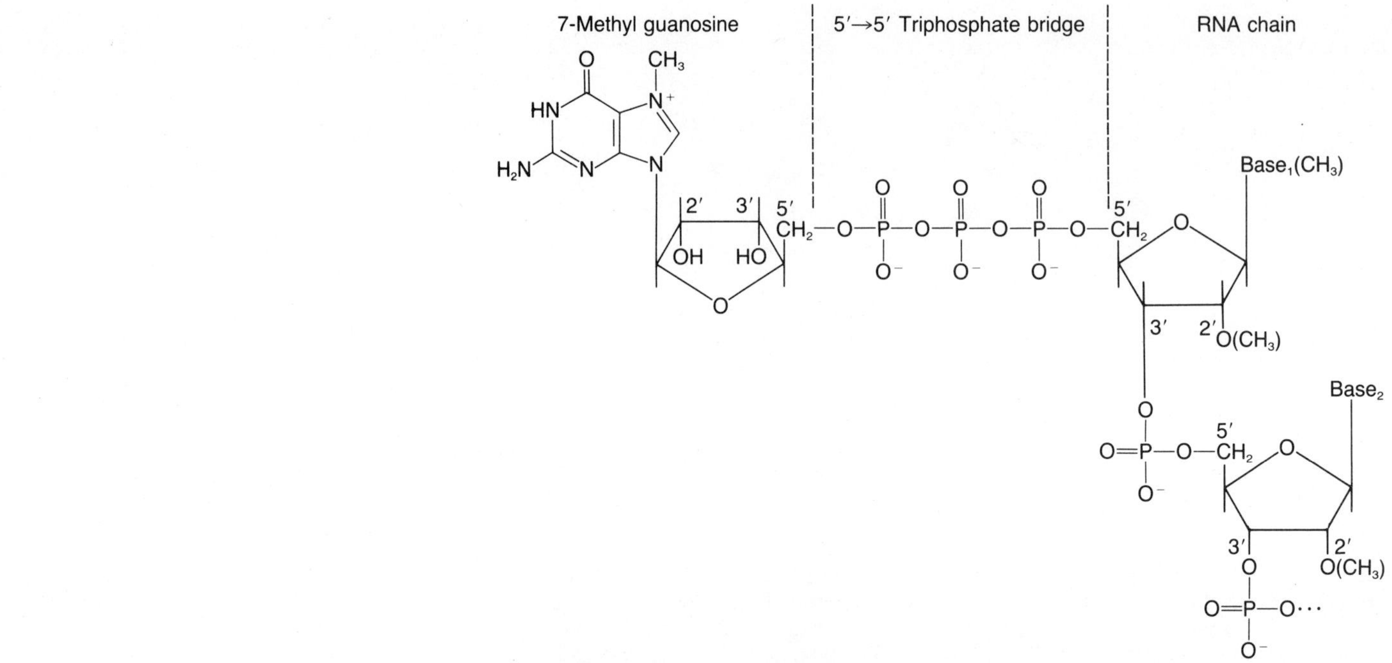

FIGURE 16.22 Structure of the cap at the 5′ end of hnRNP and ultimately the RNA product. It is unusual in possessing a 5′ → 5′ triphosphate bridge, thus leaving no terminal phosphate on this end of the molecule.

Mature mRNA, which is ultimately derived from this complex transcription fibril, contains both the poly A and the ^{7}mGppp caps, but it is much shorter than the RNA of the hnRNP complex. Thus a good portion of the hnRNA must be culled while still preserving the message and the capped ends. The manner in which this takes place is one of the most surprising molecular mechanisms used to synthesize a product.

16.5 PROCESSING OF EUCARYOTIC MESSENGER RNA

For hnRNA to become mRNA, it must be processed, a phenomenon that occurs within the nucleus before transport of the mRNA to the cytoplasm. Processing takes place on the primary transcript product as posttranscriptional events. There are two types of event: one to chemically modify or cap the ends, and a second to excise and eliminate certain sequences and splice together regions that must be retained.

The phenomenon of mRNA processing would not be necessary except for another fundamental property of the nucleus that has only recently come to light. This is the structure of the gene which, in eucaryotic cells, is commonly an interrupted or discontinuous sequence of codons, a structure now referred to as the *split gene*.

Split Genes

Procaryotic cells contain genes that consist of continuous stretches of codons without interruption. Each gene is transcribed into a primary

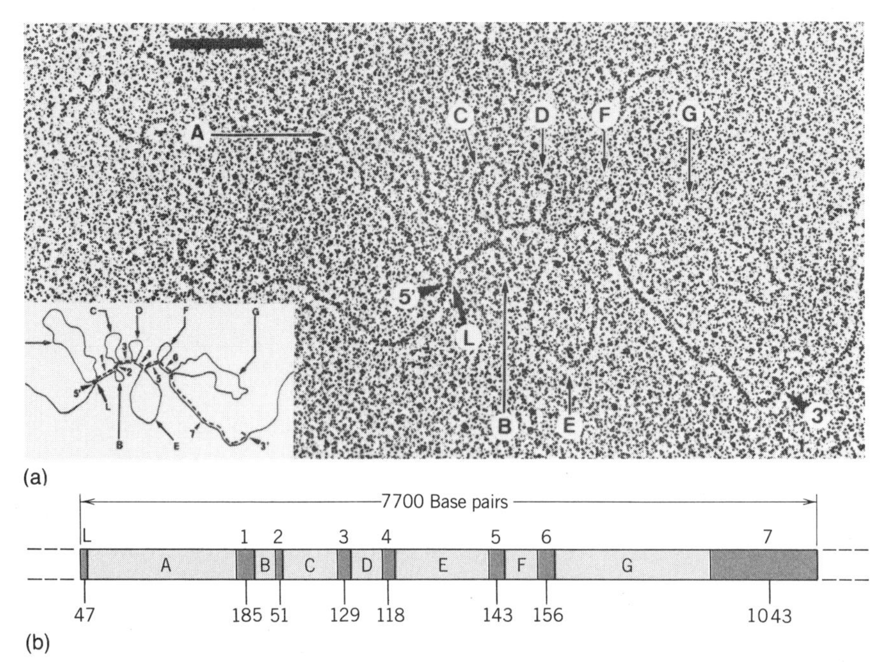

FIGURE 16.23 The split ovalbumin gene. (*a*) Electron micrograph of a single strand of DNA hybridized with mRNA, the normal transcription product of the ovalbumin gene. (Courtesy of Dr. Pierre Chambon.) The RNA binds to segments of the DNA (see dotted line of inset), leaving unbound portions of the DNA looped out. The segments of DNA that constitute the gene are numbered L, 1–7. These are exons. The loops, called introns, are designated by letters A to G in inset and in (*b*). The number of base pairs in each exon is shown. Introns range in size from 251 base pairs (B) to about 1600 base pairs (G).

transcript product, mRNA, which is the same length as the gene and specifies the sequence of amino acids assembled into a single protein. This is the principle of colinearity, which was discovered in bacteria and propagated as a molecular principle that was assumed to be universal among living systems.

The discovery in 1977 that genes in eucaryotes are *discontinuous* and that the concept of colinearity does not hold in these systems represented a sharp departure from the DNA → mRNA → protein dogma. Genes containing inserted noncodon sequences were found for such common proteins as hemoglobin,[12] ovalbumin,[13] and immunoglobulin light chains.[14]

The relation between nucleotide numbers in DNA and amino acids in ovalbumin negates the possibility of direct colinearity.

DNA	7700 base pairs
↓	
Mature RNA	1872 nucleotides
↓	
Ovalbumin	386 amino acids

One of the most dramatic examples of split genes studied so far is the gene for the major egg-white protein ovalbumin, a chain of 386 amino acids. The stretch of DNA that codes for this protein contains 7700 base pairs.[15] The length of RNA that serves as the mature messenger contains 1872 nucleotides. These numbers are obviously a denial of colinearity, for the rules of colinearity state that for every three nucleotides of DNA, three nucleotides of RNA are transcribed, and for every three nucleotides of RNA, one amino acid is incorporated into protein.

It is clear that the length of DNA encompassed by the gene is much greater than needed for the length of protein produced.

Figure 16.23*a* is an electron micrograph of the ovalbumin DNA hybridized with its mature mRNA *in vitro*. The hybrid is characterized by having unpaired loops where the two molecules have no complementary sequences. The inset shows how the loops permit the pulling together of widely spaced sequences of DNA so that the effect is a continous stretch of DNA that is complementary to the product mRNA.

Figure 16.23 also includes a representation of the split ovalbumin gene. Those regions designated L, 1, 2, 3, 4, 5, 6, and 7 are the real gene components. They have been termed *exons*. The regions designated A, B, C, D, E, F, and G are the loops for which there are no complementary RNA sequences. These regions are called *introns* or *intervening sequences*.

Capping and Splicing of Messenger RNA

The primary transcript product, which we referred to as hnRNA in the preceding section, represents an RNA that is complementary to the *entire* split gene length. This large RNA contains *both intron and exon* transcripts.

The production of the mature mRNA from hnRNA is a multistep process, and all its stages take place in the nucleus before export to the cytoplasm.

The first step in maturation is capping and polyadenylation. Capping is done at the 5′ end of the RNA and a poly A chain is added to the 3′ end. The capped end is significantly different from the natural 5′ end of an RNA, which is usually a 5′-triphosphate terminus. Although the triphosphate stretch is still present, it is sheltered by a nucleoside terminus with a free 3′-hydroxyl group (see Figure 16.22). The length of poly A on the 3′ end varies from 50 to 250 nucleotides depending on the species. The precise point in maturation at which the ends of RNA are modified in these two ways and the location and nature of the enzymes involved are still matters of investigation and discovery. But the results so far suggest that capping takes place *early* after the initiation of transcription or maybe even *directly* when it is initiated.

The second step in maturation involves a phenomenon called splicing. This is illustrated schematically for ovalbumin in Figure 16.24. This is a process that must be carried out with high precision, for if the mature mRNA that results varies in its nucleotide sequence by even one member, the message will become nonsense.

Although the specificity of splicing must be high, the details of this stage have not been worked out. But as a result of observations first reported by P. Chambon in 1978, the rule called *Chambon's rule* has emerged, namely, that splice points in mRNA may all have a common sequence: —(GU......AG)—. This rule has held true so far for nearly 100 exon–intron junctions that have been sequenced. Near these junctions there are additional "consensus" sequences in many cases that may contribute toward specificity.

In spite of the close correlation of common sequences and junction points, these may only be indirectly responsible for the specificity of cleavage. The sequences may be responsible for folding the RNA in a particular way, or they may promote an interaction with proteins. The folded or protein-associated regions may then be cleaved by endonucleases.

The roles of the caps and poly A tails on mRNA are largely a puzzle. They may protect and stabilize the message or assist in speeding up translation in eucaryotes. It is also speculated that splicing may be pre-

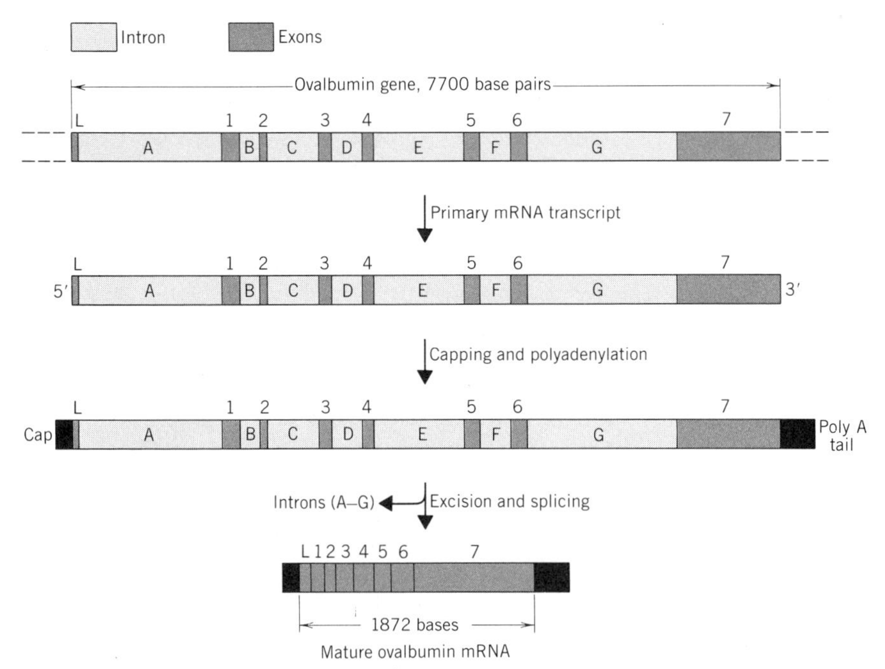

FIGURE 16.24 Stages in the maturation of mRNA from the ovalbumin gene. The entire gene, with exons and introns, is first transcribed and the primary transcript is then capped at the 5′ end and a poly A tail added to the 3′ end. Then introns are enzymatically excised and the segments spliced until only exons are left. Once the mRNA has thus matured, it is shipped off to the cell cytoplasm.

requisite for exit from the nucleus and thus an essential act for transport in the eucaryotic cell.

Splicing seems to be an extravagant and risky process, for often more RNA is discarded than is salvaged, and a one-nucleotide miss would spell disaster to the message. Whatever the reason for splicing and the mechanisms whereby it takes place and is controlled, the phenomenon does represent a unique and stimulating challenge for future research.

Having completed this chapter, it may be well to return to the beginning and read again the quotation from Terence Martin and his co-workers. It would be a mistake to think that there is nothing in this field left to discover.

Summary

The nucleus can be considered to consist of four distinct regions: the envelope, the nucleolus, condensed chromatin, and granular and fibrous material. The latter materials are made up of diffuse chromatin and a structural network called the matrix.

Purified matrix material may be obtained from isolated nuclei by a multistep procedure employing DNase and low salt extraction, extraction with high salt, solubilization with detergent, and a final DNase–RNase hydrolysis. That which remains is a nuclear skeleton of three major proteins that maintains the gross size and shape of the original nucleus.

The nuclear matrix is proposed to have functions with respect to the cell activities of DNA replication, transcription, and the posttranscriptional processing and transport of RNA products.

The nucleolus is a depot of RNA and protein that is being assembled into preribosomal subunits. It contains a fibrillar and a granular zone, the former being sites where nascent rRNA is transcribed and the latter regions where ribosomal proteins are condensing with the rRNA.

The functions of the nucleolus are transcription, processing, and assembly. Each of these functions is pertinent to the formation of ribosomes.

Outside the nucleolus, perichromatin fibrils, perichromatin granules, and interchromatin granules may be found. Perichromatin fibrils are apparently early forms of mRNA, while extranucleolar perichromatin granules are probably a storage form of mRNA, a ribonucleoprotein complex. Interchromatin granules are largely a mystery.

Messenger RNA, when first transcribed, appears as a high molecular weight heterogeneous collection of molecules complexed with protein referred to as hnRNP. Mild nuclease treatment reduces hnRNP to a homogeneous 30S unit. The high molecular weight primary transcript may contain as many as 50 proteins plus the RNA core.

The 5′ end of the hnRNP RNA has an unusual 7-methyl guanosine cap and the 3′ end contains a long stretch of poly A residues.

The conversion of hnRNA to mRNA involves excision of regions called introns and a splicing together of the remaining exons that when linearly attached, contain the message of the RNA. All these events appear to take place in the nucleus before the transport of mRNA across the nuclear envelope to its cytosolic site of function.

References

1. Structure and Function of Nuclear and Cytoplasmic Ribonucleoprotein Complexes, T. E. Martin, J. M. Pullman, and M. D. McMullen, in *Cell Biology, A Comprehensive Treatise,* vol. 4, D. M. Prescott and L. Goldstein, eds., Academic Press, New York, 1980.
2. Comparative Cytological Studies with Special Regard to the Nucleolus, T. H. Montgomery, *J. Morphol.* 15:265(1898).
3. Über totale und partielle somatische Heteropyknose, sowie strukturelle Geschlectschromosemen bei *Drosophila funebris,* E. Heitz, *Z. Zellforsch. Mikrosk. Anat.* 19:720(1933).
4. The Relation of a Particular Chromosomal Element to the Development of the Nucleoli in *Zea mays,* B. McClintock, *Z. Zellforsch. Mikrosk. Anat.* 21:294(1934).
5. Dynamic Properties of the Nuclear Matrix, R. Berezney, in *The Cell Nucleus,* vol. VII, H. Busch, ed., Academic Press, New York, 1979.
6. The Plant Nucleus, E. G. Jordan, J. N. Timmis, and A.J. Trewavas, in *The Biochemistry of Plants,* vol. 1, N. E. Tolbert, ed., Academic Press, New York, 1980.
7. The Nucleolus and Nucleolar DNA, K. Smetana and H. Busch, in *The Cell Nucleus,* vol. I, H. Busch, ed., Academic Press, New York, 1974.
8. The Ultrastructural Visualization of Nucleolar and Extranucleolar RNA Synthesis and Distribution, S. Fakan and E. Puvion, *Int. Rev. Cytol.* 65:255(1980).
9. Portrait of a Gene, D. L. Miller and B. R. Beatty, J. Cell Physiol. 74, suppl. 1:225(1969).

10. Organization of Nucleolar Chromatin, W. W. Franke, U. Scheer, H. Spring, M. F. Trendelenburg, and H. Zentgraf, in *The Cell Nucleus,* vol. VII, H. Busch, ed., Academic Press, New York, 1979.
11. Isolation of Nuclear Nucleoproteins Containing Messenger Ribonucleic Acid, O. P. Samarina, A. S. Asrijan, and G. P. Georgiev, *Proc. Natl. Acad. Sci. USSR* 163:1510(1965).
12. Intervening Sequence of DNA Identified in the Structural Portion of a Mouse β-globin Gene, S. M. Tilghman, D. C. Tiemeier, J. G. Seidman, B. M. Peterlin, M. Sullivan, J.V. Maizel, and P. Leder, *Proc. Natl. Acad. Sci. U.S.* 75:725(1978).
13. Ovalbumin Gene Is Split in Chicken DNA, R. Breathnack, J. L. Mandel, and P. Chambon, *Nature* 270:314(1977).
14. Variable and Constant Parts of the Immunoglobulin Light Chain Gene of a Mouse Myeloma Cell are 1250 Nontranslated Bases Apart, C. Brack and S. Tonegawa, *Proc. Natl. Acad. Sci. U.S.* 74:5652(1977).
15. Split Genes, P. Chambon, *Sci. Am.* 244:60(1981).

Selected Books and Articles

Books

The Nucleolus, H. Busch and K. Smetana, Academic Press, New York, 1970.

The Nucleus, A. J. Dalton and F. Haguenau, eds., Academic Press, New York, 1968.

Articles

Dynamic Properties of the Nuclear Matrix, Ronald Berezney, in *The Cell Nucleus,* vol. VII, H. Busch, ed., Academic Press, New York, 1979.

Organization of Nucleolar Chromatin, W. W. Franke, V. Scheer, H. Spring, M. F. Trendelenbrug, and H. Zentgraf, in *The Cell Nucleus,* vol. VII, H. Busch, ed., Academic Press, New York, 1979.

RNA Processing and the Intervening Sequence Problem, J. Abelson, *Annu. Rev. Biochem.* 48:1035(1979).

Split Genes, P. Chambon, *Sci. Am.* 244:60(1981).

Structure and Function of Nuclear and Cytoplasmic Ribonucleoprotein Complexes, T. E. Martin, J. M. Pullman, and M. D. McMullen, in *Cell Biology, A Comprehensive Treatise,* vol. 4, D. M. Prescott and L. Goldstein, eds., Academic Press, New York, 1980.

The Nucleolus and Nucleolar DNA, K. Smetana and H. Busch, in *The Cell Nucleus,* vol. I, H. Busch, ed., Academic Press, New York, 1974.

The Ultrastructural Visualization of Nucleolar and Extranucleolar RNA Synthesis and Distribution, S. Fakan and E. Puvion, in *Int. Rev. Cytol.,* G. H. Bourne, J. F. Danielli, and K. W. Jeon, eds., Academic Press, New York, 1980.

17

Ribosomes

Models provide a means of summarizing existing data in a way that provokes further questions and of focusing on the areas of our ignorance. Finally, they allow concentration on the interrelation between structure and function. We believe that our knowledge of ribosome structure is now at a stage where such questions can be seriously considered.

K. E. VAN HOLDE
W. E. HILL, 1974

The field today seems inundated by facts. No single investigator even attempts to keep abreast of all developments any longer. Yet, the problem is not facts per se; *it is a conceptual structure incapable of weaving the facts into a coherent, comprehensible picture.*

CARL R. WOESE, 1980

Ribosomes were first studied in the early 1930s, discovered and isolated in the early 1940s, scrutinized in the 1950s, and christened in 1958. In the 1960s they were dissociated and reconstituted, in the 1970s sequenced and studied topographically; and in the 1980s they continue to be the object of considerable research effort.

The statement that ribosomes were studied *before* they were discovered is not in error. Investigators in the 1930s noted a number of important properties of the cytoplasm that were due to ribosomes, but they were not aware of the particles that were responsible for these properties. It was noted, for example, that during spermatogenesis the amount of RNA in the cell was related to the size of the cytoplasm,[1] an early and indirect observation pointing to the cytoplasmic location of the cellular RNA. More direct studies, employing basic staining techniques that discriminated between DNA and RNA, and spectrophotometric measurements of absorption in different cell regions, confirmed that RNA is present in the cytoplasm of both plant and animal cells and suggested that DNA is found exclusively in the nucleus.

Using quantitative techniques at the close of the 1930s and the beginning of the 1940s, an enormously important observation was made concerning ribosome function. It was reported that cells were rich in RNA when they were active in protein synthesis. Secretory cells, as an example, were noted to be RNA-rich, whereas cells of other types, such as heart muscle, which makes little new protein, were relatively RNA-poor. The important conclusion drawn from these observations was that a direct correlation existed between the *rate* of protein synthesis in a cell and the *amount* of RNA it contained. RNA was thus linked to protein synthesis, an idea that was ultimately to attract hundreds of investigators into an era of research that uncovered some of the cell's greatest molecular secrets. Although these early results were obtained because of the presence of ribosomes, the organelles had not yet been discovered.

Method II

Ribosomes were not able to elude researchers for long. Around the turn of the decade Albert Claude homogenized chick and mammalian embryos and obtained a fraction containing what he called microsomes, particles of ribonucleoprotein and lipid visible with the dark field microscope. These were, of course, bits and pieces of the endoplasmic reticulum with attached ribosomes, but without electron microscopy ribosomes were still one step from being sighted as individual particles. In the mid-1950s their presence in both free and membrane-attached forms was confirmed by G. E. Palade and P. Siekevitz. Using electron microscopy, these particles were observed to be about 20 nm in diameter and ubiquitous in all tissues examined from a variety of sources.

Method III

In the 1950s yet another technique was employed to study biological systems that also gave another perspective of ribosomes. The technique was ultracentrifugal analysis. Ribosomes sedimented as discrete peaks in the 40S–76S range. When purified by centrifugation and electrophoresis, they were found to contain about half RNA and half protein.

The studies cited above, which were concerned with the location of ribosomes within the cell, their size, and composition provide the foundation for the first definitive results on ribosome function. In the mid-1950s, P. Zamecnik and his co-workers showed clearly that radioactive amino acids *first* were incorporated into proteins on ribosomes and *then* were released to the soluble portions of the cell. The definition of the ribosome as the seat of protein synthesis was thus established.

Consequently, by the middle of the twentieth century ribosomes were recognized as particles 20–30 nm in diameter, containing about half RNA and associated with protein synthesis. But they did not yet bear an official name.

In 1958 the papers presented at a meeting of the Biophysical Society at the Massachusetts Institute of Technology were published in book form.[2] R. B. Roberts edited the collection of papers and coined the name *ribosome* in his introductory comments. Thus this ribonucleoprotein particle was finally named, about 30 years after it became the subject of extensive investigation. By contrast, things nowadays are given a name, generally by the media, often long before they are seriously studied.

In the 1960s ribosomes were subjected to exhaustive electrophoretic and chromatographic procedures, this time not to purify them but to examine their parts. It soon became clear that ribosomes contain three or four kinds of RNA and scores of proteins. In the decade that followed, these components were subjected to careful compositional and structural analyses so that today we may view the primary structures of several of the components, both RNA and protein.

Present work on ribosomes is in a phase of protein and RNA topography and clarification of the relation between their three-dimensional architecture and their function. Looking back, we can see that a great deal has been accomplished by many researchers and by sophisticated instrumentation. Looking ahead, we can anticipate answers to numerous questions concerning this most complex particle on which is based the extraordinarily complicated process of protein synthesis.

In the pages that follow we will focus on the ultramolecular and molecular structures of the ribosome and the relation of these structures to their function.

Table 17.1 Major Ribosome Classes Based on Sedimentation Properties

Ribosome Type	Sedimentation Coefficient
Eucaryote (cytosol)	80S
Procaryote	70S
Eucaryote (organelle)	
Chloroplasts	70S
Mitochondria	55S–80S

17.1 THE GENERAL PHYSICAL PROPERTIES OF RIBOSOMES

There are fundamentally three classes of ribosomes within cells. Procaryotic cells possess a single class called 70S because of their sedimentation properties. Eucaryotic cells possess members of two classes. Those in the cytoplasm are 80S in their sedimentation properties, whereas within the mitochondria and chloroplasts the ribosomes are related but not identical to the 70S type of procaryotic ribosome. Many of the latter have sedimentation properties that are not strictly 70S. Table 17.1 summarizes some of the sedimentation properties of these three types of ribosome.

S stands for the Svedberg unit(s), which is a measure of sedimentation rate. The S value or sedimentation coefficient is not linear with molecular weight. The Svedberg, who developed the ultracentrifuge, is the one honored by this designation.

The most thoroughly studied and completely understood ribosome is the 70S ribosome of the bacterium *Escherichia coli*. Next in line as an object of investigation is the 80S eucaryotic ribosome. Organelle ribosomes are the least well characterized of the three types.

Procaryotic Ribosomes

Most information available regarding procaryotic ribosomes has been obtained from *E. coli*. Significant deviations from the properties of ribosomes possessed by this bacterium are not likely to be found among most other procaryotic ribosomes.

The 70S Particle

The 70S particle, which is the intact or complete ribosome, has been difficult to characterize with a high level of definition. The basic problem is that it varies somewhat from preparation to preparation, containing varying amounts of mRNA, tRNA, and other factors that are extraribosomal ingredients of the machinery of protein synthesis.

Most experimental approaches using electron microscopy have yielded dimensions for the 70S particle in the vicinity of $16 \times 18 \times 20$ nm.

This particle depends for its integrity on a magnesium ion concentration slightly greater than 2 m*M*. At higher concentrations, 70S particles dimerize to more rapidly sedimenting 100S units. At concentrations of magnesium below 2m*M*, 70S particles dissociate into 30S and 50S subunits. Thus, in response to magnesium ion concentration, 70S ribosome monomers are formed transiently from a pool of subunits (see Figure 17.1).

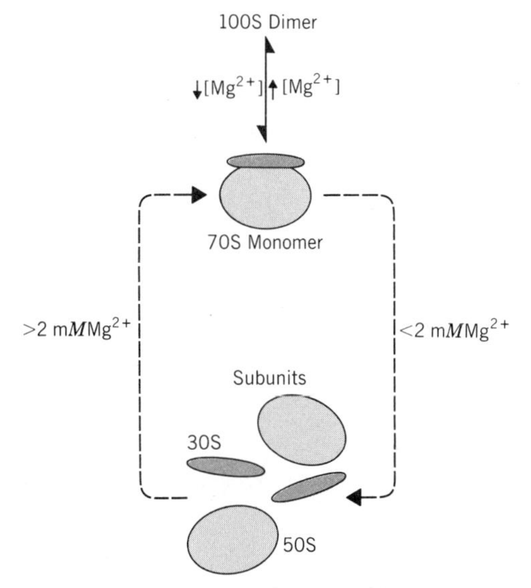

FIGURE 17.1 Mg^{2+}-dependent ribosome transformations.

These transformations are related to what occurs in the cell. When ribosomes are actively engaged in protein synthesis, they are intact 70S particles. When they are not translating RNA into protein, they dissociate into 30S and 50S subunits. Thus the 70S particle is probably rarely if ever

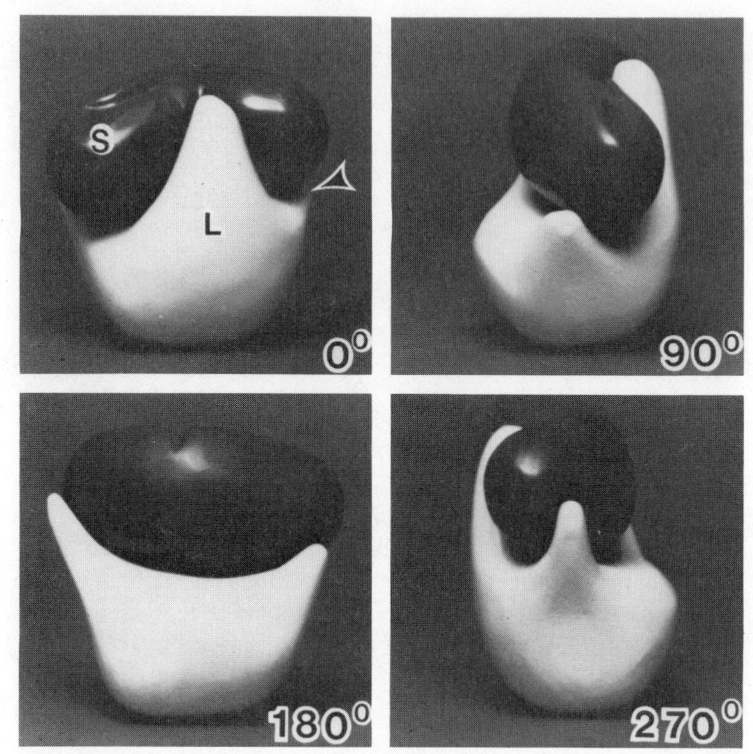

FIGURE 17.2 Three-dimensional model of 70S procaryotic ribosome. Four different views are presented; in each the 30S subunit (S) is in black and the 50S subunit (L) in white. (Courtesy of Dr. Miloslav Boublik.)

made up of the same two subunits on more than one occasion. But it makes no difference, because all 30S and all 50S subunits are structurally indistinguishable and apparently functionally equivalent.

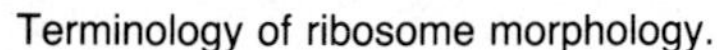
Terminology of ribosome morphology.

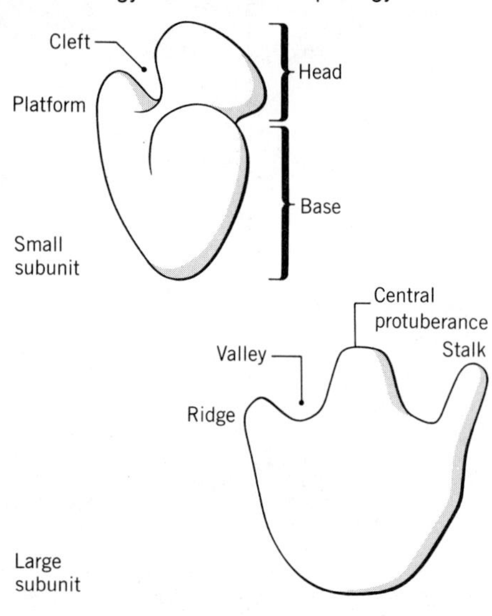

The 30S and 50S Subunits

A model of the 70S ribosome is presented in Figure 17.2. The two subunits that make up the intact particle are clearly complicated and irregular in shape. The 50S subunit has a spherical base with three protuberances. The 30S subunit is more rodlike than the 50S particle, having an axial ratio of 2:1. The subunit is deeply furrowed by a cleft, separating the rodlike structure into two lobes.

These subunits and the manner in which they fit together can be depicted conveniently in schematic form as shown in Figure 17.3. Note especially the tunnel that exists between the two subunits. A good deal of experimental evidence supports the idea that mRNA is threaded through a tunnel of this sort when it is undergoing translation.

Hydrodynamic studies on ribosomes and their subunits have yielded the molecular weight information provided in Table 17.2. The 30S subunit, when isolated under conditions described as ''unwashed,'' has a molecular weight of 1.0×10^6 daltons. When washed with NH_4Cl or precipitated

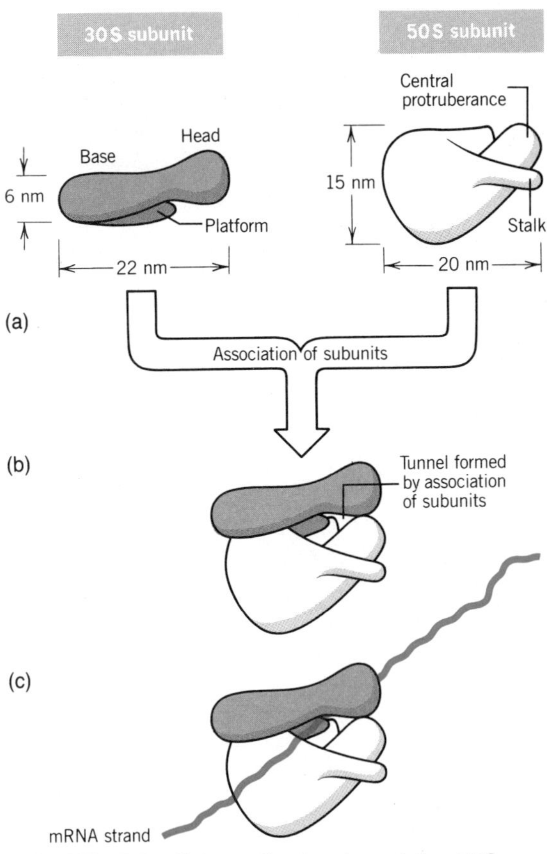

FIGURE 17.3 Schematic drawing of the 70S procaryotic ribosome. This convenieht and commonly used depiction of the two subunits shows that when they associate, a tunnellike space is left between them through which mRNA may be threaded during translation.

Table 17.2 **Sedimentation Coefficients and Molecular Weights (daltons) of Ribosome Monomers and Subunits**

	Eucaryote	Procaryote
Monomer Unit		
Sedimentation coefficient	80S	70S
Molecular weight	4.5×10^6	2.7×10^6
Small Subunit		
Sedimentation coefficient	40S	30S
Molecular weight	1.5×10^6	0.9×10^6
Large Subunit		
Sedimentation coefficient	60S	50S
Molecular weight	3.0×10^6	1.55×10^6

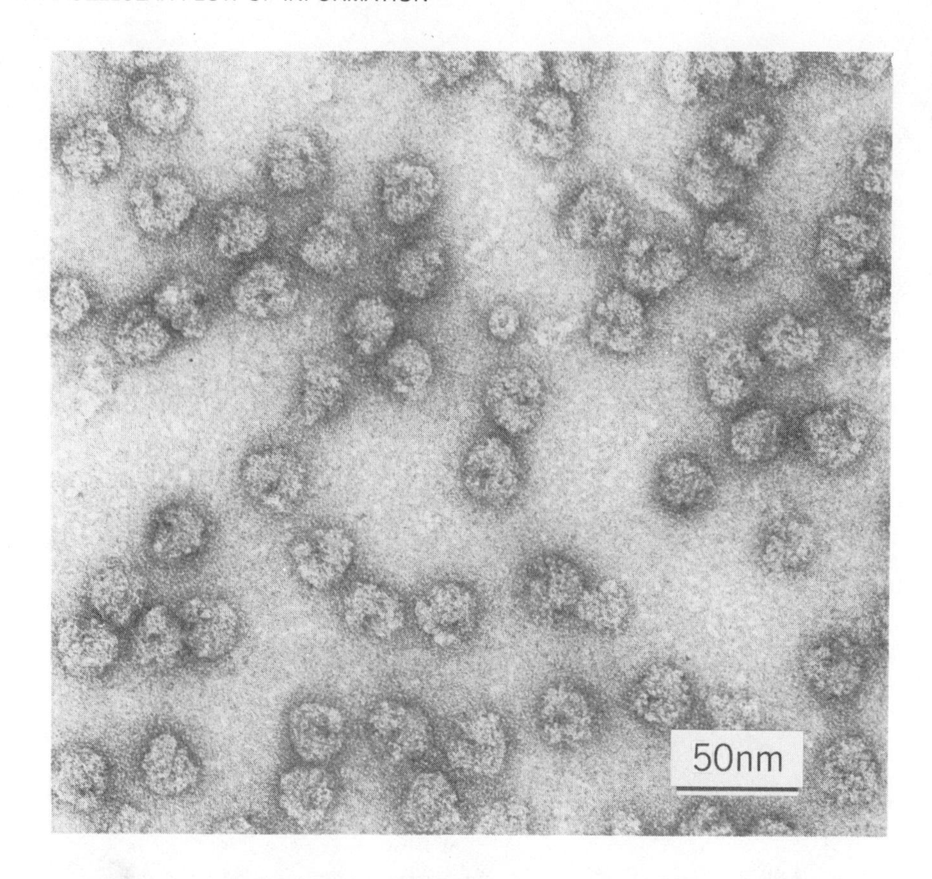

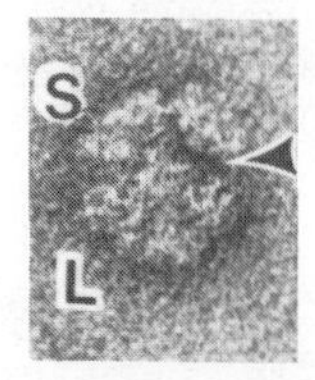

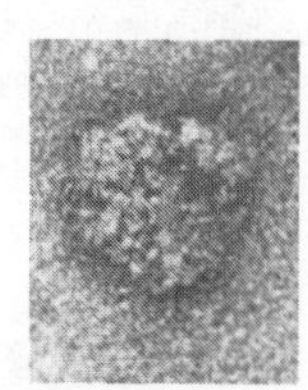
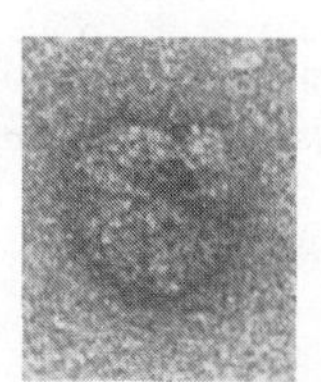
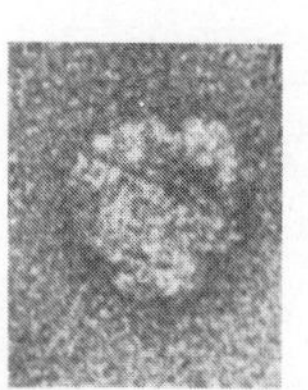
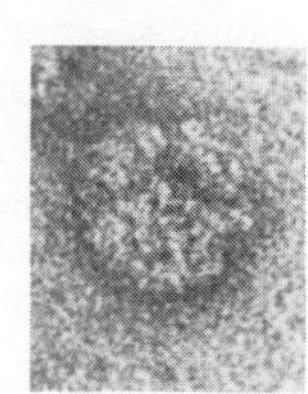

FIGURE 17.4 The 80S ribosomes from *Dictyostelium discoideum*, a eucaryotic slime mold. The lower panel contains individual ribosomes on which interfaces (arrow) between small (S) and large (L) subunits can be seen. (Electron micrographs courtesy of Dr. Miloslav Boublik.)

by $(NH_4)_2SO_4$, about 100,000 daltons of protein is removed, leaving a rather homogeneous particle of molecular weight 0.9×10^6 daltons. This latter value is generally considered to be the correct weight of the subunit, free of extraneously adsorbed proteins.

The 50S subunit also contains adsorbed protein when first isolated and must be washed to bring it to what is considered actual *in vivo* size. Unwashed particles have a molecular weight of 1.7×10^6 to 1.8×10^6 daltons; washed preparations give reproducible values of 1.55×10^6 daltons.

Eucaryotic Ribosomes

Eucaryotic ribosomes are more difficult to isolate and obtain in pure form than procaryotic ribosomes, so studying their physical properties has been a special challenge to researchers. They are larger than procaryotic ribosomes and, like their counterparts in bacterial cells, are made up of two smaller subunits. The monomer sediments near 80S, although there is considerable species variability in sedimentation properties; the subunits have sedimentation coefficients in the vicinities of 40S and 60S. Figure 17.4 shows the two-subunit structure of 80S ribosomes.

The small subunits of ribosomes from rat liver have been reported to have sedimentation coefficients of 36.9S and molecular weights of 1.44×10^6 daltons.[3] Large subunits are 56.3S particles with molecular weights of 2.9×10^6 daltons.

Eucaryotic ribosomes vary in molecular weight from about 3.9×10^6 daltons when isolated from plant sources to 4.6×10^6 daltons from animals. An interesting feature of this range is that it appears to be due to variations in mass of the 60S subunit. The 40S subunit is quite constant in size among different species.

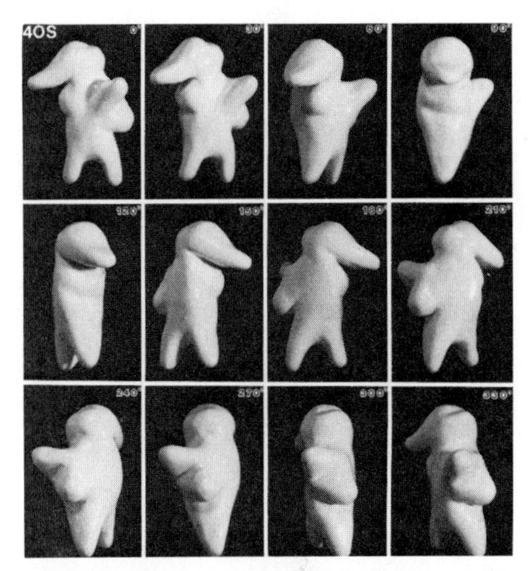

FIGURE 17.5 A model of the 40S subunit of the eucaryotic ribosome generated by computer averaging of electron micrographs. (Courtesy of Dr. Miloslav Boublik.)

In spite of significant size differences between procaryotic and eucaryotic ribosomes, their subunits appear to have roughly similar shapes. The 60S subunit is like the 50S procaryotic subunit in that it is rounded with a pointed end opposite a more blunt end, like a skiff.

The 40S subunit of the eucaryotic ribosome is an elongated particle with dimensions of about 13×25 nm. It has been subjected to close scrutiny by M. Boublik and his co-workers, who have used computer averaging of electron micrographs to arrive at their conclusions.[4] This technique has produced an alarming amount of surface detail for this subunit, which otherwise is often depicted as a simple ellipsoid. A model generated from these computerized images is shown from 12 different views in Figure 17.5. This image is enough to make one shudder, not only because of its grotesque features but because it emphasizes a degree of complexity in this subunit that is generally not appreciated. However, in spite of this topographical detail, a careful analysis of the various views reveals familiar silhouettes as they are conventionally drawn for ribosome subunits, including those used to depict procaryotic ribosomes. Compare these views with those in Figure 17.15 to see some commonly recurring structural themes.

Mitochondrial and Chloroplast Ribosomes

Organelle ribosomes vary considerably in their physicochemical properties. Chloroplast ribosomes have many structural similarities to procaryotic ribosomes, whereas the ribosomes of mitochondria differ from those of procaryotes and even differ among themselves depending on the source from which they are obtained.

Chloroplast ribosomes sediment at 70S and can be dissociated into subunits of 30S (29S–41S) and 50S (46S–54S).[5] Thus, both intact particles and their subunits are close to the size parameters of procaryotic ribosomes.

In contrast, mitochondrial ribosomes are a structurally diverse class of

Table 17.3 Sedimentation Values of Mitochondrial Ribosomes

	Monomer	Small Subunit	Large Subunit
Animals	55S–60S	30S–35S	40S–45S
Ascomycetes	70S–75S	30S–40S	50S
Ciliate protozoans (*Tetrahymena*, *Paramecium*)	80S	55S	55S
Higher plants	77S–80S	40S	60S

particles. This diversity is illustrated in the sedimentation data on monomers and subunits in Table 17.3. Four classes are distinguishable. The monomer unit in animals is small, close to the size of *E. coli* ribosomes. Its unusually low sedimentation properties (55S–60S) are due to its lower buoyant density and larger volume than *E. coli* ribosomes (70S).

Ciliated protozoans have 80S monomers uniquely characterized by their dissociation into two subunits with equal sedimentation values. The mitochondrial ribosomes of higher plants are very similar in sedimentation properties to cytoplasmic ribosomes.

17.2 RIBOSOMAL RIBONUCLEIC ACIDS

The RNAs of procaryotic ribosomes.

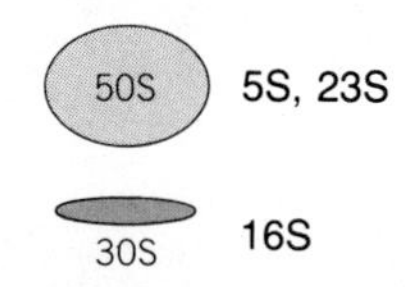

The RNAs of eucaryotic ribosomes.

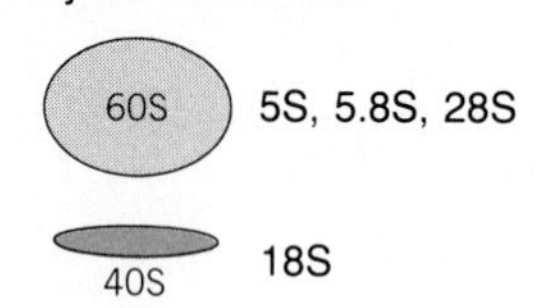

The RNAs of chloroplasts.

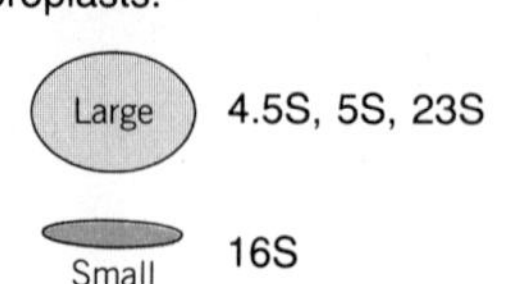

The RNAs of mitochondria.

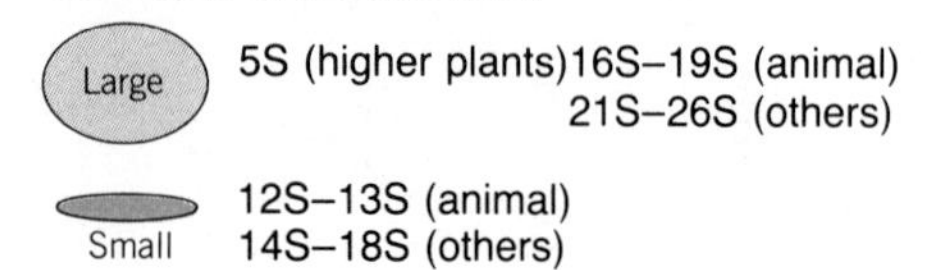

Procaryotic ribosomes possess three different RNA species, which are designated according to their sedimentation properties: 5S, 16S, and 23S. The highest and lowest molecular weight members are found in the 50S subunit, and 16S RNA is a component of the 30S subunit.

Eucaryotic ribosomes carry four types of RNA. They are designated 5S, 5.8S, 18S, and 28S. All these except the 18S RNA reside within the 60S subunit. The 18S species is a 40S subunit constituent.

Table 17.4 summarizes some of the salient physicochemical features of the various ribosomal ribonucleic acids.

The ribosomes of chloroplasts contain four species of RNA: the small subunit contains 16S RNA and the large subunit three species (23S, 5S, and 4.5S). Thus these RNAs bear sedimentation properties that resemble those of procaryotic ribosomes, but the extent to which these similarities reflect chemical and structural kinships is not yet known. The GC content of the 16S + 23S RNA for a number of angiosperm species is between 55.2 and 55.9%, whereas in *E. coli* the GC contents of 23S RNA and 16S RNA are 53.5 and 54.5%, respectively.[5] These values do not prove identity, but they do leave open the possibility of strong similarities.

The 5S RNA species of the chloroplast ribosome is apparently distinct from the 5S RNA of the cytoplasmic ribosomes of the same cell and is also different from the 5S RNA in mitochondrial ribosomes. The 4.5S chloroplast RNA is present at about the same level as 5S RNA, but it varies in size among different flowering plants.

Mitochondrial ribosomes have their own distinct set of RNAs, which are fewer in kind as compared to ribosomes of other sources. All studied so far lack a 5.8S species, and all but higher plant mitochondrial ribosomes are deficient in the 5S RNA molecule. Thus, except for the higher plants, mitochondrial ribosomes lack the lower molecular weight RNAs. The 5S

Table 17.4 Physical Properties of Ribosomal Nucleic Acids

	Eucaryote	Procaryote
Number of RNAs in monomer	4	3
Small subunit		
RNA type	18S	16S
Molecular weight (daltons)	0.7×10^6	0.6×10^6
Large subunit		
RNA types	5S 5.8S 28S	5S 23S
Molecular weight (daltons)	3.2×10^4 5×10^4 1.7×10^6	3.2×10^4 1.1×10^6

RNA of higher plants is distinct from the 5S RNA of the cytoplasmic ribosomes.

In general, mitochondrial rRNAs differ from cytoplasmic rRNAs in that they have a low GC content, little secondary structure, and a characteristic pattern of methylated nucleotides.

5S RNA

The 5S RNA was first identified in 1963 as an RNA of the 50S subunit in *E. coli* ribosomes. It has since been found in other sources, as discussed above.

Primary Structure

E. coli 5S RNA was the first RNA for which the primary structure was determined. This was reported in 1968 by Frederick Sanger and his research group.[6] Since then the sequences of many 5S RNAs have been worked out.

All 5S RNAs consist of about 120 nucleotides, the bases of which are neither methylated nor modified in other ways.[7] Strong homologies in primary structure exist between various procaryotic species. This is also true within eucaryotic species of such wide diversity as yeast, chicken, and human. There is even significant homology seen when comparing procaryotes with eucaryotes.

An interesting finding is that the two halves of the 5S RNA molecule in *E. coli* contain homologous regions. For example, the section between residues 10 through 60 contains 34 residues that have identical locations with those in a section between 61 and 110. Furthermore, 5S RNA bears many similarities to transfer RNA, a fact that has suggested to some workers that tRNA and 5S RNA have arisen from a common ancestral gene.

Regions of significant nucleotide homologies in 5S RNA.

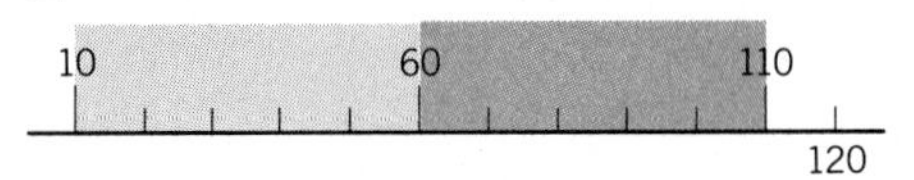

In spite of the similarities between procaryotic and eucaryotic 5S RNA, they cannot be substituted in the heterologous 50S subunit to make a functional particle. By contrast, 5S RNA in one procaryote can function in the 50S subunit of another procaryote.

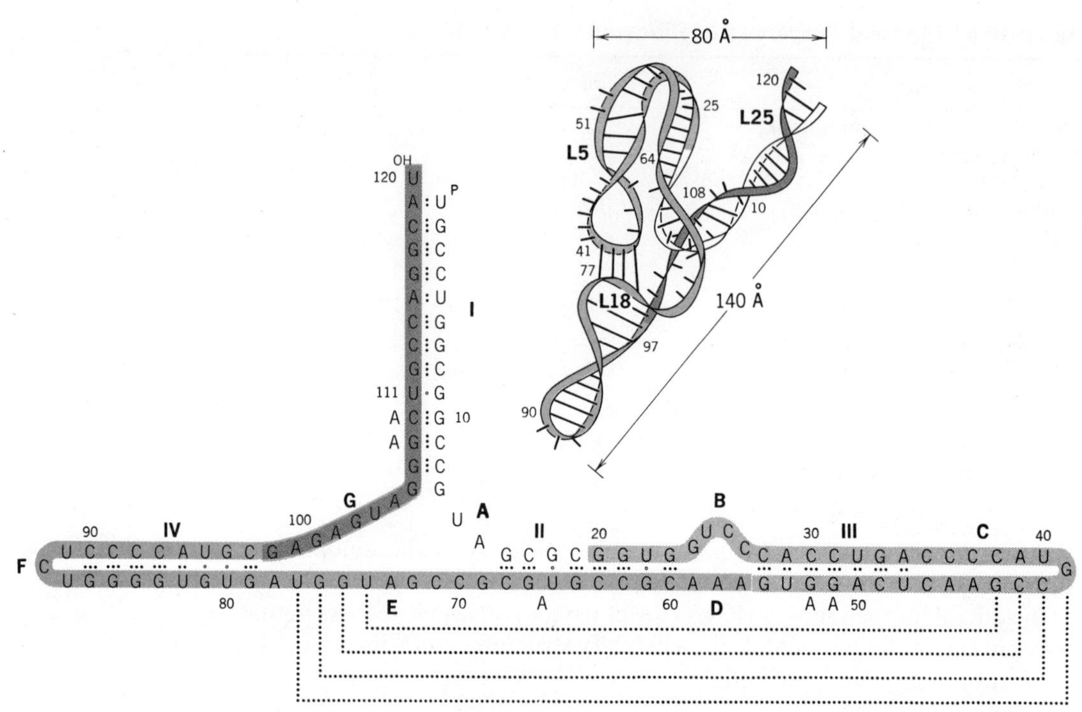

FIGURE 17.6 A secondary structure model of 5S ribosomal RNA of *E. coli*. Four helices are shown, termed the molecular stalk (I), the weak tuned helix (II), the common arm base (III), and the procaryotic loop (IV). Three proteins, L5, L18, and L25, are depicted as they are thought to relate to the RNA.

Secondary Structure and Conformation

The secondary structure of 5S RNA has not fallen into the cloverleaf pattern as neatly as for tRNA. Rather, many different models have been proposed to illustrate how folding can explain selective chemical and enzymatic modifications of the intact molecule.

A recent model of 5S RNA is presented in Figure 17.6. Two forms of the molecule, designated A and B, are often discussed. The A form is one of two stable molecular conformations that can be induced by modifying temperature and Mg^{2+} concentrations. It is considered to be the native form of the molecule within the 50S subunit. The biological significance of the B form is not known.

The overall shape of the 5S RNA molecule according to a number of studies is that of an elongated "Y".

16S RNA

Method IX

Using conventional RNA sequencing techniques, researchers at the University of Strasbourg in France were able to arrive at the primary structure of more than 90% of the 16S RNA molecule by the late 1970s. Other investigators, by using rapid RNA sequencing methods and by determining the sequence of the DNA in the cloned rrnB gene, published the complete primary structure of *E. coli* 16S RNA in 1978.[8] It has 1541 nucleotides, or is slightly more than 12 times the size of 5S RNA.

Topography

Solving the primary structure of 16S RNA has eased investigators over a major hurdle. With this information in hand, it is possible to attack problems related to the overall topography or run of the RNA molecule within the 30S subunit and the manner in which it is folded into its secondary structure. Although these problems are not yet solved completely, some very exciting lines of information are coming from several laboratories, which are clarifying both the structural and functional roles of 16S RNA within the subunit.

Electron microscopy has been one approach used to get at the question of topography. When 16S RNA is prepared under conditions normally used to reconstitute ribosome subunits, except that it is maintained in a naked state, it appears in the electron microscope as a particle with a shape and dimensions similar to those of the complete 30S subunit. The important implication of this observation is that the information for RNA folding in the 30S subunit may reside within the RNA by itself. It may therefore be the core that determines the overall shape of the subunit when proteins are bound to it.

Another approach, using nucleases that are single strand specific, has brought researchers to the conclusion that a 5′-terminal core of 500 nucleotides may act as a nucleus around which the remainder of the 30S subunit is assembled.

A very useful approach to the study of topography has been to employ single strand specific chemical modifications of bases. A reagent exceptionally well adapted for this purpose is kethoxal (α-keto-β-ethoxybutyraldehyde). This compound reacts under mild conditions to specifically modify guanine residues. Other reagents are employed that are specific for other bases.

Armed with molecular tags of this sort, the extent and positions of modifications in 16S RNA can be determined under a variety of conditions. When kethoxal is provided to the intact 30S subunit, about 25 sites in 16S RNA are modified, but not randomly. Most of the positions are from the middle to the 3′ end. In contrast, when the whole 70S ribosome is reacted, sites near the middle and the 3′ terminus are not modified. By implication, these regions of the 16S RNA must interact with the 50S subunit and are thus protected from chemical modification when the two subunits interact.

A number of base-specific chemical modifying agents are used to selectively modify RNA. Kethoxal is G-specific. It reacts under mild conditions and can be removed at will. Glyoxal is also G-specific. Bisulfite is C- and U-specific and *m*-chloroperbenzoate is A-specific.

These results and the results from many similar chemical modifications have provided details that have not yet been put together into a coherent picture for 16S RNA, but one is certain to emerge soon. Importantly, this approach has been useful in testing the credibility of using base pairing as a basis for determining the secondary structure of the molecule. Base pairing on naked RNA may lead to one model of secondary structure, whereas base pairing using information about the topography of the molecule may lead to another secondary structure.

Secondary Structure

Using all the information currently available, Harry Noller has proposed a preliminary secondary structure model for 16S RNA (Figure 17.7). According to this model, the molecule is organized into five discernible structural domains. These domains, along with their positions, associated

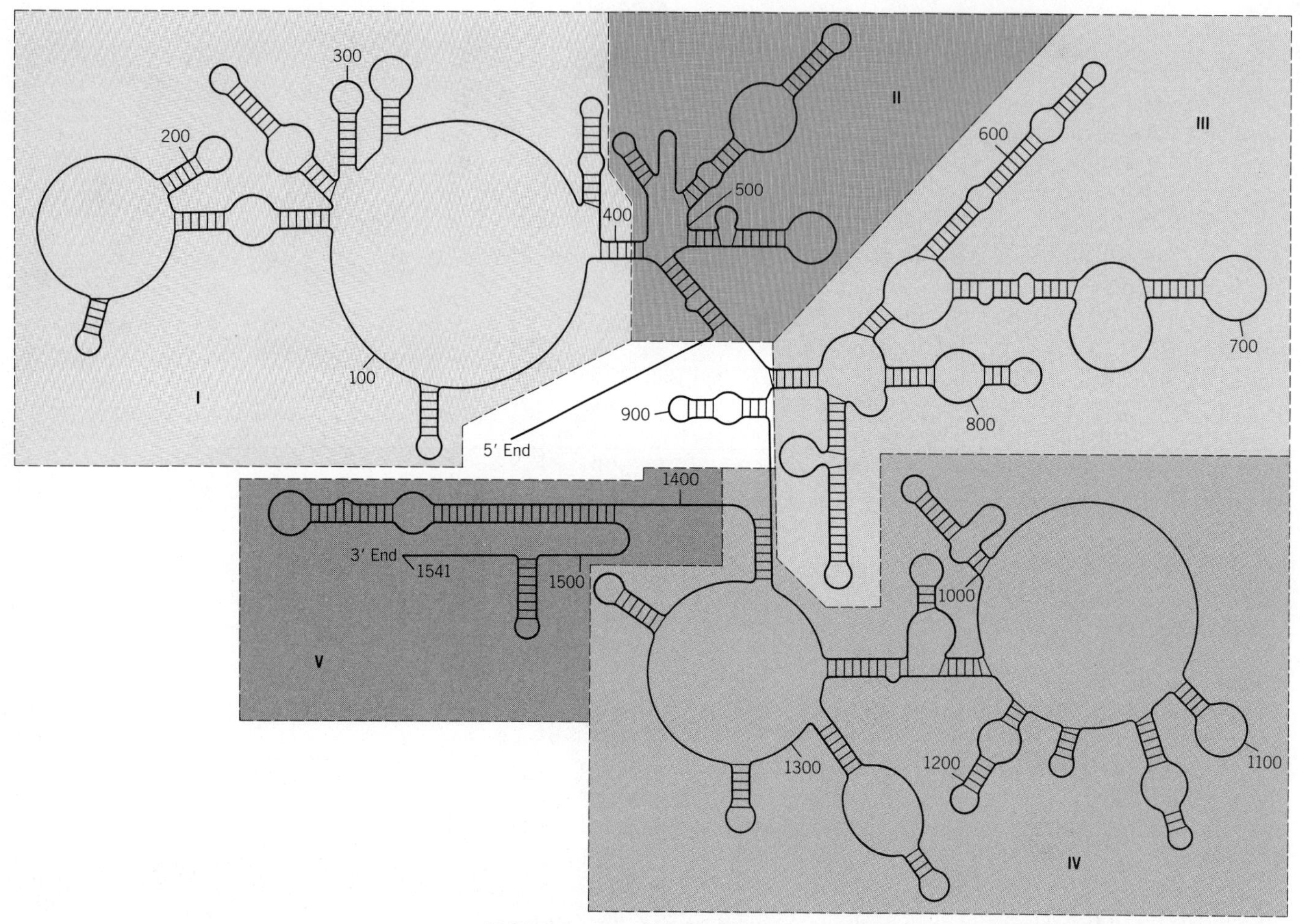

FIGURE 17.7 Secondary structure model for *E. coli* 16S ribosomal RNA. The 1541 nucleotides that make up the molecule are proposed to interact such that five different domains can be discerned (indicated approximately). The locations of these domains in the ribosome subunit and their possible functions are delineated in Table 17.5.

proteins, location in the intact particle, and possible function are summarized in Table 17.5.

Domain I is a relatively nuclease-resistant region that corresponds to the assembly core discussed above. Both domains I and II are thought to be located in the body of the 30S subunit. Domain III contains stretches of RNA that make contact with the 50S subunit. This domain is probably in the platform of the 30S subunit.

It is speculated that certain reversible interactions may take place involving domain IV. These interactions or "switches" may be between positions 100 and 1060, and 400 and 1050. The switches are closed in the 30S subunit but opened in the 70S ribosome. When closed, domains I and IV are brought together; when opened, they are apart.

It is thought that an open switch, such as is present in the 70S ribosome, may provide a site for the binding of aminoacyl-tRNA. The position of this domain relative to the intact 30S subunit is in the head region.

Table 17.5 Structural and Functional Organization of 16S RNA

Domain	Position	Proteins	Location	Function
I	25–420	S4, S20	Body	Assembly
II	420–560	S4	Body	Assembly
III	560–890	S6, S8, S15, and S18	Platform	Subunit association
IV	920–1400	S7, S9, S10, S13, and S19	Head	tRNA binding
V	1400–1541	S1 and S21	Platform	Subunit association, initiation

Domain V is a stretch on the 3′ end of the molecule. It appears to be in the platform region of the subunit. Two roles are proposed for this section of RNA. One is in binding mRNA and thus initiating translation. Another is an interaction with the 50S subunit.

23S RNA

The largest RNA of the procaryotic ribosome and a member of the 50S subunit, 23S RNA, is the most recent ribonucleic acid to be sequenced. It is about twice the size of 16S RNA, containing 2904 nucleotides. Once again, it took more than classical RNA sequencing techniques to do the job. Rather, the sequence was determined on cloned restriction fragments of the rrnB gene of *E. coli*.

Other RNAs

Eucaryotic ribosomes contain three species of RNA that have not been discussed: 5.8S, 18S, and 28S.

The sequences of 18S and 28S RNA had not been published as this was being written but may well be in print now. Genes for ribosomal constituents have been identified, and the prospects are therefore good that 18S and 23S RNA will be sequenced indirectly through the determination of gene sequences.

On the other hand, 5.8S RNA has been sequenced from a number of eucaryotic species. It bears no significant primary structure in common with procaryotic 5S RNA, although it is assumed that eucaryotic 5.8S RNA and procaryotic 5S RNA are functionally homologous.[3]

One reason for assuming functional homology is related to their biosynthetic origins from DNA. In the eucaryote, 5.8S, 18S, and 23S RNA are all transcripts coming from the same region in rDNA. In the procaryote 5S, 16S, and 23S RNA are generated from the same region in DNA. Eucaryotic 5S RNA, on the other hand, is transcribed from a separate site in DNA.

In spite of the homologies between the two 5S RNAs, they are different enough to make functional interchange impossible in *in vitro* systems. For example, eucaryotic 5S RNA has a conserved sequence YGCG that is complementary to AUCG of loop IV found only in eucaryotic initiator-tRNAs. Procaryotic 5S RNA has a conserved sequence CGAAC that is complementary to TψCG in loop IV of all elongator-tRNAs and in procaryotic initiator-tRNA.

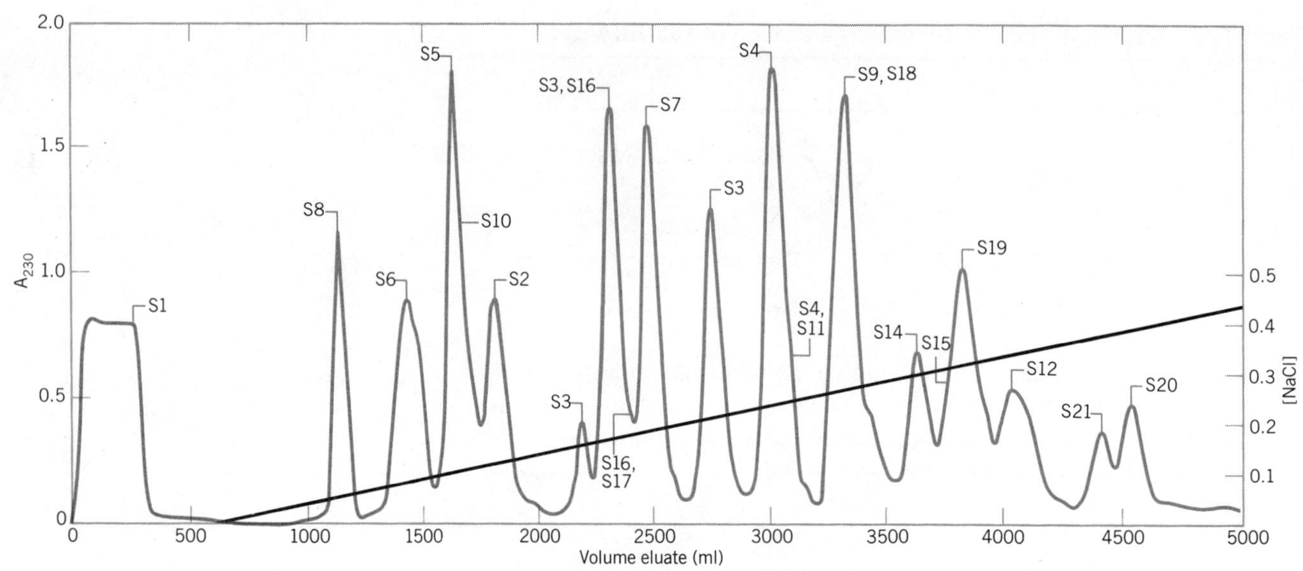

FIGURE 17.8 Elution profile of proteins extracted from the procaryotic 30S subunit on a cellulose phosphate column using a linear NaCl gradient. Although many of the 30S proteins are resolved by this technique, further purification is necessary to obtain homogeneous proteins for careful analytical studies.

The thesis that has emerged is that procaryotic 5S RNA binds elongator-tRNAs, while eucaryotic 5S RNA binds the initiator-tRNA. 5.8S RNA binds the elongator-tRNAs. Thus in the eucaryotic ribosome, 5.8S RNA is a part of the A site where it binds elongator-tRNAs. The 5S RNA is a part of the P site where it binds the initiator-tRNA.

17.3 RIBOSOMAL PROTEINS

To say that ribosomes are about half protein and half ribonucleic acid is to sketch a fairly simple picture. To point out that ribosomes contain several RNAs and scores of proteins is to add an enormous amount of detail that both clarifies the picture and provides the opportunity for a good deal of speculation regarding the significance of the ingredients. But to publish the exact number of RNAs and proteins, to give their primary, secondary, and tertiary structures, to illustrate their topography and interactions, and to discuss their functions during protein synthesis is to paint the walls and ceilings of the Sistine Chapel. Investigators are still standing on scaffolds and lying on their backs working on the project.

Extraction and Purification of Proteins

Proteins are first extracted from ribosome subunits and then purified.[9] There are several commonly employed methods of extraction. One is the use of 67% acetic acid in the presence of a high Mg^{2+} concentration. A second is to treat ribosomes with 2 *M* LiCl in the presence of 4 *M* urea. Still another is to subject unfolded ribosomes to RNase.

The precise treatment chosen depends on the type of study to which the extracted protein is to be subjected. If, for example, secondary or tertiary structure is to be studied, RNase treatment in the absence of denaturing acid or urea would be desirable. If a primary structure is to be determined, denaturation is of no consequence and other more harsh techniques can be employed.

After the proteins have been extracted from the nucleic acids in a particular ribosome subunit, they are purified by elaborate column chromatographic techniques. An elution profile of proteins from the 30S subunit is presented in Figure 17.8, which indicates both the sophistication of the technique and the complexity of the subunit.

Various chromatographic systems are useful, including cation and anion exchangers as well as gel filtration. Often one run is not sufficient to resolve all the proteins of a given subunit. Therefore eluted peaks that contain more than one component, as determined by gel electrophoresis, are rechromatographed under a different set of conditions until complete resolution is obtained.

Methods IV and V

The column is often very large in these chromatographic runs, for they must have a capacity sufficient to carry a load of proteins that can be resolved into individuals bearing enough material for continued studies. These columns are therefore preparative, not analytical columns.

Number and Nomenclature of Proteins

The number of proteins in ribosomes can be determined most readily by two-dimensional polyacrylamide electrophoresis. The ability of this technique to resolve these proteins is apparent in the results shown in Figure 17.9.

The small subunit of procaryotic ribosomes resolves into 21 spots (S1–S21) and the large subunit into 34 spots (L1–L34). Further analyses have shown that the actual number of individual proteins in the 30S subunit corresponds to the number of spots, 21, whereas spot L8 from the 50S subunit consists of three proteins: L7, L10, and L12. Each of these three also appears in its own position as well. Therefore L8 is an artifact, containing no unique proteins.

The number of different proteins in procaryotic ribosome subunits.

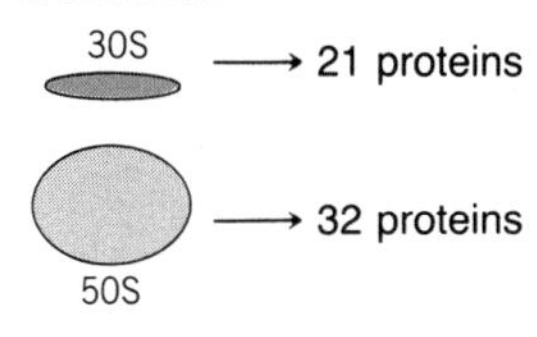

Protein L26, from the 50S subunit, is identical to S20 from the 30S subunit. This does not mean, however, that each subunit contains one molecule of this protein.

The stoichiometry is actually 0.8 S20 and 0.2 L26 per 70S ribosome, or a total of one molecule per intact ribosome. The appearance of this protein in both subunits is attributed to a distribution of it between the two subunits when they are separated. Thus, if one assumes the real position of this protein to be on the 30S subunit (S20) then the actual number of proteins in the 50S subunit is 32. This is currently taken to be the correct number in spite of the finding that two of the proteins, L7 and L12, differ only by an acetyl group at the *N*-terminus of one of them.

Table 17.6 lists the proteins as they are designated from the 30S and 50S ribosomes along with their molecular weights as determined by researchers at the Max Planck Institute in Berlin. Other research groups have carried out similar studies on proteins, namely groups in Uppsala, Madison, Geneva, and Hiroshima. Each has devised its own nomenclature, but the proteins have been sufficiently characterized so that the various designations can be correlated among laboratories.

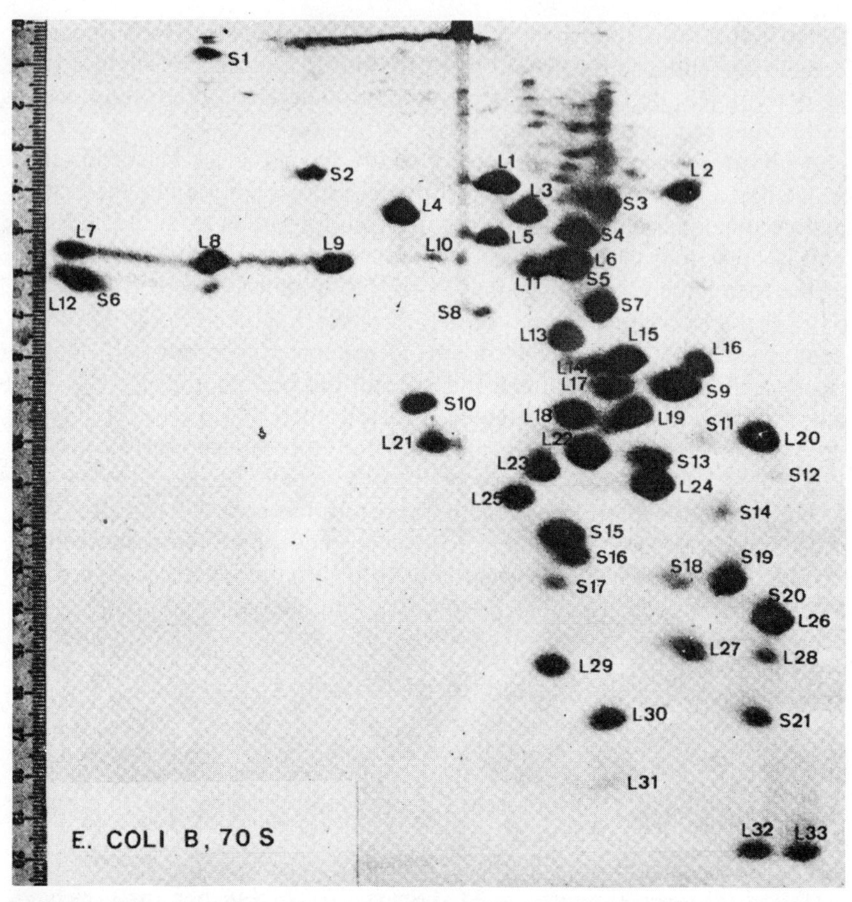

FIGURE 17.9 Two-dimensional electrophoresis of procaryotic 70S ribosome proteins on polyacrylamide. Proteins coming from the small (S) and large (L) subunits are so designated. (Courtesy of Dr. H.G. Wittmann.)

The number of individual proteins in eucaryotic ribosomes has not been determined with certainty. Rat liver ribosomes have yielded 82 proteins, 33 from the 40S subunit and 49 from the 60S subunit.[10] So for the present it is safe to say that eucaryotic ribosomes contain 70 to 80 proteins. As can be imagined, with more proteins extracted from ribosomes of different eucaryotic sources, it has been most difficult to arrive at the precise number of individuals for each subunit and to correlate the results among different eucaryotic sources and from different laboratories. Most of the work has been done on rat liver ribosomes.

The number of different proteins in eucaryotic ribosome subunits can be approximated.

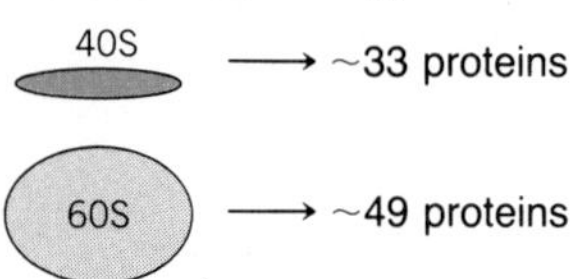

Structure of Ribosomal Proteins

Ribosomal proteins have been studied to various degrees of sophistication with respect to their primary, secondary, and tertiary structures.

Primary Structure

As of 1980, primary structures were determined for 19 proteins from the small subunit and 29 proteins from the large subunit of the *E. coli* ribo-

Table 17.6 **Comparison of Molecular Weights (daltons) of Proteins in 30S and 50S Procaryotic Subunits as Determined by SDS–Polyacrylamide Gel Electrophoresis and Sedimentation Equilibrium**[a]

	30S Proteins			50S Proteins	
	SDS–Gel	Sedimentation Equilibrium		SDS–Gel	Sedimentation Equilibrium
S1	65,000	n.d.	L1	26,700	22,000
S2	28,300	24,000	L2	31,500	28,000
S3	28,200	23,000	L3	27,000	23,000
S4	26,700	23,000	L4	25,800	28,500
S5	19,600	18,500	L5	22,000	17,500
S6	15,600	15,500	L6	22,200	21,000
S7	22,700	26,000	L7	13,400	15,500
S8	15,500	15,500	L8	17,300	19,000
S9	16,200	14,500	L9	17,300	n.d.
S10	12,400	18,000	L10	19,000	21,000
S11	15,500	n.d.	L11	19,600	19,000
S12	17,200	15,000	L12	13,200	15,500
S13	14,900	14,000	L13	17,800	20,000
S14	14,000	14,000	L14	16,200	18,500
S15	12,500	13,000	L15	17,500	17,000
S16	11,700	13,000	L16	17,900	22,000
S17	10,900	n.d.	L17	16,700	15,000
S18	12,200	10,500	L18	14,300	17,000
S19	13,100	14,000	L19	14,900	17,500
S20	12,000	12,500	L20	17,200	16,000
S21	12,200	13,500	L21	13,900	14,000
			L22	14,800	17,000
			L23	12,700	12,500
			L24	14,300	17,500
			L25	12,000	12,500
			L26	12,000	12,500
			L27	12,700	12,000
			L28	12,300	15,000
			L29	12,000	12,000
			L30	11,200	10,000
			L31	10,000	n.d.
			L32	10,500	n.d.
			L33	10,500	9,000
			L34	9,600	n.d.

[a]The designations (S1, L1, etc.) are those used by the researchers at the Max Planck Institute in Berlin. n.d.: not determined.

some.[11] By the time these lines are read it is likely that all *E. coli* proteins will have been sequenced.

Knowledge of primary structures has revealed some interesting features of these proteins that otherwise could not have been discerned. One feature is a degree of methylation not commonly seen in proteins. Protein L11, for example, is heavily methylated, containing nine methyl groups. Three of these are on the *N*-terminal alanine, giving an *N*-trimethylalanine, and the other six are evenly distributed between two internal lysines in positions 3 and 39. The resulting modified residue is a N_ϵ-trimethyllysine.

Other amino acids, such as glutamine (N^5-monomethylglutamine) and methionine (*N*-monomethylmethionine), are methylated, some are acetylated (*N*-acetylalanine and *N*-acetylserine), and some derivatized in un-

Protein L11 is methylated in ways not commonly seen among proteins.

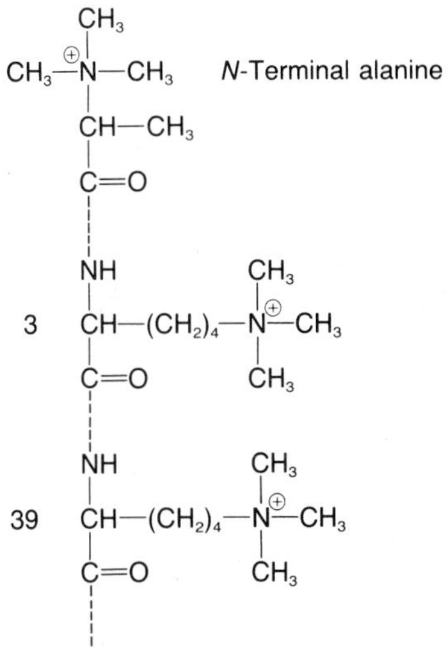

$CH_3-N^{\oplus}(CH_3)_2-CH(CH_3)-C{=}O$ — *N*-Terminal alanine

3: $NH-CH[(CH_2)_4-N^{\oplus}(CH_3)_3]-C{=}O$

39: $NH-CH[(CH_2)_4-N^{\oplus}(CH_3)_3]-C{=}O$

known ways (aspartic acid and arginine). Although the full significance of these modifications is not yet known, they are likely to be related to the special roles these proteins have in conferring structure and function to the intact ribosome.

Another feature made apparent by sequence analyses is that the different proteins in a given bacterial ribosome appear to possess no strong structural homologies, with the possible exceptions of L7/L12 and S20/L26. A lack of sequence homology suggests that these proteins are not derived from a common ancestral gene.

A final feature derived from sequence information is that ribosomes can be studied in a phylogenetic sense. Enough information on primary structures is not yet in for extensive comparisons, but with the limited information available, it looks as though the proteins of *E. coli* and *Bacillus subtilis* show significant homologies, but *E. coli* and *Artemia salina* (a brine shrimp) have little homology. A significant degree of homology is seen *among* eucaryotes such as shrimp, yeast, and rat liver. Thus ribosomal proteins are yet another example of proteins that differ between species and yet carry out the same functions for the different cells in which they are present.

Secondary Structure

Two approaches have been employed to study the secondary structure of ribosomal proteins. One has been to use the known primary structures as a basis for predicting secondary structure, and the other has been to employ optical rotatory dispersion or circular dichroism to study secondary structure experimentally. The results obtained from the two approaches have not agreed in all instances, but in general most ribosomal proteins have an α-helix content ranging between 25 and 70% as well as some β structure and random coil. All the ribosomal proteins studied thus far have unique secondary structures.

Methods VIII

Tertiary Structure

Method VIII

Techniques such as proton magnetic resonance and microcalorimetry are presently being used to study the folding and interaction of polypeptide chains in ribosomal proteins. As is true for both primary and secondary structures, most of the proteins studied so far appear to have their own discrete tertiary structures. While at the moment the results are rather difficult to interpret and to relate to ribosome structure and function, these approaches will undoubtedly yield valuable information for the eventual complete understanding of ribosomes.

Shape of Proteins

Methods III and VIII

Investigators using low angle X-ray scattering, neutron scattering, and sedimentation, viscosity, and diffusion analyses are accumulating information regarding the shapes of the various ribosomal proteins. Many of the proteins studied are elongated, with axial ratios that range from 20:1 to about 3:1. Others appear to be more globular or spherical. One of the strengths of these approaches is that the influence of RNA on protein shape can be investigated with the objective of discerning the nature of the interaction between proteins and nucleic acids in intact ribosomes.

17.4 THE RECONSTITUTION OF RIBOSOMES

As complex a particle as the ribosome is, it can be disassembled into its constitutent proteins and nucleic acids and completely reconstituted to a functional particle under *in vitro* conditions. Viruses can also be reconstituted from their constitutent protein and nucleic acid ingredients, but they are in most cases simpler than ribosomes, containing only one or a few types of protein and one species of nucleic acid.

Ribosomes, as we have already pointed out, contain several RNAs and scores of different proteins. Their reconstitution *in vitro* therefore represents a remarkable feat that provides an opportunity to study the priorities and requirements of assembly.

The 30S subunit of *E. coli* was the first to be reconstituted by M. Nomura and co-workers[12] and, being the smallest and most thoroughly understood of the subunits, has been the prototype for reconstitution studies. With considerable effort the more complex 50S subunit has also yielded to reconstitution,[13] and information is now available on both systems.

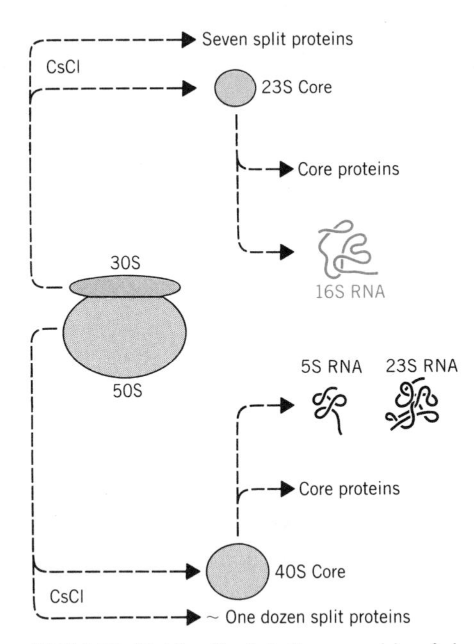

FIGURE 17.10 Partial disassembly of ribosome subunits in cesium chloride (CsCl). The 50S subunit gives rise to a 40S core particle and about a dozen split proteins. The 30S subunit similarly yields a core of 23S and seven split proteins. The RNAs of the two subunits reside in these cores along with several proteins.

Partial Reconstitution

Since the mid to late 1960s experimental techniques have been worked out to partially disassemble ribosome subunits under mild conditions and then bring about their reconstitution to intact functional units.

When ribosome subunits are centrifuged in 5 *M* CsC1 in the presence of 0.04 *M* Mg^{2+}, populations of proteins are split off from the subunit cores and separate as a band of lower density in a gradient.

Figure 17.10 illustrates this partial disassembly of subunits. Seven "split proteins" are commonly released from the 30S subunit by this technique and about a dozen from the 50S subunit. A given subunit core and its split proteins reconstitute rapidly within a few minutes at 37°C, with little sensitivity to variations in the experimental conditions.

Since total reconstitution is now workable on subunits, partial reconstitution is no longer employed. But historically, it was an important prelude to the reconstitution work that followed. In addition, partial reconstitution revealed that certain of the proteins were essential for the binding of streptomycin and chloramphenicol, for elongation factors, and for participation in the peptidyltransferase center of the subunit.

Total Reconstitution

The 30S subunits with physical and functional similarities to native subunits can be reconstituted from 16S RNA and the 21 individually purified proteins. The reconstitution in this case has requirements more stringent than those for partial reconstitution. A sharp ionic strength optimum of 0.37 is required along with a Mg^{2+} concentration of 20 m*M*. Reconstitution proceeds most rapidly between 40 and 50°C.

The 50S subunit of *E. coli* did not yield to reconstitution techniques readily, so attention was shifted for some time to *Bacillus stearothermophilus*, which contains a 50S subunit that can be reconstituted at temperatures *above* 50°C. These high temperatures produced denaturation of the 50S proteins of *E. coli*, thus blocking reassembly.

Eventually reconstitution of *E. coli* 50S subunits was accomplished by

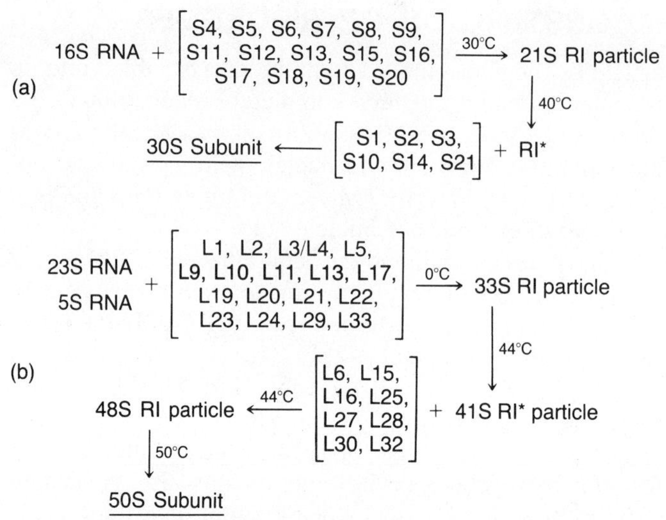

(a)

16S RNA + [S4, S5, S6, S7, S8, S9, S11, S12, S13, S15, S16, S17, S18, S19, S20] —30°C→ 21S RI particle

21S RI particle —40°C→ [S1, S2, S3, S10, S14, S21] + RI*

30S Subunit ← [S1, S2, S3, S10, S14, S21] + RI*

(b)

23S RNA + 5S RNA + [L1, L2, L3/L4, L5, L9, L10, L11, L13, L17, L19, L20, L21, L22, L23, L24, L29, L33] —0°C→ 33S RI particle

33S RI particle —44°C→ [L6, L15, L16, L25, L27, L28, L30, L32] + 41S RI* particle

48S RI particle ←44°C— [L6, L15, L16, L25, L27, L28, L30, L32] + 41S RI* particle

48S RI particle —50°C→ 50S Subunit

FIGURE 17.11 Tentative reaction schemes for the *in vitro* assembly of 30S (*a*) and 50S (*b*) ribosome subunits of *E. coli*. Assignments have not been made for proteins L7/L12, L8, L14, L18, L26, L31, and L34 in the 50S subunit.

a two-step incubation. The first incubation is carried out for 20 min in 4 m*M* Mg^{2+} at 44°C with (23S + 5S) RNA and the total proteins of the 50S subunit. This is followed by a second incubation for 90 min at 50°C in a higher concentration (20 m*M*) of Mg^{2+}. Under these conditions total reconstitution of the 50S subunit can be achieved.

Reconstitution of the 30S subunit follows first-order kinetics. The major rate-limiting step is a unimolecular rearrangement of an intermediate 21S particle that will form first at 30°C. This particle, designated RI, undergoes rearrangement to RI* at 40°C. The activated intermediate will then take on more proteins at lower temperatures until the intact 30S subunit is assembled.

For the 50S subunit, the rate-limiting steps for reconstitution are two conformational changes, each of which can be brought about by the two-step incubation conditions described above.

Figure 17.11 summarizes the reaction schemes that are proposed for reconstitution of the two subunits of *E. coli*. Reconstitution of the 30S subunit commences when 16S RNA combines with a selected group of proteins to form the intermediate particle RI. This can be converted by heating it to a state, RI*, in which it will combine with the remaining proteins to give rise to the intact 30S subunit.

As Figure 17.11 shows, there are several intermediate stages during the reconstitution of the 50S subunit. The first intermediate (33S) forms spontaneously at 0°C and the second intermediate (41S) appears when the temperature is raised to 44°C. Analysis of these two intermediates has shown that they contain identical compositions of RNA and protein and are therefore simply two different conformations of the same RNA–protein complex.

The third intermediate (48S) contains all the proteins and RNAs of the final 50S subunit, but it is conformationally different from the final product. Thus, the second and final rate-limiting step is the conversion of the 48S particle to the 50S subunit.

The Function of Components during Reconstitution

Given the above-described reconstitution conditions, a number of different experimental approaches have been used to assess the importance and roles of the proteins and nucleic acids during reconstitution. As delineated by Nomura and Held,[12] four possible roles for the components of the ribosome can be assessed experimentally.

1. *A given component may be essential for ribosome assembly but not required for any function.*
2. *A given component may be required indirectly for some ribosomal function because its presence maintains an active center in a proper configuration in the ribosome structure.*
3. *A given component may be a part of the active center, playing a direct role in a given ribosomal function.*
4. *A given component may be required for both assembly and function of the ribosome.*

The roles for all the components have not yet been assigned, but during the past decade remarkable progress has been made in matching critical functions with several of the components.

Single Component Omission Studies

One experimental approach that has been used productively is described as a *single component omission experiment*. Reconstitution experiments are performed with only one component missing, and the effect of this omission on reassembly or ribosome function is then determined.

With this approach an absolute requirement of 16S RNA for the reconstitution of 30S subunits has been determined. Thus it can be concluded that at least one function of 16S RNA is to serve as a core or initiator around which proteins aggregate.

The roles for protein components are more difficult to assign. Often the omission of a single protein will affect several functions, and each function, in turn, can be shown to depend on more than one protein. Nevertheless, the functional picture is taking form.

Proteins S1 and S6 do not appear to affect ribosome function when they are omitted singly, but S6 binds S18 during assembly and therefore may have a primary role in assembly. Protein S16 also appears to be important for assembly, since when it is omitted the rate of reconstitution is decreased. But when omitted, the 30S subunits that eventually form appear to have unimpaired functions. Several other proteins (S4, S7, S8, and S9) appear to be important for assembly.

Other proteins, when omitted, result in subunits with altered functions. Protein S12, for example, seems to be important for polypeptide chain initiation. When S11 is omitted, the resulting particles have a significantly increased frequency in translational errors.

Chemical Modification Studies

Another experimental approach that has been used is to *chemically modify subunits or their components* and then note the effect of this on reconstitution and function. For this approach to work well, reagents must be selected that are specific, or limited in action, so that the degree of modification can be controlled. Tyrosine residues, for example, can be modified with tetranitromethane, or the reagent kethoxal can be used to inactivate subunits by reacting with RNA.

Using these approaches it has been shown that tRNA will not bind to ribosomes when certain proteins are modified, suggesting that this binding depends on protein, not 16S RNA. In contrast, kethoxal does not inactivate ribosomes when tRNA is bound but will when tRNA is not in place to shelter some vulnerable site. Reconstitution experiments have revealed that kethoxal modifies 16S RNA, and when so modified activity is lost. These results suggest that tRNA binding to the ribosome depends on 16S RNA. Thus tRNA binds to *both* proteins and 16S RNA during translation.

Still other approaches are used to assign component functions. In this regard mutationally or physiologically altered ribosomes have been used with some success. Also, reconstitution experiments have been conducted with heterologous systems, such as 16S RNA from one species and ribosomal proteins from another.

As a result of applying these various approaches, several other examples could be cited to illustrate the functional assignments that have been made. But the interested reader would profit more from a study of references and books listed at the end of this chapter, since the studies cite much detail.

It should be stressed that these studies are far from simple, and the results are often difficult to interpret. A given protein, for example, may affect function only indirectly by influencing the behavior of another protein directly responsible for activity or by changing the overall shape of the ribosome subunit.

The Relation between Reconstitution and *in Vivo* Assembly

The fact that intact and fully functional ribosomes can be formed from their protein and RNA components by reconstitution experiments suggests that at least a similar if not identical process may be used by a cell *in vivo*.

But some clear differences between the *in vitro* and *in vivo* assemblies must be recognized. Assembly is more rapid and efficient *in vivo* and certainly does not require the high temperatures that are used for *in vitro* reconstitution systems.

The differences in these requirements for assembly may be related to the form of RNA within the nucleolus where ribosome precursors are generated. This RNA is higher in molecular weight than the final constituent RNAs of the ribosome (see Chapter 16), a factor that may direct a more efficient assembly of ribosomal precursors.

It is also likely that preribosomal particles are generated *during* transcription of rRNA. One can envision that as sequences of RNA are available, beginning at the 5′ end of the molecule, assembly may begin in a

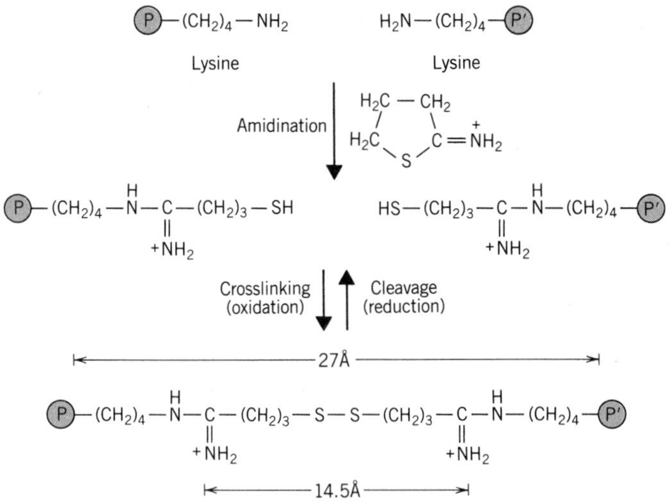

P—$(CH_2)_4$—NH_2 Lysine

H_2N—$(CH_2)_4$—P′ Lysine

Amidination (H_2C—CH_2, H_2C, S, C$=NH_2^+$)

P—$(CH_2)_4$—NH—C($=NH_2^+$)—$(CH_2)_3$—SH

HS—$(CH_2)_3$—C($=NH_2^+$)—NH—$(CH_2)_4$—P′

Crosslinking (oxidation) / Cleavage (reduction)

27Å

P—$(CH_2)_4$—NH—C($=NH_2^+$)—$(CH_2)_3$—S—S—$(CH_2)_3$—C($=NH_2^+$)—NH—$(CH_2)_4$—P′

14.5Å

FIGURE 17.12 The reagent 2-iminothiolane can be used to cross-link two adjacent proteins on the surface of a ribosome or its subunits. Reduction of the disulfide bond can then be used to free the proteins from each other.

sequential manner that is difficult to reproduce *in vitro*. Furthermore, there may be nonribosomal elements *in vivo* that are lacking in the *in vitro* reconstitution systems.

17.5 THE TOPOGRAPHY OF RIBOSOME COMPONENTS

The ultimate objective in studying ribosome structure is to be able to position each protein with respect to all others and to the RNAs. Many researchers believe that until this is accomplished there will not be a clear understanding of the ultrastructure of ribosome function.

A good deal of effort has therefore been channeled toward these goals, and many different approaches have been devised to study component topography. Of these we will look at only two: cross-linking techniques and the use of component-specific antibodies.

The Cross-linking of Components

Methods IV and VII

One of the most productive approaches to the study of component topography has been to react a given subunit with a bifunctional reagent that will form a covalent bridge between two components that are juxtaposed on the particle surface. Cross-linked products can then be analyzed by their increased molecular weights from sodium dodecyl sulfate–polyacrylamide electrophoresis followed by additional electrophoretic or immunological techniques to identify the specific components that are paired.

A typical cross-linking protocol using the bifunctional reagent 2-iminothiolane is depicted in Figure 17.12. This particular reagent has the advantage over several others in that it generates a disulfide bridge be-

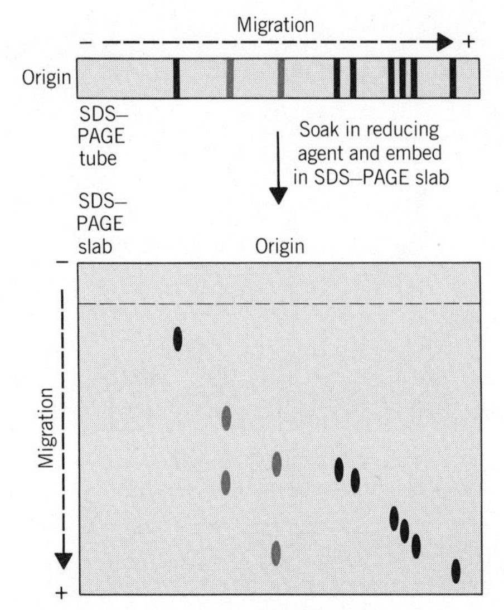

FIGURE 17.13 Electrophoretic protocol used to detect cross-linked proteins in ribosome subunits. After cross-linking, proteins are separated by electrophoresis in SDS–PAGE systems under nonreducing conditions (Top). After reduction, the tube containing the proteins is embedded in an SDS–PAGE slab and again subjected to separation (Bottom). Only proteins previously cross-linked and now reduced will move at positions off a diagonal line (color).

tween two protein components that can be reversibly reduced at will to reform the monomer units.[14]

This is taken advantage of by an ingenious application of electrophoresis shown in Figure 17.13. First, electrophoresis is carried out on the ribosome components (after cross-linking) under nonreducing conditions. The tube gel containing separated components is then soaked in a reducing agent to break the disulfide linkages between pairs.

A second electrophoresis is then conducted with the tube gel embedded in a slab. All the proteins *unmodified* by the reducing agent (and thus not cross-linked to begin with) will fall on a diagonal line across the slab. Those proteins that have been released from pairing by the reducing agent will now be *lower* in molecular weight, will possess *higher* mobilities, and will migrate *beyond* the diagonal line. A further analysis of these spots by elution and electrophoresis makes it possible to identify the specific components that have become cross-linked.

Cross-linking results of this sort have been used to formulate schemes like that shown in Figure 17.14. In Figure 17.14a the proteins of the 30S subunit are arranged in a two-dimensional projection with protein S13 represented as a central protein having nine neighbors. Three proteins, S2, S3, and S4, possess two cross-link domains designated A and B. The A domains are shown to the left and the B to the right. The two domains should be thought of as constituting different ends or regions of a given protein. Two proteins, S15 and S20, have been identified in only one cross-link pair and one protein, S16, has not been observed in a cross-link.

The 50S subunit of *E. coli,* studied in similar fashion, has yielded 54 dimers, as well as other more complex combinations of pairing. The maps are too complex to depict, but restricted mapping has been used to show the topography between certain functional proteins, such as peptidyl transferase or the 5S RNA–protein complex and other protein components.

Cross-linking studies have also provided valuable information regarding the spatial relationships of the two ribosome subunits. Figure 17.14*b* shows a scheme of cross-linking between proteins that lie at the interface of the 30S and 50S subunits of *E. coli.* Note the trans-subunit cross-links in particular, such as S9–L5, S9–L27, and S20–L26. In this figure the position of initiation factor 2 (IF 2) is also depicted as being cross-linked to the 50S proteins L7/L12 and to a group of 30S proteins that are encircled. Figure 17.14b further shows that three proteins, L5, L18, and L25 are

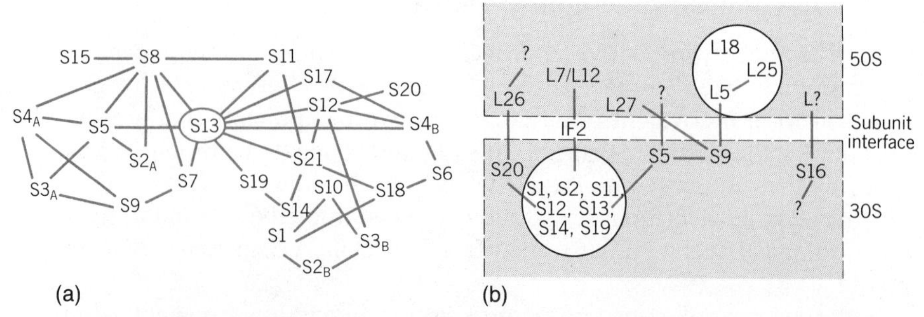

FIGURE 17.14 (*a*) Pattern of protein cross-links in 30S ribosome subunits. (*b*) Pattern of protein cross-links at the interface of the 30S and 50S subunits.

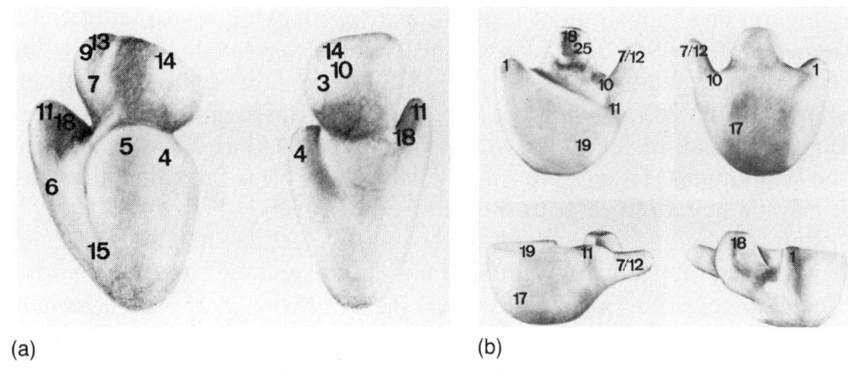

(a) (b)

FIGURE 17.15 Mapping of several proteins on the subunits of *E. coli* ribosomes. (*a*) Two views of the 30S subunit and (*b*) four views of the 50S subunit. (Courtesy of Dr. H. G. Wittmann.)

spatially related and bind specifically to 5S RNA. These proteins are close to the S9 30S subunit protein.

All these results must be understood to mean that two proteins that cross-link have domains within 15Å of each other. Their centers of mass, however, may be quite far apart, especially if they are elongated proteins with high axial ratios.

The next step in cross-linking is to determine precisely where within the known amino acid sequence of two proteins the bifunctional bridge connects the components. This task is being undertaken in some laboratories, but the data at this point are not very extensive.

Immune Electron Microscopy

Another productive approach to component topography has been realized by the use of antibodies against individual components followed by electron microscopy. First, non-cross-reacting antibodies against a certain component are bound to intact subunits. Antibodies will bind to their specific antigens and, since antibodies are bifunctional, one will bind to two subunits. These two-subunit complexes are purified and stained with uranyl acetate, and the position of the bound antibody is determined with the electron microscope.

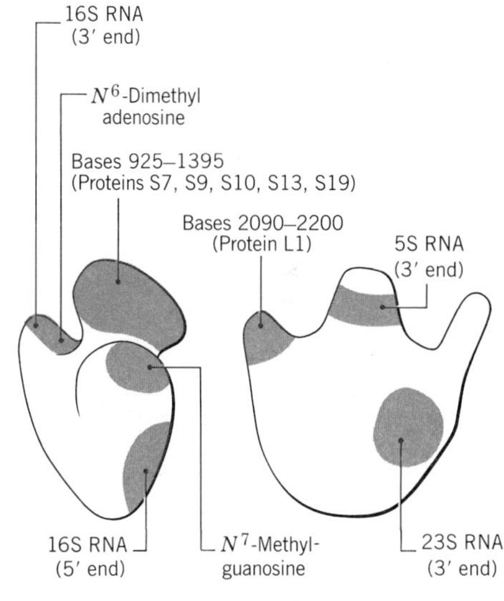

FIGURE 17.16 Locations of ribosomal RNA regions in the small (30S) and large (50S) subunits of the *E. coli* ribosome. The locations of several proteins are also shown. (Courtesy of Dr. H. G. Wittmann.)

Methodical studies of this sort have yielded models of the subunits with mapped antibody binding sites. Figure 17.15 depicts three-dimensional models of the two subunits of *E. coli* with mapped binding sites for proteins. This approach can be used to confirm the results of cross-linking as well as to give an added dimension to the details of component topography.

Method VII

Ribosomal components other than proteins are also being located on the subunit surfaces by the techniques of immune electron microscopy. Figure 17.16 depicts the surface positions that are thought to be occupied by parts of ribosomal nucleic acids. The results suggest that the nucleic acids meander throughout the subunit interiors and surface at certain identifiable points.

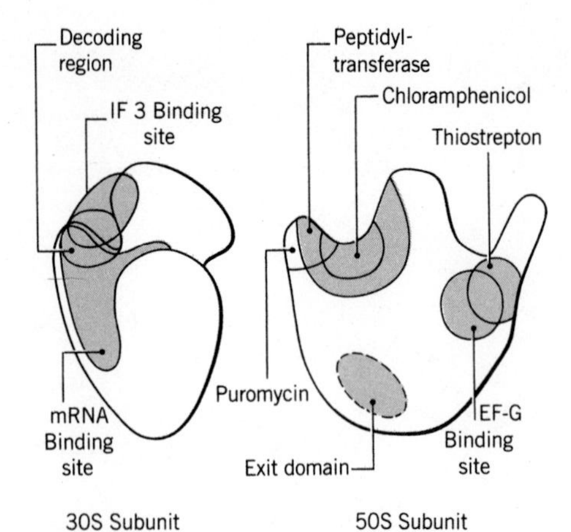

FIGURE 17.17 Several functional domains mapped on *E. coli* 30S and 50S subunits by the technique of immune electron microscopy. (Courtesy of Dr. H. G. Wittmann.)

Figure 17.17 illustrates the presumed locations of several functional domains of the 30S and 50S subunits related to protein synthesis. For example, the binding sites of mRNA and initiation factor 3 (IF 3) are shown on the 30S subunit and several places where antibiotics are presumed to bind, hence to interfere with protein synthesis, are marked on the 50S subunit. To interpret the significance of these functional domains it may be helpful to refer to the details of Figure 17.18, below.

Before leaving the topic of ribosome architecture, it is important to emphasize the speculative nature of this subject and the lack of agreement among the experts. Some researchers feel very strongly that their models of ribosome structure portray with great accuracy these ribonucleoprotein particles. Others, who feel just as strongly that we should regard model building with serious reservations, point out the difficult technical problems associated with obtaining ribosomes that are pure, unaltered, and amenable to the experimental approaches that are being used.

Thus, although several models have been presented in the foregoing discussion, we should be prepared to see models of ribosomes undergo significant change as researchers continue to obtain and interpret structural data. At this point, models are useful to the extent that they help us collect information and envision how structure *may* be related to composition and to function. But these models should not be taken as the last word in ribosome structure, and they do not represent a consensus among experts.

17.6 THE ROLE OF THE RIBOSOME DURING TRANSLATION

The ribosome is the effector of protein synthesis, a task that requires a controlled interaction between it and two RNAs, several enzymatic proteins, and nucleotides. In general terms it is necessary for the ribosome to accomplish two different but equally important functions in this role.

The first function is to deal with the genetics of the process; that is, the ribosome must be able to handle information. A proper alignment must be made between the ribosome, the mRNA, and incoming aminoacyl-tRNAs before any biochemistry can be allowed to take place. The topography of ribosomal constituents is crucial to a proper alignment between the elements of translation. It is in this area that a good deal of current research is centered.

The second function of the ribosome is to promote the biochemistry of translation. When all components are properly aligned, a peptide bond can be formed. Alignment without peptide bond formation is a futile effort. Peptide bond formation without proper alignment is also futile, for the system then loses its fidelity.

Basic Interactions of Components during Translation

It is not our objective to discuss in depth the details of protein synthesis. The reader can easily find this information in good biochemistry textbooks. Rather, we will observe briefly how the components *interact* during translation and attempt to relate certain features of component topography to the process.

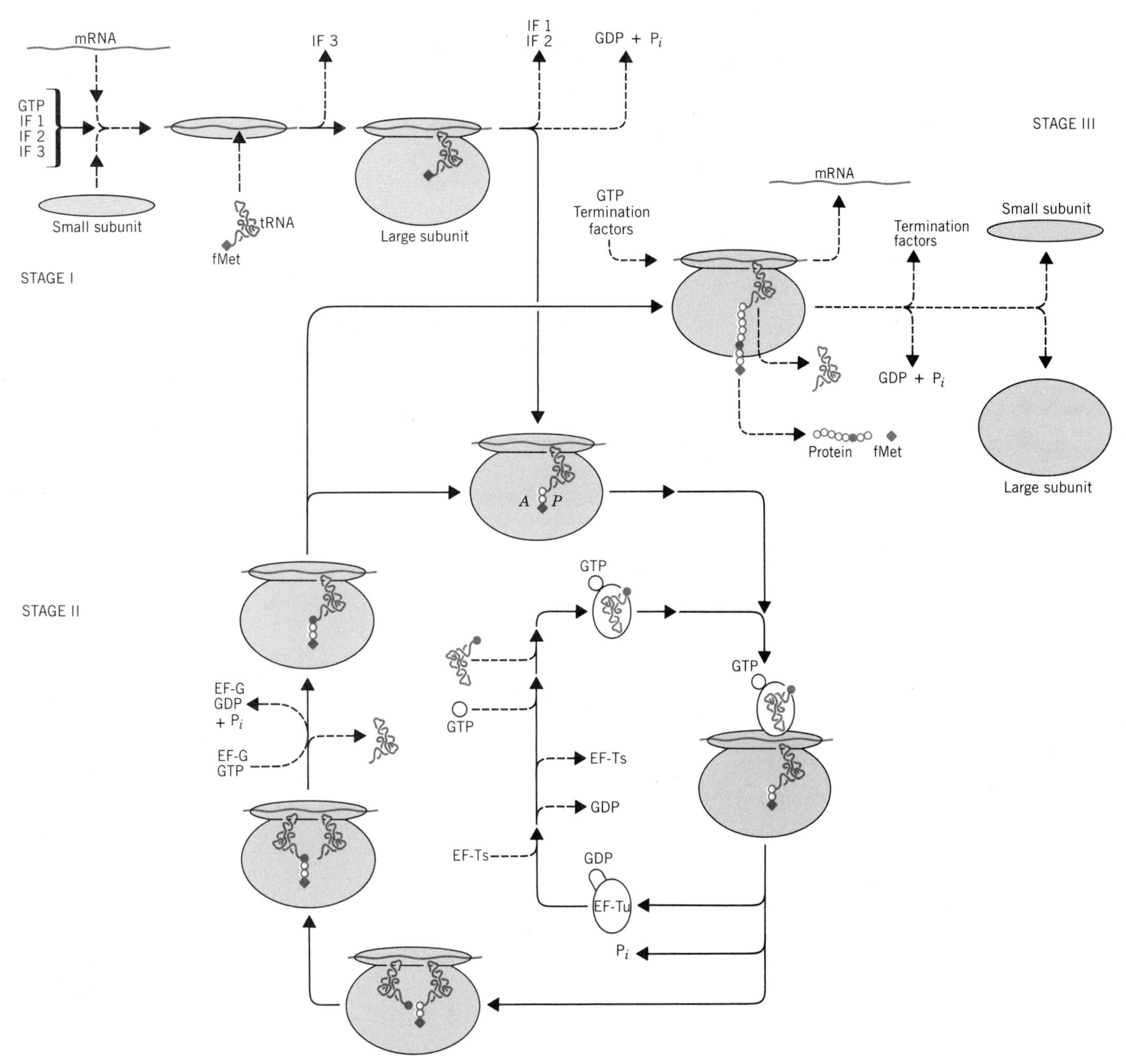

FIGURE 17.18 The three stages of protein synthesis. Stage I is termed initiation; Stage II, elongation; and Stage III, termination.

Figure 17.18 contains in abbreviated form the three stages of protein synthesis as they are conventionally portrayed. Translation is initiated when five components interact with the small subunit: mRNA, GTP, and three initiation factors. In procaryotic cells tRNA charged with a formylmethionine (fMet) then binds to its codon (AUG or GUG) and one initiation factor is released. This sets the stage for the large subunit to bind to the small, an interaction that releases hydrolyzed GTP (GDP + P_i) and the other two initiation factors.

The product of this sequence of interactions is now primed for the second stage of translation, elongation. As the nascent polypeptide chain elongates, there is a high level of coordination between two ribosome sites, A and P. The nascent peptide, covalently linked to the tRNA that in turn is bound to its codon, occupies the P site. The incoming tRNA charged with its particular amino acid comes into the A site. With the appropriate enzymatic elongation factors (EF) and sources of energy, the nascent peptide is transfered to the aminoacyl-tRNA in the A site. After ejection of the empty P-site tRNA, the whole complex moves one codon to bring the elongated peptidyl–RNA into the P site.

Protein synthesis is terminated when a stop codon (UAG, UAA, or UGA) is reached. This frees all the components of the complex, releasing the completed protein and dissociating the ribosome subunits from one another and from RNA and translation factors.

The genetic function of the ribosome is conducted by the small subunit. Here all the alignment takes place between mRNA, tRNA, and the ribosome to guarantee eventual fidelity in the protein product. The biochemical function of the ribosome is carried out by the large subunit.

It is the ultimate objective of researchers working on ribosomes to be able to explain both the genetic and biochemical functions in terms of the molecular components of the ribosome. The techniques that have been used to reveal protein and nucleic acid topography are now making it possible to address this objective in a realistic fashion.

Possible Relation between Ribosome Topography and Translational Fidelity

Calculations of errors in incorporation of amino acids and topographical studies have led James Lake and his colleagues to the conclusion that an A and a P site are not sufficient to assure the fidelity that is required for the system to manufacture proteins so free of error.[15] Simply relying on codon–anticodon interactions at the A site would, according to calculations, result in one incorrect tRNA binding for every 100 correct interactions. The actual error rate is thought to be less than one in 2000 incorporations. Thus, some factor other than simple nucleotide complementarity must be invoked.

Cross-linking studies have shown that mRNA is bound to the small subunit in the vicinity of proteins S11 and S18, which would place it in the vicinity of the cleft and platform of the subunit. Furthermore, at least two of the initiation factors can be cross-linked to proteins located in the cleft and platform regions. Based on these kinds of observation, the A and P sites are thought to be located in the vicinity of the cleft.

A cluster of proteins located on the small subunit but faced *away* from the large subunit constitutes a region called the R site by Lake. These six proteins are S3, S4, S5, S10, S12, and S14. Three of them, S4, S5, and S12, are related to the streptomycin effect—an introduction of wrong amino acids into the nascent protein when streptomycin is bound to the ribosome. The other three affect the efficiency of tRNA binding to ribosomes.

Lake proposes that *first* an incoming aminoacyl–tRNA binds to the R site with part of its structure and at the same time to the codon by its anticodon. *Next* the tRNA is flipped to the A site, with the energy for the flip coming from the hydrolysis of GTP. During this flip the tRNA is held in place *only* by its codon–anticodon interaction. The codon–anticodon unit acts like a hinge.

In this hypothesis, the flip is the test of codon–anticodon pairing. If the correct tRNA is flipped, it will withstand the strain of this movement and thus come properly into the A site. If an incorrect pairing has occurred, the base pairing will not be strong enough to withstand the motion of the flip, and the aminoacyl–tRNA will leave the ribosome.

This model provides some new insights to the mechanism of translational fidelity and points out how an understanding of component topography may be used to explain the complex interactions that take place between the various elements involved in protein synthesis. At this point it is highly speculative. But it may also be a portent of things to come in research on the structure of ribosomes and their function.

Summary

Based on their physicochemical properties, several classes of ribosomes exist, all of them carrying out the same function: protein synthesis. Procaryotic cells possess 70S ribosomes, the cytoplasm of eucaryotic cells 80S ribosomes, and organelles such as chloroplasts and mitochondria 70S and variable, respectively. All ribosomes, regardless of class, dissociate into two smaller subunits, one of which is generally larger and more complex than the other.

The 30S and 50S subunits of procaryotic ribosomes contain three RNAs, a 16S in the former and a 5S and a 23S in the latter. The 5S RNA consists of about 120 nucleotides for which the sequence is known. The 16S RNA contains 1541 nucleotides; its primary structure has also been determined. The secondary structures for these RNAs are also quite certain, as well as how the RNAs interact with various ribosomal proteins. The 23S RNA contains a known sequence of 2904 nucleotides.

The 40S and 60S subunits of eucaryotic ribosomes contain four RNAs, three (5S, 5.8S, 28S) in the 60S subunit and one (18S) in the 40S subunit. The 5S and 5.8S RNAs have been sequenced. The 5S RNA bears sequence homologies with the procaryotic 5S RNA, but the two do not appear to be functionally related. The 5.8S RNA is not structurally similar to the procaryotic 5S RNA, but these two appear to be functionally homologous. Eucaryotic 18S and 28S RNAs have not yet been sequenced.

Procaryotic ribosomes contain about 53 proteins, 21 in the 30S subunit and 32 in the 50S subunit. Eucaryotic ribosomes possess about 70 to 80 proteins. The primary structures are known for almost all the procaryotic ribosomal proteins.

Both the 30S and the 50S subunits can be completely disassembled into their protein and nucleic acid constituents and reassembled *in vitro*. Using reconstitution techniques along with chemical modifications of components, the roles of certain proteins in assembly and in ribosome function have been determined.

The components of ribosomes have been studied topographically by cross-linking techniques and immune electron microscopy. These approaches have culminated in topographical maps that are beginning to reveal how the various members of the ribosome are related three-dimensionally.

References

1. Ribosome Research: Historical Background, A Tissières, in *Ribosomes,* M. Nomura, A. Tissières, and P. Lengyel, eds., Cold Spring Harbor Laboratory, Cold Spring Harbor, N.Y., 1974.
2. *Microsomal Particles and Protein Synthesis,* R. B. Roberts, ed., Pergamon Press, New York, 1958.
3. The Structure and Function of Eukaryotic Ribosomes, Ira G. Wool, in *Ribosomes, Structure, Function and Genetics,* G. Chambliss, G. R. Craven, J. Davies, K. Davis, L. Kahan, and M. Nomura, eds., University Park Press, Baltimore, 1980.
4. Computer Averaging of Electron Micrographs of 40S Ribosomal Subunits, J. Frank, A. Verschoor, and M. Boublik, *Science* 214:1353(1981).
5. Biogenesis of Chloroplast and Mitochondrial Ribosomes, J. E. Boynton, N. W. Gillham, and A. M. Lambowitz, in *Ribosomes, Structure Function and Genetics,* G. Chambliss, G. R. Craven, J. Davies, K. Davis, L. Kahan, and M. Nomura, eds., University Park Press, Baltimore, 1980.
6. The Sequence of 5S rRNA, G. G. Brownlee, F. Sanger, and B. G. Banell, *J. Mol. Biol.* 34:379(1968).
7. 5S RNA, R. Monier, in *Ribosomes,* M. Nomura, A. Tissières, and P. Lengyel, eds., Cold Spring Harbor Laboratory, Cold Spring Harbor, N.Y., 1974.
8. Structure and Topography of Ribosomal RNA, H. F. Noller, in *Ribosomes, Structure, Function and Genetics,* G. Chambliss, G. R. Craven, J. Davies, K. Davis, L. Kahan, and M. Nomura, eds., University Park Press, Baltimore, 1980.
9. Purification and Indentification of *Escherichia coli* Ribosomal Proteins, H. G. Wittmann, in *Ribosomes,* M. Nomura, A. Tissières, and P. Lengyel, eds., Cold Spring Harbor Laboratory, Cold Spring Harbor, N.Y., 1974.
10. The Structure and Function of Eukaryotic Ribosomes, I. G. Wool, *Annu. Rev. Biochem* 48:719(1979).
11. Structure of Ribosomal Proteins, H. G. Wittmann, J. A. Littlechild, and B. Wittmann-Liebold, in *Ribosomes, Structure, Function and Genetics,* G. Chambliss, G. R. Craven, J. Davies, K. Davis, L. Kahan, and M. Nomura, eds., University Park Press, Baltimore, 1980.
12. Reconstitution of Ribosomes: Studies of Ribosome Structure, Function and Assembly, M. Nomura and W. A. Held, in *Ribosomes,* M. Nomura, A. Tissières, and P. Lengyel, eds., Cold Spring Harbor Laboratory, Cold Spring Harbor, N.Y., 1974.
13. Analysis of the Assembly and Function of the 50S Subunit from *Escherichia coli* Ribosomes by Reconstitution, K. H. Nierhaus, in *Ribosomes, Structure, Function and Genetics,* G. Chambliss, G. R. Craven, J. Davies, K. Davis L. Kahan, and M. Nomura, eds., University Park Press, Baltimore, 1980.
14. Protein Topography of *Escherichia coli* Ribosomal Subunits as Inferred from Protein Cross-linking, R. R. Traut, J. M. Lambert, G. Boileau, and J. W. Kenny, in *Ribosomes, Structure, Function and Genetics,* G. Chambliss, G. R. Craven, J. Davies, K. Davis, L. Kahan and M. Nomura, eds., University Park Press, Baltimore, 1980.
15. The Ribosome, J. A. Lake, *Sci. Am.* 245:84(1981).

Selected Books and Articles

Books

Ribosomes, M. Nomura, A. Tissières, and P. Lengyel, eds., Cold Spring Harbor Laboratory, Cold Spring Harbor, N.Y., 1974.

Ribosomes, Structure, Function and Genetics, G. Chambliss, G. R. Craven, J. Davies, K. Davis, L. Kahan, and M. Nomura, eds., University Park Press, Baltimore, 1980.

Articles

Ribosome Structure, R. Brimacombe, G. Stoffler, and H. G. Wittmann, *Annu. Rev. Biochem.* 47:217(1978).

Structure of Ribosomes, H. G. Wittmann in *Cell Compartmentation and Metabolic Channeling,* L. Nover, F. Lynen, and K. Mothes, eds., Elsevier/North-Holland Biomedical Press, Amsterdam, 1980.

The Ribosome, J. A. Lake, *Sci. Am.* 245:84(1981).

The Structure and Function of Eukaryotic Ribosomes, I. G. Wool, *Annu. Rev. Biochem.* 48:719(1979).

The Translational Machinery: Components and Mechanism, J. W. B. Hershey, in *Cell Biology: A Comprehensive Treatise,* vol. 4, D. M. Prescott and L. Goldstein, eds., Academic Press, New York, 1980.

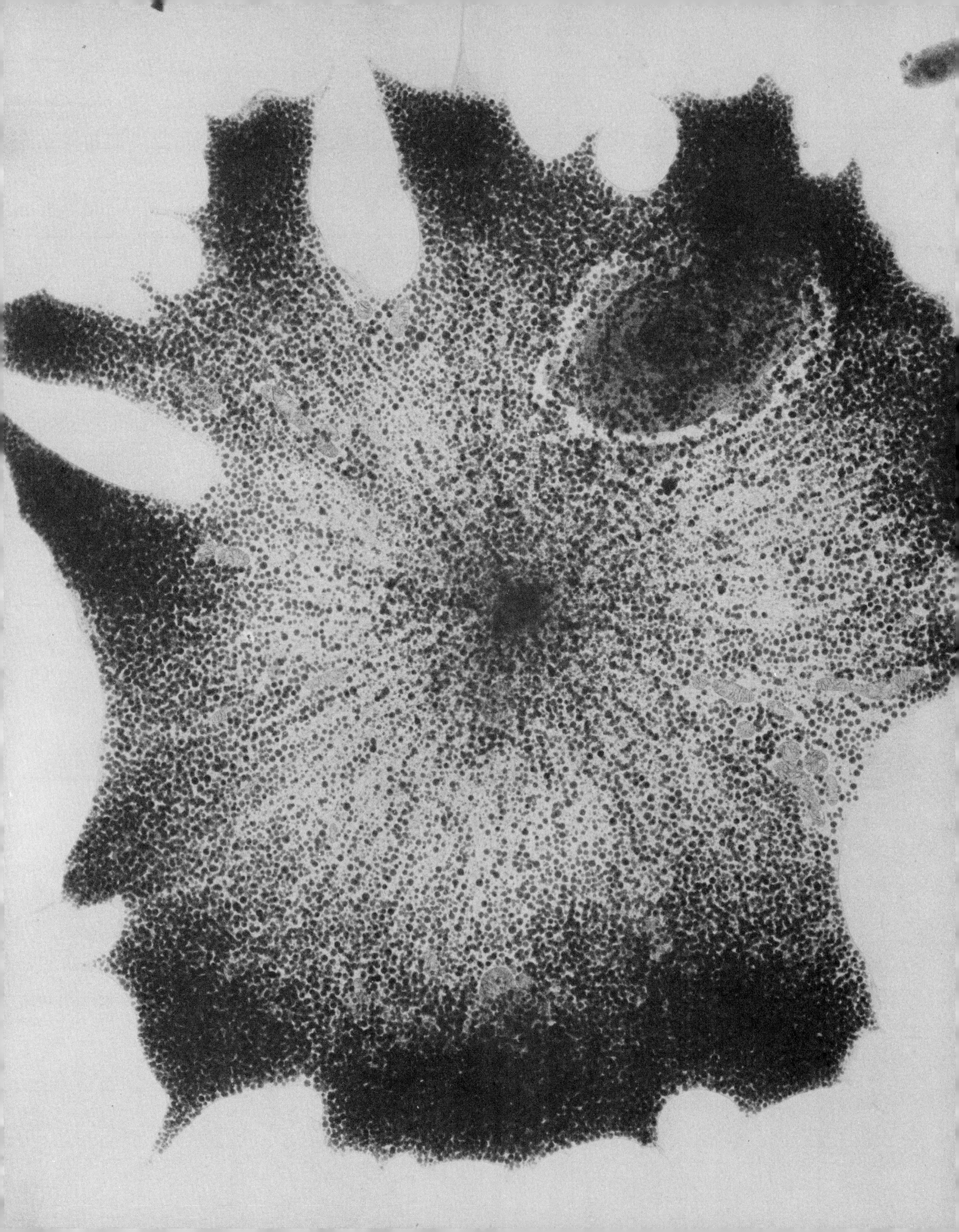

SECTION 8

THE MOLECULAR ANATOMY OF FORM AND MOVEMENT

Cells display two types of movement. One is the movement of the entire cell, a type of locomotion. The second is movement within the cell. In some cases this movement is responsible for locomotion. In other cases internal movements are related to cell replication, or to changes in cell form or to streaming of the cytoplasm.

In eucaryotes, the basis of form and movement resides in the behavior of polymer systems, which undergo polymerization–depolymerization reactions and dynamic interactions. In procaryotes the major form of movement is due to the generation of force at the base of polymeric appendages.

In the two chapters that follow we will examine the structures and molecular makeup of organelles that are responsible for movement and then review some representative types of cellular dynamics that have their basis in the behavior of these polymeric systems.

18

Components and Organelles of Cellular Dynamics

Philosophers through the ages have made the astute observation that life, in its many aspects, appears to be continuously moving. All things in the universe, from the cosmic to the atomic level, exhibit some form of movement. Getting down to earth, the capacity to move is also an essential feature of the biological world.

P. CAPPUCCINELLI, 1980

Cell motility is far from a recent discovery. In fact, the observation of cell movements and the investigation of the basis of those movements represent some of the oldest lines of biological research.

But only during the past two decades has the ultrastructural and molecular basis of cell movement come to be understood. That understanding has been accompanied by several surprises.

It was not surprising to discover that motile bacteria are propelled by appendages that use water as leverage to push and turn the organism erratically at high rates and frequencies. But it was most surprising to find a rotary motor, without precedent in the biological world, as the power source for this movement.

It was not surprising to find contractile proteins at the basis of muscle movement. But the intricate interactions between the proteins of muscle and the ubiquity of these components, even in nonmuscle cells, has been surprising to many.

It was not surprising to find that microtubules located in the centers of eucaryotic cilia and flagella are responsible for driving their activities. But it has been surprising to discover that microtubules permeate and penetrate the entire cell and are responsible for a variety of internal dynamic events that include chromosome movements, cytoplasmic streaming, amoeboid movements, and the maintenance of cell shape.

It has not been surprising to find that the cytoplasm of the cell contains organelles and particles, but to observe via high voltage electron microscopy that the cytosol, that last structureless part of the cell, contains networks of fibers and filaments resembling the structural steel and concrete reinforcement rods of a large building, makes even the simplest part of the cell surprisingly complex.

Up until the mid-1950s the cytoplasm of the cell was thought to be relatively simple and structureless, especially with respect to organelles of movement. One exception to this was the mitotic apparatus, which had been seen considerably earlier, along with cilia and flagella. It was astutely surmised around the turn of the century by E. B. Wilson[1] that these organelles may have some element in common. Now it is clear that this is true.

Microtubules were studied first in cilia and flagella, where they are quite stable and reproducibly observed. Early studies began in the late 1940s and continued into the 1950s. Cytoplasmic microtubules turned out to be much more difficult to observe because of their dynamic properties and lability. Nevertheless some important observations by S. Inoué employing polarization light microscopy in the early 1950s demonstrated organized spindle fibers in living cells, along with the ability of colchicine to eliminate this birefringence from the cell.[2,3] These were important advances because they established cytoplasmic fibers as nonartifactual and stimulated the thought that spindles could be formed from a pool of monomer units in the cell and disassembled to feed units back into the pool.

Colchicine is an alkaloid produced by various members of the family *Liliaceae*. When it is administered to dividing cells, chromosome movements are blocked at metaphase. Colchicine binds to the protein units of microtubules and prevents their polymerization.

The first detailed chemical analysis of the components that make up the microtubules of cilia was conducted by I. R. Gibbons in the early 1960s.[4] This required techniques to isolate and purify these structures which, in turn, made possible a study of their properties *in vitro*. In 1963, the same year these chemical studies were reported, the organelle was christened "microtubule" by Myron Ledbetter and Keith Porter, at Harvard University, and David Slautterback, at the University of Wisconsin.

Once the basic morphological and biochemical information was in hand, researchers turned toward relating the structures to cellular processes. It is in this area that remarkable insights were gained in the decade of the 1970s to explain the variety of cell movements that are known.

The work is by no means complete, but many things now can be reported with confidence and several theories await critical experimental proof.

When discussing organelles of movement, we have two types of movement to consider. One is a movement of the cell itself, and the other is movement within the cell. In eucaryotic cells the organelles that carry out these two types of movement are related, being of the same morphology and biochemical basis. Within procaryotic cells there are no organelle-mediated movements. But procaryotes do possess appendages that function as organelles of locomotion. We will begin with a discussion of these.

18.1 FLAGELLAR STRUCTURE AND FUNCTION IN BACTERIA

The flagella of procaryotic cells appear to be constructed from a set of plans completely different from that used for the organelles of locomotion of eucaryotes. About all they have in common is function.

Bacteria may be flagellated in a variety of ways. Some possess a single flagellum (monotrichous):

or a tuft or two (lophotrichous):

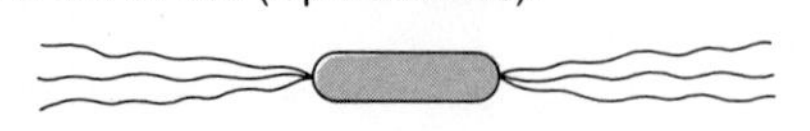

or many flagella emanating in all directions (peritrichous):

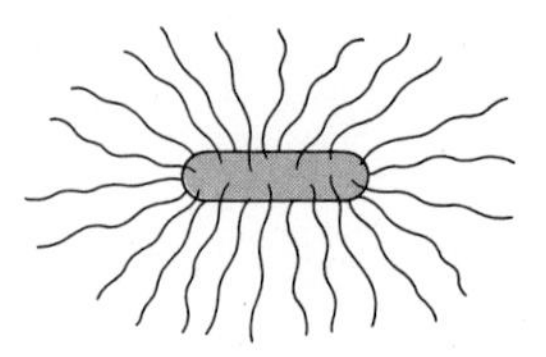

In bacteria, flagella are not essential for survival, for many species do perfectly well without them, depending only on random Brownian motion and currents to get them to sources of nutrients or to bring essentials by. But when bacteria are situated within gradients of nutrients or toxins, organisms that can move toward or away from a particular environmental agent appear to have some advantage.

Bacteria may have only a single flagellum, one or more tufts of flagella, or up to hundreds of flagella extending in all directions from the cell. Studies on flagellar movement are obviously more complicated in organisms that have peritrichous flagellation, but these can serve as the source of flagellar material for biochemical studies. Figure 18.1 contains electron micrographs of representative bacteria possessing flagella.

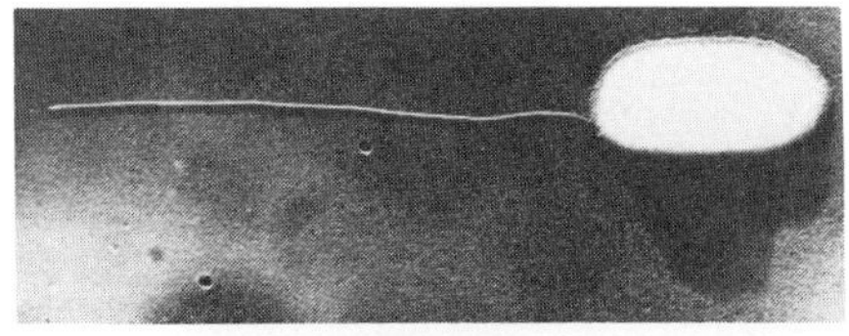

(a)

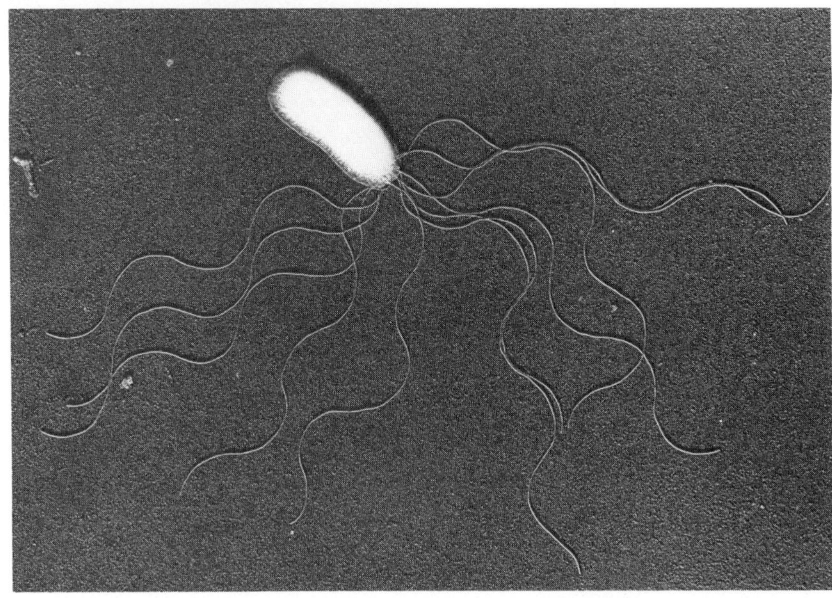

(b)

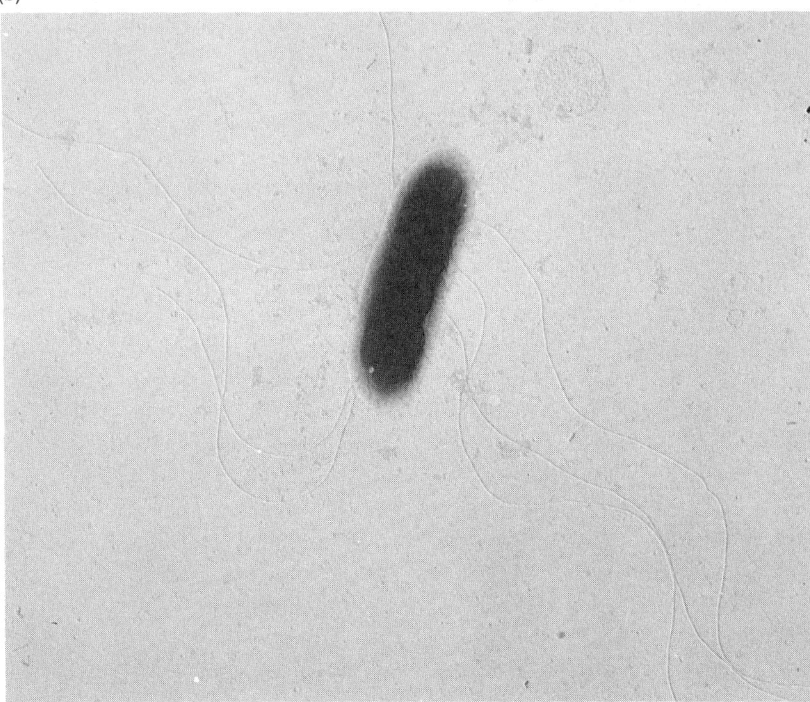

(c)

FIGURE 18.1 Three types of bacterial flagellation. (*a*) A single polar flagellum in *Pseudomonas solanacearum*. (*b*) A polar tuft in *Pseudomonas marginalis*. (*c*) Peritrichous flagellation in *Erwinia chrysanthemi*. (Courtesy of Dr. Arthur Kelman.)

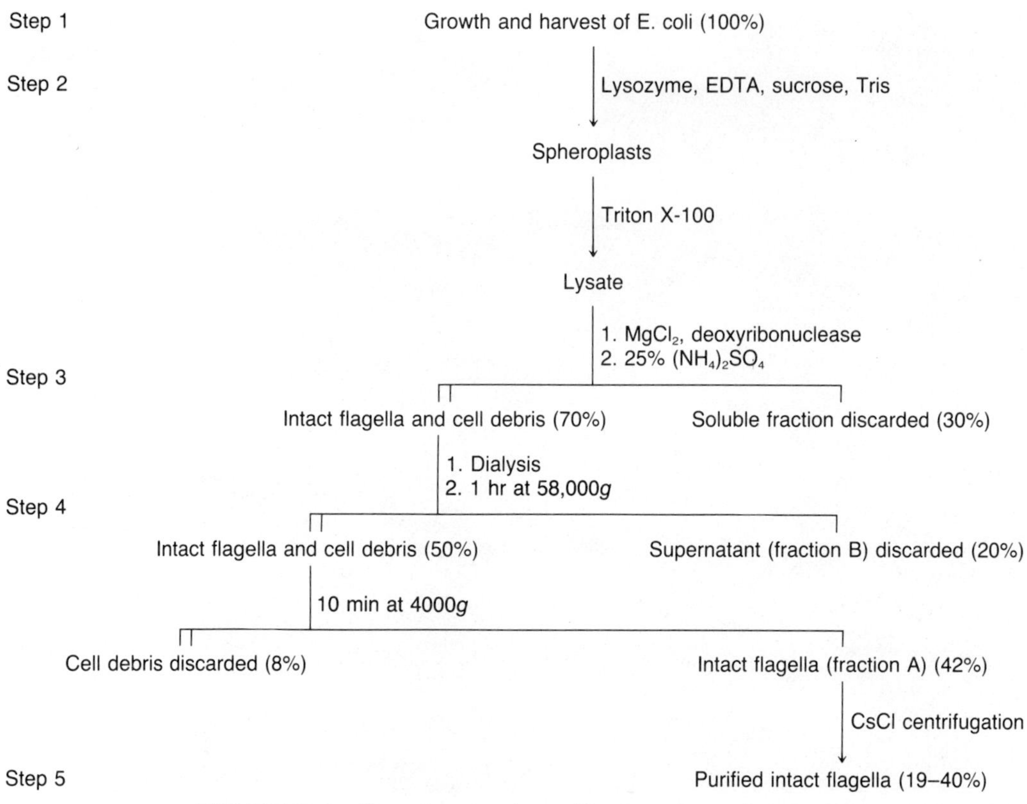

FIGURE 18.2 Flow sheet for the purification of intact flagella. The original amount of flagella, defined as 100%, would amount to 2.4 mg per 10^{12} cells. Yields are shown in parentheses for the individual steps.

The Isolation of Flagella

Bits and pieces of flagella can be obtained from bacteria by simply subjecting them to harsh treatments that shear off the organelles. Although this approach is useful for some chemical studies, it is often more desirable to obtain intact flagella.

Elegant techniques were worked out to accomplish this about simultaneously at the University of California, San Diego, and the University of Wisconsin, Madison, and published back to back in the *Journal of Bacteriology* in 1971.[5,6] An example of the protocol used to obtain intact flagella from *Escherichia coli* is presented in Figure 18.2.

The essence of the procedure is that the cell wall is enzymatically removed by lysozyme and the cell membrane lysed and partially solubilized by the detergent Triton X-100. Intact flagella are finally harvested by differential and density gradient centrifugation.

A preparation of flagella isolated in this manner is shown in the electron micrograph of Figure 18.3. Note the typical intact flagellum terminated by a hooklike structure at one end: this is the part of the flagellum that is anchored into the bacterial cell body.

Using purified intact flagella from *Escherichia coli* and *Bacillus subtilis*, M. DePamphilis and D. Adler of the University of Wisconsin[7] initiated a

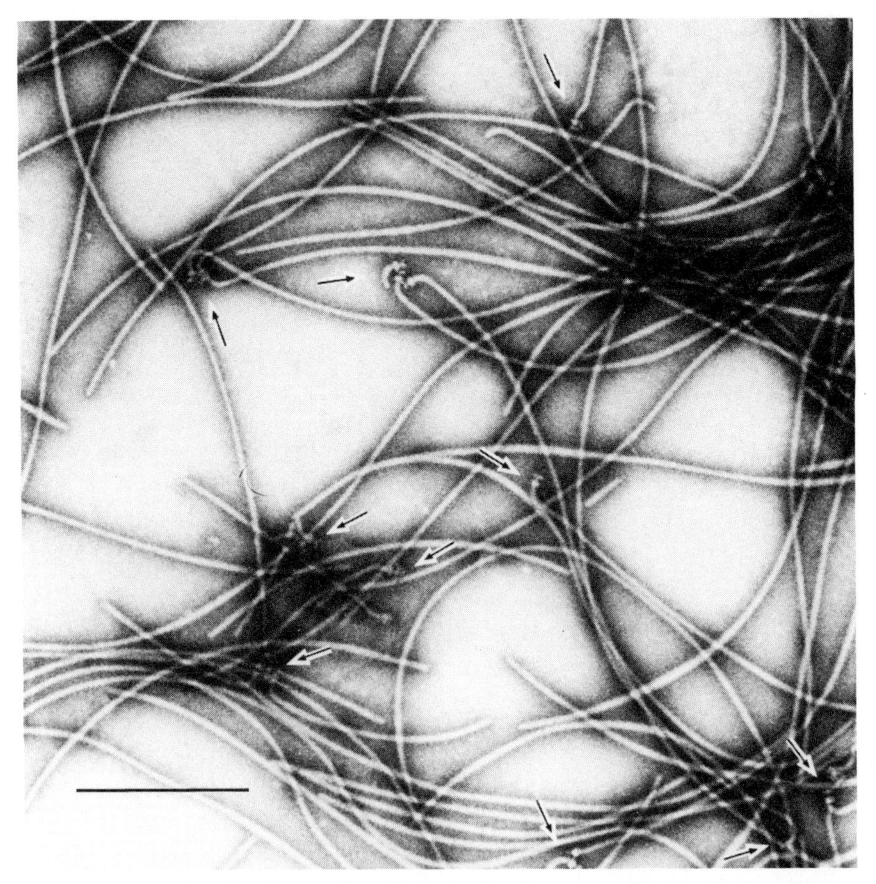

FIGURE 18.3 Purified intact flagella from the Gram-negative bacterium *Escherichia coli*. Many of the filaments have retained their basal body complexes (arrows). Bar, 0.5 μm. (Courtesy of Dr. Julius Adler.)

series of classical electron microscopic studies on the ultrastructure of flagella that have been supported and extended by other research groups. As a result, we can now discuss flagellar structure in terms of their possession of three distinct substructures: the filament, the hook, and the basal body.

The Ultrastructure of Flagella

Figure 18.4 contains electron micrographs of the three substructures of flagella as they appear in the Gram-negative organism *Escherichia coli*. Gram-positive bacteria possess flagella with a similar appearance except that they have a less complicated basal structure. These differences are related to differences in their cell wall structures.

The Filament

The filament is the portion of the flagellum that extends from the hook to the most distal end of the structure. Filaments average about 20 nm in diameter (ranging from 12 to 25 nm) and 10 to 20 μm in length. The overall form of the flagellum is a semirigid helix that on two-dimensional projec-

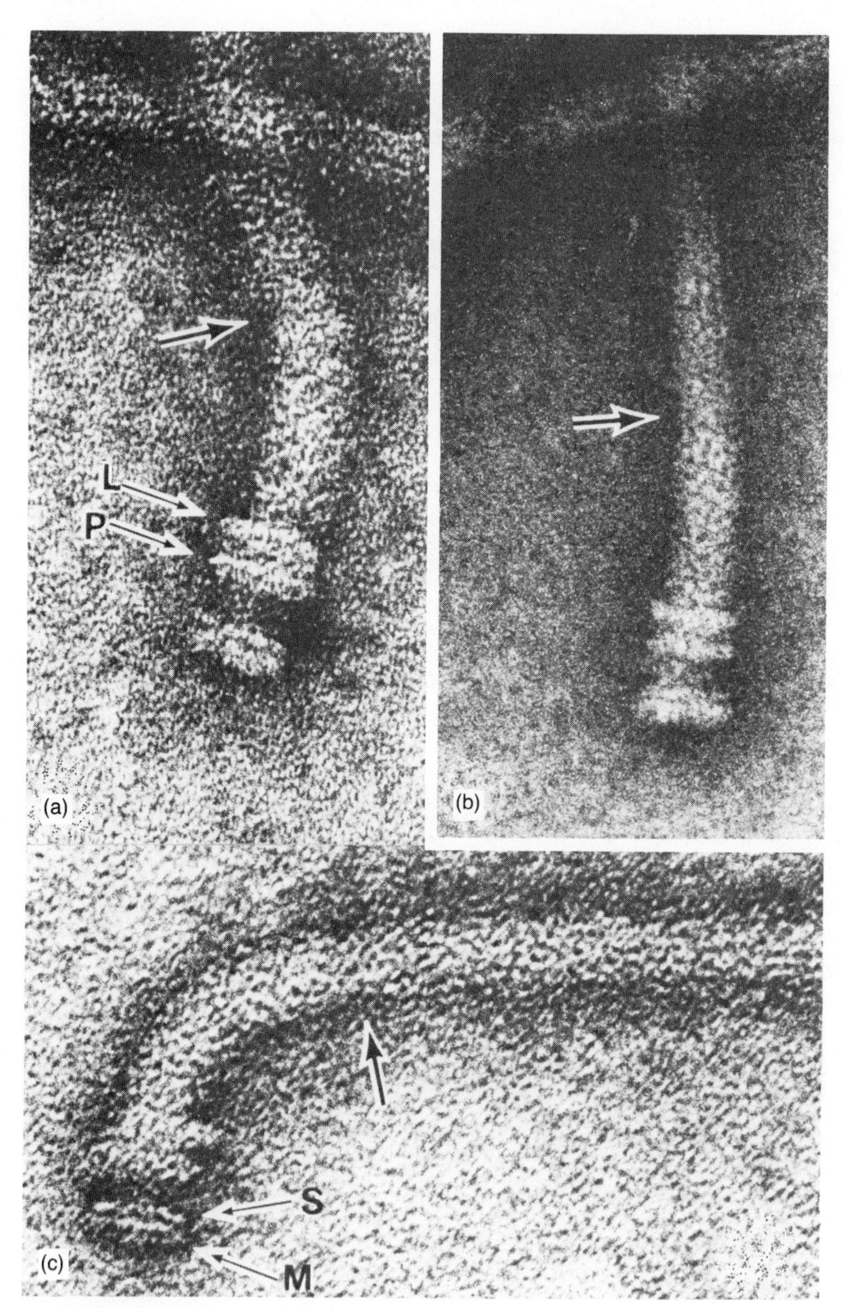

FIGURE 18.4 Highly magnified *E. coli* flagella showing their three substructures: filament, hook, and basal structure. (*a*) An arrow points to the juncture between the filament and the hook. The L and P rings are clearly seen. ×500,000. (*b*) All four rings of the basal structure are evident as well as the transition area between hook and filament. (*c*) The top edge of the S ring and the top and bottom edges of the M rings are visible. ×460,000. (Courtesy of Dr. Julius Adler.)

tions has a typical wave form of constant amplitude in a given species. In *E. coli* the wavelength is about 2.3 μm or about one tenth the length of the flagellum.

E. coli has a flagellar wave form of about 2.3 μm, one tenth the length of the flagellum. Three-dimensionally the pattern is helical. Mutants without helices or with tighter helices have been observed.

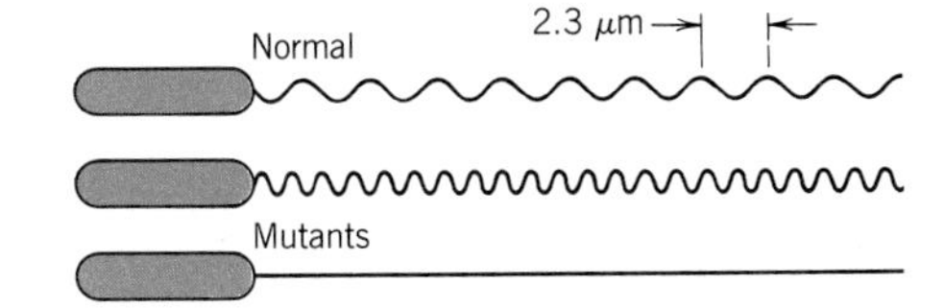

The filament is made up of a single type of protein called *flagellin*. Although identical units of flagellin make up the filament in a given species, the flagellins of different species vary in their molecular weights, chemical compositions, and manner in which they are stacked to make up the filament.

Molecular weights of flagellin ranging from 33×10^3 daltons for *Bacillus subtilis* to 51×10^3 daltons for *Salmonella* species to 60×10^3 daltons for *E. coli* have been reported.[8] The entire amino acid sequence has been determined for the flagellin of *B. subtilis*. It is a protein of 304 amino acid residues with a random distribution of hydrophobic residues but with an asymmetric charge distribution. Residues 1 to 101 possess a net charge of +6, residues 102 to 203 a net charge of −9, and residues 204 to 304, at the *C*-terminal end, a net charge of −4. The significance of this charge distribution is not known, but it is not unlikely that a relation exists between the quaternary arrangement of flagellins in a flagellum and the surface charge properties of the individual proteins.

The filament is a hollow cylinder of parallel longitudinal rows of flagellin molecules, with the number of rows constituting the circumference of the cylinder variable according to species. Between 8 and 12 rows is typical for many bacteria. The diameter of the hollow portion of the cylinder varies with the dimensions of the flagellin units and the number of longitudinal rows in the filament, but a core of 3 to 10 nm is common.

The flagellin rows are arranged in such a way that the subunits appear to be aggregated around the filament in helical fashion. The particular helical structure depends on the angle between the subunit and the filament axis. In one species of *Proteus*, for example, the flagellin molecules are wedge shaped and tilted at a 30° angle to the filament axis. This confers on the filament a characteristic helical conformation.

Filaments can be dissociated into flagellin components by high and low pH, urea, guanidine, or detergents. Remarkably, dissociated components will reassemble intact filaments when the dissociating agents are removed, suggesting that the subunits themselves possess all the necessary information for assembly of the mature filament.

This is in line with the observation that flagella grow by addition of flagellin molecules to the distal portion of the filament. Flagellins are synthesized within the bacterial cell, are immediately transported through the hollow core of the filament, and add to the growing tip of the filament.

The Hook

The hook is the section of flagellum between the filament and the basal structure. It is slightly larger in diameter than the filament and is slightly curved. It ranges in length from 70 to 90 nm and in width from 13 to 21.5 nm.

A single type of protein makes up the hook. As for the filament, hook proteins are also aggregated helically around a hollow core. These proteins, bearing no special name as yet, have molecular weights of 42×10^3 daltons in *E. coli* and 33×10^3 daltons in *B. subtilis*.

The protein is apparently neither compositionally nor structurally related to flagellin. It is antigenically distinct from it and is unaffected by low pH, urea, and other agents that dissociate flagellin units from one another.

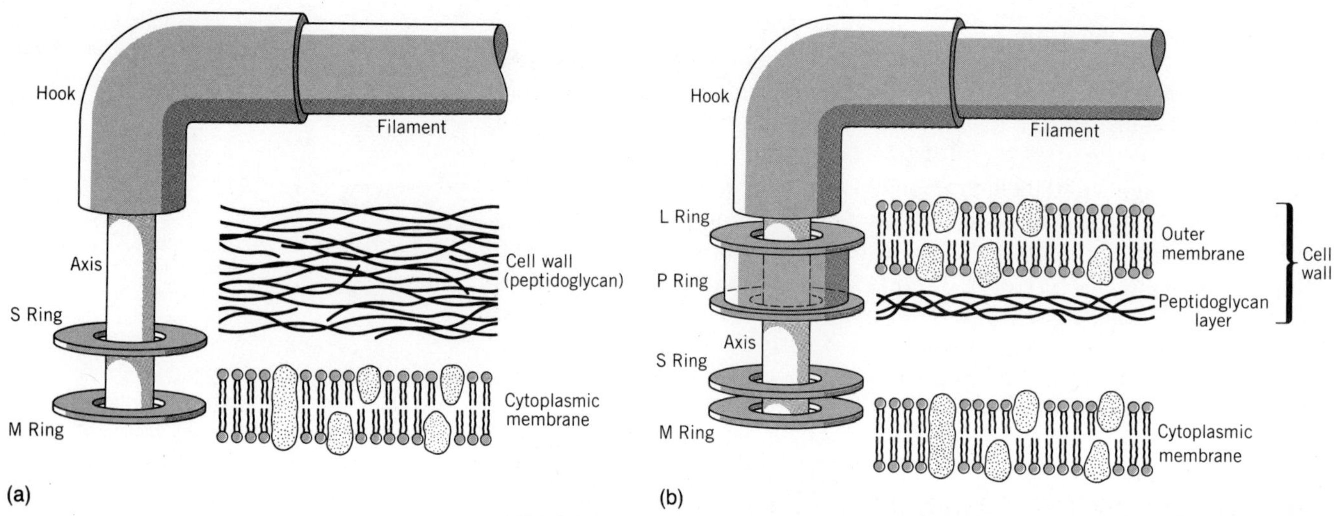

FIGURE 18.5 Schematic models of flagellar basal structures in Gram-positive (*a*) and Gram-negative (*b*) bacteria. The structures are shown as they are thought to be aligned with the plasma membrane and the major components of the cell wall.

The functional import of the hook is unknown, but two speculations are commonly made. One role may be to serve as an initiation point for flagellin polymerization, hence the growth of the filament. Another may be to serve as a "universal joint" between the rotational motion of the basal structure and the rest of the flagellum.

The Basal Structure

The basal structure at the most proximal end of the flagellum is the most complicated part of the organelle. In *E. coli* it consists of at least nine different proteins ranging in molecular weight from 9×10^3 to 60×10^3 daltons.

Electron micrographs of basal structures from Gram-positive and Gram-negative bacteria have been interpreted along the lines presented in Figure 18.5. As indicated, Gram-positive bacteria have simpler basal structures than Gram-negative organisms. But in both types, rings are fashioned around an axis or rod that is about half the diameter of the filament. Both have M and S rings, the former attached to the plasma membrane and the latter situated between the membrane and the layer of peptidoglycan. The two rings are separates by a gap of only 3 nm.

Recall from Chapter 3 that Gram-positive bacteria have walls that are peptidoglycan-rich and fairly homogeneous. Gram-negative bacteria have a thin layer of peptidoglycan overlaid by an outer membrane that contains the complex substance lipopolysaccharide.

Gram-negative bacteria have an additional pair of rings, one, termed P, associated with their relatively thin layer of peptidoglycan and the other, designated L, interacting with the layer of lipopolysaccharide characteristic of Gram-negative bacteria. In *E. coli* these rings are separated by a 9-nm gap. The two pairs are spaced by 12 nm.

Different organisms show variations on the theme of this structure, but they all look as though they belong more in the basement with the plumbing than in a cell of a living organism.

The Biomechanics of Flagellar Movement

Bacteria have the capability to move at extremely high velocities. An *E. coli* cell 2 μm long may cover a distance of 10 times its length in 1 sec. This is comparable to an automobile traveling well over the speed limit.

For some time it was thought that flagella propelled cells by a wavelike motion, with the organelle *fixed* to the cell but bending in a helical fashion. Bending this way would cause the propagation of a helical wave down the length of the filament, which would generate the force for movement. It was thought that bending was consistent with biological systems, but rotation of the flagellum at its base was viewed as implausible.

Then in 1973 H. Berg and R. Anderson proposed the implausible[9] and they and others, including M. Silverman and M. Simon, quickly moved to provide experimental support to the theory.[10] The essence of the proposal was that flagella move by a *rotary* action at their base, an actual slipping of the organelle with respect to the wall of the organism. Experimental proof came from the observation that when bacteria are tethered to a surface or to one another by way of their flagella, *the entire organism rotates*.

When bacteria are tethered to a surface or to one another, they rotate, a clear indication that their flagella rotate and do not simply bend.

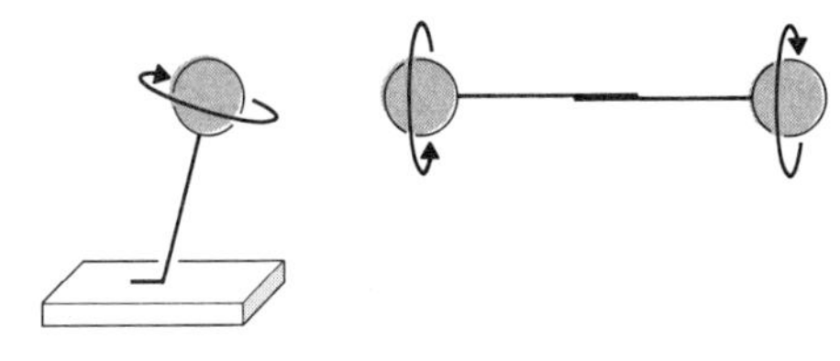

A dynamic model has been proposed to explain how the rings of the basal structure may generate the torque to turn the flagellum shaft. Figure 18.6 illustrates the principle. The M ring is assumed to be rigidly attached to the rod, functioning like a rotor in a rotary motor. The S ring is thought to be anchored to the peptidoglycan layer and not free to turn. It is functionally analogous to a stator. These two rings are the crucial components of the motor. Torque is generated between them, rotating the M ring in the plasma membrane and along with it the entire flagellum.

The P and L rings of Gram-negative bacteria do not appear to be essential for the generation of rotary motion. It is assumed rather that they serve as bushings around that portion of the shaft that projects through the complex lipopolysaccharide of the outer cell wall membrane.

An interesting find has been that the energy source for flagellar rotation is not ATP. It is instead derived from an electrochemical potential across the membrane, a protonmotive force (PMF). The PMF is made up of two components, a transmembrane potential and a transmembrane pH. If either of these components is disturbed, flagellar rotation ceases.

At this time it is not understood how PMF can drive the rotary motor. It has been hypothesized that an influx of protons may be channeled through the contact surfaces between the S and M rings to generate torque. Alternatively, protons may be only indirectly coupled to mechanical rotation.

The basal structure and its relation to flagellar rotation is one of the most surprising recent discoveries in the biological world. Nature appears to have invented and used the wheel for travel in the biological world long before it was employed to move humans and their possessions about.

During the next decade the details of this process will be significantly unraveled. It will be a most interesting area of research to observe, as it is sure to bring new insight to biological mechanisms of motion generation.

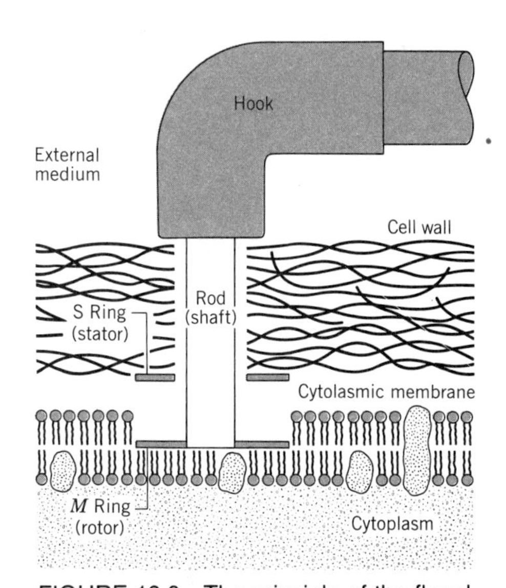

FIGURE 18.6 The principle of the flagellar rotary motor. Torque is generated between a rotating M ring and a stationary S ring. Since the rod is fixed to the M ring, it rotates with the ring. The hook may function as a universal joint between the rod and the filament.

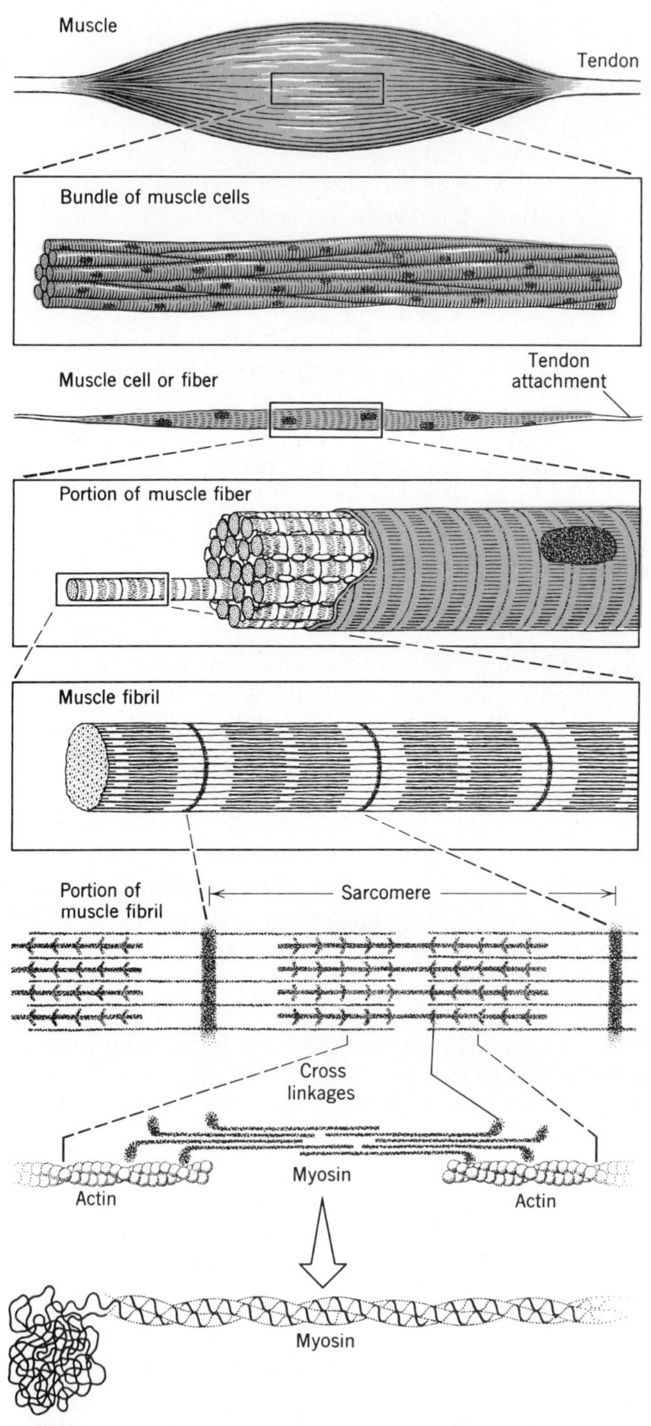

FIGURE 18.7 Levels of organization in muscle. The fundamental units are the two proteins, actin and myosin. These interact in a highly coordinated manner to form the sarcomere, the repeating unit of the muscle fibril. A collection of fibrils make up a muscle fiber or cell.

18.2 THE MUSCEL CELL ACTIN–MYOSIN SYSTEM

The bacterial flagellum is an organelle of motility that conducts its business not by contraction but by rotation. In that sense it is different from all other motility organelles. Movement in eucaryotes—whether of the whole cell or of components of the cell—is based on the activities of contractile proteins, which generally function at the expense of ATP hydrolysis.

Muscle motility has been intensively studied and is the best understood system of biological movement. Muscle cells represent a group of highly specialized cells, designed to effect supracellular contraction and little else. But they are not the only kind of cell to exhibit motility based on muscle-cell-type mechanisms. Most every kind of cell carries out some kind of internal if not external motility, so many different kinds of cell activity are presently being studied with an eye to muscle as the prototype contractile mechanism.

The term "contractile" protein is quite commonly used when discussing proteins that are involved with cellular movements. It is important to note, however, that the proteins themselves do not contract. Rather, as a result of their sliding past one another, the effect in the cell is contraction.

Since an understanding of the ultrastructure of muscle motility is so fundamental to a discussion of other types of cell movement, we will summarize its salient features.

The Microscopic Appearance of Muscle

Muscle is made up of bundles of long parallel fibers, each fiber being a multinucleate cell. Each fiber, or cell, contains fibrils that are the basic units of contraction. Fibrils, in turn, are made up of filaments of two types, which interdigitate laterally to give muscle its characteristic striated appearance. These levels of organization in skeletal muscle are delineated in Figure 18.7, which should be studied carefully before proceeding. The microscopic appearance of myofibrils is shown in Figure 18.8.

The striated appearance of muscle is due to the manner in which the myosin (thick) and actin (thin) filaments interdigitate. The wide, dark bands are called A (anisotropic) bands. The outer portions of the A band are the only regions in the myofilament where, in cross section, both actin and myosin filaments are found. The light region between the A bands is the I (isotropic) band. It contains actin filaments, but not myosin.

The I band is bisected by a narrow dark line, the Z line. The A band is discontinuous, being split by a lighter zone called the H zone, which is further bisected by a line, the M line. The H zone and the M line contain myosin but no actin. The distance between Z lines defines the sarcomere.

Microscopically observable regions of striated muscle are designated by letters derived from German words:

Z line	*zwischen*	between
H zone	*hell*	light
M line	*mittel*	middle

Gross Changes in Structure during Contraction

The structure we have been describing is typical of *extended* muscle. When muscle contracts, the sarcomere shortens by about 50%, from a length of approximately 3.5 to 2 μm. The A band maintains a constant width, while the I band narrows corresponding to the overall contraction of the sarcomere. The process of contraction is illustrated schematically in Figure 18.9.

Although the overall length of the sarcomere is shortening during contraction, neither the myosin nor the actin filaments themselves change in length.

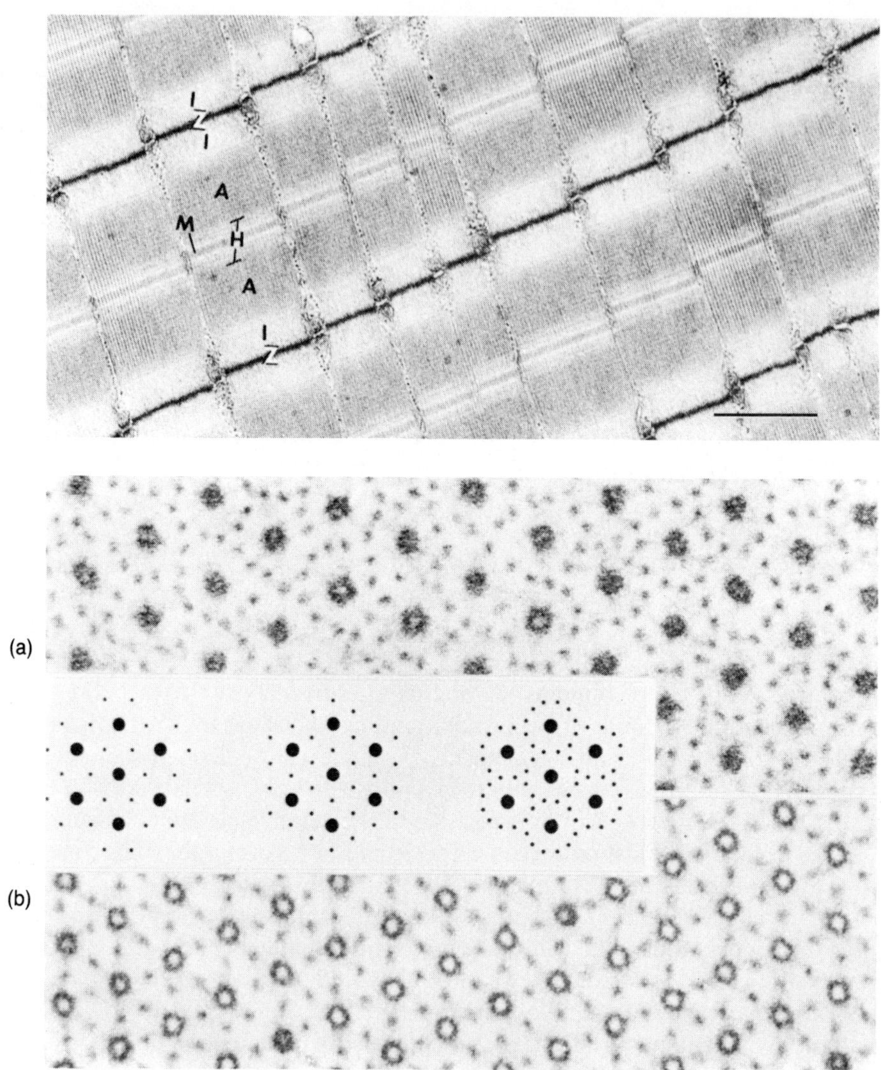

FIGURE 18.8 Electron micrographs of striated muscle specimens. (*Top*) A longitudinal section illustrating the striated appearance. Sarcomeres are clearly evident along with additional ultrastructural detail. This micrograph should be compared with Figures 18.7 and 18.9. (*Bottom*) Transversely sectioned preparation of intersegmental muscle of the cockroach (*a*) and flight muscle of the housefly (*b*). Right and center inset drawings depict these two views, respectively. Small actin filaments surround the larger myosin structures. As is apparent, the ratio of actin to myosin varies, as well as the arrangements of the two types of filaments. For comparative purposes, the left inset drawing shows the arrangement of filaments in vertebrate striated muscle. See text for discussion of nomenclature. (Courtesy of Dr. David S. Smith.)

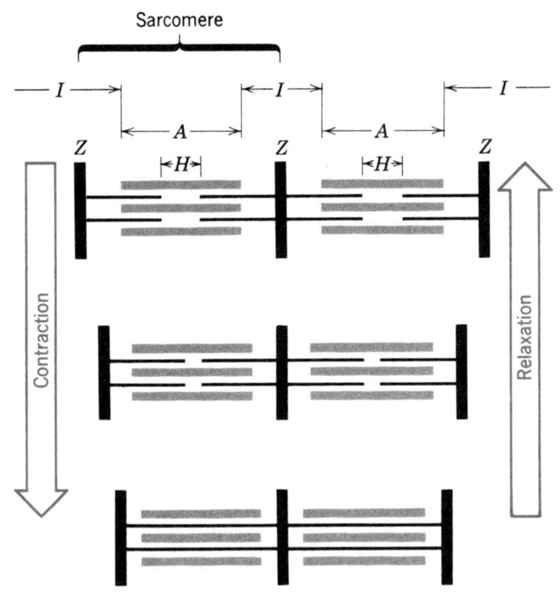

FIGURE 18.9 Schematic illustration of the process of muscle contraction emphasizing the relation between actin and myosin. Although the sarcomere shortens, the lengths of actin and myosin are unchanged during contraction. See text for discussion of nomenclature.

The Molecular Model of Muscle Contraction

Contraction of muscle is now quite well worked out at the molecular level. It is based on a movement of myosin with respect to actin, with ATP providing the energy for translocation.

The Properties of Actin and Myosin

Actin exists in two forms, a monomer globular protein of molecular weight 43,000 daltons (G-actin) and a fibrous polymer, which we have been referring to as the actin filament (F-actin).

The polymerization of G-actin to F-actin is driven by ATP hydrolysis:

$$n(\text{G-actin}) + n\text{ATP} \rightarrow \text{F-actin (actin-ADP)}_n + n\text{P}_i$$

The polymerized product contains ADP units attached to actin monomers. Although the hydrolysis of ATP is required to form the actin filament, this is not the hydrolysis that drives contraction. Once formed, actin filaments are about 7 to 9 nm in diameter and 1 μm long.

Myosin is a high molecular weight (460,000 daltons) protein consisting of two polypeptide chains twisted about each other for most of the length of the molecule. The twisted region, or tail, is rodlike, about 130 nm long and 2 nm in diameter. It is nearly completely α-helical in conformation. Each polypeptide chain terminates at its *N*-terminal end in a globular region, or head, which contains the ATPase activity of the muscle system. Each globular head has associated with it two light chains having molecular weights between 15,000 and 27,000 daltons.

Myosin and components derived from it.

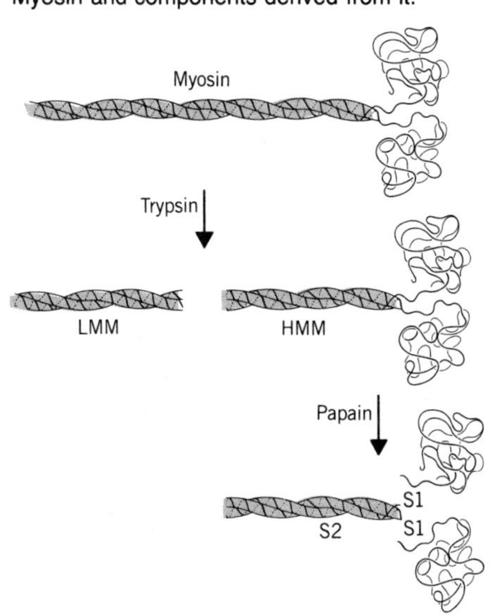

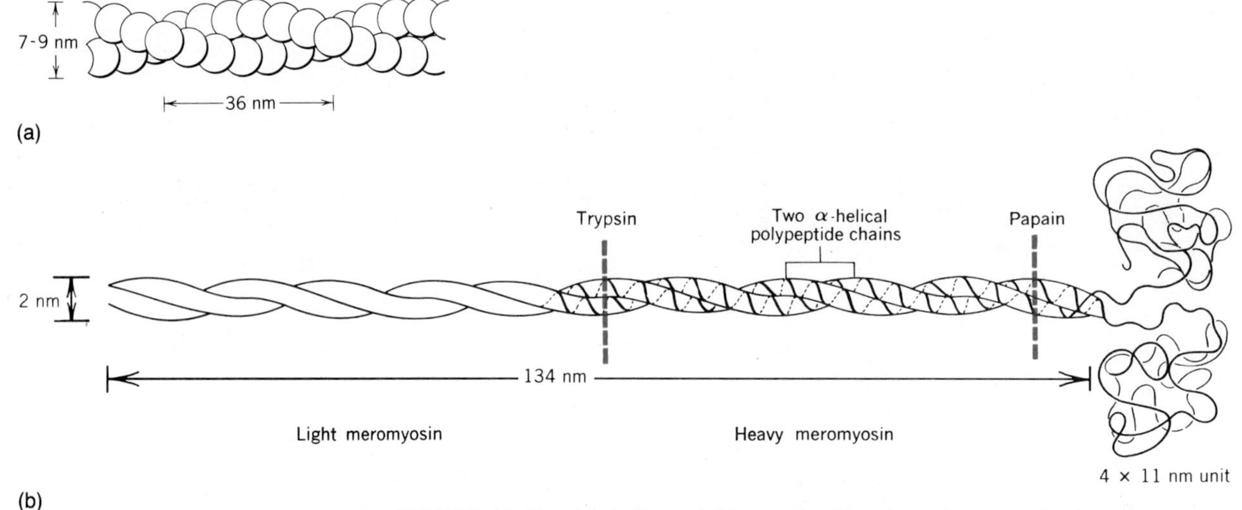

FIGURE 18.10 (*a*) Actin and (*b*) myosin. The sites of enzymatic cleavage shown can be used to produce heavy meromyosin or simply preparations of meromyosin heads.

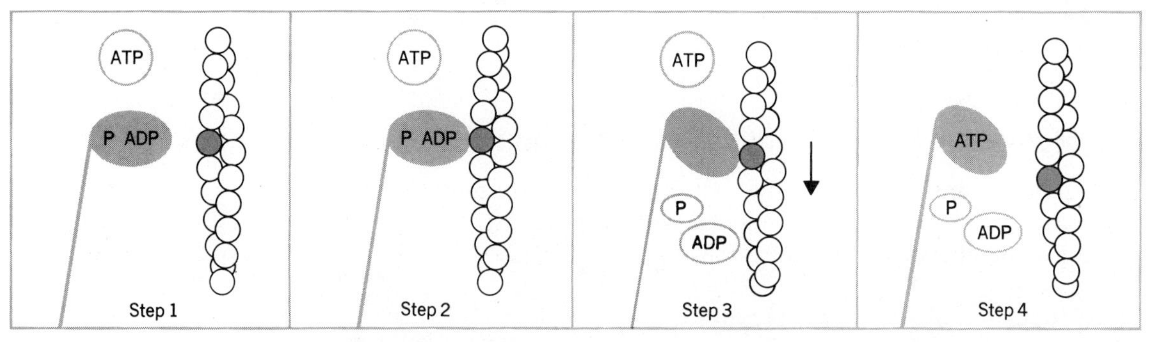

FIGURE 18.11 Stages in muscle contraction at the molecular level. Movement of actin and myosin with respect to each other occurs in step 3. The hydrolysis of ATP, which takes place in step 1, is the initial step that "cocks" the system for subsequent movement.

The myosin filament is made up of a collection of myosin molecules with overall dimensions of 15 nm in diameter and about 1.6 μm long. Myosin molecules are arranged longitudinally within the filament with the heads oriented away from the M line. Thus the filament is symmetrical with the axis of symmetry the M line.

Actin and myosin are shown in schematic form in Figure 18.10.

The Power Stroke Cycle of Contraction

Muscle contraction is based on cyclic attachment and detachment reactions between the head of myosin and the F-actin polymer. This is illustrated in Figure 18.11.

In resting or noncontracted muscle there is no contact between actin and myosin. The contraction cycle begins when actin and the head of myosin attach. Next, the orientation of the head tilts so that its long axis changes from 90° to about 45° with respect to the axis of the actin filament.

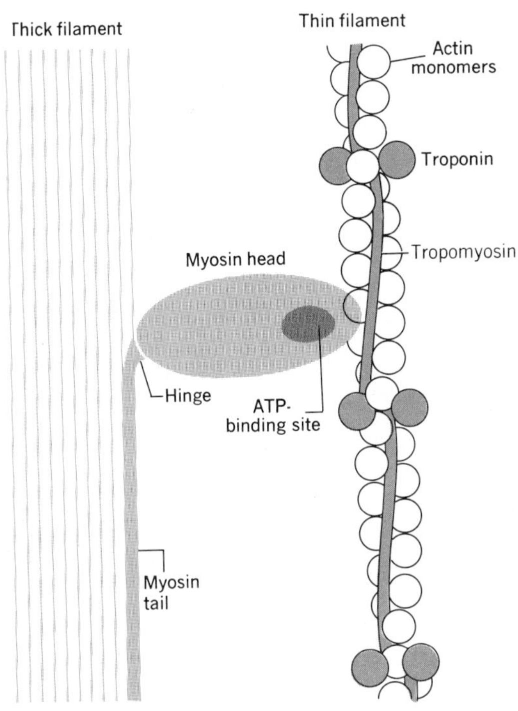

FIGURE 18.12 Additional components of the actomyosin system. Tropomyosin is situated in the groove of the actin polymer. At regular intervals troponin molecules interact with tropomyosin. These two proteins respond to Ca^{2+} and thus have a regulatory function in muscle contraction.

This action slides the filaments about 7.5 nm relative to each other. Finally, the myosin head is released from the actin filament and the power stroke is completed.

The decisive step in the cycle that brings about movement is the tilt of the globular head of myosin with respect to the actin filament. One would think that this must be the ATP-requiring step, but it is not directly so.

ATP is bound to the globular head when it is released from the actin filament (step 4). This tilted and detached configuration of the head is the low energy configuration of the system. ATP hydrolysis changes the head tilt to perpendicular (cocks it for the power stroke), and the resulting ADP and P_i remain tightly bound to the myosin head. When myosin binds to the actin filament, ADP and P_i are released. It is this *release* that causes the tilt to 45° and thus provides the power stroke for movement.

The myosin molecule has two "hinge" regions where the polypeptide chains are flexible, allowing the molecule to make the type of movement necessary for the contraction of the sarcomere.

The actomyosin system as we have described it thus far is not complete. Several other constituents have important roles in the complex. One is *tropomyosin*, a 66,000-dalton molecule made up of two α-helical subunits that coil about each other. In the actomyosin system tropomyosins lie end to end in the grooves generated by the helical configuration of actin monomers in the actin filament. Their position in the system is illustrated in Figure 18.12. Tropomyosin interacts with yet another protein complex, *troponin*, to regulate muscle contraction. This regulation is the subject of a good deal of current research.[11]

Tropomyosin and the troponin complex have essential roles in initiating the power stroke cycle. When a nerve impulse reaches the outer membrane of a muscle fiber, the membrane system of the sarcomere is depolarized. This releases Ca^{2+} to the cytoplasm where the actomyosin system is located. Calcium ion binds to the troponin complex, generating a conformational change in the complex that is transmitted to tropomyosin. Tropomyosin moves in its helical groove, unblocking access of the myosin heads to the actin polymer.

Thus the regulator of contraction is Ca^{2+}. Its release sets in motion a series of interactions between several proteins, the result being cellular contraction and organismic motility.

The actin–myosin interactions just described are applicable to nonmuscle cells as we shall see later in this chapter. Muscle and nonmuscle contractile systems are not identical, but they bear such strong similarities that what is known about muscle actomyosin and its behavior is certain to simplify studies on nonmuscle contractile activities.

18.3 MICROTUBULES

Microtubules exist singly or in higher packing orders depending on their functions in cells. Singlet or unit microtubules are commonly seen in the cell interior, while highly ordered arrangements of microtubules are characteristic of organelles of locomotion. In both instances microtubules have basically the same molecular structure.

Composition and Ultrastructure

Microtubules are made up of two types of protein subunits called alpha (α) and beta (β) tubulin. A dimer consisting of one each is the basic molecular structure on which all higher ordered structures are built.

The two proteins are very similar. They are globular, with molecular weights of about 55,000 daltons and 500 amino acid residues. Sequence studies have revealed that the two proteins are remarkably similar in primary structure, suggesting a common evolutionary origin. Not only are they similar to each other, but they show strong similarities among many species from which they have been isolated and characterized.

The α and β tubulins aggregate to form a cylinder of about 24 nm in diameter with a bore of 15 nm. The subunits are arranged in a very shallow helix: 13 tubulins typically make up one complete turn of the helix. The dimers are arranged longitudinally on the cylinder, not along the run of the helix strand. Therefore the cylinder structure may be viewed as a wall made up of 13 *protofilaments* arranged in a circle with the run of the protofilament consisting of alternating α and β tubulins. The helical architecture results from the angle between the axis of a protofilament and the positions of tubulins on adjacent protofilaments. These features are illustrated schematically in Figure 18.13.

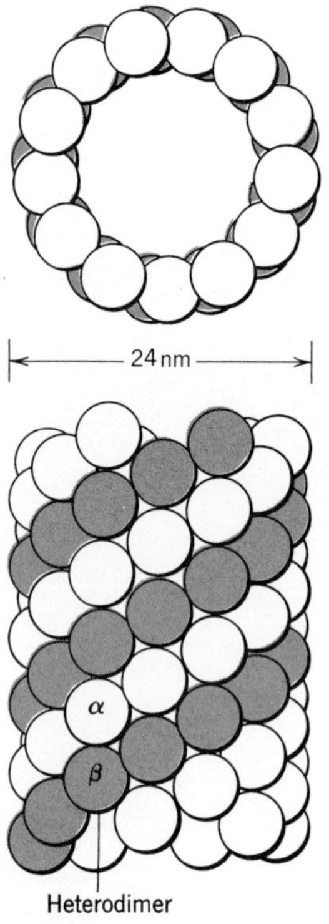

24 nm

α

β

Heterodimer

FIGURE 18.13 Helical architecture of α and β tubulins constituting microtubules. In the most typical type of microtubule 13 α, β pairs make up one complete turn of the helix. A longitudinal run of alternating α and β tubulins is called a protofilament.

The length of a given microtubule appears to be indeterminate. It may depend on the particular type of cell in which it is found and its specific function in the cell. Lengths of 10 to 25 μm can be observed within the axons of nerve cells and 5 to 200 μm within the cilia and flagella of organisms possessing those appendages.

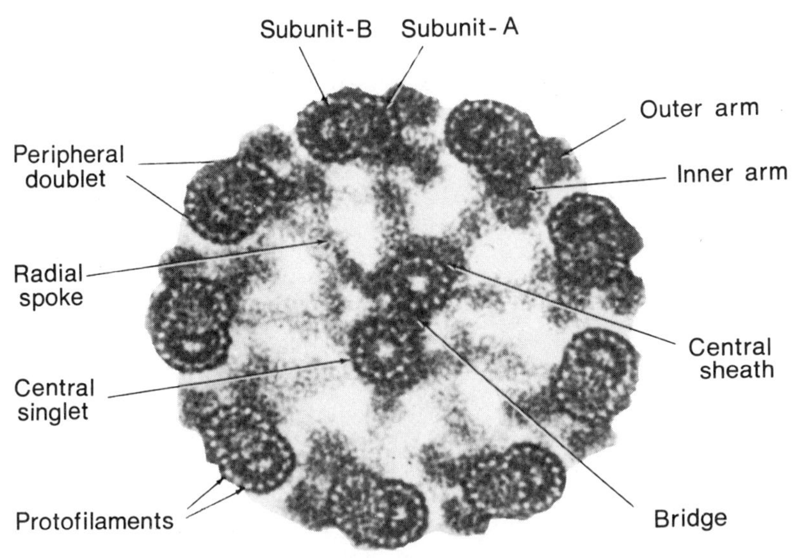

FIGURE 18.14 Electron micrograph of a ciliary axoneme showing the second level of microtubule arrangement. The classic 9 + 2 arrangement is dramatic. Compare with Figure 18.15 for a delineation of the parts of the axoneme. (Courtesy of Dr. Don W. Fawcett.)

Higher Order Packing Arrangements

In cilia and flagella, as well as in centrioles and basal bodies, microtubules are arranged with a high degree of order.

One level of order is the microtubule doublet. In this arrangement one microtubule is complete and the neighboring member of the doublet shares three of the protofilaments of the other to make it a complete cylinder.

A second level of order is the arrangement of doublets with respect to one another. In cilia and flagella there are nine doublets arranged in a circle around two singlets in the center. This arrangement, described as 9 + 2, can be seen clearly in the electron micrograph of Figure 18.14.

Microtubule doublets are not free-standing structures but are rather linked to one another and spoked to the central singlet pair. These features are illustrated in Figure 18.15.

The microtubular system and associated structures make up the *axoneme* of the cilium or flagellum. Each doublet of microtubules is skewed 10° from a tangent drawn to a circle going through all doublets. The member of the doublet closer to the center of the axoneme is designated *subfiber A*. The other member is *subfiber B*.

Subfiber A contains two arms, called *dynein* arms, extending out toward the neighboring doublet. The dynein arms point clockwise when the axoneme is viewed in the direction from tip to base. Interdoublet links (nexin), radial spokes and heads, and a central sheath and microtubule bridge make up the rest of the ultrastructure of the axoneme. The entire cilium or flagellum is encased by a membrane that is an extension of the plasma membrane.

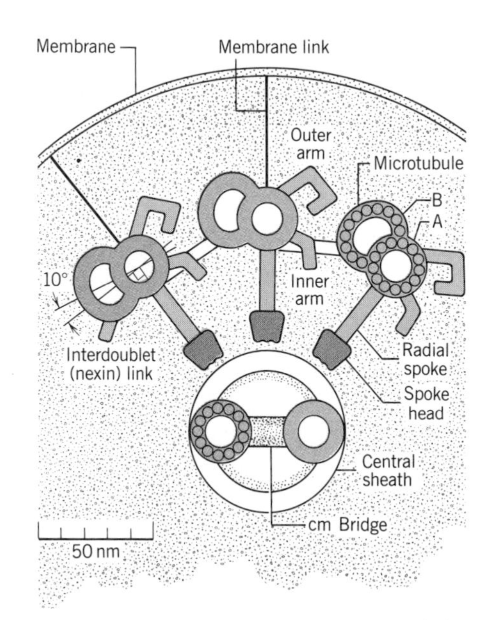

FIGURE 18.15 The components of the axoneme (Courtesy of Dr. Michael E. J. Holwill.)

Dynein, the Microtubule ATPase

The second axonemal protein, in addition to tubulin, that has been the object of intensive study is dynein. It has ATPase activity and as such is directly responsible for the hydrolysis of ATP that provides the energy for ciliary and flagellar movement. Dynein thus serves a pivotal role in microtubule-based motility.

Dynein is present in two isoenzymic forms in sea urchin sperm, a common source of material for studies on microtubules. The major form, dynein 1, accounts for about 80% of total axonemal ATPase activity. It is located in the arms on the doublet microtubules of the axoneme.

When first extracted, dynein 1 has a high molecular weight (1.5×10^6 daltons). After dialysis against low ionic strength buffers its molecular weight is cut in half. On SDS–polyacrylamide gel electrophoresis of whole anoxemes, dynein migrates as a complex band of 300,000 to 400,000 daltons.

The second isoenzymic form of dynein, dynein 2, accounts for the remaining ATPase activity of the axoneme. Its molecular weight, when purified, is around 700,000 daltons, about the same as the smaller form of dynein 1. On electrophoresis it migrates with dynein 1.

The dynein arms project obliquely from subfiber A at a repeat distance of 24 nm. In the presence of 1 mM $MgSO_4$ the arms make contact with subfiber B of the adjacent microtubule doublet. This interaction of two doublets by way of dynein cross-bridges is strikingly revealed in the electron micrograph of Figure 18.16.

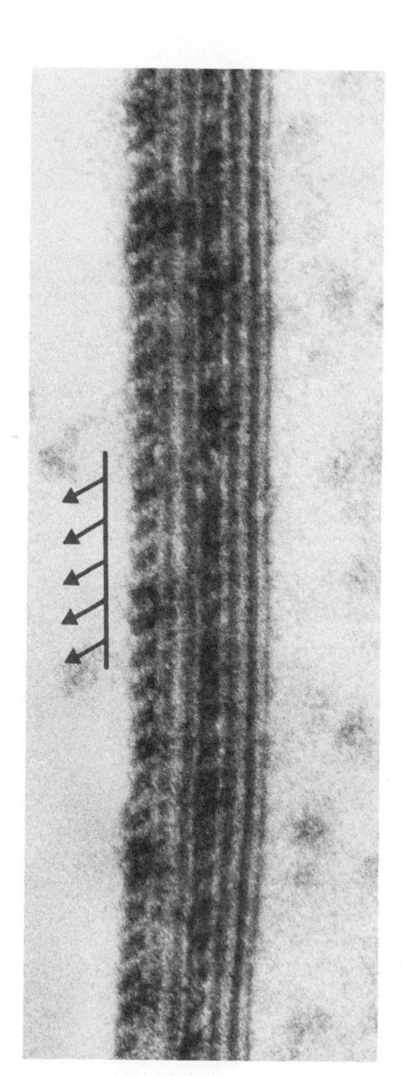

FIGURE 18.16 Two doublet microtubules and their dynein arms. These are obtained from regions of ATP-disintegrated cilia. The polarity of the arms is base directed as is clearly evident for the row projecting to the left (arrows). A careful analysis suggests that each arm consists of three subunits. ×180,000. (Courtesy of Dr. Fred D. Warner.)

At present it is believed that each arm on the A subfiber contains three subunits, at least one of which is dynein 1. The attachment of the arm to the subfiber is not covalent. Dynein can be extracted by 0.5 M KCl, and prolonged exposure to EDTA or ATP causes the arms to dissociate from the fiber. Under the appropriate experimental conditions, the arms can be restored to their positions on the A subfiber.

The Orientation of Microtubules in Cells

Microtubules in the 9 + 2 array are found in cilia and flagella as already described. In addition, they are distributed throughout the cell interior in a variety of ways in almost all eucaryotic cells.

The Marginal Band of Blood Cells

In nucleated erythrocytes carried by all vertebrates other than mammals, microtubules form an elliptical bundle encircling the cell perimeter just beneath the plasma membrane. They are also universally present in mammalian blood platelets. Marginal bands as they are seen in erthyrocytes are presented in the electron micrograph of Figure 18.17.

A similar bundle is seen in the primitive nucleated erythrocytes of mammalian embryos. In some cases the marginal band seems to be important in maintaining the elliptical shape and rigidity of the biconvex erythrocytes. In other cases the band does not maintain the shape of the mature cell, but it may be important in assembling the cell into its characteristic shape.

Parallel Longitudinal Runs in Axoplasm

Within the axon of the neuron, microtubules are found stretching longitudinally. Several of these can be seen in the micrograph of Figure 18.18, along with smaller associated filaments.

In this case the role of microtubules seems to be related more to transport than to imparting shape or rigidity to the cell. It is thought that microtubules aid in the movement of neurosecretory granules within the cell (see Chapter 19).

Spindle Fibers of the Mitotic Apparatus

The mitotic apparatus possesses two functional types of fiber. One type connects the chromosomes to a pole body, or centriole, and the other type extends from pole to pole, bypassing the chromosomes.

Both types of fiber are microtubules that are assembled before mitosis and disassembled after chromosomes are separated. Figure 18.19 is a micrograph showing the spindle orientation of microtubules in a cell during mitosis.

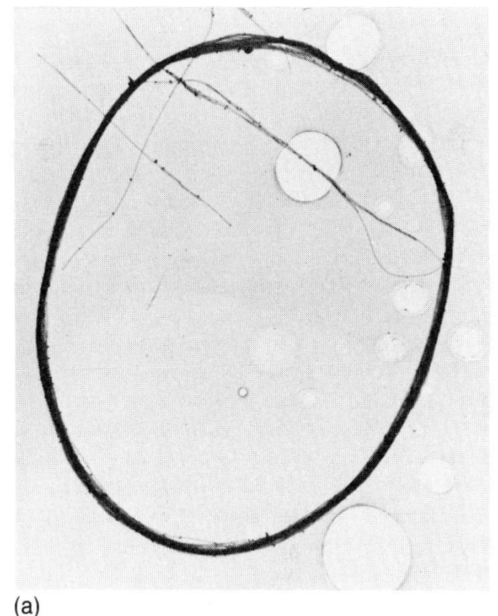

(a)

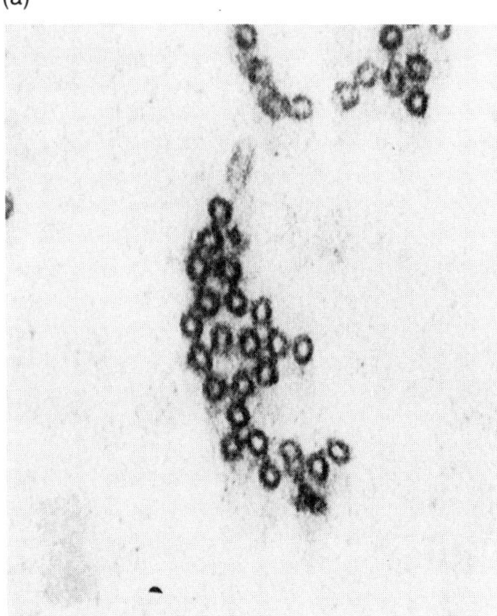

(b)

FIGURE 18.17 Electron micrographs of the marginal band of microtubules isolated from dogfish erythrocytes. The whole band (*a*) appears to be a continuous elliptical structure. In cross section (*b*) the band is clearly composed of microtubules, many of which are cross-bridged. (Courtesy of Dr. D. W. Cohen.)

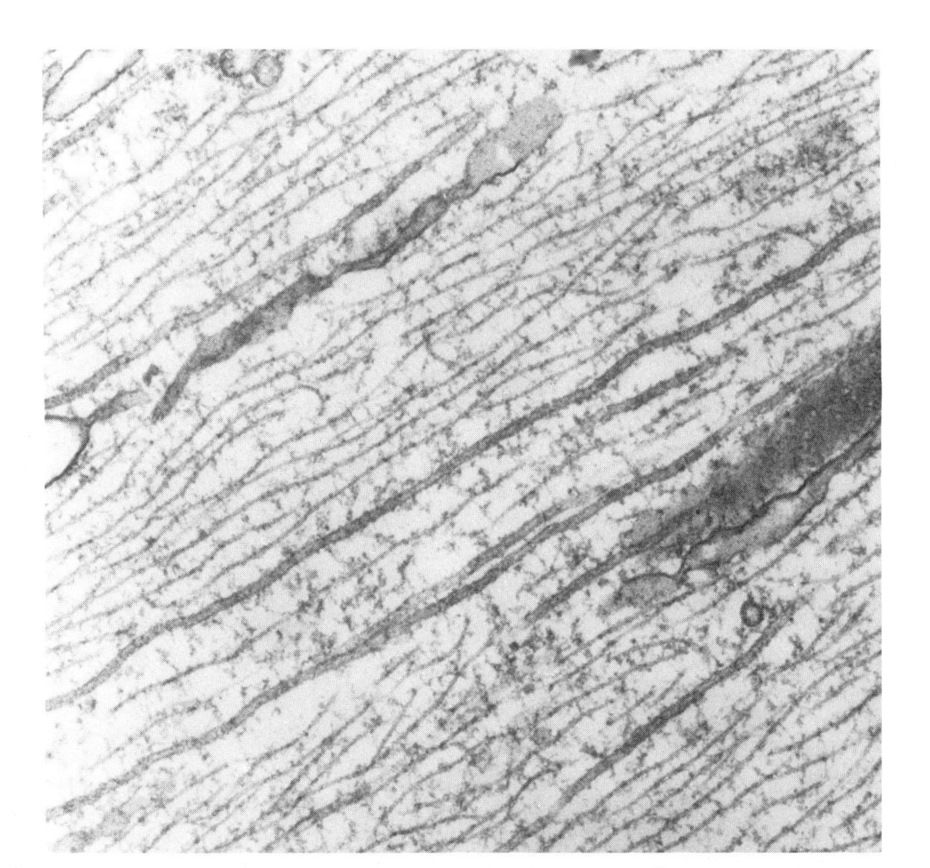

FIGURE 18.18 Axoplasm of a human nerve. The largest filaments running from lower left to upper right are microtubules. Thinner structures are neurofilaments. The background appears to contain a network of smaller filaments that interconnect the larger structures. Irregular structures are membranes of the endoplasmic reticulum. $\times 60{,}000$. (Courtesy of Dr. Pierre Dustin.)

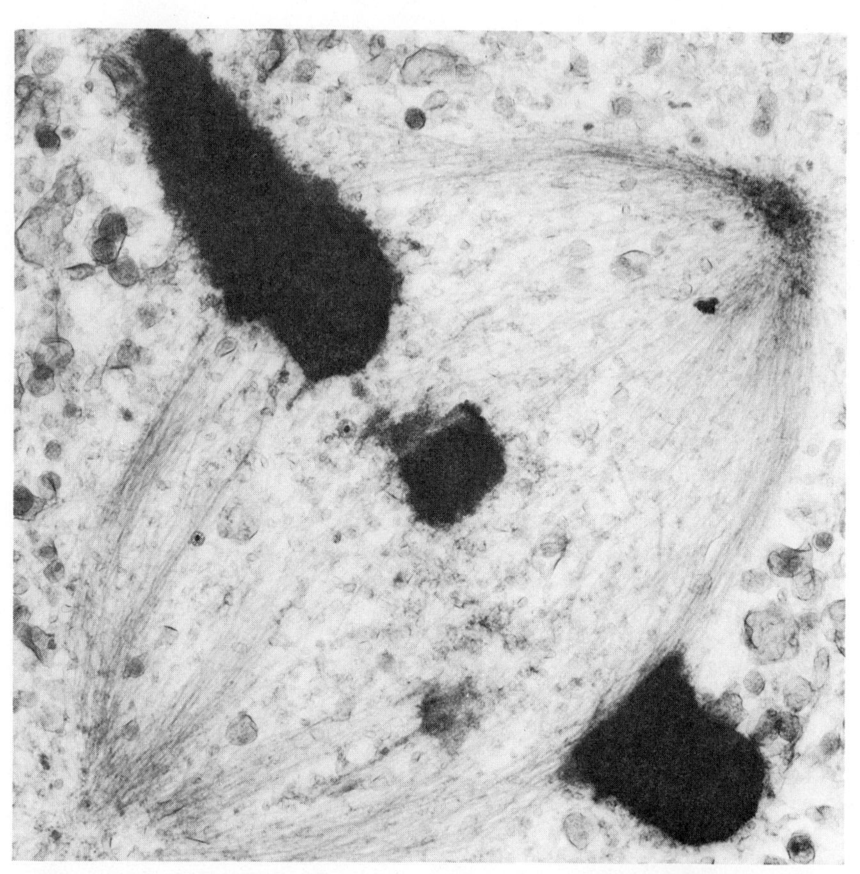

FIGURE 18.19 Electron micrograph of a section through a spindle. The microtubules are clearly oriented with the axis of the spindle, emanating from the two poles of the cell. The large dark objects are chromosomes lined up on the metaphase plate. (Courtesy of Dr. J. Richard McIntosh.)

Cytoskeleton

Method VII

By using fluorescent antibodies to tubulin it is possible to localize microtubules in any type of cell. A cell treated in this manner is depicted in Figure 18.20. In this micrograph microtubules run in a radiating orientation, forming what takes on the appearance of a cytoskeleton. In Chapter 19 we will discuss these structures, as they are revealed by immunofluorescence microscopy, in more detail.

Assembly–Disassembly Reactions

The intracellular microtubule network is highly dynamic, constantly forming and disappearing depending on cell activities. A number of different activities are known to depend on microtubules, so there must be control mechanisms to activate their assembly and direct their breakdown as the demands of the cell change.

It is not known for certain what governs microtubule assembly *in vivo,* but several factors that influence *in vitro* assembly and disassembly have been studied.

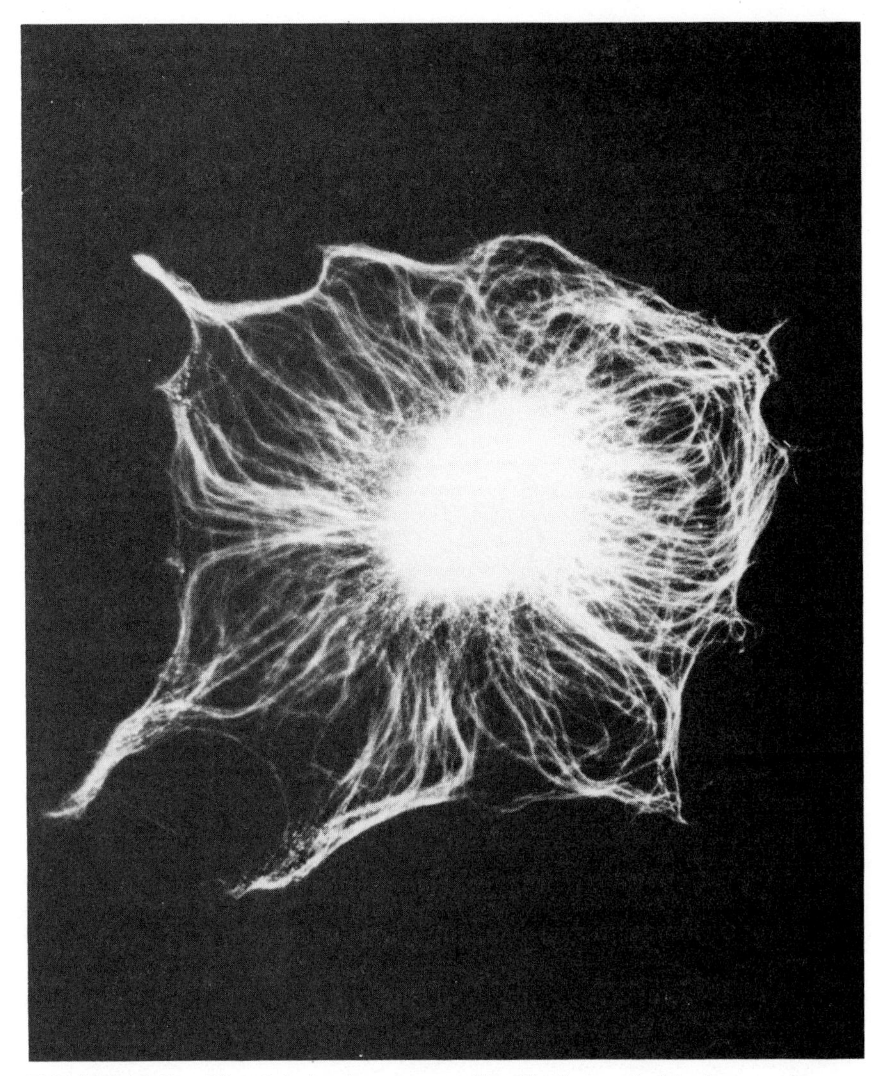

FIGURE 18.20 Microtubules in a human fibroblast revealed by immunofluorescence with antitubulin antibodies. (Courtesy of Dr. E. Lazarides.)

In the laboratory, complete microtubules can be assembled starting with α and β tubulin monomers. The conditions for assembly were worked out by Richard C. Weisenberg at Temple University.[12] Often the tubulin used in assembly experiments comes from brain tissue.

Requirements for Assembly

Self-assembly of microtubules takes place at 37°C in a pH 6.6–6.8 buffer containing 1 mM GTP and 0.5 mM $MgCl_2$. The microtubules that form are stabilized by 4 M glycerol and 1 M sucrose.

The first reaction is a dimerization of one α and one β tubulin (see Figure 18.21). Dimers aggregate further to generate rings, some of which appear to be complete and others spirals.

Electron microscopic evidence for the formation of rings is presented

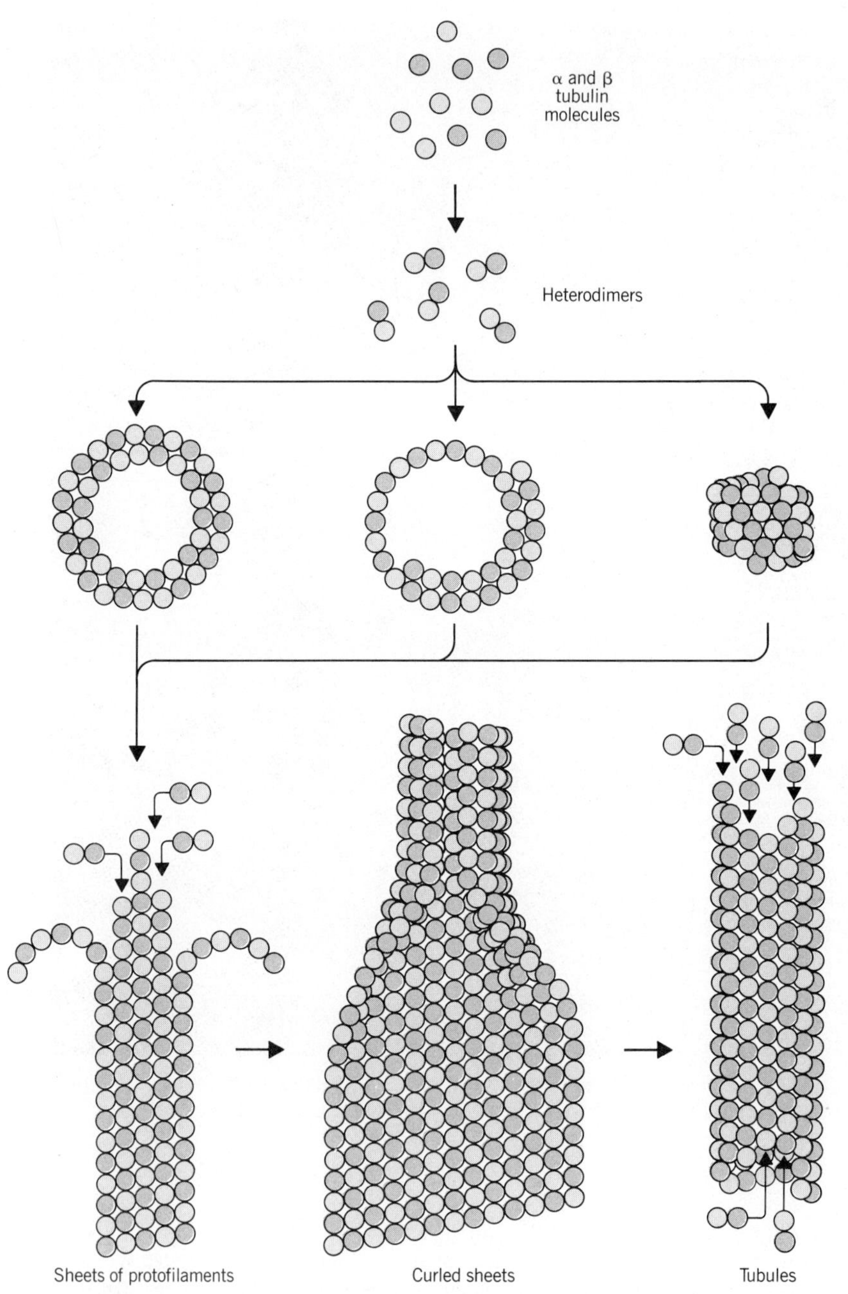

FIGURE 18.21 Proposed steps in the formation of microtubules. α and β tubulin dimers polymerize into ring structures which are converted to sheets that curl into tubes. These steps, originally formulated by Dr. M. W. Kirschner, are based on the results of electron microscopic studies such as those depicted in Figure 18.22

in Figure 18.22. These will form at 4°C, but disappear when the temperature is elevated and higher order structures appear.

The next step in assembly appears to be the formation of sheets of parallel protofilaments. The mechanism of transition from rings to sheets is not understood, but there is nevertheless electron microscopic evidence for sheet formation before the final assembly of the microtubule.

The sheet is finally thought to curl up to form a tube, which constitutes a short section of microtubule. The microtubule then grows by the addition of dimers to the ends.

GTP is required for assembly at a concentration that is at least equal to that of the tubulin dimer. The ends of growing microtubules possess GTPase activity, and this activity is responsible for the hydrolysis of the GTP that is required for assembly. However, the hydrolysis does not appear to provide the energy for polymerization. Rather, the nucleotide, perhaps in its diphosphate form, may be the essential factor that confers on tubulin the conformation that catalyzes assembly.

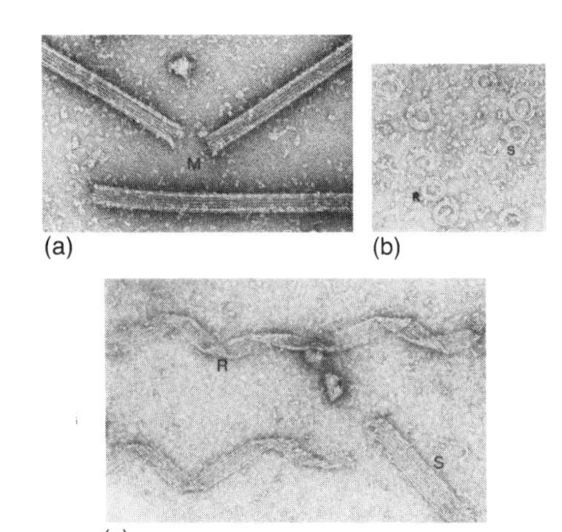

FIGURE 18.22 Electron micrographs of microtubules and possible intermediates. (*a*) Complete microtubules (M). (*b*) Double rings (R) and spirals (S). (*c*) Twisted ribbons (R) and flat sheets (S) of protofilaments. (Courtesy of Dr. Felicia Gaskin.)

Accessory Proteins

Studies on microtubule assembly have revealed that other protein factors speed up assembly and may have specific roles in the completed microtubule. These microtubule-associated proteins, (MAPs), have not been completely characterized. Several of them appear to be of relatively high molecular weight (~ 200,000 daltons) and one, designated "tau," is a 70,000-dalton protein that is found at low concentrations in microtubules. It appears to be quite important for assembly. Tau seems to be primarily involved during the formation of rings and spirals and therefore may be essential for the early stages of microtubule construction.

Another MAP, called MAP 2, appears on the completed microtubule at fixed intervals on its surface. Microtubules assembled without this protein have smooth surfaces. When present, MAP 2 projects perpendicularly from the surface at 30-nm intervals.

One type of arrangement between MAPs and the tubulins is illustrated in Figure 18.23. In this model part of the MAP runs longitudinally between protofilaments for a length of about 12 tubulin dimers. The length of a single MAP is therefore nearly 100 nm, but because of the manner in which MAPs are staggered and helically aligned, the distance between projections represents a fraction of the length of the molecule.

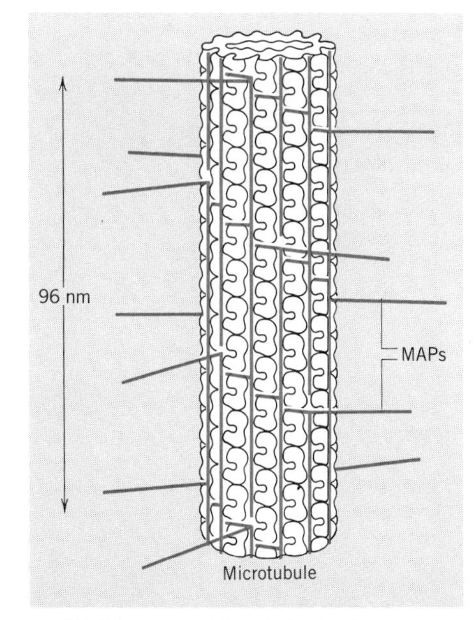

FIGURE 18.23 Model depicting possible relation between microtubule associated proteins (MAPs) and the microtubule. A part of the MAP may run between protofilaments for a length corresponding to 12 tubulin dimers (~ 96 nm). The other portion of the molecule extends outward as a filament. Side branches are thought to help stabilize the system and to assist in bringing about symmetry and a controlled diameter of the microtubule.

Factors That Influence Disassembly

Microtubules depolymerize in the cold (4°C) and are sensitive to pressure and chemicals such as colchicine and vinblastine.

It has been known since the early 1930s that colchicine, when injected into mice, causes an increase in cells at prometaphase. Although it was originally thought that colchicine stimulated mitosis, it is now clear that mitosis is *blocked* at the prometaphase stage, thus causing a pileup of cells with these characteristic chromosome arrangements.

Each α, β-tubulin dimer possesses one binding site for colchicine and one for vinblastine. These chemicals apparently compete for binding sites that are normally required for interaction between dimers or protofilaments in forming the sheet. As a result, they inhibit polymerization and microtubules are not generated.

Since the mitotic apparatus is made up of microtubules, these chemicals block the movements of chromosomes that normally take place after metaphase in mitosis.

High concentrations of divalent cations, such as Ca^{2+}, also inhibit microtubule assembly or depolymerize already assembled microtubules.

Pressure is a depolymerizing agent as well. Recall from Chapter 2 that cells do not reproduce at ocean depths where the pressure is great enough to inhibit polymerization reactions. High pressures must affect the conformation of tubulin so that aggregation is not favored.

How these factors, which influence *in vitro* assembly and disassembly, relate to the events taking place *in vivo* is largely unknown. But it is a reasonably safe guess that temperature, Mg^{2+} and Ca^{2+} concentrations, and the level of GTP in the cell all contribute toward proper assembly and disassembly.

A very interesting finding regarding microtubule assembly and disassembly is that at polymerization equilibrium, assembly and disassembly take place at opposite ends of the microtubule.[13] There is thus a polarity within the microtubule that affects its growth and breakdown. This property appears to be related to the directional activities for which microtubules are responsible within the cell. This is elaborated on in the discussion of microtubule-based movements in Chapter 19.

Calmodulin Regulation of Microtubule Dynamics

Calmodulin has been implicated as a modulator by many different enzymes and activities, including:

- Ca^{2+}-dependent phosphodiesterase
- Brain adenylate kinase
- Membrane Ca^{2+}-ATPase
- Phosphorylase kinase
- Synaptic and plasma membrane phosphorylation
- Ca^{2+} transmembrane transport
- Microtubule disassembly

During the past decade a protein called calmodulin has been discovered and studied intensively.[14] It is a Ca^{2+}-binding protein that not only is ubiquitous among eucaryotic cells but is highly conserved in its structure among all species studied.

Calmodulin stimulates a variety of enzymes directly and regulates others indirectly by way of specific protein kinases. It is beginning to appear that calmodulin is the universal acceptor of Ca^{2+} whenever this ion is employed by the cell in a regulatory role.

Calmodulin has a molecular weight of 16,790 daltons and binds up to four Ca^{2+} per molecule. Its primary structure and the proposed sites of Ca^{2+} binding are depicted in Figure 18.24. When calmodulin functions as a modulator of enzyme activity, it does so in two stages. First, Ca^{2+} binds to the protein, which results in conformation changes that vary depending on the number of Ca^{2+} ions bound. Next, the calmodulin with its Ca^{2+}-imposed conformation complexes with the enzyme it activates. Thus Ca^{2+} is not the immediate regulator of enzyme activity but operates through a calmodulin intermediary.

At the moment, the role of calmodulin in microtubule assembly and disassembly is only circumstantial. One line of evidence is the calmodulin distribution in the cell as it relates to cell cycle. During interphase, calmodulin is distributed rather diffusely throughout the cell interior. During metaphase its location is primarily in the vicinity of the pole regions of the mitotic apparatus. Later on, during anaphase and telophase, calmodulin is found at each end of the midbody of the mitotic apparatus.

Thus calmodulin has both a temporal and a spatial pattern during mitosis that suggests it to be at the cell centers where microtubules are forming and disassembling. In this regard it is speculated that calmodulin may have a dual regulatory role. On the one hand, by virtue of its Ca^{2+}-binding ability, it may maintain low Ca^{2+} levels which are necessary for microtubule assembly. On the other hand, increased calmodulin concentrations may activate tubulin depolymerization, hence microtubule disassembly.

In addition to influencing microtubule dynamics, calmodulin may regulate microtubule ATPase activity such as that which is associated with dynein.[15] The beating of cilia and flagella in eucaryotes has its basis in an

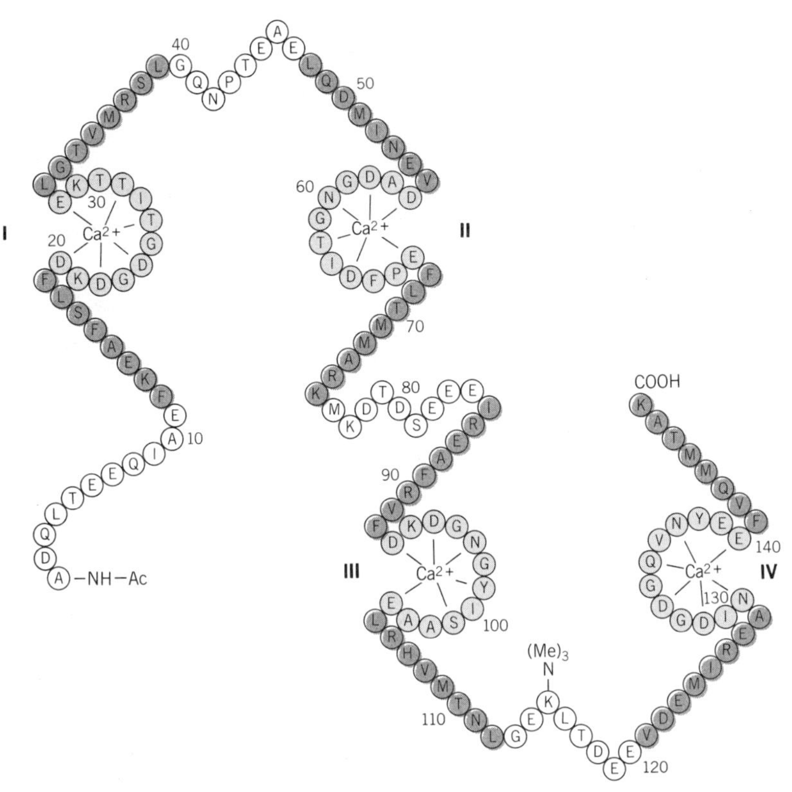

FIGURE 18.24 Calmodulin, a regulatory protein, shown with its primary structure and proposed sites of Ca^{2+} binding. The interaction between calmodulin and Ca^{2+} may be the first reaction in a complex series that results in microtubule assembly. It may also be involved in microtubule depolymerization and in dynein ATPase activity. Amino acids are designated by their one-letter symbol. Roman numerals indicate molecular domains that bind Ca^{2+}.

interaction between microtubules bridged by dynein molecules (see Chapter 19 for a more thorough treatment of this topic). The ATPase activity of dynein in the presence of Ca^{2+} is markedly enhanced when calmodulin is provided. Furthermore, dynein has been demonstrated to possess Ca^{2+}-dependent binding sites for calmodulin. Thus it appears that calmodulin is a mediator for the Ca^{2+} regulation of dynein ATPase, hence is a factor in the control of ciliary and flagellar motility.

Although the roles of calmodulin vis-à-vis microtubules have not been precisely spelled out in molecular terms, it is beginning to appear that it may have a central role with respect to a number of microtubular activities. Several research groups are presently concentrating on studies that are certain to clarify the role of this ubiquitous eucaryotic protein in regulatory processes.

Microtubule Organizing Centers

Microtubules are subject to assembly and disassembly reactions in the living cell just as they are *in vitro*. But the conditions that control their dynamics *in vivo* are not well understood.

Although microtubules are distributed throughout most of the cell, they

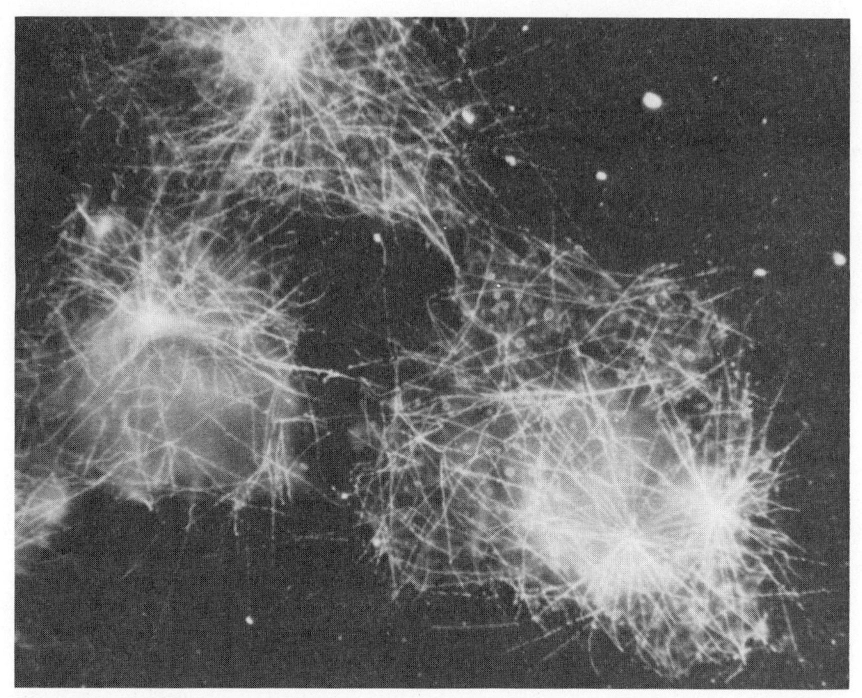

FIGURE 18.25 Microtubule pattern in cells treated to be permeable to high molecular weight materials and then incubated with pure tubulin and fluorescent antibodies to tubulin. Distinct initiation sites and radiating microtubules are apparent. (Courtesy of Dr. B. R. Brinkley.)

Centrioles and basal bodies are tiny cylindrical structures about 200 nm in diameter and 300 to 500 nm long. They resemble a section of a eucaryotic flagellum except that they possess nine triplet microtubules on the outer rim and no central microtubules. Centrioles are associated with the mitotic apparatus and basal bodies lie at the base of eucaryotic flagella and cilia.

Kinetochores are the sites at which the microtubules of the mitotic apparatus are attached to the chromosome. Little is known of their molecular anatomy.

appear to be organized and to grow out from certain centers within the cell. These are referred to as microtubule organizing centers (MTOCs). MTOCs generally have at their focus centrioles, basal bodies, or the kinetochores of metaphase chromosomes. Apparently these are nucleating centers that serve as templates for the polymerization of tubulin.

The molecular anatomy of MTOCs is a mystery, and the manner in which these centers organize and generate microtubules is equally obscure. But the fact remains that microtubules converge on the centrosome region of the cell, and when centrioles replicate and move toward opposite poles within the cell, there is a corresponding duplication and shifting of centers from which microtubules emanate.

One experimental approach that has been used to study MTOCs as nucleation sites is to treat cells with Colcemid to disrupt endogenous microtubules and then gently lyse the cells with Triton X-100.[16] The resultant cells retain their MTOCs but are extremely permeable to high molecular weight materials.

When cells thus treated are administered fluorescent antibodies to tubulin only short microtubules are seen along with spots representing MTOCs. However, when these cells are incubated under conditions that promote tubulin polymerization and then reacted with antibodies to tubulin, long microtubules are evident that extend in radial fashion from MTOCs.

Figure 18.25 contains a micrograph showing that microtubule growth occurs from distinct initiation sites within the cell. Polymerization appears

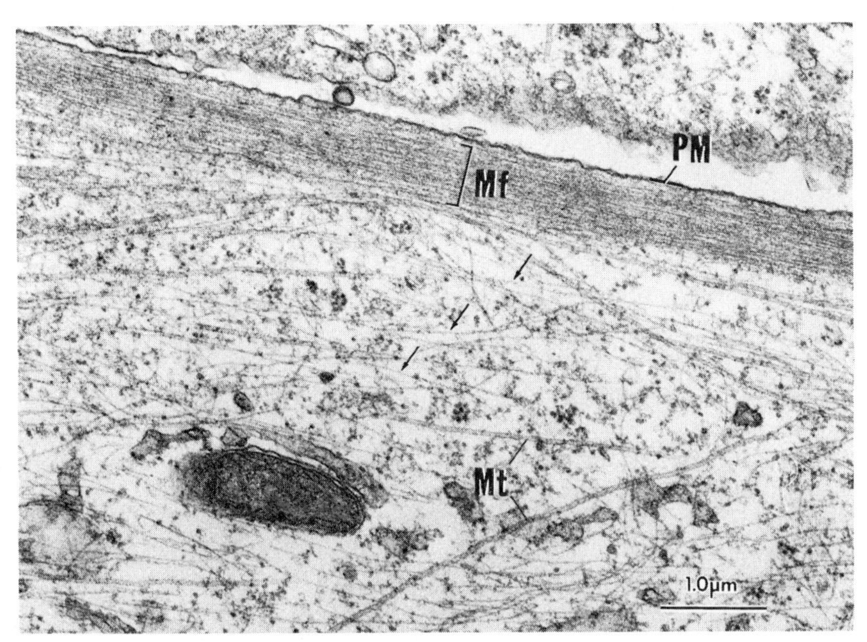

FIGURE 18.26 Cultured cartilage cell from a chick embryo showing a bundle of microfilaments (Mf) just beneath the plasma membrane (PM). Deeper in the cell microtubules (Mt) are readily apparent, and intermediate 10-nm filaments (arrows) weave through the cytoplasm. (Courtesy of Dr. Harunori Ishikawa.)

to take place on the end of the microtubule that extends away from the MTOC. This observation suggests that the MTOC governs a polarity with respect to microtubule structure. Growth does not appear to be indeterminate. Microtubules seem to terminate near the cell margins. It has been suggested that the ends of microtubules are capped as they near the cell margin, but this is only speculation. If there is a capping mechanism, hence a means of regulating microtubule length, it is likely that this reaction is controlled by something other than the MTOC. The composition and properties of the cytoplasm near the plasma membrane (where microfilaments and intermediate filaments may be found) may be important factors in controlling microtubule growth.

18.4 MICROFILAMENTS

Size and Composition

High resolution electron microscopy has uncovered a maze of filamentous structures within nonmuscle eucaryotic cells. One type of filament is the microfilament, a structure about 7 nm in diameter and of indeterminate length.

An electron micrograph containing microfilaments in a cell from a chick embryo is shown in Figure 18.26. This cartilage cell also contains microtubules and 10-nm intermediate filaments, permitting a convenient comparison between the three filamentous structures: microfilaments, microtubules, and intermediate filaments.

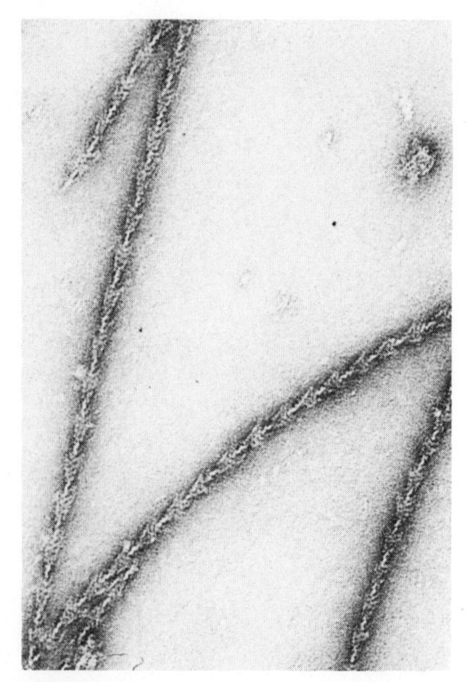

FIGURE 18.27 Electron micrograph of F-actin decorated with myosin S1 heads. The arrowheads all point in the same direction in a given strand, indicating a polarity of the actin filament. (Courtesy of Dr. James Spudich.)

Microfilaments possess about the same diameter as the actin filaments of striated muscle. In fact, actin is the main structural protein of microfilaments. The concentration of actin in nonmuscle cells is surprisingly high. It may typically account for 10% of total cell protein and in many cases is the most abundant cell protein. The actin of nonmuscle is very similar to that of muscle actin. It can be extracted and in *in vitro* settings will undergo polymerization reactions from the G-actin monomer state to F-actin. Actin from the slime mold *Physarum* has been sequenced, and the results show a striking similarity to the primary structure of rabbit skeletal actin.[17]

Actin is most commonly purified by taking advantage of its high affinity for DNase I. The principle of affinity chromatography is therefore used. DNase I is coupled to agarose beads and cell extracts are passed through columns containing this DNase I matrix. Actin is selectively adsorbed while the other ingredients of the extract pass through the column. Actin can then be eluted by application of an appropriate buffer.

Although most actins from various sources bear strong similarities, there are subtle yet significant differences. Using powerful techniques to separate actins, three types have been discerned, termed α, β, and γ.

The α form of actin is found in fully mature muscle tissue. The other two forms are more characteristic of nonmuscle cells; however, all three forms may be present in certain nonmuscle tissues.

The significance of different forms of actin in a given tissue or cell is not known. But one possibility that is being considered is that the different forms are involved in different types of cell movement.

Reaction with Myosin

In Section 18.2 we discussed the interaction between actin and myosin in muscle cells. Microfilaments, being of actin composition, also bind myosin.

In vitro and *in situ* microfilaments can be coated or "decorated" with heavy meromyosin (HMM) or S1 heads. This binding results in an arrowhead pattern to the microfilament in which the arrowheads all point in the same direction (See Figure 18.27). This pattern indicates that microfilaments possess a polarity, a property that is probably crucial to their role in mediating cell movements.

Heavy meromyosin (HMM) or S1 head decorating is a commonly employed technique to label actin-containing filaments in cells. Thus actin filaments will possess "arrows," whereas other filaments will remain smooth, or undecorated.

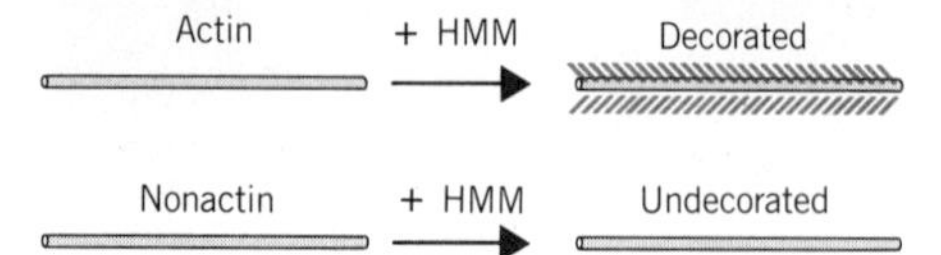

Cellular Location of Microfilaments

The HMM binding method has become a very useful procedure for identifying and localizing microfilaments in any type of cell. The electron micrograph in Figure 18.28 shows a bundle of microfilaments just beneath the plasma membrane in a flattened cultured cartilage cell. The specificity of the reaction between HMM and microfibrils is also apparent in these results. Intermediate filaments are not decorated.

Microfilaments are generally distributed in the cortical regions of the cell just beneath the plasma membrane. In contrast, intermediate filaments and microtubules are found in subcortical and deeper regions of the cell.

Microfilaments also extend into cell processes, especially where there is movement. Thus they are found in the microvilli of the brush border

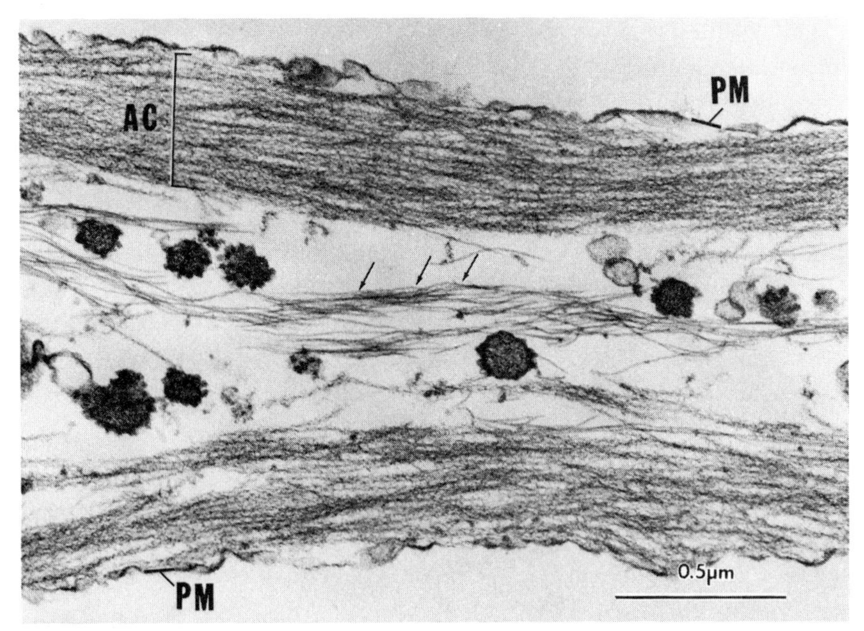

FIGURE 18.28 Part of a cultured cartilage cell treated with heavy meromyosin (HMM). Microfilaments, which lie just beneath the plasma membrane (PM), are decorated with HMM to form the arrowhead complexes (AC). Intermediate filaments remain smooth and undecorated (arrows). Compare this view with Figure 18.26, where microfilaments have not been complexed with HMM. (Courtesy of Dr. Harunori Ishikawa.)

of intestinal epithelium and in cell types or regions where amoeboid movement and cytoplasmic streaming are prominent.

Microfilaments can also be localized in cells by another technique, since they bind fluorescent-labeled antibodies to actin. Immunofluorescence micrographs (e.g., Figure 18.29) show clearly that microfilaments are oriented in bundles that appear to run the length of the cell. In cell regions where there is high motile activity, the microfilaments appear to be shorter and more randomly oriented.

Nonmuscle Myosin and Actin-Binding Proteins

In addition to actin, nonmuscle eucaryotic cells contain myosin. However the ratio of myosin to actin is much lower than that found in muscle. The two types of myosin are strikingly similar in a number of properties, including the physical parameter of shape, their content of light and heavy chains, and ATPase activity. Both types of myosin interact with actin.

A major difference between the myosins of muscle and nonmuscle cells is related to their solubilities and abilities to form thick filaments. Nonmuscle myosin does not form thick filaments easily.

The fact that nonmuscle cells contain both actin and myosin opens up the possibility that a type of muscle contraction mechanism may be used by these cells for various movements. However, since nonmuscle myosin is present in low concentrations and does not readily polymerize, actin–myosin interactions may have to be modified from that observed in muscle.

FIGURE 18.29 Actin filaments in a human skin fibroblast revealed by immunofluorescence with antibodies to actin. (Courtesy of Dr. E. Lazarides.)

The question that must be answered in this regard is whether *soluble* myosin can interact with actin to generate force. According to the results of a number of studies, the answer appears to be yes. This makes it possible to propose and test a number of schemes to explain cell movements, which are dealt with further in Chapter 19.

In 1975 a high molecular weight protein that cross-links actin to form gels was purified from rabbit lung macrophages.[18] This protein was named actin-binding protein (ABP). ABPs with similar amino acid compositions and molecular weights have since been isolated from widely divergent sources.

Method VII

Immunofluorescence microscopy has shown that ABPs are located in the cell periphery near the plasma membrane, in the vicinity of actin microfilaments. It is speculated that actin-binding proteins function in peripheral cytoplasm along with microfilaments to bring about movement.

Still other proteins are found to be associated with actin in nonmuscle cells. Among these are tropomyosin, troponin, and α-actinin. These proteins appear to be components of the F-actin form or are necessary to mediate attachments between actin and the cell membrane.

18.5 INTERMEDIATE FILAMENTS

Of the three types of filament associated with cellular movement, microtubules, microfilaments, and intermediate filaments, the latter are the most poorly understood.

Intermediate filaments lie between the other two organelles in diameter, being on the average about 10 nm wide (compared to 7 nm for microfilaments and 24 nm for microtubules). They are present singly or in bundles depending on the cell type.

Various names have been attached to these filaments that have a basis in the cell type in which they are observed. Thus, intermediate filaments in epidermal cells are called tonofilaments, in nerve cells they are referred to as neurofilaments, and in neuroglial cells they are designated glial filaments. In spite of the diverse designations, these filament types are probably all both structurally and functionally related. An example of the cellular location of one type of intermediate filament, vimentin, common to chick embryo fibroblasts, is shown in Figure 18.30.

In cross section, intermediate filaments have a tubular appearance. Each tubule appears to be made up of 4 or 5 protofilaments arranged in parallel fashion. Thus, from this perspective, intermediate filaments are scaled-down versions of microtubules, which have 13 protofilaments.

In contrast to the structures and compositions of microtubules and microfilaments, which are highly conserved among tissues and species, intermediate filaments can be immunologically and biochemically grouped into five different systems.[19,20] These can be described briefly as follows.

1. Desmin filaments. These are intermediate filaments of smooth, skeletal, and cardiac muscle cells and certain nonmuscle cells. They contain a major protein of 50,000 to 55,000 daltons called desmin or skeletin. Antibodies against desmin will not bind to intermediate filaments of cell types other than those listed above.
2. Keratin (also known as tono) filaments. These are intermediate filaments of epithelial cells containing keratin or prekeratin. Keratin is

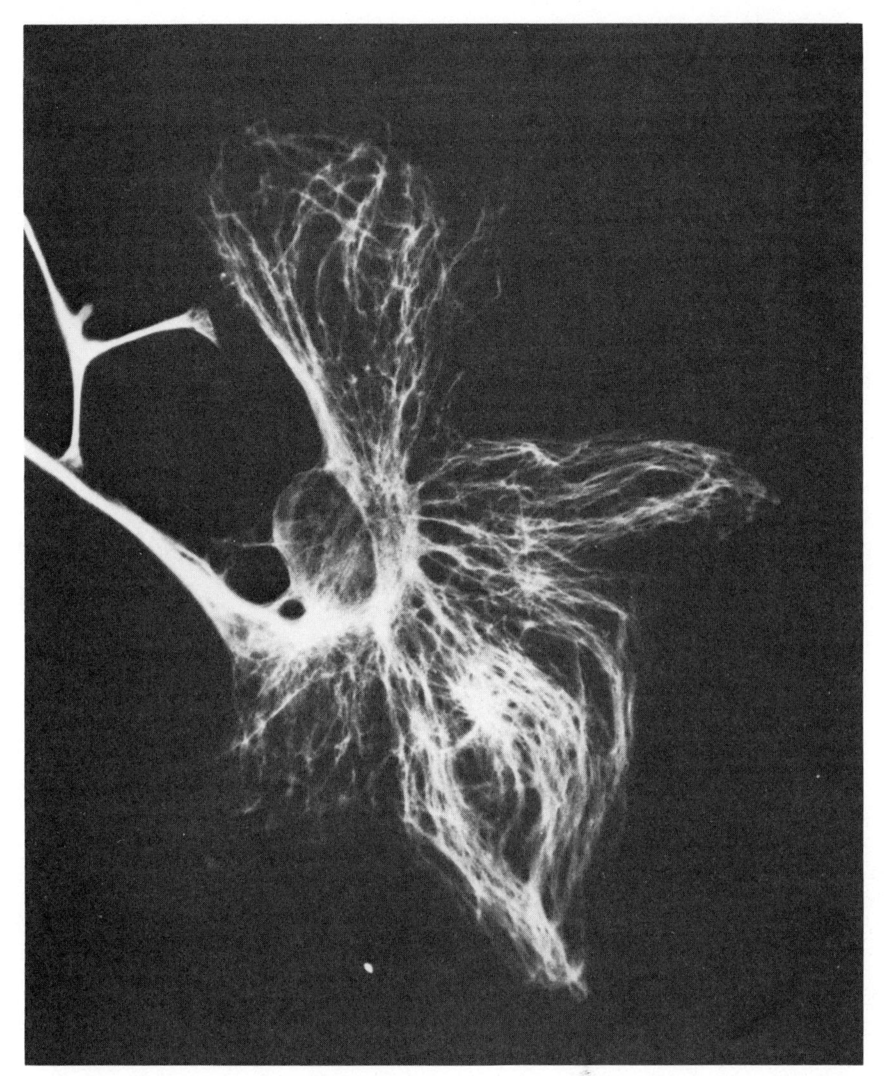

FIGURE 18.30 Intermediate filaments in a chick embryo fibroblast revealed by immunofluorescence with antibodies to vimentin. (Courtesy of Dr. E. Lazarides.)

made up of several polypeptides of molecular weight 40,000 to 65,000 daltons. Antibodies against prekeratin will not bind to intermediate filaments of brain, muscle, or various mesenchymal cells.

3. Vimentin filaments. These intermediate filaments are found in certain mesenchymal cells, especially fibroblasts, and in most differentiating cells. They contain a major polypeptide of 50,000 to 54,000 daltons called vimentin. Antibodies against vimentin will not bind to tonofilaments, brain filaments, or muscle filaments.
4. Glial filaments. Only glial cells and cells of glial origin contain these intermediate filaments.
5. Neurofilaments. These intermediate filaments are detected only in neurons.

Table 18.1 Comparison of Some Properties of Microtubules, Intermediate Filaments, and Microfilaments

Property	Microtubules	Intermediate Filaments	Microfilaments
Structure	Hollow with walls made up of 13 protofilaments	Hollow with walls made up of 4 to 5 protofilaments	Solid made up of polymerized actin (F-actin)
Diameter (nm)	24	10	7-9
Monomer units	α and β tubulin	Five types of protein defining five major classes	G-actin
ATPase activity	Present in dynein arms	None	None
Functions	Motility of eucaryotes Chromosome movements Movements of intracellular materials Contribute toward maintaining cell shape	Integrate contractile units in muscle Cytoskeletal structural function in cytoplasm	Muscle contraction Cell shape changes Protoplasmic streaming Cytokinesis

The immunological differences just stated among these five classes have been confirmed by a number of other criteria such as peptide mapping, isoelectric points, and molecular weight studies, which have shown that the major protein subunit in the various classes is distinct.

The intermediate filament desmin has been studied extensively by Elias Lazarides and colleagues who have named it from a Greek noun that means link or bond.[21] They picked this type of intermediate filament because other components of the cell in which desmins are found, namely, the contractile filaments, are quite well characterized. Therefore the possible functions of this filament may be easier to discern than those in other less well characterized cells.

When using smooth muscle as a source of material these investigators have been able to show that desmin is present in two forms, both of which have a high affinity for actin.[22] They have concluded from their studies that desmin is an insoluble, calcium-dependent polymer, some fraction of which interacts structurally with actin. They further have proposed that the function of desmin in smooth muscle is to mechanically integrate its contractile units over large distances in the cell.

Desmin in striated muscle is very similar to that of smooth muscle, but in striated muscle it may have some unique functions. Desmin is a component of the Z line in striated muscle. In this location it is thought to integrate the actin filaments from the sarcomere that terminate in the Z line. The result would be to bring the Z lines into lateral register, giving striated muscle its characteristic appearance.

Desmin is a hydrophobic protein that in muscle is also associated with the T-tubule system. In this location desmin is proposed to function in directing the invagination of the plasma membrane that eventually forms the T tubules.

Finally, desmin is proposed to link actin to the plasma membrane. Hence the name, which means link or bond.

The existence of intermediate filaments in different forms is, at the moment, quite a puzzle. We would expect their roles to be similar in cells of different kinds, and therefore the filaments themselves should be compositionally similar. Apparently their roles in different cells are related but are tailored to the specific requirements of the cell. The result is a system of filaments that are compositionally heterogeneous but morphologically similar.

Table 18.1 summarizes the major properties of microtubules, microfilaments, and intermediate filaments. All three systems consist of polymerized subunits, each system containing a predominant type of subunit. But beyond this, the three types of cellular filament display significant variations. Functionally they have both common and distinctive features. In general, they are concerned with cell shape and movement. These properties are the topic of the following chapter.

Summary

Cells demonstrate two types of movement: a movement of the cell itself and a movement within the cell.

Certain bacteria are motile by means of flagella that project from their anchoring sites in the plasma membrane through the cell wall and into the environment. The flagellum is made up of three untrastructural components. The filament, the part most distal to the cell, consists of aggregated flagellin proteins that form a hollow cylinder. The hook is also made up of a single type of protein, aggregated helically to form a curved cylinder. The basal structure consists of a rod with rings that function as bushings and as centers for the generation of rotary force. The entire flagellum moves by rotation, which appears to be energized by a proton-motive force.

Eucaryotic cell movement of either the entire cell or its internal components is based on the activities of contractile proteins or organelles that have the ability to move relative to one another.

One type of contractile system is the actin–myosin proteins of muscle. Both actin and myosin are polymeric proteins that, in the presence of ATP, make contact and undergo conformational changes that result in movement.

Another type of system involved in intracellular dynamics is the microtubule. These are hollow cylindrical structures made up of α, β-tubulin polymers. They may exist singly within cells, where they form elements of the cytoskeleton and the mitotic apparatus, or they may be grouped in pairs, as is the case in the axonemes of cilia and flagella. When paired, microtubules have attached dynein arms which possess ATPase activity and function as the basis for the movement of microtubules.

Microtubules can be assembled and disassembled *in vitro*. *In vivo* assembly appears to be governed by activities of the microtubule organizing centers of the cell.

Microfilaments are fibers of about 7 nm diameter found in nonmuscle cells; they are composed of polymerized actins. Microfilaments react with myosin, also present in nonmuscle cells, and as such appear to have roles in internal cell movements.

Intermediate filaments approximately 10 nm wide are also found widely distributed among eucaryotic cells. Although they may have roles in organizing other contractile units in the cell, their functions for the most part are unknown.

References

1. *The Cell in Development and Inheritance,* E. B. Wilson, Macmillan, New York, 1900.
2. The Effect of Colchicine on the Microscopic and Submicroscopic Structure of the Mitotic Spindle, S. Inoué, *Exp. Cell Res.* 2(Suppl.):305(1952).
3. Polarization Optical Studies of the Mitotic Spindle. I. The Demonstration of Spindle Fibers in Living Cells, S. Inoué, *Chromosoma (Berlin)*5:487(1953).
4. Studies on the Protein Components of Cilia from *Tetrahymena pyriformis,* I. R. Gibbons, *Proc. Natl. Acad. Sci. U.S.* 50:1002(1963).
5. Purification and Thermal Stability of Intact *Bacillus subtilis* Flagella, K. Dimmitt and M. Simon, *J. Bacteriol.* 105:369(1971).
6. Purification of Intact Flagella from *Escherichia coli* and *Bacillus subtilis,* M. L. DePamphilis and J. Adler, *J. Bacteriol.* 105:376(1971).
7. Fine Structure and Isolation of the Hook–Basal Body Complex of Flagella from *Escherichia coli* and *Bacillus subtilis,* M. L. DePamphilis and J. Adler, *J. Bacteriol.* 105:384(1971).
8. Flagellar Structure and Function in Eubacteria, R. N. Doetsch and R. D. Sjoblad, *Annu. Rev. Microbiol.* 34:69(1980).
9. Bacteria Swim by Rotating Their Flagellar Filaments, H. C. Berg and R. A. Anderson, *Nature* 245:380(1973).
10. Flagellar Rotation and the Mechanism of Bacterial Motility, M. Silverman and M. Simon, *Nature* 249:73(1974).
11. Regulation and Kinetics of the Actin–Myosin–ATP Interaction, R. S. Adelstein and E. Eisenberg, *Annu. Rev. Biochem.* 49:921(1980).
12. Microtubule Formation *in Vitro* in Solutions Containing Low Calcium Concentrations, R. C. Weisenberg, *Science* 117:1104(1972).
13. *In Vitro* Assembly by Microtubules, R. B. Scheele and G. G. Borisy, in *Microtubules,* K. Roberts and J. S. Hyams, eds., Academic Press, New York, 1979.
14. Calmodulin, C. B. Klee, T. H. Crouch, and P. G. Richman, *Annu. Rev. Biochem.* 49:489(1980).
15. Calmodulin Confers Calcium Sensitivity on Ciliary Dynein ATPase, J. J. Blum, A. Hays, G. A. Jamieson, and T. C. Vanaman, *J. Cell Biol.* 87:386(1980).
16. Tubulin Assembly Sites and the Organization of Cytoplasmic Microtubules in Cultured Mammalian Cells, B. R. Brinkley, S. M. Cox, D. A. Pepper, L. Wible, S. L. Brenner, and R. L. Pardue, *J. Cell Biol.* 90:554(1981).
17. The Amino Acid Sequence of *Physarum* Actin, J. Vandekerckhove and K. Weber, *Nature* 276:720(1978).
18. Actin-Binding Protein, T. P. Stossel, J. H. Hartwig, H. L. Yin, and W. A. Davies, in *Cell Motility: Molecules and Organization,* S. Hatano, H. Ishikawa, and H. Sato, eds., University Park Press, Baltimore, 1979.
19. Microfilaments, Microtubules, and Intermediate Filaments in Immunofluorescence Microscopy, K. Weber and M. Osborn, in *Cell Motility: Molecules and Organization,* S. Hatano, H. Ishikawa, and H. Sato, eds., University Park Press, Baltimore, 1979.
20. Intermediate Filaments: A Chemically Heterogeneous, Developmentally Regulated Class of Proteins, E. Lazarides, *Annu. Rev. Biochem.* 51:219(1982).
21. Immunological Characterization of the Subunit of the 100 Å Filaments from Muscle Cells, E. Lazarides and B. D. Hubbard, *Proc. Natl. Acad. Sci. U.S.* 73:4344(1976).

22. Studies of the Structure, Interaction with Actin, and Function of Desmin and Intermediate Filaments in Chicken Muscle Cells, E. Lazarides, B. D. Hubbard, and B. L. Granger, in *Cell Motility: Molecules and Organization,* S. Hatano, H. Ishikawa, and H. Sato, eds., University Park Press, Baltimore, 1979.

Selected Books and Articles

Books

Cell and Muscle Motility, vols. 1 and 2, R. M. Dowben and J. W. Shay, eds., Plenum Press, New York, 1981 and 1982.

Cell Motility: Molecules and Organization, S. Hatano, H. Ishikawa, and H. Sato, eds., University Park Press, Baltimore, 1979.

Microtubules, K. Roberts and J. S. Hyams, eds., Academic Press, New York, 1979.

Motility of Living Cells, P. Cappuccinelli, Chapman and Hall, New York, 1980.

Articles

Calmodulin, C. B. Klee, T. H. Crouch, and P. G. Richman, *Annu. Rev. Biochem.* 49:489(1980).

Flagellar Structure and Function in Eubacteria, R. N. Doetsch and R. D. Sjoblad, *Annu. Rev. Microbiol.* 34:69(1980).

How Bacteria Swim, H. C. Berg, *Sci. Am.* 233:36(1975).

In Vitro Assembly of Cytoplasmic Microtubules, S. N. Timasheff and L. M. Grisham, *Annu. Rev. Biochem.* 49:565(1980).

Microtubules, P. Dustin, *Sci. Am.* 243:67(1980).

Regulation and Kinetics of the Actin–Myosin–ATP Interaction, R. S. Adelstein and E. Eisenberg, *Annu. Rev. Biochem.* 49:921(1980).

Some Biophysical Aspects of Ciliary and Flagellar Motility, M. E. J. Holwill, in *Advances in Microbial Physiology,* vol. 16, A. H. Rose and D. W. Tempest, eds., Academic Press, New York, 1977.

19

Form, Movement, and Replication

I must say, for my part, that no more pleasant sight has ever yet come before my eye than these many thousands of living creatures, seen all alive in a little drop of water, moving among one another, each several creature having its own proper motion.

ANTONY VAN LEEUWENHOEK, 1676

Form, movement, and replication are three cell properties that appear to have little, if anything at all, in common.

All cells possess a form, usually predictable and reproducible, that is characteristic of the cell. Several factors may influence form, such as a retaining wall of cellulose or chitin or a packing arrangement between cells. But basically, when animal cells are given the freedom of a unicellular, uncrowded existence, form is maintained by components of the cell interior rather than by forces outside the plasma membrane.

With few exceptions, all cells move. They exhibit either cellular motility (i.e., movement of the whole cell itself) or movement of the complex environment of the cell interior. Thus, in certain cases movement is blatant and undisguised, such as when a flagellated protozoan streaks across the microscopic field. But in other cases the internal movements are slow and nearly imperceptible, as when pigment granules are moved to their proper locations or when a feeding amoeba cautiously sends out a probing pseudopod.

Except for certain highly differentiated cells, all cells are subject to replication, a cell behavior that is essential for the maintenance of an organ or tissue and for the continuation of the particular cell lines in the biosphere. Replication begins internally, and its processes are complex, slow, and difficult to observe in action. It culminates externally in the generation of new individuals that are capable of repeating the process over and over again.

Form, movement, and replication—on the surface three seemingly unrelated events. But they are not. All three are based on the structures and functions of dynamic internal fiber systems that have become the object of serious scientific scrutiny only during the last couple of decades.

The shafts of cilia and flagella were observed to contain one type of fiber, the microtubule, some time before it was clear that similar fibers functioned in the cell interior as organelles of structure and movement. Microtubules are now understood to play essential roles in the mitotic apparatus, hence in replication, as well as in axonemes and as a component of the cytoskeleton—the form-giving apparatus of the cell. Thus they are concerned with all three cell properties that constitute the theme of this chapter.

But microtubules are not the sole ultrastructural ingredients upon which these properties depend. Other fiber systems permeate the cell interior. Microfilaments are ubiquitous fibers that not only contribute toward the structure of the cytoskeleton but also provide it with certain of its dynamic properties and assist the internal transformations of the cell that result in such phenomena as cytoplasmic streaming and amoeboid locomotion. They may even participate in the separation of chromosomes during mitosis. Thus microfilaments too are concerned with all three cell properties: form, movement, and replication.

The cell interior possesses still other fiber systems. Intermediate filaments are plentiful in many regions of many cell types. Although their roles are not as clearly defined as for some of the others, they at times surround the mitotic apparatus or are embedded in the cell cortex, there helping to give the cell form and to stabilize the cytoskeleton.

These dynamic fiber systems of the cell, along with membranous organelles and ribosomal particles, appear to be attached to one another and positioned by an interconnecting network of fine filaments called the microtrabeculae. This network meanders through what has been conventionally considered to be the structureless cytoplasm—that one last part of the cell thought to be free from organelles and particles. But it is not. The extraorganelle cytoplasmic matrix is highly structured with this dynamic system of filaments, most of whose functions are yet to be determined. But when the last word is in on structure and function, it would be surprising to hear that the microtrabeculae are not involved in some facet of form, movement, and replication.

Thus, these three properties are not unrelated. They are all fiber-based cell features and behaviors. Without this fiber system, all cells would assume the thermodynamically favorable sphere, none would move either internally or externally, and there would be no propagation of the cell line. Therefore, the impact of these organelles on the cell ranges all the way from appearance to survival. This being the case, future studies in this area are certain to be of significant import for all of cellular biology.

19.1 THE BEATING OF CILIA AND FLAGELLA

Cilia and flagella are essentially identical surface appendages found on a wide variety of cell types. We generally think of cilia as being shorter than flagella and more numerous on cell surfaces, but these organelles possess essentially identical ultrastructural and molecular organizations.

About the most definitive distinction that can be made between cilia and flagella is in the general pattern of their beat cycles.[1] Flagella and sperm tails demonstrate a continuous propagation of relatively planar bends, whereas cilia have a three-dimensional effective recovery stroke. Except for this functional difference, the two terms can be used interchangeably. It is unfortunate that no term has been agreed on to embrace both organelles. In this chapter, unless specifically indicated, the terms are used interchangeably, with an emphasis on "cilia," simply for convenience.

The Occurrence of Cilia and Flagella

Cilia and flagella are motility organelles of eucaryotic cells that create a relative movement between a cell and its environment. Many protozoans use cilia and flagella as organelles of locomotion. Since the organism is small (unicellular), the motion created between the cell and its environment is sufficient to propel the organism through a liquid medium. Thus, these appendages are useful in the search for food and in gaining access to better environments.

The 25,000 plus species of Protozoa are divided into five classes, two of which are Mastigophora and Ciliata. The Mastigophora, commonly called flagellates, possess one or more long, whiplike appendages of locomotion, whereas the Ciliata, called ciliates, are covered by short appendages that beat in a coordinated manner. Besides the length and number of locomotive organelles, the two classes differ in certain other cellular characteristics.

Among sessile multicellular organisms, such as sponges, the beating of cilia produces currents that flow over the surface of the organism, bringing by the sought-after nutrients to sustain life. The overall effect is thus similar to that achieved by motile protozoans, even though the sessile organism must remain in its immediate environment.

A variety of mammalian tissues and cells possess cilia, which have received considerable scientific scrutiny. For example, cilia are found on the surfaces of cells that line different passages, such as the nasal, tracheal, and reproductive tracts, and are present on cells along the digestive tract. The mammalian sperm also possesses a cilium that functions as a force-generating motile tail.

Except for sperm, the cilia in mammalian systems are not organelles of locomotion. But their effect is the same, that is, to move the environment with respect to the cell surface.

The Ciliary Membrane

The entire axoneme of the cilium is surrounded by a membrane that on morphological grounds appears to be a simple extension of the plasma membrane. Thus the surface of the cilium is continuous with the surface of the cell. Figure 19.1 contains an electron micrograph of a section cut transversely through a bed of cilia, each one of which is encased by a membrane.

Some studies have revealed information suggesting that the ciliary membrane contains features distinct from the plasma membrane, even though it is physically continuous with it.[2] X-Ray diffraction results indicate that the ciliary membrane has properties similar to those of nerve myelin, which as we pointed out in Chapter 4, is a rather atypical membrane. Furthermore, when ciliated cells are freeze-fractured and examined in the electron microscope, the faces of the ciliary membrane contain relatively few particles when compared with the other surface membranes of the cell. Thus it would appear that the ciliary membrane is predominantly in lipid bilayer form with fewer than usual contributions made to the membrane by intercalated proteins.

Method VIII

Method I

Some of the proteins of the ciliary membrane, however, appear to be specific to the cilium as judged by immunological techniques. The significance of these specific and somewhat unusual membrane characteristics when compared to the plasma membrane may be related to special re-

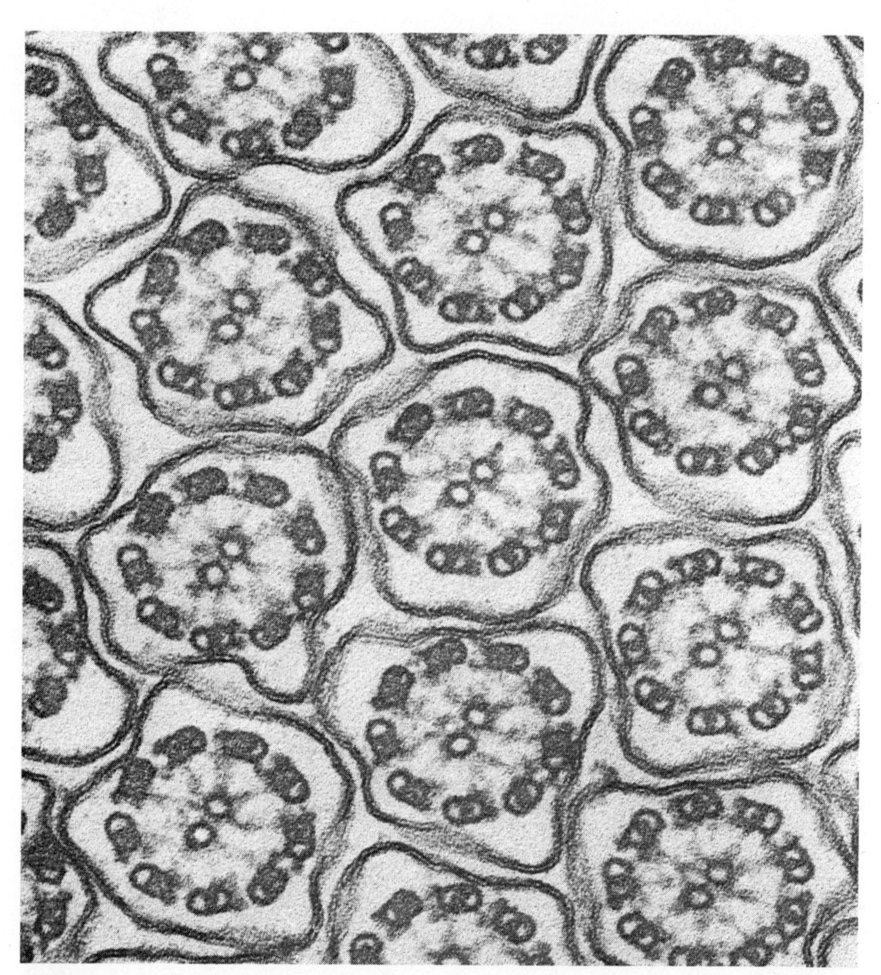

FIGURE 19.1 Cross section of mussel gill cilia. Each cilium is encased by a membrane that is an extension of the plasma membrane of the cell. The 9 + 2 arrangement of microtubules characteristic of the axoneme is nicely displayed in this section. (Courtesy of Dr. Peter Satir.)

quirements placed on these organelles. But it has been suggested that the ciliary membrane must be able to function as an especially effective barrier against the loss of ATP and certain essential ions that are required at appropriate concentrations to provide the energy for ciliary movement.

The role of this membrane in containing ATP has been demonstrated in striking fashion by treating ciliated cells with agents that dissolve the membrane, such as glycerin and Triton X-100. Membraneless cilia will continue to beat only as long as ATP is *added* to the medium. In addition, the concentration of added ATP affects the frequency of the beat.

In the intact organism ATP is generated by mitochondria, which frequently lie near the base of the cilium. ATP diffuses toward the tip of the organelle and, because of the membrane, is held at a concentration in the axoneme that provides an appropriate beat frequency.

An unusual feature of the membranes of all somatic cilia is the presence of multiple strands of particles at the base of the organelle. This feature,

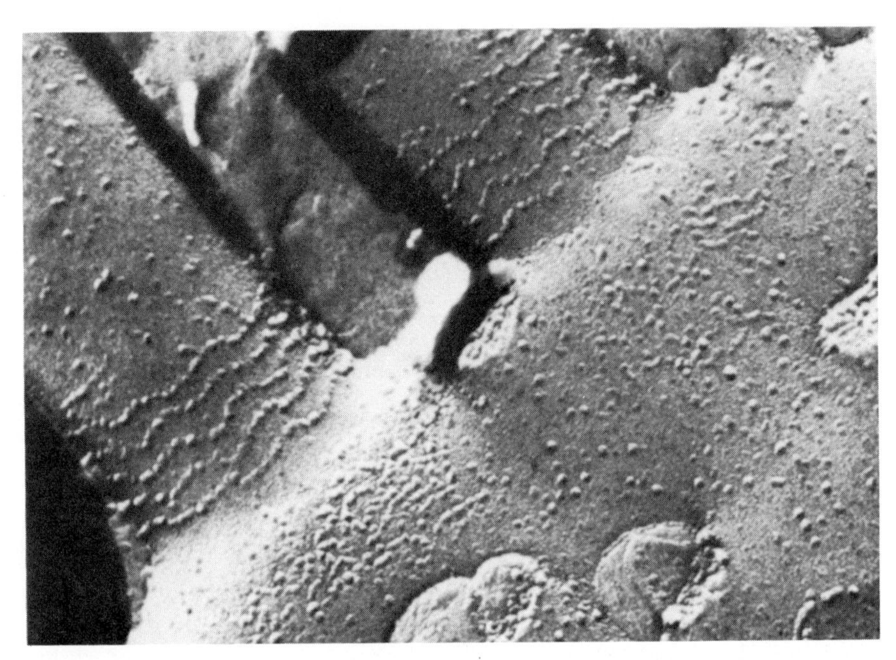

FIGURE 19.2 Multiple strands of particles, the ciliary necklace, surround the bases of these cilia in a ductus efferens of a rat. These particles appear on the P face of the membrane when freeze-fractured. (Original micrograph by Dr. Fumi Suzuki.)

termed a *ciliary necklace,* is seen in the electron micrograph of freeze-fractured material of Figure 19.2. The number of strands present reflects phylogenetic specificity, with up to 11 strands the maximum number observed.

The ciliary necklace is found at a region in the organelle where microtubules and centrioles make contact with the membrane. The precise role of the necklace in this complex is not known, but possibilities have been proposed. One function of the necklace may be to position the underlying centrioles from which the cilium is generated. A second role may be to aid in membrane differentiation of the cilium. The rings of particles may retain proteins that would otherwise diffuse out and be incorporated into ciliary membrane. Thus the membrane is encouraged to take on its rather atypical protein-poor properties.

The Ciliary Axoneme

The ultrastructure of the axoneme has been discussed in Chapter 18. This highly structured shaft, with its microtubules and interconnecting proteins held in a fixed array, is critical for the movement of the cilium.

At this point we emphasize that the axoneme contains all that is necessary for motility. Membraneless cilia beat as long as there is a sufficient supply of ATP. However, microtubules by themselves cannot generate the beat cycle, a fact that points to the significant roles played by the other components of the axoneme.

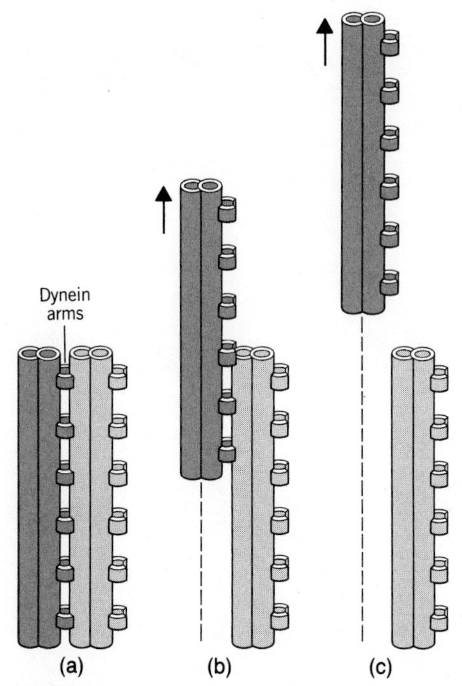

FIGURE 19.3 The sliding mechanism of microtubules. (*a*) A pair of microtubule doublets makes contact between the dynein arms of one and a microtubule of the other. (*b*) Movement of dynein generates the force for sliding. (*c*) Apart from the connecting components of the axoneme, sliding will occur until the two doublets are separated.

The Mechanism of Ciliary Motility

Ciliary and flagellar movement is a microtubule-based motility. The sliding filament model proposed by Peter Satir[3] is now widely accepted by the scientific community as the most reasonable explanation for the basis of the beat.

The ultrastructural basis of the ciliary beat is a sliding of microtubule doublets past one another by virtue of dynein arm movements. Figure 19.3 illustrates how doublets interact and slide relative to one another in response to the generation of force by dynein.

Three ingredients are essential for sliding to take place: microtubules, dynein, and ATP. Isolated axonemes from some species continue to undulate, as already discussed. However, if these axonemes are lightly trypsinized, the axoneme will not undulate. Instead, the microtubules will slide past one another until they disengage. This phenomenon points to the decisive role played by intermicrotubule connecting proteins in converting the sliding movement to a beat.

The ciliated organism *Tetrahymena pyriformis* is often used as a source of material for the study of the interaction between microtubules during sliding.[4] The properties of the axonemes in this organism are such that when isolated and exposed to ATP, they disintegrate by sliding without trypsinization.

Figure 19.4 shows the results of ATP disintegration of the axoneme in this system. The doublets have crawled by one another until they are just overlapped near their ends.

Inactive and Bridged Conformations of Dynein Arms

The axoneme as it is conventionally drawn in cross section shows the dynein arms in the inactive conformation. They are not making contact with the adjacent doublet.

In the inactive conformation the arms repeat at 24-nm intervals along subfiber A and are uniformly tilted toward the base of the cilium. All nine doublets of the axoneme show these characteristics of conformation.

The subunit composition of the dynein arms in this conformation consists of three morphological subunits lying in a common plane with a center-to-center spacing of 7.8 nm. The length of the arm is about 23.8 nm. The conformation of dynein arms when doublets are bridged is identical to that seen in the inactive state. The only distinction between the two is that in the bridged conformation the arms make contact with the adjacent subfiber B. The attachment to subfiber B appears to involve the terminal subunit of the dynein arm. If the angle of dynein inclination remains the same in both the inactive and bridged conformations, the diameter of the axoneme cylinder must and does decrease when the bridge form is assumed.

The Generation of Force

In the muscle actomyosin system force is generated when myosin undergoes a conformational change, moving with rachetlike action on the actin filament. Although the muscle concept of conformational change applies to the dynein system, the mechanism of movement appears to be quite different.

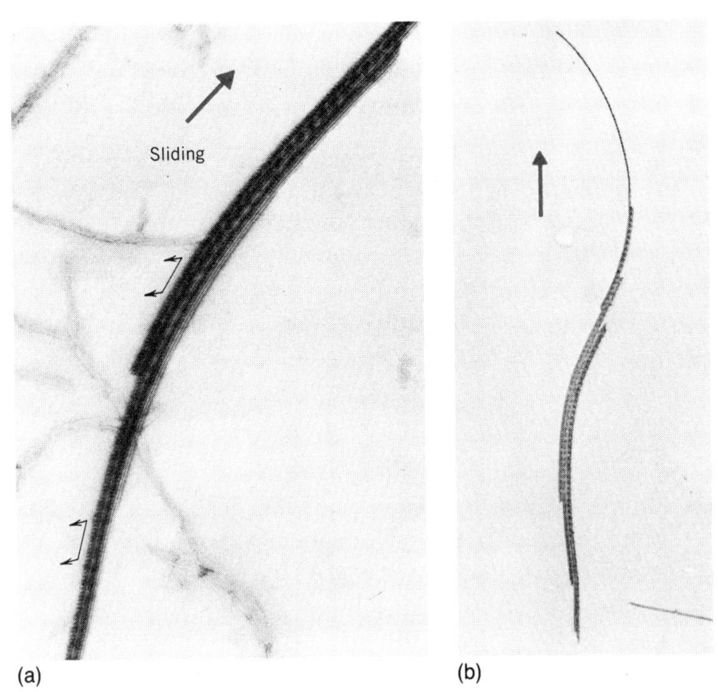

FIGURE 19.4 Isolated *Tetrahymena* cilia reactivated with ATP to cause sliding with eventual disintegration. The doublets, cross-bridged by dynein arms, become partially overlapped as they slide apart. Free dynein arms are polarized and tilt toward the base of the cilium (bracketed arrows) and away from the direction of active sliding. (*a*) ×25,000; (*b*) ×740. (Courtesy of Dr. Fred D. Warner.)

For one thing, the polarity of dynein arms is *unchanged* in both inactive and bridged conformations. This is somewhat difficult to reconcile with a change in conformation that is thought to be essential to drive the system. Also, ATP initiates bridging, which is not the case for muscle contraction (see Chapter 18).

The force that is generated during sliding always occurs in a direction toward the ciliary tip. This is opposite from the direction in which the dynein arms are inclined. The generation of force could occur simply if the dynein arms moved from their angle to a greater angle, thus sliding the adjacent subfiber B toward the ciliary tip relative to the position of the doublet containing the dynein arms.

Although this seems like a plausible explanation of sliding, the change in angle of dynein arms has not yet been observed. It may be that only the terminal subunit of the arm undergoes a conformational change accounting for the movement. This would be subtle and difficult to detect.

The Roles of Nexin Links and Radial Spokes

It is clear from the discussion above that mere microtubule sliding does not cause the cilium to bend. Rather, sliding by itself causes axoneme disintegration.

In the intact cilium or axoneme, the nexin links and radial spokes appear to play decisive roles in converting the movement of sliding to that of

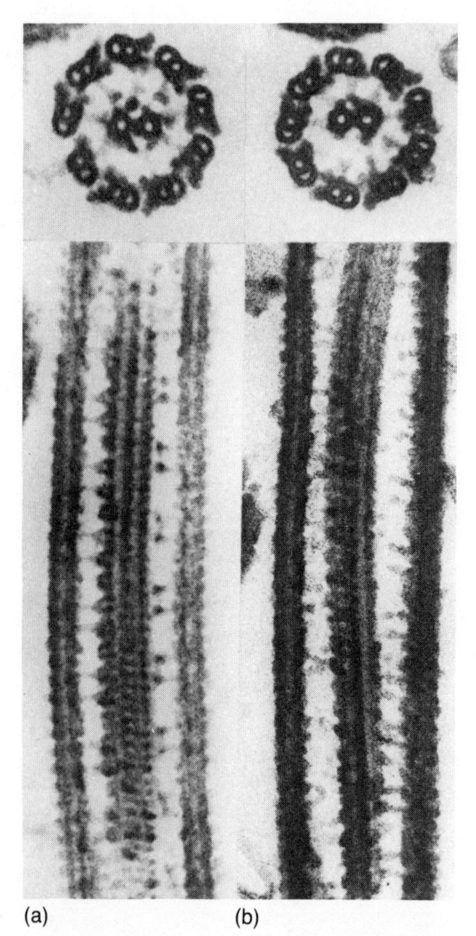

FIGURE 19.5 Electron micrographs of transverse and longitudinal sections of wild-type (*a*) and mutant (*b*) axonemes. In the wild-type organism radial spokes extend from subfiber A of the doublets toward the central pair (top) and project in pairs along the length of two outer doublets (bottom). Each spoke contains two segments, a proximal stalk and a distal spokehead. In the mutant axonemes the radial spokes project with the same periodicity but are found to extend at abnormal angles and display flexibility. The spokeheads are absent from these. ×97,500. (Courtesy of Dr. David J. L. Luck.)

bending. Figure 19.5 depicts the interactions between microtubules and radial spokes in both cross-sectional and longitudinal views of the axoneme in *Chlamydomonas* flagella. The radial spokes project in pairs between the outer nine doublets and the inner pair of microtubules.

In mutant organisms either possessing aberrant spokes or lacking spokes, motility is generally altered or the flagella are paralyzed.[5] Thus spokes clearly have a vital role in the mechanism of flagellar motility. Some of these ultrastructural features of axonemal microtubules are illustrated in Figure 19.6.

In a study of radial spoke positions vis-à-vis ciliary bending, Fred Warner and Peter Satir have shown that spokes appear to undergo detachment–reattachment cycles to convert interdoublet sliding to bending.[6]

In straight regions of a cilium, radial spokes occur in two basic conformations: parallel arrangements, with all spokes normal to subfiber A, and arrangements wherein spoke 3 is inclined. (The spokes occur in groups of three along subfiber A in the organism used for these studies—spoke 3 is closest to the tip in a cluster.) In bent regions, on the other hand, the spokes are inclined but their periodicity is not changed.

These observations are interpreted as follows. In straight regions of the cilium, spokes do not make contact with the central sheath. They remain perpendicular to the microtubule axis and are free to slide past the central sheath when microtubules slide.

In bent regions the radial spokes are attached to the central sheath. When microtubules slide relative to the central sheath, the radial spokes are forced to assume an angle from the perpendicular. The result is that the microtubule system is pulled into an arc as it moves away relative to the sheath. This generates a bend.

Perhaps the effect of this can be comprehended more clearly by imagining what happens when one tosses one hand into the air. If the member were severed from the arm, it would go straight up (microtubule sliding with no bending). But attached to the arm, the hand cuts an arc whose radius is determined by the length of arm (microtubule sliding linked to the sheath).

Figure 19.7 illustrates the relation between spokes and central doublets in bent and straight regions of cilia: the radial spoke is the transducing system in the cilium. It converts the movement of slide to that of bend.

Nexin appears to prevent excessive sliding. Evidence for this is somewhat weak, but trypsinized axonemes (in which nexin is destroyed) disintegrate in ATP by excessive sliding, whereas when nexin is intact this cannot occur, a circumstance that favors this antisliding role for nexin. It is thought that nexin remains permanently attached to microtubules and, by virtue of some elasticity, regulates the degree to which microtubules may slide past one another.

Patterns of Movement

It is in the beat pattern that the greatest distinction can be made between cilia and flagella. The difference in pattern cannot be readily related to the ultrastructure of the axoneme, although there may well be subtle differences that are ultimately reflected in patterns of movement. For the moment, the different beat patterns seem to be more pragmatic, depending

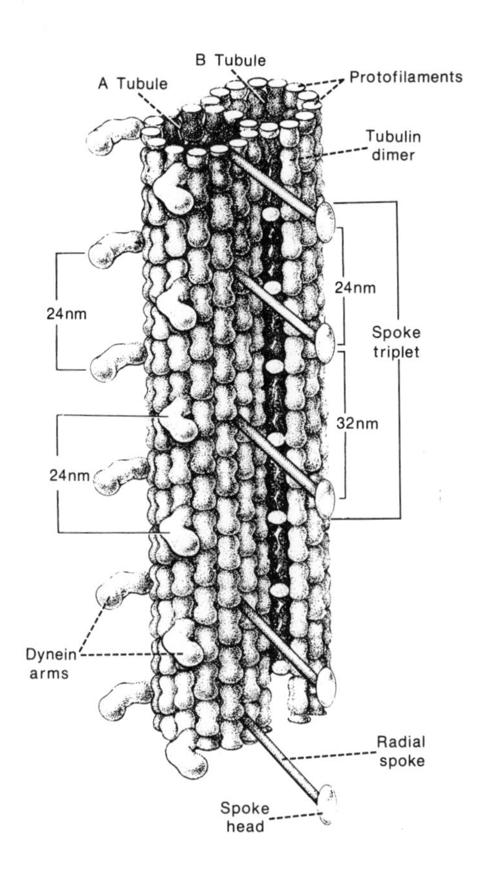

FIGURE 19.6 Model of an axonemal doublet and its appendages.

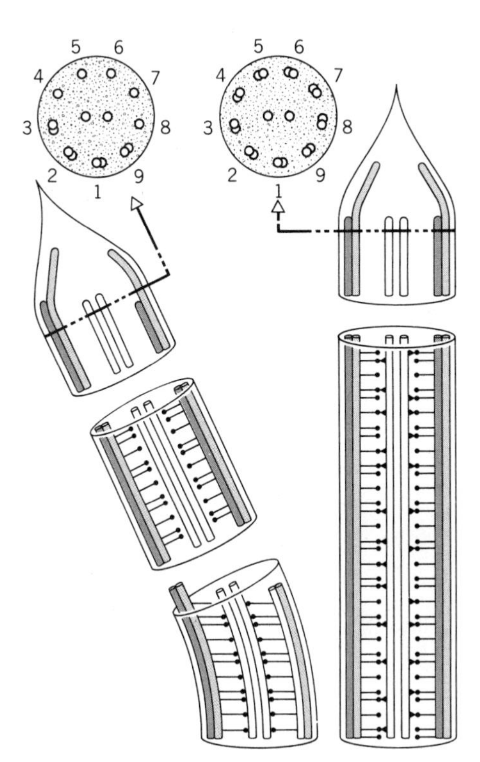

FIGURE 19.7 Drawing emphasizing the ralation between radial spokes and central doublets during ciliary movement. Note that when radial spokes make no contact with the central sheath, microtubules may slide without concurrent bending. In regions where contact is made between the two, the cilium will bend.

more on the length of the appendage and its proximity to other appendages than on axoneme structure. Longer and shorter organelles propagate different waves. Special compensations must be made in the beat pattern when the organelles are crowded closely together.

Most cells propagate the wave down the organelle from base to tip, just as the sliding force generated between microtubule doublets is from base to tip. In this case the cell normally moves in a direction opposite from that of wave propagation.

Some exceptions to this pattern have been observed. Among the trypanosomes, for example, the wave is propagated from the tip of the flagellum to its base. This direction can be reversed when the organism encounters an environment it prefers to avoid.

Another type of variation in pattern is seen among the flagellates of the genus *Ochromonas*. In these organisms cell movement follows the direction of wave propagation, from base to tip. The organism is thus pulled by the wave rather than propelled away from it. This action is due to rows of tiny rigid hairlike structures (20 nm thick and 1 μm long) called mastigonemes that line the flagellum in the plane of the wave. As a wave crest passes along the flagellum, the thrust of the mastigonemes is in the opposite direction.

These variations are summarized in Figure 19.8. Depending on the particular organism, the wave may be propagated in a single plane or in a helical manner. In the latter case the wave occupies all possible planes in the volume of space described by the helix. When the wave is prop-

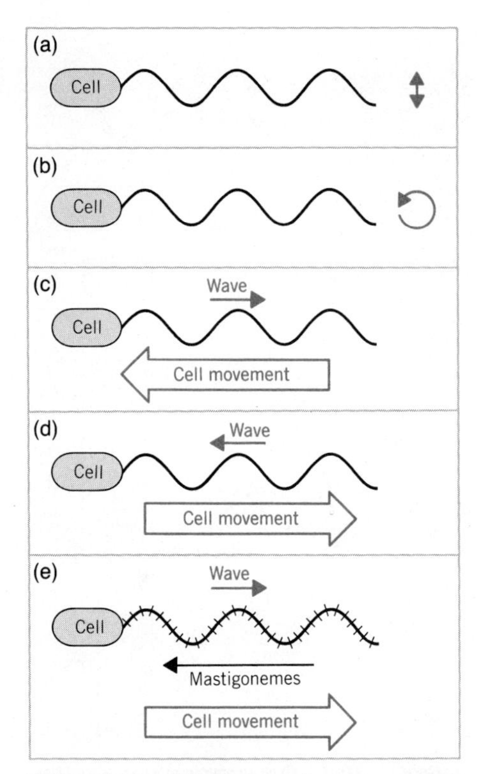

FIGURE 19.8 Patterns of flagellar and cell movement. Flagella may move in a plane (*a*) or within the volume of a cylinder cut by a helical pattern (*b*). The tips of the flagella move as shown in the insets. Cell movement is generally opposite to the direction of wave propagation (*c* and *d*) but may be the same if mastigonemes are present (*e*).

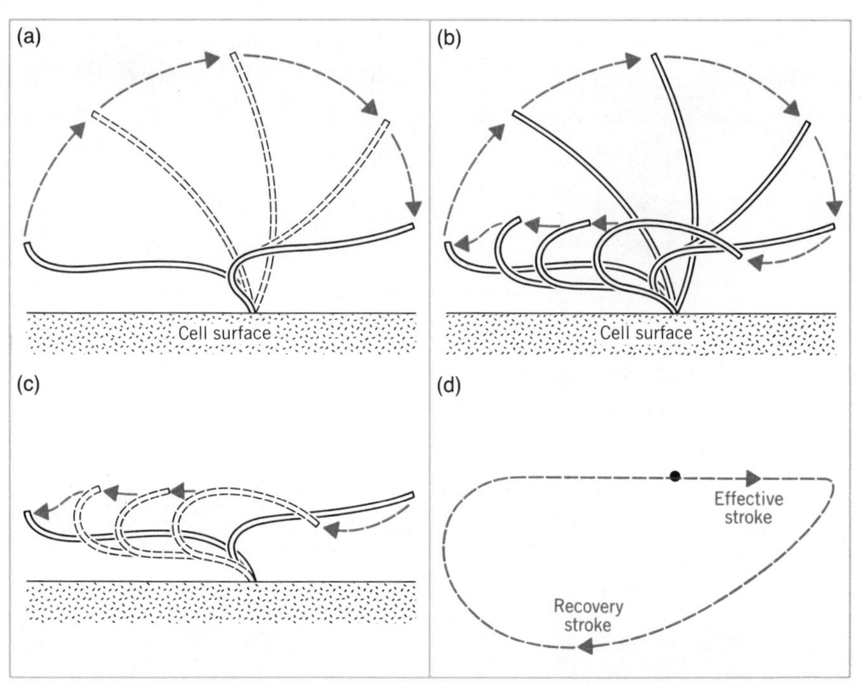

FIGURE 19.9 The basic features of the effective (*a*) and recovery strokes (*b*) of the ciliary beat. The path taken by a ciliary tip as it moves through the recovery stroke (*c*) arcs outward from its path during the effective stroke (*d*).

agated along a helical path, torque is generated that rotates the cell at some prescribed frequency, depending on the beat frequency and the hydrodynamic properties of the environment.

The beat pattern of cilia is generally described as consisting of two phases: an effective stroke and a recovery stroke (see Figure 19.9). The cilium is straight or nearly so during the effective stroke, but during recovery a bend is propagated from base to tip until the organelle is returned to its original position at the start of the effective stroke.

An interesting feature of the recovery stroke is that it does not return on the same plane as the effective stroke. The bend propagated is to the left or right of the effective stroke plane, depending on the species of organism, so that the cilium moves parallel to the surface of the cell.

Since ciliated organisms are usually covered by cilia that are spaced only about 1 μm apart, a good deal of interciliary coordination is necessary. Figure 19.10 depicts this type of coordination. The result is a metachronal wave that passes over the cell surface, with the cilia synchronized in a direction perpendicular to the direction of wave movement.

The relation between the direction of the effective stroke and the propagation of the metachronal wave varies among organisms. The type we have just described is called diaplectic metachrony.[7] Depending on the movements of an individual cilium, the two directions may be the same (symplectic) or opposite (antiplectic). In addition, two possible versions of diaplectic metachrony may be envisioned, depending on the relation between the effective stroke (right or left) and the direction of wave motion.

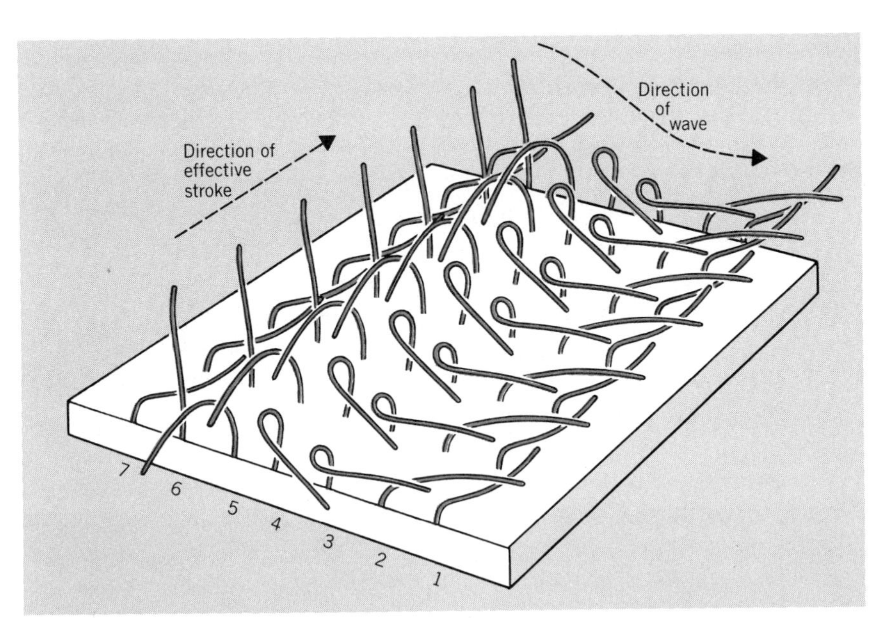

FIGURE 19.10 Illustration of the relation between the direction of effective stroke and the direction of the metachronal wave. The cilia in a given row are all in synchrony, but out of phase with the cilia in adjacent rows. Each of the cilia in a line perpendicular to the rows is participating in a different phase of the stroke cycle. The result is a wave that appears to move over the surface of the organism.

These are obviously complex phenomena that will require extensive study before they are well categorized and thoroughly understood. The beating actions of cilia and flagella thus pose a challenge for representatives of several disciplines, including investigators whose interests are mainly in molecular mechanisms as well as biophysicists who may explore the physical outcome of the molecular activities.

19.2 MICROVILLI

Microvilli are membrane-bound projections on the apical surfaces of mammalian cells of a number of types. They are commonly found on the surfaces of most epithelial cells, such as those associated with the oviduct, mammary glands, the striated border of the intestines, and the brush border of kidney tubules. They are thus found in anatomical locations similar to those of cilia.

Unlike cilia, microvilli form a rather polymorphic class of surface protuberances that are regularly packed in some tissues and loosely positioned in others. In general, they are shorter and smaller in diameter than cilia. They are commonly about 0.1 μm in diameter and range in length from a fraction of a micrometer to about 2 μm.

The Microvillar Membrane

The membranes covering microvillar projections appear to be simple extensions of the plasma membrane with possibly only minor modifications.

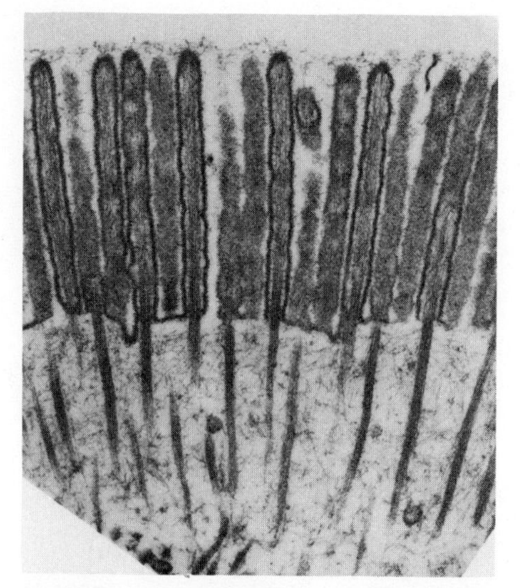

FIGURE 19.11 Isolated brush border showing each microvillus to contain a bundle of actin filaments that projects down into the terminal web region of the cell. (Courtesy of Dr. Mark S. Mooseker.)

In freeze-fracture preparations they contain particles normally interpreted to be intercalated proteins.

Cell surfaces containing microvilli are quite stable. In fact, brush borders can be isolated and studied independently from the rest of the cell. Studies on these isolated surfaces have revealed some of their enzyme activities, which may be related to the specific functions of these surfaces. Enzymes that appear to be localized in microvillar membranes, or at least preferentially located there (according to specific activities), are alkaline phosphatase, a Na^+,K^+-ATPase, and disaccharidases such as trehalase and maltase. The disaccharidases, in particular, are located uniquely in brush border preparations. The presence of these enzymes is probably related to the adsorptive and transport functions of microvilli.

The Microvillar Core

The core of the microvillus contains a bundle of 20 to 30 actin filaments, which can be seen easily in electron micrographs of isolated material (e.g., Figure 19.11). Each bundle of filaments is 50 to 60 nm in diameter, and each filament is about 7 nm wide.

At the tip of the microvillus the filaments are embedded in a densely staining region that is close to the plasma membrane. At the other end of the bundle the filaments penetrate the cortex of the cytoplasm into a zone of filaments called the terminal web. This region contains myosin filaments as well as those intermediate filaments called tonofilaments.

Within the microvillus, core actin filaments are cross-bridged to one another, often with a periodicity of 33 nm. In addition, the filaments are attached by bridges to the adjacent plasma membrane that run along the length of the microvillus. These features are apparent in Figure 19.12.

Cross-bridges are especially evident in electron micrographs when Mg^{2+} concentrations are kept high (15–40 m*M*). In the presence of Mg^{2+} and Triton X-100, the membrane can be solubilized and the bundle of filaments with associated bridges will remain intact.

Several different proteins have been identified as affiliated with the core bundle. Villin, a 95,000 dalton Ca^{2+}-dependent protein has the ability to sever actin filaments. At low Ca^{2+} concentrations it brings about actin filament bundling. Fimbrin is a 68,000 dalton protein that appears to function as a linker between actins. Four calmodulin molecules form a complex with two other proteins in the core. This complex may be the bridge seen between actin filaments and the microvillus membrane. Tropomyosin, but not troponin, is also present in the microvillus core.

A speculative model incorporating these ingredients into the microvillus and terminal web is presented in Figure 19.13. It should be viewed as a working model that is certain to take on new forms as it is subjected to further experimental analysis.

The polarity of actin filaments in the microvillus can be thought of as analogous to the actin polarity in skeletal muscle where the Z line is comparable to the villus tip.

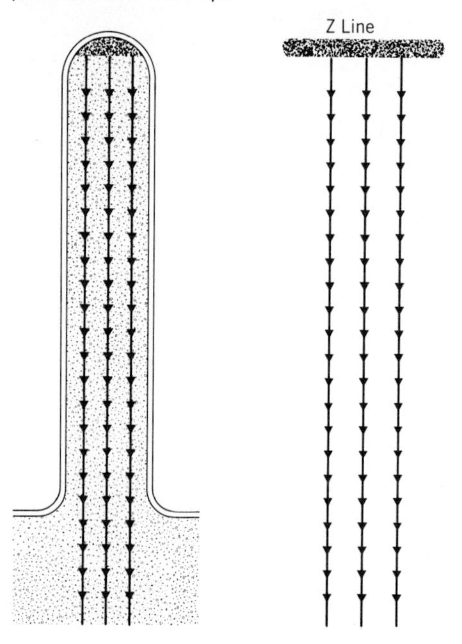

The actin filaments in the core all have the same polarity, as demonstrated by decoration with S1 of heavy meromyosin.[8] The arrowheads that result from this interaction all point away from the plasma membrane toward the terminal web. Thus the polarity of filaments is analogous to the muscle system in which the actin filaments, when decorated, have arrowheads that point away from the Z line.

From this it is apparent that the core of the microvillus has the prerequisite ingredients for a contractile system that functions along the lines of the muscle system. But the analogy breaks down when comparing the locations and roles of myosin in the two settings. In microvilli, myosin is located predominantly in the terminal web regions where it exists in monomers or dimers or, at most, in groupings of a few individuals. In muscle, many myosins aggregated to form a heavy filament with heads that function as a rachet against the actin filaments. A further puzzling feature of the microvillar myosin, distinguishing it from that of muscle, is its orientation with respect to actin. The axes of the two may be perpendicular rather than parallel, thus making it difficult to imagine how they may constitute a fully functional contractile system.

Therefore microvillus researchers are facing some very fascinating challenges. Nevertheless, the microvillus has been and continues to be an important model system that has served as a prototype for actomyosin-based intracellular movements. Certainly an important observation coming out of studies on microvilli is that actin filaments may be attached to membranes both at their ends and along their lengths. In parts of the cell other than in microvilli, actin filaments are often observed to be *parallel* to the plasma membrane rather than perpendicular. The microvillus studies have shown the potential for lateral bridges between membranes and the filament axis as a common mechanism of attachment between the two.

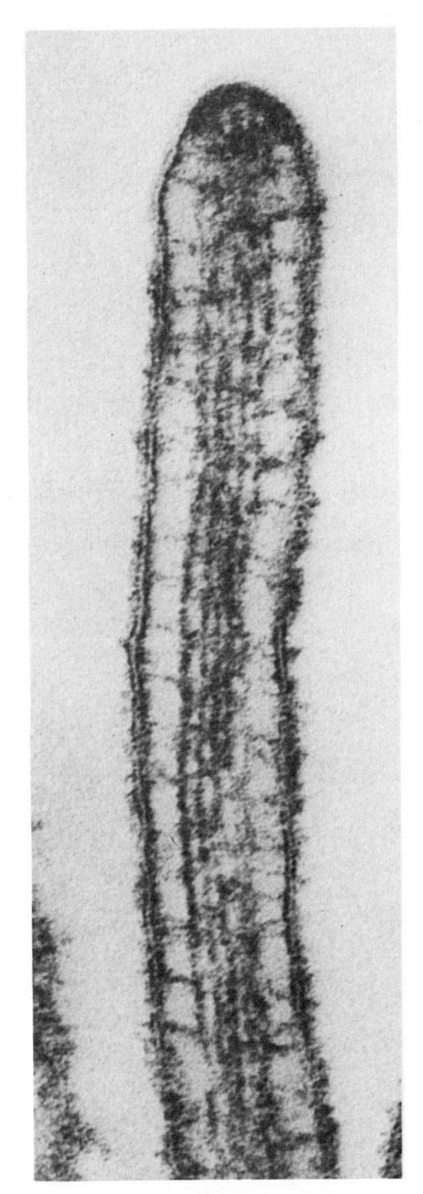

FIGURE 19.12 A highly magnified microvillus showing bridges that connect actin filaments with the plasma membrane and with each other. (Courtesy of Dr. Mark S. Mooseker.)

The Function of the Microvilli

At present, the structure of microvilli cannot be related very easily to their proposed function. But a broad range of functions have been proposed for these surface extensions that await experimental verification.

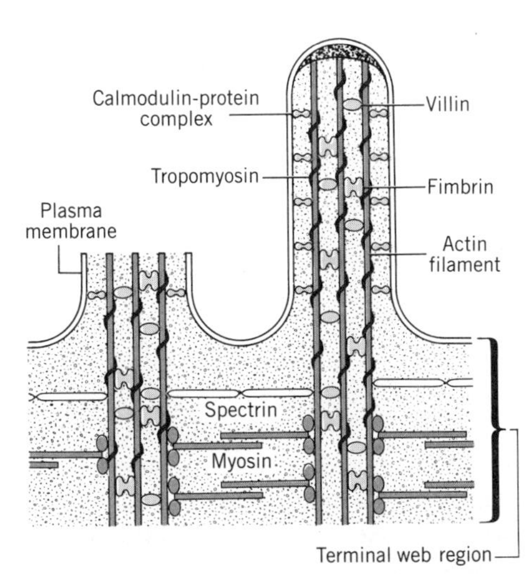

FIGURE 19.13 A speculative model illustrating the spatial relationships of the molecular members of the microvillus and terminal web region. This model, based on a drawing kindly provided by Dr. Mark S. Mooseker, incorporates the results of studies from several research groups.

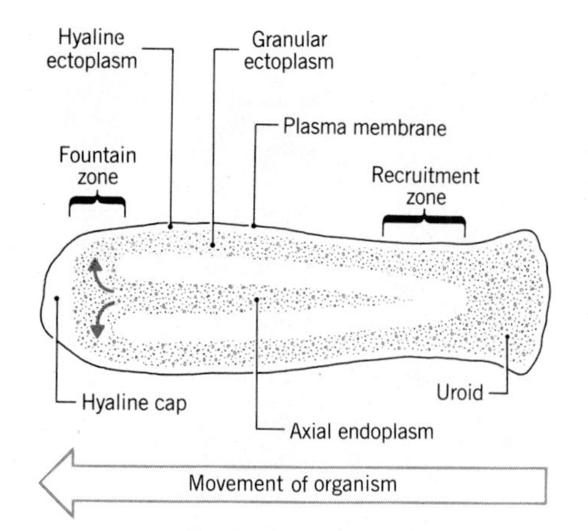

FIGURE 19.14 Terminology for cytoplasmic regions in an amoeboid cell. As illustrated, this organism would move toward the left, but inside ingredients of the endoplasm circulate.

One type of function appears to be the selective adsorption and transport of amino acids and fats. The membranes of intestinal brush borders contain the appropriate enzymes to effect a carrier–mediate type of transport so that these materials are effectively transported across the membrane in molecular fashion rather than by pinocytosis.

A second type of proposed function is to store membrane and microfilament materials. In some cell systems, such as fibroblasts, shape changes are associated with a depletion of microvilli, as though microvilli provide the raw materials for movement and changes in membrane surface areas.

Motility has also been proposed as a microvillar function. In this sense a coordinated effort by the microvilli on a surface may sweep material toward a resorptive area of the cell. The result would be very similar to that of the function of cilia.

19.3 CYTOPLASMIC STREAMING AND AMOEBOID MOVEMENTS

The observation that cells move by a mechanism described as *amoeboid* is an old observation dating back to the early use of microscopes. But the delineation of the molecular and ultrastructural basis for this form of motility is a current research area that has many loose ends.

Many theories have been advanced to explain amoeboid movement, but no one theory seems to have completely satisfied the scientific community of researchers working in this area. It is generally agreed that the basis of movement resides in the activities of the contractile proteins, namely actin and myosin. Thus the structure and dynamics of cytoplasm are important factors in any theory concerned with movement.

The terminology that designates different cytoplasmic regions in amoeboid cells is indicated in Figure 19.14.

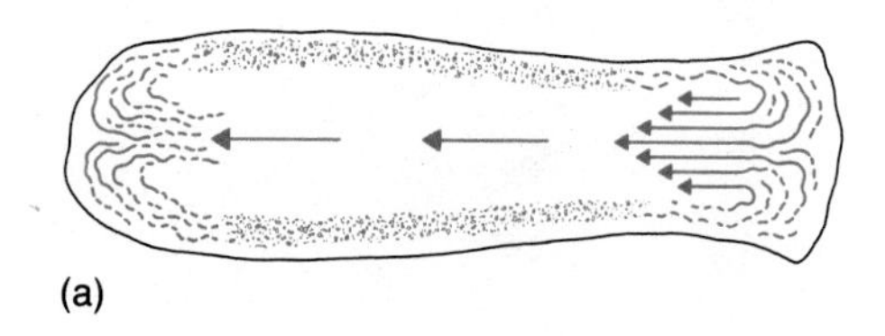

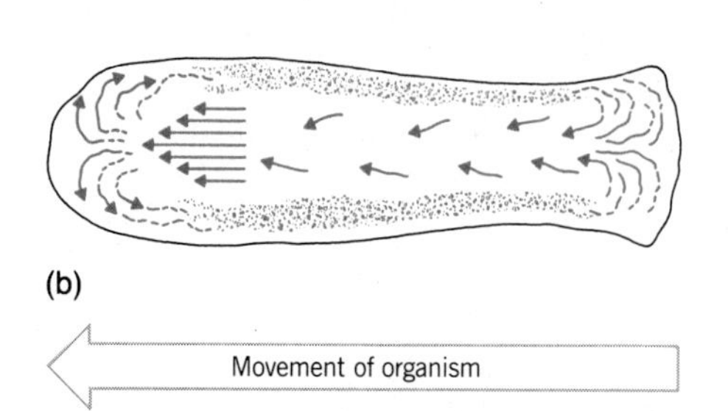

FIGURE 19.15 Depiction of the ectoplasmic tube contraction theory (*a*) and the frontal zone contraction theory (*b*) of amoeboid movement. In (*a*) the generation of force is at the opposite end of the cell from the site of pseudopod formation. In (*b*) force is generated near the region of membrane movement.

Ectoplasmic Tube and Frontal Zone Contraction Theories

Two theories that have played important roles in explaining the basis of amoeboid movements are the *ectoplasmic tube contraction theory* and the *frontal zone contraction theory*.

The first, developed by C. Pantin and S. Mast in the early 1920s, centered on contraction of the ectoplasm as the force driving the movement of endoplasm.[9,10] The endoplasm is pushed toward regions of less pressure just as a skilled entertainer maneuvers air in a balloon to create figures. Pseudopods develop as a result of endoplasm being forced into low pressure regions. In this model, the generation of force may be quite distant from the site of pseudopod formation (see Figure 19.15*a*). The endoplasm streams from high pressure to low pressure zones. This theory was developed well before nonmuscle actin and myosin filaments were recognized. Therefore, the molecular bases of the driving forces were obscure.

The frontal zone contraction theory proposed in the early 1960s by Robert Allen[11] locates the force-generating center toward the front of a moving organism in the zone of endoplasm (Figure 19.15*b*). A volume of endoplasm contracts, and in so doing pulls toward it endoplasm from the rest of the cell.

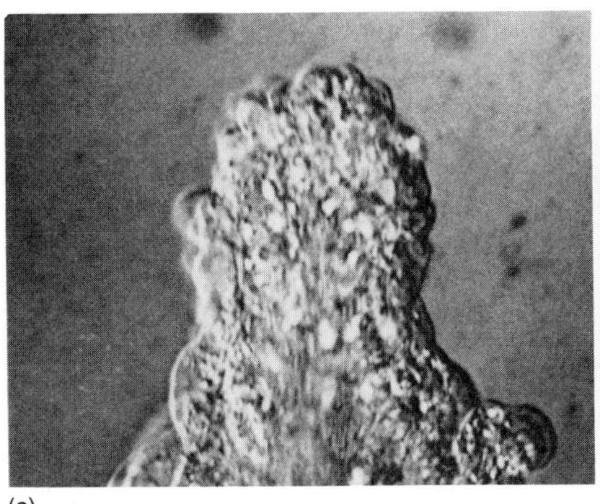

(a)

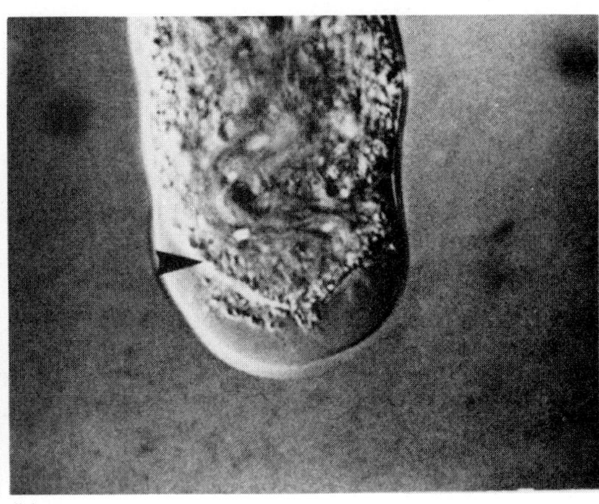

(b)

FIGURE 19.16 Nomarski micrographs of the amoeboid cell *Chaos carolinensis*. (*a*) Tail of an actively motile cell has a wrinkled appearance. (*b*) Frontal portion of an advancing pseudopod contains an organelle-free cap anterior to a plasma gel sheet (arrow) behind which organelles congregate. (Courtesy of Dr. D. Lansing Taylor.)

Cytoplasmic Movements During Amoeboid Motility

When pseudopods are forming and amoebae are moving, the cytoplasm is in a dynamic state of streaming. The amoebae most frequently used to observe these properties are *Amoeba proteus* and *Chaos carolinenis*, since they are large and easily viewed.

When amoebae move, they have a polarity related to the direction of movement. Pseudopods cannot form randomly in all directions or the organism would tear itself apart by virtue of the edges crawling away from the center. There is, rather, an anterior pole that corresponds to the tip of the advancing pseudopod and a posterior pole called the uroid, likened to the tail of the organism (refer to Figure 19.14).

The most frontal portion of the advancing pseudopod contains a thick hyaline cap, whereas the rest of the cell possesses only a thin layer of this organelle-free material. The trailing edge of the tail has a wrinkled appearance. These features are evident in Figure 19.16.

When a pseudopod advances, the hyaline cap expands in a forward direction and the axial endoplasm streams toward the tip, where it is diverted three-dimensionally like water at the top of a fountain. When diverted, the endoplasm moves into and becomes a part of the ectoplasm. The ectoplasm moves toward the tail region of the organism, where it is once more transformed or "recruited" into endoplasm. The endoplasm then moves forward in response to expansion of the hyaline cap.

The Relation between Cytoplasmic Structure and Contraction

The ectoplasm of amoeboid organisms is a gel consisting of actin held in position by cross-linked gelation factors. Myosin is also present in the ectoplasm; but as long as the actin is cross-linked, myosin can generate no force to effect contraction. The cross-linked actin of the ectoplasm

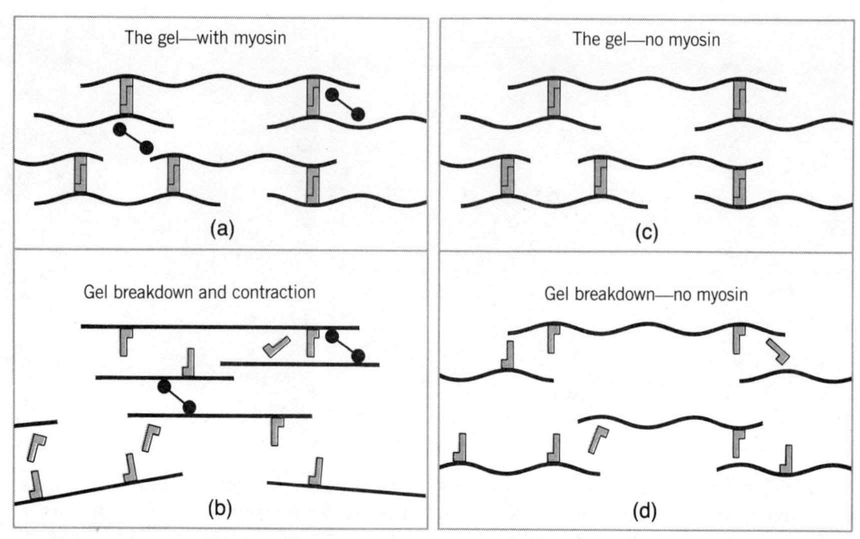

FIGURE 19.17 The solation–contraction coupling hypothesis. Actin is represented by solid lines, myosin by dumbbells, and gelation factors by blocks. If all factors are present, gelation occurs and contraction is inhibited (*a*). When cross-links are broken, contraction occurs (*b*). In the absence of myosin, gelation may occur (*c*), but contraction is not possible even when cross-linking agents are released (*d*).

can thus be viewed as a cytoskeleton that maintains the shape and rigidity of the cell.

Endoplasm contains actin also, but it is not held in cross-linked form. It may even be depolymerized to G-actin rather than F-actin.

D. Lansing Taylor and his colleagues, who have conducted a number of studies on the relation between cytoplasmic structure and contraction, have arrived at the following conclusions:[12] (1) Ectoplasm is a gel of cross-linked actin filaments. The function of the gel is to determine shape and transmit tension from regions of contraction to cell–substrate attachment sites. (2) When contractile tension is applied to optically isotropic gels, it is converted to optically anisotropic actin fibrils. (3) For free actin filaments to slide and cause contraction, cross-links between actin filaments must be locally dissociated (solation). Contraction occurs when free actin filaments, still attached to adjacent gel, pull gelled regions together. (4) Solation–gelation transformations are regulated by the level of Ca^{2+} and/or pH. Gelation is preferred at pH 6.8–7.0 and a submicromolar concentration of free Ca^{2+}. (5) In the absence of myosin, microdomains of gel solate when Ca^{2+} concentration and pH are raised above levels optimal for gelation.

These conclusions have been put together to formulate the *solation–contraction coupling hypothesis* of amoeboid movement.[13] A diagrammatic representation of the hypothesis appears in Figure 19.17.

Four states are shown. In the first condition, actin, myosin, and cross-linkers are all present. This state is a gel that cannot undergo contraction even in the presence of myosin because the cross-linking agents resist sliding of the filaments. Contraction may occur in the second state. When cross-links are broken (by appropriate levels of Ca^{2+} or pH), the actin filaments are free to slide, and contraction will ensue in the presence of myosin. In the absence of myosin (the third state), no contraction will occur. However, gelation and solation transformations (state 4) can still take place by reversible dissociations of cross-linking agents from the actin filaments.

Applying these properties of cytoplasm structure to the events taking place during amoeboid movement, the following is envisioned. The gel in the tail is induced to solate and contract by a rise in intracellular Ca^{2+} and/or pH. Adjacent gelled regions are pulled together where myosin is present, and simple solation with no contraction (recruited endoplasm) takes place in myosin-free domains. Thus, the volume of the tail decreases and endoplasm is moved forward.

The endoplasm, largely in sol form near the tail, is gradually returned to gel form in a gradient increasing from tail to pseudopod. The most anterior portion of this gradient contains the maximum extent of gelation.

To accommodate the movement that must take place for pseudopods to advance, the hypothesis predicts a local rise in Ca^{2+} and/or pH at the site of pseudopod formation. This would favor localized solation or a weakening of the gel at that point so that endoplasm would protrude into the forming pseudopod. The presence of myosin at the anterior end where localized solation has occurred may bring about contraction to pull the gelling endoplasm of the gradient forward.

The solation-contraction coupling hypothesis thus presents a favorable framework from which amoeboid movement can be viewed and tested. Before it is accorded a more widely embraced theoretical niche, however, the hypothesis deserves some careful studies. It will be important, for example, to define and quantitate the cellular locations of myosin and Ca^{2+} and to determine whether pH gradients or localized changes exist in the cytoplasm.

A major problem to be explained is the basis of regulation or control. Calcium ion at certain concentrations appears to be crucial. Is it stored, and if so where, and how is it released? What is the nature of communication between the advancing pseudopod and the contracting tail? These and other questions are currently being investigated. When the answers are in, they will likely have ramifications for other types of cell movement that are actin–myosin based, such as cytokinesis or the movements of microvilli.

19.4 THE MITOTIC APPARATUS AND CHROMOSOME MOVEMENTS

"Mitosis" and "cell division" are often thought about as interchangeable terms, but technically they are not. Cell division is a two-phase process. The first phase involves the separation of chromosomes (karyokinesis) and the second is concerned with movements of the cytoplasm (cytokinesis). Only the latter phase gives rise to individual cells. "Mitosis" and "karyokinesis" are terms that can be used interchangeably.

Both karyokinesis and cytokinesis use contractile filaments within the cell, but the former is primarily a microtubule-dependent process and the latter comes about by the action of actin and myosin microfilaments. In this section we shall concentrate on karyokinesis, or mitosis.

Mitosis was observed in living cells early in the nineteenth century, and correlations were made between chromosome movements and gene behavior in the late 1800s. The mitotic spindle was not seen, however, until near the midpoint of the twentieth century. Shortly after that it was isolated (in 1952).

Microtubules were identified as the main structural components of the spindle fibers in 1962 and in the mid to late 1960s tubulin was identified as the protein of microtubules.

Since the turn of the century, and particularly during the past decade or two, research on the mechanisms of mitosis has intensified and is continuing strong today. But, as J. R. McIntosh has pointed out, we are presently somewhat in a dilemma.

Discussions of the mitotic mechanism in the early part of this century were frustrated by a marked insufficiency of information about the process. Today we are overwhelmed by information, but even so, there is not yet sufficient detail available to describe and analyze mitosis at the molecular level. A molecular understanding of mitosis will require answers to several important questions that may be clearly defined, but it is not yet obvious how to answer them.[14]

Chromosome Movements during Mitosis

A correlation of chromosome positions with the phases of the cell cycle is demonstrated in Figure 19.18.

During interphase, chromosome form and movements are not evident, but this is a very important stage for the events about to take place in mitosis. DNA synthesis, hence chromosome replication, takes place during the S phase, and tubulin synthesis, which is necessary for the imminent construction of the mitotic apparatus, occurs during S and early G_2 phases of the cycle. Apparently the synthesis of tubulin during S and G_2 is for the long-range welfare of the cell in subsequent cell divisions. Protein and RNA synthesis can be stopped and cell division will continue for awhile, suggesting that the formation of the mitotic apparatus is not tightly coupled to protein synthesis. The cell apparently has a pool of subunits that can be drawn on to construct the spindle fibers; when they are dismantled, their subunits are returned to the stockpile.

The first obvious event of mitosis is chromosome condensation. This is manifested during prophase. During early condensation the nuclear envelope is still intact. In late prophase the primary constrictions, or centromeres, appear on the condensing chromosomes. These contain the kinetochores to which spindle fibers will be attached.

While these events are taking place in the nucleus, equally important activities are going on in the cytoplasm that are requisites for mitosis. These activities are focused around the cell center, or centrioles.

The cell center in most animal cells consists of two perpendicular centrioles surrounded by a zone of poorly defined material. During late G_1 or early S phase the centrioles duplicate while lying adjacent to the nuclear envelope. During prophase the centriole pairs move away from each other to opposite sides of the nucleus, during which time they function as centers from which mitotic spindle fibers are formed and oriented. Thus a spindle begins to form outside the nuclear envelope but often so close that it deforms the membrane to some degree. Prophase ends with the breakdown of the nuclear envelope, an event that triggers a mixing of the

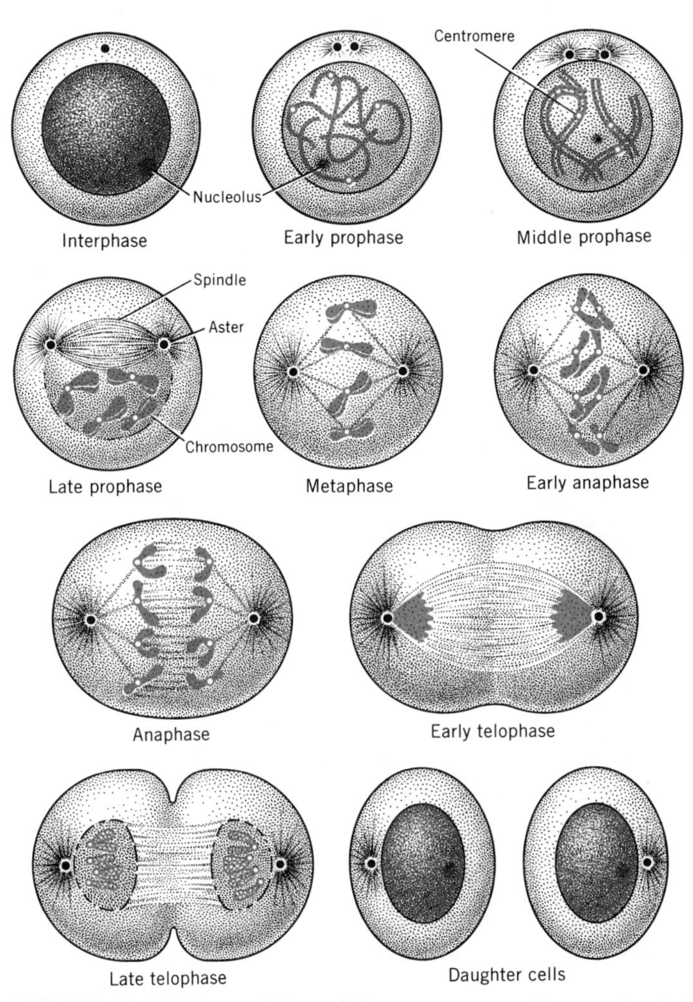

FIGURE 19.18 A correlation of chromosome positions and spindle structure with the cell cycle in an animal cell possessing four chromosomes.

cytoplasm and nucleoplasm and permits attachment of spindle fibers to chromosome kinetochores.

Two modes of spindle formation are often observed at about the time of nuclear envelope breakdown. One type results when the mitotic centers are still quite close together at the time of breakdown. When this is the case, the spindle forms between the two centers and lengthens as the centers continue to move apart to their final positions. Chromosomes scatter over the surface of the spindle during this mode of spindle assembly.

The other commonly observed mode is initiated when the mitotic centers are farther apart at the time of nuclear envelope breakdown. A radial growth of spindle fibers, called the aster, emanates from each mitotic center. The fibers elongate until they reach each other and establish a bipolar spindle. A result of this mode of spindle formation is that chromosomes are trapped in the spindle rather than spread on its surface.

The result of each mode of spindle formation is the same. Chromosomes are oriented on the metaphase plate so that the two kinetochores of a pair are directed toward opposite poles in the cell.

The spindle fibers.

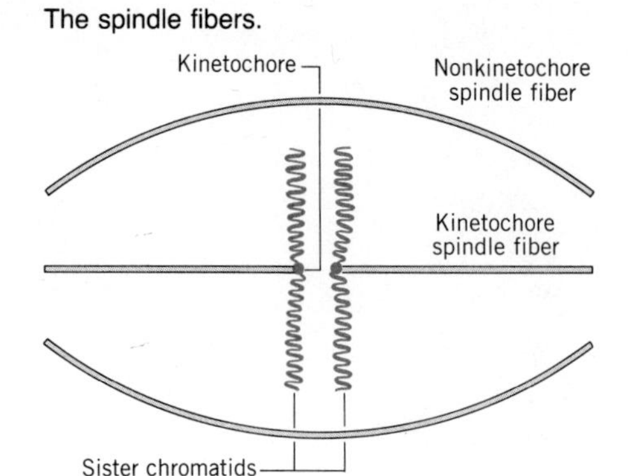

Metaphase consumes a very short period of time in the cell cycle, at least when it is defined as a condition of all chromosomes being aligned at the equator. Final alignment of all chromosomes is brief because a given chromosome may move toward and away from the metaphase plate several times before it settles down and stabilizes on the plate.

At this point two different types of spindle fiber can be identified. One type runs from pole to pole without making any contact with the chromosomes. These fibers are the most abundant and constitute most of that which is recognized as the spindle. The second type runs from kinetochore to pole. This type is approximately half the length of the other, since it extends only from the metaphase plate zone to the microtubule organizing center.

A synchronous separation of sister chromatids to opposite poles marks the onset of anaphase. The two dynamic components to this movement are at the heart of mechanisms that are proposed to explain mitosis. One component is a *movement of chromosomes toward a pole* and thus a shortening of the fiber that stretches between the two. The second is *an increase in distance between poles* and thus an extension of the pole-to-pole fiber system. Chromosome fiber shortening predominates during early anaphase, and spindle elongation is a characteristic of late anaphase.

Toward the end of anaphase the chromosomes cluster around the mitotic centers with their kinetochores approximately equidistant from the centrioles. The nuclear envelope begins to form around the cluster, and when it is complete and the chromosomes begin to disperse, anaphase is complete. Disintegration of the chromosomal fibers begins before the nuclear membrane is completely assembled, whereas remnants of the pole-to-pole spindle remain for a longer period of time. The cleavage furrow begins to form during mid- to late anaphase and the membrane at the plane of the spindle equator begins to constrict. Thus cytokinesis is begun before telophase sets in.

During telophase the chromosomes return to their interphase morphology, and the mitotic apparatus is disassembled. When cytokinesis is complete and at some point when DNA replication begins, telophase is considered to be finished.

Numerous variations on the above-described theme could be discussed, but the reader should refer to the books and articles listed at the end of this chapter to satisfy those interests.

The Microscopy of the Mitotic Apparatus

The Spindle

Materials that have indices of refraction that are not uniform in all directions are optically anisotropic. They display the property of double refraction or birefringence. The spindle with its molecular components oriented from pole to pole in the cell is birefringent.

When light microscopy is conducted with polarization optics, a birefringence can be seen that corresponds to the internal structure of the spindle. The parallel arrangement of spindle fibers, particularly those that extend from pole to pole, is responsible for the birefringence. The results of polarization microscopy do not tell us much beyond the fact that the fibers that make up the spindle are highly oriented, with a polarity that extends from pole to pole.

Methods I and VII

The use of fluorescein-labeled antibodies against tubulin has complemented polarization microscopy by revealing that the polarity and order of the spindle are due to microtubular structures. A sequence of

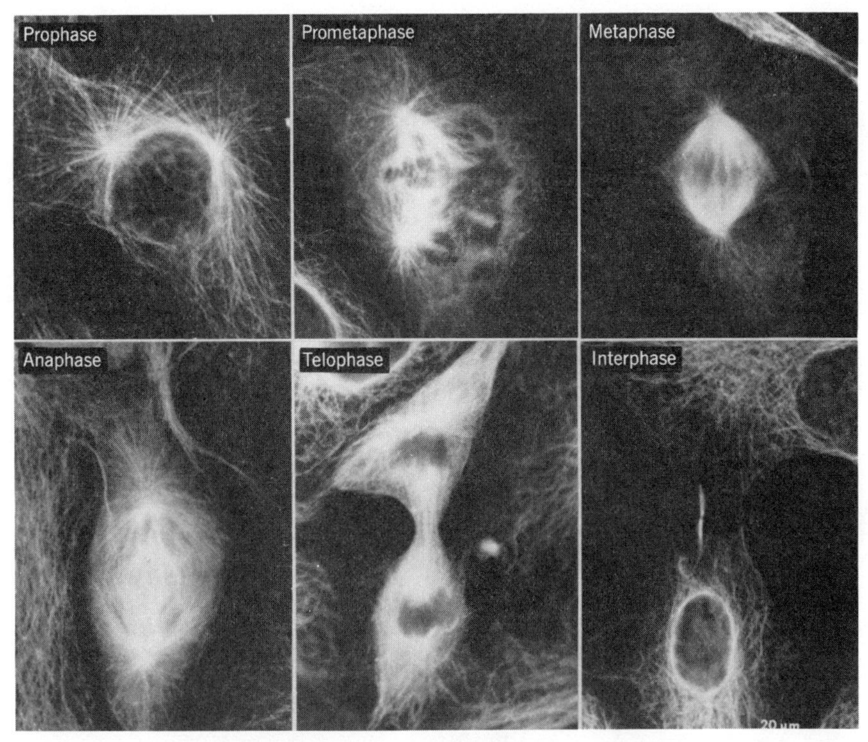

FIGURE 19.19 Indirect tubulin immunofluorescence in dividing cells. Fluorescent material that is quite amorphous during interphase becomes more and more structured as the spindle and asters are formed. (Courtesy of Dr. Hidemi Sato.)

micrographs taken during mitosis using this technique is presented in Figure 19.19.

Electron microscopy has simply confirmed the other two microscopic approaches by showing that the mitotic apparatus is a matrix of microtubules. Figures 19.20 and 19.21 show electron micrographs of spindle regions in both longitudinal and transverse planes.

The Centrioles

As noted earlier in this chapter as well as in Chapter 18, Section 3, microtubules that make up the spindle appear to emanate from a center (MTOC) at the heart of which centrioles are located. Although the precise role of centrioles vis-à-vis the production of the spindle is uncertain, their movements during mitosis suggest that their presence is not incidental to the formation and orientation of the mitotic apparatus.

The classic ultrastructural view of centrioles depicts them as cylindrical particles in pairs, with the individuals of a given pair at right angles to one another. Figure 19.22 is an electron micrograph illustrating this arrangement. Microtubules are also apparent in this micrograph. They appear to radiate from the centriole pair, but the precise relation between the pair and the microtubules is not clear.

In cross section, as can be seen in the micrograph, an individual centriole is made up of nine sets of microtubule triplets, reminiscent of the axoneme in cilia and flagella. In this case, however, each unit is made up

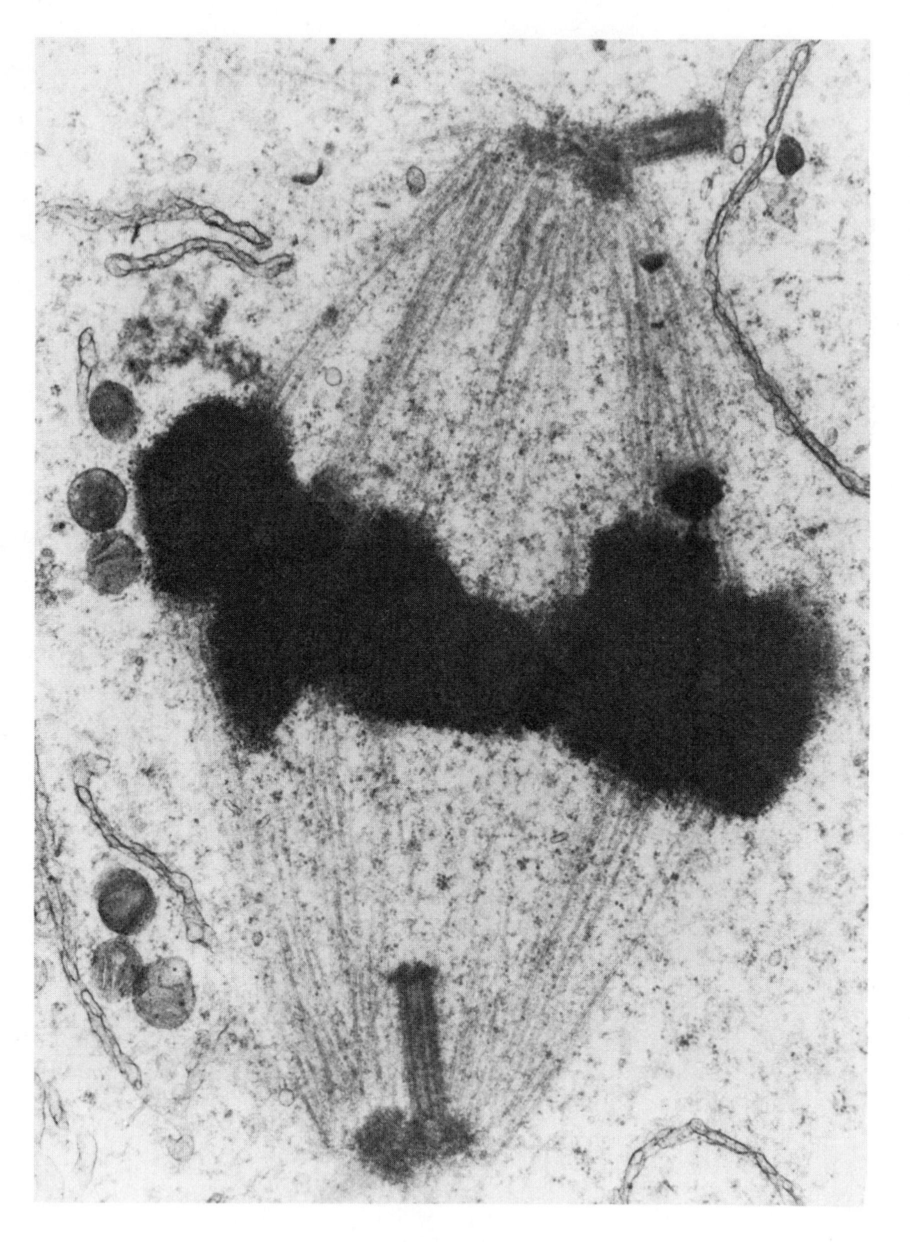

FIGURE 19.20 Longitudinal section through the mitotic apparatus. The classic spindle shape is made up of microtubules that extend between the chromosomes in the center and the pole regions of the cell where centrioles are situated. (Courtesy of Dr. J. Richard McIntosh.)

of three microtubules sharing wall structures with the central member, and there is lacking the pair of microtubules found at the center of the axoneme. On close examination, the triplet sets are seen to be connected by fibers and in some cases, depending on the point in the cell cycle, appendages stick out from the centriole cylinder.

When the two centrioles separate during mitosis, each moves with a daughter centriole growing out at right angles from the parent body. This replication is first observable near the middle of the S phase in the cell cycle (although it may begin earlier), and elongation of the daughter centriole continues until prometaphase when the two are of equal length.

Even though microtubules appear to emanate from the centriolar pair, it is presently not thought that the centriole as such is the site of microtubule formation. It is assumed, rather, that centrioles may organize the mitotic halo, a region around the pair, where tubulin units congregate and

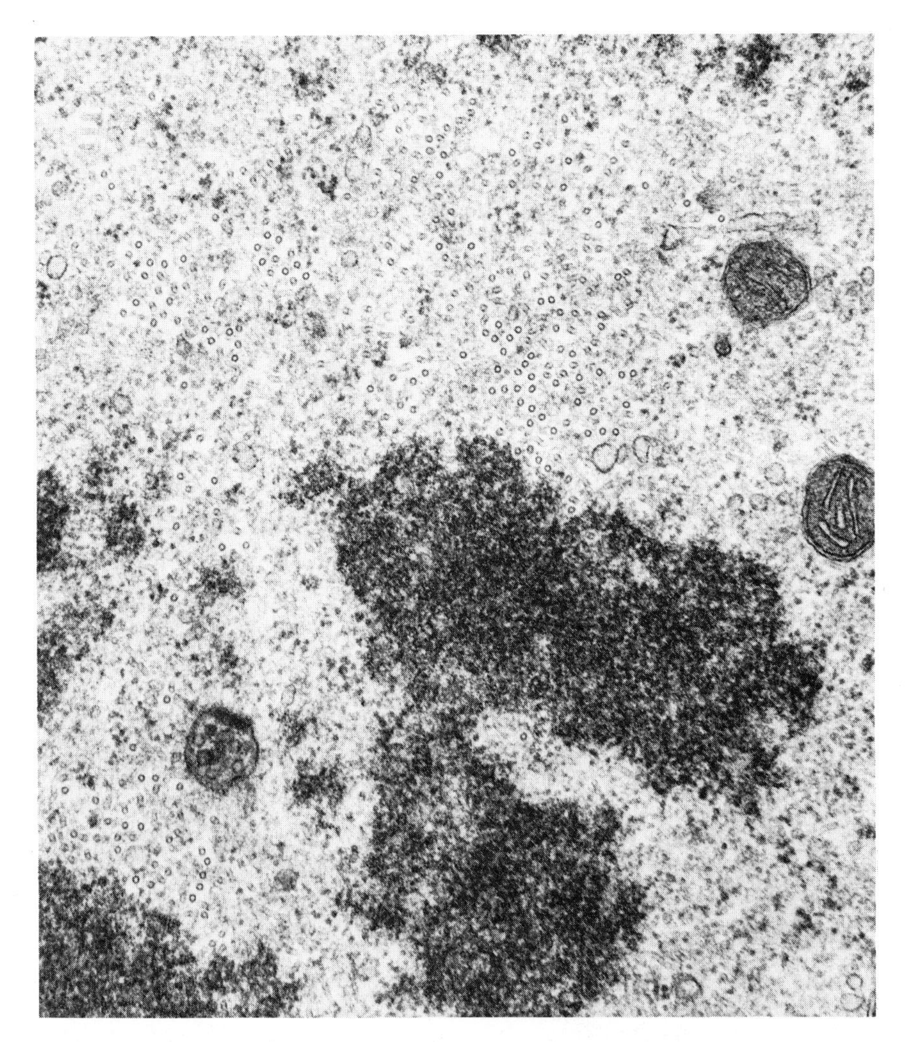

FIGURE 19.21 Transverse section through the mitotic apparatus. Hollow microtubules are clustered above the electron dense chromosome. Mitochondria and membrane fragments are also close by. (Courtesy of Dr. J. Richard McIntosh.)

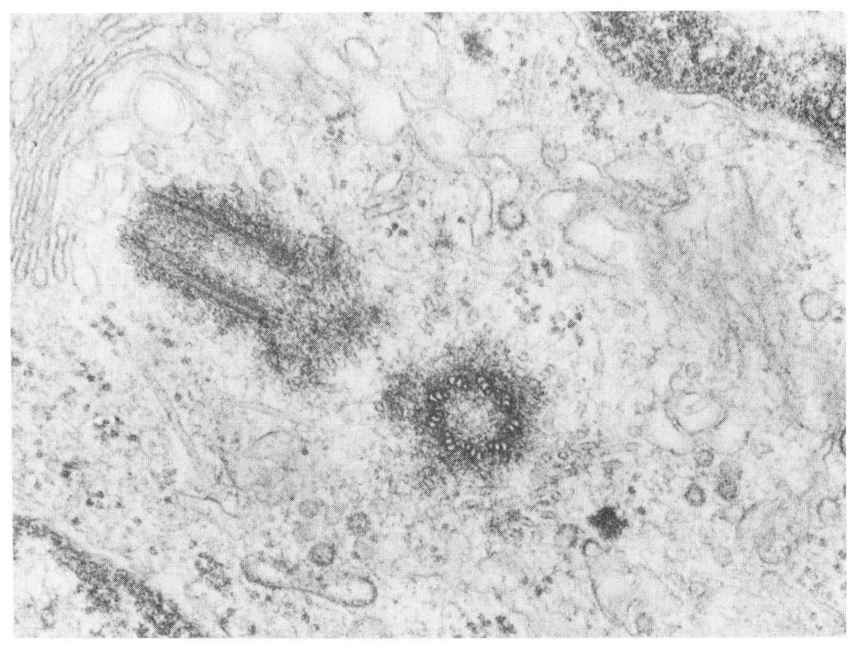

FIGURE 19.22 Electron micrograph of a thin section cut through a pair of centroles in a myelocyte from guinea pig bone marrow. Because they are at right angles to each other one is cut longitudinally and the other transversely. The transverse view shows the centriole wall to be made up of nine triplets of microtubules. (Courtesy of Dr. Don W. Fawcett.)

begin to polymerize into microtubule structures. Thus centrioles may organize the MTOC, which in turn organizes the assembly of microtubules that make up the spindle.

Hypotheses of Chromosome Movement

A number of theories dealing with mitosis have been proposed over the years. As more structural and molecular details have become available, the theories have been correspondingly modified.

At present there are three basic theoretical models of chromosome movement. Although different, they are not necessarily mutually incompatible and, in fact, the true mechanism may be a composite of aspects derived from each one.

At the heart of each theory is a consideration of how force is generated to move the chromosomes toward the poles and to extend the length of the spindle. The three types of force generation, given the ingredients of the mitotic apparatus, are (1) changes in the length of microtubules by tubulin polymerization and depolymerization, (2) microtubule sliding with dynein cross-bridges, and (3) actomyosin-mediated contractions of microtubules.

The Dynamic Equilibrium Model

The fundamental properties of the dynamic equilibrium model developed by Shinya Inoué and his co-workers in the late 1960s,[15] are illustrated in Figure 19.23. It is envisioned that a pool of tubulin exists within the cell from which are drawn the units that polymerize into growing microtubules.

Both polymerization and depolymerization must take place simultaneously to explain the movements that occur during mitosis. Kinetochore microtubules shorten to draw chromosomes toward the poles. Depolymerization is thought to proceed at the ends of microtubules in pole regions, allowing the prerequisite shortening.

At the same time, nonkinetochore microtubules are extending, pushing the poles farther apart. This is proposed to occur by adding subunits to the microtubule ends in the pole regions. There is thus some means of selection between the two types of microtubule so that subunits are added to one and removed from the other to give the desired pull–stretch effect.

Several lines of experimental evidence are supportive of this model; however, most of them are indirect. One is the effect of colchicine on microtubular systems. Colchicine is known to depolymerize microtubules *in vitro,* and at high concentrations it causes the mitotic apparatus to dismantle and blocks the anaphase movement of chromosomes. However, at lower concentrations the rate of anaphase chromosome movement is actually stepped up. This observation is interpreted to mean that depolymerization of kinetochore microtubules is enhanced, probably without an effect on the nonkinetochore microtubule polymerization reactions.

Another line of evidence consistent with the model is the effect of temperature and pressure on microtubules and mitosis. Increasing the hydrostatic pressure has roughly the same effect as colchicine both on microtubules *in vitro* and on the mitotic apparatus. This can be interpreted in the same way as the colchicine results.

The speed of chromosome movements during anaphase is directly pro-

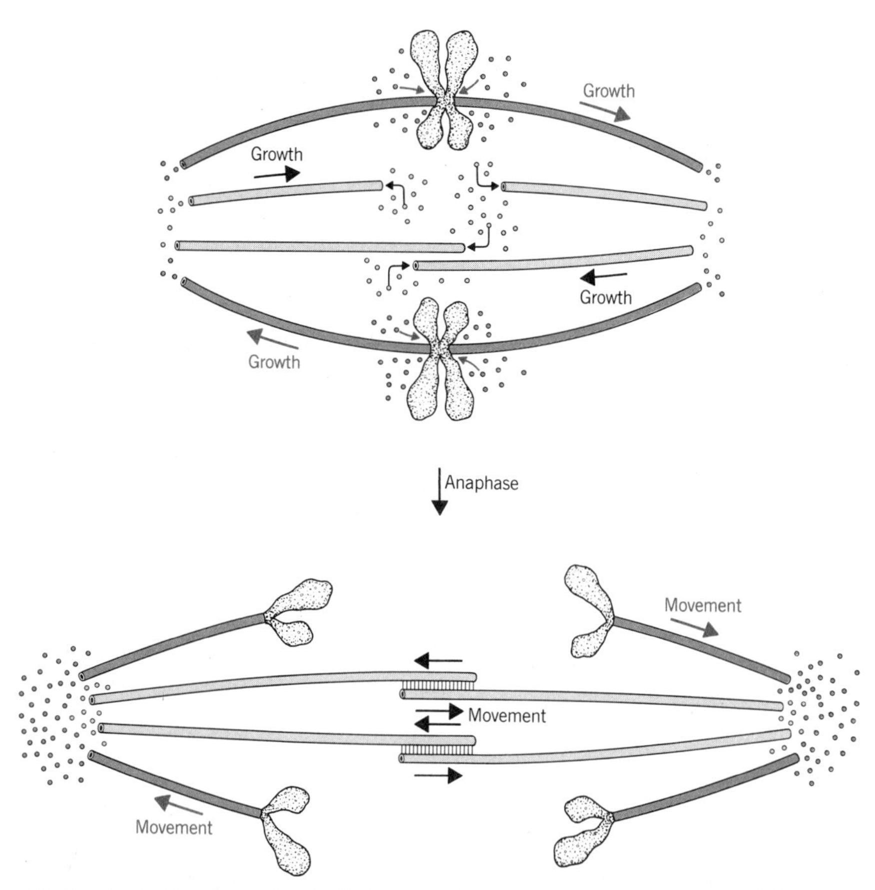

FIGURE 19.23 Dynamic equilbrium model of chromosome movement incorporating the ideas of Margolis. Microtubules grow by polymerization until kinetochore microtubules are full length and nonkinetochore microtubules overlap. The kinetochore microtubules then shorten by depolymerization and the kinetochore microtubules move apart by a force generated where they overlap. According to this model, depolymerization occurs at the poles and polymerization at ends away from the poles.

portional to temperature as long as the temperature is held within a range wherein the spindle is stable. In this case, a higher temperature favors polymerization. In the cold the mitotic apparatus disassembles. The temperature effect is most easily explained by applying it to the interpolar microtubules. At higher temperatures they elongate faster by polymerization, while polymerization is blocked or reverses at lower temperatures.

Finally, the rates of chromosome movement are consistent with rates of microtubule elongation and shortening by polymerization–depolymerization reactions.

One of the major unresolved features of the model is an explanation of how depolymerization can generate force to move the microtubules, hence the chromosomes. Anyone cutting chunks of wood off a pole soon realizes that to complete the job it is necessary to move along the length of the pole. The pole does not move toward the woodcutter. Yet this is what the model suggests takes place in the cell.

Another feature of this model that has not been sufficiently resolved experimentally is the relation between microtubule polarity and the ends to which subunits are added or from which they are subtracted. Some studies have indicated that when microtubules grow *in vitro,* subunits are added preferentially to one end only and are disassembled from the other end. Some workers have suggested that a microtubule is never a static structure but rather always in a dynamic equilibrium. A constant length may be maintained by assembly and disassembly rates that are equal but operating simultaneously on different ends of the same microtubule.

If microtubules elongate on one end, and if they are generated by opposite mitotic centers and grow toward the metaphase plate, then approaching interpolar spindles would have opposite polarities. It is not likely that they would fuse to form a continuous fiber. Even if they did, the pole ends of the fibers should be the depolymerization ends, not the ends to which subunits are added. How then can the microtubule grow to push the poles apart?

If kinetochore microtubules grow outward from the mitotic centers until they make contact with kinetochores, they are then properly aligned for disassembly at the pole end, consistent with the dynamic equilibrium hypothesis to explain chromosome movement.

Margolis and his colleagues[16] have provided a recent twist to this concept of force generation by developing a model in which all microtubules in a half-spindle are assumed to be parallel. Growth of the spindle, and thus its elongation, is envisioned to take place by an addition of subunits to the interpolar microtubules *distal* from the poles. After sufficient growth the interpolar microtubules overlap, but they are antiparallel, having grown out from opposite poles. The fibers are thought to be linked by static bridges, with a generation of force taking place in some manner between microtubules to push them apart.

Subunits are added to the chromosomal microtubules at their kinetochores. Thus the microtubule grows by being pushed toward the pole by a growing end at the kinetochore.

One clarifying feature of this concept is that both microtubule types are oriented in the mitotic apparatus so that their growth ends are at the equator and their depolymerization ends at the pole. Spindle elongation thus takes place by growth, a static interaction at the points of overlap, and force generation. The anaphase movement of chromosomes takes place as a result of depolymerization at the poles.

These concepts are obviously speculative, but they have features consistent with a number of studies and provide a framework from which continued experimental strategies will certainly arise.

The Sliding Microtubule Model

The sliding microtubule model was first proposed by J. R. McIntosh and his collaborators in 1969.[17] It has since been modified to better describe features of chromosome movement as presently understood. The basics of this model are summarized schematically in Figure 19.24.

According to this concept, force is generated between adjacent microtubules by lateral interaction that causes them to slide past each other.

Both components of the anaphase process can be addressed by sliding. The spindle will elongate when adjacent overlapping regions of interpolar

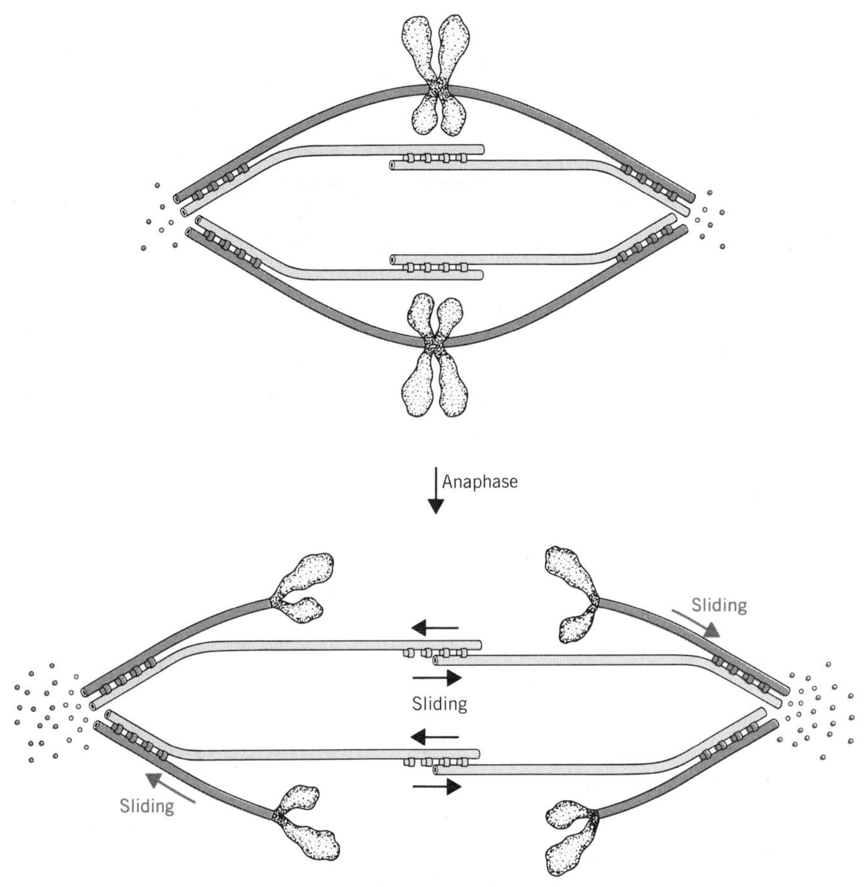

FIGURE 19.24 Sliding microtubule model of chromosome movement. Adjacent interpolar (nonkinetochore) microtubules slide past each other to effect spindle elongation. Adjacent interpolar–kinetochore microtubules slide to move chromosomes toward the poles. The generator of force for sliding is presumed to be dynein arms, positioned arbitrarily in this diagram.

fibers move in a direction that decreases the amount of overlap, thereby pushing the poles apart. Chromosomes will move toward the poles when adjacent interpolar–kinetochore microtubules slide past each other so that the overlap between the two types of fiber decreases.

This concept is supported by several lines of research. One is the discovery that dynein is a component of spindles. Thus the microtubule–dynein system of the mitotic apparatus may generate forces similar to those of the axoneme system in cilia and flagella.

The mere presence of dynein does not prove it to be the lateral component generating force in the mitotic apparatus, but other studies have shown more directly that it may be. Antisera to dynein block spindle elongation, suggesting that dynein must be free to operate during this phase of mitosis. When dynein is added to cells that have been lysed but not dismembranated, the rate of chromosome movement during mitosis is increased. Vanadate, a potent inhibitor of dynein ATPase activity, inhibits both chromosome-to-pole movements and spindle elongation. Vanadate has a similar effect on flagella; that is, it inhibits movement. Finally, ATP is required for anaphase chromosome movement.

The results above are consistent with the involvement of dynein in mitosis and thus with a mechanism of microtubule sliding akin to that proposed for cilia and flagella.

The sliding microtubule model considers sliding to result from the generation of force by conformational changes in lateral dynein bridges, but this cannot occur without depolymerization also taking place to shorten the kinetochore microtubules as they slide toward the poles. But according to this concept, depolymerization is not force generating. It is merely incidental to the process.

An apparent unresolved feature of this model is the relation between microtubule polarity and dynein-produced sliding. Assuming the polarity of the previous model—that is, growing ends positioned at the equator and depolymerization ends at the poles—the interpolar microtubules would be antiparallel at the equator, but the two types of microtubule would have similar polarity where they are found together nearer the poles. In cilia and flagella adjacent microtubules are assumed to have the same polarity.

Will antiparallel microtubules slide past each other? Does polarity affect the rate of sliding? Answers to these questions will provide some additional insight to this model, which appears to address in cogent manner most of what is known about chromosome behavior.

The Sliding Microfilament Model

The third theoretical model for the generation of force in mitosis proposes that the basis of movement is a spindle actomyosin system. As first suggested by A. Forer in 1974,[18] actomyosin was assumed to generate the force and microtubules were viewed as playing a cytoskeletal role in the mitotic apparatus. According to more recent opinions on this model, actomyosin may operate parallel to a sliding microtubule mechanism.

Figure 19.25 depicts the sliding microfilament model, which considers chromosome movement but not spindle elongation. It is assumed that F-actin is anchored to two different parts of the mitotic apparatus. One end of one polymer is attached to the kinetochore region of the apparatus. The attachment may be directly to the kinetochore or to the microtubule in the vicinity of the kinetochore. The rest of the polymer lies adjacent to the kinetochore microtubule.

One end of another actin polymer is anchored into the pole region. It too lies adjacent to the kinetochore microtubule. The role of myosin is thought to be similar to the role of dynein in flagella. It is viewed as providing the ATP-dependent cross-bridges to generate the force so that the antiparallel filaments will move past each other, imparting strain to the microtubule.

It is apparent that depolymerization of microtubules in the pole regions is also necessary for this model to work. As the kinetochore microtubules depolymerize, they can be pulled toward the poles by the force-generating actomyosin system.

This model is based primarily on the finding (by a variety of techniques) that actin and myosin are components of the spindle. Second, chromosome movement is ATP dependent, as is the actomyosin system for movement. But except for these observations, most of the experimental results are difficult to reconcile with this model. Antiserum against myosin blocks

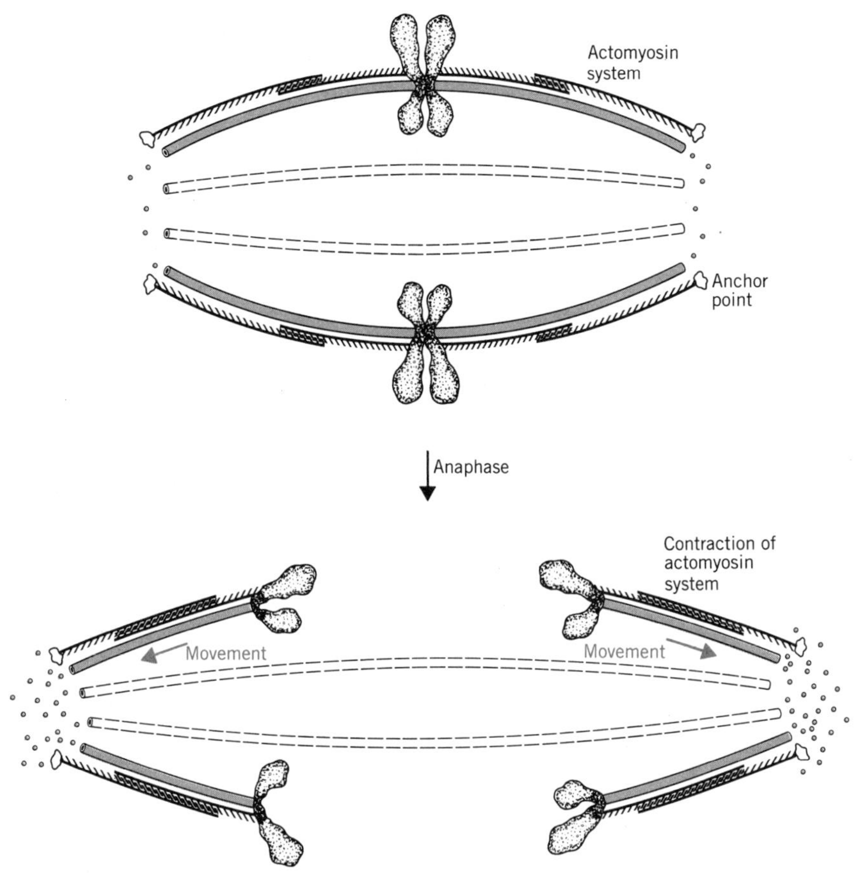

FIGURE 19.25 Sliding microfilament model of chromosome movement. Overlapping actin microfilaments are depicted as being anchored to opposite spindle regions—one to the chromosome and the other to the far end of the microtubule in the pole region. The actins are thus antiparallel and, along with myosin, generate contraction forces that move the chromosome toward the pole.

cytokinesis, but not mitosis, whereas the agents that are most effective in blocking mitosis appear to act on microtubules.

The presence of actin and myosin in the mitotic apparatus may not be incidental to anaphase chromosome movement, even if the evidence in support of the model is somewhat wanting. It may be that microtubules and actomyosin operate in concert by some mechanism that is yet to be discerned.

The Polarity of Kinetochore and Spindle Microtubules

Recent studies on the polarities of the two types of mitotic apparatus microtubule are helping to clarify the theoretical models discussed above. J. R. McIntosh has demonstrated that 90 to 95% of all the microtubules in the regions between the poles and the kinetochores in an anaphase cell are of uniform polarity with their growing ends *distal* to the poles.[19] The microtubules in this region, during anaphase, would be predominantly kinetochore microtubules (see Figure 19.26).

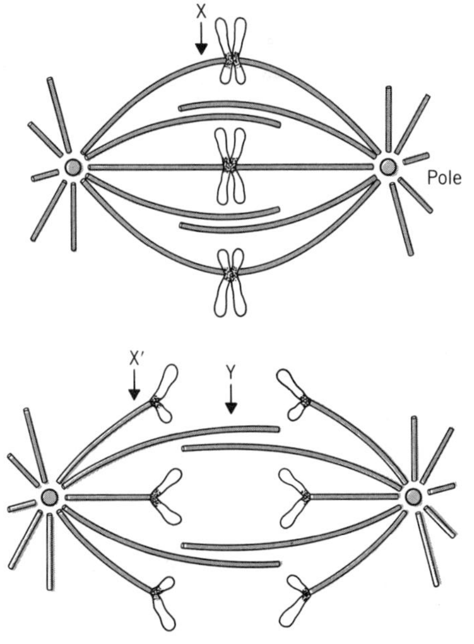

FIGURE 19.26 Schematic representation of the distribution of microtubules during metaphase and anaphase. Nonkinetochore microtubules overlap during metaphase at plane X, between kinetochores and poles. During anaphase there is no overlap at a comparable plane (X′). Sectioning through plane Y would reveal only nonkinetochore microtubules.

In these studies, microtubule polarity was determined by polymerizing hook-shaped neurotubulin appendages onto the microtubules. The orientations of the hooks, as viewed by electron microscopy, could then be used to determine and compare microtubule polarities.

Three different approaches were used. In one approach, all spindle microtubules in the zones between kinetochores and poles were analyzed. In a second, kinetochore microtubules were studied selectively by removing other microtubules with cold treatment (kinetochore microtubules are more stable than the others at low temperatures). A third approach was to identify kinetochore microtubules by tracking them to their kinetochores by serial sectioning techniques.

The finding of uniform polarities of microtubules within a half-spindle, with growth ends distal to the poles, does raise questions regarding how contact is made between microtubules and kinetochores. There appear to be two possible interpretations. One, microtubules are initiated at the pole; then they grow out and are "captured" by kinetochores. Growth continues by adding subunits in the vicinity of the kinetochore.

A second interpretation is that microtubules are initiated at the kinetochore. Growth takes place by adding subunits proximal to the kinetochore.

In another recent study, using quite a different technique, B. Telzer and L. Haimo have also shown that the half-spindle is composed of microtubules of uniform polarity.[20]

These investigators decorated meiotic spindles from the eggs of the surf clam *Spisula* with dynein isolated from the ciliary axonemes of *Tetrahymena*. Even though the systems are obviously heterologous, dynein binds at a tilt similar to that seen in the ciliary axoneme. The tilt of the dynein arms is toward the pole from which a microtubule is projecting. In an axoneme, the tilt is toward the base of the cilium. Thus, if the systems are comparable with respect to the relation between tilt and polarity, the growing ends of microtubules are distal from the poles as the growing ends of cilia are distal from the base.

Both these lines of research provide strong experimental support to the concept that microtubules within the half-spindle have uniform polarities whether they are of the kinetochore or nonkinetochore type. Furthermore, growth, or polymerization, appears to take place distal from the microtubule organizing centers.

These results lead to the implication that adjacent kinetochore–nonkinetochore microtubules are parallel in polarity, but adjacent nonkinetochore microtubules are antiparallel where they overlap at the metaphase plate.

The Mitotic Apparatus Cage

The cell interior is subjected to such extreme perturbations during mitosis that one might suspect all the ingredients within the confines of the plasma membrane to be hopelessly mixed. During prophase the elements of the cytoskeleton disperse, and later the nuclear envelope breaks down. The cell moves from a highly organized state with distinctive traits reflecting its tissue origins to a state with little apparent cytoplasmic order. After mitosis, cytoplasmic reorganization ensues, with a return to the level of order and organization extant before mitosis.

The mechanisms whereby a cell can reorganize in reproducible and predictable manner from the mitotic condition is the subject of considerable research. One observation that is coming to light is that the cell interior may not be hopelessly mixed during mitosis. Rather, the nucleus and its contents may be at least partially sequestered from the rest of the cell throughout the entire cell cycle.

The observation that lends itself to this thinking is the presence of a cage of filaments enclosing the mitotic apparatus in certain mammalian cells that have been studied.[21] When mitotic mammalian cells are extracted under conditions that stabilize the mitotic apparatus, more than 80% of the protein of the cell can be removed without removing the microtubules of the spindle. Cells thus extracted retain a fibrous cage surrounding the spindle that consists of 10-nm filaments of the vimentin type. This type of filament is designated an intermediate filament.

The biological significance of this cage is not yet entirely clear. It may serve to order the process of mitosis and the orientation and activities of the mitotic apparatus microtubules. It may function to keep the envelope-free nucleus separate from the rest of the cell. Some have speculated that the cage functions as a framework that orders the reestablishment of organization characteristic of the interphase state.

19.5 CYTOKINESIS

In late anaphase or early telophase in animal cells, after the chromosomes have been separated, the plasma membrane invaginates in the equatorial plane of the cell, constricts, and separates the cytoplasm into two approximately equal portions. This activity is referred to as *cleavage* or *cytokinesis*.

The process of cytokinesis in plants is quite different from that seen in animal cells. In plants, the cytoplasm is divided when membrane vesicles are aligned across the midsection of the cell between the new sets of chromosomes, fuse, and form a barrier of two membranes. One half of a given vesicle becomes continuous with the plasma membrane of one cell and the other half fuses with the plasma membrane of the other cell. A new wall is then generated between the two adjacent membranes in a space that is made up of the combined interiors of the vesicles that fused. The system of vesicles that align and constitute the newly forming barrier between daughter cells is called the *phragmoplast*.

Cytokinesis in animal cells takes place along significantly different lines from this. A contractile ring forms circumferentially in the *cleavage furrow*. This constricts until the membranes fuse and two independent cells are generated.

In contrast to karyokinesis, which is a microtubule-based activity, cytokinesis employs microfilaments that are composed largely of actin. But actin is not the only component of the contractile ring. Combining electron microscopy with immunological techniques as well as heavy meromyosin decoration have revealed that myosin, α-actinin, tropomyosin, and filamin are localized in this region as well.

The roles of these various components in cytokinesis are not known for certain, but it is generally agreed that the membrane is pulled down into the cleavage furrow when the underlying ring of actin filaments contracts. Thus, a contraction of an actin-containing system appears to be fundamental to cytokinesis.

Both microscopic and biochemical evidence supports this point of view. When sections from just beneath the cleavage furrow are viewed with the electron microscope, circumferentially oriented filaments, each of 5 to 7 nm diameter, are observed to form a band around the cell. These can be decorated with S1 fragments, indicative that they have an actin composition.

Agents that block the activities of actin systems, such as cytochalasin B, disrupt the ultrastructure of the contractile ring and inhibit cytokinesis. Antibodies directed against myosin also block cleavage.

Thus, the ingredients necessary for contraction, as it is understood in muscle systems, appear to be available in the contractile ring. Both actin (+ tropomyosin) and myosin are necessary to generate the force that moves the filaments in ever-tightening circles.

But unless the filaments are attached to the plasma membrane, contraction may take place without furrowing. Although experimental data are wanting in this regard, it is thought by some researchers that α-actinin may play an important role in linking the F-actin filaments to the plasma membrane. There is precedence for this in the muscle system where actin is viewed as anchored to the Z line by α-actinin.

The role of filamin during cytokinesis is even more speculative than that of α-actinin. Filamin is a high molecular weight (250,000 daltons) actin-binding protein that cross-links F-actin filaments *in vitro*. This cross-linking causes the formation of a gel.

There are several possible roles for filamin *in vivo* with respect to cytokinesis. One function it may have is to increase the cortical gel strength, which commonly occurs before cleavage. It may also regulate the interaction of actin with myosin. Or, it may organize the contractile microfilaments into parallel arrays so that they function in concerted manner in the ring and do not pull in several directions.

19.6 THE CYTOSKELETON

The cytoplasm, once thought to be amorphous and structureless, turns out to be one of the most complex environments in the cell. The term "cytosol," as it is commonly used, conjures up an image of the cytoplasm as having two components: organelles and an organelle-free solution. According to this image organelles are relatively free to move about because they are unattached, and the environment in which they float (the cytosol) permits free diffusion in all directions.

Electron microscopy, and in particular high voltage electron microscopy (HVEM), have enhanced our understanding of the extraorganelle environment of the cytoplasm, and it is now realized that essentially no part of the cell interior is structureless. This realization has opened up an area of research that is certain to be the center of much attention during the coming decade. At the moment it is so new that the relationships between structure and function are largely speculative.

Method I

High voltage electron microscopes are not the ordinary lab instrument. They weigh 22 tons and tower 32 ft. Electrons are accelerated to an energy level of one million volts, which enables the microscope to be used to view sections that are much thicker than required for ordinary transmission electron microscopy. Thus the great advantage of HVEM over conventional TEM is that it permits a three-dimensional view of the cell interior.

The Composition and Interactions of Cytoskeletal Components

The interiors of most eucaryotic cells contain, in addition to the conspicuous organelles, a network of fiber systems of three major types: microtubules, microfilaments, and intermediate filaments. These, in turn, are linked together and attached to organelles by a fourth type of filament, which forms what has been termed the "microtrabecular lattice."

All these fibers are distributed throughout the cytoplasm. Sometimes the fibers possess an orientation that reflects a particular cell activity, such as movement or cytoplasmic transport, while at other times they appear to be much more randomly arranged.

Method VII

The technique of indirect immunofluorescence microscopy has been most useful in defining the cell regions occupied by the various fibers. There are three steps to the technique. First, cells are fixed in such a way that their interior superstructures are preserved and the antigenic properties of the components are not lost. Second, the membrane is treated with an organic solvent that increases its porosity sufficiently to permit antibodies to diffuse through it and into the cell. Finally, antibodies are added.

Generally, two types of antibody are used in sequence. The first to be applied are monospecific antibodies directed against a particular ultrastructural component, such as tubulin or actin. After these antibodies have found their targets and are bound, the excess are washed out and fluorescein-labeled antibodies directed against the first type of antibody (anti-IgG) are applied. These will bind to antibodies that are bound to cytoskeletal components and, when viewed with the instruments of fluorescence microscopy, light up the cell's ultrastructural interior.

The Intracellular Locations of the Three Major Fibers

Figure 19.27 demonstrates the manner in which the three major fibers, microtubules, microfilaments, and intermediate filaments, are spiderwebbed throughout the cell interior.

Actin-containing filaments (seen in the series, *a, d, g*), the microfilaments, are oriented in parallel manner throughout the deeper interior of the cell and in bundles nearer the cell periphery. Cytochalasin B, known to depolymerize F-actin, destroys the fibrous nature of this component of the cytoskeleton.

Tubulin-containing fibers (in the series *b, e,* and *h*), the microtubules, are more randomly arranged, but in general extend radially from the cell center and then bind and circle the cell borders. Colcemid depolymerizes microtubules (Figure 19.27*h*) but leaves the other two fiber systems undisturbed.

Intermediate filaments (series *c, f,* and *i*) course through the cell (they are generally close to the cell surface) in an interlocking spiderweb pattern that is not disrupted by either of the depolymerizing agents.

All three fiber systems coexist in the intact cell, providing an internal structure that is complex and space filling.

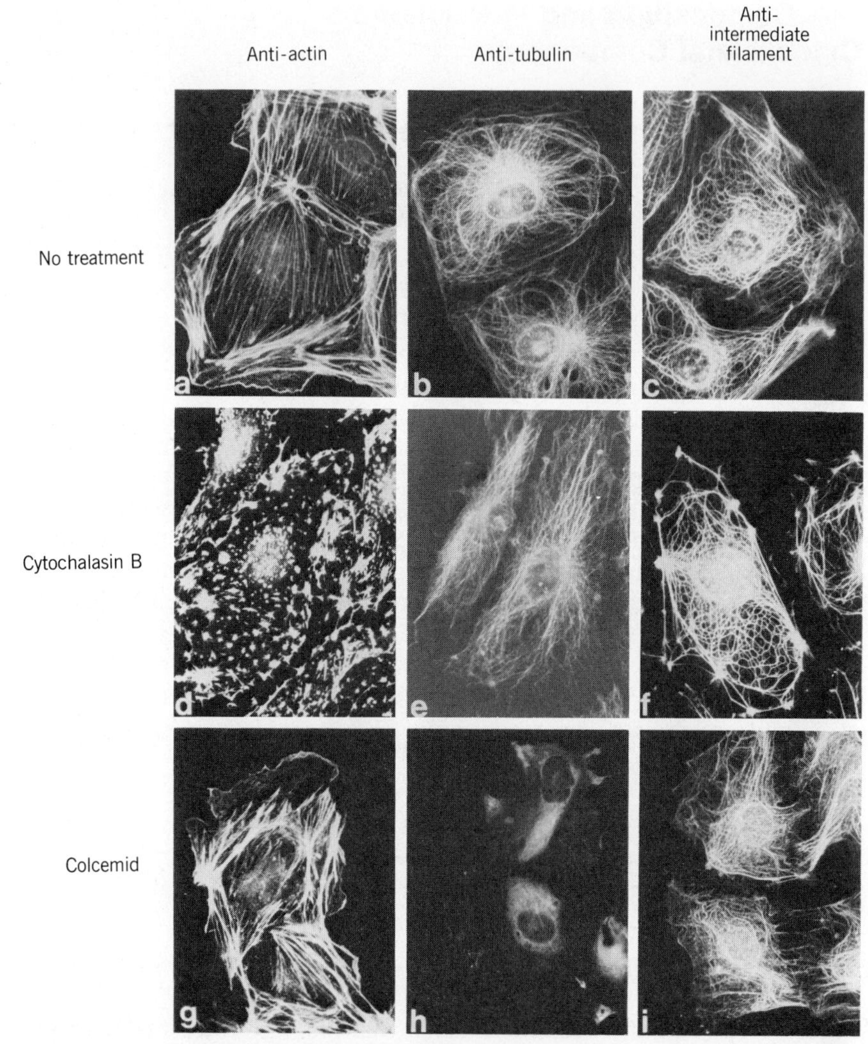

FIGURE 19.27 Immunofluorescent results of cells before any treatment (*a–c*), after treatment with cytochalasin B (*d–f*), and after treatment with Colcemid (*g–i*). The specificity of antibodies used is indicated at the top of each column. Cytochalasin B disrupts the microfilament system and Colcemid destroys microtubules. (Courtesy of Dr. Mary Osborn.)

The Microtrabecular Lattice

Yet another cytoskeletal component discovered in recent years must be added to the other three fibers to make the picture complete. These are thin filaments, 2 to 3 nm in diameter and 30 to 300 nm long, that act as linkers between the other major fiber systems. This network of filaments has been called the microtrabecular lattice by Keith Porter and co-workers, who were the first to enter into a serious study of its properties.[22]

To gain an appreciation for the relation between the microtrabeculae and the other cytoskeletal fibers, let us look first at a model and then at some supporting electron micrographs.

Figure 19.28 illustrates the relation between the components of the cytoskeleton as determined from extensive HVEM studies. The micro-

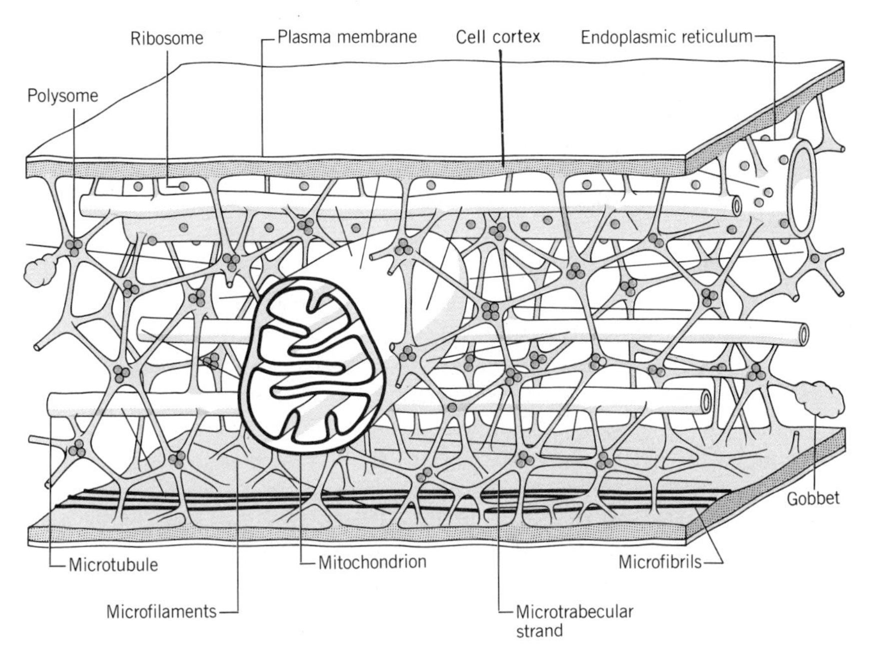

FIGURE 19.28 Model of the microtrabecular lattice. The lattice material is depicted here as being continuous with the cell cortex, surfaces of the endoplasmic reticulum, microtubules, and microfilaments. Ribosomes are found bound to the junctions of the microtrabeculae. The lattice is thought to effectively hold all organelles in place except, possibly, mitochondria. (Based on a diagram kindly provided by Dr. Keith R. Porter.)

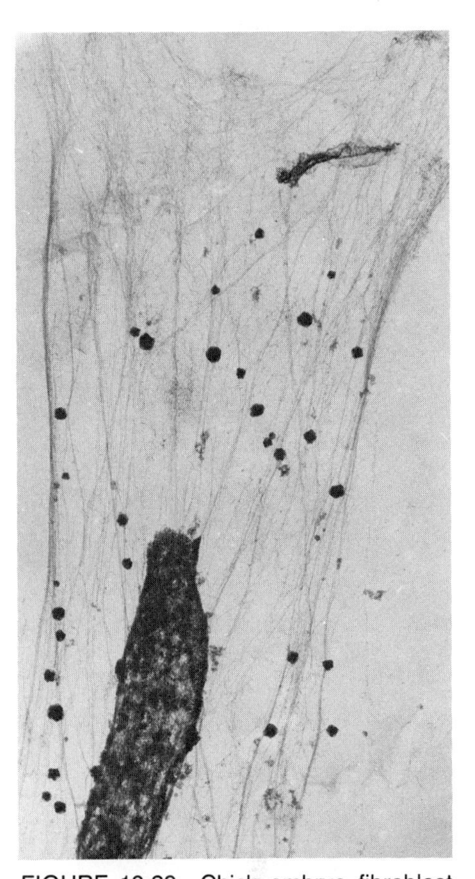

FIGURE 19.29 Chick embryo fibroblast extracted with Sarkosyl. Only actin filaments remain in the cell. They pass both over and under the remnant of the nucleus. Dense granules appear to be associated with the actin cytoskeleton. $\times 6260$. (Courtesy of Dr. Manfred Schliwa.)

trabecular filaments suspend all the other internal organelles and fibers in their places. Note the bridges formed by the microtrabeculae between the cell cortex, a layer adjacent to the plasma membrane, and the cell interior. These thin filaments interact not only with membrane-bound organelles but also with the fiber systems. They even hold clusters of ribosomes in suspension. Using HVEM and a carefully developed extraction procedure, Manfred Schliwa and Johnathan van Blerkom have been able to demonstrate with remarkable clarity the structural relationship between the microtrabecular lattice and the other major fiber systems.[23]

When cells are extracted with the ionic detergent Sarkosyl in the presence of F-actin-stabilizing phalloidin, only actin filaments remain within cells (see Figure 19.29). The filaments are oriented with the long axis of the cell and extend both over and under the nucleus. Ruffles, which appear at points at the cell edge, contain extensive networks of actin filaments.

All filaments except the intermediate filaments can be removed from the cytoskeleton by successive extraction with buffers of low and high ionic strength. The webbed pattern characteristic of intermediate filaments can be seen in the electron micrograph of Figure 19.30. Bundles of filaments are not seen in this particular cell line (BSC-1, from African green monkey kidney). However, in other cell lines bundles may be quite prominent.

Figures 19.29 and 19.30 establish the presence of F-actin and intermediate filaments throughout the cell interior. Microtubules, which are

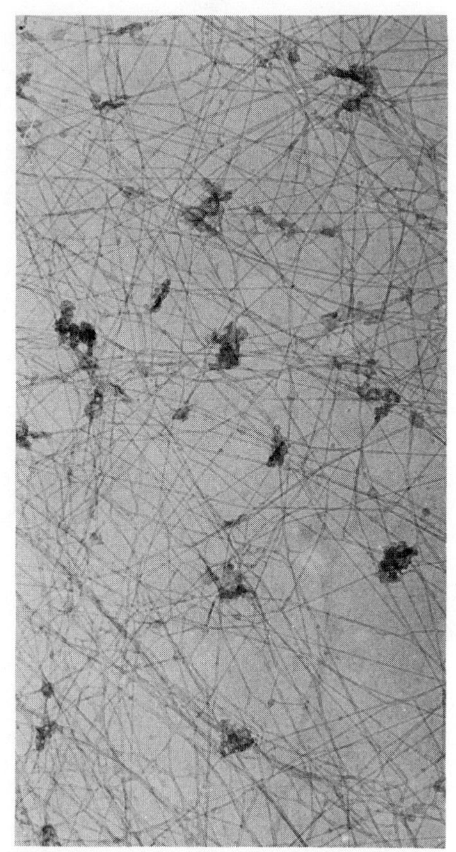

FIGURE 19.30 Monkey kidney cell extracted with low and high ionic strength buffers. Remaining are the intermediate filaments, which criss-cross the cell interior. ×34,000. (Courtesy of Dr. Manfred Schliwa.)

quite easy to distinguish from the rest of the fiber families because of their larger size, can also be seen to stretch throughout the cell interior.

Differential extraction has the disadvantage of showing only one type of fiber arrangement. It does not demonstrate the linkages that exist between fibers. However, when actin filaments are decorated with heavy meromyosin S1, the different fiber types can be seen together in the cell and can be distinguished by their characteristic dimensions. Intermediate filaments, actin filaments, microtrabeculae, and microtubules all can be distinguished in Figure 19.31, an electron micrograph that depicts the structure of the complete cytoskeleton.

An analysis of electron micrographs like those in this section has shown that the microtrabeculae are true ultrastructural linkers between the forming end-to-side contacts with other components of the cytoskeleton. Both like (e.g., actin–actin) and unlike (e.g., actin–intermediate filament; intermediate filament–microtubule) linkages have been seen. They represent a small portion of the mass of the cytoskeleton but may play a very important role in tying the other units together. The chemical nature of the microtrabeculae is presently unknown.

The Integrating Role of Actin in the Cytoskeleton

Electron microscopic studies such as those just discussed have also revealed that actin is a prominent bridging element between members of the cytoskeleton. Actin forms end-to-side contacts with all the other fiber classes as well as with itself to form Y-shaped branches.

A type of actin linkage that appears to be particularly important is that which occurs between actin and microtubules. Normally, microtubules are straight cylinders of tubulin with no tendency to bend on their own *in vitro*. When they are generated from the microtubule organizing center of the cell, they grow out radially, giving the center its aster appearance. However, as we have seen in micrographs, they bend and twist throughout the cell when they are integrated into the cytoskeleton.

Schliwa and van Blerkom have concluded that microtubules bend by interaction with other components of the cytoskeleton. Furthermore, actin may be responsible for creating the bends by placing tension on microtubules at their point of mutual contact. An electron micrograph of an actin-stressed microtubule is presented in Figure 19.32.

The Functions of the Cytoskeleton

Given the composition and structure of the cytoskeleton, it becomes apparent that in general it has two major functions: it maintains cell shape and provides for cell movement.

Except that microtubules and other filaments are splayed and interconnected throughout the cytoplasm, cells would tend to assume a more thermodynamically favorable spherical conformation. But very few cells are spherical. Rather, they take on shapes reflecting the orientation of the underlying family of fibers, as a tent assumes the shape of its supporting network of aluminum tubing.

Since the cytoskeleton is a network of dynamic fibers, it may change shape. The result is cell movement. Two types of movement are exhibited by cytoskeletal changes, but they are not necessarily mutually exclusive.

One type is a movement of the entire cell. The second type is movement within the cell interior.

During movement certain components of the cytoskeleton may have predominant roles. We have seen, for example, that actin filaments appear to be largely responsible for amoeboid movements. Microtubules, on the other hand, appear to be responsible for the movements of chromosomes during anaphase in mitosis. It is probably safe to state that any eucaryotic cell movement, either internal or external, is mediated by some component of the cytoskeleton.

We will discuss only a couple of examples of movements that are presently being investigated as cytoskeletal events. However, the reader should realize that any movement, whether amoeboid, endocytic, or cytoplasmic streaming, is based on the dynamics of the internal fiber system of the cell.

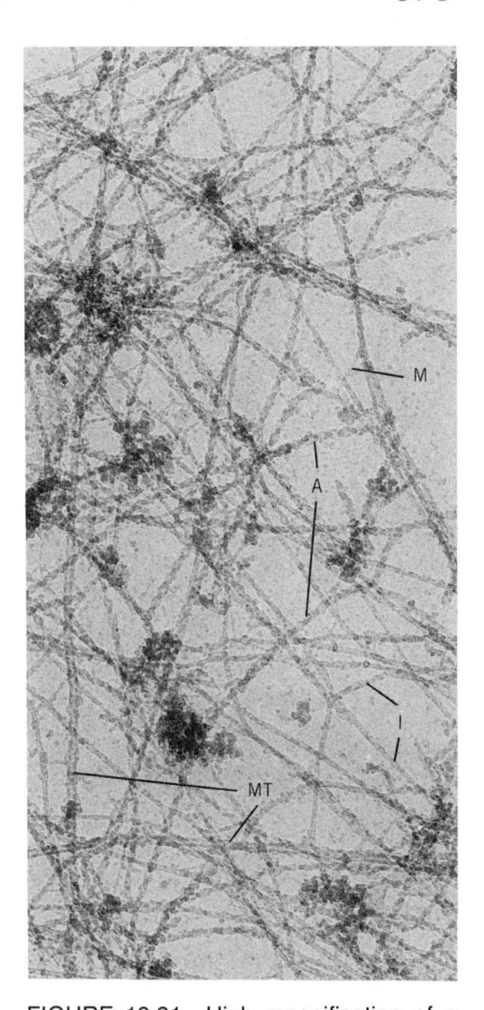

FIGURE 19.31 High magnification of a monkey kidney cell labeled with heavy meromyosin S1. Intermediate filaments are not decorated, hence smooth (I), and actin filaments are decorated (A). Microtrabeculae (M) act as linkers between other components, and microtubules (MT) are apparent. ×80,000. (Courtesy of Dr. Manfred Schliwa.)

Axoplasmic Transport

The transport of materials through the cytoplasm of the nerve cell is one of the most dramatic examples of internal cell movements. The nerve cell, or neuron, consists of a cell body from which the axon, a long cytoplasmic extension, reaches toward and makes contact with some type of receptor. The cytoplasm of the axon, termed axoplasm, is a high traffic zone for materials that are transported both away from and toward the cell body.

Two types of transport occur within the axon. One is slow and unidirectional: material moves from the cell body toward the nerve fiber terminals at a rate of only a few millimeters per day. This slow, unidirectional movement is called *axoplasmic flow*.

The second type of movement is rapid and bidirectional. Materials may be moved away from the cell body (orthograde movement) or toward the cell body (retrograte movement) at rates of 100 to 700 mm per day. This type of transport is called *axonal transport*.

Axoplasmic Flow. Axoplasmic flow and axonal transport are fundamentally different internal movements involving different mechanisms and generating different accomplishments.[24] Axoplasmic flow is a movement of bulk axoplasm along the length of the axon.

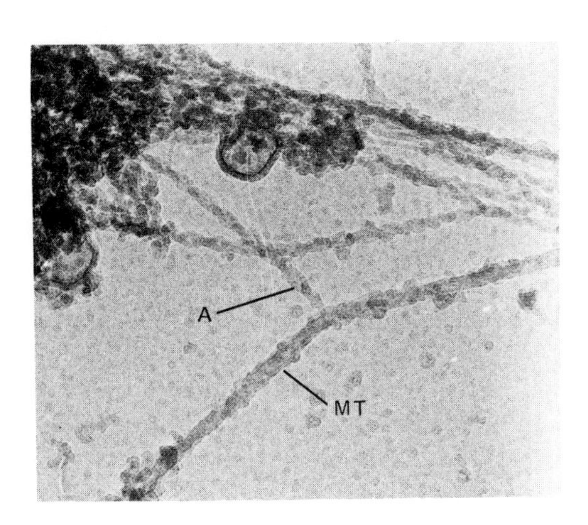

FIGURE 19.32 High magnification of an actin-stressed microtubule. The actin filament (A) attaches by its end to the side of a microtubule (MT) causing it to kink; S1 arrowheads on the actin point toward the site of contact. ×120,000. (Courtesy of Dr. Manfred Schliwa.)

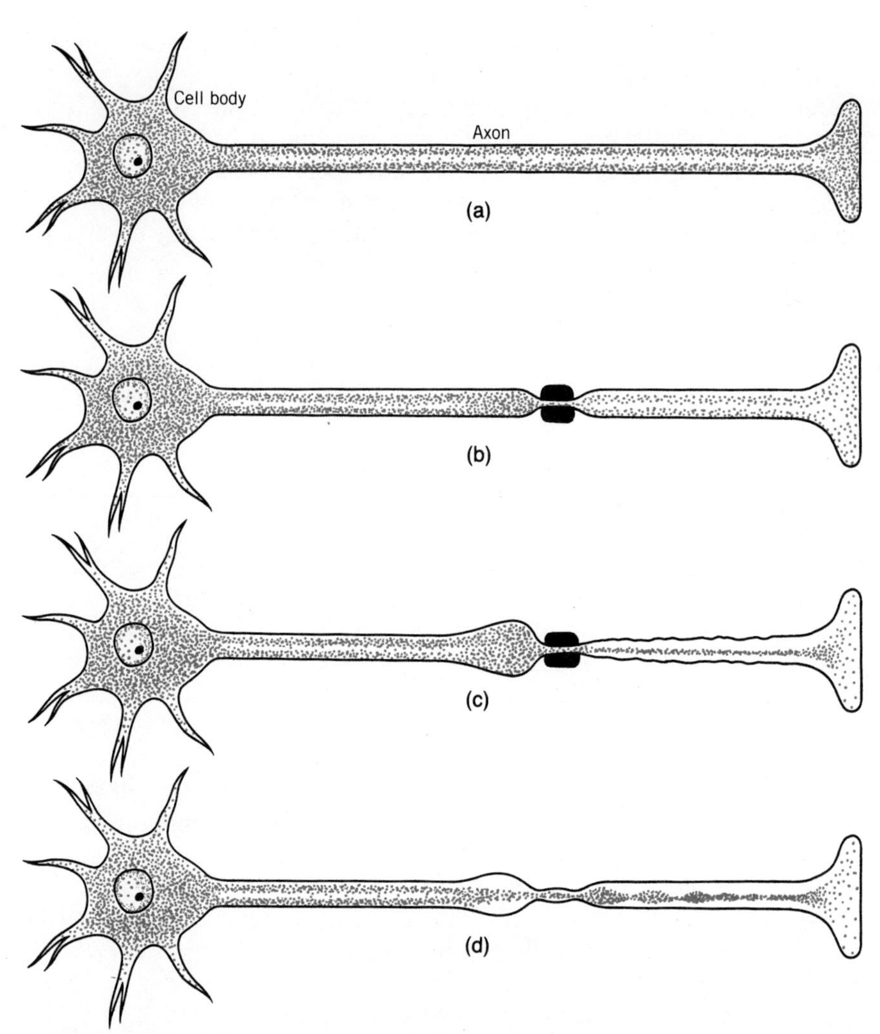

FIGURE 19.33 Schematization of constriction experiment conducted by Paul Weiss. A mature neuron (*a*) was subjected to a cuff on its axon (*b*). After several weeks the axon on the side of the cuff proximal to the cell body swelled, whereas on the other side it atrophied (*c*). This was due to a damming up of axoplasm. Upon removal of the cuff, normal flow and morphology were gradually restored (*d*).

Paul Weiss and his co-workers studied this slow transport in the late 1940s by constricting branches of the sciatic nerve *in vivo* and noting morphological changes on both sides of the constriction site with time (Figure 19.33). Within a few weeks, shape changes take place above and below the constriction. Above the constriction the axon distends; below it deterioration sets in.

An analysis of axoplasm reveals that about 80% of the dry weight of axoplasm is protein, and 20% of the protein is cytoskeletal, namely microtubules, microfilaments, and neurofilaments (intermediate filaments). During axoplasmic flow, 80% of the proteins that move, as determined by Raymond Lasek and Masanori Kurokawa, are components of the cytoskeleton. These investigators injected radioactive amino acids into

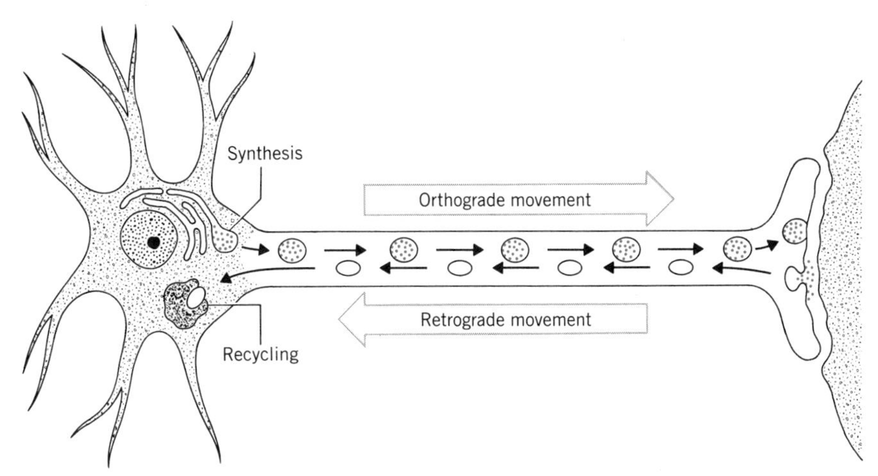

FIGURE 19.34 The bidirectional movement of materials during axonal transport. Material moves rapidly from the cell body to the synapse and there functions in transmission to a receptor. Membrane vesicles return, again by rapid transport, to the cell body.

dorsal root ganglia and followed the appearance of labeled cytoskeletal elements by gel electrophoresis with time along the length of the axon.

This axoplasmic flow appears to represent a movement of the *entire cytoskeleton* outward from the cell body, a movement that is presumed to have two functions. When neurons are damaged, this movement appears to be a mechanism of nerve regeneration. The cytoskeleton moves the regenerating axon in the direction of its receptor site until contact with another nerve or muscle is made. In an intact, healthy neuron, on the other hand, axoplasmic flow is apparently a mechanism to continuously renew and replace axoplasm.

The mechanism by which the cytoskeleton advances is not known. It appears to move more as a unit than as free subunits. Certainly the basis of its movement lies in the fact that it is a dynamic structure made up of contractile and polymerizing–depolymerizing protein systems.

Axonal Transport. Axonal transport is a rapid, bidirectional movement of materials along the axon (Figure 19.34). Membrane-bound vesicles of several kinds form in the cell body and move rapidly toward the synapse. There, by means of exocytosis, the contents of the vesicles are delivered into the synapse to carry out the transmitter activities in that region.

Movement in the direction of the synapse, termed orthograde movement, is only one part of the transport picture. Vesicles are generated at the synapse surface and return (retrograde movement) to the cell body to complete the cycle of transport.

The type of material moved in orthograde transport consists of protein, sulfated macromolecules, and calcium. The membranes of the vesicles resemble those of the smooth endoplasmic reticulum. It is thought that certain of the vesicles are used to maintain the plasma membrane of the axon while others carry transmitter substances to the synapse.

Retrograde movement begins with endocytosis at the presynaptic membrane. It is not certain whether the membrane patches that are retrieved

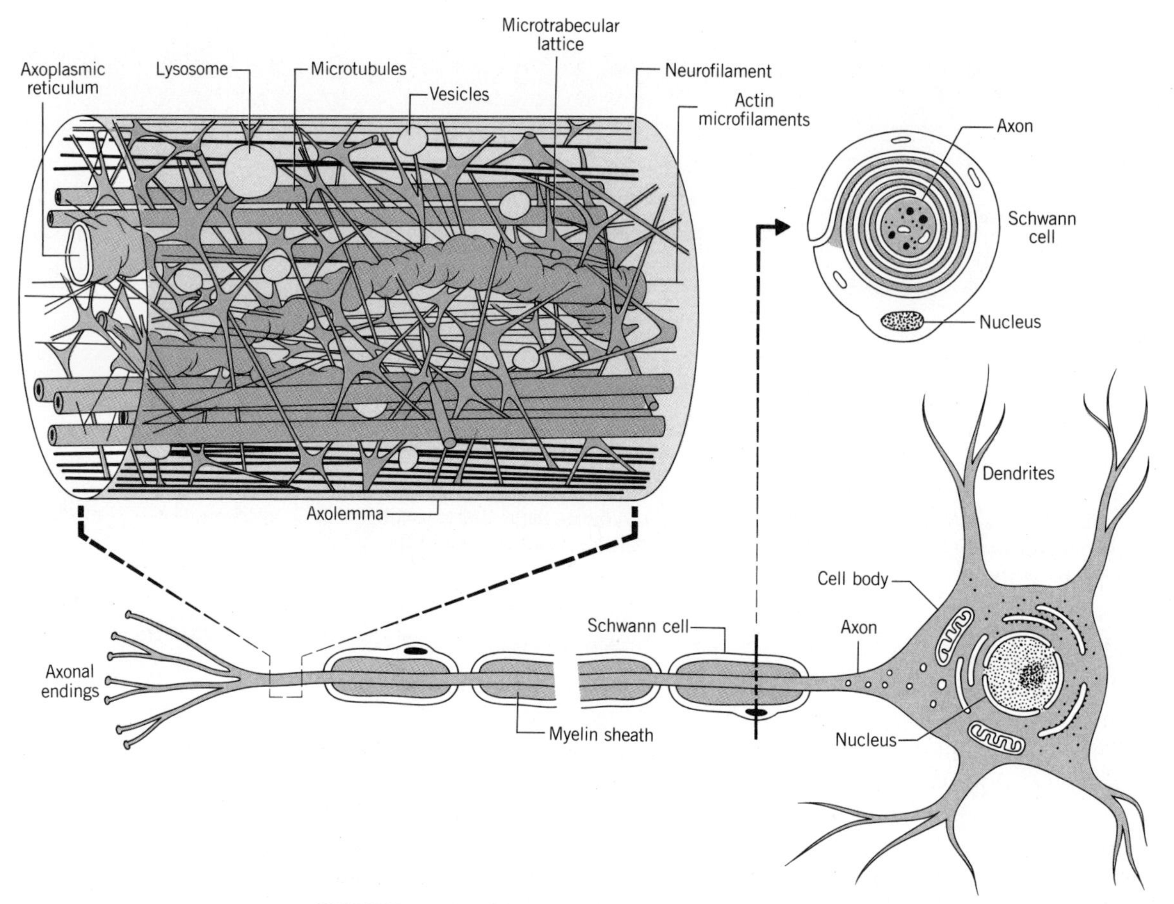

FIGURE 19.35 Model of the nerve cell emphasizing the structural detail of the axoplasm.

in this way are randomly selected or the same membranes involved in orthograde transport are returned to the cell body. The latter could occur if exocytosis were followed immediately by endocytosis of an incompletely emptied vesicle.

In any event, one of the characteristics of retrograde vesicles is association with lysosomes. The function of retrograde movement, therefore, appears to be returning and recycling the membranes and any materials picked up from the synapse.

Characteristics and Mechanisms of Axonal Transport. The characteristics of axonal transport cannot be accounted for by growth of the axoplasm, as is the case for axoplasmic flow, or by simple diffusion. The vesicles move by discontinuous jumps, independently of one another. The movement is temperature dependent, and once the vesicles are generated in the cell body and delivered to the axon, they continue to move down the axon independent of the cell body, provided there is a supply of energy.

Two types of hypothesis have been put forth to explain axonal transport. The type proposed first suggested that vesicles are moved passively along the axon either by a peristaltic contraction of axoplasmic walls or by ciliary movements. There is little direct experimental support for this hypothesis.

High voltage electron microscopy has shown the axon to consist of a maze of microtubules and microfilaments, along with the interconnecting elements of the microtrabecular lattice. Figure 19.35 models this complex system. Given this internal structure, a second hypothesis seems more plausible, namely, that microtrabecular cross-linkages make transient contact with the membrane-bound vesicles. If a cross-link makes contact with a vesicle with one end and maintains contact with a stationary component of the axoplasm (such as a microtubule) with its other end, it may then contract and pull the vesicle ahead. According to this hypothesis, microtubules function as a track on which the vesicles move, propelled by a dynamic interaction with cross-bridging elements. An electron micrograph demonstrating cross-bridging elements to axonal vesicles is presented in Figure 19.36.

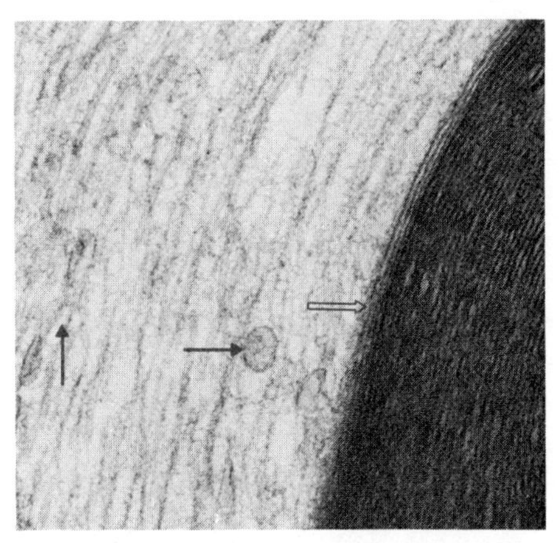

FIGURE 19.36 High voltage electron micrograph of a longitudinal thick section through a myelinated peripheral nerve axon. Careful scrutiny reveals linkages between the lattice and the plasma membrane (open arrow) and membrane-bound vesicles of the smooth endoplasmic reticulum (solid arrows). (Courtesy of Dr. Mark H. Ellisman.)

At the moment, the factors that dictate the polarity of movement are a puzzle. What determines whether movement should be orthograde or retrograde?

There is evidence that orthograde and retrograde vesicles are morphologically different and that the polarity of microtubules within a given axon is uniform.[25] This would suggest that the basis for direction in movement must reside in the vesicle rather than in the axoplasmic superstructure. Different vesicles may recognize different components of the axoplasm (e.g., microtubules or neurofilaments), and therefore movement in one direction may be guided by one type of filament while another type may be responsible for movement in the reverse direction.

These matters are, of course, speculative. Therefore some most interesting results are certain to be generated in this area of research within the next few years. Answers to the mechanisms of axoplasmic transport will be helpful in understanding intracellular movements in general.

Pigment Dispersion in Chromatophores

Chromatophores are pigment-containing cells that are common in certain amphibians, fishes, and reptiles. The distribution of pigment granules within the cells is under hormonal and nervous control to enable the organism to control the intensity of its color an even the color shade.

The distribution of pigment granules is generally in a radial pattern from the cell center, as is also true for the arrangement of microtubules. This type of distribution is obvious in the micrographs of Figure 19.37.

High voltage electron micrographs show that the microtubules run parallel to the linear alignment of pigment granules (Figure 19.38). The granules are associated with the microtrabecular lattice, which may link them together and to the microtubules.

In the presence of ATP the granules are fully dispersed throughout the cell, and the microtrabeculae appear to be fully extended. Dispersion is relatively slow, taking several seconds for completion. In the dispersed state the system appears to store potential energy that is converted to kinetic energy when ATP is removed. The granules then move toward the cell center much more rapidly than they were dispersed.

As the granules aggregate, they move along the tracks defined by the radially positioned microtubules. The microtrabeculae shorten and sever their connections with the granules.

The molecular mechanism underlying pigment dispersion and aggregation is not yet worked out, but it is apparent that elements of the

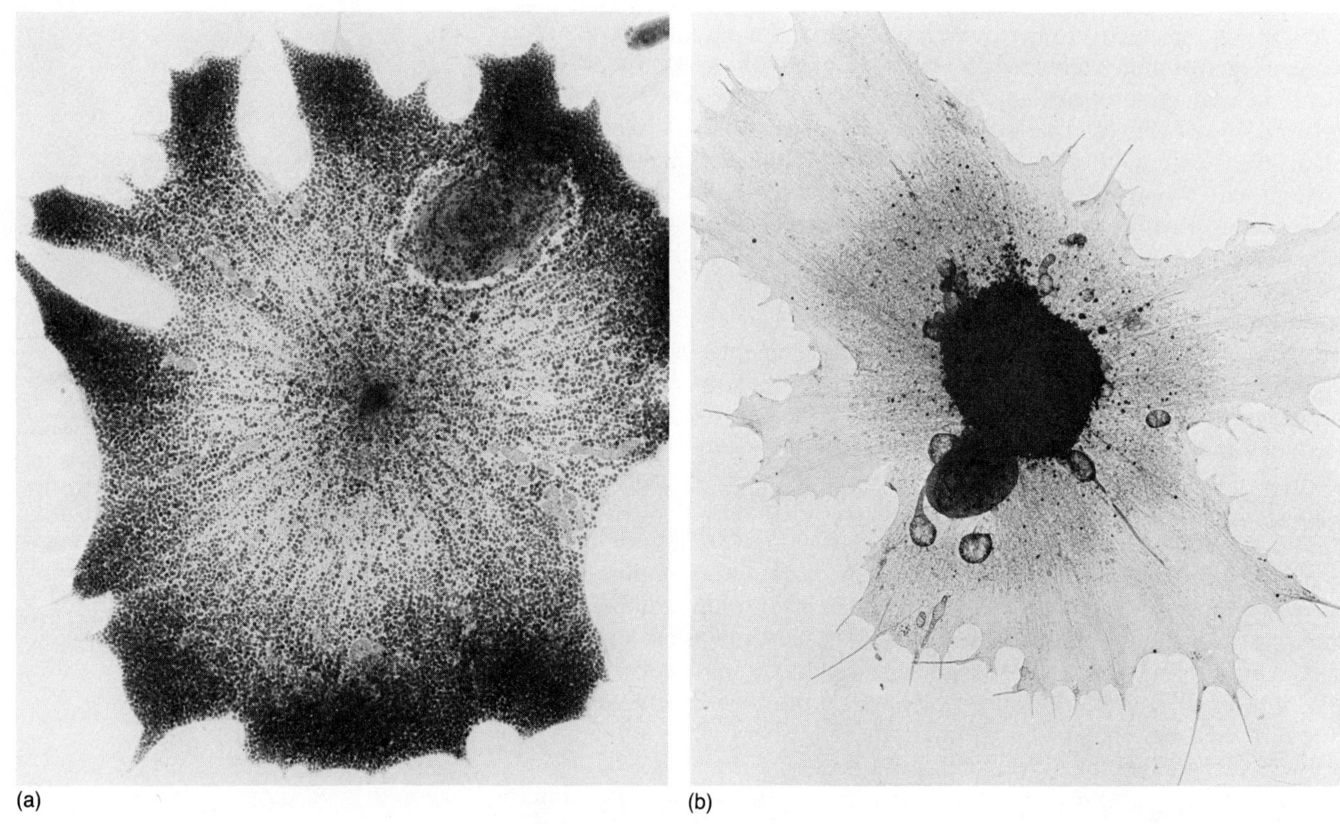

FIGURE 19.37 High voltage electron micrograph of whole cells of the erythrophore, or red pigment cell, of the squirrelfish. Pigment granules are dispersed radially from the cell center (*a*) along tracks that would be identical to those of microtubules. Color is diminished in these cells when the pigment granules aggregate (*b*) into a clump at the center of the cell. The large dark shape in each micrograph is the cell nucleus. (Courtesy of Dr. Keith R. Porter.)

cytoskeleton are involved. Microtubules form the tracks along which the granules move. Microtrabeculae shorten and lengthen in response to ATP and make contact with the granules. The ATP dependence demonstrated by this system suggests that contractile proteins, such as actin, have an important role in granule movement. The possibility therefore exists that microtrabeculae have actin cores that form the molecular basis for this dynamic cell activity.

Summary

One of the most obvious motility behaviors of cells is that which results from the beating of cilia and flagella. The basis of this beating is described by the sliding filament model. Microtubule doublets arranged circumferentially in the axoneme slide past one another by virtue of dynein arm movements. Dynein, activated by ATP, extends from the A subfiber and makes contact with the B subfiber of an adjacent doublet. Force is generated when dynein arms move. The movement of sliding is converted to bending by virtue of radial spokes that bridge each outer doublet to the inner pair of microtubules.

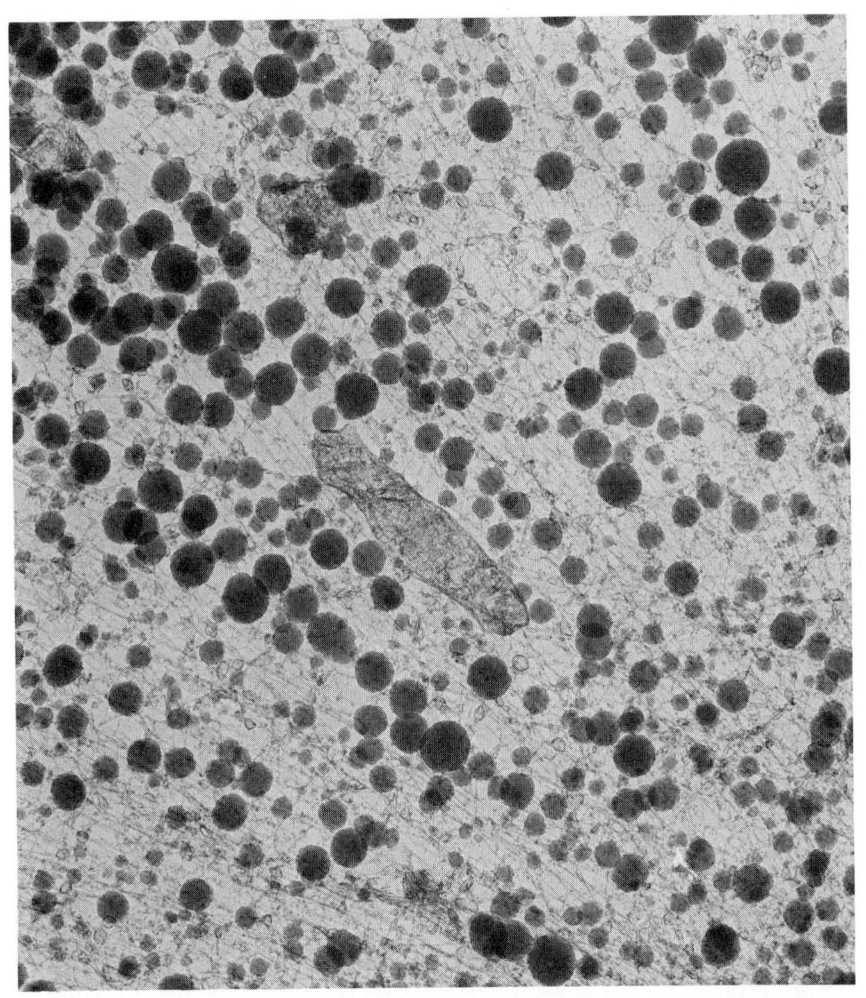

FIGURE 19.38 Spherical pigment granules embedded in the microtrabecular lattice. Microtubules extend through this portion of the cell from lower right to upper left. The granules appear to be coated with material of the lattice and on close examination can be seen to be connected to the microtrabeculae. ×29,000. (Courtesy of Dr. Keith R. Porter.)

The wave that is generated by sliding is propagated down the organelle from base to tip, with the cell generally moving in a direction opposite from that of wave propagation.

In cilia, the beat pattern has two phases: an effective stroke and a recovery stroke.

Microvilli are membrane-bound projections on certain surface cells that have actin filaments in their cores. The actin filaments from adjacent microvilli extend into a region called the terminal web, where they are cross-bridged by myosin. Several other proteins are affiliated with the actin filament core. Their roles within this surface projection are presently unsolved puzzles.

One type of intracellular cytoplasmic streaming is that which is related to and forms the basis of amoeboid movements. Several theories have emerged to explain amoeboid movement. All propose that movement is related to properties and activities of polymer systems, most specifically actin. Both sol–gel processes and contractile reactions are proposed to lie at the basis of movement.

The mitotic apparatus is a microtubule-containing structure whose activities bring about the concerted movement of chromosomes to daughter cells in mitosis. Three major theories of chromosome movements have been advanced.

The dynamic equilibrium model proposes that kinetochore microtubules shorten by depolymerization reactions and that nonkinetochore microtubules lengthen by the addition of subunits to growing ends. There is thus an equilibrium between polymerized and depolymerized states of tubulin, the dynamic state of which provides the basis for chromosome movements.

The sliding microtubule model suggests that force is generated for movement when adjacent microtubules slide by one another by virtue of dynein activities. Thus nonkinetochore microtubules slide by one another in a direction that results in a lengthening of the spindle and kinetochore microtubules slide so that they are moved toward the poles.

The sliding microfilament model proposes that the basis of movement resides in contraction reactions of actin with microtubules viewed as serving a skeletal role in the mitotic apparatus. As the microtubules are pulled, they depolymerize to allow for the necessary shortening that takes place.

Microtubules within a given half-spindle appear to have uniform polarity. The growing or polymerizing ends are distal from the microtubule organizing centers.

The entire mitotic apparatus in certain mammalian cells is surrounded by a cage of intermediate filaments. This may serve to keep the material within the mitotic apparatus relatively free from cytoplasmic components in the cell during mitosis.

Cytokinesis is the separation of a dividing cell into two individuals. In plants a phragmoplast forms of membranous vesicles, which then fuse to separate the daughter cells. In animals the plasma membrane in the midsection invaginates into a cleavage furrow, which constricts to the point of generating two independent cells. Furrowing appears to have its basis in contractile activities of actin microfilaments, which form a bundle of rings around the circumference of the cell midsection.

The cell interior contains three major fiber systems: microtubules, microfilaments, and intermediate filaments. These, in turn, are linked by a dynamic network of 2 to 3-nm filaments designated the microtrabecular lattice. The entire network of filaments constitutes the cytoskeleton, which gives form to the cell and serves as the basis of cellular movements and intracellular transport.

Several types of transport are fiber based. Two examples of this in which the transport is vivid are axoplasmic transport and the dispersion of pigment granules in chromatophores.

References

1. Cilia and Flagella: Microtubule Sliding and Regulated Motion, F.D. Warner, in *Microtubules*, K. Roberts and J.S. Hyams, eds., Academic Press, New York, 1979.

2. Microvilli and Cilia: Surface Specializations of Mammalian Cells, P. Satir, in *Mammalian Cell Membranes,* vol. 2, G.A. Jamieson and D.M. Robinson, eds., Butterworths, London, 1977.
3. Studies on Cilia. III. Further Studies on the Cilium Tip and a "Sliding Filament" Model of Ciliary Motility, P. Satir, *J. Cell Biol.* 39:77(1968).
4. Structural Conformation of Ciliary Dynein Arms and the Generation of Sliding Forces in *Tetrahymena* Cilia, F.D. Warner and D.R. Mitchell, *J. Cell Biol.* 76:261(1978).
5. Radial Spokes of *Chlamydomonas* Flagella: Genetic Analysis of Assembly and Function, B. Huang, G. Piperno, Z. Ramanis, and D.J.L. Luck, *J. Cell Biol.* 88:80(1981).
6. The Structural Basis of Ciliary Bond Formation. Radial Spoke Positional Changes Accompanying Microtubule Sliding, F.D. Warner and P. Satir, *J. Cell Biol.* 63:35(1974).
7. Some Biophysical Aspects of Ciliary and Flagellar Motility, M.E.J. Holwill, *Adv. Microb. Physiol.* 16:1(1977).
8. Organization of an Actin Filament–Membrane Complex. Filament Polarity and Membrane Attachment in the Microvilli of Intestinal Epithelial Cells, M.S. Mooseker and L.G. Tilney, *J. Cell Biol.* 67:725(1975).
9. C.F.A. Pantin, *J. Mar. Biol. Assoc.* 13:24(1923).
10. Structure, Movement, Locomotion, and Stimulation in Amoeba, S.O. Mast, *J. Morphol. Physiol.* 41:347(1925).
11. A New Theory of Ameboid Movement and Protoplasmic Streaming, R.D. Allen, *Exp. Cell Res.* 8(Suppl.):17(1961).
12. The Solation–Contraction Coupling Hypothesis of Cell Movements, D.L. Taylor, S.B. Hellewell, H.W. Virgin, and J. Heiple, in *Cell Motility: Molecules and Organization,* S. Hatano, H. Ishikawa, and H. Sato, eds., University Park Press, Baltimore, 1979.
13. The Contractile Basis of Amoeboid Movement. VI. The Solation–Contraction Coupling Hypothesis, S.B. Hellewell and D.L. Taylor, *J. Cell Biol.* 83:633(1979).
14. Cell Division, J.R. McIntosh, in *Microtubules,* K. Roberts and J.S. Hyams, eds., Academic Press, New York, 1979.
15. Cell Motility by Labile Association of Molecules. The Nature of Mitotic Spindle Fibers and Their Role in Chromosomal Movement, S. Inoué and H. Sato, *J. Gen. Physiol.* 50:259(1967).
16. Mitotic Mechanism Based on Intrinsic Microtubule Behavior, R.L. Margolis, L. Wilson, and B.I. Kiefer, *Nature* 272:450(1978).
17. Model for Mitosis, J.R. McIntosh, P.K. Hepler, and D.G. van Wie, *Nature* 224:659(1969).
18. Possible Roles of Microtubules and Actin-Like Filaments During Cell Division, A. Forer, in *Cell Cycle Controls,* G.M. Padilla, I.T. Cameron, and A.M. Zimmerman, eds., Academic Press, New York, 1974.
19. Structural Polarity of Kinetochore Microtubules in PtK Cells, U. Euteneuer and J.R. McIntosh, *J. Cell Biol.* 89:338(1981).
20. Decoration of Spindle Microtubules with Dynein: Evidence for Uniform Polarity: B.R. Telzer and L.T. Haimo, *J. Cell Biol.* 89:373(1981).
21. Isolation and Partial Characterization of a Cage of Filaments That Surrounds the Mammalian Mitotic Spindle, G.W. Zieve, S.R. Heidemann, and J.R. McIntosh, *J. Cell Biol.* 87:160(1980).
22. Microtrabecular Lattice of the Cytoplasmic Ground Substance. Artifact or Reality, J.J. Wolosewick and K.R. Porter, *J. Cell Biol.* 82:114(1979).
23. Structural Interaction of Cytoskeletal Components, M. Schliwa and J. van Blerkom, *J. Cell Biol.* 90:222(1981).
24. The Transport of Substances in Nerve Cells, J.H. Schwartz, *Sci. Am.* 242:151(1980).

25. Microtrabecular Structure of the Axoplasmic Matrix: Visualization of Cross-linking Structures and Their Distribution, M.H. Ellisman and K.R. Porter, *J. Cell Biol.* 87:464(1980).

Selected Books and Articles

Books

Cell Motility, H. Stebbing and J.S. Hyams, Longman, New York, 1979.

Cell Motility: Molecules and Organization, S. Hatano, H. Ishikawa, and H. Sato, eds., University Park Press, Baltimore, 1979.

Microtubules, K. Roberts and J.S. Hyams, eds., Academic Press, New York, 1979.

Motility of Living Cells, P. Cappuccinelli, Chapman and Hall, New York, 1980.

Prokaryotic and Eukaryotic Flagella, Symposia of the Society for Experimental Biology, No. XXXV, Cambridge University Press, Cambridge, 1982.

Articles

Centrioles in the Cell Cycle. I. Epithelial Cells, I.A. Vorobjev and Yu. S. Chentsov, *J. Cell Biol.* 93:938(1982).

Cytoplasmic Structure and Contractility in Amoeboid Cells, D. Lansing Taylor and J.S. Condeelis, *Int. Rev. Cytol.* 56:57(1979).

How Cilia Move, P. Satir, *Sci. Am.* 231:44(1974).

Microtrabecular Structure of the Axoplasmic Matrix: Visualization of Cross-linking Structures and Their Distribution, M.H. Ellisman and K.R. Porter, *J. Cell Biol.* 87:464(1980).

Microvilli and Cilia: Surface Specialization of Mammalian Cells, P. Satir, in *Mammalian Cell Membranes,* vol. 2, G.A. Jamieson and D.M. Robinson, eds., Butterworths, London, 1977.

Some Biophysical Aspects of Ciliary and Flagellar Motility, M.E.J. Holwill, *Adv. Microb. Physiol.* 16:1(1977).

Structural Interaction of Cytoskeletal Components, M. Schliwa and J. van Blerkom, *J. Cell Biol.* 90:222(1981).

The Ground Substance of the Living Cell, K.R. Porter and J.B. Tucker, *Sci. Am.* 244:56(1981).

The Molecular Basis of Cell Movement, E. Lazarides and J.P. Revel, *Sci. Am.* 240:100(1979).

The Transport of Substances in Nerve Cells, J.H. Schwartz, *Sci. Am.* 242:152(1980).

APPENDIX

METHODS IN CELL BIOLOGY

Microscopy

The microscope has retained a top position as the instrument of choice for cell studies since it was first used in the 1600s, even though biochemical researchers have made enormous contributions toward our understanding of the cell by techniques that largely disrupt its structural integrity. There is something to the adage that seeing is believing, even among scientists who have developed an image of a particular cell structure through other physical or chemical means. To actually see something generally makes one a bit more confident of its existence.

Many different versions of the microscope are in use today, ranging all the way from the simple light microscope used by students early in their academic careers to the enormously sophisticated high voltage electron microscope available to only a handful of research scientists. Each instrument has its particular niche in cell studies and each also has certain limitations.

The following brief description of several types of microscopy emphasizes information that will be of help in understanding this text.

BRIGHT FIELD LIGHT MICROSCOPY

The most common microscope, used by all students of science, is the bright field light microscope. It is an extremely useful tool for observing basic sizes and shapes of stained cells and beyond that for noting the presence of nuclei and nucleoli.

The bright field microscope is quite limited in its power of resolution and in its ability to detect slight refractive index differences within the cell such as those produced by membrane-containing organelles.

The limit of resolution is defined by the equation:

$$R = \frac{0.61\lambda}{\text{NA}}$$

where λ is the wavelength of light employed and NA stands for numerical aperture, a property of the lens system.

Under the best of conditions, using an objective lens with NA of 1.40 and a violet filter (λ = 400 nm), a resolving power of about 175 nm, or 0.175 μm, can be obtained. In practice, this means that closer than this distance, the microscope cannot see as separate two objects, or that any object with a diameter smaller than 0.175 μm cannot be usefully magnified.

Since membranes have a width of 6 to 10 nm and ribosomes are 20 to 30 nm in diameter, it is easy to see why the light microscope cannot be used to study these organelles. Bacteria, possessing diameters in the vicinity of 1 μm or slightly less, as well as mitochondria, approximately the same size, are about at the practical limit of usefulness for the light microscope.

PHASE CONTRAST LIGHT MICROSCOPY

Living or unstained cells are most often studied by types of interference microscopes. Both phase contrast and Nomarski microscopy are examples of interference microscopy.

Basically, interference microscopy enhances the contrast between cells and their environments and between internal organelles and their surroundings with a specially designed optical system that can alter the phase of light waves.

One can imagine that two light waves, when passing side by side through a cell, will emerge in phase if they have encountered the same refractive index and thickness of medium:

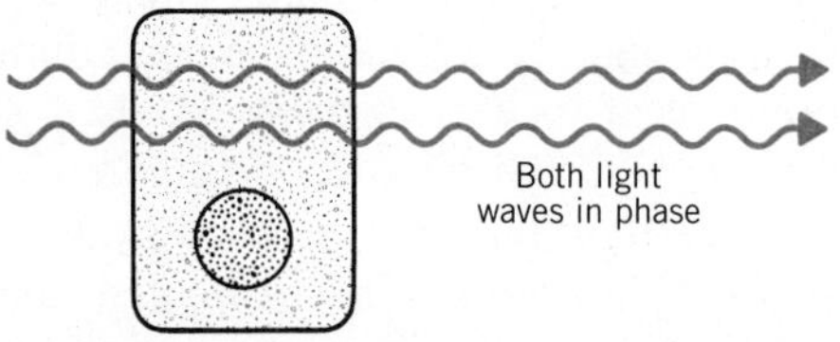

If, however, one light wave encounters an organelle not passed through by the other, it will be retarded, and will emerge slightly out of phase:

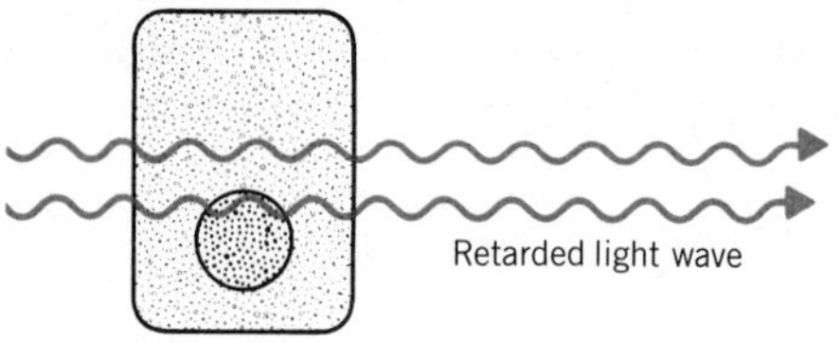

Normally the phase shifts are on the order of 0.25 of a wavelength (0.25 λ), which results in partial destructive interference of the light (Figure 1). The phase contrast microscope is equipped to selectively retard the out-of-phase light wave with respect to the other another 0.25λ so that a condition of complete destructive interference is achieved. When this occurs, the two waves combine to cancel each other out. The visual effect is darkness.

Thus the contrast at the edge of a cell or at organelle boundaries is enhanced from that seen with ordinary light microscopy, enabling one to see mitochondria, mitotic chromosomes, nucleoli, and other organelles quite clearly in living cells (pp. 394, 545).

Nomarski optical systems make use of the rate of change of refractive index in the cell. Thus, where the rate of change is greatest, such as at the edges of cells or their organelles, the contrast is greatest. The visual effect of this is three-dimensional, permitting the cell and its organelles to take on the appearance of depth (see Figure 12.5).

POLARIZATION MICROSCOPY

Polarization microscopy is a special adaptation of ordinary light microscopy in which the incident light is controlled by polarizing filters.

Light, before it impinges on the specimen, is passed through a polarizing filter. Thus the light that strikes the specimen is plane polarized; that is, the light waves are vibrating in only one plane. If the specimen is made up of material that is randomly oriented, the plane-polarized light will pass through it and emerge as it entered—still polarized in the same plane. If this light is then passed through a second filter that has polarizing properties perpendicular to the first, the light will be stopped and nothing will be seen in the microscope.

If, on the other hand, the specimen possesses structures that are aligned in some manner, plane-polarized light will be altered and a portion of it will pass through the second polarizing filter and will appear bright in the microscope.

With this technique oriented structures can be observed within a cell even if the structures themselves are beyond the resolving power of the microscope. The mitotic spindle, for example, is made up of microtubules that cannot be resolved with the light microscope, but because they are oriented from pole to pole in the cell, a brightness can be seen that has an outline of the spindle (p. 656).

In the same manner certain plant cell wall components are amenable to study by polarization light microscopy because of their ordered orientation.

ELECTRON MICROSCOPY

The development of the electron microscope has provided the cell biologist with one of the most important tools of the trade. Two major types of electron microscope are employed in biological studies, the transmission electron microscope (TEM) and the scanning electron microscope (SEM).

Transmission Electron Microscopy

The TEM is the most commonly used electron microscope. Its basic features are compared to those of a light microscope in Figure 2. The energy source is an electron beam which emanates from a tungsten filament functioning as a cathode in the instrument. Electrons are accelerated by a high voltage applied between the cathode and an anode such that they are projected down a column from which interfering air has been removed. The beam of electrons is controlled by electromagnets so located in the column that the beam is properly focused on the specimen placed in its path.

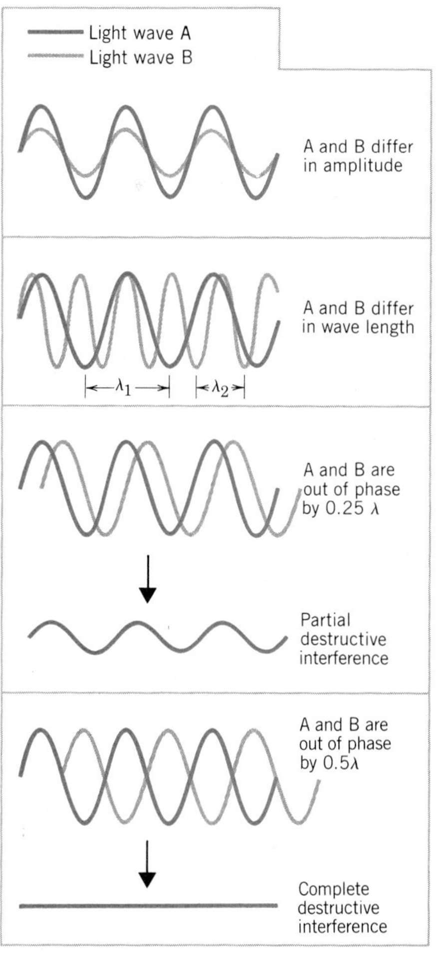

FIGURE 1 Properties of light waves that result in partial and complete destructive interference.

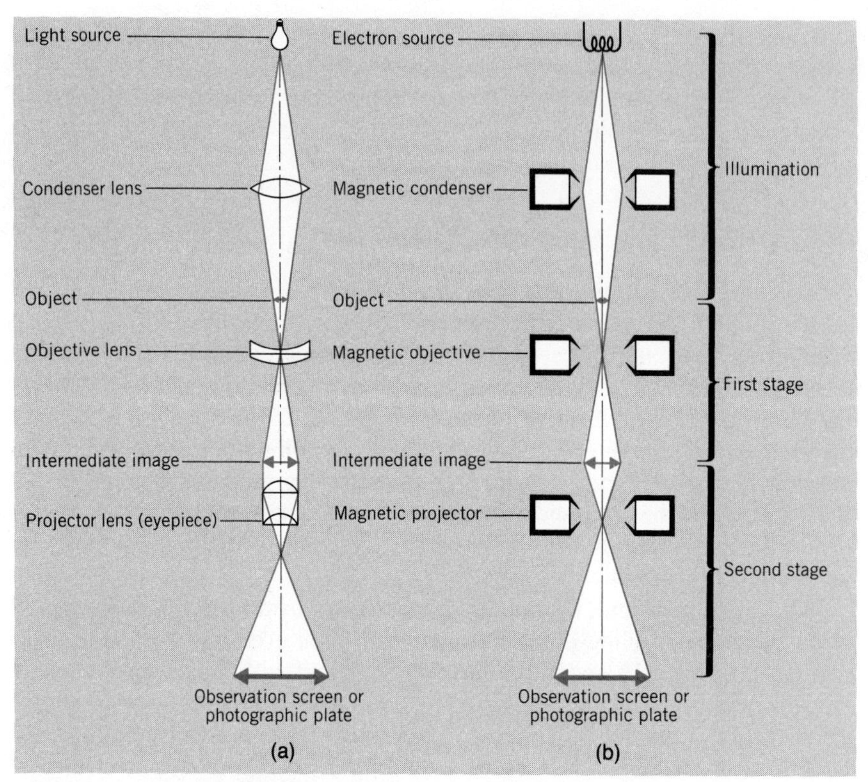

FIGURE 2 Basic features of the electron microscope (*b*) as compared to the ordinary light microscope (*a*).

Electrons that are transmitted through the specimen strike a viewing screen, which is coated with a fluorescent material. The electrons excite this material to emit visible light, which can be viewed directly by the operator. Electrons that are not transmitted by the specimen because they encounter electron-dense regions leave correspondingly dark regions on the viewing plate. By lifting the viewing plate, the operator can capture the electron image directly on film.

The wavelength of electrons is a reflection of the speed at which they are traveling. This speed is controlled by the voltage applied to the microscope. Commonly, between 50,000 and 100,000 V is used, resulting in a resolving power that approaches 1 Å. In practice, resolving powers somewhat less than this are obtained, but limits in the vicinity of 10 Å are common.

Because of this high resolving power very high magnifications can also be achieved. Routinely, magnifications between 5000 and 50,000 diameters are obtained by the instrument directly, with values up to about 300,000 times possible given ideal conditions. These images can be photographically enlarged further to give magnifications on the order of 1×10^6.

Chemical Fixation and Staining

Biological materials cannot be viewed in a living state by electron microscopy, nor can they normally be viewed unless properly fixed and cut into extremely thin sections.

Several types of fixative are used to preserve the structures of cells and to prepare them for subsequent sectioning and staining procedures. In practice, a small sample of tissue is immersed in a solution of fixative to capture the molecular anatomy of the cell in its most native configuration.

Among the various fixatives employed, osmium tetroxide (OsO_4), potassium permanganate ($KMnO_4$), and glutaraldehyde are very commonly used. Each possesses its own peculiar advantages in fixing and preserving cell components.

Tissue that is fixed must be embedded in a resin that polymerizes to a hardened state to support the tissue as it is sectioned. The embedded tissue is then subjected to ultrathin sectioning to produce sections 90 nm or less in thickness. These are placed on a grid that has the structure of a miniature screen. The grid can be coated with an electron-transparent film if one is studying small particles such as ribosomes that otherwise would fall through the holes in the grid.

Because chemical fixing and staining may introduce artifacts into tissues, other approaches have been explored. One, quite commonly employed now, is the use of various freezing processes.

Freezing and Fracturing

Techniques have been worked out for the rapid freezing of tissues with a minimum of ice crystal formation, a phenomenon that may cause damage and distortion. The freezing can be done in liquid nitrogen (boiling point −198°C) or in a variety of other media. After freezing, and while the specimen is still cold, substitution solvents may be added to stain and prepare it for embedding and sectioning.

The technique of freeze-etching has provided valuable insight regarding the presence and nature of particles in membranes and the properties of organelle surfaces. First, a rapidly frozen tissue, held in the cold at a high vacuum, is fractured with a knife in such a way that a surface is exposed. Often the fracture bisects a membrane to reveal the membrane interior. The surface is then left exposed briefly in the vacuum, which sublimes some of the ice, thus "etching" the surface or causing the nonsublimed areas to be raised above the regions that have lost material. This surface is then shadowed with an evaporated metal and coated with a layer of evaporated carbon. The result is a shadowed replica of the cut, sublimation-etched surface. The replica is detached from the underlying tissue by dissolving the latter, picked up on a grid, and examined in the electron microscope. Figure 3 portrays some of the salient features of this process.

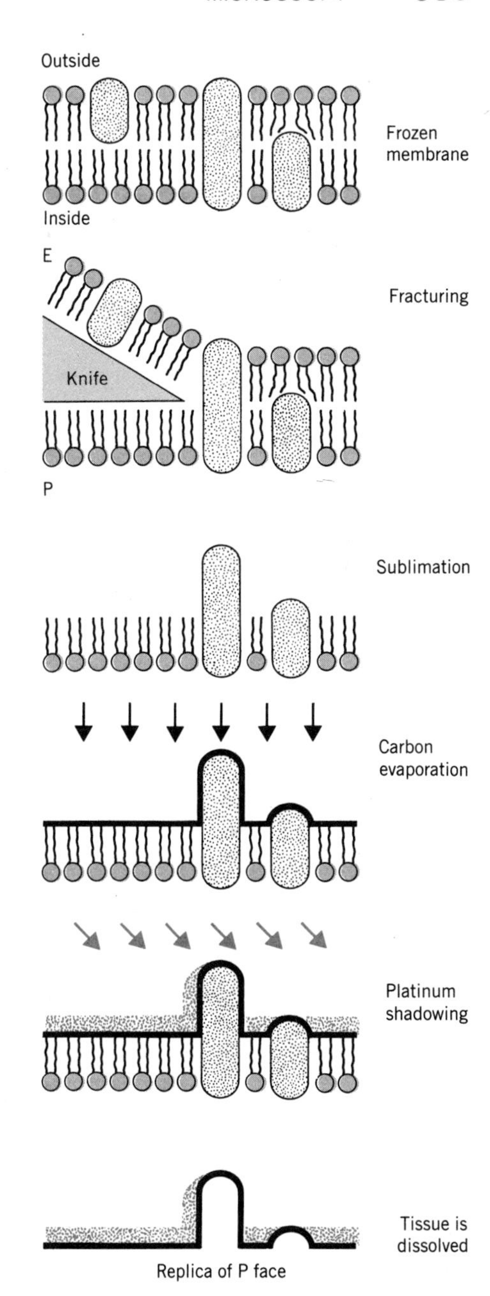

FIGURE 3 Steps involved in freeze-fracturing, etching, and making a replica of a tissue surface for observation by transmission electron microscopy.

We have referred to the technique of freeze fracturing many times in this text—for example, when discussing the structure of gap junctions (p. 172), the surface membranes of ciliary cells (p. 648), and the interior of the plasma membrane (see Figure 4.16).

Scanning Electron Microscopy

The scanning electron microscope is an instrument that is especially useful for examining the surface of a specimen. A spot of electrons is focused on the specimen and is then scanned over the surface. Some of the electrons may be reflected from the surface, or they may excite secondary electrons from the specimen. These electrons are attracted to a detector,

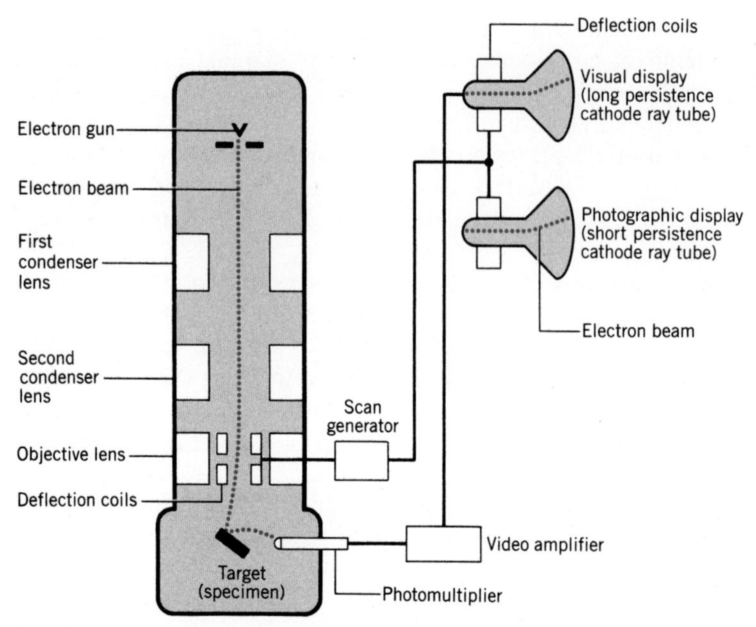

FIGURE 4 A scanning electron microscope.

which gives rise to a flash of light in a solid scintillator. This light output is amplified, and the resulting electrical signal governs the brightness of a display spot that scans synchronously with the probe spot. An image of the surface is thus obtained on a display tube; photographs of such images are called scanning electron micrographs. Figure 4 contains a diagram of the SEM.

Biological specimens are prepared by dehydrating them in liquid carbon dioxide or a similar cryogenic medium and then placing them in a vacuum evaporator. A thin film of gold or another heavy metal is condensed on the surfaces.

Resolving power in SEM is limited by the diameter of the probe spot. Resolving powers of 50 Å are obtainable, but SEM is more commonly used under conditions that do not require such high resolution.

An example of the kind of result obtainable by SEM is seen in the electron micrograph of Figure 5.14.

High Voltage Electron Microscopy

A relative newcomer to the field of microscopy is the high voltage electron microscope (HVEM). This mammoth and costly instrument employs one million volts as an energy source.

With a voltage of this magnitude, thick sections of tissue may be examined rather than the conventional thin sections. This has enabled research scientists to view the cell interior in considerable depth as well as with extremely high resolution. As a result of employing this instrument, a new level of structural sophistication has been observed in the cytosol, giving rise to the concept of the microtrabecular lattice, as discussed in Chapter 19.

Tissue and Cell Disruption

For many types of study in cell biology it is important that tissues or cells remain intact. A cell functions as a cell only when it is intact, and the nature of the interactions between cells specifies the properties and functions of a particular tissue.

But many studies cannot be carried out on intact systems. It is then necessary to disrupt the integrity of a tissue or cell and begin to examine its components.

Several different techniques are available to disrupt cells. The technique chosen depends on the resistance of the tissue to disruption and on the nature of the study to be carried out.

In general, animal tissues are easier to disrupt than plant tissues. Liver is especially amenable to disruption and, partly for this reason, has gained a prominent place among tissues used for investigation by the cell biologist and biochemist.

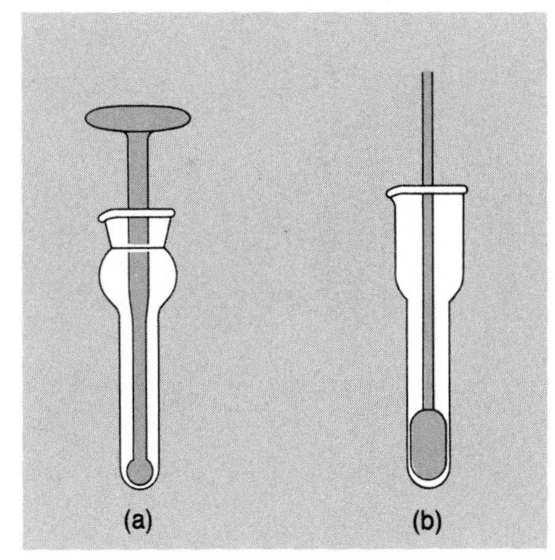

FIGURE 5 Tissue homogenizers. (*a*) Dounce and (*b*) Potter-Elvehjem.

TISSUE HOMOGENIZATION

One of the most frequently employed ways to disrupt animal tissue, such as liver, is to subject it to the action of the tissue homogenizer. Several different types are available, but two common ones used are the Dounce and the Potter-Elvehjem homogenizers (see Figure 5). Both have glass tube-shaped mortars, which are fashioned to provide a very small clearance between their surfaces and the surface of the pestle when placed together.

The Dounce homogenizer has a glass pestle terminated by a ball that permits a clearance of 0.001 to 0.006 mm between it and the inside of the mortar. It is operated by placing small pieces of tissue in an appropriately buffered medium into the mortar and forcing the pestle down until all the tissue has been forced up past the ball. A few strokes are sufficient to completely disrupt liver tissue. This homogenizer is especially useful for studying nuclei, since they are generally not disrupted as they are squeezed between the two surfaces.

The Potter–Elvehjem homogenizer is operated in basically the same way, except that the pestle, generally made of a rounded Teflon tip mounted on a stainless steel shaft, is attached to a stirring motor. As the pestle is lowered into the tissue-containing medium, it is also turning, which has the effect of very efficiently disrupting cells.

Normally homogenization is conducted in an isotonic medium so that the particles and organelles that result are not further affected by osmotic changes. Sucrose is frequently used as an osmotic protectant at a concentration of 0.25 *M*.

Following homogenization the disrupted preparation is normally fractionated by centrifugation to procure the desired cell components (see Method III).

We have referred to tissue homogenization several times in this text, as for the preparation of microbodies (p. 224), lysosomes (p. 286), mitochondria (p. 394), nuclei (p. 477), and the microsomal fraction (p. 198).

SONIC OSCILLATION

When cells suspended in an aqueous environment are subjected to high frequency sonic oscillation, they break open. The sonic oscillation produces cavitation in the solution, which is the immediate cause of cell breakage.

This technique is most applicable to unicellular systems, such as microorganisms or red blood cells, or to tissues that are first treated to eliminate the tight interactions between cells. It is also a procedure that can be used to effectively disrupt sealed organelles, such as lysosomes and mitochondria. As discussed in this text, it has been successfully employed to shake the outer membrane from the inner compartment of mitochondria (p. 400).

A number of other techniques can be employed to disrupt tissues, such as grinding, mechanical shaking, pressure methods, and osmotic shock. We have emphasized here only those that are most pertinent to this text.

III

Centrifugation

Centrifugation was one of the early physical techniques that was used to fractionate cells and separate and purify their components. Without its development, much of what we know of the molecular anatomy of the cell would still be far beyond our reach.

The basic equipment of a centrifuge is very simple. It consists of a rotor in which samples are placed and a mechanism, usually a motor, to spin the rotor. A sample, in a spinning rotor, is subjected to an outward directed force F, the magnitude of which depends on the angular velocity in radians per second ω, and the radius of rotation r, expressed as:

$$F = \omega^2 r$$

When compared to the earth's gravitational force, the force on a sample is called the relative centrifugal force (RCF), expressed as a number times g, as $10{,}000g$. This value is derived quantitatively from:

$$\text{RCF} = \frac{\omega^2 r}{980}$$

Since ω can be expressed in revolutions per minute (rpm), one can calculate RCF knowing the radius of the sample from the point of rotation by applying:

$$\omega = \frac{\pi(\text{rpm})}{30}$$

Beyond the basics of a rotor and a drive mechanism, centrifuges can be purchased in a variety of sizes with a raft of accessories to assist the operator in accomplishing the desired task.

Although there are many ways to carry out a sedimentation experiment, we will consider them as belonging to two major types: high speed centriguation, and ultracentrifugation.

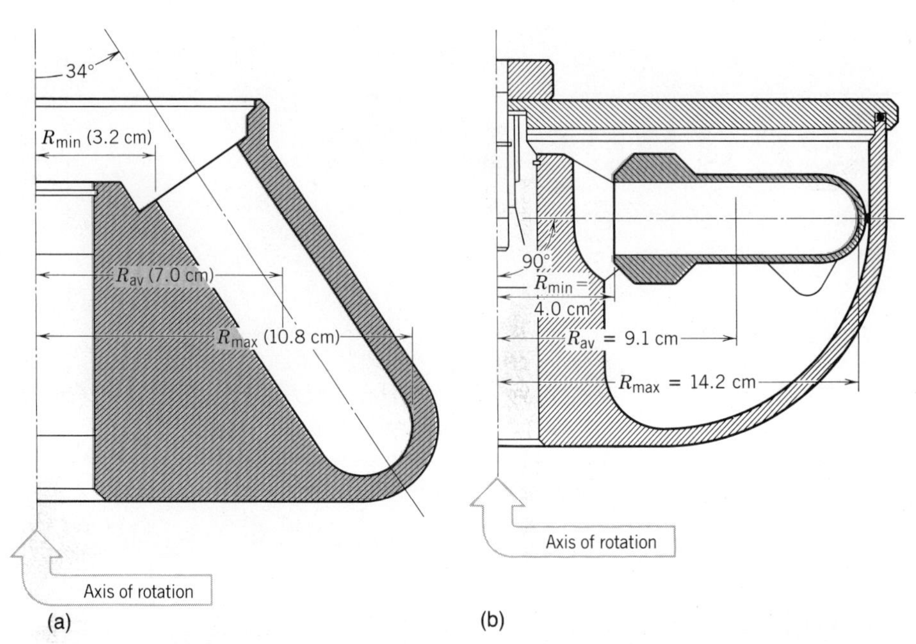

FIGURE 6 Specifications of typical fixed-angle (*a*) and swinging bucket (*b*) rotors. This swinging bucket is constructed with an outer aluminum shell to reduce the effects of wind during rotation. (Courtesy of Beckman Instruments, Inc.)

HIGH SPEED CENTRIFUGATION

Most students of science are familiar with the use of some type of preparative centrifuge. The desk-top clinical centrifuge is the simplest example of this, for it is often used to sediment down precipitates or red blood cells at speeds that reach about 3000 rpm maximally. Since the objective is to *prepare* material for additional study, the technique is called *preparative centrifugation*. The clinical centrifuge is not considered to be high speed and has only limited value in cell fractionation.

Researchers more commonly use a somewhat more sophisticated version of the clinical centrifuge for cell fractionation and sedimentation in which the temperature can be controlled, the duration of the run automatically timed and terminated, and the rotor driven at speeds in the vicinity of 20,000 rpm. These centrifuges commonly generate forces up to about 50,000g at which most microorganisms, cell debris, and cellular organelles will sediment.

By subjecting a tissue homogenate to centrifugation at a relatively low force, a *pellet* containing cell debris and nuclei is deposited at the bottom of the tube, overlaid by a *supernatant solution* of nonsedimentable material. This upper solution can be decanted and resedimented at a higher force, generating a new pellet and a second supernatant solution lacking the material of the new pellet. By repeating this procedure, a homogenate can be sedimented into fractions until only low molecular weight nonsedimentable material remains in the supernatant solution. This technique is referred to as *differential centrifugation* and is a commonly used preparative technique to separate out enriched *but not pure* fractions of cell components.

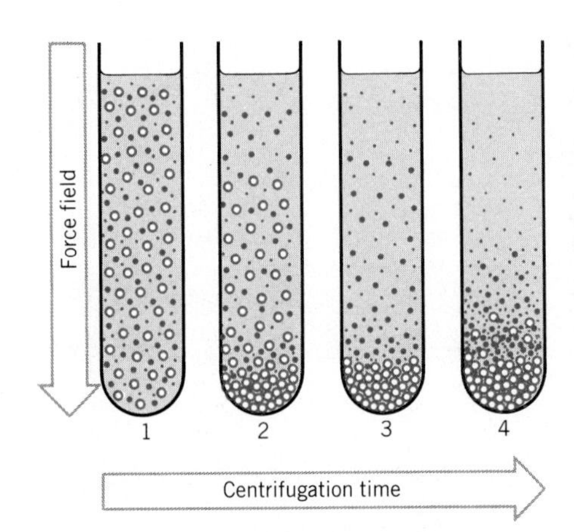

FIGURE 7 The principle of differential centrifugation, or pelleting.

Preparative centrifuges employ rotors in which the samples may be placed at a fixed angle or at an angle that changes with rotor speed. Figure 6 illustrates typical fixed-angle and swinging bucket assemblies. An advantage of the swinging bucket system is that the lines of force are parallel to the sides of the tube containing the sample, thereby eliminating the kind of turbulence and mixing that occur in a fixed-angle rotor as material scuffs along the edge of the sample wall.

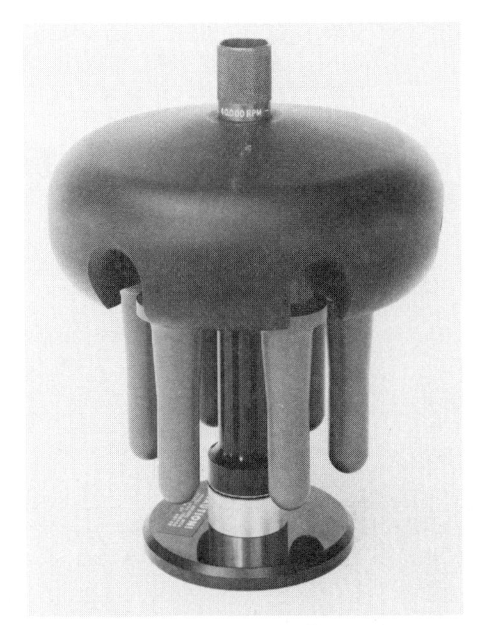

FIGURE 8 A modern swinging bucket rotor used for density gradient centrifugation. (Courtesy of Beckman Instruments, Inc.)

Ultracentrifugation

Centrifuges capable of operating at extremely high rpm (75,000) and force (500,000*g*) were first developed and put into operation in the early 1920s by The Svedberg. This development represented a major technological breakthrough, for with these high centrifugal forces it became possible to pellet lower molecular weight materials such as viruses and nucleic acids and to analyze them in gradients.

Both fixed-angle and swinging bucket rotors may be used in ultracentrifuges, depending on the type of application desired.

Differential Centrifugation

Differential centrifugation is perhaps the most common separation made in preparative ultracentrifuges. The result, as discussed earlier, is a pellet and a supernatant solution. Figure 7 illustrates the principle of pelleting as it would take place during differential centrifugation.

Density Gradient Centrifugation

The method of density gradient separation was developed in the late 1940s. It required the design and construction of a special breed of rotors with swinging buckets that could operate at extremely high centrifugal forces. Figure 8 shows a modern swinging bucket rotor.

The method involves filling the centrifuge tubes in such a way that the density of the fluid decreases in magnitude from the bottom of the tube to its top. This gradient can be formed in steps, in which case it is referred to as a discontinuous gradient, or linear gradients can be prepared (i.e., the change in density per unit of distance in the tube is constant throughout).

Many different materials can be used to form gradients, but sucrose and cesium chloride are two of the most commonly used media for these experiments.

Rate Zonal Centrifugation

One type of density gradient centrifugation is termed rate zonal. The sample is layered on top of a preformed gradient as depicted in Figure 9. Upon centrifugation, the material sediments through the gradient as a zone with a rate that depends on its density as compared to the density of the gradient. Materials with different densities will therefore separate from one another and move as separate zones through the gradient. The run is stopped before the zones reach the bottom of the tube.

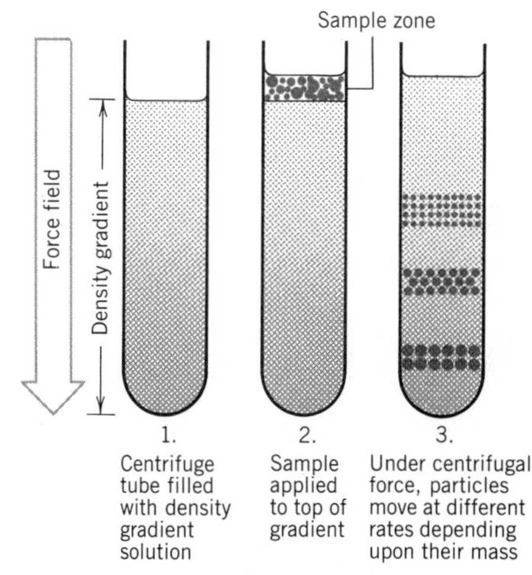

FIGURE 9 The principle of rate zonal centrifugation.

In this technique the densities of the sedimenting materials must be greater than the density at any point in the gradient. If the density of the

Table 1 Characteristics of Sedimentation Velocity and Sedimentation Equilibrium Centrifugation

	Sedimentation Velocity	Sedimentation Equilibrium
Synonym	Zone centrifugation	Isopycnic, density equilibration
Gradient	Shallow, stabilizing—maximum gradient density below that of least dense sedimenting species	Steep—maximum gradient density greater than that of most dense sedimenting species
Centrifugation	Incomplete sedimentation, short time, lower speed	Complete sedimentation to equilibrium position, prolonged time, high speed

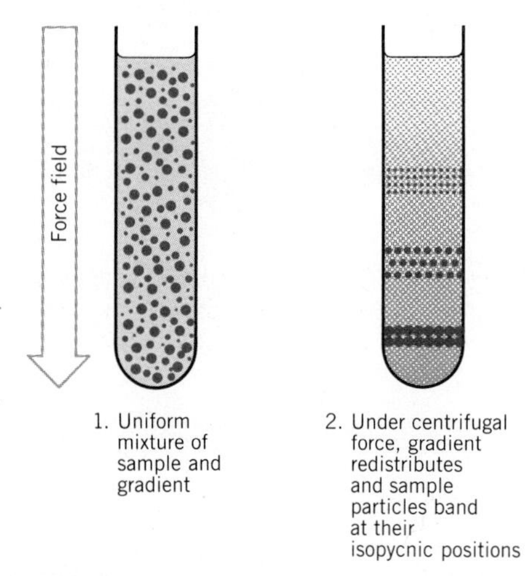

FIGURE 10 The principle of isopycnic centrifugation.

gradient is higher at any point than the density of any sedimenting component, that component will cease moving toward the tube bottom and a pile-up of zones could result at that point.

Isopycnic Centrifugation

A second type of density gradient technique, called isopycnic, employs a gradient that encompasses the densities of all sedimenting materials in a sample. When these conditions are met, a given material will come to rest in the gradient at a position where its density is the same as the density of the gradient. Materials of different densities will therefore separate into different zones in the gradient.

Using this technique it is possible to start with a preformed gradient or to start with material that will form a gradient automatically under high centrifugal force. Cesium chloride is often used to accomplish this. Thus, a uniform mixture of sample in cesium chloride, when under high forces, will resolve into a gradient with separated zones of sample, as illustrated in Figure 10.

Recovery and Analysis of Results

After materials have been separated by density gradient techniques, they must be removed from the tube for analysis and sometimes for further work.

One method of recovery is to remove the tube from the rotor and puncture the bottom with a device that will permit a controlled withdrawal of fluid into fractions. The fractions can be analyzed by spectrophotometric, chemical, or other techniques that may be useful in detecting the material separated. Once a particular component has been found and isolated, it can be subjected to additional studies. Thus density gradient centrifugation can be used as a preparative procedure as well as one that may give an indication of homogeneity.

Some of the characteristics of density gradient centrifugation and the nomenclatures commonly employed are summarized in Table 1.

Analytical Ultracentrifugation

The king of the centrifuges is the analytical ultracentrifuge. This instrument contains a rotor with sample wells through which a light beam can be projected. The quartz-windowed sample container and the optical system of the instrument permit the researcher to measure the rate at which a material sediments during the run. If only one sedimenting component is present, there will be generated a boundary of sedimenting material that moves from the top of the sample container to the bottom. This boundary can be seen by the optical system because there is a refractive index difference in the sample on the two sides of the boundary.

A sedimentation coefficient for any material can be calculated from the equation:

$$s = \frac{dx/dt}{\omega^2 x}$$

where x is the distance of the sedimenting boundary from the center of rotation in centimeters, t is the time in seconds, and ω is the angular velocity of the rotor in radians per second. Sedimentation coefficients range between 1×10^{-13} and several hundred $\times 10^{-13}$ sec as denoted in Table 2. A sedimentation coefficient of 1×10^{-13} sec is called a Svedberg unit and is designated S. Thus, ribosomes, for example, which have a sedimentation coefficient of 70×10^{-13} sec, are called 70S particles.

Determining Molecular Weights By Sedimentation–Diffusion

Given the sedimentation coefficient of a material, it is possible to calculate its molecular weight, provided information is obtained on its diffusion properties. A diffusion coefficient can be experimentally determined that is a function of the size and shape of the molecule and the frictional resistance of the medium in which it is measured.

The rate of diffusion is given by Fick's first law of diffusion, which states that the amount of material ds diffusing across an area A in a period of time t is proportional to the concentration gradient dc/dx at that point:

$$\frac{ds}{dt} = -DA\frac{dc}{dx}$$

where D is the diffusion coefficient, which is the quantity of material diffusing per second across a surface area of 1.0 cm^2 when there is a concentration gradient of unity. Diffusion coefficients range from 11×10^{-7} cm^2/sec for small proteins to 0.5×10^{-7} cm^2/sec for large particles such as viruses.

Given a diffusion coefficient and a sedimentation coefficient, the molecular weight of a component can be calculated from

$$M = \frac{RTs}{D(1 - \bar{v}\rho)}$$

where R is the gas constant [8.31×10^7 ergs/(mol)(K)], T the absolute temperature, $\bar{v}$ the partial specific volume of the material, ρ the density of solvent, and s and D the sedimentation and diffusion coefficients.

As an example of the values above, human hemoglobin has a sedimentation coefficient of 4.46×10^{-13} sec, a diffusion coefficient of 6.9×10^{-7} cm^2/sec, and a molecular weight of 64,500 daltons.

Table 2 Sedimentation Coefficients (*S*) of Biological Molecules and Particles

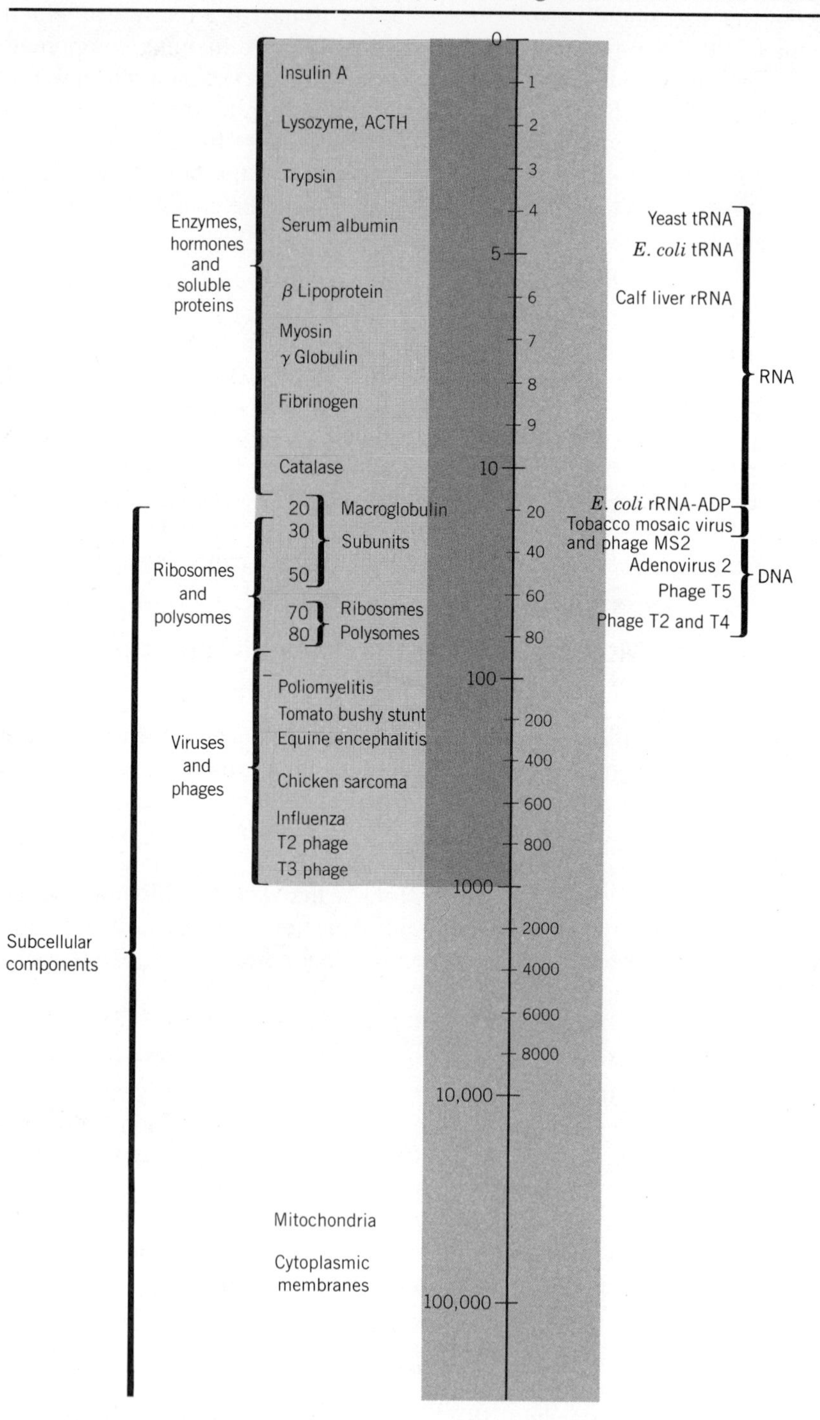

We have made numerous references to the use of centrifugation in this text. As examples, both differential and density gradient centrifugation were employed to purify lysosomes (pp. 287 and 289), the Golgi complex (p. 361), mitochondria (p. 395), and nuclear envelopes (p. 477). Ribosomes are often characterized according to their sedimentation coefficients (p. 569). Chromatin can be pelleted from lysed nuclei by simple preparative centrifugation (p. 505).

Many other examples are encountered in the chapters. As is apparent, in most cases centrifugation is used as an early step in the isolation and purification of a cell component that is subsequently subjected to additional scrutiny.

Electrophoresis

Electrophoresis is one of the most important techniques used for the separation of complex mixtures. It is useful for the analysis of numbers and types of components in a mixture as well as for isolating and purifying molecules.

The principle of electrophoresis is to propel charged molecules through some porous buffered support medium by application of an electric field to the system. By varying the pH of the buffer and by varying the properties of the support medium, separation can be effected by differences in molecular weight as well as differences in charge densities.

Many different types of support medium are used, including paper, cellulose polyacetate, starch gels, and polyacrylamide gels. In all these the experimental apparatus varies but the principle is the same: molecules possessing a charge move when placed in an electric field at a rate that reflects their charge densities and their sizes and shapes.

POLYACRYLAMIDE GEL ELECTROPHORESIS

Polyacrylamide gel electrophoresis (PAGE) is perhaps the most versatile and widely employed electrophoretic technique known today.

Polyacrylamide gels are made by the free-radical polymerization products of the acrylamide monomer

$$CH_2{=}CH{-}\overset{\overset{\displaystyle O}{\|}}{C}{-}NH_2$$

and the cross-linking comonomer *N,N'*-methylene-bis-acrylamide

$$CH_2{=}CH{-}\overset{\overset{\displaystyle O}{\|}}{C}{-}NH{-}CH_2{-}NH{-}\overset{\overset{\displaystyle O}{\|}}{C}{-}CH{=}CH_2$$

according to the scheme shown in Figure 11. Polymerization is carried out in tubes to generate cylinders of gels or on flat surfaces to form slabs.

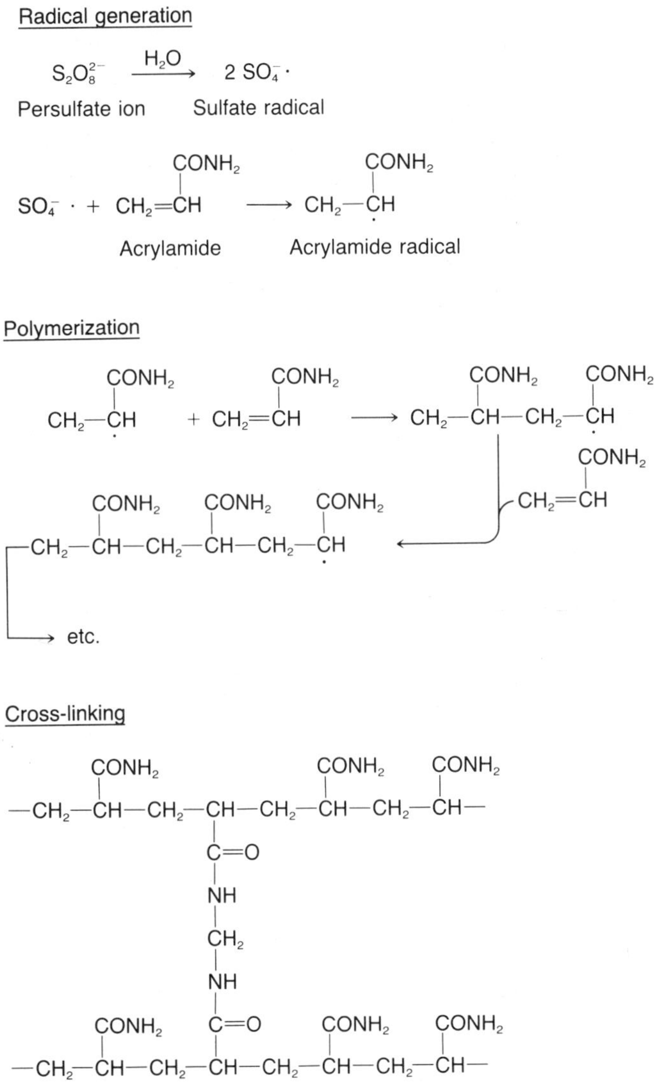

Radical generation

$$S_2O_8^{2-} \xrightarrow{H_2O} 2\,SO_4^{-}\cdot$$

Persulfate ion — Sulfate radical

$$SO_4^{-}\cdot + CH_2{=}CH(CONH_2) \longrightarrow CH_2{-}\dot{C}H(CONH_2)$$

Acrylamide — Acrylamide radical

Polymerization

$$CH_2{-}\dot{C}H(CONH_2) + CH_2{=}CH(CONH_2) \longrightarrow CH_2{-}CH(CONH_2){-}CH_2{-}\dot{C}H(CONH_2)$$

$$\xrightarrow{+\ CH_2{=}CH(CONH_2)} {-}CH_2{-}CH(CONH_2){-}CH_2{-}CH(CONH_2){-}CH_2{-}\dot{C}H(CONH_2) \longrightarrow \text{etc.}$$

Cross-linking

$${-}CH_2{-}CH(CONH_2){-}CH_2{-}CH{-}CH_2{-}CH(CONH_2){-}CH_2{-}CH(CONH_2){-}$$

$$|$$

$$C{=}O{-}NH{-}CH_2{-}NH{-}C{=}O$$

$$|$$

$${-}CH_2{-}CH(CONH_2){-}CH_2{-}CH{-}CH_2{-}CH(CONH_2){-}CH_2{-}CH(CONH_2){-}$$

FIGURE 11 Reactions involved in the production of polymeric material for polyacrylamide gel electrophoresis.

A wide variety of pore sizes can be obtained by varying the concentrations and ratios of monomer and cross-linking comonomer.

A normal run on PAGE is conducted at a pH chosen such that all the species possess the same charge but have different charge densities (charge to mass ratio). This is normally the case when proteins are separated. Since the acrylamide gel may slow down the large molecules, separation is effected by the combined action of charge density and sieving.

Nucleic acids have identical charge to mass ratios but are still separable on PAGE if a gel is selected that will sieve the molecules according to size.

It is possible to separate proteins on the basis of size alone if before electrophoresis they are solubilized in the detergent sodium dodecyl sulfate (SDS) and a reducing agent to break disulfide linkages. This causes proteins to lose their particular three-dimensional shapes and they are

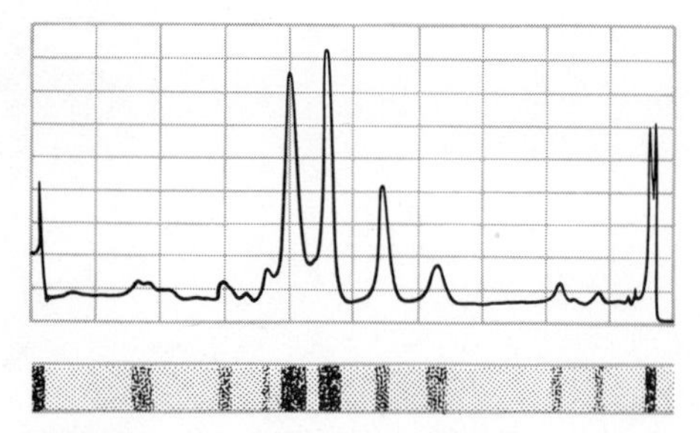

FIGURE 12 Relation between bands produced by staining electrophoretically separated proteins and a graph generated by a density scan of the gel.

converted to rodlike structures. SDS, a negatively charged molecule, binds to these proteins with a constant weight ratio, thus imparting to the proteins a constant charge per unit weight. This being the case, electrophoretic mobility becomes strictly a function of molecular weight or size governed by the sieving effects of the gel. This technique, designated SDS–PAGE, is now commonly employed to determine the molecular weights of unknown proteins by comparing their mobilities to those of standard proteins.

When the electrophoretic separation of proteins or nucleic acids is completed, the components are generally revealed by staining. Pictures can be taken showing the bands, or the gels can be subjected to an instrument that will scan the material for band densities and convert these to peaks, the heights of which reflect the density and thus the amount of material in the band (see Figure 12).

Polyacrylamide electrophoresis has had wide application in cell biology. Examples of its use are in the determinations of the makeup of the electron transport complexes of mitochondria (p. 407) and the ATPase of chloroplasts (p. 444). SDS–PAGE has been used in studies on the nucleus, including chromosome scaffolds (p. 517) and matrix proteins (p. 542). A complete separation of the proteins in ribosomes has been achieved by two-dimensional PAGE (see Figure 17.9), and the technique has also been helpful in studies on protein topologies of ribosomes (p. 589). Microtubules have yielded many of their anatomical secrets to researchers using SDS–PAGE (p. 618).

CARRIER-FREE CONTINUOUS ELECTROPHORESIS

In contrast to PAGE, where molecules are propelled through a porous support medium, carrier-free electrophoresis employs only a buffer through which molecules migrate. This requires a specially designed apparatus. One type used consists of two sheets of glass separated by a small space in which the buffer is contained. The sample is fed in between the sheets at one edge along with buffer, which is pumped toward the opposite edge. The electric current is applied perpendicular to this flow. Thus, molecules will emerge from the edge of the sheet at different points where they are collected in fractions.

This rather specialized application of electrophoresis has been used in cell biology for the separation of organelles from one another, such as in the purification of lysosomes (p. 290).

V

Chromatography

The term "chromatography" embraces a wide variety of related laboratory techniques that are employed to separate different molecular species. The underlying principle in all these techniques is a differential migration of components of a mixture such that they separate out from one another. For this to occur, there must be some force that impels these molecules to move, and there must be some selective impedance causing some species to move more slowly than others.

The force that propels molecules in chromatographic systems is a moving liquid or gas phase. These phases can be pumped through a column containing the impeding substance, or they may move by gravity or capillary action. The substance providing selective impedance may do so by monitoring charge, size, affinity, or partition differences in the components. The major distinctions between the types of chromatography are in the impedance substances.

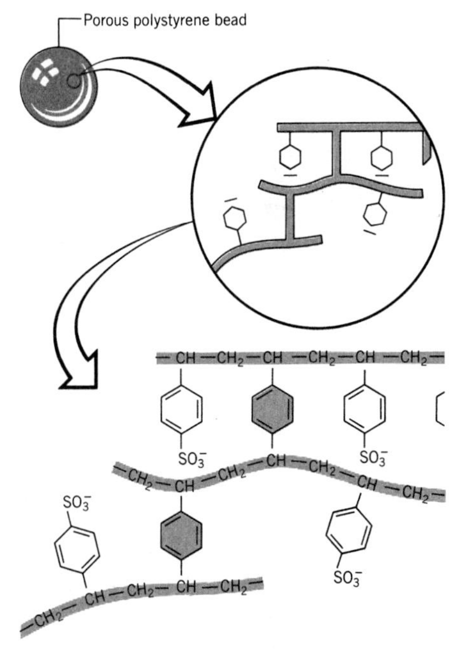

FIGURE 13 The basic features of the matrix used for ion exchange chromatography.

ION EXCHANGE CHROMATOGRAPHY

In ion exchange chromatography molecules are separated by their differences in charge. The experimental setup for this most often consists of a column into which a charge-containing medium is packed and through which a buffer is pumped. The medium commonly consists of porous polystyrene beads chemically derivatized with negative or positive substituents (see Figure 13). A negatively charged medium would have mobile counterions associated with it. Their places can be taken over by other cations, such as those present in a mixture to be separated. Thus there may be an exchanging of cations on this column, hence the name cation exchange chromatography. In like manner, a positively charged medium is an anion exchange medium.

When a mixture of molecules possessing different charges is passed through a column, molecules possessing an opposite charge will be se-

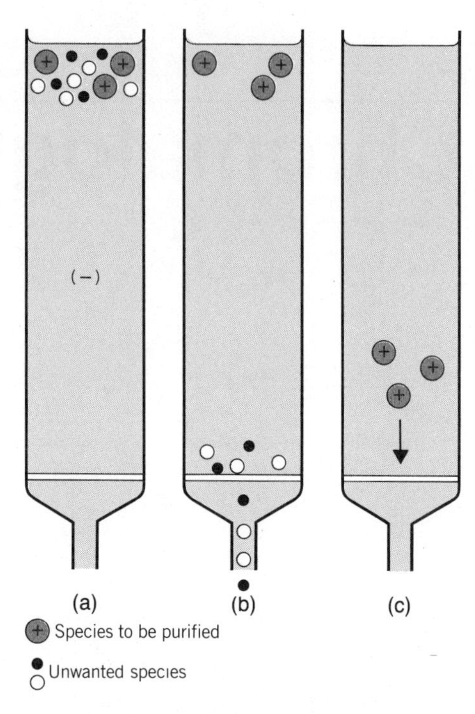

FIGURE 14 The principle of sorption chromatography. A column containing a negatively charged matrix (*a*) will retain the desired positively charged species until it is free of all unwanted species (*b*). Then it is swept off the column (*c*).

lectively impeded, while those with no charge or the same charge as the medium will pass through. If the compound of interest is retained—or sorbed—the technique is referred to as sorption chromatography. After all unwanted species have passed through the column, the retained molecules can be eluted by increasing the concentration of counterion in the buffer to exchange with the retained molecules and/or by changing the pH of the eluent to neutralize the charge differences between the column and the molecules of interest. The species will then sweep off the column (Figure 14 illustrates this principle).

Alternatively, the compound of interest may be eluted from the column and all undesirable species be retained.

A slightly different application of these principles results in fractionation of the mixture into its various parts, all of which may be recovered for further studies (see Figure 15). First, all species are sorbed by electrostatic interaction with the column. The eluent is then gradually changed in composition with time such that weakly charged species are eluted first, followed successively by species bearing increasing charges.

Ion exchange chromatography is useful for separating charged substances such as amino acids and proteins, which can be induced to bear different degrees of the same charge by changing pH. Ribosomal proteins, as an example, have been separated from one another by this technique (p. 580).

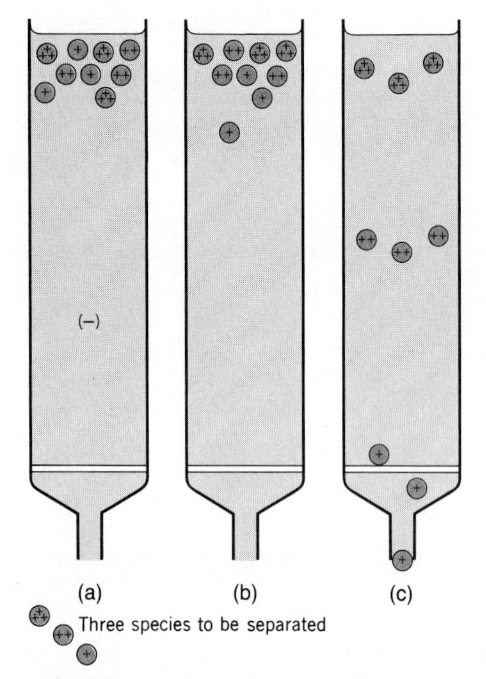

FIGURE 15 The principle of fractionation chromatography. A mixture of positively charged species will bind to a negatively charged column (*a*). By changing the pH and ionic environment in the column, the species with the lowest net charge will be released first (*b*), followed in sequence by the others until they sweep off the column separated (*c*).

AFFINITY CHROMATOGRAPHY

The experimental setup used for affinity chromatography may be identical to that used for ion exchange chromatography. The main difference between the techniques resides in the manner in which the matrix material in the column is derivatized.

Generally some small ligand molecule (such as the substrate of an enzyme) or a specific protein (such as an antibody) is chemically linked to the porous matrix of the column. Upon passage of a mixture of molecules through the column, only those that form a biospecific bond with the matrix-bound molecule will be retained; all others will wash through the column. The bound molecules may then be eluted by increasing the ionic strength of the eluent or by adding the ligand to the buffer percolating through the column. Since the biospecific reaction between the retained molecule and the column is reversible, the free ligand will displace the column-bound ligand and the retained molecule will move off the column, as illustrated in Figure 16.

Many different kinds of molecule may be used to derivatize the matrix material. Thus this type of chromatography is not only highly specific for the isolation of the material in question but it is also highly versatile. The system is capable of being used for the isolation of any kind of molecule as long as it can be made to interact with the bound ligand.

Antibodies, which can be prepared with a very narrow specificity against other molecules, are often bound to the matrix. Thus a particular antigen can be plucked from a mixture and, after being made free of contaminants, can be released and recovered in pure form. In other cases the interaction between bound ligand and specific molecule may be much more fortuitous based on unique complementary interactions.

Affinity chromatography can be used to isolate and purify biological materials that may be present in very low amounts in cells or are for various reasons hard to separate from other cell components. Nonmuscle actin, a protein present in eucaryotic cells, has been isolated in this way (p. 628).

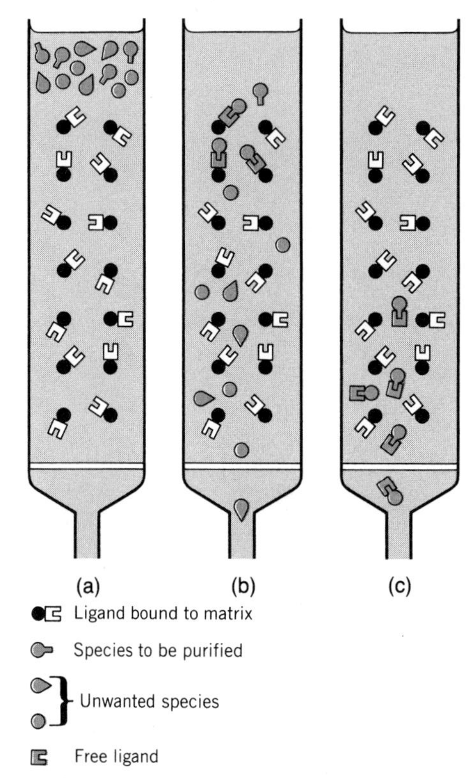

FIGURE 16 The principle of affinity chromatography. (*a*) Original sample. Only the desired species binds to the ligand and is retarded (*b*). After undesirable species have been removed, the species to be purified is eluted (*c*) by excess ligand or changing the conditions in the column.

PARTITION AND ADSORPTION CHROMATOGRAPHY

Every student who has taken organic chemistry is familiar with the principle of separation effected by the separatory funnel. When a substance is shaken with two mutually immiscible liquid phases, it will partition between the phases with a higher concentration in one phase than in the other. If one phase is made to move with respect to the other, the material will also move depending on its partition coefficient. If it prefers the mobile phase, it will move rapidly. If it prefers the stationary phase, its movement will be impeded.

If a material is impeded in its movement in a mobile liquid or gas phase by virtue of attraction by adsorption to a stationary solid phase, the process is called adsorption chromatography. In this case, depending on the intensity of adsorption, the material is either strongly or slightly impeded, and mixed materials can be separated from one another if their adsorption characteristics differ (see Figure 17).

The principles of partition and adsorption chromatography underlie many different techniques, including common paper chromatography (partition), thin layer chromatography (partition and/or adsorption), and gas chromatography (generally partition).

In paper chromatography the paper serves mainly as an inert support, which becomes hydrated. The hydrated phase is the stationary phase. An organic phase moves past this aqueous phase and partitioning molecules move at rates that depend on their partition coefficients.

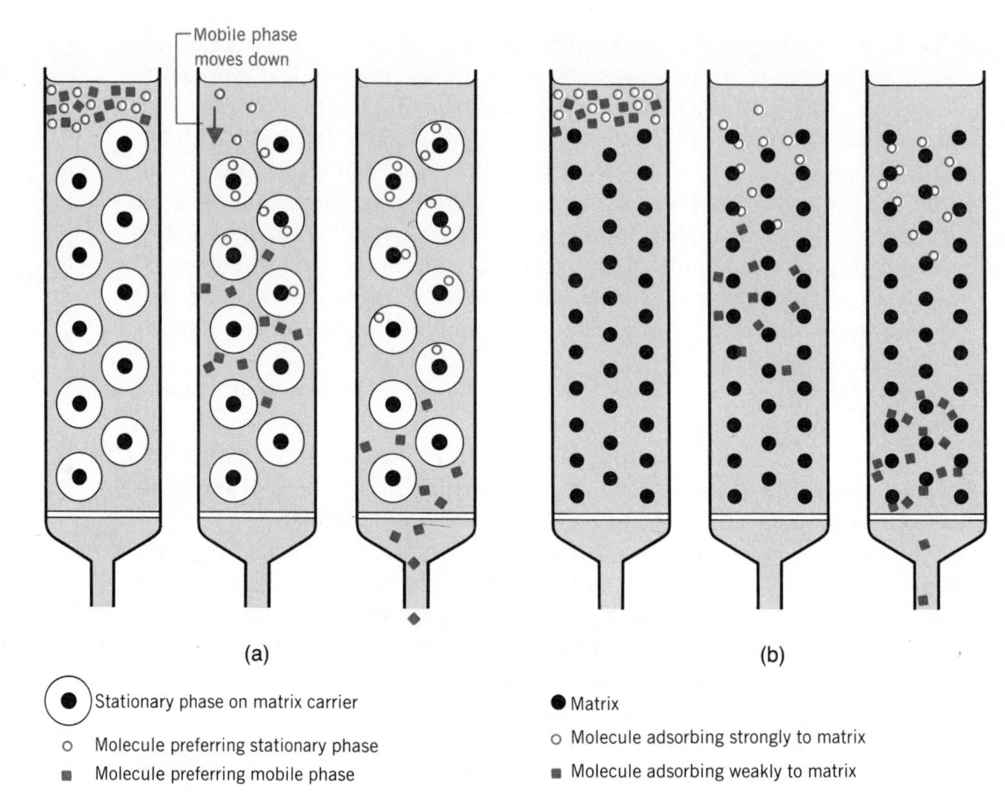

FIGURE 17 The basic features of partition (*a*) and adsorption (*b*) chromatography.

In gas chromatography materials may partition between a moving gas phase and a stationary nonvolatile phase that coats the matrix material of a column. This partitioning is generally temperature dependent, and therefore a gradual temperature increase during a run may sequentially push different materials into the moving or gaseous phase. Alternatively, adsorption rather than partition may function in the technique of gas chromatography.

Either the principle of partition or of adsorption chromatography may be employed in thin layer chromatography. In this technique the matrix material is applied as an extremely thin layer to an inert plastic or glass backing, and in this respect the technique is much like paper chromatography. However, many different substances can be applied to this backing, thereby making the technique more versatile than paper. Because of the extreme thinness of this matrix layer, moreover, the technique is generally more sensitive and rapid than paper.

Chromatography has been one of the major workhorses of the cell biologist interested in isolating and purifying materials, ranging from plant pigments to amino acids to lipids. Information can be gained on both composition and quantities of materials in a mixture. In addition, the chromatographic methods are useful both for preparative as well as analytical purposes.

GEL FILTRATION CHROMATOGRAPHY

Gel filtration chromatography is also called gel permeation chromatography and molecular sieving. The principle of this technique is to separate molecules according to their size and shape differences.

This is accomplished by using a column matrix material consisting of beads of carefully controlled porosity. A matrix bead of a particular porosity is chosen depending on the type of separation that must be achieved. Generally, a porosity is chosen such that one species may enter the bead through its pores while the other species is excluded from the bead completely.

Thus when a mixture of molecules is passed through a column, molecules larger than the pore diameters will pass by the beads and will be rapidly eluted from the column. Molecules that enter the beads have access to a larger volume in the column and are therefore retarded in their movement (see Figure 18).

A common use for gel chromatography is to desalt a sample or to move a large molecule into a new buffer. Proteins, for example, are rapidly and gently separated from low molecular weight salts by this technique. When the protein emerges from the column, it is immersed in whatever buffer the column contains. Thus rapid buffer exchange is possible.

By carefully choosing the pore size of beads, mixtures of high molecular weight materials, such as proteins or nucleic acids, may be fractionated. If all members in a mixture have access to the pores but to different degrees, the protein with least access will emerge first from the column.

Gel filtration chromatography is obviously an extremely useful tool for the cell biologist. It has been of enormous help in the isolation of many different molecular components of the cell. An example of its use is in the isolation of ribosomal proteins (p. 581).

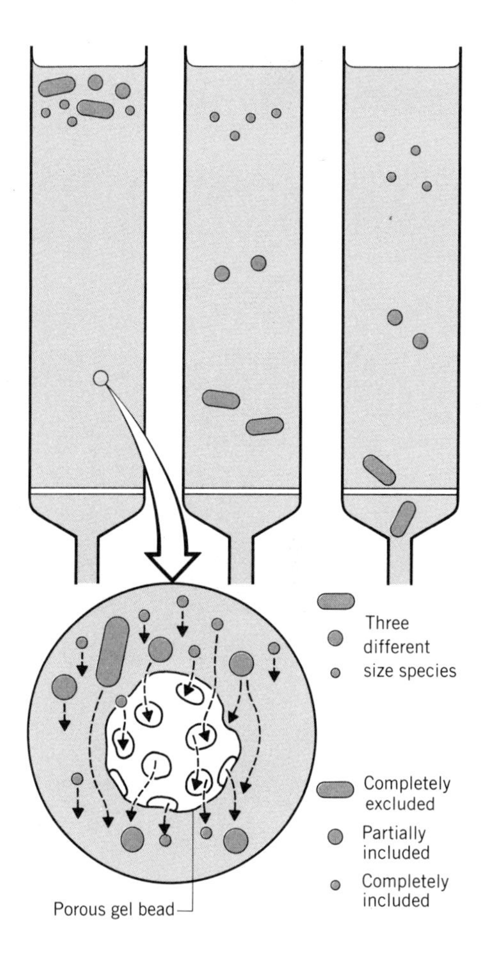

VI Radioactive Labeling

Radioactive isotopes have proved to be powerful tools for monitoring cellular events and for localizing these events and their participants within the cell. The isotopes most commonly used are ^{3}H, ^{14}C, ^{32}P, ^{35}S, ^{125}I, and ^{131}I, since these are unstable and release a high energy electron, or β particle, as they decay.

Generally, substrate molecules or building blocks of the cell's macromolecules are made radioactive by incorporating a radioactive element into their structures. The isotopes can be administered to intact cells or to a cell fraction, and the fate of the radioactive species can be followed by very sensitive monitoring devices.

The approach often used by the biochemist is to provide a radioactive precursor to a cell or cell fraction and trace the fate of the radioactive component with time. An example of this type of study is to follow the incorporation of [^{3}H]-thymidine into various forms of DNA in the nucleus (p. 543). The radioactive material is administered to the system being studied; then at various time intervals aliquots are removed and subjected to fractionation, and the fate of the radioactivity is noted. In the example used here, the radioactivity is found to move to different fractions with time, thus indicating the manner in which DNA is handled by the nucleus. For this type of approach, radioactivity is normally monitored by a liquid scintillation spectrometer, which can detect the radioactive component and also measure the amount of component present.

The complex metabolic pathways used by cells for energy retrieval and biosynthesis and the pathway of carbon during carbon dioxide fixation were worked out using radioactive precursors in this manner (p. 431).

A very widely used adaptation of radioactive tracer technology is called *autoradiography*. The principles of this technique are outlined in Figure 19. Cells are first exposed to a radioactive precursor, which becomes incorporated into a structure or functional molecule of the cells. The cells are then washed free of unincorporated precursor and fixed in a medium that kills the cells and suspends them in a state that maintains their natural structure. The suspended cells are embedded in wax or plastic by conventional histological techniques and placed on a slide. A thin film of photographic emulsion is coated over the cells, usually by dipping the slide in a liquid emulsion in the dark. Slides are subsequently stored in the dark for a period of days or weeks during which time the β particles emitted from the radioactive isotope activate silver halide crystals in the photographic emulsion. The slide is then processed as one would process a photographic film. Crystals activated by the radiation develop into silver grains, which appear as dark spots in light microscopy.

The technique of autoradiography is applicable to electron microscopy as well as to light microscopy. For electron microscopy the cells are placed on a grid rather than a slide and are then processed in a manner similar to that for light microscopy. When viewed in the electron microscope silver grains are opaque, and thus are readily observed when in contrast with a more electron-transparent background.

One of the important uses of autoradiography is in "pulse–chase" experiments by which the movement with time of a material in a cell can be observed. Tissues or cells are exposed for a brief period of time (pulse) to a radioactive precursor and are then washed free of radioactivity. Immediately a continuing supply of the nonradioactive form of the same precursor is provided (chase). Thus the radioactivity that enters into a stream of movement in the cell has been both preceded and followed by nonradioactive material. By taking aliquots with time and examining them with biochemical or autoradiographic techniques, both the biochemical transformations of this material and its movement in the cell can be monitored.

Many references have been made in this text to the use of autoradiographic techniques. Good examples of the use of the pulse–chase technique appear in the discussions of the secretory pathway (p. 328), the role of the Golgi complex in intracellular transport (p. 370), and in the activities of chromatin in the nucleus (p. 556).

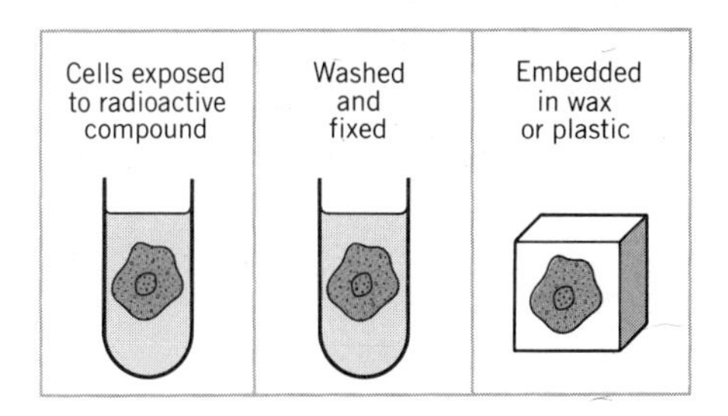

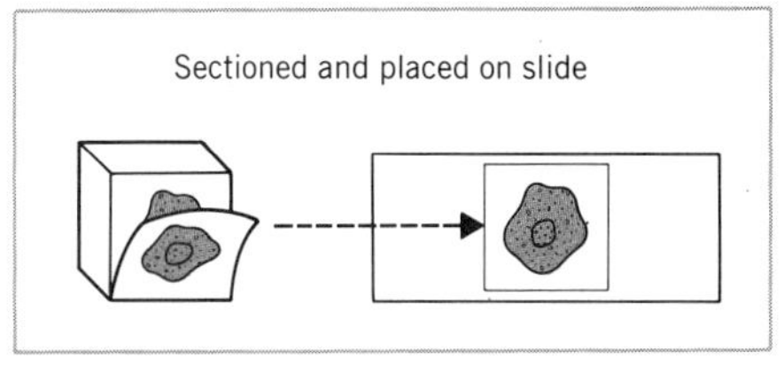

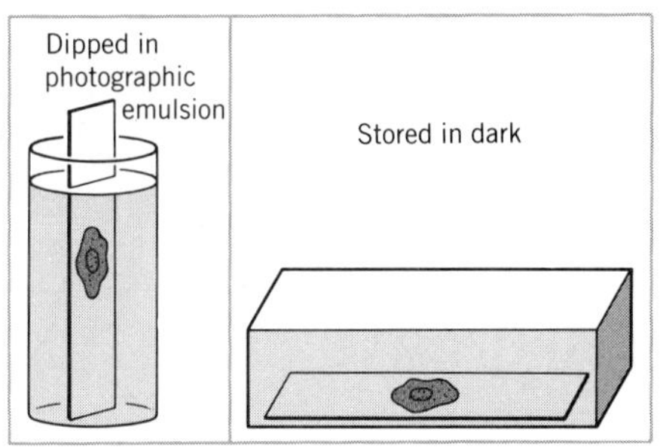

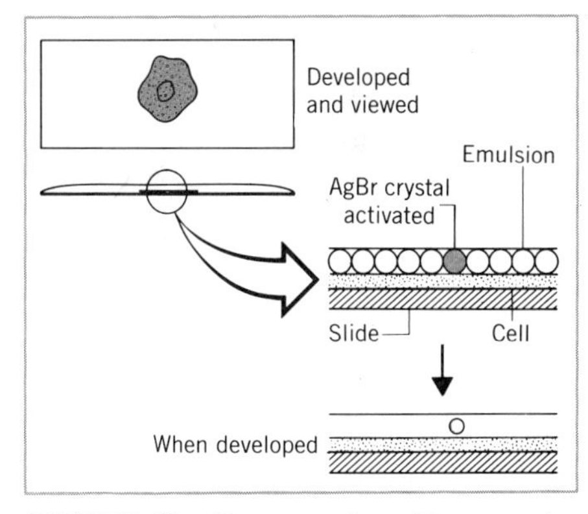

FIGURE 19 Steps employed in preparing tissue for autoradiography.

Immunochemical Techniques

The use of antibodies has many applications in the study of the cell. Since antibodies can be produced having a specificity against individual cell components, they can be used to ferret out a particular organelle or component in the midst of all other molecular species.

The first task, when using antibodies in a study, is to produce antibodies with the desired specificity, such as against tubulin. This means that tubulin from a given tissue or organism must be isolated and purified. It is then injected into an animal in which the protein is foreign or at least dissimilar enough to the animal's own protein to induce the production of antibodies. The antibody fraction of the animal's blood is then isolated and, in our hypothetical case, it would be enriched in antibodies against tubulin.

Once antibodies are isolated, they can be used in several different ways to probe the cell. We will discuss just a few of these that are especially pertinent to the material discussed in this text.

FLUORESCENT ANTIBODIES

Certain kinds of small organic molecule can be electronically excited when they are placed in ultraviolet light. These molecules lose their excitation energy by emitting a quantum of light as fluorescence. Thus, they absorb invisible light (ultraviolet) and reemit visible light (i.e., they fluoresce).

Organic molecules capable of fluorescence can be covalently attached to antibodies, thereby imparting to antibodies the property of fluorescence. Once a fluorescent antibody is available, it can be used in two different ways: in direct or indirect immunofluorescence microscopy.

The direct technique is very straightforward. Fluorescent-labeled antibodies are reacted with thin sections of cells or with whole cells treated to permit the antibodies to penetrate the cell membrane. These preparations are then examined in a microscope equipped with an ultraviolet light source. The only cell components visible will be those that fluoresce, and they will be seen brightly illuminated against a dark background. Fluorescent antibodies directed against tubulin would attach only to microtubules, and their locations would be clearly evident in the cell. The principles of this technique are outlined in Figure 20.

Indirect immunofluorescence microscopy has a wider application than the direct technique, although the final results and the microscopic setup employed are the same. Two types of antibody are used in this technique. The first—produced, for example, in a rabbit—is directed against the cellular component to be studied (see Figure 21). It is not labeled with fluorescent material. The second type of antibody is produced with a specificity against the first type. Thus a goat may be used to make antibodies against rabbit antibodies. The goat antibodies are then made to be the recipient of the fluorescent label.

Equipped with these two types of antibody, the researcher first reacts a cell section with unlabeled rabbit antibodies (directed against a specific component such as tubulin) and, after washing out excess unbound antibodies, applies the fluorescent-labeled antibodies. These bind only to the rabbit antibodies, which then become sandwiched in between the tubulin and the tagged goat antibodies. The result is microtubules that have fluorescent labeling.

The advantage of the indirect method over the direct technique is that only one type of antibody containing a fluorescent label need be prepared. This one preparation can be used to light up any cell component that has been reacted with its homologous antibody.

Fluorescent immunochemical techniques have provided dramatic evidence of the intracellular locations of different cell components, such as microtubules (p. 621), and actin proteins (p. 670). Micrographs resulting from these techniques have given us a new image of the cell interior.

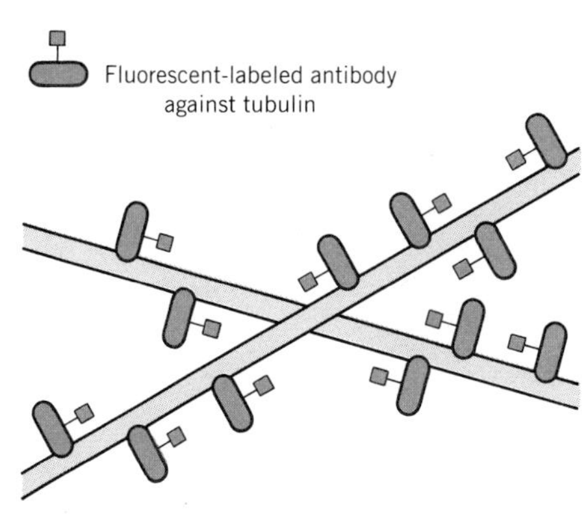

FIGURE 20 Direct fluorescent-labeled antibody technique.

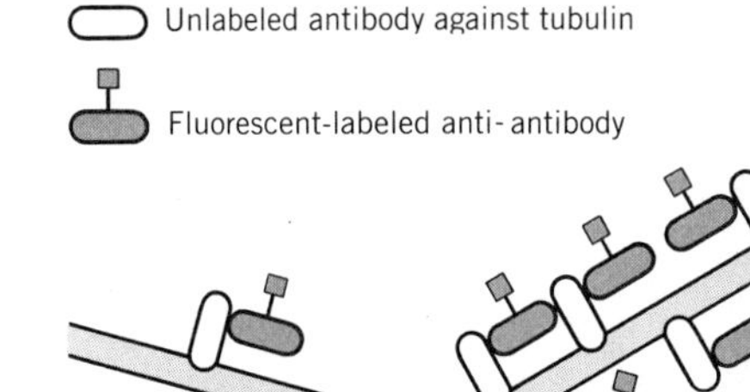

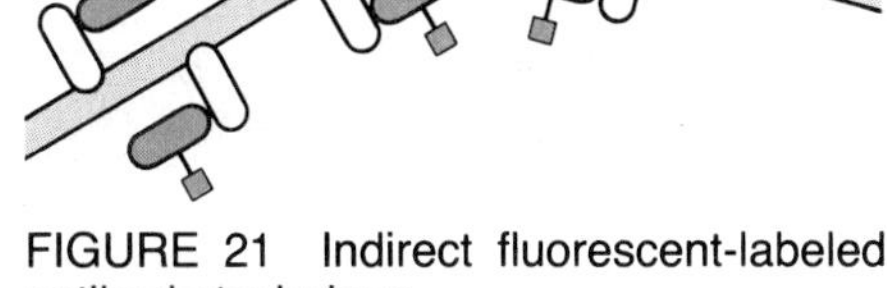

FIGURE 21 Indirect fluorescent-labeled antibody technique.

FERRITIN-LABELED ANTIBODIES

Ferritin is a high molecule weight iron–protein complex that functions as a carrier or iron in many organisms, including the human. The complex contains 17 to 23% iron as a core of hydrated ferric hydroxide around which 24 subunits of protein aggregate to form a protein shell. A complex that carries a complete load of iron contains as many as 2000 atoms of iron.

Because of its composition of iron, ferritin cannot be penetrated by electrons and is therefore opaque in the electron microscope. By itself, it is useless unless it can be specifically attached to different cell components. This can be accomplished by attaching ferritin to antibodies and applying these ferritin-labeled antibodies to a tissue or cell under study. The antibody will provide the desired specificity and the ferritin provides electron opacity.

Numerous studies have been carried out using ferritin-labeled antibodies, including studies on the cell surface (p. 131). Ferritin itself is taken into erythroblasts by the process of micropinocytosis (see Figure 8.16), where the iron is apparently used in the synthesis of hemoglobin. This process can be followed with electron microscopy because of the opacity of the ferritin.

Since lectins bind to cell surfaces with a high specificity, these can also be used in the place of antibodies as an object of ferritin labeling. These ferritin-labeled lectins are useful for a variety of investigations on the cell surface (p. 134).

RADIOACTIVE ANTIBODIES

Another adaptation of the use of antibodies is to label them with a radioactive isotope such as ^{3}H, ^{14}C, or ^{131}I. An organic compound containing one of these isotopes can be covalently linked to the antibody, or the antibody can be directly iodinated with isotopic iodine.

Radiolabeled antibodies can then be used in conjunction with autoradiography to localize various components in the cell. As an example, the fate of the plasma membrane during endocytosis has been studied using labeled antibodies against membrane components (p. 272). After reaction, cells or cell fractions can be analyzed for their content of radioactivity or a section of the cell may be subjected to autoradiography (see Method VI).

IMMUNE ELECTRON MICROSCOPY

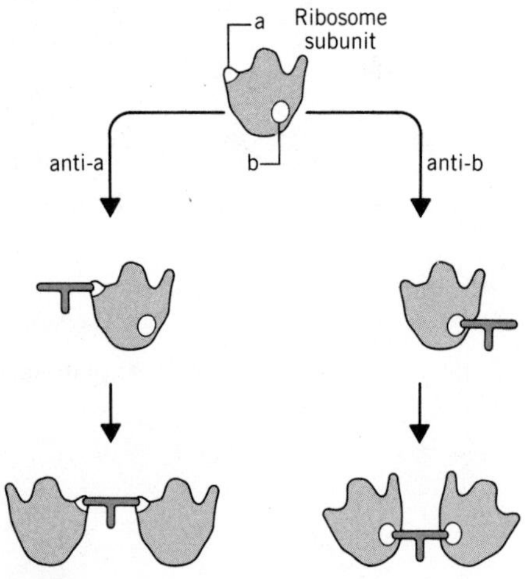

FIGURE 22 The principle of immune electron microscopy as applied to ribosomes.

The technique of immune electron microscopy is a relatively new approach to studying the topology of cell or organelle components by use of the electron microscope. The application of this technique to the study of ribosomes has yielded dramatic and insightful results.

Antibodies, directed against a specific ribosomal protein, are first of all reacted with ribosomes or their subunits. Since antibodies are bivalent, one antibody will cross-bridge two ribosomal particles, linking identical proteins on the two particles. These dimers can be examined directly in the electron microscope (see Figure 22). A careful analysis of the ribosome morphology and the point of contact made by the antibody bridge will reveal where on the surface of the ribosome a particular component is located (p. 591).

Somewhat more complicated adaptations of this approach have been used to localize where tRNA is bound to the ribosome. First, tRNA is covalently attached to the ribosome by irradiation. The tRNA is one that is derivatized with a dinitrophenyl (DNP) group on the amino acid it carries. Next, the ribosome–tRNA complex is reacted with anti-DNP antibodies. Monomers and dimers will be formed that can be examined in the electron microscope. Once again, by way of a careful analysis of shapes and bridgepoints, the site of binding of tRNA to the ribosome can be determined. These approaches have yielded the dramatic binding maps such as those appearing in Figures 17.15, 17.16, and 17.17.

VIII

Spectroscopy

"Spectroscopy" is a blanket term that encompasses a variety of techniques. Basically, it is the study of the interaction that takes place between electromagnetic radiation and matter. By using techniques that monitor these interactions, information may be gained regarding structure and bonding as well as intra- and intermolecular phenomena.

There are many forms of electromagnetic radiation. Light is but one form. Others are microwaves, infrared rays, and X-rays. Wave theory is quite complicated, and we shall not discuss the many theories that surround the study of electromagnetic radiation. Suffice it to say that an electromagnetic wave consists of two components: an electric field and a magnetic field. The two fields oscillate in space perpendicular to each other and in a direction perpendicular to the direction of wave propagation.

An additional property of electromagnetic radiation is that when emitted it possesses small, discrete amounts of energy rather than energy values within a continuous range. These discrete amounts are referred to as quanta of energy. Thus light, for example, is viewed as consisting of particles called light quanta or photons of energy. Light has both the properties of waves and the properties of particles.

All types of spectroscopy can be placed into one of two categories: absorption or emission. Matter may absorb electromagnetic radiation, in which case its energy will be correspondingly boosted, as by exciting an electron from a lower to a higher energy level. Alternatively, matter may emit electromagnetic radiation, in which case there is an energy loss of the matter that corresponds to the amount of energy emitted. This emission may take place spontaneously, as when an excited molecule loses an amount of energy in the form of radiation. In other cases, the emission may be induced by applying external irradiation to the matter.

It is not difficult to imagine that when a molecule is exposed to electromagnetic radiation, its atoms will undergo certain transitions in response to the irradiation that will depend on the atomic nature of the material and its environment or the interactions in which it is involved. By measuring absorption or emission and changes in electronic or nuclear properties, a good deal of information can be ascertained regarding structure.

The material that follows summarizes the various kinds of spectroscopy referred to in this text, emphasizing the kind of information that can be derived from a particular technique.

NUCLEAR MAGNETIC RESONANCE SPECTROSCOPY

Nuclear magnetic resonance (NMR) spectroscopy is a technique that is concerned with the interaction with atomic nuclei of the magnetic component of electromagnetic radiation. This approach is possible because of the spin properties of protons and neutrons in a nucleus. If the spins of all the particles are paired in a nucleus, there will be no net spin. If, however, there is one net unpaired spinning particle, this will impart a magnetic moment to the nucleus. When such a nucleus is placed in a magnetic field and an appropriate amount of energy is applied, the nuclei will absorb the energy and become aligned in the magnetic field.

Several atomic nuclei found in biological molecules are suitable for NMR study, such as ^{1}H, ^{13}C, ^{14}N, ^{17}O, and ^{31}P. An instrument can be set for a magnetic field energy that interacts with any of these isotopes, such as ^{1}H. When ^{1}H is the nucleus under study, the technique is frequently referred to as proton magnetic resonance (PMR). The energy levels vary slightly, depending on the environment in which the protons are found. Thus, an adjustment in the magnetic field may be necessary for absorption to occur, and this adjustment can be used to interpret the nature of the chemical environment in which the nucleus under study is located.

The basic features of an NMR spectrometer include a source of radiation, a receiver coil to detect the absorption of energy, a dc magnetic field, and a recorder or an oscilloscope to display the signals. In practice, the frequency of the electromagnetic radiation is kept constant and the externally applied magnetic field is varied until absorption, or resonance, is achieved.

An NMR or PMR spectrum for a compound is a special fingerprint that is useful for depicting the environment or interaction in which the molecule participates. This has been useful, for example, in studying the environment of molecules in the plasma membrane (p. 133) and for determining the nature of polypeptide chain folding in ribosomal proteins (p. 584).

ELECTRON SPIN RESONANCE SPECTROSCOPY

Electron spin resonance (ESR), also called electron paramagnetic resonance (EPR) spectroscopy, is a study of the interaction of the magnetic component of electromagnetic radiation with the electrons of molecules. In theory it is similar to NMR except that the spinning species of concern is the electron. A spinning electron generates a magnetic field. When placed in an external magnetic field a certain amount of energy will cause an unpaired spinning electron to become aligned in the field, which can be detected when the energy is absorbed.

In most molecules electrons occur in pairs with opposite spins. These molecules are not subject to ESR studies because the spins within a pair are mutually neutralized. A few molecules, such as O_2, NO, and NO_2 and transition metal ions like Fe^{3+}, Mn^{2+}, and Ca^{2+}, possess one or more unpaired electrons and thus give observable ESR spectra.

ESR is useful for identifying in certain biological processes participants that may be present at very low levels. This technique has been helpful to gain some insight into the nature of the electron acceptor in photosynthesis, which is not a readily identifiable or easily isolated molecular species (p. 440).

OPTICAL ROTATORY DISPERSION AND CIRCULAR DICHROISM

When plane-polarized light is passed through an optically active material, the plane of polarization will be rotated through a certain angle. This optical rotation is wavelength dependent, or, in other words, the amount of rotation changes as the wavelength changes. The rate of change of rotation with wavelength is called optical rotatory dispersion (ORD), and a plot of molar rotation versus wavelength gives an ORD curve.

Certain optically active substances may absorb right and left circularly polarized light differently. Whenever this occurs, the resulting radiation is no longer plane polarized; rather, it becomes elliptically polarized. The technique for determining this polarization is called circular dichroism (CD).

Both ORD and CD are extremely useful tools for studying the conformation of biological molecules in solution. Proteins, for example, contain two contributors toward optical activity: the L-amino acids and the α-helical folding of the polypeptide chains. By studying the optical rotation of proteins, the percentage of helical structure in the molecule can be estimated. This approach has been very useful in delineating the helical nature of the proteins that are present in the plasma membrane (p. 115) and in studying the secondary structure of ribosomal proteins (p. 584).

X-RAY DIFFRACTION

Of all the methods that are used to study biological molecules, X-ray diffraction is the only one that will give insight regarding the three-dimensional positions of atoms in the molecule.

The experimental setup is quite simple, consisting only of an X-ray source and a detector, usually a photographic film. The sample to be studied is placed between the source and the detector as illustrated in Figure 23.

For the determination of a three-dimensional atomic structure, the sample should be in a crystalline state to avoid some of the geometrical distortions that would otherwise be present. Fibers, which have a degree of order that approaches crystallinity, are also good subjects for X-ray diffraction. The collagen fiber is an example of this. But even films and powders of samples can reveal the symmetry of the material or even be useful for identification.

Let us suppose that the sample to be analyzed is a collagen fiber. It is mounted between the source and the detector, and X-rays of a single wavelength are impinged on it. Most of the X-ray beam will completely penetrate the sample, passing through unchanged. Some of the beam,

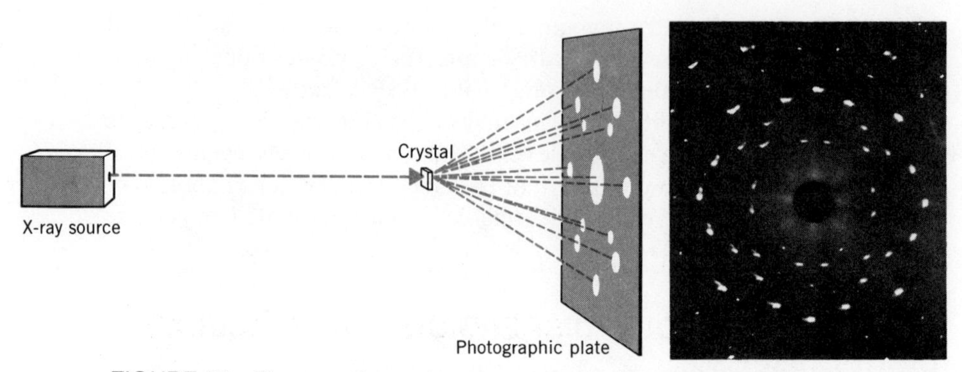

FIGURE 23 The experimental setup for X-ray diffraction analysis.

however, will be scattered, or diffracted, as it interacts with electrons concentrated at each atom in the molecule.

If there is no order in the molecule, the scattered X-rays will emerge at all angles and uniformly darken the photographic film. If, on the other hand, there is a degree of atomic periodicity in the molecule, the scattered X-rays will emerge with a particular pattern of constructive interference (which will darken the detector) or destructive interference (which will give no detectable scattering).

The resulting photograph shows an array of spots, which is subjected to a complex mathematical treatment of intensities and positions that requires the use of high speed computers with large memories.

Even the simplest cell is too complex to be studied intact by X-ray diffraction. But certain of its components are amenable to these analyses. Collagen has a structural periodicity that can be seen both by electron microscopy as well as by X-ray diffraction (see Figure 2.12). Several macromolecules from the cell interior, such as myoglobin, have yielded complete three-dimensional pictures by X-ray diffraction. At a somewhat more complex level, membranes have been studied to gain insight into their order by way of this technique (see pp. 132 and 380).

INFRARED SPECTROSCOPY

When molecules absorb light in the infrared region (10^3–10^4 nm), they undergo transitions between vibrational levels of their ground states. These vibrational levels are generated by bond stretching and bending of various functional groups, such as methyl, carbonyl, and amide groups. Since the modes of vibration of the functional groups are very sensitive to environment, information can be gained about the interactions between functional groups in macromolecules.

Infrared spectroscopy is useful in the determination of secondary structure in proteins, such as the amount of α helix in histones (p. 511). This is possible because the amide bond has different values depending on whether it is involved in the α-helical structure or is present as a random coil, and has not hydrogen bonded.

RAMAN SPECTROSCOPY

Normally light is scattered without a change in frequency. This is termed "elastic scattering." A small fraction of light in the infrared region scatters with a frequency shift, or inelastic scattering. This type of scattering is referred to as "Raman scattering." The frequency change that occurs is by excitation of the molecule to a higher vibrational level, which results in a loss of energy and a frequency decrease.

Raman spectroscopy examines these vibrational transitions and gives information regarding biological molecules in several different ways. The content of α-helical structure in proteins can be determined by this technique as well as the number of S—S bonds in proteins. The number of paired and unpaired bases in RNA can also be ascertained as well as differences between adenosine mono-, di-, and triphosphates and their ionization forms.

Although this technique probably has not been fully exploited in biological systems, in recent years it has, with the development of lasers, become more prominent in structural assays.

Nucleic Acid Sequencing

Very effective and rapid techniques have been worked out to determine the amino acid sequence of proteins, beginning with the landmark accomplishment of the insulin sequence by Frederick Sanger in 1953. Elucidating the sequence of nucleotides in nucleic acids has proven to be a more intractable problem until the past decade, during which time some remarkably rapid and accurate techniques were developed for sequencing both RNA and DNA.

Several approaches have been used, and the ones in use today are still evolving. We therefore do not concentrate on the details of the technology, since these are likely to keep changing, but emphasize certain of the principles involved in nucleic acid sequencing.

Success in nucleic acid sequencing depends on several important advances that have taken place in recent years. One is the availability of polyacrylamide gel electrophoresis systems that resolve polynucleotides that differ in length by only one nucleotide. PAGE is the analytical system on which determination of the sequence ultimately depends after all the preparative chemical and enzymatic work has been completed.

Another advance, which has had enormous implications in many areas of molecular biology, is the discovery and preparation of enzymes that cleave nucleic acids with high specificity. Foremost among these is a group called restriction endonucleases, which recognize specific base sequences and cleave both strands of the DNA in these sequence regions. These enzymes have their utility in sequence work in that they can be used to break giant DNA molecules into fragments of workable size, a major objective in nucleic acid sequencing.

Beyond the use of restriction endonucleases, a variety of chemical modifications of nucleic acids have been devised so that a polynucleotide can be *chemically* cleaved at any of its four bases. By establishing conditions that cause partial cleavage at each base, a set of products is yielded that represents the number of times a particular base appears in the nucleotide. Thus, a polynucleotide with the following sequence can be cleaved at its four different bases to give the indicated products.

^{32}P-AGTGATCA

Cleavage at A yields	^{32}P-AGTG
	^{32}P-AGTGATC
Cleavage at G yields	^{32}P-A
	^{32}P-AGT
Cleavage at C yields	^{32}P-AGTGAT
Cleavage at T yields	^{32}P-AG
	^{32}P-AGTGA

An analysis of the entire complement of products reveals that the products differ from one another in length by *only one nucleotide*. Thus they are all separable on PAGE. To detect these separated products, the original polynucleotide is first made radioactive by enzymatically attaching a ^{32}P to its 5′ end. Thus all products will be radioactive, and upon electrophoresis they will resolve as depicted in Figure 24.

The sequence of the polynucleotide can be read *directly from the gel*, the band moving farthest from the origin being the smallest fragment. Each successive band in ascending order represents a position in the polynucleotide. Using similar principles, RNA can also be rapidly sequenced.

In some cases the sequence of a particular RNA is more conveniently determined by sequencing its DNA template. One advantage of this approach is that with the aid of restriction endonuclease enzymes a given DNA segment can be incorporated into a bacterial genome and many copies of it made. Thus large amounts of DNA segments can be harvested, once again using restriction endonucleases, and these can be subjected to sequence analysis. The RNA sequence in question can then be inferred from the known DNA sequence.

Sequencing techniques of this sort have been used in elucidating the structure of ribosomal 16S and 23S RNAs (see pp. 576 and 579).

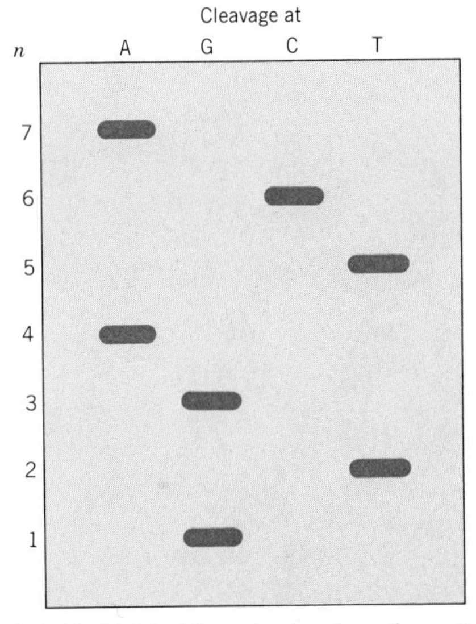

FIGURE 24 The electrophoretic pattern obtained on four samples of an oligonucleotide by cleaving at its four bases. The sequence can be read directly from bottom (the smallest nucleotide) to the top (the largest).

Glossary

Acrosome A lysosomelike organelle found in the sperm of animals and having a hydrolytic role in fertilization.
Actin A protein that exists in monomeric (G-actin) and polymeric (F-actin) forms; the latter constitutes the basic structure of microfilaments in muscle and nonmuscle cells.
Active Transport The energy-requiring movement of solutes against a concentration gradient.
Actomyosin A term that describes the complex of actin and myosin when these proteins interact to form a contractile system in the cell.
Adenosine Triphosphatase (ATPase) An enzyme that hydrolyzes ATP to adenosine diphosphate (ADP) and inorganic phosphate (P_i).
Affinity Chromatography A technique to separate a molecular species from a mixture based on employing a matrix for which the species has a selective affinity (see Method V).
α-Actinin A protein that is especially abundant in the Z line of muscle fibers (molecular weight, 95,000 daltons).
Amphipathic Describing a molecule possessing both a hydrophilic and a hydrophobic region.
Anaphase A stage in mitosis during which the chromosomes move toward opposite ends of the spindle.
Angstrom (Å) A unit of length commonly used to describe molecule dimensions (1 Å = 10^{-1} nm = 10^{-10} m).
Antibody A protein produced by plasma cells that is capable of specifically binding to the antigen that triggered its production.
Anticodon The group of three bases on a transfer RNA molecule that is complementary to the three-base codon of messenger RNA.
Antigen A substance that is capable of stimulating the production of an immune response when injected into a vertebrate.
Aster The region at a pole of a dividing cell where microtubules radiate out from a pair of centrioles.
Autophagic Vacuole A type of secondary lysosome contrived by fusion of a lysosome with a cytosegresome, a vesicle containing certain of the cell's own organelles.
Autophagy The intracellular digestion of endogenous material of the cell involving lysosomes.
Autoradiography A method used to locate radioactive materials in cells by microscopy of exposed photographic film developed after a period of contact with labeled cells (see Method VI).
Axoneme The microtubular system and associated structures that make up the interior of a cilium or flagellum in eucaryotes.
Azurophil Granule A type of primary lysosome in neutrophils containing myeloperoxidase in addition to hydrolases.

Basement Membrane A specialized form of connective tissue made up of collagen-rich extracellular matrices.
Bimolecular Leaflet The bilayer of amphipathic lipids that makes up the lipoid portion of the membrane.
Bivalent Describing an associated pair of homologous chromosomes as seen in meiosis.

Calmodulin A protein that binds calcium ion (Ca^{2+}) and possesses a regulatory role for a number of cell processes.
Centriole A cylindrical microtubule-containing organelle found at the center of spindle poles in dividing cells and at the bases of cilia and flagella.
Centromere The primary constriction region of the chromosome where the spindle fibers attach to induce movement of chromosomes to the poles during anaphase.
Chemiosmotic Hypothesis A unidirectional flux of protons across a membrane generates a gradient that consists of differences both in hydrogen ion concentration and in electrical potential.
Chitin The major wall component in most filamentous fungi, consisting of linear polymers of *N*-acetylglucosamine units.
Chlorophyll A family of photosynthetic light-capturing pigments located in chloroplast thylakoid membranes or in procaryotic photosynthetic membranes.
Chloroplast The chlorophyll-containing photosynthetic organelle in eucaryotic cells.
Chromatid One half of a newly replicated chromosome, generally joined to its counterpart at the centromere region.
Chromatin The deoxyribonucleoprotein material of the nucleus making up the chromosomes.
Chromosome The gene-containing structure in cells made up of DNA (in procaryotic cells) or DNA plus protein (in eucaryotic cells).
Ciliary necklace Multiple strands of particles that appear in the membranes at the base of somatic cilia.
Cilium An organelle of locomotion possessed by organisms of the phylum *Ciliophora*.
Codon A sequence of three nucleotides in messenger RNA that codes for one amino acid in a protein or for chain initiation or termination.

Colchicine A drug derived from a flowering plant that is capable of disrupting microtubules or blocking their formation.
Collagen The most ubiquitous and abundant fibrous protein of the extracellular matrix in animals.
Concanavalin A (Con A) A type of bifunctional protein called a lectin; it can bind to certain cell surfaces.
Coupling Factor; F_1 Factor The headpiece of the protein complex on mitochondrial inner membranes that conducts ATP synthesis during oxidative phosphorylation.
Cristae Infolded regions of the mitochondrial inner membrane where electron transport and oxidative phosphorylation take place.
Cyclosis A streaming phenomenon involving the cytoplasmic materials in cells of plant leaves.
Cytocavitary Network The dynamic system of interior cell membranes that separates the cell interior into a cytoplasmic and an intracavitary compartment.
Cytochrome Oxidase The terminal enzyme system of aerobic respiration that transfers electrons to molecular oxygen.
Cytochrome *P*-450 An integral transmembrane protein of the endoplasmic reticulum functioning with its reductase in hydroxylation reactions.
Cytochromes Heme-containing proteins that transport electrons by way of iron atom valency changes.
Cytoplasm The extranuclear protoplasmic content of the cell.
Cytosegresome A membrane-bound vesicle containing organelles from the cell interior.
Cytosol The extraorganelle portion of the cytoplasm, in which are found most of the soluble molecules of the cell.

Dalton Unit of mass approximately equal to the weight of a hydrogen atom; used interchangeably with molecular weight.
Dark reactions Light-independent photosynthetic reactions during which carbon dioxide is reduced to carbohydrates.
Desmosome An adhering junction between cells characterized as having a seven-layered region where the cell membranes of adjacent cells are parallel, forming either a spot or a belt.
Dictyosome That part of the Golgi complex where the cisternae are highly stacked, also used to describe the Golgi complex of plants.
Diploid A cell having two sets of homologous chromosomes in the nucleus.
Dynein The protein arm of the microtubule doublet subfiber A of cilia and flagella that possesses ATPase activity.

Electron Transport Chain A group of electron transport molecules that transfer electrons from a donor to an acceptor, usually within the confines of some membrane structure.
Electrophoresis A method of separating molecules based on their charge and mass differences (see Method IV).
Endocytosis The bulk transfer of solutes or particles into a cell by an infolding and pinching off of a portion of plasma membrane.
Endoplasmic Reticulum (ER) The major membrane system distributed throughout the cytoplasm of eucaryotic cells.
Eucaryotes (Greek: true nucleus) Organisms having their genetic material enclosed within a membrane and generally containing a number of membranous organelles in their cytoplasm.
Euchromatin Chromosomes or chromosome regions that are not condensed and are considered active.
Exocytosis The bulk transfer of solutes and particles out of the cell by fusion of an internal vesicle with the plasma membrane followed by the expulsion of its contents to the cell exterior.
Exon A segment of DNA in eucaryotes containing a portion of a gene, separated from other portions of the same gene by introns; an "expressed" region.

Facilitated Diffusion The transport of molecules across the plasma membrane down a concentration gradient with the assistance of membrane proteins.
Fibronectin A protein that serves to anchor cells to an extracellular collagen matrix.
Flagellum An organelle of locomotion in both procaryotes and eucaryotes. There is essentially no structural similarity between the flagella of procaryotes and eucaryotes.
Fluid Mosaic Model A model of cell membranes depicting proteins as floating in a bilayer of lipid molecules.
Fluorescent Antibody Technique The detection of selected molecules in cells by interacting them with a specific antibody conjugated with a fluorescent dye (see Method VII).
Freeze-fracture Method A technique for preparing tissues or cells for electron miscroscopy by rapid freezing and fracturing to expose surface structures (see Method I).
Frets Chloroplast thylakoids that stretch through the stroma between grana.
Furrowing A constriction of the plasma membrane as a ring that promotes the formation of the daughter cells from the parent cell in animal systems.

Gap Junction A site of communication between cells characterized structurally as a disc-shaped domain containing particles where the two membranes of adjacent cells are separated by a gap.
GERL *G*olgi-associated *e*ndoplasmic *r*eticulum involved in the formation of *l*ysosomes.
Glucans Glucose polymers found in an array of forms in fungal cell walls.
Glycolate Pathway A sequence of reactions that takes place in leaf peroxisomes in conjunction with the carbon cycle of chloroplasts.
Glycolipids Lipids that contain covalently bound carbohydrates.
Glycolysis The anaerobic oxidation to pyruvic acid of monosaccharides such as glucose.
Glycoprotein Proteins that contain one or more covalently bound carbohydrate residues.
Glyoxylate Cycle A metabolic pathway that involves part of the Krebs cycle plus two unique enzymes, malate synthetase and isocitrate lyase.
Glyoxysome A class of microbody found in plant endosperm that contains enzymes of the glyoxylate cycle in addition to catalase and oxidases.
Golgi Complex A region of specialized

membranes that functions in processing and packaging proteins prior to their secretion from the cell or their incorporation into membranes.

Haploid Describing a cell having a single set of chromosomes.

HeLa Cells A line of human cervical carcinoma (cancer) cells that grow readily in tissue culture and are extensively used for research purposes.

Hemicellulose A varied group of polysaccharides in plant cell walls characterized as branched heteropolymers not possessing charged groups.

Heterochromatin Highly condensed chromosomes or chromosome regions found in the interphase nucleus.

Heterophagy Intracellular digestion of exogenous materials involving lysosomes.

Histone A family of basic proteins of the chromosome, especially rich in arginine and lysine.

Hydrophilic Describing molecules or their parts that are attracted to and stable in water.

Hydrophobic Describing molecules or their parts that are repelled by water and not stable in it.

In situ (Latin: "in place") Describing an organelle or component that is in its original place in the cell or organism.

In vitro (Latin: "in glass") Describing experiments done on isolated cells, tissues, or cell-free extracts.

In vivo (Latin: "in life") Describing experiments done on or within intact living organisms.

Integral Membrane Protein A protein embedded in the lipid bilayer of membranes, often protruding at one or both surfaces.

Intermediate Filaments Hollow cylindrical intracellular fibers 10 nm in diameter that differ somewhat in composition depending on their tissue source: desmin, tonofilaments, vimentin, glial filaments.

Interphase The "resting" state of the eucaryotic nucleus between the stages of mitosis or meiosis, consisting of G_1, S, and G_2 periods in the cell cycle.

Intron A segment of DNA in eucaryotes that separates the DNA of a given gene into pieces; an untranslated intervening sequence.

Kinetochore Site of attachment of the spindle microtubule to the chromosomal centromere.

Krebs Cycle Pathway for oxidative metabolism of acetyl groups resulting in their "conversion" to carbon dioxide and water and a reduction of coenzymes; also called the citric acid cycle or the tricarboxylic acid cycle.

Lampbrush Chromosomes Very large bivalent chromosomes in amphibian oocyte nuclei containing paired regions of looped-out chromatin fibers, characteristic of the diplotene stage of meiosis I prophase.

Lectins Bifunctional proteins that can attach to cell surfaces and agglutinate cells.

Light Reactions The light-dependent reactions in photosynthesis by which radiant energy is converted to the chemical bond energy of ATP and the reducing power of nicotinamide adenine dinuclotide phosphate (NADPH).

Liposomes Single bilayer or multilamellar vesicles formed by dispersing polar lipids in water.

Loculus The interior of a chloroplast thylakoid.

Lysosome A membrane-bound cytoplasmic organelle containing a wide variety of hydrolytic enzymes capable of breaking down most types of biological molecule.

Mannan Component The major polysaccharide in most yeast with linear and branched polypyrannose structures.

Matrix General term describing the interiors of organelles where the medium is assumed to be relatively unstructured.

Meromyosin, Heavy (HMM) A portion of the myosin molecule containing ATPase activity and an ability to bind calcium ion (Ca^{2+}); it may be cleaved from the intact myosin molecule by trypsin and used to "decorate" actin, with which it binds.

Metaphase The stage of mitosis during which chromosomes are highly coiled and compacted and are aligned along the equatorial plane of the mitotic spindle before moving to opposite poles in the cell.

Microbody A membrane-bound cytoplasmic organelle containing catalase and a variety of oxidase enzymes; consists of two classes: peroxisomes and glyoxysomes.

Microfilaments Intracellular fibers of 7 nm diameter and indeterminant length made up of polymerized actin (F-actin). Microfilaments function in the maintenance of cell structure and movement.

Micrometer (μm) A unit of length commonly employed to describe cellular dimensions, equal to 10^3 nm or 10^4 Å.

Micropinocytosis The engulfment by cells of tiny amounts of fluid into vesicles approximately 70 nm in diameter.

Microsome A functional term describing the endoplasmic reticulum-rich fraction of a tissue produced by homogenization and harvested by centrifugation.

Microtrabecular Lattice A network of filaments 2 to 3 nm thick that act in the cytoplasm as linkers between the other major fiber systems.

Microtubule A cylinder of 24 nm in diameter and indeterminant length consisting of aggregated α and β tubulin molecules; the basic structural unit of spindle fibers, ciliary subfibers, and centriole subfibers.

Microtubule Organizing Center (MTOC) A location in the cell containing a centriole or basal body where microtubules appear to become organized and grow.

Microvilli Projections of the plasma membranes of animal epithelial cells that considerably increase the surface area for absorptive purposes.

Mitochondrion (Greek: *mito*–a thread; *chondros*—a grain) A cytoplasmic organelle bounded by two membranes, the inner one of which is infolded (cristae) and contains the components of electron transport and oxidative phosphorylation; the matrix contains the enzymes of the Krebs cycle.

Mitosis The events that take place in the nucleus to separate the chromosomes into two daughter nuclei.

Mononuclear Phagocyte System A system of cells found in a variety of tissues that bear the relationship of morphology, origin, and function (phagocytosis).
Myosin A protein that, along with actin, constitutes the contractile system of both muscle and nonmuscle cells, containing ATPase activity.

Nonhistone Proteins A very heterogeneous group of proteins associated with chromosomes that are concerned with DNA replication, transcription, and regulation, as well as with other cell functions.
Nuclear Cortex A fibrous, electron-dense material that coats the nuclear side of the inner membrane of the nuclear envelope.
Nuclear Envelope The double membrane system that surrounds the chromosomes in eucaryotic cells.
Nucleoid A region in procaryotic cells wherein the DNA is segregated (but not membrane bound) and is discernible by microscopic techniques.
Nucleolar Organizing Region (NOR) The specific region on a chromosome containing the genes of ribosomal RNA and constituting the site at which the nucleolus is located.
Nucleolus A depot region of the nucleus in which ribosomal RNA is synthesized and ribosomal subunits assembled in conjunction with a specific region of a chromosome.
Nucleoplasm The unstructured nonchromatin matrix portion of the nucleus, comparable to the cytosol in the cytoplasm.
Nucleosome A structural unit of chromatin in eucaryotes consisting of the core particle (146 base pairs of DNA plus an octamer of histones) plus the spacer DNA (54 base pairs) and histone H1.
Nucleus The chromosome-containing compartment of the eucaryotic cell, bounded by a double membrane system.

Organelle A component of the cell that is normally distinguishable from all other cell parts both structurally and functionally.
Oxidative Phosphorylation A type of ATP synthesis coupled with the flow of electrons from a reduced substrate to molecular oxygen.

Pectins A complex family of polysaccharides in plant cell walls characterized in part by their content of branched chains and acidic groups.
Peptidoglycan The basic structural unit of the bacterial cell wall, containing *N*-acetylglucosamine and peptide units.
Peripheral Protein Term designating a protein that interacts only with the hydrophilic surface of the lipid bilayer of a membrane, hence is easily extractable from it.
Peroxisome A class of microbody (containing catalase and oxidases) found in both plants and animals.
Phagocytosis (Greek: *phagein*—to eat; *kytos*—hollow or cell) Endocytosis that involves the engulfment of particulate material.
Photophosphorylation A type of ATP synthesis coupled with the flow of electrons from water to a coenzyme, driven by radiant energy.
Photorespiration Light-stimulated production of carbon dioxide involving chloroplasts and leaf peroxisomes.
Photosynthesis The light-dependent and light-independent reactions that culminate in the production of carbohydrate from carbon dioxide and water.
Photosystem I (PS I) A reaction system in photosynthesis that transfers electrons from PS II to nicotinamide adenine dinucleotide phosphate (NADPH) without the evolution of molecular oxygen.
Photosystem II (PS II) A reaction system in photosynthesis that transfers electrons from water to PS I, during which process molecular oxygen is evolved.
Pinocytosis (Greek: *pinein*—to drink; *kytos*—hollow or cell) Endocytosis that involves the engulfment of soluble rather than particulate materials.
Polyacrylamide Gel Electrophoresis (PAGE) A type of electrophoresis using polyacrylamide gel as the support through which molecules possessing charge move and are separated (see Method IV).
Polysome A string of aggregated ribosomes, connected by messenger RNA.
Polytene Chromosomes Giant chromosomes generated during early larval development in cells of *dipterans* as a result of cessation of mitotic divisions without a halt in DNA replication.
Primary Cell Wall The first wall to form around plant cells, having randomly oriented microfibrils.
Procaryote (Greek: primitive nucleus) Small, comparatively simple unicellular organisms in which the genetic material is not surrounded by a membrane: bacteria and cyanobacteria.
Procollagen A hydroxylated and glycosylated precursor of collagen that is secreted by the cell before it is processed.
Prophase The first stage of mitosis during which chromosomes thicken and become visible by light microscopy.
Protoplast A cell without any structural wall material exterior to its plasma membrane.
Puff Generally describing an expanded region of giant polytene chromosomes where transcription is taking place.

Registration Peptides Describing regions of the α chains of collagen that promote their pairing to form the procollagen molecule.
Residual Bodies Aged varieties of lysosomes that contain undigested materials.
Reticuloendothelial System A widely divergent collection of phagocytic cells in a variety of tissues.
Ribophorins Integral membrane proteins of the rough endoplasmic reticulum that appear to be receptor and binding sites for ribosomes.
Ribosomal DNA (rDNA) The genes present in the nucleolar organizing region of chromosomes that code for ribosomal RNA.
Ribosomal RNA (rRNA) The collection of ribonucleic acids that along with proteins make up the ribosome structure.
Ribosomes Complex ribonucleoprotein particles that, in conjunction with messenger and transfer RNA and several other factors, constitute the site of protein synthesis in procaryotic and eu-

caryotic cells and in chloroplasts and mitochondria.

Rough Endoplasmic Reticulum (RER) Endoplasmic reticulum with ribosomes attached.

S Period The stage in the cell cycle during which DNA replication occurs.

Scanning Electron Microscope (SEM) An electron microscope that is particularly useful for observing the surface of a structure; the results give a three-dimensional effect (see Method I).

Secondary Cell Wall A highly ordered wall, containing parallel microfibrils, that surrounds mature plant cells.

Secretory Pathway A route taken by secretory proteins that includes the endoplasmic reticulum where they are formed, the Golgi complex, and their sites of storage before leaving the cell.

Sedimentation Coefficient A quantitative measure of the rate of sedimentation of a given substance in a centrifugal field, expressed in Svedberg units (S).

Signal Hypothesis A proposal pertaining to the mechanism whereby proteins destined to be secreted are sequestered into channels of the endoplasmic reticulum because they contain a unique sequence of amino acids that signals an interaction between the growing protein and the membranes of the endoplasmic reticulum.

Smooth Endoplasmic Reticulum (SER) Endoplasmic reticulum that does not have attached ribosomes.

Specific granule A type of granule in neutrophils containing hydrolases having neutral pH optima.

Spindle A characteristic grouping of microtubules that occurs during nuclear division to align and move the chromosomes at metaphase and anaphase.

Spindle Fibers Microtubules that make up the mitotic apparatus, extending either from pole to pole or from one pole to a chromosome.

Stroma The unstructured matrix material of the chloroplast outside of the grana and stroma thylakoids.

Teichoic Acids Polymers of derivatized ribitol or glycerol phosphates attached to the peptidoglycan of certain Gram-positive bacteria.

Teichuronic Acid A heteropolysaccharide of *N*-acetylmannosaminuronic acid found attached to the peptidoglycan of certain Gram-positive bacteria.

Thylakoid (Greek: baggy trousers) A disc-shaped or elongated enclosed membrane space within chloroplasts where the light-dependent reactions of photosynthesis take place.

Tight Junction A type of contact between cells where a true fusion of cell membranes takes place to form an impermeability barrier between cells.

Transcription The process by which a particular base sequence of DNA is converted into a complementary base sequence of RNA.

Translation The process by which a particular base sequence in messenger RNA is used for the construction of a correspondingly specific sequence of amino acids in a protein.

Tritosomes Lysosomes that have taken up the detergent Triton WR-1339.

Tropocollagen The building block molecule of collagen, arising from procollagen in the extracellular matrix by enzymatic processing.

Tropomyosin A protein that interacts with troponin and actin in striated muscle to regulate contraction.

Troponin A regulatory protein in striated muscle that binds calcium ion (Ca^{2+}), causing conformational changes in the actomyosin system to affect contraction.

Tubulins Globular protein subunits of two types, α and β, that polymerize into a regular helix to form microtubules.

Unit Membrane A concept of membrane structure consisting of a bilayer of polar lipids coated with extended proteins.

Vacuole A membrane-enclosed sac in the cell cytoplasm holding a fluid that may contain dissolved molecules and small particles.

Vesicle A small, spherical, membrane-bound organelle containing a fluid with dissolved molecules.

Zymogen Granule A membrane-enclosed compartment containing digestive enzyme precursors, especially prominent in secretory cells such as the pancreatic acinar cells.

CREDITS

SECTION OPENERS

Section 1, page 18 Electron micrograph courtesy of Drs. Eva Frei and R. D. Preston.

Section 2, page 106 Electron micrograph courtesy of Dr. Daniel Branton.

Section 3, page 190 Electron micrograph courtesy of Dr. Robert P. Bolender. Reproduced from R. P. Bolender, D. Paumgartner, G. Losa, D. Muellener, and E. R. Weibel, *J. Cell Biol.* 77:565(1978) by copyright permission of The Rockefeller University Press, New York.

Section 4, page 246 Electron micrograph courtesy of Dr. Daniel Friend. Reproduced from *The Cell,* D. W. Fawcett, copyright 1981, by permission of W. B. Saunders Co., Philadelphia.

Section 5, page 316 Electron micrograph courtesy of Dr. Richard L. Ornberg. Reproduced from R. L. Ornberg and T. S. Reese, *J. Cell Biol.* 90:40(1981) by copyright permission of The Rockefeller University Press, New York.

Section 6, page 386 Electron micrograph courtesy of Dr. Harry T. Horner.

Section 7, page 470 Electron micrograph courtesy of Dr. Gunter F. Bahr.

Section 8, page 608 Electron micrograph courtesy of Dr. Keith R. Porter.

CHAPTERS

Chapter 1

Figure 1.1 Models reproduced from *Biology, A Contemporary Perspective,* G. C. Stephens and B. B. North, copyright 1974, by permission of John Wiley & Sons, Inc., New York.

Figure 1.2 Reproduced from *The Microbial World,* R. Y. Stanier, M. Doudoroff, and E. A. Adelberg, copyright 1970, by permission of Prentice-Hall, Englewood Cliffs, New Jersey.

Figure 1.3a Reproduced from F. Morel, R. Baker, and H. Wayland, *J. Cell Biol.* 48:91(1971) by copyright permission of The Rockefeller University Press, New York.

Chapter 1 Marginal Comments

Page 4 and page 5, top Electron micrographs courtesy of Dr. Harry Horner.

Page 6, top Reproduced from H. Shio and P. B. Lazarow, *J. Histochem. Cytochem.* 29:1263(1981) by permission of Elsevier North Holland, New York and by copyright permission of The Histochemical Society, Inc., New York.

Page 7, center Reproduced from *The Cell,* D. W. Fawcett, copyright 1981, by permission of W. B. Saunders Co., Philadelphia.

Page 7, bottom Electron micrograph courtesy of Dr. T. E. Weier.

Chapter 2

Figure 2.1 Reproduced from *Biology of Microorganisms,* Thomas D. Brock, 3rd ed., copyright 1979, p. 250, by permission of Prentice-Hall, Inc., Englewood Cliffs, N.J.

Figure 2.2 Reproduced from L. M. Santo, H. R. Hohl, and H. A. Frank, *J. Bacteriol.* 99:824(1969), by permission of the American Society for Microbiology, Washington, D.C.

Figure 2.4 Reproduced from *Cell Physiology,* Arthur C. Giese, 5th ed., copyright 1979, W. B. Saunders Co., by permission of Holt, Rinehart and Winston, CBS College Publishing, New York.

Figure 2.5 Adpated from The Effects of Pressure on Organisms: A Summary of Progress, J. A. Kitching, in *The Effects of Pressure on Organisms,* M. A. Sleigh and A. G. MacDonald, eds., copyright 1972, by permission of Cambridge University Press.

Figure 2.6 Reproduced from Protein and Nucleic Acid Synthesis in *Escherichia coli:* Pressure and Temperature Effects, J. V. Landau, *Science* 153:1273(1966), by permission of the American Association for the Advancement of Science, Washington, D.C.

Figure 2.7 Reproduced from The Adaptation of Enzymes to Pressure in Abyssal and Midwater Fishes, P. W. Hochachka, T. W. Moon, and T. Mustafa, in *The Effects of Pressure on Organisms,* M. A. Sleigh and A. G. MacDonald, eds., copyright 1972, by permission of Cambridge University Press.

Figure 2.9 Reproduced from H. Freeze and T. D. Brock, *J. Bacteriol.* 101:541(1970) by copyright permission of the American Society for Microbiology, Washington, D. C.

Figure 2.10 Reproduced from T. A. Langworthy, *Biochim. Biophys. Acta* 487: 37(1977) by copyright permission of Elsevier North Holland Biomedical Press, Amsterdam.

Figure 2.11 Reproduced from *Cell Biology,* P. Sheeler and D. Bianchi, copyright 1980, by permission of John Wiley, & Sons, Inc., New York.

Figures 2.15 and 2.16 Adapted from The biosynthesis of Procollagen, M. E. Grant and D. S. Jackson, in *Essays in Biochemistry,* vol. 12, P. N. Campbell and W. N. Aldridge, eds., copyright 1976, by permission of the Biochemical Society (London).

Figure 2.17 Reproduced from J. C. Murray, G. Stingl, H. K. Kleinman, G. R. Martin, and S. I. Katz, *J. Cell Biol.* 80:197(1979) by copyright permission of The Rockefeller University Press, New York.

Tables 2.1, 2.4, and 2.5 Reproduced from *Thermophilic Microorganisms and Life at High Temperatures,* Thomas D. Brock, copyright 1978, by permission of Springer-Verlag, New York.

Table 2.2 Reproduced from *Cell Physiology,* Arthur C. Giese, 5th ed., copyright 1979, W. B. Saunders Co., by permission of Holt, Rinehart, and Winston, CBS College Publishing, New York.

Table 2.3 Reproduced from Effects of Deep-Sea Pressures on Microbial Enzyme Systems, C. E. Zobell and J. Kim, in *The Effects of Pressure on Organisms,* M. A. Sleigh and A. G. MacDonald, eds., copyright 1972, by permission of Cambridge University Press.

Table 2.6 Reproduced from H. K. Kleinman, R. J. Klebe, and G. R. Martin, *J. Cell Biol.* 88:473(1981) by copyright permission of The Rockefeller University Press, New York.

Tables 2.9 and 2.10 Reproduced from N. A. Kefalides in *Mammalian Cell Membranes,* vol. 2, G. A. Jamieson and D. M. Robinson, eds., Butterworths, London, 1977, by copyright permission of G. A. Jamieson.

Chapter 3

Figure 3.4 Reproduced from Biogenesis of the Wall in Bacterial Morphogenesis, H. J. Rogers, in *Adv. Microb. Physiol.* vol. 19, copyright 1979, by permission of Academic Press, Inc. (London) Ltd.

Figure 3.10 Adapted from Composition and Structure of Bacterial Lipopolysaccharide, S. G. Wilkinson, in *Surface Carbohydrates of the Prokaryotic Cell,* I. Sutherland, ed., copyright 1977, by permission of Academic Press, Inc. (London) Ltd.

Figure 3.12 Reproduced from P. J. Mühlradt, V. Wray, and V. Lehmann, *Eur. J. Biochem.* 81:193(1977), by permission of Springer-Verlag, New York.

Figure 3.13 Reproduced, with permission, from V. Braun and K. Hantke in the Annual Review of Biochemistry, vol. 43, copyright 1974 by Annual Reviews, Inc., Palo Alto, CA.

Figure 3.14 Reproduced from A. C. Steven, B. Heggler, R. Müller, J. Kistler, and J. P. Rosenbusch, by copyright permission of The Rockefeller University Press, New York.

Figure 3.15 Reproduced from D. L. Dorset, A. Engel, M. Häner, A. Massalski, and J. P. Rosenbusch, *J. Mol. Biol.* 165:701(1983) by copyright permission of Academic Press, New York.

Figures 3.16, 3.17, and 3.18 Reproduced from The Actions of Penicillin and Other Antibiotics on Bacterial Cell Wall Synthesis, J. L. Strominger, in *Biochemistry of Cell Walls and Membranes,* C. F. Fox, ed., copyright 1975, by permission of Butterworths, London.

Figure 3.19 Reproduced from Teichoic Acids, M. Duckworth, in *Surface Carbohydrates of the Prokaryotic Cell,* I. Sutherland, ed., copyright 1977, by permission of Academic Press, Inc. (London) Ltd.

Figure 3.20 Reproduced from T. E. Rohr, G. N. Levy, N. J. Stark, and J. S. Anderson, *J. Biol. Chem.* 252:3460 and N. J. Stark, G. N. Levy, T. E. Rohr, and J. S. Anderson, *J. Biol. Chem.* 252:3466(1977), by copyright permission of the American Society of Biological Chemists, Inc., Bethesda, Maryland.

Figure 3.21 Adapted from Biosynthesis of Peptidoglycan, J. M. Ghuysen and G. D. Shockman, in *Bacterial Membranes and Walls,* L. Leive, ed., copyright 1973, by permission of Marcel Dekker, New York.

Figures 3.22, 3.23, and 3.24 Reprinted from D. J. Tipper and J. L. Strominger, *Proc. Natl. Acad. Sci.* 54:1133(1965) by courtesy of the National Academy of Science, U.S.A.

Figure 3.26 T. J. Beveridge, University of Guelph/Biological Photo Service.

Figure 3.27 Adapted from H. J. Rogers, *Adv. Microb. Physiol.* 19:1(1979), by permission of Academic Press, Inc. (London) Ltd.

Figure 3.28 Reproduced from an article by A. P. J. Trinci in *Science Progress* 65:75(1978), by copyright permission of Blackwell Scientific Pub. Ltd., Oxford.

Figure 3.29 Reproduced from S. Bartnicki-Garcia and E. Reyes, *Biochim. Biophys. Acta* 165:32(1968), by copyright permission of Elsevier Biomedical Press, Amsterdam.

Figures 3.31 and 3.32 Reproduced from Structure and Biosynthesis of the Mannan Component of the Yeast Cell Envelope, C. Ballou, in *Adv. Microb. Physiol.,* vol. 14, copyright 1976, by permission of Academic Press, Inc. (London) Ltd.

Figure 3.34 Reproduced from Biosynthesis of Cell Walls of Fungi, V. Farkaš, *Microbiol. Rev.,* vol. 43, copyright 1979, by permission of the American Society of Microbiology, Washington, D.C.

Figure 3.38 Reproduced from Biosynthesis of Cell Wall Polysaccharides in Glycoproteins, M. C. Ericson and A. D. Elbein, in *The Biochemistry of Plants,* vol. 3, copyright 1980, by permission of Academic Press, New York.

Figure 3.39 Reproduced from Plant Cell Walls, W. D. Bauer, in *The Molecular Biology of Plant Cells,* H. Smith, ed., copyright 1977, by permission of Blackwell Scientific Pub. Ltd., Oxford.

Figures 3.40, 3.41, and 3.42 Reproduced from *Surface Carbohydrates of the Eukaryotic Cell,* G. M. W. Cook and R. W. Stoddart, copyright 1973, by permission of Academic Press, Inc. (London) Ltd.

Figure 3.43 Reproduced from The Primary Cell Walls of Flowering Plants, A. Darvill, M. McNeil, P. Albersheim, and D. P. Delmer, in *The Biochemistry of Plants,* N. E. Tolbert, ed., copyright 1980, by permission of Academic Press, New York.

Figures 3.45 and 3.48 Based on a model by A. W. Robards. Redrawn with his permission.

Figure 3.47 Adapted from Plant Cell Walls, W. D. Bauer, in *The Molecular Biology of Plant Cells,* H. Smith, ed., copyright 1977, by permission of Blackwell Scientific Publications, Ltd., Oxford, England.

Figures 3.51, 3.52, 3.53, and 3.54 Reproduced from T. H. Giddings, D. L. Brower, and L. A. Staehelin, *J. Cell Biol.* 84:327(1980), by copyright permission of The Rockefeller University Press, New York.

Table 3.1 Reproduced from Chemistry of Cell Wall Polysaccharides, G. O. Aspinall, in *The Biochemistry of Plants,* vol. 3, J. Preiss, ed., copyright 1980, by permission of Academic Press, New York.

Chapter 4

Figure 4.1 Reproduced from J. F. Danielli and E. N. Harvey, *J. Cell. Physiol.* 5:491(1935) by copyright permission of Alan R. Liss, Inc., New York.

Figure 4.2 Reproduced from J. F. Danielli and H. Davson, *J. Cell. Physiol.* 5:498(1935) by copyright permssion of Alan R. Liss, Inc., New York.

Figures 4.4 and 4.16 Reproduced from *The Cell,* D. W. Fawcett, copyright 1981, by permission of W. B. Saunders Co., Philadelphia.

Figure 4.5 Reproduced from J. D. Robertson, *Ann. N. Y. Acad. Sci.* 137:421(1966) by copyright permission of the New York Academy of Sciences, New York.

Figure 4.7 Adapted from The Fluid Mosaic Model of the Structure of Cell Membranes, S. J. Singer and G. L. Nicolson, *Science* 175:720(1972), by copyright permission of the American Association for the Advancement of Science.

Figure 4.8 Reproduced from M. P. Sheetz and S. J. Singer, *J. Cell Biol.* 73:638(1977) by copyright permission of The Rockefeller University Press, New York.

Figure 4.10 and Table 4.1 Reproduced from *The Molecular Biology of Cell Membranes,* P. J. Quinn, copyright 1976, by permission of Macmillan Press, Ltd., London.

Figure 4.15 Reproduced by permission from *Models of Cell Membranes,* A. D. Bangham, Hospital Practice 83:79(1973) and from *Cell Membranes: Biochemistry, Cell Biology and Pathology,* G. Weissman and R. Claiborne, eds., H. P. Publishing Co., New York, 1973.

Figure 4.17 Adapted from S. J. Singer, *J. Supramolec. Struc.* 6:313(1977) by permission of Alan R. Liss, Inc., New York.

Figure 4.23 Reproduced from R. M. Williams and D. Chapman, *Progr. Chem. Fats and Other Lipids,* 11:1(1970), by permission of Pergamon Press, New York.

Figure 4.24 Reproduced from B. D. Ladbrooke, R. M. Williams, and D. Chapman, *Biochim. Biophys. Acta* 150:333(1968), by permission of Elsevier Biomedical Press, Amsterdam.

Figure 4.27 Adapted from Membrane Fluidity and Cellular Functions, S. J. Singer, in *Control Mechanisms in Development,* R. H. Meints and E. Davies, eds., by permission of Plenum Publishing Corp., New York, 1975.

Figure 4.28 Adapted from *Biochemistry,* L. Stryer, copyright 1981, by permission of W. H. Freeman and Company, San Francisco.

Figure 4.31 Adapted, with permission, from S. J. Singer in the Annual Review of Biochemistry, vol. 43, copyright 1974 by Annual Reviews, Inc., Palo Alto, CA.

Figure 4.34 Adapted from Synthesis and Segregation of Secretory Proteins: The Signal Hypothesis, G. Blobel, in *International Cell Biology,* B. R. Brinkley and K. R. Porter, eds., copyright 1977, by permission of The Rockefeller University Press, New York.

Figure 4.36 Adapted, with permission, from W. Wickner in the Annual Review of Biochemistry, vol. 48, copyright 1979 by Annual Reviews, Inc., Palo Alto, CA.

Figure 4.37 Reproduced from D. D. Sabatini, G. Kreibach, T. Morimoto, and M. Adesnik, *J. Cell Biol.* 92:1(1982) by copyright permission of The Rockefeller University Press, New York.

Tables 4.2 and 4.3 Reproduced, with permission, from S. J. Singer in the Annual Review of Biochemistry, vol. 43, copyright 1974 by Annual Reviews, Inc., Palo Alto, CA.

Table 4.4 Reproduced, with permission, from W. Wickner in the Annual Review of Biochemistry, vol. 48, copyright 1979 by Annual Reviews, Inc., Palo Alto, CA.

Chapter 5

Figure 5.1 Plates 6 and 7 from Paul Ehrlich's "On Immunity with Special Reference to Cell Life" Croonian lecture given to the Royal Society, March 22, 1900.

Figures 5.2 and 5.3 Reproduced from The Complex Carbohydrates of Mammalian Cell Surfaces and Their Biological Roles, R. C. Hughes, in *Essays in Biochemistry,* vol. 11, P. N. Campbell and W. N. Aldridge, eds., copyright 1975, by permission of the Biochemical Society, London.

Figures 5.7 and 5.8 Adapted from Cell Surface Receptors, a Biological Perspective, M. F. Greaves, in *Receptors and Recognition,* P. Cuatrecasas and M. F. Greaves, eds., copyright 1976, by permission of Chapman and Hall, London.

Figures 5.9, 5.10, and 5.11 Reproduced from Cell-Surface cAMP Receptors in *Dictyostelium,* P. C. Newell and I. A. Mullens, in *Cell-Cell Recognition,* A. Curtis, ed., copyright 1978, by permission of Cambridge University Press.

Figure 5.13 Reproduced from *Cell Biology,* P. Sheeler and D. Bianchi, copyright 1980, by permission of John Wiley & Sons, Inc. New York.

Figures 5.15, 5.16, and 5.17 Adapted from The Complex Carbohydrates of Mammalian Cell Surfaces and Their Biological Roles, R. C. Hughes, in *Essays in Biochemistry,* vol. 11, P. N. Campbell and W. N. Aldridge, eds., copyright 1975, by permission of the Biochemical Society, London.

Figures 5.18, 5.20, 5.22, and 5.24 Reproduced from Junctions Between Cells, N. B. Gilula, in *Cell Communication,* R. P. Cox, ed., copyright 1974, by permission of John Wiley & Sons, Inc. New York.

Figures 5.23 and 5.28 Reproduced from *Cell Biology,* P. Sheeler and D. E. Bianchi, 2nd ed., copyright 1983, by permission of John Wiley & Sons, Inc. New York.

Figure 5.26 Adapted from Junctions Between Living Cells, L. Andrew Staehelin and B. E. Hull, *Sci. Am.* 238:140(1978) by copyright permission of Scientific American, Inc., New York.

Figure 5.27 Adapted from a model in *Cell* 28:441(1982) by permission of P. Pinto da Silva and the Massachusetts Institute of Technology.

Figures 5.29 and 5.30 Adapted from H. K. Kleinman, R. J. Kleve, and G. R. Martin, *J. Cell Biol.* 88:473(1981) by copyright permission of The Rockefeller University Press, New York.

Figure 5.31 Adapted from *Biochemistry,* L. Stryer, copyright 1981, by permission of W. H. Freeman and Company, San Francisco.

Table 5.1 Reprinted from H. K. Kleinman, R. J. Kleve, and G. R. Martin, *J. Cell Biol.* 88:473(1981) by copyright permission of The Rockefeller University Press, New York.

Chapter 6

Figures 6.2 and 6.6 Reproduced from R. P. Bolender, O. Paumgartner, G. Losa, O. Muellener, and E. R. Weibel, *J. Cell Biol.* 77:565(1978) by copyright permission of The Rockefeller University Press, New York.

Figure 6.3 Reproduced from *Cell Fine Structure,* Thomas L. Lentz, copyright 1971, by permission of W. B. Saunders Co., Philadelphia.

Figure 6.5 Reproduced from A. K. Christensen and D. W. Fawcett, *J. Biophys. Biochem. Cytol.* 9:653(1961) by copyright permission of the Rockefeller University Press, New York.

Figure 6.9 Reproduced from T. H. Giddings and L. A. Staehlin, *J. Cell Biol.* 85:147(1980) by copyright permission of The Rockefeller University Press, New York.

Figure 6.10 Adapted from J. J. Geuze, M. F. Kramer, and J. C. H. deMan in *Mammalian Cell Membranes,* vol. 2, G. A. Jamieson and D. M. Robinson, eds., copyright 1977, Butterworths, London, by copyright permission of G. A. Jamieson.

Figure 6.11 Adapted from R. W. Estabrook, J. Wenningloer, B. S. S. Masters, H. Jonen, T. Matsubara, R. Ebel, D. O'Keeffe, and J. A. Peterson in *The Structural Basis of Membrane Function,* Y. Hatefi and L. Djavadi-Ohaniance, eds., copyright 1976, by permission of Academic Press, New York.

Figures 6.12 and 6.13 Reproduced from Multiple Forms of Cytochrome P-450: Criteria and Significance, E. F. Johnson, in *Reviews in Biochemical Toxicology,* vol. 1, copyright 1979, by permission of Elsevier Science Pub. Co., New York.

Figures 6.16 and 6.17 Reproduced from *Biochemistry,* D. E. Metzler, copyright 1977, by permission of Academic Press, New York.

Figure 6.20 Adapted from The Intracellular Organization of Protein Synthesis, H. N. Munro and P. M. Steinert, in *MTP International Review of Science: Biochemistry,* Series 1, vol. 7, copyright 1975, by permission of Butterworths, London.

Table 6.1 Reproduced, with permission, from J. W. DePierre and L. Ernster in the Annual Review of Biochemistry, vol. 46, copyright 1977 by Annual Reviews, Inc.

Table 6.2 Reproduced from Hepatic Cytochrome P-450, A. J. Paine, in *Essays in Biochemistry,* P. N. Campbell and R. D. Marshall, eds., vol. 17, copyright 1981, by permission of the Biochemical Society, London.

Chapter 7

Figures 7.1, 7.11, 7.12, and 7.13 Reproduced from H. Shio and P. B. Lazarow, *J. Histochem. Cytochem.* 29:1263(1981) by permission of Elsevier North Holland, New York and by copyright permission of The Histochemical Society, New York.

Figures 7.2 and 7.3 Reproduced from S. E. Frederick and E. H. Newcomb, *J. Cell Biol.* 43:343(1969), by copyright permission of The Rockefeller University Press, New York.

Figure 7.4 Reproduced from F. Leighton, B. Poole, H. Beaufay, P. Baudhuin, J. Coffee, S. Fowler, and C. deDuve, *J. Cell Biol.* 37:482(1968) by copyright permission of The Rockefeller University Press, New York.

Figure 7.5 Reproduced from P. Baudhuin, H. Beaufay, and C. deDuve, *J. Cell Biol.* 26:219(1965) by copyright permission of The Rockefeller University Press, New York.

Figure 7.9 and Table 7.2 Reproduced, with permission, from N. E. Tolbert, in the Annual Review of Plant Physiology, vol. 22, copyright 1971 by Annual Reviews, Inc.

Figures 7.10 and 7.14 Reproduced from Biogenesis of Peroxisomes and the Peroxisome Reticulum Hypothesis, P. B. Lazarow, H. Shio, and M. Robbi, in *Biological Chemistry of Organelle Formation,* T. Bücher, W. Sebald, and H. Weiss, eds., copyright 1980, by permission of Springer-Verlag, Berlin.

Figure 7.15 Adapted, with permission, from H. Beevers, in the Annual Review of Plant Physiology, vol. 30, copyright 1979 by Annual Reviews, Inc., Palo Alto, CA.

Table 7.1 Reproduced from N. E. Tolbert and R. P. Donaldson, in *Mammalian Cell Membranes,* vol. 2, G. A. Jamieson and D. M. Robinson, eds., copyright 1977, Butterworths, London, by copyright permission of G. A. Jamieson.

Chapter 8

Figure 8.2 Reproduced from *Cell Fine Structure,* Thomas L. Lentz, copyright 1971, by permission of W. B. Saunders Co., Philadelphia.

Figure 8.4 Reproduced from Factors Thought to Contribute to the Regulation of Egress of Cells from Marrow, M. A. Lichtman, J. K. Chamberlain, and P. A. Santillo, in *The Year in Hematology,* A. S. Gordon, R. Silber, and J. LoBue, eds., vol. 2, copyright 1978, by permission of Plenum Press, New York.

Figures 8.5 and 8.7 Adapted from *Phagocytic Engulfment and Cell Adhesiveness,* C. J. van Oss, C. F. Gillman and A. W. Neumann, copyright 1975, by permission of Marcel Dekker, New York.

Figure 8.6 Reproduced from Kinetics of the Phagocytic Function of Reticulo-endothelial Macrophages *in vivo,* C. Stiffel, D. Mouton, and G. Biozzi, in *Mononuclear Phagocytes,* R. van Furth, ed., copyright 1970, by permission of Blackwell Scientific Publications, Ltd., Oxford.

Figure 8.8 Reproduced from D. F. Bainton, *J. Cell Biol.* 58:249(1973) by copyright permission of The Rockefeller University Press, New York.

Figure 8.9 Reproduced from J. Boyles and D. F. Bainton, *Cell* 24:905(1981) by copyright permission of the Massachusetts Institute of Technology.

Figures 8.10, 8.11, 8.12, and 8.13 Reproduced from E. D. Korn and R. A. Weisman, *J. Cell Biol.* 34:219(1967) by copyright permission of The Rockefeller University Press, New York.

Figure 8.14 Reproduced from *Fundamental Concepts of Biology,* G. E. Nelson, G. G. Robinson, and R. A. Boolootian, copyright 1974, by permission of John Wiley & Sons, Inc. New York.

Figure 8.15 Reproduced from *The Cell,* D. W. Fawcett, copyright 1981, by copyright permission of W. B. Saunders Co., Philadelphia.

Figure 8.16 Reproduced from D. W. Fawcett, *J. Histochem. Cytochem.* 13:75(1965) by copyright permission of the Histochemical Society, New York.

Figure 8.17 Adapted from Endocytosis, T. P. Stossel, in *Receptors and Recognition,* P. Cuatrecasas and M. F. Greaves, eds., copyright 1977, by permission of Chapman and Hall, London.

Table 8.1 Reproduced from *The Macrophage,* B. Vernon-Roberts, copyright 1972, by permission of Cambridge University Press, Cambridge.

Tables 8.2 and 8.3 Reproduced from *Mononuclear Phagocytes,* R. van Furth, ed., copyright 1970, by permission of Blackwell Scientific Publications, Ltd., Oxford.

Table 8.4 Reproduced from Chemotaxis of Mononuclear and Polymorphonuclear Phagocytes, E. Sorkin, J. F. Borel, and V. J. Stecher, in *Mononuclear Phagocytes,* R. van Furth, ed., copyright 1970, by permission of Blackwell Scientific Publications, Ltd., Oxford.

Tables 8.5, 8.6, 8.7, and 8.8 Reproduced from *Phagocytic Engulfment and Cell Adhesiveness,* C. J. van Oss, C. F. Gillman, and A. W. Neumann, copyright 1975, by permission of Marcel Dekker, New York.

Table 8.9 Reproduced from Endocytosis and Intracellular Digestion, Z. A. Cohn, in *Mononuclear Phagocytes,* R. van Furth, ed., copyright 1970, by permission of Blackwell Scientific Publications, Ltd., Oxford.

Chapter 9

Figures 9.1 and 9.2 Reproduced from The Lysosome in Retrospect, C. de Duve, in *Lysosomes in Biology and Pathology,* vol. 1, J. T. Dingle and H. B. Fell, eds., copyright 1969, by permission of Elsevier Biomedical Press, Amsterdam.

Figure 9.3 Reproduced from P. Baudhuin, H. Beaufay, and C. de Duve, *J. Cell Biol.* 26:219(1965) by copyright permission of The Rockefeller University Press, New York.

Figure 9.5 Reproduced from *The Cell,* D. W. Fawcett, copyright 1981, by permission of W. B. Saunders Co., Philadelphia.

Figure 9.8 Reproduced from W. Halperin, *Planta (Berlin)* 88:91(1969) by copyright permission of Springer-Verlag, New York.

Figures 9.9, 9.10, 9.11, and Table 9.2 Reproduced from Methods for the Isolation of Lysosomes, H. Beaufay, in *Lysosomes—a Laboratory Handbook,* J. T. Dingle, ed., copyright 1972, by permission of Elsevier Biomedical Press, Amsterdam.

Figure 9.12 Adapted from Lysosomal Membranes, J. A. Lucy, in *Lysosomes in Biology and Pathology,* vol. 2., J. T. Dingle and H. B. Fell, eds., copyright 1969, by permission of Elsevier Biomedical Press, Amsterdam.

Figure 9.13 Adapted from Nature of Lysosomal Enzymes, J. M. W. Bouma, in *Enzyme Therapy in Lysosomal Storage Diseases,* J. M. Tager, G. J. M. Hooghwinkel, and W. Th. Daems, eds., copyright 1974, by permission of Elsevier Biomedical Press, Amsterdam.

Figure 9.14 Reproduced from Lysosomal Enzymes, A. J. Barrett, in *Lysosomes—a Laboratory Handbook,* J. T. Dingle, ed., copyright 1972, by permission of Elsevier Biomedical Press, Amsterdam.

Figure 9.15 Reproduced from A. B. Maunsbach, *J. Ultrastruct. Res.* 15:197 (1966) by copyright permission of Academic Press, New York.

Figure 9.16 and Table 9.4 Reproduced from Primary Lysosomes of Blood Leukocytes, D. F. Bainton, B. A. Nichols, and M. G. Farquhar, in *Lysosomes in Biology and Pathology,* vol. 5, J. T. Dingle and R. T. Dean, eds., copyright 1976, by permission of Elsevier Biomedical Press, Amsterdam.

Figure 9.17 Reproduced from D. F. Bainton, *J. Cell Biol.* 58:249(1973) by copyright permission of The Rockefeller University Press, New York.

Figure 9.18 Adapted from Mechanisms of Cellular Autophagy, J. L. E. Ericsson, in *Lysosomes in Biology and Pathology,* vol. 2, J. T. Dingle and H. B. Fell, eds., copyright 1969, by permission of Elsevier Biomedical Press, Amsterdam.

Figure 9.19 Adapted from Functions of Lysosomes in Kidney Cells, A. B. Maunsbach, in *Lysosomes in Biology and Pathology,* vol. 1, J. T. Dingle and H. B. Fell, eds., copyright 1969, by permission of Elsevier Biomedical Press, Amsterdam.

Figures 9.20 and 9.26 Adapted from *Lysosomes and Cell Function,* D. Pitt, copyright 1975, by permission of Longman Group Ltd., London.

Figure 9.21 Reproduced from Lysosomes and the Cellular Physiology of Bone Resorption, G. Vaes, in *Lysosomes in Biology and Pathology,* vol. 1, J. T. Dingle and H. B. Fell, eds., copyright 1969, by permission of Elsevier Biomedical Press, Amsterdam.

Figure 9.22 Reproduced from Hormonal Regulation of Metamorphosis, J. R. Tata, in *Sym. Soc. Exp. Biol.* XXV, copyright 1971, by permission of Cambridge University Press, Cambridge.

Figures 9.23, 9.24, and 9.25 Reproduced from Lysosomal Enzymes in Mammalian Spermatozoa, D. B. Morton, in *Lysosomes in Biology and Pathology,* vol. 5, J. T. Dingle and R. T. Dean, eds., copyright 1976, by permission of Elsevier Biomedical Press, Amsterdam.

Figure 9.27 Adapted from Catabolism of Glycoproteins, G. Gregoriadis, in *Lysosomes in Biology and Pathology,* vol. 4, J. T. Dingle and R. T. Dean, eds., copyright 1975, by permission of Elsevier Biomedical Press, Amsterdam.

Figure 9.30 Adapted from Lysosomotropic Cancer Chemotherapy, A. Trouet, D. Deprez-de Campeneere, A. Zenebergh, and G. Sokal, in *Activation of Macrophages,* W. H. Wagner and H. Hahn, eds., copyright 1974, by permission of Elsevier, New York.

Table 9.3 Reproduced from *Cell Biology,* P. L. Altman and D. D. Katz, eds., *Fed. Am. Soc. for Exp. Biol.* copyright 1976, Bethesda, Maryland.

Table 9.5 Reproduced from Endocytosis and Intracellular Digestion, Z. A. Cohn, in *Mononuclear Phagocytes,* R. van Furth, ed., 1970, by permission of Blackwell Scientific Publications, Ltd., Oxford

Table 9.6 Reproduced, with permission, from E. F. Neufeld, T. W. Lim, and L. J. Shapiro, in the Annual Review of Biochemistry, vol. 44, copyright 1975, by Annual Reviews, Inc., Palo Alto, CA.

Chapter 10

Figure 10.1 Reproduced from *Cell Fine Structure,* Thomas Lentz, copyright 1971, by permission of W. B. Saunders Co., Philadelphia.

Figures 10.4, 10.16, and 10.26 Reproduced from G. Palade, *Science* 189:347(1975) by copyright permission of The Nobel Foundation.

Figure 10.5 Adapted from Synthesis and Segregation of Secretory Proteins: The Signal Hypothesis, G. Blobel, in *International Cell Biology,* B. R. Brinkley and K. R. Porter, eds., copyright 1977, by permission of The Rockefeller University Press, New York.

Figure 10.7 Reproduced from J. D. Jamieson and G. E. Palade, *J. Cell Biol.* 34:597(1967) by copyright permission of The Rockefeller University Press, New York.

Figure 10.8 Adapted from Membranes and Secretion, J. D. Jamieson in *Cell Membranes,* G. Weissmann and R. Claiborne, eds., copyright 1975, by permission of HP Publishing Co., New York.

Figure 10.9 Reproduced from D. A. Brodie, *J. Cell Biol.* 90:92(1981) by copyright permission of The Rockefeller University Press, New York.

Figures 10.10, 10.11, 10.12, and 10.13 Reproduced from A. B. Novikoff, M. Mori, N. Quintana, and A. Yam, *J. Cell Biol.* 75:148(1977) by copyright permission of The Rockefeller University Press, New York.

Figure 10.14 Reproduced from A. R. Hand and C. Oliver, *J. Cell Biol.* 74:399(1977) by copyright permission of The Rockefeller University Press, New York.

Figure 10.15 Reproduced from Membranes and Secretion, J. D. Jamieson, in *Cell Membranes,* G. Weissman and R. Claiborne, eds., copyright 1975, by permission of HP Publishing Co., New York.

Figures 10.17, 10.21, and 10.22 Reproduced from D. Lawson, M. D. Raff, B. Gomperts, C. Fewtrell, and N. B. Gilula, *J. Cell Biol.* 72:242(1977) by copyright permission of The Rockefeller University Press, New York.

Figures 10.18 and 10.19 Reproduced from P. Pinto da Silva and M. L. Nogueira, *J. Cell Biol.* 73:161(1977) by copyright permission of The Rockefeller University Press, New York.

Figure 10.20 Adapted from P. Pinto da Silva and M. L. Nogueira, *J. Cell Biol.* 73:161(1977) by copyright permission of The Rockefeller University Press, New York.

Figure 10.23 Reproduced from Modification of the Secretory Granule During Secretion in the Rat Parotid Gland, Z. Selinger, Y. Sharoni, and M. Schramm, in *Advances in Cytopharmacology,* vol. 2, B. Ceccarelli, F. Clement, and J. Meldolesi, eds., copyright 1974, by permission of Raven Press, New York.

Figures 10.24 and 10.25 Reproduced from R. L. Ornberg and T. S. Reese, *J. Cell Biol.* 90:40(1981) by copyright permission of The Rockefeller University Press, New York.

Figure 10.28 and Table 10.2 Reproduced from Secretory Mechanisms in Pancreatic Acinar Cells. Role of the Cytoplasmic Membranes, J. Meldolesi, in *Advances in Cytopharmacology,* vol. 2, B. Ceccarelli, F. Clementi, and J. Meldolesi, eds., copyright 1974, by permission of Raven Press, New York.

Figure 10.29 Adapted from Redundant Cell-Membrane Regulation in the Exocrine Pancreas Cells After Pilocarpine Stimulation of the Secretion, M. F. Kramer and J. J. Geuze, in *Advances in Cytopharmacology,* vol. 2, B. Ceccarelli, F. Clementi, and J. Meldolesi, eds., copyright 1974, by permission of Raven Press, New York.

Figures 10.30, 10.31, and 10.32 Reproduced from Proteolytic Cleavage in the Posttranslational Processing of Proteins, D. F. Steiner, P. F. Quinn, C. Patzelt, S. J. Chan, J. Marsh, and H. S. Tager, in *Cell Biology—A Comprehensive Treatise,* vol. 5, D. M. Prescott and L. Goldstein, eds., copyright 1980, by permission of Academic Press, New York.

Table 10.1 Reproduced from Production of Secretory Proteins in Animal Cells, J. D. Jamieson and G. E. Palade, in *International Cell Biology,* B. R. Brinkley and K. R. Porter, eds., copyright 1977, by permission of The Rockefeller University Press, New York.

Chapter 11

Figure 11.1 Reproduced from The Discovery of the Golgi Apparatus by the Black Reaction and Its Present Fine Structural Visualization, C. Inferrera and G. Carrozza, in *Golgi Centennial Symposium: Perspectives in Neurobiology,* M. Santini, ed., copyright 1975, by permission of Raven Press, New York.

Figure 11.5 and Tables 11.3 and 11.4 Reproduced from Membranes of the Golgi Apparatus, P. Favard, in *Mammalian Cell Membranes,* vol. 2, G. A. Jamieson and D. M. Robinson, eds., copyright 1977, by permission of Butterworths, London.

Figure 11.6 Reproduced from D. S. Friend and M. J. Murray, *Am. J. Anat.* 117:135(1965) by copyright permission of Alan R. Liss, Inc., New York.

Figure 11.7 Reproduced from A. Rambourg, W. Hernandez, and C. P. Leblond, *J. Cell Biol.* 40:395(1969) by copyright permission of The Rockefeller University Press, New York.

Figure 11.8 Reproduced from The Golgi Apparatus: Form and Function, G. M. W. Cook, in *Lysosomes in Biology and Pathology,* vol. 3, J. T. Dingle, ed., copyright 1973, by permission of Elsevier Science Publishing Co., New York.

Figures 11.9, 11.25, 11.26, 11.27, and Tables 11.1 and 11.6 Reproduced from Membrane Flow and Differentiation: Origin of Golgi Apparatus Membranes from Endoplasmic Reticulum, D. J. Morré, T. W. Keenan, and C. M. Huang, in *Advances in Cytopharmacology,* vol. 2, B. Ceccarelli, C. Clementi, and J. Meldolesi, eds., copyright 1974, by permission of Raven Press, New York.

Figure 11.12 Reproduced from Role of the Golgi Apparatus in Terminal Glycosylation, C. P. Leblond and G. Bennett, in *International Cell Biology,* B. R. Brinkley and K. R. Porter, eds., copyright 1977, by permission of The Rockefeller University Press, New York.

Figure 11.13 Reproduced from P. Whur, A. Herscovics, and C. P. Leblond, *J. Cell Biol.* 43:289(1969) by copyright permission of The Rockefeller University Press, New York.

Figures 11.15 and 11.16 Reproduced from *Cell Biology,* P. Sheeler and D. Bianchi, copyright 1980, by permission of John Wiley & Sons, Inc., New York.

Figures 11.17, 11.18, and 11.28 Adapted from Membranes of the Golgi Apparatus, P. Favard, in Mammalian Cell Membranes, vol. 2, G. A. Jamieson and D. M. Robinson, eds., coyright 1977, by permission of Butterworths, London.

Figure 11.19 Reproduced from Glycosylation of Glycoproteins During Intracellular Transport of Secretory Products, H. Schachter, in *Advances in Cytopharmacology,* vol. 2, B. Ceccarelli, C. Clementi, and J. Meldolesi, eds., copyright 1974, by permission of Raven Press, New York.

Figure 11.20 Reproduced from G. Whaley, M. Dauwalder, and J. Kephart, *J. Ultrasturc. Res.* 15:169(1966) by copyright permission of Academic Press, New York.

Figure 11.24 Reproduced from C. E. Bracker, S. N. Grove, C. E. Heintz, and D. J. Morré, *Cytobiologie* 4:1(1971) by copyright permission of Wissenschaftliche Verlagsgesellschaft, Stuttgart.

Table 11.2 Data collected from *Cell Biology,* P. L. Altman and D. D. Katz, eds., copyright 1976, by permission of the Federation of American Societies for Experimental Biology, Bethesda, MD.

Table 11.5 Reproduced from T. W. Keenan and D. J. Morré, *Biochemistry* 9:19(1970) by copyright permission of the American Chemical Society, Washington, D. C.

Chapter 12

Figure 12.3 Reproduced from W. A. Anderson, *J. Histochem. Cytochem.* 18:201 (1970) by copyright permission of The Histochemical Society, Inc., New York.

Figures 12.5 and 12.6 Reproduced from *Mitochondria,* B. Tandler and C. Hoppel, copyright 1972, by permission of Academic Press, New York.

Figures 12.7 and 12.12 Reproduced from *The Cell,* D. W. Fawcett, copyright 1981, by permission of W. B. Saunders Co., Philadelphia.

Figure 12.8 Reproduced from *Biology, A Contemporary Perspective,* G. C. Stephens and B. B. North, copyright 1974, by permission of John Wiley & Sons, Inc., New York.

Figure 12.11 (top) Reproduced from *Muscle,* D. S. Smith, copyright 1972, by permission of Academic Press, New York.

Figure 12.11 (bottom) Reproduced from D. S. Smith, *J. Cell Biol.* 19:115(1963) by copyright permission of The Rockefeller University Press, New York.

Figure 12.13 Reproduced from G. D. Pappas and P. W. Brandt, *J. Biophys. Biochem. Cytol.* 6:85(1959) by copyright permission of The Rockefeller University Press, New York.

Figure 12.16 Reproduced from G. L. Sottocasa, B. Kuylenstierna, L. Ernster, and A. Bergstrand, *J. Cell Biol.* 32:415(1967) by copyright permission of The Rockefeller University Press, New York.

Figure 12.17 Reproduced from C. Schnaitman and J. W. Greenawalt, *J. Cell Biol.* 38:158(1968) by copyright permission of The Rockefeller University Press, New York.

Figures 12.18, 12.21, and 12.23 Reproduced from The Structure of Mitochondrial Membranes, R. A. Capaldi, in *Mammalian Cell Membranes,* vol. 2, G. A. Jamieson and D. M. Robinson, eds., copyright 1977, by permission of Butterworths, London.

Figure 12.25 Reproduced from *Cell Biology,* P. Sheeler and D. Bianchi, copyright 1980, by permission of John Wiley & Sons, Inc., New York.

Figure 12.28 Adapted from The Structure of Mitochondrial Membranes, R. A. Capaldi, in *Mammalian Cell Membranes,* vol. 2, G. A. Jamieson and D. M. Robinson, eds., copyright 1977, by permission of Butterworths, London.

Figure 12.30 Reproduced from How Cells Make ATP, P. C. Hinkle and R. E. McCarty, *Sci. Am.* 238:104(1978) by copyright permission of Scientific American, Inc., New York.

Figure 12.31 Reproduced, by permission, from R. L. Cross, in the Annual Review of Biochemistry, vol. 50, copyright 1981, by Annual Reviews, Inc., Palo Alto, CA.

Figure 12.32 and Table 12.5 Reproduced, by permission, from E. Racker, in the Annual Review of Biochemistry, vol. 46, copyright 1977, by Annual Reviews, Inc., Palo Alto, CA.

Figures 12.33 and 12.34 Reproduced from *Biochemistry,* A. L. Lehniger, copyright 1975, by permission of Worth Publishers, New York.

Figure 12.35 Reproduced from L. Pikó, D. Blair, A. Tyler, and J. Vinograd, *Proc. Nat. Acad. Sci.* 59:838(1968) by courtesy of the National Academy of Science.

Figures 12.37 and 12.38 Reproduced from W. J. Larsen, *J. Cell Biol.* 47:373(1970) by copyright permission of The Rockefeller University Press, New York.

Figure 12.39 Adapted from The Assembly of Mitochondria, J. Saltzgaber, F. Cabral, W. Birchmeier, C. Kohler, T. Frey, and G. Schatz, in *International Cell Biology,* B. R. Brinkley and K. R. Porter, eds., copyright 1977, by permission of The Rockefeller University Press, New York.

Tables 12.1, 12.2, 12.3, and 12.4 Reproduced from *The Structure of Mitochondria,* E. A. Munn, copyright 1974, by permission of Academic Press Inc. (London) Ltd.

Tables 12.6, 12.7, and 12.8 Reproduced from Structure and Function of Mitochondrial DNA, P. Borst, in *International Cell Biology,* B. R. Brinkley and K. R. Porter, eds., copyright 1977, by permission of The Rockefeller University Press, New York.

Table 12.9 Reproduced from Transcription and Translation in Mitochondria, T. W. O'Brien, in *International Cell Biology,* B. R. Brinkley and K. R. Porter, eds., copyright 1977, by permission of The Rockefeller University Press, New York.

Chapter 13

Figure 13.9 Reproduced from R. Emerson and C. M. Lewis, *Amer. J. Botany* 30:165(1943) by copyright permission of the American Journal of Botany, Miami, FL.

Figures 13.10, 13.11, and 13.12 Reproduced from Photosystem I Photoreactions, J. R. Bolton, in *Primary Processes of Photosynthesis,* vol. 2, J. Barber, ed., copyright 1977, by permission of Elsevier Biomedical Press, Amsterdam.

Figure 13.14 Reproduced from Mechanisms of Oxygen Evolution, R. Radmer and G. Cheniae, in *Primary Processes of Photosynthesis,* vol. 2, J. Barber, ed., copyright 1977, by permission of Elsevier Biomedical Press, Amsterdam.

Figures 13.16 and 13.19 Reproduced from Biochemistry of the Chloroplast, R. G. Jensen in *The Biochemistry of Plants,* vol. 1, N. E. Tolbert, ed., copyright 1980, by permission of Academic Press, New York.

Figure 13.17 Reproduced from J. M. Anderson, *FEBS Letters* 124:1(1981) by copyright permission of the Federation of European Biochemical Societies.

Figure 13.18 Reproduced from B. Andersson and J. M. Anderson, *Biochim. Biophys. Acta* 593:426(1980) by copyright permission of Elsevier North-Holland Biomedical Press, Amsterdam.

Figure 13.20 Adapted from CO_2 Metabolism and Plant Productivity, R. H. Burris and C. C. Black, eds., copyright 1976, by permission of University Park Press, Baltimore.

Figure 13.21 Reproduced from *Cell Biology,* P. Sheeler and D. E. Bianchi, copyright 1983, by permission of John Wiley & Sons, Inc., New York.

Figure 13.22 Reproduced from P. Seyer, K. V. Kowallik, and R. G. Herrmann, *Current Genetics* 3:189(1981) by copyright permission of Springer-Verlag, New York.

Figure 13.23 Adapted from Plastid DNA—The Plastome, R. G. Herrmann and J. V. Possingham, in *Chloroplasts,* J. Reinert, ed., copyright 1980, by permission of Springer-Verlag, New York.

Figure 13.24 Reproduced from Types of Plastids: Their Development and Interconversions, E. Schnepf, in *Chloroplasts,* J. Reinert, ed., copyright 1980, by permission of Springer-Verlag, New York.

Table 13.1 Reproduced from RNA and Protein Synthesis in Plastid Differentiation, R. Wollgiehn and B. Parthier, in *Chloroplasts,* J. Reinert, ed., copyright 1980, by permission of Springer-Verlag, New York.

Chapter 14

Figures 14.2, 14.6, and Table 14.2 Reproduced from The Nuclear and the Cytoplasmic Pore Complex: Structure, Dynamics, Distribution, and Evolution, G. G. Maul, in *Internation Review of Cytology,* Suppl. 6, G. H. Bourne, J. F. Danielli, and K. W. Jeon, eds., copyright 1977, by permission of Academic Press, New York.

Figures 14.3 and 14.4 Adapted from An Enzyme Profile of the Nuclear Envelope, I. B. Zbarsky, in *International Review of Cytology,* vol. 54, G. H. Bourne, J. F. Danielli, and K. W. Jeon, eds., copyright 1978, by permission of Academic Press, New York.

Figure 14.5 Reproduced from D. W. Fawcett and H. E. Chemes, *Tissue and Cell* 11:147(1979) by copyright permission of Longman Group Ltd., London.

Figures 14.7 and 14.20 Reproduced from Structures and Functions of the Nuclear Envelope, W. W. Franke and U. Scheer, in *The Cell Nucleus,* vol. 1, H. Busch, ed., copyright 1974, by permission of Academic Press, New York.

Figure 14.8 Reproduced from A. C. Fabergé, *Cell Tiss. Res.* 151:403(1974) by copyright permission of Springer-Verlag, New York.

Figure 14.9 Reproduced from R. H. Kirschner, M. Rusli, and T. E. Martin, *J. Cell Biol.* 72:118(1977) by copyright permission of The Rockefeller University Press, New York.

Figures 14.10, 14.13, 14.14, and 14.15 Reproduced from G. Schatten and M. Thoman, *J. Cell Biol.* 77:517(1978) by copyright permission of The Rockefeller University Press, New York.

Figures 14.11, 14.12, and 14.17 Reproduced from P. N. T. Unwin and R. A. Milligan, *J. Cell Biol.* 93:63(1982) by copyright permission of The Rockefeller University Press, New York.

Figure 14.16 Reproduced from The Plant Nucleus, E. G. Jordan, J. N. Timmis, and A. J. Trewavas, in *The Biochemistry of Plants,* vol. 1, N. E. Tolbert, ed., copyright 1980, by permission of Academic Press, New York.

Figure 14.19 Reproduced from B. J. Stevens and H. Swift, *J. Cell Biol.* 31:55(1966) by copyright permission of The Rockefeller University Press, New York.

Tables 14.1 and 14.3 Reproduced from The Nuclear Envelope in Mammalian Cells, D. J. Fry, in *Mammalian Cell Membranes,* vol. 2, G. A. Jamieson and D. M. Robinson, eds., copyright 1977, by permission of Butterworths, London.

Chapter 15

Figure 15.2a Reproduced from A. K. Kleinschmidt, A. Burton, and R. L. Sinsheimer, *Science* 142:961(1963) by copyright permission of the American Association for the Advancement of Science, Washington, D.C.

Figure 15.2b Reproduced from D. Friefelder, A. K. Kleinschmidt, and R. L. Sinsheimer, *Science* 146:254(1964) by copyright permission of the American Association for the Advancement of Science, Washington, D.C.

Figures 15.4 and 15.47 Reproduced from *Principles of Genetics,* E. J. Gardner and D. Peter Snustad, copyright 1981, by permission of John Wiley & Sons, Inc., New York.

Figures 15.7 and 15.8 Reproduced from *General Virology,* S. E. Luria, J. E. Darnell, D. Baltimore, and A. Campbell, copyright 1978, by permission of John Wiley & Sons, Inc., New York.

Figure 15.9 Reproduced from J. Vinograd, J. Lebowitz, R. Radloff, R. Watson, and P. Laipis, *Proc. Nat. Acad. Sci.* 53:1104(1965) by courtesy of the National Academy of Science.

Figure 15.10 Adapted from J. Vinograd, J. Lebowitz, R. Radloff, R. Watson, and P. Laipis, *Proc. Nat. Acad. Sci.* 53:1104(1965) by courtesy of the National Academy of Science.

Figure 15.11 Reproduced from K. Mizuuchi, L. M. Fisher, M. H. O'Dea, and M. Gellert, *Proc. Nat. Acad. Sci.* 77:1847(1980) by courtesy of the National Academy of Sciences and the permission of Dr. Martin Gellert.

Figure 15.12 Reproduced from R. Kavenoff and B. C. Bowen, *Chromosoma (Berl.)* 59:89(1976) by copyright permission of Springer-Velag, New York.

Figures 15.14 and 15.15 and Table 15.2 Reproduced from Chemical, Physical, and Genetic Structure of Prokaryotic Chromosomes, D. E. Pettijohn, and J. O. Carlson, in *Cell Biology, A Comprehensive Treatise,* vol. 2, D. M. Prescott and L. Goldstein, eds., copyright 1979, by permission of Academic Press, New York.

Figure 15.16 Reproduced from *Fundamental Concepts of Biology,* G. E. Nelson, G. G. Robinson, and R. A. Boolootian, copyright 1974, by permission of John Wiley & Sons, Inc., New York.

Figure 15.18 and Table 15.3 Reproduced from Nucleosomes: Composition and Substructure, R. R. Rill, in *Molecular Genetics, Part III Chromosome Structure,* J. H. Taylor, ed., copyright 1979, by permission of Academic Press, New York.

Figure 15.21 Reproduced from Chromosomes and Chromatin Structure, G. F. Bahr, in *Molecular Structure of Human Chromosomes,* J. J. Yunis, ed., copyright 1977, by permission of Academic Press, New York.

Figures 15.23 and 15.26 Reproduced from *Genes,* B. Lewin, copyright 1983, by permission of John Wiley & Sons, Inc., New York.

Figure 15.24 Adapted, with permission, from J. D. McGhee and G. Felsenfeld, in the Annual Review of Biochemistry, vol. 49, copyright 1980, by Annual Reviews Inc., Palo Alto, CA.

Figure 15.25 Reproduced from J. T. Finch, L. C. Lutter, D. Rhodes, R. S. Brown, B. Rushton, M. Levitt, and A. Klug, *Nature* 269:29(1977) by copyright permission of Macmillan Journals Ltd., London.

Figure 15.27 Adapted from Mammalian Chromosome Structure, D. E. Comings, in *Chromosomes Today,* vol. 6, A. de la Chapelle and M. Sorsa, eds., copyright 1977, by permission of Elsevier Biomedical Press, Amsterdam.

Figure 15.28 Reproduced from A. L. Bak, J. Zeuthen, and F. H. C. Crick, *Proc. Nat. Acad. Sci.* 74:1595(1977) by courtesy of the National Academy of Science.

Figure 15.29 Reproduced from Higher-Order Structure of Mitotic Chromosomes, A. Leth Bak and J. Zeuthen, in *Cold Spring Harbor Symposium on Quantitative Biology,* vol. 42, copyright 1978, by permission of Cold Spring Harbor Laboratory, New York.

Figure 15.30 Adapted from the reference for Figure 15.29.

Figures 15.31, 15.32, 15.33, and 15.34 Reproduced from Metaphase Chromosome Structure: The Role of Nonhistone Proteins, U. K. Laemmli, S. M. Cheng, K. W. Adolph, J. R. Paulson, J. A. Brown, and W. R. Baumbach, in *Cold Spring Harbor Symposium on Quantitative Biology,* vol. 42, copyright 1978, by permission of Cold Spring Harbor Laboratory, New York.

Figure 15.36 Reproduced from J. G. Gall, *Brookhaven Symp. in Biology* 8:18(1955).

Figures 15.37 and 15.41 Reproduced from Organization of Transcriptionally Active Chromatin in Lampbrush Chromosome Loops, U. Scheer, H. Spring, and M. F. Trendelenburg, in *The Cell Nucleus,* vol. 7, H. Busch, ed., copyright 1979, by permission of Academic Press, New York.

Figure 15.38 Reproduced from O. L. Miller and B. R. Beatty, *J. Cell Physiol.* 74, Suppl. 1:225(1969) by copyright permission of Alan R. Liss, Inc., New York.

Figure 15.39 Reproduced from H. G. Callan, *Proc. Roy. Soc. Lond. B* 214:417(1982) by copyright pemission of The Royal Society, London.

Figure 15.43 Reproduced from E. J. DuPraw and P. M. M. Rae, *Nature* 212:598(1966) by copyright permission of Macmillan Journals, Ltd., London.

Figure 15.46 Reproduced from *Principles of Genetics,* E. J. Gardner, copyright 1975, by permission of John Wiley & Sons, Inc., New York. Based on the work of W. Beermann and U. Clever.

Table 15.1 Reproduced from The DNA and RNA of Bacterial and Viral Chromosomes, M. G. Smith, in *MTP International Review of Science,* vol. 6, K. Burton, ed., copyright 1974, by permission of Butterworths, London, and MTP Medical and Technical Publishing Co., Ltd., Lancaster.

Chapter 16

Figures 16.1, 16.5, 16.6, 16.7 and Table 16.1 Reproduced from R. Berezney and D. S. Coffee, *J. Cell Biol.* 73:616(1977) by copyright permission of The Rockefeller University Press, New York.

Figures 16.2, 16.3, 16.4, and 16.8 Reproduced from Dynamic Properties of the Nuclear Matrix, R. Berezney, in *The Cell Nucleus,* vol. 7, H. Busch, ed., copyright 1979, by permission of Academic Press, New York.

Figure 16.12 Reproduced from *The Cell,* D. W. Fawcett, copyright 1981, by permission of W. B. Saunders Co., Philadelphia.

Figures 16.13, 16.17, 16.18, 16.19, and 16.20 Reproduced from The Ultrastructural Visualization of Nucleolar and Extranucleolar RNA Synthesis and Distribution, S. Fakan and E. Puvion, in *International Review of Cytology,* vol. 65, G. H. Bourne, J. F. Danielli, and K. W. Jeon, eds., copyright 1980, by permission of Academic Press, New York.

Figure 16.14 Reproduced from M. F. Trendelenberg, H. Spring, U. Scheer, and W. W. Franke, *Proc. Nat. Acad. Sci.* 71:3626(1974) by courtesy of the National Academy of Sciences, Washington, D. C.

Figure 16.21 Reproduced from Structure and Function of Nuclear and Cytoplasmic Ribonucleoprotein Complexes, T. E. Martin, J. M. Pullman, and M. D. McMullen, in *Cell Biology—A Comprehensive Treatise,* vol. 4, D. M. Prescott and L. Goldstein, eds., copyright 1980, by permission of Academic Press, New York.

Table 16.2 Reproduced from The Nucleolus and Nucleolar DNA, K. Smetana and H. Busch, in *The Cell Nucleus,* vol. 1, H. Busch, ed., copyright 1974, by permission of Academic Press, New York.

Chapter 17

Figure 17.2 Reproduced from M. Boublik, W. Hellmann, and A. K. Kleinschmidt, *Cytobiologie* 14:293(1977) by copyright permission of Wissenschaftliche Verlagsgesellschaft, Stuttgart.

Figure 17.3 Reproduced from *Cell Biology,* P. Sheeler and D. Bianchi, copyright 1983, by permission of John Wiley & Sons, Inc., New York.

Figure 17.6 Reproduced from T. Pieler and V. A. Erdmann, *Proc. Nat. Acad. Sci.* 79:4599(1982) by courtesy of the National Academy of Sciences, Washington, D. C.

Figure 17.8 Reproduced from S. J. S. Hardy, C. G. Kurland, P. Voynow, and G. Mora, *Biochemistry* 8:2897(1969) by copyright permission of the American Chemical Society, Washington, D. C.

Figure 17.9 Reproduced from E. Kaltschmidt and H. G. Wittmann, *Proc. Nat. Acad. Sci.* 67:1276(1970) by courtesy of the National Academy of Sciences, Washington, D. C.

Figure 17.11a Adapted from Reconstitution of Ribosomes: Studies of Ribosome Structure, Function, and Assembly, M. Nomura and W. A. Held, in *Ribosomes*, M. Nomura, A. Tissieres, and P. Lengyel, eds., copyright 1974, by permission of Cold Spring Harbor Laboratory, New York.

Figure 17.11b Adapted from Analysis of the Assembly and Function of the 50S Subunit from *Escherichia coli* Ribosomes by Reconstitution, K. H. Nierhaus, in *Ribosomes—Structure, Function and Genetics*, G. Chambliss, G. R. Craven, J. Davies, K. Davis, L. Kahan, and M. Nomura, eds., copyright 1980, by permission of University Park Press, Baltimore.

Figures 17.12 and 17.14 Reproduced from Protein Topography of *Escherichia coli* Ribosomal Subunits as Inferred from Protein Crosslinking, R. R. Traut, J. M. Lambert, G. Boileau, and J. W. Kenney, in *Ribosomes: Structure, Function, and Genetics*, G. Chambliss, G. R. Craven, J. Davies, K. Davis, L. Kahan, and M. Nomura, eds., copyright 1980, by permission of University Park Press, Baltimore.

Figures 17.15, 17.16, and 17.17 Reproduced, with permission, from H. G. Wittmann, in the *Annual Review of Biochemistry*, vol. 52, copyright 1983, by Annual Reviews, Inc., Palo Alto, CA.

Table 17.3 Reproduced from Biogenesis of Chloroplast and Mitochondrial Ribosomes, J. E. Boynton, N. W. Gillham, and A. M. Lambowitz, in *Ribosomes: Structure, Function and Genetics*, G. Chambliss, G. R. Craven, J. Davies, K. Davis, L. Kahan, and M. Nomura, eds., copyright 1980, by permission of University Park Press, Baltimore.

Table 17.5 Reproduced from Structure and Topography of Ribosomal RNA, H. F. Noller, in *Ribosomes: Structure, Function and Genetics*, G. Chambliss, G. R. Craven, J. Davies, K. Davis, L. Kahan, and M. Nomura, eds., copyright 1980, by permission of University Park Press, Baltimore.

Chapter 18

Figures 18.2 and 18.3 Reproduced from M. L. DePamphilis and J. Adler, *J. Bacteriol.* 105:376(1971) by copyright permission of the American Society for Microbiology, Washington, D. C.

Figure 18.4 Reproduced from M. L. DePamphilis and J. Adler, *J. Bacteriol.* 105:384(1971) by copyright permission of the American Society for Microbiology, Washington, D. C.

Figure 18.5 Adapted from *Motility of Living Cells*, P. Cappuccinelli, copyright 1980, by permission of Chapman and Hall, Methuen, Inc., New York.

Figures 18.7, 18.9, 18.10, 18.11, 18.12, and 18.13 Reproduced from *Cell Biology*, P. Sheeler and D. Bianchi, copyright 1980, by permission of John Wiley & Sons, Inc., New York.

Figure 18.8 Reproduced from *Muscle*, D. S. Smith, copyright 1972, by permission of Academic Press, New York.

Figure 18.14 Reproduced from *The Cell*, D. W. Fawcett, copyright 1981, by permission of W. B. Saunders Co., Philadelphia.

Figure 18.15 Reproduced from M. Holwill, *Symp. Soc. Gen. Microbiol.* 30:278(1980), by copyright permission of Cambridge University Press, New York.

Figure 18.16 Reproduced from F. D. Warner and D. R. Mitchell, *J. Cell Biol.* 76:261(1978) by copyright permission of The Rockefeller University Press, New York.

Figure 18.17 Reproduced from W. D. Cohen, D. Bartelt, R. Jaeger, G. Langford, and I. Nemhauser, *J. Cell Biol.* 93:828(1982) by copyright permission of The Rockefeller University Press, New York.

Figure 18.21 Reproduced from *Cell Biology*, P. Sheeler and D. Bianchi, copyright 1983, by permission of John Wiley & Sons, Inc., New York, based on an original model by M. Kirschner, L. Honig, and R. Williams in *J. Mol. Biol.* 99:263(1975).

Figure 18.22 Reproduced from Microtubules and Intermediate Filaments, F. Gaskin and M. L. Shelanski, in *Essays in Biochemistry*, vol. 12, P. N. Campbell and W. N. Aldridge, eds., copyright 1976, by permission of Academic Press, New York.

Figure 18.23 Reproduced from Structure of Microtubules, L. A. Amos, in *Microtubules*, K. Roberts and J. S. Hyams, eds., copyright 1979, by permission of Academic Press, New York.

Figure 18.25 Reproduced from B. R. Brinkley, S. M. Cox, D. A. Pepper, L. Wible, S. L. Brenner, and R. L. Pardue, *J. Cell Biol.* 90:554(1981) by copyright permission of The Rockefeller University Press, New York.

Figures 18.26 and 18.28 Reproduced from Identification and Distribution of Intracellular Filaments, H. Ishikawa, in *Cell Motility: Molecules and Organization*, S. Hatano, H. Ishikawa, and H. Sato, eds., copyright 1979, by permission of University of Tokyo Press, Tokyo.

Chapter 19

Figure 19.2 Reproduced from *The Cell*, D. W. Fawcett, copyright 1981, by permission of W. B. Saunders Co., Philadelphia.

Figures 19.3, 19.7, 19.8, and 19.15 Adapted from *Motility of Living Cells*, P. Cappuccinelli, copyright 1980, by permission of Chapman and Hall, Methuen, Inc., New York.

Figure 19.4 Reproduced from F. D. Warner and D. R. Mitchell, *J. Cell Biol.* 89:35(1981) by copyright permission of The Rockefeller University Press, New York.

Figure 19.5 Reproduced from B. Huang, G. Piperno, Z. Ramanis, and D. J. L. Luck, *J. Cell Biol.* 88:80(1981) by copyright permission of The Rockefeller University Press, New York.

Figure 19.6 Reproduced from Advances in the Ultrastructural Analysis of the Sperm Flagellar Axoneme, R. W. Linck, in *The Spermatozoan*, D. W. Fawcett and M. Bedford, eds., copyright 1979, by permission of Urban and Schwarzenberg, Inc., Baltimore.

Figure 19.10 Reproduced from How Cilia Move, P. Satir, in *Sci. Am.* 231:44(1974) by copyright permission of Scientific American, Inc., New York.

Figures 19.11 and 19.12 Reproduced from M. S. Mooseker and L. G. Tilney, *J. Cell Biol.* 67:725(1975) by copyright permission of The Rockefeller University Press, New York.

Figure 19.16 Reproduced from D. L. Taylor, Y. -L. Wang, and J. M. Heiple, *J. Cell Biol.* 86:590(1980) by copyright permission of The Rockefeller University Press, New York.

Figure 19.17 Adapted from S. B. Hellewell and D. L. Taylor, *J. Cell Biol.* 83:633(1979) by copyright permission of The Rockefeller University Press, New York.

Figure 19.18 Reproduced from *Principles of Genetics*, E. J. Gardner and D. P. Snustad, copyright 1981, by permission of John Wiley & Sons, Inc., New York.

Figure 19.19 Reproduced from Assembly and Disassembly of the Mitotic Spindle, H. Sato, Y. Ohnuki, and Y. Sato, in *Cell Motility: Molecules and Organization*, S. Hatano, H. Ishikawa, and H. Sato, eds., copyright 1979, by permission of University of Tokyo Press, Tokyo.

Figure 19.20 Reproduced from J. R. McIntosh, *J. Cell Biol.* 61:166(1974) by copyright permission of The Rockefeller University Press, New York.

Figure 19.27 Reproduced from M. Osborn, W. W. Franke, and K. Weber, *Proc. Nat. Acad. Sci.* 74:2490(1977) by courtesy of the National Academy of Sciences, Washington, D. C.

Figures 19.28 and 19.35 Reproduced from *Cell Biology*, P. Sheeler and D. Bianchi, copyright 1983, by permission of John Wiley & Sons, Inc., New York. Figure 19.28 is based on an original model by Dr. K. R. Porter.

Figure 19.36 Reproduced from M. H. Ellisman and K. R. Porter, *J. Cell Biol.* 87:464(1980) by copyright permission of The Rockefeller University Press, New York.

METHODS IN CELL BIOLOGY

Figure 6 Courtesy of Beckman Instruments, Palo Alto, CA.

Figure 8 Courtesy of Beckman Instruments, Palo Alto, CA.

Figure 23 Reproduced from *Fundamentals of Chemistry*, J. E. Brady and J. R. Holum, copyright 1981, by permission of John Wiley & Sons, New York.

Table 1 Reproduced from *The Tools of Biochemistry*, T. G. Cooper, copyright 1977, by permission of John Wiley & Sons, Inc., New York.

Table 2 Courtesy of Beckman Instruments, Palo Alto, CA.

Index